D1162694

Physiology

Chapter 20 on *Immunology* contributed by Assistant Professor Vinay Likhite, Biology Department, New York University, Washington Square

Chapter 25 on *Muscle and Bone as a Functional Unit* contributed by Professor Günter Friedebold, Director of the Orthopedic Clinic and Polyclinic of the Free University, Berlin, Germany

FLEUR L. STRAND

PROFESSOR OF BIOLOGY
New York University Washington Square

Physiology

A Regulatory Systems Approach

MACMILLAN PUBLISHING CO., INC.
New York

COLLIER MACMILLAN PUBLISHERS
London

A portion of this material has been adapted from *Modern Physiology: The Chemical and Structural Basis of Function* copyright © 1965 by Fleur L. Strand.

Macmillan Publishing Co., Inc.
866 Third Avenue, New York, New York 10022

Collier Macmillan Canada, Ltd.

Library of Congress Cataloging in Publication Data

Strand, Fleur L
Physiology.

Bibliography: p.
Includes indexes.
1. Human physiology. 2. Biological control systems.
I. Title. [DNLM: 1. Physiology. QT104 S897p]
QP34.5.S8 612 77-933
ISBN 0-02-417670-2

Printing: 4 5 6 7 8 Year: 0 1 2 3

Cover photograph of scanning electromicrograph courtesy of J. LeBlanc, Laboratory of Embryology, New York University.

Acknowledgments for Poems

Many of the poems chosen to introduce the chapters of this book were found in a fine anthology by Helen Plotz: *Imagination's Other Place. Poems of Science and Imagination.* Compiled by Helen Plotz. Copyright 1955 by Thomas Y. Crowell Company.

p. vii. "The Secret Sits" by Robert Frost. From *The Poetry of Robert Frost* edited by Edward Connery Lathem. Copyright 1942, © 1962 by Robert Frost. Copyright © 1969 by Holt, Rinehart and Winston. Copyright © 1970 by Lesley Frost Ballantine. Reprinted by permission of Holt, Rinehart and Winston, Publishers.

Chapter 1. "One's Self I Sing" by Walt Whitman. From *Leaves of Grass: Selected Prose of Walt Whitman,* edited by John Kouwenhoven, Random House, Inc., New York, 1950.

Chapter 2. From *The Poems of Emily Dickinson,* edited by Martha Dickinson Bianchi and Alfred Leete Hampson. Copyright 1929 by Martha Dickinson Bianchi. Little, Brown and Company, Boston, 1929.

Chapter 3. "The High Cost of Being Human" by H. J. Morowitz. From *The New York Times,* Feb. 11, 1976, p. 41. © 1976 by The New York Times Company. Reprinted by permission.

Chapter 4. "Honey and Salt" by Carl Sandburg. From *Honey and Salt,* Harcourt Brace Jovanovich, Inc., New York, 1963.

Chapter 5. "Measurement" by A. M. Sullivan. From *Selected Lyrics and Sonnets,* Thomas Y. Crowell Company, Inc., New York, 1970.

Chapter 6. "Regulation of Bioelectric Events" by A. M. Sullivan. From *Selected Lyrics and Sonnets,* Thomas Y. Crowell Company, Inc., New York, 1970.

Chapter 7. "A Reflex" by Robert Frost. From *The Poetry of Robert Frost* edited by Edward Connery Lathem. Copyright 1942, © 1962 by Robert Frost. Copyright © 1969 by Holt, Rinehart and Winston. Copyright © 1970 by Lesley Frost Ballantine. Reprinted by permission of Holt, Rinehart and Winston, Publishers.

Chapter 8. "The Dinosaur" by Bert Leston Taylor. From *The Chicago Tribune.*

Chapter 9. "Men Say They Know Many Things" by Henry Thoreau. From *Collected Poems of Henry Thoreau,* edited by Carl Bode, The Johns Hopkins University Press, Baltimore, Md., 1964.

Chapter 10. "Little Gidding" by T. S. Eliot. From "Little Gidding" in *Four Quartets,* copyright, 1943, by T. S. Eliot, copyright 1971 by Esme Valerie Eliot. Reprinted by permission of Harcourt Brace Jovanovich, Inc.

Chapter 11. "Exotic" by A. R. Ammons. Reprinted from *Collected Poems,* by A. R. Ammons. By permission of W. W. Norton & Company, Inc. Copyright © 1972 by A. R. Ammons.

Chapter 12. "Song at Sunset" by Walt Whitman. From *Leaves of Grass: Selected Prose of Walt Whitman,* edited by John Kouwenhoven, Random House, Inc., New York, 1950.

Chapter 13. "Surgeons Must Be Very Careful" by Emily Dickinson. From *The Poems of Emily Dickinson,* edited by Martha Dickinson and Alfred Leete Hampson. Copyright 1929 by Martha Dickinson Bianchi. Little, Brown and Company, Boston, 1929.

Chapter 14. "Protozoa" by A. M. Sullivan. From *Selected Lyrics and Sonnets,* Thomas Y. Crowell Company, Inc., New York, 1970.

Chapter 15. "What Am I, Life?" by John Masefield. Reprinted with permission of Macmillan Publishing Co., Inc., from *Sonnets* by John Masefield. Copyright 1916 by John Masefield, renewed 1944 by John Masefield.

Chapter 16. "Song of Myself" by Walt Whitman. From *Leaves of Grass: Selected Prose of Walt Whitman,* edited by John Kouwenhoven, Random House, Inc., New York, 1950.

Chapter 17. "Quandary" by Robert Frost. From *The Poetry of Robert Frost* edited by Edward Connery Lathem. Copyright 1942, © 1962 by Robert Frost. Copyright © 1969 by Holt, Rinehart and Winston. Copyright © 1970 by Lesley Frost Ballantine. Reprinted by permission of Holt, Rinehart and Winston, Publishers.

Chapter 18. "A Man Said to the Universe" by Stephen Crane. From *The Collected Poems of Stephen Crane,* Alfred A. Knopf, Inc., New York, 1930.

Chapter 19. "Auguries of Innocence" by William Blake. From *William Blake, Poet and Mystic* by P. Berger, Haskell House Publishers, Ltd., Brooklyn, N.Y., 1914, Reprint 1969.

Chapter 20. "The Staff of Aesculapius" by Marianne Moore. Courtesy of Abbott Laboratories' *What's New,* North Chicago, Ill.

Chapter 21. "Unsaid" by A. R. Ammons. Reprinted from *Collected Poems,* by A. R. Ammons. By permission of W. W. Norton & Company, Inc. Copyright © 1972 by A. R. Ammons.

Chapter 22. "No Single Thing Abides" by Titus Lucretius Carus, 95–52 B.C. Translated by W. H. Mallock. Dodd, Mead & Company, Inc., New York.

Chapter 23. "God's First Creature Was Light" by Winifred Welles. From *Blossoming Antlers* by Winifred Welles. Copyright 1933 by Winifred Welles. All rights reserved. Reprinted by permission of The Viking Press, Inc., New York.

Chapter 24. "With a Daisy" by Emily Dickinson. From *The Poems of Emily Dickinson,* edited by Martha Dickinson Bianchi and Alfred Leete Hampson. Copyright 1929 by Martha Dickinson Bianchi. Little, Brown and Company, Boston, 1929.

Chapter 25. "Song at Sunset" by Walt Whitman. From *Leaves of Grass: Selected Prose of Walt Whitman,* edited by John Kouwenhoven, Random House, Inc., New York, 1950.

Chapter 26. Margaret Mead. From *The New York Times,* Oct. 30, 1966.

Chapter 27. "Heredity" by Thomas Hardy. From *Collected Poems* by Thomas Hardy, copyright 1925 by Macmillan Publishing Co., Inc.

To my husband and daughter
For the precious hours taken from family life

We dance around in a ring and suppose
But the Secret sits in the middle and knows.

From "The Secret Sits," in *The Poetry of Robert Frost,* edited by Edward Connery Lathem, New York: Holt, Rinehart and Winston, 1971.

Foreword

Too much of physiology is a description of analytical experiments on some system isolated from the whole organism. This work is often very elegant, and also provides in its limited field definitive solutions. Wonderful as these isolated systems are, and challenging as they are to the scientist, it is important that a text for undergraduate students should be directed more to the integrative processes whereby all of the diverse organs and systems are eventually oriented to the performance of the complete organism. This synthesis should not be established by some integrative chapters at the end of the analytically based chapters. Instead, as in this book, each phase of the synthesis should follow immediately on the analysis. This arrangement demonstrates to the students the significance of the regulatory mechanisms and makes the physiological text more relevant and interesting. As biologists, we must view the organism as being evolved by the dual processes of genetic creativity and selectionist rejection of unfavorable mutants. It is an emergent process that results in design and purpose in the immensely complex hierarchy of regulatory controls.

This book is remarkable for the emphasis that is placed on such controls at all levels—cellular, nervous, and endocrine. A description of these controlling mechanisms is presented first to give the student a basic understanding of the mechanisms for interaction of all parts of the body. Then follows the treatment of the various systems and organs. The text is remarkable in that each system is treated as a problem situation that has been studied to give solutions, but not final solutions. It is no more than the nearest approach to truth that we have yet achieved, and doubtless it will be superseded. In a sense, the scientific enterprise can be regarded as a search for absolute values, not only of truth, but also of beauty. Do we not recognize the beauty as well as the truth of some wonderful new insight of great generality and simplicity, as for example, the double helix of the genetic code? In this text, the student becomes a participator in the scientific enterprise. The text is no dull assembly of known facts to be remembered and regurgitated in examinations. Rather, it is a living experience replete with the romance of discovery.

The immense growth of literature makes physiology, in toto, beyond the comprehension of any mind if the aim is to present the whole story at a level satisfactory for the experts in one or another field. There are many such large compendiums or handbooks with multitudes of authors writing very limited chapters with even some later chapters attempting synthesis. But for the undergraduate student these books are very unattractive and even misleading. Such books should be for reference after a good overall knowledge of physiology has been attained. A truly readable text must be the expression of the thoughts of one scientist with a wide range of interests and with the drive to study, criticize, and assimilate many secondary sources as well as the most significant primary sources. Professor Strand has done just that in a concentrated effort of many years spent in completely reconstructing her earlier text *Modern Physiology.* A new title has been chosen for what is essentially a new book. She has achieved a lucid presentation even of complex phenomena. In order to emphasize current ideas in physiology, there has been a different selection from that of a typical undergraduate textbook of physiology. Special emphasis has been placed on regulatory activities of the hypothalamus, both neural and endocrine. This makes the book more readable, pertinent, and interesting. However, it requires the elimination of older, less important material that is found in standard texts. This book is thus not a piling of new findings on top of the old. Rather, it is a judicious selection of classical physiology together with recent discoveries right up to the current time.

I have been delighted that Professor Strand has incorporated into her text many selections from my book *The Understanding of the Brain,* where I have tried to present to the student my ideas on specially chosen aspects of brain physiology. It gives me great pleasure to express my enthusiasm concerning this truly remarkable book by Professor Strand. She has devoted herself over the years to the art of clear communication to student audiences of all ages, not only to the students of her own university department, but also in her admirable Sunrise Semester Series on the Columbia Broadcasting System some years ago. I had the pleasure of a joint presentation with her in the last program of the series. I conclude by stating that this book is truly a work of dedication to the student audience, and to them I recommend it enthusiastically.

Sir John Eccles
Locarno, Switzerland
May 1976
Nobel Laureate in Physiology and Medicine

Preface

THIS TEXT HAS been written to enable the student to master, and then apply, the basic concepts of physiological regulation from the level of the cell to the level of the integrated, intact organism. Consequently, Part I emphasizes cellular regulation and is followed by an introduction to neural, hormonal, and neuroendocrine regulatory systems in Part II. From my teaching experience, I have found that if a student has grasped the fundamental processes by which cells are regulated, then any organ system can be studied with relative ease. Learning then proceeds in a logical and intellectually satisfying manner, rather than through a formal memorization of accumulated facts. This permits the selection of subsequent chapters according to the amount of time available for the course: the student has the scientific vocabulary and scientific approach to master them in almost any order.

Because physiology is a dynamic, changing science, the basic concepts presented are supplemented wherever feasible with up-to-date experimental evidence. Controversial ideas and hypotheses are included to illustrate that present ideas often have to be modified as new work is published. On the other hand, it is often startling to realize how many physiological concepts are based on studies originating at the beginning of this century, or even earlier, by scientists who used their imagination as well as the available facts to create workable hypotheses. These hypotheses could then be tested by newer techniques available to modern investigators and modified, extended, or corrected as necessary. Although current ideas in physiology are emphasized in this text, homage is paid to some of the classical studies, to give a sense of the continuity of science and our debt to previous generations of scientists.

Equal rights are also given to women, who of course have a physiology apart from their reproductive organs. To the physiological values usually calculated for the "typical" 70-kg male have been added, wherever available, the equivalent values for the female.

Sufficient anatomy is included in this book for an adequate understanding of physiological processes. Although comparative animal studies are frequently mentioned, the orientation is toward human physiology. This makes the text pertinent to the interests of a wide range of students, from biology majors and premedical students to nurses, and perhaps even an occasional nonscience major, fascinated by the glorious complexities of the living body.

To suit such diverse needs and interests, additional reading material has been listed at the end of each chapter. *Cited references* follow the specific references mentioned in the chapter. *Additional readings* are much wider in scope and have been divided into a section on books and, for most chapters, a section listing pertinent articles. The book sections include general reference texts, reviews, and handbooks. These texts contain much useful information, and the reader is strongly advised to consult them.

As I came to the final, difficult stages of this book, I realized how much help my editors had given me. I would particularly like to thank Mr. Woodrow Chapman for his quiet, calm reassurance when I most needed it. I would also like to thank Mrs. Elaine Wetterau for her patient attention to important detail and unusual ability to coordinate the host of problems—and their solutions—that arise during production. My gratitude is also due to Mr. Russell Peterson and to Mr. Andrew Mudryk who so skilfully and artistically interpreted my sketches to produce the many excellent illustrations that form an essential part of this text.

I would especially like to thank Sir John Eccles for his help and critical reading of this book, and for the beautifully written foreword. I am also greatly indebted to Professor Albert S. Gordon for friendly encouragement over many years and, in particular, for his helpful comments and criticisms on the chapter on blood. Writing this book will have been well worth the effort if it can inspire my students the way in which he has inspired me.

Fleur L. Strand
New York University
Washington Square

Contents

Cellular Regulation

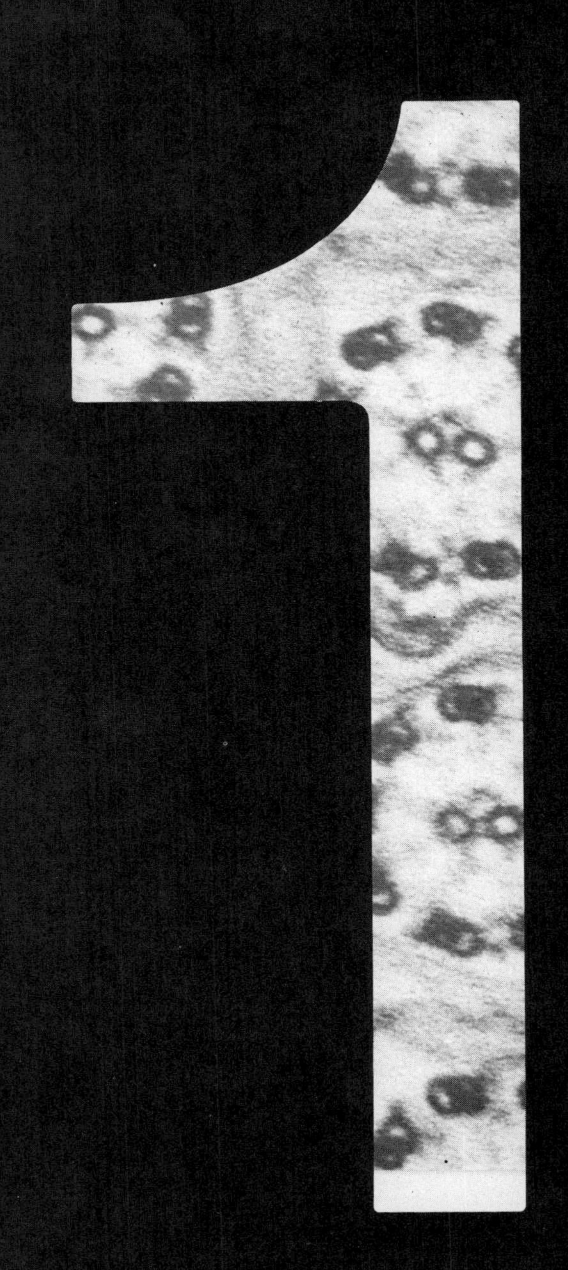

Chapter 1

Physiological Regulation

Of physiology from top to toe I sing,
Not physiognomy alone nor brain alone is worthy for the Muse,
I say the form complete is worthier far,
The Female equally with the Male I sing.

Of Life immense in passion, pulse, and power,
Cheerful, for freeset action form'd under the laws divine,
The Modern Man I sing.

Walt Whitman, "One's-Self I Sing"

1-1 Concepts of Physiological Regulation and Homeostasis

Physiology in its true sense encompasses all aspects of biology, for it is the study of function which, in turn, is inseparable from structure. Both structure and function must be studied on all levels, from the molecule to the intact organism, and in some cases should even be extended to the social groupings of organisms. It is obvious that in this book we must make a drastic choice in both the topics to be covered and the aspect from which they will be approached. As the requirements of living organisms, from a physiological standpoint, are astonishingly similar, and often seem to be modifications of a basic pattern, this book will attempt to select the most important requirements for the life of the organism and show their development in the mammalian, and particularly the human, organism.

The unifying point of view will be the manner by which the regulatory systems of the mammalian organism are integrated to provide the optimal environment for the basic unit of living matter, which is the cell. All the various systems—respiratory, digestive, circulatory—which are individually designated as though they existed for their own sake and dignity, serve in a subsidiary role to fulfill the biologic requirements of the cell. This set of essential conditions is the *internal environment* or, in the more elegant French phrase of the physiologist Claude Bernard (1878), *le milieu interne*. The respiratory system, usually described with emphasis on the lungs and the passages leading to them from the outside, is of significance only in that it brings the oxygen necessary for fuel combustion to the cells and removes from the tissues the accumulating carbon dioxide that acts as a poison when present in high concentrations. This internal respiration at the level of the tissues is the ultimate significance of the entire respiratory system, although obviously any disturbance in the lungs or bronchi would affect respiration as seriously as a disturbance at the tissue level. Today, the heart-lung machine can replace the work of these organs for limited periods of time, circulating aerated blood to the tissues and removing the carbon dioxide from the blood, supplying glucose when necessary, and totally bypassing the heart and lungs, except as they themselves may be supplied with oxygen and glucose by the machine.

Similarly, the digestive system may be considered to be an elaborate assembly line, speeded up or slowed down by the nervous system and by chemical regulators that act upon the foodstuffs passing along this line. This process transforms the appetizing yet nonutilizable gross food particles into the smaller and simpler molecules that can penetrate through the walls of the digestive system and then pass to the tissues by way of the blood, to be ultimately used by the cell for energy or for structure.

The excretory system is perhaps the most accurately attuned to changes that might affect the constancy of the internal environment. The ureters, bladder, and urethra are again the physical instruments for removing superfluous fluids and ions from the blood. The precisely adjustable activity of the cells of the kidney tubules is responsible for the constancy of the composition of blood, and thereby for the internal environment of all the cells of the body.

Even the reproductive system contributes to the well-being of the cell, particularly in the adult, because the hormones secreted by the normally functioning gonads regulate many metabolic processes and, together with the secretions of the other members of the endocrine system, act as coregulators of the functioning and adjustment of the body.

Rapid coordination of internal organ systems and response to the external environment are provided by the complex nervous system. Interactions between the nervous and endocrine controlling systems are precisely adjusted by intricate mechanisms, the significance of which has only just begun to be appreciated and has given rise to the new term *neuroendocrine regulation*.

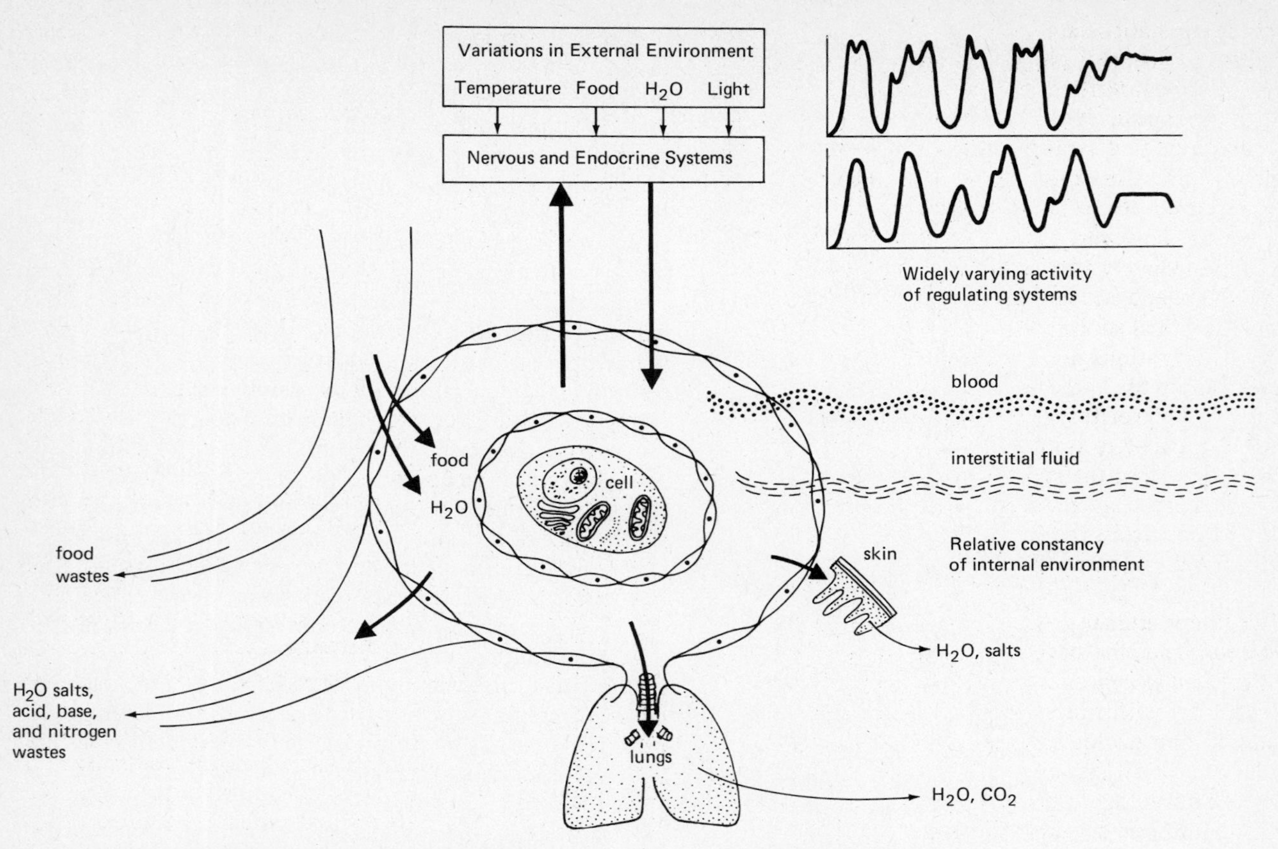

Figure 1-1. **Through regulatory adjustments of different organ systems (nervous, endocrine, digestive, kidney, skin, lungs), the blood and interstitial fluid that form the internal environment for the cell remain relatively constant despite wide fluctuations in the components of the external environment, including food and water intake.**

Figure 1-1 should be studied carefully. It shows the circulatory system acting as a mediator between the delicate cell and the harsh external environment. The circulating blood must be under constant regulation through the activities of the lungs, skin, kidneys, digestive organs, and neurohumoral secretions. Thus, provided that a cell is protected and nourished, and the waste products removed, it should survive and perhaps even function. The science of tissue culture, or the growth and reproduction of cells outside the organism, has enabled us to determine to a large extent the basic conditions necessary to the life of a cell. This permits us to look at the various systems of the body from a somewhat different perspective. Each system is enormously complicated, it is true, but if we keep in mind that each one functions properly only insofar as it cooperates with the others in maintaining the proper environment for the cells, then we can usually deduce what changes will occur in the human body (the patient) when the normal function (*physiology*) is interfered with as a result of disease, malfunction, or removal of a part (*pathology*).

This leads us logically to the way in which the constancy of the components of the immediate environment of the cell is maintained. This is called *homeostasis* and differs from the concept of the constant internal environment only in that the latter describes a condition, the former the active processes by which this condition is achieved. It is particularly important not to think of homeostasis as the maintenance of

Figure 1-2. **Normal fluctuation in blood glucose levels in a healthy individual.**

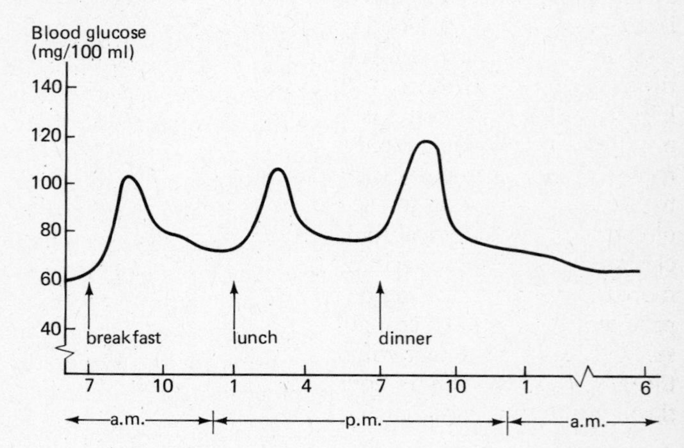

an unchanging, or static, state. Biology is life; life is change and the ability to adapt. Biological systems show this to a marked degree. The limiting factor is the extent of the range within which life can survive. This is the *physiological range,* and only variations that stray beyond these realms are incompatible with life. For example, average blood sugar level is usually reported to be 80 mg glucose/100 ml blood. Before breakfast, however, the level may be down to 60 mg/100 ml; immediately after a hearty meal it may shoot up to 120 mg. Neural and endocrine controlling mechanisms in the healthy individual soon return blood sugar levels to the norm. These fluctuations are physiologically tolerable, for they set into motion the homeostatic (perhaps *homeodynamic* would be the better word) regulations that adjust the circulating level of glucose to optimal (Fig. 1-2).

We shall be studying the mechanisms by which the controlling systems are triggered and the ways in which they effect their actions, from the molecular level to that of the integrated animal. A great deal of information, not yet properly correlated, may be obtained from *comparative studies* of different animal groups to show the evolution of homeostatic controls, the development of these controls in the individual (*ontogeny*), and the gradual failure of homeostatic mechanisms with illness (*pathology*) and age (*gerontology*). Just as the geological stages of the earth, or the evolutionary stages of life, are continuous and continuing processes, so embryology, physiology, pathology, and gerontology flow into one another in the life and death of the individual.

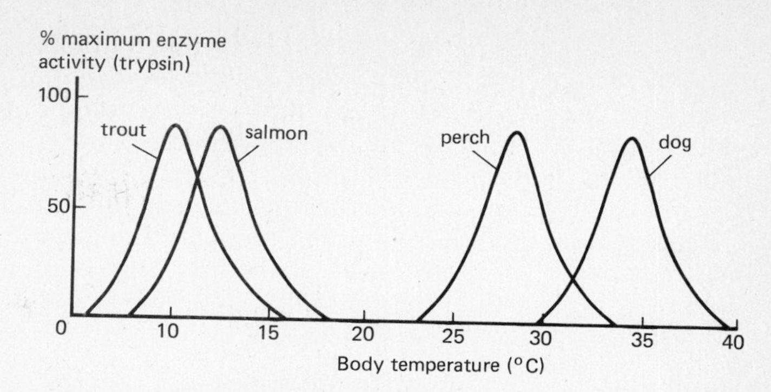

Figure 1-3. **The optimum temperature for the enzyme trypsin corresponds to the body temperature of the organism. Trout and salmon are cold-water fish, whereas perch are warm-water fish.**

1–2 Physiological Control Systems

Adaptation

Living systems are constantly adjusting and reacting to changes in the environment. These changes are *stimuli,* and the reactions of the organism are its *responses.* We must differentiate between the immediate response of an organism and the effects of long-lasting environmental changes. For example, an increase in blood pressure in response to a painful stimulus, and its return to the normal steady state as a result of homeostatic mechanisms, is a brief and short-lived phenomenon. If blood pressure rises because of exercise, and the exercise is maintained for a certain length of time, the blood pressure, too, will be held at a new and higher, steady state, better adapted to supply the exercising muscles with blood at a rapid rate under high pressure. This means that the homeostatic mechanisms have been "reset" to facilitate adaptation to the new circumstances. There is, of course, a physiological limit to this. The exercising muscles ultimately produce acid metabolites, such as lactic acid, more rapidly than they can be buffered or oxidized, and this produces permeability changes in muscle cells, causing the soreness and fatigue that will stop the strenuous exercise. If the animal or human is constantly subjected to exercise through training, basic changes in the biochemistry of the muscle occur; there is an increase in the number of open capillaries supplying the muscle, consequently an increase in blood supply, contractile proteins, and the oxygen-storing protein, myoglobin. The muscle has adapted to the new environment by *hypertrophy,* or increase in size of the individual cells. The opposite response occurs with disuse; the muscle *atrophies* as the individual cells lose their contractile protein and myoglobin content.

This individual adaptation of the tissues is of no evolutionary significance unless there is a change in the genes of the germ cells (ova or sperm). Immediate tissue adaptation has to utilize the available biochemical machinery of the cells, whereas in evolutionary adaptation the biochemical material itself changes, due to the production of new or different enzymes, which will then direct cell metabolism along new metabolic pathways. An interesting example of this can be seen in the great difference found between the temperature optima of enzymes from different species corresponding to the body temperature of the organism (Fig. 1-3).

Ontogeny of Regulatory Processes

When and how do the physiological mechanisms that regulate homeostasis occur in the individual? Although satisfactory evidence from the fetus is hard to come by, careful studies on newborn and very young mammals indicate that most systems mature at different rates. Many enzymes are missing or nonfunctional at birth, so that physiological regulation is only gradually attained by the developing infant. In some cases, the difference in metabolism between the neonate and the adult represents a beneficial adaptation to the prenatal environment: the fetus gets oxygen through the placenta at levels that are roughly equivalent to the oxygen content of maternal venous blood. This relatively anaerobic environment is adequate for the developing fetus, which functions largely on anaerobic metabolic pathways. Fetal hemoglobin also has a greater affinity for oxygen than adult hemoglobin, which permits it to take up large volumes of oxygen at low oxygen pressures; consequently, newborn mammals can survive without oxygen longer than the adult.

This resistance to anoxia is particularly evident in the brain and is probably a protective mechanism against the asphyxia of birth. Circulatory and respiratory adjustments must take place rapidly and are quite adequate by the first 2 weeks of life for the human, but the delicate regulatory mechanisms of the kidney take much longer to perfect, probably as long as 2 years. Statistics on infant mortality show the highest incidence of death occurs in the first 24 hr after birth.

Feedback Systems

Negative Feedback Systems

To maintain a biological system in a steady state, there must be an integrative system that can measure and compare the input with the output, and then be able to do something about any discrepancy. One of the most important integrative systems is a small area of the brain, the hypothalamus, which receives both endocrine and neural input and responds by both secretory and neural output. If the temperature of the blood circulating near the hypothalamus is increased one or two degrees, certain hypothalamic cells become active, as shown by an increase in their firing rate from 7 to 15 impulses/sec. Through the connections of these cells with many neural circuits, a whole chain of events is initiated, resulting in increased circulation through the skin and sweat glands and a consequent loss of heat by the body to the environment. The cooled blood circulating to the hypothalamus slows down the activity of the neurons there, stopping the process when the blood is the "right" temperature (37.1°C). These temperature-sensitive neurons respond to slight differences in temperature, adjusting their activity accordingly and acting like a precision thermostat with a specific set point. Here the output is regulated by the input, but the input in turn is inversely affected by the output. In other words, the more effective the neurons are in lowering the temperature of the blood, the more rapidly will their activity be decreased. This *negative feedback* tends to restore the original physiological state.

On-Off Control Systems. In an engineering control system, such as the thermostat that controls room temperature, the temperature is the *controlled variable,* which fluctuates within a relatively narrow range. An *error detector* compares the actual room temperature with the desired temperature, the *set point.* The error detector also activates a control mechanism that corrects any deviation in the room temperature. For example, when the room temperature drops, a switch turns the furnace on, which raises the temperature of the room. This is a typical negative feedback system, with a *closed loop control,* and it is diagrammatically illustrated in Fig. 1-4.

This type of *on-off control system* inevitably has oscillations in the temperature, for it depends upon the furnace's being turned on or off, without any intermediate adjustments. There are more complex control systems that are able to provide a relatively constant output of the controlled variable.

Proportional Control Systems. Proportional control systems adjust the output in proportion to the degree of fluctuation of the controlled variable. A physiological example is seen in the relationship between respiratory rate and depth (ventilation) and the carbon dioxide content of the blood. If the set point for the carbon dioxide content of arterial blood at rest is $P_{CO_2} = 40$ mm Hg, then if the P_{CO_2} is increased by strenuous exercise, the respiratory centers in the central nervous system are stimulated and ventilation increases. The increased ventilation is directly proportional to the increased P_{CO_2}, lowering the blood carbon dioxide accordingly.

This simple proportional system will still have minor errors that can be corrected by an *integral control system.* In an integral control system, the *rate of change of the output* is proportional to the input.

Regulatory systems in physiology are really much more complex than any of these controls described, for physiological set points can be varied according to diurnal cycles and metabolic state. There are many inputs feeding information into the system. In addition, physiological regulation depends to a large extent on simultaneous stimulation and inhibition.

The many complexities and wonders of regulatory systems will be described throughout this text. The development of new experimental techniques has shown that our earlier explanations of physiological regulation may have been too simplistic. We are still very far from unravelling all the intricacies that simultaneously adjust physiological responses to the continuous changes of the internal and external environments.

Figure 1-4. **Components of a negative feedback control system represented by a room thermostat.**

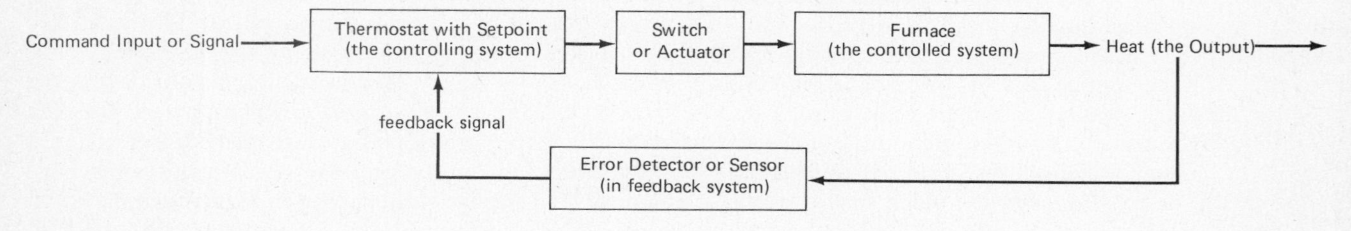

Positive feedback systems

In some systems, the greater the output, the more the controlling system is activated. This *positive feedback* mechanism does not happen very often in biological systems. In mammals, the best examples are shown by behavioral reflexes, where food gathering and body grooming evoke pleasurable sensations that reinforce the original stimulus and act to perpetuate the feeding or cleaning behavior. An example on a cellular level is the series of events that occurs when a nerve membrane is stimulated. The stimulus causes a change in membrane permeability, which results in an influx of sodium, partially depolarizing the membrane. This increases its permeability to sodium still further and permits more sodium to flow inward until the peak of the impulse is reached.

Feedback mechanisms are seen for neural, hormonal, and enzymatic controls. Sometimes they work directly on the central stimulating mechanism, sometimes they involve many components in a linked chain, but biological efficiency requires that each control itself be controlled. It is when these mechanisms no longer function that we appreciate their significance. In aging individuals, there are sharp shifts in the relative amounts of hormones secreted, with consequent impairment of the functions these hormones regulate. As the nervous system is profoundly affected by hormone levels, the neural regulatory mechanisms become desynchronized. The aging process is a gradual one, varying for different parameters in different individuals, and it has been suggested that the nucleoproteins of the genes themselves may be altered with age, thus affecting enzyme synthesis and enzymatic direction of homeostatic mechanisms.

Cited Reference

Bernard, C. *Leçons sur les Phénomènes de la Vie Commune aux Animaux et aux Vegetaux,* Vol. 1. Ballière, Paris, 1873.

Additional Readings

BOOKS

Adolph, E. F. *Origins of Physiological Regulations.* Academic Press, Inc., New York, 1968.

Adolph, E. F., ed. *The Development of Homeostasis.* Academic Press, Inc., New York, 1960.

Jones, R. W. *Principles of Biological Regulation; An Introduction to Feedback Systems.* Academic Press, Inc., New York, 1973.

Milhorn, H. T., Jr. *The Application of Control Theory to Physiological Systems.* W. B. Saunders Company, Philadelphia, 1966.

Wolstenholme, G. E. W., and J. Knight, eds. *Homeostatic Regulators.* J. A. Churchill Ltd., London, 1969.

Yamamoto, W. S., and J. R. Brobeck, eds. *Physiological Controls and Regulations.* W. B. Saunders Company, Philadelphia, 1965.

ARTICLES

Machin, K. E. "Feedback Theory and Its Application to Biological Systems." In *Homeostasis and Feedback Mechanisms. Symp. Soc. Exp. Biol. 18.* Academic Press, Inc., New York, 1964.

Tustin, A. "Feedback." *Sci. Am.,* Sept. 1952.

Chapter 2

The Living Cell

"Faith" is a fine invention
When Gentlemen can see—
But Microscopes are prudent
In an Emergency.

Emily Dickinson

SINCE THE EARLIEST description of the unit of living matter as a "cell" or cavity circumscribed by a wall, like the cells of a honeycomb, much biological endeavor has been directed toward determining the nature of such a unit and the factors controlling its growth and behavior. One of the most important discoveries was that the "cavity" is no empty space; on the contrary, it is filled with a fluid of variable viscosity, called *cytoplasm*. The circumscribing wall is also of vital importance because in animal cells it consists of an invisible, living membrane, the *plasma membrane*, that controls the rate and type of substance getting into and leaving the cell. In plant cells the cell membrane is covered by a thicker, protective coat, which is the microscopically visible *wall*. Animal cells do not have a cell wall, and the delicate plasma membrane is protected only by a thin layer of cementing material that binds neighboring cells together. Some eggs are protected by a layer of jelly or even by a shell, but these may be regarded as specializations.

Cytoplasm is an unbelievably complex structure. One can analyze its physical and chemical constituents, but the intricacies of its organization cannot be duplicated yet in a laboratory. This indicates that the properties of cellular life are intimately bound to the organization of the cell's molecular structure rather than to the mere presence of the various molecules. One of the most fundamental discoveries of modern biology is that basic physiological processes are common to both the plant and animal kingdoms. The differences in the molecular organization of the chromosomes, which control the inheritance and functioning of the cell, result in the evolution of the infinite number of variations, found not only between widely different animal groups but even between individuals of the same species.

The chromosomes in most forms are found within a specialized body inside the cell, the *nucleus*. Plant and animal cells with true nuclei are called *eukaryotic cells*. Bacteria and blue-green algae have their nuclear material scattered throughout the cytoplasm; those cells that lack a nuclear membrane are *prokaryotic cells*. If the nucleus of a eukaryotic cell is removed, or if the chromosomes are damaged or destroyed, the entire cell disintegrates and dies. Thus the life of the cell depends on the functional integration of the nucleus and the cytoplasm surrounding it. It can be immediately appreciated that the distribution of this controlling material, the chromosomes, to new cells formed is of vital importance to ensure the continuation of the same type of molecular and structural organization of the daughter cells as was characteristic of the mother cell. The *cell theory* asserts that all cells come from pre-existing cells, and consequently that the cell is a vital intermediate in the perpetuation of life.

2–1 Methods of Investigating Cell Structure and Function

Our ideas of cell structure and function depend very much on our senses and the tools we use to magnify their sensitivity. The cell appears different to the cell physiologist, the cytologist, the biochemist, and the molecular biologist. Some of the methods used to study the cell are outlined in the following discussion.

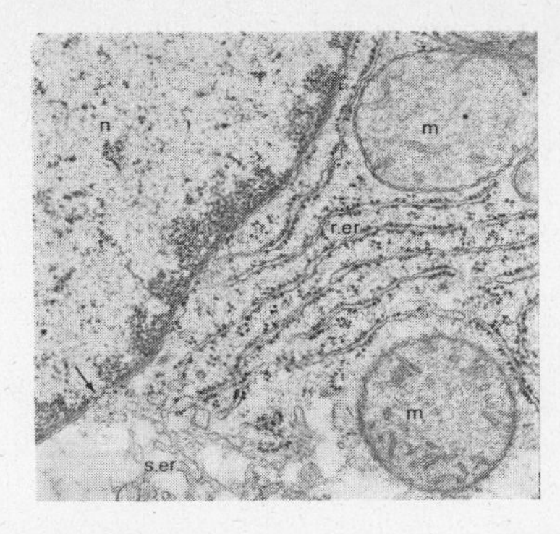

Figure 2-1B. **Electronmicrograph of part of a liver cell, showing some of the structures seen in Fig. 2-1A. n = nucleus, m = mitochondria, r. er = rough endoplasmic reticulum, s. er = smooth endoplasmic reticulum. The arrow indicates a pore in the nuclear membrane. Dark clumps on either side of the pore are chromatin. ×20,000. (Courtesy of J. Gennaro, Laboratory of Cellular Biology, New York University.)**

Microscopy

The magnification of an object under the *electron microscope* may be 100 times as great as under the *light microscope,* but magnification is not the only criterion for obtaining a satisfactory enlarged image. Unless the lens can distinguish minute details, merely increasing the size of the image gives us no more information. The ability of a lens to distinguish detail is its *resolution.* The maximum resolution of the unaided eye is about 0.1 mm. That means that two points, 0.1 mm apart, can just be distinguished as distinct entities. If the distance between them is decreased, the eye perceives them as one point. The resolution of the light microscope is about 0.2 μm; that of the electron microscope is 3 to 5 Å, or 0.3 to 0.5 nm. (See the Appendix for measurements.)

Figure 2-1A. **A diagram of a generalized cell.**

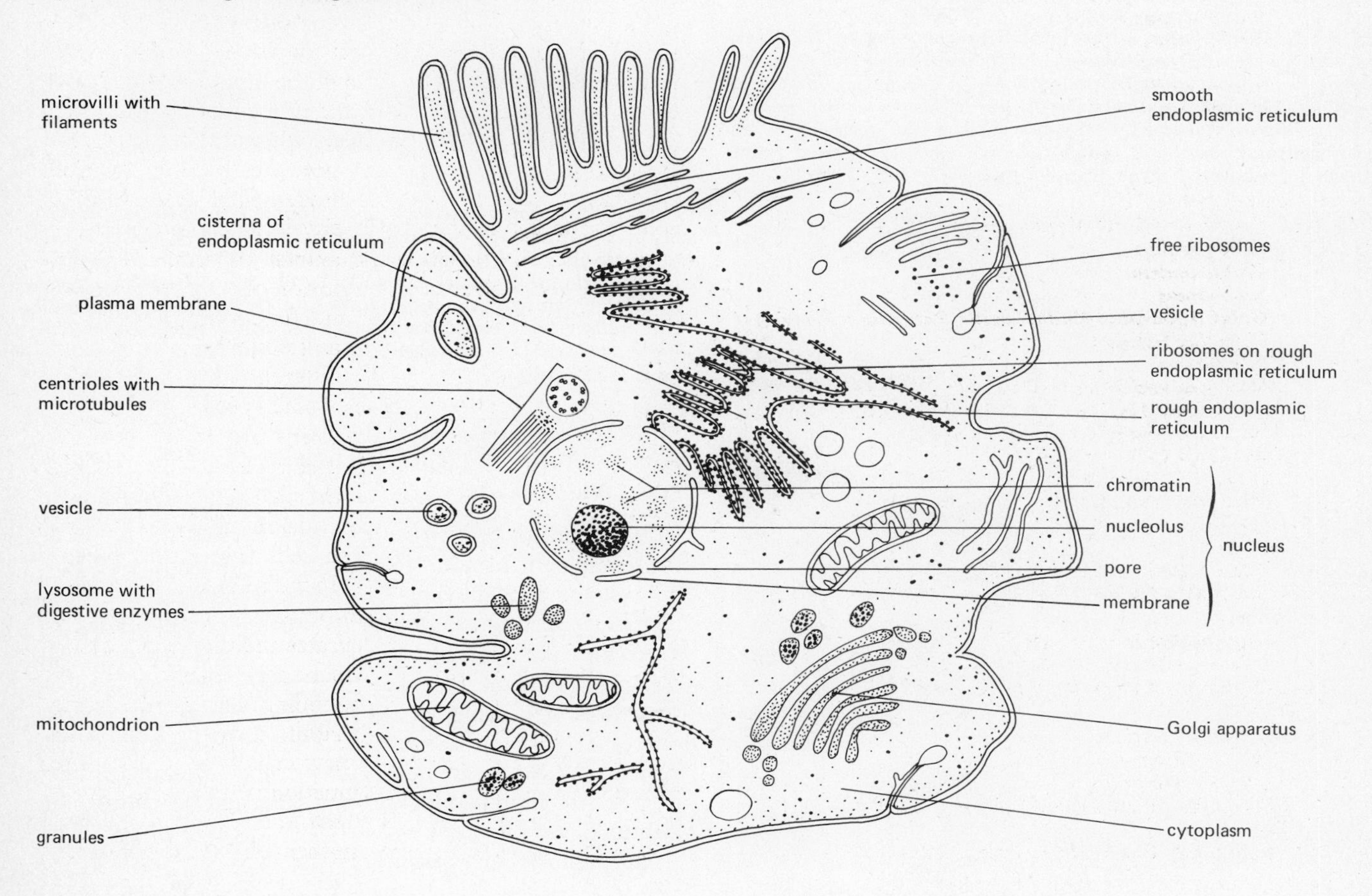

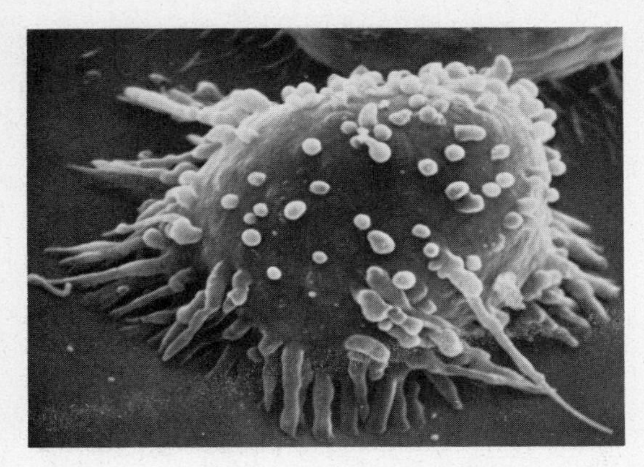

Figure 2-1C. **This scanning electronmicrograph shows the three-dimensional structure of an epidermal (skin) cell in culture. The long processes growing out from this embryonic cell attach it to the glass coverslip on which it is being cultured. ×1200. (Courtesy of J. M. LeBlanc, Laboratory of Cellular Biology, New York University.)**

When the *scanning electron microscope* is used, a three-dimensional view of the cell is obtained. This technique has been used for relatively few types of cells because of technical problems, but the three-dimensional structure is beautifully explicit. Figures 2-1A, B, and C show the cell as seen diagrammatically and under transmission electron and scanning electron microscopes.

Figure 2-2. **Cells from rat peritoneal fluid as seen by Nomarski differential-interference-contrast optics. m = mast cell, e = eosinophilic leukocyte, mp = macrophages, n = nuclei, and v = vacuoles. The nuclei and vacuoles appear as "depressions," in contrast to highly refractive structures such as the granules in the mast cell and eosinophil, which appear in marked raised relief. [From J. Padawer, *J. Roy. Microsc. Soc.* 88(3): 305 (1968).]**

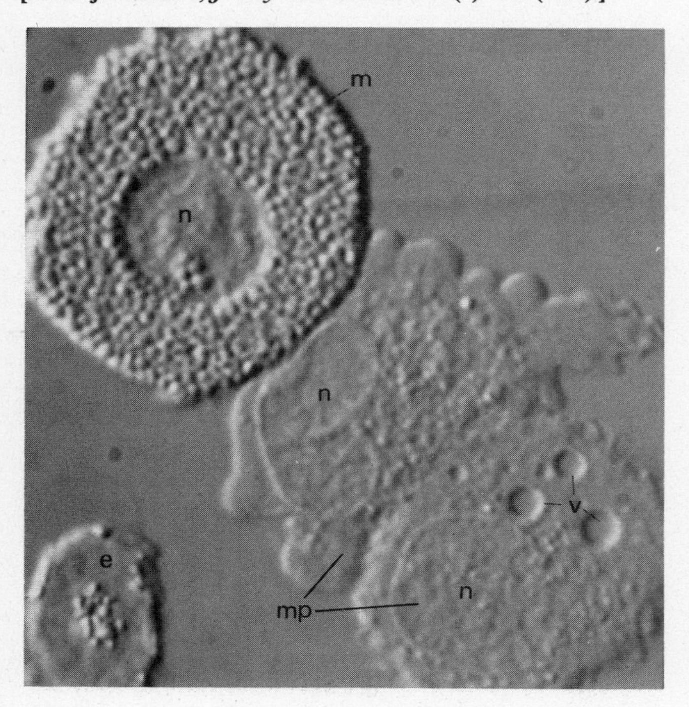

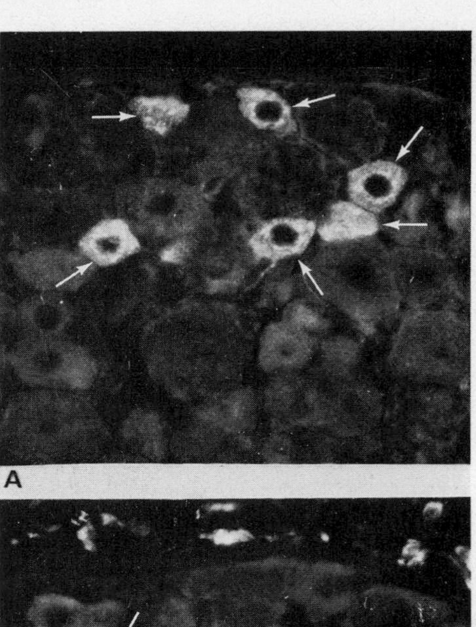

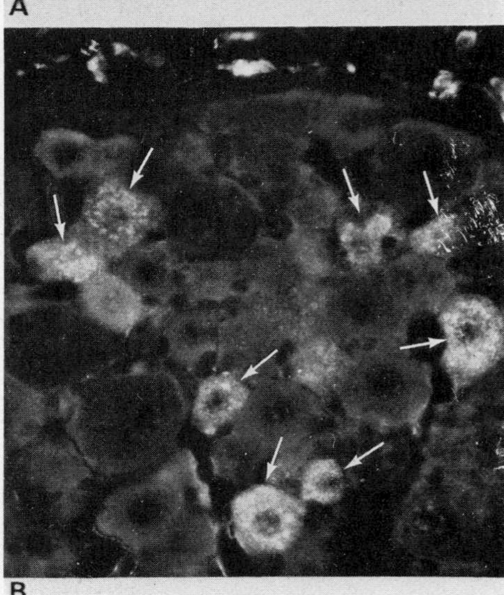

Figure 2-3. **Immunofluorescence techniques permit the localization of the small peptides, somatostatin and substance P, in nerve cells of spinal ganglia. (A) Following incubation of the cells with antiserum to somatostatin, several small cell bodies (arrows) fluoresce, indicating that they contain somatostatin. (B) Following incubation with antiserum to substance P, other cell bodies fluoresce, indicating the presence of substance P. [From T. Hökfelt et al., *Neuroscience* 1: 131 (1976).]**

Adaptations of Light Microscopy for Living Cells

A number of techniques has been devised for rendering the transparent structures within cells visible under the light microscope and allowing the study of living cells. These include *phase contrast, dark field, interference,* and *Nomarski differential interference microscopy*. These techniques in general rely upon the fact that light passes through different structures at varying speeds (i.e., that the structures have different indices of refraction) so that objects either scatter light (dark field) or are outlined by the resulting interference between the light passing through them and the background illumination (Fig. 2-2).

Polarization microscopy uses the fact that the speed of

polarized light passing through some objects varies with the direction of polarization; this is especially useful for studying filamentous structures such as the mitotic apparatus.

Fluorescence microscopy involves the irradiation of a cell with ultraviolet light. If the cell contains a substance that can absorb this light and emit it as light at a different wavelength, this is seen as a fluorescent emission. The fluorescent substance may be normally present within the cell, such as norepinephrine within certain neurons (Fig. 2-3), or it may be artificially introduced into the cell as a marker. An example of the latter is the administration of antibodies tagged with a fluorescent substance: the fluorescent antibodies will be visible wherever there are cell antigens.

USE OF STAINS TO STUDY CELLS UNDER LIGHT MICROSCOPY

The transparent cellular structures can be made visible by *staining* with dyes. The cells are usually killed (*fixed*) first, and then exposed to one or more dyes. When the dye used is specific for the particular cell structure, additional information can be obtained. A number of dyes can be classified as acid or basic, and thus they will differentially color either the *nucleus* (which will react with basic dyes such as hemotoxylin) or the *cytoplasm* (which generally stains with a more acid dye, such as eosin).

Using specific stains for different structures (*histochemistry*), a great deal of information can be obtained about the composition of cell organelles and the distribution of enzymes. With special techniques, proteins, fats, glycogen, and nucleic acids can be identified and localized in the cell.

PREPARATION OF CELLS FOR ELECTRON MICROSCOPY

Materials are usually prepared for viewing under the electron microscope by techniques very similar to those used to prepare killed materials for light microscopy. The stains used for electron microscopy are usually *heavy metals,* such as lead, osmium, or uranium, which scatter electrons well. This use of electron microscopy is called *transmission electron microscopy*.

Cell Replicas. Sometimes cells are coated with a metal film, called a *replica,* which reproduces their surface contours. This replica can then be seen under the electron microscope, giving a three-dimensional picture of the original surface. In one variation of this technique, the specimen is rapidly frozen and then shattered by a knife along natural cleavage planes. A replica is then made of the exposed surface. This technique is called *freeze-fracture*.

The breaks in the freeze-fractured preparation tend to follow membranes, peeling them into two halves along the center of the lipid bilayer and exposing the protein-containing structures within the membrane. These structures appear as bumps, or *particles*. This technique provides a way to see possible pathways through which substances penetrate the membrane. This is explained in more detail in Chapter 4 and illustrated in Fig. 4-20.

In *freeze-etch* microscopy, a thin film of platinum and carbon is evaporated onto the surface of the fracture, which can then be examined by electron microscopy.

SCANNING ELECTRON MICROSCOPY

Scanning electron microscopy utilizes a narrow beam of illuminating electrons that scans back and forth across the specimen; the resulting image is viewed on a television screen. Replicas are usually made so that the three dimensional surface structure of large, complex specimens can be studied (Fig. 2-1C).

Autoradiography

Autoradiography is one of the most useful techniques for studying the localization of biochemical processes within the cell. If a cell or a tissue takes up a substance preferentially, and that substance can be made radioactive, then this radioactivity can be localized if a section of the cell or tissue is placed in contact with a photographic emulsion. The emulsion contains silver bromide granules, which are sensitized by the radiation and so produce an image corresponding to the site of radioactive atoms in the cell. For example, if radioactive iodine, ^{131}I, is administered to a rat, it will be metabolized in the same manner as the nonradioactive isotope and will be concentrated preferentially in the thyroid gland. If thin sections of the gland are prepared at different time intervals, the changes in the radioactive areas show the movement of the ^{131}I into the space in the center of the ring of thyroid cells (follicle) in which the thyroid hormone is stored (Figs. 2-4A and B).

Figure 2-4A. **In this light microscope radioautograph of a thyroid gland of a rat treated with radioactive iodine (^{125}I), the small dark dots represent silver grains darkened by the radiation. These dots are localized throughout the colloid (col) in the spaces or follicles surrounded by the thyroid cells (c). ×200. (From C. Simon and B. Droz, in *Current Topics in Thyroid Research,* C. Cassano and M. Andreoli, eds., Academic Press, Inc., New York, 1965.)**

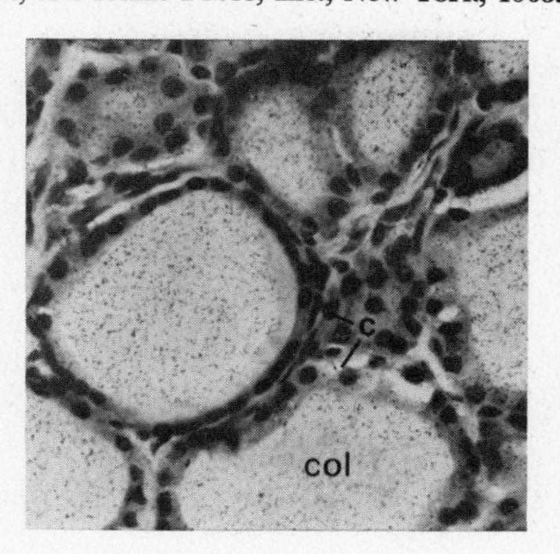

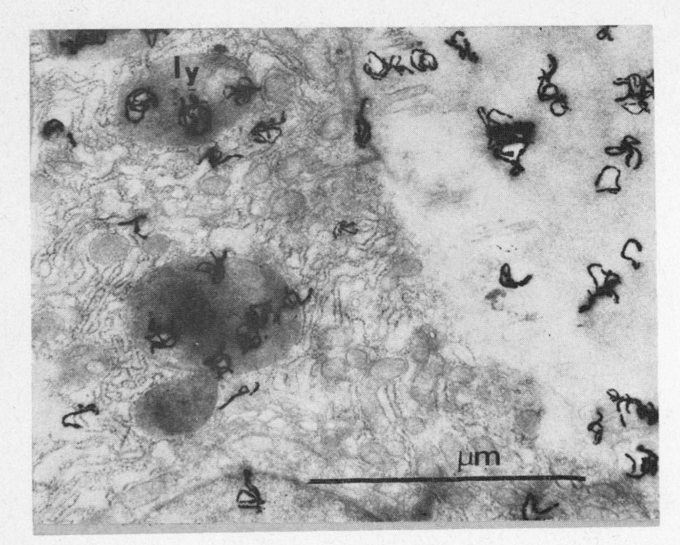

Figure 2-4B. **Electron microscope radioautograph of a thyroid follicle prepared from a rat that had been given ^{125}I in its drinking water for 55 days. In this figure, as in Fig. 2-4A, the amount of radioactivity in the thyroid slice is directly proportional to the amount of iodine in that area. Most of the silver grains are in the follicular space (light area on the right), where the thyroid hormones are stored as colloid. Those grains seen in the cells are restricted to the lysosomes (ly), where the iodine-rich hormones are prepared for secretion into the follicles. [From C. Simon and B. Droz, *J. Physiol. (Paris)*, 66: 65 (1973), Masson, S.A., Paris.]**

Table 2-1. **Eagle's Medium**

	Milligrams per 1000 ml
L-Arginine	105
L-Cystine	24
L-Histidine	31
L-Isoleucine	52
L-Leucine	52
L-Lysine	58
L-Methionine	15
L-Phenylalanine	32
L-Threonine	48
L-Tryptophan	10
L-Tyrosine	36
L-Valine	46
L-Glutamine	292
Choline	1
Nicotinic acid	1
Pantothenic acid	1
Pyridoxal	1
Riboflavine	0·1
Thiamine	1
i-Inositol	2
Folic acid	1
Glucose	2000
Phenol red	20
Penicillin	0.5
NaCl	8000
KCl	400
$CaCl_2$	140
$MgSO_4 \cdot 7H_2O$	100
$MgCl_2 \cdot 6H_2O$	100
$Na_2HPO_4 \cdot 2H_2O$	60
KH_2PO_4	60
$NaHCO_3$	350

Tissue Culture

If the *milieu interne* of the body can be adequately reproduced, the cell survives, grows, and reproduces in vitro. This technique of tissue culture was initiated by Harrison (1907), who took pieces of undifferentiated frog embryo and placed them in sterile clots of frog lymph. Not only did the cells survive but they even started the development of nerve fibers, showing that specialization can occur in tissue culture. Ideally, the medium for tissue culture should be chemically defined, but this has not been completely achieved. Animal cells require a natural medium or a synthetic one supplemented with a natural product, such as serum. The complexity of Eagle's medium is shown in Table 2-1. Even so, it is necessary to add serum to it, to ensure cell viability.

Isolation of Cell Organelles

In order to have large enough amounts of cellular organelles to study and analyze, the organelles must be separated from the cell and remain relatively undamaged in the process. It is usually necessary to destroy the plasma membrane first by *homogenization* in a blender. The force used must be enough to rupture the cell with as little damage to the organelles as possible. To prevent inactivation of enzymes, both the tissue and the medium are kept cold during this process. The homogenate now consists of a suspension of cellular components of different size and density. Through the use of *differential centrifugation,* in which different speeds of centrifugation separate the components, or through *density gradient centrifugation,* in which the medium used has a pre-established density gradient, a fairly sharp separation of the cell elements can be obtained.

2-2 The Fundamental Cell

The cell that can be best described in general terms is probably the embryonic cell, because it has not yet become specialized, or differentiated, and its general structure is fairly representative of the basic plan of all animal cells (Fig. 2-1A). The embryonic cell divides much more rapidly than adult cells do and also possesses the remarkable potentiality for developing into many other kinds of cells, such as those of muscle, blood, or connective tissue.

Cell Membranes

All cells are surrounded by a membrane, the *plasma membrane,* which is capable of invaginations into the cell, and evaginations to the exterior. This gives the plasma membrane a contiguity with the intracellular membranes, which form a complex network throughout the cytoplasm. This intracellular membranous system is the *endoplasmic reticulum* (ER), a system of extraordinary importance for the functioning of the cell. Under the electron microscope, the structure of the plasma membrane and of the intracellular membranes appears as two thin, dark lines, separated by a clear space (Fig. 2-5). This trilaminar structure has been called the *unit membrane* by Robertson (1964) and has an average thickness of about 7.5 to 10 nm, although the thickness varies considerably. The ER is modified in certain regions within the cell by a covering of granules, *ribosomes,* on the outer layer of the unit membrane. Because of the granular thickening, this portion of the ER is called the *rough ER* and is associated with protein synthesis; the nongranular regions are referred to as the *smooth ER,* which acts as a transport and storage system. The endoplasmic reticulum divides the cytoplasm into compartments so that the cytoplasm cannot be considered a homogeneous medium. Rather, it is a series of highly organized areas in which specific chemical reactions can take place—either separated physically, or spatially organized to co-ordinate with one another, perhaps on or in the membrane itself.

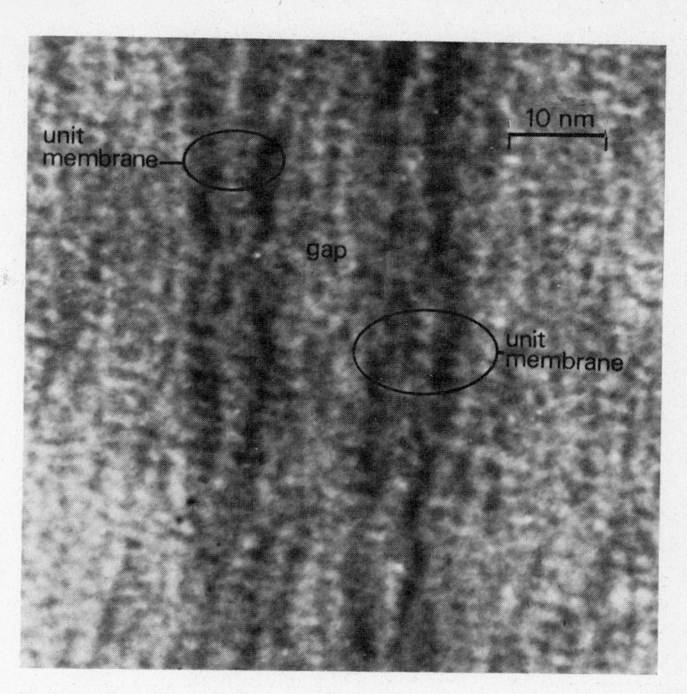

Figure 2-5. **Electronmicrograph of the boundary between two smooth muscle cells. Each cell membrane consists of a double layer that forms the unit membrane. ×500,000. (From J. B. Robertson, in *Molecular Biology,* D. Nachmansohn, ed., Academic Press, Inc., New York, 1960.)**

Functions of cell membranes

The study of cell membranes is one of the most active fields in biology at the moment, for cell membranes are considered to be involved in many disparate phenomena: the distribution of ions and solutes between the cell and its environment; the rate at which substances enter or leave the cell and consequently the regulation of cellular metabolism; bioelectric potentials; immunochemical reactions; and cell recognition, the way in which cells recognize foreign bodies or cells. Contact inhibition, a most important mechanism by which cell division and growth are normally inhibited when similar cell membranes come in contact, appears to be a membrane function that goes awry when cancerous cells multiply uncontrollably.

During certain stages of cell development, the plasma membrane, the ER, the membranes surrounding the cell nucleus, and the membranes of many of the cell organelles, such as the mitochondria, appear to be continuous. This continuity suggests that they may all have a common origin, possibly in the invaginations of the plasma membrane. It does not, however, follow that all cell membranes are structurally or functionally the same. In fact, many of the important specializations of cells and cell organelles can be attributed to modifications of the unit membrane. These vital modifications will be discussed in detail in Chapter 4.

Cell Organelles

Cell organelles include mitochondria, ribosomes, the Golgi-endoplasmic reticulum-lysosome complex, granules, vacuoles, and structures associated with cell division (not always present in the nondividing cell).

Mitochondria

These organelles look like granules under the light microscope, but under the electron microscope they are seen as highly organized structures (Fig. 2-6). Their size is variable according to cell type, averaging about 0.5 μm wide and 3 μm in length. The more active a cell, the greater the number of mitochondria it is likely to possess. A mitochondrion is basically a fluid-filled vessel with an involuted wall, consisting of a *double unit membrane,* each about 6 nm thick. The inner membrane is deeply folded into *cristae* and covered with small spheres (Fig. 2-7). The cristae or crests increase the surface area enormously, and the complexity of these infoldings varies in different cells. Some are simple folds; others are so numerous they look like stacks of coins.

The mitochondria are referred to as the powerhouse of the cell, for they are capable of extracting energy from the chemical bonds of cell nutrients by oxidation. This energy is then stored in the form of high-energy phosphate bonds in a compound found in all cells, *adenosine triphosphate* (*ATP*). This compound, the universal energy carrier of the cell, will be discussed in more detail in Chapter 5. The enzymes that are responsible for these *oxidative phosphorylation* reactions are arranged in an orderly manner in the inner membrane of the mitochondrion. If the mitochondria are subjected to sonic oscillation, the inner membrane, together with its spheres, may be separated from the rest of the organelle, but these particles still retain their capacity for oxidative phosphorylation.

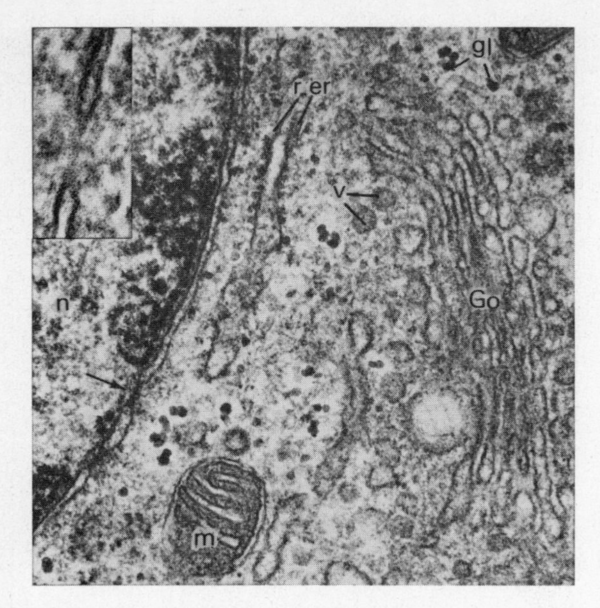

Figure 2-6. **Cell organelles and the nucleus of a liver cell as seen under the electron microscope. Go = Golgi apparatus, gl = glycogen granules, m = mitochondrion, and v = vesicles. ×25,000. Part of the nucleus (n) is shown on the left, and the arrow points to a nuclear pore. The insert shows the pore in the double nuclear membrane at a higher magnification (×40,000). (Courtesy of M. Yoder, New York University.)**

Ribosomes

These particles are considerably smaller than the mitochondria, and they are usually attached to the endoplasmic reticulum, forming the regions of rough ER (Figs. 2-6 and 2-10). They are composed of ribonucleic acid (RNA) in combination with protein. Each ribosome includes three kinds of RNA molecules of different size, which can be separated on the basis of the speed with which they sediment when subjected to centrifugation. (This speed is measured in Svedburg units or *s*, so that, e.g., the molecules can be identified as 70*s*, 50*s*, and 30*s*. The actual numbers vary according to the source of the ribosomes).

Figure 2-7. **Diagrammatic representation of the three-dimensional structure of a mitochondrion that has been cut longitudinally.**

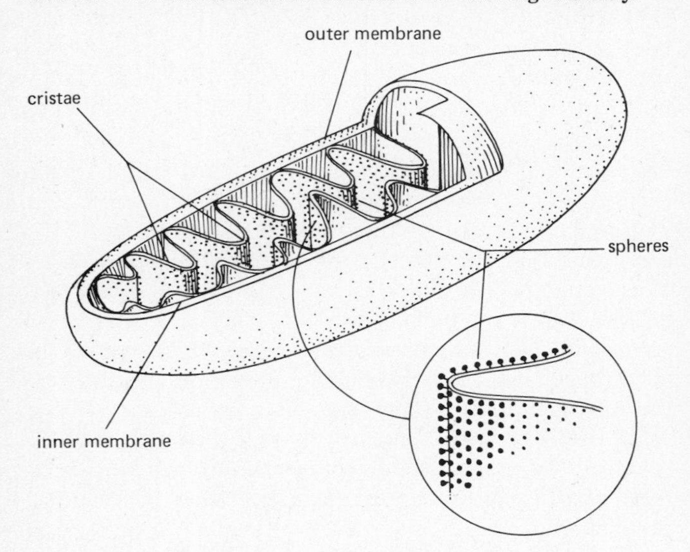

The ribosomes are the site of protein synthesis in the cell, and they receive their instructions from the molecular organization of the genetic material encoded in the genes of the chromosomes. The energy for the synthetic processes is obtained from the reactions occurring in the mitochondria. (See Chap. 5 for details on the synthesis of protein.)

Golgi Apparatus–Endoplasmic Reticulum–Lysosome Complex (GERL)

The Golgi apparatus and the lysosomes are closely associated with the ER and may be considered to form a functional unit (Fig. 2-8).

The *Golgi apparatus* appears to be a specialized region of the ER. It has no ribosomes but consists of smooth mem-

Figure 2-8. **Golgi apparatus-endoplasmic reticulum-lysosome complex (GERL). The parallel, smooth membranes of the Golgi apparatus (Go) can be seen to give rise to buds (b) that are freed as vesicles (v) filled with the substances (mostly lipids and polysaccharides), which are synthesized on the Golgi membranes. In close association is a mitochondrion (m), which provides the energy for synthesis, and the rough endoplasmic reticulum (r. er), where protein synthesis occurs. ×25,000. The insert shows a lysosome that is sometimes found in conjunction with this complex. (Golgi apparatus photograph courtesy of D. Rutherford, Laboratory of Cellular Biology, New York University. Lysosome insert courtesy of F. G. Bauman, Medical School, New York University.)**

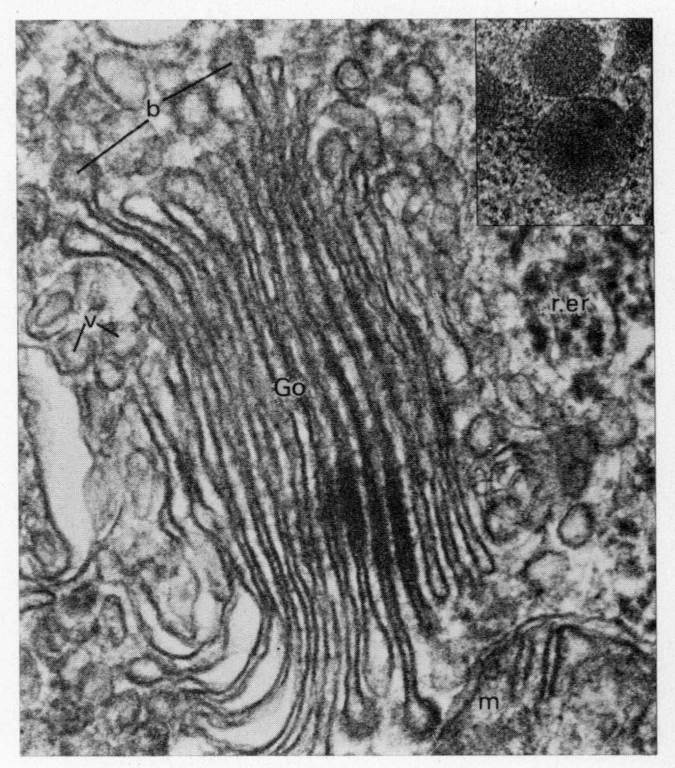

branes lying parallel to one another and enclosing sacs or cisternae, about 6 to 9 nm wide. Associated with these cisternae are vesicles and vacuoles, and the entire system seems to be an assembly area for the production and storage of different cellular secretions, particularly lipids and complex polysaccharides, and in some cells, proteins. After the discharge of its secretions, the Golgi apparatus shrinks considerably, showing a rhythmic pattern of activity. Strangely enough, the Golgi apparatus is sometimes found in cells that have no apparent secretory activity. Part of the function of the Golgi apparatus is the assembly of membranes that have specific characteristics, due to the type of proteins and polysaccharides incorporated in them. These specialized membranes are then transported to different parts of the cell, and it is inferred that the characteristics of the plasma membrane and the intracellular membranes may be dependent in part on membrane specialization in the Golgi apparatus.

The *lysosomes* are vesicles surrounded by a single membrane and containing *enzymes,* most of which are acid hydrolases. These hydrolases are released when the membrane bursts, permitting them to digest cellular structures and to act as "suicide bags." This is an oversimplification, however: the enzymes are released in an apparently controlled manner during the normal metabolism of the cell and are involved in the resorption of bone, thyroid hormone function, disposal of transient structures during embryonic development, and defense mechanisms against bacterial and viral infections.

Lysosomes are highly important organelles, the control of which is still very poorly understood. The release of such potent digestive enzymes must be meticulously regulated to prevent accidental cell injury or death. Part of the control appears to lie in the specificity of the lysosomal membrane and the membranes with which it can fuse, prior to releasing its enzymes. The Golgi apparatus produces these "surface-recognizing" membranes and probably packages the lysosomal hydrolases inside them. An interesting example of lysosomal control is seen in the manner by which the phagocytes, scavenger blood cells, destroy foreign particles without committing hara-kiri in the process. The phagocytes send out cytoplasmic processes or pseudopodia, which surround and engulf the bacterium, so that it lies within a pouch or phagosome. The lysosome of the phagocyte approaches, and its membrane fuses with the pouch membrane, permitting the direct liberation of lysosomal enzymes into the pouch. This results in the bacterium's being safely digested within a vesicle, analogous to the way in which higher animals digest their food in a stomach, without destroying their cells (Fig. 2-9).

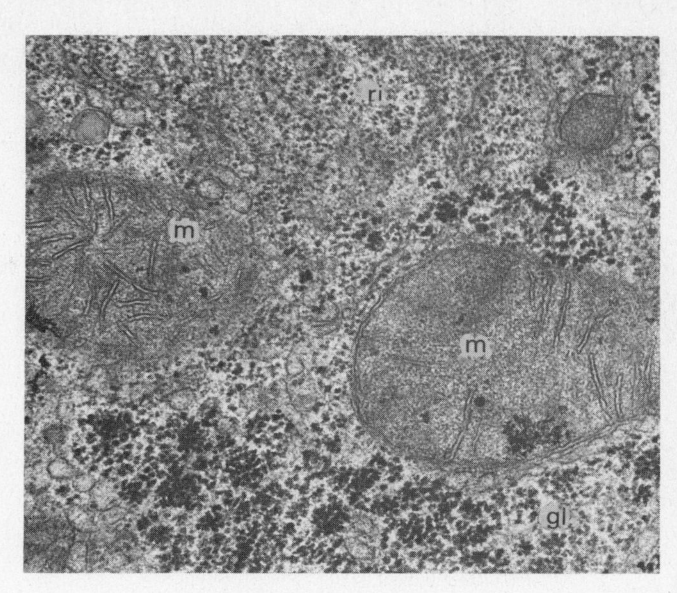

Figure 2-10. **The glycogen granules (gl) are the numerous dark inclusions in this liver cell. The smaller, lighter structures are the ribosomes (ri); mitochondria = m ×30,000. (From J. Rhodin, *An Atlas of Ultrastructure,* W. B. Saunders Company, Philadelphia, 1963.)**

Figure 2-9. **Lysosomal digestion of a bacterium. Fusion of the lysosomal and phagosomal membranes (C) empties the lysosomal digestive enzymes into the phagosome.**

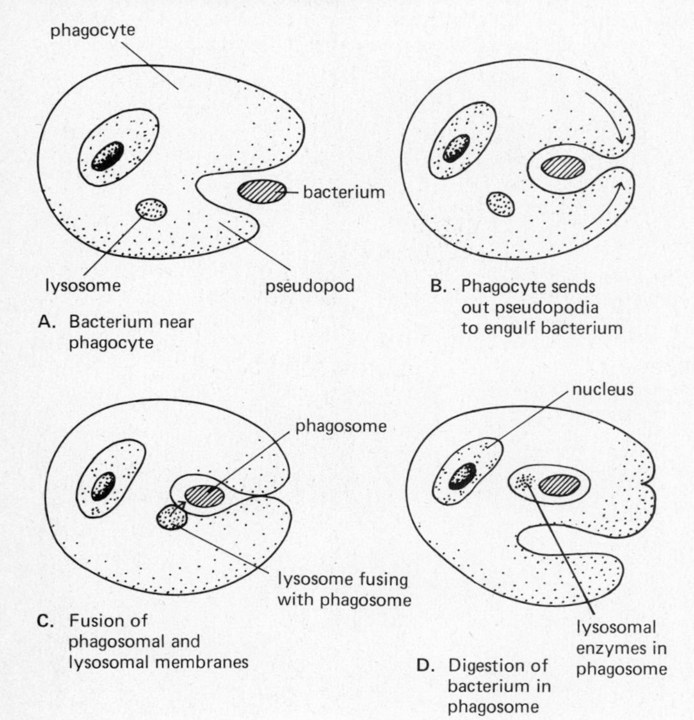

Granules

Various types of granules represent the stored secretion of many cells (Figs. 2-10 and 2-11). They may consist of protein (enzymes, hormones), carbohydrate (glycogen), fats (lipids), or even various kinds of pigment. In some cells, the pigment granules are melanin, a black-brown substance that gives the cell, and the organ of which it is part, its typical color. The skin of various races, for example, varies in color

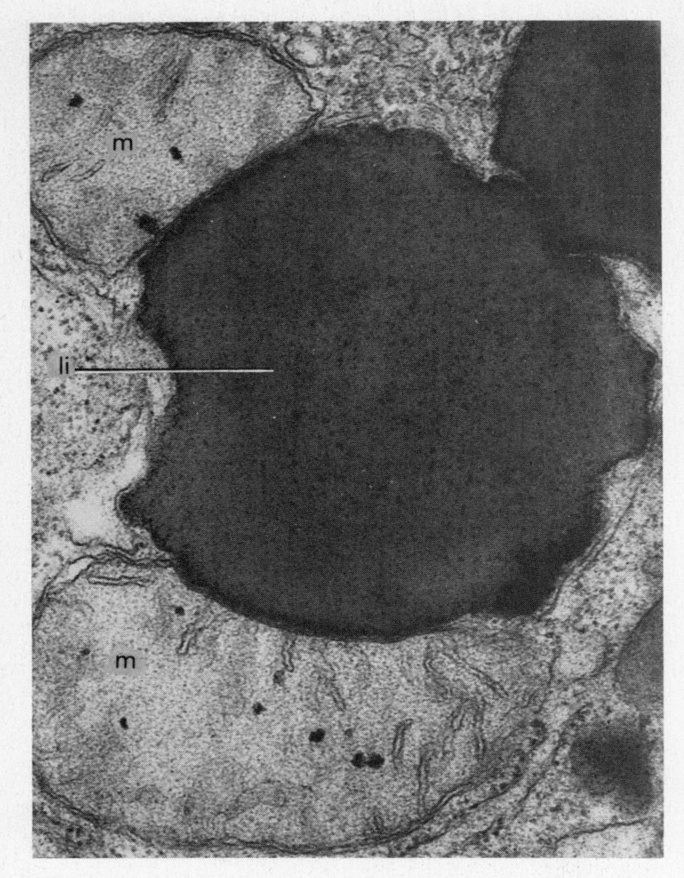

Figure 2-11. **Lipid granules (li) are also found in liver cells, usually close to the mitochondria (m). ×46,000. (From J. Rhodin, *An Atlas of Ultrastructure,* W. B. Saunders Company, Philadelphia, 1963.)**

in accordance with the concentration and distribution of melanin granules within certain layers of the epidermis.

Vacuoles and microbodies

Vacuoles represent regions of the cytoplasm, screened off by a membrane and containing the secretions of the cell, usually in a liquid or semiliquid form, like fat droplets. *Microbodies,* like lysosomes, are vesicles containing enzymes, especially oxidative ones. Because many microbodies contain oxidases that produce hydrogen peroxide, they are also called *peroxisomes.* Peroxisomes also contain the enzyme *catalase,* responsible for the breakdown of hydrogen peroxide to water.

Structures associated with cell division: centrioles, mitotic spindle, and microtubules

The centriole is a small, rod-shaped body near the nucleus (Fig. 2-1A), which divides to form two centers for spindle fiber formation during cell division. The centriole is made of microtubules, hollow, straight cylinders about 24 nm in

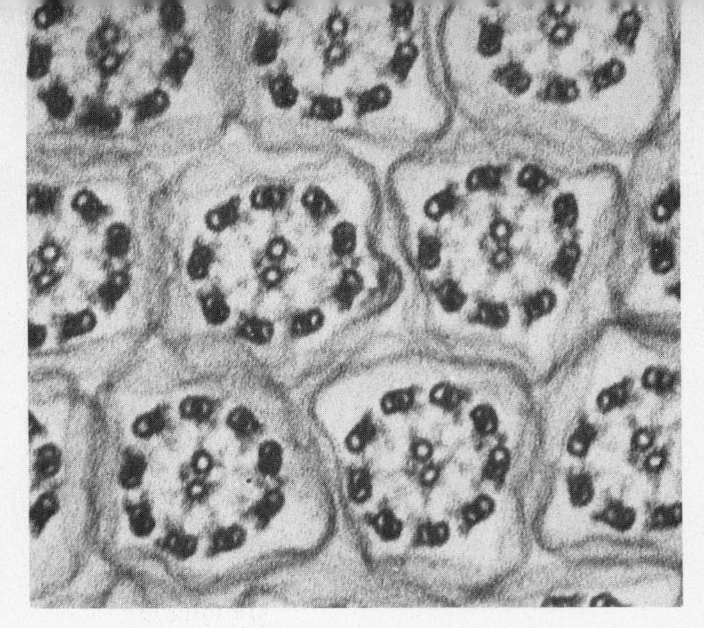

Figure 2-12. **Cross sections of gill cilia. The arrangement of the filaments is identical to that in the sperm tails in Fig. 2-13. [Reprinted by permission of the Rockefeller Institute Press, from I. R. Gibbons, *J. Biophys. Biochem. Cytol.* 11: 179, (1961). The photographs here reproduced from printed copy inevitably show a loss of detail, and quality of the results is not representative of the originals.]**

diameter. The microtubules are arranged in nine bundles, each of which has three microtubules. The fibers in the mitotic spindle are also microtubules. They are made of a protein tentatively called *tubulin.*

Figure 2-13. **Cross sections of spermatozoan tails. In the central portion of the tail are nine pairs of peripheral filaments (pf) around one central pair (cf). ×71,000. (From J. Rhodin, *An Atlas of Ultrastructure,* W. B. Saunders Company, Philadelphia, 1963.)**

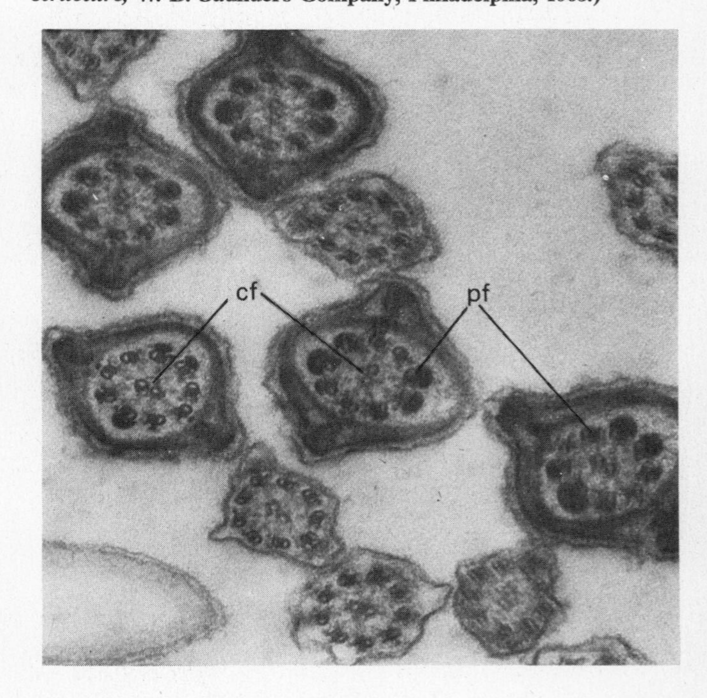

Specialized Cell Surfaces and Protrusions

CILIA AND FLAGELLA

Cilia and flagella are fine protrusions of the cell, continuous with the cell membrane. If the protrusion is long and whiplike, like the tail of a sperm, for example, it is a flagellum; if there are many short, hairlike processes they are cilia, which characteristically move back and forth like oars. The basic structural plan is the same whether the cilia are from mammalian cells in the respiratory tract or from the gill of a fresh water mussel; whether the flagellated sperm are human or from a moss plant (Figs. 2-12 and 2-13). They are all made of the centriolar units of nine bundles of *microtubules,* attached to the cell by a *basal plate, basal bodies,* and *rootlets* (Fig. 2-13). If the structure is modified for movement, as are the cilia and flagella, there are two central bundles of microtubules in addition to the nine peripheral ones (Fig. 2-14). The rod cells of the retina of the eye are highly modified cilia, in which specific ionic permeabilities have been retained. Contraction has been lost, so that only the nine peripheral bundles remain. The stalks have been widened into an outer segment into which infoldings of the cell membrane move as disks from the base to the tip, giving rise to well-organized *lamellae,* on which molecules of the light-sensitive pigment rhodopsin are geometrically spaced for optimal response to light (Fig. 2-15).

MICROVILLI

Microvilli are multitudinous microscopic projections of the cytoplasm, covered with the cell membrane. A single

Figure 2-14. **A cilium, showing its microtubules.**

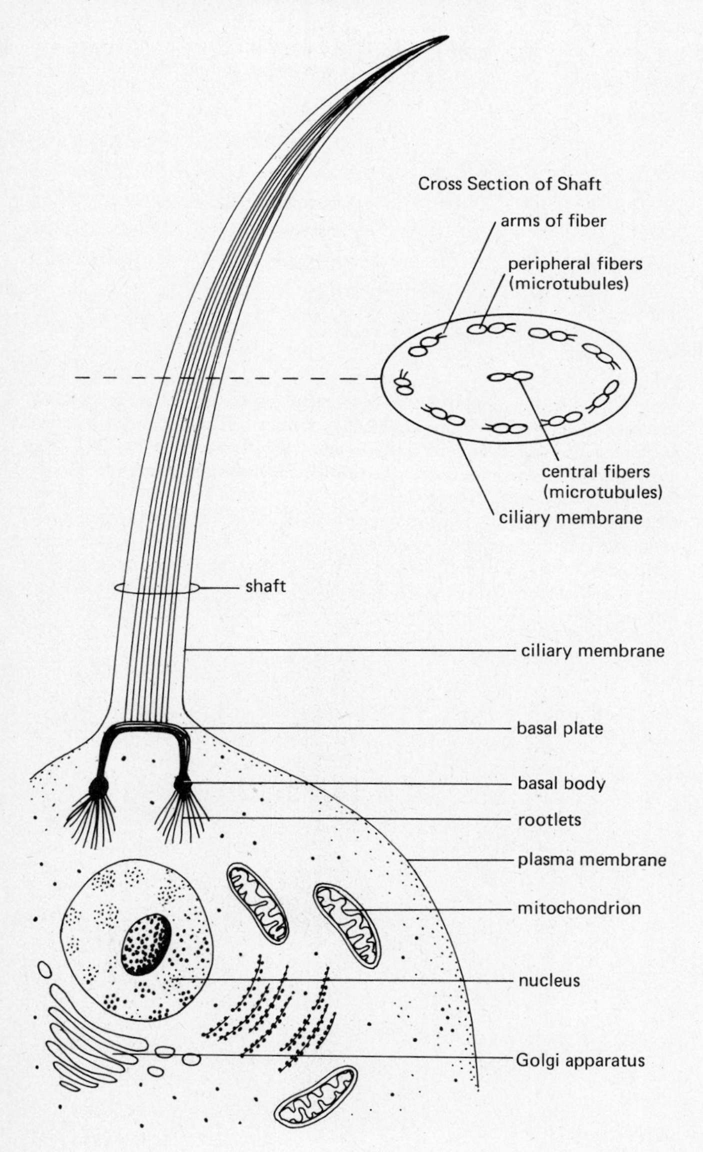

Figure 2-15. **A rod cell of the retina, showing its modification from a cilium.**

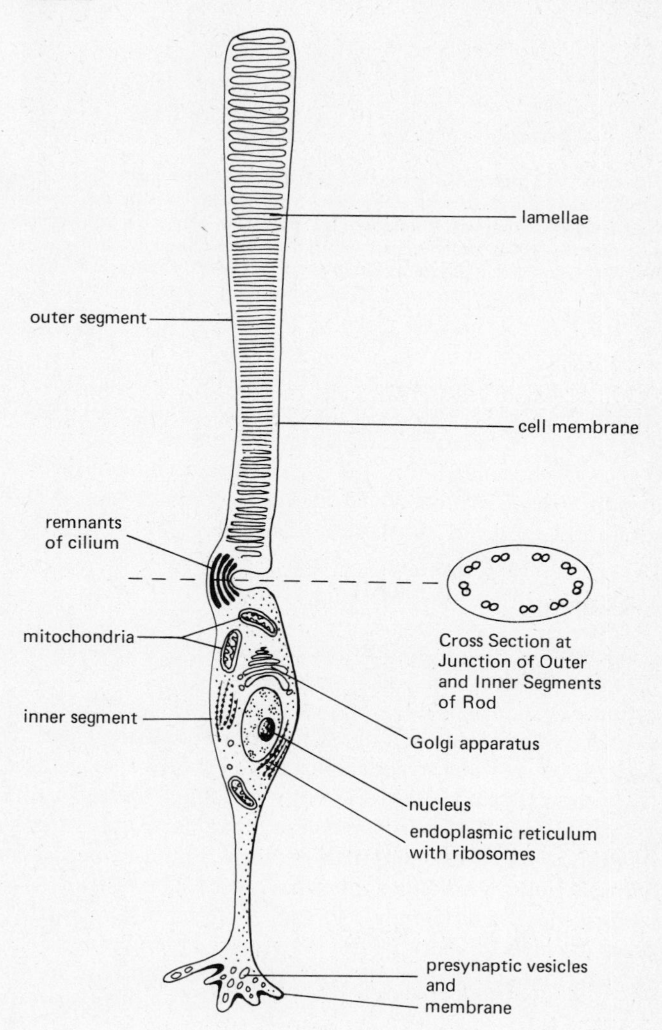

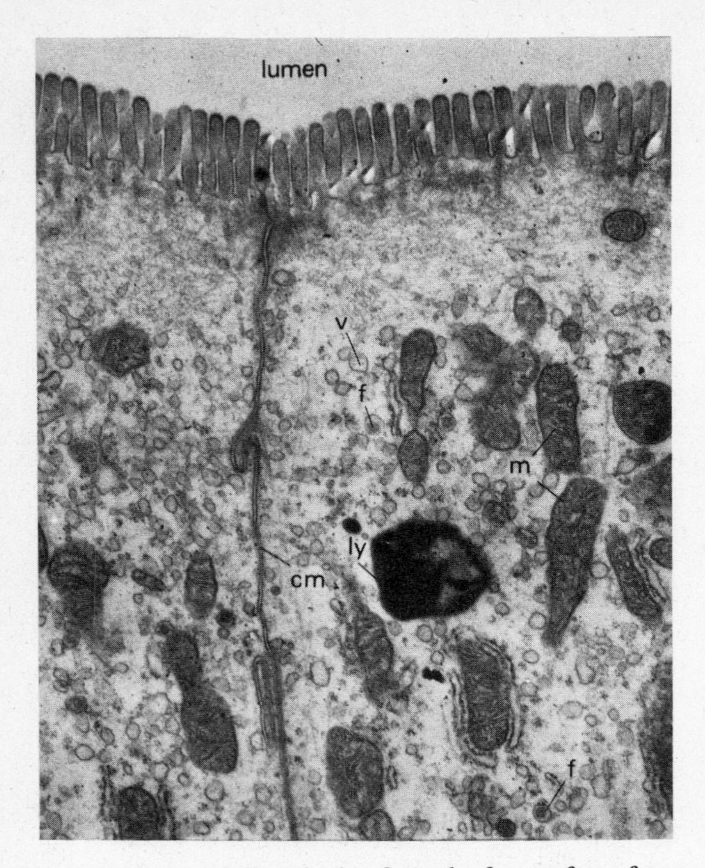

Figure 2-16. **Microvilli projecting from the free surface of two intestinal surface cells into the lumen of the digestive tract. Small vesicles (v) transport absorbed fat particles (f); also shown are a lysosome (ly), a vesicle containing digestive enzymes, mitochondria (m), and the cell membrane (cm). ×15,000. (From J. Rhodin, *An Atlas of Ultrastructure,* W. B. Saunders Company, Philadelphia, 1963.)**

epithelial cell may have several thousand microvilli projecting from its free surface into the lumen of the digestive tract (Fig. 2-16). A microvillus is about 100 nm in diameter and 600 to 900 nm in length. This increases the available surface area of a cell enormously, an important consideration in the rapid absorption of substances from the gut.

Membrane junctions

Membrane junctions are specialized regions of contact between adjacent cells; they serve as a means of communication. Two plasma membranes may come very close together, with a space of only 2 to 4 nm separating them to form a *gap junction,* which may act as a barrier to substances from the exterior as well as firmly attaching the cell to its neighbor (Fig. 2-17). It is presumed that electrotonic transmission of electrical potentials of nerve impulses occurs at gap junctions. This is not the usual mechanism of transmission of electrical potentials in higher vertebrates. Usually, there is a cleft of about 20 nm between the membrane of the exciting cell and that of the responding cell, and both membranes are modified for chemical transmission of the electrical impulse. The two modified membranes and the separating cleft form the *synapse* (Fig. 2-18), which is discussed in detail in Chapter 7.

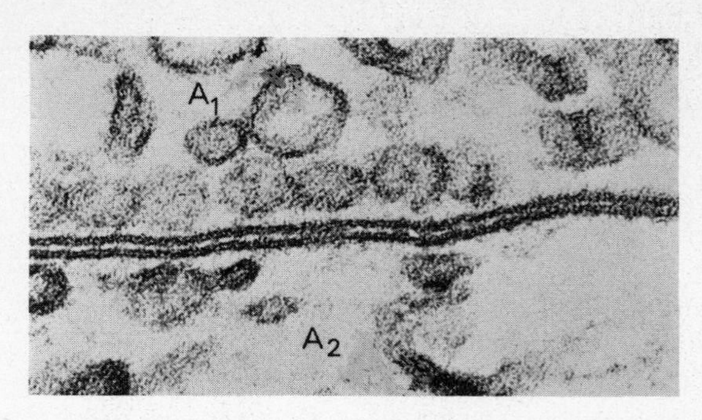

Figure 2-17. **In this very thin electronmicrograph of a crayfish synapse, the two membranes that form the gap junction are separated by a narrow gap of approximately 2 nm. The overall thickness of the membranes and gap is about 18 nm. There are vesicles in both the axons (A_1, A_2), but their function is unknown because this is a synapse where electrical transmission occurs. ×130,000. (From G. D. Pappas and S. G. Waxman, in *Structure and Function of Synapses,* G. D. Pappas and D. P. Purpura, eds., Raven Press, New York, 1972, pp. 1–43.)**

Sometimes the space between two adjacent plasma membranes may be widened and filled with a disk-shaped structure. The plasma membranes are thickened, and dense filaments radiate out into the cytoplasm. This structure is a *desmosome,* and is primarily involved in cell-to-cell adhesion, rather than in cell communication (Fig. 2-19).

Figure 2-18. **Chemical synapse between a nerve and muscle in a frog shows the specialized presynaptic nerve membrane (n) and postsynaptic muscle membrane (m). Upper arrows show the active zones of the presynaptic membrane, where the vesicles (v) that contain the chemical transmitter that stimulates the muscle are released. The muscle membrane is thrown into folds in this synaptic region, and the thickened areas that contain the receptors for the transmitter are seen. c = synaptic cleft, f = muscle contractile filaments, and S = Schwann cell fingers wrapping around the nerve. ×25,000. (Courtesy of D. Rutherford, Laboratory of Cellular Biology, New York University.)**

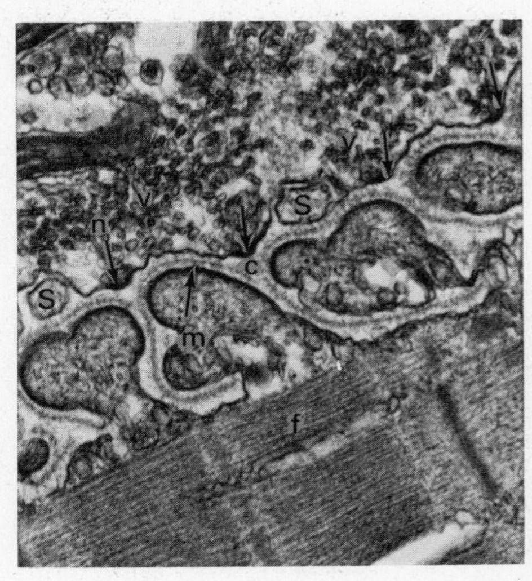

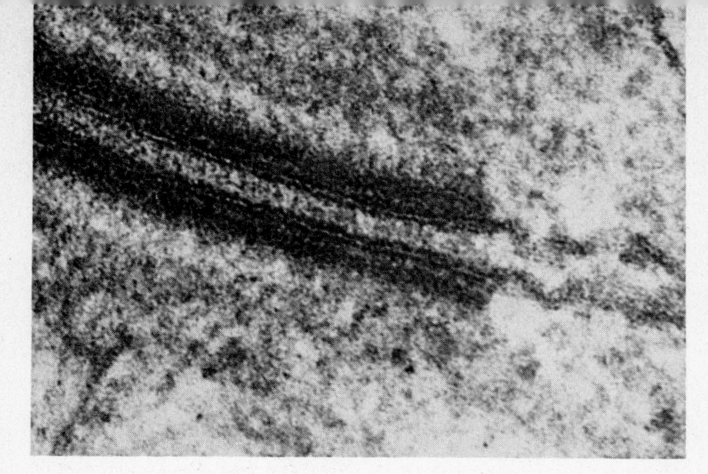

Figure 2-19. **Desmosomes firmly holding epithelial cells. ×212,000. (Electronmicrograph courtesy of M. Yoder, New York University.)**

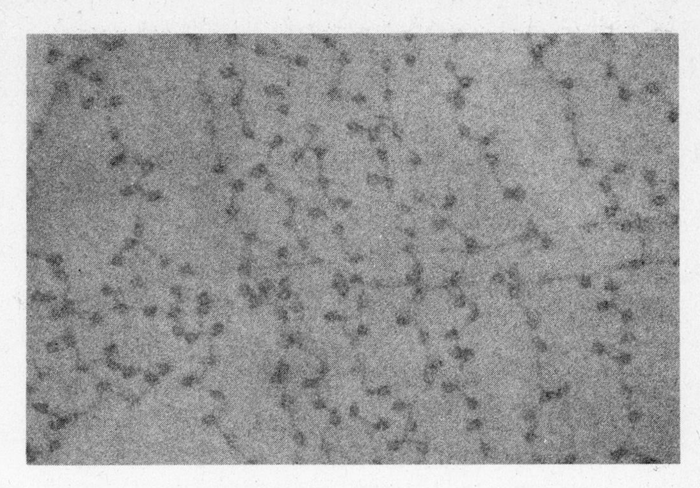

Figure 2-21. **Chromatin fibers streaming out of the nucleus of a chicken red blood cell. The chromatin looks like beads on a string, with each bead having a diameter of about 6.9 nm. The distance between the beads is approximately 14 nm. [From D. E. Olins and A. L. Olins, *Science* 188: 1097 (1975). Copyright 1975 by the American Association for the Advancement of Science.]**

Nucleus

The nucleus is a large round or oval body that moves within the cytoplasm to the site of greatest metabolic activity. Removal or destruction of the nucleus by microdissection (Fig. 2-20) or localized irradiation results in the rapid death of the cell, for the nucleus contains not only the genetic material, deoxyribonucleic acid (DNA), but also ribonucleic acid (RNA) that controls cellular activities such as protein synthesis, including enzyme synthesis, and consequently cell metabolism and growth.

The nucleus is separated from the cytoplasm by the *nuclear membrane,* a modified unit membrane with large pores, about 15 nm in diameter (Fig. 2-6). This results in an incomplete barrier between the nucleoplasm and cytoplasm; nevertheless, differences in the chemical composition of these regions are maintained. There is no clear evidence that the nuclear membrane connects directly with extracellular space, although the nuclear pores are believed to empty into the cisternae of the ER and consequently permit the passage of substances between the nucleus and cytoplasm.

Figure 2-20. **Enucleation of a cell. The cell is kept in place by three microneedles that can be accurately controlled by a micromanipulator. A fourth needle is used to puncture the nucleus and pull it out of the cell.**

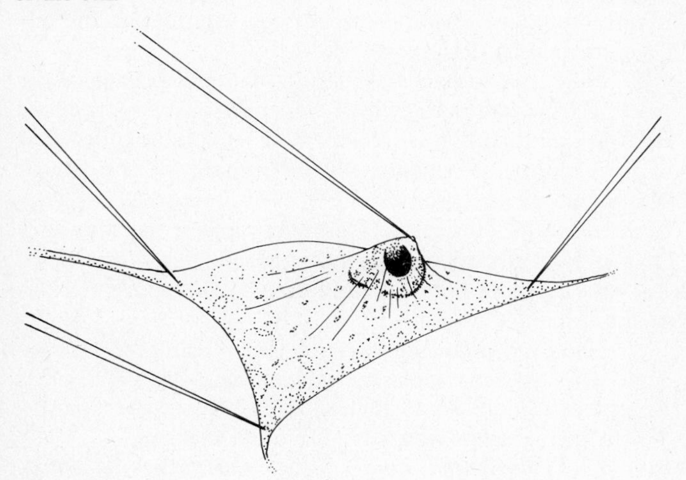

The nuclear membrane encloses the darkly staining *chromatin granules,* which is the indefinite form taken by the *chromosomes* in the cell not undergoing division. These granules consist of DNA and a special type of protein known as a histone; together they form complexes that are regularly arranged like beads on a string (Fig. 2-21). In this beaded form, the DNA takes up only one sixth of the length it would be if it were extended. Although the chromosomes are difficult to visualize as individual structures at this time, they are extremely active, for it is in the times between divisions, as well as in the very early stages of division, that they synthesize the protein, RNA, and DNA of which they are composed.

In addition to the chromosomes, the nucleus usually contains one or more spherical, darkly staining bodies, the *nucleoli,* which are made of granules of protein and RNA very similar to the composition of the cytoplasmic ribosomes. The nucleoli are probably the site of ribosome synthesis, the newly formed ribosomes entering the cytoplasm through the nuclear pores.

2–3 The Molecular Nature of the Gene

The vital role of the chromosomes, as has been discussed above, is twofold: first, to *regulate the synthetic reactions* occurring in both the nucleus and the cytoplasm; and second, to *transfer the characteristics* of one cell to its daughter cells and thus ensure the continuity of specific cell types through countless generations. This controlling information is contained in specific sites, the *genes,* along the length of the chromosome. The genes are discrete segments of the long molecules of DNA and are combined in specific numbers to form the individual chromosome. The number of genes varies greatly from chromosome to chromosome within a nucleus, but the general distribution is the same throughout a species. Humans possess 23 pairs of chromosomes whose lengths differ because of the different number of genes in each pair. This results in individually recognizable chromosomes, and any alteration in the order or the number of the genes along the length of the structure will result in a specific genetic alteration, provided that the change is within the chromosomes of the reproductive (germ) cells.

A change in the actual structure of the DNA is a *mutation* and can be induced by radiations of various types, such as ultraviolet, X-ray, and ionizing radiations. Changes in the genes of the germ cells (sperm and egg) are changes that will be reflected in succeeding generations. Any genetic alteration in the cells forming the tissues of the other organs of the body, the *somatic cells,* will be reflected only as an alteration in the cells of that particular individual. This distinction between the somatic cells, which give rise only to subsequent cell generations in one individual, and the germ cells, which are capable of forming an entirely new individual, is of great basic importance.

Under certain circumstances, however, it is possible to inherit the characteristics of a somatic cell. In frogs, if the nucleus is removed from an unfertilized ovum and replaced, through microsurgical techniques, by the nucleus of an intestinal cell, the egg will develop into an individual that is the genetic replica of the donor of the nucleus. All frogs produced in this way will be identical. This technique of producing identical individuals is known as *cloning.*

2–4 Duplication and Specialization of Cells

Mitosis

The development of many cells from the single fertilized egg, which is now called the *zygote,* and the subsequent cellular specialization into tissues involve many processes, the chief of which is an increase in the number of cells. Fundamental to this process is the continuous synthesis or duplication of the genetic material within the chromosomes, which is then meticulously divided between the two daughter cells formed by the mother cell. This duplicating ability is the unique contribution of DNA, and the entire nuclear content of DNA is doubled just prior to the initiation of cell division. The DNA, together with intranuclear RNA, apparently also is responsible for synthesizing the protein with which it is combined. These syntheses are followed by an intricate series of steps that permit accurate distribution of the chromosomes to the daughter cells. The entire process is called mitosis.

Figure 2-22. **The phases of the cell cycle for a typical mammalian cell.**

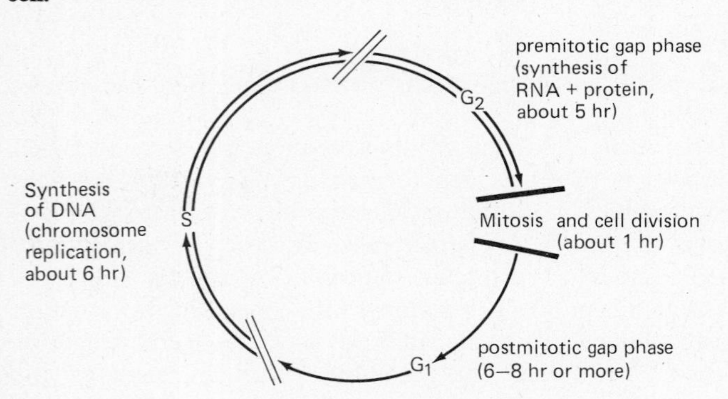

Preparation for mitosis

During the interval between cell divisions, *interphase,* little structural change takes place in the cell. As the cell prepares to divide, it begins the orderly synthesis, first of RNA and protein (the G_1 or postmitotic gap phase), followed by a period during which only DNA is synthesized (the S phase). This cycle is completed by the premitotic gap phase (G_2), during which time only RNA and protein are synthesized (Fig. 2-22). By the time the cell is ready to enter mitosis, it has doubled its content of DNA, RNA, and protein. The centrioles divide even earlier than this, forming a pair of new centrioles, at right angles to one another.

The four phases of mitosis

The four phases of mitosis are shown in Fig. 2-23.

1. *The First Phase of Mitosis—Prophase.* The nuclear membrane disappears so that the contents of the nucleus and the surrounding cytoplasm are no longer separated. The chromosomes are now distinct organelles, and each separates longitudinally into identical halves, the *chromatids.* The two halves of the centriole separate, migrating to the poles of the cell; fine fibers, or microtubules, are formed between them, giving rise to the *spindle,* along which the chromosomes will ultimately travel, as though along guide lines. The beginning of spindle formation can be seen even before the nuclear membrane disappears.

2. *The Second Phase of Mitosis—Metaphase.* Each of the pairs of chromosomes, which had previously reproduced itself longitudinally, is attached to a spindle fiber and aligned along the equator of the spindle. The point of attachment of the chromosome is the *centromere,* a constriction that divides the chromosome into two arms. This very characteristically appearing phase represents the line-up of the chromosomes before distribution to the opposite poles of the cell.

3. *The Third Phase of Mitosis—Anaphase.* The chromosome pairs are separated, one half of each pair being pulled along the spindle fibers toward each pole of the spindle. Which half goes to which pole is quite random.

4. *The Fourth and Final Phase of Mitosis—Telophase.*

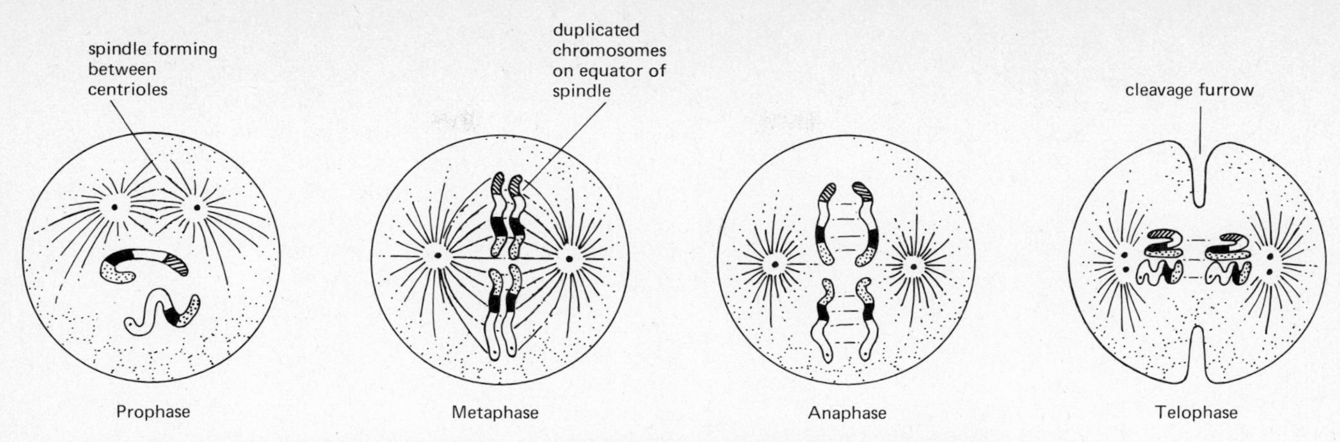

Figure 2-23. **Mitosis in an animal cell, showing the duplication of the DNA of the chromosomes and its subsequent equal distribution to each of the daughter cells.**

The entire complement of chromosomes reaches the poles of the spindle, which then slowly disappears. The nuclear membrane re-forms around each group of chromosomes, the nucleolus reappears, and a *cleavage furrow* indents the cytoplasm, distributing roughly equal amounts of cytoplasm to each daughter cell around the accurately divided nuclear material within the newly formed nucleus. Two daughter cells have been produced, each of which now receives the same type of instructions from the DNA within its chromosomes as was transmitted to the ancestral cells.

2–5 Organization of Cells into Tissues and Organs

The Germ Layers

The mitotic production of new cells continues until the zygote looks like a ball of cells with a cavity in the center. This stage is called the *blastula.* The blastula is no bigger than the original ovum because no growth has occurred yet. In the human, this stage is called the blastocyst because this is the time when the blastula encysts in the wall of the receptive uterus, burrowing in to establish its vital nutritive connections with the tissues of the mother. The tiny human blastocyst does not have stored food in the form of yolk that the eggs of birds and reptiles have, so that its ability to obtain nourishment rapidly from the maternal uterus is essential to its further development (Fig. 2-24).

An important distinction between cells has already been made at this time, for the outermost layer of the blastocyst is set off from an *inner cell mass* (Fig. 2-25). The outer layer, the *trophoblast* or "feeding layer," will never form a part of the actual embryo but is destined to participate in the formation of the *placenta,* the efficient organ formed by both embryo and uterus to nourish and protect the embryo during its sojourn within the mother. From the rather scanty information available about very young human embryos, the trophoblast seems to grow into the tissues of the uterus about 7 to 10 days after the ovum has been released from the ovary (*ovulation*). As the fingerlike processes of the trophoblast penetrate the lining of the uterus, they destroy and digest part of the uterine lining and the blastocyst feeds parasitically on the debris. The uterus reacts by surrounding the invader with a protective wall of cells, and from this time on the uterus and the trophoblast cooperate to form the placenta.

Figure 2-24. **Young human blastocyst. (From E. Blechschmidt, *The Stages of Human Development Before Birth,* W. B. Saunders Company, Philadelphia, 1961. S. Karger, Basel and New York.)**

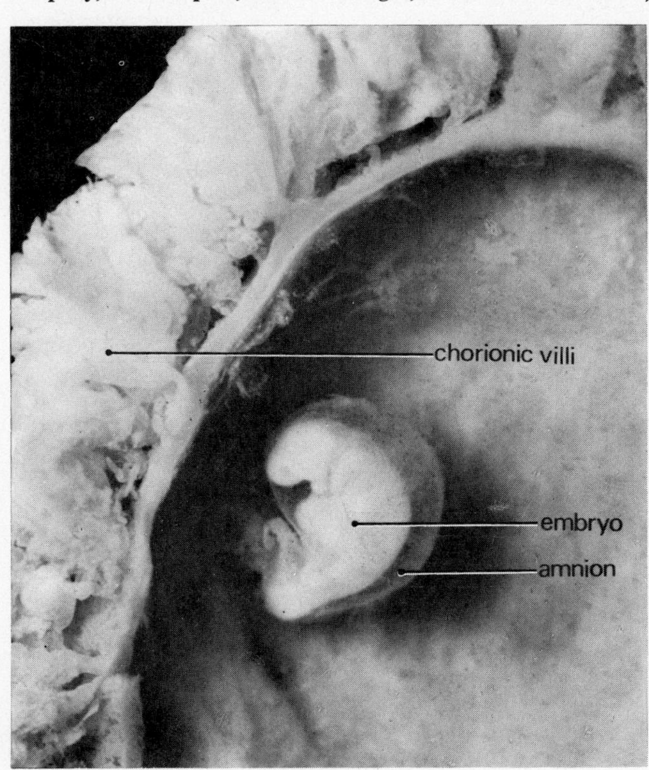

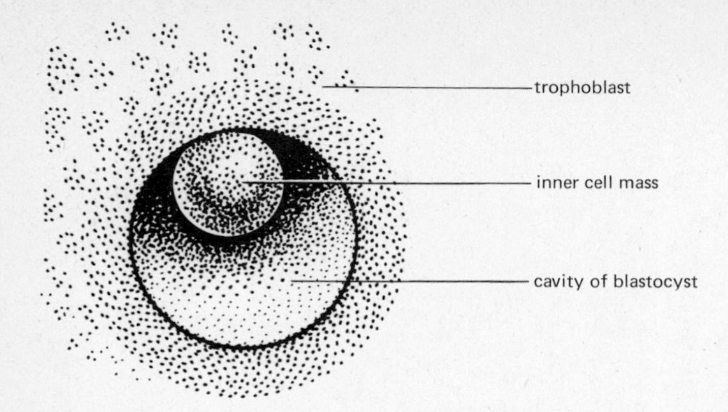

Figure 2-25. **Formation of the inner cell mass and trophoblast in the human embryo.**

The inner cell mass again distinguishes between those cells that will form the embryo and those extra-embryonic cells that exist to feed and protect it. The inner cell mass forms a figure 8, with the top and bottom circles of the eight separated from one another by a thin plate of cells, the *embryonic disk.* The top circle develops into a fluid-filled sac, the *amnion,* in which the embryo floats and from which it also obtains nourishment. The lower circle forms a small, empty sac attached to the belly side of the embryo called the *yolk sac* (Fig. 2-26). In human embryos it is quite without yolk, but it develops in the same way and is in the same position as the functional yolk sac of lower animals.

The amnion grows rapidly and surrounds the embryo and the yolk sac everywhere except at the thick umbilical cord, through which run the blood vessels connecting the embryo with the placenta.

The embryonic disk, the future embryo, now appears to split into two layers. The top layer of cells is the *ectoderm,* the lower layer the *endoderm.* These are the first two of the three *germ layers,* each of which will give rise to specific tissues in the developing embryo (Fig. 2-26). The embryo is now double-layered, and the process which forms the double-layered organism is *gastrulation.* Some very primitive animals, like the hydra, never develop a proper third layer, but in all higher animals a middle layer appears, the *mesoderm,* which separates the ectoderm and endoderm. The development of the three germ layers is the first stage in tissue differentiation, and although there are variations in the plan, each germ layer gives rise to the same type of tissue in all animals. Ectoderm, for example, is always the tissue from which the nervous system is formed.

Now the groundwork for the tissues and organs has been laid.

In the embryo

1. The outermost layer, the *ectoderm,* forms the protective outer layers of the skin as well as the nervous system and the sense organs.
2. The innermost layer, the *endoderm,* forms the lining of the digestive, respiratory, and urinary tracts.
3. The middle layer, the *mesoderm,* spreads throughout the embryo to give rise to the supporting tissues of the body—bone, muscle, connective tissue, and blood—and also the kidneys and the reproductive organs.

Outside the embryo

The extra-embryonic mesoderm is mesoderm that has spread out to line the trophoblast. It soon separates from its inner surface to develop into a thick, double-layered membrane, the *chorion.* The chorion develops into a shaggy sac around the embryo and all its other membranes, and at the site where it is in contact with the lining of the uterus, the chorion sends deep projections into the rich substance of

Figure 2-26. **The three germ layers of the human embryo.**

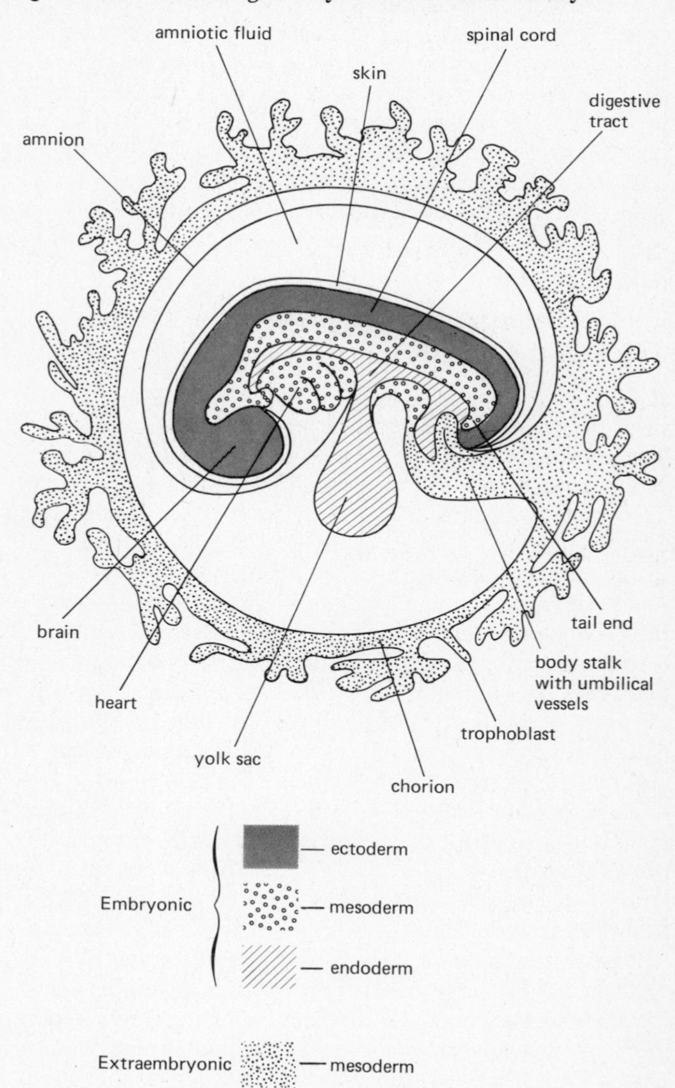

this tissue to form the *embryonic part of the placenta* (Fig. 2-26).

Differentiation and Specialization

Mitosis accounts for the increase in number of cells but not for their specialization. During the development of the embryo, the various areas of the rapidly dividing cells come under the influence of chemicals secreted by cells that have already specialized, or differentiated. These chemicals are known as *embryonic organizers* and will organize cells and tissues to develop into a definite structure, such as nerve, muscle, or eye, according to the location on the embryo. The way in which the instructions are followed by the cell is not understood, but this complex action can be demonstrated by removing a small piece of tissue from one area, the part destined to become a leg, for example, and placing it in the head area. If the transplanted tissue has already received its instructions and become determined, it will continue this development to become a supernumerary limb attached to the head; but if these cells are still quite undetermined, they will follow the instructions of the new organizer in the head area and become incorporated as part of the head (Fig. 2-27). This is the next step to be explained in terms of the messenger RNA carrying the instructions from the genes to the ribosomes in the cytoplasm, with the organizer undoubtedly affecting the type or transmission of the code. When we understand this mechanism better, a major part of the mystery of biological development will have been explained.

Tissues

A tissue is composed of many cells, all with the same structure and function. The embryonic cell, endowed with the *properties of life*—including the ability to *grow* and *reproduce,* to *metabolize,* to *receive and respond to stimuli,* and to show *movement* and *adaptation*—is not specialized for any one of these functions, except perhaps for reproduction in terms of repeated cell division. From this increase in cell number, the tissues of the organism are formed, each tissue being composed of thousands of cells modified to perform a specific function in the coordinated body more efficiently than could an undifferentiated cell. In this process, the infinite ability of the undifferentiated cell to divide is lost. Cells specialized to the extent of muscle and nerve cells practically never undergo mitosis. Cell development in tissues where regeneration is possible, as in the liver, ends with each specialized cell, and the new cells must go through the entire process of differentiation and specialization again.

The main tissues of the body are epithelial, muscle, nerve, and connective (Fig. 2-28).

Figure 2-27. **Tissue from the leg area of a newt transplanted to the head develops into an extra leg if the cells of the leg area have already been determined.**

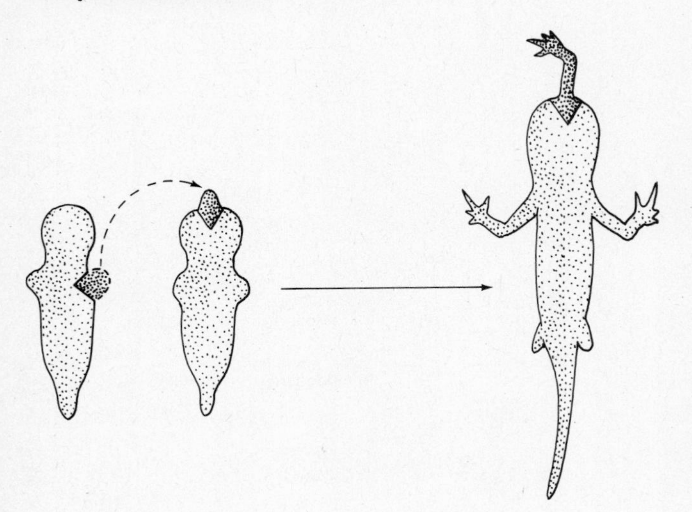

EPITHELIAL TISSUE—PROTECTIVE, LINING, AND SECRETORY (ECTODERM OR ENDODERM)

Epithelial tissue is extremely variable in form and function, ranging from that consisting of very *thin, flat* cells only one layer thick, such as the thin covering or serosa of the viscera, to the many-layered outer portion of the skin, the epidermis. It includes secretory cells that range in height from *cuboidal* to high *columnar,* and it is found lining the digestive, respiratory, and urogenital tracts (Fig. 2-28).

The epithelial cells themselves may have specialized structures within them, as do the ciliated cells of the respiratory tract. The constant movement of these very fine processes keeps this tract clear of small foreign particles. Sometimes the layers of epithelial cells are arranged so that they can be stretched, as in the lining of the urinary bladder. When the bladder is empty, this *transitional* epithelium seems to consist of many layers of cells; but when the bladder is distended with urine, it appears to be only two or three layers deep.

MUSCLE TISSUE—SPECIALIZED CONTRACTILITY (MESODERM)

All protoplasm is contractile, but muscle cells possess, in addition, special contractile fibrils that increase their contractile efficiency enormously. There are three main types of muscle, which differ in their structure and thus their function in certain ways.

Skeletal Muscle. As the name implies, this is muscle attached to the bones and responsible for moving the body or parts of it. It is under the control of the voluntary nervous system and contracts the fastest of the three types of muscle. There are two types of *contractile filaments;* one is made of the protein *myosin,* and the other is composed of three proteins, *actin, tropomyosin,* and *troponin.* The distribution of these filaments into bands gives skeletal muscle a

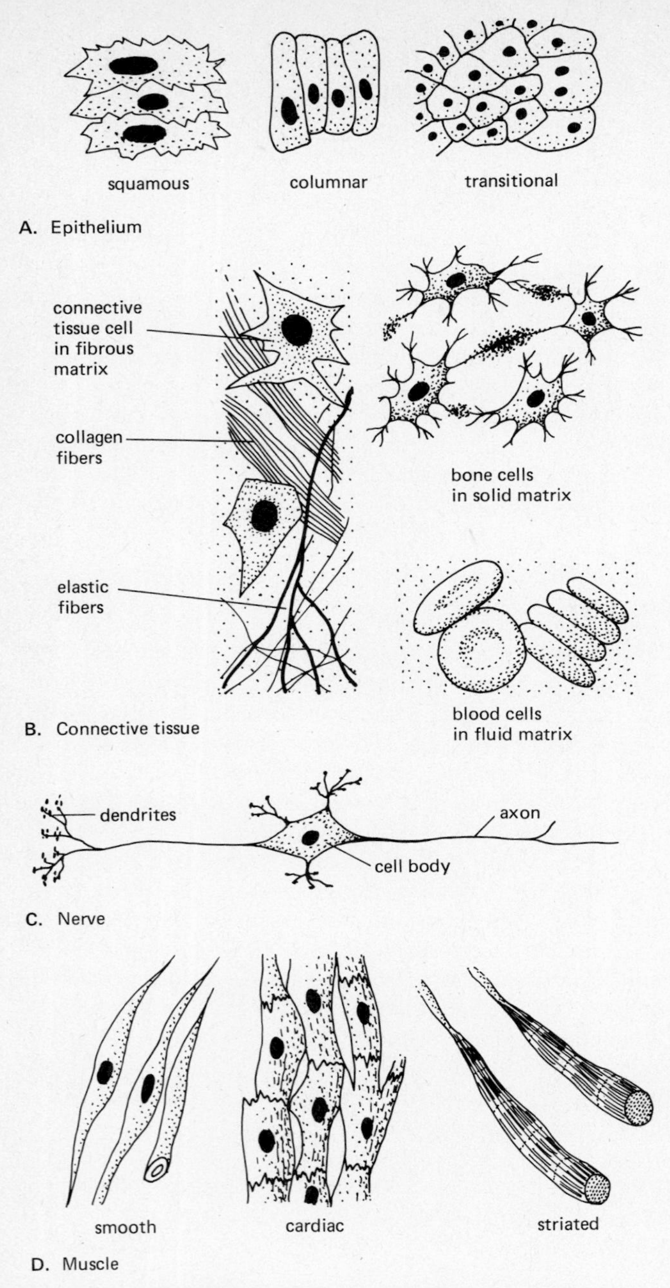

Figure 2-28. **The cells that make up the main tissues of the body.**

striped or striated appearance under the microscope. The significance of the molecular structure of these filaments is discussed in Chapter 23.

Smooth Muscle. This muscle has no striations and is not under voluntary control. It makes up the muscle of the viscera and the blood vessels, and sometimes rather isolated groups of such muscle cells are found in connective tissue. It contracts much more slowly than skeletal muscle, but it has an inherent rhythmicity that is modified by the action of the involuntary nervous system.

Cardiac Muscle. The muscle of the heart is made up of cells that are essentially similar in composition to striated muscle. These muscle fibers branch but are separated into individual cells by continuations of the plasma membrane, *the intercalated disks.* Contractile filaments are present, but the cross-striations are not very clear due to the granular nature of the *sarcoplasm* (muscle cytoplasm).

NERVE TISSUE—CONDUCTING (ECTODERM)

This tissue is specialized for the *conduction and transmission of electrical impulses,* and the organization of these nerve

Figure 2-29. **Organization of tissues (epithelial, connective, and muscle) into an organ, the esophagus.**

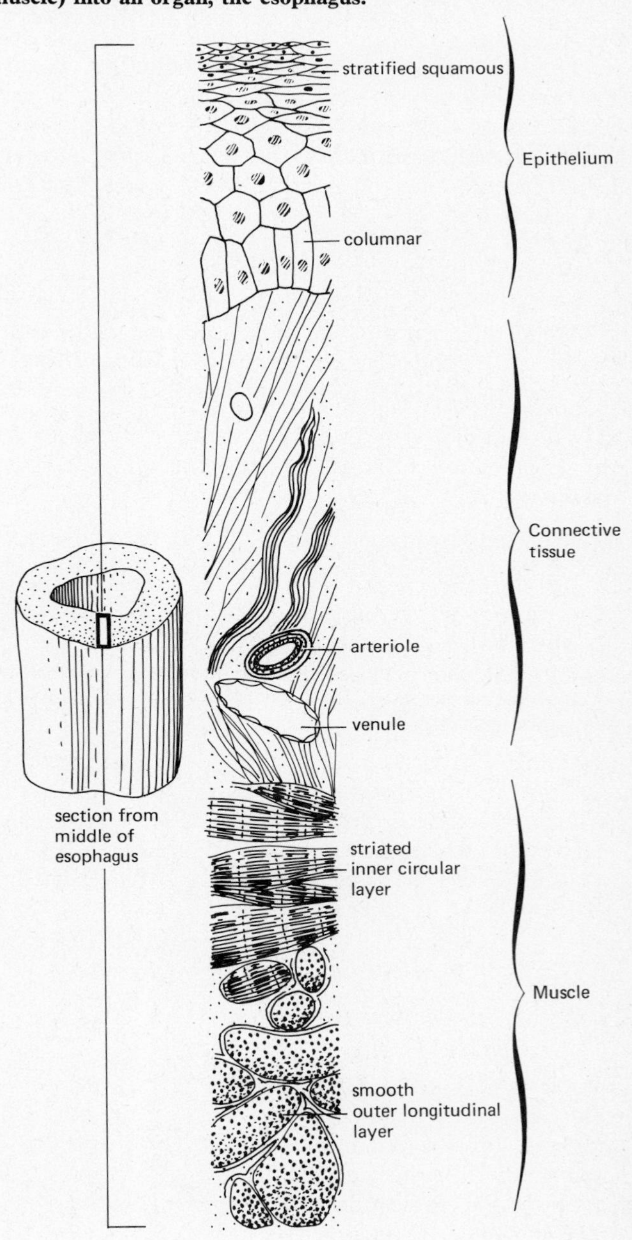

cells or *neurons* is the most complex of that of any of the tissues. The neuron consists of a *cell body* that contains the nucleus and the other organelles typical of a cell with very high metabolic activity, such as many mitochondria and ribosomes. The neuron is further specialized by the possession of *processes,* which connect it through *synapses* to other neurons, making a long chain of conducting tissue that links the various parts of the body. The processes that *conduct information toward the cell body* are usually quite numerous, receiving synapses from many other cells, and like the twigs upon a tree are called *dendrites.*

Usually, a single process *conducts impulses away from the cell body.* These processes are called *axons;* frequently they branch profusely so as to contact many other nerve cells to form the synapse, or they may terminate peripherally in a wide variety of specialized endings, in a muscle, or in another organ.

CONNECTIVE TISSUE—SUPPORTING, CONNECTING, AND TRANSPORTING (MESODERM)

This is a term covering a wide variety of cell types, having in common the possession of more *intercellular material,* or *matrix,* than cells. Such tissues are blood, bone, and cartilage, and the elastic and fibrous connective tissues found holding the epithelium firmly to the underlying tissues (Fig. 2-29). All these have cells that are specialized in different ways but are all widely separated by the matrix, which may range in consistency from the very fluid plasma containing the blood cells to the extremely rigid matrix separating the cells of bone. (For more details about each type of specialized cell, refer to the chapters on blood, skin, and bone.)

Usually, connective tissue is interspersed among the cells forming other tissues, and blood vessels, nerves, and some type of protective covering are always found in relation to a tissue. Often, several tissues are organized anatomically to form a functional *organ,* like the uterus. This is made up of an inner lining of epithelial tissue, capable of enormous development in pregnancy, and several layers of strong muscle tissue, all surrounded by a thin serous membrane. A great deal of connective tissue appears between the muscle cells, and the entire structure is plentifully supplied with blood vessels. These tissues are all essential for the integrated function of the organ. (The tissues of the esophagus are shown in Fig. 2-29.)

2–6 From Molecule to Gene; From Gene to Cell; From Cell to Tissue, Organ, and Integrated Organism

Modern physiology is finding the explanation of many of the mysteries of living organisms in terms of the molecular structure and organization of the cell. The unique self-duplicating ability of the cell has been found to reside in the long molecules of DNA within the nucleus. These molecules are responsible for the organization of the functional cell. This orderly, directive process is continuous, resulting in the multiplication of cells and their subsequent organization into specialized tissues. These tissues are combined in many ways to form the still more highly specialized organs, which in turn are functionally integrated by nervous and hormonal means to result in the harmoniously balanced body of the physiologically normal animal. An understanding of these physiological processes always involves a return to the nature of the *molecular reactions* occurring within the cells of the organism, but this must then be reassociated with the *integration of these processes with the intact organism.*

Cited References

HARRISON, R. G. "Observations on the Living Developing Nerve Fiber." *Proc. Soc. Exper. Biol. Med.* **4:**14, 1907.

ROBERTSON, J. D. "Unit Membranes: A Review with Recent New Studies of Experimental Alterations and a New Subunit Structure in Synaptic Membranes." In *Cellular Membranes in Development,* M. Locke, ed. Academic Press, Inc., New York, 1964, p. 1.

Additional Readings

BOOKS

BLOOM, W., and D. W. FAWCETT. *A Textbook of Histology,* 10th ed. W. B. Saunders Company, Philadelphia, 1975.

DEROBERTIS, E. D. P., W. W. NOWINSKI, and F. A. SAEZ. *Cell Biology,* 4th ed. W. B. Saunders Company, Philadelphia, 1965.

KESSEL, R. G., and C. Y. SHIH. *Scanning Electron Microscopy in Biology.* Springer-Verlag New York, Inc., New York, 1974.

PATTEN, B. M. *Human Embryology,* 3rd ed. McGraw-Hill Book Company, New York, 1968.

PEASE, D. C. *Histological Techniques for Electron Microscopy,* 2nd ed. Academic Press, Inc., New York, 1964.

POLLISTER, A. W., ed. *Physical Techniques in Biological Research.* Academic Press, Inc., New York, 1966.

RHODIN, J. A. G. *An Atlas of Ultrastructure.* W. B. Saunders Company, Philadelphia, 1963.

RUGH, R. *Vertebrate Embryology.* Harcourt Brace Jovanovich, Inc., New York, 1964.

ARTICLES

BRACHET, J. "The Living Cell." *Sci. Am.,* Sept. 1961.

EPHRUSSI, B., and M. C. WEISS. "Hybrid Somatic Cells." *Sci. Am.,* Apr. 1969.

MAZIA, D. "The Cell Cycle." *Sci. Am.,* Jan. 1974.

NEUTRA, M., and C. P. LEBLOND. "The Golgi Apparatus." *Sci. Am.,* Feb. 1969.

NOMURA, M. "Ribosomes." *Sci. Am.,* Oct. 1969.

NOVIKOFF, A. B., E. ESSNER, and N. QUINTANA. "Golgi Apparatus and Lysosomes." *Fed. Proc.* **23:**(5):1010 (1964).

WESSELS, N., and W. J. RUTTER. "Phases in Cell Differentiation." *Sci. Am.,* Mar. 1969.

Chapter 3

Cellular and Extracellular Constituents

Another annual cycle inevitably passed and the pain was eased by a humorous birthday card. The front bore the caption, 'According to biochemists the materials that make up the human body are only worth 97¢. . . . I decided to make a thorough study of the entire matter. I started by sitting down with a catalogue from a biochemical company and began to list the ingredients. Hemoglobin was $2.95 a gram, purified trypsin was $36 a gram, and crystalline insulin was $47.50 a gram. I began to look at slightly less common constituents such as acetate kinase at $8,860 a gram. The real shocker came when I got to follicle stimulating hormone at $4,800,000 a gram, clearly outside the reach of anything that Tiffany's could offer. For the really wealthy there is prolactin at $17,500,000 a gram, street price. Not content with a brief glance at the catalogue, I averaged all the constituents over the best estimate of their percentage in the composition of the human body and arrived at $245.54 as the average price of a gram dry weight of human being. . . . The next computation was done with a great sense of excitement, I had to multiply the price per gram by my dry weight. The number literally jumped out at me—$6,000,015.44. I was a Six Million Dollar Man! . . .

"We must somehow reconcile the 97-cent figure and the $6 million figure. The answer is at the same time very simple and very profound: Information is much more expensive than matter. . . . We are, at the molecular level, the most information-dense structures around, surpassing by many orders of magnitude the best that computer engineers can design or even contemplate by miniaturization.

H. J. Morowitz, "The High Cost of Being Human"

IT IS, PERHAPS, a dangerous venture to list the chemical components of a cell; this might imply that the cell is a static structure, with a constant, unchanging composition. Nothing could be further from the truth. Substances are always entering or leaving the cell; the cell is producing, using up, and storing; the cytoplasm and its contents are in constant movement. There are, however, certain basic components on which the cell is dependent, and these components are present, inside and outside the cell, in what we have called the *physiological range* (Chapter 1).

3–1 Water

General Significance of Water in Living Systems

We are all conscious of the essential role that water plays in our diet. From the slight discomfort we have experienced with thirst and from our knowledge of the agony of men deprived of water for long periods of time, we know how vital it is to prevent dehydration of the body. This dehydration can come from fluid loss from unchecked external or internal hemorrhage; it can result from inadequate fluid intake or from too great a loss of fluid from the kidneys, skin, or lungs. It can be aggravated by high temperatures, both external and internal (fever). All these varied circumstances have as a result the alteration of the internal environment, and thus the life of the cell is endangered.

Functions of Water

Living tissues are composed of a small amount of solids and a varying proportion of water, depending on the tissue age and type, from 50 to 90 per cent. Very rigid tissues such as bone have less water than fluid tissues such as blood, but in all cases, water is the chief constituent of tissue, particularly in young animals. The dehydration that comes with age probably has a great deal to do with decreased efficiency of tissue metabolism and is one of the outstanding problems in geriatrics. All chemical reactions that take place in the body do so in an aqueous medium, and usually the water molecules are actually involved in the reaction. The principal function of water in the body, then, is to provide a suitable *medium for chemical reactions.*

Because water is a poor conductor of heat, it fulfills the role of an *insulator,* making it possible for the heat-regulating mechanisms to adjust to changes in the external temperature gradually. An animal with these heat-regulating devices experimentally destroyed still would not show the same uptake of heat from the sun as would a piece of stone. The water content of animal tissue prevents heat from being absorbed or lost too rapidly.

On the other hand, water with ions dissolved in it can *conduct an electrical current.* This is an important characteristic of many physiological phenomena, such as the conduction of nerve impulses. Water has a *high surface tension,* a factor that plays a fundamental role in the formation of cell membranes and the concentration of substances upon these surfaces, which facilitate chemical reactions.

Water is of great importance in maintaining the *proper size and shape of cells and tissues.* The amount of water inside and outside the cell will determine its turgidity: a swollen cell in danger of bursting or a shrunken cell cannot function efficiently. When this maldistribution of water is extended to the level of the tissues, it results in *edema* (excess fluid in the tissue spaces) or *dehydration* (lack of fluid in the tissue spaces). Either can lead to severe disturbances in the blood supply and thereby the nutrition of the affected areas.

Water can also act as a *fluid medium for the removal of excreted wastes.* In the human, metabolic wastes secreted by way of the kidneys are dissolved in the urine, and the concentration of these waste products is determined largely by the simultaneously excreted volume of water. This is not a universal biological property—many other animals excrete such wastes in solid or semisolid form—but it is nevertheless essential for the proper functioning of the mammalian kidney.

Various other mechanical processes are *lubricated* by watery fluids containing mucus or salts; for example, chewing, swallowing, and talking are severely handicapped by insufficient amounts of saliva; the eyes are kept free from dirt particles by the tears; sexual intercourse is possible only when sufficient amounts of lubricating fluid are secreted by the accessory sex glands to permit penetration.

To summarize the general functions of water:

1. Maintenance of cellular size and form.
2. Medium for chemical reactions.
3. Insulation from temperature extremes.
4. Electrical conduction.
5. Facilitation of surface reactions.
6. Fluid medium for excreted wastes.
7. Lubrication.

Molecular Structure and Chemical Reactivity

Water, as we refer to it in everyday life, is a compound containing, apart from microscopic biologic forms, many common salts and gases. It varies in composition and concentration depending on its source, its proximity to salt water, the metal of the pipes through which it passes, and many other factors. Water itself, however, always has the same chemical structure, regardless of where it is found, and all other components are contaminations of this compound.

The smallest particle of this compound that still retains its properties is the *molecule* of water, H_2O. Water is referred to as a compound because its molecules are made up of atoms from more than one kind of *element,* the hydrogen (H) and the oxygen (O) atoms. Separately, these *atoms* have completely different properties from the molecules of water. The most obvious difference is that at the same temperature and pressure the hydrogen and oxygen atoms are gases, whereas the water molecules are liquid. One can also have molecules that are composed of two or more identical atoms. For example, gaseous oxygen usually exists in the molecular form; that is, two atoms of oxygen form the molecule of oxygen. Despite the fact that molecular oxygen is made up of only one kind of atom, it has different reactive properties from atomic oxygen.

The *definition of a molecule* is still not complete. We must say that a molecule is the smallest part of a substance, compound or simple, that still retains the properties of that substance and is composed of atoms in very specific proportions. H_2O is water. H_2O_2 is a very different compound, hydrogen peroxide. Yet these two compounds are composed of the same types of atoms, the hydrogen and the oxygen atoms. Under certain circumstances, the one compound can be transformed into the other.

$$H_2O_2 + H_2O_2 \xrightarrow{\text{light energy}} H_2O + H_2O + O_2$$

2 molecules of hydrogen peroxide — 2 molecules of water, 1 molecule of oxygen

This is more usually written:

$$2H_2O_2 \xrightarrow{\text{light energy}} 2H_2O + O_2$$

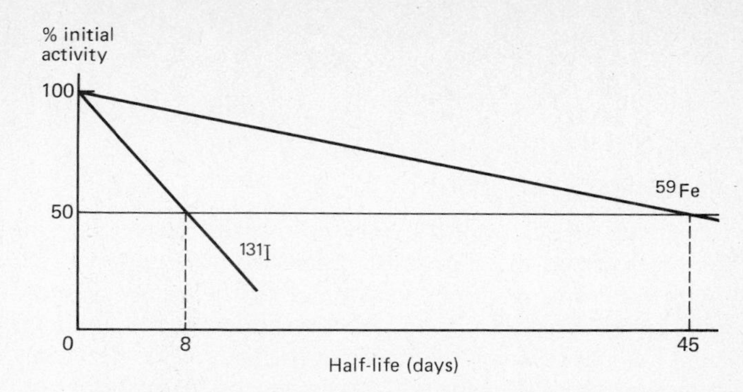

Figure 3-1. **Half-lives of two radioactive elements, iodine (^{131}I) and iron (^{59}Fe).**

This transformation is obviously caused by a *change in the proportion of the atoms* forming the respective molecules. Thus the chemical reaction indicated above is a shorthand for the initial disruption of the forces holding the atoms together in the one type of molecule and the re-formation of these atomic forces to yield a new molecule. The *forces that hold the atoms together* to form a molecule are very similar to those that hold the atom itself together, only the latter are so much stronger that extremely powerful reactors are required to split them. To understand these binding forces, it is necessary to know a little about the basic structure of an atom.

Structure of the atom

An atom is composed of a *nucleus* and *orbiting electrons.* The nucleus contains positively charged particles, the *protons,* which are equal (in the uncharged state) to the number of negatively charged particles, or *electrons,* which give the atom its atomic number and its chemical properties. In reacting molecules, or in ions, some electrons may be lost or gained. Ions are discussed later in this section and in Sec. 3-2.

In addition, the nucleus contains uncharged particles, the *neutrons,* which have mass but no electrical charge. Elements with the same number of protons, and therefore of planetary electrons, but with different numbers of neutrons, are *isotopes.*

Because the nucleus of the atom contributes mainly mass, and the planetary electrons determine entirely the chemical properties of the element, it is obvious that since isotopes have the same number of electrons, they have the same chemical properties, differing only in their mass. They can be substituted experimentally for the naturally occurring element in the living body and be traced through their metabolic pathways because they eventually can be detected through their weight difference. If the heavy isotope is unstable, that is, if the nucleus tends to disintegrate as its particles escape, it is called *radioactive:* the rate of disintegration varies with the type of nucleus and is usually expressed in terms of the length of time for half the number of nuclei to disintegrate, the *half-life of the element* (Fig. 3-1). These isotopes can then be traced by their radioactivity. Two of the isotopes of hydrogen are stable or nonradioactive; a third one (*tritium*) is radioactive and has a mass number of 3. The heavy stable isotope of mass number 2 is called *deuterium,* as opposed to the light one of mass number 1, which is the commonly occurring form (*protium*) (Fig. 3-2). When deuterium is incorporated into water, the resulting substance is called heavy water; it may be given to experimental animals as a tracer to determine the fate of water in the body. The results of these experiments will be discussed later.

Chemical reactivity of the atom

The planetary electrons are arranged in orbits around the nucleus. The innermost orbit requires two electrons (one electron pair) for stability; the next orbit requires eight (four electron pairs); and the next ones require varying numbers of negatively charged particles for completion. Depending on whether acceptance or donation of electrons will most readily complete the outermost orbit, the combining power, or *valence,* of an atom is designated by a minus (−) or a plus (+). The hydrogen atom donates an electron, losing one negative charge and thereby becoming positively charged. Its valence is 1+. Oxygen accepts two electrons, resulting in a net negative charge of 2, and a valence of 2−. It is important to remember that the positive or negative charge is a way of expressing *ability to donate or accept electrons, which are always negatively charged particles.* When atoms combine to form molecules, they do so in various ways, one of which is by sharing their incompleted orbits of planetary electrons and thus satisfying their valence (number of positive or negative charges). These shared electrons (one contributed by each atom) form *covalent bonds.* Figure 3-3 shows the covalent bonds in a molecule of water. In some molecules, electrons are shared in groups of four, rather than in pairs. These are *double bonds*

H:C̤::C̤:H
  H   H

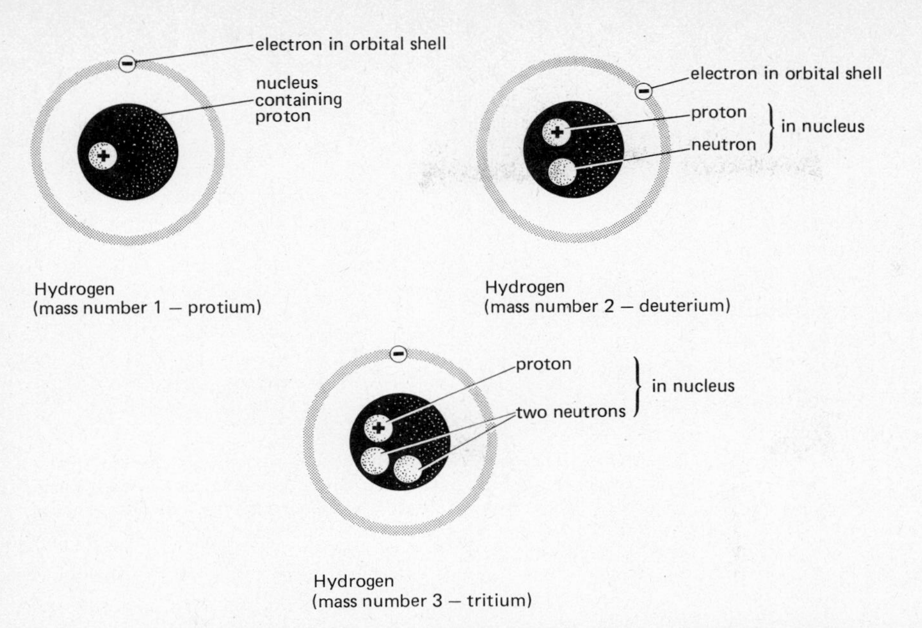

Figure 3-2. **The isotopes of hydrogen. The first two isotopes, protium and deuterium, are stable; the third isotope is the unstable, radioactive tritium.**

Figure 3-3. **Atomic structure of a molecule of water, showing the shared electrons of the covalent bonds.**

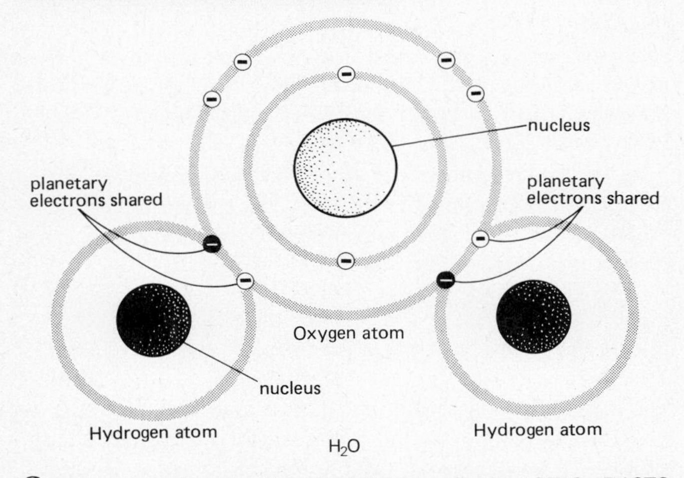

Chemical reactivity of water: ions, acids, bases, and the pH of blood

Ions. Water is important in biological chemical reactions for two reasons: first, it participates itself in many reactions, being either incorporated or released; second, it dissolves other substances so that they can react chemically. What does this actually mean? Water facilitates the breakdown of loosely associated molecules into their component electrically charged atoms, or groups of atoms. These electrically charged particles are called *ions.* Ions may be negatively charged (*anions*) or positively charged (*cations*) and may have single or multiple charges (H^+; Ca^{2+}; K^+; OH^-; O^{2-}).

Water itself is a *dipole,* having both positive and negative charges, for it is composed of two hydrogen atoms, which donate electrons easily, and one oxygen atom, which accepts electrons ravenously

$$\begin{matrix} +H \searrow \\ \quad O^- \\ +H \nearrow \end{matrix}$$

Only molecules that are polar themselves (or are charged) are highly soluble in water.

Acids and Bases. An *acid* is defined as a compound that on ionization yields hydrogen ions, or protons, plus ions capable of combining with protons. Those ions that can combine with protons are *bases.*

$$\underset{\text{acid}}{\boxed{HCl}} + H_2O \rightleftharpoons H^+ + \underset{\text{base}}{\boxed{Cl^-}} + (H^+ + OH^-)$$

An *alkali* is defined as a substance that on ionization yields hydroxyl ions (OH^-).

$$\underset{\text{alkali}}{\boxed{NaOH}} + H_2O \rightleftharpoons Na^+ + \boxed{OH^-} + (H^+ + OH^-)$$

The *strength of an acid* is dependent on the relative ease with which its associated base can combine with protons. Hydrogen cyanide (HCN) is a weak acid because its base (CN^-) combines firmly with protons. Hydrochloric acid (HCl) is a strong acid because its base (Cl^-) combines very loosely with protons. The strength of an acid or base is expressed in terms of the pH scale.

pH. *The pH scale* is an expression of the relationship between the concentration of hydrogen ions and hydroxyl ions released by a substance on ionization. The number of these ions released is so large that it is quite unwieldy for practical use so the relationship is simplified mathematically by being expressed as the *negative logarithm of the hydrogen ion* (H^+) *concentration.* The pH scale indicates the degree of acidity of a solution, but it is in inverse magnitude to the number of H^+ released.

A solution of hydrochloric acid (HCl) that on dissociation yields 1×10^{-2} g mol per liter has a pH of

$$\boxed{-\log H^+ \text{ concentration}} = \log \frac{1}{H^+} \text{ concentration}$$

$$\text{or } pH = \log \frac{1}{10^{-2}} = \log 10^2 = 2$$

For convenience this solution is said to have a pH of 2. The useful range of the scale is

pH	1	2	3	4	5	6	7	8	9	10	11	12	13	14
	Very acid (Most H^+ released)						Neutral ↑						Very basic (Most OH^- released)	

Water dissociates into H^+ and OH^-, but only about one molecule of water in 550 million molecules is dissociated into ions at ordinary room temperatures. This number increases greatly if the temperature is increased.

Water must be neutral as it dissociates (although slightly) to yield equal numbers of hydrogen and hydroxyl ions. Physiological solutions usually have a pH that varies only minimally around neutrality, despite the complexity of their composition. This means that the homeostatic mechanisms controlling this delicate adjustment are very involved. The pH of blood is of great importance: it determines the pH of the cellular environment and consequently affects the cell itself. Living cells are extraordinarily sensitive to changes in pH. The normal pH of blood is 7.4, that of the tissue fluids between 7.3 and 7.4, and the pH of the interior of the cell varies somewhat around 7.1. These values are all just slightly alkaline. When metabolic wastes are discharged rapidly into the circulation as a result of strenuous exercise, the pH of the blood may sink to 7.2 (more acid), but controlling mechanisms prevent any further variation. These important homeostatic mechanisms will be considered in detail in the chapter on blood.

3–2 Important Biological Ions

The ions found in the tissues and in the blood of animals bear a very interesting relationship to the theory of evolution of modern forms from an original sea-water environment (Fig. 3-4). The mechanisms of water metabolism and the maintenance of ionic concentrations are different in animals from fresh-water, salt-water, and terrestrial environments because the problems of water conservation are so different. The structure of the kidney varies accordingly in these forms and will be discussed with kidney function in Chapter 15.

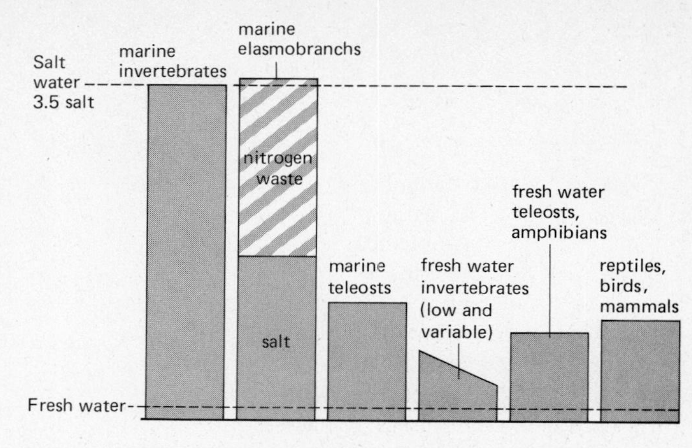

Figure 3-4. **The osmotic concentrations of the body fluids of various groups of animals in relation to the concentration of sea water. (Adapted from Knut Schmidt-Nielsen, *Animal Physiology,* 2nd Edition, © 1964, p. 51. Reprinted by permission of Prentice-Hall, Inc., Englewood Cliffs, New Jersey.)**

In animals, the most important positively charged ions are sodium, potassium, hydrogen, magnesium, calcium, and iron (Na^+, K^+, H^+, Mg^{2+}, Ca^{2+}, Fe^{3+}); these are all positively charged *cations.* The most important negatively charged *anions* are the hydroxyl, chloride, bicarbonate, sulfate, phosphate, and carboxyl ions (OH^-, Cl^-, HCO_3^-, SO_4^{2-}, PO_4^-, and COO^-).

In addition to these inorganic ions, which are present in an approximate concentration of 200 mEq/liter, charged organic molecules, such as proteins, are important in many physiological processes.

3–3 Macromolecules

Macromolecules are large molecules, polymers made up of many subunits and varying in size from a molecular weight of about 4000 to several million. The significance of this to biological systems cannot be underestimated. While polymers of sugars are relatively simple, tending to be made of chains of the same subunit (e.g., glycogen is made of linked glucose molecules only), protein macromolecules are composed of up to 20 different amino acids. As these are three-dimensional structures, the isomeric possibilities are immense. Nucleic acids are similarly complex macromolecules. The entire genetic code is transcribed in the DNA molecules, which not only can reproduce themselves but can also direct the reproduction of other molecules, a necessity for the survival of the species and, parenthetically, the basis for biological variation and evolution.

3-3.1 Proteins

General significance of protein in living systems

Proteins make up the bulk of the organic material within a cell, amounting to 12 to 14 percent of the cell. These macromolecules vary in molecular weight (Fig. 3-5) from several thousand (myoglobin) to more than a million (lipoprotein). Their importance is signified by the Greek origin of the word, protein, from *proteios,* meaning of first importance. Proteins are prime contributors to cellular *structure,* whether in the form of fibrils, or networks, that increase the liquid *sol* state of cytoplasm to the more viscous *gel* state, or in combination with lipids to form complex cell membranes. The protein, *collagen,* is the major fibrous constituent of connective tissue; the *elastin* fibers of the blood vessel walls are protein, as is *fibrinogen,* the precursor of the fibrin that forms in blood clots. Protein is in intimate contact with the nucleic acid, DNA, within the chromosomes and with the cytoplasmic nucleic acid, RNA, in the ribosomes. Some organisms consist almost completely of protein and nucleic acid: the structure of the poliomyelitis virus is due to its protein component; its enormous ability to reproduce is due to its single strand of RNA.

Not only do proteins give structure to the cell, they are closely concerned with the *metabolic reactions* occurring within it. The *enzymes* catalyzing these reactions are proteins, often in combination with other chemical groups. Many *hormones* are proteins, for example, insulin and the hormones secreted by the anterior pituitary gland. The *distribution of water* between the blood and tissue fluid is partly regulated by the osmotic pressure generated by the nondiffusing proteins in the plasma (Sec. 4-2). The efficiency of *transportation of oxygen and carbon dioxide* through the blood is immensely enhanced by the chemical properties of the iron-containing protein, hemoglobin.

All living material, plant and animal alike, contains protein, but the type of protein varies considerably, not only from species to species but even between individuals. The early difficulties in the attempts to transfuse blood from one individual to another came from a lack of understanding of the specificity of the protein molecule, in particular the blood groups. Proteins of the skin seem to be even more specific, and skin grafts, unless they are donated by the "host" himself or his identical twin, are rejected with extreme rapidity. This ability of protein molecules to distinguish differences in the conformation of similar protein molecules is based on small variations at the surface of a protein molecule, involving perhaps a surface patch of a few amino acids. Years of work have been devoted to the unravelling of the internal structure of the protein molecule. In 1958 Sanger won the Nobel prize for his analysis of the two-dimensional structure of insulin. Since then, Perutz and Kendrew have been similarly honored for their description of the three-dimensional structure of the proteins hemoglobin and myoglobin.

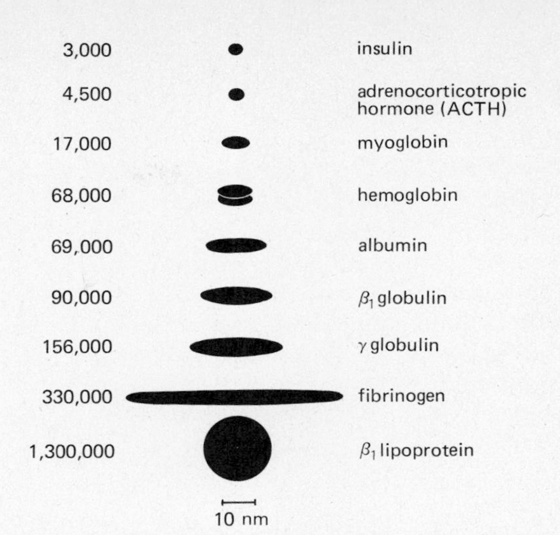

Figure 3-5. **The relative sizes of protein molecules and their molecular weights.**

Molecular structure and reactivity

Proteins are made of the elements carbon, hydrogen, oxygen, and nitrogen, with variable amounts of other elements, such as phosphorus (P), sulfur (S), magnesium (Mg), and iron (Fe).

Amino Acids—Building Blocks for Proteins. The basic building block of the large protein molecule is the relatively small molecule, the *amino acid.* There are more than 20 different amino acids, and a protein molecule may consist of many hundreds of these, linked together in a specific formation that gives that protein molecule its specific attributes. When one considers the enormous variety of words one can form from the 26 letters of the alphabet, where the length of the word is limited to about 30 letters in most languages, and where the linkage of the letters is in only one direction, the infinite possibilities of the amino acid chains, with three-dimensional linkages, are awe-inspiring.

If we take the simple seven-letter word *protein* and try to see how many arrangements we can make from it, we soon give up for lack of space—and patience:

PROTEIN NIETORP PRTIENO PORTIEN POTRINE
ETNIRTP EONIRTP OTEINPR EINTROP NTPIERO

There are 10^{65} ways of arranging 20 amino acids into a protein 50 amino acids long.

Structure of amino acids. Amino acids possess both acidic and basic qualities. The basic qualities are derived from the presence of one group of atoms that can *accept hydrogen ions,* the *amino radical:*

$$\cdots NH_2^+ + H^+ \rightleftharpoons \cdots NH_3$$

The acidic qualities come from the *carboxyl group,* which

can *donate hydrogen* ions. This radical dissociates to yield

$$\cdots COOH \rightleftharpoons \cdots COO^- + H^+$$

The amino and carboxyl radicals are attached to the same carbon atom, designated as the *alpha carbon atom.* This C_α also carries a hydrogen atom and the rest of the amino acid. All amino acids possess this grouping around the C_α and vary only in the nature of the rest of the molecule. Since there are 20-odd amino acids, each having an amino and carboxyl group attached to a C_α, the residue, R, of each obviously must differ.

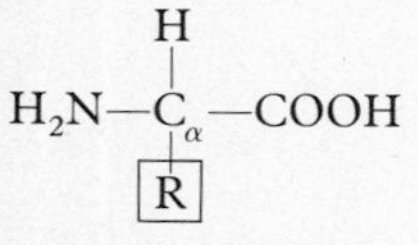

$H_2N{-}C_\alpha(H)(\boxed{R}){-}COOH$ General formula for an amino acid

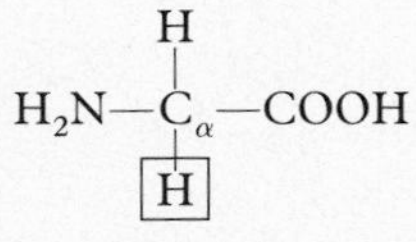

$H_2N{-}C_\alpha(H)(\boxed{H}){-}COOH$ Simplest amino acid (glycine)

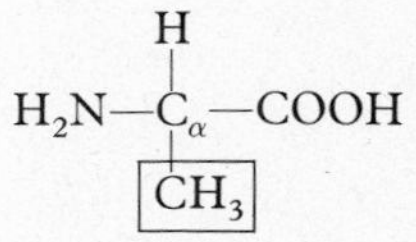

$H_2N{-}C_\alpha(H)(\boxed{CH_3}){-}COOH$ Alanine

If R is a methyl group (CH_3), the amino acid, alanine, is formed. Glycine is the only amino acid with two identical chemical groups (two hydrogen atoms) attached to the alpha carbon; its mirror images are superimposable and it is said to be symmetrical. All other amino acids are asymmetrical; their mirror images are not superimposable. They are called *enantiomorphs.* If a beam of polarized light is passed through a crystal of an asymmetric amino acid, the light may be rotated to the left, levo (l), or to the right, dextro (d), depending on the asymmetry of the amino acid. These optical isomers are distinguished by the cell, and only the L amino acids are metabolized. On the other hand, only the D configuration of most sugars can be utilized.

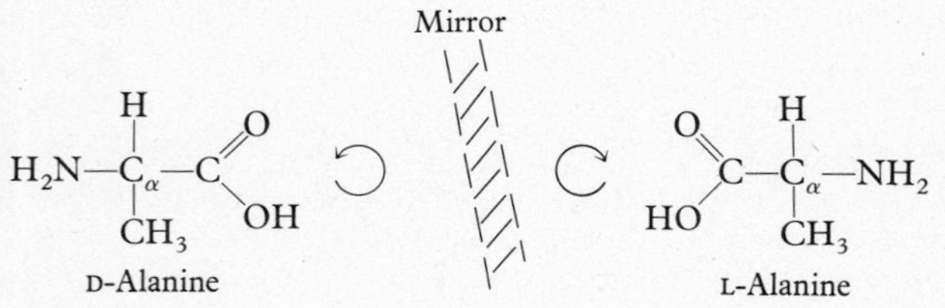

Mirror

$H_2N{-}C_\alpha(H)(CH_3){-}C(=O)OH$ D-Alanine

$HO(O=)C{-}C_\alpha(H)(CH_3){-}NH_2$ L-Alanine

(The direction of the rotation of light is indicated by the small d and l letters; D and L indicate the structural configuration of the enantiomorphs.)

The Oparin-Haldane hypothesis (1964 and 1954) of chemical evolution suggests that complex organic molecules were formed from simple molecules, present in the primitive atmosphere of the earth or other planets before the appearance of life. The classic laboratory experiment performed by Miller in 1953 showed that amino acids could be formed by the discharge of an electric spark through a mixture of methane, ammonia, and water. Later experiments by Harada and Fox (1965), among others, have shown that various atmospheres containing carbon dioxide, carbon monoxide, formaldehyde, and hydrogen sulfide can give rise to an assortment of amino acids if sufficient energy (comparable to a bolt of lightning) is applied in the absence of oxygen.

Support for the extraterrestrial synthesis of amino acids comes from an analysis by Ponnamperuma of amino acids found on the Murchison meteorite, which fell on Australia in 1969. There were equal amounts of D and L amino acids; if they had been contaminated by terrestrial amino acids they would have been preponderantly of the L form. These molecules, the product of naturally occurring chemical evolution, apparently stopped at this point and did not proceed on to life. The fact that we know that organic molecules can be formed from simple molecules, provided that energy is available, and that these organic molecules may proceed to build protein macromolecules, does not mean that we know how the first cell was formed, nor the first tissues or multicellular organisms. Like "Humpty Dumpty," it is always much easier to take things apart than to put them together again.

FUNCTIONS OF AMINO ACIDS. Amino acids are important in many metabolic pathways, apart from protein synthesis (Fig. 3-6). If the methyl group of alanine is replaced by a benzene ring, a ring structure made of six CH groups with alternating single and double bonds, the amino acid, *phenylalanine,* is formed. If one hydroxyl group is added to phenylalanine, *tyrosine* results. Tyrosine, in turn, is the precursor for the hormones of the thyroid gland, the chief of which is *thyroxine.* Two hydroxyl groups added to phenylalanine result in the formation of *dihydroxyphenylalanine* (*DOPA*). In turn, DOPA gives rise to the chemical transmitter *dopamine,* which is essential for the normal function of certain parts of the nervous system. Lack of dopamine is associated with Parkinson's disease, a serious abnormality of the motor system, which can be treated by the administration of relatively large amounts of L-DOPA. Dihydroxyphenylalanine is also on the metabolic pathway for the synthesis of other neurotransmitters, such as norepinephrine, and in addition is one of the precursors for the pigment melanin, found in the skin, eyes, and hair. If there is a metabolic defect in the beginning of this long pathway, caused by the lack of the enzyme necessary to metabolize phenylalanine to tyrosine (*phenylalanine hydroxylase*), then instead of these useful molecules being synthesized, phenylalanine is broken down to ketones. These ketones accumulate in the tissues, being particularly toxic to nervous tissue. If untreated, mental retardation inevitably occurs. However, a simple test of the blood or urine of the newborn can indicate the presence of these ketones (the PKU test for *phenylketoneurea*); appropriate dietary therapy can prevent the retardation but not reverse it subsequently.

PEPTIDE BONDS. Amino acids are linked to form protein macromolecules through *peptide bonds.* The molecule formed when 2 amino acids are joined is a *dipeptide;* 3 form a *tripeptide,* several form a *polypeptide,* and more than 50 are usually defined as a protein. If a protein is broken down in the presence of water (hydrolyzed), either in the digestive system through enzymatic action or in a test tube with heat, polypeptides are obtained first, followed by dipeptides and individual amino acids.

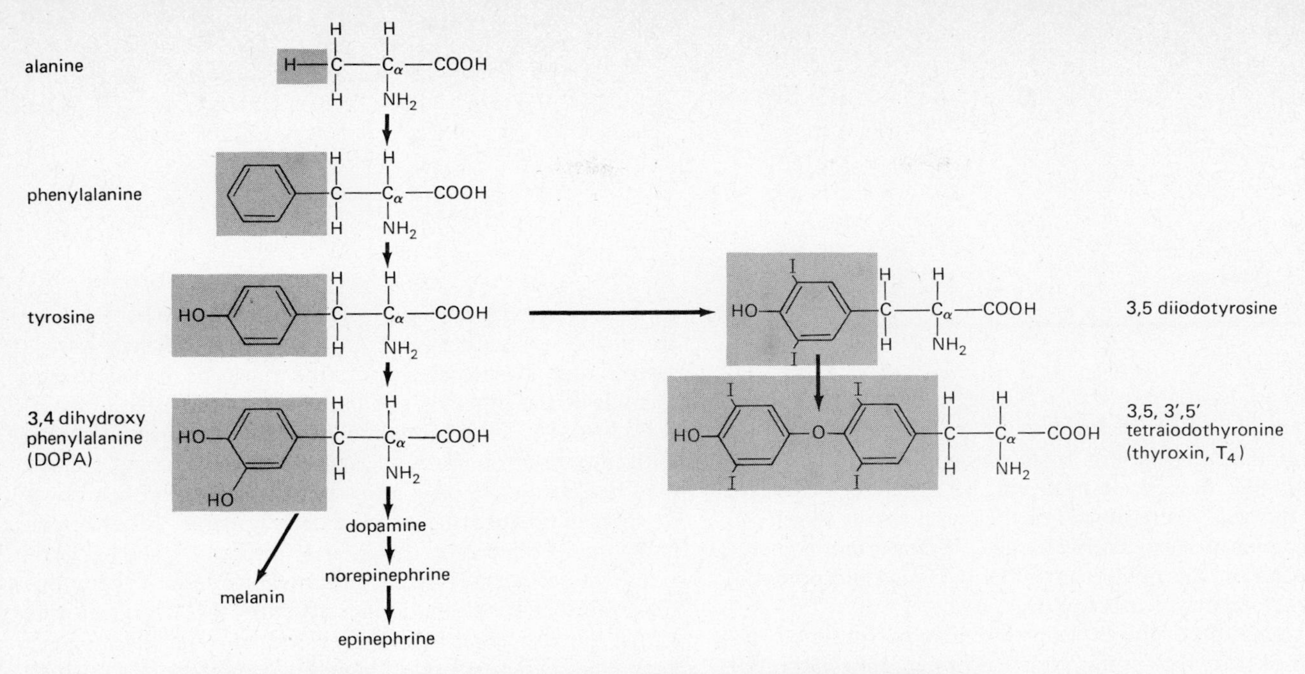

Figure 3-6. **Some of the metabolic pathways of an important amino acid, alanine. Melanin is a pigment; dopamine, norepinephrine, and epinephrine are neurotransmitters; and epinephrine is also a hormone, as is thyroxin.**

The ability of amino acids to form peptide bonds depends on their ampholytic qualities. *Ampholytes* can either accept a proton from a strong acid or donate a proton to a strong base. In other words, amino acids behave as dipolar ions with both acidic and basic properties. The pH of the medium will determine the net charge of the molecule.

$$H_2N{-}R{-}COOH + H^+ \longrightarrow {}^+H_3N{-}R{-}COOH$$

(acid solution)

R $^+$ = net charge of molecule

$$H_2N{-}R{-}COOH \longrightarrow H_2N{-}R{-}COO^- + H^+$$

(basic solution)

Ⓡ $^-$ = net charge of molecule

Some amino acids have more than one free basic or acidic group; consequently, there will be variable numbers of amino and carboxyl groups ionized at a particular pH. If equal numbers of acidic and basic groups are ionized, the molecule will have a net charge of zero and will not migrate to either pole when placed in an electric field. The pH at which the amino acid, or other dipolar molecule, is electrically neutral is known as its *isoelectric point* (p*K*).

$${}^+H_3N{-}R{-}COO^- \qquad Ⓡ^{\pm} = \text{net charge of molecule}$$

There are sharp changes in the characteristics of a molecule at its isoelectric point; the viscosity, solubility, electrical conductance, and so forth, are decreased. This is particularly important in the case of proteins, for their behavior will depend largely on their amino acid composition and the pH of the environment (Fig. 3-7 and Table 3-1). If the pH of the medium is on the acid side of the p*K* for a particular protein, that protein will be positively charged. If the pH of the medium is on the basic side of the p*K*, the protein will carry a negative charge. Thus, if

Figure 3-7. **Solubility of egg albumin (●—●) and hemoglobin (○—○) as a function of pH. The charge of these proteins is also altered by the pH of the medium. Note that at a pH of 5.5, egg albumin (p*K* 4.6) is negatively charged, whereas hemoglobin (p*K* 6.8) is positively charged.**

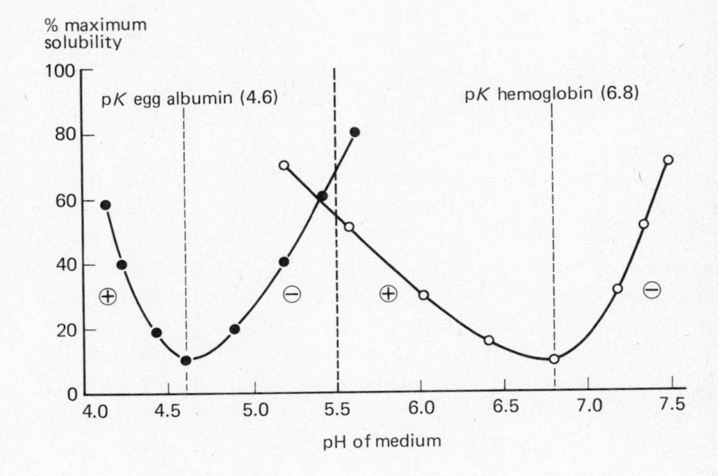

Table 3–1. Isoelectric Points (pK) of Some Proteins

Protein	pK
Pepsin	1.0
Egg albumin	4.6
Adrenocorticotropin (ACTH)	4.7
Insulin	5.3
Hemoglobin	6.8
Ribonuclease	7.8
Cytochrome *c*	10.7

hemoglobin, with a pK of 6.8, and egg albumin, with a pK of 4.6, are placed in the same medium, which has a pH of 5.5, then the hemoglobin molecule will have a positive charge, while the egg albumin will be negatively charged. Within the cell, where the pH of the cytoplasm is about 7.2, most proteins are negatively charged, an important contributing factor in the establishment of a bioelectric potential across the plasma membrane (Sec. 6–2).

The formation of the peptide bond results from the combination of an amino group attached to the alpha carbon of one amino acid, with the carboxyl group attached to the alpha carbon of a second amino acid. This union is accomplished through the elimination of a molecule of water.

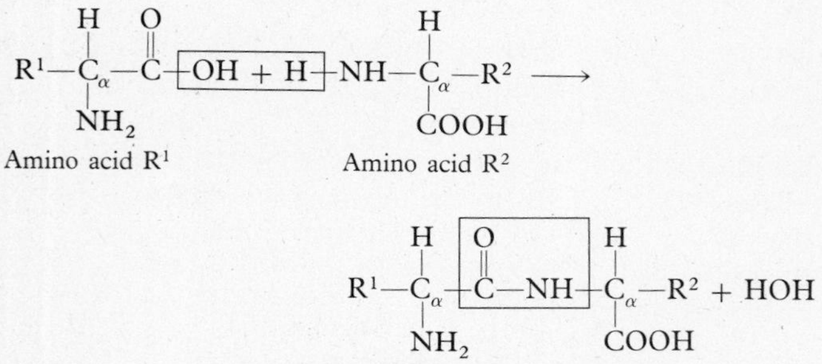

$$R^1{-}C_\alpha H(NH_2){-}C({=}O){-}\boxed{OH + H}{-}NH{-}C_\alpha H(COOH){-}R^2 \longrightarrow$$

Amino acid R[1]   Amino acid R[2]

$$R^1{-}C_\alpha H(NH_2){-}\boxed{C({=}O){-}NH}{-}C_\alpha H(COOH){-}R^2 + HOH$$

Peptide bond of dipeptide

The plane of the peptide bonds can be seen in this *tripeptide:*

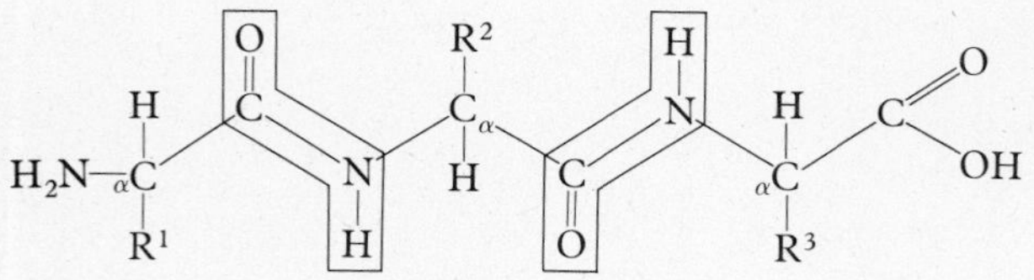

$$H_2N{-}{}_\alpha CH(R^1){-}\boxed{C{=}O}{-}\boxed{N{-}H}{-}C_\alpha H(R^2){-}\boxed{C{=}O}{-}\boxed{N{-}H}{-}{}_\alpha CH(R^3){-}C({=}O)OH$$

The long chains of amino acids, linked to form polypeptides, can fold in a number of ways, and in many cases additional cross-links help to stabilize these folds. The *disulfide bond* (S–S) of the amino acid *cystine* permits one cystine molecule to enter into two separate polypeptide chains, linking them together or causing a single chain to fold back on itself to form a ring. These disulfide bridges can be broken by relatively strong chemical or physical forces that reduce them to the sulfhydryl form (-SH), yielding two *cysteine* molecules.

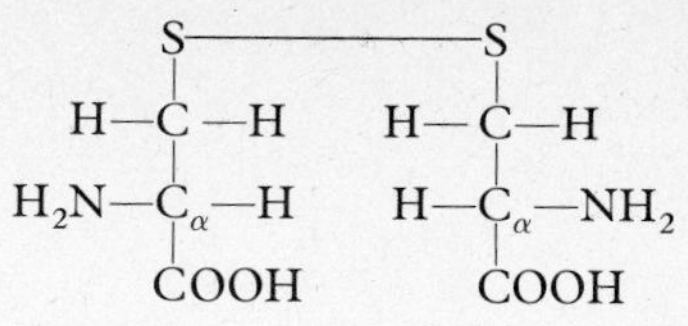

$$HOOC{-}C_\alpha H(NH_2){-}CH_2{-}S{-}S{-}CH_2{-}C_\alpha H(NH_2){-}COOH \qquad HS{-}CH_2{-}C_\alpha H(NH_2){-}COOH$$

Cystine 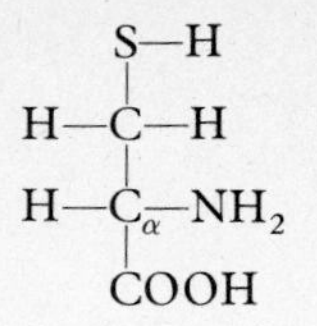   Cysteine

Cystine is an essential amino acid for premature babies, who are unable to synthesize it. Cow's milk is deficient in this amino acid; consequently, cystine must be added to the formula if the infant is not being breast-fed.

PEPTIDES. There are many naturally occurring peptides with important biological activities. Many proteins, in fact, may become active only after they have been broken down to shorter polypeptide or peptide fragments. Most of the hormones produced by the anterior pituitary gland are peptides (the rest are proteins), and the hormones of the hypothalamus, the neuroendocrine structure that regulates this gland, are also peptides. Many vasodilator substances have been identified as peptides, including *substance P* (so named because it was first isolated as a powder). Substance P will produce a marked effect on smooth muscle contractile activity and is an extremely potent vasodilator. Fragments of *adrenocorticotropic hormone* (*ACTH*), a polypeptide consisting of 39 amino acids produced by the anterior pituitary gland, are suspected of influencing neurotransmission and the excitability of the central nervous system. A similar neuroexcitatory role is attributed to *MSH* (*melanocyte stimulating hormone*), the better-known function of which is to stimulate the synthesis of the pigment, melanin, in mammals, and to cause melanin dispersal in the melanophores of cold-blooded vertebrates.

Two of the hypothalamic hormones, *vasopressin* and *oxytocin*, are built on the same basic plan: a ring structure of amino acids, with a tail. The substitution of one or two amino acids, in the third and eighth positions, is responsible for both qualitative and quantitative differences between these two octapeptides. Although these hormones are made up of eight amino acids, there are nine positions, for the first and the sixth are the identical halves of a single cystine molecule, linked to form a ring.

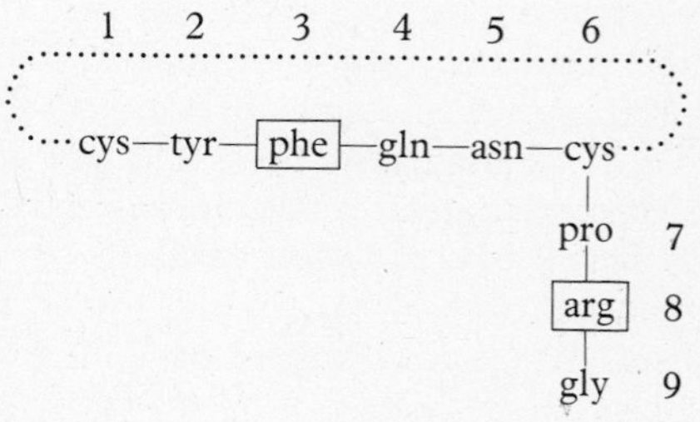

1 cys — 2 tyr — 3 [phe] — 4 gln — 5 asn — 6 cys (1 and 6 linked in a ring)
6 cys — 7 pro — 8 [arg] — 9 gly

cys = cystine
tyr = tyrosine
phe = phenylalanine
gln = glutamine
asn = asparagine
pro = proline
arg = arginine
gly = glycine
iso = isoleucine
leu = leucine

Vasopressin

Vasopressin not only has marked vasopressor effects on the circulation, but it causes water retention by the kidney; hence its other name, *antidiuretic hormone* (*ADH*). If isoleucine replaces phenylalanine in the third position, and leucine substitutes for arginine in the eighth position, a hormone with quite different properties is formed. This is oxytocin, which causes milk ejection from the mammary glands, or breasts, and contraction of the uterus at parturition. Oxytocin may have very mild vasopressin effects.

```
  1    2    3     4     5    6
  cys—tyr—[iso]—gln—asn—cys
                         |
                        pro  7
                         |
                       [leu] 8
                         |
                        gly  9
```

Oxytocin

If isoleucine is retained in position 3, but arginine is in position 8, an octapeptide is formed with properties somewhere in between those of vasopressin and oxytocin. It is *vasotocin*, found in lower vertebrates like the frog, and it probably was the precursor of the two hormones of the higher vertebrates.

```
  1    2    3     4     5    6
  cys—tyr—[iso]—gln—asn—cys
                         |
                        pro  7
                         |
                       [arg] 8
                         |
                        gly  9
```

Vasotocin

Despite the obvious fact that a frog does not eject milk nor does it have a uterus to contract, frog vasotocin will cause these effects when injected into mammals. This leads to the fascinating inference that the hormones were produced first, before the appropriate tissues were developed, early in the evolution of animal phyla. Only later were the hormones adapted to the use of terrestrial animals. This is supported by the observation that vasopressin, the antidiuretic hormone, is present in fresh-water fish, which certainly have no need to conserve water.

LEVELS OF PROTEIN STRUCTURE

The individuality of a protein molecule lies not only in its composition of amino acids, usually a selection of from about 15 to 20 different ones, but also the amino acid sequence in the polypeptide chain. These two factors result in the *primary structure* of the protein. The *secondary structure* is created by the helical twisting of the amino acid chains, which average about 3.6 amino acids for each twist of the helix. The helices, in turn, are coiled into a complex three-dimensional conformation, which gives the protein its *tertiary structure,* with its exposed areas for enzyme action, cell recognition, and other specific parameters.

Methods of Separating Proteins. To *extract* a protein from a cell, various methods that will fragment the cell are used. These include grinding, shaking with glass beads, freezing and thawing, and disintegration through sonic vibration. The proteins then are extracted from the homogenate with appropriate solvents and *separated* from this mixture on the basis of their *solubility.* This usually involves the careful adjustment of the pH of the medium, for the solubility of each protein varies with its isoelectric point (Fig. 3-7). Separation of proteins by *ultracentrifugation* on the basis of their molecular weight is another useful technique.

Proteins can be separated by *electrophoresis,* a method based on the migration of charged particles to the poles of an electric field. As a protein's net charge varies with pH, two proteins with different isoelectric points will have different mobilities. If a protein mixture is placed in a U tube, together with a buffer, a negatively charged protein molecule will move toward the anode in one arm of the tube, away from the cathode in the other arm (Fig. 3-8). There are many refinements of this technique. One can measure changes in optical density as the protein concentration changes at the boundary between it and the buffer: the electrophoresis may take place on *starch strips,* on *agar,* or on polyacrylamide, which will slow the movement of the protein molecules and permit easier separation (Fig. 3-9).

Proteins can also be separated on the basis of their size and diffusion constant by using a column of inert polysaccharide called *Sephadex,* which can be prepared like a series of sieves, each with different size holes. Smaller molecules can penetrate the appropriate Sephadex column, while the larger ones remain in the solvent surrounding the column.

Primary Structure. The sequence of amino acids in a chain makes up the primary protein structure. The

Figure 3-8. **Separation of proteins by electrophoresis in a liquid medium. The negatively charged proteins (anions) migrate toward the anode (positive electrode) in one arm of the U tube; the positively charged cations move toward the cathode (negative electrode).**

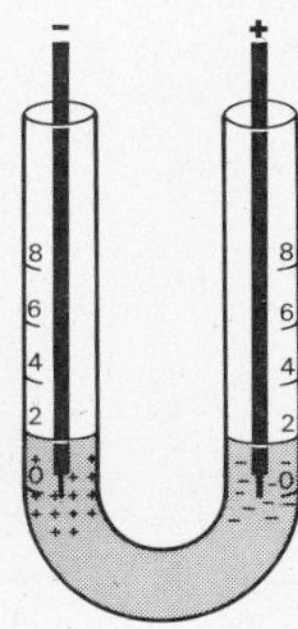

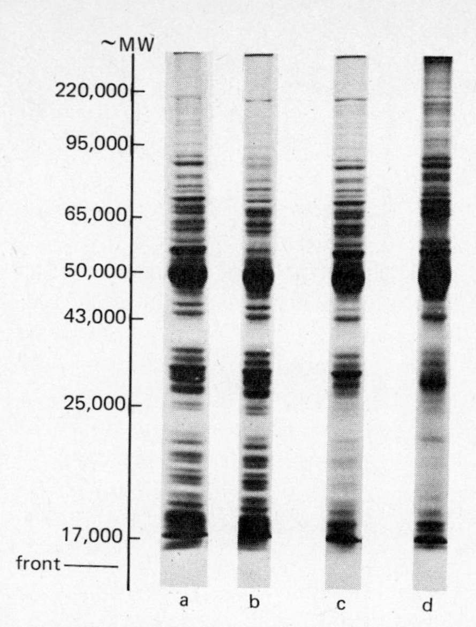

Figure 3-9. **Proteins can be separated by electrophoresis on polyacrylamide gels. These proteins are from different microsomal subfractions of rat liver. The gel patterns are formed when the proteins migrate at different speeds in an electric field. a = rough microsomes (rm), b = rm after removal of the cisternal content, c = rm after removal of the membrane-bound ribosomes, d = smooth microsomes, and mw = molecular weight. (Courtesy of G. Kreibich, Department of Cell Biology, New York University Medical School.)**

binding of two polypeptide chains by the disulfide bridge of a cystine molecule is part of the primary structure. Hydrolysis of the peptide bonds with strong acid at 100°C will free the amino acids, which then can be separated by various means. One of the simplest is *paper chromatography.* This technique is based on the different solubility of amino acids in a particular solvent and consequently the more rapid rate at which the soluble amino acids will be carried along with the solvent as it moves up a paper column by capillary action (Fig. 3-10). Under constant conditions, each amino acid will move a specific distance up the column, where it becomes adsorbed. The paper then can be dried and sprayed with ninhydrin. This chemical reacts with the individual amino acids, identifying the location of each by a spot of color on the paper. Better separation of complex mixtures can be obtained by rotating the paper 90° and repeating the chromatography, using a different solvent to get a *two-dimensional chromatogram.* An *electric field* may be placed around the system to speed up the passage of the substances on the basis of their charge, or *radioactive amino acids* can be located by processing the strip through a Geiger-Müller tube.

The same principle can be utilized using a *column of starch or silica gel.* In this case, the organic solvent and water are added at the top and allowed to penetrate down through the column, then the protein hydrolysate and more solvent are added. The amino acids separate out along the column, according to their respective solubility in the organic solvent and water, and their individual adsorption on the starch. If the differences between the amino acids are great enough, complete separation can be obtained. The use of columns containing ions (ion exchange resins) that react with charged groups of the amino acids is a further development of this technique, *ion exchange chromatography.*

However, our knowing the amino acid composition of a protein does not tell us the amino acid sequence. Sanger's brilliant investigations on insulin, as reported in 1956, were based on his use of *end-group analysis,* through which he was able to identify the amino acids with a free alpha amino or carboxyl group; that is, those at the end of the chain. By subsequent enzymatic hydrolyses, the terminal amino acid was split off, permitting the identification of the new terminal amino acid. Sanger determined the entire amino acid sequence of insulin, showing that this protein is made of two polypeptide chains, one of 21 amino acids (the A chain), the other of 30 amino acids (the B chain). These chains are linked by disulfide bonds (Fig. 3-11).

Secondary Structure. Several forces, other than the disulfide and peptide bonds of the primary structure, are involved in the conformation of a protein molecule. The most important is the *hydrogen bond,* a fairly weak bond resulting from the tendency of hydrogen atoms to share electrons with two other atoms, usually oxygen atoms. In the polypeptide chain, hydrogen bonds are usually between the nitrogen and the double-bonded oxygen of different peptide bonds.

$$
\begin{array}{ccc}
RC_{\alpha}H & & HC_{\alpha}{-}R \\
\quad N{-}\boxed{H\cdots\cdots O}{=}C & & \\
O{=}C & & NH
\end{array}
$$

Although individual hydrogen bonds are weak, the great number of them occurring in the large protein molecule helps to secure the folds of the helices together in a regular, repeating stable conformation. This is characteristic of fibrous and contractile proteins such as keratin of hair, fibrin of blood clots, myosin of muscle, and collagen and elastin of

Figure 3-10. **A two-dimensional chromatogram obtained by separating amino acids by paper chromatography, then rotating the paper 90° and using a different solvent to further separate the mixture.**

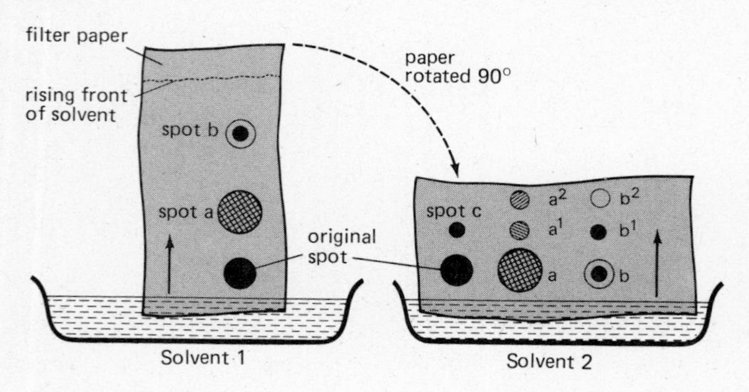

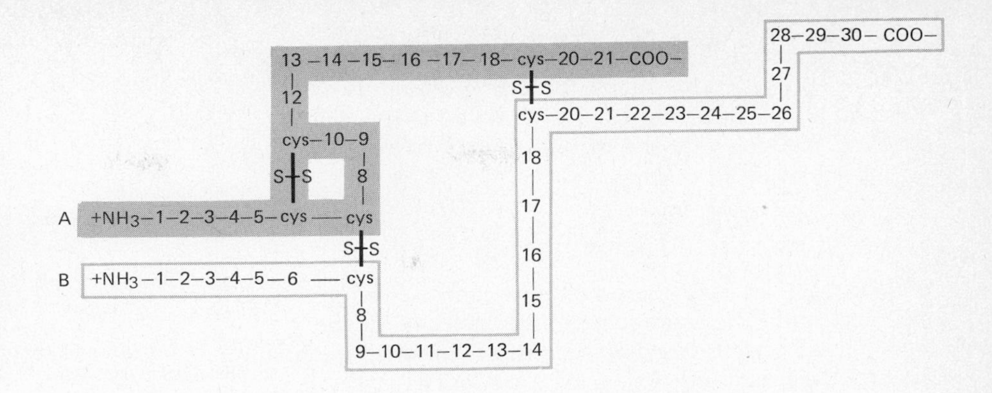

Figure 3-11. **Diagrammatic representation of the A and B polypeptide chains of the insulin molecule. The numbers represent amino acids. There are two disulfide (S—S) linkages connecting the chains and one intrachain linkage in the A chain.**

connective tissue. Fibrous proteins can exist in a stretched-out form (α *form*), or in a folded, and consequently shorter form (β *form*).

Tertiary Structure. The helix is folded still more, by other types of noncovalent bonds between the side chains of amino acids, to give it its tertiary structure. The chain with its helical regions is wrapped around itself, often forming a compact globular molecule. Hemoglobin and myoglobin are globulins. Figure 3-12 shows the three-dimensional structure of hemoglobin.

This tertiary conformation is a very stable form of the molecule, in a thermodynamic sense. Its integrity is essential for the specific activity of the protein, particularly for enzymes. If secondary and tertiary structures are destroyed by strong acids or bases, or by heat, radiation, or other physical forces, the protein loses its biological activity and is *denatured.* If denaturation has caused the unfolding of the polypeptide chains, it is irreversible.

The determination of the three-dimensional structure of proteins was made possible through the use of *X-ray diffraction,* which shows the scattering of electrons as they pass through a protein crystal. Painstaking analysis of the data shows the position of the atoms in the molecule. The recent development of electronic computers, and their application to this problem, has facilitated these investigations considerably.

Figure 3-12. **The three-dimensional structure of hemoglobin as determined by X-ray diffraction studies. The two chains nearest the reader are darker. The true and approximate axes of symmetry are indicated by the solid and broken lines. (From R. E. Dickerson and I. Geis, *The Structure and Action of Proteins,* W. A. Benjamin, Inc., Menlo Park. Copyright 1969 by Dickerson and Geis.)**

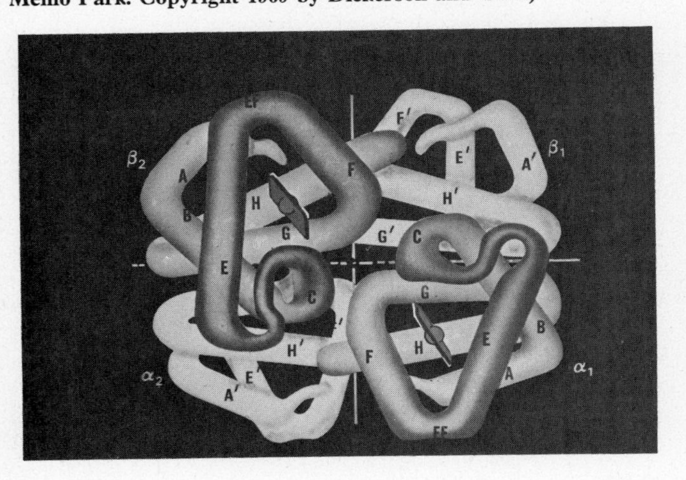

Quaternary Structure. Some macromolecules consist of aggregations of subunits. The quaternary structure of hemoglobin is formed by two pairs of polypeptide chains, each with a molecular weight of 17,000, forming the functional hemoglobin molecule, with a weight of 68,000. These subunits can be separated without denaturing the protein. Several enzymes, including ribonuclease, also have subunits, forming a quaternary structure.

3-3.2 Nucleic Acids

General significance

The nucleic acids, DNA and RNA, are long-chain molecules, which together are responsible for the inheritance of genetic information to subsequent generations of daughter cells and the control of the cellular metabolism of the parent cell.

The genetic information carried in the chromosomes is localized in particular regions, known as *genes.* Each gene represents a specific biochemical portion of the long molecule of DNA; each gene carries the coded information for the synthesis of a single protein or group of related proteins. This code is expressed in the linear arrangement of the units, called *nucleotides,* that made up the nucleic acid. This linear arrangement of the code units is ideal for information storage and transfer. On the other hand, the complex coiling of the protein molecule makes it quite unsuitable for coding, but gives it the tremendous structural variation needed for processing and control.

For the information that is coded in DNA to be effective, the following processes must occur:

1. The information must be accurately and reliably transmitted to daughter cells during cell division. This means that the DNA of the chromosomes must be

precisely replicated to provide a full complement of genetic information to each daughter cell.

2. The encoded instructions for cell metabolism and control must be delivered to the cytoplasm. The mediating molecules are the various forms of RNA, discussed below.

Structure of nucleic acids and nucleotides

Nucleic acids are long-chain molecules composed of units called nucleotides. The difference in the nature of the individual nucleic acids is based on the manner in which these nucleotides are arranged. This is very similar to the way in which a protein is built up by amino acids, organized in a specific linear sequence, only instead of a choice of 20 amino acids, the nucleic acids are made up of four of a possible five nucleotides. The DNA molecule is always formed from the nucleotides *thymine, adenine, cytosine,* and *guanine.* In the RNA molecule, the thymine is replaced by the fifth type of nucleotide, *uracil* (Table 3-2).

A nucleotide is a monomer, consisting of three parts (Fig. 3-13):

1. A five-carbon sugar (pentose). In RNA, this sugar is ribose. In DNA, the sugar has one less oxygen atom and is called deoxyribose (Sec. 3-3.4).
2. A phosphate group (PO_4).
3. One of five possible nitrogen-containing bases.

The nucleotides join end to end, to form the polymeric nucleic acids, DNA or RNA. The linkage is between the phosphate group of one unit and the sugar of another, forming a pattern of a phosphate-sugar backbone. The variation in this backbone lies in the particular sequence of bases sticking out from the backbone (Figs. 3-14 and 3-15).

DNA. Deoxyribonucleic acid is a macromolecule composed of the sugar, deoxyribose, phosphate groups and the nitrogen-containing bases, thymine, adenine, cytosine, and guanine. The total length and molecular weight of most forms of DNA is not known, but the smallest seems to be 1.2×10^8, and the largest has a length of about 400 μm and a molecular weight of 10^9. Watson and Crick, in 1953, suggested that the DNA molecule is formed by two long chains of nucleotides. These two chains twist in opposite directions, so that they fit like two pieces of rope twisted

Table 3–2. **Nucleotide Composition of DNA and RNA**

DNA	RNA
Phosphate	Phosphate
Deoxyribose sugar	Ribose sugar
Base	Base
Thymine	Uracil
Adenine	Adenine
Cytosine	Cytosine
Guanine	Guanine

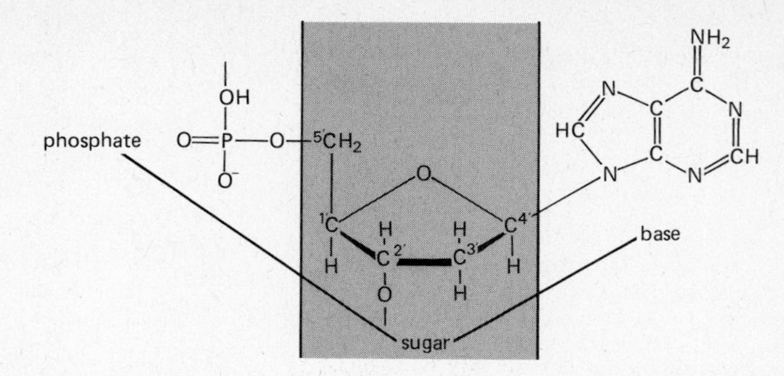

Figure 3-13. **Structure of a nucleotide of the nucleic acid DNA. In this figure, the sugar is deoxyribose and the base is adenine.**

together (Fig. 3-14). This is the *double helix* that Watson describes so vividly in his book of that name.

How are the two strands held together? The bases from one strand pair up with the bases from the other strand, *but* the base pairing is always fixed, so that cytosine always teams up with guanine, thymine with adenine. Consequently, the sequence of bases in one strand of the double helix determines the sequence in the other: This will be a *complementary sequence.*

A T G G A will pair up with
T A C C T

Replication of DNA. The *replication* of DNA consists of two main steps:

1. The molecule begins to unwind its two helical strands, separating them out and so exposing them to free nucleotide bases in the surrounding medium.
2. Each strand picks up its complementary bases, so that there now are two DNA molecules where there was only one. Each new molecule is an exact replica of the original, which has acted as a *template* on which the new molecule has been synthesized (Fig. 3-15).

RNA. Ribonucleic acid is present in small quantities in the nucleus and in much larger amounts in the cytoplasm. Although RNA is a macromolecule, it is considerably smaller than DNA. It contains the sugar, ribose, and the nitrogenous bases, uracil, adenine, cytosine, and guanine. It is generally single-stranded and does not have the helical structure of DNA. The cell contains three quite different types of RNA, with different properties and functions. There is a fourth type of RNA found in viruses.

Messenger RNA (*m*RNA). Messenger RNA has molecular weights of up to 5×10^6. It is synthesized in the nucleus and moves into the cytoplasm, where it attaches to the ribosomes of the rough ER to direct the synthesis of a single protein or group of related proteins. The *transcription* of the information necessary for this occurs when the *m*RNA is synthesized on a portion of a DNA molecule used as a template. The pairing of the bases for RNA with those of DNA is the same as for the replication of DNA, so that

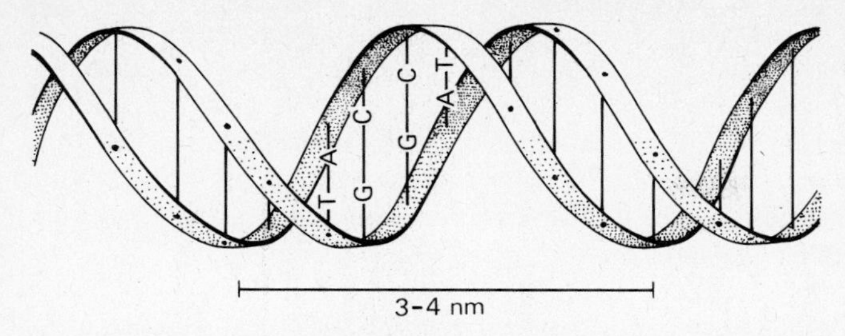

Figure 3-14. **The twisted helix of the DNA molecule as suggested by Watson and Crick in 1953. The chains are joined by hydrogen bonding between the base pairs. A = adenine, C = cytosine, G = guanine, and T = thymine.**

the information content of *m*RNA is identical to that of the DNA on which it is made. Because only a small part of the DNA molecule is transcribed to form the relatively small RNA molecule, the latter carries only a fraction of the total information coded into DNA.

TRANSFER RNA (*t*RNA). Transfer RNA is a fairly small macromolecule with a molecular weight of about 25,000. Its role is to pick up amino acids from the cytoplasm and bring them to the surface of the ribosome-*m*RNA complex. Before the amino acids can bind to *t*RNA, they have to be activated through reactions involving ATP. Each *t*RNA is coded to

Figure 3-15. **The replication of DNA. The top diagram shows the sequence of bases in the two twisted strands of the molecule. (A = adenine, C = cytosine, G = guanine, and T = thymine). The lower diagram shows that the strands have separated and each one acts as a template to synthesize a replica of its previous partner. Note that the original strands are shown with heavy lines and their bases in dark boxes. The newly synthesized strands are indicated by broken lines and their bases are shown in open boxes.**

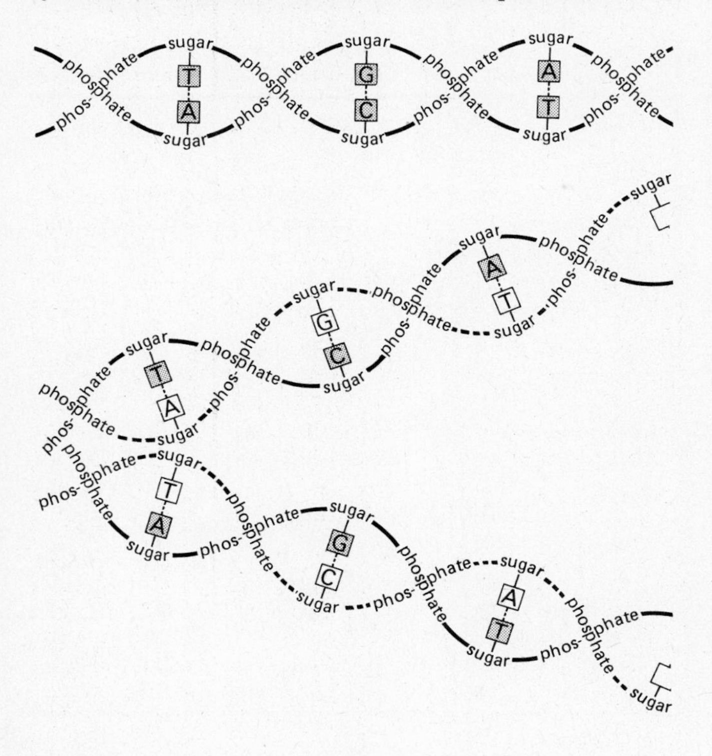

pick up a specific activated amino acid and transfer it to the correct order for the particular polypeptide chain it is destined to produce. This translation of information into protein synthesis is the function of *t*RNA.

RIBOSOMAL RNA (*r*RNA). The 70*s* ribosome may be broken into two subunits, a large 50*s* and a small 30*s*, which cannot be separated without destroying the ribosome. The large subunit contains 34 different protein molecules and two molecules of ribosomal RNA. The small subunit is made up of 21 proteins and only one RNA molecule. The smaller *r*RNA subunit has a molecular weight a little less than a million; the larger *r*RNA subunit has a molecular weight of about 1.6 million. The ribosomes interact with *m*RNA and bring together the other molecules required for the orderly elongation of the protein chain.

THE GENETIC CODE

Tremendous advances have been made in recent years in our understanding of how DNA can serve as genetic material in protein synthesis. This work has been rewarded with several Nobel prizes to scientists: among them are Nirenberg, Ochoa, Jacob, and Monod. We have no idea, however, of the code(s) used for the synthesis of the contents of the entire cell, but knowing the details of genetic control of protein synthesis is of great importance. It has been suggested that because all enzymes are proteins, and enzymes direct cellular metabolism, then if the enzymes are synthesized in the correct order and amounts, other syntheses should follow in a controlled manner. Most of this pioneering work has been done on microorganisms, such as *Escherichia coli*, but the genetic code appears to be applicable to higher organisms as well.

The symbols of the code that will determine the amino acid sequence of a protein are the individual nucleotide bases. This nucleotide code is a *triplet code*, made up of three bases. These base triplets are crowded along the DNA molecule, with neither overlapping nor spaces, so that for the code to be read properly, it must be started at the right place.

The first step in the transcription of the code is the synthesis of *m*RNA on the DNA template. The base pairing of *m*RNA includes just enough triplets to provide the right number of amino acids for a single protein, which can be an enormous number, but still represents only a fraction of the information in the DNA.

In turn, *m*RNA then serves as a template for this specific protein. The *m*RNA attaches to several ribosomes, with which it then functions (Fig. 3-16). Now *t*RNA molecules will bring their specific amino acids to the correct place on the *m*RNA molecule. In order for such geometric recognition to occur, *t*RNA must have a mirror-image code (anticodon) of the specific triplet (codon) on the *m*RNA. For example, the codon in *m*RNA composed of three uracyl molecules (UUU) will code one amino acid, phenylalanine; the anticodon for UUU would be on the *t*RNA. The ultimate DNA codon would be three adenine molecules, AAA.

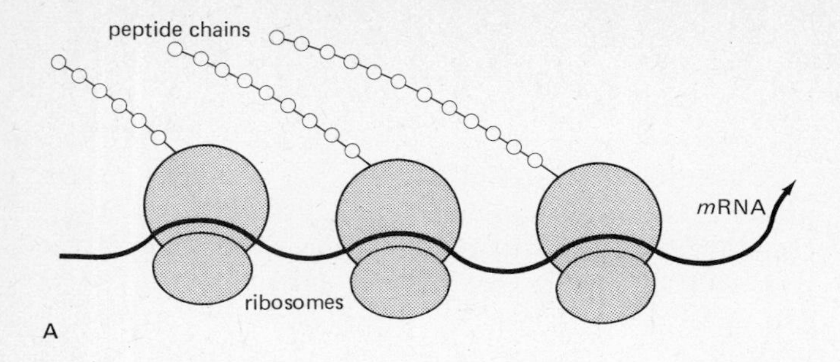

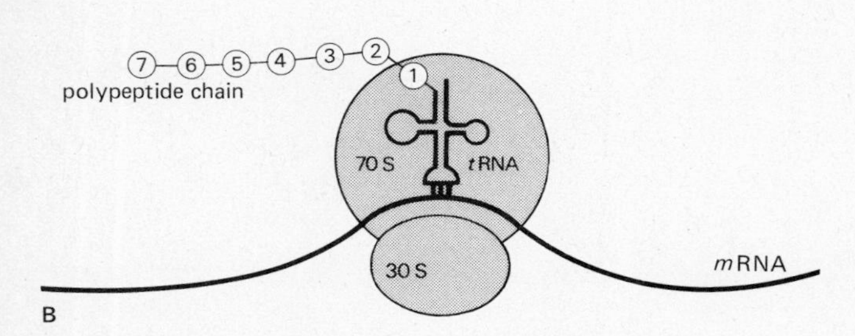

Figure 3-16. **Relation of ribosomes to *m*RNA and *t*RNA. (A) Three ribosomes are seen scanning a single *m*RNA strand with polypeptide chains of increasing length issuing from them. (B) A *t*RNA molecule bearing amino acid 1 is already attached to the growing peptide chain by enzymic formation of a peptide bond. (From J. L. Howland, *Cell Physiology,* Macmillan Publishing Co., Inc., New York, 1973.)**

Some of the codons of the genetic code are shown in Table 3-3. This coding system would permit more information to be stored than is actually used. There are 64 possible triplet combinations, but only 20 of them spell out known amino acids. There is also a certain amount of redundancy: alanine has three codons, GCU, GCC, and GCA, for example. The remaining codes have been called *nonsense codes,* but in fact they do serve a purpose. They are like the STOP used in a telegram to separate sentences, only these STOPS separate polypeptide chains and cause them to be released from the ribosomes. They are then transported through the endoplasmic reticulum to different parts of the cell.

Physiological regulation at the level of *m*RNA—induction and repression

Some enzymes can be synthesized only if the cell has been exposed previously to the substance on which the enzyme normally acts, its *substrate*. The enzymes are said to be *induced enzymes* and are of particular interest because they indicate the influence of environmental factors on the genetic code.

Repression of enzyme synthesis may be considered a check on induction. Repression occurs when the substance that an enzyme would normally help synthesize is already present in adequate amounts in the medium. Further production of this substance is unnecessary, and economy dictates that the process be stopped several steps back; hence, the enzyme is no longer produced.

The *operon model* is a neat explanation proposed by Jacob and Monod (1961) for these regulative processes. Genes that respond to induction-repression control are grouped together in clusters called *operons*. An operon consists of four parts, the *structural gene* that is responsible for protein synthesis, and its three controlling genes, the *operator, promotor,* and *regulator genes* (Fig. 3-17).

1. The operator gene permits the structural gene to function. If the product of the repressor gene (a protein) binds to the operator, the structural gene is inactive.
2. If the operator gene is free of repressor, the structural gene can synthesize protein. This is probably the mechanism of induction. The inducer (substrate) binds to the repressor, thus removing the inhibiting factor and permitting the operator to drive the structural gene.
3. The promotor gene lies between the regulator and operator genes and facilitates the synthesis of *m*RNA by the structural genes.

The *complete synthesis* of an artificial gene that functions in a living cell has recently been announced by Nobel laureate Khorana and his associates at the Massachusetts Institute of Technology (1976). Khorana's technique makes it possible to change gene sequences at will. Although the particular gene that has been synthesized is one that already exists in and is essential to living cells, it would theoretically be possible to synthesize entirely new genes. The hazards and advantages of such techniques are discussed in Sec. 27-12.

Table 3-3. **The Genetic Code. The Specific Triplet of Bases (Codon) Will Determine the Amino Acid Synthesized. The Nonsense Codes, Which Do Not Code for Any Amino Acid, Are Shown in Boxes.**

UUU	Phe	UCU	Ser	UAU	Tyr	UGU	Cys
UUC	Phe	UCC	Ser	UAC	Tyr	UGC	Cys
UUA	Leu	UCA	Ser	UAA	STOP	UGA	STOP
UUG	Leu	UCG	Ser	UAG	STOP	UGG	Trp
CUU	Leu	CCU	Pro	CAU	His	CGU	Arg
CUC	Leu	CCC	Pro	CAC	His	CGC	Arg
CUA	Leu	CCA	Pro	CAA	Gln	CGA	Arg
CUG	Leu	CCG	Pro	CAG	Gln	CGG	Arg
AUU	Ile	ACU	Thr	AAU	Asn	AGU	Ser
AUC	Ile	ACC	Thr	AAC	Asn	AGC	Ser
AUA	Ile	ACA	Thr	AAA	Lys	AGA	Arg
AUG	Met	ACG	Thr	AAG	Lys	AGG	Arg
GUU	Val	GCU	Ala	GAU	Asp	GGU	Gly
GUC	Val	GCC	Ala	GAC	Asp	GGC	Gly
GUA	Val	GCA	Ala	GAA	Glu	GGA	Gly
GUG	Val	GCG	Ala	GAG	Glu	GGG	Gly

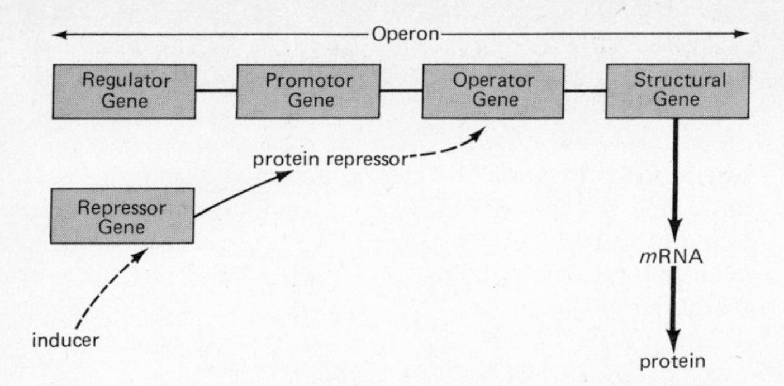

Figure 3-17. **The four parts of an operon. The activity of the structural gene is enhanced by the combined action of the regulator, operator, and promotor genes, but inhibited by the product of the repressor gene. An inducer probably acts by binding with, and inhibiting, the repressor gene. Broken arrows indicate inhibition.**

3–3.3 Lipids

GENERAL SIGNIFICANCE

The layer of adipose tissue (fat) that covers the human form, sometimes so gracefully and sometimes so grossly, is an evolutionary development seen first in those animals that maintain a constant body temperature. It represents stored food, taken in excess over the amount utilized by the body and deposited in certain sites that have a "predilection" for fat deposition. This fat blanket is not only an inert insulating layer: it is constantly being metabolized, and energy and heat are actively generated from it. Hibernating animals possess a special type of *brown fat* that is metabolized as the animal begins to come out of hibernation: the heat liberated by the oxidation of the fat raises the body temperature to the level needed for normal physiological function.

The areas where fat is deposited are determined genetically and regulated by the endocrines; the rate at which it is metabolized is controlled by both the nervous and the endocrine system. In our prosperous society, excessive fat becomes a social and medical problem: our standards of physical beauty demand slimness in women, a lean line in men. This is unlike the standards of primitive societies, where food is scarce; the accumulation of fat indicates wealth, and the epitome of female beauty may be the 280-lb chief's wife, barely able to move herself around. Apart from physical appearance, however, overweight, which is almost always due to excessive fat, has been shown to decrease life expectancy, probably due to the great strain put on the heart. Fat deposition in the major arteries is associated with *atherosclerosis,* which often leads to circulatory failure. This type of fat deposition is related to the accumulation in the blood of certain types of fats, the *saturated* fats, which accounts for the great emphasis today on the use of polyunsaturated fats in the diet.

Phospholipids and *cholesterol* are found throughout the body as part of cell membranes and the many structures that are formed from membranes, such as the endoplasmic reticulum, the mitochondria, the myelin sheath of nerves, and the nuclear membrane. Many important compounds, such as the steroid hormones and the bile salts, are also derived from cholesterol. All these types of fats are found in blood in combination with protein as lipoproteins, which act as a means of transport for the lipids.

TYPES OF LIPIDS

Lipids are a heterogeneous group of fatlike substances, which are water insoluble but generally fairly soluble in organic solvents like ether, benzene, or chloroform. From this large group, our discussion will be limited to naturally occurring substances made up of long chains of fatty acids (simple and compound lipids) and the ring-based steroids.

Simple Lipids. Simple lipids consist of three molecules of fatty acid, attached by an ester linkage

$$-\overset{\displaystyle O}{\overset{\|}{C}}-O-$$

to a molecule of glycerol (Fig. 3-18). This large molecule is a *triglyceride,* or *neutral fat.* Most of the fat of the normal diet is neutral fat and any excess is stored in adipose tissue in this form. In the adult mammal, more than 10 per cent of the body weight may be stored triglyceride.

Compound Lipids. Compound lipids contain other elements, such as sulfur, phosphorus, or nitrogen, in addition to the carbon, hydrogen, and oxygen of the simple lipids.

Phospholipids are compound lipids, containing two fatty acids (diglycerides) attached to the glycerol molecule. Instead of the third fatty acid, they have a phosphate molecule,

Figure 3-18. **Synthesis of a simple lipid. Triglycerides are formed by esterification of three molecules of fatty acid with one molecule of glycerol. The ester linkage is shown in the dark boxes.**

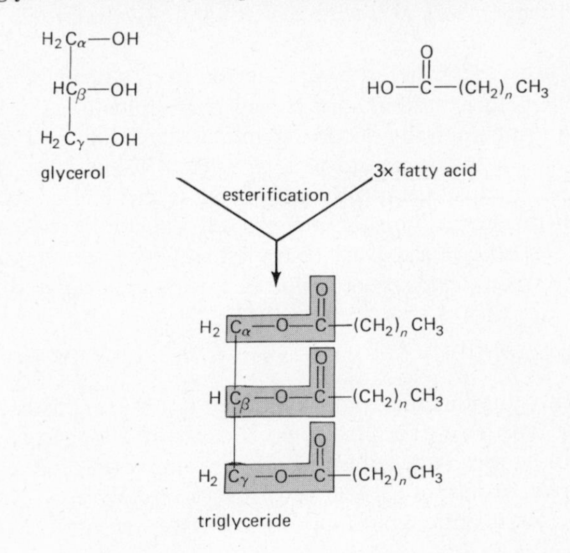

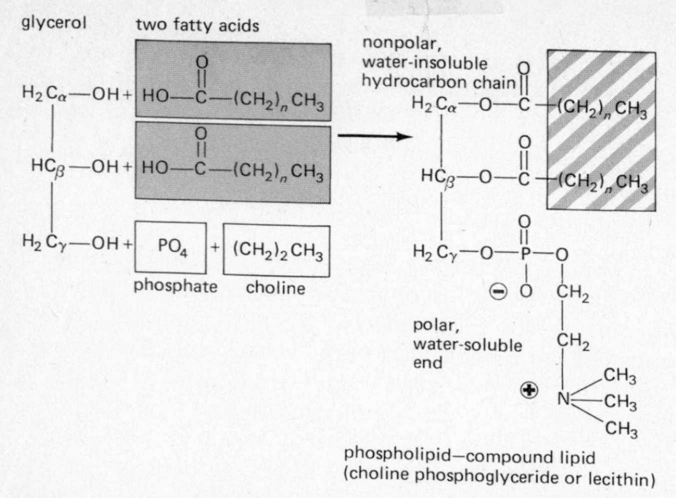

Figure 3-19. **Synthesis of a compound amphipathic lipid, choline phosphoglyceride (lecithin), a phospholipid. Whereas the C_α and C_β of glycerol combine with fatty acids, C_γ binds to phosphate and choline.**

usually with an additional water-soluble molecule attached to it. This chemical constitution makes the phospholipid molecule *amphipathic,* with a nonpolar, water-insoluble part (the fatty acids) and a highly polar, water-soluble part, the rest of the molecule (Fig. 3-19). This structural asymmetry determines their orientation in cell membranes (Secs. 4-3 and 4-5). In cell membranes there are five main phospholipids, which differ according to the type of water-soluble molecule attached to the phosphate. The most abundant form is *lecithin,* in which this additional molecule is choline, a vitamin B. Its newer name is *choline phosphoglyceride.* If the phosphate group has no attached molecules, the phospholipid is slightly less polar. This is *phosphatidic acid,* and it has been suggested that it functions as a movable molecule within the cell membrane, acting as a carrier for ions, whereas the other phospholipids are bound into the structure of the membrane.

Compound lipids also include lipid complexes with protein (*lipoproteins*) and with carbohydrate (*glycolipids*).

Steroids. Although they are derived from a phenanthrene ring structure, steroids are classified as lipids (Fig. 3-20). The most commonly occurring animal steroid is *cholesterol,* which is a major constituent of normal tissues, for it is present in most cell membranes. It is the metabolic precursor of the steroid hormones of the adrenal cortex (aldosterone, cortisol), of the ovary (estradiol and progesterone), and of the testis (testosterone). The bile acids are also steroids.

Fatty acids

The common unit from which these lipids are constructed is the fatty acid. This molecule consists of a long chain of carbon atoms, attached to a methyl group at one end and a carboxyl group at the other end. The carboxyl group gives it its acidic qualities, the long carbon chain its fatty, hydrophobic (nonpolar) properties. Almost all the naturally occurring fatty acids have an even number of carbon atoms.

Saturated Fatty Acids. Saturated fatty acids have only single bonds between the carbon atoms of the chain. Palmitic acid (C_{16}) is the most abundant saturated fatty acid found in animal food, stearic acid (C_{18}) the next most plentiful. Saturated fatty acids usually form solid fats, such as margarine or lard.

Structure of palmitic acid (saturated)

H_3C—$(CH_2)_{14}$—COOH

Methyl group — Hydrocarbon chain — Carboxyl group

Unsaturated Fatty Acids. Unsaturated fatty acids contain one or more double bonds between the carbon chain atoms. This gives rise to a geometric isomerism known as *cis-trans isomerism,* because the four valence bonds of the double-bonded carbon atoms lie in one plane

$$>C=C<,$$

and there is no freedom of rotation about the axis of the double bond.

Cis (naturally occurring form) — *Trans*

As a result of this, there is a greater rigidity in the structure of these unsaturated fatty acids, an important determinant in cell membrane structure and function (Sec. 4-5).

The greater the number of double bonds in a fatty acid, the more unsaturated it is. Highly unsaturated fatty acids are found in fish oils and the liver of grass-eating animals. Beef liver is rich in unsaturated fatty acids, whereas pork liver is not. A commonly found important unsaturated fatty acid is *linoleic acid,* found in vegetable seed oils.

Essential and Nonessential Fatty Acids. Some fatty acids can be produced by the body, but there are two or three that must be provided in the diet: these are the essential fatty acids (EFA), one of which is linoleic acid. From the EFA, which are unsaturated, the remaining fatty acids (nonessential) can be synthesized by the cell by hydrogenation.

Chemical Reactivity of Fatty Acids

1. They can be combined in synthetic reactions to form triglycerides and phospholipids (Fig. 3-18). Degrada-

tion products of fatty acids are used to synthesize cholesterol.
2. They can be oxidized to yield energy. Both (1) and (2) are discussed in detail in Sec. 18-4.
3. They can be saturated by hydrogenation. The addition of hydrogen to the unsaturated fatty acid destroys the double bonds and converts the essential fatty acid to the nonessential form. This process is used commercially to produce margarine from oils.

3-3.4 Carbohydrates

GENERAL SIGNIFICANCE

Carbohydrates are the most prevalent energy source for humans, whether they take the form of the rice diet of the Chinese or the staple potatoes of the American diet. Energy can be obtained from these substances only after their digestion and breakdown to molecules small enough to be absorbed by the cells and, through a series of complex metabolic reactions, ultimately oxidized to yield energy.

In humans, the large carbohydrate molecules of starch and glycogen are hydrolyzed to yield simple sugars, the most important of which is glucose. One of the most carefully regulated constituents of the internal environment is the level of circulating blood glucose, which in turn is the result of the amount of glucose ingested and absorbed, the

Figure 3-20. **The steroid cholesterol and some of the hormones derived from it. The steroid (phenanthrene) ring is shaded, and the carbon atoms that form the structure are represented by numbers. Only the basic carbon and hydrogen structure is shown for the three classes of steroid hormones. Details are depicted in later chapters.**

C_{27} cholesterol

C_{19} androgens (testosterone)

C_{21} progesterone, corticosteroids

C_{18} estrogens

amount stored in the form of glycogen in the tissues, and the amount oxidized by the tissues for energy.

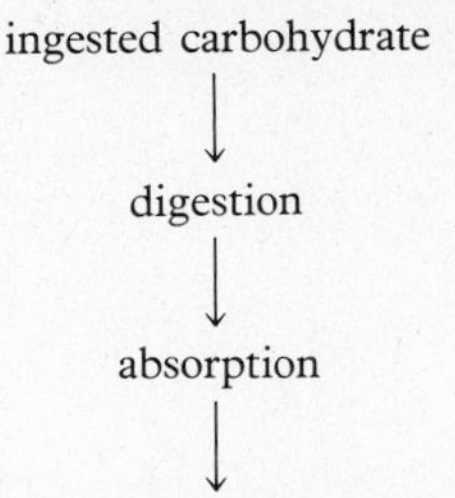

Although the blood glucose will vary normally for short periods following a heavy meal or a fast or after strenuous exercise, the homeostatic mechanisms of the liver, adrenal medulla, and pancreas rapidly return the level to the average 100 mg glucose per 100 ml blood. The efficient reabsorbing action of the kidney tubule prevents any loss of glucose into the urine unless the blood glucose level is unusually high.

Carbohydrates also form important structural and storage elements in both plants and animals. *Cellulose* and *starch* are plant polysaccharides, macromolecules made up of glucose subunits, linked by *glycosidic bonds.* Animals store carbohydrate as *glycogen,* which is a polymer of glucose molecules linked in the same manner. Although polysaccharides are built from sugar subunits as proteins are built from amino acids, the polysaccharides do not carry a definite number of monomer units; consequently, they vary in molecular weight and other physical properties.

MOLECULAR STRUCTURE

Carbohydrates contain the elements carbon, hydrogen, and oxygen. They have the chemical structure of either polyhydroxy-aldehydes (—CHO attached to one of a series of hydroxylated carbon atoms) or polyhydroxy-ketones

$$-\overset{|}{C}=O$$

attached to a similar series of carbon atoms.

L-Glyceraldehyde	Dihydroxyacetone (a ketone)
CHO	CH_2OH
HO—C—H	C=O
CH_2OH	CH_2OH

Monosaccharides. The simplest of carbohydrates, monosaccharides are made up of from three to eight carbon atoms. They form the basic subunit from which the larger carbohydrate molecules are built. The three carbon monosaccharides are *trioses,* formed by the fission of the six-

glucose

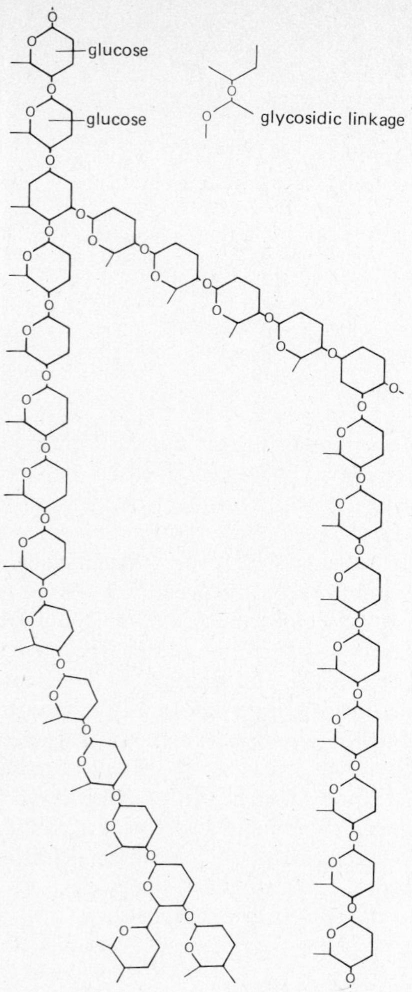

glucose

glycosidic linkage

Figure 3-21. **A fragment of a glycogen molecule. The glucose units are joined by glycosidic linkages.**

carbon sugar, glucose, during glucose degradation in animals to yield energy (Sec. 5-12).

The other most important monosaccharides in biological systems are the *pentoses* (C_5) and *hexoses* (C_6). *Pentoses* occur widely in all living systems as intermediates in metabolism and as part of some macromolecules, such as nucleic acids and coenzymes. The five-carbon sugar *ribose* is associated with RNA; the deoxygenated form of this pentose, *deoxyribose,* is the sugar moiety of DNA.

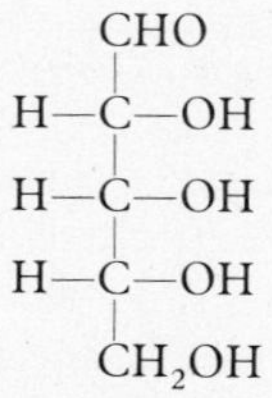

CHO
H—C—OH
H—C—OH
H—C—OH
CH_2OH
Ribose

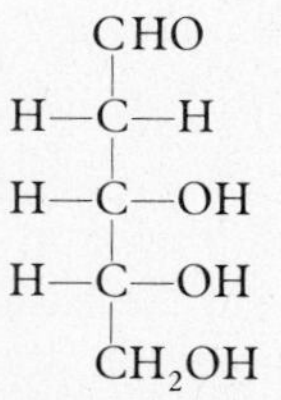

CHO
H—C—H
H—C—OH
H—C—OH
CH_2OH
Deoxyribose

Hexoses are six-carbon sugars in which the ratio of hydrogen to oxygen is that of H_2O, with the general formula of $C_6H_{12}O_6$. This formula represents about 16 different hexoses, which differ in the arrangement of the CHOH group in space. This type of isomerism is *stereoisomerism,* and the small structural variations result in sugars that differ in their chemical and physical properties and in their metabolic rates. As a biological source of energy, the most important hexose is *glucose,* but *fructose, galactose,* and *mannose* also are involved in metabolism.

There are several ways to indicate the structure of a sugar. The simplest is to show the linear structure of the six-carbon chain, but a more accurate representation is the ring form. The carbon atoms are considered to form a closed ring structure due to the reaction between the aldehyde group at carbon 1 and the hydroxyl group at carbon 5.

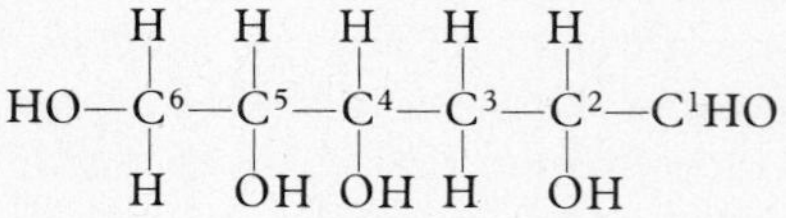

$$HO—C^6H_2—C^5H(OH)—C^4H(OH)—C^3H(H)(OH)... $$

H H H H H
HO—C^6—C^5—C^4—C^3—C^2—C^1HO
H OH OH H OH

Open-chain linear form of D-glucose

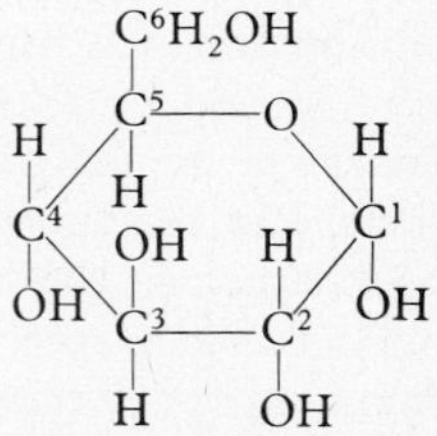

C^6H_2OH
C^5—O
H H
C^4 H C^1
OH H
OH C^3—C^2 OH
H OH

Ring form of D-glucose

Disaccharides. Disaccharides contain two monosaccharide subunits linked by a *glycosidic bond.* The linkage is between a hydroxyl group of one sugar unit and the carbonyl carbon

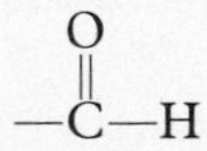

O
‖
—C—H

of another sugar unit. Depending upon the steric configuration at the carbon 1, the glycosidic linkage is called α or β.

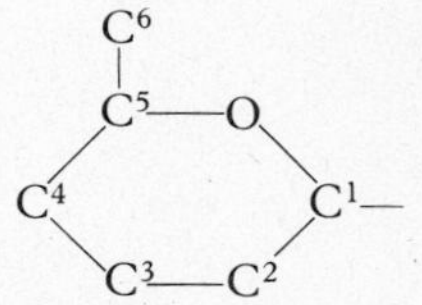

C^6
C^5—O
C^4 C^1—
C^3—C^2
Glucose unit

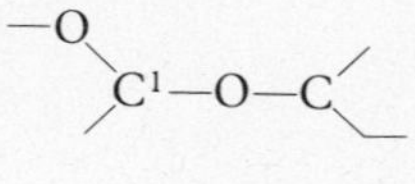

—O
C^1—O—C
Glycosidic linkage

One can compare the glycosidic linkage binding monosaccharides into larger molecules to the peptide bond joining amino acids into polypeptides, or to the ester linkage between fatty acids. The most common *disaccharides* are *sucrose* and *lactose.* Sucrose is formed by a glycosidic linkage between glucose and fructose, whereas lactose is made of glucose and galactose. *Maltose* is another disaccharide formed by two molecules of glucose. In the process of forming a disaccharide from two simple sugars, a molecule of water is removed, giving the disaccharide the general formula of $C_{12}H_{22}O_{11}$.

Glucose — Glycosidic linkage — Fructose = Sucrose

Sucrose is obtained from sugar cane or beets but is produced by most photosynthetic plants, which are capable of utilizing solar energy to synthesize organic material from inorganic sources.

Polysaccharides. Polysaccharides are simply long chains of sugars linked by glycosidic bonds. They vary individually according to the number and type of monosaccharides in the chain, the amount of branching, and the nature of the glycosidic bond (α or β). *Cellulose* and *starch* are the two most important plant polysaccharides and contribute structure and energy storage, respectively. *Glycogen* is the storage form of hexoses in animals. It is made of glucose subunits linked by α glycosyl bonds. It is a much larger macromolecule than starch, is highly branched, and has estimated molecular weights of from 270,000 to 1,000,000,000. Figure 3-21 shows a fragment of a glycogen macromolecule. Another animal polysaccharide is *chitin,* the hard covering of insects and crustaceans. Some complex polysaccharides are formed with protein and lipid components (*glycoproteins* and *glycolipids*), and it is believed that these substances form part of the outer zone of the plasma membrane and perhaps play a role in molecular recognition processes (Sec. 4-5).

For the utilization of polysaccharides by the organism, the glycosidic bonds have to be enzymatically hydrolyzed (digested), freeing the disaccharides and ultimately the monosaccharides. Only the latter are small enough to pass through cell membranes and thus become available for energy release (Secs. 17-6 and 18-3).

Cited References

HALDANE, J. B. S. *The Origin of Life.* New Biological Series No. 16. Penguin Books, Ltd., London, 1954.

HARADA, K., and S. W. FOX, "Thermal Polycondensation of Free Amino Acids with Phosphoric Acid." In *The Origins of Prebiologic Systems,* S. W. Fox, ed. Academic Press, Inc., New York, 1965, p. 289.

JACOB, F., and J. MONOD, "Genetic Regulatory Mechanisms in the Synthesis of Proteins." *J. Mol. Biol.* **3:** 318 (1961).

KENDREW, J. C. "Myoglobin and the Structure of Proteins." *Science* **139:** 1259 (1963).

MILLER, S. L. Formation of Organic Compounds on the Primitive Earth." *Science* **117:** 528 (1953).

OPARIN, A. I. *Life, Its Nature and Development.* Academic Press, Inc., New York, 1964.

PERUTZ, M. F. "X-Ray Analysis of Hemoglobin." *Science* **140:** 863 (1963).

SANGER, F. "The Structure of Insulin." In *Currents in Biochemical Research,* D. Green, ed. Interscience Publishing, Inc., New York, 1956.

WATSON, J. D., and F. H. C. CRICK, "Molecular Structure of Nucleic Acids—A Structure for Deoxyribose Nucleic Acid." *Nature* **171:** 737 (1953).

Additional Readings

BOOKS

ALLEN, J. M., ed. *Molecular Organization and Biological Function.* Harper & Row, Publishers, New York, 1967.

BALDWIN, E. *An Introduction to Comparative Biochemistry,* 5th ed. Cambridge University Press, New York, 1967.

BENNETT, T. P., and E. FRIEDEN. *Modern Topics in Biochemistry.* Macmillan Publishing Co., Inc., New York, 1966. (Paperback.)

DICKERSON, R. E., and I. GEIS. *The Structure and Action of Proteins.* Harper & Row, Publishers, New York, 1969.

HALDANE, J. B. S. *The Origin of Life.* New Biology Series No. 16. Penguin Books Ltd., London, 1954. (Paperback.)

MCELROY, W. D. *Cell Physiology and Biochemistry,* 3rd ed. Foundation of Modern Biology Series. Prentice-Hall, Inc., Englewood Cliffs, N.J., 1971. (Paperback.)

MCGILVERY, R. W. *Biochemical Concepts.* W. B. Saunders Company, Philadelphia, 1975.

TERNAY, A. L. *Contemporary Organic Chemistry.* W. B. Saunders Company, Philadelphia, 1976.

WATSON, J. D. *Molecular Biology of the Gene,* 2nd ed. W. A. Benjamin, Inc., Menlo Park, Calif., 1970. (Paperback.)

WATSON, J. D. *The Double Helix.* Atheneum Publishers, New York, 1968.

ARTICLES

BROWN, D. B. "The Isolation of Genes." *Sci. Am.,* Aug. 1973.

CLARK, B. F. C., and K. A. MARCKER. "How Proteins Start." *Sci. Am.,* Jan. 1968.

CRICK, F. H. C. "The Genetic Code." *Sci. Am.,* Oct. 1966.

GREEN, D. E. "The Synthesis of Fats." *Sci. Am.,* Aug. 1960.

KORNBERG, A. "The Synthesis of DNA." *Sci. Am.,* Oct. 1968.

MILLER, O. L., JR. "The Visualization of Genes in Action." *Sci. Am.,* Mar. 1973.

MIRSKY, A. E. "The Discovery of DNA." *Sci. Am.,* June 1968.

NIRENBERG, M. W. "The Genetic Code II." *Sci. Am.,* Mar. 1963.

PALADE, G. "Intracellular Aspects of the Process of Protein Synthesis." *Science* **189:** 347 (1975).

PTASHNE, M., and W. GILBERT. "Genetic Repressors." *Sci. Am.,* June 1970.

SPECTOR, D., and D. BALTIMORE. "The Molecular Biology of Poliovirus." *Sci. Am.,* May 1975.

YANOFSKY, C. "Gene Structure and Protein Structure." *Sci. Am.,* May 1967.

Chapter 4

The Plasma Membrane as a Regulatory Organelle

How long does love last?
..
...love might last as
six snowflakes, six hexagonal snowflakes,
six floating hexagonal flakes of snow
or the oaths between hydrogen and oxygen
in one cup of spring water

Carl Sandburg, "Honey and Salt"

THE PLASMA MEMBRANE is the barrier that particles and water must penetrate in order to enter or leave the cell. It determines the composition of the cell and protects the cell from changes in the internal environment, but only within a physiological range. It acts as a homeostatic organ in this respect, maintaining an optimal concentration of substances within the cell, and controlling the volume of fluid within the cell, the distribution of ions, the passage of oxygen and carbon dioxide, of small and large molecules. Practically all biochemical phenomena can be brought ultimately to a membrane dimension, whether it is the interaction of cancer cells, the transduction of energy, cell division, or immunochemical reactions. How the plasma membrane does this, what its chemical and molecular structure may be, has been the object of intensive investigation since the early part of this century. Concepts of cell membrane structure are being more hotly contested today, perhaps, than ever before. The chronological development of these ideas is interesting because many such ideas grew out of the investigative techniques available at the time.

4–1 The Cell as an Osmometer

The classical studies on membrane permeability involved the rate of penetration of different substances into and out of the cell and the consequent changes in cell volume. The passive movement of particles (*solutes*) through the plasma membrane may be by *diffusion;* subsequent inequalities in particle distribution will cause the movement of water by *osmosis,* which may cause the cell to swell or shrink. A mechanical (hydrostatic) pressure may be exerted to force fluids and small particles through membranes; this is *filtration.* Energy may be expended by the cell to get substances through the membrane by *active transport* or by *pinocytosis.* These last two mechanisms will be considered later in this chapter.

Diffusion

In 1827, the English botanist Robert Brown noticed that pollen grains of certain plants had particles which, when observed under the microscope, exhibited a constant, irregular, vibratory movement. He ascribed this to some vital force, but all types of material, living or inanimate, show this *Brownian movement,* provided the particles are small enough and the medium not too viscous. All particles in dilute solution are in constant motion due to their kinetic energy, and unless their movement is impeded by external forces, any given particle will move at random in any direction. Such a particle moves in jumps with the energy derived from collisions with other particles. One cannot determine the speed of a given particle by following its exact path, because it probably changes its path 10 million times a second, but Einstein showed that its *mean displacement* will depend on its molecular weight. The larger the particle, the less its mean displacement or the distance it will travel per unit time.

If there is a difference in concentration of particles in two parts of the solution, there will be a *net* movement of particles (solute) from where they are more concentrated to where they are less concentrated. This is because there are statistically more solute particles to move from the region of higher concentration; the movement is still random. One can also consider the solvent molecules (usually water, in biological systems) as moving from where they, the water molecules, are more concentrated to the region where they are less concentrated, as illustrated in Fig. 4-1.

Diffusion may be defined as the net movement of particles from a region of higher concentration to one of lower concentration. The steeper the gradient, the greater the net movement or flux. The larger the particle, the slower the movement. The higher the temperature, the more rapid the Brownian movement. These relationships are quantitatively expressed in the Fick equation:

$$\frac{ds}{dt} = -DA\frac{dc}{dx}$$

where

$\frac{ds}{dt} = \frac{\text{amount of substance in moles}}{\text{distance per unit time}}$ = rate of transport

A = cross section of the area of membrane through which particles pass

$\frac{dc}{dx}$ = the concentration gradient

D = the diffusion coefficient

Figure 4-1. **Diffusion of sugar molecules in water. Both sugar and water molecules move from their respective regions of highest to lowest concentrations, through random movement, until they are equally distributed.**

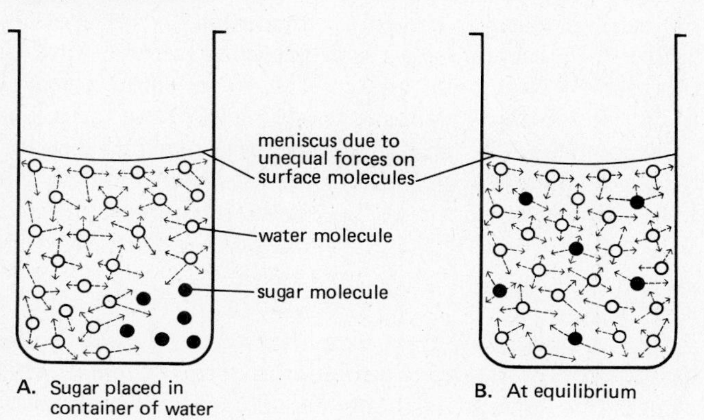

Table 4-1. **Diffusion Coefficients (*D*) and Molecular Weights of Selected Substances**

Substance	Molecular Weight	Diffusion Coefficient at 20°C (cm²/sec)
O_2	32	1.9×10^{-5}
CO_2	44	1.8×10^{-5}
Urea	60	1.2×10^{-5}
NaCl	58	1.6×10^{-5}
KCl	75	1.9×10^{-5}
$CaCl_2$	111	1.3×10^{-5}
Glucose	180	6.7×10^{-6}
Hemoglobin	68,000	6.9×10^{-7}
Urease	480,000	3.5×10^{-7}
Collagen	345,000	6.9×10^{-8}
DNA	6,000,000	1.3×10^{-8}

Sources: H. S. Harned and B. B. Owen, *The Physical Chemistry of Electrolytic Solutions,* 3rd ed., Van Nostrand Reinhold, New York, 1958; and C. Tanford, *Physical Chemistry of Macromolecules,* Wiley, New York, 1961.

This equation must be modified for substances that dissociate (electrolytes), because dissociation increases the number of particles involved.

The *diffusion coefficient* is useful for characterizing macromolecular movement in biological systems. It varies inversely with the molecular weight. Larger molecules with smaller diffusion coefficients usually move more slowly, as shown in Table 4-1. However, the shape of the molecule is important and, because the asymmetry of large molecules will hamper their movement, urease, a globular protein, diffuses more rapidly than the smaller, fibrous collagen molecule.

Diffusion is generally a very slow process; in cooking we speed up the movement of particles by heating or stirring the solution. In the living organism, where temperatures are restricted to a rather narrow range, diffusion is a significant force only because it is restricted to very short distances, measured in nanometers. Where the distances are greater, other forces are involved, especially the pressure generated in a closed circulation.

The simplest driving force for the movement of materials across the cell membrane is the *concentration gradient.* If the membrane is extremely permeable to both the solvent and solute, it will impede the movement only imperceptably. Very permeable membranes, such as those lining the liver capillaries or the alveoli (air spaces) of the lungs, hardly slow down the diffusion process at all. It is thus of advantage to the cell to maintain a concentration gradient, but this means that the diffusing substance must constantly be replenished on one side and removed from the other side.

Osmosis

By far the most abundant substance to diffuse across the cell membrane is water, the biological solvent. Approximately 100 times the cell volume passes across the cell membrane

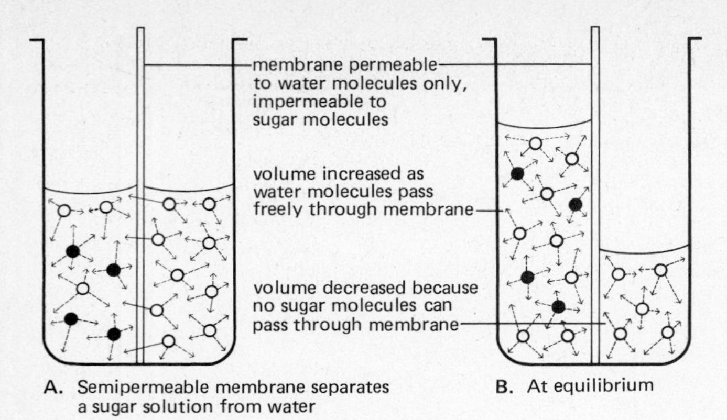

Figure 4-2. **Osmosis. (A) Movement of water molecules (open circles), through a membrane, from their region of highest concentration to their region of lowest concentration. Sugar molecules (dark circles) are unable to pass through the membrane. (B) This results in an uneven distribution of fluid. The height to which fluid in the left compartment can rise by osmotic pressure is limited by hydrostatic pressure in the reverse direction.**

every second, yet under normal circumstances there is no net change in cell volume, so delicately are these forces balanced.

A concentration gradient can be set up for water, just as for solute particles, and water will move from its region of highest concentration to where it is least concentrated, until an equilibrium is established. If the system is made more complex by separating the regions of variable concentrations by a membrane that is permeable to water but impermeable to the solute particles, water will pass through the membrane with the net flux in the direction of the solution with lower water concentration—that is, higher solute particle concentration—thus increasing the volume of this compartment (Fig. 4-2). If one of the compartments is a cell, then the volume of the cell will change. If there are more particles within the cell that cannot penetrate the cell membrane than there are nonpenetrating particles in the outside solution, then more water will enter the cell than leaves it. The cell will swell, and if it is an animal cell (which does not have the supporting wall of plant cells) it will burst or *lyse.* Figure 4-3 shows that the solution that causes lysis is one

Figure 4-3. **Lysis of an animal cell in a hypotonic medium.**

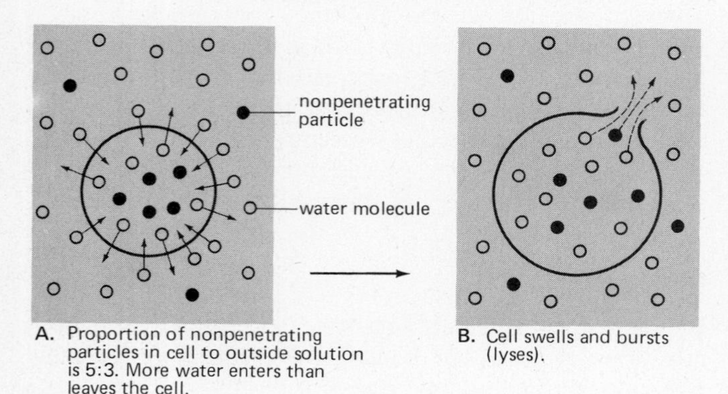

with fewer nonpenetrating particles than the cell and is *hypotonic* to the cell.

nonpenetrating particle
water molecule
A. Proportion of nonpenetrating particles in cell to outside solution is 5:3. More water enters than leaves the cell.
B. Cell swells and bursts (lyses).

If the solution has a greater number of nonpenetrating particles than the cell, the solution is *hypertonic* to the cell and will cause it to shrink. If a cell shows no change in volume when placed in a solution, that solution is said to be *isotonic.* It is important to realize that there is *no one* isotonic solution for all cells. The tonicity of a solution depends on the relative number of nonpenetrating particles inside and outside the cell; this, in turn, is dependent on whether the solute particles penetrate the plasma membrane.

Many of the studies on plasma membranes have utilized *erythrocytes.* These *red blood cells,* placed in a hypertonic solution, shrivel up or become *crenated.* If they are placed in a hypotonic solution they swell and burst, or *hemolyse* (Fig. 4-4). The stroma containing hemoglobin is liberated, and the remaining cell membranes are precipitated. These precipitated membranes, or "ghosts," have been widely used for studies on the chemical nature of plasma membranes, which have been shown to consist chiefly of protein and lipid. Actually, changes in cell volume are not always so dramatic. If a substance penetrates fairly slowly, it is "nonpenetrating" at first, causing a temporary shrinkage of the cell, then as it penetrates, an equilibrium is established and, if the change has not been too abrupt, the cell regains its original volume. Figure 4-5 shows the volume changes in erythrocytes exposed to solutions with particles that penetrate at different rates, or do not penetrate at all.

Osmosis is the movement of water through a semipermeable or selectively permeable membrane, in response to inequalities of solute distribution. The term is restricted completely to *water movement.*

Osmotic pressure

Osmotic pressure is the force that moves water in such a system; it can be measured in terms of the pressure necessary to prevent the movement of water, or to retain the original volume of the system. Osmotic pressure is expressed in millimeters of Hg or centimeters of water pressure. Osmotic pressure may be measured also by determining the hydraulic pressure developed in the capillary tube of an osmometer filled with the solution to be studied, when it is separated by a semipermeable membrane from a volume of distilled water (Fig. 4-6). Obviously, the semipermeability of the membrane is the determining factor: it must be permeable to water and impermeable to the solute particles in the osmometer. This is the reasoning for using the cell as an osmometer; the characteristics of the membrane will determine the passage of solvent particles, which in turn will tell us about the nature of the plasma membrane.

Osmotic activity

The osmotic activity is expressed as the *osmolarity,* or osmotic concentration, of a solution in terms of *osmols.* One osmol is the amount of solute, dissolved in a liter of water,

Figure 4-4A. **Hole torn in the membrane of a red blood cell during hemolysis. ×24,000. [From D. Dannon, *J. Cell Comp. Physiol.* 57:115 (1961).]**

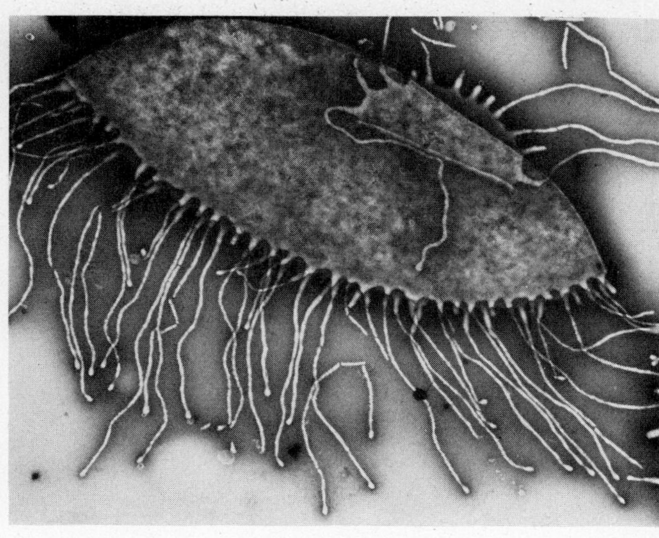

Figure 4-4B. **Human red blood cell, hemolyzed with a hypotonic salt solution. (Courtesy of Richard F. Baker, University of Southern California.)**

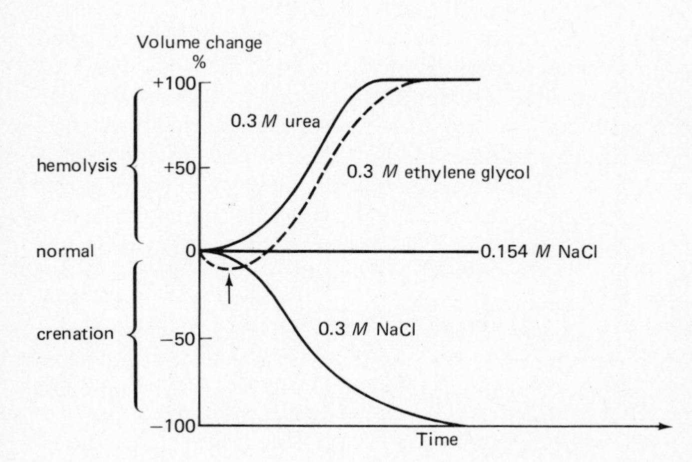

Figure 4-5. **Volume changes of erythrocytes exposed to particles that penetrate at different rates. NaCl does not penetrate at all; urea penetrates very rapidly. Ethylene glycol penetrates somewhat more slowly, so that there is a brief period during which it is hypertonic to the cell, which shrinks (arrow).**

Figure 4-6. **A simple osmometer. (A) A sac is made by attaching a semipermeable membrane (permeable to water, impermeable to sucrose) to a capillary tube. The sac is filled with a sucrose solution and immersed in a container of distilled water. Because water is more concentrated outside the sac than in it, the net flow of water is into the sac. (B) This increase in sac volume is seen as a rise of the fluid in the capillary tubing. At equilibrium, the hydrostatic pressure developed by the weight of the column of fluid in the tube equals the osmotic pressure forcing the water into the sac.**

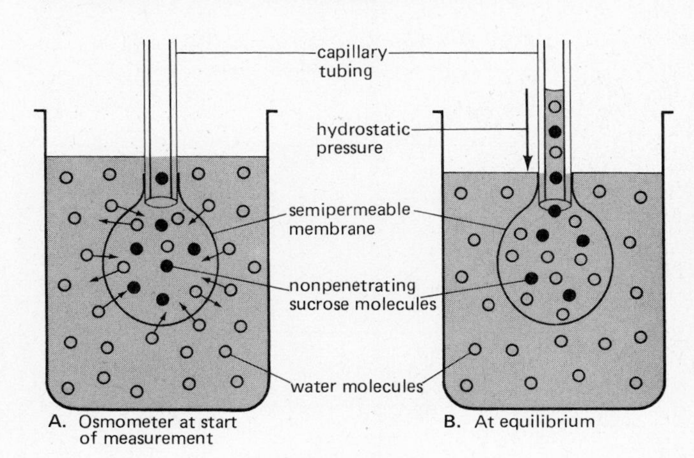

that would exert the same osmotic pressure as one mole of an ideal nonelectrolyte. For example, a 0.2 molar glucose solution (a nonelectrolyte) is also 0.2 osmolar, and osmotically equivalent to a 0.1 molar NaCl solution. The latter dissociates, and so is also a 0.2 osmolar solution.

Comparison of osmolarity with osmoticity

It is important to realize that although *osmoticity* may be used to compare the osmotic concentrations of solutions (isosmotic solutions have exactly the same osmotic concentrations), such a comparison yields no information about the permeability of the plasma membrane or the effect of the solution on cell volume. It simply tells us how many osmotically active particles the solution contains, not whether they can penetrate the plasma membrane. Consequently, an isosmotic solution is not necessarily isotonic to the cell. For example, if a cell is placed in a 0.15 *M* NaCl solution and there is no volume change, this solution is isotonic to the cell. If the same cell is placed in a 0.30 *M* urea solution and the urea penetrates the cell, the cell will burst. The *urea solution is hypotonic* to the cell. Both solutions have the same osmolarity (0.3 osmolar) and are isosmotic to each other and to the cell; but one is isotonic, the other hypotonic, to the cell (Fig. 4-7A).

If, on the other hand, a different type of cell is placed in these solutions, and there is *no volume change* in 0.15 *M* NaCl, *or* in the 0.3 *M* urea solution, this tells us that *0.3 M urea is isotonic* to this particular cell (Fig. 4-7B). This cell membrane is impermeable to urea, whereas the first cell membrane permitted urea to penetrate. Both cells have membranes that are impermeable to NaCl. To maintain the normal volume of isolated cells, it is necessary to place them in an isotonic solution. This is equally important when blood is replaced by infusions of saline or other fluids: all such fluids must be isotonic to the cells.

4–2 Distribution of Water in the Body

One of the earliest experiments on water distribution is a report on the amount of water in the entire body of an executed criminal who was ground up finely (after being

Figure 4-7. **The cell as an osmometer. In (A) the cell is impermeable to 0.15 M NaCl and permeable to an isosmotic solution of 0.3 *M* urea. The NaCl solution is isotonic to the cell; the urea solution is hypotonic. In (B) another cell type is shown to be impermeable to both the urea and NaCl solutions. These isosmotic solutions are both isotonic to this cell.**

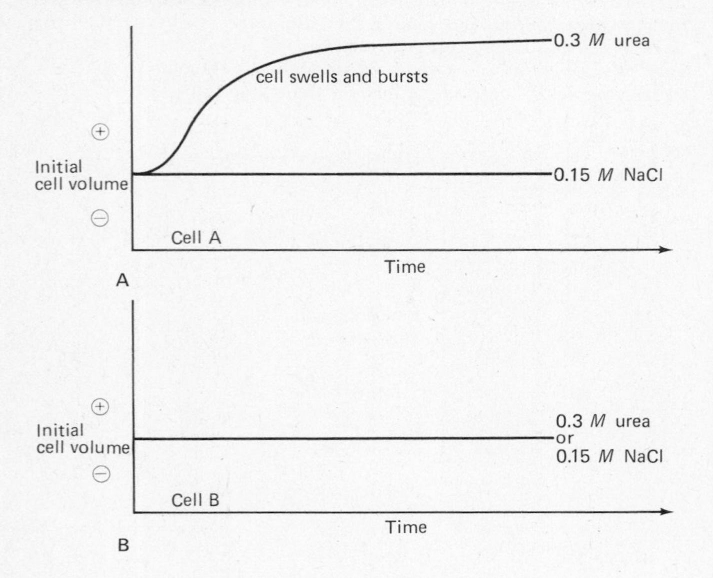

Table 4–2. **Comparison of Permeabilities of Cells to Water***

Sea urchin egg (unfertilized)	0.1
Sea urchin egg (fertilized)	0.3
Fresh water ciliated protozoa	0.1
Erythrocyte (human)	3.0
Erythrocyte (ox)	2.5

Source: Davson, H. and J. F. Danielli, *The Permeability of Natural Membranes,* 2nd ed., Cambridge University Press, New York, 1952.

* Cubic micra of water that pass through 1 square micron of cell surface per min. Constant temperature and pressure.

hanged) so that the proportion of water to solid tissue could be determined. The results showed that 75 per cent of the entire body was water. Since then, somewhat more refined techniques, such as the use of heavy water, have been used to show that this enormous percentage of fluid is distributed in various compartments in the body. Either deuteriated (D_2O) or tritiated water (HTO) may be used; D_2O is determined on the basis of its weight and HTO by its radioactivity. The fluid contained within the cells themselves is called *intracellular;* the fluid within the blood and interstitial spaces forms the *extracellular* compartment. Fluid exchange between blood and the interstitial spaces is so rapid as to permit their being considered as one compartment, and approximately 17.5 liters of water (lean, adult male) or 13.2 liters (lean, adult female) may be found in this extracellular compartment. Intracellular fluid amounts to about 22.5 liters for males, 16.8 liters for females. This distribution of water is essential for tissue turgidity and blood volume. But what prevents the water from leaving one compartment and passing into another? The answer is, of course, that water passes freely in all directions, but there is a *difference in the rate of passage* and a very minimal variation in the final balance of water distribution. This is particularly surprising when one considers the multitude of factors that influence the passage of water through the body compartments and their consequent interaction on water metabolism.

The rate of water passage through cell membranes is determined not only by the *unequal distribution of osmotically active particles,* but by the *changeable permeability* of the cell membrane itself. Different physiological states will profoundly affect membrane permeability: one way in which hormones exert their effects on cells is by changing membrane permeability (Sec. 10-8). Vasopressin (antidiuretic hormone) increases the water permeability of kidney collecting duct cells, thus permitting water to leave the lumen of the duct and become reabsorbed into the circulation and thereby decreasing urine output (Sec. 15-4). Fertilization of the egg increases its permeability to water, and there are differences in the permeability of cells from different species to water (Table 4-2). The mechanisms underlying water transport through cell membranes are poorly understood. In addition to osmosis and filtration pressure (discussed below), special pores in the membrane have been postulated. Other possibilities include the establishment of an electrical gradient across a membrane to speed up osmosis (electro-osmosis) and the active transport of ions to change the concentration gradient along which water then passively follows.

Exchange of Fluid on the Capillary Level

The physical forces that determine whether fluid will move into or out of a capillary, and consequently out of or into the interstitial fluid, are *osmotic pressure* and *hydrostatic pressure.* This two-way transfer can occur only at the level of the capillaries, the smallest of the blood vessels, because only the capillary walls are thin enough to act as membranes through which water and small particles can pass. Because the number of capillaries in the body is enormous, the total capillary bed represents an area of semipermeable membrane tremendous in extent, through which fluid exchange takes place (see also Sec. 12-11).

The endothelial cells of the capillary walls are so thin in some places that they have been thought to be holes or pores. Although there is practically no cytoplasm in these regions of the wall, the plasma membrane is stretched across them, completely closing the "pore." This means that the capillary pores are not an easy route for large particles to pass through the capillary wall. All particles must pass through either the thick part of the endothelial cell (including the cytoplasm and plasma membrane) or through the thin part, where only the plasma membrane is present. This *capillary pore* should not be confused with the *molecular pores within the membrane itself.* The capillary pore represents a space between cells closed off by a membrane, whereas the membrane pore is real space between molecules (Figs. 4-8 and 4-18).

Figure 4-8. **The closed pores of the capillary wall. [From J. Rhodin, *J. Ultrastructure Research* 6:171 (1962).]**

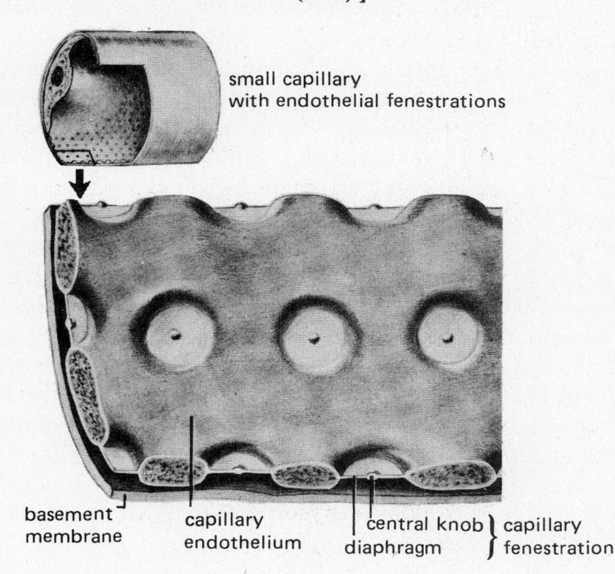

Table 4-3. **Permeability of Capillaries to Lipid-Insoluble Molecules of Different Molecular Weights**

Substance	Molecular Weight	Capillary Permeability $(cm^3)(sec)/100$ g tissue
Water	18	3.7
Urea	60	1.83
Glucose	180	0.64
Sucrose	342	0.35
Myoglobin	17,000	0.005
Hemoglobin	68,000	0.001
Serum albumin	69,000	0.000

Table 4-3 shows the relative permeability of capillary pores to *different size molecules.* Smaller molecules like glucose pass through much more easily than the large albumin molecules. Molecules larger than 17,000, the molecular weight of myoglobin, cannot get through these pores, which have a radius of about 3 nm. However, there is a limited passage of very large molecules, including plasma proteins, which may get out through a few *extra large pores,* or "leaks," or may be extruded by evaginations of the plasma membrane (exocytosis). There is also great variation in pore size in blood capillaries from different organs. The liver has very large pores, especially in the hepatic sinusoids; skeletal muscle capillary pores are small; and those in the intestinal capillaries seem to be of both sizes. Figure 4-9 shows the results of an experiment by Mayerson (1960) in which radioactive albumin (molecular weight 69,000) was injected into an anesthetized dog. The rate at which radioactivity can be picked up in the lymph of various organs was measured. The lymph represents the excess interstitial fluid that is drained off into this special part of the circulation (Sec. 13-8).

The *osmotic pressure* that pulls water back into the capillary is caused by the retention of most of the plasma proteins in the capillary, as they are too large to pass through the capillary membrane with ease. A relatively small amount of protein does escape and generates a small osmotic force in the reverse direction. The other force that influences the return of water to the blood is the pressure of fluid in the tissues; this *tissue pressure* is of significance only under pathological conditions in which the accumulation of water in the tissues is abnormal (*edema*).

The main force that drives water out of the capillaries is the *hydrostatic pressure* generated by the heart pumping blood through a closed circulation. Although it was once believed that the pressure at the arterial end of the capillary is considerably higher than at the venous end, more modern techniques indicate very little pressure difference. However, a rhythmic dilation and constriction of the arterioles that control the capillary bed (vasomotion) alternately increase and decrease capillary pressure, so that water exchange takes place along the entire length of the capillary. Glucose, salts, ions, gases, and the other constituents of blood that are small

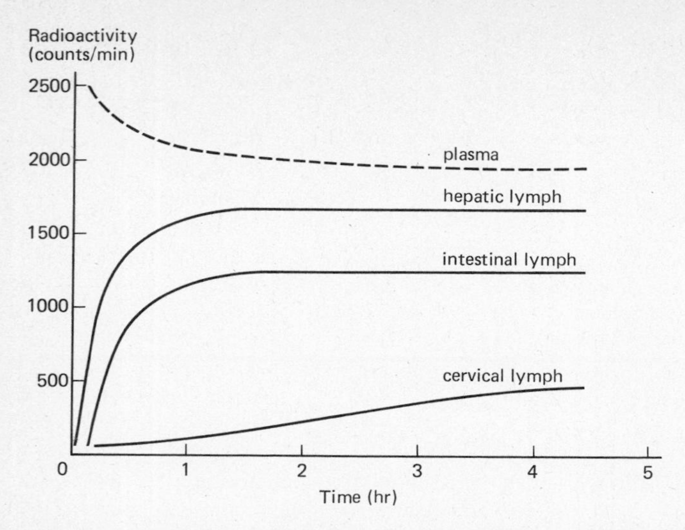

Figure 4-9. **Relative permeability of the lymphatics of different organs to albumin is shown by the rate of appearance of radioactive albumin in the lymph following injection into the plasma. (Redrawn from H. S. Mayerson, in *Handbook of Physiology,* Sec. 2, Vol. II, The American Physiological Society, Bethesda, Md., 1963, p. 1035.)**

enough to pass through the capillary membrane will move in both directions at a rate depending on their individual concentration gradients. This modification of Starling's original hypothesis (1896) of fluid exchange simply balances capillary and osmotic pressures on an alternating basis, rather than depending on an arterial/venous gradient.

The final distribution of water between the blood and the tissues is thus an equilibrium between water passing out of the blood because of the *filtration pressure* engendered by the heart and closed circulation, and the water reentering the blood due to the *osmotic pressure* generated by the plasma proteins, assisted by the *tissue pressure.*

$$\underset{35\text{ mm Hg}}{\text{capillary pressure}} \xrightleftharpoons{\text{opposed by}} \underset{22\text{ mm Hg}}{\text{osmotic pressure}} + \underset{5\text{ mm Hg}}{\text{tissue pressure}}$$

$$\textit{effective filtration pressure} = 35 - (22 + 5)\text{ mm Hg} = 8\text{ mm Hg}$$

Figure 4-10. **The relationship between the forces regulating fluid passage between the blood capillaries, interstitial fluid, and the tissue cells. The effective filtration pressure is about 8 mm mercury. The accumulated interstitial fluid formed is drained off into the lymphatic capillaries.**

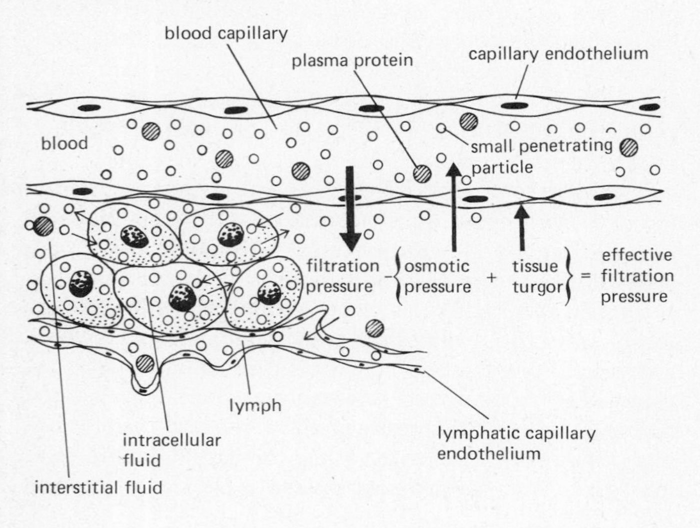

Lymphatic System

This difference of 8 mm Hg *effective filtration pressure* results in more water leaving the capillary than reenters it directly. The overflow is drained by the *lymphatic capillaries,* which have a pressure of −2 mm Hg. These lymphatic capillaries lead into a fine system of ducts which join to form the large lymph vessels that ultimately return the tissue fluid, or *lymph,* to the blood circulation. These ducts, together with the widespread chain of lymph nodes or glands that filter the lymph and localize infection, make up the *lymphatic system.* The fluid exchange between the capillaries, the interstitial fluid, the cells and the lymphatics is diagrammed in Fig. 4-10. The lymphatic system is discussed in more detail in Secs. 13-8 and 14-2.

Edema and Dehydration

Any change in any one of these balancing factors will obviously upset the normal water distribution. If blood pressure rises markedly, the filtration pressure will rise accordingly, resulting in more fluid in the tissues than can be drained off by the lymph ducts. This accumulation of fluid in the tissues is called *edema.* Similarly, if the lymph ducts themselves are blocked in any way—for example, by parasitic worms (nematodes) in the tropical disease *filariasis*—fluid will accumulate in the tissues to such a degree that marked swelling of the affected part occurs. If this be an extremity, the resulting deformity may be grotesque enough to be named *elephantiasis.* In starvation, in which the concentration of plasma proteins falls because of the low protein intake, bloating of the tissues is typical. Any factors that severely lower blood pressure, on the other hand, will lower filtration pressure and result in *dehydration* of the tissues. The most obvious example is hemorrhage. The loss of blood volume within the closed circulation causes a fall in blood pressure. This volume can be replenished by simple intravenous administration of isotonic saline; however, because the hemorrhage results in loss of the plasma proteins as well, this fluid will bring immediate relief to the dehydrated tissues but will not be able to maintain the normal circulating volume of the blood. Fluid will seep out into the tissues because the depletion of plasma proteins diminishes the osmotic pull of water back to the circulation. Unless the osmotic balance is restored by plasma transfusions, or still better, by whole blood, the relief is very temporary.

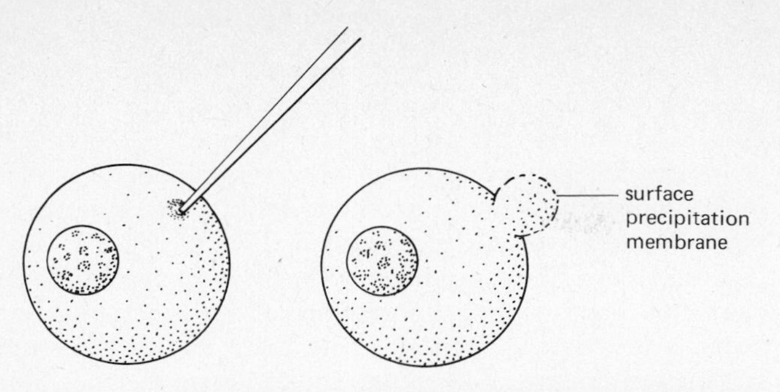

Figure 4-11. **Formation of a surface precipitation membrane following the rupture of the plasma membrane of a cell in an aqueous solution. The new membrane is formed by surface forces orienting proteins and lipids at the cytoplasmic-water interface.**

4–3 Studies on Surfaces and Surface Films

How does one know there is a plasma membrane? Using the technique of microdissection, if one punctures an *Arbacia* (sea urchin) egg, after having removed the protective jelly and vitelline membrane, the cytoplasmic contents flow out. If the tear is extensive, the cell will disintegrate. If the hole is small, however, a new film appears and seals the puncture, allowing the cell to maintain its integrity. This represents the formation of a new plasma membrane, called a *surface precipitation membrane.* It is caused by surface forces orienting the amphipathic proteins and lipids of the cytoplasm at the interface between the torn cytoplasm and the aqueous environment (Fig. 4-11).

Characteristics of Surfaces and Surface Tension

Molecules at the surface of a liquid are subjected to unequal forces: on one side the cohesive forces of molecules within the body of the liquid act as restraining forces, whereas the other side is relatively free of such attraction (Fig. 4-12). A surface, or area between two immiscible phases such as liquid/liquid (oil and water), liquid/gas (water and air), liquid/solid (cytoplasm and granules) is an *interface.* A fundamental property of an interface is that work must be performed to increase its area. *Surface tension* (σ) is a measure of the energy required to expand a surface, bringing molecules from the interior to the surface. It bears some relation to the heat of vaporization, indicating the amount of energy required to free surface molecules from a liquid and permit them to escape into the vapor phase.

Water has a high surface tension. At body temperature (37°C) the surface tension of pure water is about 70 dynes/cm (a dyne is the force required to accelerate a one-gram mass one centimeter per second). If the temperature is raised to 60°C, the surface tension of water falls to 65 dynes/cm. The surface tension of blood plasma and tissue fluid at 37°C is about 50 dynes/cm. This decrease in surface tension is caused by the surface active substances present in these fluids.

Figure 4-12. **Surface molecules at an interface between water and air are subjected to unequal forces. The cohesive forces of the water molecules in the body of water are much greater than those in air as water vapor.**

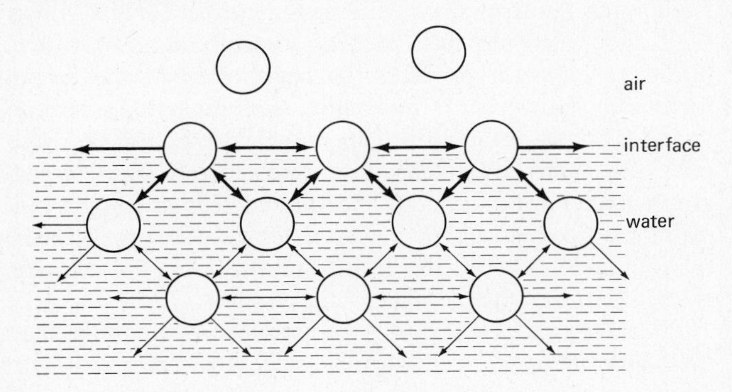

Surface-Active Substances

A surface-active substance concentrates at a surface (*adsorption*) and lowers surface tension. This is because the substance is *amphipathic;* that is, it has both polar and nonpolar groups and so can orient itself between two phases, at the interface. The amphipathic molecules have weak forces of attraction for each other, and they act to separate the molecules of other species at the surface, lowering their cohesive forces. *Detergents* are familiar surface-active agents. Acting as a bridge between oil and water, they penetrate and disperse oily substances, surrounding each oil droplet with a layer of negative charges as the polar ends (projecting into the aqueous phase) dissociate. This prevents coalescence and the accumulation of grease film (see Fig. 4-13). This breakdown of fat into small droplets is *emulsification,* physiologically an important function of the bile salts, resulting in the exposure of a greatly increased surface area of the fats to the digestive action of the appropriate enzymes (lipases).

Protein and Lipid Films

Proteins are surface-active molecules. They lower the surface tension of the surface on which they are adsorbed. Protein molecules in dilute solution can be made to spread out on a pure water surface in a monomolecular film in a Langmuir trough. The thickness of the protein monolayer is 0.9 to 1.0 nm; the molecules are oriented at the air-water interface with their polar groups (amino and carboxyl) pointed into the water, whereas the nonpolar hydrocarbon groups project into the air (Fig. 4-14). One way of measuring the surface tension of these films is through the use of a Maxwell frame. This consists of a U-shaped wire frame, with a separate bar across the open end. The film pulls at the cross bar: the force needed to prevent the bar from moving

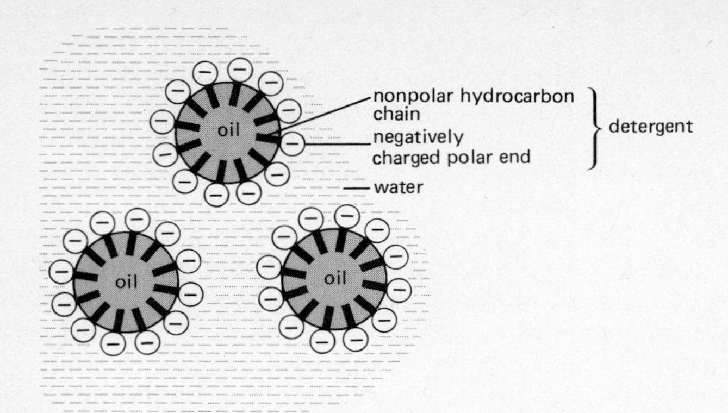

Figure 4-13. **Emulsification of oil droplets by a surface-active substance, for example, a detergent or bile salts. The negatively charged polar ends of the detergent are oriented into the aqueous phase, preventing coalescence of the oil droplets.**

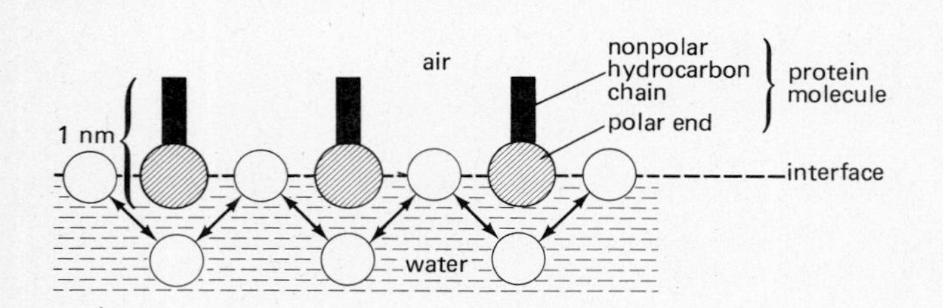

Figure 4-14. **Orientation of protein molecules on a water surface to form a monomolecular film.**

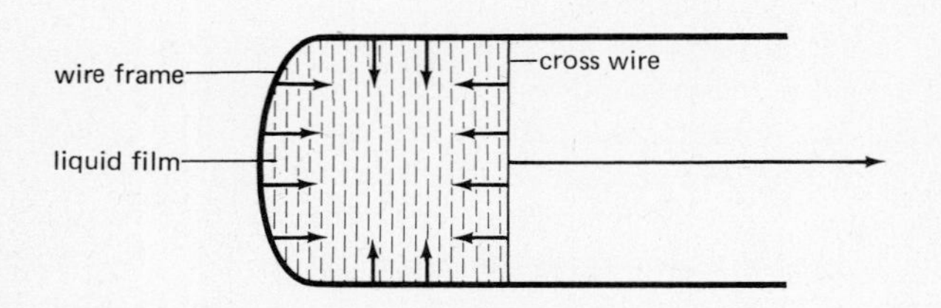

Figure 4-15. **Measurement of surface tension with a Maxwell frame. The force needed to prevent the thin cross wire from moving toward the bottom of the U-shaped frame is proportional to the surface tension of the liquid film within the frame.**

down toward the bottom of the U is proportional to the surface tension (Fig. 4-15).

Protein monolayers retain their biological properties, such as their enzymatic characteristics and their ability to react with specific antibodies. The surface forces only partially open the tertiary structure of the protein; they are not severe enough to destroy it. This mild change in the native structure of the protein is called *surface denaturation.*

Lipid molecules will also form *monolayers* in this manner, but the surface tension of lipid films is higher than that of protein films. Measurements of the surface tension of erythrocytes (red blood cells) by Cole in 1932 and of marine invertebrate eggs by Danielli and Harvey in 1935 showed cells to have a low σ of between 0.03 and 1.0 dynes/cm. Oil droplets have a surface tension of from 1.0 to 10.0 dynes/cm. The low value for cell membranes is due to the proteins in the membrane. The technique used for the determination of the surface tension of a cell is to expose the cell to centrifugal force (g) while observing it under a special microscope; the minimal g needed to split the cell into two is proportional to the surface tension.

The normal lung produces a surface active substance (*surfactant*), probably a lipoprotein, which is essential for keeping the alveoli open. It coats them and reduces the surface tension from 40 dynes/cm to about 2 dynes/cm. Extracts from the lungs of premature infants who succumbed to hyaline membrane disease (*atelectasis*) do not lower surface tension below 18 dynes/cm. The airways collapse with each exhalation, and the infant has to exert enormous muscular force to breathe. Apparently, a certain degree of cell maturity is necessary to produce sufficient amounts of stable surfactant. The most promising therapy at the moment appears to be simply "blowing up the baby"—that is, using a positive pressure respirator to inflate the baby's lungs—until such time when enough surfactant can be produced to lower lung σ to proper levels.

4–4 The Nature of the Penetrating Particle

Lipid Solubility

Studies by Overton in the early part of this century, on the rate of penetration of substances into plant cells, led to the generalization that the more *lipid soluble* a particle, the more readily it would enter. This may be expressed also in terms of the *oil/water partition coefficient,* which is a measure of relative solubility in oil and water. A high oil/water partition coefficient indicates high lipid solubility. Table 4-4 shows that methyl alcohol penetrates plant cells so rapidly that the actual rate cannot be determined, whereas the more water-soluble ethylene glycol, glycerol, and erythritol penetrate more slowly. The rate of penetration decreases with increasing polarity, because of increased numbers of hydroxyl groups.

Size of Particle

Looking at the formulas of the substances in Table 4–4, we see clearly that there is an increase in molecular size, which might be expected to deter the penetration of the larger molecules. However, if molecules of comparable size but very different lipid solubility are compared, the lipid soluble molecules penetrate much more rapidly. Propionamide and glycerol have approximately the same molecular weight, but propionamide penetrates the erythrocyte 400 times faster than glycerol. The greater lipid solubility of propionamide is the dominant factor here.

If one compares a series of homologous compounds that gradually increase in molecular size, one finds that the smaller molecules penetrate more rapidly, with the plasma

Table 4–4. **Correlation Between Permeability and Lipid Solubility for Leaf Cells (*Chara ceratophylla*)**

	Substance	Relative Permeability* of Cell	Oil/Water Partition Coefficient
Methyl alcohol	CH_2OH	0.99	0.78
Ethylene glycol	$CH_2OH \cdot CH_2OH$	0.043	0.049
Glycerol	$CH_2OH \cdot CHOH \cdot CH_2OH$	0.00074	0.007
Erythritol	$CH_2OH\ (CHOH)_2CH_2OH$	0.000046	0.003

Source: Data from R. Collander. *Physiol. Plant,* **2:**300 (1949).

* Complete permeability is 1.00.

membrane apparently acting as a *molecular sieve.* It has been calculated that the pores of biological membranes vary in diameter, but on an average measure about 0.7 nm. A pore this size should easily pass both the hydrated sodium ion (0.5 nm) and the hydrated potassium ion (0.4 nm), but permeability to these two ions varies dramatically and rapidly depending on the state of excitation of the cell; consequently, it is unrealistic to think of these pores as being static or permanent. There is also a considerable difference in the relative permeability of different cells to these two ions. Again, the hydrated chloride ion (0.39 nm) is approximately the same size as the hydrated potassium ion, but penetrates the red blood cell considerably faster, because of its negative charge.

Consequently, one can tentatively state that small, lipid soluble particles, either uncharged or with a negative charge, will penetrate more rapidly than large, polar molecules, especially those with a positive change. However, membranes are far more complex than this simple analysis would indicate; membranes other than those of red blood cells do not show the preferential permeability to anions, and, in fact, are relatively impermeable to salts. Danielli (1962) has shown that the permeability of the mouse erythrocyte to erythritol or glycerol is more than one hundred times the permeability of the ox erythrocyte to these substances. One has to consider, then, the variable nature of the plasma membrane, *varying among species,* varying with physiological condition, and even varying, like a *mosaic,* from one place on the membrane to another. A good example of membrane heterogeneity is seen in the membrane of an innervated cell. The membrane of this postsynaptic cell can be excited electrically everywhere *except* at the region where the nerve comes in apposition to it, forming the synapse. In the synaptic region, the postsynaptic cell can be excited only by chemical transmitters, which are normally released by the nerve endings (Fig. 4-16). This indicates a basic difference in membrane structure at the synaptic and nonsynaptic areas of the same cell.

Figure 4-16. **Differences in membrane structure and function at synaptic and nonsynaptic regions of a skeletal muscle cell.**

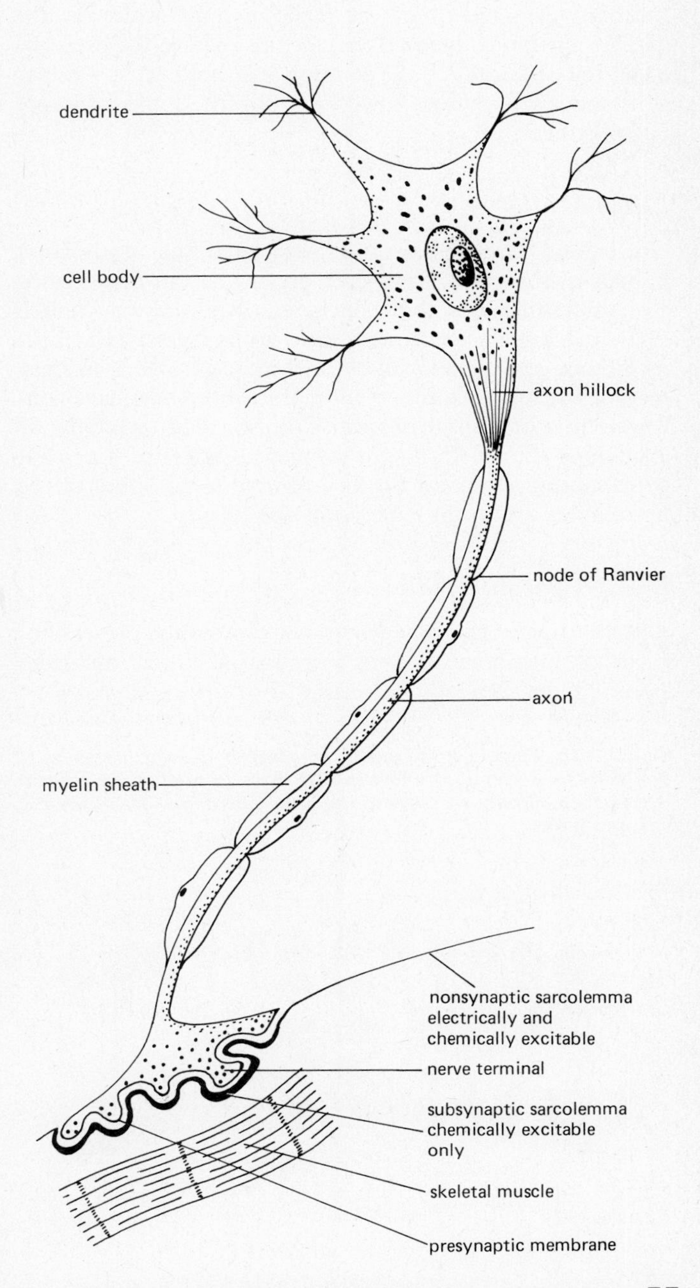

4–5 Current Concepts of Membrane Structure

Clearly emerging from the early studies on the plasma membrane were the concepts that its characteristic of per-

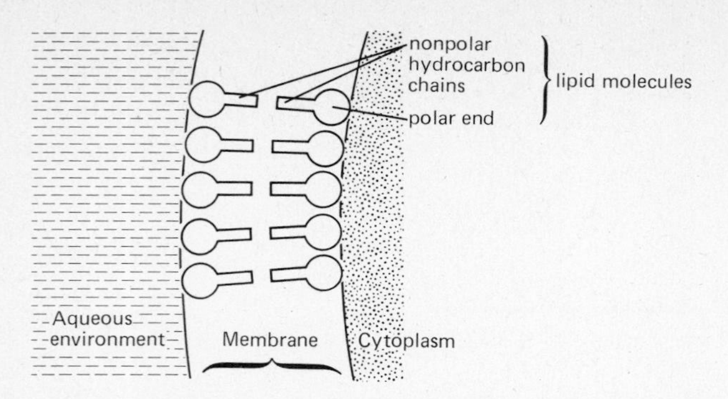

Figure 4-17. **Bimolecular lipid membrane of a red blood cell as postulated by Gorter and Grendel in 1925.**

meability required that it be made of lipid, while its low surface tension and sievelike structure require the presence of protein. Basically, these concepts still hold: what is under continuous speculation, however, is how the components are put together.

Lipid Model

Gorter and Grendel, in 1925, extracted the lipids from human erythrocytes; from data obtained from compressed monomolecular files, they concluded that there was enough lipid to form a *bimolecular layer* around each cell. This membrane structure is shown in Fig. 4-17. More modern methods of lipid extraction and film compression have confirmed this organized bilayer of lipids. The technique of examining the electron spin resonance (ESR) of paramagnetic substances shows further that the more liquid of the lipid layers are in the interior of the bilayer.

Lipid-Protein Model

A purely lipid membrane would not explain the presence of pores or membrane charge, nor would it give sufficient stability to the structure. Davson and Danielli proposed in 1935 that the *lipid bilayer* was sandwiched in between *two layers of adsorbed proteins,* for proteins will form monomolecular films at interfaces, as previously discussed. This membrane model assumes that the lipid bilayers are oriented with their polar groups facing outward, binding the protein molecules at both surfaces (Fig. 4-18). This model also can explain the presence of charged pores. The total thickness of the membrane is calculated at between 7.5 and 10 nm. This view of the membrane is particularly appealing when correlated to the three-layer structure of the unit membrane as seen under the electron microscope, where the inner lipid bilayer is seen as a light zone between two dark zones, the protein layers (Fig. 2-5).

Figure 4-18. **Protein-lipid model of plasma membrane as suggested by Davson and Danielli in 1935. A layer of globular protein is adsorbed on both surfaces of the lipid bilayer and also lines the charged pores.**

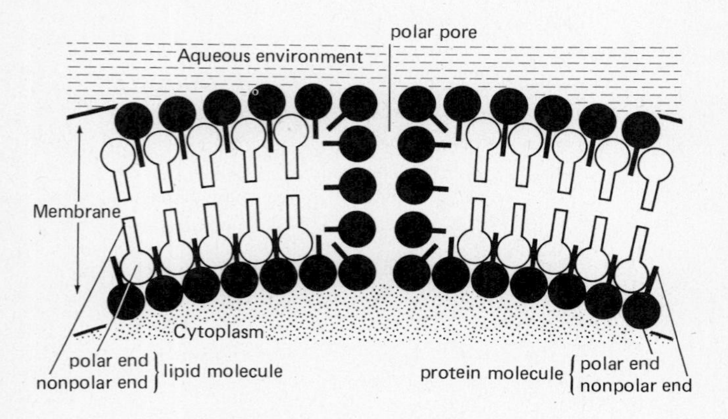

Appealing though the Davson-Danielli model is, it requires that the lipid-protein ratios of cells be fairly constant, so that there would be just a sufficient amount of protein to cover the lipid bilayer on either side. Table 4-5 shows clearly that these ratios vary considerably. The most intensely investigated system is myelin, and one finds that there is only enough protein to cover 43 per cent of the lipid area. Similarly, there is enough protein in the erythrocyte ghost to provide the cell with a monomolecular layer 2.5 times the area of the lipid. Most bacterial membranes seem to have enough protein to cover five times the area occupied by the lipid. The particular lipid and protein composition also varies considerably in different membranes, so one must conclude that there is a wide spectrum of membrane structure and composition, all perhaps variations on the unit membrane theme. It is not surprising that membranes that have to carry out energy transduction, transport of small molecules, protein biosynthesis, and many other intricate metabolic tasks should show modifications of structure to permit them to perform these specialized functions.

Fluid Mosaic Models

Current concepts of membrane structure bear little relationship to the neat and regular bilayer models conceived from earlier work. Cell membranes are now visualized as dynamic fluid structures, with globular proteins floating like icebergs in a moving sea of lipid (Singer and Nicolson, 1972).

MEMBRANE PROTEINS

Intrinsic Proteins. In the fluid model of the cell membrane, those proteins that are incorporated into the lipid are the *intrinsic proteins.* They are amphipathic and structurally asymmetrical, with one highly polar end and one nonpolar end. The polar region emerges from the surface of the membrane, whereas the nonpolar portion sinks into the hydrophobic lipid interior. Drastic treatment with protein denaturants or organic solvents is required to dissociate these intrinsic proteins from the membrane.

There are many possible arrangements of the protein molecules in the lipid matrix. Some may extend only partially through the lipid bilayer; other proteins may penetrate

Table 4-5. **Protein and Lipid Composition of Different Membranes (Molar Ratio*)**

Membrane	Amino Acid	Phospholipid	Cholesterol	Area Ratio (Protein:Lipid)
Myelin	264	111	75	0.43
Erythrocyte	500	31	31	2.5
Bacillus megaterium (a bacterium)	520	23	0	5.4

Source: Data from E. D. Korn, *Science* **153:**1491 (1966).

* Data are calculated from the percentage compositions using the appropriate molecular weights.

through both membrane surfaces. In addition, several proteins may be arranged into a functional unit (Fig. 4-19).

Freeze-etch studies and studies of fractured membranes (Figs. 4-20, 7-8, and 7-9) show a granulated structure in which the granules (about 5 to 8.5 nm in diameter) probably represent protein aggregates. About 10 to 20 per cent of the protein within a cell is physically associated with membranes. Experiments with electron spin resonance techniques have shown that the globular proteins and the surface antigens float within the membrane, rather than being rigidly fixed in place.

Proteins within the membrane are intimately associated with *membrane transport,* acting as *enzymatic carriers* or as peptide cages or channels (Sec. 4-6). Membrane proteins may also act as *surface antigens,* with specific chemical groups exposed to the outer surface of the plasma membrane (antigen-antibody reactions are discussed in Sec. 20-2). Proteins are the selective *receptors* for many types of information carrying molecules, such as hormones and neurotransmitters (see Secs. 10-8 and 7-5).

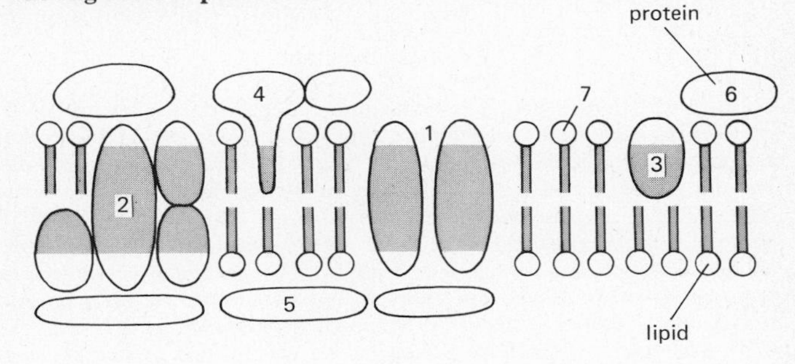

Figure 4-19. **Fluid mosaic model of membrane structure as suggested by Singer and Nicholson in 1972, and modified by other investigators since then. The protein molecules (large shapes) float in the lipid matrix, with their nonpolar regions (shaded areas) oriented toward the nonpolar lipid center of the membrane. Proteins occupy different positions in the membrane according to their structure and function. (1) Hydrophilic channel between adjacent protein molecules; (2) complex of proteins forming a functional unit; (3) protein acting as an enzymatic carrier; (4) surface antigen protein; (5) inner surface protein; (6) protein receptor for hormone or neurotransmitter, and so forth; and (7) lipid molecule in the area lacking surface protein.**

Extrinsic Proteins (Peripheral Proteins). Extrinsic proteins are associated only with the surfaces of the membrane. These proteins do not penetrate to any marked degree into the lipid matrix and so may be relatively easily dissociated from the membrane without destroying its structural integrity. Some extrinsic proteins can be visualized with the electron microscope: the enzyme *ATPase* on the *inner surface* of the mitochondrial membrane, for example, and the polypeptide *spectrin,* which looks like a fuzzy lining on the inside of the red cell ghost. Extrinsic proteins (often glycoproteins) on the *outer surface* of the cell membrane include those that are associated with *surface recognition* phenomena of cells.

Membrane lipids

The degree of fluidity of the membrane depends primarily on its lipid composition: the greater the proportion of unsaturated fatty acids to saturated fatty acids, the more fluid will the membrane be. Transport is considerably faster in fluid membranes, which are rich in unsaturated fatty acids, than in the more rigid membranes with a greater proportion of saturated fatty acids.

Many animals, in preparing for winter, are able to change fatty acid proportions, thus keeping cell membranes and other lipid containing structures fluid despite the cold. It is as though they change from summer to winter oil.

Summary of Membrane Structure Concepts

One must conclude that the irreducible minimum for a cell membrane is a continuous lipid bilayer, with polar regions oriented outward and apolar regions oriented toward the center of the membrane. This lipid matrix has the consistence of gasoline at body temperature. The lipid is interrupted by protein structures and pores, most of which, like the lipid molecules, are constantly moving within the membrane. The heterogeneity of cell membranes can be accounted for by variations in the specific lipids and proteins incorporated into the membrane and by differences in the structural organization of the protein molecules in the lipid.

4-6 Dynamic Nature of the Plasma Membrane

Active Transport

Biological membranes do not behave as a purely passive, porous barrier to the passage of substances. If this were so,

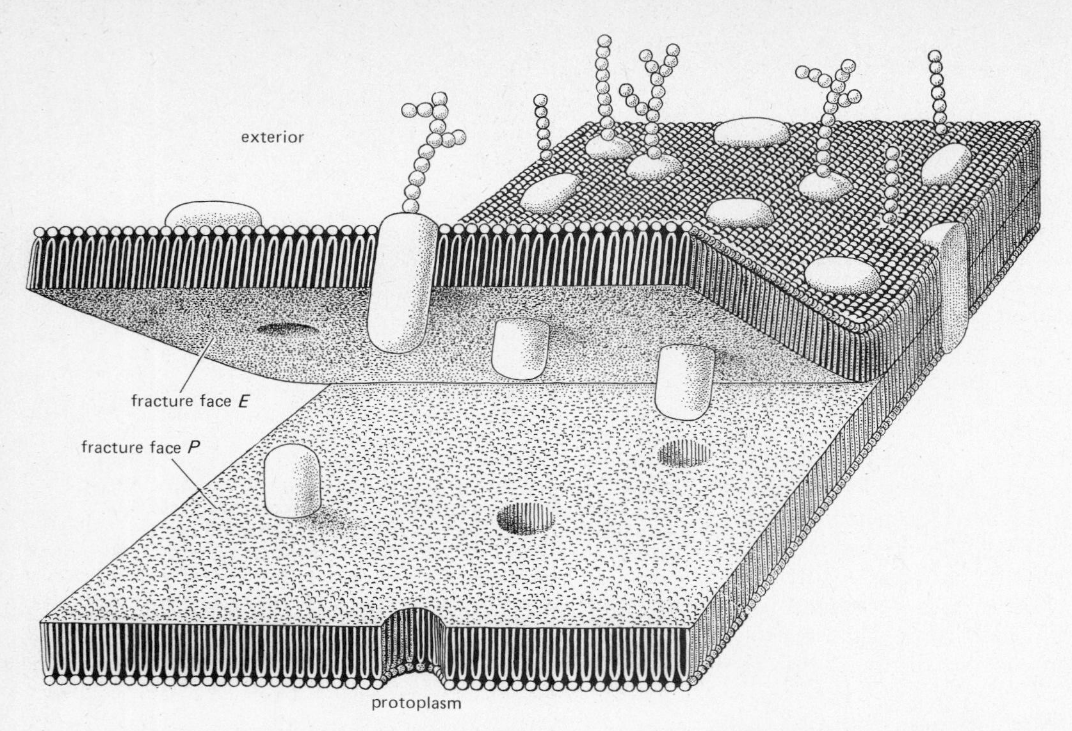

Figure 4-20. **Freeze-fracture technique is depicted as it is employed to elucidate the structure of a membrane. The intact membrane appears at the upper right. It consists of a double layer of lipid molecules, which have water-soluble heads and water-insoluble tails. The heads face outward and the tails inward, end to end. Certain proteins, shown here as large elliptical objects, can either penetrate the membrane or interact with its two outer surfaces. Proteins and lipids facing the exterior surface are often combined with sugars, shown as straight and branched, linked chains. When the membrane is frozen and fractured, it comes apart at the plane between the lipid tails. The penetrating proteins are not cleaved and hence appear on a fracture face as protuberances and depressions. (From "The Final Steps in Secretion," by B. Satir, *Scientific American*, Oct. 1975, p. 28. Copyright © 1975 by Scientific American, Inc. All rights reserved.)**

no transport could occur against a concentration gradient, and electrochemical and concentration gradients could not be efficiently maintained. Yet marked concentration gradients are seen in all living systems. The plant cell *Nitella* maintains a K^+ concentration within the cell cytoplasm that is more than 1000 times as much as the K^+ outside the cell. The K^+ concentration in the erythrocyte is 150 meq/liter while the plasma K^+ concentration is only 5.35 mEq/liter. In general, cells maintain a very high internal K^+ concentration relative to the external environment, but a relatively low Na^+ concentration. In the case of the erythrocyte, the external Na^+ concentration is approximately 12 to 15 times greater than that of intracellular Na^+.

This ability of biological membranes to establish and maintain steep concentration gradients is also seen in the intestine, where the cells will transport glucose from the intestinal lumen into the circulation even if the blood glucose is higher than the gastric glucose. This is obviously of significance in salvaging all available glucose for the organism, even under starvation conditions. The homeostatic functions of the kidney, too, depend to a large extent on its ability to reabsorb substances that have been filtered by the nephrons and return them, against a concentration gradient if necessary, to the blood. This type of transport requires the expenditure of metabolic energy and is termed *active transport*. It is profoundly influenced by certain hormones, an important physiological regulatory device discussed in Secs. 10-8 and 15-4.

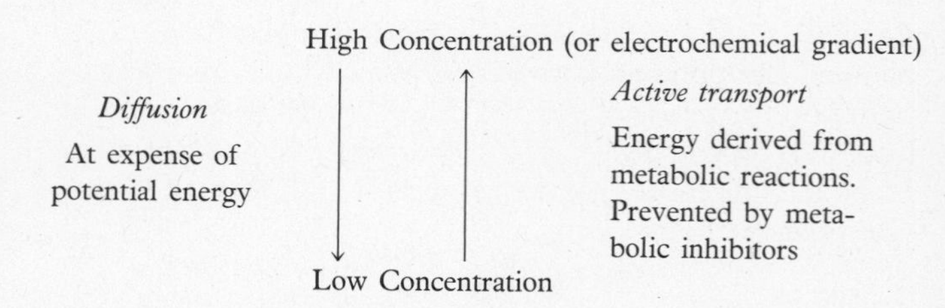

Figure 4-21 is an arbitrary plot showing the effect of a respiratory poison on the active transport of a substance into a cell. Before the poison is administered, the substance accumulates within the cell interior with time, until it finally levels off. This point of equilibrium is far above the concentration of the substance in the external medium. If the poison is given before equilibrium is reached, there will be a rapid fall in the rate at which the solute is passing into the cell; if it is administered later, the cell will be unable to maintain the concentration gradient it had already established.

Mechanisms of active transport

The kinetic data favor the concept of a carrier molecule mechanism, but they do not prove it. If one compares the slope of the curves in Fig. 4-22, one sees that if the cell membrane is acting purely as a mechanical barrier there is a linear relationship between the amount of solute transported and the concentration gradient. If there is a carrier system involved, however, the curve rises rapidly and then flattens out, for a maximum rate (V_{max}) is reached when all the carrier molecules are being utilized. The mechanism is said to be *saturated.*

Carrier molecules

For a molecule to be a carrier, it must have the following characteristics:

1. It must be able to pick up the substance to be carried across the membrane at one side and be able to release it at the other.
2. It must have some structural or electrical characteristic that will enable it to select the particular molecule or ion it is to carry.
3. It must be a part of the cell membrane yet be able to move through it.

One contender for the role of a carrier molecule is *phosphatidic acid,* a phospholipid that lacks the water-soluble moiety attached to the other phospholipids (Sec. 3-3.3). Consequently, it will tend to stay in the lipid phase of the membrane and not pass into the aqueous phase on either side. Phosphatidic acid has an anionic charge that could bind monovalent cations such as Na^+ and K^+. The energy for this ion transport can be shown to come from the diphosphorylation of ATP. The active transport of sodium ions out of cells and the inward transport of potassium ions are sometimes called the sodium and potassium *pumps.* Table 4-6 shows the K^+ flux for erythrocytes of different animals.

Table 4-6. **Potassium Flux for Erythrocytes of Various Animals**

	K^+ *Influx* μ Equivalents (cm/sec)	K^+ *Efflux* μ Equivalents (cm/sec)
Frog	1.1×10^{-8}	0.19×10^{-9}
Cat	8.9×10^{-9}	3.4×10^{-9}
Human	6.7×10^{-9}	0.24×10^{-9}

Proteins may also act as *carrier molecules for charged molecules,* providing a protected polar passageway or "*peptide cage*" that traps the cation on one side of the membrane. The ion enters the cage and is stripped of its bound water; this changes the conformation of the protein shield, closing the cage (Fig. 4-23). The protein swings around within the membrane so that the closed gate now faces the inner side of the membrane; the gate opens and the hydrated ion escapes, restoring the protein to its original conformation.

Proteins are also responsible for the *enzymatic transport* of large molecules, like sugars and amino acids, across cell membranes. These proteins are enzymes called *transferases;* being intrinsic proteins, they pick up their substrate at one side of the membrane and shuttle it across to the other side. The enzymes themselves are hydrophobic and cannot enter the aqueous environment on either side of the membrane.

The classic experiments by Cohen and Monod (1957) at the Pasteur Institute in Paris were performed on bacteria (*Escherichia coli*) in which the sugar (*lactose*) transport system develops, or is *induced,* only by the addition of the

Figure 4-21. **Effect of a respiratory poison on active transport. Active transport results in a concentration of solute within the cell considerably greater than that of the medium. A respiratory poison administered before equilibrium is reached (1) immediately decreases transport. Administration of the poison after equilibrium (2) prevents maintenance of the concentration difference.**

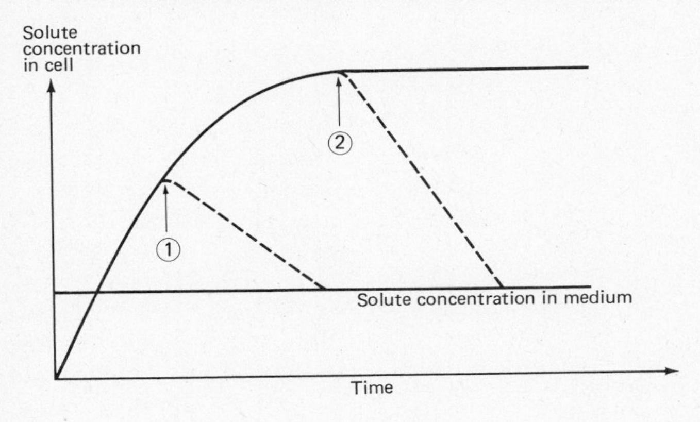

Figure 4-22. **Comparison of kinetic data for diffusion and active transport of solute into a cell. The solid line shows a linear relationship for diffusion. Active transport (broken line) reaches a maximal rate (V_{max}), indicating a possible carrier mechanism that is saturated at this point.**

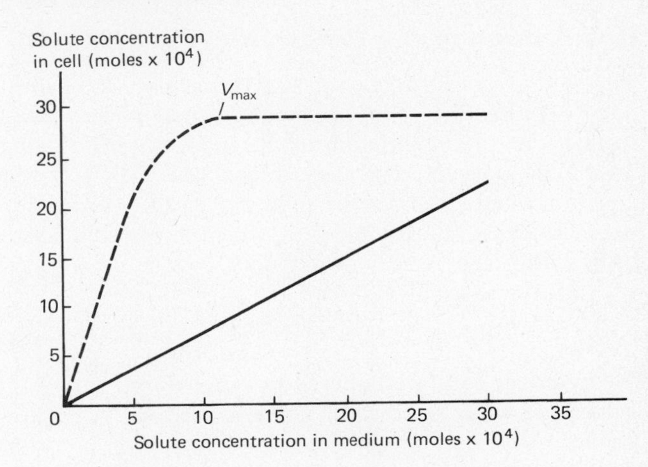

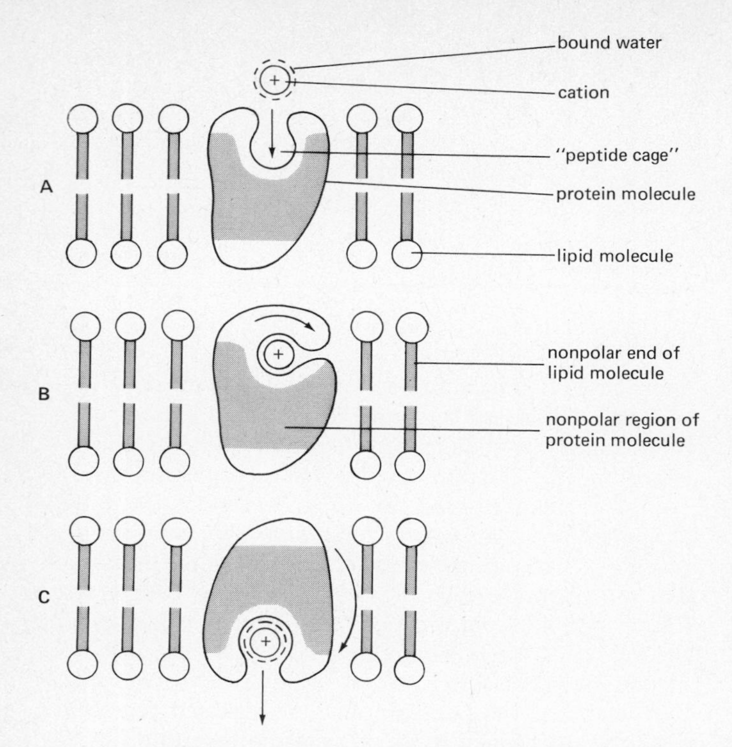

Figure 4-23. **Passage of a cation through a hydrophilic "peptide cage" of a protein molecule (A). The cation is stripped of its water shell within the cage (B), changing the conformation of the protein molecule, which swings around to the inner surface of the membrane. Rehydration of the cation releases it from the cage (C).**

substrate. After a brief delay, during which time enzyme induction occurs, the substrate is transported across the membrane even against a concentration gradient of 1:500. This type of induced enzyme transport system is known as a *permease system.*

The characteristics of the rate at which lactose is transported are those of enzymatic reactions (Sec. 5-1), and the system is blocked by metabolic poisons. This enzyme transport system, once formed, functions only to bring the substrate through the barriers of the cell membrane but is quite separate from the enzymes that metabolize the sugar.

Modification of the Penetrating Particle

What often appears to be an active transport mechanism may be caused by a change in the actual concentration gradient due to an alteration of the substance penetrating. Carbon dioxide entering a red blood cell (RBC) is very quickly converted, under the influence of the enzyme *carbonic anhydrase,* to carbonic acid, which in turn dissociates to the bicarbonate and hydrogen ions.

$$\begin{array}{ll}\text{Plasma} & \text{RBC} \\ CO_2 \longrightarrow & CO_2 + H_2O \rightleftharpoons H_2CO_3 \rightleftharpoons HCO_3^{+} + H^{+}\end{array}$$

The hydrogen ions are buffered by hemoglobin within the erythrocyte while the bicarbonate ions diffuse out into the plasma, where they are buffered by sodium ions, forming $NaHCO_3$, part of the alkali reserve of the body. The net result is that, by these reactions, the CO_2 is removed almost as fast as it enters, so that the concentration gradient always favors the entry of CO_2.

Many enzymes are located on the outside of the cell membrane, in the fuzzy coat. This is particularly true of the lipases of the gastric mucosa cells. Fat droplets may be partially digested by these extracellular enzymes before they can enter the cell.

Pinocytosis (Cell Drinking)

Pinocytosis is a rather neat device whereby large particles may evade the membranous barrier by adhering to special sites on the membrane, which are then activated and invaginate, forming a vesicle with the particle inside. Often, a small amount of extracellular fluid is incorporated into the vesicle, which then penetrates through the cytoplasm, leaving membranous channels to mark its pathway (Fig. 4-24). The vesicle may then become detached from the cell membrane to form part of the cytoplasm or to be attacked by lysosomal enzymes, which destroy the vesicular membrane, liberating the engulfed particle.

Figure 4-24. **Pinocytosis. A large extracellular particle is bound to a special site on the plasma membrane, which then invaginates, trapping the particle in a vesicle. The vesicle may later be freed from the invaginated membrane and digested by lysosomal enzymes.**

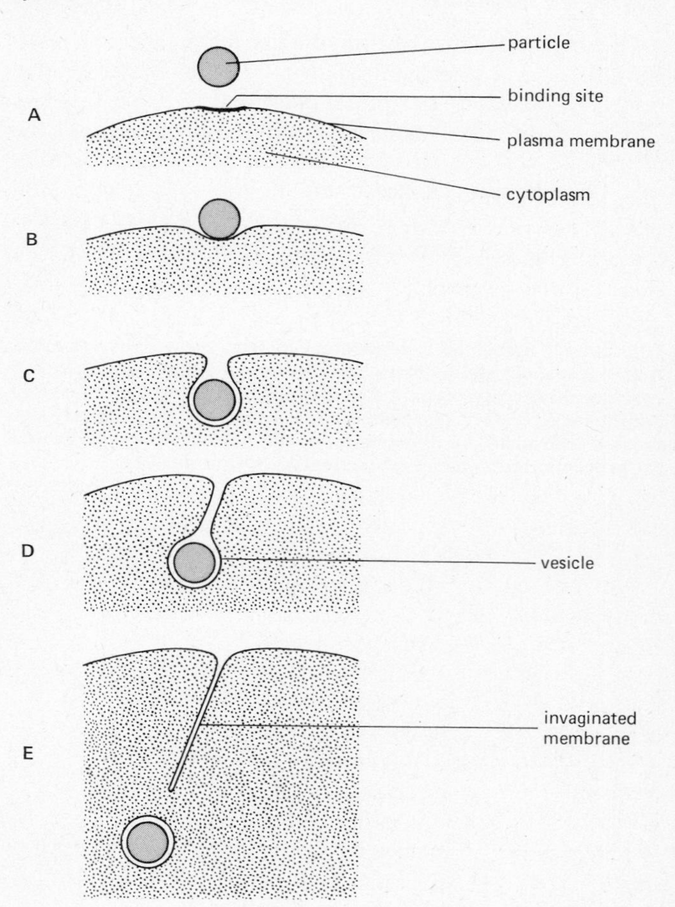

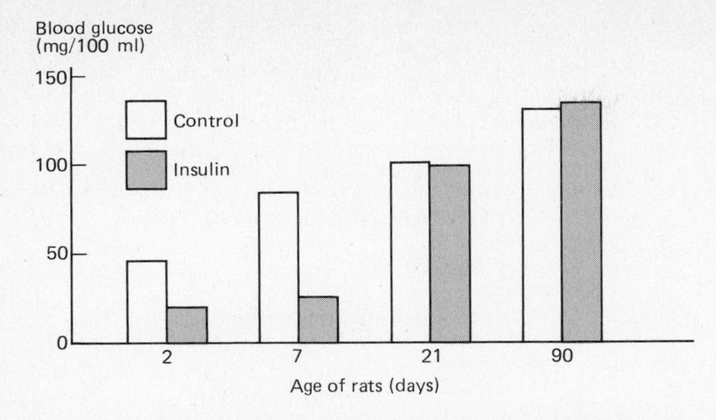

Figure 4-25. **Effect of orally administered insulin (40 International Units/100 g) on blood glucose levels in immature and mature rats. There is a marked fall in blood glucose in the immature animals, but by 3 weeks of age orally administered insulin is ineffective.**

Pinocytosis may be demonstrated under certain physiological circumstances only. A hungry ameba will greedily engulf protein molecules, marked with fluorescent dyes or radioactive traces. A well-fed ameba will not respond. (The engulfing of large particles, such as bacteria, is *phagocytosis,* or cell eating.) Pinocytosis is inhibited by metabolic inhibitors such as dinitrophenol, carbon monoxide, cyanide, and low temperatures, as well as by many drugs. Thus pinocytosis is yet another indicator of the dynamic state of the cell membrane.

Effect of Maturation on Active Transport and Pinocytosis

Pinocytosis may be considered a primitive form of digestion and occurs in the intestinal cells of the immature mammal; it is a property that is lost with maturity. Suckling infants can absorb the large protein antibodies in the mother's milk, presumably by pinocytosis. This will confer a temporary *passive immunity* on the young infant. Similar studies in newborn rats show that insulin administered orally will lower blood sugar markedly, but it will not cause any significant change in older rats, which are unable to absorb protein molecules from the digestive tract (Fig. 4-25). The interpretation of this experiment is complicated by the fact that there is very little protein digestion in the infant stomach.

Although pinocytotic activity decreases with maturation, active transport mechanisms increase. This is particularly marked for glucose transport which reaches a peak at the time of weaning. Prior to weaning, the milk sugar (lactose) is digested to galactose which is readily absorbed, and glucose, which is poorly absorbed. Once a solid diet has been instituted, the main monosaccharide is glucose, and the more mature intestine now preferentially and actively transports glucose. It is interesting that this intestinal maturation is not dependent on the diet, for it will occur even if the infant is maintained on a milk diet.

Cited References

COHEN, G., and J. MONOD, "Bacterial Permeases." *Bact. Rev.* **21:** 169 (1957).

COLE, K. S. "Surface Forces of the Arbacia Egg." *J. Cell. Comp. Physiol.,* **1:** 1 (1932).

DANIELLI, J. F. "Structure of the Cell Surface." *Circulation* **26:** 1163 (1962).

DANIELLI, J. F., and H. A. DAVSON. "A Contribution to the Theory of Permeability of Thin Films." *J. Cell. Comp. Physiol.* **5:** 495 (1935).

DANIELLI, J. F., and E. N. HARVEY. "The Tension at the Surface of Mackerel Egg Oil with Remarks on the Nature of the Cell Surface." *J. Cell. Comp. Physiol.* **5:** 483 (1935).

GORTER, E., and F. GRENDEL. "On Biomolecular Layers of Lipids on the Chromocytes of the Blood." *J. Exp. Med.* **41:** 439 (1925).

MAYERSON, H. S., C. G. WOLFRAM, H. H. SHIRLEY, and K. WASSERMAN. "Regional Differences in Capillary Permeability." *Amer. J. Physiol.* **198:** 155 (1960).

MONOD, J., and F. JACOB. "Genetic Regulatory Mechanisms in the Synthesis of Proteins." *J. Mol. Biol.* **3:** 318 (1961).

SINGER, S. J., and G. L. NICOLSON. "The Fluid Mosaic Model of the Structure of Cell Membranes." *Science* **175:** 720 (1972).

STARLING, E. H. "On the Absorption of Fluid from the Connective Tissue Spaces." *J. Physiol. (London)* **19:** 312 (1896).

Additional Readings

BOOKS

DAVSON, H., and J. F. DANIELLI. *The Permeability of Natural Membranes,* 2nd ed. Cambridge University Press, New York, 1952.

KORN, E. D. *Transport.* Plenum Publishing Corporation, New York, 1975.

KOTYK, A., and R. JANACEK. *Cell Membrane Transport,* 2nd ed. Plenum Publishing Corporation, New York, 1974.

PACKER, L. *Biomembranes: Architecture, Biogenesis, Bioenergetics and Differentiation.* Academic Press, Inc., New York, 1975.

PONDER, E. *Hemolysis and Related Phenomena.* Grune & Stratton, Inc., New York, 1948.

STEIN, W. D. *The Movement of Molecules Across Cell Membranes.* Academic Press, Inc., New York, 1967.

WEISSMAN, G., and R. CLAIBORNE, eds. *Cell Membranes: Biochemistry, Cell Biology and Pathology.* Hospital Practice Publishing Co., Inc., New York, 1975.

ARTICLES

CAPALDI, R. A. "A Dynamic Model of Cell Membranes." *Sci. Am.,* Mar. 1974.

FOX, F. "The Structure of Cell Membranes." *Sci. Am.,* Feb. 1972.

HODGKIN, A. L., and R. D. KEYNES. "Active Transport of Cations in Giant Axons from *Sepia* and *Loligo.*" *J. Physiol. (London)* **128:** 61 (1955).

KORN, E. D. "Structure and Function of the Plasma Membrane: A Biochemical Perspective." *J. Gen. Physiol.* **50:** 257 (1968).

LODISH, H. F., and R. D. KEYNES. "The Assembly of Cell Membranes." *Sci. Am.,* Jan. 1979.

ROBERTSON, J. D. "The Unit Membrane and the Davson-Danielli Model." In *Intracellular Transport,* K. B. Warren, ed., Academic Press, Inc., New York, 1965, p. 1.

RUSTAD, R. C. "Pinocytosis." *Sci. Am.,* Apr. 1961.

SINGER, S. J., and G. L. NICOLSON. "The Fluid Mosaic Model of the Structure of Cell Membranes." *Science* **175:** 720 (1972).

SOLOMON, A. K. "The State of Water in Red Cells." *Sci. Am.,* Feb. 1971.

Chapter 5

Enzymes and Metabolic Regulation

Stars and atoms have no size,
They only vary in men's eyes.

Men and instruments will blunder,
Calculating things of wonder.

A seed is just as huge a world
As any ball the sun has hurled.

Stars are quite as picayune
As any splinter of the moon.

Time is but a vague device;
Space can never be precise;

Stars and atoms have a girth,
Small as zero, ten times Earth.

There is by God's swift reckoning
A universe in everything.

A. M. Sullivan, "Measurement"

Enzymes

Enzymes are biological catalysts that not only speed up the many biochemical reactions of the organism, but also direct and select metabolic pathways. Enzymes, in turn, are modified in their activity or concentration by hormones that transmit information about situations external to the responsive cell. A chain of biochemical information and response is set up so that a change in one part of the organism will catapult a series of responses elsewhere in the organism. A carbohydrate-rich meal, for example, will, through the hormonal action of insulin, result in the activation of glycogen synthetase, an enzyme that builds up liver glycogen from glucose. A fall in blood glucose, through epinephrine and glucagon action, will activate another series of enzymes to cause glycolysis, the breakdown of glycogen to glucose. These interconvertible enzyme systems, functioning at a level of cellular homeostasis, are switched through hormonal integrative mechanisms to processes that serve the homeostatic requirements of the entire organism.

5–1 Enzymes as Catalysts

Catalysis speeds up a chemical reaction without altering its final equilibrium. The biochemical reactions accelerated by enzymes could occur in the absence of these protein catalysts, but so slowly as to be of little use to the organism. If this rate were not so low, many substances would disintegrate spontaneously.

An enzyme, like other catalysts, does not influence the equilibrium of the reaction ($A + B \rightarrow C$), but it combines with the *substrate,* the substance on which it acts, to form an unstable intermediate complex, the *enzyme-substrate complex*. In this manner, it lowers the *activation energy* of the reaction. This is the energy barrier that must be overcome

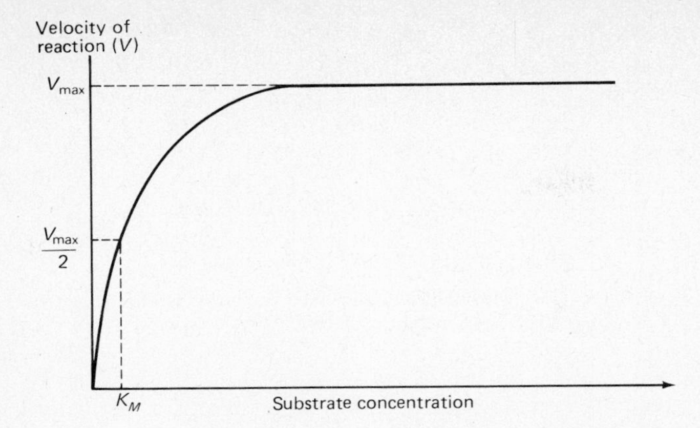

Figure 5-1. **The rate of an enzyme-catalyzed reaction and substrate concentration. The maximum rate of the reaction (V_{max}) is reached when the substrate concentration "saturates" the enzyme; K_M is the substrate concentration at half the maximum reaction velocity.**

before the reaction can occur. The activation energy may be supplied by the internal energy of the molecules as they collide (a very slow process at physiological temperatures), or it may be speeded up by adding external energy in the form of heat. This increases the rate of movement of the molecules and the likelihood of their colliding and reacting—not a very practical mechanism for living systems, because heat rapidly denatures proteins and destroys cells.

In biological systems, the activation energy is lowered by the formation of the intermediate enzyme-substrate complex. As formulated originally by Michaelis and Menten, the enzyme (E) combines with the substrate (S) to form the enzyme-substrate complex (ES) from which the enzyme is then regenerated essentially unchanged.

$$E + S \rightleftharpoons \boxed{ES} \longrightarrow E + P$$

P is the product of the reaction, which may be rate limiting if it accumulates, thus inhibiting the enzyme and controlling its own rate of synthesis.

Under controlled conditions of temperature and pH the rate of enzyme catalysis increases with increased substrate concentration up to a limiting value: further increase in substrate concentration produces no additional effect. This is explained by the "saturation" of the enzyme with the substrate to form ES. At this point, the velocity (V) of the reaction is maximal (V_{max}). Increasing the substrate concentration beyond this cannot increase the amount of ES, so that the rate of the reaction is now independent of substrate concentration (Fig. 5-1).

The Michaelis constant (K_M) is the substrate concentration at half maximum velocity ($V_{max}/2$). A high Michaelis constant indicates that the enzyme has a low affinity for a substrate and will preferentially bind another substrate with a lower K_M.

5–2 Enzyme Structure

All enzymes are proteins; with the exception of the digestive enzymes pepsin and trypsin, and some other enzymes, they consist of a protein part (the *apoenzyme*) conjugated with a nonprotein portion. If the nonprotein moiety is an organic molecule, easily separated from the apoenzyme, it is called a *coenzyme;* if it is firmly attached it is usually referred to as a *prosthetic group*. The atoms of copper, zinc, and iron that are bound to certain enzymes are prosthetic groups; many coenzymes are closely related to vitamins. The coenzymes also

Figure 5-2. **Great structural similarity (A) permits the substrate to fit exactly into the binding pocket of the active site of the enzyme to form the enzyme-substrate complex (B). This pulls the catalytic site side-chains of the protein (enzyme) into the correct alignment to cleave the substrate and free the products (C).**

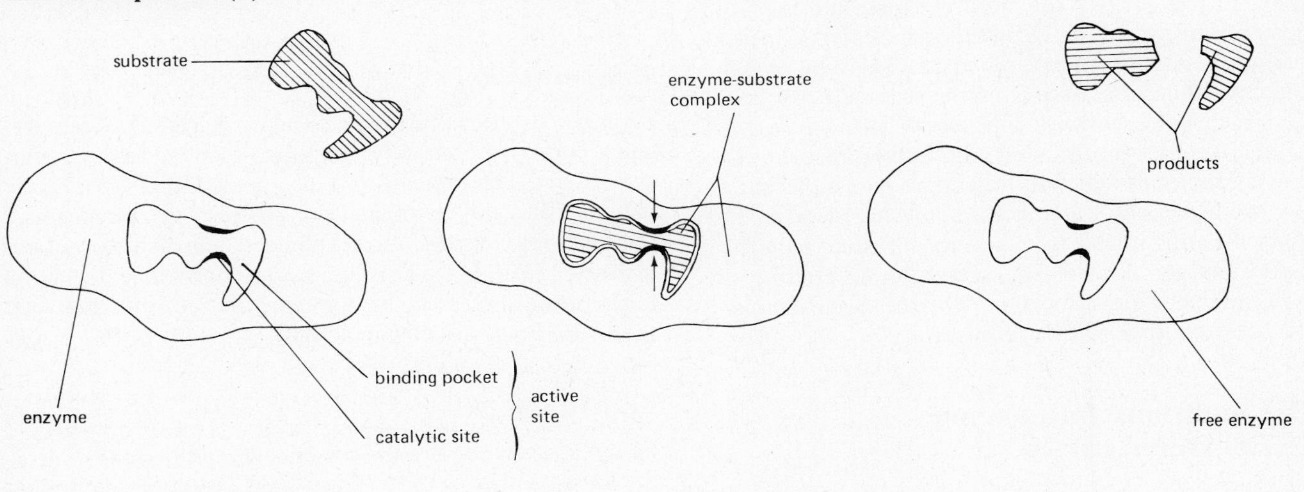

function as catalysts; therefore they, like enzymes, are needed only in very small amounts, for they are not used up in the reaction.

Specificity

The great specificity of enzyme is caused by their protein structure. They are formed from polypeptide chains, tightly folded into a compact globular form. A specific geometric arrangement of chemical groups on a small patch of the surface of the molecule assures that the enzyme will combine only with its particular substrate. This small patch where the enzyme and substrate are bound to each other and where catalysis takes place is the *active site* (Fig. 5-2).

Enzymes can distinguish between highly similar substrates, and the proteases that split the peptide linkages between adjacent amino acids exemplify this well. The digestive enzyme trypsin will cleave a bond only between the amino acids serine and lysine; on the other hand, chymotrypsin, another digestive protease, will act only on a phenylalanine-serine bond. How can an enzyme discriminate between these amino acids? It appears that enzyme and substrate must be bound together at several points. Dickerson (1964) and his colleagues believe that there is a "pocket" in the surface of the enzyme, near or in the active site, that has very specific chemical properties. In the case of trypsin, this pocket is negatively charged, forming a receptor for the positively charged side chains of its amino acid substrates, which are neatly fitted into it. Electrostatic attraction now holds these side chains in place; this orients the substrate on the active site at the precise configuration needed for catalysis (Fig. 5-2).

Induced Fit

One of the theories suggested to explain enzyme action is that enzymes are "flexible," changing their shape slightly to provide a perfect or "induced" fit of the enzyme to its substrate. In this process the substrate molecule is placed under strain, distorting its structure sufficiently to permit the catalytic groups of the enzyme to attack it. The changed substrate molecule is released from the enzyme as the product (P), freeing the enzyme, which returns to its original shape, able to pick up another substrate molecule. The number of molecules of substrate that one molecule of enzyme can act upon in 1 min is its *turnover number*. For most enzymes this ranges from several substrate molecules to 10,000, but for catalase and acetylcholine esterase the turnover numbers are astonishingly large, approaching a million substrate molecules a minute.

5–3 Conditions Influencing Enzyme Activity

Because enzymes are proteins, they are greatly influenced by any factors that alter their charge or structure. These factors

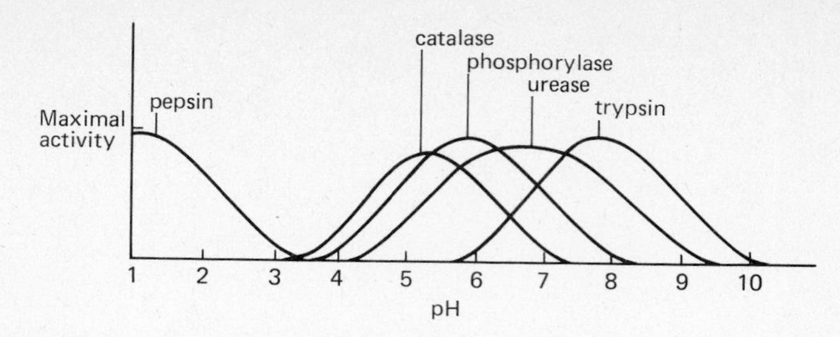

Figure 5-3. **The optimal pH of different enzymes is shown by the pH at which they have maximal activity.**

include pH, temperature, and a wide variety of activators and inhibitors.

Optimal pH

There is an *optimal pH* at which each enzyme exhibits its greatest activity. For most enzymes this optimum is at a neutral or slightly acid pH. Physiologically, the range is quite narrow, for the pH of the blood and tissues remains constant around 7.4 and 7.3 (Sec. 3–1). A slight increase in acidity seems to increase the efficiency of some of the enzymes involved in muscle contraction and is supposedly one of the factors involved in "warming up" before an athletic feat, as the active muscles produce acid metabolites.

Some enzymes show extremes of pH optima. The optimal pH for pepsin is very acid (1.5–2.0), whereas that for trypsin is very alkaline (8–11). These requirements are reflected in the low pH of gastric juice, where pepsin is active, and in the alkaline pH of the duodenum, where trypsin activity occurs. One enzyme that is functional only in infancy is rennin. It is found in the stomach, and in the adult is completely nonfunctional at the low pH characteristic of gastric juice; in the infant, however, the amount of acid secreted is considerably less, and the rennin of early infancy is probably involved in the curdling of milk prior to digestion. Figure 5-3 illustrates the effect of pH on the activity of various enzymes.

Temperature

A similar curve can be drawn to depict the effect of temperature on enzyme activity. In warm-blooded animals, enzymes have their *optimal temperature* around 40°C (slightly higher than body temperature), whereas the temperature optimum for enzymes of cold-blooded animals is considerably lower (Fig. 5-4). This figure also shows that increasing temperature increases enzyme denaturation, thus reducing the overall rate of enzyme activity.

Activators

Many enzymes require the presence of monovalent or divalent ions for their activity. Muscle ATPase, an enzyme that hydrolyzes ATP, requires both Mg^{2+} and Ca^{2+}. Most coenzymes and prosthetic groups are activators; the enzyme

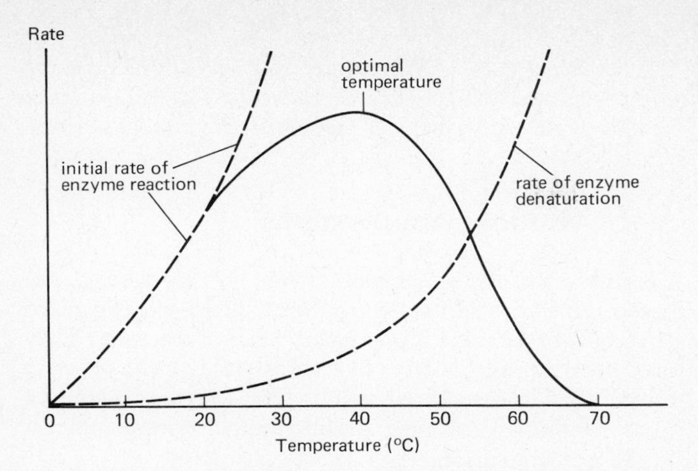

Figure 5-4. **The optimal temperature for enzyme activity is a composite of the increase in reaction rate with increased temperature, offset by the gradual denaturation of the enzyme by heat.**

has no biological activity in their absence. Another enzyme may be the activator, converting the inactive form of an enzyme to the active one. This important form of regulation is seen when protein kinases, enzymes present in all cells, are first activated by the cyclic AMP mechanism (Sec. 10-8), then in turn activate other enzyme systems to produce the biochemical response characteristic of that particular tissue.

Inhibitors

There are many types of enzyme inhibitors. Some compete with the substrate for the active site; others distort the active site by binding elsewhere and changing the structural geometry of the enzyme. Some inhibitors play a role in drug action; for example, antibiotics inhibit several enzyme systems necessary for bacterial growth. Heavy metals, such as cyanide, arsenic, silver, mercury, and copper, act as poisons that inhibit enzyme action completely. Other inhibitors are metabolic products, the accumulation of which stops the reaction.

Figure 5-5. **Reaction rates distinguish competitive from noncompetitive enzyme inhibition. As the competitive inhibitor competes with the substrate for the enzyme, increasing substrate concentration ultimately reverses the inhibition. This dependence on substrate concentration is not seen in noncompetitive inhibition.**

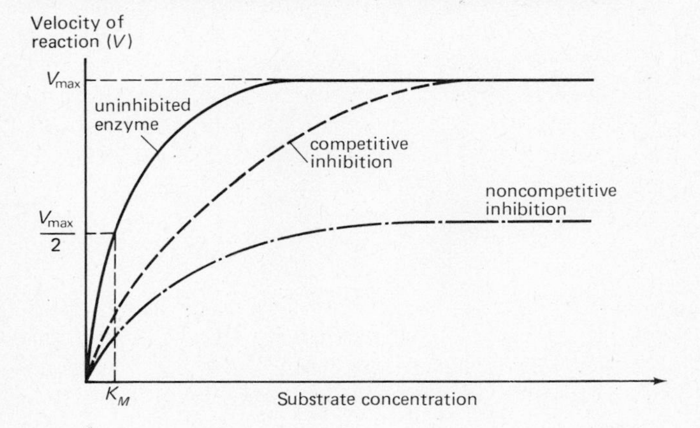

COMPETITIVE INHIBITORS

Organic molecules that compete with the proper substrate for the active site on the enzyme are called competitive inhibitors. Such competition requires a structural similarity between substrate and inhibitor, but also sufficient dissimilarity so that the inhibitor is not catalyzed by the enzyme but only forms a loose complex with it. Increasing the substrate concentration displaces a proportional number of inhibitor molecules; consequently, one of the identifying characteristics of competitive inhibition is that it can be reversed by increasing substrate concentration (Fig. 5-5).

A classical example of competitive inhibition is that by malonic acid of succinic acid, an important intermediary in the Krebs cycle (Sec. 5-12) for the enzyme succinic acid dehydrogenase. Although malonic acid has a structure very similar to that of succinic acid, it cannot be dehydrogenated by the enzyme and so blocks the entire cycle.

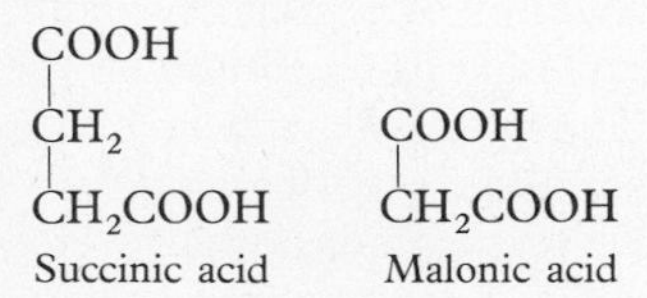

NONCOMPETITIVE INHIBITORS

Noncompetitive inhibitors do not compete with the substrate for the active site of the enzyme. There are two main types: allosteric inhibitors and inactivators of reactive groups.

Allosteric Inhibitors. These molecules bind with the enzyme at a site other than the catalytic one (*allosteric* = "other site"). They are not involved themselves in the chemical reaction but affect enzyme activity by altering its conformation. The binding at the allosteric site causes a progressive distortion of the enzyme that eventually involves the catalytic site, altering its shape and preventing the substrate from binding properly or producing misalignment of the catalytic groups (Fig. 5-6). Some enzymes may have a special binding site for the end product of the reaction, so that the end product acts as an inhibitor of the reaction that gave rise to it (*end-product inhibition*).

Inactivators of Reactive Groups. Some inhibitors form covalent bonds with important chemical groups at or near the active site. Nerve gases (organophosphates) form a complex with one or more of the —SH groups at the active site of the enzyme acetylcholine esterase, preventing the binding of the substrate acetylcholine and having profound effects on nerve and muscle. Increasing substrate concentration cannot reverse noncompetitive inhibition.

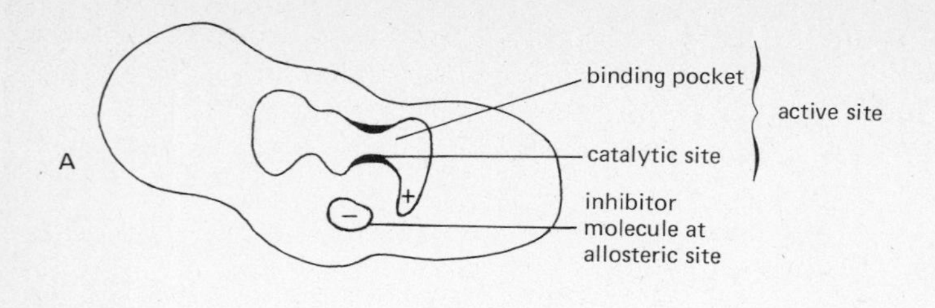

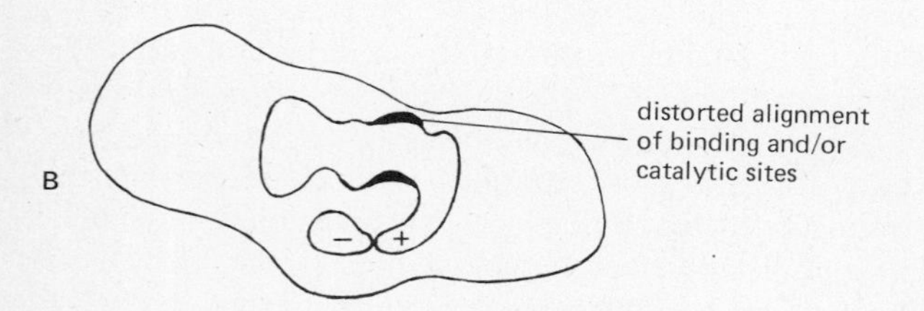

Figure 5-6. **Binding of a noncompetitive inhibitor at an allosteric site distorts the flexible enzyme so that the substrate cannot be properly bound and catalyzed at the active site.**

Regulatory molecules

Regulatory molecules may be either activators or inhibitors of enzyme action. A regulatory molecule binds at an allosteric (or regulatory) site, distorting the enzyme. If the induced change in enzyme shape results in a correct fit of the substrate to the catalytic site, the regulatory molecule "turns on" or activates the enzyme. If the reverse occurs and the catalytic function is disrupted, the enzyme is inhibited or "turned off." Hormones are among the most important regulatory molecules, turning enzymes on and off according to the metabolic requirements of the organism (Sec. 10-8).

Enzyme induction and repression

The concentration of an enzyme may be increased (induced) or decreased (repressed) by agents that affect protein synthesis within the cell. These include hormonal, nutritional, genetic, and pharmacological factors. In bacteria, these agents clearly affect the cell at the level of the gene, resulting in an increase in *m*RNA synthesis. Present evidence indicates this to be true also for mammalian systems because inhibitors of *m*RNA synthesis prevent the increase in enzyme concentration characteristic of some hormones and drugs.

5–4 Isozymes

Some enzymes may exist in several forms; they are very similar chemically, physically, and in the reactions they catalyze, but slight differences in their structure permit them to be separated by electrophoresis (Fig. 5-7). Lactic acid dehydrogenase exists in five forms, with proportions of the isozymes present in different organs. So far, this discovery has been a useful tool for studying genetic variability but its biological significance is uncertain. It appears that, in general, different tissues use a given enzyme under different conditions and that poikilotherms (cold-blooded animals) need enzymes with different properties in different seasons.

5–5 Multienzyme Systems

Enzymes that act in sequence, with the product of one becoming the substrate for the next, often are organized physically to facilitate this complex series of reactions. They may be arranged linearly, like an assembly belt, or more often in a well-structured sphere. The enzyme systems of tissue respiration are organized into multienzyme complexes in the mitochondrion.

5–6 Classification of Enzymes

In an attempt to systematize the random nomenclature of enzymes, many of which have been known for years by trivial names (pepsin, trypsin, ptyalin), an international commission has established six main classes according to the reaction catalyzed and the type of bond affected. The suffix *-ase* is attached to the name of the reaction. In the following list, the examples in parentheses have been chosen from enzymes discussed in this text.

1. *Dehydrogenases.* Enzymes that remove hydrogen (succinic dehydrogenase).
2. *Transferases.* Enzymes that transfer chemical groups from one substance to another (choline acetyl transferase).
3. *Hydrolases.* Enzymes that hydrolyze their substrate. Hydrolases form several subclasses according to the substrate affected: (a) esterase—cleaves ester linkages (acetylcholine esterase); (b) glycosidase—cleaves gly-

Figure 5-7. **Lactic acid dehydrogenase isozymes separated by starch gel electrophoresis; arrow shows the direction of migration to positive pole. The isozymes were obtained from mouse (a) muscle, (b) tumor, and (c) kidney.**

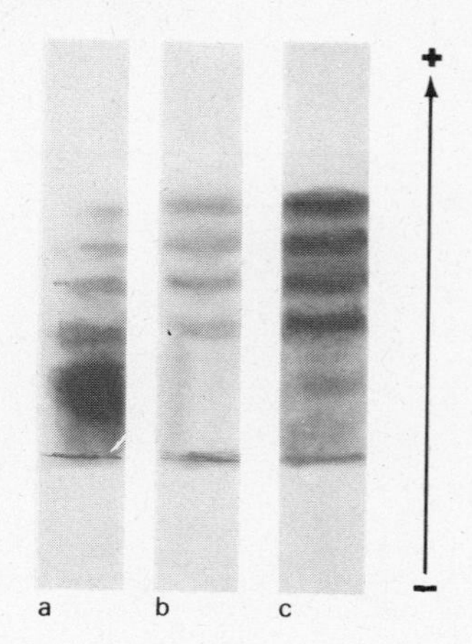

cosides (amylase); (c) peptidase (protease)—cleaves peptide bonds (trypsin, chymotrypsin).

4. *Lyases.* Enzymes that degrade or break down the substrate (pyruvate decarboxylase).
5. *Isomerases.* Enzymes that catalyze an internal change in molecular structure to form the isomer (phosphohexoisomerase).
6. *Ligases.* Enzymes that catalyze synthetic processes (amino acid activating enzyme).

Bioenergetics

Energy for biological processes comes ultimately from the sun. Plant cells are transducers of light energy, which is absorbed by their chlorophyll pigments and transformed into chemical energy. This chemical energy is stored in the compound *adenosine triphosphate* (ATP),[1] which is found in all cells, plant and animal. In plants, the energy derived from ATP hydrolysis is subsequently used to reduce carbon dioxide to glucose, which is stored as starch and cellulose.

Because animals cannot transduce light energy for synthetic reactions (photosynthesis), they must depend on plant products, or on other animals that have ingested plants, for their primary energy source. The products of carbohydrate, fat, and protein metabolism can all be oxidized enzymatically to yield chemical energy, but carbohydrate is usually the preferred fuel.

The degradative oxidation of food molecules liberates energy, partly as heat and partly as chemical energy stored in the ATP molecule. As a result of these reactions, ATP is readily available to all cells to supply energy for the synthetic, mechanical, and ion transport work of the cell. It is the universal common denominator: its energy can be used for all energy-requiring reactions, but in the process ATP is hydrolyzed to adenosine diphosphate (ADP). It is the energy released from food oxidation that permits the rephosphorylation of ADP, a phosphate acceptor, to form ATP, a phosphate donor.

The significance of this ATP-ADP system is that it occupies an intermediate position in the enzymatic phosphate transfer reactions in the cell, accepting phosphate groups from high-energy donors and passing them on to low-energy acceptors, raising the energy level of the latter. The net flow of phosphate is always from high-energy to low-energy compounds (Fig. 5-8). In this system of energy flow, ATP is an obligatory link: there are no enzyme systems that can transfer phosphate groups directly from high-energy donors to low-energy acceptors without their first being transferred to ATP.

The many chemical steps involved in oxidation and phosphorylation are catalyzed by enzyme systems that are carefully and systematically coupled and synchronized. A great many of these steps have been worked out in detail by biochemists; a great many more remain to be clarified.

[1] There are several other energy-rich compounds, similar in structure to ATP (*guanosine triphosphate,* for example), which are of significance in energy-requiring processes, but which are not discussed here).

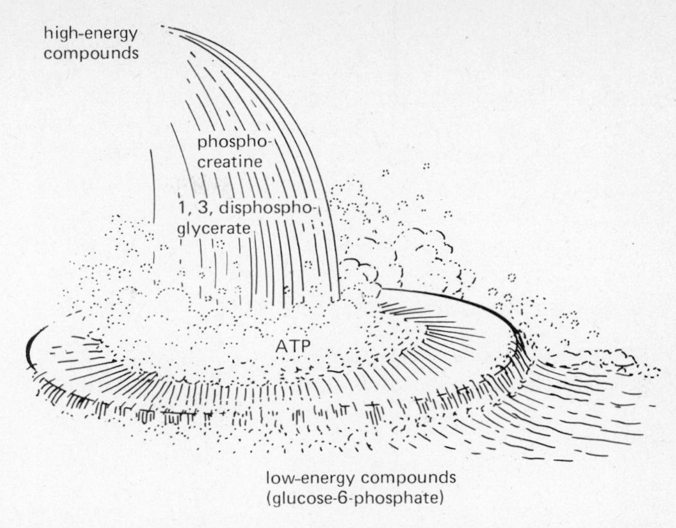

Figure 5-8. **The flow of phosphate is from high-energy to low-energy compounds, with ATP an essential link.**

5–7 High-Energy Compounds

High-energy compounds are characterized by one or more high-energy bonds that have a large free energy of hydrolysis. Free energy is the maximum useful work that can be obtained from a chemical reaction. These high-energy bonds are symbolized by a wriggle bond (~); the phosphate bond energy is transferred to molecules that utilize it directly, or to other storage molecules, such as *creatine phosphate* (*CP*). Whereas energy-rich compounds are primarily phosphate compounds, they may be derivatives of complex sulfur compounds such as *acetyl coenzyme A,* an important intermediate in food oxidation.

It is important to realize that the formation of high-energy bonds requires the *input of energy.* In the reaction

$$\textit{reactant} \rightleftharpoons \textit{product} + \textit{released energy}$$

the reactant is a higher-energy compound than the product if a high-energy bond is broken and energy is released.

If energy is released during the formation of the product, the reaction will go spontaneously to the right and energy is required to remake the reactant.

5–8 Structure and Properties of ATP and Related Compounds

Adenosine triphosphate (ATP) was first isolated from muscle, where it was found to be involved in the energy for muscle contraction. It is one of a series of phosphorylated organic compounds that store and distribute energy in cells.

This energy is easily transferred from one to another compound in the series in the presence of the proper enzyme. These phosphorylated compounds are distinguished from each other by the number of phosphate groups and the type of phosphate linkage to the rest of the molecule (Fig. 5-9). They are all *nucleotides,* compounds that consist of a nitrogenous base (adenine), a five-carbon sugar (ribose), and one or more phosphate groups (Sec. 3-3.2).

Adenosine Monophosphate (AMP)

Adenosine monophosphate has only one phosphate group, which is attached by an ester linkage to the 5′ position on

$$-\overset{|}{\underset{|}{C}}-O-\overset{O}{\underset{|}{\overset{/\!/}{P}}}-$$

the ribose molecule. This is not a high-energy phosphate bond.

Adenosine Diphosphate (ADP)

Adenosine diphosphate is a nucleotide with two phosphate groups, the second one being attached in anhydride linkage

$$-\overset{O}{\overset{/\!/}{P}}-O\sim\underset{|}{P}-$$

with the 5′ phosphate group of AMP. This second linkage is a high-energy phosphate bond.

Adenosine Triphosphate (ATP)

Adenosine triphosphate has a third phosphate group in linear, anhydride linkage, giving it two energy-rich phosphate bonds.

Cyclic Adenosine Monophosphate (Cyclic AMP)

Cyclic adenosine monophosphate is derived from ATP, but it has its single phosphate group esterified in a cycle through the condensation of two hydroxyl groups in the same molecule. This compound is not involved in energy transfer but is a most important "second messenger" between a hormone and its effects on enzyme systems. Its functions are discussed in Sec. 10-8.

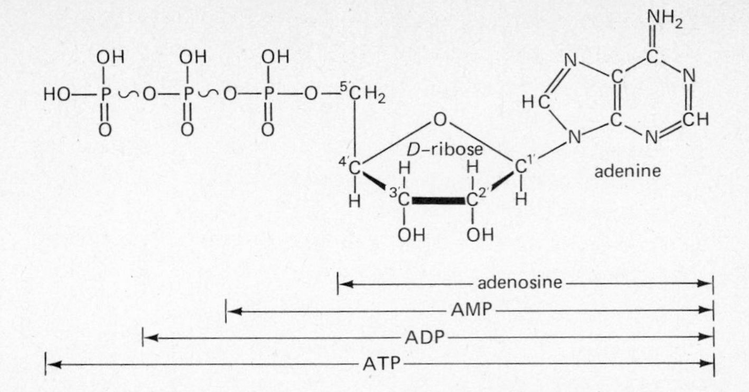

Figure 5-9. **The structure of the nucleotides, adenosine monophosphate (AMP), adenosine diphosphate (ADP), and adenosine triphosphate (ATP). The symbol ∼ indicates high-energy bonds.**

5-9 Hydrolysis of ATP

Depending on which bond reacts in hydrolysis, one of the following reactions will occur:

1. Most commonly, when ATP is enzymatically hydrolyzed, the terminal phosphate group is transferred to water, and ADP and inorganic phosphate (Pi) are released. The free energy derived from this reaction can be coupled with energy-requiring reactions. Enzymes catalyzing this type of hydrolysis are adenosine triphosphatases, or *ATPases.* Those enzymes that transfer the phosphate group from ATP to another substrate are *kinases.*

$$\text{ATP} + H_2O \longrightarrow \text{ADP} + \text{Pi} + \text{energy}$$

2. ATP can be enzymatically hydrolyzed and both energy-rich phosphate bonds removed to yield AMP.

$$\text{ATP} + H_2O \longrightarrow \text{AMP} + \text{Pi} + \text{energy}$$

3. ATP can be enzymatically hydrolyzed to cyclic AMP under the influence of the enzyme *adenyl cyclase.*

$$\text{ATP} + H_2O \longrightarrow \text{cAMP} + \text{Pi} + \text{energy}$$

5-10 Conservation of Energy of Oxidation

The energy present in food molecules is liberated through a series of oxidation-reduction reactions that produce compounds with high-energy phosphate bonds which are then donated to ATP. This occurs in two steps:

1. *The substrate,* for example, an aldehyde derived from glucose degradation, *is oxidized* (with the addition of inorganic phosphate) *to a phosphorylated acid.*

$$\underset{\text{Aldehyde}}{R-CHO} + \text{Pi} \longrightarrow \underset{\text{Phosphorylated acid}}{R-\overset{O}{\overset{\|}{C}}-O-\overset{O}{\overset{\|}{\underset{\underset{OH}{|}}{P}}}-OH}$$

2. The second step *transfers the phosphate group* from the phosphorylated acid to ADP to form ATP.

$$R-\overset{\overset{O}{\|}}{C}-O-\underset{\underset{OH}{|}}{\overset{\overset{O}{\|}}{P}}-OH + ADP \longrightarrow ATP + R-COOH$$

The shorthand that we shall use throughout this text for coupled reactions is:

substrate red → substrate ox, coupled with ADP + Pi → ATP

red = reduced
ox = oxidized

5–11 Oxidation-Reduction Reactions

Oxidation is the addition of oxygen to a substance or, more frequently in biological systems, oxidation is the enzymatic removal of hydrogen (or loss of electrons) from a substrate. The electrons must be accepted by an oxidizing agent, which becomes reduced. These two reactions are usually coupled.

Ferric Ion

If a ferric ion, Fe^{3+}, is the electron acceptor, it is reduced to the ferrous ion Fe^{2+}:

$$Fe^{3+} + 2e^{-} = 2Fe^{2+}$$

Ferric ions are important electron acceptors in the respiratory enzymes known as the *cytochromes.*

Molecular Oxygen

Molecular oxygen can act as an oxidizing agent by picking up four electrons:

$$O_2 + 4e^{-} = 2O^{2e^{-}} \xrightarrow{+4H^{+}} 2H_2O$$

Molecular oxygen is the final acceptor of electrons in the electron transport chain in tissue respiration and water is formed in the process.

Coenzymes

Coenzymes function as electron carriers, carrying electrons from one reaction to the next, alternately being oxidized and reduced in the process.

NICOTINAMIDE ADENINE DINUCLEOTIDE

An important coenzyme is nicotinamide adenine dinucleotide (NAD). It is the nicotinamide portion of the molecule that accepts electrons and becomes reduced and nicotinamide is formed from niacin, one of the B vitamins. If this vitamin is not present in the diet of humans and other mammals, the deficiency disease pellagra results (Sec. 21–7).

In the anaerobic breakdown of glucose (*glycolysis*), vertebrate tissues reduce pyruvic acid to lactic acid; in the process reduced NAD is oxidized and thus becomes available to accept hydrogen from other substrates. The enzyme involved is *lactic acid dehydrogenase.* This occurs in muscle when there is not enough oxygen immediately available, and energy in small amounts for a short time can be released through anaerobic reactions.

$$\underset{\text{Pyruvic acid}}{\begin{matrix} CH_3 \\ | \\ C{=}O \\ | \\ COOH \end{matrix}} + NAD\cdot 2H \longrightarrow \underset{\text{Lactic acid}}{\begin{matrix} CH_3 \\ | \\ CHOH \\ | \\ COOH \end{matrix}} + NAD$$

In yeast, this reduction is preceded by *decarboxylation,* the removal of a carboxyl group. NAD · 2H is the coenzyme for the lyase *pyruvate decarboxylase.* The product of anaerobic glycolysis in yeast (fermentation) is, of course, ethyl alcohol.

$$\underset{\text{Pyruvic acid}}{\begin{matrix} CH_3 \\ | \\ C{=}O \\ | \\ COO{\mid}H \end{matrix}} + NAD\cdot 2H \longrightarrow \underset{\text{Ethyl alcohol}}{\begin{matrix} CH_3 \\ | \\ CH_2OH \end{matrix}} + NAD + CO_2$$

FLAVIN ADENINE DINUCLEOTIDE

Another coenzyme that accepts electrons is *flavin adenine dinucleotide* (FAD). Part of this molecule is made of the vitamin riboflavin, or B_2. The enzyme *succinic dehydrogenase* contains FAD as its coenzyme.

UBIQUINONE

Ubiquinone is an electron carrier between the flavoproteins and the cytochromes. Its functional group (hydroquinone) becomes oxidized to quinone. It is found in almost all cells (hence the name ubiquinone), and is classified among the fat-soluble vitamins.

5–12 Glycolysis and Respiration

There are two main enzymatic pathways involved in the process of extracting energy from food molecules, one that does not require oxygen (*glycolysis*) and one that does (*respiration*). In anaerobic glycolysis, glucose is converted to lactic acid, and in the process about 7 per cent of the available energy is retrieved. Much of the early work on glycolysis was done on isolated muscle, and it was found that although muscles can function under strictly anaerobic conditions, lactic acid accumulates and causes early fatigue.

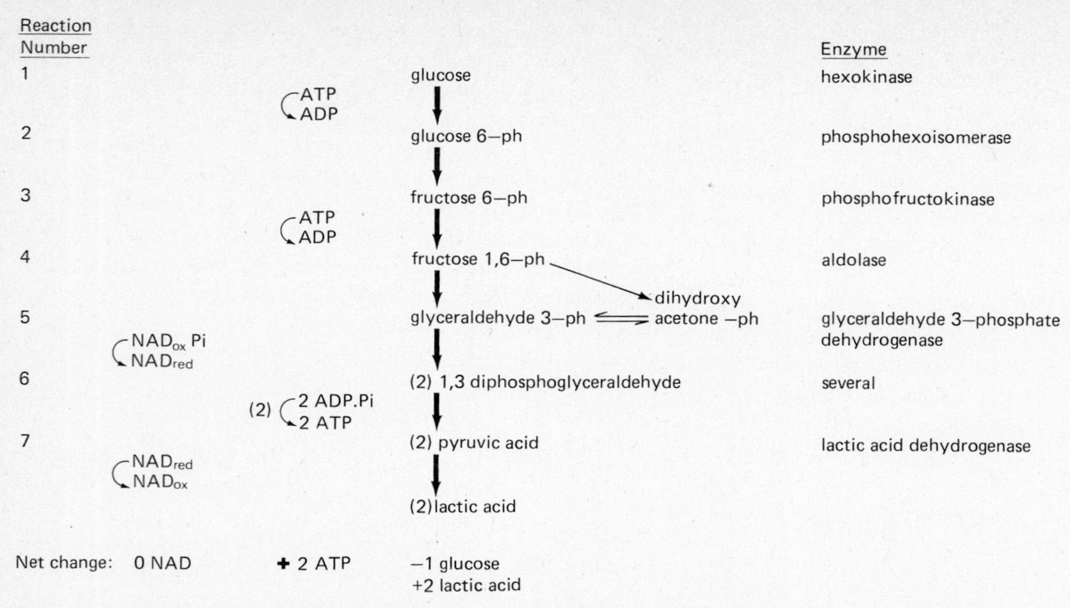

Figure 5-10. **The glycolytic balance sheet (not all reactions are included). As a result of anaerobic glycolysis, there is a net gain of two molecules of ATP from one molecule of glucose.**

Normally, there is an adequate oxygen supply to muscle, and under these conditions lactic acid is not formed; glycolysis takes place only up to the stage of pyruvic acid, which is then oxidized completely to $CO_2 + H_2O$ and most of the remaining 93 per cent of the energy extracted.

Glycolysis

The anaerobic breakdown of the 6C glucose molecule to two 3C lactic acid molecules is inseparably linked to the formation of ATP. For every molecule of glucose broken down to lactic acid, 2 molecules of ATP are generated.

glucose	$C_6H_{12}O_6$	2ADP + 2Pi
lactic acid	$2CH_3CHOH{\cdot}COOH$	2ATP

This overall reaction involves 11 separate enzymes, each acting on a specific chemical reaction but each linked to the preceding step. Only some of the steps will be discussed here. Further details may be found in any modern biochemistry text.

These reactions occur in the soluble part of the cytoplasm, without the highly structured arrangement that is typical of the enzymes associated with respiration. As the plasma membrane is permeable to glucose and its end product, lactic acid, but not to the highly charged phosphate intermediates, a free flow of fuel and product through the cell ensures the continuation of glycolysis without a corresponding loss of ATP and ADP.

The glycolytic balance sheet

The numbers of the following reactions refer also to those in Fig. 5-10.

1. Glucose can enter the cascade of anaerobic reactions only after it has been phosphorylated to glucose 6-ph and its energy level appropriately raised. This uses up one molecule of ATP.
2. Glucose 6-ph is isomerized to fructose 6-ph, or phosphorylated fructose can enter the pathway at this point.
3. Fructose 6-ph undergoes another phosphorylation to fructose 1,6-ph, using up a second ATP molecule.
4. The 6C fructose molecule, with its two phosphate groups, is split into two 3C molecules (trioses), each phosphorylated. These isomers are interconvertible and enter the subsequent reactions as two molecules of glyceraldehyde 3-ph.
5. Each of these molecules of glyceraldehyde 3-ph undergoes phosphorylation, which is linked with the reduction of NAD.
6. Subsequent oxidation of these two molecules results in the formation of four molecules of ATP (intermediate reactions are omitted here).
7. In this last step, the reduction of pyruvic acid to lactic acid, NAD is oxidized. *The net energy gain* from these glycolytic reactions is two molecules of ATP.

Respiration

Almost all of the energy for aerobic cells comes from respiration, a process which is dependent ultimately on molecular oxygen. The enzymes necessary for these complex oxidative reactions are assembled in the mitochondria and function as multienzyme complexes.

There are three main processes to consider:

1. The major food molecules—carbohydrates, fats, and proteins—must be prepared to enter oxidative pathways by being degraded to a 2C compound, acetic acid, which is always found in combination with coenzyme A, as the compound *acetyl CoA*.
2. Acetyl CoA enters a cyclic pathway of tricarboxylic acids (acids with three carboxyl groups), and as it spins through the cycle, $CO_2 + H_2O$ are split off. In these reactions, hydrogen atoms are removed by dehydrogenases. This is the *Krebs tricarboxylic cycle* (or the citric acid cycle), and is shown in simplified form in Fig. 5-11.
3. For every pair of hydrogen atoms removed, two electrons are passed down a linear series of electron carriers that form the *electron transport chain of respiration.* As they pass down the chain, these electrons lose much of their energy, which is captured in the form of ATP. This phosphorylation in the respiratory chain is *oxidative phosphorylation.* The electrons are passed on finally to molecular oxygen, which becomes reduced to water.

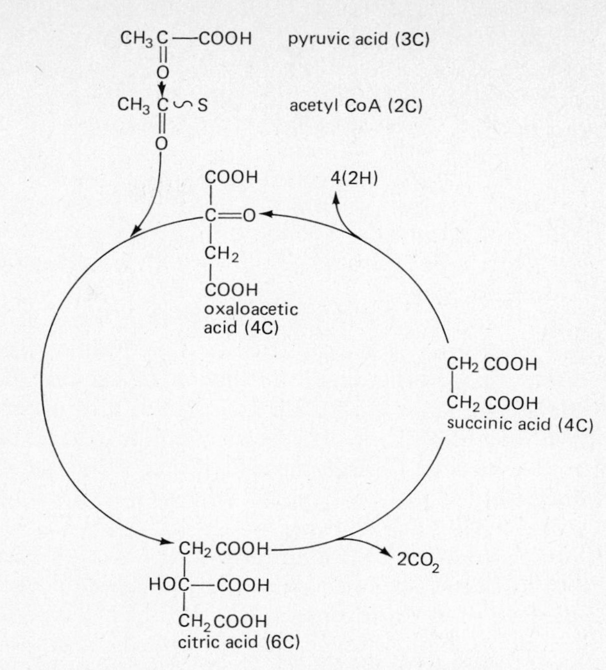

Figure 5-11. **Some reactions of the Krebs tricarboxylic acid (citric acid) cycle.**

The flow sheet for the degradation of food molecules in glycolysis to their entry into the Krebs cycle as acetyl CoA, and the subsequent passage of electrons along the respiratory chain, is diagrammatically shown in Figs. 5-11 and 5-12.

Formation of acetyl coenzyme A

In the discussion on glycolysis it was stated that pyruvic acid is usually the end product of glycolysis. The next step, before pyruvic acid enters the Krebs cycle, is the enzymatic removal of a carboxyl group accompanied by oxidation (in yeast, as mentioned earlier in this chapter, decarboxylation occurs anaerobically, and ethyl alcohol is produced). The acetic acid formed appears as a compound of coenzyme A, acetyl CoA. The thiol group (—SH) of CoA binds the acetic acid through a thioester linkage, which is a high-energy bond.

$$\underset{\text{Pyruvic acid}}{CH_3COCOOH} + \text{CoA-SH} \longrightarrow \underset{\text{Acetyl CoA}}{CH_3\,CO{\sim}S\text{—}CoA} + CO_2$$

The free energy of hydrolysis of acetyl CoA is a little higher than that of ATP, so that this molecule is in an activated form; only in this way can it enter the tricarboxylic cycle.

The Krebs tricarboxcylic acid cycle

This cycle is named after a Polish biochemist, later a British citizen and peer, who demonstrated the metabolic interrelationship between certain tricarboxcylic acids in the aerobic oxidation of carbohydrate (Krebs and Kornberg,

Figure 5-12. **Oxidative phosphorylation along the electron transport chain produces three molecules of ATP and one molecule of water. Molecular oxygen is needed only for the last reaction in this chain.**

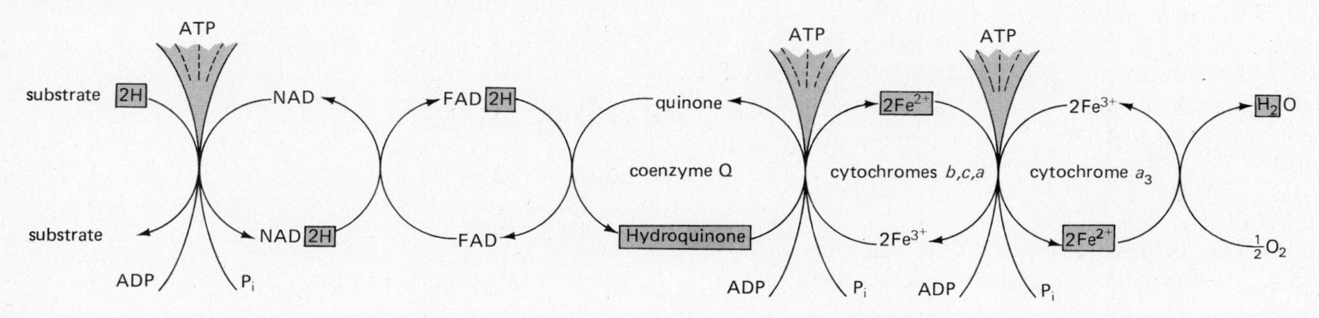

1957). He showed that poisoning of any one of the enzymes involved in this cycle blocks the entire sequence. Only some of the steps involved are outlined here and in Fig. 5-11; further details can be found in biochemistry texts.

1. The activated 2C acetyl CoA enters the cycle, where its acetyl group is transferred to a 4C dicarboxylic acid, *oxaloacetic acid,* forming a 6C compound, *citric acid* (a tricarboxylic acid). In this reaction, coenzyme A is liberated.
2. Two molecules of CO_2 are split off from citric acid and its derivatives during a series of enzymatic reactions, which result in the formation of another 4C compound, *succinic acid.* Succinic acid in turn undergoes a series of four dehydrogenations to regenerate *oxaloacetic acid.* Oxaloacetic acid, which is essential to draw acetyl CoA into the cycle, can now start a second cycle.
3. During the stepwise dehydrogenation of succinic acid to oxaloacetic acid, four pairs of hydrogen atoms are removed by dehydrogenases. For each pair of hydrogen atoms produced, one pair of electrons enters the respiratory chain.

THE ELECTRON-TRANSPORT CHAIN OF RESPIRATION

The electron-transport chain (Fig. 5-12) is a system for efficiently coupling the production of water to the phosphorylation of ATP, for the energy liberated by the reaction

$$2H_2 + O_2 \rightleftharpoons 2H_2O$$

is used to drive the series of reactions in the electron-transport chain.

For each revolution of the tricarboxylic acid cycle there are four dehydrogenations, for three of which NAD_{ox} is the electron acceptor, becoming NAD_{red} in the process. The electron acceptor for the fourth dehydrogenation is FAD. These coenzymes form a chain along which electrons are passed to yet another acceptor, ubiquinone, from which the electrons finally reach the series of iron-containing enzymes known as the *cytochromes.*

The active group of the cytochromes contains an atom of iron that in its oxidized state, Fe^{3+}, can accept an electron to become reduced to Fe^{2+}. The cytochromes can transport only one electron at a time, whereas NAD and FAD can carry two. The last in this series of cytochromes (*b, c, a*) is *cytochrome oxidase* (a_3), the only cytochrome that can give up its electrons directly to molecular oxygen. One of the most potent poisons known is cyanide, which blocks this last step in the respiratory chain.

Oxidative Phosphorylation

When a pair of electrons passes down the transport chain from NAD_{red} to molecular oxygen, three molecules of ATP are formed from ADP and phosphate. This process is oxidative phosphorylation. *The respiratory chain is a device by which the large amounts of energy produced by each oxidation in the chain are delivered in small packages and made available to the cell as ATP by the mitochondria.*

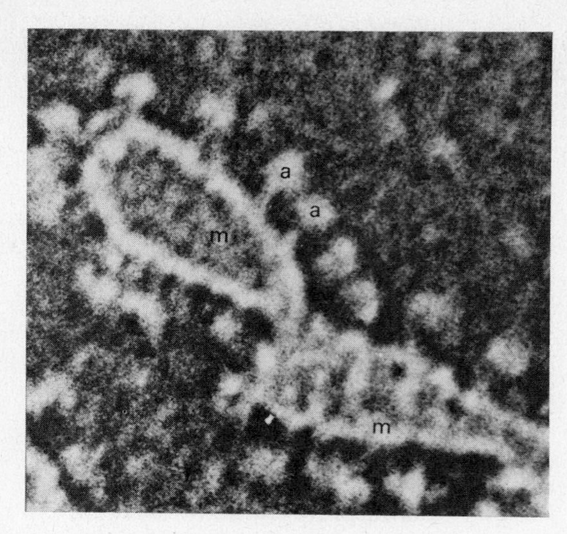

Figure 5-13. **This electronmicrograph shows ATPase molecules (a) lined up on two mitochondria (m). (From H. Fernandez-Moran, *Ciba Symposium 31,* Elsevier Excerpta Medica, Amsterdam, Holland, 1976.)**

The overall equation for oxidative phosphorylation may be shown as:

$$NAD_{red} + 3ADP + 3Pi + 2H^+ + \tfrac{1}{2}O_2 \longrightarrow NAD_{ox} + 3ATP + 4H_2O$$

It is possible to separate the two processes of electron transport and the phosphorylation of ADP to ATP through the use of certain poisons. One of these is 2,4 dinitrophenol, which *uncouples* oxidation and phosphorylation and consequently has been a most useful tool to study these reactions. Another ingenious research technique is to separate the particles of the inner mitochondrial membrane that contain the phosphorylating enzymes from the rest of the inner membrane. When this is done, the inner membrane can transport only electrons. The ability of the inner membrane to produce ATP is restored if the phosphorylating enzymes are replaced. Figure 5-13 shows the globular ATP molecules lined up along the edges of the mitochondria that have produced them.

Cited References

DICKERSON, R. E. "X-Ray Analysis and Protein Structure." In *The Proteins,* 2nd ed., Vol. 2, H. Neurath, ed. Academic Press, Inc., New York, 1964, p. 603.

KREBS, H. A., and H. L. KORNBERG. *Energy Transformations in Living Matter.* Springer-Verlag, Berlin, 1957.

Additional Readings

BOOKS

BERNHARD, S. A. *The Structure and Function of Enzymes.* W. A. Benjamin, Inc., Menlo Park, Calif., 1968. (Paperback.)

BOYER, P. D., H. LARDY, and K. MYRBACK. *The Enzymes,* Vols. 1-3. Academic Press, Inc., New York, 1959-1961.

CHANCE, B., R. ESTABROOK, and J. R. WILLIAMSON, eds. *Control of Energy Metabolism.* Academic Press, Inc., New York, 1965.

LEHNINGER, A. L. *Bioenergetics,* 2nd ed. W. A. Benjamin, Inc., Menlo Park, Calif., 1971. (Paperback.)

ARTICLES

ERNSTER, L. and C. P. LEE. "Biological Oxidoreductions." *Ann. Rev. Biochem.* **33:** 729 (1964).

HINKLE, P. C., and R. E. MCCARTY. "How Cells Make ATP." *Sci. Am.,* March 1978.

KOSHLAND, D. E. "Correlation of Structure and Function in Enzyme Action." *Science* **142:** 1533 (1963).

KOSHLAND, D. E. "Protein Shape and Biological Control." *Sci. Am.,* Oct. 1973.

LEHNINGER, A. L. "How Cells Transform Energy." *Sci. Am.,* Sept. 1961.

PHILLIPS, D. C. "The Three-Dimensional Structure of an Enzyme Molecule." *Sci. Am.,* Nov. 1966.

RACKER, E. "The Inner Mitochondrial Membrane: Basic and Applied Aspects." *Hosp. Pract.,* Feb. 1974.

STROUD, R. M. "A Family of Protein-Cutting Proteins." *Sci. Am.,* July 1974.

Regulation by the Nervous and Endocrine Systems

Chapter 6

Regulation of Bioelectric Events

Here is the test for unbelievers
Who cannot see the electric beavers

Ferrying the metal flakes across
Anode and cathode without a loss

Of one electron in furious tides
Where worlds are joined and a world divides;

Copper and zinc, nickel and brass,
Gold and silver and iron pass

Through eyes of the atom where the beavers trade
The base and the precious in a masquerade.

A. M. Sullivan, "Electrolysis"

REGULATION AND INTEGRATION within the multicellular organism have developed along two main lines, *chemical* and *nervous*. Neurons, the specialized cells of the nervous system, rapidly conduct information to other cells, to which they are connected by specific anatomical pathways. Chemical regulation in higher animals usually refers to the effects of the secretions (hormones) of the endocrine glands. These hormones reach all cells of the body through the circulation, and the specificity of their action depends upon the ability of certain cells to bind the hormone. Hormonal actions are slower and longer lasting than neural effects; they are discussed in Chapter 10, and subsequently in appropriate places in more detail throughout the text.

The distinction between chemical and neural regulatory processes is more a matter of convenience than of accuracy, for the two methods of integrating the activities of the multitudinous cells of the body are closely interwoven. In vertebrates, electrical impulses are generally transmitted from the conducting neuron to the receiving cell by chemical transmitters. Even when the concept of chemical regulation is confined to the hormones of the endocrine glands, the interdependence of the nervous and endocrine systems is very close, as we shall emphasize when we discuss neuroendocrine regulation (Chap. 11).

The functional unit of the nervous system is the *neuron*, organized into highly complex pathways through synaptic contact with other cells. The variations in the complexities of these pathways distinguish humans from lower animals, yet the way in which information is conducted to other cells is basically the same throughout the animal kingdom.

6–1 The Neuron

Neurons are cells specialized to respond to stimuli, and to express this response in the form of coded electrical impulses conducted to other cells. The speed with which the impulses can be conducted makes nerve systems invaluable for the rapid integration of function, especially in complex organisms.

Structure of the Neuron

A neuron consists of a cell body, the *soma*, with many fine branches or *dendrites*, and usually one large fiber, the *axon* (Sec. 2-5). Many sensory neurons are bipolar, with two axons (Fig. 6-1). The axon projects from the soma of the cell at the *axon hillock*. The *nerve terminals* or *boutons* of the axon synapse with the next cell in the anatomical pathway. Through the use of a special staining technique, the Golgi method, which for some unknown reason stains only a few cells in a neuronal population, it is possible to see the many variations on the basic neuronal form. More recently, greater selectivity in cell staining has been achieved through injecting cells with procion yellow dye. Other staining techniques make use of the fact that some substances are picked up by the nerve terminals and transported back along the

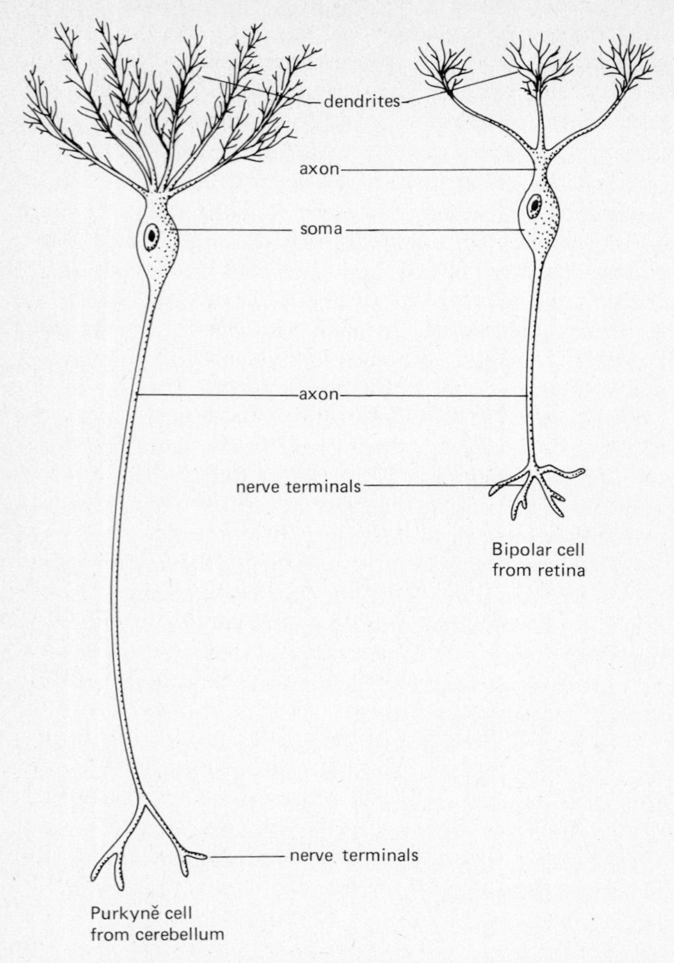

Figure 6-1. **Two types of neurons. The Purkyně cell has only one long axon, whereas the retinal cell has a long axon and a short axon. The cerebellum is part of the brain (see Chap. 8), and the retina is the sensory portion of the eye (see Chap. 22).**

Figure 6-2. **Neurons visualized by different techniques. (A) A motoneuron from the spinal cord after being injected with procion yellow dye. (B) A Golgi cell from the cerebellum, stained by the Golgi technique of silver impregnation.**

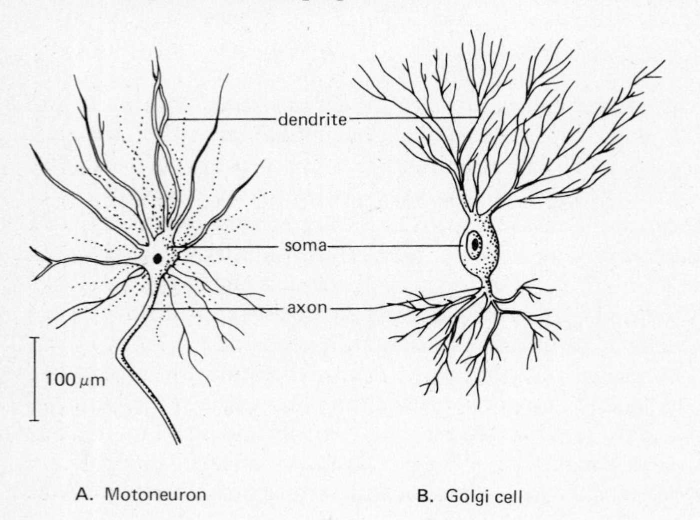

axon to the cell body (*retrograde transport*). If the transported substance is the enzyme horseradish peroxidase, the enzyme reacts with a substrate in the cell body, resulting in the deposition of a dense end product.

Radioactive amino acids can be injected near nerve terminals. These amino acids are incorporated into labeled protein that can be detected by radiographic techniques. Figures 6-2 and 11-1 show the appearance of neurons visualized by these different techniques.

Neuronal Organelles

Nerve cells possess all the organelles characteristic of highly active cells: mitochondria, endoplasmic reticulum, Golgi apparatus, and so forth, in addition to the nucleus and nucleolus. Around the nucleus and in the dendrites are deeply staining particles, called *Nissl bodies*. When these are magnified under the electron microscope, it can be seen that they are clumps of ribosomes attached to the endoplasmic reticulum, and that they are consequently associated with protein synthesis in the neuron. When an axon is damaged or cut, these Nissl bodies disappear. If the neuron recovers, the granules return within a few weeks as indicators of cell vitality. This has been used as a technique to trace the cell bodies of damaged axons and determine nerve pathways. Motoneurons in the spinal cord that have been selectively attacked by the poliomyelitis virus show this disappearance of the Nissl bodies, which in most cases is irreversible (Fig. 6-3).

Specialized Regions of the Neuron

Each nerve cell must be capable of receiving messages, integrating them, and preparing suitable coded responses to pass on to the next cell. There are three specialized areas for these functions (Fig. 6-1).

1. *A region of a neuron specialized to respond to a stimulus.* This is generally a dendrite, or in some sensory cells it may be an ending of an axon. This receptive region is especially sensitive to a given type of stimulus—for example, to light or to a particular chemical. The response to the stimulus is a change in voltage (*depolarization*) across the cell membrane at this region. If the voltage falls enough, it can trigger an action potential (nerve impulse) which travels along the axon, the next region of neuron specialization.

   Chapter 7 discusses specializations of these receptive regions with reference to signals from other neurons; Chapter 22 is concerned with receptive regions for external stimuli, such as light, sound, and taste.
2. *A region of a neuron specialized to conduct an action potential.* This consists of the nerve axon and its surrounding sheath. If the voltage in the receptive area reaches a certain critical level (*threshold*) it initiates the action potential. The *action potential is a wave of depolarization that is conducted along the axon without*

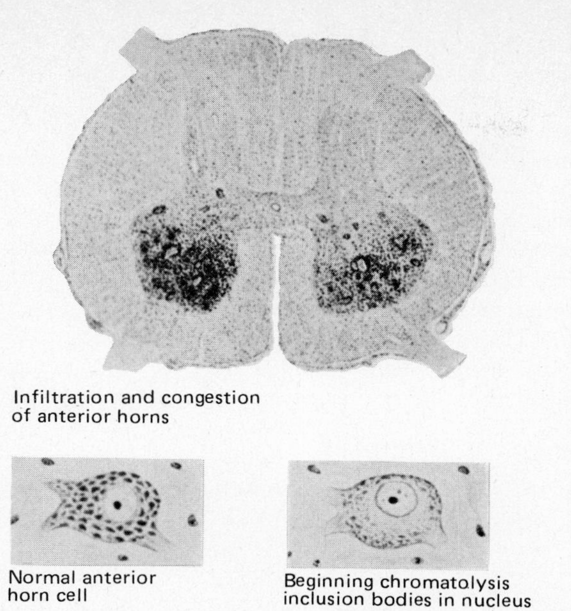

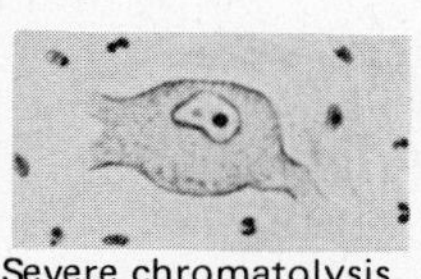
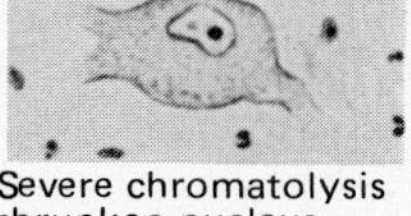

Figure 6-3. **Motoneurons in the spinal cord showing the destruction of the Nissl bodies by the poliomyelitis virus. (Adapted from an original painting by Frank H. Netter, M.D. From The CIBA Collection of Medical Illustrations, copyright by CIBA Pharmaceutical Co., Division of CIBA-GEIGY Corporation.)**

loss in amplitude. The size of the action potential is constant for the given physiological conditions: once the events in the receptive region have evoked the action potential, the action potential is full sized. This is an example of the "all-or-nothing" law, although it is always rash to talk of laws in biological systems. Nerve conduction, based on circulating electric currents, is very rapid, up to 120 meters per second.

3. *A terminal region.* This is the area of the nerve terminals, where the chemical transmitter is stored in vesicles. When the action potential reaches the terminals, a series of reactions is set off that culminates in the release of the transmitter and its diffusion across the synaptic cleft (Secs. 7-2, 7-4, and 7-5) to the receptive membrane of the next cell. Transmission, being a chemical phenomenon, is considerably slower than conduction.

6–2 Membrane Potentials

History of "Animal Electricity"

In the study of the nerve impulse, the earliest workers concentrated on the conducting zone or the axon, for the logical reason that it is the easiest part on which to work. We shall consider all three zones, but shall commence with the fundamental discoveries of the nature of membrane potentials and the production of an action potential on a nerve axon.

One of the first recorded experiments that purported to show "animal electricity" was described by Galvani in 1791. He was fascinated by electricity and had acquired a machine for generating it electrostatically. Because he was also a professor of anatomy, frogs' legs were routinely dissected and studied, often at the same time that the electric machine was used. This led to the accidental discovery that the frogs' muscles would contract if certain parts were stimulated electrically. Galvani interpreted this to mean that animal cells contained the same kind of electricity as his machine and that changes in this electricity caused by "induction at a distance" would cause the muscles to twitch. Galvani also noticed the twitching of frogs' legs that were hung by brass hooks and then placed on an iron rod. Although he misinterpreted this as being brought about by inherent "animal electricity," his work drew much attention to the sensitivity of nerve and muscle fibers to electric currents. These observations were the basis for Volta's invention of the battery that is based on bimetallic electricity.

The amount of electrical energy produced by living systems is, in general, very slight, but it is associated with all types of protoplasmic activity, from the formation of pseudopodia to mitotic spindle movements to nerve and muscle action potentials. A few bony fish and elasmobranchs possess electric organs that can generate large amounts of electricity—even enough to kill. The electric organs of eels are really highly specialized neuromuscular junctions placed in series.

Demonstration of Membrane Potentials

For more than 100 years it has been known that the cell maintains a difference in electrical potential between its inside and the outside medium, that is, across the cell membrane. This is the *"resting potential"* of nerve and muscle or the *"membrane potential"* of cells in general. Its existence, however, was not directly measured at first; the techniques available were too crude. Rather, it was inferred from experiments that showed that, if one electrode is placed on the surface of a nerve or muscle and another on a damaged part (considered to represent the inside axoplasm or sarcoplasm), the injured portion is electronegative with respect to the healthy region. This *injury potential* is about 30 mV and was considered to represent the resting potential. It is much too low, actually, because of the short-circuiting between the electrodes under these conditions. DuBois Reymond (1852) used a galvanometer to measure the injury current that flowed between these two regions, but much more sensitive techniques were needed before accurate determinations could be made. Great progress was made in the 1920s by Gasser and Erlanger (1922), who first used the cathode ray oscilloscope to give an accurate time course of nerve action potentials, but this was for bundles made up of large numbers of nerve fibers.

In the mid-1930s, Osterhout was able to insert an electrode directly into the sap of the large plant cell *Nitella* and to record the potential across the cell membrane as 80 mV. A report by Young in 1936 that the giant axon of the squid measured about 500 μm in diameter led Hodgkin and Huxley (1952) to repeat Osterhout's work on the animal cell. They succeeded in this difficult task, using the newly developed cathode follower amplifier and the cathode ray oscilloscope. They found the resting potential to be about -50 mV (the voltage across the cell membrane from inside to outside). The action potential of this giant axon was studied in detail and the ionic mechanism upon which it is based was elucidated in the 1950s.

The cathode ray oscilloscope

This instrument is capable of following instantaneous potential changes, but it is relatively insensitive and requires considerable amplification to pick up the small signals from living cells, which are often much less than 1 mV. The cathode ray oscilloscope consists basically of a tube that ejects electrons in a steady stream against a special screen, which becomes fluorescent where the electrons hit it. The stream of electrons is displaced up or down when passed through an appropriate electrical field. The electrical fields are produced by a pair of vertical deflection plates, appropriately charged. Powerful amplifiers magnify the electrical changes in the tissue sufficiently to affect the direction of the electron stream and thus deflect the spot of light up or down.

A pair of horizontal deflection plates is used to sweep the spot across the screen at the desired speed so that the response can be spread out and photographed and its time base measured. Figure 6-4 shows the dramatic difference in potential between the recording electrode and the indifferent electrode when the recording electrode penetrates the membrane. Prior to penetration, with both electrodes on the cell surface, there is zero potential difference; on penetration there is an immediate change of as much as -80 mV.

Figure 6-4. **The intracellular resting potential can be measured by inserting a recording electrode inside the cell and connecting it by a voltage meter to a reference electrode on the surface of the cell. The cell is immersed in a suitable salt solution. In (A) both the recording and the reference electrodes are on the surface of the cell and the extracellular potential is zero. In (B) the recording electrode has been inserted into the cell, and there is an abrupt change in potential (-80 mV), demonstrating the inside of the cell to be negative with respect to the outside.**

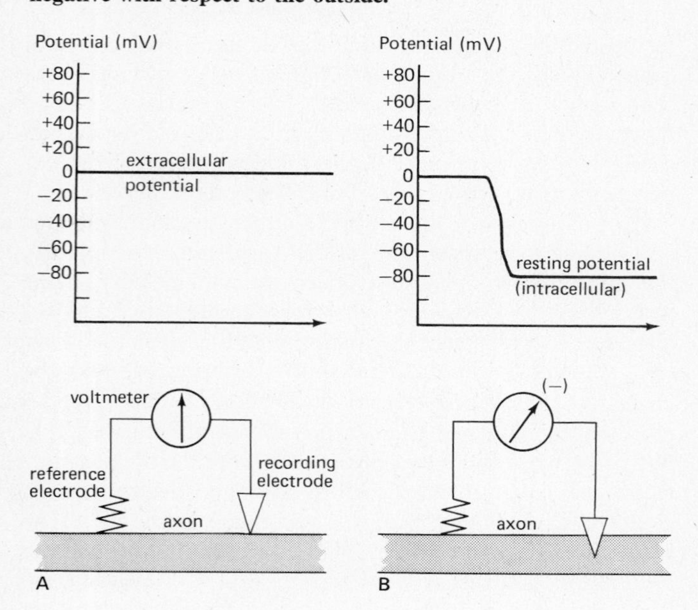

Establishment and Maintenance of the Membrane Potential

The charge across a cell membrane depends upon the difference in concentration of certain ions on either side of the membrane: this in turn is dependent on the structure and characteristics of the cell membrane and its components. The external concentrations for mammalian cells are very much the same as for a protein-free filtrate of blood plasma. The internal concentrations for sodium and chloride are lower on the inside than on the outside, the ratios being about 8 : 1 and 14 : 1. Potassium ions are more than 30 times more concentrated inside the cell than outside.

Factors influencing ion movements

The passage of ions across a membrane is dependent on the *electrochemical gradient* for each ion species and the *permeability* of the membrane to the ions. In living cells, these forces are modified by the active transport of specific ions by "pumps," especially the *sodium-potassium pump*, which is an *ATP-ase* specific for the transport of sodium and potassium across the membrane. Recently, it has been shown that chloride pumps are also very important in special sites. The unequal distribution of charged ions across a membrane will give rise to a potential difference that can be calculated by the Nernst equation. In its simplified form the equation at 37°C for univalent ions is

$$E = 61 \log \frac{C_1}{C_2}$$

where E is the potential in millivolts and C_1 and C_2 are the concentrations of the given ion species on either side of the membrane.

The Nernst equation gives the electrical potential necessary to balance the concentration difference of charged particles, so that they diffuse at equal rates in both directions across the membrane. The Nernst equation is valid only for an ion free to distribute passively. This accounts for the very different ratios for sodium and potassium with a single potential across the membrane.

THE POTASSIUM EQUILIBRIUM POTENTIAL

Radiotracer studies have shown that the resting nerve cell is 75 times as permeable to K^+ as to Na^+, thus permitting K^+ to traverse the membrane far more rapidly than Na^+. However, the concentration gradient of the high internal potassium would tend to move K^+ out of the cell, down the electrochemical gradient, because the concentration difference is not fully compensated for by the membrane potential (Fig. 6-5).

It is not known exactly which anions balance the high internal potassium. Certain proteins are important because, at the pH of the cell, they are anions. Other anions would be glutamate, bicarbonate, and chloride. The electrochemical difference for potassium is 20 mV (−70 mV for the membrane potential and +90 mV from the concentration gradient, as calculated from the Nernst equation). This outward transport of potassium is balanced by an inward potassium pump.

THE SODIUM EQUILIBRIUM POTENTIAL

The electrochemical potential difference for sodium ions is very large (130 mV). This is derived from the concentration difference of −60 mV as calculated from the Nernst equation and −70 mV for the membrane potential. This results in the inward diffusion of sodium ions 100 times more rapidly than outward. Balancing this inrush, the cell membrane is far less permeable to sodium than to potassium, and the sodium pump very efficiently pumps sodium ions out of the cell. Both pumps, the sodium and the potassium, are coupled and dependent on the same metabolic processes.

THE CELL MEMBRANE

The characteristics of the cell membrane are prerequisites for the establishment and maintenance of the membrane potential. As we have seen (Chap. 4), the cell membrane is basically a bimolecular leaflet of phospholipid molecules, with the hydrophilic polar groups exposed on both surfaces. Structural proteins stabilize the membrane, and specific proteins may extend partially or completely through it. Receptor sites of many types are found on the outer surface (Fig. 4-19).

Pores and channels penetrate the membrane. There are separate *channels* for the sodium and potassium ions, and they are controlled by *gates*. The channels are probably protein-containing systems extending across the membrane. These systems specifically recognize and allow facilitated transport of sodium and potassium. These are two quite different systems, each with its own distinct properties. The poison *tetrodotoxin* specifically blocks the sodium gates when it is applied to the outside of the membrane. When the sodium currents are reduced to zero, in the absence of sodium and by blocking the sodium gates with tetrodotoxin,

Figure 6-5. **Various conditions across the membrane of a cat motoneuron: (A) gives approximate ionic compositions inside and outside and the respective equilibrium potentials; (B) is a formal electrical model for an average motoneuron as measured by a microelectrode in the soma; and (C) shows ionic fluxes across the membrane for K^+ and Na^+ ions under resting conditions. These fluxes are in part diffusional down the electrochemical gradients as indicated, and in part due to specific ionic pumps driven by metabolism. The fluxes due to diffusion and the operation of the pump are distinguished by crosshatching and the magnitudes are given by the respective widths of the channels. (From J. C. Eccles, *The Physiology of Nerve Cells,* The Johns Hopkins University Press, Baltimore, 1957.)**

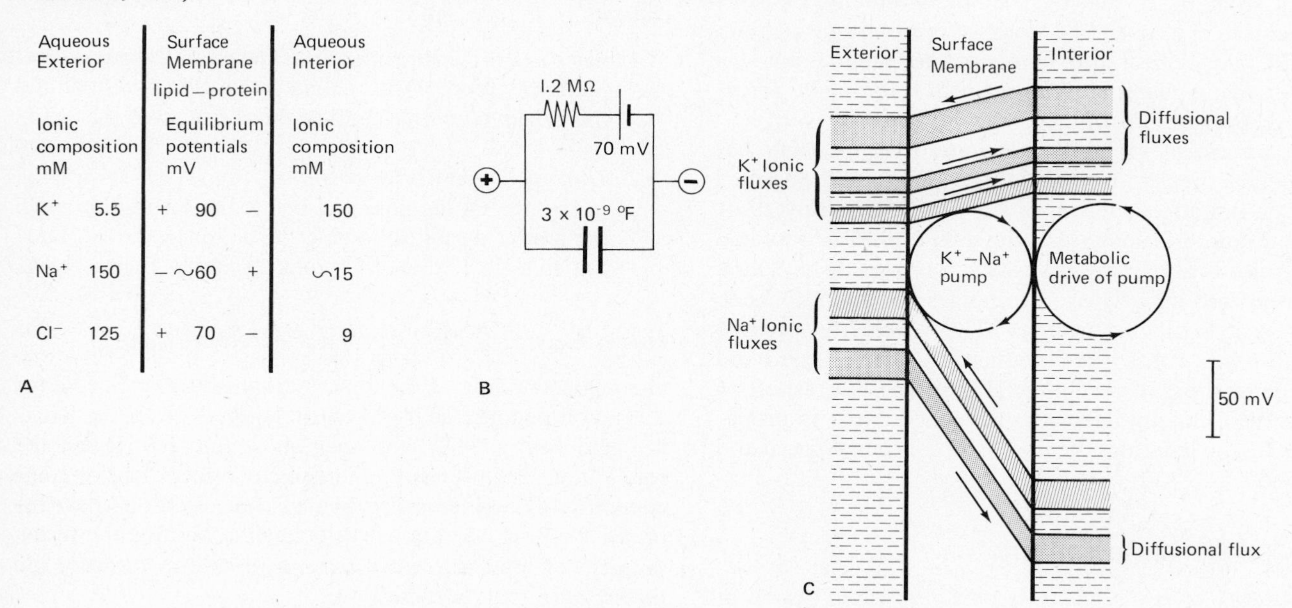

the opening or closing of the sodium gates generates brief currents (*gating currents*) in the opposite directions across the membrane. The number of sodium gates has been estimated to be almost 500/cm^2, which is very close packing indeed. There is as yet no evidence for a related process for the potassium gates.

These characteristics of the cell membrane, as well as the difference in size of the hydrated ions (the hydrated potassium ion is considerably smaller than the hydrated sodium ion), partially explain the selective treatment of sodium and potassium ions by the membrane. Once the concentration difference has been established, the resulting electrical potential is maintained to a certain degree by the membrane, which has a *high electrical resistance* and a *large capacitance*. The *capacitance* of a membrane indicates its ability to hold a potential; the thinner the membrane the greater its capacitance, and the cell membrane is only 7 nm thick.

This membrane potential is by no means stable. Unless free energy from metabolism is constantly applied to maintain the sodium and potassium pumps, it tends to "leak" or run down. Its dependence on energy yielding metabolic reactions is demonstrated by the use of metabolic inhibitors such as cyanide or dinitrophenol. These inhibitors prevent the formation of ATP and so switch off the linked sodium-potassium pump; the passive flow of sodium into the nerve cell along its concentration gradient can then be observed, using a radioactive isotope of sodium. The sodium pump can be started up again by the intracellular injection of energy-yielding substances, such as ATP.

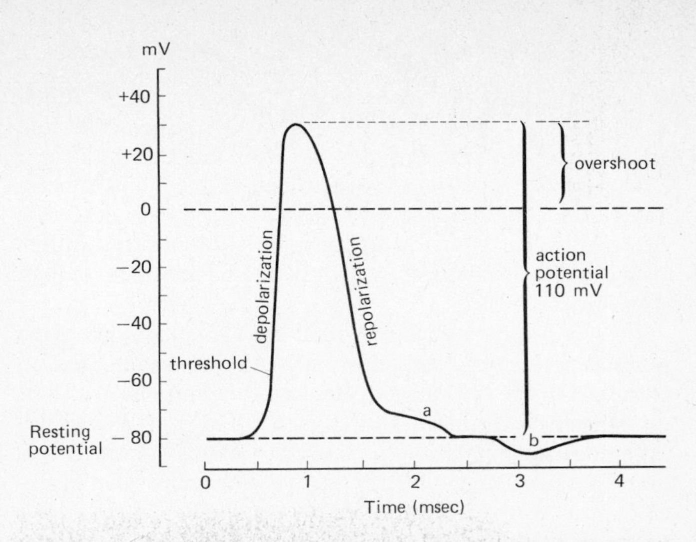

Figure 6-6. **The action potential of a nerve cell measured with the recording electrode inside the cell. The different phases of the action potential are shown with respect to their time course. a = depolarizing after-potential, and b = hyperpolarizing after-potential.**

6–3 The Action Potential

Initiation of the Action Potential

When an axon is stimulated by an electrical current of sufficient strength, so that the membrane potential is suddenly reduced or depolarized from −70 to −50 mV (inside to outside), this critical drop in membrane potential initiates a more dramatic potential change, which is the *action potential,* or *spike* (Fig. 6-6).

Once the critical level of depolarization, the *threshold,* has been reached, further increases in the strength of the applied current do not affect the size of the spike. It is truly all or nothing. Figure 6-6 shows that the total change in potential of the spike is about 110 mV. As the change in potential crosses the zero line (where the charges inside and outside are equal), it is moving from −70 mV *inside* to +40 mV *inside*. In other words, the potential has been reversed and the inside of the cell membrane is positive with respect to the outside. Once initiated, the action potential is propagated along the length of the fiber with a constant speed and amplitude.

Phases of the Action Potential

1. *Upstroke or rising phase.* In this very rapid period of change, the cell loses its negative resting potential, becomes depolarized (zero potential), and reverses the membrane potential so that the inside of the cell is briefly positive.
2. *The overshoot.* The short positive phase is called the overshoot and is usually about 30 to 40 mV.
3. *Repolarization phase.* The downstroke of the potential is the repolarization, a slightly slower procedure than the initial depolarization.

After-potentials

In some but not all cells, there are small potentials at the end of, or following, the repolarization phase; they reflect changes in the excitability of the cell.

1. *Depolarizing after-potentials.* The membrane potential is slightly more positive than the resting potential and the cell is therefore slightly more excitable than normal.
2. *Hyperpolarizing after-potentials.* Some cells show a fall in the membrane potential below the resting potential for a brief period following the action potential. During this time, the cell is less excitable than normal.

Ionic Basis of the Action Potential

Our understanding of the ionic mechanisms responsible for the action potential is based on the work of Hodgkin, Huxley, and Katz (1952), and that of Keynes (1951), on the squid giant axon. Using radioactive tracers and voltage clamping of the membrane (which controls the transmembrane potential at a known value and so permits the measurement of ionic currents), they were able to measure ion fluxes with great accuracy.

Another ingenious method perfected by Baker, Hodgkin,

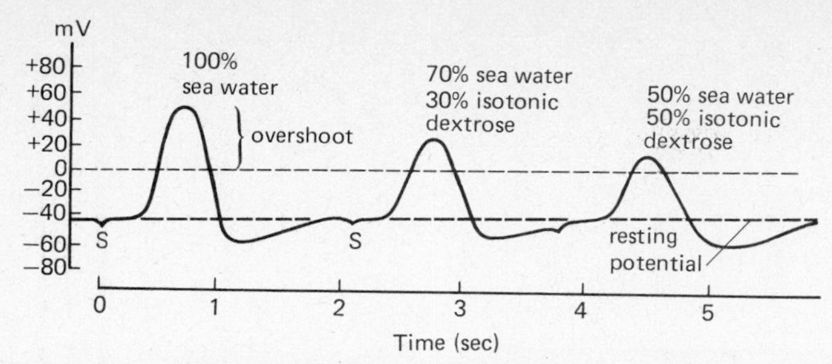

Figure 6-7. **Reduction in the Na^+ concentration gradient across a nerve membrane decreases the amplitude of the action potential without affecting the resting potential. In this figure the external Na^+ concentration is reduced, but the same effect can be achieved by increasing the internal Na^+ and keeping the external Na^+ constant. S = time of stimulation.**

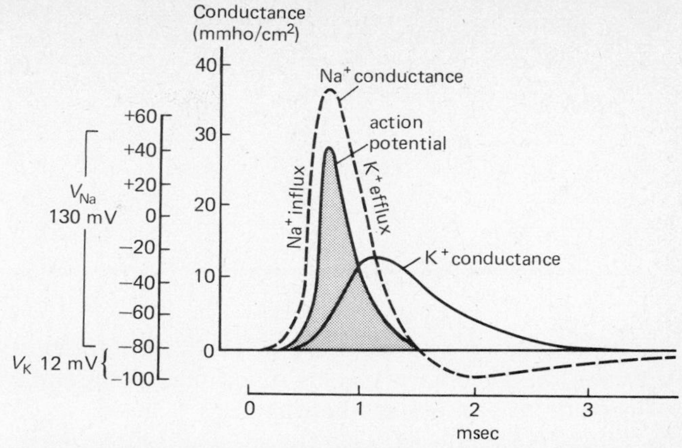

Figure 6-8. **Correlation of the changes in sodium and potassium conductance with phases of the action potential, where V_{Na} and V_K are the equilibrium potentials for sodium and potassium across the membrane. [Adapted from A. L. Hodgkin and A. F. Huxley, *J. Physiol. (London)* 117:23 (1952).]**

and Shaw (1962) is to extrude mechanically most of the axoplasm of the squid axon and to reinflate it with an artificial solution of potassium salts. The axon is quite able to produce normal action potentials for several hours. This technique makes it possible to vary the internal salt concentration as well as the external, and to determine the effects of changes in concentration gradient obtained in this manner on the resting and action potentials (Fig. 6-7).

As a result of these experiments, the *different phases of the action potential* can be correlated with the following changes in *ionic fluxes* (Fig. 6-8):

1. The initial depolarization of the membrane leads to an increase in the permeability of the membrane to sodium (the sodium conductance).

2. The sodium conductance rises very steeply by a self-propagating (positive feedback) mechanism, because the more sodium that enters (*influx*) the greater the depolarization and the greater the increase in sodium conductance, up to the peak of the impulse. This explains the all-or-nothing nature of the action potential: once it has started it utilizes the energy of the ions running down their electrochemical gradients. During this period the sodium gates are open.

3. The potassium gates open a little later than the sodium gates and stay open longer. Consequently, the increase in potassium conductance begins a little later and lasts longer. The outward flow (*efflux*) of the potassium ions slows the rise of the potential, then causes it to fall to its initial level (*repolarization*) and even briefly reverses it.

Figure 6-9. **The absolute and relative refractory periods of nerve. (A) During the absolute refractory period, which starts immediately after the first stimulus (arrow) to reach the threshold, the nerve is completely inexcitable. In the relative refractory period, the threshold is higher than normal and only stronger stimuli can excite the nerve. In (B) two action potentials are evoked as a result of six separate stimuli, administered to the same nerve as in (A). The relative strength of the stimuli is indicated by the length of the arrows. At 1 msec, the nerve is still absolutely refractory following the stimulus administered in (A). At 2 msec, a stimulus twice the strength of the original one is needed to evoke an action potential. At 4 msec, the nerve is again in the absolute refractory period. At 5 and 6 msec, the original strength of stimulation is inadequate to overcome the relative refractoriness of the nerve, but by 7 msec the nerve has regained its original threshold and the initial low stimulus strength results in an action potential.**

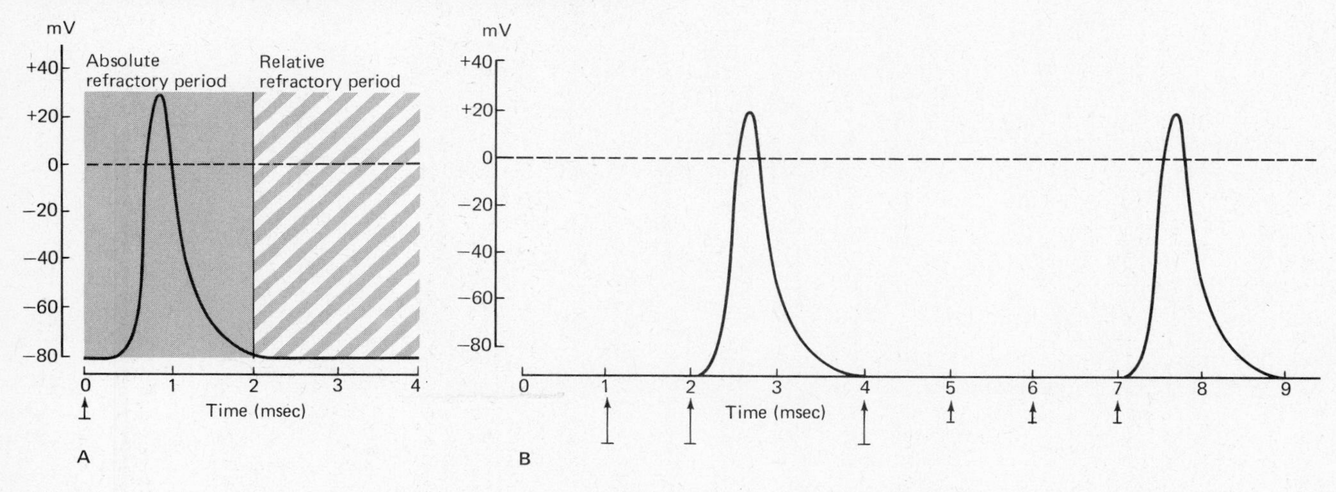

Table 6–1. **Relationship Between Nerve Fiber Diameter, Myelination, and Speed of Conduction**

	Diameter (μm)	% Myelin	Conduction Velocity (msec)	Temperature (°C)
Squid giant axon	650	None	25	20
Cat sensory nerve				
A fibers	4	30–50	25	38
C fibers	1	None	2	38

As the potential returns to its original level (negative feedback), the membrane regains its original permeability and is ready to conduct another impulse.

4. There is a very brief period of less than 1 msec, during which time the sodium gates are closing and the potassium gates are still open. During this time the nerve fiber is unresponsive to a depolarizing current and consequently cannot conduct an impulse. This is the *absolute refractory period* (Fig. 6-9). Because this interval is extremely short (about 2 msec), these fibers can carry very fast trains of impulses. The absolute refractory period is followed by a recovery of excitability during which time the threshold of the nerve is higher than normal, and so only stimuli of greater strength can evoke a propagated impulse, which is itself smaller and slower. This recovery time is called the *relative refractory period* (Fig. 6-9); it lasts another 2 msec after the end of the absolute refractory period.

Conduction of the Action Potential

CONDUCTION AND THE EVOLUTION OF THE MYELIN SHEATH

An important evolutionary advance is the *increase in velocity* with which nerve fibers conduct electrical impulses. Invertebrates developed giant fibers, often through the fusion of many smaller ones, thus decreasing the ratio of membrane capacity to the internal electrical resistance and so speeding up the impulse. However, increasing the diameter of the nerve fiber 16 times, as from the crab axon to squid axon, results in only a fourfold increase in conduction velocity. Vertebrates escaped the dilemma of a tremendous increase in the bulk of nerve fibers to achieve quickness of response by wrapping the axon with concentric layers of *myelin,* a lipid sheath. Specialized connective tissue cells, which in the peripheral nervous system are called *Schwann cells,* revolve around the axon during neurogenesis, laying down the lipoprotein lamellae that form the myelin sheath (Fig. 6-10). These axons are called *myelinated;* those that lack the sheath are thinner, *unmyelinated* fibers and conduct much more slowly. Recent electronmicrographs indicate, however, that many of these "unmyelinated" nerves have a thin, myelinated sheath (Fig. 6-11). An important fact to note is that the sheath of the myelinated fibers is interrupted along the length of the axon at regular intervals (usually of 1–2 mm), the *nodes of Ranvier* (Figs. 4-16 and 6-11). The addition of the myelin sheath permits an enormous increase in conduction velocity with a relatively small increase in fiber diameter. This is discussed further in the following section.

Table 6–1 compares the speed of conduction of unmyeli-

Figure 6-10. **Formation of the myelin sheath (m) by Schwann cells. (From J. D. Robertson, in *Molecular Biology,* D. Nachmansohn, ed., Academic Press, Inc., New York, 1960.)**

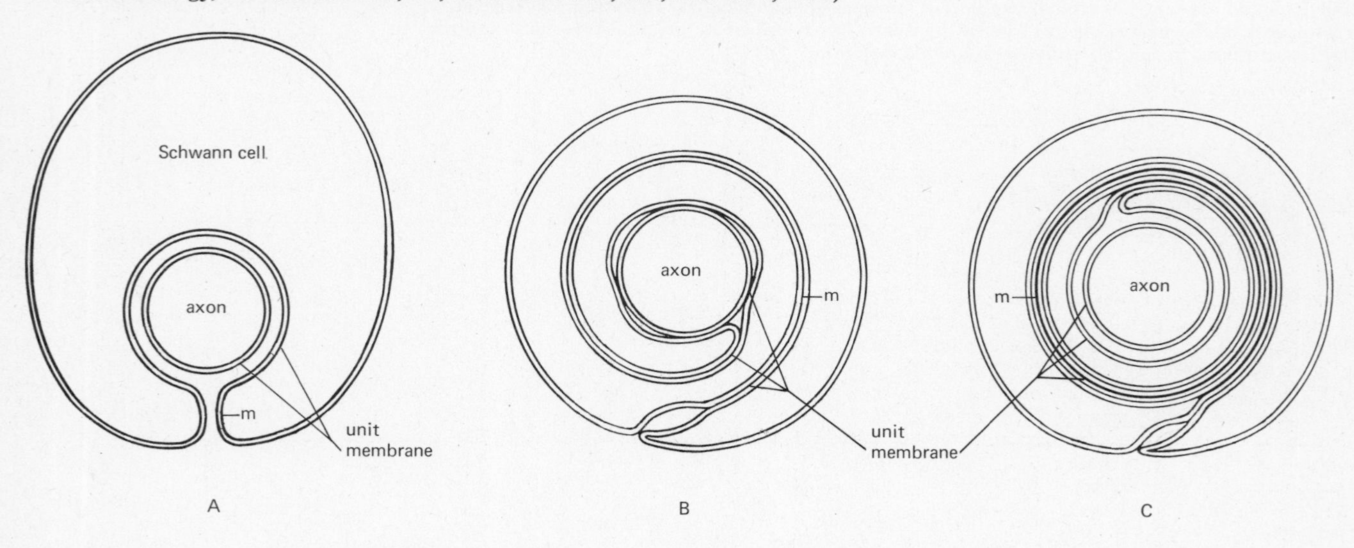

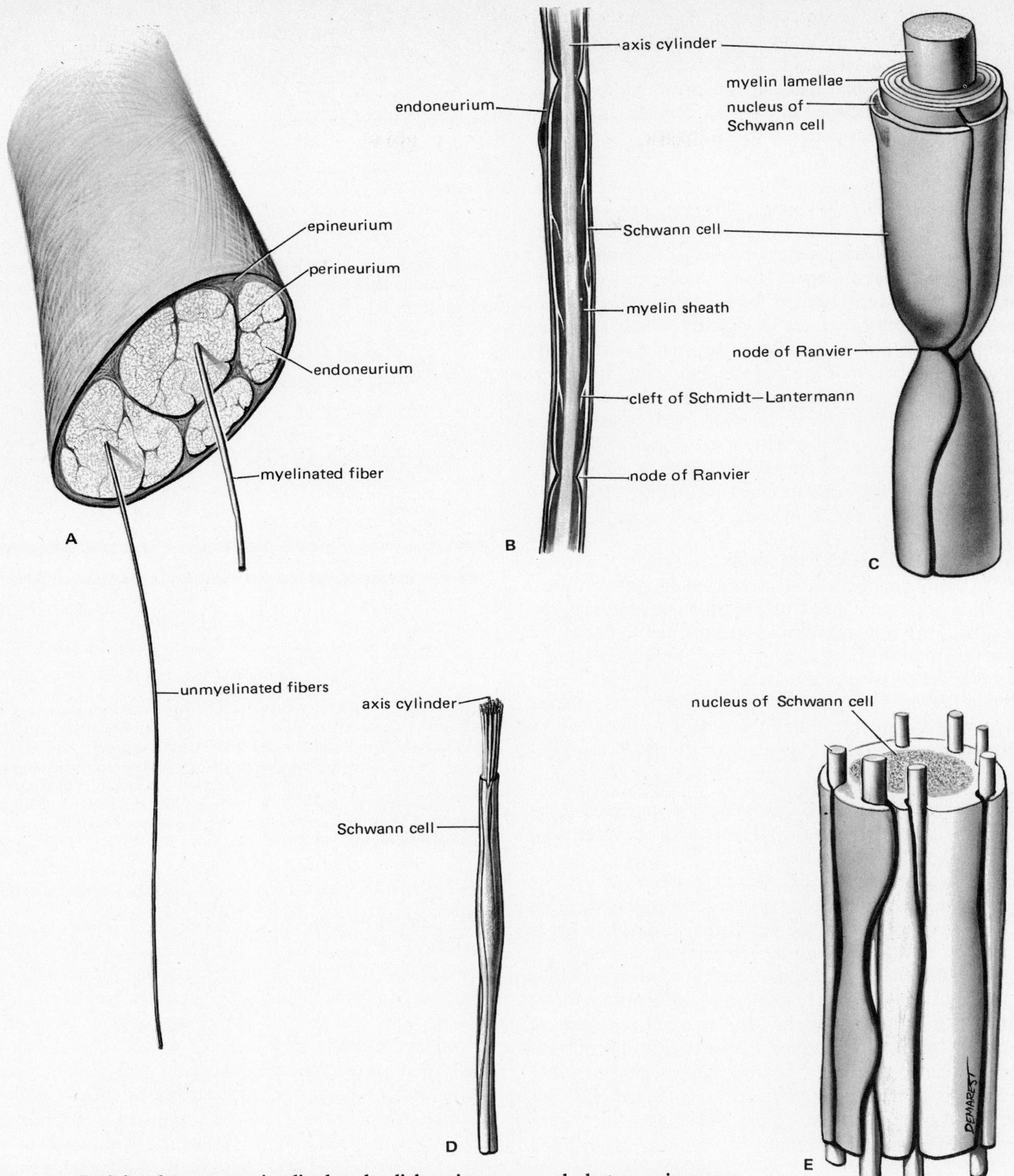

Figure 6-11. **Peripheral nerve as visualized under light microscope and electron microscope magnifications. (A) A myelinated nerve fiber and several unmyelinated nerve fibers extending out of the peripheral trunk. (B) Myelinated nerve fiber as visualized with light microscope. (C) Myelinated nerve fiber as reconstructed from electronmicrographs. The helically laminated myelin sheath (jelly roll) is continuous with the cell membrane of the neurolemma cell. (D) Several unmyelinated nerve fibers as viewed with the light microscope. One neurilemma cell ensheaths several nerve fibers. (E) Several unmyelinated nerve fibers ensheathed by one neurilemma cell, as reconstructed from electronmicrographs. (From The *Human Nervous System,* 2nd ed., by C. Noback and R. J. Demarest, McGraw-Hill Book Company, New York, 1975. Used with permission of McGraw-Hill Book Company.)**

nated, giant fibers of the squid with that of myelinated and unmyelinated fibers of different sizes in the cat. The practical significance of myelination is that it permits the thousands of nerve fibers that make up a nerve trunk to take up comparatively little space, and yet to conduct nerve impulses at velocities reaching up to 120 m/sec.

SALTATORY CONDUCTION

The elongated nerve axon acts like a poorly conducting electric cable with a leaky sheath, the surface membrane. In vertebrate myelinated nerves these inefficient electrical characteristics are compensated for (1) by the wrapping of the axon in concentric layers of myelin, which act as an insulating sheath that increases the resistance and greatly lowers the capacitance of the surface; and (2) by the insertion of "boosters" at approximately 1-mm intervals to lift the attenuated signals. These intervals correspond to the interruptions in the myelin sheath at the nodes of Ranvier. When there are defects in the myelin sheath, such as occur in the disease *multiple sclerosis,* there is a failure of impulse conduction although the axon itself is unaffected.

Saltatory Conduction Compared to Cable Conduction. In cable conduction, electric currents flow outward from the membrane, ahead of the impulse, through the surrounding medium and then reenter the membrane at the point of the impulse. They complete the circuit by returning up the conducting core of the axon (Fig. 6-12). Cable conduction in nerves is slow; in myelinated nerves it is speeded up by having the impulse jump from node to node. There are three important factors involved in this *saltatory conduction:*

1. The myelin sheath acts as an *insulator* preventing the flow of current across the membrane of the internodal regions.
2. The myelin sheath *greatly reduces the electrical capacity* between the axis cylinder and its surrounding medium so that much less of the very brief inward current is lost by leakage through the capacitance.
3. When the cablelike spread of current reaches the *nodal region,* it *depolarizes the membrane* sufficiently to open the sodium gates and to start the self-regenerative process of sodium conduction that results in the action potential at this nodal region and so generates the currents to depolarize the next node (Fig. 6-12). Because this is an all-or-nothing mechanism, the action potential remains constant in size as it travels along the nerve (Fig. 6-13).

A further biological advantage to saltatory conduction is that it requires a minimum of metabolic energy: just that amount necessary to recover the ionic changes in the small area of exposed membrane at the nodes, through the use of the sodium and potassium pumps. This means that a nerve can conduct action potentials tirelessly, with very little energy consumption.

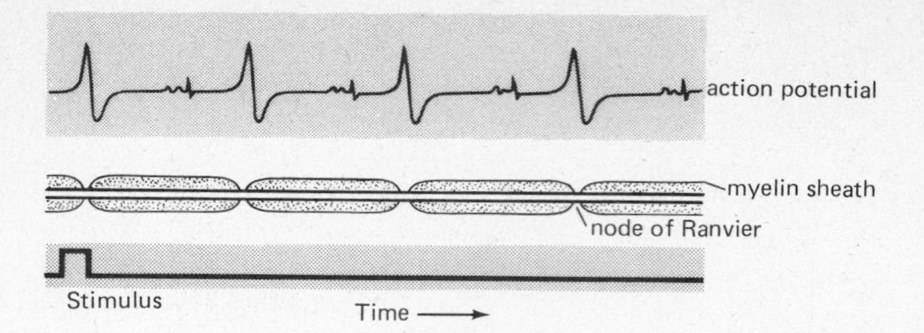

Figure 6-12. **Saltatory conduction along a myelinated nerve fiber. The action potential is recorded without diminution in amplitude at the nodal areas, as it travels along the fiber from the point of stimulation. Only very small currents can be recorded at the internodal regions. The membrane current "jumps" from active to inactive node, as shown in Fig. 6-13.**

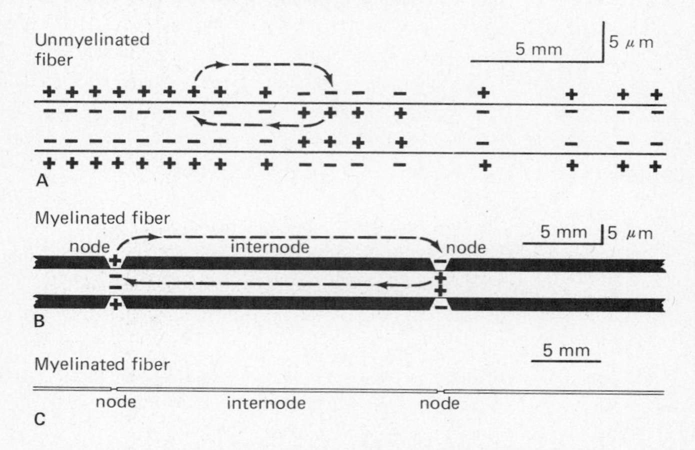

Figure 6-13. **Conduction of an action potential along a nerve fiber. (A) shows the current flow in an unmyelinated fiber. In (B) the propagation is in a myelinated fiber with the current flow restricted to the nodes. The dimensions in (B) are transversely exaggerated, as shown by the scale but are correctly shown in (C). (From *The Understanding of the Brain* by J. C. Eccles, McGraw-Hill Book Company, New York, 1973. Used with permission of McGraw-Hill Book Company.)**

Monophasic and Diphasic Action Potentials

When one electrode is placed inside the cell and the other is outside, one can measure the *potential change across the membrane* at the point of insertion of the intracellular electrode. When these changes are properly amplified and led into a cathode ray oscilloscope, they can be seen on the screen as the *monophasic action potential* (Fig. 6-14). It is often more convenient when working with intact animals to record from the surface of the nerve or muscle with two electrodes, and under these circumstances a *diphasic action potential* is recorded. With surface electrodes one is comparing the potential between *two regions of the cell membrane surface;* the difference in potential is first recorded at the one electrode, and then a mirror image is obtained as the action potential reaches the second electrode (Fig. 6-15).

The Compound Action Potential

In a nerve trunk, where fibers of many sizes are present (Fig. 6-16), the action potential is quite complex, with

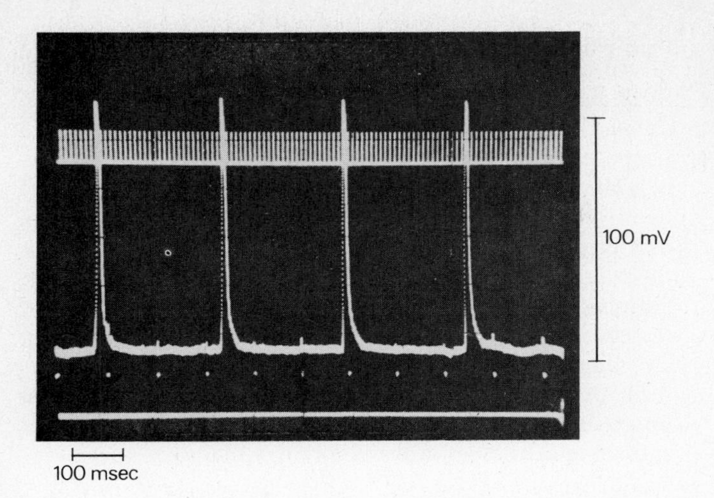

Figure 6-14. **A monophasic action potential is obtained when the recording electrode is inside the cell and the reference electrode is on the cell surface. The cell in this record is from the atrial region of the heart. The upper trace shows the rate of electrical stimulation. (Courtesy of J. Piedilato, Physiology Laboratory, New York University.)**

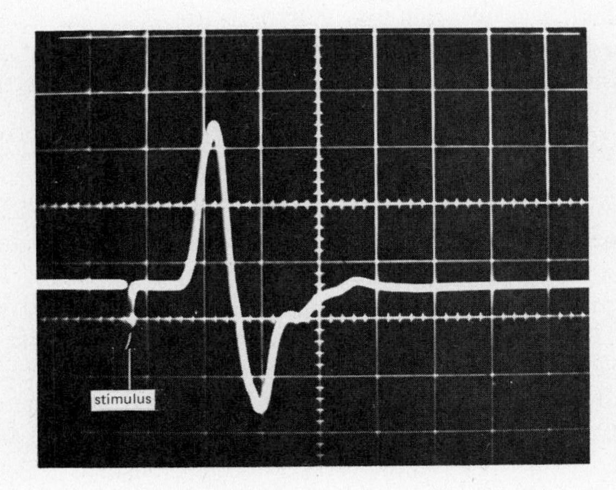

Figure 6-15. **A diphasic action potential is obtained when both electrodes are on the surface of the cell so that the difference in potential is recorded first at one electrode and then, as the action potential travels along the cell membrane, at the second electrode. Vertical scale, 1 cm = mV; horizontal scale, 1 cm = msec.**

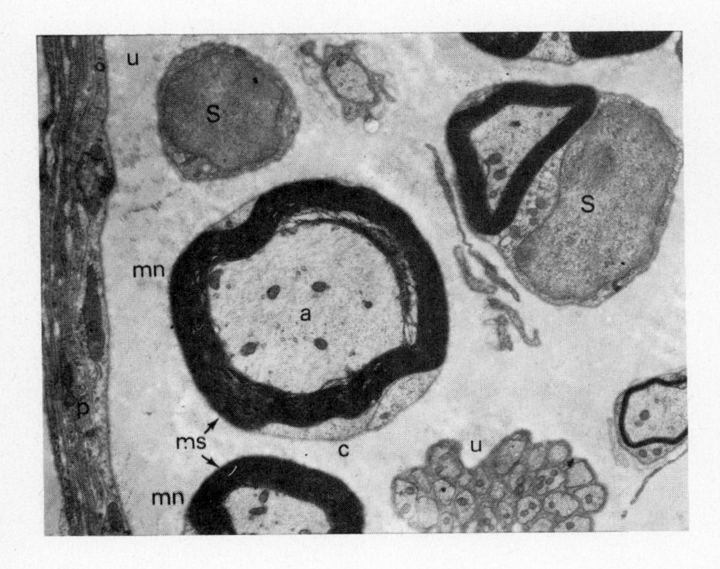

Figure 6-16. **Transverse section of nerve fibers in the sciatic nerve. The darkly staining myelin sheath (ms) covers the single axon (a) of the myelinated nerve fiber (mn). The sciatic nerve also contains unmyelinated fibers (u). S = Schwann cell. c = collagen fibers in a fine connective tissue sheath, the endoneurium. To the left is a heavy connective tissue layer, the perineurium (p). ×9000. (From J. Rhodin, *An Atlas of Ultrastructure,* W. B. Saunders Company, Philadelphia, 1963.)**

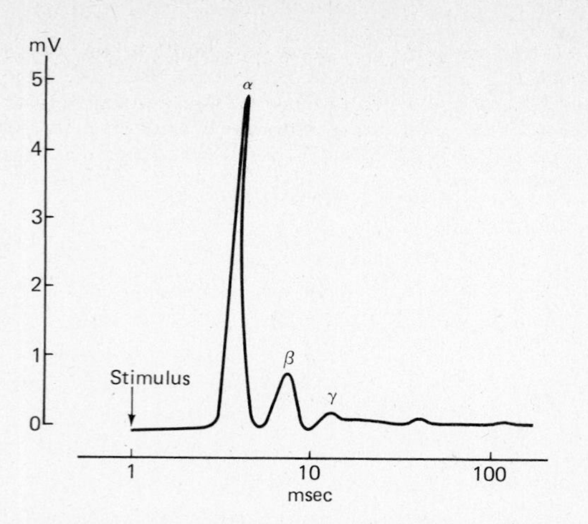

Figure 6-17. **Compound action potential recorded from the sciatic nerve of a frog. The different peaks represent nerve fibers of different diameters conducting at different velocities. The α fibers conduct most rapidly.**

Figure 6-18. **Erlanger and Gasser demonstrated in 1937 that the α and β peaks of the compound action potential represent different fiber groups conducting at different velocities and that the velocity of conduction is constant. The numbers on the left represent the distance from the site of stimulation to the recording electrodes. S = time of stimulation.**

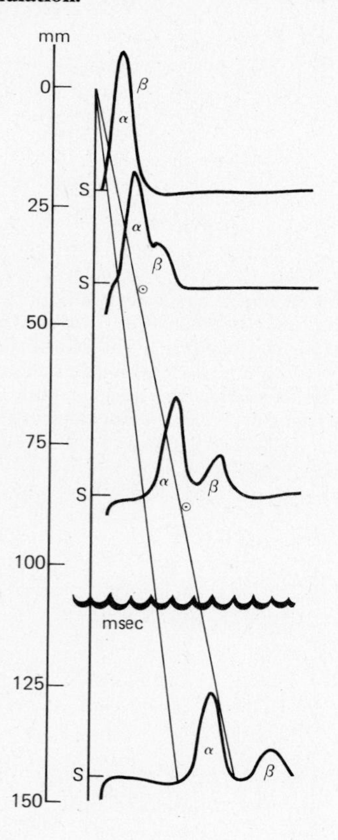

several peaks caused by the difference in conduction velocity of the various fiber components (Figs. 6-17 and 6-18).

The *larger the diameter* of the fiber, the *faster it will conduct* an action potential, so that the largest myelinated fibers, carrying impulses from the spinal cord to the cerebellum and also down the spinal cord, conduct at speeds of up to 140 m/sec. Unmyelinated fibers from 0.2 to 1.0 μm in diameter conduct impulses at about 0.2 to 2 m/sec. Because there are many more of these fine fibers, the size of the peripheral nervous system would have to be tremendously increased if they were large enough to conduct impulses at the same speed as the large fibers. The myelinated fibers seem to have been chosen for the conduction of the most urgent information, such as that concerned with movement and some of the pathways involving cutaneous sense.

6–4 Two-Way Conduction

A nerve fiber is capable of conducting an impulse in both directions, away from the cell body (*orthodromically*) or toward the cell body (*antidromically*). This two-way conduction is caused by the spread of the cable current in both directions if the axon is stimulated along its length. Physiologically, this is not common, and synaptic excitation usually generates impulses in or close to the soma.

Antidromic conduction is sometimes seen when an action potential travels up one branch of a sensory axon and down the other terminal branches of that axon (Fig. 6-19). This *axon reflex* results in an unusually widespread response to a simple stimulus, like a skin scratch. There is a *triple response;* a red line along the scratch, a flare spreading around it and a white wheal extending for several centimeters, which is caused by vasodilation resulting from the axon reflex.

6–5 The Stimulus

A stimulus is any change that can alter the energy state of a tissue sufficiently to depolarize the membrane and initiate an action potential. A nerve may be stimulated by tapping (mechanical), heating (thermal), placing a little table salt on it (chemical-osmotic), or by electrical stimulation. These various stimuli are converted, or transduced, by the nerve to their fundamental denominator, a change in energy level, to which it gives an electrical response, an action potential.

The most easily regulated and least damaging type of stimulation to use experimentally is electrical. It is consequently the type of stimulation most frequently utilized to investigate the responses of nerve and muscle.

6–6 Excitability

Excitability may be defined as the ability of a cell to respond to a stimulus with an action potential.

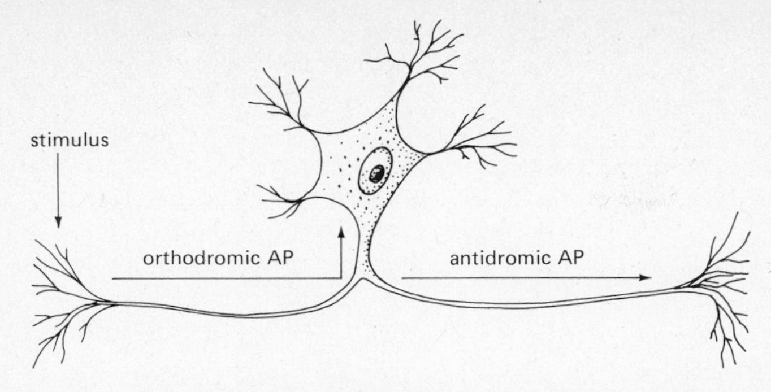

Figure 6-19. **The axon reflex is caused by impulses from skin receptors travelling centrally along a sensory fiber and also being conducted peripherally by other branches of the same nerve. Orthodromic conduction for sensory neurons is toward the cell body, the reverse of the normal conduction direction for motoneurons.**

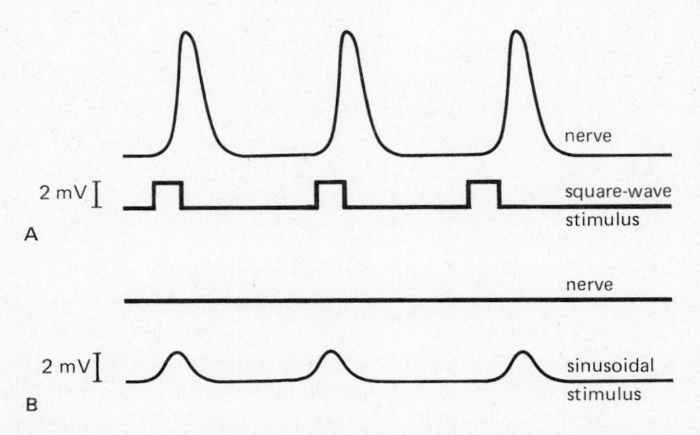

Figure 6-20. **A square-wave pulse that reaches its maximum rapidly evokes a response from an excitable cell, such as a nerve (A). A sinusoidal stimulus, which reaches its maximum much more slowly, results in accommodation, or increase in cell threshold, so that the same strength of stimulus that was used in (A) has no effect in (B).**

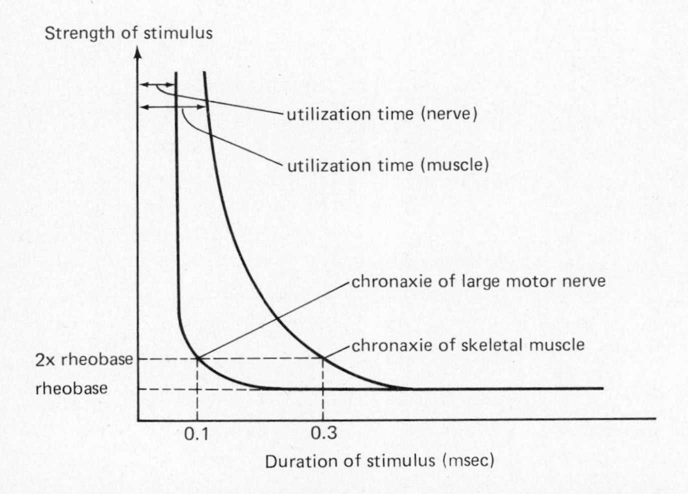

Figure 6-21. **The strength-duration curve shows that a minimum duration (the utilization time) is essential for even the strongest stimulus to evoke a response. The chronaxie of a large motor nerve is considerably less than that of skeletal muscle, demonstrating the greater excitability of the nerve.**

Excitability and Parameters of the Stimulus

The excitability of a tissue may be measured if one knows the conditions that a stimulus must fulfill to evoke a response. The energy change required involves the following parameters:

1. Strength of the stimulus.
2. Duration of the stimulus.
3. Rate of rise of the stimulus.

The current strength that is just enough to produce a response is the *threshold of excitation*. But the strength necessary will vary with the other two parameters, so for controlled physiological experiments it is usual to keep the duration constant (at under 1 msec) and use *square-wave* or *rectilinear pulses,* which reach their maximum strength rapidly. This type of stimulation is much more effective than one that reaches its maximum slowly—a sinusoidal pulse, for example (Fig. 6-20). The use of square-wave pulses prevents *accommodation,* which is a progressive increase in the threshold of excitability during the process of stimulation. One would have to increase the strength of the stimulus constantly to excite an accommodating tissue. Accommodation is also seen to other types of stimuli, such as pressure, light, smell, and noise.

The Strength-Duration Curve

The relationship between the strength of the stimulus and its duration is an important one. It is demonstrated in the *strength-duration curve* in Fig. 6-21. It can be seen that a certain minimum duration, the *utilization time,* is needed for even the strongest stimulus to evoke a response. The weakest stimulus that will excite the tissue, provided it is applied for a long enough time, is called the *rheobase*. In order to compare excitability of different fibers, a stimulus of twice the strength of rheobase is used to test the length of time needed to excite a response. This time is known as *chronaxie*. The chronaxie of nerve fibers varies according to the size and type of the fiber (as well as with its ability to accommodate), but it serves as a rough indicator of excitation time. It is used as a clinical device to determine whether the innervation of a muscle is intact. If it is intact, the chronaxie will be compound because of the difference in excitability between the nerve and muscle fibers; if the nerves have been destroyed, a simple chronaxie will be obtained, corresponding only to the muscle fibers (Fig. 6-21).

Cited References

Baker, P. F., A. L. Hodgkin, and T. I. Shaw. "Replacement of the Axoplasm of Giant Nerve Fibers with Artificial Replacements." *J. Physiol.* (*London*) **164:** 330 (1962).

Baker, P. F., A. L. Hodgkin, and T. I. Shaw, "The Effects of Changes in Internal Ionic Concentrations on the Electrical Properties of Perfused Giant Axons." *J. Physiol.* (*London*) **164:** 355 (1962).

DUBOIS-REYMOND, E. *On Animal Electricity.* Trans. Bence Jones. Churchill Publishers, London, 1952.

GALVANI, L. *De Biribus Electricitatus in Moto Musculari, Commentarius.* Typographia Instituti Scientarum, Bologna, Italy, 1791.

GASSER, H. S., and J. ERLANGER. "A Study of the Action Currents of Nerve with a Cathode Ray Oscilloscope." *Amer. J. Physiol.* **62:** 496 (1922).

HODGKIN, A. L., and A. F. HUXLEY. "A Quantitative Description of Membrane Currents and Its Application to Conduction and Excitation in Nerve." *J. Physiol.* (*London*) **117:** 500 (1952).

HODGKIN, A. L., and A. F. HUXLEY. "Currents Carried by Sodium and Potassium Ions Through the Membrane of the Giant Axon of *Loligo*." *J. Physiol.* (*London*) **116:** 449 (1952).

HODGKIN, A. L., and A. F. HUXLEY. "The Components of Membrane Conductance in the Giant Axon of *Loligo*." *J. Physiol.* (*London*) **116:** 473 (1952).

HODGKIN, A. L., and A. F. HUXLEY. "The Dual Effect of Membrane Potential on Sodium Conductance in the Giant Axon of *Loligo*." *J. Physiol.* (*London*) **116:** 497 (1952).

HODGKIN, A. L., A. F. HUXLEY, and B. KATZ. "Measurement of Current-Voltage Relations in the Membrane of the Giant Axon of *Loligo*." *J. Physiol.* (*London*) **116:** 424 (1952).

KEYNES, R. D. "The Ionic Movements During Nervous Activity." *J. Physiol.* (*London*) **114:** 119 (1951).

YOUNG, J. Z. "The Structure of Nerve Fibers and Synapses in Some Invertebrates." *Cold Spring Harbor Symp. Quant. Biol.* **4:** 1 (1936).

Additional Readings

BOOKS

A. Of General Interest to Chapters 6, 7, 8, and 9

BARNES, C. D., and C. KIRCHNER. *Readings in Neurophysiology.* John Wiley & Sons, Inc., New York, 1968.

BOURNE, H. H. *The Structure and Function of Nervous Tissue,* Vols. 1–6. Academic Press, Inc., New York, 1968–1972.

ECCLES, J. C. *The Physiology of Nerve Cells.* The Johns Hopkins University Press, Baltimore, 1957.

ECCLES, J. C. *The Understanding of the Brain,* 2nd ed. McGraw-Hill Book Company, New York, 1977.

KUFFLER, S. W., and J. G. NICHOLLS. *From Neuron to Brain.* Stamford, Conn.: Sinauer Associates, Inc., Sunderland, Mass., 1976.

OCHS, S. *Elements of Neurophysiology.* John Wiley & Sons, Inc., New York, 1965.

SCHADÉ, J. P., and D. H. FORD. *Basic Neurology,* 2nd ed. American Elsevier Publishing Co., Inc., New York, 1973.

SCHMIDT, R. F., *Fundamentals of Neurophysiology.* Springer-Verlag New York, Inc., New York, 1975.

TOWER, D. B., ed. *The Nervous System,* Vol. 1. In *The Basic Neurosciences,* R. O. Brady, Vol. ed. Raven Press, New York, 1975.

B. Of Specific Interest to Chapter 6

BURÉS, J., M. PETRÁN, and J. ZACHAR. *Electrophysiological Methods in Biological Research.* Academic Press, Inc., New York, 1962.

HODGKIN, A. L. *The Conduction of the Nerve Impulse.* Charles C Thomas, Publisher, Springfield, Ill., 1964.

KATZ, B. *Nerve, Muscle, and Synapse.* McGraw-Hill Book Company, New York, 1966. (Paperback.)

NASTUK, W. L., ed. *Physical Techniques in Biological Research.* Academic Press, Inc., New York, 1964.

ARTICLES

BAKER, P. F. "The Nerve Axon." *Sci. Am.,* Mar. 1956.

BODIAN, D. "The Generalized Vertebrate Neuron." *Science* **137:** 323 (1962).

BRAZIER, M. "The Historical Development of Neurophysiology." In *Handbook of Physiology,* Sec. 1, Vol 1: 1. American Physiological Society, Washington, D.C., 1959.

FUHRMAN, F. A. "Tetrodotoxin." *Sci. Am.,* Aug. 1967.

HODGKIN, A. L. "The Ionic Basis of Nerve Conduction." *Science* **145:** 1148 (1964).

HUXLEY, A. F., and R. STAMPFLI. "Evidence for Saltatory Conduction in Peripheral Nerve Fibers." *J. Physiol.* (*London*) **108:** 315 (1949).

KATZ, B. "Quantal Mechanism of Neural Transmitter Release." Nobel Prize Lecture. *Science* **173:** 123 (1971).

KEYNES, R. D. "Ion Channels in the Nerve-Cell Membrane." *Sci. Am.* March 1979.

KEYNES, R. D. "The Nerve Impulse in the Squid." *Sci. Am.,* Dec. 1958.

WEBSTER, H. DEF. "Peripheral Nerve Structure." In *The Peripheral Nervous System,* D. B. Tower, ed. Plenum Publishing Corporation, New York, 1974, p. 3.

Chapter 7

Chemical Regulation at the Synapse

Hear my rigmarole
Science stuck a pole
Down a likely hole
And he got it bit.
Science gave a stab
And he got a grab.
That is what he got.
"Ah," he said, "Qui vive,
Who goes there and what
ARE we to believe?
That there is an It?"

Robert Frost, "A Reflex"

THE FUNDAMENTAL CONCEPT that has emerged from the physiological investigations that began at the beginning of this century is that most neurons communicate with other cells through chemical transmitters. These transmitters are released as the result of the action potential (AP) arriving in the presynaptic nerve terminals: the transmitter diffuses across the synaptic cleft that separates the cells to combine with specific receptors on the postsynaptic membrane. Depending on the nature of the transmitter, it will open one or another type of the highly distinctive gates in the membrane, causing either excitation or inhibition of the postsynaptic cell.

7–1 The Neuronal Doctrine

The anatomical separation of each individual neuron was first described by Ramon ý Cajal (1909–1911) on the basis of his histological studies using Golgi's silver staining technique [Fig. 6-2(B)]. Although these two investigators shared a Nobel prize for their work, they disagreed violently on the interpretation, Golgi maintaining that the nerves were continuous with one another (1898). Sherrington's studies on the reflex arc (1906) gave support to the *neuronal doctrine* of Cajal, and the tremendous development of electrophysiological and neurochemical techniques in recent years has unequivocally proved that each neuron is anatomically separate, with the area of cellular contact, the synapse, involving many specializations that affect cellular activities.

7–2 Types of Synapses

Chemical Synapses

The development of unidirectional pathways between neurons, with connections that channel impulses along selected routes to excite or inhibit other cells, is associated with the formation of a highly specialized area of communication between the axon terminal and the cell with which it communicates. This region is called the *synapse,* a term introduced by the physiologist Sherrington (1906). The word synapse is derived from a Greek word meaning "to clasp tightly." Sherrington was a scientist of extraordinary vision; not only did he emphasize the integrative nature of the nervous system in his book of that title (1906), but he also trained and inspired some of our finest neurophysiologists and Nobel Prize laureates, among them Sir John Eccles and Professor Ragnar Granit.

In the nervous system of vertebrates, and in many invertebrates, transmission at the synapse is chemical, dependent upon the secretion and release of a neurotransmitter from the terminals of the presynaptic cell, across the narrow space, the *synaptic cleft,* that separates it from the surface of the *postsynaptic cell* (Fig. 7-1). The region of the postsynaptic membrane immediately under the presynaptic terminal is known as the *subsynaptic membrane.* Because only the *presynaptic terminals* possess this chemical transmitter, which is released when the nerve impulse reaches the terminals, transmission is *unidirectional.* Some neurotransmitters excite the postsynaptic cell, others inhibit it. These two kinds of synapse can be distinguished in electromicrographs because of fine differences in structure and in the shape of the vesicles in which the transmitters are stored (Fig. 7-2).

There is no diminution in excitation in synaptic systems. Impulses are transmitted from cell to cell, often to many other cells, with the synapse acting as an *amplifier.* However, the synapse is subject to a host of physiological variables that control the amount of transmitter released and inactivated and the state of excitation of the postsynaptic membrane. A neuron within the central nervous system may be covered by thousands of synaptic knobs from other neurons; these synapses may be on dendrites, on the soma, or on the axon. Combinations of synapses form *serial synapses* (Fig. 7-3), some of which may be excitatory, others inhibitory. Consequently, the development of synaptic systems bestows an enormous variability of response upon the central nervous system of higher animals.

Figure 7-1. **Nerve terminals (from different neurons) forming synapses (chemical). Note that the presynaptic membranes are thickened where the vesicles are lined up opposite the postsynaptic membrane, which is also thickened. Many mitochondria are present in the presynaptic terminals. Differences in the shape and contents of the vesicles indicate different transmitters; however, only one kind of transmitter is produced by any one neuron.**

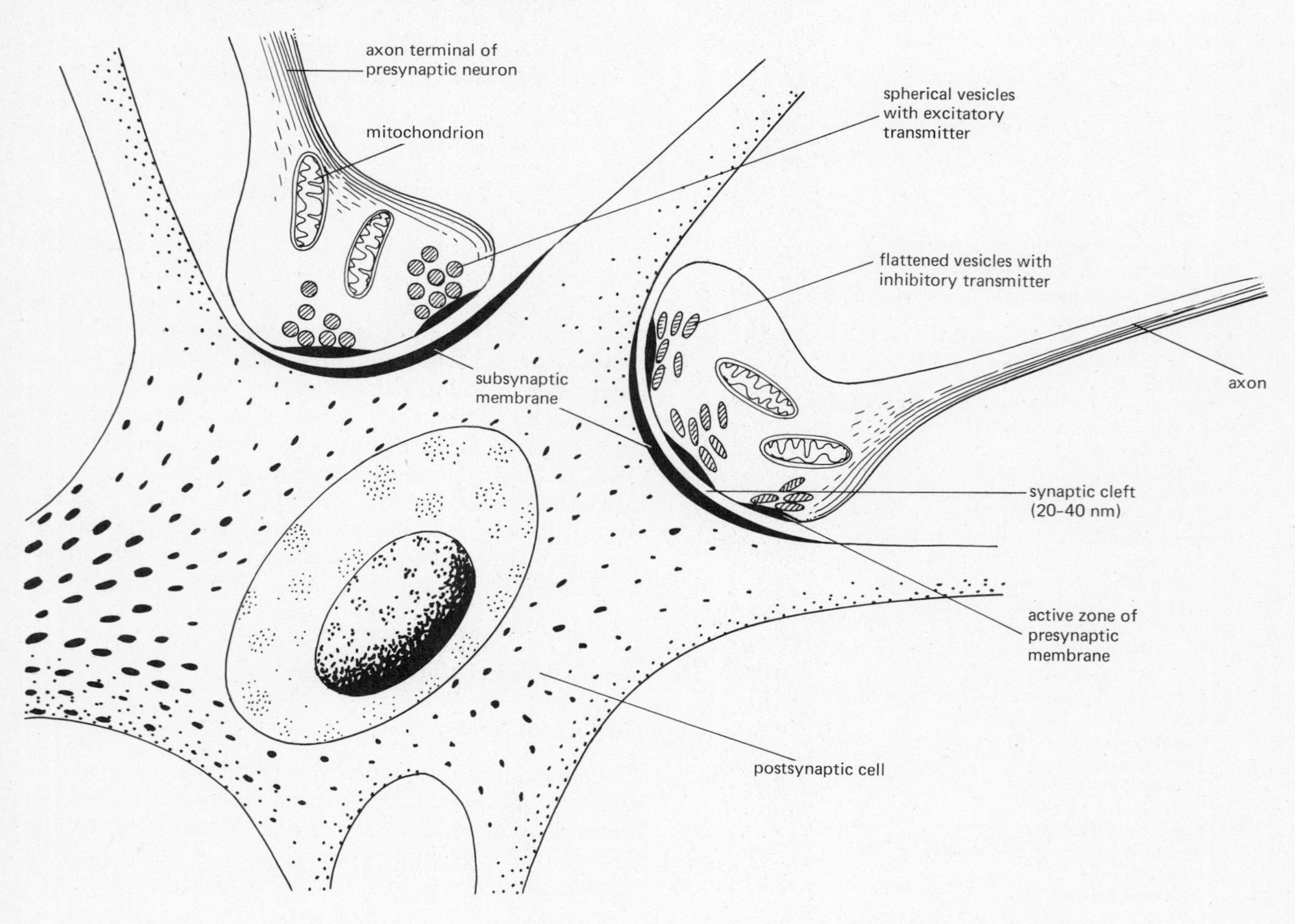

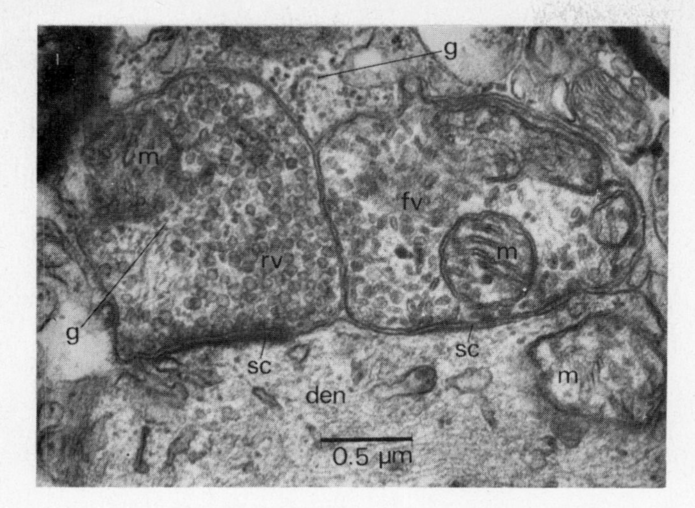

Figure 7-2. **Electronmicrograph of two axodendritic synapses on a fish motoneuron. The left knob has round vesicles (rv) (excitatory), whereas the flattened vesicles (fv) containing inhibitory transmitter are found in the right knob. g = granules, m = mitochondria, den = dendrite, and sc = synaptic cleft. (From E. G. Gray, *Cellular Dynamics of the Neuron,* S. H. Barondes ed., Academic Press, Inc., New York, 1969.)**

Electrical Synapses

If the cleft between synapsing cells is reduced from the 20 to 40 nm characteristic of the chemical synapse, to about 2 nm, this is called a *gap junction;* transmission across this type of junction is electrical. In another type of electrical synapse, there is no gap between the cell membranes, which are fused to form a *tight junction* (Figs. 7-4A and B).

The nerve impulse can be transmitted electrically across this tight junction often in either direction. In these electrotonic junctions, transmission can be polarized if one of the membranes is rectified so that the electrical resistance is considerably higher in one membrane than in the other. Electrical synapses are far more frequently found in invertebrates than in vertebrates. Some vertebrates have electrotonic synapses in specialized areas: neurons controlling the electric organs of fish; the giant Mauthner cells at the base of the brain in teleost fish where transmission is electrical; the presence of both electrical and chemical transmission in the ciliary ganglion of the chick. A mixture of these types of transmission is probably more widespread than we realize.

Figure 7-3. **Different types of chemical synapses. Note the serial synapse in the upper part of the diagram.**

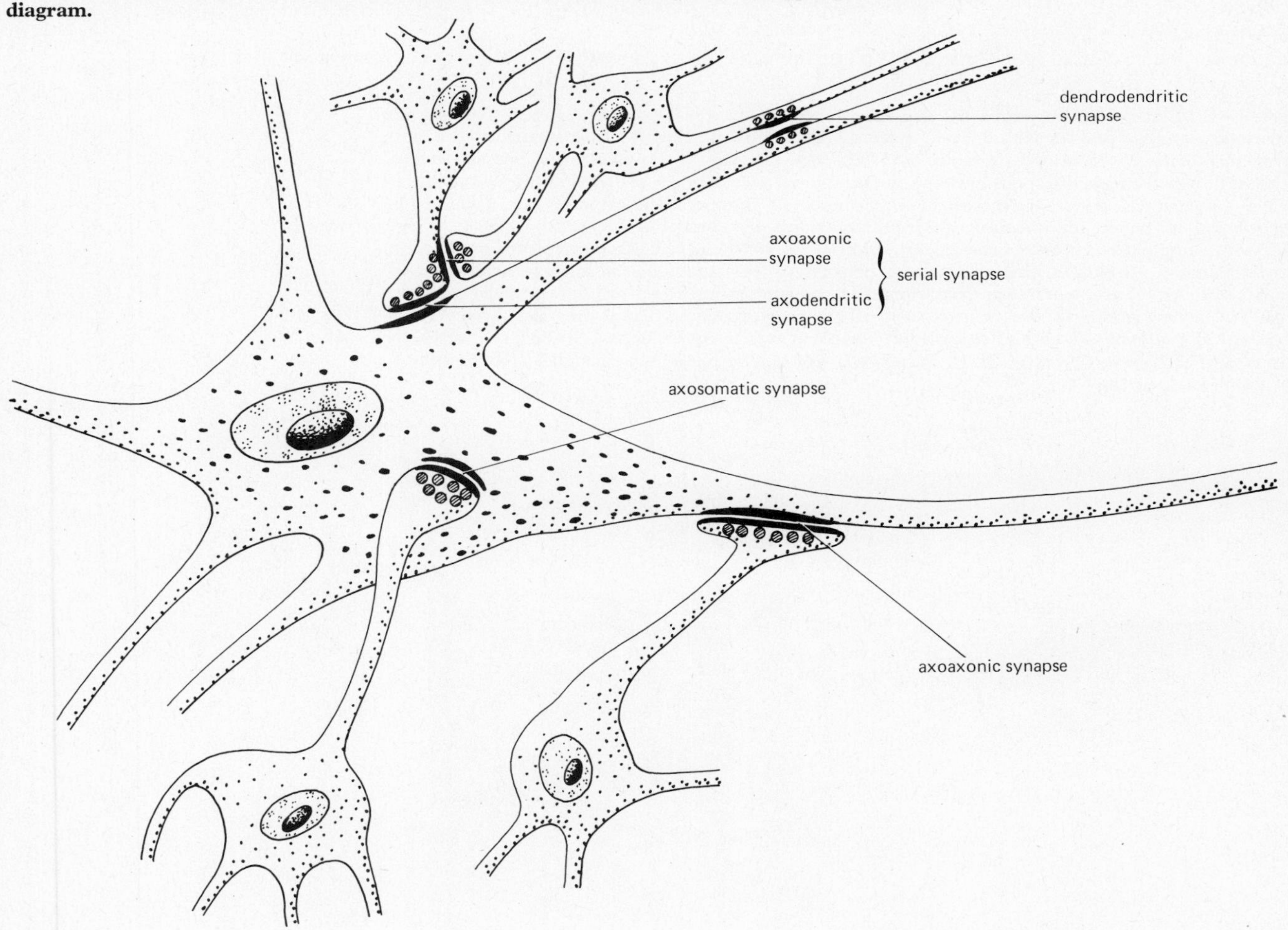

7–3 History of Chemical Transmission

In 1921 Otto Loewi used the perfused frog heart to demonstrate that when its inhibitory nerve, the vagus, is stimulated and the heart stops beating, some chemical appears in the perfusate which is able to stop the contractions of another, denervated heart (Fig. 7-5). With scientific caution, Loewi named the chemical in the perfusate "Vagusstoff"; it was later shown to be acetylcholine (ACh). The structural formula of acetylcholine is

$$(CH_3)_3N^+ - CH_2CH_2\text{—}O\text{—}\overset{\overset{\displaystyle O}{\|}}{C}\text{—}CH_3$$

Loewi repeated this experiment, stimulating the acceleratory nerves to the heart and found that the application of the resulting perfusate to a second, unstimulated heart speeded up its contractions. The substance released from the sympathetic nerves was not identified as norepinephrine until 1946, when von Euler isolated and identified it as the chemical transmitter at the endings of the postganglionic fibers of the sympathetic nervous system (reviewed by von Euler, 1967). The detailed studies of Axelrod (1971) have shown norepinephrine to be a transmitter in the brain as well as in the peripheral nervous system and have clarified its synthesis, storage, and inactivation (Sec. 9–5).

Another type of *peripheral synapse,* the *neuromuscular junction,* has been extensively studied; much of the information obtained has been applied to *central synapses,* which are considerably more difficult to analyze. In 1936, Dale demonstrated that when motor nerves are stimulated, acetylcholine is released. The application of small doses of ACh close to the muscle cause rapid contractions, similar to the effect of nerve stimulation. In 1952, Fatt and Katz showed that ACh is released at motor nerve endings in discrete units or *"quanta"* consisting of many thousands of molecules. This correlated beautifully with the electronmicrographs showing vesicles clustered in the nerve terminals and led to the idea that ACh is stored in these vesicles.

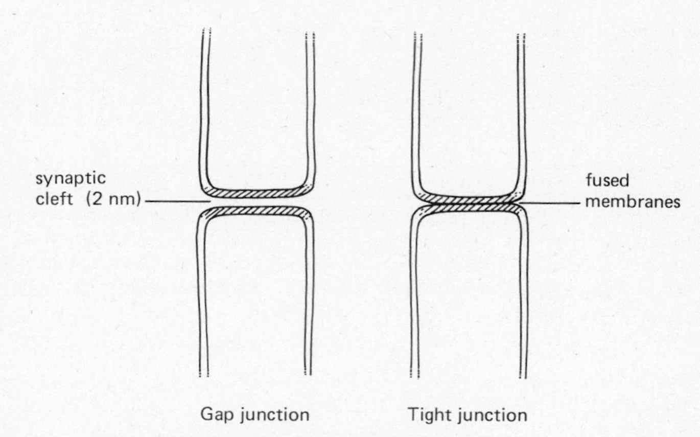

Figure 7-4A. **Diagram of electrotonic synapses.**

Figure 7-4B. **Electronmicrographs of electrotonic synapses. (1) Two membranes forming an electrotonic synapse are separated by a space of 2 to 4 nm (at arrows) that separates the unit membranes of this typical gap junction. ×177,000. (2) The use of the electron-dense substance, lanthanum, which enters the space between the junctional membranes, shows that the extracellular space is interrupted by a honeycomb of interconnected channels. Lanthanum was applied extracellularly; its penetration outlines the regular arrays of subunits (indicated by arrow) where intercytoplasmic pathways may exist in gap junctions. ×40,000. (3) In this electrotonic synapse in the midbrain of a toadfish, the presynaptic process (upper area) forms a structurally mixed synapse: vesicles (v) are seen in one area where the apposing membranes are separated by a 20-nm space. The arrow points to the characteristically narrow space of a gap junction. There is no physiological evidence for both chemical and electrical transmission here. ×36,500. (From G. D. Pappas, in *The Nervous System,* D. B. Tower, ed., Vol I., *The Basic Neurosciences,* Raven Press, New York, 1975, p. 19.)**

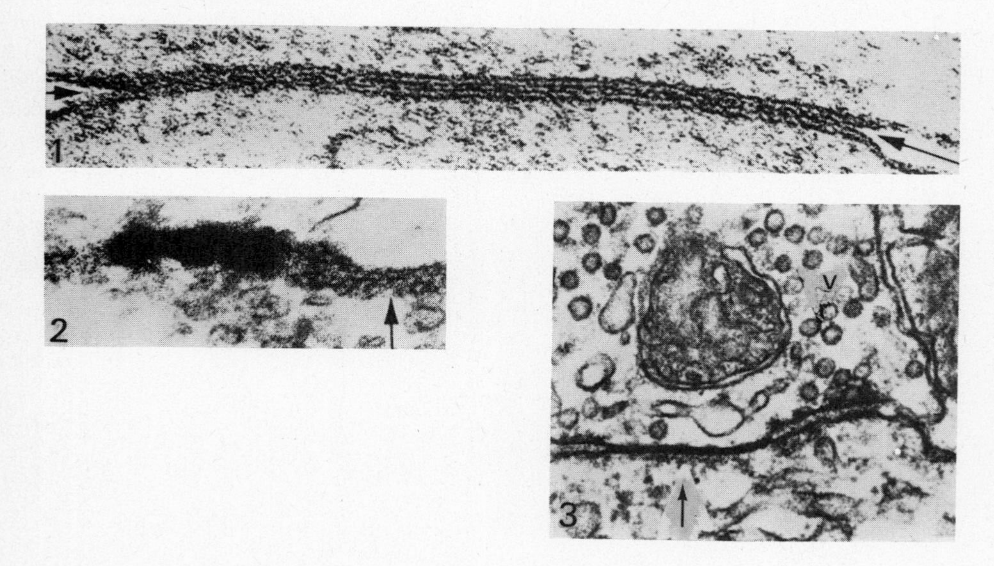

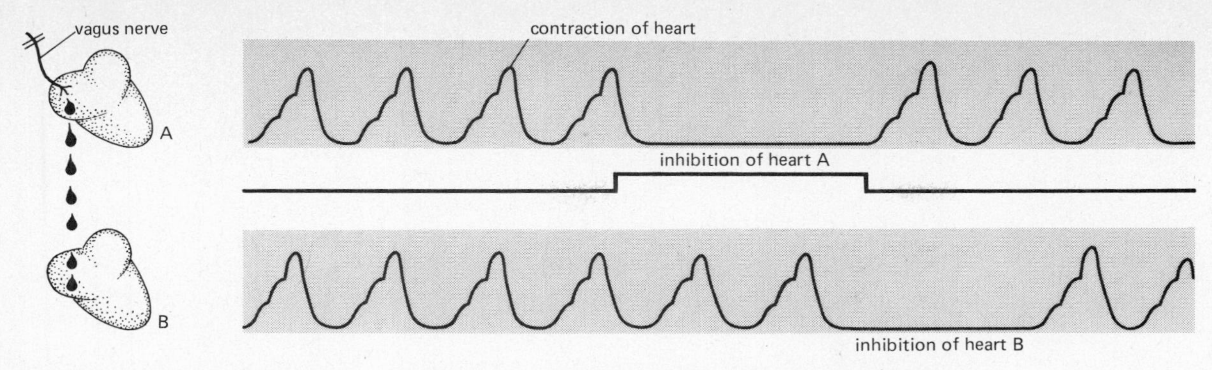

Figure 7-5. **Loewi's experiment to show the production of an inhibitory substance (Vagusstoff) on stimulation of the vagus nerve to the frog heart. Heart A stops contracting when its vagus is stimulated, and after a brief latent period the perfusate from A which drips over heart B causes the inhibition of the denervated heart B. Note that the first systole after recovery from inhibition is larger than the preinhibition systole as a result of increased cardiac filling. The elevation in the signal line between the contraction records of hearts A and B indicates the duration of stimulation of the nerve.**

Electronmicrographs show that all healthy nerve terminals contain vesicles, which may be flattened or spherical, granular or clear, but that only one type is found in any of the branches of a particular neuron (Figs. 7-1 and 7-2). This is excellent support for the concept put forward many years ago by Dale (called *Dale's principle* by Eccles), that each neuron produces only one kind of chemical transmitter. The histochemical "mapping" of different transmitters throughout the nervous system is based upon this specificity of distribution.

All of the investigators mentioned in this section—Loewi, Dale, von Euler, Axelrod, and Katz—have been honored by Nobel Prizes for their work on chemical transmission.

7–4 The Neuromuscular Junction or Synapse

All movements are composites of the contractions of *motor units,* the motor nerve (motoneuron), its axon, and all the muscle fibers it innervates (Fig. 7-6). Just as each nerve action potential is all-or-nothing, so the resulting contraction of each muscle fiber of the motor unit is all-or-nothing; increases in the strength of muscle contraction are obtained through the recruitment of greater numbers of motor units.

Each muscle fiber receives only one of the many terminal branches of the nerve fiber. The nerve terminal has a characteristic form in different vertebrates, being highly compact in mammals and much simpler and more spread out in the frog.

Microscopic Structure of a Neuromuscular Synapse

Presynaptic structures

The axon terminates in little knobs on the surface of the muscle fiber but does not fuse with it. Within the terminals of the axon are the spherical *synaptic vesicles* (40–200 nm in diameter) containing ACh, and the many mitochondria needed for the active synthetic processes occurring in the terminals (Fig. 7-7).

Freeze-fracture studies, in which the presynaptic membrane is split longitudinally as shown in Figs. 7-8 and 7-9, give a beautiful three-dimensional view of the inner and outer surfaces of this membrane, as well as of the fracture

Figure 7-6. **A motor unit consists of a motoneuron in the spinal cord, its axon and nerve terminals, and the skeletal muscle fibers it innervates.**

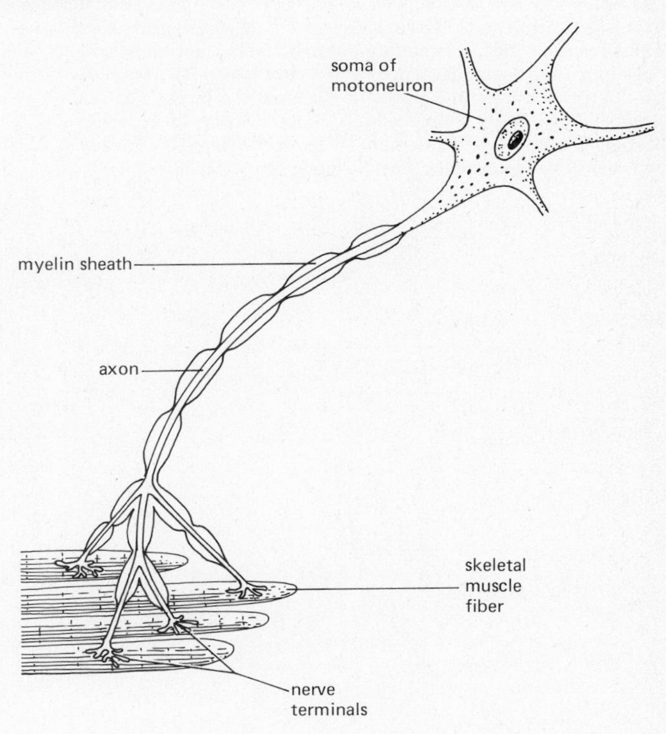

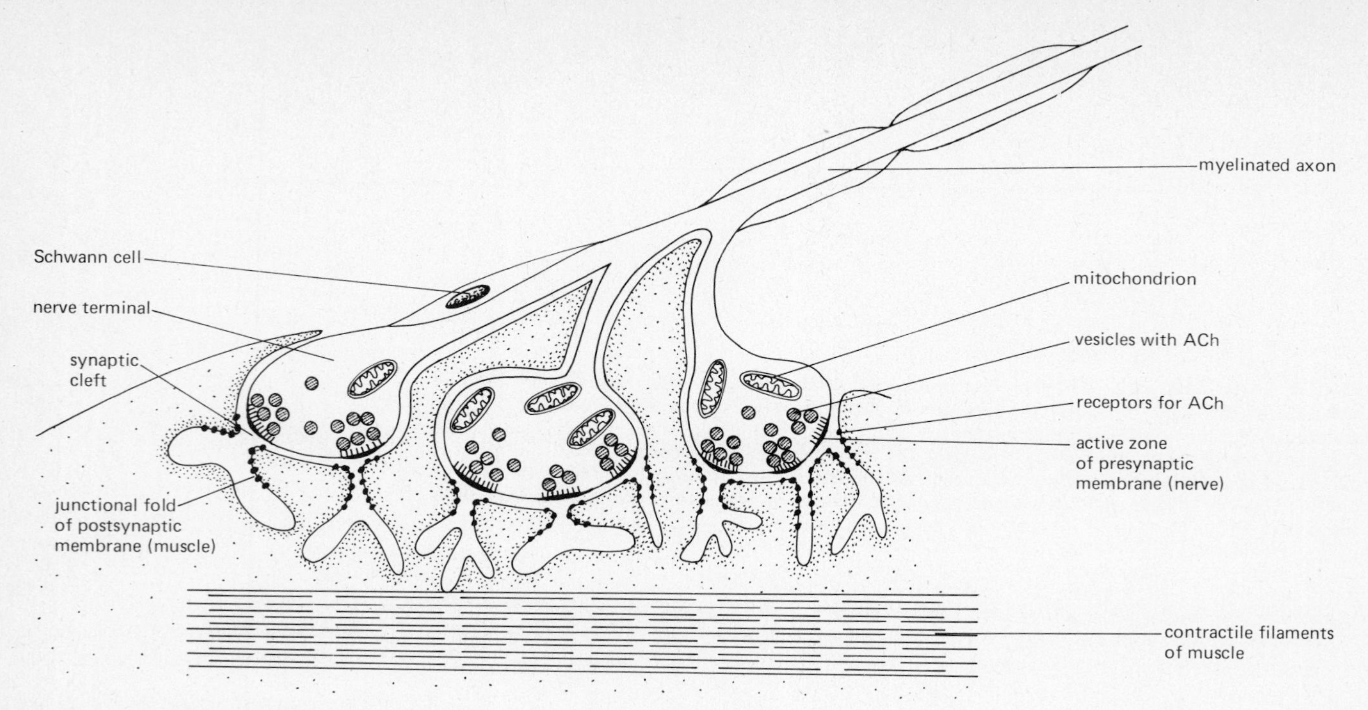

Figure 7-7. **The neuromuscular synapse. The motor nerve terminals lose their myelin sheath and lie naked on indentations (synaptic gutters) in the muscle fiber. Vesicles in the presynaptic terminals are lined up near release areas, known as active zones. The deep junctional folds of the postsynaptic membrane contain many particles, concentrated especially in the areas opposite the presynaptic active zones. These particles are believed to be ACh receptors.**

Figure 7-8. **Three-dimensional diagram of presynaptic and postsynaptic structures of the neuromuscular junction as observed upon freeze fracturing. The plasma membranes are split (at arrows in B) and shown in C. The cytoplasmic half of the presynaptic membrane at its active zone shows, on its fracture face, protruding particles whose counterparts are seen as pits on the fracture face of the outer membrane leaflet. Vesicles that fuse with the presynaptic membrane give rise to characteristic protrusions and pores in the fracture faces. The fractured postsynaptic membrane in the region of the folds shows a high concentration of particles (ACh receptors) on the cytoplasmic leaflet. (From U. J. McMahan, in *From Neuron to Brain,* S. W. Kuffler and J. G. Nicholls, eds., Sinauer Associates, Inc., Sunderland, Mass., 1976.)**

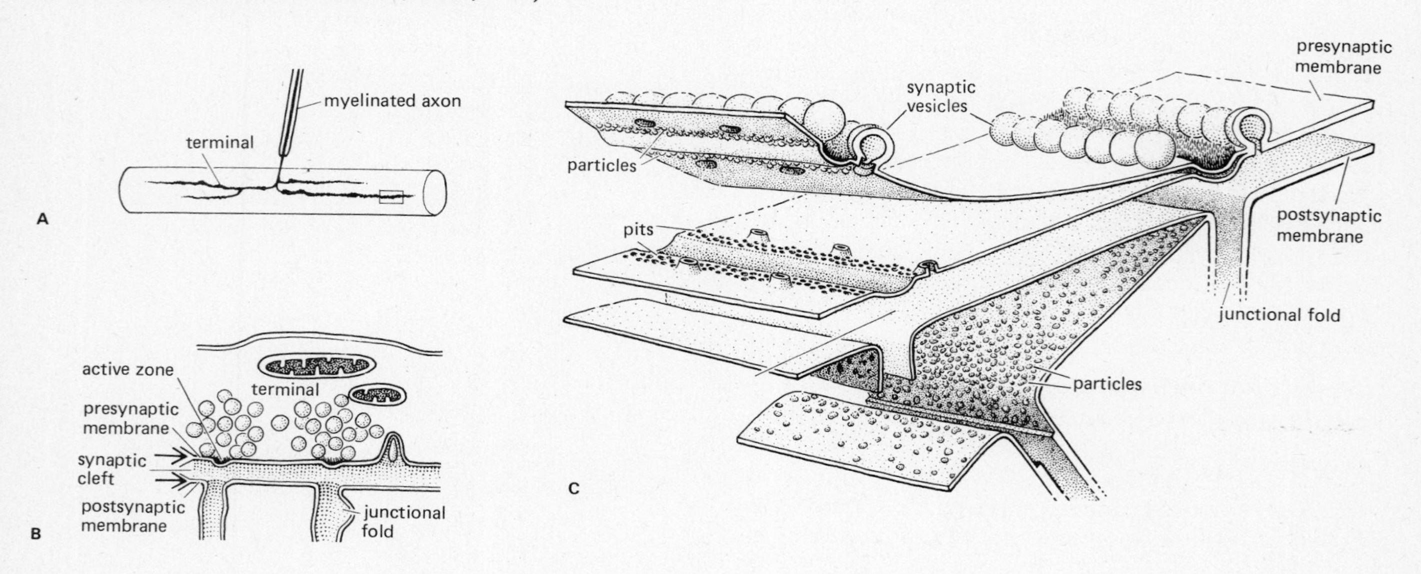

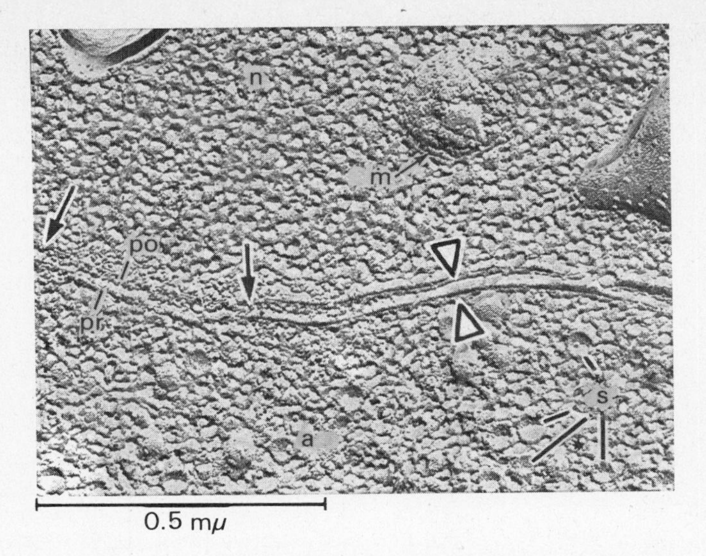

Figure 7-9. **Freeze-etched axosomatic synapse. The arrows point to the active zone. The presynaptic terminal (pr) is below; the postsynaptic density (po) is seen above it. The double leaflet structure of the cell membrane can clearly be seen (arrow heads). n = neuron, a = axon terminal, m = mitochondria, and s = synaptic vesicles. The vesicles are often associated with a single large particle in the concavity, which may be a calcium binding site. (From K. Akert and C. Sandri, in *Excitatory Synaptic Mechanisms*, P. Andersen and J. K. S. Jansen, eds., Universitetsforlaget, Oslo, Norway, 1970, p. 27.)**

faces. Figure 7-8 also shows the orderly line-up of *synaptic vesicles* in a double row along a central band of dense material, to which, presumably, they are normally attached. These vesicles are on the *cytoplasmic surface* of the presynaptic membrane. Each row of vesicles represents the bottom line of a hexagonal pile of vesicles.

The corresponding *fracture face of the cytoplasmic half* of the presynaptic membrane is covered with small particles, which follow the line-up of the vesicles on the cytoplasmic surface of this membrane.

The *fracture face* of the *outer layer* of the presynaptic membrane shows *pits* that neatly correspond to the particles on the other side of the fracture. These presynaptic regions of vesicle and particle accumulation are the *active zones* of the presynaptic membrane, where transmitter release presumably occurs. What cannot be seen even with the electron microscope is the molecular structure of the presynaptic membrane with its selective ionic gates. These are diagrammatically illustrated in Fig. 7-10.

The synaptic cleft

Separating the axon terminal from the muscle fiber membrane (the sarcolemma) is a gap of about 40 nm. This is wide enough to permit the flow of currents generated by the action of the chemical transmitter on the postsynaptic membrane, but too large for any detectable electrical transmission of the nerve impulses to the muscle fiber.

Postsynaptic structures

There is an enlargement of the sarcoplasm of the muscle fiber at the junctional area, which is called the *end plate.* The end plate is the postsynaptic region where depolarization occurs to give rise to the *end-plate potential* (EPP), which is described in the following discussion of the physiology of the neuromuscular synapse. The postsynaptic surface area is markedly increased by deep *junctional folds* (*postsynaptic folds*).

Freeze-fracture studies of the postsynaptic membrane show a very high concentration of particles at the edges of the synaptic folds, just opposite the active zones of the presynaptic membrane (Fig. 7-8). It is believed that these particles are the ACh receptors.

The postsynaptic membrane is both structurally and physiologically different from the rest of the cell membrane, for this region directly under the terminals of the presynaptic nerve is *electrically nonexcitable.* The postsynaptic region will respond only to chemical stimulation or inhibition.

McMahan and Kuffler (1971), using Nomarski differential interference microscopy, were able to study the neuromuscular junction in the living muscle fiber. This important advance was the basis for experiments designed to show the exquisite correlation between structure and function.

Now that it had become possible to visualize the nerve terminals in the living preparation, Kuffler and Yoshikama

Figure 7-10. **Chemical transmitter (T) released from the presynaptic nerve terminals opens separate Na^+ and K^+ gates and their associated channels. This results in the influx of Na ions and the efflux of K ions, according to their respective electrochemical gradients, and depolarizes the postsynaptic membrane. The chloride gates remain closed in the presence of an excitatory transmitter but are opened by inhibitory transmitters.**

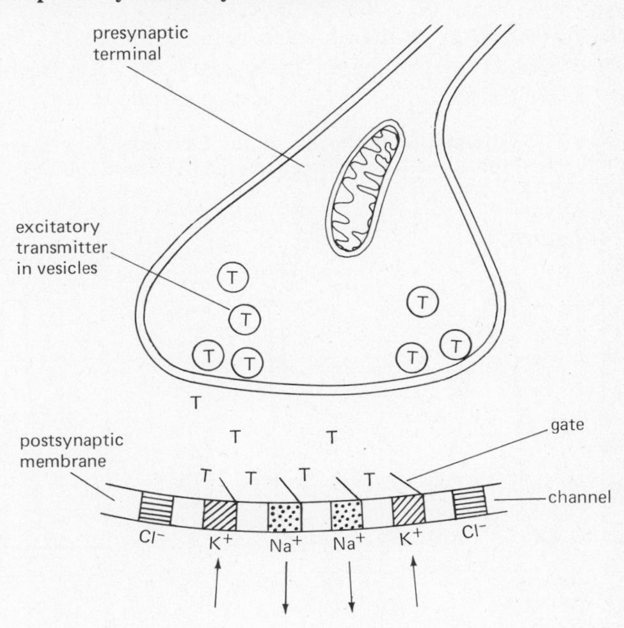

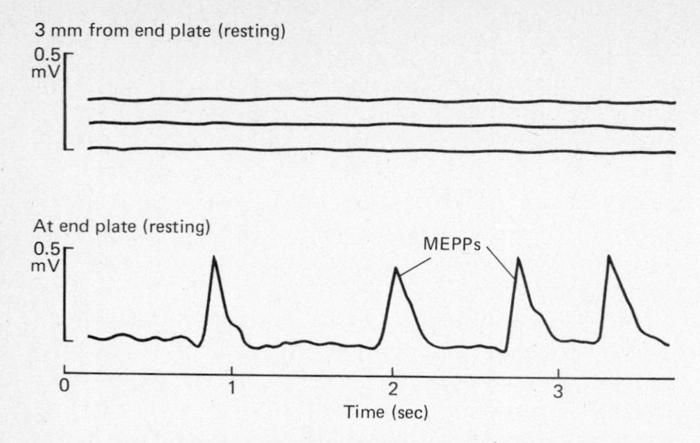

Figure 7-11. **MEPPs (miniature end-plate potentials) recorded intracellularly from a resting neuromuscular synapse. The bottom trace shows these random depolarizations to be approximately 0.5 mV, occurring about 1/sec. The top trace indicates that this "biological noise" is present only at the end-plate region and not found a short distance away.**

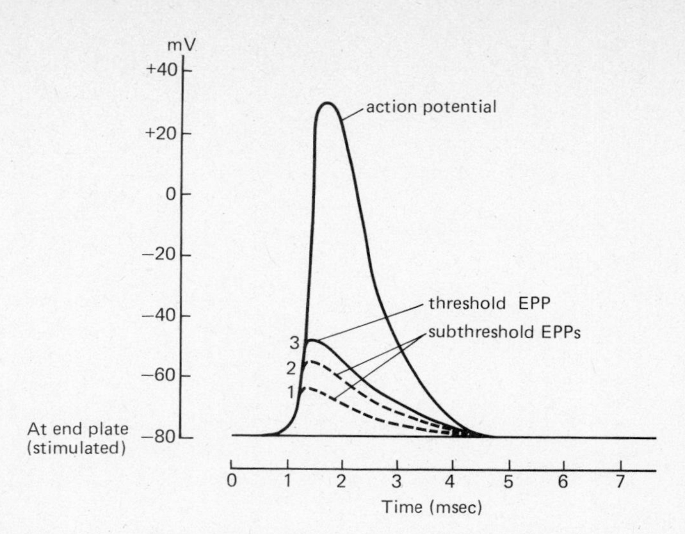

Figure 7-12. **The graded EPP (end-plate potential) can be summated by increasing the strength of stimulation (1, 2, and 3). When the EPP reaches a critical level of depolarization at 3, an action potential is evoked.**

(1975) were able to apply ACh iontophoretically to either of two parallel nerve terminals, for these run longitudinally in the frog. If the ACh is applied close to either terminal, a large depolarization of the muscle membrane is obtained. If ACh is applied between the terminals, there is little or no response, clearly indicating that it is only the muscle surface membrane under the nerve terminal that is sensitive to ACh.

Physiology of the Neuromuscular Synapse

The important discovery of the quantal release of transmitters by Fatt and Katz, mentioned above, was almost serendipitous. While studying muscle fiber potentials with intracellular electrodes, they observed small, random depolarizations at a frequency of about 1/sec and an amplitude of about 0.5 mV. These small potentials are spontaneous and found only at the end-plate region (Fig. 7-11). They first described this finding as "biological noise." On further study they discovered that these *miniature end-plate potentials* (MEPPs) are due to the expulsion of brief jets of ACh, acting on the end plate of the muscle fiber. These jets correspond to a quantal release of ACh, and it can be shown by various experimental and mathematical methods that when the nerve fiber is stimulated ACh is released in quanta, which build up to form an EPP. The EPP is graded in size, and at a critical level of depolarization—about 50 mV—it triggers the impulse, which can be recorded as it travels along the muscle membrane (Fig. 7-12). The relationship between the graded EPP and the all-or-nothing action potential (AP) is clearly shown when the neuromuscular synapse is poisoned with *curare,* the poison used on arrow tips by the South American Indians. Today curare (or one of its many derivatives, such as *d*-tubocurarine) is used clinically,

Figure 7-13. **Diminution in the size of the EPP caused by curare poisoning does not change the amplitude of the all-or-none action potential (AP) until the EPP drops below the threshold. At this point no AP is evoked.**

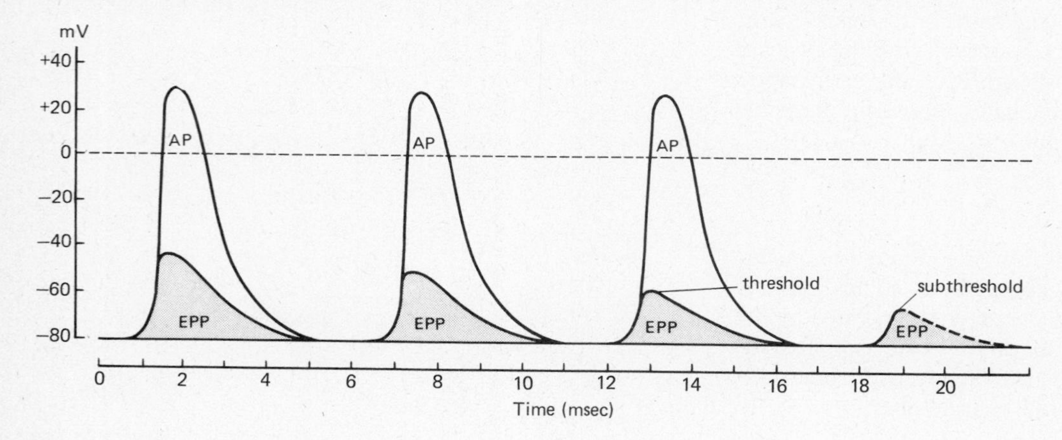

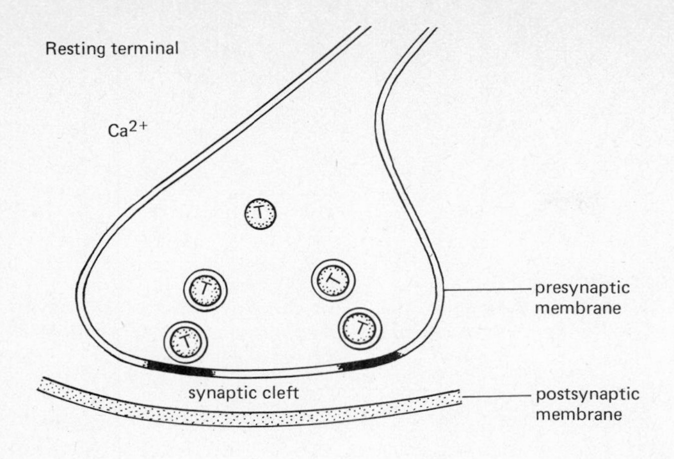

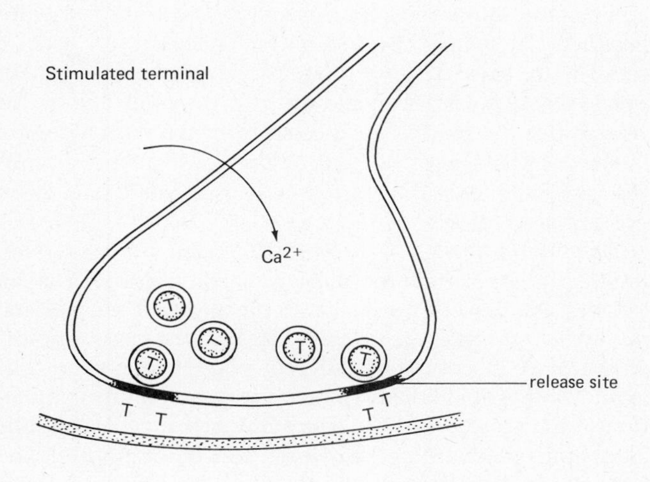

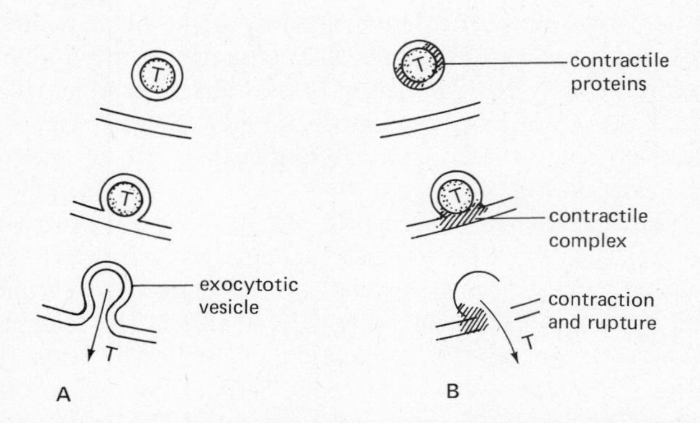

Figure 7-14. **Release of transmitter from synaptic vesicles is absolutely dependent upon the entry of Ca^{2+}, which occurs when the resting nerve terminal is depolarized (stimulated). One theory, A, holds that the transmitter (T) is released by exocytosis as the vesicle fuses with the presynaptic membrane. Another theory, B, suggests that proteins of the presynaptic and vesicle membranes form a contractile complex (crosshatching), which on contraction pulls the vesicle open, freeing its contents.**

under controlled conditions, to obtain muscular relaxation without excessively deep anesthesia.

Figure 7-13 shows that when recordings are made from the end-plate region of an isolated nerve muscle preparation, to which curare is added, the initial EPP that fires the impulse gradually becomes smaller and smaller until no AP is evoked. The AP appears full size or is completely absent.

Experiments with curare also have helped to elucidate the mechanism of action of ACh on the postsynaptic membrane.

7–5 Acetylcholine

Mechanism of Action of Acetylcholine

Release

The AP reaching the nerve terminals depolarizes the membrane sufficiently (about 30 mV) to open the calcium gates, permitting the inrush of calcium ions down their steep electrochemical gradient. This triggers the release of ACh from the synaptic vesicles (Fig. 7-14). It has been calculated that four Ca^{2+} ions are needed to burst one vesicle (probably by changing the negative charge on the inner surface of the vesicle membrane). There is an absolute need for extracellular Ca^{2+} in a minimum concentration of about 10^{-4} *M* in order for depolarization to release ACh from nerve terminals. Removal of Ca^{2+} from the external medium or blockage of Ca^{2+} action by the addition of Mg^{2+} depresses transmitter release. Recently, Miledi (1973) has been able to inject Ca^{2+} electrophoretically into a presynaptic terminal pretreated with tetrodotoxin. The calcium injection causes an enormous quantal release of transmitter.

The ejection of ACh from the burst vesicle is probably accomplished by exocytosis, which presumes that the vesicle has fused with the presynaptic membrane prior to its rupture (Fig. 7-14A). Another suggestion is that there is a contractile complex formed between proteins of the vesicle membrane and the presynaptic membrane, similar to the actomyosin complex of muscle. The inrush of Ca^{2+} causes this complex to contract, pulling the vesicle apart (Fig. 7-14B).

Recycling of vesicles

Although there is some evidence that the vesicle membrane is *recycled* into the presynaptic membrane, studies on the composition of these two types of membrane have shown considerable differences between them. It is more likely that the disrupted vesicles are taken up by cisternlike structures in the nerve terminals. In these structures, the old membranes are modified, and new vesicles are pinched off and filled (Fig. 7-15). Although it is certain that transmitters are stored in these presynaptic vesicles, it is also possible that some transmitter may be stored in the cytoplasm, outside the vesicles.

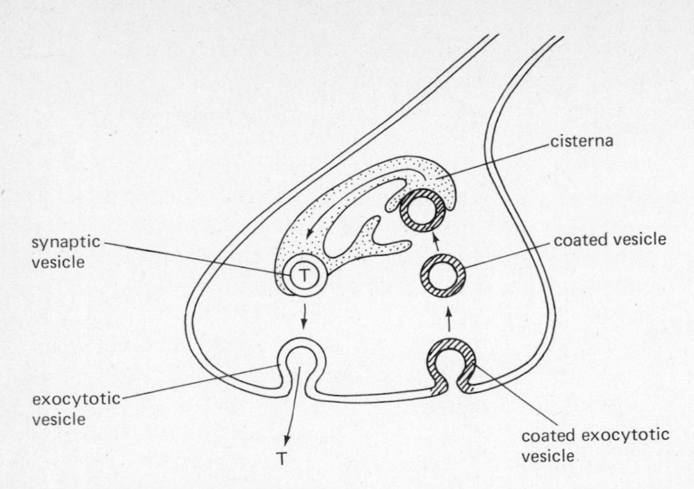

Figure 7-15. **Recycling of synaptic vesicles may involve modification of the vesicle membrane. The membrane appears to become coated, then processed through a cisterna (made of fused vesicles), which then frees an uncoated, transmitter-filled vesicle.**

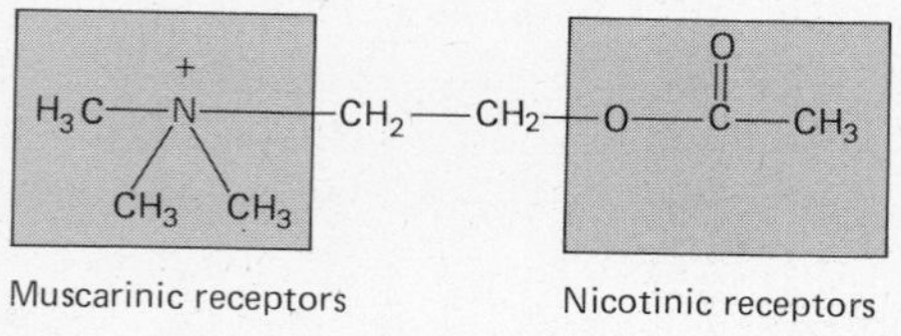

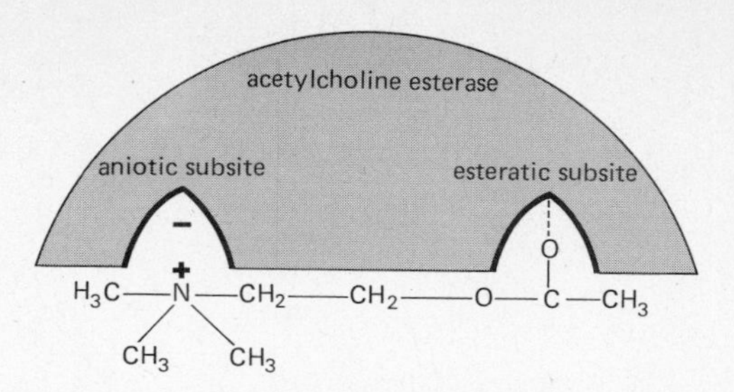

Figure 7-17. **Binding of ACh to the catalytic center of the enzyme acetylcholine esterase occurs at an anionic and an esteratic subsite.**

ACh ACTIVITY AT THE END PLATE

At the motor end plate, ACh undergoes two different types of reactions: it combines with a membrane receptor that results in the opening of ionic gates to cause depolarization, and it combines with a hydrolytic enzyme, *acetylcholine esterase* (*AChE*), which rapidly inactivates it.

ACh Receptor. Most evidence indicates that the ACh receptor is a protein, built into the structure of the membrane in a highly ordered way. A change in the structural geometry of this protein occurs when ACh binds to it, resulting in the opening of the ionic gates and a change in permeability. Junctional inhibitors such as curare also bind to the receptor protein, but alter it to an inactive form, which does not result in depolarization. When the receptor site is occupied by curare, ACh is unable to activate it. Snake venom contains toxins (e.g., α bungarotoxin) that bind very tightly and specifically to ACh receptors, and studies using these toxins have contributed greatly to our knowledge of these receptors. There is a very high density of receptors (3×10^7) per end plate, which is enough for the 10^4 quanta of ACh liberated, as estimated by Katz and Miledi (1967). There are between 12,000 and 21,000 molecules of ACh per quantum, which can easily be packed into a vesicle. Because the synaptic gap is so small, there is an excellent chance for most of the ACh molecules to collide with a receptor site within a few milliseconds of their release and cause depolarization within the requisite time for proper transmission.

There are quite separate ionic gates and transmembrane channels for sodium, potassium, and calcium. For each of these ions there are two types of gates, one that can be opened by depolarization and one that responds only to the transmitter. As might be expected from the morphological studies of the synapse described previously, the transmitter-sensitive sites with their ionic gates are found only in the postsynaptic region.

DIFFERENT TYPES OF ACh RECEPTORS. The classical differentiation of ACh receptors was established by Dale in 1914 on the basis of the effects of the alkaloids *nicotine* and *muscarine* on cholinergic junctions. Both drugs mimic the action of ACh but act at different postsynaptic sites. The application of small doses of nicotine results in the stimulation of the receptors of the neuromuscular junction and the autonomic ganglia: larger amounts will block transmission just as high concentrations of ACh will. Traditionally, these cholinomimetic effects are called *nicotinic*, and these ACh receptors are blocked by *curare* and related compounds.

Muscarine affects the ACh receptors in the target tissues of the parasympathetic nervous system (vagus nerve endings on heart muscle and the smooth muscle of the intestine, for example), and these *muscarinic* effects are blocked by *atropine*. The nicotinic receptors apparently interact with the carboxyl side of ACh, and the muscarinic with its methyl side (Fig. 7-16).

The character of the receptor seems to depend primarily on the type of cell on which it is found, but more than one type of receptor can be present on the same cell. Gardner and Kandel (1972) have demonstrated this in the sea snail *Aplysia*. Iontophoretic application of ACh will result in either depolarization or hyperpolarization of the same cell, depending on which receptor is stimulated. The two types of receptors respond to differences in concentration of ACh. Certain spinal cord cells and sympathetic ganglion cells in the mammal also appear to possess two types of receptors.

Inactivation of ACh. The concentration of ACh at the end-plate region remains high only for a very brief moment,

Figure 7-16. **Proposed sites of interaction of nicotinic and muscarinic receptors with acetylcholine.**

$H_3C—\overset{+}{N}—CH_2—CH_2—O—\overset{O}{\overset{\|}{C}}—CH_3$ (with two CH_3 groups on N)

Muscarinic receptors | Nicotinic receptors

for it is hydrolyzed rapidly by AChE into the inactive products choline and acetate. Histochemical staining shows that the enzyme is highly concentrated in the end-plate region and localized on the postjunctional membrane and in its infoldings. AChE is also found on the surface of the presynaptic membrane and in the cleft, although this may be due to diffusion.

Nachmansohn (1959) has shown that ACh is bound to the enzyme at a catalytic center made up of two subsites: an anionic site that has a negative charge, and an esteratic site (Fig. 7-17). Inactivation of ACh occurs in two steps after the formation of the *enzyme substrate complex:*

$$\text{AChE} + \text{ACh} \rightleftharpoons \text{AChE-ACh}$$

1. The *ester linkage is broken* and the *enzyme becomes acetylated,* freeing choline:

$$\text{AChE-ACh} \rightleftharpoons \text{acetyl-AChE} + \text{choline}$$

2. The *acetylated enzyme is rapidly hydrolyzed,* producing free enzyme for further activity, and acetate:

$$\text{acetyl-AChE} + H_2O \rightleftharpoons \text{AChE} + \text{acetate}$$

When AChE is inhibited, the accumulation of ACh desensitizes the membrane receptors, which after an initial period of repetitive firing can no longer respond so that neuromuscular transmission fails. A nerve gas (diisopropylphosphofluoridate, DFP) is an organophosphate that combines irreversibly with AChE, preventing it from binding with its physiological substrate, ACh, and resulting in massive accumulation of ACh that paralyzes the responses of the muscles to nerve impulses. The respiratory muscles and the central nervous system are also affected, and death results.

Anticholinesterases with short-term effects, such as eserine and prostigmine, are used clinically to alleviate the muscle weakness characteristic of *myasthenia gravis,* where there is a decrease in ACh effectiveness due to the production of antibodies that block junctional receptors.

Synthesis, Transport, and Storage of ACh

The early experiments of Nachmansohn (1959) led to the discovery that an acetylating cofactor is needed for the

Figure 7-18. **The resynthesis of ACh in nerve terminals involves its Ca^{2+}-dependent release, its inactivation by AChE, and the reuptake of choline into the terminals. The site of action of some drugs that interfere with the release, binding, breakdown, or resynthesis of ACh is indicated by the broken arrows.**

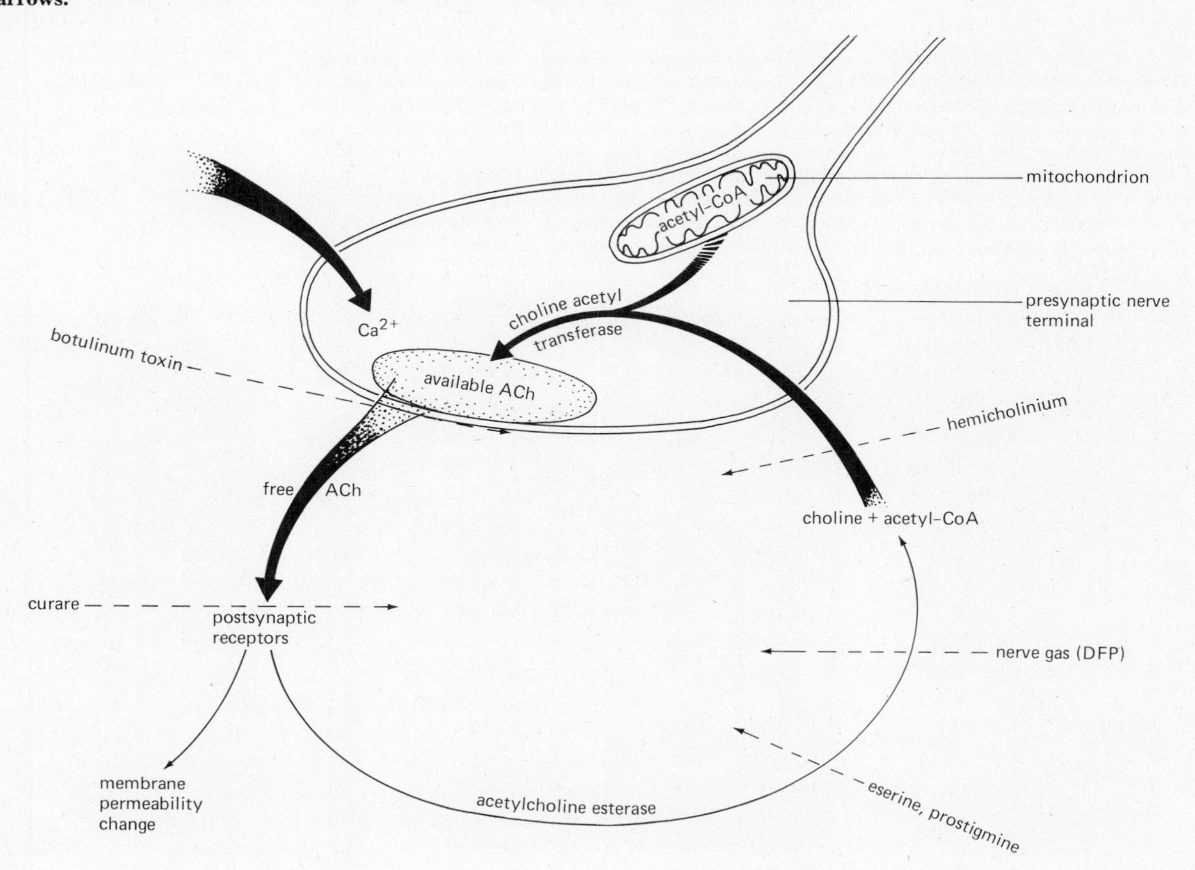

synthesis of ACh. This is Coenzyme A (CoA), which is discussed in Chapter 5. Acetyl CoA is one substrate, choline is another, and the enzyme is choline acetyltransferase.

$$\text{acetyl CoA} + \text{choline} \xrightleftharpoons[]{\text{choline acetyl transferase}} \text{ACh} + \text{CoA}$$

Neurons, like other cells, obtain acetyl CoA from oxidative metabolism in the mitochondria. Choline is taken up from the extracellular fluid by a carrier mechanism which can be blocked by the drug hemicholinium, effectively inhibiting the biosynthesis of ACh (Fig. 7-18).

Most ACh is synthesized in the *bodies of the motoneurons* and transported by *axonal flow* to the nerve terminals, where the largest fraction is stored in the synaptic vesicles. This process takes about 10 days and is too slow to replenish the amounts of transmitter used on rapid stimulation. However, the *nerve terminals* themselves synthesize ACh, with the rate of synthesis being closely linked to neuronal activity. Stored ACh is released continually and spontaneously; stimulation increases the rate of release. At the same time, choline is being formed rapidly as a result of the hydrolysis of ACh by AChE, and more than half of this choline is taken up by the nerve terminals and resynthesized to ACh.

Not all the ACh in the nerve terminals is available readily. Some appears to be *stationary,* some is in a depot that is being constantly *replenished,* and a small fraction is *available immediately.* Radioactive tracer studies indicate that this last fraction is made up of the newly synthesized ACh and that it is released twice as rapidly as the preformed stores. These various compartments for ACh storage are in equilibrium with one another (Fig. 7-19A) and assure that the nerve terminals will not be depleted of ACh, giving a large safety factor to neuromuscular transmission. The junction is sensitive, however, to many poisons, especially the deadly *Botu-*

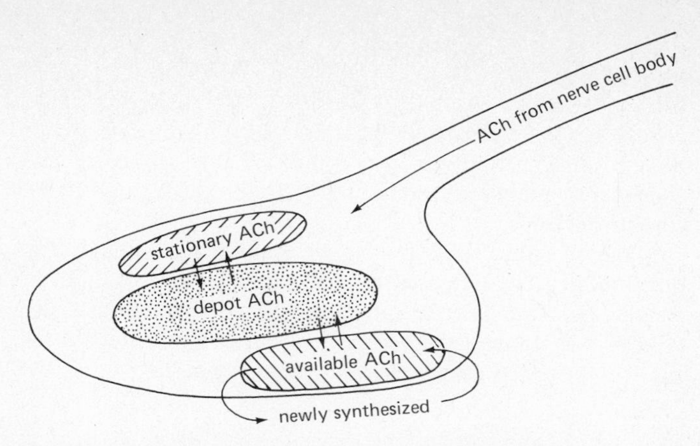

Figure 7-19A. **The various compartments for ACh storage are in equilibrium with one another. Replenishment from ACh synthesized in the cell body requires about 10 days, but depletion is prevented by the very rapid resynthesis of ACh in the nerve terminals themselves. The newly synthesized ACh appears to be preferentially used.**

Figure 7-19B. **Cycle of depletion and resynthesis of synaptic vesicles in frog neuromuscular junction during stimulation and recovery. (1) Unstimulated preparation showing nerve terminals filled with synaptic vesicles. (2) Depletion of synaptic vesicles following 1-hr stimulation in a high K^+ medium. The large areas of invaginated membrane (im) probably represent incorporation of vesicle membranes into the presynaptic membrane. (3) Recovery after 20-min rest in normal Ringer's solution. Vesicles have accumulated behind the active zones. n = nerve terminal, v = vesicles, c = synaptic cleft, m = muscle junctional folds (postsynaptic membrane), f = muscle contractile filaments, and S = Schwann cell fingers wrapping around the nerve. Upper arrows indicate active zones. (Courtesy of D. Rutherford, Laboratory of Cellular Biology, New York University.)**

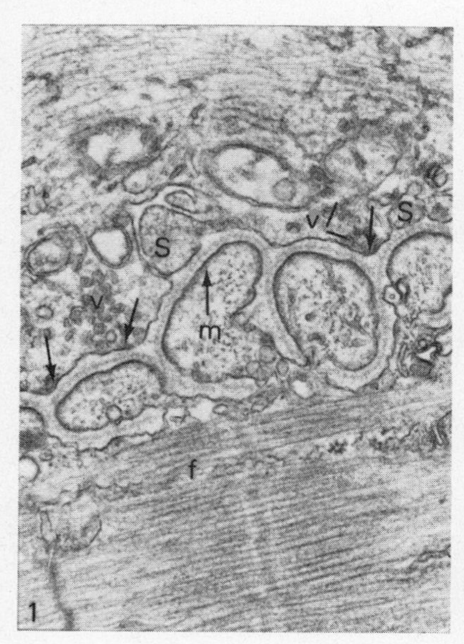

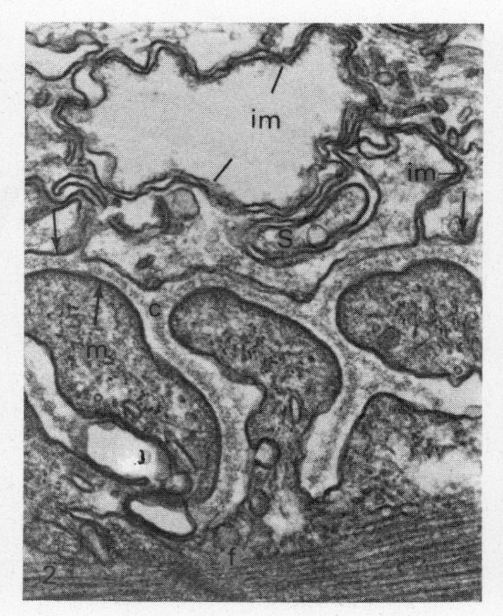

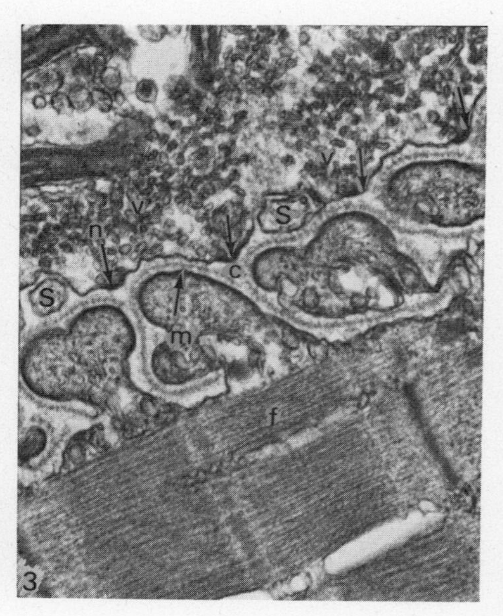

linum toxin, which prevents the release of ACh from nerve endings and causes death by respiratory paralysis.

Normally, the rapid resynthesis of ACh at neuromuscular junctions during continuous stimulation prevents the depletion of transmitter stores. If, however, resynthesis is prevented by immersion in a high K^+ medium, the almost complete disappearance of synaptic vesicles can be seen. The membrane of the stimulated presynaptic terminal appears to increase enormously in surface area, infolding upward between the active zones, presumably through incorporation of the membranes of the empty vesicles. If the nerve-muscle preparation is allowed to recover in normal Ringer's solution, vacuoles appear and become the transmitter-filled vesicles. This *cycle of depletion and resynthesis* is beautifully demonstrated in the electronmicrographs prepared by Rutherford, Nastuk, and Gennaro (Fig. 7-19B).

7–6 Central Synapses

The central nervous system is composed of an enormous number of individual nerve cells, organized into functional systems by the synaptic contacts they make with one another (Chaps. 8 and 9). Figure 7-20 shows a drawing of the synaptic contacts of eight neurons from the frontal cortex of a child. It is extremely difficult to study changes at a synaptic level in such complex systems, and much of our knowledge of central synapses has come from almost 100 years of intensive study of spinal reflexes.

Organization of Reflexes

The term *reflex,* as Sherrington (1906) originated it, involved the idea that information from sensory receptors traveled along specific nerve pathways to motoneurons in the CNS to produce or prevent movement. The simplest reflexes are stereotyped responses to stimuli, below the level of consciousness. The simplest *reflex arc* is that of the stretch reflex, where only two neurons are involved: a sensory neuron that makes synaptic contacts with a motoneuron (Fig. 7-21). The knee jerk in humans is an example of a simple stretch reflex: a tap on the tendon at the knee joint will elicit the sharp extension of the lower leg, provided that the stimulated leg is held in a relaxed position, usually crossed over the supporting knee of the other leg. This reflex can be inhibited easily by the subject, for reflexes are not independent entities functioning in isolation. Many reflex pathways are interconnected, and most spinal reflexes are under very effective control by motor centers in supraspinal structures, which can either inhibit or facilitate them (Chap. 8).

Except for the stretch reflex, reflex arcs always include at least one interneuron between the sensory and motor neurons, and also make synaptic contacts with tract neurons projecting to higher levels, as do stretch reflexes. Sherrington's concept that the reflex is the unit of nervous integration still holds true, but it must be emphasized that these

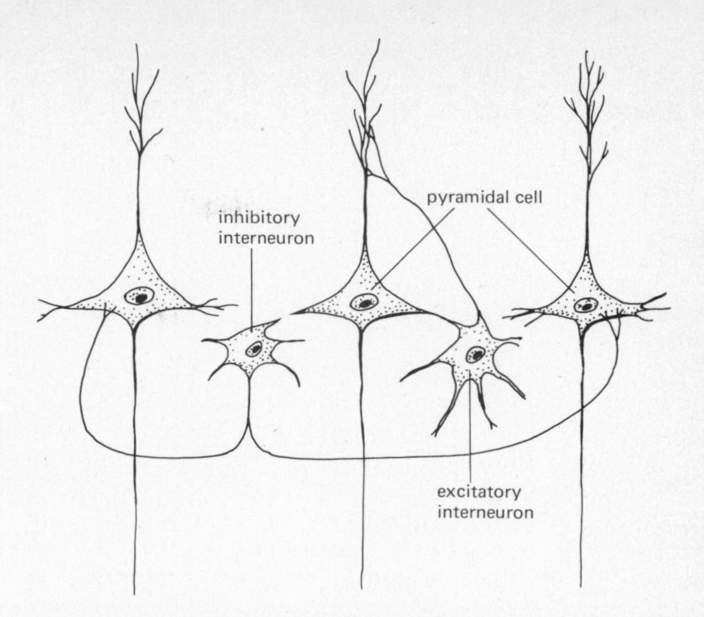

Figure 7-20. **Synaptic contacts of a few of the many cell types of the cerebral cortex.**

units are built into a tremendous variety of neural pathways to provide for the coordination of responses. There is also a continuous feedback of afferent information during the execution of a movement, so that the output of the muscles is changed by the changing sensory input. This feedback mechanism is dependent on information coming from stretch receptors within the muscles themselves, the *muscle spindles.* These basic reflexes are worth considering in detail, for they provide us with our fundamental concepts of synaptic function.

The Stretch Reflex

Muscles contain not only the strong twitch fibers which contract and cause movement or tension, but, in addition, special complex receptors called muscle spindles, which are located between and parallel to the twitch fibers (*extrafusal fibers*) (Fig. 7-22). These receptors consist of sensory nerve endings (*annulospiral endings*) wrapped around delicate muscle fibers within the spindle (*intrafusal fibers*). The intrafusal fibers also receive motor innervation from nerves arising from small cells in the ventral horn of the spinal cord (Sec. 8–6). These are the gamma cells; their axons are the efferent *gamma fibers.* (About one third of the motor nerves that leave the spinal cord are gamma fibers. The rest are the large, myelinated efferent alpha fibers that come from the ventral horn motor cells and innervate the twitch fibers, forming the *motor units* that were described in this chapter.)

The large sensory nerves of the muscle spindle are classified as 1a afferents: they are myelinated and conduct impulses extremely rapidly. The whole apparatus of intrafusal fibers and their motor and sensory nerves is contained within a

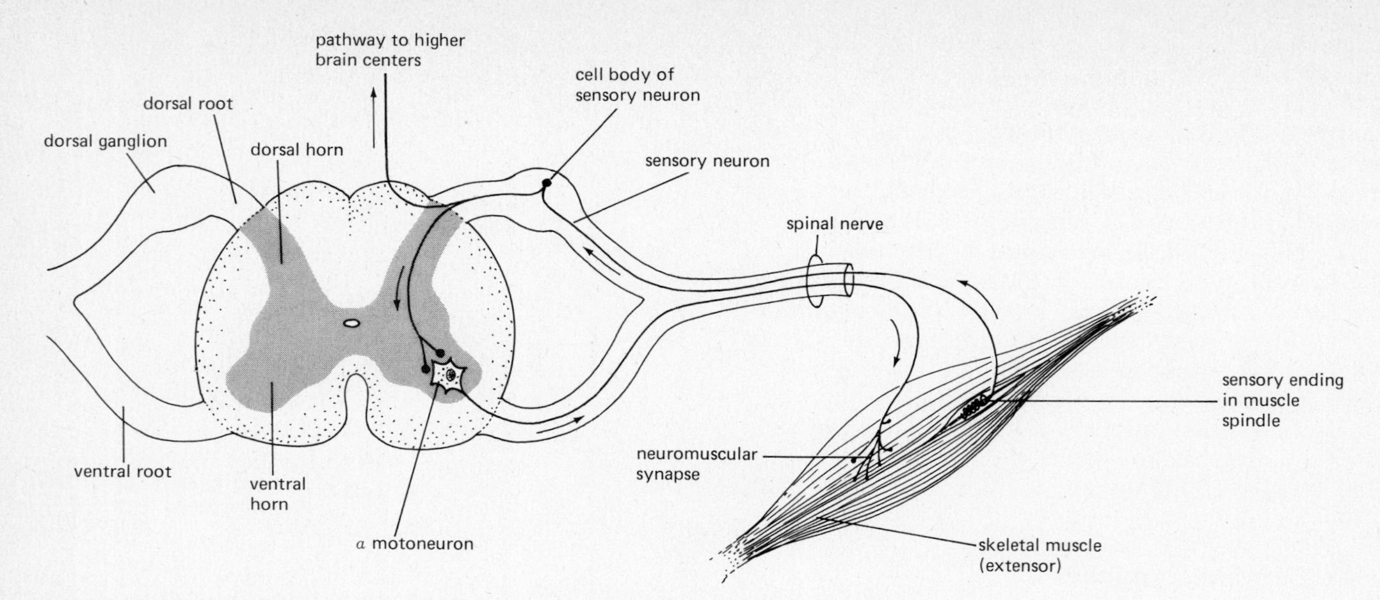

Figure 7-21. **The stretch reflex is a two-neuron, monosynaptic pathway. Stretching of the sensory ending in the muscle stimulates the sensory nerve to fire an impulse that causes the release of an excitatory transmitter (ACh) directly on the alpha motoneuron in the ventral horn of the spinal cord. The motoneuron fires and all the muscle fibers of its motor unit contract. The direction of nerve impulses along the reflex pathway is indicated by arrows.**

Figure 7-22. **Receptors and effectors of the stretch reflex. The muscle spindle is shown in greater detail in this diagram than in Fig. 7-21. Note the dual motor innervation of the muscle: the fast pathway to twitch muscle fibers via the alpha nerve fibers, and the slightly slower pathway to the intrafusal fibers via the gamma nerve fibers.**

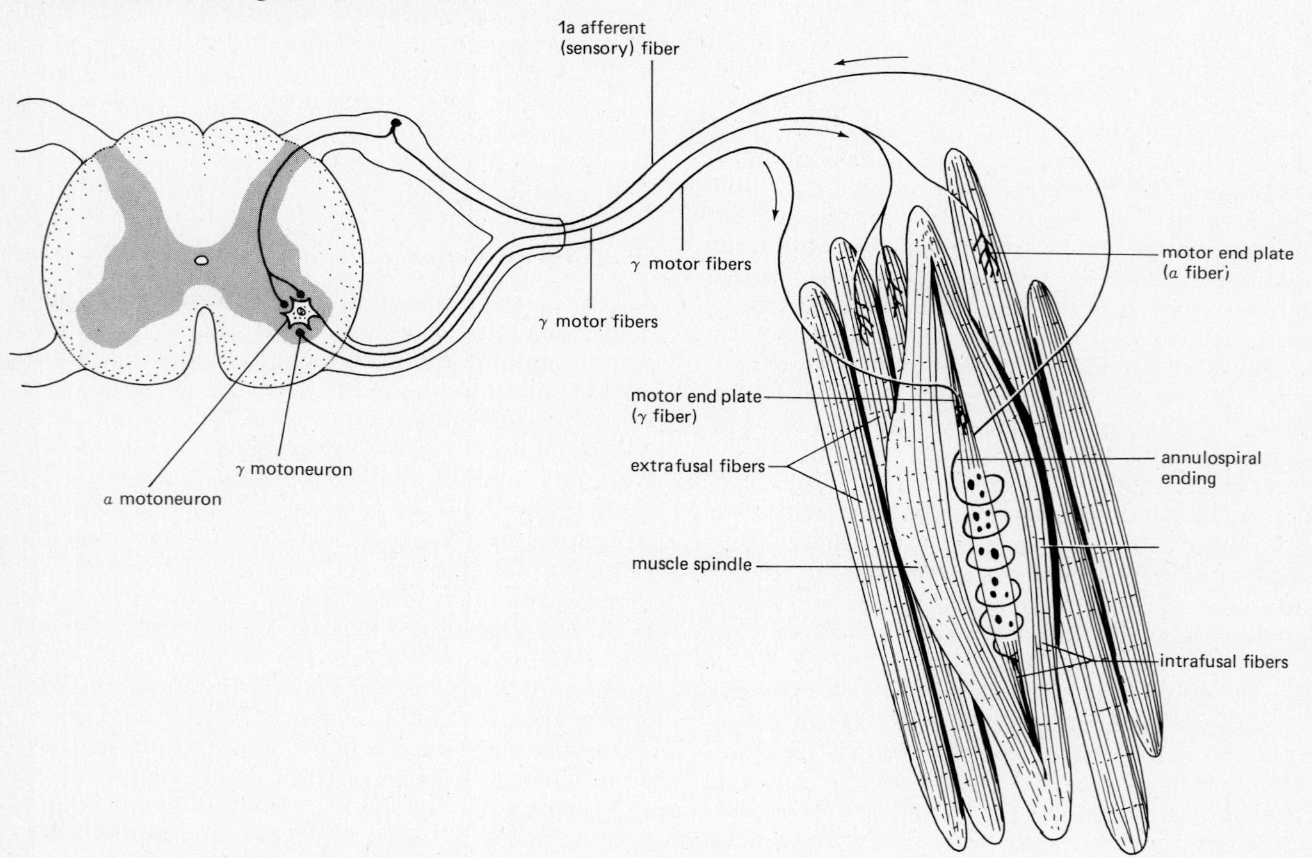

spindle-shaped capsule filled with lymph. The muscle spindle is an extremely sensitive device for delivering information about the length of a muscle and the rate of change in length; its intricacies will be discussed in greater detail in Chapter 23.

The afferent nerves of the muscle spindle can be stimulated in two ways: peripherally, through stretching the muscle; and centrally through higher levels of the nervous system.

Peripheral control

The normal stretch placed on extensor muscles is caused by gravity. As the twitch fibers are stretched, so are the parallel intrafusal fibers. The spirals of the annulospiral nerve endings are pulled apart. They respond by firing impulses which reach the ventral horn motor cells by a monosynaptic pathway and cause the motoneurons to fire, which results in the contraction of the twitch fibers. This shortens the muscle and relieves the stretch on the muscle spindles, which become quiescent. As a result, the muscle as a whole relaxes, to be stretched again by gravity. The whole cycle recommences, keeping the extensor muscles of the body in a constant state of *tonic contraction* as they act against gravity to keep the body upright. In bipeds, such as humans, the antigravity muscles are the *muscles of the trunk* and the *extensor muscles* of the leg, which keep the knee joint in a locked position. The antigravity muscles of the sloth, which spends its life hanging upside down from a branch, are the flexor muscles of its four limbs.

Central control

The control over the muscle spindles is too accurately tuned to be merely a reaction to gravity. The degree to which the 1a afferents are stimulated is controlled centrally through the gamma efferents, which in turn are regulated through higher motor centers in the midbrain, cerebellum, and cerebral cortex, in response to information coming in from the muscle spindles and other peripheral receptors. The integrated regulation of muscle contraction is responsible for the resulting smoothness of the contraction, a topic that is discussed in detail in Chapter 23.

The Flexor Reflex

This familiar reflex is a rapid, unconscious withdrawal from an unpleasant or harmful stimulus and results from the contraction of the flexor muscles, pulling the stimulated area away from the stimulus. A simple example is stepping on something sharp: you withdraw your leg by contracting your flexor muscles. This is the result of stimulation of many touch and pain receptors which branch out profusely to reach the motoneurons through interneurons. Figure 7-23 shows a simplified version of a flexor reflex pathway. At the same time, tract fibers are stimulated and conduct impulses

Figure 7-23. **The flexor reflex is a three-neuron, disynaptic pathway. Stimulating the pain receptors in the skin causes nerve impulses to travel along the sensory neuron, which synapses with an interneuron in the dorsal horn. The interneuron then stimulates the alpha motoneuron in the ventral horn, which results in contraction of the twitch muscle fibers. All these synapses are excitatory. Note the pathway by which pain sensation travels to the brain.**

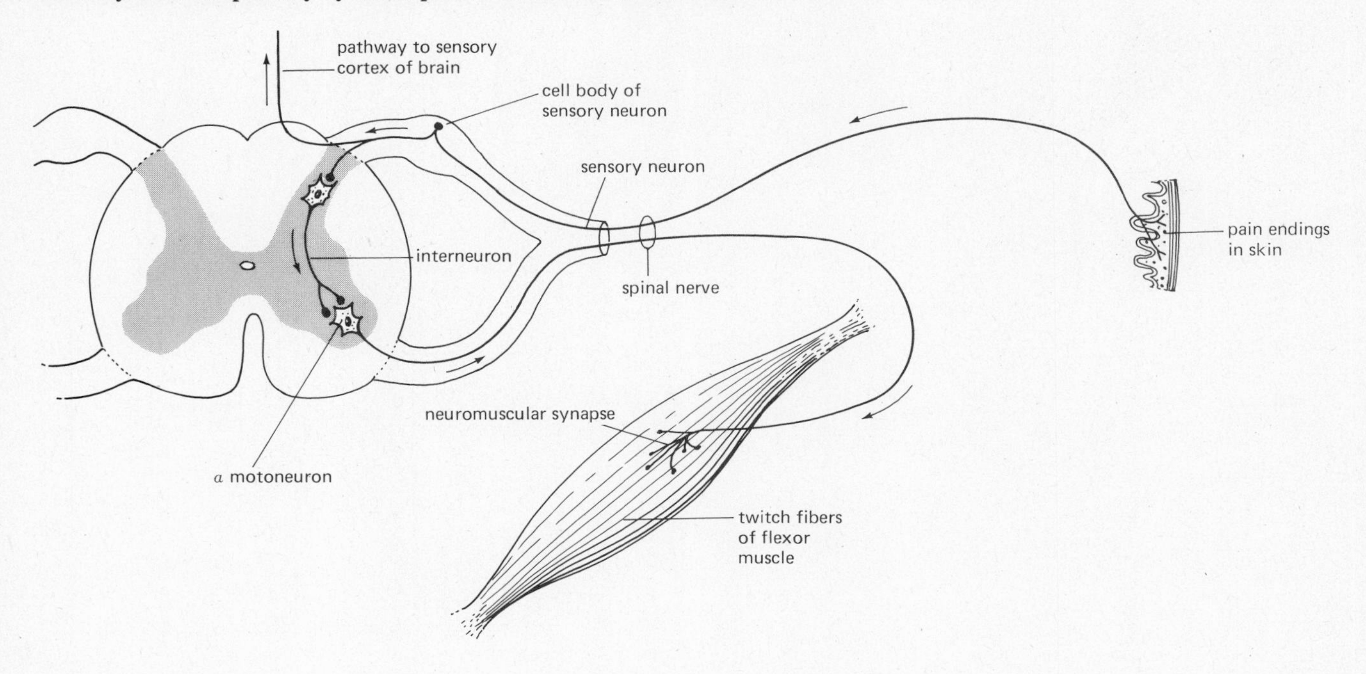

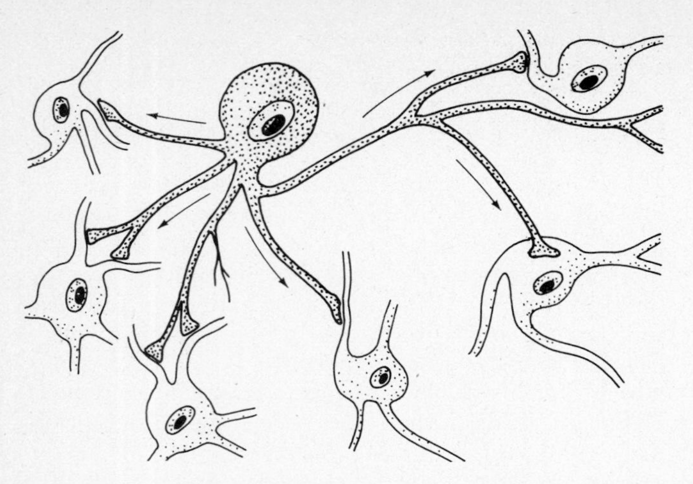

Figure 7-24. **Divergence of impulses from one neuron to many other neurons. All impulses from one neuron are either inhibitory or excitatory.**

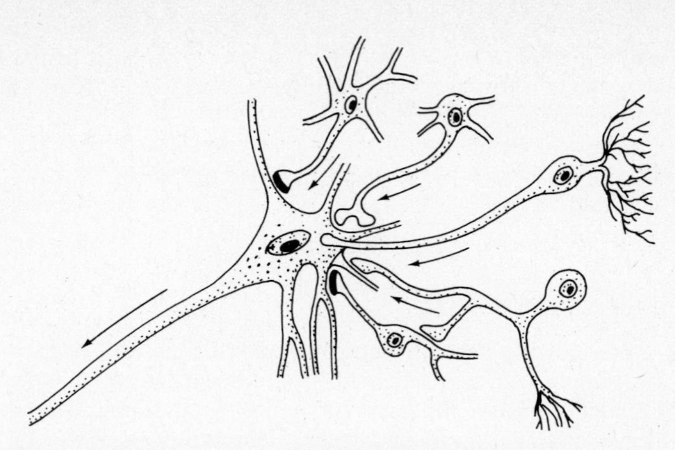

Figure 7-25. **Convergence of impulses from many neurons on to one neuron. Some of these may be excitatory (open nerve terminals), whereas others may be inhibitory (solid nerve terminals).**

Figure 7-26. **Reciprocal innervation and the crossed extensor reflex. On the right side of the diagram, an ipsilateral flexor reflex is shown. Note that stimulation of the sensory neuron results in simultaneous stimulation of the right flexor motoneuron (F) via interneuron 1, and inhibition of the right extensor motoneuron (E) via interneuron 2. On the left side of the diagram, the crossed extensor reflex is shown. The same sensory neuron now stimulates interneuron 3, which crosses to the other side of the spinal cord to stimulate the left extensor motoneuron. Interneuron 3 also synapses with an inhibitory interneuron 4, which inhibits the left flexor motoneuron. Inhibitory interneurons are shown as dark structures.**

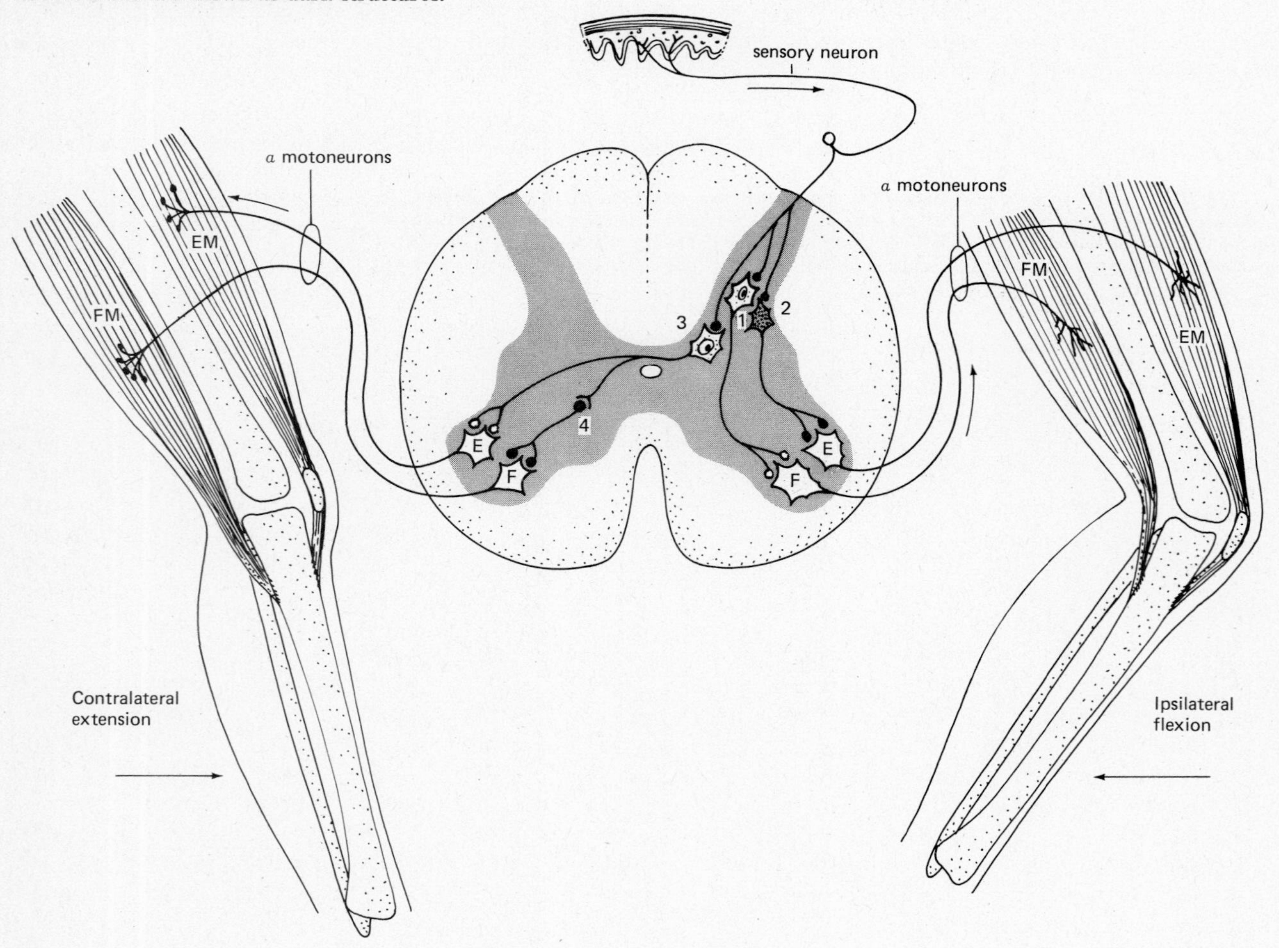

to the sensory cortex, so that one is aware of the pain a fraction of a second after the withdrawal has occurred. This time difference is caused by the slower conduction of the thin, myelinated fibers along which pain impulses travel to the brain. There are also several synapses en route, which delay the impulses further.

Concepts of Reflex Pathways

These two relatively simple reflexes illustrate several important concepts, for the description of which Sherrington received the Nobel Prize in 1932.

Divergence

One afferent fiber coming into the spinal cord delivers its message to many pathways, activating many nerve cells (Fig. 7-24).

Convergence

Each neuron in the CNS is the site of convergence for many pathways. For example, the motoneurons involved in the stretch reflex receive impulses from many other receptors, as well as from higher levels in the CNS (Fig. 7-25).

Reciprocal innervation

For movement to occur at a joint, one muscle must be stimulated to contract while its antagonistic muscle is inhibited. In the stretch reflex, when the afferent fiber from the muscle spindle is excited, it not only excites its own motoneuron (E) but sends out a branch to an interneuron. This interneuron sends its axon to the antagonist motoneuron (F) and forms inhibitory synapses on it. There is a similar neural pathway resulting in the inhibition of the extensor muscles when the flexors are stimulated (Fig. 7-26).

Crossed extensor reflex

This is a logical extension of the principles already discussed. If you step on a sharp nail, the immediate flexor withdrawal is compensated for by the extension of the opposite limb, which takes the weight of the body and prevents it from falling. This shift in equilibrium is due to the contraction of the flexors on the stimulated (ipsilateral) side, while the extensors contract on the compensating (contralateral) side, with appropriate inhibition of the antagonistic muscles of both sides (Fig. 7-26).

These reflexes form the units from which our postural and locomotor reflexes and patterns of movement are built. They are discussed in more detail in Chapter 23. At the moment, our interest is concentrated on the synaptic mechanisms involved in the stimulation and inhibition of neurons in the central nervous system.

Excitatory Synaptic Action

In the discussion of the stretch reflex, it was claimed that motoneuron E is excited to fire, whereas motoneuron F is inhibited. What evidence do we have to support this? In the early 1950s Eccles and his colleagues succeeded in inserting an extremely fine micropipette, the tip of which was only 0.5 μm in diameter, into the soma of a motoneuron (diameter about 70 μm). The delicacy of this technique can be appreciated when one remembers that the diameter of the squid giant axon is 500 μ. This micropipette is filled with 3 M KCl, an electrically conducting salt solution. The cell membrane seals itself around the glass, and the impaled cell functions normally for hours. It is then possible to measure the potential changes across the surface membrane as a result of synaptic action, a much more difficult task than measuring these changes at the neuromuscular junction.

If a group of 1a afferent fibers from a muscle spindle is electrically stimulated by stimuli of gradually increasing strength, more and more nerve fibers fire single impulses up the spinal cord on to the motoneuron. Figure 7-27 shows the *depolarization* of the motoneuron: the depolarization rises rapidly and decays slowly, looking very much like the prolonged end-plate potentials produced by neuromuscular synapses. These potentials in the depolarizing direction are called *excitatory postsynaptic potentials* (EPSPs) and like EPPs are the result of the ionic changes induced on the postsynaptic membrane by the excitatory transmitter. The events that occur here are essentially the same as at the neuromuscular synapse; that is, the nerve terminals are depolarized and the calcium gates are opened, resulting in the release of transmitter; the transmitter is briefly bound to the steric receptor, opening the sodium and potassium gates and depolarizing the postsynaptic membrane.

Not much is known about the excitatory transmitters of the central nervous system. Acetylcholine certainly excites some central synapses; glutamate and the catecholamines are probably excitatory in action at some other synapses.

Coding of information

As the excitatory transmitter is released in quanta, the resulting EPSP is graded in size, depending upon the number of quanta of transmitter binding and briefly distorting the postsynaptic receptors. The additive effect of increased amounts of transmitter may be achieved by either spatial or temporal summation.

1. *Spatial summation.* The increase in the size of the EPSP shown in Fig. 7-27 is caused by the convergence onto the motoneuron of impulses from more excited 1a afferents. As the stimulus increases in strength it reaches the threshold of more 1a afferents, exciting them to fire. This summated effectiveness of impulses from *several converging neurons* is called *spatial summation.* When the EPSP reaches a critical level as a result of this summation, the motoneuron fires an action potential. The critical level in the CNS varies with different types of neurons from about 10 to 20 mV,

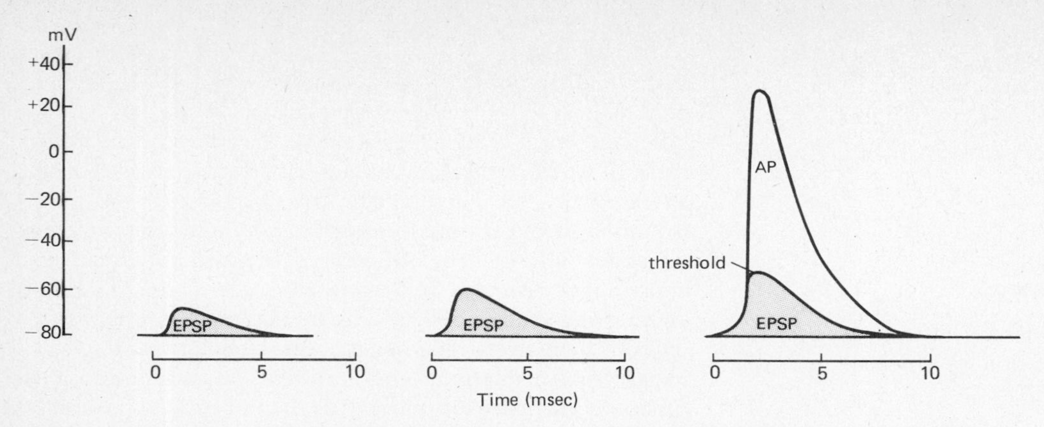

Figure 7-27. **Graded depolarization of the motoneuron results in excitatory postsynaptic potentials (EPSP) that summate with increasing strength of stimulation until the threshold of the motoneuron is reached and the all-or-none action potential (AP) is evoked.**

considerably lower than at the neuromuscular synapse. The action potential is, of course, all or nothing in amplitude and travels along all the branches of the axon to the synapses on the muscle fibers that make up the motor unit.

It is perhaps another example of biological efficiency that on an average only one quantum of transmitter is liberated per impulse at an excitatory synapse on a motoneuron, as compared to 200 to 300 at the neuromuscular synapse, for there is a tremendous convergence of synapses on the motoneuron, permitting the summation of their individual excitatory effects within a fraction of a millisecond. Because inhibitory effects can also be summated, and because the neuron will respond according to the algebraic sum of excitation and inhibition, the central synapse allows tremendous flexibility in contrast to the "on" or "off" response of the neuromuscular junction.

2. *Temporal summation.* The pattern of firing in the nervous system is not usually a single synchronized volley of impulses but rather a sustained, repetitive discharge. The depolarization caused by *repetitive* impulses is very effectively summated to fire a neuron: this is temporal summation and permits the *coding of information* concerning the strength of the stimulus. As the action potential is all-or-nothing, the only language available to the neuron to indicate the strength of the stimulus is a change in the frequency of discharge: the stronger the stimulus, the more rapidly the neuron fires (Fig. 7-28). Many neurons in the CNS can be driven to fire at 500 to 1000 impulses per second.

Figure 7-28. **Increasing the strength of the stimulus increases the frequency of firing of a neuron. Within a certain range, this increase is linear. The broken line indicates the threshold of excitation of the neuron.**

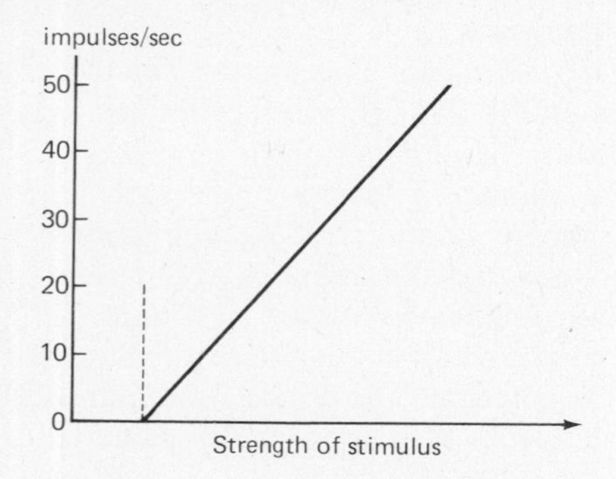

In the integrated nervous system, a strong stimulus will excite more nerve fibers and cause them to fire at higher frequencies than will a weak stimulus, so that both spatial and temporal summation are involved simultaneously to regulate the rate of firing of neurons. Both spatial and temporal summation depend upon a rapid accumulation of transmitter from the presynaptic nerve terminals; this must occur within a fraction of a millisecond, so that transmitter accumulation is faster than transmitter inactivation.

Inhibitory Synaptic Action

INHIBITORY POSTSYNAPTIC POTENTIALS

Using the same careful micropipette techniques they had used to show that stimulation of the 1a afferents from the muscle spindle caused a depolarization and consequent stimulation of the extensor motoneuron (E), Eccles and his co-workers studied the charge on the membrane of the motoneuron F of the antagonistic muscle. Figure 7-29 shows that following graded stimulation of these afferent fibers, there is an increased charge, or *hyperpolarization,* on the surface of the F motoneuron. This hyperpolarization is the *inhibitory postsynaptic potential* (IPSP). Like the EPSP, it is variable in size, depending on the number of converging inhibitory synapses. Unlike the EPSP, it cannot give rise to an action potential but slowly decays to restore the membrane potential to its initial level.

In terms of synaptic connections on the motoneuron, inhibitory synapses appear to be concentrated on the soma, whereas the excitatory synapses are on the dendrites. The actual charge across the neuronal membrane will be an arithmetical balance between the graded IPSPs and EPSPs,

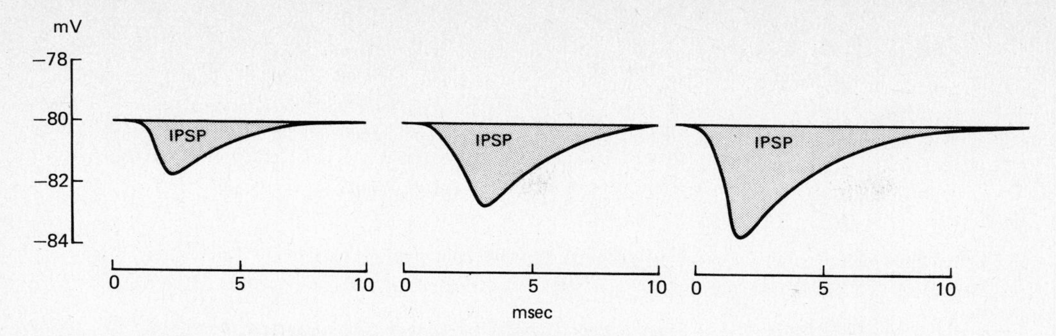

Figure 7-29. **Hyperpolarization of the postsynaptic neuronal membrane increases with increased amounts of inhibitory transmitter to give rise to the graded inhibitory postsynaptic potential (IPSP). The time scale indicates the duration of stimulation of the inhibitory presynaptic neuron.**

and will determine whether the motoneuron will fire or not (Fig. 7-30). Eccles (1957) postulates that all inhibitory pathways in the mammalian central nervous system must include an interneuron that produces an inhibitory transmitter. Because all afferent fibers entering the spinal cord appear to be excitatory in their action, an interneuron is essential for central inhibition. Using the stretch reflex as an example, the monosynaptic pathway which excites motoneuron E must be enlarged to include a disynaptic pathway, by way of an interneuron, in order to inhibit motoneuron F (Fig. 7-26). This postulate is strongly based on the early ideas of Dale (1936), who suggested that *only one type of chemical transmitter is produced at the terminals of a single neuron*. Otherwise we would have to assume that the same transmitter would stimulate E and inhibit F during the stretch reflex, then reverse these effects during flexion, a highly improbable assumption.

Ionic mechanisms of postsynaptic inhibitory synapses

The inhibitory transmitters in the mammalian central nervous system are the amino acid *glycine* and an amino acid derivative, *gamma aminobutyric acid* (GABA). They are probably packaged in the flattened vesicles characteristic of inhibitory synapses and released in a quantal manner into the synaptic cleft to cause hyperpolarization of the postsynaptic membrane. Glycine inhibition has been demonstrated mainly in the spinal cord, although glycine is also found in high concentrations in the brain. Gamma aminobutyric acid is concentrated in inhibitory axons in the spinal cord and in the brain, especially in the cerebellum. The structural formula for GABA is

$$HOOC—CH_2—CH_2—CH_2—NH_2$$

The structural formula for glycine is

$$CH_2(NH_2)COOH$$

Until recently, it had been thought that Cl^- ions were in equilibrium across the membrane and that inhibition involved the opening of both Cl^- and K^+ gates in order to achieve the hyperpolarization of the IPSP. However, the chloride pump normally pumps Cl^- ions outward across the membrane, reducing the internal concentration of chloride ions $(Cl^-)_i$ and so maintaining the equilibrium potential at about −80 mV, or at least 10 mV in the hyperpolarizing direction.

The inhibitory transmitter attaches to a receptor site on

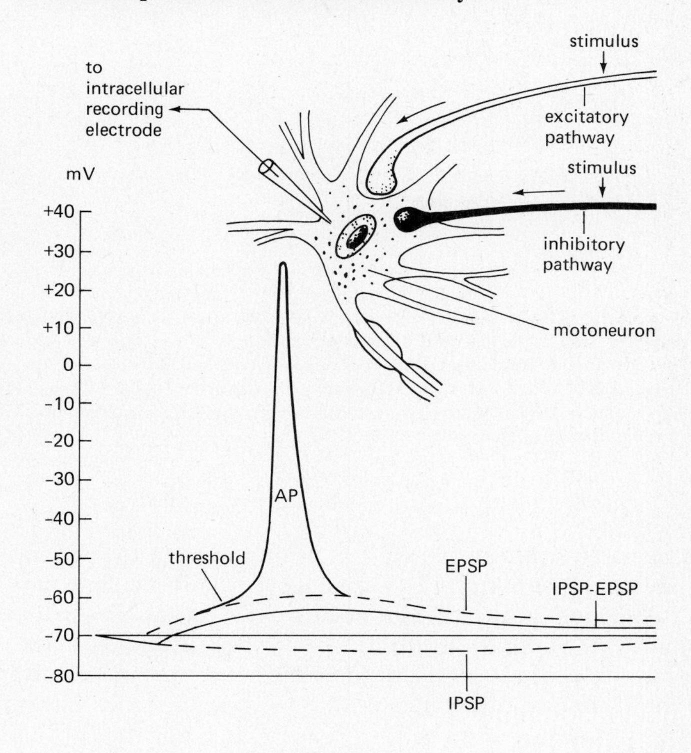

Figure 7-30. **The charge across the neuronal membrane is an arithmetical balance between the graded IPSPs and the EPSPs (IPSP − EPSP). In this figure, the EPSP alone evokes an AP, but when an IPSP is simultaneously produced the resultant charge is below threshold and no AP results. The inset in the upper right represents the experimental arrangement: the motoneuron is stimulated through excitatory and inhibitory neurons and the membrane potential recorded intracellularly.**

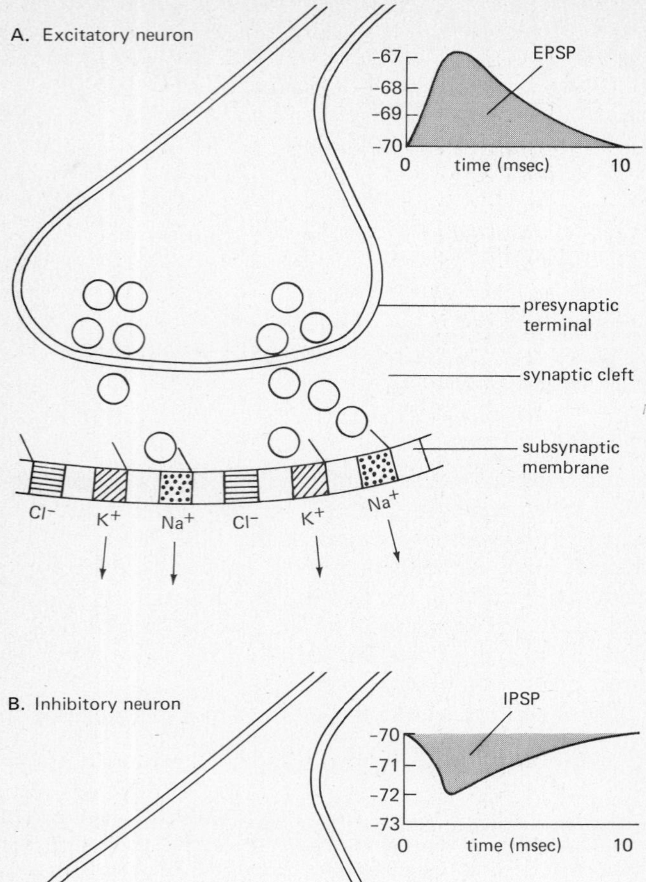

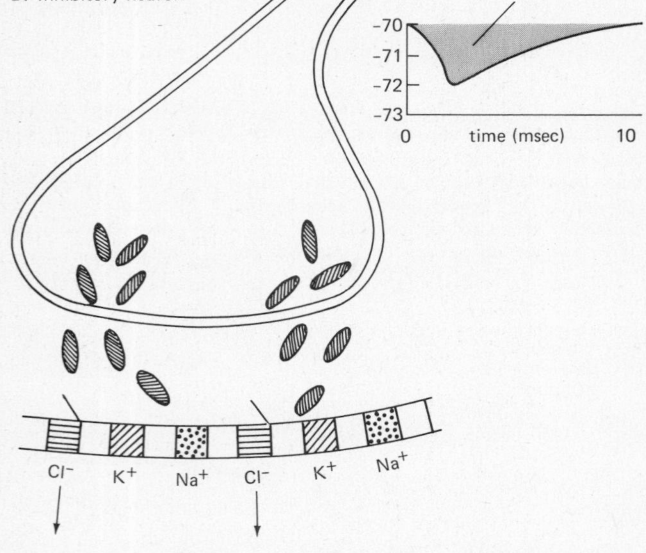

Figure 7-31. **The excitatory transmitter (open circles) in (A) opens the Na^+ and K^+ gates and associated channels in the subsynaptic membrane, depolarizing the membrane and causing current to flow inward (EPSP). In (B), the inhibitory transmitter (solid ellipses) opens only the Cl^- gates and channels and the current flows out, hyperpolarizing the membrane (IPSP).**

the postsynaptic membrane and distorts it briefly to open an ionic gate, permitting Cl^- ions to move by diffusion into the opened ionic channel. This results in the subsynaptic current that generates the IPSP (Fig. 7-31). Even when the chloride pump is specifically inhibited by ammonium ions, the inhibitory transmitter is able to open the ionic gates.

The ionic channels for inhibition are surprisingly nonspecific, depending purely on pore size: all anions smaller than a critical size in the hydrated state (0.29 nm) pass through. Some inhibitory synapses depend on cation (K^+) flux, as does the inhibition of the heart, but most others depend on chloride movement.

Feedback inhibitory pathways

Renshaw Cell Pathway (Recurrent Inhibition). We owe practically all our fundamental information about inhibitory synapses and the significance of inhibition in the functions of the nervous system to the magnificent work and writings of the neurophysiologist Sir John Eccles (1957, 1964). One of his earliest contributions was the study of special inhibitory cells, Renshaw cells, in the ventral horn of the spinal cord. Renshaw cells are stimulated by the terminals of recurrent or collateral branches of a motoneuron axon returning to the spinal cord. The axon of the Renshaw cell, in turn, forms inhibitory synapses on groups of motoneurons (Fig. 7-32). The existence of the Renshaw cell has been demonstrated clearly, both by electrophysiological recordings and by injecting the responding cells with a dye, procion yellow, which beautifully delineates their form and synaptic contacts.

When a motoneuron fires intensely, the impulse not only travels along its main axon and branches to the muscle fibers, but also travels antidromically to excite the Renshaw cells. Renshaw cells must be excited by a large number of motoneurons before they are activated. Acetylcholine is the transmitter at all these motoneuron terminals (*Dale's principle*).

The response of the activated Renshaw cell is to fire with a long burst of high-frequency impulses, liberating the inhibitory transmitter glycine, and resulting in the formation of an IPSP on the motoneuron. As a result, the more intensely a motoneuron fires, the more it turns itself off by negative feedback through the Renshaw cells (Fig. 7-32). This eliminates any weakly discharging motoneurons in the pool, whereas more strongly firing ones continue to respond, thus sharpening the selected movement.

Pharmacological tests also prove that different transmitters are required for inhibition and excitation. Curarelike compounds, which prevent the action of acetylcholine, prevent the firing of the Renshaw cell following motoneuron stimulation, demonstrating that acetylcholine is the excitatory transmitter. Strychnine does not affect these synapses, permitting the Renshaw cell to discharge but preventing its inhibitory action on the motoneuron.

Disinhibition. This is the removal of inhibition, resulting in excitation, and can again be demonstrated using the Renshaw cell. Not only does this cell form inhibitory synapses on motoneurons, but it also synapses on inhibitory interneurons, decreasing their inhibitory activity and so facilitating excitation (Fig. 7-32).

Cerebellar Inhibitory Pathways. There are many inhibitory pathways in the CNS, but the cerebellar cortex is unique in that its entire output is inhibitory. The cerebellum is a part of the hindbrain discussed in detail in Chapter 11. The largest cells of the cerebellar cortex, the Purkyně cells, which form the only efferent pathway from the cerebellum, are inhibitory in their action and exert this inhibition on the cells in Deiter's nucleus (a collection of large nerve cells in the brain stem) and the intercerebellar nuclei. The Purkyně cells receive their excitatory input from two types of fibers, the climbing fibers and the mossy fibers, and they are inhibited by other cells in the cerebellar cortex (Fig. 7-33).

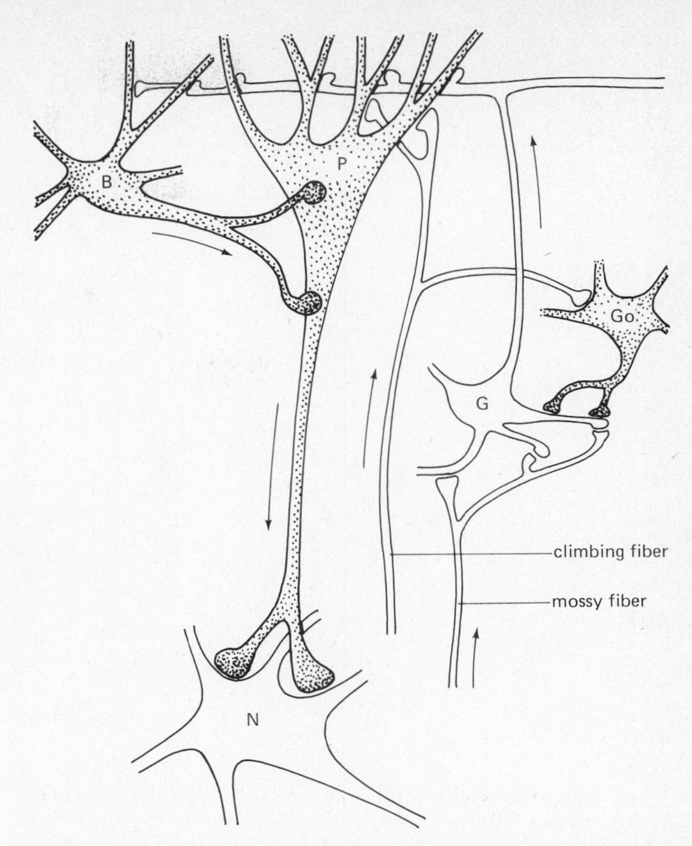

Figure 7-33. **The inhibitory output of the Purkyně cells (P) of the cerebellum is carefully processed through neuronal circuitry involving excitatory afferents and interneurons (clear structures) and inhibitory neurons and interneurons (dark cells). The afferent fibers are the mossy fibers and climbing fibers. The mossy fibers synapse with excitatory interneurons, the granule cells (G), which in turn excite the Purkyně cells and the basket cells (B). The climbing fibers excite the Purkyně cells and the Golgi cells (Go). Purkyně, basket, and Golgi cells are all inhibitory neurons; N is a large nerve cell in the brain stem.**

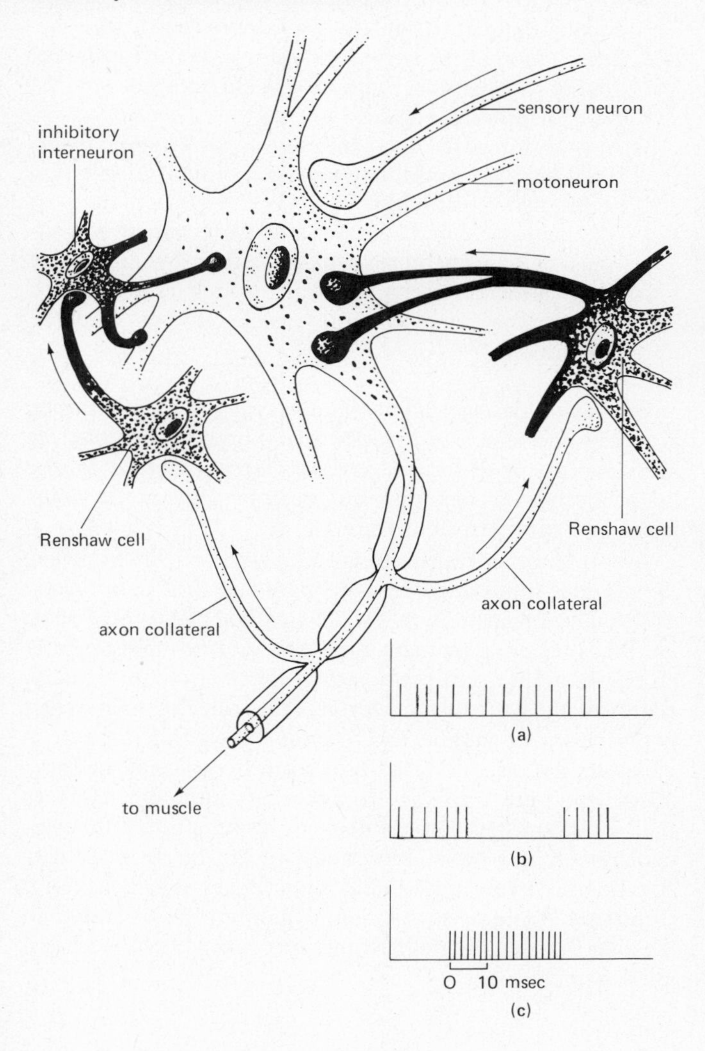

Figure 7-32. **Recurrent inhibition occurs in the spinal cord when axon collaterals of the motoneuron form excitatory synapses on Renshaw cells. The fast-firing Renshaw cells release glycine, which inhibits the motoneuron despite the excitatory action of stimulated sensory fibers. Excitatory synapses are open, inhibitory synapses are dark. (a) Firing of stimulated sensory nerve; (b) firing of motoneuron, inhibited when the Renshaw cell firing rate (c) reaches a certain level. Disinhibition is shown on the left of the drawing, where a Renshaw cell inhibits an inhibitory interneuron, decreasing inhibitory action on the motoneuron.**

The unusual circuitry of the cerebellum routes the input from the mossy fibers to the Purkyně cells through excitatory synapses involving other cerebellar cells. Any of these synapses can be blocked by inhibitory feedback circuits, including recurrent inhibition. This carefully processes the information allowed to reach the Purkyně cell, balancing the strength of the IPSPs and EPSPs on it and determining whether or not it will discharge.

Presynaptic inhibition

Presynaptic inhibition is an evolutionary development that decreases the amount of neural information bombarding excitatory synapses and protects the organism against the flood of sensory impulses received. Presynaptic inhibition is achieved by depressing the release of the excitatory transmitter from presynaptic terminals and so preventing the formation of EPSPs on the postsynaptic membrane.

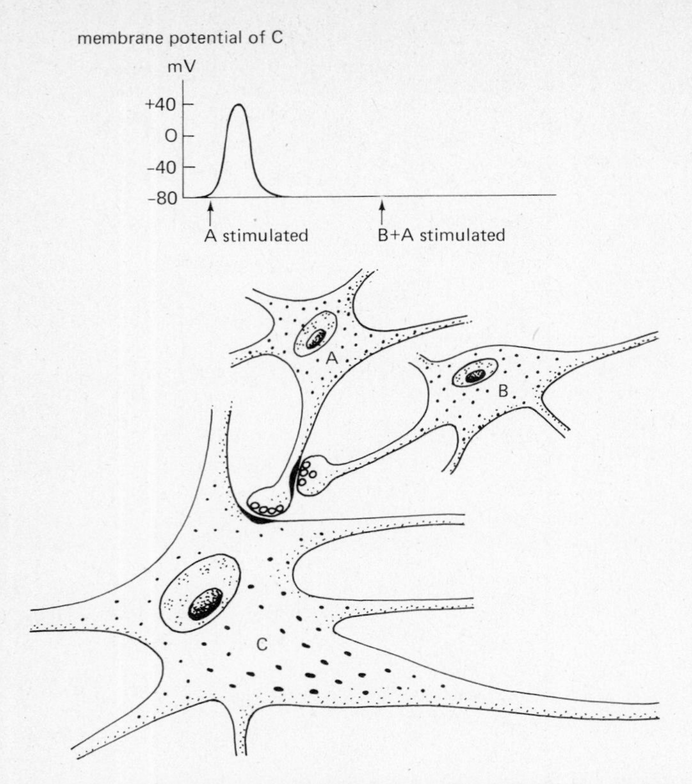

Figure 7-34. **Presynaptic inhibition occurs through the prevention of the release of an excitatory transmitter. Stimulation of cell B prevents cell A from releasing ACh. The inset shows that stimulation of cell A normally depolarizes cell C, but if B is also stimulated, no transmitter is released from A so that no membrane change occurs in C.**

Unlike postsynaptic inhibition, which involves the release of an inhibitory or hyperpolarizing transmitter from presynaptic endings, there is no hyperpolarization of the postsynaptic membrane, the electrical characteristics of which remain unaffected (Fig. 7-34).

Presynaptic inhibition is especially important at lower levels of the brain and spinal cord, where it exerts a widespread suppressor influence. It is not found at the highest levels of the mammalian brain—the cerebellar cortex and the cerebral cortex—where postsynaptic inhibition dominates.

GABA as a Depolarizing Agent in Presynaptic Inhibition. GABA is the transmitter involved in presynaptic inhibition. It depolarizes the presynaptic terminals of primary afferent fibers by opening the Cl^- gates and pumping Cl^- ions outward. This reduces the size of the presynaptic spike potentials and so reduces the output of excitatory transmitter. The equilibrium potential for this ionic mechanism is about -40 mV, which indicates that there is an inward Cl^- pump in the presynaptic fibers that causes the E_{Cl} to be about 30 mV in the depolarizing direction from the resting potential.

GABA as a Hyperpolarizing Agent in Postsynaptic Inhibition. When GABA acts as a postsynaptic inhibitory transmitter resulting in IPSPs, it does so by affecting the same ionic gates, that is, the Cl^- gates, but in this case it pumps Cl^- ions in, causing hyperpolarization. It appears that in the mammalian nervous system, a particular transmitter always acts by opening the same ionic gates, although the direction of ionic flux may vary.

Characteristics of Neurotransmitters

Many stringent criteria must be satisfied before a substance may be classified as a neurotransmitter. So stringent are the requirements that they have been fulfilled completely by only two chemicals, acetylcholine and norepinephrine. Some of these characteristics are

1. Nerves should have the enzymatic machinery needed to synthesize the substance, and it must be found in adequate quantities in the presynaptic terminals.
2. Stimulation of the nerves should release the substance, which should react with a specific receptor on the postsynaptic membrane.
3. Application of the substance to the postsynaptic membrane should result in the same response as stimulation of the nerve terminals.
4. Mechanisms should be available to terminate the actions of the chemical rapidly, either by enzymatic inactivation or by its removal through reuptake into the terminals and synaptic vesicles and/or into the circulation.

Enzymatic destruction at the postsynaptic membrane by acetylcholinesterase is the dominant process involved in acetylcholine inactivation, whereas active uptake is the major mechanism removing norepinephrine from the vicinity of the postsynaptic membrane.

Although other transmitters fulfill some of these criteria, they do not fulfill them all and so are classified temporarily as "putative transmitters." They include the amino acids GABA, glycine, glutamate, and aspartic acid. Substance P, an eleven amino acid peptide, may be the excitatory transmitter of the 1a sensory neurons. The catecholamines epinephrine and dopamine, and the amine serotonin (5 hydroxytryptamine), are classified as transmitters. Many of these transmitters are active in the brain and although much of their biochemistry and metabolism, especially of the catecholamines, has been elucidated, we are far from understanding their role either on a cellular level or in behavioral responses. Some of these transmitters will be discussed in Chapter 9, when the higher functions of the nervous system are considered.

Cited References

AXELROD, J. "Noradrenaline: Fate and Control of Its Biosynthesis." *Science* **173:** 598 (1971).

CAJAL, R. S. y. *Histologie du Systé Nerveux de l'Homme et des Vertébrés.* 2 vols. A. Maloine, Paris, 1952. Reprinted by Institute Ramón y Cajal, Madrid.

DALE, H. H., and M. VOGT. "Release of Acetylcholine at Voluntary Motor Nerve Endings." *J. Physiol.* (*London*) **86:** 353 (1936).

ECCLES, J. C. *Physiology of Nerve Cells.* The Johns Hopkins University Press, Baltimore, 1957.

ECCLES, J. C. *The Physiology of Synapses.* Springer-Verlag, Berlin, 1964.

FATT, P., and B. KATZ. "Spontaneous Subthreshold Activity at Motor Nerve Endings." *J. Physiol.* (*London*) **117:** 109 (1952).

GARDNER, D., and E. KANDEL. "Diphasic Postsynaptic Potential: A Chemical Synapse Capable of Mediating Conjoint Excitation and Inhibition." *Science* **176:** 675 (1972).

GOLGI, C. "Sur la Structure des Cellules Nerveuses." *Arch. Ital. Biol.* **30:** 60 (1898).

KATZ, B., and R. MILEDI. "The Release of Acetylcholine from Nerve Endings by Graded Electric Pulses." *Proc. Roy. Soc. London Biol.* **16:** 23 (1967).

KUFFLER, S. W., and D. YOSHIKAMI. "The Distribution of Acetylcholine Sensitivity at the Post-Synaptic Membrane of Vertebral Skeletal Twitch Muscles: Iontophoretic Mapping in the Micron Range." *J. Physiol.* (*London*) **251:** 465 (1975).

LOEWI, O. "Über humorale Übertragbarkeit der Herznervenwirkung." *Pflüg. Arch. ges. Physiol.* **189:** 239 (1921).

MCMAHAN, U. J., and S. KUFFLER. "Visual Identification of Synaptic Boutons on Living Ganglion Cells and of Varicosities in Post-Ganglionic Axons in the Heart of the Frog." *Proc. Roy. Soc. London Biol.* **181:** 421 (1971).

MILEDI, R. L. "Transmitter Release Induced by Injection of Calcium Ions into Nerve Terminals." *Proc. Roy. Soc. London Biol.* **183:** 421 (1973).

NACHMANSOHN, D. *Chemical and Molecular Basis of Nerve Activity.* Academic Press, Inc., New York, 1959.

SHERRINGTON, C. S. *Integrative Action of the Nervous System.* Yale University Press, New Haven, Conn., 1906.

VON EULER, U. S. "A Specific Sympathomimetic Ergone in Adrenergic Nerve Fibers (Sympathin) and Its Relations to Adrenaline and Noradrenaline." *Acta Physiol.* (*Scand.*) **12:** 73 (1946).

VON EULER, U. S. "Adrenal Medullary Secretion and Its Neural Control." In *Neuroendocrinology,* Vol. 2, L. Martini and W. F. Ganong, eds. Academic Press, Inc., New York, 1967, p. 283.

Additional Readings

BOOKS

BACQ, Z. M. *Chemical Transmission of Nerve Impulses: A Historical Sketch.* Pergamon Press, Inc., Elmsford, N.Y., 1975.

BENNETT, M. V. L., ed. *Synaptic Transmission and Neuronal Interaction.* Society of General Physiologists Series, Vol. 28. Raven Press, New York, 1974.

ECCLES, J. C. *The Physiology of Synapses.* Springer-Verlag, New York, Inc., New York, 1964.

HALL, Z. W., J. G. HILDEBRAND, and E. A. KRAVITZ. *Chemistry of Synaptic Transmission: Essays and Sources.* Chiron Press, Portland, Oreg., 1974. (Paperback.)

HORRIDGE, G. A. *Interneurons.* W. H. Freeman and Co., San Francisco, 1968.

HUBBARD, J. I., ed. *The Peripheral Nervous System.* Plenum Publishing Corporation, New York, 1974.

MCLENNAN, H. *Synaptic Transmission.* W. B. Saunders Company, Philadelphia, 1970.

PAPPAS, G. D., and D. P. PURPURA, eds. *Structure and Function of Synapses.* Raven Press, New York, 1972.

ROBERTS, E., T. CHADE, and D. B. TOWER, eds. *GABA in Nervous System Function.* Raven Press, New York, 1975.

The Synapse. Cold Spring Harbor Symposia on Quantitative Biology, Vol. 11. Cold Spring Harbor Laboratory, Cold Spring Harbor, N.Y., 1976.

WASER, P. G. *Cholinergic Mechanisms.* Raven Press, New York, 1975.

ARTICLES

BIRKS, R. I., H. E. HUXLEY, and B. KATZ. "The Fine Structure of the Neuromuscular Junction of the Frog." *J. Physiol.* (*London*) **150:** 134 (1960).

COUTEAUX, R. "Morphological and Cytochemical Observations on the Postsynaptic Membrane at Motor Endplates." *Exp. Cell. Res.* (*Suppl.*) **5:** 294 (1958).

DALE, H. H. "Acetylcholine as a Chemical Transmitter of the Effects of Nerve Impulses. *J. Mount Sinai Hosp.* **4**(5): 401 (1937).

DELORENZO, A. J. "Electron Microscopy: Tight Junctions in Synapses of the Chick Ciliary Ganglion." *Science* **152:** 76 (1966).

DEROBERTIS, E. D. P., A. PELLEGRAINO DE IRALDI, G. RODRIGUEZ, and C. J. GOMEZ. "On the Isolation of Nerve Endings and Synaptic Vesicles." *J. Biophys. Biochem. Cyto.* **9:** 229 (1961).

ECCLES, J. C. "Ionic Mechanism of Postsynaptic Inhibition." *Science* **145:** 1140 (1964).

ECCLES, J. C. "The Synapse." *Sci. Am.,* Jan. 1965.

FATT, P., and B. KATZ. "Spontaneous Subthreshold Activity at Motor Nerve Endings." *J. Physiol.* (*London*) **115:** 109 (1952).

FLOREY, E. "Comparative Physiology: Transmitter Substances." *Ann. Rev. Physiol.* **23:** 501 (1961).

KATZ, B. "Microphysiology of the Neuromuscular Junction." *Johns Hopkins Hosp. Bull.* **102:** 275 (1958).

KATZ, B., and S. THESLOFF. "On the Factors Which Determine the Amplitude of the 'Miniature End-Plate Potentials.'" *J. Physiol.* (*London*) **137:** 267 (1957).

KRNJEVIC, K. "Chemical Nature of Synaptic Transmission in Vertebrates." *Physiol. Rev.* **54:** 418 (1974).

NACHMANSOHN, D. "Role of Acetylcholine in Neuromuscular Transmission." *Ann. N.Y. Acad. Sci.* **135:** 136 (1966).

NASTUK, W. L. "Some Ionic Factors That Influence the Action of Acetylcholine at the Muscle End-Plate Membrane." *Ann. N.Y. Acad. Sci.* **81:** 317 (1959).

PALAY, S. L. "The Morphology of Synapses in the Central Nervous System." *Exp. Cell. Res.* (*Suppl.*) **5:** 275 (1958).

THESLEFF, S. "The Mode of Neuromuscular Block Caused by Acetylcholine, Nicotine, Decamethonium, and Succinylcholine." *Acta Physiol.* (*Scand.*) **34:** 218 (1955).

VON EULER, U. S. "Adrenergic Neurotransmitter Function." *Science* **173:** 202 (1971).

WHITTAKER, V. P., and E. G. GRAY, "The Synapse: Biology and Morphology." *Brit. Med. Bull.* **18:** 223 (1962).

WILSON, V. "Inhibition in the Central Nervous System." *Sci. Am.,* May 1966.

Chapter 8

Integrated Nervous Regulation

Behold the mighty dinosaur
Famous in prehistoric lore,
Not only for his weight and length
But for his intellectual strength.
You will observe by these remains
The creature had two sets of brains—
One on his head (the usual place),
The other at his spinal base.
Thus he could reason "a priori"
As well as "a posteriori."
No problem bothered him a bit:
He made both head and tail of it.
So wise he was, so wise and solemn
Each thought filled just a spinal column.
If one brain found the pressure strong
It passed a few ideas along;
If something slipped his forward mind
'Twas rescued by the one behind.
And if in error he was caught
He had a saving afterthought,
As he thought twice before he spoke
He had no judgments to revoke;
For he could think without congestion,
Upon both sides of every question.

Bert Leston Taylor, "The Dinosaur"

THE ULTIMATE UNDERSTANDING of the human brain is an extraordinarily complex and ambitious goal, which is far beyond our reach at the moment. We can strive, however, to reach lesser goals on the way and to synthesize and integrate information derived from the many fields of investigation that make up neurobiology. We have fascinating leads from areas such as neuroembryology, neurophysiology, neurochemistry, neuroendocrinology and neurocybernetics. A complete barrage of investigative tools is being used to try to unravel some of the mysteries of the brain, that organ which is infinitely more than the sum of its parts, with which we react to both our internal and external environment, with which we think, learn, remember, and create—with which, above all, we are aware of ourselves and our world.

This chapter will provide a brief introduction to evolutionary trends in the nervous system of significance to its function. It will also describe the basic organization of the nervous system, introducing functional subdivisions that will be described in greater detail in later chapters.

8–1 Evolutionary Trends in the Nervous System

Once the *specialized nerve cell* had evolved, its pattern of connections to other cells, and its mode of information transmission to these connecting cells, assumed the utmost functional importance. The development of *complex chemical synapses* bestowed enormous variability and selectivity of response (Chap. 7, and also Chaps. 22 and 23). *Saltatory conduction,* made possible by the development of myelinated

Figure 8-1. **The brain, spinal cord, and spinal nerves.**

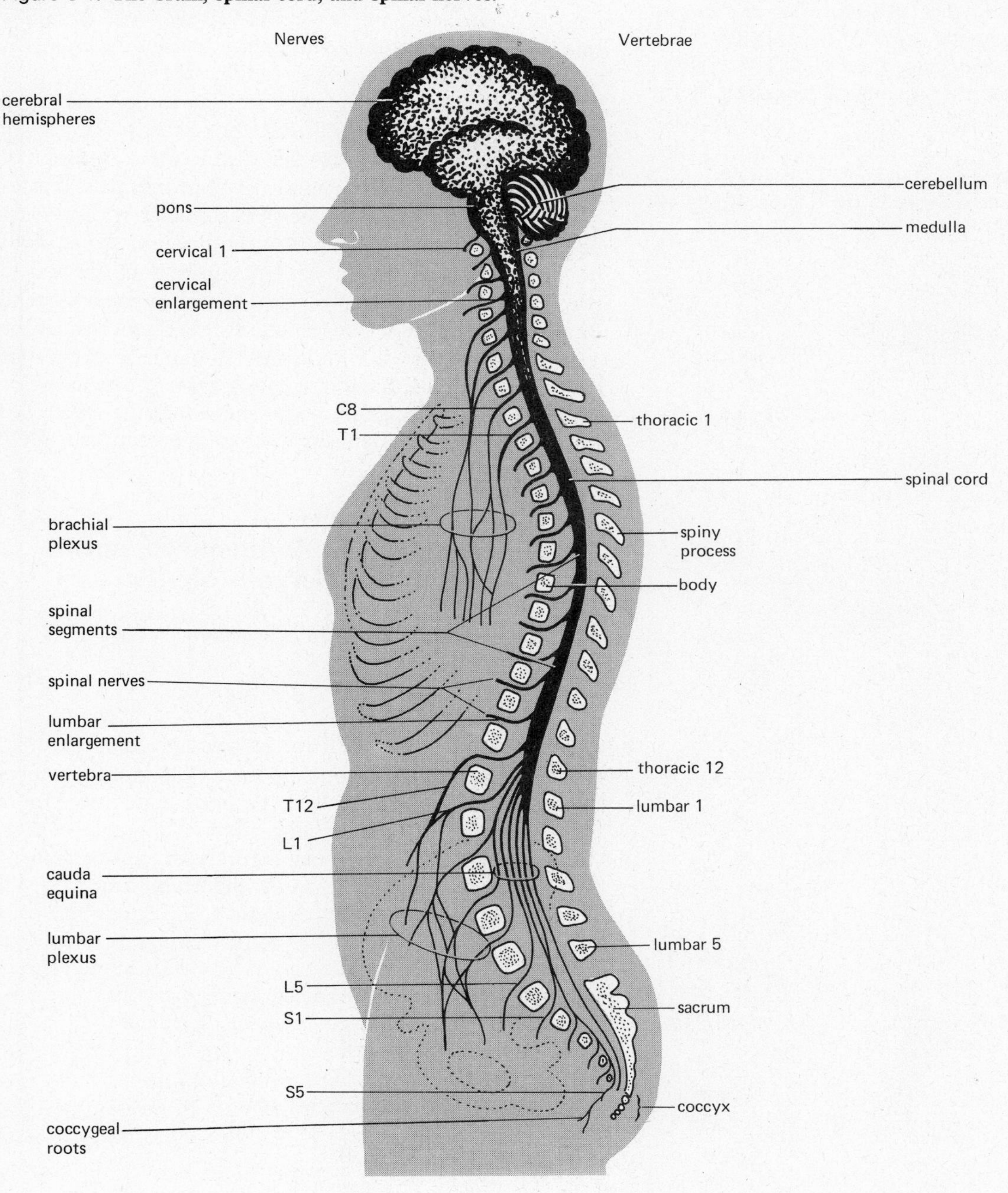

nerve fibers, greatly accelerated the speed by which nerve impulses can travel, permitting the extremely rapid response to a stimulus characteristic of higher organisms.

Centralization and *cephalization* are two additional evolutionary advances in the nervous system of considerable significance to the ability of the animal to regulate and integrate its responses to internal and external stimulation.

Centralization and Cephalization

Central and peripheral neurons

As nerve pathways become more complex, there is an evolutionary tendency for nerve cells to move inward to a more protected position in the body, forming the *central nervous system,* which is made up of the elongated *spinal cord* and its bulbous anterior end, the *brain.* These structures are protected further by three membranes and by the bony vertebral column and skull that encase them (Fig. 8-1).

Some neurons remain *peripheral,* forming groups of nerve cells outside the CNS, called *ganglia.* An exception to this is found in the sensory cells of the retina and the olfactory mucosa, but these cells are really extensions of the forebrain. Although the ganglionic cells are outside the central nervous system, they lie very close to it or in very protected regions.

Figure 8-2. **Comparison of the brain of the dogfish, sheep, and human. Note the tremendous increase in the relative size of the cerebral hemispheres (forebrain in the dogfish).**

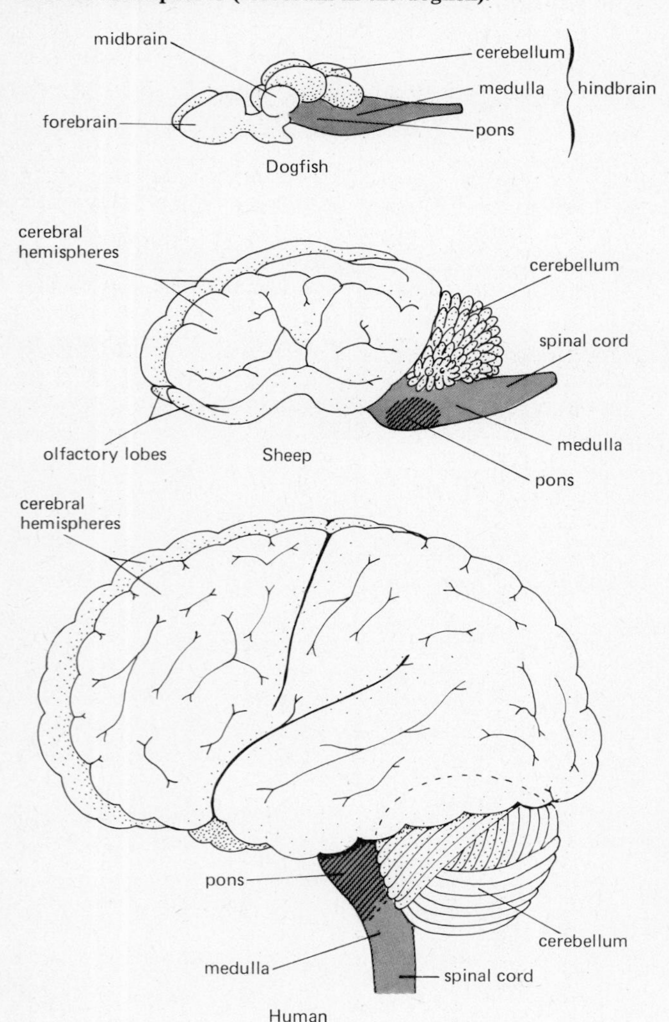

Cephalization

The anterior, or cephalic, end of most vertebrates develops more rapidly than the posterior structures. This process of cephalization is seen clearly in the development of the nervous system, in which the brain grows more rapidly than the spinal cord and exerts a progressively more dominant role over spinal functions. The anterior part of the brain, the cerebral hemispheres, grows so rapidly that it doubles over and conceals much of the brain stem. This is most marked in primates and reaches a peak in the brain of humans (Fig. 8-2). Yet it is more than mere size that distinguishes the brain of humans from that of other highly developed animals. It is far beyond our ability to relate the characteristics of consciousness, personality, and thought to the material structure of the brain. We shall have to confine ourselves in this text to the physiological aspects of the central nervous system. Remembering these severe limitations, we shall consider the central nervous system with its untold numbers[1] of neurons as a giant, computerized, integrating center, amplifying or filtering out information and channeling the processed results as efferent impulses along the appropriate nerve pathways.

8–2 Development, Maintenance, and Aging of the Nervous System

Factors Influencing Development and Maintenance

Genetic control

Neurons are derived from embryonic ectoderm that thickens, invaginates, and finally forms a long hollow tube, the *neural tube.* The cephalic or head end grows more rapidly and forms the bulging brain while the smaller, caudal portion forms the spinal cord; the process of cephalization is quite clear (Fig. 8-3). As the neural tube closes, it separates from the overlying epidermis, and ectodermal cells on either side are squeezed away, forming a long, thin column of cells, the *neural crest.* These cells migrate laterally to give rise to the cells of the sympathetic and parasympathetic ganglia, the sensory cells of the spinal ganglia, the Schwann cells, and some of the supporting cells of the nervous system. The pigment-containing cells of the skin and eyes, the melanocytes, are also of neural crest derivation.

[1] One estimate is 50–100 billion neurons.

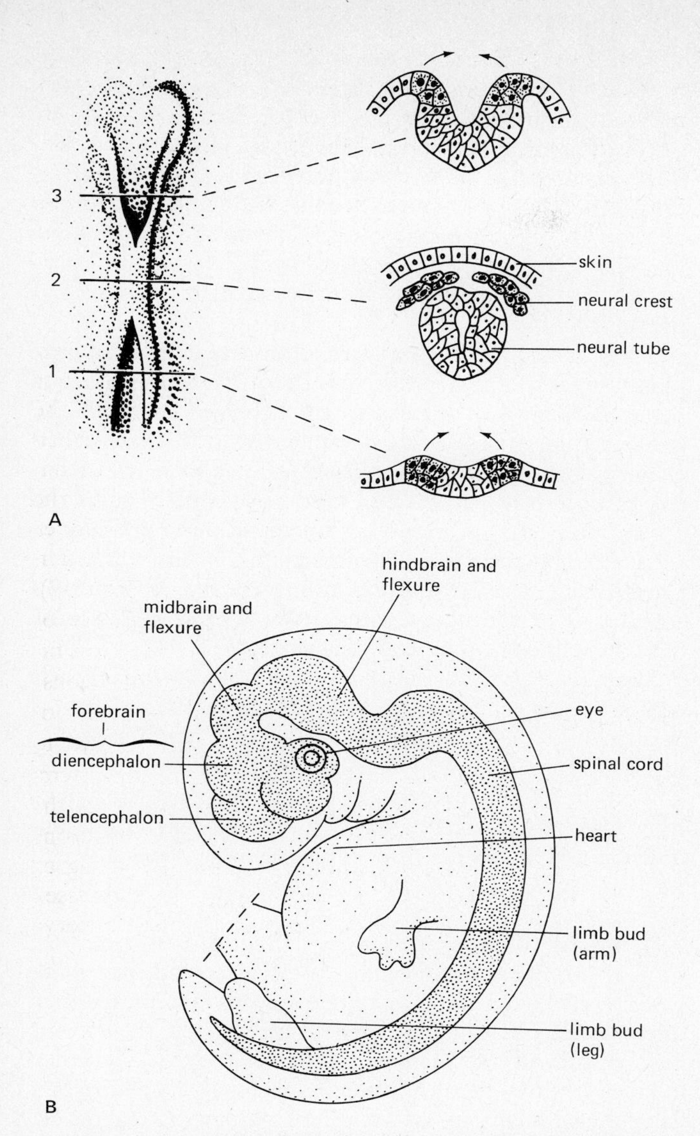

Figure 8-3 **(A) Stages in the closure of the neural tube. Left: (1) open posterior end, (2) closed middle section, and (3) partially open anterior end. On the right are cross sections of the neural tube at each stage. (B) Development of the neural tube into the brain and spinal cord, showing the marked development of the brain, and the hindbrain and midbrain flexures.**

The cells of the neural tube divide mitotically to produce an enormous population of primitive germinal cells, which divide to form neuroblasts. These cells have lost forever their ability to divide and develop eventually into nerve cells; they possess in their genome the necessary information to form the distinctive and highly organized structure characteristic of many parts of the brain, especially the cerebellum. Eccles and his associates have made a brilliant study of the organization of the cerebellum, that part of the brain responsible for equilibrium and the integration of movement. In his book *The Understanding of the Brain* (1977), he makes some general statements about neurogenesis:

1. "At no stage in development are the neurons of the brain connected together in random order.
2. The organization of the brain is highly specific, involving not only connections between specific neurons but also the number and location of synaptic knobs on that cell and the precise location of its own synaptic on other cells."

This appears to be incredibly accurate planning, and many ingenious experiments have been devised to determine how the neurons get connected and how they know where to go. At one time it was believed that use of the structures brought organization to a previously random arrangement. This is not so: the specific nerve connections have been found to be ready in their final form before use. A kitten opens its eyes only several days after birth, yet at birth the retina is already connected to the visual cortex of the brain in the detailed topographical pattern of the adult. If the kitten is kept in the dark for several weeks after birth, however, this organization is permanently destroyed.

Sperry (1944, 1945) has suggested that neurons are guided and organized through the use of intricate chemical codes under genetic control, and that, as the searching fiber tips of the growing neuron reach a choice of pathways, they are guided by a chemical touch system, which functions in three dimensions. This hypothesis is supported by experiments on fish and amphibia, in which nerve regeneration occurs readily. If the optic nerve is cut and allowed to regenerate, it will grow back to connect with the brain, making its original, point-to-point connections between the retina and the optic tectum, or lobe (Fig. 8-4).

Figure 8-4. **Point-to-point regeneration of neurons occurs in the visual system of the frog. On the right side of the diagram, the scar tissue of the cut and regenerated optic nerve is seen. The nerve fibers grow back and connect up with the same cells in the brain with which they originally synapsed.**

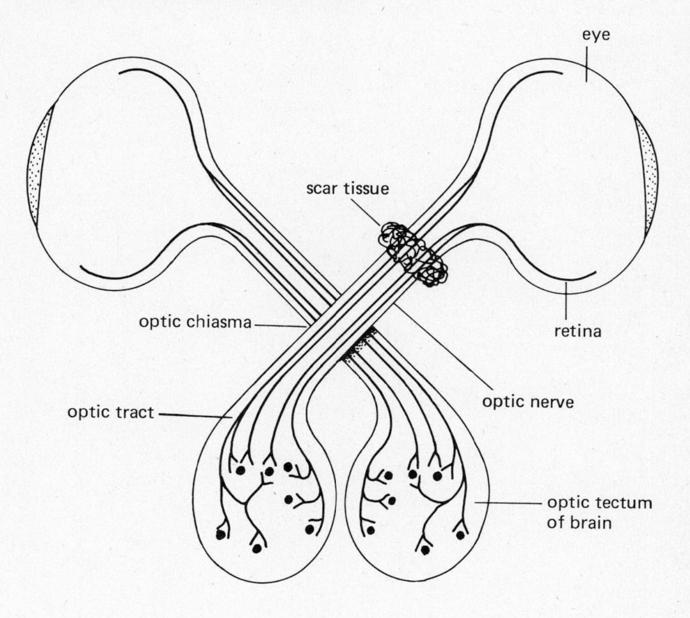

Glial cells (neuroglia)

These are the supporting cells of the central nervous system. Their name is derived from the Greek word for glue, and one of their functions is to glue the central nervous system together. They tightly surround the neurons and consequently are between the nerve cell and the capillaries (Fig. 8-5). There are several types of glial cells, and many roles have been ascribed to them, including transport of nutrients from the capillaries and wastes back to the capillaries, maintenance of the highly selective blood-brain barrier (discussed in Sec. 8-5), and formation of myelin in the central nervous system by one type of neuroglia.

During neurogenesis, the glial cells probably play an important role in guiding the neurons to their correct destination. In the cerebellum and cerebral cortex, glial cells become organized as long parallel strands at right angles to the surface of the brain. The growing nerve fibers then glide along these strands, perhaps being guided by chemical recognition. We really have no idea, however, as to what forces organize the glial cells originally.

Hormonal and other extracellular influences

Hormones. The growth of neural circuits and their final integration into complex patterns of behavior are influenced strikingly by hormones. Timing is of crucial importance. *Thyroid hormone* appears to be the most important hormonal moderator; there is a critical stage in mammalian brain development, including that of humans, during which lack of this hormone will cause irreversible damage. This is seen as severe mental retardation in cretins.

The rat has proved to be a particularly suitable experimental animal for studies of the effects of hormones on the developing nervous system, for maturation of the central nervous system, which in humans takes place *in utero,* continues after birth in the rat. Hamburgh (1971) has shown that the inability of thyroxine-deficient rats to learn is caused by a decrease in myelination of axons and a failure to develop the right synaptic connections at the proper time. He calls thyroxine the "time clock" of the developing nervous system. Thyroxine also increases protein synthesis and the metabolic rate.

Figure 8-5. **Glial cells surround the neurons in the central nervous system. The processes of the glial cells form a compact layer around the capillaries.**

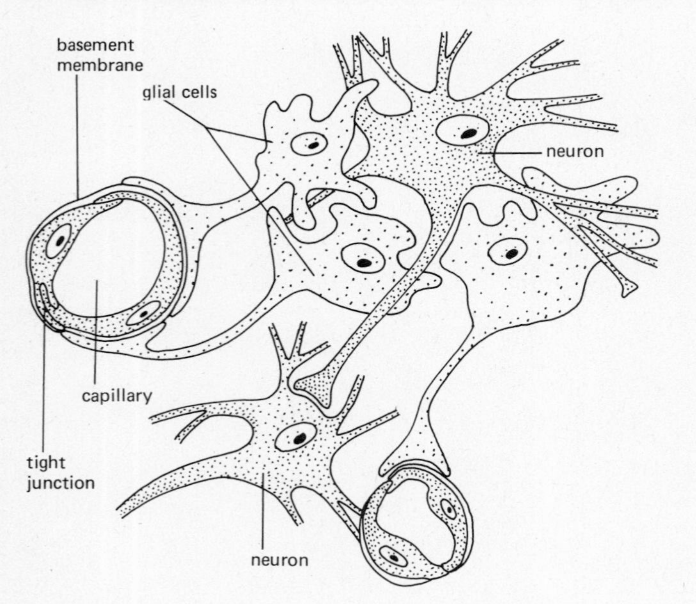

There is a similar critical period during which sex hormones organize the hypothalamus into the male pattern. If neonatal (newborn) male rats are deprived of *androgens* (male sex hormones) a few days after birth (by surgical or chemical castration) the hypothalamus never changes from the basic female pattern of cyclic activity to that of the acyclic male. Administration of male sex hormones later in life is not able to induce adequate male copulatory activity. Control rats, castrated 2 weeks after birth and subsequently supplied with androgens will display typical male sexual behavior although they will, of course, be sterile.

Similarly, a single injection of androgens during this critical period in the female will change the hypothalamus permanently to the male, acyclic type. These "androgenized" females do not ovulate and consequently are sterile. They also develop malelike copulatory and ejaculatory patterns later in life. The patterns of sexual behavior are laid down at a very specific time in neurogenesis, during which they are irrevocably affected by the presence or absence of androgens.

A number of *other hormones,* including corticosterone, growth hormone, and insulin have widespread, but more general, effects on the developing nervous system, for they affect levels of available energy supplies and the rate of metabolism. Any deficiency in *nutritional supply*—whether indirectly, through a hormonal inadequacy or directly, through malnutrition—will interfere with the normal development and function of the nervous system.

Nerve Growth Factor. Nerve growth factor (NGF) is a protein that promotes the growth, development, and maintenance of certain parts of the nervous system, mainly the sympathetic and spinal ganglia. NGF resembles insulin in both its structure and function, for it stimulates the synthetic and energy-producing activities of its target cells, just as insulin does (see Chap. 19). Levi-Montalcini (1975) has shown that NGF stimulates several metabolic pathways in the target nerve cells, acting perhaps at the level of transcription. The submaxillary (salivary) glands of male mice are the best source of NGF, producing 10 times more than the submaxillary glands of female mice and up to 1000 times more than other tissues. It is possible that the submaxillary glands store NGF as well as synthesizing it.

The sympathetic nervous system is particularly sensitive to NGF, atrophying in its absence. If antisera to NGF, which inactivate it, are administered, devastating effects on

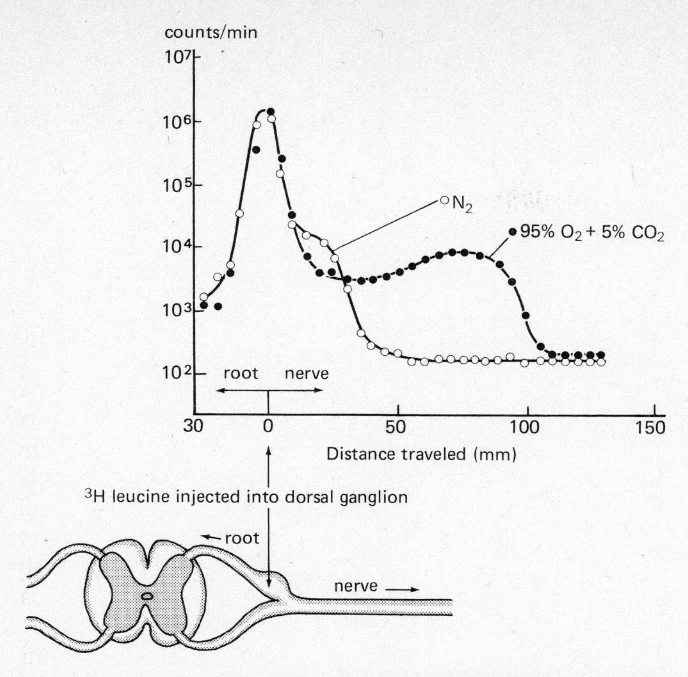

Figure 8-6. **Fast axonal transport and anoxia. Following the injection of the dorsal root ganglion with a radioactive amino acid (^{3}H-leucine), counts of the radioactivity were made under aerobic (95% O_2 + 5% CO_2) and anaerobic (N_2) conditions. The rapid rate of transport of the ^{3}H-leucine was drastically reduced after 2 hr of anaerobia. The diagram below the graph shows the site of ^{3}H-leucine injection and that axoplasmic transport takes place in two directions. [Redrawn from S. Ochs, *Science* 176:252 (1972). Copyright 1972 by the American Association for the Advancement of Science.]**

the sympathetic ganglia are seen and the animal is in effect "immunosympathectomized." It seems unlikely that only these two regions of the nervous system should be blessed by a trophic growth factor. There may be others of which we are not aware.

Trophic Factors of Neurons. If most of the synapses impinging on a neuron degenerate, the cell will die. It appears that synapses provide a mechanism of transfer for important macromolecules to cross from one neuron to the next and that these macromolecules are vital for the life of the neuron. Grafstein (1969) has shown that radioactively labeled amino acids injected into the eye of a mouse will be synthesized into macromolecules and then travel along the visual pathways, crossing synapses along the way until they reach the visual cortex. This is a remarkable observation, for it means that there must be some mechanism by which these large molecules can get through synaptic membranes. These molecules are carried along the nerve axons by a highly organized transport system that is responsible for *axonal flow.* It has been suggested by Ochs (1972) that the molecules attach to the neurofibrils and neurotubules that run the length of the axon and then travel along as though on a conveyor belt. This *fast transport* rate is about 400 mm a day (Fig. 8-6). There is also a *slower transport* mechanism in nerves and even a *retrograde transport* that carries substances back up to the nerve cell. These transport mechanisms are dependent on aerobic metabolism. Axonal transport is discussed in greater detail in Sec. 11-1.

Aging

There is a progressive and marked decrease in the number of neurons in the central nervous system with age. This neuronal death starts just after birth and, because neurons neither divide nor regenerate, it has been estimated that by the age of 90 there has been a loss of 50 per cent of the neurons of the frontal cortex of the human male. Despite the fact that there is probably a great redundancy in the number of neural connections in the brain, the consequent slowing of responses with age is hardly surprising.

Even though neurons of the adult mammal do not regenerate if they are damaged, it appears that regeneration does occur on a microscale, for lost synapses are replaced.

Figure 8-7. **Plasticity of central nervous system neurons is indicated by the changes that occur in the small dendritic spines of neurons in the visual cortex of mice. Growth of the spines is postulated to occur by hypertrophy (B) or branching (C), whereas disuse atrophy occurs with visual deprivation (D). (From *The Understanding of the Brain,* 2nd ed., by J. C. Eccles, McGraw-Hill Book Company, New York, 1974. Used with permission of McGraw-Hill Book Company.)**

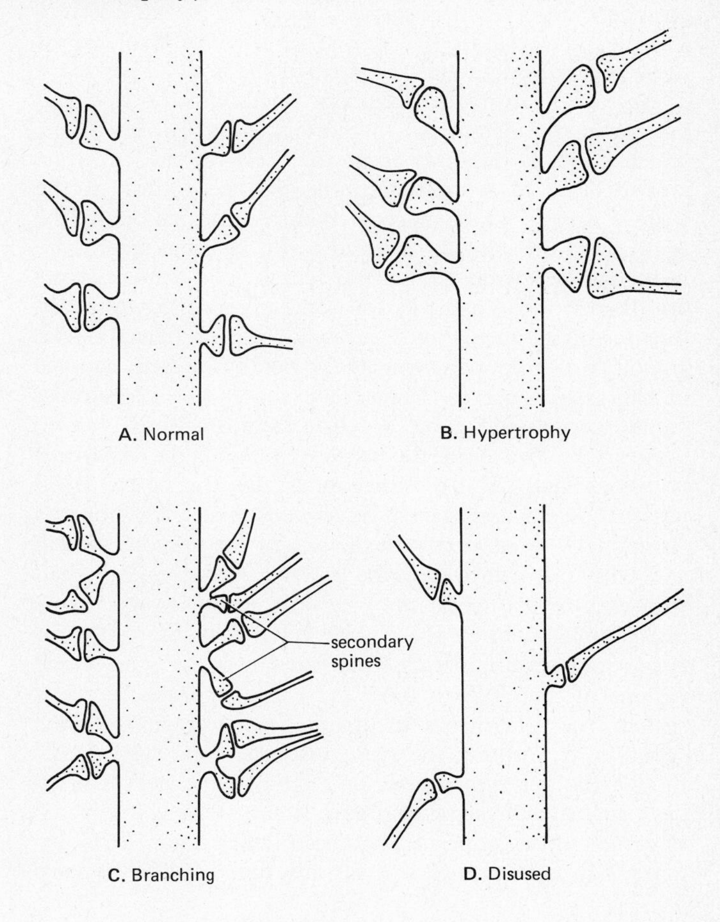

There is a *plasticity* of the mammalian central nervous system hitherto unsuspected (Fig. 8-7). If nerve pathways in the brain of an adult rat are cut, collateral fibers sprout from intact axons to occupy vacant synaptic sites. This can occur only over very small distances, implying that:

1. There is some kind of chemical recognition that normally prevents such sprouting.
2. When the synapse degenerates and is devoured by a glial cell, a neighboring axon is stimulated in some way, probably by chemical signals from the glial cells, to send out a sprout to the empty synapse.

These interpretations are purely hypothetical, but they do suggest that the same microregenerative processes may occur during aging, so that the effects of the marked decrease in neuron numbers are somewhat ameliorated.

8–3 Organization of the Nervous System: Central, Peripheral, and Autonomic

The nervous system is usually divided into the *central nervous system* (CNS), comprising the brain and spinal cord, and the *peripheral nervous system.* The latter consists of ganglia and the fiber bundles that form the 31 pairs of spinal roots (Fig. 8-1). These *spinal nerves* connect the spinal cord to the peripheral receptors (sensory endings) and to the effectors (muscles and glands) (Figs. 7-21 and 7-23). Branches of these nerves reach almost all parts of the body. They carry *afferent fibers* from the periphery (including the skin and viscera in this definition) to the CNS, and *efferent fibers,* which conduct impulses from the CNS to the effectors.

The brain is connected to the periphery through the *cranial nerves*. These nerves conduct information to the brain from the special senses of sight, hearing, smell, and taste, and also from the general senses. The cranial nerves are the peripheral pathway for nerve impulses to voluntary muscles of the face, mouth, eyes, tongue, and larynx. The cranial part of the parasympathetic nervous system consists of the cranial nerves. Figure 8-8 shows the 12 pairs of cranial nerves, and Table 8-1 lists their functions.

A further division of the nervous system is the *autonomic nervous system* (ANS), which innervates the heart, blood vessels, viscera, and glands. Its name is misleading, for it is functionally integrated with the rest of the nervous system and with the neuroendocrine system. However, we shall start with this simplification, for anatomical reasons. Most organs innervated by the ANS receive a double supply of nerves, one set originating from neurons in the *craniosacral* regions of the CNS to form the *parasympathetic* nervous system, the other from the *thoracolumbar* regions of the spinal cord, forming the *sympathetic* nervous system. Although the parasympathetic and sympathetic nerves often have opposite effects on an organ, their activity is integrated to assure the appropriate response of that organ. It is important to realize that the various parts of the nervous system must function as a harmonious unit to obtain optimal physiological function; when this harmony is disrupted, widespread malfunctions occur.

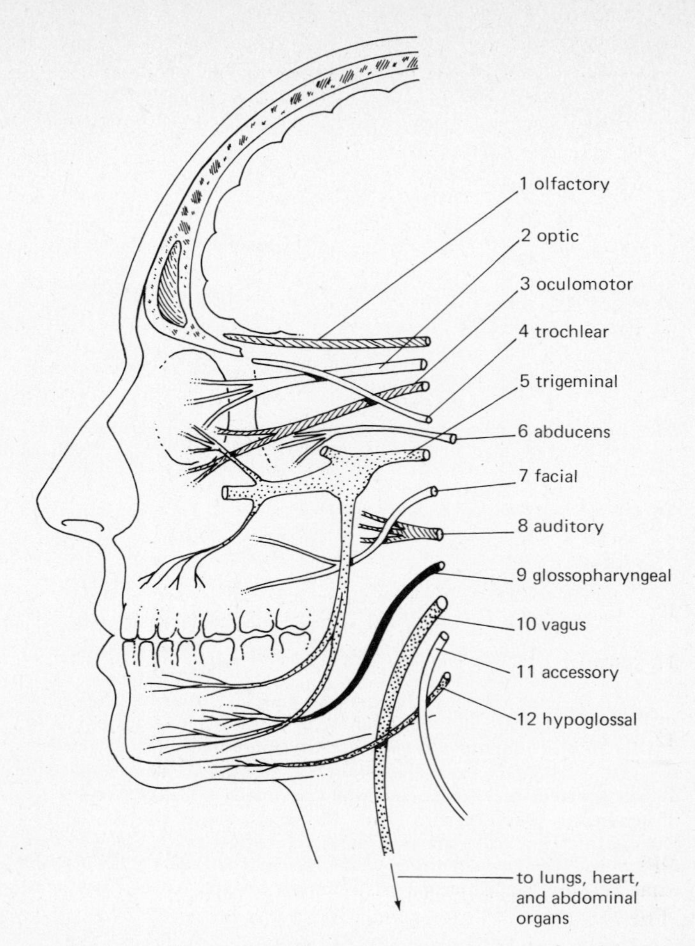

Figure 8-8. **The 12 cranial nerves (on one side of the head). (From *General Physiology,* 5th ed., by P. Mitchell, McGraw-Hill Book Company, New York, 1956. Used with permission of McGraw-Hill Book Company.)**

8–4 Levels of the Central Nervous System

The central nervous system is often divided rather arbitrarily into levels, according to the relative position of the elongated nervous system, as developed from the neural tube. The neural tube continues to thicken as the cells grow and proliferate, and by the fourth week, the human embryo shows the development of the three primary divisions of the brain into *fore-, mid-,* and *hindbrain,* as well as the elongated thinner *spinal cord,* already closed along its full length [Fig. 8-3(B)]. This is an extremely critical time of development, and the failure of even a tiny portion of the neural tube to close properly can result in the death or malformation of the embryo. The entire embryo at this time is only about 5 mm in length.

Table 8–1. **The Functions of the 12 Pairs of Cranial Nerves**

Cranial Nerve	Organ Innervated	Type of Fiber	Function
1. Olfactory	Olfactory mucous membrane	Sensory	Smell
2. Optic	Retina	Sensory	Vision
3. Oculomotor	Four of the six eye muscles	Motor	Eye movement
4. Trochlear	Superior oblique eye muscle	Motor	Eye movement
5. Trigeminal	Skin and mucous membrane of head	Sensory	Sensation
	Muscles of jaw	Sensory	Proprioception
6. Abducens	Lateral rectus eye muscle	Motor	Eye movement
7. Facial	Muscles of face, scalp, neck	Motor	Movement
	Salivary glands (sublingual and submaxillary)	Motor	Secretion of saliva
	Taste buds at back of tongue	Sensory	Taste
8. Auditory			
Vestibular	Semicircular canals, utricle, saccule	Sensory	Equilibrium
Cochlear	Organ of Corti	Sensory	Hearing
9. Glossopharyngeal	Pharynx, back of tongue	Sensory	Taste, sensation
	Parotid gland	Motor	Secretion of saliva
10. Vagus	Pharynx, larynx, trachea, esophagus, thoracic and abdominal viscera	Sensory and motor	Visceral reflexes
11. Accessory	Thoracic and abdominal viscera	Sensory and motor	Visceral reflexes
	Muscles of neck and shoulder	Motor	Movement
12. Hypoglossal	Muscle of tongue	Motor	Movement

From this time on, the brain of the human embryo develops far more rapidly and in much greater complexity than that of other vertebrates, although the basic plan is the same. The *forebrain* is divided into an *anterior* portion, the *telencephalon*, and a posterior *diencephalon* (Fig. 8-9). The telencephalon continues growing at the most rapid rate, especially those parts that are necessary for adaptation to the external environment. These are the specializations for the eyes, nose, ears, and mouth. The foremost portion of the telencephalon forms the two cerebral hemispheres, structures that are extremely well developed in humans.

The levels of the central nervous system are conceptualized as follows, from the highest to the lowest:

Forebrain	Telencephalon	—cerebral hemispheres corpus striatum
	Diencephalon	—thalamus hypothalamus pineal gland
Midbrain	Mesencephalon (Brain stem)	—corpora quadrigemina (roof) red nucleus substantia nigra cerebral peduncles (floor)
Hindbrain	Rhombencephalon (Brain stem)	—cerebellum pons medulla oblongata
Spinal cord		—cervical thoracic lumbar sacral

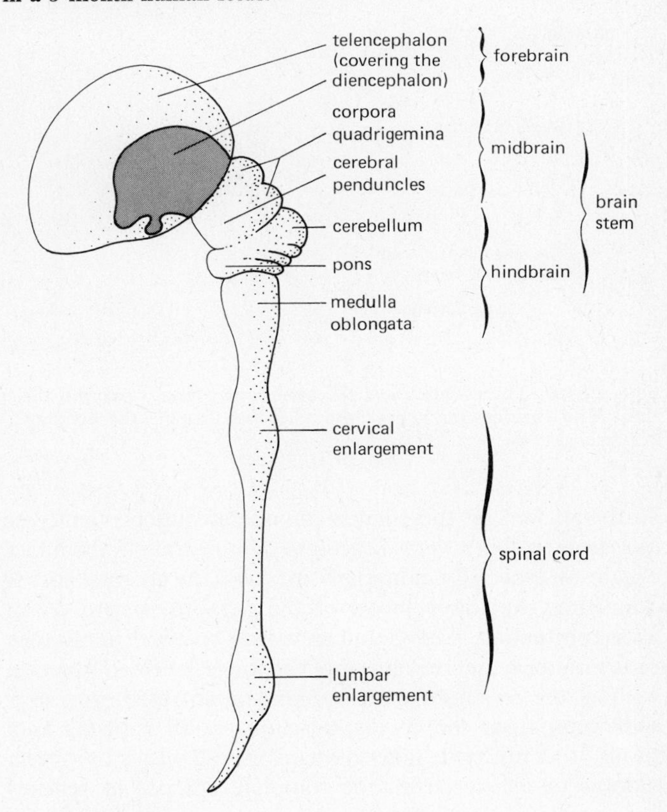

Figure 8-9. **The levels of the central nervous system as developed in a 3-month human fetus.**

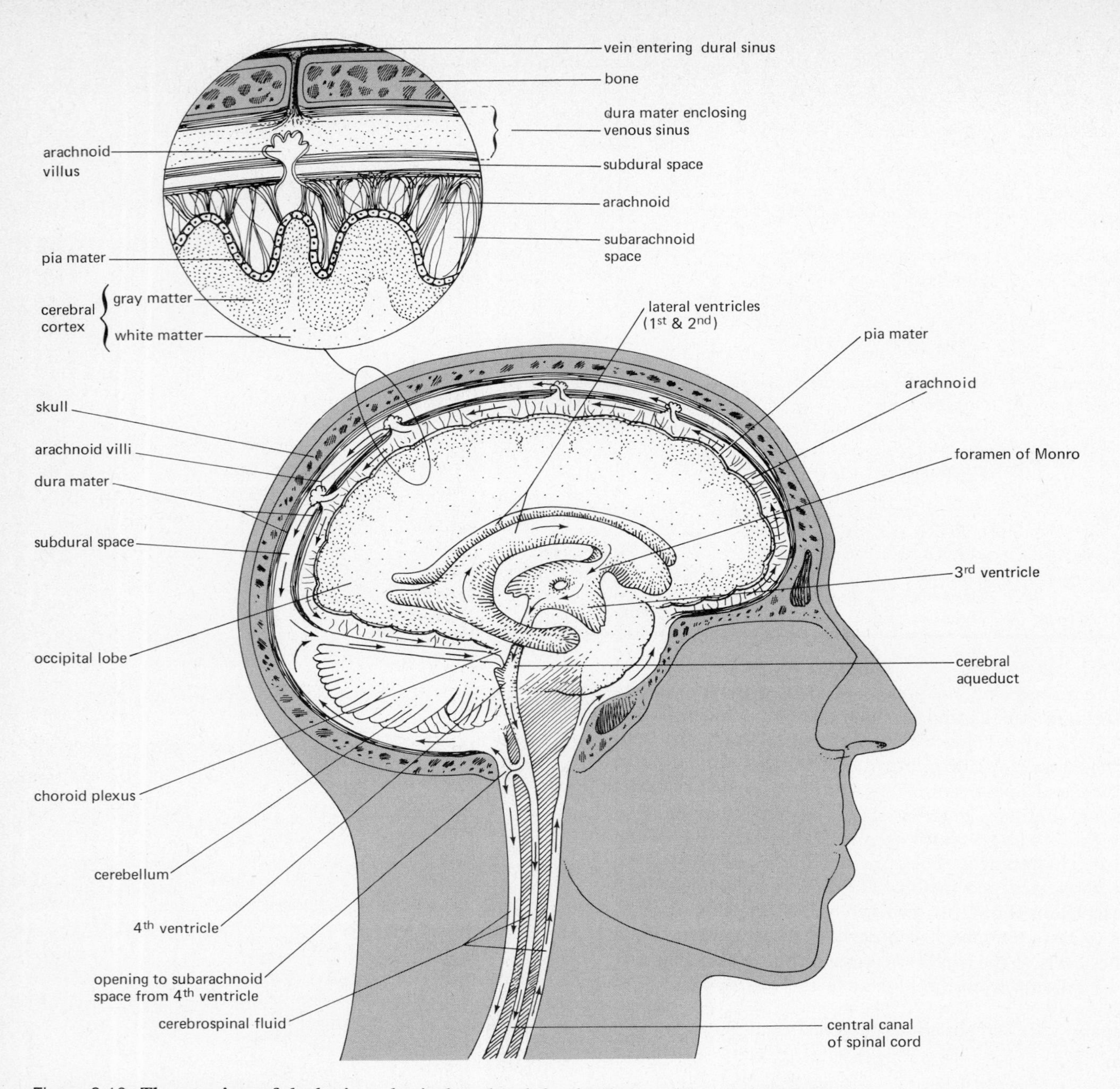

Figure 8-10. **The coverings of the brain and spinal cord and the circulation of the cerebrospinal fluid. The top drawing represents a section through the layers of the skull, membranes, and the cerebral cortex.**

Although some of these levels can operate independently of one another, this is very difficult to demonstrate in the intact organism, especially in humans, in whom the cerebral cortex dominates the lower levels of the nervous system to an enormous extent. This dominance may be either inhibitory or facilitatory, and the ability of the lower levels to function without the cerebral cortex can be demonstrated only after separation from the cortex, as occasionally happens as a result of an injury. It takes humans from 2 weeks to several months to recover from the complete absence of reflexes resulting from such a transection. The period during which almost all reflexes are absent is termed *spinal shock,* and if the patient is carefully nursed through this time, many reflexes[2] that are under the control of the spinal cord come back, often much more marked than before, due to the lack of cortical inhibition. Apparently, it takes the neurons in the spinal cord this much time to recover from the abrupt loss

[2] Vasomotion, evacuation of the rectum and urinary bladder, sexual reflexes, and postural and rhythmic reflexes may return but frequently do not.

of impulses from the higher centers. In animals in which cerebral dominance is less marked, recovery from spinal shock is much more rapid, taking only about 2 min in the frog and several days in the cat and the dog.

8–5 Protective Devices of the Nervous System

Coverings of the Brain and Spinal Cord

The entire central nervous system is surrounded by a protective layer of bone (Fig. 8-10). This bone forms the *skull* or *cranium,* covering the brain, and the segmented *vertebral column* that protects the spinal cord. The vertebral column is composed of vertebrae that vary in size and structure through the length of the column, being modified in the *cervical* (neck) region for the attachment of the ribs and pectoral (shoulder) girdle, and in the *lumbar* (abdominal) region for the torsion and strain of the weight of the upper body and its movements. The last vertebrae are fused to form the *sacrum,* to which the pelvic (hip) girdle is attached, and the *coccyx* (Chap. 24). The nerves that grow out from the central nervous system must penetrate through special canals or *foramina* (singular, *foramen*) between these bones to reach the other organs of the body. The pathways of the nerves from the brain in their course through the skull to reach the muscles of the eye, face, and jaw are especially complex.

Between this hard covering of bone and the delicate nervous tissue of the brain and spinal cord are three protective and nourishing membranes, the *meninges.* The innermost, most fragile membrane, which covers all the convolutions of the outer surface of the brain and cord, is the highly vascular *pia mater,* extending even into the deeper portions of the brain in certain areas, bringing the blood capillaries into close contact with the central cavity of the neural tube (now the *ventricles*). The pia mater is in close contact with a spidery membrane, the *arachnoid,* and the space between them, the *subarachnoid space,* is filled with *cerebrospinal fluid.* Surrounding the arachnoid is the thick fibrous *dura mater,* which adheres tightly to the skull in many places. Through the dura mater run the large meningeal arteries that supply the meninges and the cranial bones. The dura mater is a double membrane, one layer lining the skull and the inner layer separating parts of the brain into definite compartments. Even more important is the separation of these two layers of the dura mater in certain places to form blood channels, the *venous sinuses,* through which most of the copious blood supply of the brain drains before returning through the large veins to the heart. The space between the arachnoid and the dura has very little fluid in it, but projections of the arachnoid, the *arachnoid villi,* bulge into the dura mater, especially in the sinuses, and appear to act as valves for the circulation of the cerebrospinal fluid.

The Ventricular System and the Cerebrospinal Fluid

THE VENTRICLES

The entire neural tube, including the brain, retains its original central cavity, which becomes enlarged and modified in the brain to form the ventricles (Fig. 8-10). These spaces are continuous with one another and with the central cavity of the spinal cord. They also connect to the subarachnoid space through three openings from the fourth ventricle in the hindbrain. Thus the cerebrospinal fluid (CSF) contained within these cavities also circulates through the subarachnoid space and provides a cushioning layer of fluid around the delicate nervous tissues of the brain and cord.

The *four ventricles* of the brain correspond to the early development of the forebrain, midbrain, and hindbrain. The first and second ventricles are in the telencephalon of the forebrain, extending laterally in each of the cerebral hemispheres. The third ventricle is in the diencephalon and is connected to the junction between the first two ventricles by a small opening, the foramen of Monro. A narrow channel through the midbrain, the cerebral aqueduct (aqueduct of Sylvius), connects the third ventricle with the wide, diamond-shaped fourth ventricle in the hindbrain.

THE CEREBROSPINAL FLUID

Functions. The CSF provides the delicate brain with protection from mechanical trauma so that it does not bump

Figure 8-11. **The choroid plexus is formed by the penetration of the pia mater with its many blood capillaries into the ependyma of the ventricles.**

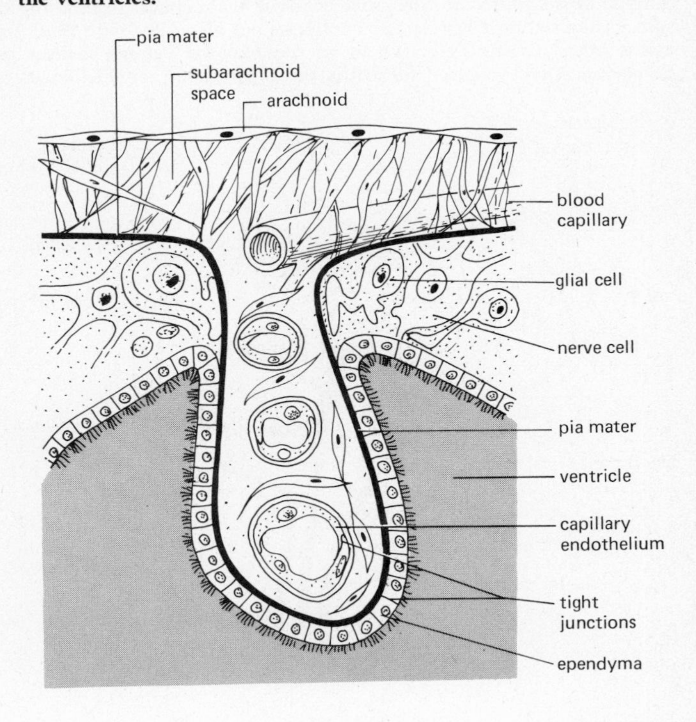

against the hard skull; it provides the nerve tissue, which it bathes, with the optimal physiological fluid environment for nerve function; it serves as a "sink" for brain metabolites and as a relief mechanism for the increase in intracranial pressure that occurs with each arterial pulse of blood to the brain.

Formation. The entire system of the central canal and the ventricles is lined with a thin layer of ciliated cells, the ependyma. At certain places in the brain, the pia mater penetrates down to the ependyma, carrying with it a complex network of relatively large blood capillaries (up to 15 μm in diameter) called the *choroid plexus* (Fig. 8-11). Each ventricle is supplied with these networks, which are the chief site of formation of the CSF. Recent experimental evidence has shown that the ventricular walls and the subarachnoid pial surface also contribute to CSF production. The cells of the choroid plexus are secretory, and if enzymatic inhibitors are administered, the production of CSF stops. The rate of secretion of CSF has been estimated at about 0.3 to 0.4 ml/min, about one third the rate of urine formation. By using sophisticated perfusion techniques, it is possible to demonstrate that the rate of CSF formation remains fairly constant, despite moderate changes in intracranial pressure.

Composition. Cerebrospinal fluid is an almost cell-free fluid, slightly alkaline and isotonic to blood plasma. It contains very little protein. Table 8-2 compares the composition of CSF and plasma. Cerebrospinal fluid is an ideal physiological solution for the brain, providing it with the fluid environment optimal for the conduction of nerve impulses: low potassium, and high sodium and magnesium.

Table 8–2. **Composition of Plasma and Cerebrospinal Fluid (CSF) mEq/liter**

Substance	Plasma	CSF
Protein (mg/100 ml)	7000.0	20.0
Glucose (mg/100 ml)	83.0	60.0
Lactate	9.9	2.6
Na^+	145.0	150.0
K^+	4.3	2.9
Ca^{2+}	9.5	2.3
Mg^{2+}	2.0	2.3
Cl^-	103.0	130.0
HCO_3^-	27.4	21.0

Through the use of isotopic tracers, it has been shown that there are separate transport systems for certain anions and cations, which result in the regulation of CSF composition within very narrow limits. Amino acids are selectively transported, probably by active mechanisms that regulate their movement in both directions, so that CSF may act briefly as either a store or sink for these molecules. This control of amino acid levels of the brain is of considerable significance, for they are directly or indirectly involved in synaptic transmission in the CNS.

Figure 8-12. **Oscillation of cerebrospinal fluid with each arterial pulse. In (A), the pressure builds up in the brain as blood is forced into it during contraction of the heart. Pressure is relieved by the bulging of the brain into the subarachnoid brain space, forcing the CSF into the spinal canal and spinal subarachnoid space. The ventricles are also filled with fluid. In (B), during relaxation of the heart, blood drains from the brain through the venous sinuses and the volume of the brain decreases. Cerebrospinal fluid fills the brain subarachnoid space and the ventricles are small.**

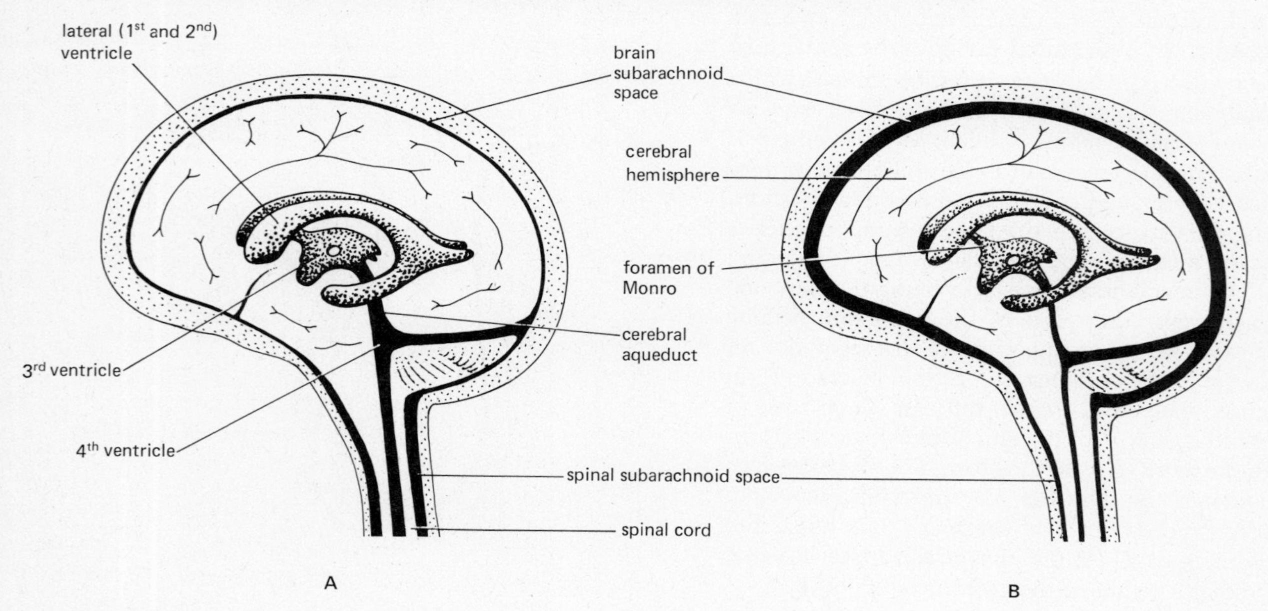

The transport of sugars is highly stereospecific, with glucose being moved in either direction, most probably by facilitated diffusion, for although its transport is blocked by structural analogues it is not affected much by anoxia. These selective transport mechanisms between blood and CSF form the *blood-CSF barrier.*

Circulation. There are two important concepts to be considered here: one is the basic movement of CSF out of the ventricular spaces into the subarachnoid spaces; the other is the continuous oscillation of CSF within the ventricular system with each arterial pulse (corresponding to a contraction of the heart).

MOVEMENT OUT OF THE VENTRICULAR SYSTEM. This circulation begins in the lateral ventricles, where the movement of the ciliated ependymal cells and the arterial pulsations of the choroid plexus may be of aid in propelling the fluid into the third ventricle and thence into the fourth ventricle. Through the apertures in the fourth ventricle, CSF seeps into the subarachnoid spaces over the cerebral hemispheres and the spinal cord (Fig. 8-10). It is absorbed into the blood at several sites, through the villi and granulations of the large venous sinuses in the skull and at the nerve root sheaths in the spinal column. The villi act as one-way valves, permitting CSF flow from the subarachnoid spaces into the blood sinuses.

OSCILLATIONS WITHIN THE VENTRICULAR SYSTEM. With each arterial systole there is a momentary increase in the contents of the intracranial cavity, causing the ventricular system and the choroid plexuses to swell. As the brain is enclosed in the rigid skull, there must be relief from this pressure. This is provided not only by the flow of CSF out of the ventricular system into the subarachnoid spaces, but also by its flow down from the *cranial* subarachnoid spaces to the *spinal* subarachnoid spaces.

This flow is reversed at the end of systole as blood continues to drain away from the brain, decreasing the intracranial contents, and fluid returns to the cranial subarachnoid spaces and the ventricles (Fig. 8-12). If there is an interruption in this to-and-fro movement, the intracranial pressure is increased and the ventricles swell, causing the condition known as *hydrocephalus,* which can be extremely damaging to brain tissue. If the pressure falls, through the removal of only a few milliliters of fluid, as is sometimes necessary for the analysis of CSF, very painful headaches result.

The Blood-Brain Barrier

The stable internal environment so necessary for the CNS of mammals is further provided for by a series of special transport mechanisms that protect it from small variations in the composition of the blood. These form the blood-brain barrier (BBB). The first demonstration of the existence of the BBB was in 1882 by Ehrlich, who injected animals with a vital dye, trypan blue, and noted that all tissues of the body rapidly became blue, with the exception of the brain and spinal cord, which remained white. However, if he injected a small amount of the dye into the subarachnoid space or directly into a ventricle, the brain and cord quickly took up the color. It is now known that the dye attaches to albumin in the blood and that the dye-albumin complex cannot penetrate the BBB.

MORPHOLOGICAL BARRIER

From a morphological point of view, the barrier appears to be the very constricted tight junctions between the continuous cells of the brain capillaries, which do not permit the passage of molecules with weights even as low as 2000 (Figs. 8-5 and 8-11). Consequently, proteins and molecules that bind to proteins cannot enter the brain. Lipid-soluble substances usually can penetrate readily: these include alcohol and the steroid hormones.

PHYSIOLOGICAL BARRIER

Physiologically, special transport systems permit the passage of selected molecules into the brain and exclude others. Glucose, amino acids, and ions are transported by carrier systems. Fructose and L-glucose cannot enter, but D-glucose penetrates freely. There is great variation in the rate of movement of different amino acids into the brain, and competitive inhibition of their transport is easily demonstrated. Essential amino acids not synthesized by the brain are rapidly taken up, whereas those that are readily synthesized from glucose metabolites are excluded. Simple charged ions of the blood exchange with brain ions, although they do so more slowly than with other tissues.

PATHOLOGICAL BARRIER

The BBB may be broken down pathologically by any injury to the brain, resulting in subsequent diffusion of proteins into brain tissues. Administration of a fluorescent dye bound to a protein can be used to localize the site of injury. The treatment of certain spine and brain disorders is impeded by the inability of many drugs to pass the BBB.

REGIONAL DIFFERENCES

Regional differences in the blood-brain barrier are interesting. It is absent in the median eminence of the hypothalamus and the pineal gland. Both these brain regions "sample" the contents of the blood and adjust homeostatic regulatory mechanisms accordingly. The function of these regions depends on variations in blood composition, whereas the function of the rest of the CNS requires essential stability in the internal environment.

8–6 Structure of the Nervous System

The Spinal Cord

MACROSCOPIC STRUCTURE

This cylindrical mass of nervous tissue represents the original neural tube, much modified in its development. The bony vertebral column grows more rapidly than the spinal cord which, being firmly anchored anteriorly to the bulging brain, is pulled up during development and in the adult reaches only to the level of the first or second lumbar vertebrae. Because the lumbar, sacral, and coccygeal nerves still leave the vertebral column through their original foramina, their roots are greatly elongated to form a thick bundle within the canal, called the *cauda equina* (horse's tail). (A spinal puncture, which requires the removal of fluid from the spinal cord, must enter above this level, for the accessible cerebrospinal fluid is contained within the subarachnoid space surrounding the true spinal cord.)

The adult spinal cord is about 40 to 45 cm in length, and is surrounded by the three meninges (Sec. 8–5) and a rather thick cushion of adipose tissue containing a plexus of veins between the dura mater and the surrounding vertebrae. The spinal cord is somewhat flattened ventrodorsally and has two swellings or enlargements; the anterior swelling represents the origin of the nerves innervating the arms and forming the *brachial plexus,* whereas the posterior swelling contains the cell bodies of the neurons innervating the legs through the *lumbar plexus* (Fig. 8-1). (The dinosaurs had disproportionately large hindlimbs, and their lumbar enlargements were equal in size to their brains.)

Figure 8-13. **The dermatomes of the ventral surface of the hand. The differences in shading represent areas of the skin innervated by different sensory nerves. Note the overlapping of the fields.**

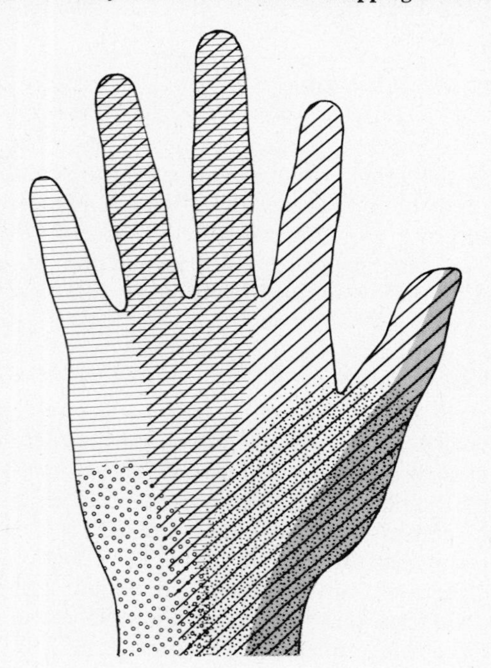

The spinal cord terminates in a narrow filament, which is surrounded by the roots of the nerves forming the cauda equina. Throughout its length, the spinal cord shows an obscure *segmentation,* reminiscent of the embryonic development of nerves and muscles in groups and seen also in the segmentation of the vertebrae. In humans, there are 7 segments in the cervical section, 12 in the thoracic, 5 in the lumbar, 5 in the sacral, and 1 in the coccygeal. This corresponds (with the exception of 8 cervical roots) to the number of roots leaving the spinal cord in each region, to total 31 pairs of roots. The corresponding muscle and skin areas that are innervated are known as *myotomes* and *dermatomes* (Fig. 8-13); because of the shifting position of the limbs during embryonic development, the distribution of nerves to these areas of muscle and skin is complicated but regular. Although it is not possible to trace the pathways of these nerves by dissection, they have been demonstrated clinically by the areas of paralysis or anesthesia following lesions of nerve roots within the spinal cord.

MICROSCOPIC STRUCTURE

The spinal cord consists of nerve cell bodies and their processes or fibers. The cell bodies, together with some unmyelinated fibers and blood vessels, are massed together in columns, which in cross section have the form of the letter H. This is known as the *gray matter,* because of its appearance, as opposed to the *white matter* that surrounds it. The surrounding white matter is formed by the axons of the cells within the H and also the axons of the sensory cells that lie outside the central nervous system, sending their processes into the cord to synapse with cells in the gray matter or to travel directly to the brain. The cord is therefore equipped to send and to receive impulses. The myelination of most of the axons gives the white matter its characteristic color.

The crossbar of the H is pierced by the central canal of the spinal cord. This canal is lined with the ependymal cells and filled with cerebrospinal fluid. Between the neurons and their fibers are the specialized connective cells of the nervous system called neuroglia. These are of different origin from the connective tissue found elsewhere in the body, and are perhaps derived from the ependymal cells.

The gray matter is composed mainly of longitudinal columns of cell bodies connected with each other through a large number of association neurons. These connections, or synapses, are on various levels of the spinal cord and extend up to different levels in the brain and also across both sides of the central canal. A *localization of special nerve cell bodies* (the largest cells, small cells, and medium-sized cells) occurs in three main areas of the gray matter (Fig. 8-14).

The Largest Cells. The largest and most easily recognizable are the *ventral horn alpha motor cells,* found in the

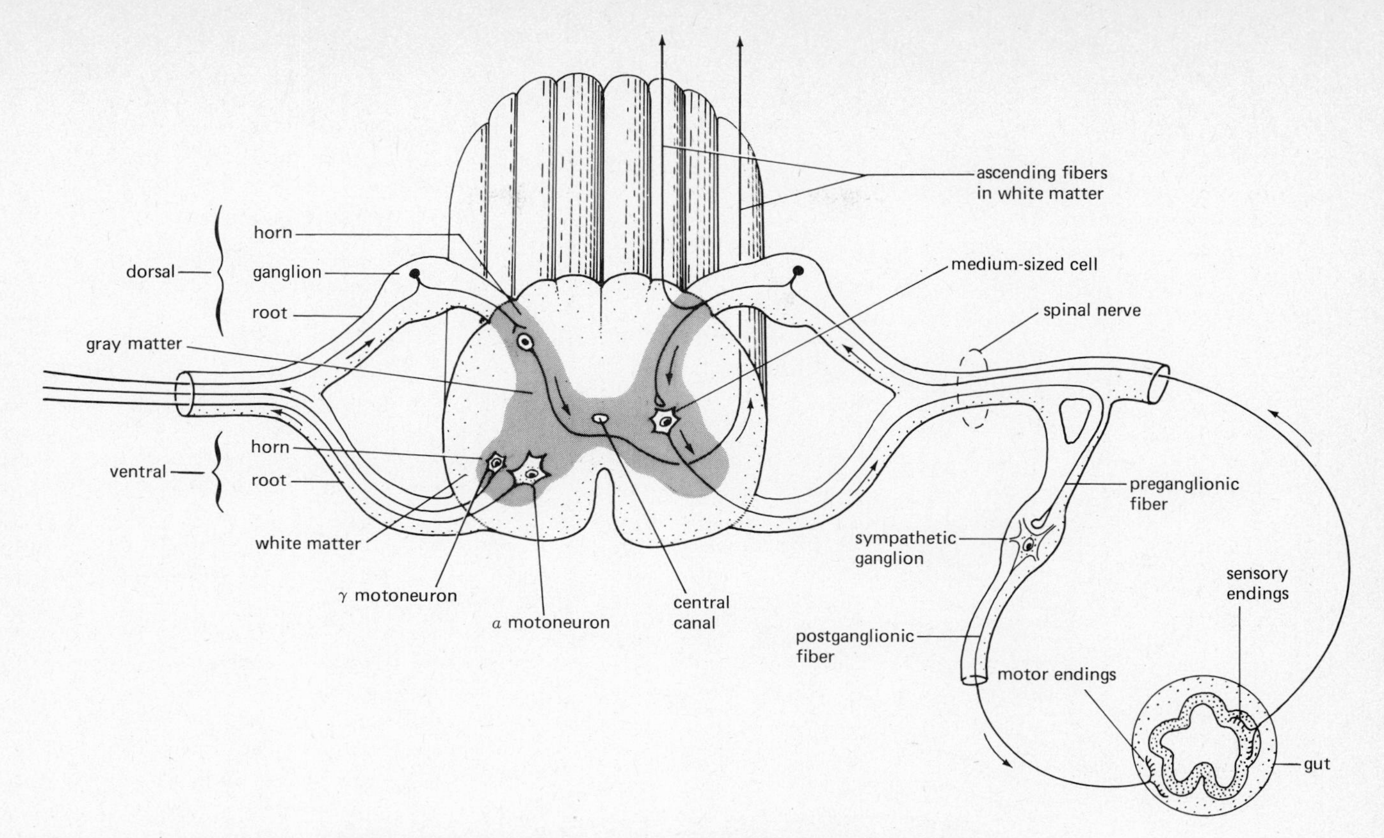

Figure 8-14. **Cross section of the spinal cord, showing localization of cell types. On the right is shown a reflex pathway involving the autonomic nervous system: the sensory neuron synapses on a medium-sized cell in the gray matter between the dorsal and ventral horns. The motor pathway involves a presynaptic and a postsynaptic fiber, with a synapse in the sympathetic ganglion outside the spinal cord. On the left are shown motoneurons in the ventral horn and a small cell in the dorsal horn. The axon of the small cell crosses to the other side of the cord and travels up the ventrolateral white matter to the cerebellum. A similar ascending fiber is shown as a branch of the sensory neuron of the autonomic reflex: this branch travels up a different part of the white matter to sensory areas of the brain.**

ventral column of the H of the gray matter. These cells are the motor cells that form the *final common path of all impulses going to the skeletal muscles.* Their axons pass out of the spinal cord by way of the ventral root to end as motor end plates in striated muscle. The smaller *gamma motoneurons* are also in the ventral horn.

Small Cells. In the *dorsal column of gray matter* (dorsal horn) small cells are important for connecting the impulses that come into the spinal cord with different levels of the nervous system on both sides.

Medium-Sized Cells. Medium-sized cells lie in the column of gray matter laterally between the ventral and dorsal horns. These cells form the first part of a two-neuron pathway for the innervation of smooth muscle, cardiac muscle, and glands. The first neuron is the one lying in the gray matter of the spinal cord; its axon also leaves the cord through the ventral root, but, instead of directly innervating the muscle or gland, it synapses with neurons in the *autonomic ganglia* that lie outside the spinal cord. The axon of this first neuron is called the *preganglionic fiber;* that of the second neuron is the *postganglionic fiber.* The postganglionic fiber ends in the muscle or gland and reaches its destination either by joining the motor fibers in the spinal nerve or else by forming an independent visceral nerve (Fig. 8-14).

The Autonomic Ganglia

THE SYMPATHETIC GANGLIA: THORACOLUMBAR LEVEL

All the preganglionic fibers leaving the spinal cord from the thoracic and lumbar levels enter a chain of ganglia that lies on either side of the spinal cord (Figs. 8-14, 8-15, and 8-16). These *sympathetic ganglia* are connected to one another, forming the sympathetic trunk. The preganglionic sympathetic nerve terminals secrete *acetylcholine.* In contrast, the postganglionic fibers of the sympathetic neurons secrete *norepinephrine,* or *noradrenaline* as it is often called, and consequently are termed *adrenergic nerves.*

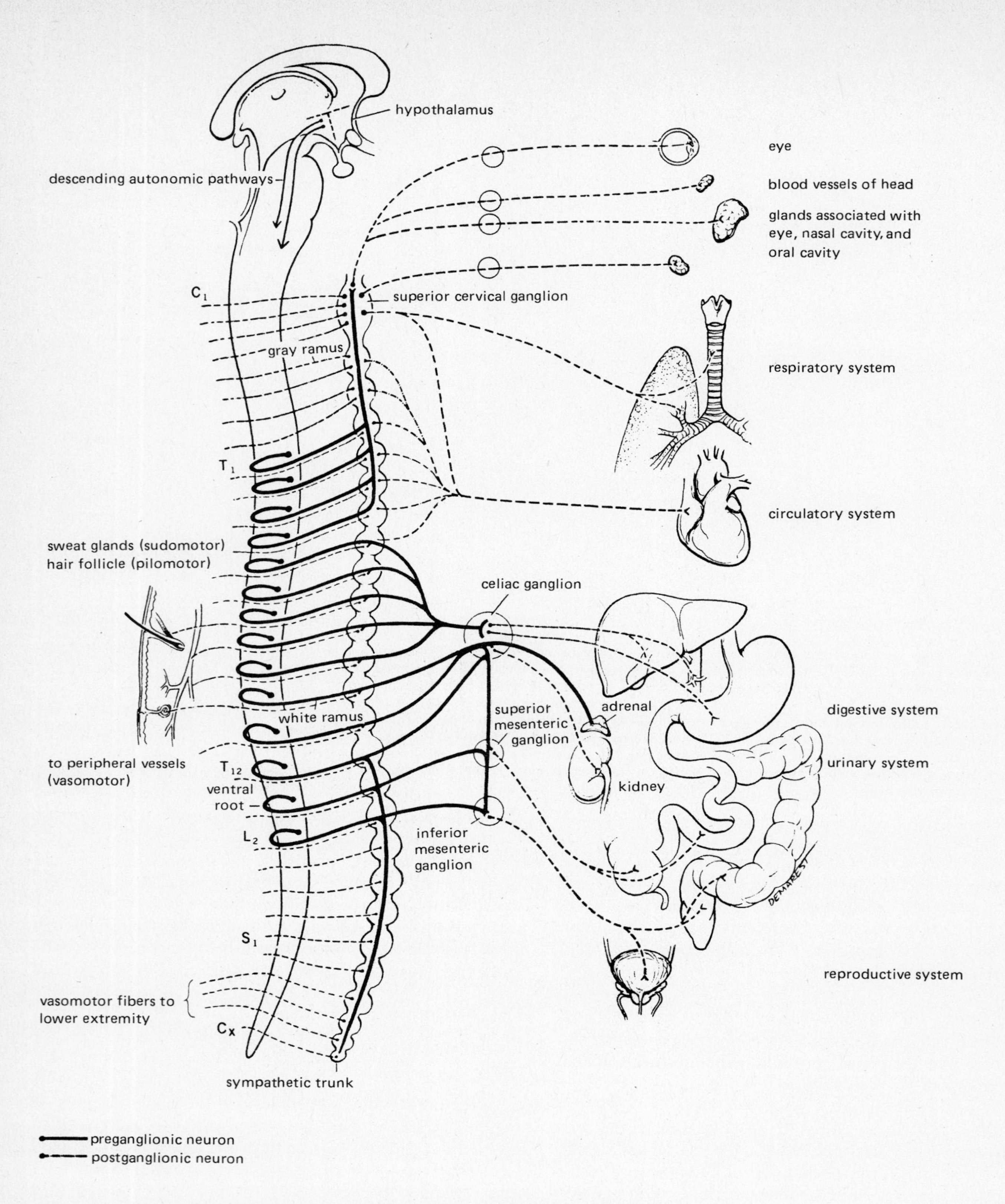

Figure 8-15. **The sympathetic nervous system. (From *The Human Nervous System,* 2nd ed., by C. R. Novack and R. J. Demarest, McGraw-Hill Book Company, New York, 1975. Used with permission of McGraw-Hill Book Company.)**

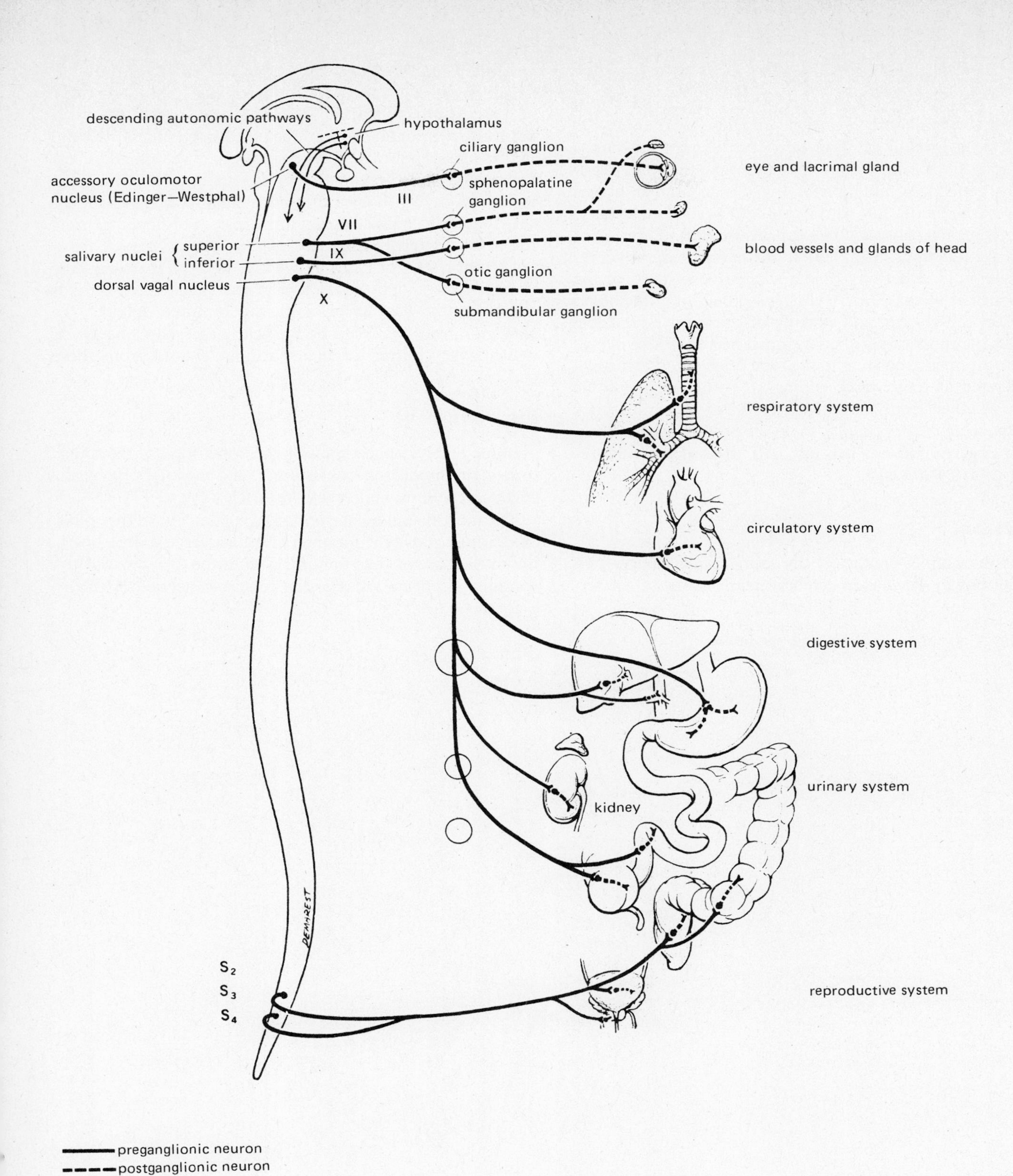

Figure 8-16. **The parasympathetic nervous system. (From *The Human Nervous System,* 2nd ed., by C. R. Novack and R. J. Demarest, McGraw-Hill Book Company, New York, 1975. Used with permission of McGraw-Hill Book Company.)**

The parasympathetic ganglia: craniosacral level

Those preganglionic fibers that leave the central nervous system at the level of the brain or sacrum synapse in ganglia very close to, or even within, the end organ that they regulate. In general, stimulation of the thoracolumbar nerves prepares the body for activity in emergency—increased heart rate, respiration, and blood pressure, and inhibition of vegetative functions. Stimulation of the craniosacral nerves usually results in activity of the vegetative systems of the body, such as increased peristalsis, vasodilation of the blood vessels in the viscera, and slowing of the heart rate. Table 8-1 lists the functions of the 12 cranial nerves. These nerves are also grouped according to the type of chemical transmitter secreted at their nerve endings. All the preganglionic fibers, whether from the parasympathetic or sympathetic systems, secrete *acetylcholine,* as do the postganglionic fibers of the parasympathetic neurons. All these nerves are referred to as *cholinergic.*

The Brain

The brain occupies practically the entire cranial cavity, with the enormously developed cerebral hemispheres of the telencephalon covering almost all the dorsal surface. We shall describe the human brain, however, as though it were stretched out longitudinally, because the neural pathways are formed through the connections between the lower and higher levels in terms of the embryonic development of the neural tube. The gray matter in the hindbrain and midbrain is formed by cell bodies in groups called *nuclei,* which are distributed irregularly through the masses of fibers forming the white matter. In the *telencephalon,* there is a more regular arrangement of gray and white matter. However, the arrangement is opposite to that of the spinal cord, because the outer thin gray layer of the cerebral hemispheres, the *cerebral cortex,* is formed by the cell bodies; the inner thick white layer consists of fibers leading to and from these cortical cells.

The hindbrain

Immediately connecting with the upper level of the spinal cord is the hindbrain, consisting of the medulla oblongata, pons, cerebellum, and fourth ventricle (Fig. 8-17).

The medulla oblongata, a thickening just above the spinal cord, is just above the foramen magnum (large space), opening in the base of the cranium. Within the 2 to 5 cm of the medulla oblongata are the vital centers concerned with *res-*

Figure 8-17. **Sagittal section through the brain.**

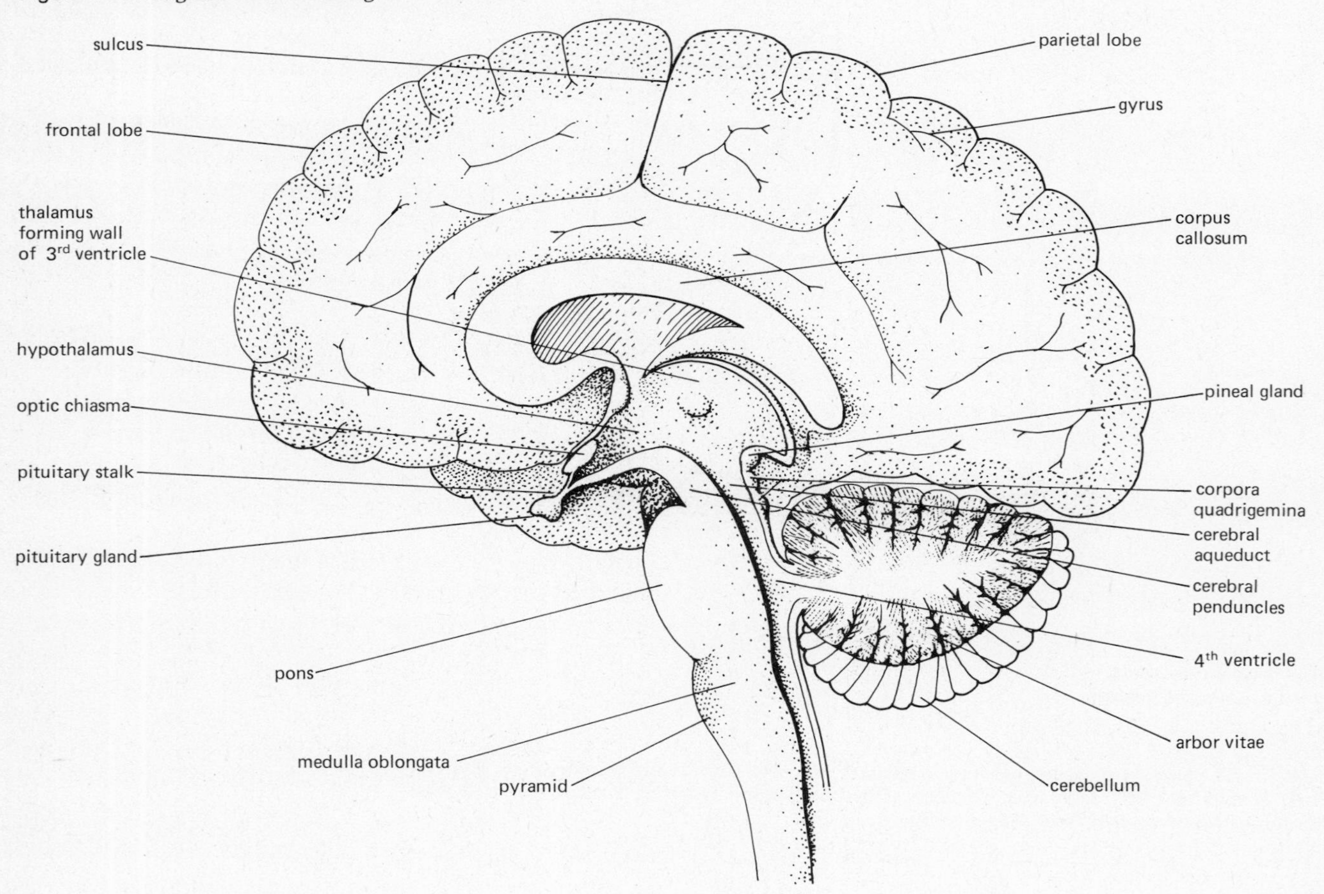

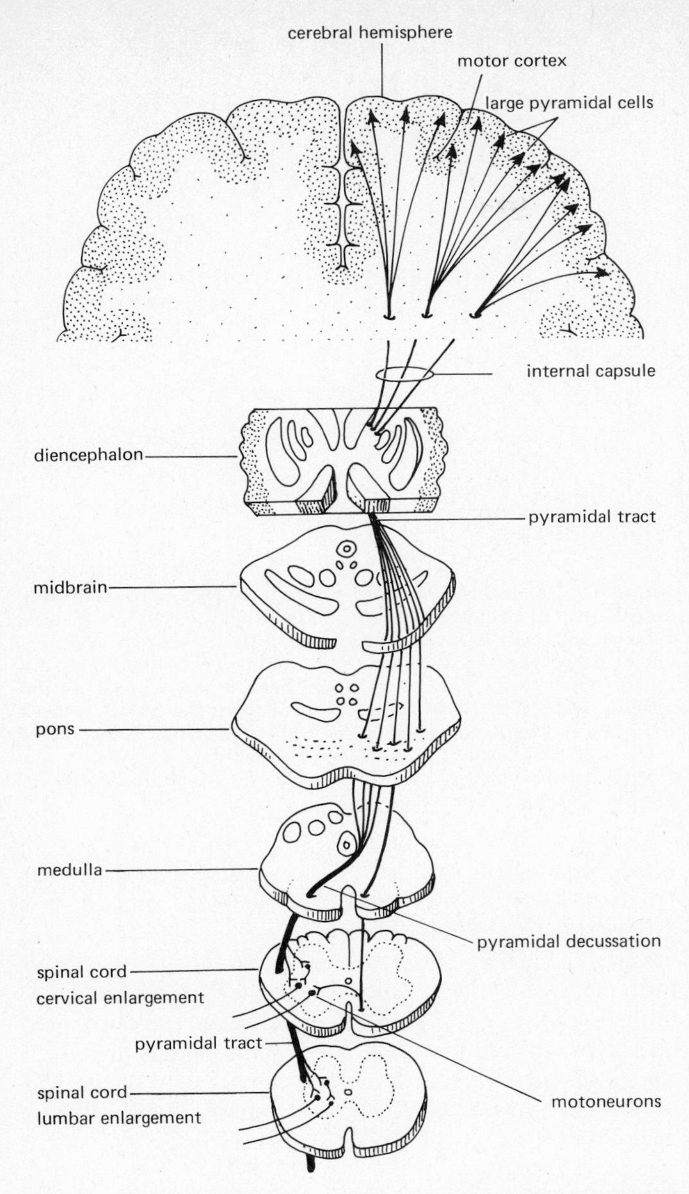

Figure 8-18. **Pyramidal (corticospinal) tract from the right motor cortex of the brain to the ventral horn motoneurons in the spinal cord. Most of the fibers cross over in the medulla (pyramidal decussation), but some continue down uncrossed to cross lower in the cord. (From *The Understanding of the Brain* by J. C. Eccles, McGraw-Hill Book Company, New York, 1974. Used with permission of McGraw-Hill Book Company.)**

piration, heart rate, and *circulation* and *emetic* (vomiting) *control,* so that a small lesion in this area can easily prove fatal. Here are also the *cochlear nuclei,* which transmit auditory (sound) information to the auditory cortex. As the medulla joins the spinal cord, two bundles of fibers cross over the median plane, forming the *decussation of the pyramids* (Fig. 8-18). These contain the axons of motor neurons mainly from the motor area of the cerebral cortex; in humans, most synapse directly on the ventral horn motor cells of the other side of the spinal cord. In humans, not all the fibers cross over in the medulla, but some continue down on the same side to cross over much farther down in the cord (Fig. 8-18).

The *pons* forms the ventral part of the hindbrain above (superior to) the medulla oblongata. Its nuclei form pathways to the two cerebellar hemispheres, and it also contains ascending and descending fiber tracts that connect it to different levels of the CNS. Very important nuclei within the pons are associated with the regulation of respiration and cardiovascular mechanisms.

The cerebellum

The cerebellum is a highly convoluted organ dorsally situated over the medulla and pons. It is a very specialized component of the central nervous system and is particularly well developed in mammals. The title of the book by Eccles, Ito, and Szentágothai, *The Cerebellum as a Neuronal Machine* (1967), indicates the ability of this organ to handle extremely complex neuronal calculations, permitting the mastery of intricate movements. The more intricate the movements of which an animal is capable, the more developed is the cerebellum. Eccles compares the cerebellum to a computer, able to coordinate information from the periphery coming through receptors in muscles and joints, and from the visual, auditory and equilibrium receptors, with instructions coming from the cerebral cortex, or higher brain.

Macroscopic Structure. The primitive cerebellum of early vertebrates has become greatly enlarged in the progression from the simple lamprey to fish, amphibia, birds and mammals, but its size is variable, depending apparently on the complexity of the adaptation of the animal to its external environment rather than its particular classification. For example, most fish have relatively simple cerebella, but that of the electric fish is enormously hypertrophied and capable of handling the vast amount of data derived from its lateral line organs (sensory detecting system) and from the ampullae of Lorenzini (electrical sensing organs).

Anatomically, the mammalian cerebellum can be described in several ways, but the simplest system is the division into a central *vermis* (wormlike in appearance), more laterally on each side the *pars intermedia,* and still

Figure 8-19. **The cerebellum viewed from below.**

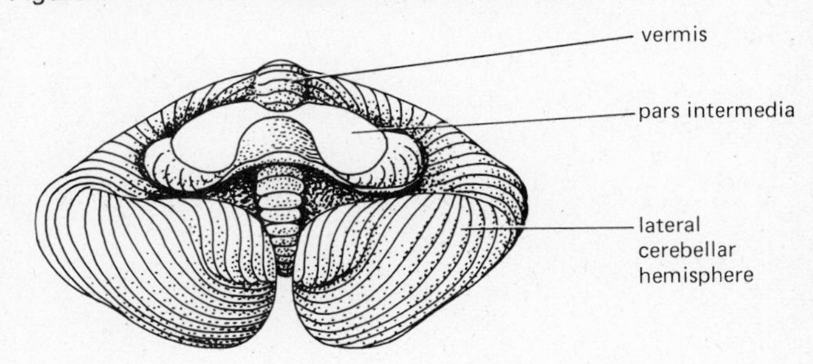

more laterally the two large *cerebellar hemispheres.* On each side the vermis and pars intermedia are divided into an *anterior* and a *posterior lobe.* These are the phylogenetically older structures (*paleocerebellum*), and are concerned with posture and basic equilibrium during movement. The lateral hemispheres are newer (*neocerebellum*) and have undergone extensive development correlated to the integration of finely coordinated motor skills, which in humans includes speech (Figs. 8-17 and 8-19).

Microscopic Structure. The cerebellum is joined to the rest of the nervous system by a wealth of afferent and efferent fibers that form the *superior, middle,* and *inferior peduncles.* They compose the central white matter of the cerebellum. Deeply imbedded within this white matter are the *cerebellar nuclei,* collections of nerve cells that mediate the output of the Purkyně cells of the cerebellar cortex.

The *cerebellar cortex* is tightly packed into minimal space through deep infoldings which form a series of *folia.* Although the basic plan of a folium is remarkably similar in most vertebrates, there is a tremendous elaboration in the structure of the principal cell of the cortex, the *Purkyně cell,* in higher forms. In the lamprey, each Purkyně cell has only a few branches, whereas in humans it may have many hundreds (Fig. 8-20). Yet there is little overlapping of territory of the individual Purkyně cells, so that their innumerable synaptic contacts are well defined and contained. The axons of these cells provide the *only efferent path* from the cerebellar cortex, and their axons end as *inhibitory* synapses on neurons of the deep cerebellar nuclei and other nuclei of the brain stem.

There are *two afferent pathways,* the *climbing fibers* and the *mossy fibers.* The climbing fibers climb along the Purkyně cells to make as many as 2000 synapses with a single cell. The mossy fibers achieve a logarithmic increase in synaptic contacts by branching out and synapsing with hundreds of *granule cells,* each of which has three to five dendrites. The axons of the granule cells grow up towards the layer of Purkyně cells, where they bifurcate to form the *parallel fibers,* which excite spines on the dendrites of the Purkyně cells (Fig. 8-21). In addition, there are three other types of cells that are excited by the parallel fibers and which, in turn, have an inhibitory output (Sec. 7-6). The function of this beautiful neuronal circuitry is discussed in Sec. 23-13.

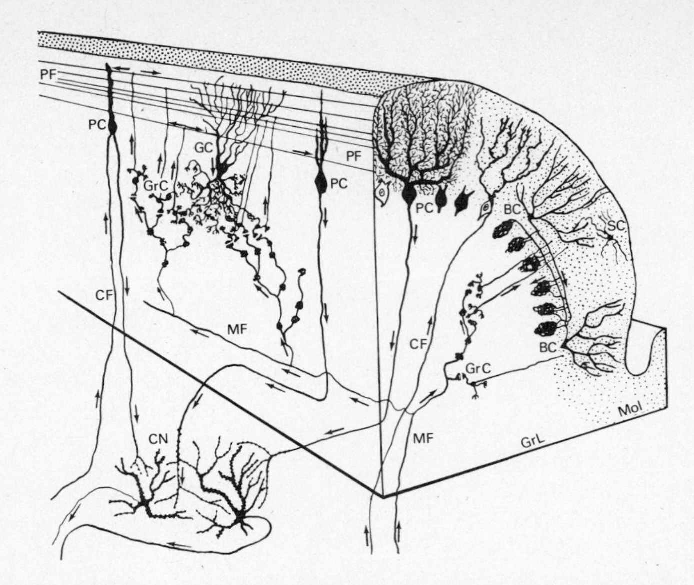

Figure 8-21. **Drawing of a three-dimensional section of a cerebellar folium. The principal neurons are the Purkyně cells (PC) which synapse with cells in the cerebellar nuclei (CN). Afferent impulses enter the cerebellum through the climbing fibers (CF) and the mossy fibers (MF). The mossy fibers branch and synapse on the granule cells (GrC), the axons of which form the parallel fibers (PF). These fibers synapse with the basket cells (BC), stellate cells (SC), and Golgi cells (GC). (From Elizabeth C. Crosby, et al., *Correlative Anatomy of the Nervous System,* Macmillan Publishing Co., Inc., New York, 1962, p. 193.)**

Figure 8-20. **Purkyně cells from a lamprey and a human cerebellum.**

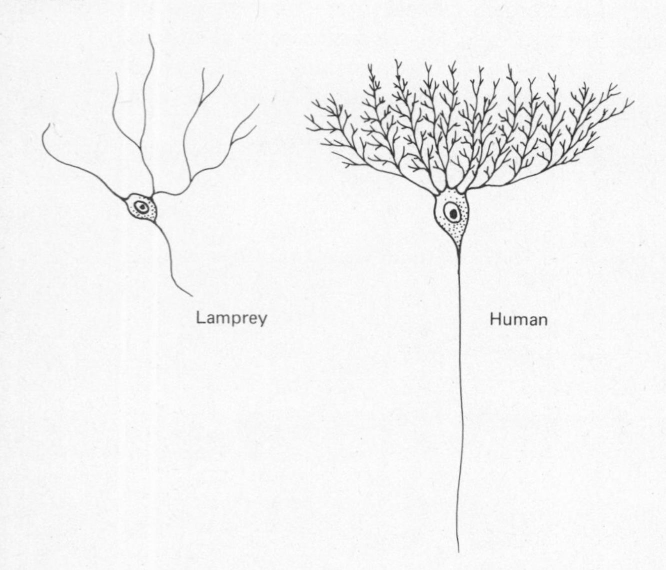

The Midbrain

This is a short section connecting the hindbrain with the diencephalon and consists of the corpora quadrigemina, the nucleus ruber (red) and the substantia nigra (black), and the cerebral peduncles.

The *roof* (tectum) of the midbrain in mammals is formed by four bumps, the *corpora quadrigemina.* The two anterior bumps are the *anterior colliculi* (little hills) involved in visual reflexes. The two *posterior colliculi* are associated with auditory reflexes. Lower vertebrates possess only the two anterior colliculi, which are the end stations for all impulses from the optic tracts. Mammals have developed the two posterior colliculi for hearing, in association with the development of the elegant spiral cochlea.

The *middle core* of the midbrain (tegmentum) contains reticular material: interlacing longitudinal and transverse fibers between which are embedded small masses of gray matter; the *nucleus ruber* and the *substantia nigra,* which are associated with movement; the nuclei of some of the *cranial nerves;* and some *reticular nuclei.*

The *basal zone* of the midbrain is made of massive motor fiber tracts that connect the cerebral cortex to lower levels. These form the *cerebral peduncles* (Fig. 8-22).

The reticular formation of the brain stem

As its name indicates, the *reticular formation* (RF) consists of a network of nerve cells and fibers connecting the medulla, pons, and mesencephalon with each other and with higher and lower levels of the CNS. Aggregations of cells of different types and sizes are present, but the reticular formation does not include the specific sensory and motor nuclei of these regions.

The *general function* of the RF may be viewed as the *regulation of neuronal excitability,* from the level of the cerebral cortex to that of spinal motoneurons. Its discharge of impulses to the higher levels of the brain helps to maintain consciousness and the waking state; the withdrawal of its stimulating effect on the cerebral cortex is one of the factors involved in sleep. The reticular formation receives a constant flow of impulses from the muscles and also from the main sensory pathways, so that changes in the position of the body, noise, and light all tend to keep us awake and alert. When we relax and when external stimulation is at a low level (e.g., dim light and quiet), the amount of activity in the reticular formation and consequently in the cerebral cortex diminishes, and sleep is possible. Destruction of the reticular system results in chronic somnolence or coma.

Stimulation of *specific areas* of the RF in the medulla (bulbar region) *facilitates the response of spinal motoneurons* to stimulation from the motor cortex, causing exaggerated reflexes; stimulation of other bulbar areas of the RF powerfully *inhibits these motoneurons.* As a result of the diffuse and widespread anatomical connections between the RF and the motor cortex, the cerebellum, and the spinal motoneurons, the RF is in an excellent position to "tune" the responses of CNS neurons and modulate motor activity (Figs. 8-22 and 9-1).

The forebrain: diencephalon and the limbic system

The diencephalon consists of the *thalamus,* the *hypothalamus,* and the *pineal gland* (Fig. 8-17). The *optic nerves* and the *retinae* develop from it. It *connects* the midbrain to the cerebral hemispheres and contains many *nuclei* which form a feedback system between the diencephalon and the cerebral

Figure 8-22. **The midbrain and reticular formation. The midbrain is shaded darker than the rest of the brain stem. The reticular formation consists of cells and fibers linking the medulla, pons, and midbrain with each other and with higher and lower centers, including the thalamus, cerebral cortex, cerebellum, and spinal cord. See also Fig. 9-1.**

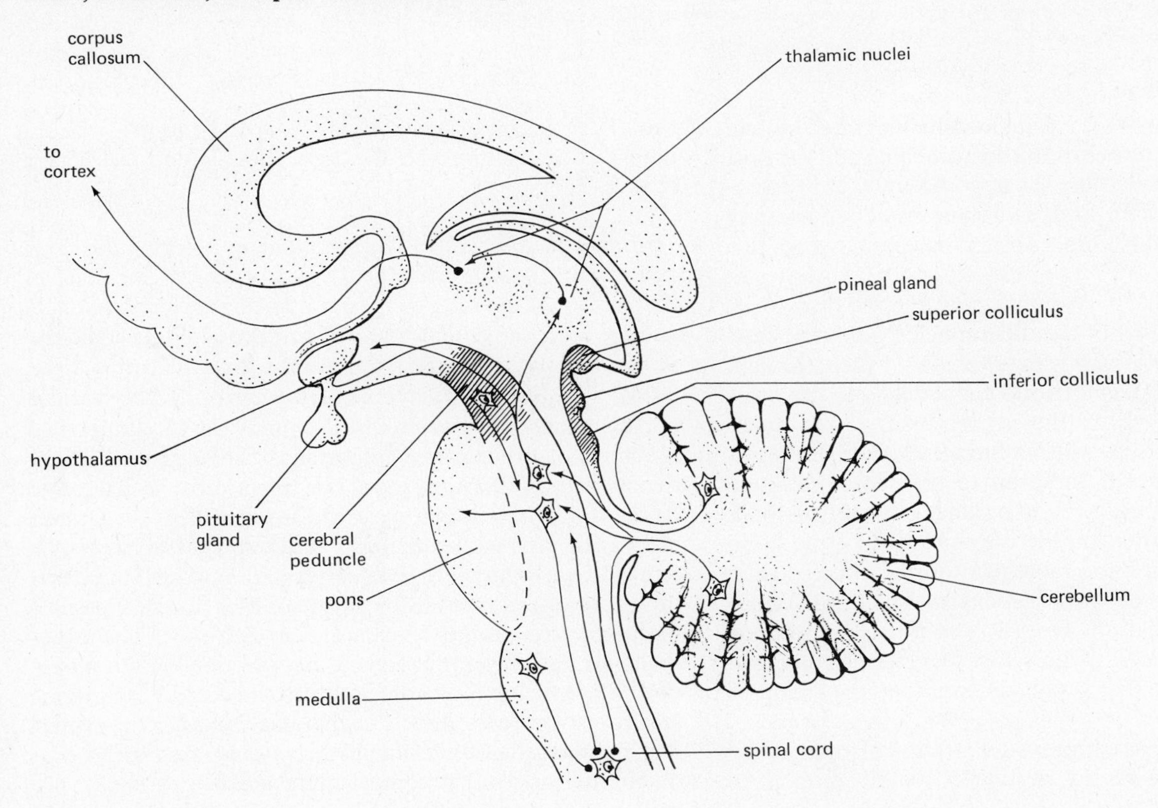

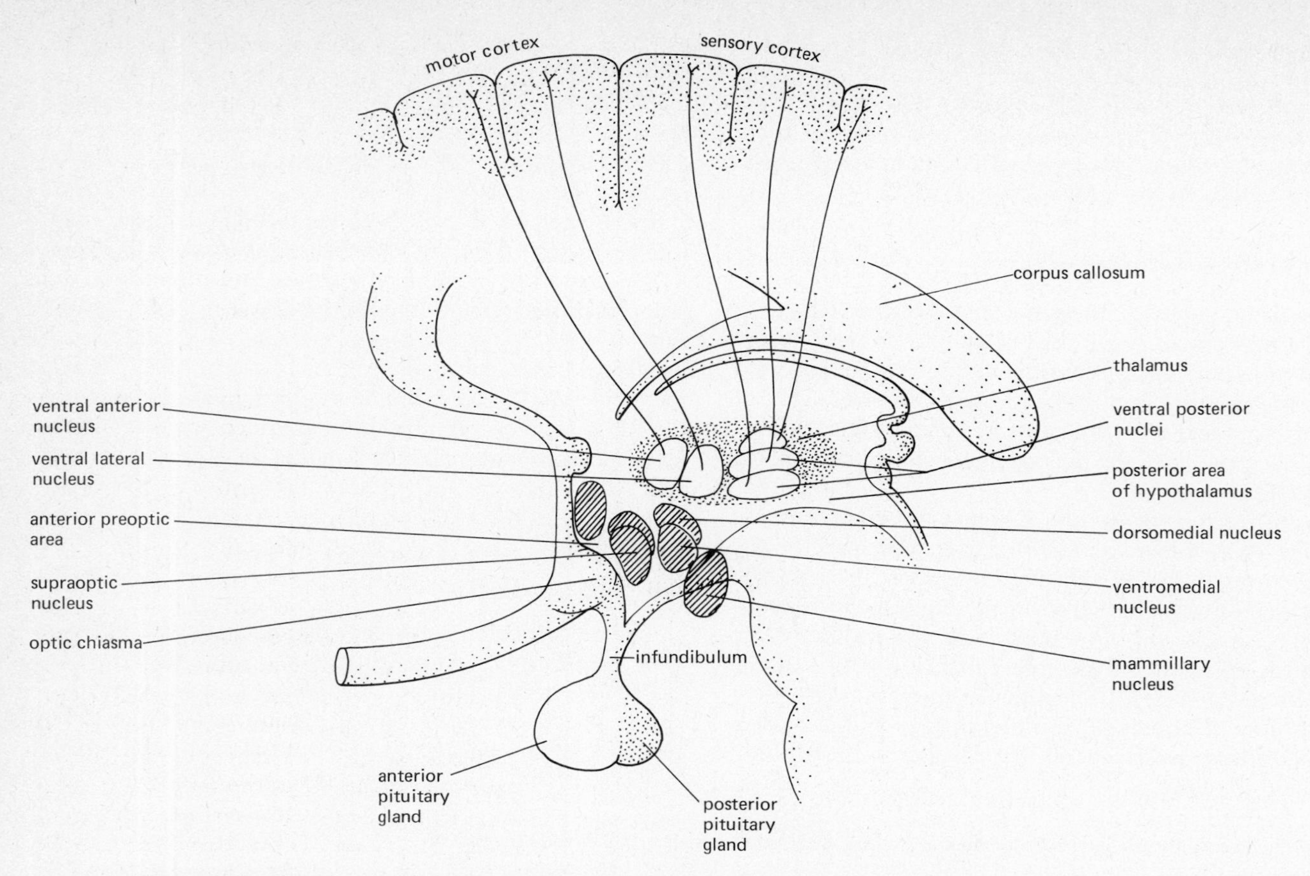

Figure 8-23. **The nuclei of the thalamus and hypothalamus. Some of the connections of the thalamus to the motor and sensory cortex are shown. The nuclei of the hypothalamus are indicated by striping (not all nuclei are shown). The optic chiasma is the crossing of the optic tracts.**

cortex. The *limbic system* is a functionally integrated system involving neural structures in the diencephalon and the base of the telencephalon. Because the hypothalamus is an essential unit of this system and its functions are intricately associated with it, the limbic system is discussed in this section.

The Thalamus. Most of the diencephalon—its walls and part of the floor—is formed by the *thalamus,* which, through its specialized nuclei, has intricate connections to the cerebral cortex, particularly to the *sensory areas of the cortex* (Fig. 8-23). It integrates sensory impulses from the eyes, ears, and skin and sends these on to the sensory cortex. Lesions in the thalamus have the peculiar effect of not only causing loss of sensation on the opposite side of the body but often resulting in unbearable pain on the affected side; shaving the face may evoke sensations as though the entire skin were being peeled off. Destruction of the connections between the thalamus and the sensory cortex is often the only way to stop pain that is not susceptible to the action of drugs.

There are also nerve pathways from the thalamus to the *motor cortex,* with the ventral lateral (VL) and ventral anterior (VA) thalamic nuclei acting as relay stations between both the cerebellum and the basal ganglia and the cortex (Fig. 8-23).

The Hypothalamus. The hypothalamus forms the floor and part of the side walls of the diencephalon. Included in this area is the crossing of the optic tracts (the *optic chiasma*) and the *posterior pituitary gland,* suspended by its stalk, the *infundibulum,* from the base of the hypothalamus (Fig. 8-23). The rest of the hypothalamus is formed by various nuclei, which control an astonishingly large number of functions of fundamental importance. The activities of the hypothalamus are influenced by practically every other organ of the body and it, in turn, affects them. It is anatomically associated with the higher centers of the nervous system, and with the *limbic system* which is involved in behavioral responses essential for survival. The limbic system is discussed later in this section and in Sec. 9-4. The hypothalamus has nuclei that control *appetite* and *body temperature;* it is involved in *sleep,* and it is the central integrating area for response to *stress.* The hypothalamus has the centers that regulate the *autonomic nervous system* and nuclei producing hormones that control the *endocrines.*

Hypothalamic control of the *endocrine system* is achieved through a complex of stimulating and inhibiting hormones that regulate the secretions of the *anterior pituitary gland.* The pituitary gland is suspended by a stalk, the *infundibulum,* from the floor of the diencephalon, and the anterior portion of this gland produces the *tropic hormones* that regulate most of the other endocrine glands. There is a *special circulation* between the hypothalamus and the anterior pituitary, which conveys the hypothalamic regulating hormones to the anterior pituitary (Fig. 10-3).

There are *direct neuronal connections* between the hypothalamus and the *posterior* portion of the *pituitary gland,* which acts as an endocrine storage organ for the secretions of special hypothalamic cells. Hypothalamic control of the endocrine system is discussed in detail in Chapters 10 and 11.

Both the autonomic nervous system and the endocrine system are intimately involved in regulating the response of the organism to *stress* (any harmful, extreme stimulus). The hypothalamus integrates the responses of these systems: for example, emotional responses characteristic of rage or excitement involve the *autonomic nervous system,* the *endocrine system,* and *higher cortical centers.* All the characteristics of a *stimulated sympathetic nervous system* are seen in rage, including increased heart rate and blood pressure, increase in rate and depth of respiration, and dilation of the pupils. Vasopressin is released from the hypothalamo-posterior pituitary stores and increases blood pressure: a specific hormone from the hypothalamus stimulates the endocrine response to stress by way of the anterior pituitary gland and the adrenal cortex (the *alarm reaction,* which is discussed in Chap. 11).

The integrated role of the cerebral cortex in such emotional displays was first shown by Sherrington (1906) in animals from which the cerebral cortex had been removed. In such a *decorticate* animal, a caress from a well-known animal keeper will cause it to show all evidence of rage; this is called *sham rage* and only occurs if the hypothalamus has not been destroyed along with the cerebral hemispheres. The cerebral cortex probably acts as a constant inhibitor of the emotional expression of the hypothalamus in the intact individual.

Similarly, there is a center for the regulation of *parasympathetic activity* (craniosacral); electrical stimulation of this center will cause slowing of the heart, contraction of the visceral muscles, and other evidence of a shift to activity of a more passive and visceral nature.

Figure 8-24. **Some of the reciprocal connections of the limbic system. The hippocampus, amygdala, and the olfactory lobe form the limbic lobe.**

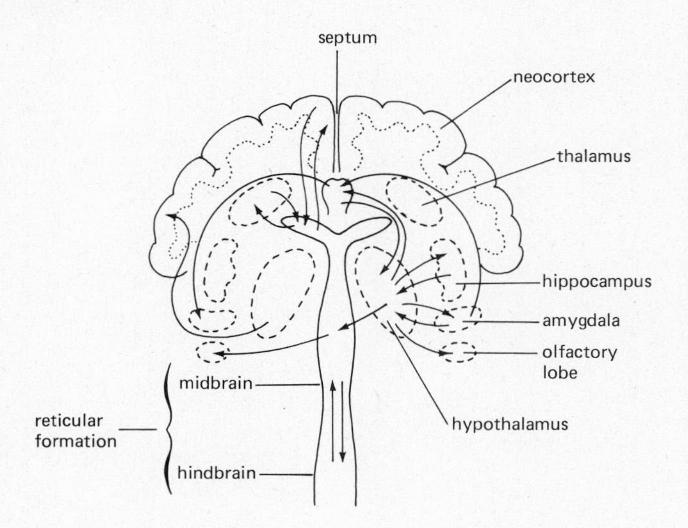

The Limbic System. The limbic system is a functionally integrated system involving neural structures in the *diencephalon* and the *base of the telencephalon.* The *telencephalic* portion is called the *limbic lobe,* and is formed by the phylogenetically ancient cortex which surrounds the brain stem and has become buried through infoldings of the greatly developed newer cortex. In higher forms most of it lies in the *hippocampus,* in the inferomedial part of the temporal lobe. There is a close evolutionary relationship between the limbic lobe and olfaction (which humans appear to have lost), but it has many functions apart from the olfactory. Some of the other telencephalic structures involved in the limbic system are the *amygdala* and the *basal ganglia,* discussed later in this section. The *diencephalic* structures that contribute to the limbic system are the *hypothalamus,* part of the *thalamus,* and the *preoptic area.* Figure 8-24 shows the limbic system as being a laterally extending neural circuit connecting the midbrain, diencephalon, and limbic lobe.

Through the use of stereotaxic instruments, modified from the type first introduced by Horsely and Clark in 1908, it is possible to place lesions accurately in specific regions of the brain. When lesions are placed in the limbic system, with minimal damage to surrounding areas, a marked loss of emotional responsiveness occurs in experimental animals. The extent of the behavioral change is correlated to the site and size of the lesion.

If specific regions of the limbic system are electrically stimulated, pleasant or unpleasant emotions can be evoked. This can be shown in experimental animals through the use of electrodes implanted in the brain. The implantation is performed under anesthesia, which is not required subsequently, permitting tests to be made on free-moving rats. Using this technique, Olds and Milner (1954) have shown the existence of *pleasure centers* and *punishment centers* in the limbic system. Their experiments are unique enough to deserve description. The electrodes were implanted in the brain and the animal placed in an experimental box. Whenever the animal went to one specific corner of the box, it received a mild shock from the electrodes. To the surprise of the investigators, it was found that the animal immediately returned to that corner; this could be repeated and repeated, with the rat always returning for more stimulation (Fig. 8-25).

The experimental conditions were then changed somewhat, and the rats were put into a "do-it-yourself" situation,

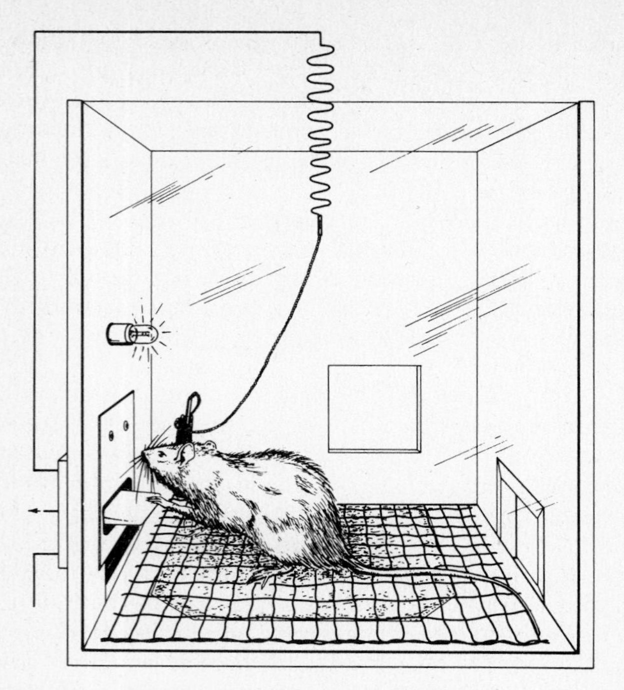

Figure 8-25. **Self-stimulation circuit. When the rat presses on the bar, it receives an electric stimulus to the brain. (Redrawn from "Pleasure Centers in the Brain," by J. Olds, *Scientific American*, Oct. 1956, p. 108. Copyright © 1956 by Scientific American, Inc. All rights reserved.)**

in which they received a shock every time they pressed a lever. After the few minutes needed to learn this, the rats stimulated their own brains for hours, about once every 5 sec. When the electrodes were in the area of the hypothalamus and certain limbic nuclei, the animals would stimulate themselves up to 5000 times/hr. This appears to be the pleasure center or reward center, and it is also the center for digestive, sexual, and excretory processes. Stimulation of the pleasure center appeared to be much more satisfying to the animals than food, for even starved rats would choose self-stimulation rather than food.

Similarly, the existence of a punishment center in the same general subcortical areas was found, and the rat would avoid self-stimulation of this center even at the cost of having to work to avoid the shock. Because a reward will speed up learning and punishment will retard it, the *limbic system is considered to be involved in motivation and learning.* It also appears to *select the sensory input* relayed to the cortex, eliminating distracting cues (such as external noise) and permitting the cortex to concentrate on important cues (the lecturer's voice). In this manner, the limbic system prepares the cortex for learning.

Rather complex behavioral patterns that are necessary for the *preservation of the species* are controlled through the limbic system. These include pleasure and grooming reactions, which form the preliminary social behavior conducive to copulation and reproduction, as well as penile erection and ovulation. Behavior that is concerned with the *survival of the individual* (licking, chewing, retching, sniffing, and searching for food, as well as attack or defense) are all controlled to some extent by the limbic system.

Practically all the behavioral alterations following destruction or stimulation of parts of the limbic system can be seen in humans after trauma, tumors, or infection of this region. Terror, rage, anxiety, or euphoria have all been described, varying with the site of damage. Whether in experimental animals or in humans, it has been demonstrated clearly that important biological functions such as hunger, thirst, sleep, sex, and behavior patterns essential for survival are dependent on the hypothalamus and the rest of the limbic system.

The Pineal Gland. The pineal gland is a small pine-cone-shaped mass of tissue in the center of the brain, rostral to the corpora quadrigemina of the midbrain. It loses all nerve connections with the brain shortly after birth, but it is extensively innervated by nerves from the sympathetic nervous system (Fig. 8-26). Its function has changed markedly through the phyla. In amphibia and primitive reptiles, the pineal cells are photoreceptors, acting as a "third eye." The amphibian pineal also secretes *melatonin,* an extremely potent skin lightening factor in amphibia. In mammals, however, the pineal photoreceptors are absent, and melatonin has evolved into a *gonad-inhibiting substance,* the production of which is abolished by exposure of the animal to light (Sec. 11-5).

THE FOREBRAIN: TELENCEPHALON

The telencephalon consists of the cerebral hemispheres, which are divided into the *frontal, parietal, occipital,* and

Figure 8-26. **The pineal gland and its sympathetic innervation. The broken line indicates a nerve pathway from the retina involving synapses in the brain stem and spinal cord, which ultimately inhibits its neurons in the superior cervical ganglion.**

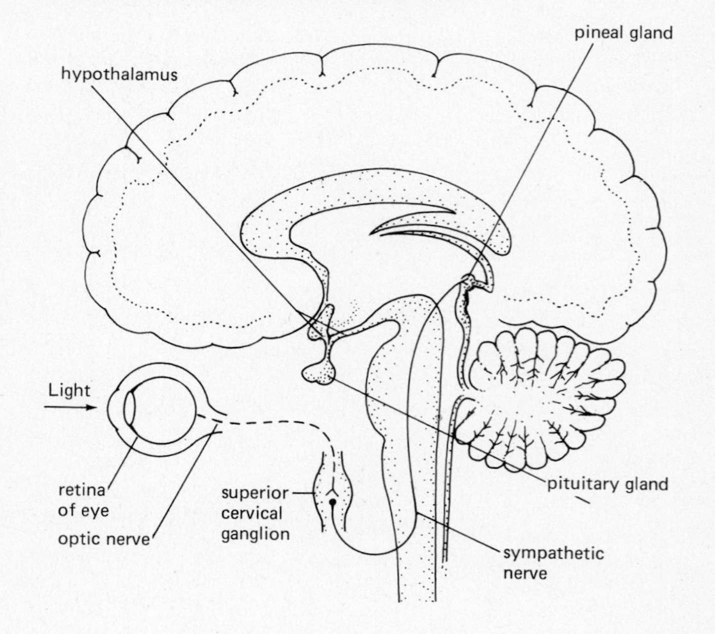

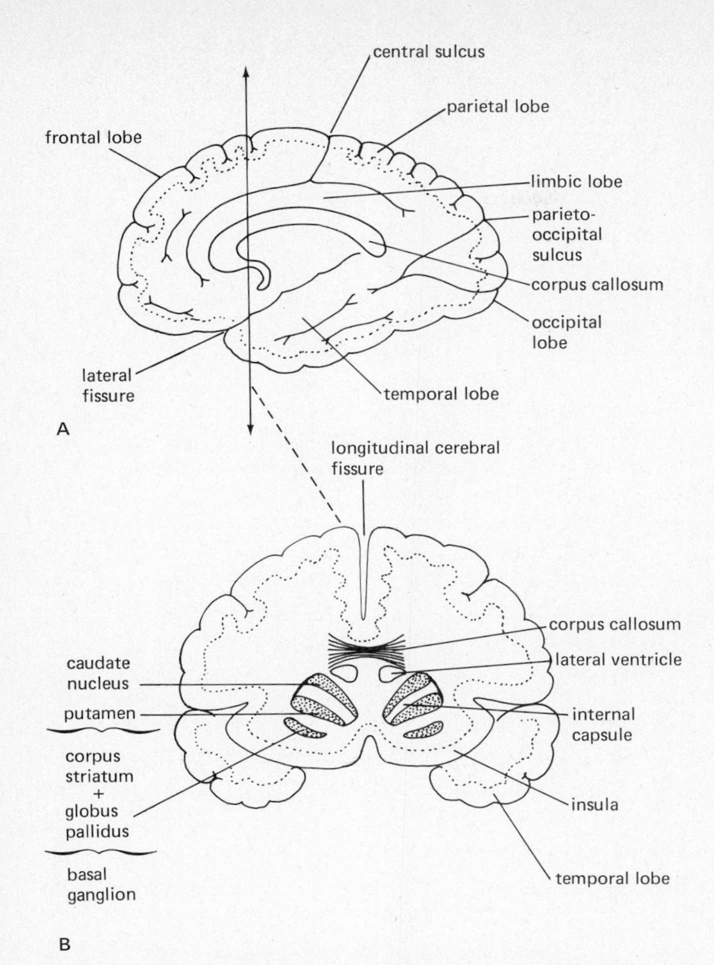

Figure 8-27. **(A) Median surface of the cerebral hemispheres, showing the lobes. The insula is buried deep in the temporal lobe and is shown in the cross section of the brain in (B). (B) Cross section as indicated by the arrow in (A). The insula and basal ganglia are shown.**

temporal lobes, and a fifth lobe which we have described as the *limbic lobe* [Fig. 8-27(A)]. Another lobe is the *insula,* which is buried deep in the lateral (Sylvian) fissure. Even deeper in the telencephalon are the *basal ganglia,* masses of gray matter that are part of the more primitive sensorimotor system that predominates in lower species. The insula and basal ganglia are shown in Fig. 8-27(B).

The exaggerated size of the cerebral hemispheres in humans involves not only an increased functional ability of the cortex, but also a great increase in the number of feedback circuits between the cortex and the subcortical elements. The ability of primates, especially humans, to perform highly skilled tasks is dependent not only on the remarkable development of the cerebellum but also on the phylogenetically new and direct connection between the cortex and the motoneurons of the spinal cord (the *corticospinal tract*) (Fig. 8-18). The capacity to discriminate fine differences in forms, shape, and textures also derives from the greater development of the sensory cortex.

The Development of the Cerebral Cortex. During the third month of the embryonic development of the human being, nerve cells from the inner layers of the forebrain migrate to the marginal zone, to give rise to the external layer of gray matter that forms the cerebral cortex. This cortical layer continues to grow at a very rapid pace so that it is thrown into folds or *gyri,* separated by *fissures and sulci,*[3] which rest upon the inner white matter. All the larger mammalian brains show well-developed gyri, and the degree of convolution of the cortex is a rather reliable indicator of the evolutionary stage of development of the brain (Figs. 8-2 and 8-27).

THE LOBES OF THE CEREBRAL HEMISPHERES

Macroscopic Structure. Although the phrenologists have no basis at all for their claim to tell character by "bumps" on the skull (which are far more likely to correspond to pressures exerted on the infantile skull during birth than to later development of personality traits), there is some correlation between function and the division of the cerebral cortex into lobes by the recognizable sulci and fissures (Fig. 8-28).

These lobes have been mapped out initially by their superficial appearance, by the study of diseased brains, and through electrophysiological studies. Because of the spread of electrical excitement through various parts of the brain, including circuits to and from lower areas of the nervous system, it is very difficult to pinpoint function in the intact, healthy brain. The correspondence is most clearly marked for the occipital lobe, where visual functions are markedly localized, but there is so much electrical intercommunication among the other lobes of the brain that the functional division is considerably less specific. The divisions are very useful, however, as points of reference and as landmarks. By referring to Figs. 8-27(A) and 8-28, the general areas assigned to the frontal, parietal, temporal, and occipital lobes can be recognized through the boundary lines of the lateral fissure and the central sulcus. The deepest lobe, the insula, is hidden by the temporal and frontal lobes. The limbic lobe has already been discussed in this section.

Microscopic Structure. The basic cell structure of the cerebral cortex is organized into six layers of varying thickness. Some differences exist in the cell types in different areas, especially in the motor cortex of the frontal lobe. In this area are found the giant *pyramidal cells* (cells of Betz), the axons of which pass down the corticospinal tracts to synapse directly on the ventral horn motor cells in the spinal cord. The large size of these cells is analogous to that

[3] The central sulcus running dorsolaterally separates the frontal lobe from the parietal; the lateral fissure (fissure of Sylvius) separates the long, tongue-shaped temporal lobe from the more dorsal frontal and parietal lobes. Other fissures partly separate the triangular occipital lobe from the parietal. The longitudinal cerebral fissure separates the right and left cerebral hemispheres.

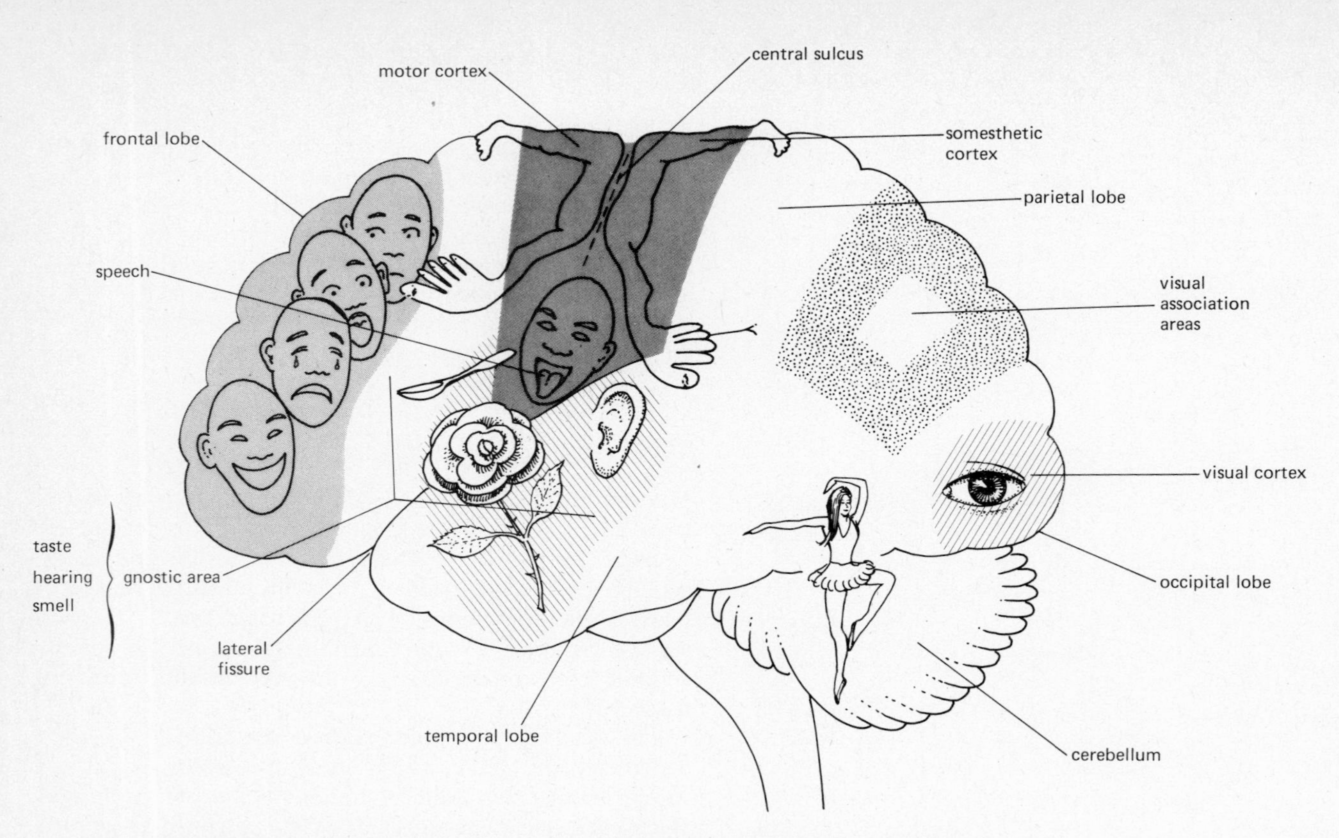

Figure 8-28. **The functions of the lobes of the cerebral hemispheres, schematically illustrated.**

of the ventral horn motor cells, which act as the final common path for so many neurons. However, many fibers of the pyramidal tract come from moderately sized pyramidal cells. The pyramidal cells, too, receive impulses from a great number of cells that synapse upon them from all directions (Fig. 8-29). In the sensory areas of the cortex, the cells are considerably smaller and closer together.

Motor Areas Immediately anterior to the central sulcus, in the *frontal lobe,* is an area which on stimulation results in movement of specific muscles on the other side of the body. This is the *primary motor area* (Figs. 8-18, 8-28, and 8-30). Electrical stimulation of the upper part of this region causes movement of the muscles of the lower limb; stimulation in its middle region moves the trunk and arms; and the lowest part of this primary motor area controls the muscles of the neck and face. Destruction of the motor area results in paralysis of voluntary muscle on the side of the body opposite to the damaged area.

The primary motor cortex does not initiate movement. Ingenious experiments on human subjects have shown that there is a change in brain potential as long as 0.8 msec before the onset of movement. This *readiness potential* first involves a negativity of a wide area of the cerebral cortex then narrows down to that area of the motor cortex concerned in the movement. It seems that complex patterns of

Figure 8-29. **Pyramidal cell of the motor cortex (largest cell) receives impulses from many cells that synapse with it. [Redrawn from P. K. Anokhin, *Ann. N. Y. Acad. Sci.* 92:899 (1961).]**

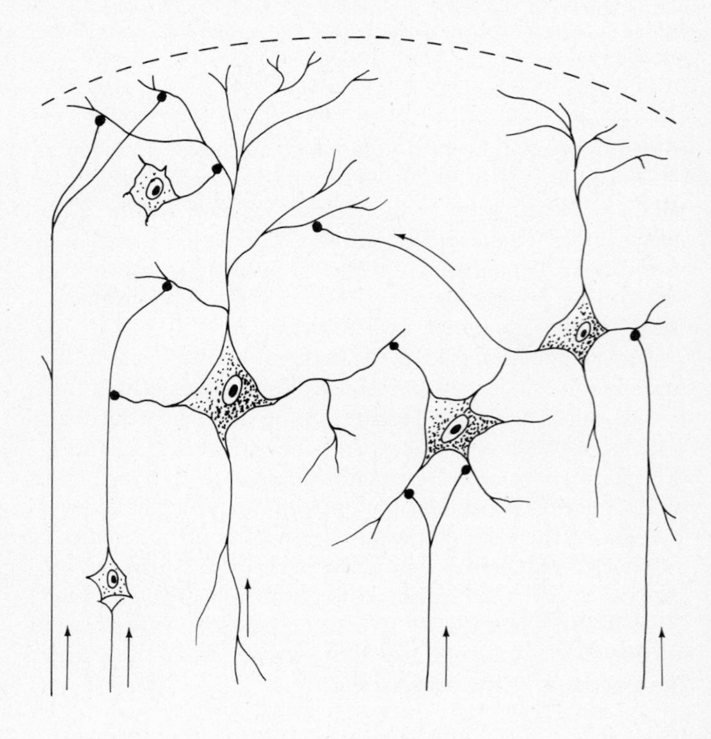

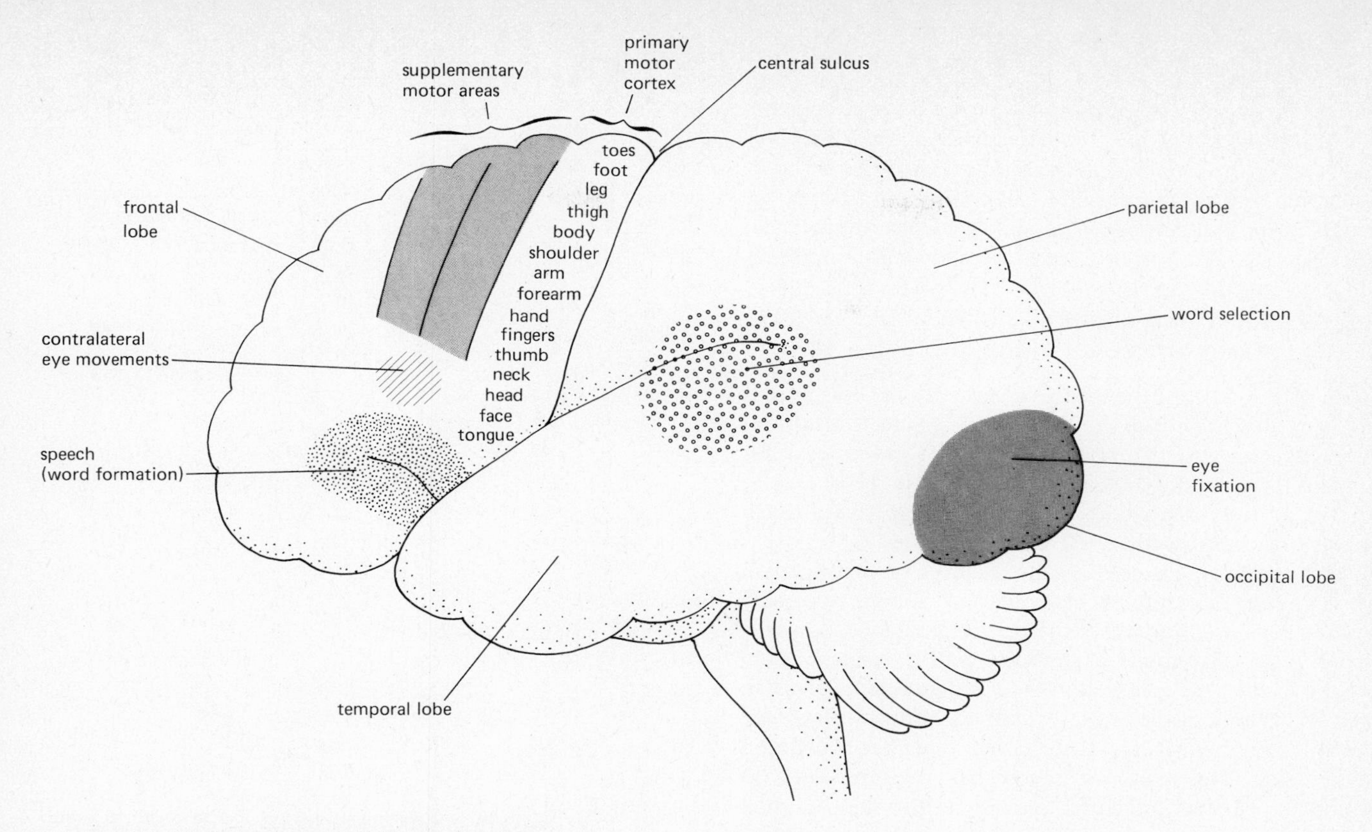

Figure 8-30. **The motor areas of the cerebral cortex.**

neuronal discharge eventually activate the correct motor cortical areas. Nor does the primary motor center alone control voluntary movement. In addition to stimulating and inhibiting impulses reaching it from the thalamus, the cerebellum, and the reticular formation, the activity of the motor cortex may be suppressed by several other cortical areas. Electrical stimulation of such *suppressor areas* causes the relaxation of muscular contraction. Other areas of the cortex have been described as *supplementary motor areas* because they cause bilateral movement, whereas stimulation of the primary motor cortex results in movement on one side of the body only.

Although simple movements can be elicited by stimulating the primary motor cortex, no coordinated, complex movements can be produced in this manner. Other areas of the frontal lobes, near the motor cortex, are necessary for integrated movements. Disturbances in motor function indicate the many processes that contribute to the successful execution of a movement like forming a letter of the alphabet. Depending on which part of the frontal lobe has been damaged, the individual may be able to move his arm and hand adequately but not be able to direct his pen to form the letter; he may be able to copy the letter but not to comprehend the meaning of it; he may be quite unable to execute these movements in their correct order, trying perhaps to scratch the paper against the pen. The inability to perform useful movements correctly is called *apraxia.*

Sensory Areas. The *parietal, temporal,* and *occipital lobes* all have areas in which specific sensations are localized. A large area at the junction of these lobes is responsible for the appreciation and interpretation of all sensations (the *gnostic area*). Picking and eating a peach produce a variety of sensations: the feel of its texture, the color of its skin, the scent of its ripeness, and finally the taste of the flesh. Destruction of the gnostic area prevents the simultaneous appreciation and understanding of these separate sensations.

It has been shown by recording the electrical activity of the cortex that more than one area is excited by each type of sensation, so that destruction of large areas of the sensory cortex still permits some kind of discrimination of sensation.

The *somesthetic cortex,* immediately behind the central sulcus in the *parietal lobe,* has the same pattern of representation of the parts of the body as the primary motor cortex in front of the central sulcus. The one side of the body is represented in the cortex of the other hemisphere. Stimulation of the somesthetic cortex gives rise to general sensations of touch, pressure, and temperature change. One of the most important functions of this area is the appreciation of spatial relationships, which permits the localization of sensation. Damage to the parietal lobe still permits awareness of touch or pressure or temperature stimulation, but the sensation cannot be localized accurately. This loss of accurate discrimination has been attributed to a general inability to concentrate, which also occurs after damage to the frontal lobes.

The representation of *sound frequency* discrimination in the *temporal lobe* and of *visual discrimination* in the *occipital lobe* is discussed in more detail in Chapter 22 in connection with hearing and vision. The perception of *taste* involves large areas of the *temporal cortex* and the *insula*. Removal or destruction of these regions does not completely destroy the sense of taste but decreases the ability to discriminate between flavors. The sensory areas are shown in Fig. 8-28.

The dominant and minor cerebral hemispheres

One of the cerebral hemispheres, usually the left one, is dominant over the other, the minor hemisphere. This dominance increases with the number of learned processes accumulated. As a result, the right side of the body is usually more adept in its movements than the left, right-handedness being more common than left-handedness.

There are far more subtle consequences of this division into dominant and minor hemispheres. Speech and self-consciousness are the factors that give dominance. There is often a clear anatomical asymmetry in the human brain, with the *speech area* enlarged in the dominant hemisphere in about 70 per cent of the brains examined. Human infants are born with this asymmetry, and the dominance is usually on the left side. No such asymmetry has been reported in nonhuman primates, a fact that corresponds to their linguistic inability. There is also a difference in the higher functions of the two hemispheres, as has been shown by the remarkable experiments on the split brain by Sperry (1970). This will be discussed in Sec. 9-3.

The basal ganglia

The basal ganglia are deeply placed masses of gray matter within the cerebral hemispheres. They are formed by the *caudate nucleus,* the *putamen,* and the *globus pallidus.* As we shall see, the first two nuclei are separated from each other by the fibers of the internal capsule, so that alternating bands of white and gray material give the area a striped appearance (corpus striatum) [Figs. 8-27(B) and 8-31]. These structures have many connections to one another and to different parts of the brain. They are closely associated with the nucleus ruber and the substantia nigra in the mesencephalon.

The globus pallidus is the main output; it sends impulses to the VL and VA thalamic nuclei, and so back to the cerebral cortex. Impulses from the VL reach the motor cortex; those from the VA are sent to the association cortex.

Most of our information about the function of the basal ganglia comes from studies on patients in whom these structures are diseased. Such patients show a wide variety of uncontrolled movements, including rigidity, tremor, and rapid and aimless movements such as seen in Huntington's chorea and other bizarre motor disturbances. Neurochemical studies of the brains of patients with Parkinson's disease have demonstrated a defect in the amount of the neurotransmitter dopamine in the basal ganglia and the substantia niger. The beneficial action of administered L-DOPA[4] confirms the functional relationship between the deficiency in

Figure 8-31. **The basal ganglia. The fibers of the internal capsule separate the caudate nucleus from the globus pallidus. The putamen is the third component of the basal ganglia.**

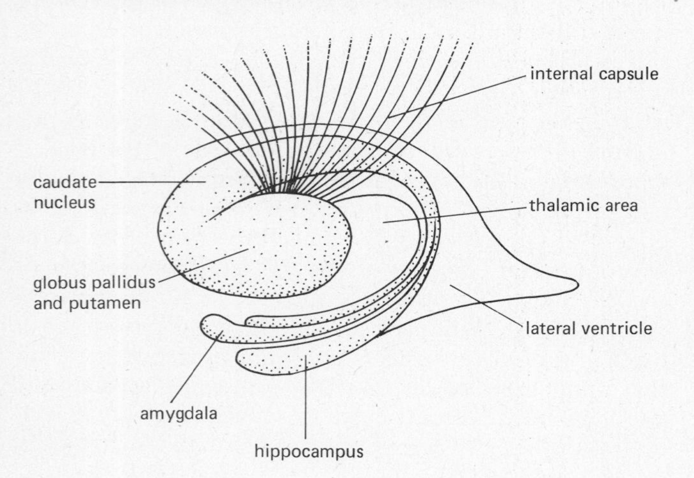

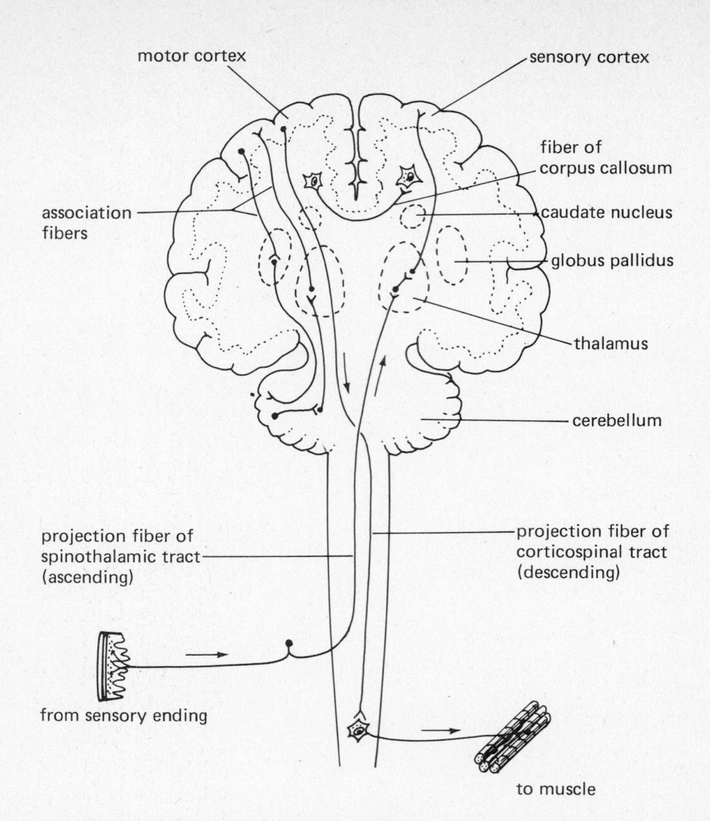

Figure 8-32. **Types of connecting pathways in the cerebral hemispheres.**

[4]L-DOPA rather than dopamine is given, for this *precursor* is able to cross the blood-brain barrier, whereas dopamine cannot (see dopamine synthesis, Sec. 9-5).

dopamine and the lack of coordination of motor function seen in these patients.

CONNECTING PATHWAYS

Fibers Connecting Different Areas of Each Cerebral Hemisphere. These *association fibers* vary greatly in length and complexity of synaptic connections. They appear to be arranged in circuits rather than in longitudinal formation. They form the anatomical basis for the concept of reverberating feedback systems that involve almost all of the areas of the cortex in any given situation (Fig. 8-32).

Fibers Connecting the Two Cerebral Hemispheres. The largest bundle of these commissural fibers forms the solid mass of white matter called the *corpus callosum* (Figs. 8-17, 8-23, and 8-32). This strong band can be seen at the base of the longitudinal fissure separating the two hemispheres, and it connects almost all areas of the left and right hemispheres in a mirror-image linkage. The corpus callosum is a most important communication system for the transmission of all events in one hemisphere to the other hemisphere (see Sec. 9-3).

Fibers Connecting the Cerebral Cortex with Lower Levels of the Nervous System. These *projection fibers* vary greatly in length. They may be the short fibers connecting the cortex with the thalamus or with the cerebellum, for example, or they may form part of the very long *tracts* that connects the cortex with the lower part of the spinal cord. These projection fibers bring impulses to the cortex (*afferent*) and conduct impulses away from the cortex (*efferent*).

Cited References

ECCLES, J. C. *The Understanding of the Brain,* 2nd ed. McGraw-Hill Book Company, New York, 1977.

ECCLES, J. C., M. ITO, and J. SZENTÁGOTHAI, *The Cerebellum as a Neuronal Machine.* Springer-Verlag New York, Inc., New York, 1967.

GRAFSTEIN, B. "Axonal Transport: Communication Between Soma and Synapse." In *Advances in Biochemical Psychopharmacology,* Vol. 1, E. Costa and P. Greengard, eds. Raven Press, New York, 1969.

HAMBURGH, M., L. A. MENDOZA, J. F. BURKART, and F. WEIL. "The Thyroid as a Time Clock in the Developing Nervous System." In *Cellular Aspects of Neural Growth and Development,* D. C. Pease, ed. UCLA Forum Med. Sci. No. 14. University of California Press, Los Angeles, 1971.

LEVI-MONTALCINI, R. "NGF: An Uncharted Route." In *The Neurosciences: Paths of Discovery.* The M.I.T. Press, Cambridge, Mass., 1975, p. 245.

OCHS, S. "Fast Transport of Materials in Mammalian Nerve Fibers." *Science* **176:** 252 (1972).

OLDS, J., and P. MILNER. "Positive Reinforcement Produced by Electrical Stimulation of Spatial Area and Other Regions of the Rat Brain." *J. Comp. Physiol. Psychol.* **47:** 419 (1954).

SHERRINGTON, C. S. *Integrative Action of the Nervous System.* Yale University Press, New Haven, Conn., 1906.

SPERRY, R. W. "Optic Nerve Regeneration with Return of Vision in Anurans." *J. Neurophysiol.* **7:** 57 (1944).

SPERRY, R. W. "Perception in the Absence of the Neo-cortical Commissures." In *Perception and Its Disorders,* D. A. Hamburg, K. H. Probram, and A. J. Stunkard, eds. The Williams & Wilkins Company, Baltimore, 1970, p. 123.

SPERRY, R. W. "Restoration of Vision After Crossing of Optic Nerves and After Contralateral Transplantation of Eye." *J. Neurophysiol.* **8:** 15 (1945).

Additional Readings

BOOKS

BERNHARD, C. G., and J. P. SCHADÉ, eds. *Developmental Neurology.* American Elsevier Publishing Co., Inc., New York, 1967.

BRAZIER, M., ed. *Growth and Development of the Brain: Nutritional, Genetic, and Environmental Factors.* Raven Press, New York, 1975.

CSERR, H. F., J. D. FENSTERMACHER, and V. FENCL. *Fluid Environment of the Brain.* Academic Press, Inc., New York, 1975.

DRACHMAN, D., ed. *Trophic Functions of the Neuron.* Ann. N.Y. Acad. Sci., 1974.

ECCLES, J. C., M. ITO, and J. SZENTÁGOTHAI. *The Cerebellum as a Neuronal Machine.* Springer-Verlag New York, Inc., New York, 1967.

HESS, W. R. *The Functional Organization of the Diencephalon.* Grune & Stratton, Inc., New York, 1958.

HYDÉN, H., ed. *The Neuron.* American Elsevier Publishing Co., Inc., New York, 1967.

ISAACSON, R. L. *The Limbic System.* Plenum Publishing Corporation, New York, 1974.

JACOBSON, M. *Experimental Neurobiology.* Holt, Rinehart and Winston, New York, 1970.

NOBACK, C. R., and R. J. DEMAREST. *The Human Nervous System,* 2nd ed. McGraw-Hill Book Company, New York, 1975.

RAPOPORT, S. I. *Blood-Brain Barrier in Physiology and Medicine.* Raven Press, New York, 1976.

SKOK, V. I. *Physiology of Autonomic Ganglia.* Igaku Shoin, Ltd., Tokyo, 1973.

ARTICLES

CLEMENTE, C. D., and M. H. CHASE. "Neurological Substrates of Aggressive Behavior." *Ann. Rev. Physiol.* **35:** 329 (1973).

DELGADO, J. M. R. "Recent Advances in Neurophysiology." *Excerpta Medica, Int. Congress Ser. No.* **180:** 36 (1968).

FRENCH, J. D. "The Reticular Formation." *Sci. Am.,* May 1957.

GAZZANIGA, M. S. "The Split Brain in Man." *Sci. Am.,* Aug. 1967.

GRAFSTEIN, B. "The Eyes Have It: Axonal Transport and Regeneration in the Optic Nerve." In *The Nervous System,* Vol 1: *The Basic Neurosciences,* D. B. Tower, ed. Raven Press, New York, 1975, p. 147.

LEVIN, S. "Stress and Behavior." *Sci. Am.,* Jan. 1971.

OCHS, S. Axoplasmic Transport–Energy Metabolism and Mechanism." In *The Peripheral Nervous System,* J. I. Hubbard, ed. Plenum Publishing Corporation, New York, 1974, p. 47.

OLDS, J. "Hypothalamic Substrates of Reward." *Physiol. Rev.* **42:** 554 (1962).

SNIDER, R. S. "The Cerebellum." *Sci. Am.,* Aug. 1958.

SPERRY, R. W. "The Eye and the Brain." *Sci. Am.,* May 1956.

WATSON, W. E. "Physiology of Neuroglia." *Physiol. Rev.* **54:** 245 (1974).

Chapter 9

Regulation by Higher Centers of the Brain

Men say they know many things;
But lo! they have taken wings,—
The arts and sciences,

And a thousand appliances;
The wind that blows
Is all that any body knows.

Henry Thoreau, "Men Say They Know Many Things"

THE HIGHER FUNCTIONS of the brain include the ability to learn, to think and remember, and for the human brain the awareness of self (state of consciousness) and the unique facility of speech. All of these phenomena depend on the *state of wakefulness,* a prerequisite for the brain to be able to attend, to select from and filter out superfluous sensory impulses, and to integrate important information into motor responses or to retain it as memory.

9-1 Wakefulness and Sleep

Wakefulness

THE RETICULAR ACTIVATING SYSTEM

The reticular formation extends through the upper medulla, pons, and midbrain, with profuse ascending connections to the thalamus and cerebral cortex and descending connections to the spinal cord. The reticular formation receives sensory input from all the peripheral receptors, including those of the muscles and joints, pain nerve endings, and visceral receptors, and the visual and auditory input from the eye and ear. Consequently, this collection of neural subcenters is in an excellent position to monitor all incoming and outgoing information (Fig. 9-1).

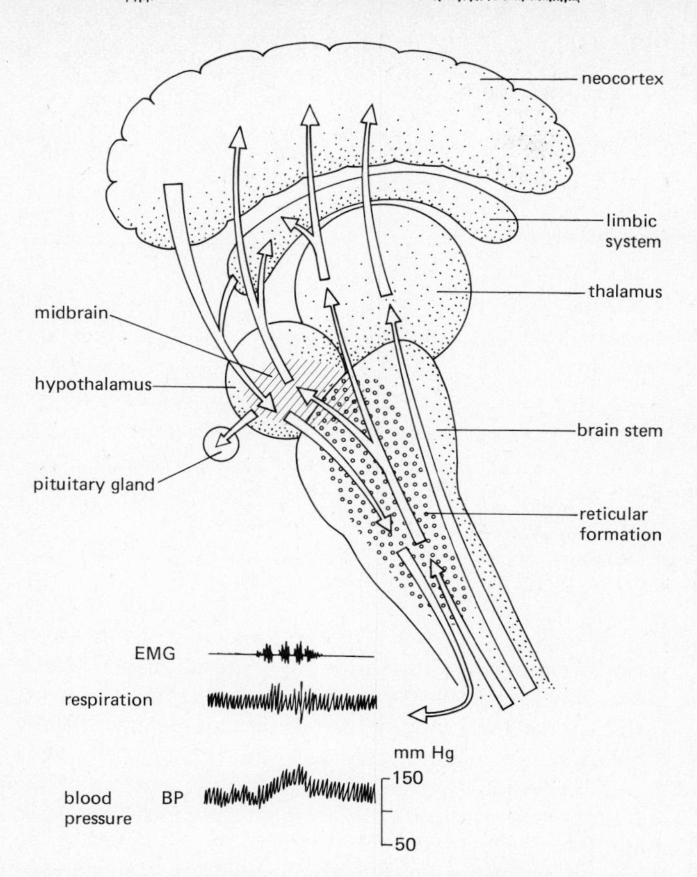

Figure 9-1. **The reticular formation receives sensory input from receptors and relays this information to the midbrain, limbic system, and neocortex. The response of the neocortex can be seen in the electroencephalogram (EEG). Muscle responses are recorded on the electromyogram (EMG). Visceral changes can be seen in the recordings of respiration and blood pressure. (Adapted from A. Zanchetti, in *The Neurosciences, A Study Program,* G. C. Quarton, T. Melnechuk, and F. O. Schmitt, eds., The Rockefeller University Press, New York, 1967.)**

Stimulation of the reticular formation activates the cerebral cortex, and so it is often referred to as the reticular activating system (RAS). Electrical stimulation of the RAS will immediately and gently wake a sleeping animal: a cat will go through its normal, lazy waking procedures without reacting, as though it had been subjected to a generalized shock treatment. The physiological mechanisms for RAS stimulation are through the profuse sensory input. Pain receptors are particularly effective for eliciting the arousal response; this is probably an evolutionary mechanism of benefit for survival.

Several positive feedback mechanisms are involved in the RAS. The aroused cerebral cortex further stimulates the RAS, particularly through the barrage of impulses from the motor cortex; this results in further RAS stimulation of spinal motoneurons (Sec. 23-13). The increased muscle tone and increased level of autonomic activity feed back through peripheral afferent pathways to maintain the level of excitement of the RAS. Movement is a particularly effective stimulus for the RAS, and wakefulness can be maintained by muscle activity, as evinced by the wriggling of sleepy, bored students.

Attention. Although impulses from the brainstem portion of the RAS cause generalized excitation through the cerebral cortex, more specific mechanisms involving the thalamic portion of the RAS stimulate selected areas of the cortex, permitting us to concentrate on the desired aspects of the sensory input and to ignore extraneous information.

Sleep

Sleep is an age-associated phenomenon. Sleep length decreases rapidly in the first year of life, then more slowly until the adult sleep-waking pattern of 7 to 8 hr of sleep, alternating with a long waking period, is achieved. Age disturbs this pattern, and older people enjoy frequent naps. However, individual variations are considerable, and sleep patterns are modifiable to some extent.

To ensure a wakeful state and alertness, the brain-stem *reticular formation* must be active. The onset of normal sleep is brought about by a reduction in sensory stimuli to the RAS. Yet sleep is not considered to be a passive state opposed to wakefulness; it can be induced by electrical stimulation of the thalamus or by chemical stimulation of various parts of the brain. The neurotransmitters are particularly effective in this regard: injection of epinephrine into the ventricles of the brain will induce sleep, as will the injection of acetylcholine or serotonin into various brain structures, including the caudate nucleus and limbic midbrain circuit.

The primary sleep-inducing areas of the brain appear to be in the *pons* and *upper medulla.* Section of the brain stem at the midpontine level permanently prevents sleep because it separates the RAS and its neural connections to the cerebral cortex, from the sleep-inducing nuclei posterior to the section.

The two states of sleep

Sleep is composed of two regularly alternating states, the first of which is slow wave sleep, the second being rapid eye movement sleep (REM sleep). REM sleep is also called paradoxical sleep, because of the sporadic jerky movements of some muscles during this stage (Fig. 9-2).

Slow Wave Sleep. A cat, which spends about 60 per cent of its life asleep, is a favorite experimental animal for sleep studies. A cat in slow wave sleep is recognizable by its curled posture (dependent on some muscle tonus, especially of the neck muscles), by its slow and steady breathing, and by the lack of eye movements behind the closed eyelids. If recordings are made of brain waves at this stage of sleep, the electroencephalogram (EEG) shows slow cortical waves of 11 to 16 cycles per second (cps). These brain waves are of

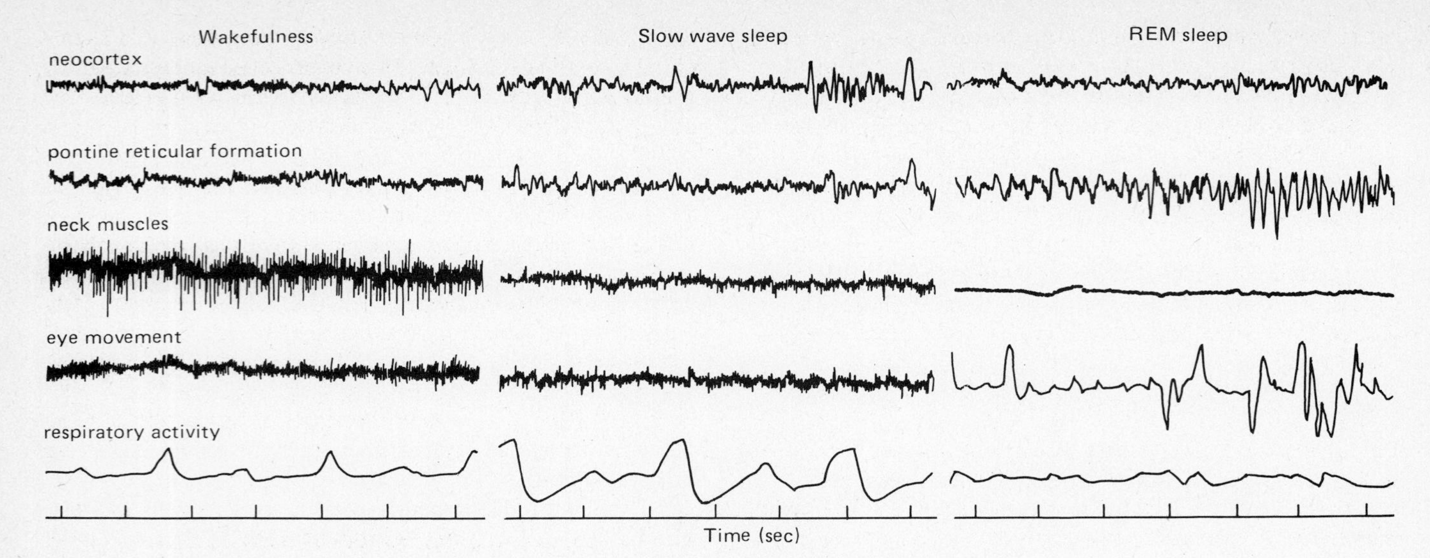

Figure 9-2. **Recordings made from the neocortex, the reticular formation and neck and eye muscles during wakefulness, slow wave sleep, and REM sleep. Note the rapid eye movements characteristic of REM sleep and the loss of muscle tonus that occurs during this period.**

large amplitude and are found mainly in the frontal and associated cortical areas. The normal activity in the midbrain region decreases, causing the general decrease in muscle tone characteristic of sleep. This decreased muscle tone is due to the removal of much of the facilitatory action of the reticular formation on spinal motoneurons.

REM Sleep. As the cat passes into REM sleep, the cortical activity becomes fast and irregular. It is accompanied by the *rapid eye movements* that give this stage its name. *Muscle tonus* almost completely disappears. The cat loses its curled-up position, and the head droops on the paws. While the antigravity muscles, especially the neck muscles, lose their tone, other twitching and jerking movements of the tail, limbs, and whiskers burst through the generalized tonic inhibition. In humans, REM sleep is accompanied by *dreaming,* and in males, by periodic penile erections. Rapid eye movement sleep appears to be deeper than slow wave sleep, for the arousal threshold is higher. Although prolonged deprivation of REM sleep results in hallucinations, aberrant behavior, and memory impairment, many individuals have been deprived of REM sleep by drugs for as long as a year without any significant resultant problems. REM sleep may be associated with the consolidation of memory needed for long-term memory (Sec. 9-3). No real unifying biological function for REM sleep has been discovered. The inference is that it is a restorative process, and that consequently there must be a need for it.

Figure 9-3. **Brain structures needed for sleep include the raphe nuclei, which, through the production of serotonin, depress the arousal effect of the reticular formation. The norepinephrine-producing cells of the locus ceruleus are needed for REM sleep.**

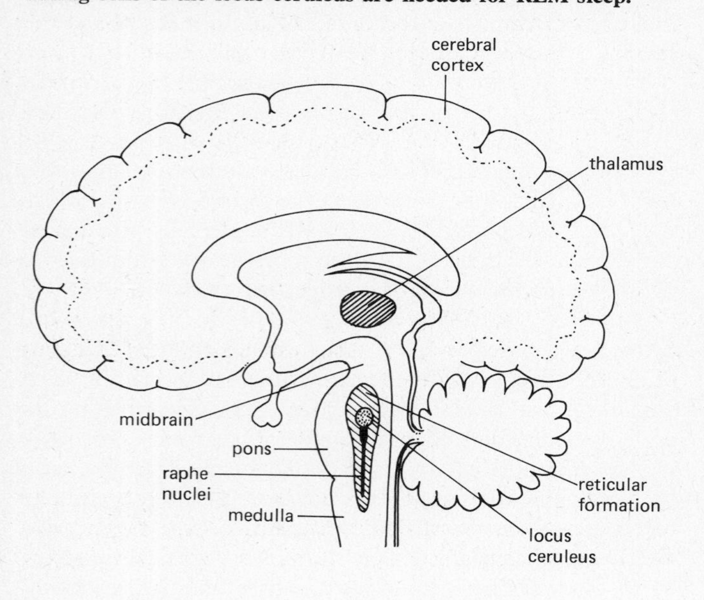

Brain structures needed for sleep

Surgical lesions of different parts of the brain have shown that REM sleep and slow wave sleep depend on different structures (Fig. 9-3). REM sleep is dependent on clusters of neurons in the roof (tegmentum) of the pons, forming the *nucleus locus ceruleus.* These neurons have been shown by histochemical fluorescent techniques to contain *norepinephrine;* bilateral destruction of these nuclei impairs REM sleep but not slow wave sleep. Slow wave sleep is abolished, however, if the cerebral cortex is removed.

Another distinct group of neurons implicated in sleep is located in the midline of the pons. These cells form the *raphe nuclei,* and they contain *serotonin.* If these raphe nuclei are destroyed, a permanent state of insomnia results.

Jouvet (1969) has proposed a linkage between the norepinephrine-containing neurons of the locus ceruleus and the raphe nuclei, which contain serotonin. He suggests that the serotonin-containing neurons trigger the norepinephrine neurons to induce REM sleep (Fig. 9-4). This is a rather complicated mechanism, involving cholinergic interneurons, but it may explain why REM sleep will only appear after a certain level of slow wave sleep has been attained (equivalent to more than 15 per cent of the day).

PHYLOGENESIS AND ONTOGENESIS OF SLEEP

Phylogenetically, REM sleep appears first in birds, where it is very short, only about 10 sec. All mammals show the two states of sleep, but REM sleep is much longer in newborn mammals, where the cortex is poorly developed and slow sleep consequently short. As the animal matures, the cortex develops; slow sleep lengthens, and REM sleep becomes correspondingly shorter.

A *basic rest-activity cycle* (BRAC) for the functioning of the nervous system has been proposed by Kleitman (1939). There is an alternation of high-voltage, low-frequency activity (rest) with low-voltage, mixed-frequency (REM) activity. Correlated to these electrical cycles are cycles in heart rate, respiration, rapid eye movements, and so forth. The length of this period has been called the *biological hour;* it increases in duration in proportion to body size and maturation. In the rat, the biological hour is about 10 min, in the cat about 30, in humans 85 to 90 min, and in the elephant 120 min. In humans, the BRAC lengthens with maturation from 50 min in the infant, to 70 min in the five-to-six-year-old, until it reaches the 90-min biological hour in the adult.

The BRAC operates during waking hours as well as in sleep and may be reflected in the attention span, for example, 90-min work followed by a coffee break and another 90-min work before lunch. It may also be locked into our internal biological clock, with a circadian rhythm of about 24 hr.

Slow sleep and REM sleep appear to be two different processes based on different structures and neurochemical mechanisms. Sleep laboratories are actively investigating the physiological and behavioral correlates of sleep and sleep deprivation, and the effects of drugs on these functions. The conclusion of most investigators in this difficult and interesting field appears to be that while the phenomena of the two states of sleep and some of their mechanisms are beginning to be understood, almost nothing is known about their function.

Figure 9-4. **Dependence of REM sleep upon a minimal amount of daily slow wave sleep.**

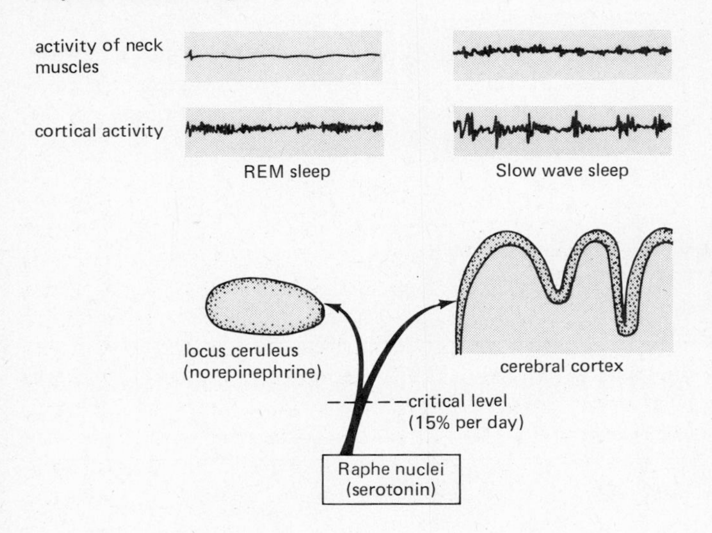

9–2 Electrical Activity of the Brain

The electrical activity of the brain has been studied extensively since the discovery of rhythmic brain waves by Hans Berger in 1924. These bioelectric phenomena can be studied by placing microelectrodes into specific brain regions or by recording from surface electrodes placed on the skull. The intensity and pattern of brain activity are determined largely by the degree of excitation from the reticular activating system and by specific sensory input. The record of the brain waves is the *electroencephalogram (EEG).*

Evoked Potentials

The brain responds with an electrical change, the evoked potential, to stimulation of a peripheral sense organ, a sensory nerve, or a point on the sensory pathway. Because these afferent impulses usually synapse with neurons in the thalamus, the resulting stimulation of the cortical cell is a blend of the activity of the first and second neurons in the pathway. Despite this complication, stimulation of the peripheral sense organ and the resultant location of an area of electrical excitement in the cortex do give an indication of the localization of function in the cortex. Our knowledge of the sensory areas of the cortex has been obtained to a large extent by this type of experimentation.

Intrinsic Potentials

The most prominent of the brain waves are the alpha waves, but the normal individual also shows delta, beta, and theta rhythms to varying degrees. Figure 9-5 shows these different brain waves in a normal adult.

ALPHA RHYTHMS

During *mental relaxation,* when there is a minimum sensory input, alpha waves can be recorded. These are about 8 to 13 cps and approximately 50 mV in amplitude. Alpha

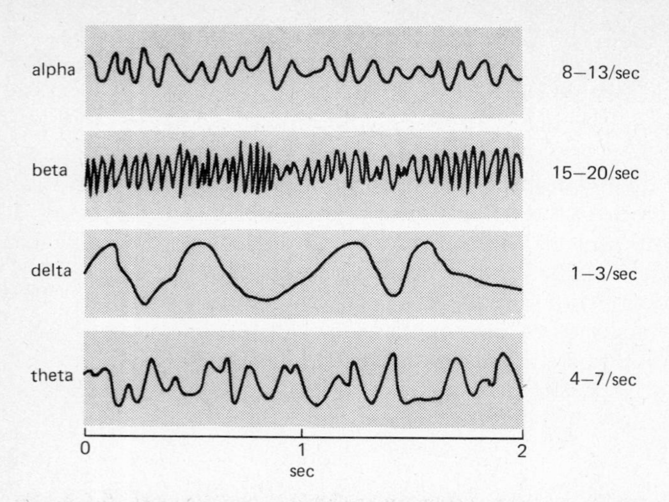

Figure 9-5. **The electroencephalographic waves of normal individuals.**

waves arise in the occipital lobe and spread over the rest of the brain. These waves disappear completely during sleep, and they are reversibly blocked out by visual attention (opening the eyes) or mental concentration (Fig. 9-6).

Beta rhythms

Beta rhythms have frequencies higher than 14 cps and are generally associated with *activation* and *tension*. They probably influence alpha activity to some extent.

Delta rhythms

Delta rhythms are slow, from 1 to 3 cps, and are associated with *deep sleep* in adults. They appear in infants and in patients with organic brain disease during periods of wakefulness. When delta rhythms persist beyond the age of 10 to 12 years, it is usually suggestive of mental immaturity.

Theta rhythms

Theta rhythms, with a frequency of 4 to 7 cps, are also associated with *childhood*. They appear in young people experiencing feelings of disappointment and frustration.

Abnormal Potentials

Paroxysmal discharges associated with epilepsy

The electrical manifestations of epilepsy are remarkably correlated with the clinical loss of consciousness and convulsions involving the skeletal musculature. The EEG shows a discharge of convulsive waves, which are generalized over the scalp or restricted to only a part of the scalp depending on the extent of the involvement (Fig. 9-7).

Generalized Discharge. In *grand mal* there is a discharge of spikes at from 8 to 12 cps, the amplitude of which increases as the frequency decreases, and these can be correlated to the contraction and relaxation of the muscles. It is accompanied by a loss of consciousness and lasts for more than 1 min.

In *petit mal,* the loss of consciousness may not be complete and lasts only a few seconds. The muscular contractions are slow, at 3 cps, hardly discernible, and are preceded by a rhythmic spike discharge of 3 cps.

Partial Discharge. These seizures are manifested by mental, sensory, or motor symptoms, including hallucinations and the gestures in response to them. It is impossible to locate an exact area of the brain to which the origin of each of these symptoms may be ascribed. Of course, a partial seizure may become generalized if there is a diffuse spread of excitement over the entire brain.

Causes of epilepsy

There are two main categories of epilepsy. One is associated with a lesion in the brain and accounts for 95 per cent of all cases. If the lesion is widespread, severe neurological or psychiatric disturbances between seizures may be associated with the epilepsy. This type is often called *organic epilepsy*. The second category, *functional epilepsy,* is only encountered in about 5 per cent of cases and is always manifested by seizure of either the *grand mal* or *petit mal* variety. It results from a fault in the functioning of the brain, often inherited as a "predisposition."

Figure 9-6. **The brain waves of the inactive brain (eyes closed) are alpha waves of approximately 10/sec. When the brain is active, as in problem solving or when the eyes are open, the alpha waves are replaced by the much smaller, faster beta waves.**

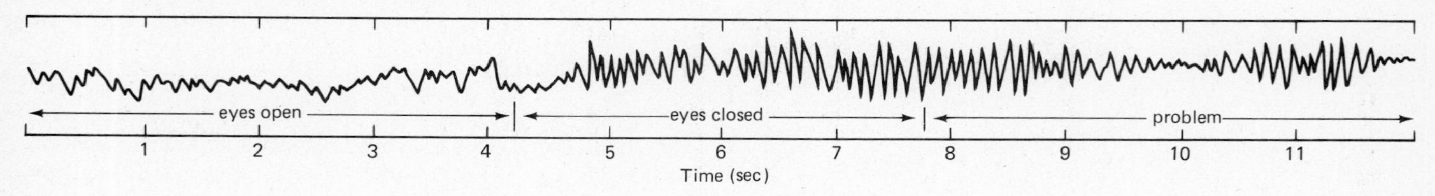

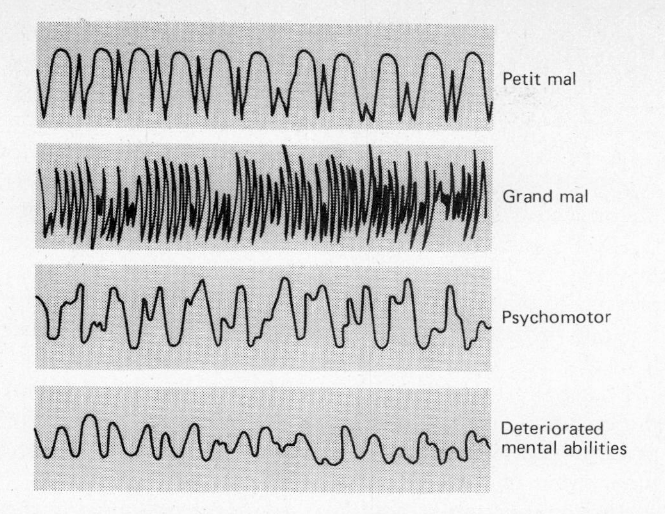

Figure 9-7. **Electroencephalograms in different types of epileptics. The neurons involved are hyperactive, and this hyperactivity appears to be autonomous. (From A. Guyton, *Textbook of Medical Physiology,* 5th ed., W. B. Saunders Company, Philadelphia, 1976.)**

Theory of mechanism of discharge

The present theory is that a localized discharge in the cerebral cortex originates from a hypersensitive focus and spreads through the cortex and subcortical centers, which in the epileptic individual are also hypersensitive and have a lowered threshold for electrical and chemical stimulation. It is fairly easy to evoke an epileptic seizure by applying a strong enough electrical stimulus to the whole of the brain or by the injection of certain drugs. It is very difficult, however, to cause a seizure by the localized electrical stimulation of the brain in the normal individual, so that the propagation of the electrical excitement from a localized area in the epileptic is assumed to be associated with an increase in excitability of the neurons.

9–3 Modern Theories of Learning, Memory, and Thought

The human brain is unique in its ability to add to its stock of information and to utilize this information. This is *learning,* and the retention and retrieval of this information is *memory.* Although we really know very little about the way in which we learn and remember, whatever evidence we do have clearly shows these phenomena to be based upon neurons and nerve pathways within the brain. The brain has often been compared to a computer, but it contains far more units (neurons) than an electronic computer and can respond in a far more varied manner. Sir Charles Sherrington's description (1906) of the brain as "an enchanted loom where millions of flashing shuttles weave a dissolving pattern, always a meaningful pattern though never an abiding one" shows that more than 70 years ago physiologists were visualizing the activity of the brain to be basically electrical in nature. Today it seems that not only are electrical impulses of vital importance in the activity of the brain, but also that chemical changes in the RNA and protein composition of the cells may be the way in which information is imprinted.

Learning

The classical concept of learning is that it is based upon the *conditioned reflex.* This is still the philosophy of the Russian physiologists adhering firmly to the dictates of Pavlov (1960), whose brilliant experiments first showed clearly the importance of this type of response. Pavlov showed that a simple reflex response to a physiological stimulus can be evoked when the stimulus is gradually replaced by a new and quite unphysiological one. For example, the physiological stimulus for salivation is the presence of food in the mouth. If food is seen and then placed in the mouth, the animal or human learns to associate the sight of the food with its later presence in the mouth, and finally the sight of the food alone is sufficient to cause salivation. This is now a conditioned reflex. It can be conditioned further if a bell is rung every time the animal sees the food; eventually the ringing of the bell will result in salivation. The animal has learned that when the bell is rung, it will be fed. This learning process has always been considered to involve only the cerebral cortex.

Incentive

Learning is accelerated or inhibited according to the incentive, and it is quite clear that learning is better if some motivation and reward are present. The experiments of Olds and Milner (1954), described in Chapter 8, have shown the existence of pleasure and punishment centers in the limbic system of rats. While learning takes place in the cortex, it appears to be directed toward stimulating the pleasure centers and away from the punishment centers, so that learning involves several levels of cortical and subcortical interaction.

Brain structures involved in learning

During conditioned reflex training, widespread changes in brain electric potentials occur, indicating the involvement of many brain structures. Yet ablation studies show that destruction of a brain structure usually affects only a certain type of learning. Figure 9-8 shows that in maze learning, the entire cortex is equipotential, but that for visual discrimination the visual area is essential. Complete destruction of this area renders the rat incapable of learning to discriminate between a black square and a white circle, for example. However, if only one sixtieth of the visual area is left intact, the rat is still able to learn form discrimination. This indi-

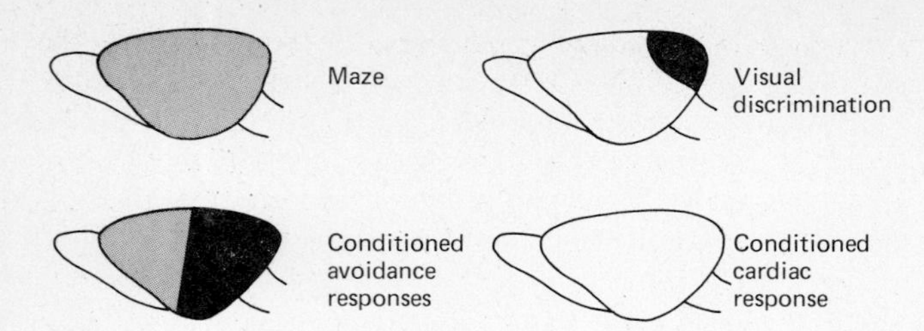

Figure 9-8. **Graph showing differential effects of cortical lesions on various learning tasks in rats. In maze learning, the entire cortex is equipotential. In visual form discrimination, only the visual area is critical. In conditioned avoidance response to visual stimuli, the entire cortex is involved, but lesions of the posterior half are more detrimental than lesions of the anterior half. In conditioned cardiac response to visual stimulus, complete decortication does not affect the learning or retention of this task. (From K. L. Chow, in *The Neurosciences: A Study Program,* The Rockefeller University Press, New York, 1967.)**

cates that a very small piece of brain tissue can take over the function of the whole area.

Ablation studies usually show the effect of the lesion to be temporary. The learning functions that have been lost can be recovered by extensive retraining, demonstrating the enormous potential of the brain to take over new functions.

Learning and the split brain

The vertebrate brain is a bilaterally symmetrical organ. The two cerebral hemispheres can be separated by sectioning the connecting fibers of the corpus callosum (Fig. 9-9), leaving two separate, functional half-brains, each with a separate realm of consciousness and awareness, and each with separate sensing, perceiving, thinking, and remembering systems. The *split brain* provides an intriguing technique for investigating the different properties of the dominant (left) and minor (right) cerebral hemispheres. It has been produced in cats and in monkeys, and in more than 20 human subjects who were suffering from uncontrollable epileptic seizures. The rigorously controlled experimental methods developed by Sperry (1974) have permitted the separate testing of each half of the brain.

Figure 9-9. **Cutting through the corpus callosum effectively splits the cerebral hemisphere into two separately functioning brains.**

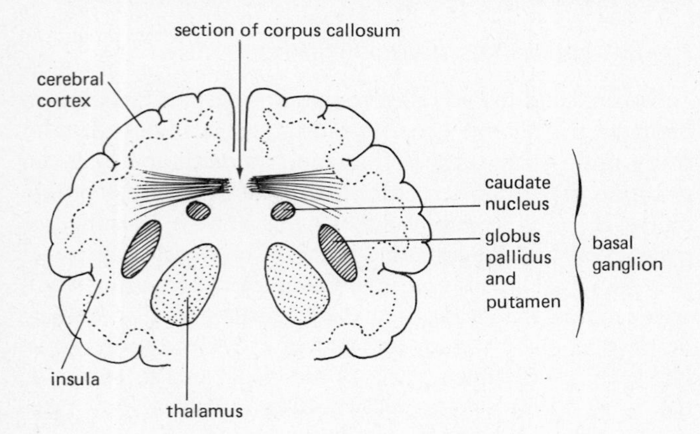

To understand the results, it is necessary to refer to Fig. 9-10, which shows how the left and right visual fields are projected onto the right and left visual cortices, respectively. Because there is a partial decussation (crossing over) in the optic chiasma, the right visual field for both eyes projects to the visual cortex in the left hemisphere (note the letter R in the left visual cortex). Similarly, the left visual fields for both eyes project to the right visual cortex.

Figure 9-10 also shows that there is a strictly unilateral projection of smell, and that the fiber tracts for hearing are predominantly crossed. The crossed motor and sensory innervation of the hands is also indicated.

For the split-brain individual, an object presented to the left-half visual field is recognized and remembered by the right hemisphere only. If the same object is subsequently presented to the left hemisphere (through the right-half visual field) it is new and unfamiliar. This is true also for objects identified by touch. *Memory traces* (*engrams*) laid down in one hemisphere are not accessible to the other hemisphere. This applies both to *short-term memory* (seconds to hours) and to *long-term memory* (days to years, often the lifetime of the animal).

Using special techniques, split-brain monkeys may be trained to learn two mutually contradictory performances. One hemisphere can learn exactly the reverse of what the other is learning, with no functional interference evident. These findings confirm that learning and memory can be confined to one hemisphere and that it is probably the neocortical structures that are involved.

Lateralization of Language. In tasks that involve language, the special complexity of the human brain becomes evident. One of the most striking symptoms in split-brain patients is their inability to describe in speech or writing anything presented to the left hand or to the left-half visual field. The speech centers and the engrams laid down for speaking and writing seem to be confined exclusively to the dominant hemisphere. The *left hemisphere* is the *talking hemisphere.* The right hemisphere is almost mute and can express itself only through simple motor responses, like pointing or signaling.

The *right hemisphere* has its own specialties, however. It has strongly developed *spatial abilities* and *pictorial and pattern sense,* far surpassing the dominant hemispheres in this regard. The minor hemisphere is also *musical.* The dominant hemisphere is more analytical; it is an excellent arithmetical computer, but it leaves geometry, with its spatial requirements, to the minor hemisphere. The eminent neurophysiologist Eccles (1977), believes that only the dominant hemisphere is in contact with the conscious self. The functions of the two hemispheres appear to be complimentary, each influencing the neural input individually.

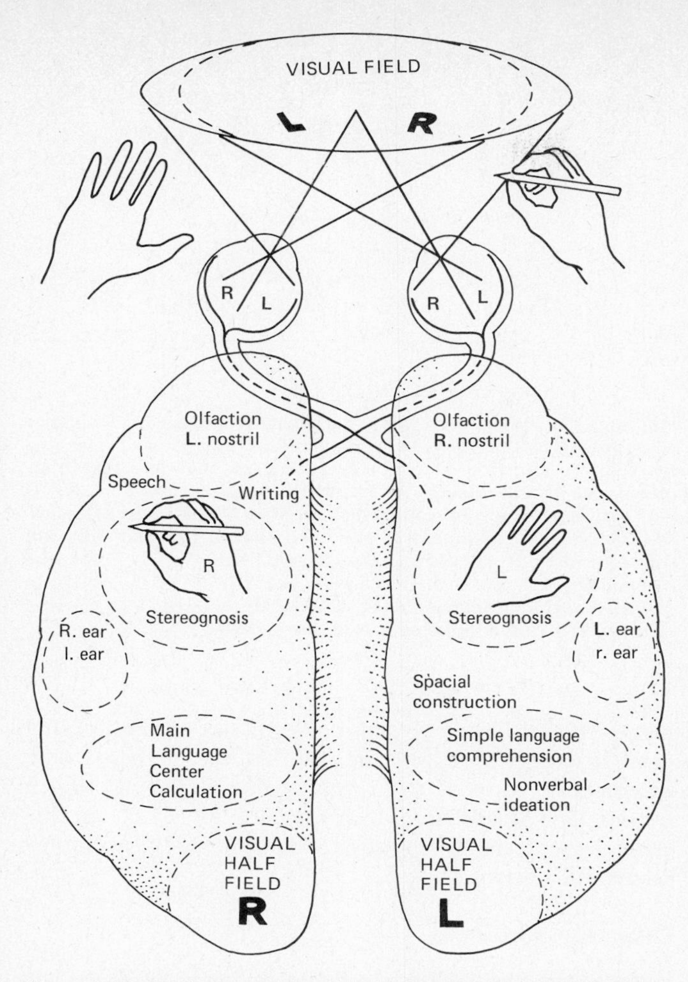

Figure 9-10. **Projection of the visual fields onto the visual cortices. The way in which the left and right visual fields are projected onto the right and left visual cortices, respectively, is due to the partial decussation in the optic chiasma. The diagram also shows the contralateral projection from the limbs to the sensory cortices. Olfaction, however, is ipsilateral. (Redrawn from R. W. Sperry, in *Perception and Its Disorders,* Res. Publ. Vol. 48, Association for Research in Nervous and Mental Disease, Inc., 1970.)**

Memory

SHORT-, INTERMEDIATE-, AND LONG-TERM MEMORY

The storage of memory that is established immediately on learning continues for hours and even days after learning. A biological analysis of memory must consider the link between the electrical and chemical events in the brain and the mechanisms by which new functional interneuronal relationships are created over a relatively long period of time.

There appear to be three interconnected steps in memory formation. The first, immediate step is *short-term memory,* which occurs in a matter of seconds. Then there is an *intermediate-term memory* of about 30 min to 3 hr, the amount of time needed for the establishment of true *long-term memory.*

THEORIES OF MEMORY FORMATION

Neuronal Connections and Functions. Circuits involving many different regions of the cerebral cortex are essential for the establishment of long-term memory. These circuits include reverberating circuits from the hippocampus and other limbic structures. The importance of the integrity of these brain circuits is demonstrated by one tragic instance in which a patient suffered bilateral lesions of the hippocampus (Milner, 1962). He completely lost all ability to establish memories for any experience occurring after the lesion. This *amnestic syndrome* means that the patient can remember only incidents of a few seconds' duration. For example, he cannot find his way to the bathroom in his new home, whereas he still has total recall for his old home. This also indicates that although the hippocampus and/or its associated circuits are involved in the consolidation of memory, memory storage does not occur in the hippocampus. The *memory trace,* or *engram,* is probably located in the neocortex.

Modification of Synapses. One concept of memory is that its structural basis in the brain lies in *modifications of synapses,* for there is no growth or change in mammalian neuronal pathways after their original formation. It is possible that repeated use of a particular synaptic system will cause hypertrophy of the synapses, or the budding of additional synapses. This is quite conjectural, but there is convincing histological evidence that visual deprivation in mice will result in loss of spine synapses in the visual cortex. Figure 8-7 shows postulated changes in dendritic spine synapses with varying degrees of use.

The response of synapses in frequent use may be *potentiated.* This has been demonstrated in the hippocampus, an area of the temporal lobe that appears to be a relay in the laying down of *long-term memory.* When hippocampal cells are subjected to mild stimulation for very brief periods (15 sec) repeated once every half-hour, there is a tremendous increase in the amplitude of their spike responses. Possible explanations for this vary from increased transmitter efficiency to receptor sensitivity and changes in cellular metabolism.

Potentiation of neuronal function has also been demonstrated in the *cerebellum.* When the input to the cerebellum from the *visual pathways* (by way of the climbing fibers) is superimposed upon the input from the *vestibular pathways* (by way of the mossy fibers), a change occurs so that the information from the vestibular pathways becomes more effective in controlling eye movements. The Purkyně cells appear to be modified in such a way that they respond more effectively to the mossy fiber input.

Integration of Functional Neuronal Groups. Studies on trained chimpanzees seem to indicate that each task that is learned forms neurons into groups that respond together. The ability to perform a particular task then depends on the particular response pattern generated in the brain. The correctness or incorrectness of motor acts by these chimpanzees can be accurately predicted by analyzing the brain waves appearing in a specific part of the cortex: one pattern emerges just prior to a correct response, a slightly different pattern when the chimpanzee is about to make a mistake.

Biochemical Changes. Behavioral studies in goldfish have shown that *protein synthesis* is necessary for long-term memory formation. When goldfish are treated immediately after training with puromycin, an inhibitor of protein synthesis, they are unable to remember the task they have just learned, which is to swim over a hurdle before the onset of an electric shock. Puromycin-treated goldfish can retain old memory tasks learned several hours or days before administration of the inhibitor, but they cannot retain memory of a new task.

Working on a cellular level, Hydén (1965) has shown that the *synthesis of RNA* by hippocampal cells is markedly increased following learning, and that this increase in RNA is the trigger for changes in protein synthesis. These biochemical changes that take place during learning are assumed to be prerequisites for the formation of long-term memory, because inhibition of protein synthesis by cells in the hippocampus prevents memory formation. In some way, the newly synthesized protein must influence the assemblage of neurons into a *functional pattern* that results in a particular behavior. We know practically nothing about the transduction of these biochemical changes into electrical changes.

The importance of the *sensory input* for the development and maintenance of proper synaptic contacts and for the synthesis of brain RNA and protein cannot be underestimated. *Sensory deprivation,* especially in young animals and children, results in structural and biochemical degradation of brain neurons. On a philosophical basis one might infer that the judicious selection of the sensory input to which young children are subjected may be of considerable significance in determining their subsequent behavior.

Pituitary Polypeptides and Memory. The pituitary polypeptide hormones *ACTH, MSH,* and *vasopressin* (antidiuretic hormone) are all released during stress; all, to varying degrees, *affect the performance of behavioral tasks.* The behavioral effects of ACTH are not mediated through the adrenal cortex, for they can be seen in *hypophysectomized animals* (animals in which the pituitary gland—hypophysis—has been removed), and, most convincingly, after the administration of peptide fragments of ACTH that do not stimulate adrenal cortical secretion (Fig. 9-11).

The long-lasting action of the posterior pituitary hormone vasopressin is of great interest. Rats that are *genetically deficient in vasopressin* suffer from *diabetes insipidus,* a disease characterized by the excretion of large volumes of dilute urine (Sec. 15-4), and although they are able to learn, their ability to retain information is severely affected. De Wied (1975) has shown that a single injection of vasopressin improves retention for as long as 120 hr; he has

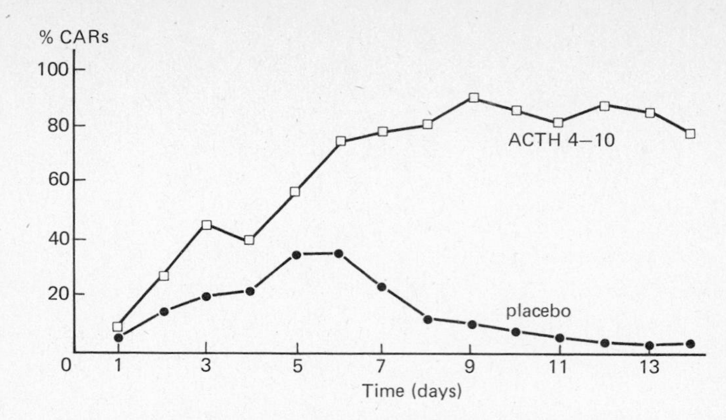

Figure 9-11. **The rate at which rats learn conditioned avoidance responses (CAR's) may be considered a measure of learning (avoidance acquisition). These graphs show that hypophysectomized rats given a placebo do not learn. Administration of ACTH 4–10 enables hypophysectomized rats to learn avoidance behavior. ACTH 4–10 is a fragment of the hormone ACTH that does not stimulate the adrenal cortex, showing this behavioral effect to be independent of adrenocortical hormones. (From D. de Wied, in *Frontiers of Neuroendocrinology,* W. F. Ganong and I. Martini, eds., Oxford University Press, New York, 1969.)**

suggested that these polypeptide hormones released during stress may play an important adaptive role, perhaps being enzymatically broken down into smaller fragments, which are rapidly released and which have specific effects on different parameters of nervous function, including complex behavioral responses.

Creative Thought

Some of the hypotheses outlined above may be applied to thinking. The brain may be compared to a supercomputer, possessing a multitude of available pathways for responses. The cortex may be considered as being continuously stimulated from the periphery and from the subcortical centers and responding in varied manner to this bombardment. Or a slow building of specific RNA and protein molecules within the neurons may open up new pathways and facilitate passage across old ones. It may be all these processes or none, but at the moment these ideas are in themselves creative enough to provide the outline for many more investigations into this difficult but fascinating field.

Eccles (1970) describes the characteristics of a brain that exhibits creative imagination. . . . "certain general statements can be made, though their inadequacy is all too apparent. There must, firstly, be an adequate number of neurones, and, more importantly, there should be a wealth of synaptic connection between them, so that there is, as it were, the structural basis for an immense range of patterns of activity. It is here that the inadequacy of explanation is so evident. There is but a poor correlation between brain size and intelligence, but in this assessment one is assuming a proportionality of brain size and neurone population. Furthermore, a chimpanzee brain may have a neurone population as high as 70% of a human brain, yet it displays almost no creative imagination. Secondly, there should be a particular sensitivity of the synapses to increase their function with usage so that memory patterns or engrams are readily formed and are enduring. Both of these properties will ensure that eventually there is built up in the brain an immense wealth of engrams of highly specific character. If added to this there is a peculiar potency for unresting activity in these engrams so that the spatio-temporal patterns are continually being woven in most complex and interacting forms, the stage is set for the deliverance of a "brain child" that is sired, as we say, by creative imagination."

9–4 Limbic and Reticular Systems and Behavior

The centers of emotional behavior have migrated, according to scientific belief, to many anatomical regions, including the heart and the cerebral cortex. They are now considered to reside in the *hypothalamus* and the *limbic structures* (Chap. 8). Limbic emotional behavior is modified by the brain-stem reticular formation, which mediates both *emotional experience and the expression of that emotion.* The RAS provides the background state of excitation that is necessary for the various types of behavior to occur. Emotion is manifested by both *motor* and *visceral reactions,* and these result mainly from discharges from the hypothalamus and midbrain through the descending reticular formation. The RAS monitors the main sensory pathways, and so is able to organize the complex phenomena of perception, dreams, and images by its *sensory filtering* and control of attention.

9–5 Biogenic Amines and Behavior

The main biogenic amines are the neurotransmitters norepinephrine (NE), dopamine (DA), and serotonin (5 hydroxytryptamine—5HT). Epinephrine (E) is also found in certain areas of the brain, in rather low concentrations, and may play a role as a neurotransmitter in the central nervous system. There is considerable evidence that the levels of these amines in the brain are altered in the affective psychoses, such as depression and schizophrenia. Depression is associated with low levels of NE, and it is possible that schizophrenia may be correlated with high levels of DA, with DA-NE systems closely interrelated. Profound behavioral changes take place in animals and in humans following alterations in the levels of these amines. There are, of course, no adequate animal models for the disturbed behavior of psychiatric patients, but studies using drugs that affect the action and metabolism of neurotransmitters are effective probes of the neurochemistry of behavioral disorders.

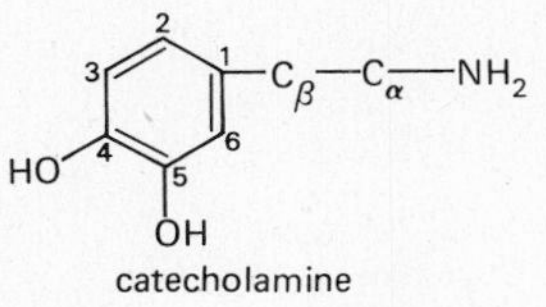

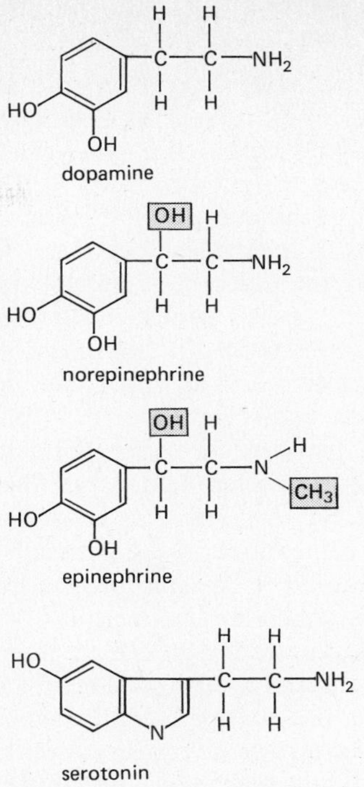

Figure 9-13. **Structures of the monoamines. Dopamine, norepinephrine, and epinephrine are catecholamines. Serotonin is an indolamine.**

Figure 9-12. **Structure of a catecholamine.**

Monoamines: Catecholamines and Serotonin

Norepinephrine, epinephrine, dopamine, and serotonin are *catecholamines,* compounds that contain a *catechol nucleus* (a benzene ring with two adjacent hydroxyl groups) and an *amine* group, as shown in Fig. 9-12. These monoamine neurotransmitters are all derived from the amino acid tyrosine. Serotonin, (5 hydroxytryptamine, 5HT) is an indolalkylamine derived from the amino acid tryptophan. The structural formulas of these biogenic amines are shown in Fig. 9-13. They are all classified as *monoamines* because they all possess a single amine group.

CATECHOLAMINES

Distribution of Catecholamines. Special techniques have been developed to visualize nerve cells and axons that contain catecholamines. Fluorescence histochemistry (Sec. 2-1), combined with electron microscopy, has been a very successful tool, permitting the mapping of *monoaminergic pathways* (nerve pathways containing monoamines) in the brain (Fig. 9-14). The monoamines are concentrated in the swellings, called *varicosities,* of these nerve terminals. Unlike the clear vesicles of ACh seen in cholinergic neurons, vesicles filled with monoamines are characterized by an electron dense core when viewed under the electron microscope (Fig. 9-15).

Other helpful techniques for the study of monoamines include sensitive radiochemical assays and fluorometric assays that permit the analysis of minute quantities of the catecholamines in various tissues and in blood.

CATECHOLAMINES IN THE PERIPHERAL NERVOUS SYSTEM. The pioneer work of von Euler in Sweden, first published in 1946 and reviewed in 1956, showed clearly a fine correlation between the postganglionic fibers of the sympathetic nervous system and nerve fibers with a high concentration of NE. As adrenaline is the British and European term for epinephrine, these sympathetic nerves are called *adrenergic nerves,* as contrasted to the nerve fibers containing ACh, which are cholinergic.

The highest concentrations of NE are found in the splenic nerve (a nonmyelinated sympathetic nerve) and in the sympathetic ganglia. The adrenal medulla, which is considered to be a modified sympathetic ganglion, contains the greatest concentration of epinephrine of all mammalian tissues, for the adrenal medulla is the tissue where significant amounts of the enzyme *N-methyl transferase* are found. This enzyme is essential for the conversion of NE to E. These reactions are described below and shown in Fig. 9-16.

Norepinephrine fulfills the requirements for a neurotransmitter in the peripheral nervous system, as it is synthesized and stored in sympathetic neurons and released in significant amounts when the sympathetic nerves are stimulated. In addition, the application of NE to the effector organs produces the same response as stimulation of these organs through the sympathetic nerves. Blocking agents that prevent the effects of sympathetic stimulation also prevent the effects of exogenous NE.

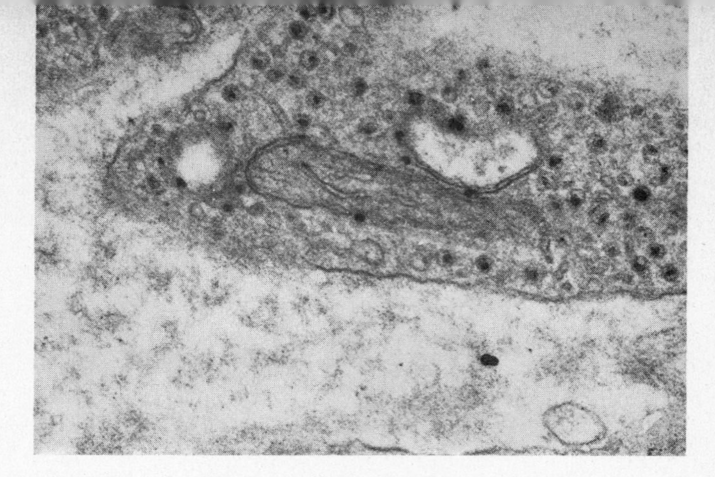
Figure 9-15. **Varicosities on the nerve endings from a rat pineal gland contain dense-core vesicles filled with norepinephrine. (Courtesy of F. Bloom, Salk Institute.)**

The role of DA in the peripheral nervous system is far less clear, and epinephrine functions almost exclusively as a hormone, except, as mentioned in the following discussion, for a possible role in limited regions of the central nervous system.

CATECHOLAMINES IN THE CENTRAL NERVOUS SYSTEM. *Norepinephrine.* Most of the NE in the mammalian nervous system is concentrated in the limbic area of the brain, especially in the hypothalamus and other brain areas that control the sympathetic nervous system [Fig. 9-14(A)]. Very

Figure 9-14. **(A) Distribution of norepinephrine-containing nerve terminals in the brain (striped areas). (B) Distribution of dopamine-containing nerve terminals in the brain (striped areas).**

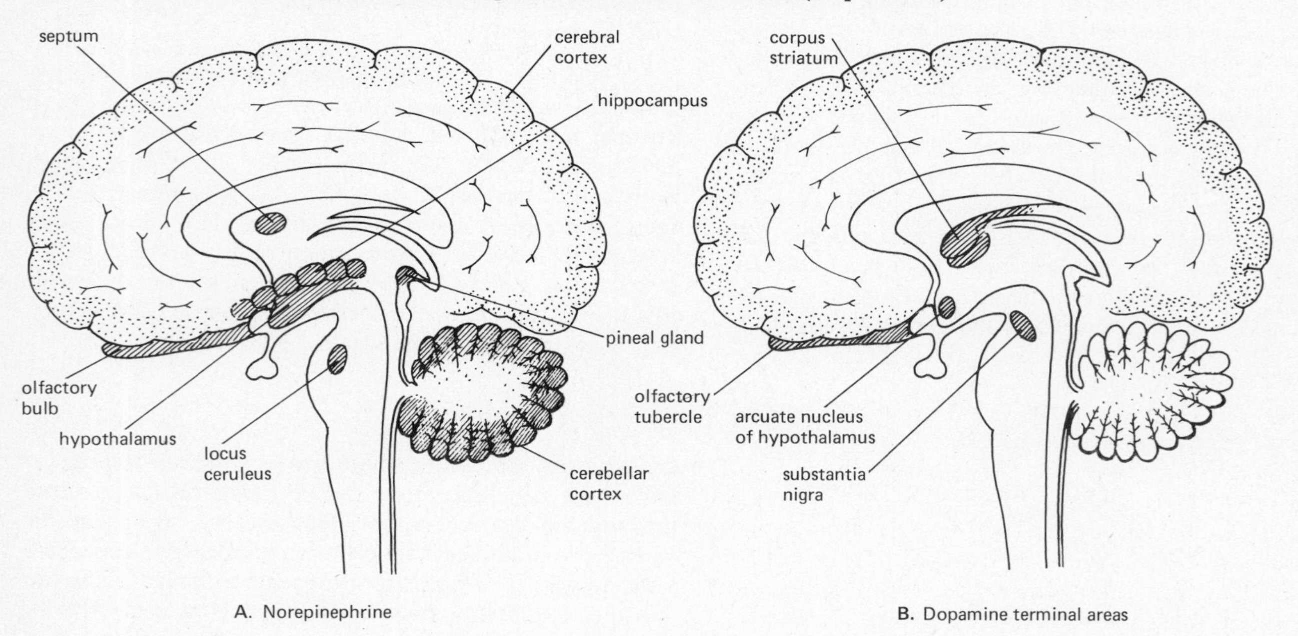

high concentrations of NE are found in the locus ceruleus in the pons, as mentioned earlier in this chapter. The connections of these NE neurons are quite diffuse and probably account for the ability of these neurons to affect various moods and degrees of wakefulness.

Dopamine. Brain DA is distributed quite differently from the NE-containing areas, indicating that while DA is a precursor for NE synthesis [Fig. 9-14(B)], DA is also a full-fledged neurotransmitter in its own right. More than 50 per cent of the total central nervous system catecholamine content consists of DA. Most of this DA is found in the *substantia nigra* and its projections to the *neostriatum,* a part of the *basal ganglia,* and in the olfactory lobe of the brain. The depletion of DA in the basal ganglia is associated with the motor disturbances characteristic of Parkinson's disease, a topic discussed in Secs. 8-6 and 23-13. Dopamine-containing neurons are also found in the *hypothalamus,* where they are involved in hypothalamic pituitary control (Sec. 11-3).

Dopamine is also found in the *superior cervical ganglion,* a sympathetic ganglion which has been shown to have significant amounts of three types of neurotransmitters, NE, DA, and ACh. Each of these transmitters is, of course, contained in a different type of neuron.

Epinephrine. The concentration of E in the central nervous system is very low, only about 10 per cent of the NE content. Like the adrenal medulla, the olfactory structures of the brain appear to have the enzyme *N*-methyl-transferase necessary for the formation of E from NE.

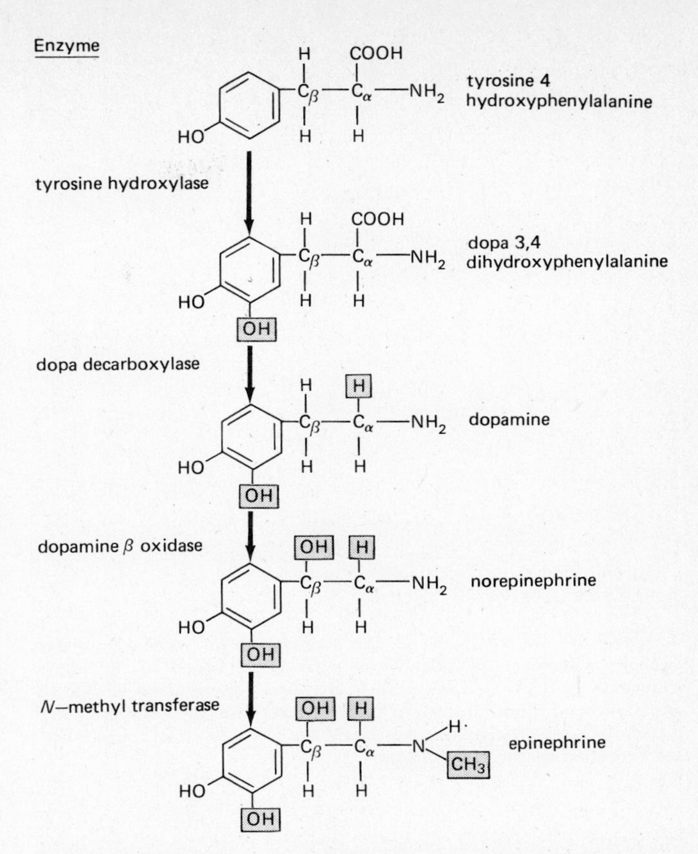

Figure 9-16. **Enzymatic synthesis of the catecholamine neurotransmitters.**

Synthesis of Catecholamines. The catecholamines DA, NE, and E are synthesized from the amino acid tyrosine through a series of enzymatic reactions shown in Fig. 9-16. While the nerve terminals are able to synthesize the transmitters, the proteins that are packaged with the transmitters in the vesicles must be sent down from the nerve cell body by axonal transport, since the terminals have no ribosomes for protein synthesis.

The steps in catecholamine synthesis include the following reactions:

1. *Tyrosine* (4 hydroxyphenylalanine) is *hydroxylated* to 3,4 dihydroxyphenylalanine (DOPA) by the rate-limiting enzyme, tyrosine hydroxylase.

2. *DOPA is decarboxylated* to form the monoamine neurotransmitter *dopamine* which can cross the blood-brain barrier. This reaction is catalyzed by the enzyme DOPA decarboxylase. Thus if L-DOPA is administered to patients with Parkinson's disease, L-DOPA can be converted in the basal ganglia to dopamine.

The formation of DA is the last synthetic reaction of the series in dopaminergic neurons.

3. *Dopamine is hydroxylated* at the β carbon in NE-containing neurons. This reaction requires the enzyme dopamine-β-oxidase. Dopamine-β-oxidase, ATP, and a soluble protein are all found inside the vesicles of the adrenergic nerve terminals where the conversion of DA to NE occurs.

4. *Norepinephrine is converted to E* in the adrenal medulla (and to a very much lesser extent in certain areas of the brain). The enzyme *N*-methyl transferase adds a methyl group to the amine of the molecule of NE.

Release of Catecholamines. Stimulation of the sympathetic nerves, especially the splenic nerve, causes an immediate release of NE. This release, like the release of ACh, is completely dependent upon the influx of Ca^{2+}. Exocytosis is probably involved in the release of the transmitter from the vesicles, although other mechanisms may also be important (Sec. 7-5). Again, like ACh, NE is probably stored in different pools within the nerve terminals, with the most recently synthesized transmitter released first.

It is very difficult to deplete the stores of NE within the nerve terminals, because of the efficient system of synthesis and uptake of released NE by these terminals.

Inactivation of Catecholamines. UPTAKE OF NOREPINEPHRINE. Almost 80 per cent of the NE released upon stimulation of the postganglionic sympathetic nerve fibers is taken back into the nerve terminals by a process Iversen (1973) has called *Uptake I* (Fig. 9-17). This is an active transport process and functions against a concentration gradient. Uptake I has been demonstrated by the injection of radioactive NE into the circulation: it is taken up almost

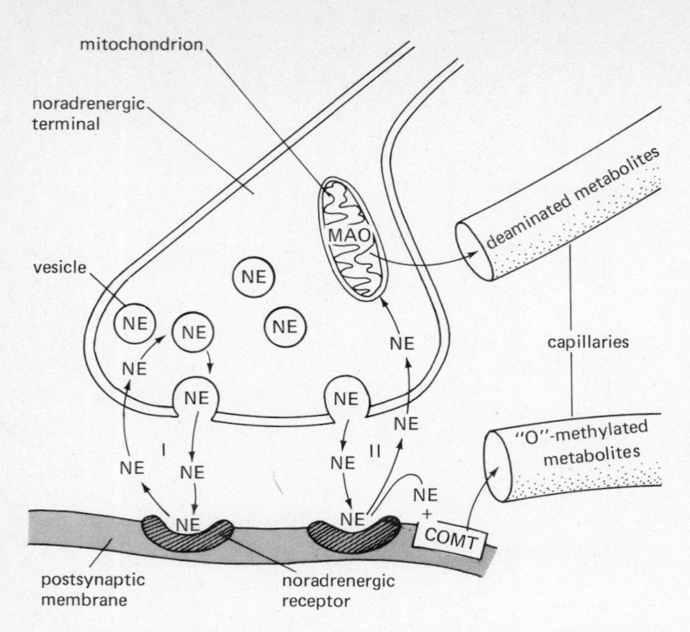

Figure 9-17. **Inactivation of the catecholamine norepinephrine (NE) by Uptake I and II. NE is actively returned to the nerve terminals by Uptake I. Uptake II involves inactivation of NE by intraneuronal monoamine oxidase (MAO) or by extraneuronal catechol-O-methyltransferase (COMT). The resulting metabolites are returned to the circulation.**

exclusively by those tissues that are rich in sympathetic nerve fibers. If the sympathetic nerves to these tissues are cut, the tissues are unable to concentrate significant amounts of NE. This rapid re-uptake of NE is a very effective and efficient way of inactivating the transmitter by removing it from the synaptic area.

ENZYMATIC INACTIVATION OF NOREPINEPHRINE. The remaining 20 per cent of the NE released into the synaptic cleft is inactivated enzymatically, and the resulting metabolites are taken up by the circulation. This *extraneuronal* re-uptake of NE is *Uptake II* (Fig. 9-17).

Figure 9-18. **Inactivation of a monoamine by the enzyme monoamine oxidase (MAO).**

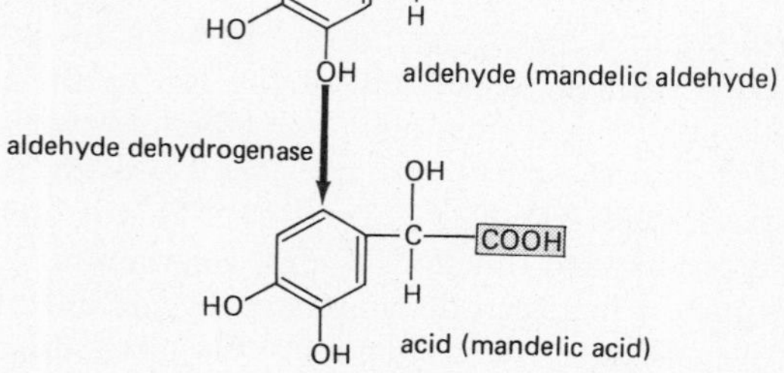

Enzyme

MAO

aldehyde dehydrogenase

monoamine (norepinephrine)

aldehyde (mandelic aldehyde)

acid (mandelic acid)

There are two enzymes involved in the metabolism of NE to its biologically inactive metabolites. *Monoamine oxidase* (MAO) is a relatively nonspecific enzyme that oxidizes most monoamines; MAO deaminates NE, E, and DA to their corresponding aldehydes, which are then rapidly oxidized to acids (Fig. 9-18). Monoamine oxidase is found in the *mitochondria* of the nerve terminals as well as in most tissues. The NE released by the vesicles into the cytoplasm of the terminals, and NE that is taken back up into the terminals by Uptake I, are subjected to deamination by MAO. The deaminated metabolites are released into the circulation (Uptake II).

The second enzyme involved in the inactivation of NE is *catechol-O-methyltransferase* (COMT), which is found *outside the neuron,* in the synaptic cleft, probably closely associated with the adrenergic receptors. The enzyme COMT was discovered by Axelrod (1959). Axelrod, von Euler, and Katz shared a Nobel Prize for their work on neurotransmitters: the first two investigators having studied catecholamine distribution and metabolism, while Katz was honored for his work on acetylcholine.

The enzyme COMT inactivates NE by transferring a methyl group to the hydroxy group on the catechol nucleus (the metahydroxy group). This reaction requires the presence of Mg^{2+}. The product is normetanephrine, which is then deaminated to an aldehyde (mandelic aldehyde) by MAO (Fig. 9-19). These reactions have been studied mainly in the liver, and it is assumed that similar reactions occur in neuronal tissue.

Mode of Action of Catecholamines. Catecholamine neurotransmitters combine with specific protein receptors on the postsynaptic membrane, initiating a series of reactions that culminate in the particular response of the affected cell. All catecholamine-triggered reactions affect the level of a "second messenger" cyclic AMP within the cell; cyclic AMP is responsible for the subsequent metabolic events in the cell. This topic is discussed in detail in Sec. 10-8.

Norepinephrine increases the heart rate and contractility, and causes contraction of the smooth muscles of the blood vessels (Sec. 12-5). The effects of NE on central neurons are less well understood. Microiontophoretic application of NE and of DA invariably causes a depression of spontaneous activity of these neurons.

SEROTONIN

Serotonin, 5-hydroxytryptamin (5-HT), is an indolealkylamine, the structure of which is shown in Fig. 9-12. It is similar, because of its indole component, to the psychedelic drug lysergic acid diethylamide (LSD), which has led to the concept, partially supported by experimental evidence, that the hallucinogens may affect behavior through serotinergic neurons.

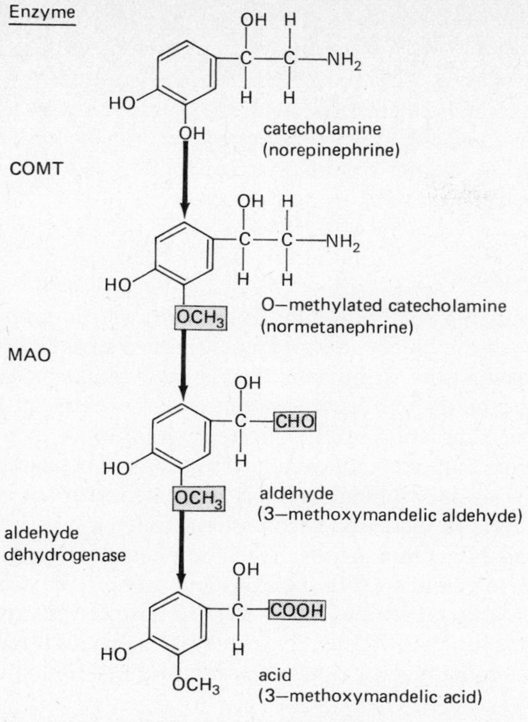

Enzyme

catecholamine (norepinephrine)

COMT

O–methylated catecholamine (normetanephrine)

MAO

aldehyde (3–methoxymandelic aldehyde)

aldehyde dehydrogenase

acid (3–methoxymandelic acid)

Figure 9-19. **Inactivation of a catecholamine by the enzyme catechol-O-methyltransferase (COMT), followed by MAO inactivation.**

Distribution of Serotonin. Serotonin is found in blood and in certain cells of the intestinal mucosa. Serotonin causes contraction of smooth muscle organs and is involved in blood coagulation. It is also a putative neurotransmitter in the central nervous system.

In the brain, fluorescence histochemistry has shown that most of the serotonin-containing neurons are concentrated in the raphe nuclei of the pons, where they are functionally involved in the regulation of sleep and wakefulness. As in other monoamine-containing cells, serotonin is found in electron-dense synaptic vesicles.

The *pineal gland,* which was described in Sec. 8-6, is an especially efficient site of serotonin synthesis, containing more than 50 times more serotonin per gram than the rest of the brain (except for the raphe nuclei). The pineal utilizes serotonin as a precursor for the pineal hormone *melatonin.*

Synthesis of Serotonin by the Brain

1. *Uptake of tryptophan.* Serotonin is synthesized from the amino acid tryptophan, which is actively taken up by the brain from plasma tryptophan. If the diet is deficient in this amino acid, brain serotonin falls to seriously low levels, so that dietary deficiencies can affect brain and behavior through this mechanism.
2. *Tryptophan is hydroxylated* at the fifth position to form form 5 hydroxytryptophan. This reaction is catalyzed by the enzyme tryptophan hydroxylase. Figure 9-20 shows this reaction and the subsequent stages in serotonin synthesis and metabolism.
3. *5 hydroxytryptophan is decarboxylated* by amino acid decarboxylase to form the monoamine serotonin (5-hydroxytryptamine).

Metabolism of Serotonin. Serotonin is inactivated in the brain by MAO by transformation to an aldehyde, which is rapidly oxidized to 5-hydroxy indole acetic acid.

In the pineal gland, which has the necessary enzymes for the synthesis of melatonin, serotonin is first acetylated to *N*-acetyl serotonin by 5-HT *N*-acetylase; *N*-acetyl serotonin is then *O*-methylated to form melatonin. The enzyme required for this last reaction is found only in the pineal gland and is 5-hydroxy indole *O*-methyl transferase (HIOMT).

LIGHT REGULATION OF PINEAL ENZYMES. A particularly interesting aspect of these enzymatic reactions responsible for serotonin and melatonin synthesis in the pineal gland is their regulation by light. Wurtman (1968) has shown that light entering the central nervous system through the eyes will stimulate the sympathetic nerves that innervate the pineal gland. This causes the release of NE, which, through cyclic AMP, increases the production of tryptophan hy-

Figure 9-20. **Synthesis and metabolism of serotonin. The pathway for melatonin synthesis in the pineal is also shown.**

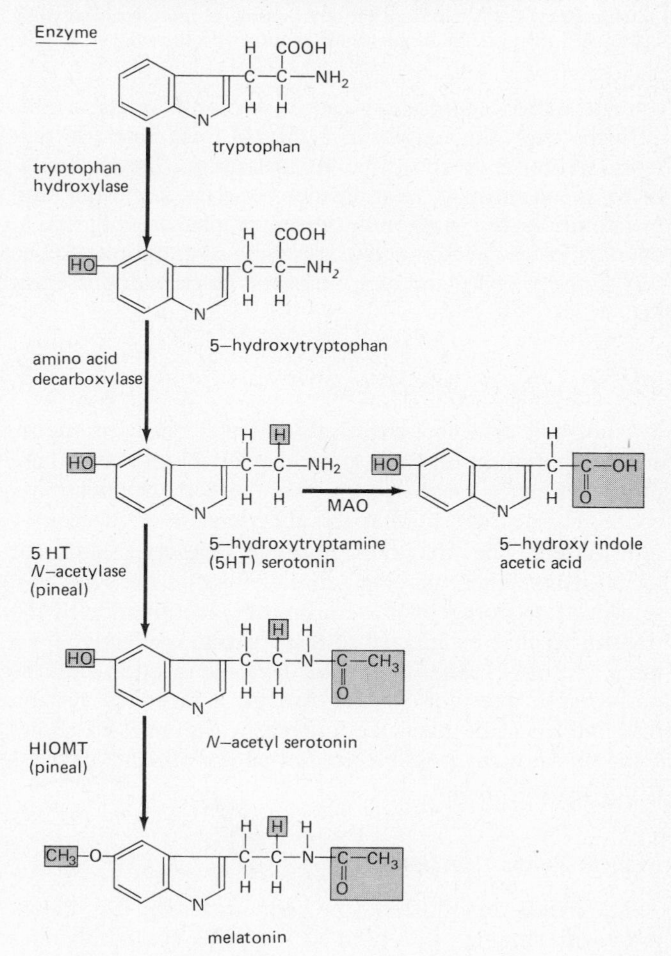

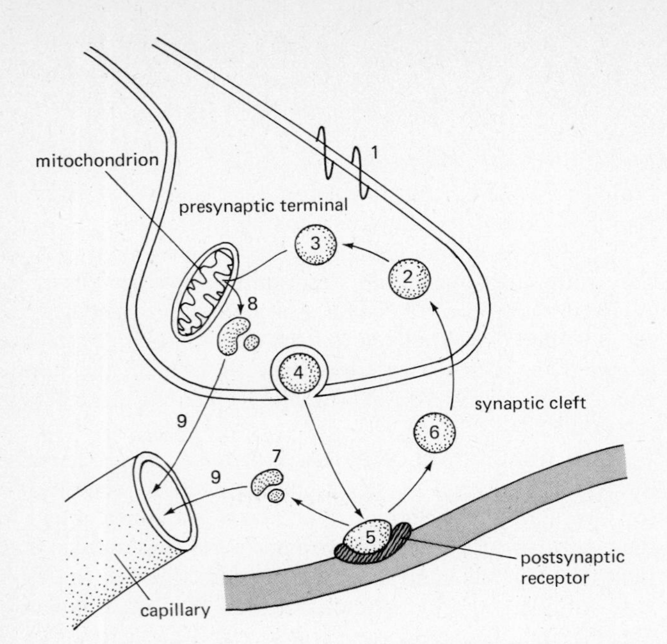

Figure 9-21. **Possible sites of action of drugs at a sympathetic nerve terminal: (1) nerve impulse, (2) synthesis of transmitter, (3) transfer and storage of transmitter, (4) release of transmitter, (5) combination of transmitter with postsynaptic receptor, (6) uptake of transmitter into neuron, (7) inactivation of transmitter in cleft by COMT, (8) inactivation of transmitter intraneuronally by MAO, and (9) passage of metabolites into circulation.**

droxylase and consequently causes serotonin levels to rise.

In the dark, the activity of 5-HIOMT increases, so that more serotonin is converted to melatonin. Consequently, there is an inverse relationship between serotonin and melatonin levels; serotonin concentrations are highest at noon, whereas melatonin levels are highest at midnight. The role of the pineal gland as a biological clock is discussed in Sec. 11-5.

The Psychotropic Drugs

Psychotropic drugs are chemicals that alter behavior, mood, and perception in humans, and behavior in animals. These drugs appear to react in some way with the monoamine-containing neurons to alter the effectiveness of the neurotransmitters. They may do so through a variety of different mechanisms. There may be a direct effect on the synthesis, transport, or storage of the transmitter. Inactivation of the transmitter may be affected so that it remains effective for a longer period of time. There may be some reaction with the postsynaptic receptor which, through a feedback system, may influence the presynaptic neuron. Figure 9-21 shows some of the many possible sites of action of some psychotropic drugs.

Major tranquilizers

The major tranquilizers (the phenothiazines, e.g., chlorpromazine) interact with both NE- and DA-containing neurons. The DA neurons are the most affected, and there is a marked increase in DA turnover, but the exact action at the DA synapse is still controversial. Some tranquilizers cause abnormal movements involving the extrapyramidal system, similar to those movements characteristic of Parkinson's disease, indicating the involvement of DA neurons in the nigro-striatal system.

Stimulants

Stimulants, such as amphetamine, a drug structurally very similar to the brain catecholamines, induce a state of behavioral excitability in humans that is very similar to schizophrenia and may be caused by an excessive release of DA or by an accumulation of DA through inhibition of its uptake. In animals, amphetamines induce stereotypic behavior, such as gnawing and sniffing, which is associated with DA but the characteristic increased locomotor activity more probably involves NE. On a cellular level, amphetamines have been shown to stimulate NE neurons but inhibit the firing of dopaminergic neurons (Fig. 9-22). Because DA neurons are inhibitory in their effects, the excitatory behavioral action of amphetamine may be due to the removal of DA inhibition.

Antianxiety drugs

Antianxiety drugs, such as meprobamate (Miltown) and benzodiazepine derivatives (Valium and Librium) appear to interact with NE neurons and to block the stress-induced increase in NE turnover.

Antidepressants

There are two major classes of antidepressants: the *monoamine oxidase inhibitors* (MAOI) and the *tricyclic antidepressants.* MAOI inhibit the enzyme monoamine oxidase, which oxidatively deaminates catecholamines (see Fig. 9-18). These inhibitors unselectively increase the endogenous levels of both DA and NE in the brain, resulting in behavioral stimulation. Because of the toxicity of this class of drugs, they are little used clinically.

Figure 9-22. **(A) Amphetamine accelerates the firing rate of neurons in the raphe nucleus, whereas LSD depresses their firing. The deflections are calibrated as spikes per minute. (B) Amphetamine depresses the firing rate of dopaminergic neurons.**

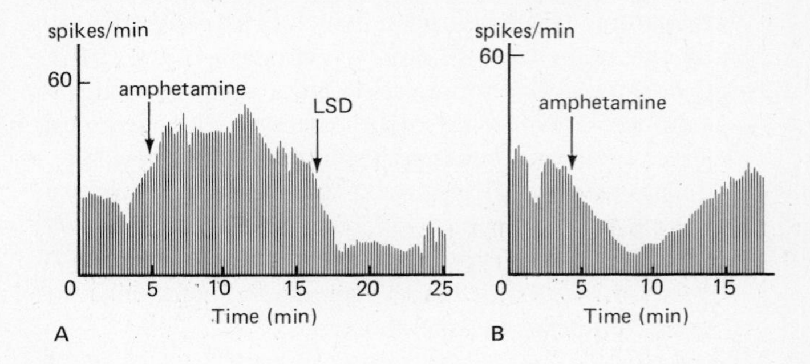

The *tricyclic antidepressants* (*imipramine, amitriptyline*) are widely used for the treatment of depression. Some of them are very potent inhibitors of NE uptake but have only minimal effect on DA. Other tricyclic antidepressants have less effect on NE uptake but are very effective in blocking serotonin uptake.

Hallucinogens

Hallucinogens, such as *cannabis* (*marijuana*) and *lysergic acid diethylamide* (*LSD*), may exert their central effect on brain amines, particularly NE and 5-HT. LSD enters the brain quickly and is removed slowly. Humans are susceptible to very low doses (20–25 μg); most animals, however, appear to be resistant to the effects of this drug, except in very high concentrations. The most significant biochemical effect of LSD is to increase 5-HT levels and decrease 5-HT turnover. Electrophysiological studies show that LSD inhibits the firing of the *raphe nuclei* that convey most of the 5-HT to the rest of the brain. Marijuana in very large doses has a similar effect. The relationship of these neurophysiological findings to the behavioral effects is not understood.

The link between hallucinogens, 5-HT, and the raphe nuclei sleep system in the brain is particularly interesting. If the sensory modulating system in the RAS is disrupted, by electrical stimulation or by drugs such as LSD, abnormal sensory phenomena dominate; there is a tremendous increase in the span of awareness, and sensory inputs are altered so that insignificant things become monumentally meaningful. Proprioceptive sensory input is no longer scanned and filtered properly, and strange alien feelings arise in the body, together with bizarre sensory disturbances.

Opiate Receptors and Naturally Occurring Brain Opiates

A relatively new and fascinating discovery is that the brain has *opiate* receptors, which have the ability to specifically bind opiates such as morphine and heroin. Snyder and Matthysse (1975) have shown that opiate receptors are concentrated in those areas of the brain concerned with the perception of pain, including the limbic system (pain is discussed in Sec. 22-9). These regions are also where opiates have their two main effects, *analgesia* (inhibition of pain sensation) and *euphoria* (elevation of mood).

Morphine does not occur naturally in the brain, but a number of different peptides found in the brain have opiate activity. The smallest of these are the *enkephalins,* pentapeptides for which the entire amino acid sequence is known. The *endorphins* are much larger polypeptides, the most active of which is *β-endorphin,* which is structurally identical to β-lipotropin (61-91), produced by the pituitary gland. It is significant that only the amino acid fragment 61-69 of β-lipotropin has opiate activity, and that the sequence 61-65 is identical to that of one of the enkephalins (met-enkephalin). As β-lipotropin also contains the amino acid sequence (β-lipotropin 41-58) common to the pituitary hormones ACTH and MSH, β-lipotropin may be the parent molecule that gives birth to an array of peptides that influence the brain and behavior. This concept is discussed in more detail in Sec. 10–7 and shown in Table 10-5.

Opiate antagonists

Opiate antagonists also bind to opiate receptors. Some of these antagonists are useful in treating overdoses of heroin because they competitively inhibit heroin action within a few minutes. However, many opiate antagonists also induce euphoria and addiction. The ideal drug would be one with sufficient opiate strength to be effective against pain but with adequate antagonist activity to prevent addiction.

Behavioral effects of endorphins

Profound behavioral changes are evoked by the injection of small amounts of β-endorphin into the brain or cerebrospinal fluid of rats. Within 30 min, prolonged muscular rigidity and immobility are induced, similar to the *catatonia* seen in human schizophrenics (Bloom, Segal, Ling, and Guillemin, 1976). The rigidity is so severe that a rat can be supported at its neck and the base of the tail by the thin edges of two metal bookends. This catatonia, which lasts for hours, can be reversed in seconds by the administration of naloxone, a specific morphine antagonist.

The questions that now arise are exciting. Because relatively large doses of enkephalins and endorphins are needed to produce analgesia, are they physiologically active? If endorphins are not of physiological significance, why does the brain have special opiate receptors? Can endorphins or enkephalins be used as nonaddictive pain killers to replace morphine clinically? Can certain mental illnesses be due to alterations in the regulation of endorphin metabolism? These are the far-reaching implications that are being investigated in many laboratories.

Cited References

Axelrod, J. "Metabolism of Epinephrine and Other Sympathomimetic Amines." *Physiol. Rev.* **39:** 751 (1959).

Berger, H. Über das Elektrenkephalogram des Menschen. *Arch. f. Psychiat.* **87:** 527 (1929).

Bloom, F., D. Segal, N. Ling, and R. Guillemin. "Endorphins: Profound Behavioral Effects in Rats Suggest New Etiological Factors in Mental Illness." *Science* **194:** 630 (1976).

de Wied, D., B. Bohus, and Tj. B. van Wimersma Greidanus. "Memory Deficit in Rats with Hereditary Diabetes Insipidus." *Brain Res.* **85:** 152 (1975).

Eccles, J. C. *Facing Reality: Philosophical Adventures by a Brain Scientist.* Springer-Verlag New York, Inc., New York, 1970, p. 129.

Eccles, J. C. *The Understanding of the Brain,* 2nd ed. McGraw-Hill Book Company, New York, 1977.

Hydén, H. "Activation of Nuclear RNA of Neurons and Glia in Learning." In *Anatomy of Memory.* D. P. Kimble, ed. Science and Behavior Books, Palo Alto, Calif., 1965, p. 178.

IVERSEN, L. L. "Catecholamine Uptake Processes." *Brit. Med. Bull.* **29:** 130 (1973).

JOUVET, M. 1969. "Biogenic Amines and the States of Sleep." *Science* **63:** 32.

KLEITMAN, N. *Sleep and Wakefulness.* University of Chicago Press, Chicago, Ill., 1963.

MILNER, B. "Les Troubles de la Mémoire Accompagnant des Lésions Hippocampiques Bilaterales." In *Physiologie de l'Hippocampe.* Colloques Internationaux No. 107. C.N.R.S., Paris, 1962.

PAVLOV, I. P. 1960. *Conditioned Reflexes.* G. V. Anrep., ed. and transl. Dover Publications, Inc., New York, 1960, p. 31.

SHERRINGTON, C. S. *Integrative Action of the Nervous System.* Yale University Press, New Haven, Conn., 1906.

SNYDER, S., and S. MATTHYSSE. *Opiate Receptor Mechanisms.* The M.I.T. Press, Cambridge, Mass., 1975.

SPERRY, R. W. "The Split Brain." In *The Neurosciences 3rd Study Program.* F. O. Schmidt and F. G. Wordon, eds. The Rockefeller University Press, New York, 1974.

VON EULER, U. S. *Noradrenaline.* Charles C Thomas, Springfield, Ill., 1956.

WURTMAN, R. J., J. AXELROD, and D. E. KELLY. *The Pineal.* Academic Press, Inc., New York, 1968.

VALDMAN, A. V., ed. *Pharmacology and Physiology of the Reticular Formation.* Progress in Brain Research 20. American Elsevier Publishing Co., Inc., New York, 1967.

WEITZMAN, E. D., ed. *Advances in Sleep Research,* Vol 2. Spectrum Publications, Inc., New York, 1976.

Additional Readings

BOOKS

AKERT, K., C. BALLY, and J. P. SCHADÉ, eds. *Sleep Mechanisms.* Progress in Brain Research 18. American Elsevier Publishing Co. Inc., New York, 1965.

BLACK, P., ed. *Physiological Correlates of Emotion.* Academic Press, Inc., New York, 1970.

CHASE, M. H., ed. *The Sleeping Brain.* Brain Research Institute, University of California, Los Angeles, 1972. (Paperback.)

COOPER, J. R., F. E. BLOOM, and R. H. ROTH. 1974. *The Biochemical Basis of Neuropharmacology,* 2nd ed. Oxford University Press, New York, 1974. (Paperback.)

COSTA, E., and P. GREENGARD, eds. *Advances in Biochemical Psychopharmacology,* Vols. 1-15. Raven Press, New York, 1969-1976.

ECCLES, J. C., ed. *Brain and Conscious Experience.* Springer-Verlag New York, Inc., New York, 1966.

ECCLES, J. C. *Facing Reality.* Springer-Verlag New York, Inc., New York, 1970. (Paperback.)

KALES, A., ed. *Sleep, Physiology and Pathology.* J. P. Lippincott Co., Philadelphia, 1969.

KARCZMAR, A. G., and J. C. ECCLES, eds. *Brain and Human Behavior.* Springer-Verlag New York, Inc., New York, 1972.

MECHOULAM, R. *Marijuana: Chemistry, Pharmacology, Metabolism, and Clinical Effects.* Academic Press, Inc., New York, 1973.

PRIBAM, K. H., ed. *Mood, States and Mind. Brain and Behaviour 1, Penguin Modern Psychology UPS 21.* Penguin Books, Middlesex, England, 1969. (Paperback.)

ARTICLES

AGRANOFF, B. W. "Memory and Protein Synthesis." *Sci. Am.,* June 1967.

AKISKAL, H. S., and W. T. MCKINNEY, Jr. "Depressive Disorders: Towards a Unified Hypothesis." *Science* **182:** 20 (1973).

DE WIED, D., B. BOHUS, and TJ. B. VAN WIMERSMA GREIDANUS. "Memory Deficit in Rats with Hereditary Diabetes Insipidus." *Brain Res.* **85:** 152 (1975).

FERNSTROM, J. D., and R. WURTMAN. "Nutrition and the Brain." *Sci. Am.,* Feb. 1974.

GONATAS, N. K. "Axonic and Synaptic Lesions in Neuropsychiatric Disorders." *Nature* **214**(5086): 352 (1967).

HORN, G., S. P. R. ROSE, and P. P. G. BATESON. "Experience and Plasticity in the Central Nervous System." *Science* **181:** 506 (1973).

JOUVET, M. "The States of Sleep." *Sci. Am.,* Feb. 1967.

KANDEL, E., and W. A. SPENCER. "Cellular Neurophysiological Approaches in the Study of Learning." *Physiol. Rev.* **48**(1): 65 (1968).

MILLER, N. E. "Chemical Coding of Behavior in the Brain." *Science* **148:** 328 (1965).

MORGANE, P. J. "Anatomical and Neurobiochemical Bases of the Central Nervous Control of Physiological Regulations and Behaviour." In *Neural Integration of Physiological Mechanisms and Behaviour.* G. J. Mogenson and F. R. Calaresu, eds. University of Toronto Press, Toronto, Canada, 1975, p. 24.

NATHANSON, J. A., and P. GREENGARD. "Second Messengers in the Brain." *Sci. Am.,* Aug. 1977.

PERT, C., M. J. KUHAR, and S. H. SNYDER. "Opiate Receptor: Autoradiographic Localization in Rat Brain." *Proc. Natl. Acad. Sci. USA* **73:** 3729 (1976).

SINSHEIMER, R. L. "The Brain of Pooh: An Essay on the Limits of Mind." In *Contemporary Readings in Biology,* G. E. Nelson and J. D. Ray, Jr., eds. John Wiley & Sons, Inc., New York, 1973.

SNYDER, S. H., S. P. BANERJEE, H. I. YAMAMURA, and D. GREENBERG. "Drugs, Neurotransmitters and Schizophrenia." *Science* **184:** 1243 (1974).

SNYDER, S. H., R. SIMANTOV, and G. W. PASTERNAK. "The Brain's Own Morphine, "Enkephalin": A Peptide Neurotransmitter?" In *Neurotransmitters, Hormones and Receptors: Novel Approaches.* The Society for Neuroscience, Bethesda, Md., 1976.

TSUKAHARA, N., H. HULTBORN, F. MURAKAMI, and Y. FUJITO. "The Response of the Dentate Gyrus to Partial Deafferentation." In *Golgi Centennial Symposium 1973: Perspectives in Neurobiology.* Raven Press, New York, 1975, p. 295.

UNGAR, G. 1971. "Chemical Transfer of Information." In *Handbook of Neurochemistry,* Vol. 6, A. Lajtha, ed., Plenum Publishing Corporation, New York, 1971, p. 241.

Chapter 10

Hormonal Regulation

We shall not cease from exploration
And the end of all our exploring
Will be to arrive where we started
And know the place for the first time.
Through the unknown, remembered gate
When the last of earth left to discover
Is that which was the beginning;
At the source of the longest river
The voice of the hidden waterfall
And the children in the apple-tree
Not known, because not looked for
But heard, half-heard, in the stillness
Between two waves of the sea.

T. S. Eliot, "Four Quartets"

10–1. Development of Humoral Control Systems

Integration within the organism has developed along two main lines, chemical and nervous. Chemical integration in its most primitive form utilizes chemicals that are produced as a result of metabolism and that affect almost all cells; the regulatory effect of carbon dioxide is an example of nonspecific chemical integration. More complex humoral control came with the development of the *endocrine glands,* or glands of internal secretion. These form an interdependent system of organs that produce chemical substances, *hor-*

Table 10–1. **Amino Acid Sequences of Vasopressin and Oxytocin**

	1	2	3	4	5	6	7	8	9
Vasopressin (arginine)	*cys-	tyr-	phe	-gln-	asn-	cys-	pro-	arg	-gly-NH_2
Oxytocin	*cys-	tyr-	ile	-gln-	asn-	cys-	pro-	leu	-gly-NH_2

* The two half-cystine residues are connected by a disulfide bridge to form a five-membered ring, to which a side chain of three amino acids is attached.

Abbreviations:

cys = cystine	asn = asparagine
tyr = tyrosine	arg = arginine
phe = phenylalanine	ile = isoleucine
gln = glycine	leu = leucine

mones, by active secretion and liberate them directly into the circulation. They are transported to various parts of the body, where they regulate and direct metabolic reactions. The tissues and organs upon which the hormones act are the *target organs.*

Hormones are *informational molecules,* carrying limited information to their target organs, which can interpret this information only if they have the right *receptors* to bind the circulating hormone. As not all cells have receptors for all hormones, the neat fit between a hormone and the receptors of its target organ forms a complex code for the regulation of a particular tissue by the appropriate hormone. Because hormones are much simpler molecules than the nucleic acids, the amount of information they carry is much less, but their effects on the nucleus and on enzyme systems in the cell membrane makes them capable of directing and selecting metabolic pathways.

In vertebrate evolution there have been only minor changes in the chemical structure of hormones, but these changes often result in profound differences in biological response. A change in two amino acids in the phylogenetically ancient hormone oxytocin can give it the quite different properties of vasopressin (Table 10–1). But the evolution of hormonal control seems to have concentrated on changes in response of the target organs so that the same hormones are used for quite different purposes in different species. Prolactin, which is a phylogenetically old hormone, will cause milk secretion in mammals, has growth hormone-like properties in pigeons, and is involved in water balance in some fish and amphibia. Melanocyte-stimulating hormone (MSH) causes darkening of the skin in lower vertebrates, which acts as a vital link in protective camouflage; in humans it appears to have a totally different role, probably influencing the excitability of the nervous system and improving its efficiency during stress.

Robert Gaunt (1967) points out that the endocrine glands were the first specialized apothecaries known to the vertebrates: "In many respects their history parallels that of applied pharmacology. Their chemical products, the hormones, like manufactured drugs, had a fairly good but by no means perfect quality control. Every so often an enzyme went berserk and an unexpected variant of an old hormone poured out—as for instance in the various forms of congenital adrenal hyperplasia. There was no federal agency to force its recall from the blood stream, but Darwinian natural selection served the purpose. If the product was bad enough to be lethal, the record was lost along with the line that produced it. If not so bad, it could hang around indefinitely and become a raw material of evolution."

10–2 Integrative Action of Hormones

In lower animals, especially evident in invertebrates where the nervous system is not so greatly developed, hormonal regulation coordinates not only the internal organs, but also the activities of the animal. Even processes as complex as insect metamorphosis are dependent almost entirely on hormonal control. In higher animals, as the nervous system has developed and gained dominance, the two integrating systems supplement each other in controlling both the coordination of the various internal organs and the animal's response to the external environment. The nervous system in humans is of more immediate importance in regulating reactions to changes in the external world, for speed and accuracy of response are characteristic of the electrically conducted nerve impulse. The endocrine response, chemical in propagation, is slower and often longer lasting, continuing the coordinated response after the initial nervous activity has abated.

The separation of integrative responsibility into nervous and endocrine systems is by no means as distinct as it was once believed to be. *The functions of the two systems not only overlap but are themselves integrated.* The endocrine system is regulated by a part of the nervous system, the hypothalamus, which in turn is regulated not only by different levels of the nervous system but to a very large extent by the levels of circulating hormones. The study of the complex interactions between the nervous and the endocrine systems is *neuroendocrinology* and is the subject of Chapter 11.

10–3 Types of Chemical Messengers

Many chemicals that are carried through the circulation affect distant cells and yet are *not* classified as hormones. *Metabolites* such as carbon dioxide, which are produced by all cells and affect almost all cells, do not fit the classification because they are too generalized. Nerve cells produce chemicals (e.g., norepinephrine and acetylcholine) that act locally and then are inactivated or taken up into the nerve endings or back into the circulation. Usually, the amount that returns to the circulation is too small to be of significance. Norepinephrine also is produced by the adrenal medulla, the cells of which are embryologically derived from neural crest and consequently can be considered modified nerve cells. The adrenal medulla liberates norepinephrine into the circulation; thus, from this source, norepinephrine may be a hormone, whereas from the adrenergic nerve endings it is a *neurotransmitter.* But this distinction is somewhat pedantic, for only small amounts of norepinephrine are produced by the medulla: most of its endocrine activity results in the production of epinephrine, a hormone.

One other type of chemical messenger should be mentioned: *pheromones.* These are secretions that are liberated externally to affect other individuals. They are usually perceived only by other members of the species or subspecies and are effective in extremely small amounts. Many insects, including ants, lay trails for other members of their species to follow by secreting pheromones as they travel. Some are sexual attractants, such as bombykol, produced by the female silkworm moth; her bombykol activates chemoreceptors on the antennae of male moths, but females are insensitive to it. Although this hormone is similar in many respects to our definition given above, it affects other individuals rather than other groups of cells in the same individual. It is quite likely that humans produce pheromones, perhaps accounting for sudden "chemical attractions," and some recent experiments confirming this hypothesis have been reported.

10–4 Characteristics of Hormones

Hormones are distinguished from other chemicals in that they:

1. Are produced by a specialized gland or by a localized group of cells and are secreted directly into the blood stream.
2. Exert their effects on distant parts of the body, not where they are locally produced.
3. Act upon their target cells by regulating the rates of specific metabolic reactions; this is accomplished through the regulation of appropriate enzyme activity.
4. Are required—because they are not utilized to provide energy—in very small amounts and may be excitatory or inhibitory in their effect, depending on their concentration and the physiological state of the responding tissue.

10–5 Demonstration of Hormonal Activity

The first clear demonstration of hormonal activity was made in 1849 by Berthold, who transplanted a testicle from a cock into a capon and showed that this implantation induced development of the retarded crest of the capon. This experiment was later misinterpreted and distorted into the "monkey-gland" craze, especially in Europe; it was believed that transplants of testicles from monkey to man would increase and improve man's sexual activity. This assumption was soon shown to be false: sexual activity in man is a highly complex pattern of activities, affected by the nervous system, by training and social environment, by personality, and by many other nebulously understood processes. Any transitory benefits the transplants may have had were purely psychological, for tissues from the monkey implanted in man are not viable. However, Berthold's work demonstrated a basic premise for the new science of endocrinology (the study of the endocrine glands): the absence of a specific secretion from an endocrine organ can be replaced by the secretion of a similar, but active, gland from another animal.

Techniques Used to Investigate Endocrine Function

Removal of endocrine gland and replacement therapy

The activity of an endocrine gland may be most effectively demonstrated by implanting it into an animal lacking that gland. Such an animal is called a "test" or "assay" animal, and the extent of its response is the basic quantitative measurement of hormonal activity. *Transplantation* of the missing endocrine gland is the oldest method. Because innervation is not essential but circulation is, the transplanted organ is usually placed into an area rich in blood vessels, like the anterior chamber of the eye if the organ is small enough. This has an added advantage in that the development of the gland can be seen. The development of follicles and corpora lutea can actually be seen in the different stages of the estrus cycle in a mouse ovary transplanted to the eye in this manner. As yet, transplants of whole organs on a large scale have not become feasible in humans, because of the antibody reactions developed in the host against the foreign tissue.

The administration of *purified extracts* from endocrine glands or of *synthetic hormones* (some of which are considerably more potent than the naturally occurring hormone) will correct the altered metabolism of the deficient host. Extracts may be made from urine, which has a high concentration of the metabolites of some hormones. For some strange reason, the urine of stallions is very rich in estrogens, so that this has become a commercially valuable source of this sex steroid.

Bioassay. The bioassay method tests the activity of a hormone in a living animal or on living tissue; for example, insulin lowers blood glucose whereas the anti-inflammatory activity of cortical hormones reduces the size of a fibrotic deposit produced by an irritant. For each bioassay, the normal source of hormone must be removed; in the last example, adrenalectomized rats would be used. *Radioimmunoassay,* the most sensitive technique for measuring hormone levels, is based on the amount of labelled hormone that will bind to its antibody.

Administration of selected chemicals

Certain chemicals will selectively destroy the cells of a particular endocrine gland. Radioactive iodine, ^{131}I, is picked up specifically by the cells of the thyroid; if enough is administered, the radioactivity will kill the thyroid cells and prevent thyroxine production. Thiouracil will block the synthesis of thyroxine. The β cells of the pancreas, which produce insulin, are selectively destroyed by alloxan, producing an experimental diabetes. The secretion of the anterior pituitary hormone, adrenocorticotropic hormone (ACTH), is prevented by the administration of dexamethasone, a synthetic glucocorticoid.

Diseases of the endocrine glands

Pathological processes that destroy the gland have the same effect as surgical removal of that gland, and tumors that produce excess hormone will evoke the same responses as the administration of large amounts of the hormone. Adrenal insufficiency is seen in patients with Addison's disease, where the adrenal cortex is atrophied, and adrenal hyperfunction in patients with tumors of the adrenal cortex. Destructive tumors of the anterior pituitary gland will be reflected by regressive changes in most of the other endocrine organs dependent upon it. Hyperfunctioning tumors of the anterior pituitary will stimulate the thyroid, adrenals, and gonads to greater than normal activity. Correlation of these metabolic and pathological changes has provided us with valuable information but there is far more variation than in the results from controlled experiments.

Isotopic tracers

Isotopic tracers are extensively used to show which cells pick up radioactive hormone or hormonal precursors and their metabolic fate.

Perfusion techniques

In the perfusion technique an endocrine gland is perfused with a fluid of known composition, containing perhaps some precursors of the hormone, and the perfusate analyzed after it has flowed through the gland. One can then determine the factors that regulate the synthesis and release of the hormone. This may be done in the intact animal (in vivo) or in an isolated preparation (in vitro).

Figure 10-1. **The endocrine glands.**

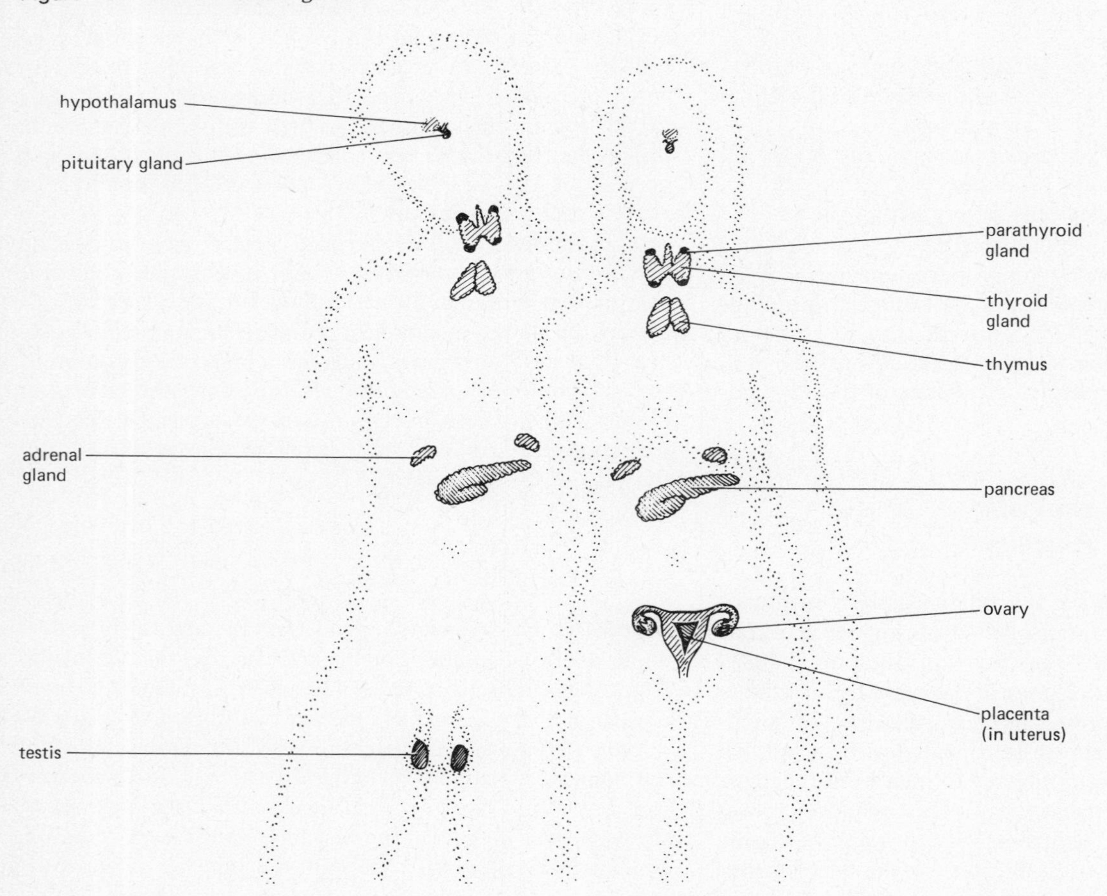

10–6 The Endocrine System

Types of Endocrine Glands

The endocrine system consists of a group of endocrine glands, most of which are anatomically unrelated, but which are coordinated in their activities through hormones reaching them through the circulation. The most important are the pituitary, thyroid, adrenals, gonads, pancreas, parathyroids, and, in pregnancy, the placenta. The position of these glands in the human body is shown in Fig. 10-1. Some endocrine glands produce several hormones; in fact, an endocrine gland may be formed from two quite different embryological germ layers. The *adrenal gland,* for example, consists of the inner *medulla,* an endocrine gland of nervous tissue origin, which produces two hormones, epinephrine and norepinephrine, which are amino acid derivatives. The outer *cortex* of the adrenal is of mesodermal origin and secretes a variety of cortical steroids. These two parts of the adrenal appear to be independent of one another. As a matter of fact, they are quite separate in some fish, the cell types form a mixed tissue in amphibia, and only in mammals are the tissues segregated into medulla and cortex. However, the epinephrine-rich medullary blood does flow through the cortex and may have some influence on cortical activity. The thyroid gland produces *thyrosine* and *triiodothyronine,* hormones that influence cell metabolism, and *thyrocalcitonin,* a hormone involved in the regulation of calcium balance.

The *pancreas* is a gland of mixed endocrine and exocrine function. The endocrine portion is formed by the islets of Langerhans, which consist of three types of cells, the function of two of which is known; the β cells produce the hypoglycemic hormone *insulin,* and the α cells produce the hyperglycemic hormone *glucagon.* The exocrine portion of the pancreas produces a large number of digestive enzymes which are passed into the duodenum.

The *pituitary gland* is another gland, of mixed embryological origin, which produces many different hormones. To demonstrate its central role in the control of the endocrine system, the pituitary and the adrenals will be described in this chapter, whereas the other endocrine glands and their secretions are discussed in connection with their regulatory effects on the appropriate metabolic reactions.

The Pituitary Gland

The pituitary gland (hypophysis) is an unpaired organ recessed in the floor of the skull (Figs. 10-2 and 10-3). It is

Figure 10-2. **The pituitary gland and its neural and circulatory connections to the hypothalamus. Four hypothalamic nuclei are shown. The paraventricular and supraoptic nuclei have large cells, the axons of which extend into the posterior lobe of the pituitary gland. Cells of the dorsomedial and ventromedial nuclei deposit their neurosecretions in the region of the median eminence, where they enter the portal system that leads to the anterior pituitary.**

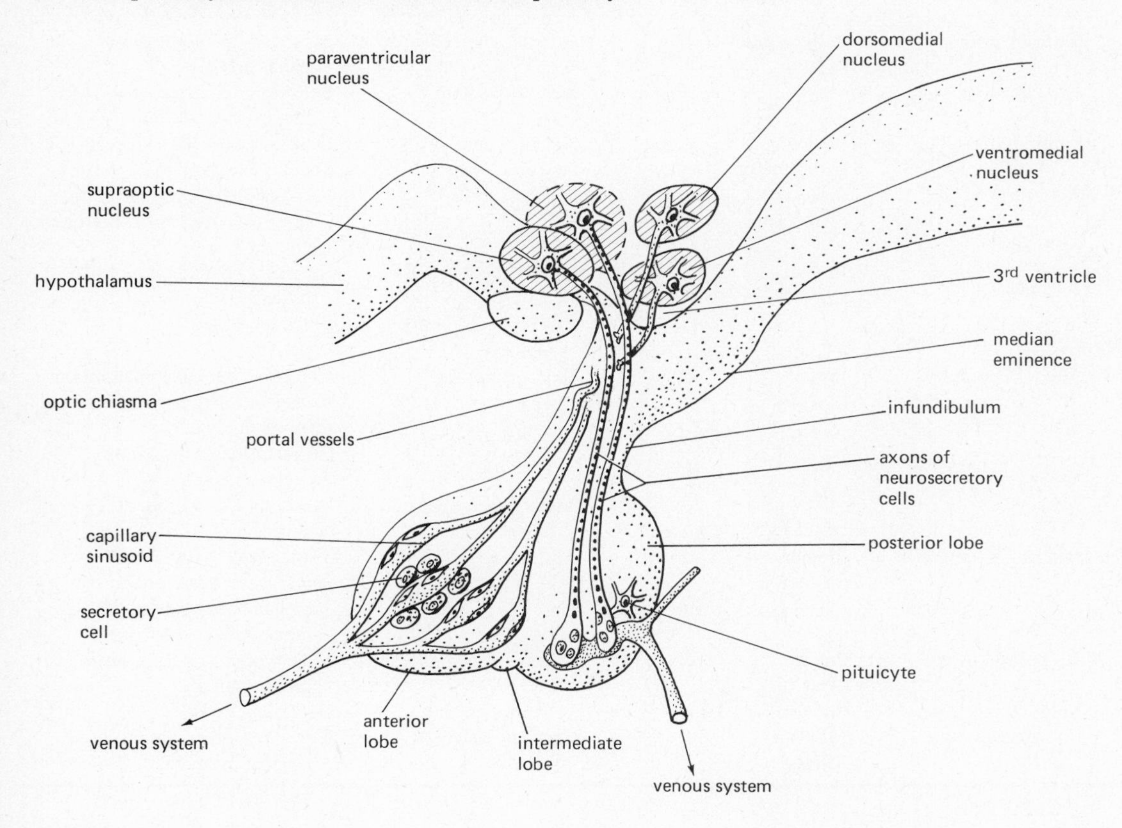

connected to the *hypothalamus,* a part of the brain, by a *stalk* (the *infundibulum*). The junctional region between the stalk and the hypothalamus is the *median eminence*. The pituitary weighs about 0.5 to 0.8 g in the human, being slightly larger in the female and increasing in size with pregnancy due to its increased secretion of the gonadotropins FSH (follicle stimulating hormone) and LH (luteinizing hormone). It is derived from ectoderm from two sources; therefore it has two anatomically and functionally different lobes, the anterior lobe (adenohypophysis) and the posterior lobe (neurohypophysis). The intermediate lobe is a small area, considered part of the neurohypophysis; in amphibia it produces *melanocyte stimulating hormone* (MSH), a hormone that causes darkening of the skin. In mammals, the function of the intermediate lobe is unknown, and MSH is produced by the anterior lobe.

THE ANTERIOR LOBE

The anterior lobe is derived from epithelium from the pharyngeal region of the embryo. The secretory cells are arranged in irregular cords around large capillary sinusoids; this copious blood supply is important, for secretion is regulated by circulating hormones reaching the cells from the other endocrine glands and from the hypothalamus. This regulation of the pituitary gland by the hypothalamus is discussed in Chapter 11.

There are at least six different cell types in the anterior pituitary, each associated with a specific hormonal secretion. These cells can be differentiated by various histochemical procedures and their activity correlated to changes in physiological conditions such as pregnancy and lactation, as well as to pathological states related to thyroidectomy, dwarfism, and so on.

Tropic Hormones. The hormones of the anterior pituitary exert a *stimulating effect* on the growth and secretion of the *other endocrine glands,* which are their target organs. In 1927, P. E. Smith conclusively demonstrated that removal of the anterior gland, also called the hypophysis (*hypophysectomy*), prevents growth in young rats and results in atrophy of the adrenal and thyroid glands, the gonads, and consequently of the accessory reproductive structures. These effects can be prevented by the administration of pituitary extracts. The "tropic" effects of the pituitary on the target endocrine organs are limited to some but not all of the hormones of the anterior pituitary. Table 10–2 lists these tropic hormones and the effects they evoke from their target

Table 10–2. **Site of Synthesis and Effects of Pituitary Protein and Polypeptide Hormones**

Hormone (Alternate name)	Synthesized by	Type of Molecule (Molecular weight)	Target Organ	Effect
Adrenocorticotropic hormone (ACTH) (corticotropin)	Anterior pituitary	Polypeptide (4,500)	Adrenal cortex	Synthesis and release of glucocorticoids
			Nervous system	Behavioral changes and threshold modulation
Lipotropin (LPH) (prohormone)	Pituitary and brain	Polypeptide (LPH 9,500)	Brain	Opiate effects of fragments
			Adipose tissue	Fat mobilization (?)
Thyroid stimulating hormone (TSH) (thyrotropin)	Anterior pituitary	Protein (28,000)	Thyroid gland	Synthesis and secretion of thyroxine and triiodothyronine
				Lipolysis
Follicle stimulating hormone (FSH)	Anterior pituitary	Protein (29,000)	Ovary (follicles)	Initiation of maturation of follicles
			Testis (seminiferous tubules)	Spermatogenesis
Luteinizing hormone (LH) (interstitial cell stimulating hormone (ICSH)	Anterior pituitary	Protein (28,000)	Ovary (follicles)	Complete follicular maturation Estrogen secretion Ovulation Corpus luteum formation Progesterone secretion
			Testis (interstitial cells)	Synthesis and secretion of androgens
Growth hormone (GH) (somatotropin STH)	Anterior pituitary	Protein (21,500)	All tissues	Tissue growth, especially marked in long bones, fat metabolism, and protein metabolism
Prolactin	Anterior pituitary	Protein (23,000)	Mammary glands	Milk production from hormone-primed glands
Melanocyte-stimulating hormone (MSH)	Anterior pituitary Intermediate lobe	Polypeptide MSH (α) MSH (β)	Melanocytes Nervous system	Darkening of skin Threshold of excitation in nervous system

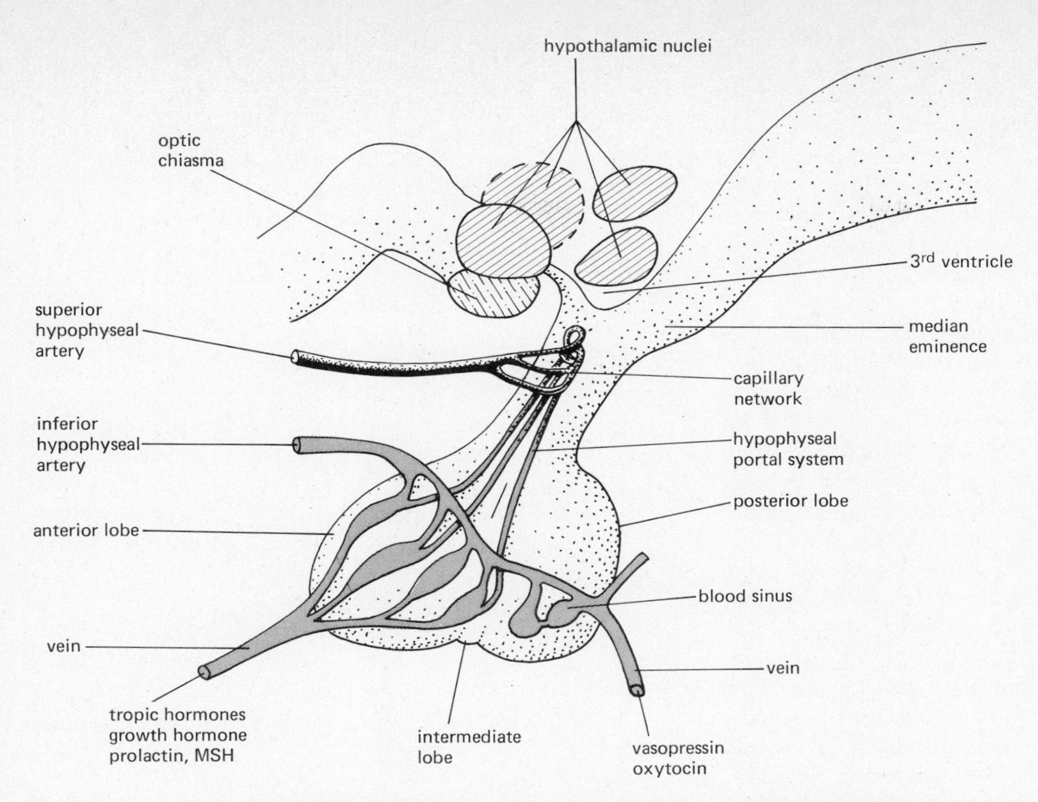

Figure 10-3. **The double arterial circulation of the anterior pituitary gland. The capillaries formed in the median eminence by the superior hypophyseal artery are connected by the relatively straight hypophyseal portal vessels to the blood sinuses in the anterior lobe. These sinuses, and those of the posterior lobe, are formed by the inferior hypophyseal artery.**

organs. Note that growth hormone, prolactin, and MSH do not stimulate other endocrine glands.

Special Blood Supply. The anterior lobe has a *double arterial circulation* (Fig. 10-3). It receives arterial blood from a branch of the internal carotid artery, the inferior hypophyseal artery, which forms the capillary sinusoids that bathe the secretory cells. It also receives blood from the superior hypophyseal artery, which enters at the median eminence where it forms another capillary network. These two capillary beds are connected by straight vessels running through the stalk, the *hypophyseal portal system.* It is called a portal system because it begins and ends in capillaries, and it is the main route through which the regulating hormones of the hypothalamus reach the anterior pituitary. The significance of this portal system was first demonstrated by Harris (1955); severance of these vessels prevents ovulation in the rabbit, which normally ovulates 10 hr after mating. In other words, mating (in the fecund rabbit) stimulates the hypothalamus reflexly, which causes the release of hypothalamic hormones that pass down into the hypophyseal portal system. This evokes the release of the gonadotropic hormones from the anterior pituitary, which result in ovulation (Fig. 10-4).

Figure 10-4. **Ovulation in the rabbit is triggered by mating. Destruction of the portal blood vessels between the hypothalamus and the anterior pituitary obliterates the response.**

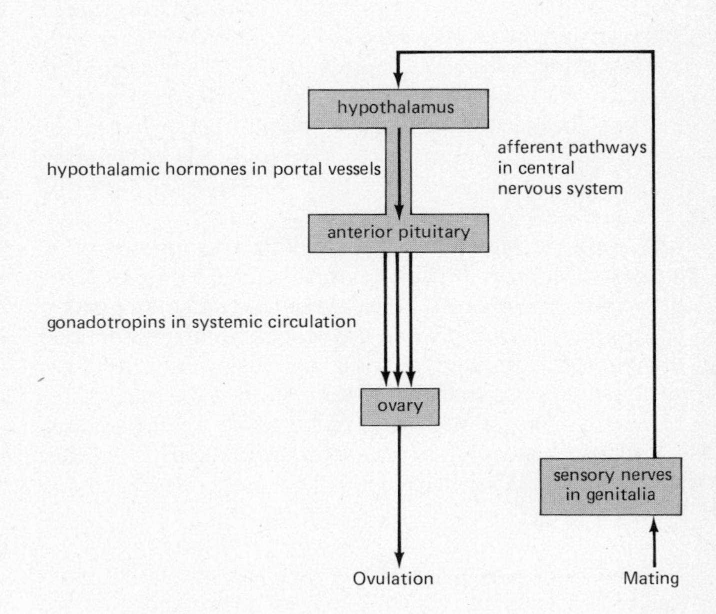

The posterior lobe

The posterior lobe arises from an outgrowth from the floor of the third ventricle of the brain and is connected by the stalk to the hypothalamus. Passing through this stalk are the axons of neurons, the *bodies* of which are in the *supraoptic* and *paraventricular nuclei* of the hypothalamus. The *terminals* of these neurons lie in the posterior lobe and are surrounded by small cells, the *pituicytes,* and by a rich supply of capillaries (Fig. 10-2).

The posterior pituitary is completely dependent on the hypothalamus for its supply of the two octapeptides, *antidiuretic hormone* (which is also called *vasopressin*) and *oxytocin.* (The structure of these hormones is discussed in Secs. 3-3.1.) Both of these hormones are synthesized in the hypothalamus. Antidiuretic hormone is produced by the magnocellular (large-cell) elements that form the supraoptic nucleus, whereas the synthesis of oxytocin occurs in the paraventricular nucleus. These large neurosecretory cells have their bodies in the hypothalamic nuclei, their axons in the infundibular stalk, and their secretory endings in the posterior pituitary gland (Fig. 10-2).

Studies using radioactive cysteine show that the octapeptides first are produced in association with a large polypeptide, *neurophysin.* This complex is transported down the axons of the neurosecretory cells to the posterior pituitary gland, where most of it is stored in neurosecretory granules. These granules are released by exocytosis directly into the circulation upon appropriate stimulation (Sec. 11-3). To be biologically active, the small octapeptides must be split off from the neurophysin molecule. It is not clear just where this process occurs, for neurophysin can be found in the circulation as well in the hypothalamus and posterior pituitary.

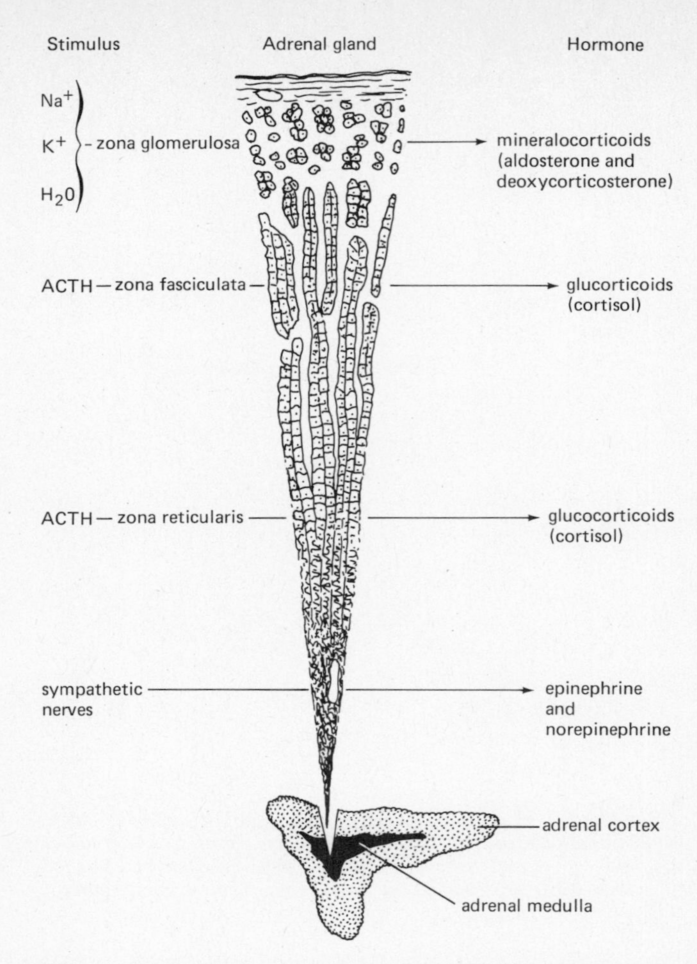

Figure 10-5. **Hormones secreted by different regions of the adrenal gland. On the left of the diagram are listed the stimuli that usually evoke the secretory response. Epinephrine and norepinephrine are produced by the adrenal medulla. The corticoids are secreted by the adrenal cortex.**

The adrenal glands

The adrenal glands are two small organs, each of which is embedded in the fat above the kidney. The outer layer of the adrenal gland is the *cortex.* Although the cortex is copiously supplied with blood from the adrenal and renal arteries, most of the nerve fibers that penetrate the cortex merely pass through it en route to the adrenal *medulla,* the central portion of this gland.

The Adrenal Medulla. The medulla is composed of secretory cells, really modified sympathetic ganglion cells that liberate their secretions, *epinephrine* and small amounts of *norepinephrine,* directly into the venous circulation draining the adrenals. Stimulation of the adrenal medulla results in essentially the same effects as stimulation of the sympathetic nervous system. The role of epinephrine in carbohydrate metabolism, and of norepinephrine as a neurotransmitter, are considered elsewhere in this text (Secs. 18-3; and 7-6, 8-6, and 9-5).

The Adrenal Cortex. The cortex is made up of three zones, the cells of which differ in form and function (Fig. 10-5). The outermost *zona glomerulosa* is formed by closely packed, low columnar cells, and it secretes hormones concerned with water and electrolyte metabolism. These hormones are called *mineralocorticoids,* the most important of which are *aldosterone* and *deoxycorticosterone.* The secretion of these hormones is mainly regulated by negative feedback through plasma sodium and potassium levels, but there is some evidence that ACTH may also be involved in their regulation.

The thick, middle *zona fasciculata* is formed by cords of cells filled with *lipid,* the amount of which varies with the secretory activity of these cells.

The innermost zone of the cortex is the *zona reticularis,* and both the middle and inner zones are associated with the secretion of *glucocorticoids,* hormones that influence carbohydrate and protein metabolism. *Cortisol* (*hydrocortisone*) is the most important glucocorticoid, and its secretion is regulated by ACTH.

The *X zone* is a region of the zona reticularis, just next to the medulla. The X zone is found only in the fetus and infant, and it produces androgenlike steroids. It normally atrophies soon after birth but sometimes develops into a hormone-producing tumor, resulting in premature virilism in males, or in masculinization of females.

The adrenal cortex is a superbly developed organ for the regulation of homeostasis. It is essential for life: in its absence life can be maintained by administration of the mineralocorticoids, but normal glucose and protein metabolism, and the ability to withstand stress, require the administration of glucocorticoids as well.

Levels of Control of the Endocrine System

The coordination of the activities of the different endocrine glands and of the nervous and endocrine systems may be categorized into the following four levels.

Blood constituents

In some cases, the level of activity of the endocrine gland is regulated by the concentration in the blood of the substance that it is regulating. Insulin secretion by the pancreas is increased almost immediately by a rise in blood glucose. This causes a rapid fall in blood glucose levels which slows the rate of insulin secretion.

Tropic hormones of the anterior pituitary gland

The anterior pituitary gland produces a series of protein and polypeptide hormones that are liberated into the circulation and appropriated by the target organs that have the necessary receptors. In this way, a tropic hormone acting specifically on its target organ causing it to secrete its own hormone. Thyroid-stimulating hormone (TSH), or *thyrotropin,* specifically acts upon the cells of the thyroid gland (target organ) to cause the secretion of thyroid hormones (Fig. 10-6).

Figure 10-6. **Regulation of the thyroid gland by thyrotropic hormone (TSH) of the anterior pituitary gland. On the left are the actively secreting cells in the thyroid of a rat that had received injections of TSH. Note the loss of colloid (stored thyroid hormone) from the follicles. On the right is the thyroid of a rat 6 months after the removal of the pituitary gland. The cells are inactive and the follicles distended with colloid. (From C. D. Turner, in *General Endocrinology,* W. B. Saunders Company, Philadelphia, 1955.)**

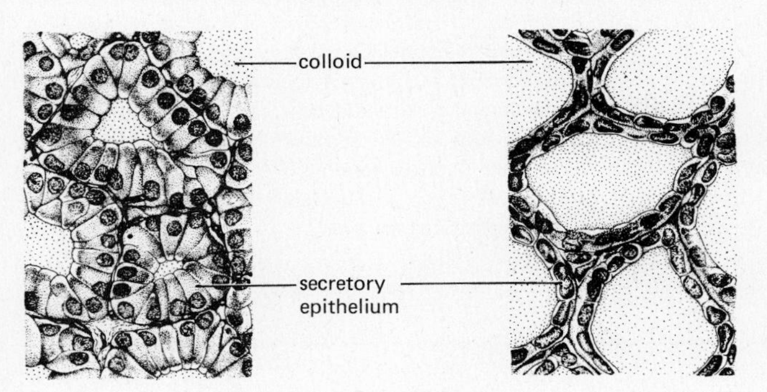

There is a reciprocal relationship between the secretions of tropic hormones and those of the target organs. The amount of circulating thyroid hormone, for example, will affect the release of TSH from the pituitary by a negative feedback mechanism, so that as thyroxine levels rise they inhibit the secretion of TSH. A similar and even more important relationship exists between target organ hormones and the hypothalamic releasing hormones (Sec. 11-3).

Considerable evidence exists to show that, apart from their effects on target endocrine glands, ACTH, TSH, and FSH may influence the activity of nonendocrine tissues, but a feedback control mechanism does not seem to be involved.

Hormones of the hypothalamus

The regulating hormones[1] of the hypothalamus are of two kinds: *releasing hormones* (RH), which stimulate the release of tropic hormones from the anterior pituitary, and hormones that *inhibit release* (IH). The neurons that secrete these hormones (neurosecretory cells) pass their secretions down their axons and liberate them from their nerve terminals. The hormones appear to be liberated as granules in the region of the median eminence and pass directly into the hypophyseal portal system, the short vascular pathway between the hypothalamus and the anterior pituitary.

Recently, it has been shown that hypothalamic hormones also reach the pituitary through the cerebrospinal fluid in the third ventricle, and that anterior pituitary hormones may reach the brain through this route.

Both negative and positive feedback mechanisms operate among these three levels of control; they are discussed in detail in Chapter 11 as examples of the intricate relationship between the nervous and endocrine systems.

Direct nervous control of the endocrine glands

Direct innervation of the endocrine glands plays a very small role in their regulation except for the adrenal medulla, which markedly reduces its secretion of epinephrine after denervation. The adrenal medulla normally responds to sympathetic stimulation by increased epinephrine secretion, but it is also affected by other factors: changes in the titers of blood glucose will modify the amount of epinephrine released by the medulla.

The nerve fibers that enter the posterior pituitary from the hypothalamus deposit their secretions in that gland and in this manner may be considered to affect it, but the posterior lobe is acting only as a storage area and really should not be classified as an endocrine gland.

[1] Two other hypothalamic hormones are *vasopressin* (*antidiuretic hormone*) and *oxytocin*. Vasopressin may stimulate the release of ACTH from the anterior pituitary, but its main function is to regulate the permeability of the kidney tubules (Sec. 15-4). Oxytocin causes uterine contractions in the pregnant female and milk ejaculation in the lactating mammal. These functions are discussed in Sec. 26-15.

10–7 Synthesis, Metabolism, and Excretion of Hormones

Hormones are modulators of metabolic reactions; therefore, their output must be regulated. The control of the rate of secretion is delicately adjusted by both the nervous and the endocrine systems (Chap. 11). The amount of hormone present in the blood will depend on the rate of secretion and release, the rate of inactivation or metabolism, and the rate of excretion. Accurate measurement of hormonal blood concentrations is not always possible, for laboratory methods are often neither sensitive nor specific enough to distinguish small changes. However, the recent development of radioimmunological assays provides an extremely accurate, if laborious, technique for the assay of protein and polypeptide hormones, and to a lesser extent of some nonprotein hormones.

Many hormones are metabolized by the tissues upon which they act. This usually results in the inactivation of the hormone, but sometimes it is transformed into a more active substance once it has entered its target cell. Testosterone, the main sex steroid produced by the testes, is metabolized to dihydrotestosterone (HT) by certain of its target cells. It is this HT molecule that then binds with the intracellular receptor to influence the genome.

The liver is the site of inactivation for most hormones. It may form complexes of the metabolites, which may still have some hormonal effects. The metabolites are excreted by the kidneys, and measurement of the urinary metabolites affords some indication of the functional activity of the endocrine glands.

Hormones may be divided into three main groups, according to their chemical structure:

1. *Steroids,* produced by the ovaries, testes, placenta, and adrenal cortex.
2. *Proteins and polypeptides,* produced by the anterior pituitary gland and the endocrine portion of the pancreas.
3. *Amino acid derivatives,* produced by the thyroid, adrenal medulla, and the pineal.

The most important hormones, the glands that synthesize them, and their effects are shown in Tables 10–2, 10–3, 10–4, and 10–6. These tables should be used purely for reference at this point, for *the functions of the specific hormones will be considered in detail at the appropriate place later in the text.* This chapter attempts to describe basic characteristics of hormones: their synthesis, metabolism, and excretion, and the possible mechanisms of their action on the metabolism of other cells.

Steroid Hormones

SYNTHESIS

The most commonly occurring animal steroid is cholesterol, found in all cell membranes, in the myelin sheath of nerves, and in many lipids. It is also the metabolic precursor of most steroid hormones. Its structure (Fig. 10-7) shows the phenanthrene nucleus (rings A, B, and C) and the attached cyclopentane ring (D) characteristic of all steroids. Cholesterol gives rise to the three basic types of steroid hormones:

1. Androgens (C_{19} compounds).
2. Estrogens (C_{18} compounds).
3. Progesterones and the corticosterones (C_{21} compounds).

The basic steroid nuclei for each of these three classes are illustrated in Fig. 10-8.

For the synthesis of steroid hormones, cholesterol is hydroxylated at C_{20} and C_{22}, and the six carbon atoms of the side chain are split off. This leaves *pregnenolone,* which can be oxidized readily to *progesterone* (Fig. 10-9). Either of these C_{21} derivatives is the precursor of all the other steroids produced by the adrenals, ovaries, and testes. The first step in this complex series of reactions, the conversion of cholesterol to pregnenolone, is dependent on cell stimulation by the tropic hormones of the pituitary: ACTH stimulation of the adrenals and gonadotropin stimulation of the gonads.

Synthesis of Androgens. The androgens are the simplest of the steroid hormones and are produced from preg-

Table 10–3. **Site of Synthesis and Effects of Steroid Hormones**

Hormone	Synthesized by	Number of Carbon Atoms	Target Organ	Effect
Androgen (testosterone)	Testis (interstitial cells)	19	Most cells	Development and maintenance of male secondary sex characteristics and behavior
Estrogen (estradiol)	Ovary (follicles)	18	Most cells	Development and maintenance of female secondary sex characteristics and behavior
Progesterone	Ovary (corpus luteum)	21	Uterus Mammary glands	Development and maintenance of uterine endometrium Mammary duct formation
Hydrocortisone Corticosterone	Adrenal cortex (zona fasciculata and reticularis)	21	Most cells	Carbohydrate, fat and protein metabolism Anti-inflammatory action
Aldosterone Deoxycorticosterone	Adrenal cortex (zona glomerulosa)	21	Kidney tubule	Reabsorption of Na^+ into blood

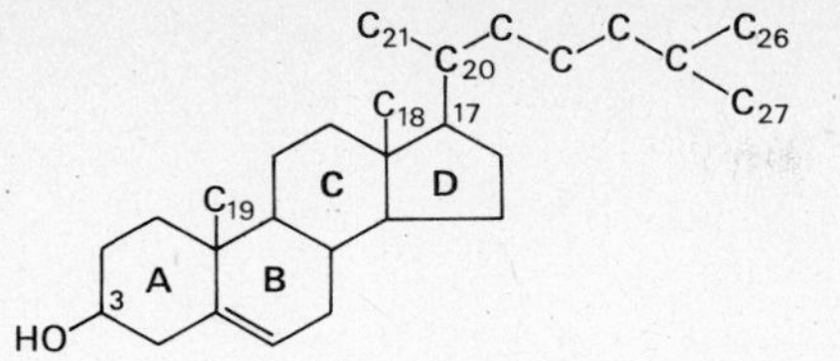

Figure 10-7. **The structure of cholesterol. The rings that form the steroid nucleus are lettered. There are side chains at C17, C18, and C19; C26 and C27 are methyl groups. Some of the other important carbon atoms are numbered.**

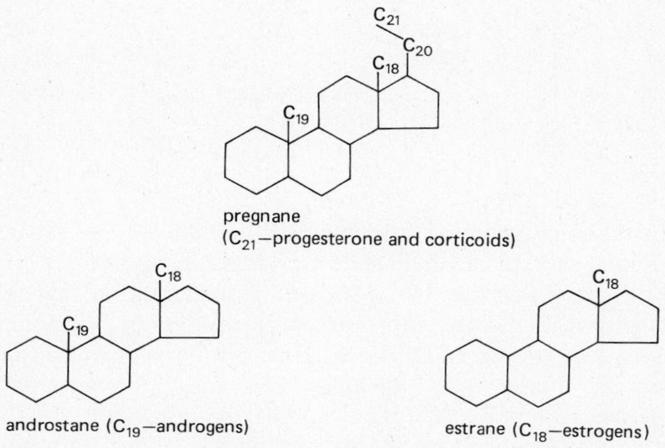

pregnane
(C_{21}—progesterone and corticoids)

androstane (C_{19}—androgens)

estrane (C_{18}—estrogens)

Figure 10-8. **The three basic steroid nuclei, C_{21}, C_{19}, and C_{18}.**

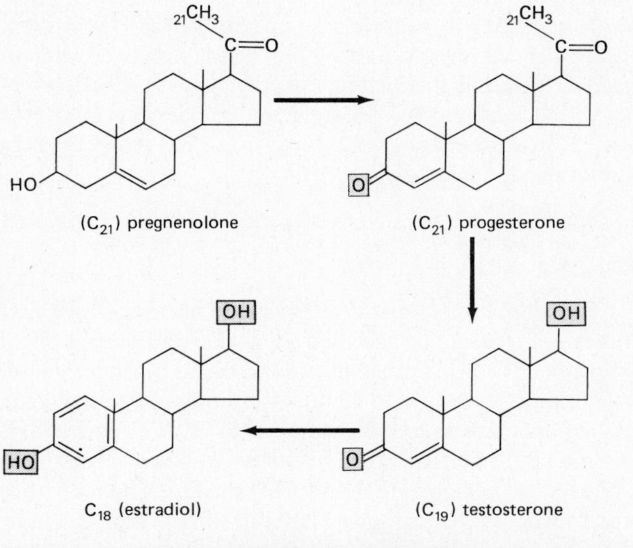

(C_{21}) pregnenolone

(C_{21}) progesterone

C_{18} (estradiol)

(C_{19}) testosterone

Figure 10-9. **Structural relationship and probable order of synthesis of pregnenolone, progesterone, testosterone, and estradiol.**

nenolone or progesterone principally by the interstitial cells of the testes. The most active biologically occurring androgen is *testosterone,* but some synthetic androgens are more potent, especially when administered orally, probably because the synthetic molecule is more easily absorbed by the intestine. The structure of testosterone is shown in Fig. 10-9.

Synthesis of Estrogens. Studies with radioactive tracers have shown conclusively that estrogens are formed primarily from androgens produced by the ovary (Fig. 10-9). The androgen precursor, testosterone, loses the C_{19} methyl group and its A ring is changed (aromatized) to a phenolic ring, which is a benzene ring with a hydroxyl group at C_3. The most active naturally occurring estrogen is *estradiol,* which has 2 hydroxyl (diol) groups. Estradiol is the main hormone secreted by the ovary, but the placenta, the adrenals, and even the brain can convert testosterone to estrogens, a fact which may be of some significance during the embryological development of the brain (Sec. 11-2). Figure 10-9 shows the slight differences in chemical structure between the most potent of the male and female sex hormones. As the French say, "Vive la différence."

Synthesis of Progesterone. Progesterone is one of the simplest C_{21} steroids. It is secreted mainly by the corpus luteum and by the placenta during pregnancy. It may be produced in small amounts by the adrenal glands. As we have seen, it is important as a precursor for the synthesis of the other steroid hormones. Synthetic progesterones are several thousand times more potent than the naturally occurring hormone in their ability to delay ovulation and are widely used as orally active contraceptives. This is discussed in Sec. 26-9. A combination of estrogen and progesterone appears to be the most effective "Pill," perhaps because of the inhibition of pituitary gonadotropins, which are essential for follicle maturation and ovulation.

Synthesis of Corticosteroids. The *hormones of the adrenal cortex* are usually divided into two main classes:

1. Predominantly *glucocorticoid* activity, affecting carbohydrate metabolism, such as hydrocortisone (cortisol), corticosterone, and small amounts of cortisone.
2. Predominantly *mineralocorticoid* activity, affecting water and salt metabolism, such as aldosterone and deoxycorticosterone.

There are many other cortical steroids with varying degrees of activity, overlapping both classes.

The adrenal steroids originate from cholesterol, by way of pregnenolone or progesterone. The main changes are additions of hydroxyl groups (Fig. 10-10). The addition of one hydroxyl group to progesterone at C_{21} forms *11 deoxycorticosterone,* a mineralocorticoid. Further hydroxylation at C_{11} produces a glucocorticoid, *corticosterone.* If C_{17} is also hydroxylated, the most potent, naturally occurring glucocorti-

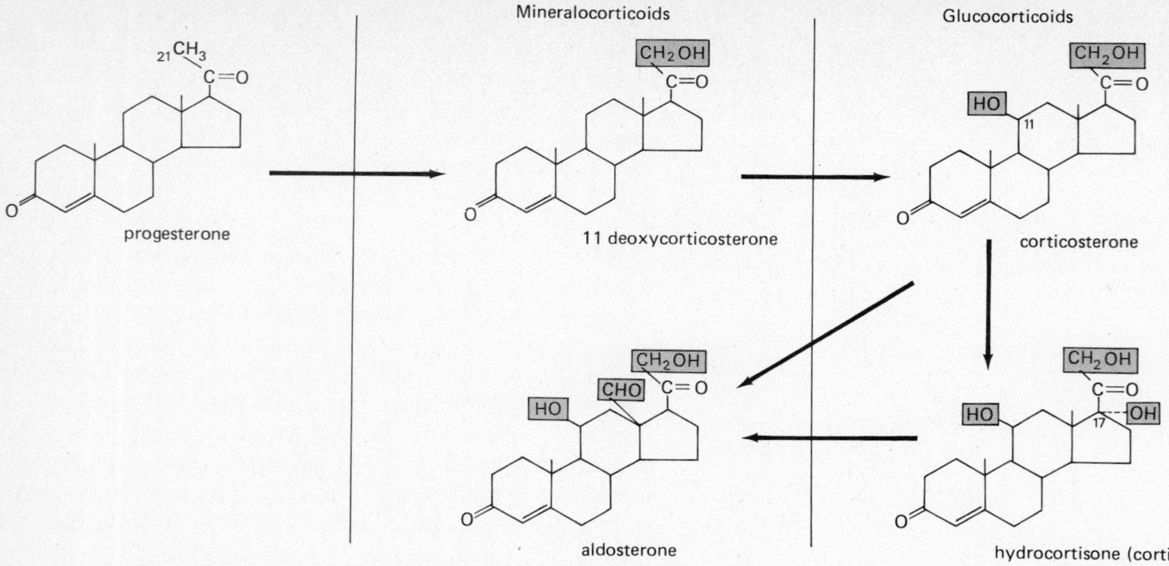

Figure 10-10. **Interrelationships of the adrenocortical hormones.**

coid is formed, *hydrocortisone.* A rather more complex pathway is needed for the synthesis of *aldosterone,* by far the most important mineralocorticoid synthesized by the body.

Metabolism and excretion

Because so many of the steroid hormones are interconvertible, many important metabolic reactions have already been considered in the section on their biosynthesis. The details of their inactivation by the peripheral tissues and by the liver are not completely clear, but the liver can convert hydroxysteroids to water-soluble derivatives, which are easily excreted. It is able also to reduce or oxidize the steroids. If the hormones are oxidized at C_{17}, 17 ketosteroids are formed and are subsequently found in the urine. Certain diseases, such as the adrenogenital syndrome (Cushing's disease) and tumors of the testes, are characterized by a change in urinary 17 ketosteroids. Depending upon the type of 17 ketosteroid present, it can be determined whether the defect lies in the testes or in the adrenal cortex. Because there are great variations in the steroid excretion levels at different times of the day and night, caused by diurnal variation in endocrine secretion, a 24-hr specimen must be analyzed.

Protein and Polypeptide Hormones

Synthesis

The synthesis of polypeptides and proteins and the control of these processes by the genome have been considered in Chapter 3. The *specific hormone* that a cell will synthesize will depend on its genome and the biochemical machinery available to it in its cytoplasm—that is, the particular enzymes, substrate concentrations, and cooperating ions (Table 10-4). The *rate* at which a cell will synthesize polypeptide and protein hormones will be affected by instructions it receives through the two integrating systems, nervous and endocrine. In both cases, information is through chemical mediators, either neurotransmitters or hormones. *Secretion of ACTH* by the anterior pituitary, for example, can be increased by *norepinephrine* and by *corticotropin releasing hormone* (*CRH*) from the hypothalamus. In the case of norepinephrine there is convincing evidence to show that protein synthesis is accelerated through the activation of enzyme systems first in the cell membrane and then in the cytoplasm, involving *cyclic AMP.* This mechanism of hormonal regulation is discussed in Sec. 10-8. CRH has been isolated only recently, and its mode of action is not yet known. It is a peptide, and so it may act in a manner similar to norepinephrine, which is an amino acid derivative. *Vasopressin,* another polypeptide, also increases ACTH secretion.

Steroid hormones, on the other hand, influence protein synthesis through their effects on the *genome.* Corticosterone, which inhibits ACTH secretion, apparently interferes with DNA-dependent RNA synthesis, and consequently inhibits protein synthesis.

Prohormones

It is probable that many polypeptide hormones are first synthesized as proteins and then the active part is split off. β lipotropin (βLPH), a substance found in the pituitary gland, has 90 amino acids (Table 10-5). Within this large molecule is a heptapeptide sequence common to ACTH and MSH. It seems likely that LPH may be a *prohormone,* a relatively inactive large hormone that splits off smaller sequences that comprise the active hormones. LPH itself has been shown to have a fat-mobilizing effect, the physiological significance of which is unknown. The relationship between LPH and the endorphins and enkephalins was discussed in Sec. 9-5.

Table 10–4. **Site of Synthesis and Effects of Protein and Polypeptide Hormones (Nonpituitary)**

Hormone (Alternate name)	Synthesized by	Type of Molecule (molecular weight)	Target Organ	Effect
Insulin	Pancreas	Polypeptide (5734)	All cells	Carbohydrate, fat, and protein metabolism Hypoglycemia
Glucagon		Polypeptide (3485)	Liver	Hyperglycemia
Thyrocalcitonin (calcitonin)	Thyroid	Polypeptide (3500–4000)	Bone	Increases Ca^{2+} deposition in bone Decreases serum Ca^{2+} Mg^{2+}, urinary PO_4, and hydroxyproline
Parathyroid hormone	Parathyroid	Polypeptide (8447)	Bone (osteoclasts)	Resorption of bone Increases serum Ca^{2+}, Mg^{2+}, urinary PO_4, and hydroxyproline
Relaxin	Ovary	Polypeptide (9000)	Pelvic ligaments Uterus	Separation of pelvic bones in parturition Dilation of cervix
Erythropoeitin	Kidney	Glycoprotein (50,000–60,000)	Bone marrow	Increased erythrocyte production
Renin	Kidney	Protein (40,000)	Adrenal cortex Vascular smooth muscle (via formation of angiotensin)	Aldosterone secretion Hypertension
Antidiuretic hormone (ADH) (vasopressin)	Hypothalamus	Octapeptide (1084) (arginine ADH)	Kidney Arteries	Reabsorption of water Contraction of smooth muscle
Oxytocin	Hypothalamus	Octapeptide (1007)	Uterus Mammary glands	Contraction of smooth muscle Milk ejection

The 39 amino acid ACTH may itself be considered a prohormone, for there are many polypeptide fragments of ACTH that have greater activity than the 39 amino acid parent molecule. Perhaps in this analogy, the LPH would be a grandparent molecule.

Carrier proteins

The polypeptides found in the posterior pituitary gland, antidiuretic hormone and oxytocin, are *synthesized in the hypothalamus* simultaneously with the carrier protein neurophysin. This mechanism may provide for storage of many small hormones with different, specific functions, which can be released rapidly as a result of any of several stimuli.

Metabolism and excretion

Hardly anything is known of the metabolic fate of ACTH and the other pituitary hormones, except that they are taken up by the kidney and the target organs. They have a fairly short half-life in the blood (from 2–30 min). Little or none of these hormones can be found in the urine.

Table 10–5. **Amino Acid Sequence of Some Peptide Hormones Showing Their Similarities. The Shaded Areas Indicate Sequences with Marked Effects on Behavior.**

Hormone	Sequence	Residue numbers
β Lipotropin	pro-tyr-lys-met-glu-his-phe-arg-trp-gly-ser-pro-pro-lys-asp----- gln	47 … 53 … 90
ACTH*	H-ser-tyr-ser-met-glu-his-phe-arg-trp-gly-lys-pro-val-----phe-OH	1 2 3 4 5 6 7 8 9 10 11 12 13 … 39
ACTH 4-10	H-met-glu-his-phe-arg-trp-gly	
αMSH	CH_3-C(=O)-ser-tyr-ser-met-glu-his-phe-arg-trp-gly-lys-pro-val-NH_2	
βMSH (horse)	tyr-lys-met-glu-his-phe-arg-trp-gly-----asp-OH	5 6 7 8 9 10 11 12 13 … 18
Vasopressin (lysine)	cys-tyr-phe-glu-asp-cys-pro-lys-gly-NH_2 (cys–cys S—S bridge)	

* Natural ACTH. The synthetic hormone is ACTH 1-24.
Amino acid abbreviations as in Table 10-1, plus pro = proline trp = tryptophan met = methionine val = valine his = histidine

Table 10–6. **Site of Synthesis and Effects of Hormones Derived From Amino Acids**

Hormone	Synthesized by	Type of Molecule	Target Organ	Effect
Thyroxine, Triiodothyronine	Thyroid (follicular cells)	Iodinated derivatives of tyrosine	Most cells	Increased metabolic rate; Growth
Norepinephrine, Epinephrine	Adrenal medulla	Catecholamines derived from tyrosine	Cardiac muscle	Increased heart rate and force
			Arteriolar smooth muscle	Vasoconstriction
			Liver	Glycolysis
			Neurons	Neurotransmission
Melatonin	Pineal	O-methylated indole derived from tryptophan (via serotonin)	Melanocytes	Skin darkening (dispersion of melanin)

Amino Acid Derivatives

In this category are several very important hormones: the hormones of the thyroid gland[2] (*thyroxine* and *triiodothyronine*), those of the adrenal medulla (*epinephrine* and *norepinephrine*) and *melatonin*, the hormone of the pineal gland (Table 10–6). Each employs a very specialized metabolic pathway and the thyroid hormones have been selected arbitrarily to represent this group. Epinephrine biosynthesis is discussed in Sec. 18–3.

SYNTHESIS OF THYROID HORMONES

The *thyroid gland* is shaped like an H, with the cross-bar on the surface of the trachea, just below the larynx. The four small parathyroid glands are on its back surface, but despite this close proximity, there is no functional connection between the thyroid and the parathyroids. The thyroid gland, like all endocrine glands, consists of secretory cells, buttressed by connective tissue and well supplied with blood vessels (Figs. 10-11 and 10-6). The thyroid is unique, however, in the arrangement of its secretory cells into rings or follicles, within the lumen of which its secretions are stored. The thyroid hormones *thyroxine and triiodothyronine*, are stored in the form of *thyroglobulin*, a large iodinated protein with a molecular weight of 660,000. The story of its synthesis and release is a fascinating one that is not yet complete.

Role of Iodine in the Formation of Thyroid Hormones

CONTENT OF SOIL AND WATER. The iodine necessary for the formation of the thyroid hormones is obtained from the diet, chiefly in the form of the inorganic iodides. The concentration of these, in turn, is dependent on the iodide content of the soil or water. In areas where the soil is characteristically lacking in iodine, such as the Alps and Pyrenees in Europe, the Himalayas in Asia, and the Great Lakes area in the United States, the consequence is the endemic occurrence of *goiter* (thyroid enlargement) (Fig. 10-12) and *cretinism* (mental retardation; see also Sec. 11–2). In Japan, where the iodine content of the soil is very low, these deficiency diseases are not often encountered because the Japanese are extremely fond of the iodine-rich seaweed and seafoods.

The classical experiments of Marine and Kimball in Akron, Ohio (1917) resulted in the widespread use of iodized salt to counteract iodine deficiencies in the diet. This was a study of about 7000 schoolgirls over a period of from 1 to 2½ years. They were divided into two groups, one of which received sodium iodide daily, the other serving as a control group. There was such clear evidence of reduction of any existing goiters and low incidence of development of new goiters in the experimental group that the use of iodized table salt has been highly recommended and proven to be extremely effective (1 part of sodium iodide is added to 100,000 parts sodium chloride).

Figure 10-11. **The thyroid gland and its secretory follicles.**

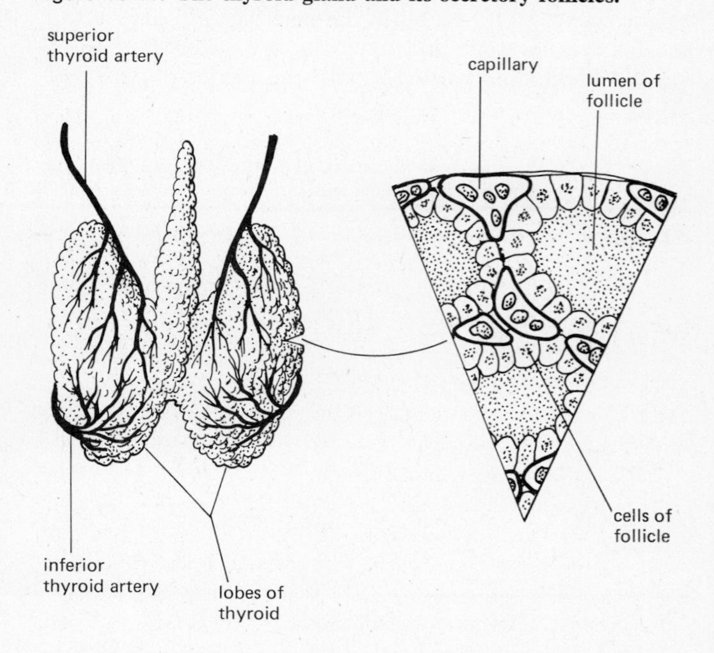

[2] Another thyroid hormone is the polypeptide *calcitonin*, which is involved in the regulation of calcium metabolism. Calcitonin is discussed in Sec. 24–7.

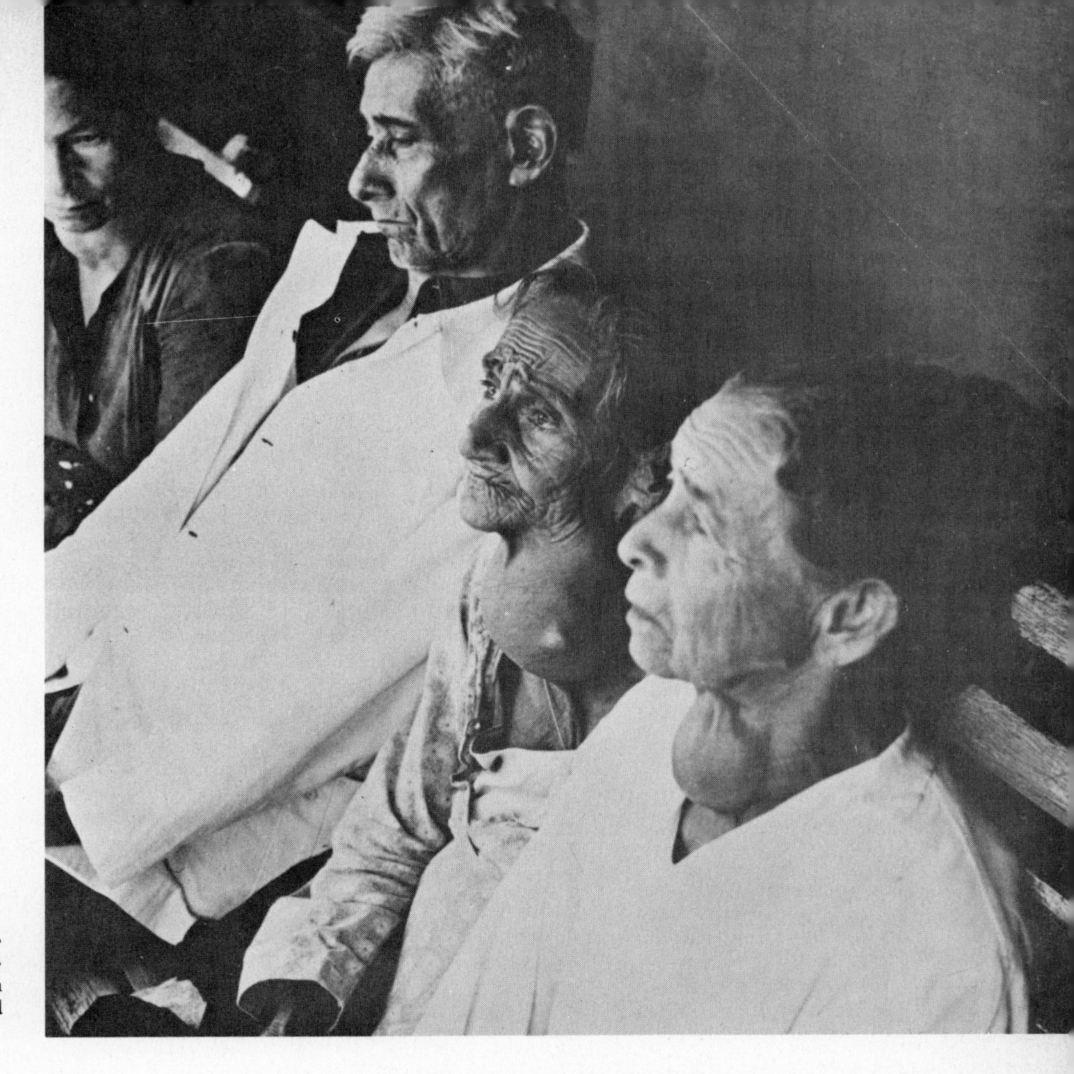

Figure 10-12. **Goiter caused by iodine deficiency is prevalent in Paraguay. (Photograph by Paul Almasy, courtesy of Pan American Health Organization, World Health Organization.)**

Goiter produced by a deficiency of iodine is characterized by large follicles filled with colloid, which flattens the follicular cells. Goiter can be due also to an overactive thyroid, caused by excessive TSH secretion. In this case, the follicles are small, contain little colloid, and have high columnar cells. The active hormones are released as soon as they are formed, so that very little remains (Fig. 10-6). This type of goiter is called *exophthalmic goiter,* because of the accompanying protrusion of the eyes.

Agents that lead to the formation of goiter are called *goitrogens*. Some block the iodination process so that uniodinated thyroglobulin is produced. Cabbage and yellow turnips contain such goitrogens. There are also several synthetic goitrogens, such as propylthiouracyl, that can be used to treat hyperthyroidism.

ACCUMULATION AND OXIDATION OF IODIDE. The ingested iodide is rapidly and efficiently absorbed through the digestive tract and avidly taken up by the thyroid gland, which concentrates the iodide it obtains from the blood about 10,000 times (Fig. 2-4). The amount of radioactive iodide taken up by the thyroid gland after 24 hr is a good indication of the activity of this gland (Fig. 10-13). This *"iodide pump"* has the same dependence on oxidative metabolism as the sodium and potassium "pumps" discussed in Sec. 6-2, and it is also blocked by metabolic inhibitors such as cyanide and dinitrophenol. The trapped iodide has to be oxidized before it can be incorporated into organic molecules. This is accomplished by a peroxidase enzyme system (which may be lacking in certain types of cretins).

Both thyroid hormones are iodinated derivatives of the amino acid tyrosine. Oxidation appears to be coupled with iodination and the first product is *monoiodotyrosine* (MIT), followed by the production of *diiodotyrosine* (DIT) (Fig. 10-14). Two molecules of diiodotyrosine condense to form *thyroxine* (T_4), a synthesis that is speeded up by TSH, the thyroid-stimulating hormone of the anterior pituitary gland. The most active of all the thyroid hormones, however, contains three atoms of iodine and is *triiodothyronine* (T_3). The hormone T_3 is probably formed by the coupling of one molecule of MIT and one of DIT, or by the loss of MIT from T_4. Iodination probably occurs after the peptides are linked to the globulin molecule.

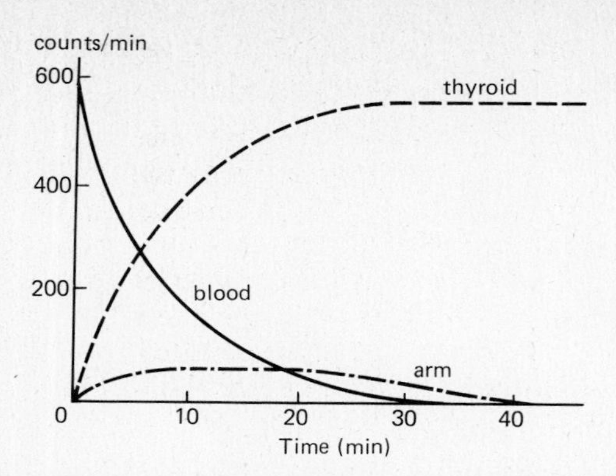

Figure 10-13. **Uptake of ^{131}I by the thyroid gland after an intravenous injection of the radioactive iodine at 0 min. Compare with the low accumulation by tissues of the arm and the loss of ^{131}I from the blood. (From *Thannhauser's Textbook of Metabolism and Metabolic Disorders,* 2nd ed., N. Zöllner and S. Estren, eds., Georg Thieme Verlag, Stuttgart, Germany, 1962.)**

Figure 10-14 **The synthesis of thyroid hormones. Before the iodide can be incorporated, it must be oxidized (I^0) by a peroxidase system of enzymes, which is under the influence of TSH.**

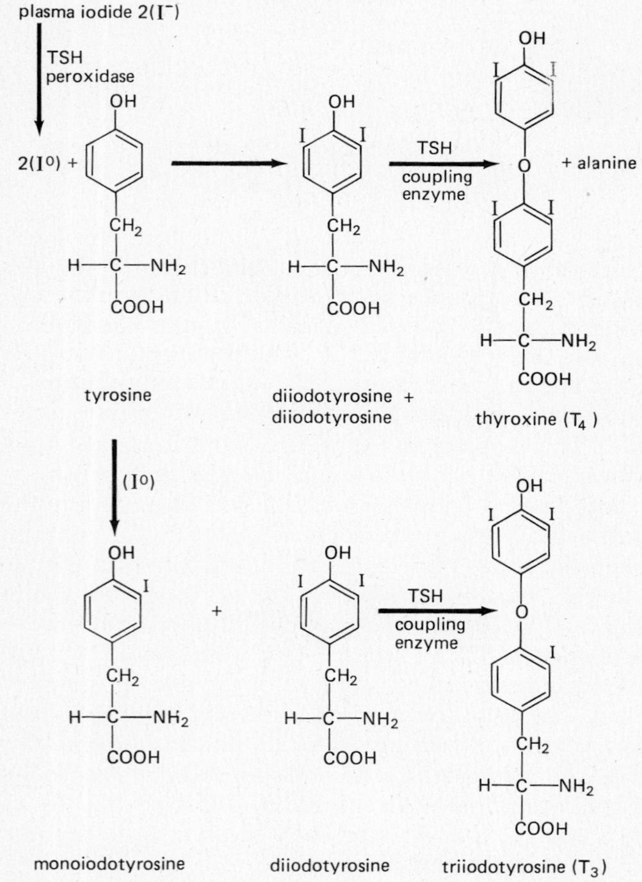

plasma iodide $2(I^-)$
TSH
peroxidase
$2(I^0)$ + tyrosine → diiodotyrosine + diiodotyrosine —TSH, coupling enzyme→ thyroxine (T_4) + alanine
tyrosine —(I^0)→ monoiodotyrosine + diiodotyrosine —TSH, coupling enzyme→ triiodotyrosine (T_3)

Release of thyroid hormones

Thyroglobulin is usually broken down to its constituent T_3 and T_4 components, as well as to other iodinated compounds, by a proteolytic enzyme before it is released from the follicle into the blood. This reaction, like all the others involving the synthesis of thyroid hormones, is stimulated by TSH. The common denominator is probably an increase in cyclic AMP, for the addition of cyclic AMP has the same effect as TSH stimulation.

Most of the thyroid hormone in the blood is bound firmly to plasma proteins, and the concentration is expressed in terms of protein-bound iodide (PBI). Present in the plasma of hyperthyroid individuals, but not in normal ones, is a *long-acting thyroid stimulator* (LATS), an immunoglobulin secreted by lymphocytes. It seems to act to stimulate the thyroid in a manner similar to TSH (through cyclic AMP) but is much slower. It is not certain whether LATS is the cause or the result of hyperthyroidism.

Metabolism and excretion

Thyroxine and triiodothyronine remain in the plasma much longer than most hormones. If they are labelled with radioactive iodide, thyroxine can be shown to disappear from the blood of a normal individual with a half-life of 6 to 7 days, T_3 with a half-life of 1 day. Most other hormones have half-lives measurable in minutes.

The thyroid hormones can be inactivated by practically all tissues by deiodination. The iodide may be excreted in the urine or be reabsorbed and reused to make more hormone. The liver forms conjugates of the hormones and excretes them into the bile and then into the intestine. Once in the intestine they may be economically reabsorbed into the circulation or excreted in the feces.

10–8 Mechanisms of Hormone Action

Hormones direct cellular metabolism, usually through their influence on selected enzyme-catalyzed reactions and by their influence on the transport of solutes across membranes. This brings the necessary substrates into the cell for the enzymes to work on and accelerates the removal of the end products from the cell and into the circulation. Two questions that naturally arise are

1. Why do only certain cells respond to the circulating hormones?
2. How do the hormones affect the enzymes within the responsive cells?

Hormones Acting Through Second Messenger—Cyclic AMP

Partial answers to both of these important questions have been provided by the Nobel Prize-winning work of Earl W.

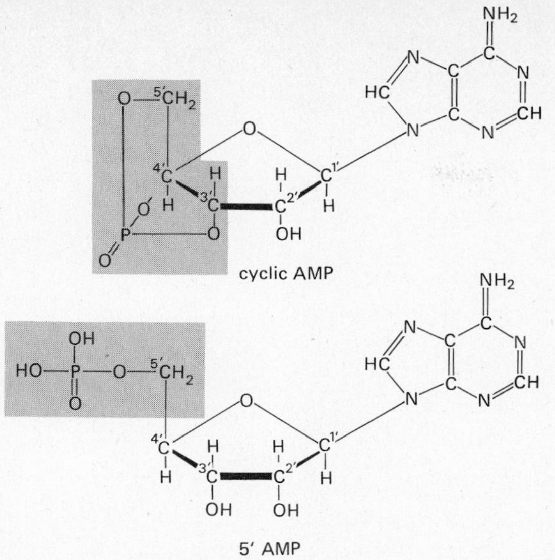

Figure 10-15. **Cyclic AMP and 5′ AMP.**

Sutherland (1972). It was his discovery of a small, heat-stable molecule called cyclic 3′ 5′ adenosine monophosphate (cyclic AMP) that demonstrated a universality of biological action: this unique molecule mediates the action of many hormones and consequently regulates cellular activity in all forms of life, including higher plants and microorganisms. Compare the structure of cyclic AMP with that of its biologically inactive form, 5′ AMP (Fig. 10-15).

FUNCTIONS OF CYCLIC AMP

Basic Functions. One function of cyclic AMP is to *indirectly make energy available* to the cell through the acceleration of glycogenolysis and lipolysis. The hormones epinephrine and glucagon increase cellular levels of cyclic AMP, which in turn affects both the rate of enzyme action and also, through direct effects of cyclic AMP on the genome, the amount and kind of enzyme synthesized.

Specialized Functions. Cyclic AMP has many *highly specialized functions.* Some of these follow.

RELEASE OF HORMONES FROM THE ENDOCRINE GLANDS. In response to ACTH, for example, cyclic AMP levels in adrenocortical cells increase, and adrenocortical hormones are synthesized and released from this tissue.

INCREASED PROTEIN SYNTHESIS. Cyclic AMP increases the production of enzymes necessary for protein synthesis. It may induce the synthesis of different types of enzymes appropriate for the particular physiological state. In starvation, when the normal substrate, glucose, is absent, cyclic AMP can direct the synthesis of enzymes required for the metabolism of alternate substrates, such as lactose.

Protein synthesis can be affected at various levels, depending upon the type of cell involved.

1. Energy production in the cell may be increased through glycogenolysis and the transport of the necessary amino acids and sugars enhanced by changes in membrane permeability.
2. Cyclic AMP may influence transcription of different regions of the genome. This will affect the rate and the selection of proteins to be assembled.
3. Cyclic AMP can influence translation by activating protein kinases (Sec. 5-8).

CHANGES IN PERMEABILITY. Changes in permeability of the kidney tubules affect the osmolarity of the urine. In response to vasopressin, cyclic AMP levels in kidney tubule cells increase, and water is reabsorbed, increasing the tonocity of the urine.

PIGMENT DISPERSAL IN MELANOCYTES. In response to MSH, cyclic AMP levels in these cells increase and the melanin granules are dispersed, causing skin darkening (Fig. 10-16). This action can be mimicked by ACTH, which shares a common amino acid sequence with MSH (Table 10-5).

SYNAPTIC TRANSMISSION IN THE NERVOUS SYSTEM. Epinephrine, norepinephrine, and serotonin, all of which act as transmitters in the nervous system, change cyclic AMP levels at the synapse and consequently change the synaptic

Figure 10-16. **The MSH increases the level of cyclic AMP in amphibian melanocytes. This causes the dispersal of the melanin granules throughout the cell, darkening the skin.**

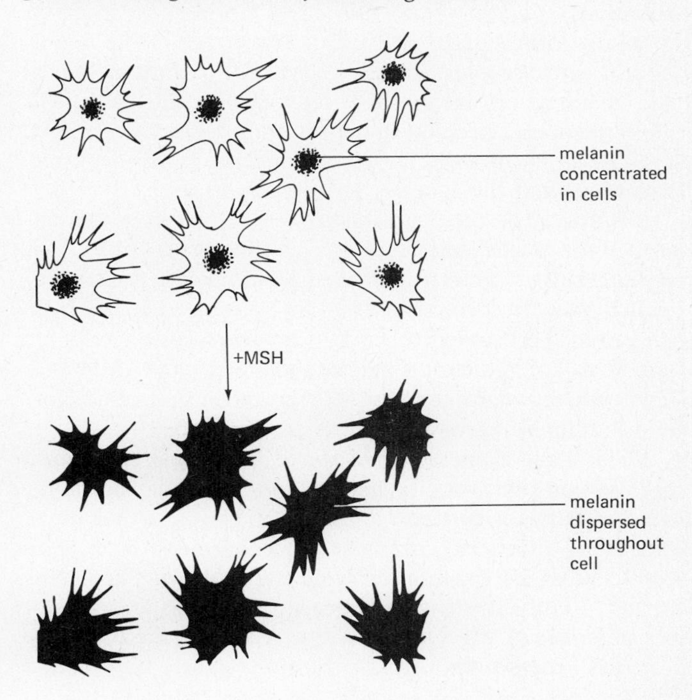

Table 10–7. **Hormonal Effects Involving Cyclic AMP**

Hormone	Tissue	Response
Adrenocorticotropin (ACTH)	Adrenal cortex	Hydrocortisone secretion
	Fat	Lipolysis
Melanocyte-stimulating hormone (MSH)	Frog skin melanocytes	Pigment dispersal—skin darkening
Vasopressin	Kidney tubules	Permeability change—water reabsorption
Glucagon	Liver	Glycogenolysis
	Fat	Lipolysis
	Pancreas	Insulin release
Thyroid-stimulating hormone (TSH)	Thyroid follicle	Thyroid hormone secretion
Luteinizing hormone (LH)	Ovary—corpus luteum	Progesterone secretion
Parathyroid hormone	Bone	Calcium reabsorption
	Kidney tubule	Phosphate excretion
Epinephrine	Liver	Glycogenolysis
	Muscle	Glycogenolysis
	Fat	Lipolysis
	Heart	Increased contractility
Norepinephrine	Cerebellum	Firing of Purkÿne cells
	Synapses	Change in synaptic potentials (hyperpolarization?)
	Heart	Increased heart rate
Serotonin (hormone?)	Synapses in CNS	Change in synaptic potentials (hyperpolarization)
	Liver	Phosphofructokinase activation

membrane potential. ACTH and related polypeptides affect synaptic transmission and the nervous system, change cyclic AMP levels at the synapse, and consequently change the synaptic membrane potential; they also affect synaptic transmission and nervous excitability in both the central and the peripheral nervous system. It is presumed that changes in cyclic AMP are involved, probably influencing intracellular calcium levels, and so indirectly altering the membrane potential.

INHIBITION OF SOME CELLULAR ACTIVITIES. The inhibition of some cellular activities, such as the immunological and inflammatory responses of leukocytes, controls the immune response, keeping it within homeostatic levels. The increase in cyclic AMP levels is triggered by epinephrine, histamine, and the E series prostaglandins (Sec. 10-9).

INHIBITION OF CELLULAR GROWTH. This can be demonstrated for normal and cancerous cells in tissue culture; it is presumed that cyclic AMP normally prevents the uncontrolled growth characteristic of malignancies. It has been suggested that, among other abnormalities, cancer cells are unable to accumulate proper amounts of cyclic AMP. It is unknown to what extent growth hormone and other hormones influencing tissue growth are implicated.

Each of the examples cited above shows that specialized cells respond in a specific manner to a particular hormone. This interaction between hormone and cyclic AMP is an intricately fitting scheme, where the hormone acts as an extracellular information-carrying molecule, or "first messenger." This information is brought from distant parts of the organism to affect levels of intracellular cyclic AMP, the "second messenger" that transmits the information throughout the cell. The cell interprets it according to its own language, that of available molecular components. Depending upon the particular biochemistry of the cell, the response will be glycogenolysis (liver or muscle cells), hormone secretion (endocrine glands), water reabsorption (kidney tubules), skin darkening (melanocytes), or changes in synaptic potential (synapses). Table 10-7 shows some of the hormonal effects that are known to be mediated by cyclic AMP.

It is of interest to note that all the hormones that utilize cyclic AMP, as far as we know today, are polypeptides (ACTH, vasopressin, MSH, glucagon, etc.) or amino acid derivatives (epinephrine, norepinephrine, and serotonin). The mechanism of action of steroid hormones is discussed later in this chapter.

HORMONE RECEPTORS

The answer to the first question raised, why only certain cells respond to a particular hormone, is based on the existence of specific, high-affinity receptors in the plasma membrane. These receptors are proteins that can recognize and bind small amounts of the appropriate hormone despite a vast excess of other types of molecules. Through the use of labelled hormone the formation of specific hormone-receptor complexes can be demonstrated. Only those cells capable of binding a hormone can respond to it.

Many hormones *regulate the activity* of their own and other hormone receptors. For example, high levels of insulin in the blood, characteristic of obesity, decrease the number of insulin-binding receptors. This may be a regulatory mechanism that prevents cells from over-responding to high hormone concentrations. Excess thyroid hormone may regulate TSH secretion by causing a loss of TRH receptors in the pituitary gland. At the same time, high thyroid hormone levels increase the number of catecholamine receptors on heart cells, which may explain the rapid heart beat and palpitations seen in patients with hyperthyroidism. Another application of receptor regulation is the ability of hormones to act in sequence. The gonadotropin FSH acts first on the ovary to cause development of the ovarian follicles and then prepares the way for the action of the second gonadotropin LH by increasing the number of LH receptors.

The change in the number of active receptors may be due to their inactivation or activation through conformational changes, or related to their position on the cell surface. "Lost" receptors may sink into the membrane. Some receptors may be effectively blocked by antibodies, as in myasthenia gravis (Sec. 7-5). Thus variations in responses to hormones may be partially caused by changes in the target cells themselves.

Hormone-stimulated cyclic AMP activity

Sutherland's brilliant work answered many parts of the second question, as to the way in which hormones change enzyme levels within a cell. It is a cyclic AMP-mediated mechanism, starting with the binding of the hormone to its membrane receptor. The immediate effect is to increase the activity of an enzyme, adenyl cyclase, which is bound to the cell membrane, presumably on its inner surface. Elements within the membrane act as "communicators" to convey the information from the hormone receptor to adenyl cyclase (Fig. 10-17). What these elements are is not known for certain. It has been proposed that calcium ions, prostaglandins, and guanosine triphosphate (GTP) may be involved, for many hormone-influenced reactions require the presence of one or more of these substances to activate the cyclase.

Locked into this increased adenyl cyclase activity is the next step, a rise in cyclic AMP levels within the cell. Cyclic AMP is produced by the catalytic action of adenyl cyclase on ATP:

$$\text{ATP} \xrightarrow{\text{adenyl cyclase}} \text{cyclic AMP and 2 Pi}$$

The cyclic AMP formed diffuses readily throughout the cell, acting as a "second messenger" instructing the cell to respond in its own inherent manner (Fig. 10-17).

Figure 10-18 shows the chain reaction triggered by the binding of a molecule of epinephrine to a receptor site on a liver cell membrane. Through the communicating elements within the cell membrane, adenyl cyclase bound to the inner side of the membrane is activated. As a result, ATP is hydrolyzed, and cyclic AMP is formed. A cascade of additional enzyme activations results:

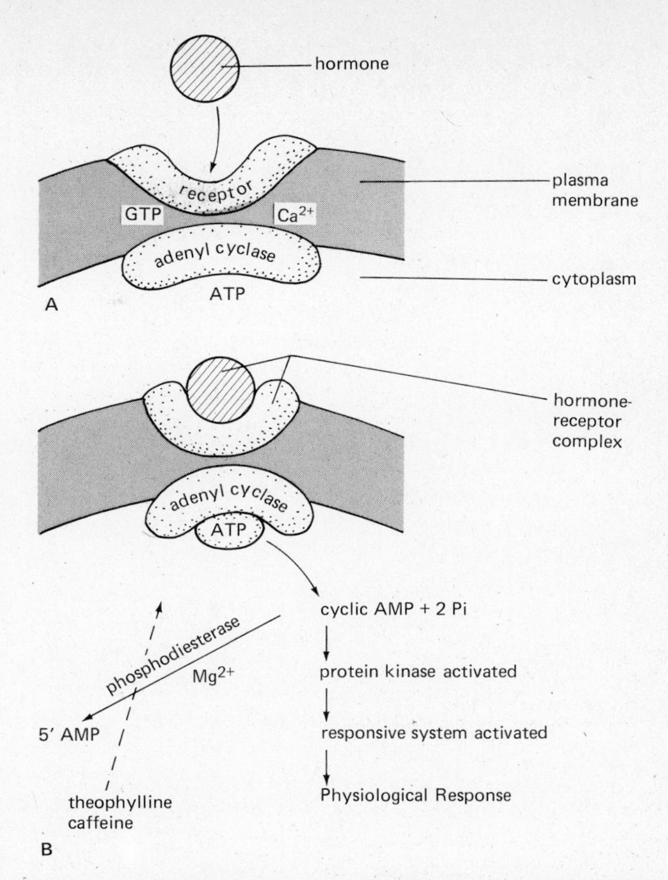

Figure 10-17. **(A) Hormones acting through cyclic AMP bind first to a receptor protein in the plasma membrane. "Communicators" such as Ca^{2+} and guanosine triphosphate (GTP) convey the change in receptor conformation to the membrane-bound adenyl cyclase. (B) The activated adenyl cyclase catalyzes the formation of cyclic AMP from ATP. This initiates the sequence of reactions shown as a generalized response of a cyclic AMP-sensitive cell. Cyclic AMP is inactivated by phosphodiesterase, an enzyme that is inhibited (broken arrow) by theophylline or caffeine.**

1. Protein kinase is activated, which then activates
2. Phosphorylase kinase, which then activates
3. Glycogen phosphorylase *b* to its active form, glycogen phosphorylase *a*.

Now glycogen becomes involved, because active phosphorylase initiates stepwise degradation of glycogen to glucose-1-phosphate. This molecule is isomerized to glucose-6-phosphate. Liver possesses the enzyme necessary for the production of free (nonphosphorylated) glucose, which is rapidly released into the circulation. Refer to Sec. 5-12 for details. This cyclic AMP cascade process has the great advantage of amplifying the effect of the hormone, because each enzyme molecule activated affects many other molecules.

A concomitant synthesis of glycogen by the liver cell at this time would be most uneconomical; such a synthesis is prevented by the inactivation of glycogen synthetase by other cyclic AMP molecules.

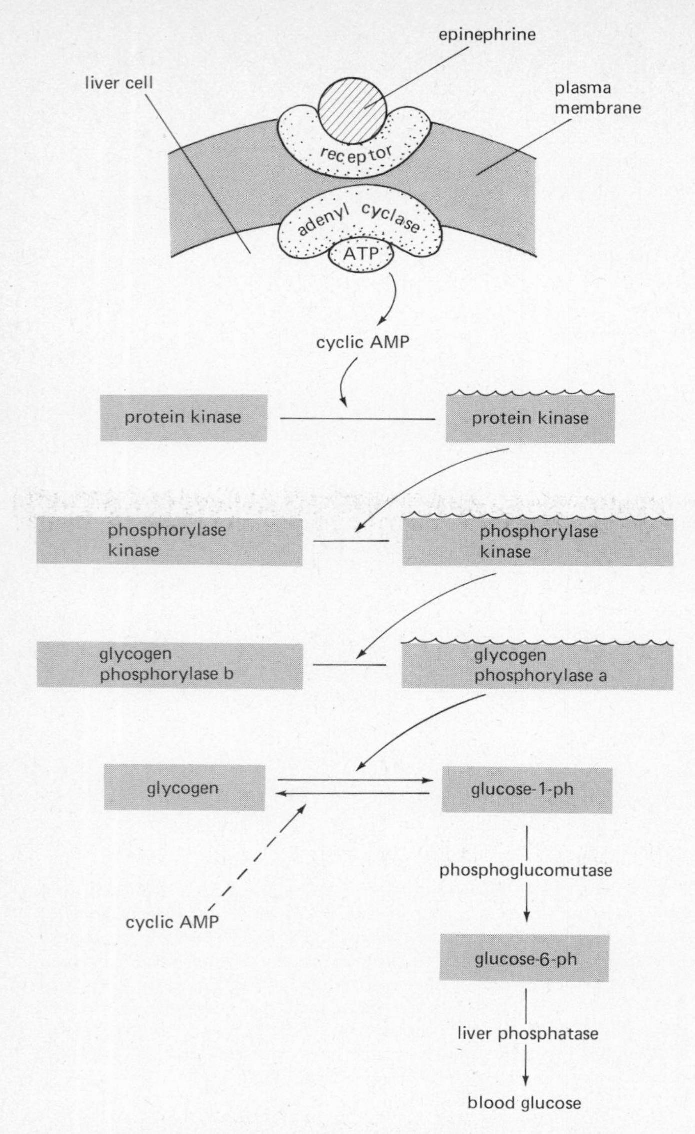

Figure 10-18. **The specific response of a liver cell to epinephrine is the breakdown of glycogen to glucose, a complex response mediated through cyclic AMP. The reactions that would cause the synthesis of glycogen are simultaneously inhibited by cyclic AMP (broken arrow). —— to ⁀⁀ indicates the change from the inactive to the active form.**

Turning off cyclic AMP activity

Once the hormone-activated reactions have been completed, cyclic AMP is hydrolyzed to 5′ AMP by an enzyme, phosphodiesterase, a reaction that requires Mg^{2+} (Fig. 10-17). All cyclic AMP–dependent reactions are turned off, because 5′ AMP is ineffective on these targets. Controlling the level of phosphodiesterase is obviously important: again many factors are involved and not all of them are understood. Calcium levels are significant, and phosphodiesterase can be inhibited experimentally by a class of compounds called *methylxanthines,* which may affect calcium levels. One of these is *theophylline* and another is *caffeine.* Presumably some of the stimulating effects of caffeine are due to the increased Ca^{2+} levels which inhibit phosphodiesterase and thus raise cyclic AMP levels. However, caffeine also raises Ca^{2+} concentrations in muscle, thus enhancing contractility through the activation of mysosin ATPase, but cyclic AMP has not been implicated.

The turning off of these cyclic AMP-stimulated reactions also requires the removal or inactivation of the hormone from the membrane receptor and the inactivation of adenyl cyclase, but these processes are still poorly understood.

Cyclic GMP

Another cyclic nucleotide, guanosine 3′, 5′ monophosphate (cyclic GMP) is present in all living systems, but in much smaller amounts than cyclic AMP. It is consequently much more difficult to measure and study cyclic GMP. The role of cyclic GMP appears to be to promote reactions opposed by cyclic AMP in such cells where these opposing reactions are synchronized or integrated.

One example of this is in liver cells, in which epinephrine stimulates *glycogenolysis* by increasing the cellular concentration of cyclic AMP: in the same liver cells the hormone insulin may spur *glycogenesis* through increasing cyclic GMP concentrations. Another important example is in the nervous system. In some peripheral ganglia, cyclic GMP may be the mediator for one neurotransmitter (*acetylcholine*), whereas cyclic AMP mediates the effects of another type of transmitter (*norepinephrine* or *dopamine*). Depending then on the relative concentrations of these two cyclic nucleotides within the cell, the cell membrane may be *depolarized* (stimulated) or *hyperpolarized* (inhibited).

Hormones Acting on the Genome

Unlike hormones that utilize cyclic AMP as a mediator (most polypeptide and amino acid derivatives, as discussed above), steroid hormones enter the cell, where they may enter the *nucleus* directly or bind with intracellular protein receptors in the soluble part of the *cytoplasm.* The formation of the hormone-receptor complex apparently facilitates the entry of the hormone into the nucleus, where it reacts with DNA (Fig. 10-19). This leads to changes in protein synthesis in the cytoplasm of the cell, appropriate to the biochemical organization of that specific cell. For example, testosterone entering a muscle cell will, through its action on the DNA of the genome, influence protein synthesis on the ribosomes to increase muscle protein. Progesterone will influence the growth of the cells lining the uterus. The specificity of the action of a hormone requires first that a cell have the requisite receptors in the plasma membrane or cytoplasm and also the necessary biochemical machinery to respond to the changes thus initiated. The mode of action of polypeptide and steroid hormones is diagrammed in Fig. 10-20.

The proteins to which these hormones bind are consid-

Figure 10-19. **Steroid hormone receptors in the rat brain can be located by administering corticosterone labelled with tritium. One or 2 hours after administration of the labelled hormone, frozen sections of the animal's brain are placed in contact with photographic emulsion and stored for several months. The radioactive atoms make black dots in the emulsion, revealing the presence of cells containing receptors for corticosterone. This autoradiogram shows that nerve cells in the hippocampus are heavily labelled. (Courtesy of B. McEwen, The Rockefeller University.)**

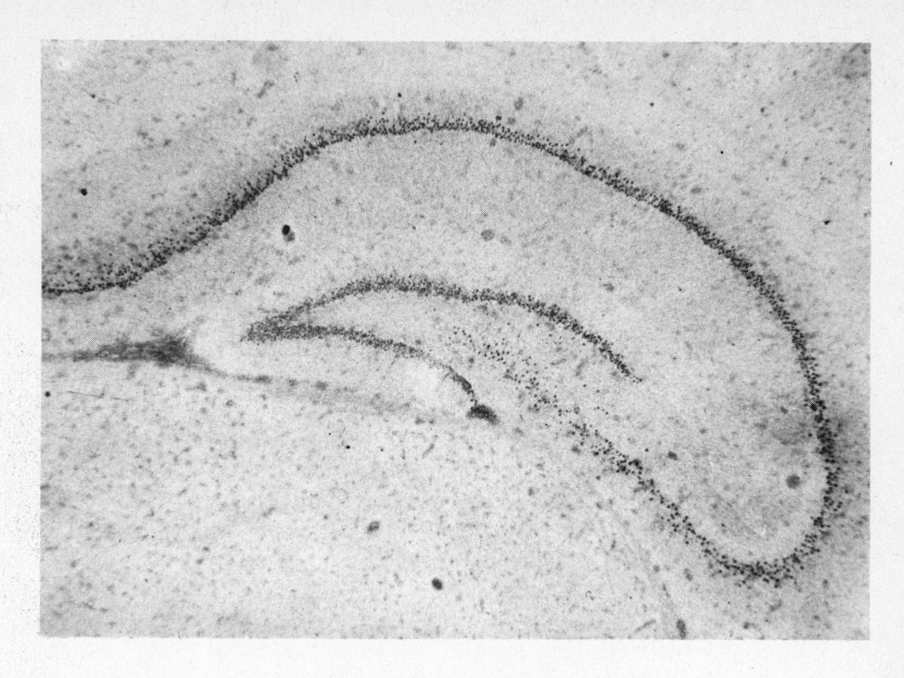

ered receptors because they can be isolated from the appropriate target tissues and not from unresponsive tissues. Additional support comes from the observation that where target tissues do not respond appropriately, the binding proteins are deficient. This can be seen in certain strains of mutant mice that are insensitive to androgens: the binding proteins for dihydrotestosterone are deficient in these animals.

Some steroids are metabolized in the target cells before combining with the cytoplasmic receptor. This occurs with testosterone, which can be metabolized to the biologically more active molecule 5 dihydrotestosterone (DHT) or to estrogen, as occurs in the brain. These routes are shown in Fig. 10-20.

Localization of steroids within the brain

McEwen (1976) has shown that different regions of the brain preferentially accumulate and retain specific steroid hormones. Binding of the female sex hormone estradiol is highest in the hypothalamus, while corticosterone is found in greater concentration in the hippocampus and amygdala, components of the limbic system. This provides a neat mechanism for the hormonal control of nervous activity.

Figure 10-20. **Mechanism of action of polypeptide and steroid hormones. See text for details. [From F. L. Strand, *Bioscience* 25 (9):568 (1975).]**

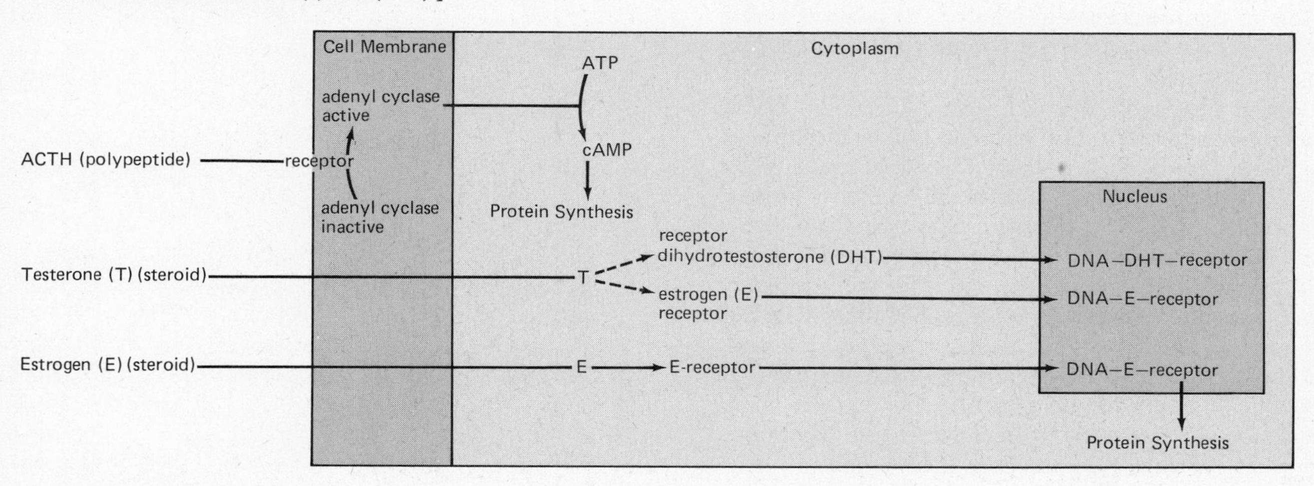

Hormones as directors of cellular metabolic pathways

Once inside the nucleus, the hormone acts on the genome, possibly by switching on a repressed gene. This results in a changed level of protein synthesis, permitting the cell to fluctuate from the production of enzymes associated with carbohydrate metabolism, for example, to fat metabolism. In this way, *hormones are truly directors of cellular metabolic pathways.*

The regulation of protein synthesis by these hormones may occur on several levels:

1. It has been suggested that the hormones activate genes by acting as *inducers*. The intracellular protein receptor normally functions as a repressor, controlling *m*RNA synthesis. Combination with the hormone alters its function, presumably through allosteric effects, and it no longer represses a specific gene. This results in increased *m*RNA synthesis.
2. The effect may be on *translation*, the process that programs the ribosomes for the synthesis of specific protein molecules according to the information received from *m*RNA molecules.
3. A direct effect of these hormones has also been postulated. The *ribosomes themselves* may be sensitive to the hormone; consequently their synthesis, turnover, and activity may be affected.

10–9 Prostaglandins

Structure and Synthesis

Prostaglandins form a group of compounds with surprisingly diverse effects. They are often classified as hormones, although they are produced by almost all tissues and cells. They are modified *fatty acids*, all variants of a basic 20-carbon chain; like the fatty acids they have a carboxyl (acid) group, but unlike them prostaglandins incorporate a *cyclopentane ring* (Fig. 10-21). Differences in the position of three oxygen atoms within the molecule create different members of the prostaglandin family. *The primary prostaglandins are PGE and PGF;* they are the precursors for the other prostaglandins.

Prostaglandins are bound to plasma proteins and to cell membranes and appear to be synthesized by membranes, using phospholipids as a precursor. They are found in most tissues in infinitesimal amounts, apparently because they are broken down rapidly by catabolic enzymes; they have a half-life of less than a minute. However, fresh semen is very rich in prostaglandins and was the source for their discovery and their naming by von Euler (1935). They are synthesized by the seminal vesicles, part of the male reproductive tract that contributes to the semen, but they were at first thought to originate in the prostatic secretion, hence their name. Prostaglandins can now be synthesized in the laboratory, which greatly facilitates research on their action.

Functions

Uterine contractility

Prostaglandins are among the most potent biological molecules, and their wide range of effects includes their remarkable ability to stimulate uterine contractility. Karim (1970), in Uganda, showed that prostaglandins are present in the amniotic fluid and venous blood of women during labor. They act synergistically with (enhance the action of) oxytocin, and it is logical to infer that they are involved in normal parturition. The administration of prostaglandins to pregnant women is now used to induce labor or abortion (Fig. 10-22). The role of prostaglandins in semen is unknown, perhaps acting to regulate the emptying of the male glands. African women have been reported to swallow semen as a contraceptive device (see also Sec. 26–18), although present evidence indicates that prostaglandins are less effective as contraceptives than they are as abortive agents.

Circulation and secretion

One prostaglandin (PGF_2-alpha) elevates blood pressure by causing arteriolar constriction, whereas PGE_2 has the reverse effect, lowering blood pressure sharply. Through their effects on blood flow, and through other mechanisms discussed below, prostaglandins regulate many physiological

Figure 10-21. **Prostaglandin structure. Differences in the position of oxygen atoms on the cyclopentane ring, attached to the fatty acid chain, differentiate members of the prostaglandin family, PGE, PGF, and PGA.**

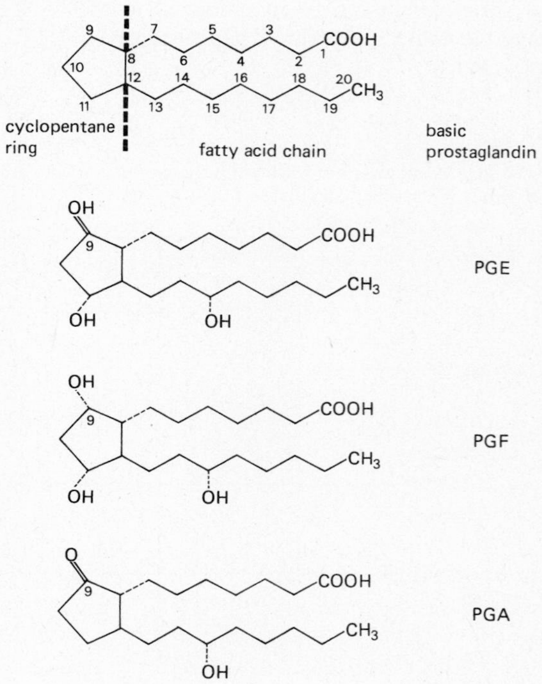

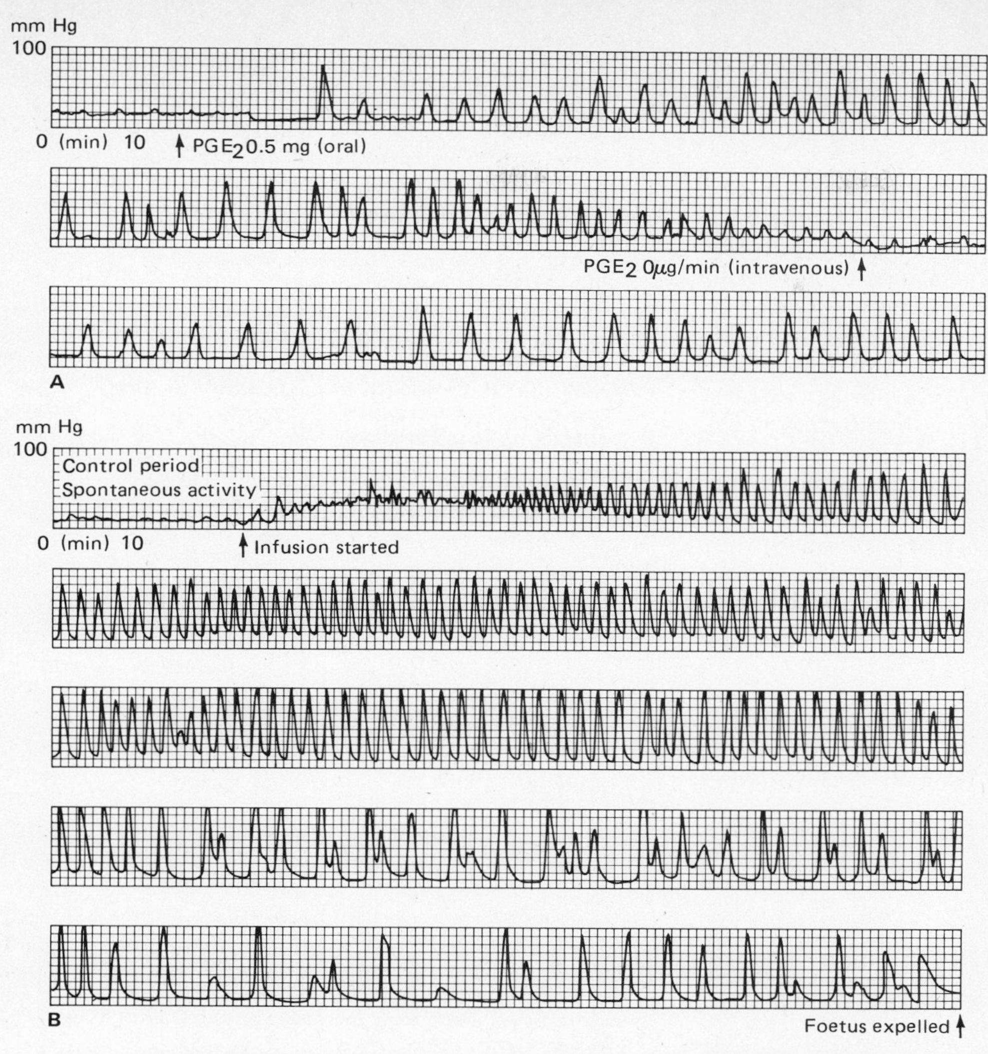

Figure 10-22. **(A) Induction of labor at term with oral and intravenous administration of a prostaglandin, PGE_2. The deflections indicate the contractions of the pregnant human uterus at 42 weeks. (B) Therapeutic abortion with intravaginal administration of PGE_2. The deflections indicate the contractions of the pregnant uterus at 14 weeks. [From S. Karim, *Ann. N.Y. Acad. Sci.* 180:483 (1971).]**

parameters, including secretion in both exocrine and endocrine glands. Prostaglandins of the E group inhibit gastric secretion and may help protect the stomach wall against ulceration. In the thyroid and corpus luteum, prostaglandins mimic the action of TSH and LH by increasing the secretion of thyroxine and progesterone, respectively.

Nervous system

Prostaglandins are the first known, naturally occurring substances that inhibit transmission in the nervous system in very low concentrations. They prevent the release of norepinephrine from stimulated sympathetic nerves, a finding of great significance: it indicates that they play a modulatory role in sympathetically innervated tissue.

Inflammatory response and blood clotting

Prostaglandins are released in trauma and shock and in other types of cell damage, such as allergic eczema. They cause the formation of wheals and flares, probably by increasing vascular permeability, and the action of aspirin and other anti-inflammatory agents may be due to their blocking action on prostaglandin synthesis. A prostaglandin PGI_2 affects blood clotting by preventing platelet aggregation and may be used therapeutically to prevent abnormal clots that lead to heart attacks and strokes.

Antilipolytic action

Prostaglandins regulate metabolism, especially fat metabolism, by counteracting the effects of hormones on fat breakdown. This antilipolytic action can be elicited by very low prostaglandin concentrations.

Mechanism of Action

To explain such a diversity of physiological effects, one must look for a common mode of action. Just as the basic mode of action for nerve transmission is a change in membrane permeability that results in excitation or inhibition, just as hormones influence enzyme systems, there must be some basic biological phenomenon to which prostaglandin action can be related. In this new field of research, it appears that cyclic AMP, and perhaps other cyclic mononucleotides, may be the mediator for prostaglandins. Stimulating effects of prostaglandins are associated with increased tissue levels of cyclic AMP and increased adenyl cyclase activity. Conversely, inhibitory effects are associated with a decrease in cyclic AMP formation. The significance of the cell membrane is brought to the fore: it is the common site of action for neurotransmitters, and hormones, and for this intermediate group, the prostaglandins.

Cited References

GAUNT, R. Comparative Endocrinology: Hormones and Receptor Response. *Fed. Proc.* **26**(4): 1192 (1967).

HARRIS, G. W. *Neural Control of the Pituitary Gland.* Edward Arnold & Co., London, 1955.

KARIM, S. M. M., and G. M. FILSHIE. "Therapeutic Abortion Using Prostaglandin F_2." *Lancet* **1:** 157 (1970).

MARINE, D., and O. P. KIMBALL. "The Prevention of Simple Goiter in Man: A Survey of the Incidence and Types of Thyroid Enlargement in the Schoolgirls of Akron, Ohio, from the 5th to the 12th Grades, Inclusive; The Plan of Prevention Proposed." *J. Lab. Clin. Med.* **3:** 40 (1917).

MCEWEN, B. "Interactions Between Hormones and Nerve Tissue." *Sci. Am.,* July 1976.

SMITH, P. E. "The Disabilities Caused by Hypophysectomy and Their Repair." *J. Amer. Med. Assoc.* **88:** 158 (1927).

SUTHERLAND, E. W. "Studies on the Mechanism of Hormone Action." *Science* **177:** 401 (1972).

VON EULER, U. S. "Über die spezifische blutdrucksenkende Substanz des menschlichen Prostat- und Samenblasensekretes." *Klin. Wochenschr.* **14:** 1182 (1935).

Additional Readings

BOOKS

BARRINGTON, E. J. W. *An Introduction to General and Comparative Endocrinology.* Oxford University Press, New York, 1975. (Paperback.)

BENTLEY, P. J. *Comparative Vertebrate Endocrinology.* Cambridge University Press, New York, 1976. (Paperback.)

DORFMAN, R. I., ed. *Methods in Hormone Research,* Vols. 1-5. Academic Press, Inc., New York, 1966-1969.

FREEDMAN, M. A., and S. N. FREEDMAN. *Introduction to Steroid Biochemistry and Its Clinical Application.* Harper & Row, Publishers, New York, 1970. (Paperback.)

FRIEDEN, E., and H. LIPNER. *Biochemical Endocrinology of the Vertebrates.* Foundations of Modern Biochemistry Series. Prentice-Hall, Englewood, N.J., 1971. (Paperback.)

RAMWELL, P., and J. E. SHAW, eds. "Prostaglandins." *Ann. N.Y. Acad. Sci.* **180:** (1971).

ROBISON, G. A., R. W. BUTCHER, and E. W. SUTHERLAND. *Cyclic AMP.* Academic Press, Inc., New York, 1971.

TEPPERMAN, J. *Metabolic and Endocrine Physiology,* 3rd ed. Year Book Medical Publishers, Chicago, 1973.

TURNER, C. D. *General Endocrinology,* 5th ed. W. B. Saunders Company, Philadelphia, 1971.

VILLEE, D. B. *Human Endocrinology: A Developmental Approach.* W. B. Saunders Company, Philadelphia, 1975.

ARTICLES

ATKINSON, R. C., and R. M. SCHRIFFRIN. "The Control of Short-Term Memory." *Sci. Am.,* Aug. 1971.

GILLIE, R. B. "Endemic Goiter." *Sci. Am.,* June 1971.

GUILLEMIN, R., and R. BURGUS. "The Hormones of the Hypothalamus." *Sci. Am.,* Nov. 1972.

MCEWEN, B. S., C. F. DENEF, J. L. GERLACH, and L. PLAPINGER. "Chemical Studies of the Brain as a Steroid Hormone Target Tissue." In *The Neurosciences, Third Study Program.* The M.I.T. Press, Cambridge, Mass., 1974, p. 599.

O'MALLEY, B. W., and W. T. SCHRADER. "Receptors of Steroid Hormones." *Sci. Am.,* Feb. 1976.

PASTAN, I. "Cyclic AMP." *Sci. Am.,* Aug. 1972.

PIKE, J. E. "Prostaglandins." *Sci. Am.,* Nov. 1971.

STONE, T. W., D. A. TAYLOR, and F. E. BLOOM. "Cyclic AMP and Cyclic GMP May Mediate Opposite Neuronal Responses in the Rat Cerebral Cortex." *Science* **187:** 845.

TOMKINS, G. M. "The Metabolic Code." *Science* **189:** 760 (1975).

Chapter 11

Neuroendocrine Regulation

Science outstrips
other modes &
reveals more of
the crux of the matter
than we can calmly
handle

A. R. Ammons, "Exotic"

ONE OF THE basic tenets of physiology is that the optimal function of cells and tissues is obtained through the maintenance of an optimal internal environment. This environment may fluctuate within a physiological range, but extreme deviations result in pathological changes and death. Included in the concept of the internal environment are the nutrients, respiratory gases, salts, tonicity and pH of blood and the body fluids, and also the temperature and pressure with which these fluids are delivered to the cells. Consequently, a most delicate yet firm integration of regulatory functions is essential. The chief regulators are the nervous and endocrine systems, which through a series of negative and positive feedback mechanisms regulate not only the internal environment, but also one another.

11–1 Development of Integrative Systems

Metabolites

Nature is in general a most thrifty housewife: one sees again and again how simple molecules manufactured by almost all tissues become modified and specialized for different functions. In *unicellular animals,* such as the ameba, there is neither a nervous nor an endocrine system, yet the many activities of the cell are well integrated—presumably through a series of metabolic and electrophysiological cues. The local metabolic alterations result in bioelectrical potentials that are conducted over the entire surface of the organism.

In *higher organisms* carbon dioxide and other products of cellular metabolism also are put to good use in integrating organ systems. Carbon dioxide acts as a potent stimulator of the respiratory center, initiating an increase in the rate and depth of respiration as the level of CO_2 rises in the blood. Nerve cells in these centers increase their rate of firing with increases in blood carbon dioxide pressure, up to a point; then, like other cells, they succumb to its toxic effects. The physiological range has been exceeded.

Neurotransmitters

Nerve cells are not only cells specialized for the conduction of electrical impulses; they also produce special chemicals that are essential for the transmission of these impulses across the gap that separates the nerve terminals from the following cell. These chemicals are neurotransmitters, and many of them are formed from simple amino acids, such as tyrosine. *Tyrosine* is the precursor for the neurotransmitters norepinephrine, epinephrine, and dopamine (see Fig. 9-16 for the metabolic pathways involved). Another amino acid, *tryptophan,* produces the transmitter serotonin (Fig. 9-20). The amino acid, glycine, and gamma amino butyric acid or

GABA (derived from another amino acid, *glutamate*), are both inhibitory transmitters of the central nervous system. The chemical structures of these neurotransmitters are shown in Sec. 7-6.

The relatively simple ion *choline* is combined with *acetyl coenzyme A* by all animal tissues to form acetylcholine, a highly important neurotransmitter in the peripheral and central nervous systems (see Sec. 7-5).

Neurotransmitters are synthesized both by the cell bodies of the neurons and in the nerve terminals. Transmitter made by the soma of the neuron is *transported* by *axonal flow* to the nerve endings, where it is *packaged*, usually in combination with a protein, inside membrane-bound vesicles. This appears to be the storage form of the transmitter, but, when it is being rapidly utilized and inactivated, the terminals have the appropriate enzyme systems to resynthesize the transmitter. This saves the long delay that axonal transport requires (from 1 to 3 days, depending on the length of the nerve) and makes the nerve terminals practically inexhaustible.

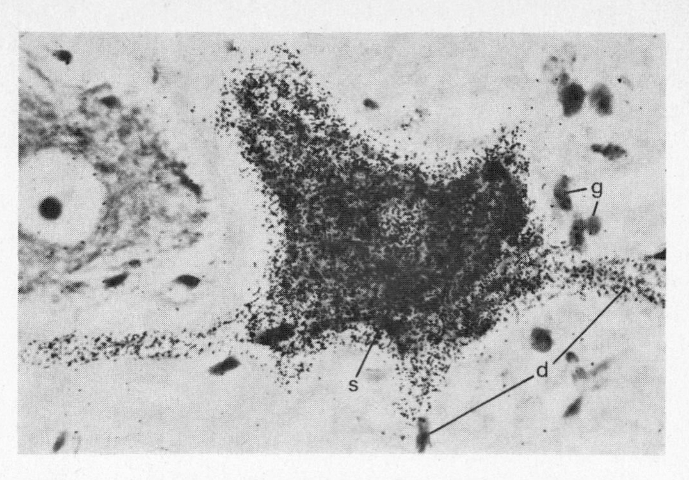

Figure 11-1. **Incorporation of ^{3}H-glycine, about 30 min after intracellular application. Labelling is confined to the soma (s) and proximal dendrites (d) of the injected neuron. No silver grains exceeding background level are seen over the neighboring glial (g) cells, nor the adjacent neuron in the left corner. Autoradiography stained by toluidine blue. ×670. (From P. Schubert and G. W. Kreutzberg, in *Advances in Neurology,* Vol. 12, G. W. Kreutzberg, ed., Raven Press, New York, 1975, p. 255.)**

Axonal Transport

There are three main types of transport of materials along nerve axons. A *slow axonal transport* moves soluble proteins and various structural proteins along the axon at a rate of about 1 to 3 mm per day. This supplies the nerve terminals with substances they are unable to synthesize because of the absence of ribosomes in the terminals. It is probable that the microtubules of the axon are involved in slow axonal transport (see also Sec. 8-2).

The *fast axonal transport* system moves other particles including cell organelles much more rapidly down the axon, at rates estimated, from experiments involving labelled amino acids, as being between 10 to 3000 mm per day (Fig. 8-6).

The *retrograde axonal transport,* as demonstrated by the use of horseradish peroxidase, was discussed in Sec. 6-1. This mechanism brings materials from the terminals back to the cell body. It is remarkable to watch this two-way transport in nerve axons with time lapse photography.

In addition, Grafstein (1969) has also shown by radiotracer techniques that labelled macromolecules are transported across synapses in the vertebrate central nervous system. This *transneuronal transport* has been shown most clearly in the visual system. Radioactive amino acids injected into the *retina* are incorporated into labelled protein that can be followed as it is transported along the optic nerves, to *synapses* in the lateral geniculate body (see Sec. 22-5), and then to the areas of the *visual cortex* that normally receive the visual input from the injected eye.

Significance of Axonal Transport. Through axonal transport the entire neuron is supplied with materials synthesized by the cell body and with materials needed for the turnover of constituents in the nerve terminals. This *nutritive role* is probably the function of the slow transport system.

The fast transport system brings *enzymes or neurosecretory materials* from the cell body to the terminals and so is involved in the function of neurons secreting transmitters, and in those producing neurosecretions, such as the neurosecretory cells of the hypothalamus.

Another very important function is the transport along the nerve axon of materials that are responsible for the development and function of the postsynaptic cell. These *trophic* materials have been demonstrated most clearly in their absence: denervated muscle, for example, undergoes specific atrophic changes that involve depletion of its protein stores as well as changes in the distribution of its acetylcholine receptors (Sec. 23-3).

Dendritic transport

Most of the substances needed in the neuronal processes have to be synthesized in the cell body or soma. This is true of the dendrites as well as of the axonal terminals and a similar transport system exists to maintain the structure and the metabolic activity of the dendrites. This can be demonstrated by the injection of radioactive amino acids into a neuron: within 4 min. after such intracellular application, autoradiographically demonstrable labelling is seen over the cell body. This is interpreted as due to the incorporation of the amino acids into proteins by the cell body. Thirty minutes after the intracellular injection, the labelled proteins can be seen in the dendrites but not over the neighboring neurons or in the closely associated glial cells. The transport rate to the dendrites appears to be at least 3 mm/hr (Schubert and Kreutzberg, 1975), as illustrated in Fig. 11-1.

Localization of neurotransmitter action

Unlike the general metabolic effects exerted by carbon dioxide, the effects of the neurotransmitters are *localized.* The discharge of the transmitter from the vesicle in which it is stored is initiated by the arrival of the nerve impulse or action potential at the nerve endings. This means that the *release of the chemical mediator* is dependent on *information from the nervous system.* The transmitter diffuses across the synaptic cleft to the next cell, where it causes either stimulation or inhibition, but in either case its effect is confined to a very small area.

Endocrine Secretions—Hormones

The chemicals secreted by the endocrine glands (discussed in Chap. 10) are released into the circulation to be distributed to all organs of the body. They will affect only those cells that have the specific receptors that permit them to bind the hormone (Sec. 10-8), but nevertheless hormones have a much more general distribution than neurotransmitters. They also reach their target cells more slowly. For example, the tropic hormone of the anterior pituitary gland, thyroid-stimulating hormone (TSH), has to be released into the veins that drain the pituitary gland, return to the heart, and pass through the pulmonary system into the systemic arterial circulation before it can reach its target endocrine gland, the thyroid. Once TSH is bound by thyroid cell receptors, a series of chemical reactions is initiated that results in the production and release of thyroxine. This hormone now enters the venous circulation, and it, too, has to reach the arterial circulation before it is ultimately distributed to all tissues of the body, where it causes an increase in metabolic rate.

Neurosecretions

Neurons may produce other chemical mediators apart from the neurotransmitters previously discussed. Some nerve cells are specialized to secrete hormones; these cells are called neurosecretory cells. Many of the cells of the hypothalamus are neurosecretory cells. They synthesize a specific hormone in the cell body, combine it with a protein, and send the hormone-protein package along the axon to the nerve terminals. Some of these terminals end directly in another gland, the posterior pituitary, where the hormones are deposited and stored. This is true for the peptide hormones vasopressin and oxytocin. Other hypothalamic hormones are deposited into a short vascular system, the hypothalamic-hypophyseal portal system, which drains the hypothalamus and empties into the anterior pituitary gland (Figs. 10-2 and 10-3). These hypothalamic hormones now either stimulate or inhibit the release of tropic hormones from the pituitary. Here we see neural control of endocrine secretion through a rather complex pathway.

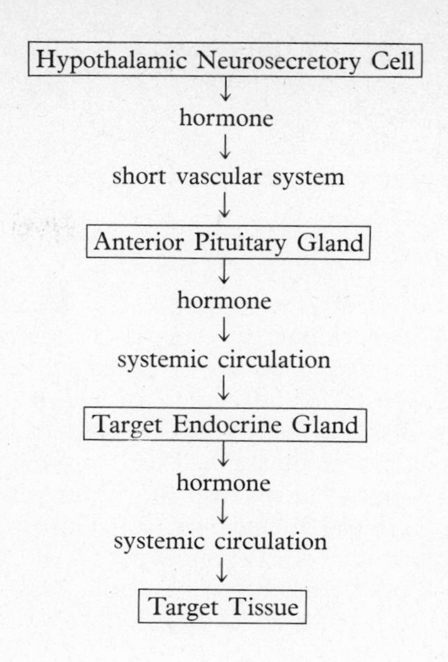

11-2 Neuroendocrine Regulation

In the Fetal or Neonatal Mammal

Neuroendocrine regulation is the control exerted by the nervous system over the endocrines. However, hormones play an equally important role in the *development and function of the nervous system* in mammals. The growth of neural circuits and their final integration into complex patterns of behavior are influenced strikingly by hormones at different levels of development. If *thyroid hormone* is lacking early in development, irreversible brain damage occurs. In the human, this default is characterized by cretinism, a form of mental retardation which can be ameliorated only if the retardation is slight and the treatment is begun very early. It has been suggested by Hamburgh (1971) that thyroxine acts as a time clock in the developing nervous system; if the clock stops during the critical period when synaptic contacts are being laid down, then the timing is off and the necessary pathways do not develop in the correct sequence.

The *sex hormones,* too, play a most important role in determining the patterns of development of the central nervous system. In female rats and monkeys, a single injection of male sex hormone in *prenatal* or *neonatal life* permanently changes the hypothalamus to the male type, which is acyclic or nonrhythmic. Consequently, these androgenized females do not ovulate and, of course, are sterile. Similarly, males castrated just after birth retain the basic pattern of the female, cycling hypothalamus. Hormones administered later in life cannot evoke typical male sexual behavior. These males tend to behave like females sexually. This means that the newly born male rat (which continues development after birth that corresponds to the latter part of development *in utero* in humans) or the unborn human male must secrete enough androgen in a certain *critical period of development* to affect permanently the organization of the hypothalamus and thus ensure male sexual activity.

In the Adult

In most *adult vertebrates,* with the possible exception of humans, the sex hormones are essential for complete mating behavior. This presupposes the normal development of the appropriate type of hypothalamus. In addition, thyroxine, the sex hormones, and many of the polypeptide hormones of the pituitary gland modulate the excitability of the nervous system and change adaptive behavior patterns. This is a new and exciting area of biological investigation, and to underscore the importance of the regulatory effects of hormones in the nervous system, I have called it *endocrineurology* (Strand, 1974).

11–3 Neuroendocrine Feedback Systems

For any energy-requiring system to function efficiently, there must be some relationship between the input of the system and its output. If these two parameters can be measured by some kind of sensing device (usually chemical, electrical, or thermal in biological systems), then it is feasible for the output to influence the input either negatively or positively (see Chap. 1). The hypothalamus is a biological sensing device par excellence. It is not protected by the blood-brain barrier, and so it is exposed to all the physical and chemical variables of the blood. It has membrane receptors that permit it to sense these changes and to transduce them into electrical (nerve) or endocrine (hormonal) responses.

Levels of Control of Endocrine Secretion

NEGATIVE FEEDBACK WITHOUT A SECOND ENDOCRINE GLAND

The *original concept* of endocrine gland regulation was that of an organ secreting hormones into the circulation, affecting distant target organs and through this action altering the composition of the blood so as to inhibit any further release of the hormone. Insulin secretion by the endocrine portion of the pancreas, for example, is stimulated by increased blood glucose levels. Insulin facilitates the passage of glucose into cells, causing blood glucose to drop and thus sharply limiting further insulin secretion. While other factors do play a role in the secretion of insulin, blood glucose is the decisive one. This is a negative feedback control mechanism that does not require an intermediary endocrine gland.

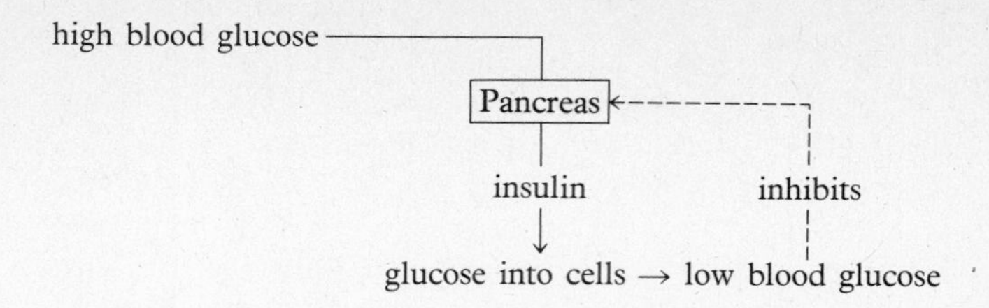

ANTERIOR PITUITARY AS AN INTERMEDIARY

For most other hormone systems, it becomes necessary to include an *intermediary,* the *anterior pituitary gland,* which was found to secrete tropic hormones that were supposed to exert their effects only on specific target organs, other endocrine glands. All of these tropic hormones are proteins or polypeptides. One of them is adrenocorticotropic hormone (ACTH), a polypeptide of 39 amino acids. ACTH is synthesized by the anterior pituitary gland, and in turn it stimulates the adrenal cortex to produce glucocorticoids. It has very little influence on the secretion of another adrenal cortical hormone, aldosterone, which is involved in water and salt metabolism. Other tropic hormones of the anterior pituitary are thyroid-stimulating hormone (TSH); the gonadotropic hormones (FSH and LH); prolactin; growth hormone (somatotropin) and melanocyte-stimulating hormone (MSH). The functions of all these hormones have been discussed in Chapter 10. With the exception of growth hormone and MSH, the mechanism inhibiting the release of these tropic hormones was attributed solely to the rising titers of hormone released by the target endocrine gland. Thus as ACTH secretion by the pituitary gland increased the output of corticoids by the adrenal cortex, the glucocorticoids in turn would act directly on the anterior pituitary to inhibit further ACTH release.

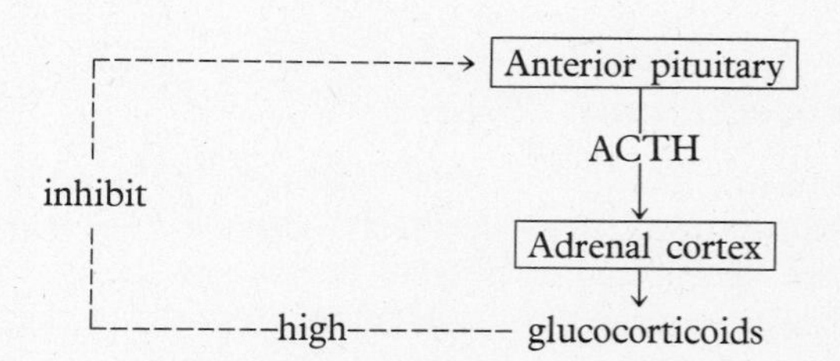

HYPOTHALAMIC CONTROL OF THE ANTERIOR PITUITARY

The last 20 or 30 years have seen the endocrine regulating system placed one step higher in the brain, in a region known as the *hypothalamus.* Although the hypothalamus in humans comprises less than 5 per cent of the entire weight of the brain, it is credited with an enormous number of vital functions. It is part of the limbic system of the brain, which is a phylogenetically older system coordinating physiological and behavioral responses. This integrative role is possible because the hypothalamus receives an abundant input of information, both neural and humoral, and regulates in turn

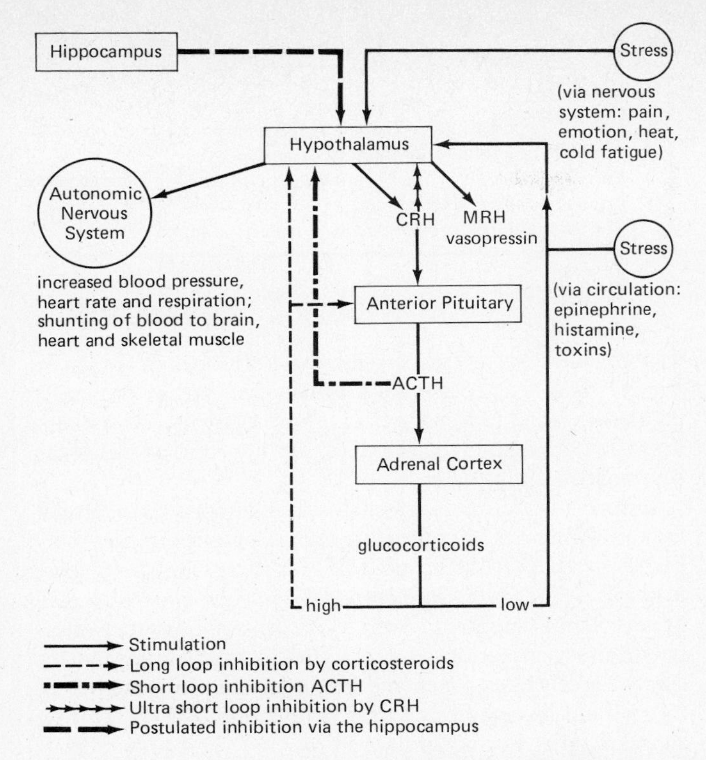

Figure 11-2. **Secretion of releasing hormones by the hypothalamus may be inhibited by various negative feedback mechanisms: a long loop, a short loop, and an ultrashort loop. [From F. L. Strand: *Bioscience* 25 (9):568 (1975).]**

both the endocrine and autonomic nervous systems. The importance of this neuroendocrine regulation is seen in the 1977 Nobel Prize award to R. Guillemin and A. Schally, and to a woman scientist, R. Yalow, who developed the radioimmunoassay that made their work possible.

Hormonal Control of Hypothalamic Neurons That Influence the Anterior Pituitary Gland

We now have to include the hypothalamus in the system of negative feedback mechanisms controlling tropic hormone secretion. It secretes a series of five or six or releasing hormones and three release-inhibiting hormones. These hypothalamic hormones are probably all polypeptides, several of which have been isolated and synthesized. The releasing hormones and their functions are shown in Tables 11-1 and 11-2.

THE CONTROL OF HYPOTHALAMIC HORMONES

The feedback between CRH, ACTH, and the glucocorticoids is used as an example and illustrated in Fig. 11-2.

Long-Loop Negative Feedback. CRH stimulates the release of ACTH from the pituitary, which stimulates the release of glucocorticoids from the adrenal cortex. When these steroid hormones reach a critical level in the blood, they inhibit the hypothalamus, thus cutting off the supply of CRH and, indirectly, of ACTH.

Short-Loop Negative Feedback. High levels of circulating glucocorticoids inhibit anterior pituitary secretion of ACTH *directly*. There may be a physiological regulating device that permits ACTH to act as its own control. High levels of circulating ACTH may directly inhibit CRH release from the hypothalamus and ACTH release from the pituitary.

Ultrashort-Loop Negative Feedback. CRH also inhibits its hypothalamic production of CRH itself.

Other Areas in the Brain. The hippocampus appears to be sensitive to levels of corticosteroids. When these hormones accumulate in the hippocampus, the firing of the hippocampal neurons inhibits the hypothalamus through direct nerve pathways.

HYPOTHALAMIC RELEASE-INHIBITING HORMONES

It is of particular interest that the three inhibiting hypothalamic hormones are for those pituitary hormones that do not evoke endocrine secretions from their target organs. In other words, although growth hormone-releasing hormone (GRH) stimulates the anterior pituitary to produce growth hormone (GH), which accelerates growth of most tissues, no hormone is secreted by these target tissues. Consequently,

Table 11-1. **Hypothalamic Release-Stimulating Hormones**

Release-Stimulating Hormone		Function	
(Adreno) corticotropic-releasing hormone	(CRH)	Stimulates production and release of ACTH	by anterior pituitary gland
Thyroid-stimulating hormone-releasing hormone	(TRH)	Stimulates production and release of TSH	
Follicle-stimulating hormone-releasing hormone	(FRH)	Stimulates production and release of FSH	
Luteinizing hormone-releasing hormone	(LRH)	Stimulates production and release of LH	
Growth hormone-releasing hormone	(GRH)	Stimulates production and release of GH	
or	or	or	
Somatotropic hormone-releasing hormone	(SRH)	STH	
Melanocyte-stimulating hormone-releasing hormone	(MRH)	Stimulates production of MSH from intermediate lobe of pituitary gland in some species, from anterior lobe in other species.	

Table 11–2. **Hypothalamic Release-Inhibiting Hormones**

Release-Inhibiting Hormone		Function
Growth hormone-release inhibiting hormone or Somatostatin or Somatotropic hormone-release-inhibiting hormone	(GIH) (SIH)	Prevents secretion of GH and TSH by anterior pituitary. Inhibits release of insulin and glucagon from pancreas.
Melanocyte-stimulating hormone-release-inhibiting hormone	(MIH)	Prevents secretion of MSH by anterior pituitary (or intermediate lobe in certain species).
Prolactin release-inhibiting hormone	(PIH)	Prevents production of prolactin by anterior pituitary gland.

there is *no built-in mechanism for negative feedback* from the periphery. Growth does not continue rampant, however, for the hypothalamus produces an inhibiting hormone, *GIH,* or *somatostatin,* which inhibits pituitary release of GH. There are several synonyms for these hormones influencing growth; they are included in Tables 11-1 and 11-2.

Somatostatin has several effects apart from its inhibiting action on the release of growth hormone. It has been shown to prevent the release of TSH from the anterior pituitary, and of insulin and glucagon from the pancreas. The release of gastrin, a digestive hormone produced by the gut, is also inhibited by somatostatin. The significance of these effects is still obscure but indicates once more that a hormone may evoke many different actions.

The long-loop negative feedback that controls the level of hypothalamic secretion of releasing hormones is shown in Fig. 11-3. When the level of hormone secreted by the target endocrine gland (adrenal cortex, thyroid, ovary, or testis) is high, hypothalamic secretion of the appropriate releasing hormone is inhibited.

Figure 11-3 also shows that although there is a similar hypothalamic set of releasing and release-inhibiting hormones for MSH (MRH and MIH) as there is for GH, there is *only a release-inhibiting hormone for prolactin* (*PIH*). The hypothalamus appears to exert a continuous inhibitory effect on pituitary secretion of prolactin. In pregnancy, the increased levels of estrogen and progesterone produced by the ovaries and placenta inhibit the production of PIH. During lactation, this hormonal inhibition of PIH production is replaced by nerve impulses reaching the hypothalamus from

Figure 11-3. **Inhibition of hypothalamic release-stimulating hormones, CRH, TRH, FRH, and LRH, occurs when the circulating levels of the hormones of the appropriate target organ glands (adrenal cortex, thyroid, ovary, and testis) are high. Only the long-loop negative feedback is illustrated. The short loop would involve the inhibition of anterior pituitary tropic hormones as well. Note that MSH, GH, and prolactin act without the intermediary secretion of a target endocrine organ, and consequently there is no negative feedback on the hypothalamus. Instead, there are both release-stimulating and release-inhibiting hormones MRH and MIH, for MSH; GRH and GIH, for GH; but only a release-inhibiting hormone, PIH, for prolactin. Broken arrows indicate inhibition.**

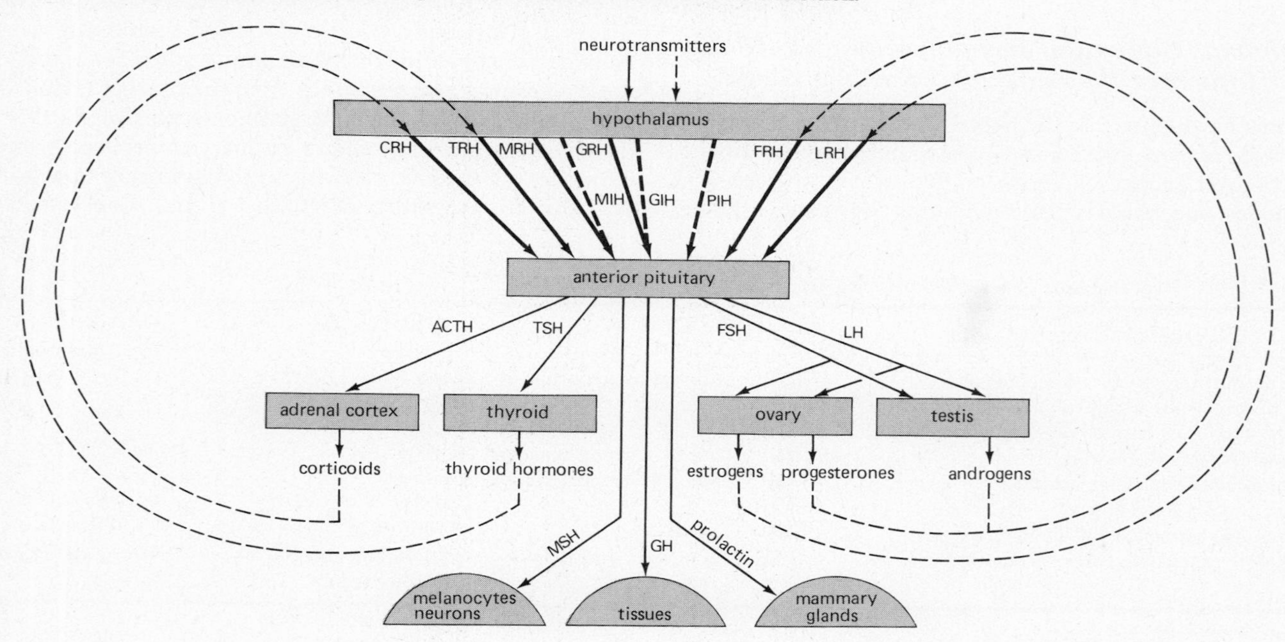

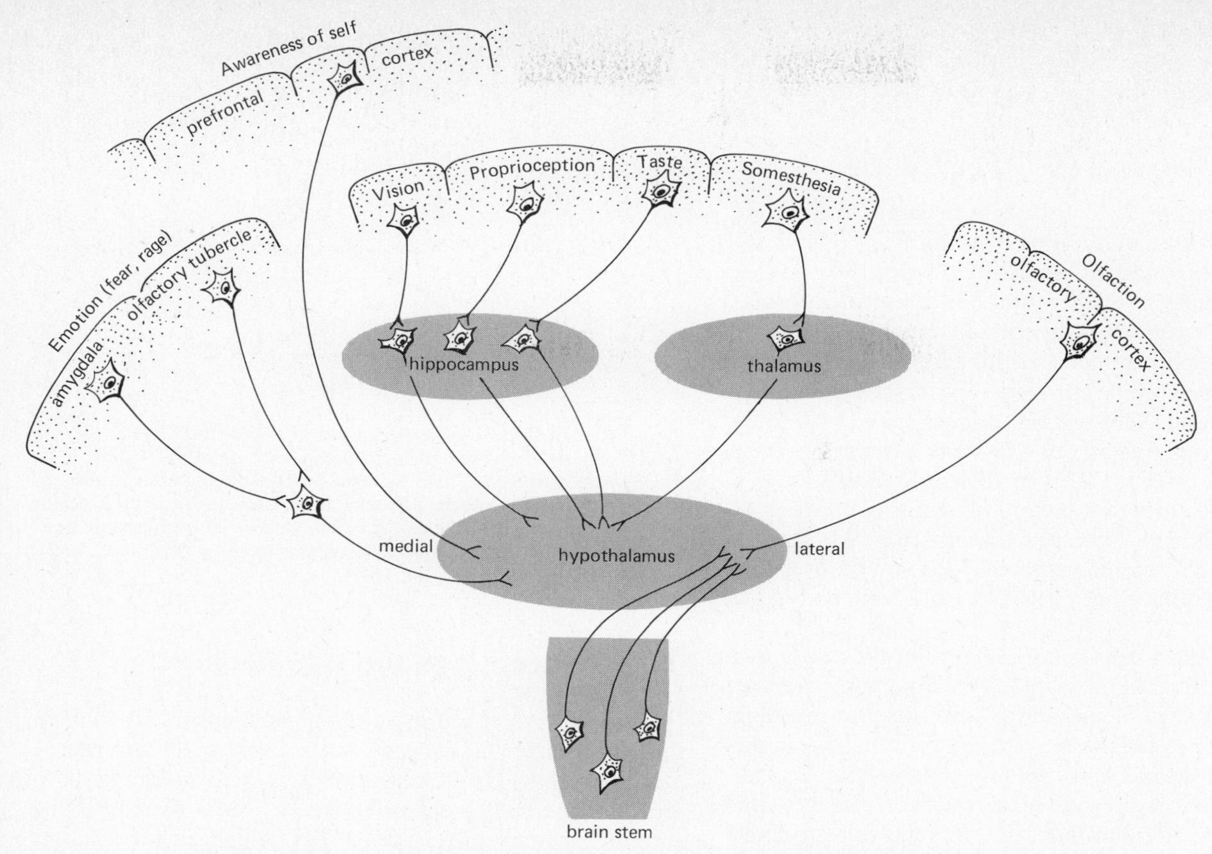

Figure 11-4. **The hypothalamus receives input from the sensory systems via several pathways. The olfactory cortex has a direct pathway to the lateral hypothalamus. Other olfactory connections, together with pathways from the amygdala, synapse finally in the medial hypothalamus. Pathways mediating vision, proprioception, and taste are shunted through the hippocampus to the anterior hypothalamus and preoptic area. Fibers from the sensory cortex via the thalamus also project to the hypothalamus. These several routes from the neocortex to the hypothalamus are important pathways by which affective states influence autonomic and endocrine responses. Ascending pathways connect the reticular formation and raphe nuclei in the brain stem with the hypothalamus.**

sensory endings in the nipples, which are stimulated by the suckling young. As the young suckle less frequently, the level of PIH rises, and soon lactation ceases (Sec. 26-15).

Sensory Control of Hypothalamic Neurons

The hypothalamus acts as an *integrative center for stimuli coming from the external environment,* especially *light.* As the days become longer in spring and early summer, light stimulates the hypothalamus to produce more FRH and LRH, resulting in increased gonadotropin secretion by the pituitary and consequently increased sex hormone secretion by the gonads. This accelerates the maturation of the ovaries or testes and also the development of secondary sexual characteristics. In birds and lower mammals increased activity and mating behavior result; in humans "spring fever" perhaps may be explained on the same basis.

One of the most important sensory inputs to the hypothalamus is through the *olfactory system,* which is the only sensory system with a direct route to the hypothalamus (Fig. 11-4). Although humans pay little attention to this sense, most other animals are keenly aware of specific odors. One of the most interesting discoveries of natural birth control depends on the stimulation of the hypothalamus through the olfactory tract. It is called the *Bruce effect* after its English discoverer (1966). The common house mouse is usually protected from other males by her mate, who keeps invaders out of their territory. If she is mounted by a strange male mouse within four days of conception by her own mate, she will abort spontaneously. She need not actually be impregnated by the stranger, she need only see or smell him and she will abort. This is what happens when there are too many mice. Apparently, these impulses to the hypothalamus through the *visual, olfactory,* or *proprioceptive tracts* prevent the release of the hypothalamic hormones that control gonadotropin secretion; consequently, the sex hormones necessary to maintain implantation of the embryo are not secreted in a normal manner.

Control of Hypothalamic Neurons Influencing the Posterior Pituitary Gland

ANTIDIURETIC HORMONE (ADH) OR VASOPRESSIN

The neurosecretory cells of the hypothalamus that secrete ADH and deposit it in the posterior pituitary gland are grouped together to form the paired supraoptic nuclei. These cells respond to both neural and humoral information. They are sensitive to changes in the plasma osmotic pressure and respond to increased tonicity with an increase in firing rate, which can be correlated to an increase in ADH secretion and release. This was first demonstrated by Verney (1947); although he was not able to show a direct response of hypothalamic cells, he inferred their role indirectly as a result of decreased urine formation following an injection of hypertonic NaCl into the carotid artery (Fig. 11-5). The same amount of hypertonic saline, injected into the systemic circulation, would have been too diluted to have had much effect.

Verney called these cells *osmoreceptors,* but they react in a similar manner to changes in blood volume and pressure. For example, a fall in blood volume and pressure resulting from hemorrhage stimulates nerve endings of the vagi in the right atrium and carotid sinus, resulting in a reflex release of ADH. This causes increased reabsorption of water by the kidney, increases arteriolar tone, and thus acts to restore the volume and pressure of the blood. These mechanisms are discussed in detail in the chapters on maintenance of blood pressure and of water balance (Chaps. 12 and 15).

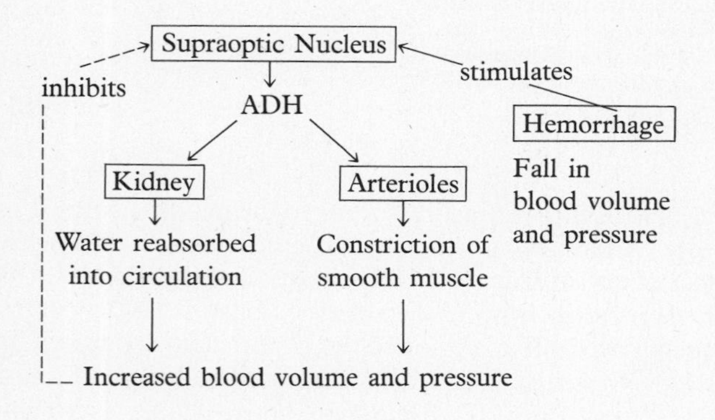

OXYTOCIN

The production and release of oxytocin respond to a *positive feedback system* in most mammals. As pregnancy nears its term, the decrease in progesterone secretion causes a secretion of oxytocin, which induces parturition. Distention of the cervix as the fetus is pushed toward the cervix of the uterus *reflexly* stimulates further secretion of oxytocin. By a similar positive feedback process, oxytocin, which is essential for the ejection of milk from the lactating mammal, is secreted as a reflex response to suckling (Fig. 11-6). This is discussed further in Sec. 26-15.

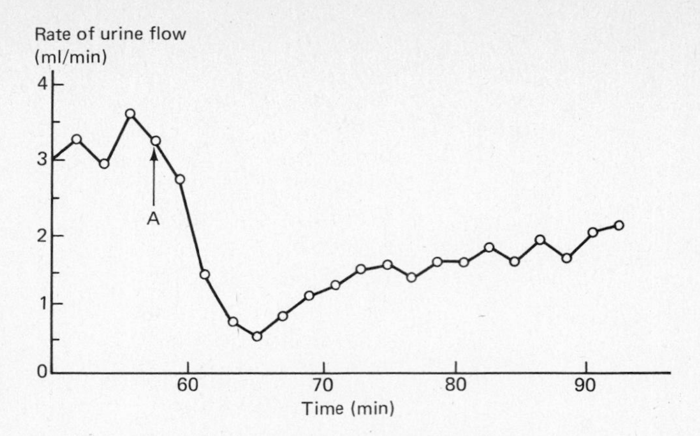

Figure 11-5. **Injection of hypertonic NaCl into the internal carotid artery causes a marked drop in urine output. Initial rapid urine flow (water diuresis) is first induced by introducing a large volume of water into the stomach. The antidiuretic effect of the 1.5 per cent NaCl (arrow at A) is due to the reflex release of antidiuretic hormone from the hypothalamus-posterior pituitary. [From E. B. Verney, *Lance:* 781 (Nov. 30, 1946).]**

11-4 Stress and the Hypothalamus

In 1946 Selye invoked the pituitary-adrenal axis to explain the response of the body to stress, in what he termed the *alarm reaction.* He described it as a generalized "call to arms" of the body's defensive forces. Stress stimulates the anterior pituitary to release ACTH, which, in turn, causes the secretion of glucocorticoids from the adrenal cortex. This is the endocrine version of Cannon's original concept of the adaptation of an animal for "flight or fight" during stress (1932). Cannon's ideas involved primarily the sympathetic nervous system and the hormone of the adrenal medulla, epinephrine (which may be considered as part of the sympathetic nervous system). We now realize that the hypothalamus is involved in integrating both endocrine and

Figure 11-6. **Positive feedback mechanisms controlling oxytocin secretion. Oxytocin increases uterine contractions during parturition and causes milk ejection from the mammary glands in lactation.**

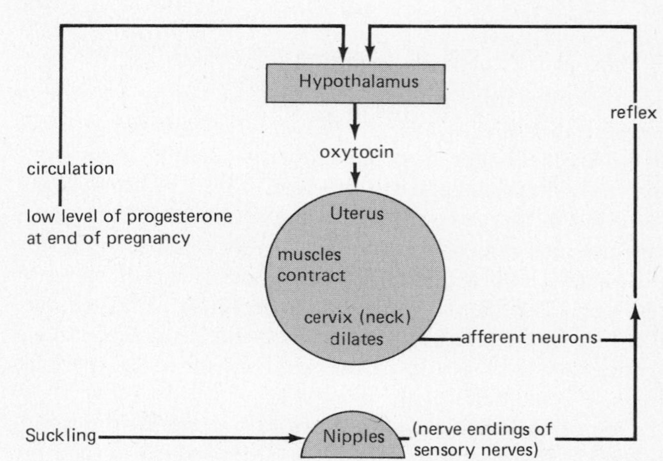

neural responses to stress. It acts as a neurosecretory organ in that it controls the anterior pituitary gland; as a neurosecretory organ in that it deposits oxytocin and vasopressin in the posterior hypothalamus, and as a purely nervous center in that it regulates the sympathetic and parasympathetic autonomic systems.

Figure 11-2 shows the response of the hypothalamus to a wide range of disturbing changes, collectively called "stress." These may include extremes of temperature, severe exercise, exposure to bacterial toxins, surgical operations, and emotional stress. The hypothalamus responds in three ways:

1. It secretes at least two releasing hormones, CRH and MRH, into the vascular system in the median eminence, or perhaps directly into the ventricular fluid.
2. It causes the release of the polypeptides MSH, vasopressin, and, to a lesser extent, oxytocin into the circulation.
3. It causes the stimulation of the sympathetic nervous system and the inhibition of the parasympathetic system.

As a result of CRH release, the anterior pituitary gland releases ACTH, which causes the secretion of glucocorticoids by the adrenal cortex. The glucocorticoids mobilize the metabolic stores of the animal and by their action on the thymus gland and other lymphatic tissues produce antibodies to fight against toxins and foreign bodies.

As a result of the release of the other polypeptides, the general excitability of the nervous system is enhanced, blood pressure rises, and water is retained by the kidney, contributing to the elevated blood pressure.

As a result of sympathetic stimulation, blood pressure rises, respiration increases in rate and depth, there are shunts in blood to the actively exercising muscles and away from the viscera, the metabolic rate rises, and the animal is prepared to fight or run for its life.

The *specific effects* of the hormones described in this chapter are discussed in detail in appropriate sections of the text.

11-5 The Pineal Gland— A Neuroendocrine Transducer

The pineal gland, like the adrenal medulla, is a fine example of a *neuroendocrine transducer,* changing neural information derived from sympathetic neurons into an endocrine response. In the case of the pineal gland, the hormone produced is *melatonin.* Melatonin, a methoxylated indole whose biological activity requires a methyl (CH_3) group attached to an oxygen atom, is derived from an amino acid precursor, 5 hydroxytryptophan, by way of serotonin (Fig. 11-7). Melatonin was first extracted and isolated by Lerner (1958) from cattle pineals. He demonstrated its blanching effect on amphibian skin and also the lack of such an effect on mammalian skin.

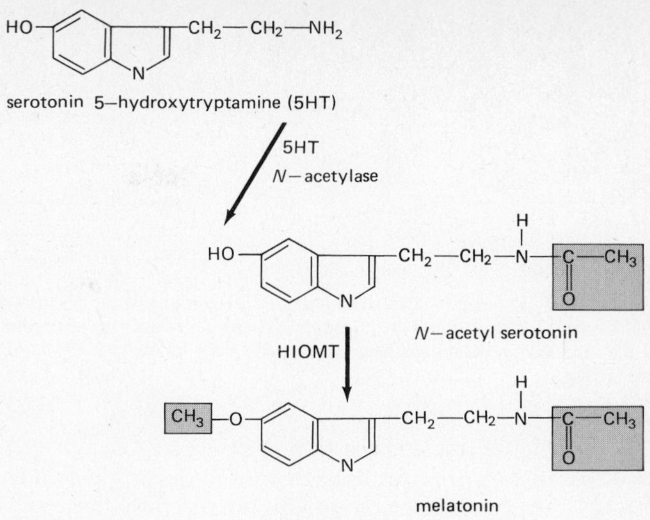

serotonin 5–hydroxytryptamine (5HT)

5HT *N*–acetylase

N–acetyl serotonin

HIOMT

melatonin

Figure 11-7. **Synthesis of melatonin from serotonin.**

Functions of Melatonin

In mammals, melatonin acts to *inhibit gonadal development and function.* If female rats are exposed to continuous light, their ovaries increase in weight and the estrus cycle is accelerated. At the same time, pineal weight decreases. If the pineal glands are removed (pinealectomy), ovarian weight increases as though the rats were exposed to continuous light. The administration of pineal extracts, or of purified melatonin, decreases ovarian weight. These experimental observations show that the pineal gland secretes melatonin, which inhibits ovarian growth, and that light inhibits the inhibiting effect of the pineal. Recently it has been demonstrated that melatonin inhibits ovulation by preventing the release of luteinizing hormone (LH).

Regulation of Melatonin Synthesis by Light

Meticulous studies over many years by Wurtman and Axelrod (1968) have shown that the synthesis of melatonin is dependent on a rate-limiting enzyme (hydroxy-indole-O-methyl-transferase or HIOMT). The production of HIOMT, in turn, is inhibited by light. This is a fascinating observation, indicating that external influences can affect the rate of enzyme (protein) synthesis.

THE "BIOLOGICAL CLOCK"

The route for the *transduction of light energy,* first into electrical nerve impulses and then into a chemical response, originates in the retinal cells. This information is conducted to the pineal cells by way of the sympathetic nerves (Sec. 8-6). Blinding or sympathectomy (cutting the sympathetic nerves or destroying the superior cervical ganglion) completely removes the responsiveness of HIOMT forma-

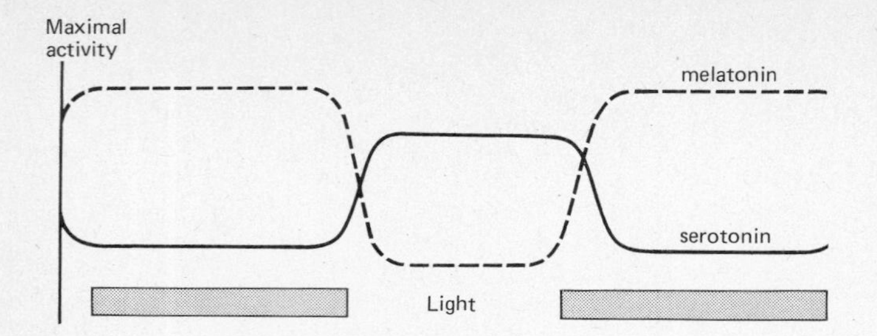

Figure 11-8. **Serotonin and melatonin secretion rhythms are mirror images of one another. Light increases serotonin production but inhibits melatonin synthesis. The solid bars indicate periods of darkness.**

tion to light. Light acts as a timing device or "Zeitgeber," resulting in a 24-hr (circadean) rhythm in the production of HIOMT and consequently of melatonin, with the levels highest at midnight and lowest at noon.

Serotonin, the precursor of melatonin (Fig. 11-7), also demonstrates a circadean rhythm in the pineal. Serotonin rhythm is the mirror image of the melatonin rhythm, being highest at noon and lowest at midnight (Fig. 11-8). This is due to two important effects of light on enzyme systems. First, norepinephrine, which is secreted at the sympathetic nerve endings, is increased by light, and through the action of norepinephrine on cyclic AMP, tryptophane hydroxylase, an enzyme needed for serotonin synthesis (Fig. 11-9), is increased. Consequently, serotonin synthesis is accelerated.

The second effect of light is inhibition of HIOMT activity. As there is no melatonin being synthesized, the precursor, serotonin, accumulates. Denervation of the pineal suppresses this rhythm, indicating that it is normally generated by the diurnal release of norepinephrine secreted at sympathetic nerve terminals in response to light. Going one step further, destruction of a part of the *hypothalamus,* or of the connections of the sympathetic nerves to the hypothalamus, also abolishes this rhythm. It appears that the origin of these circadean rhythms is a "biological clock" in the hypothalamus, which is modified by the inhibitory influence of external light.

Figure 11-9. **Effect of light on serotonin and melatonin synthesis. Intermediate reactions are not included; they are shown in Figs. 9-20 and 11-7.**

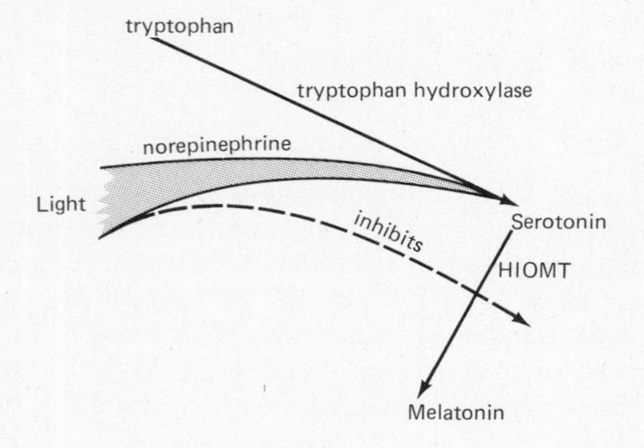

Cited References

BRUCE, H. M. "Smell as an Exteroceptive Factor." *J. Animal Sci.* (*Suppl.*) **25:** 83 (1966).

CANNON, W. B. 1932. *The Wisdom of the Body.* Kegan Paul, London, 1932.

GRAFSTEIN, B. "Axonal Transport: Communication Between Soma and Synapse." In *Advances in Biochemical Psychopharmacology,* Vol. 1, E. Costa and P. Greengard, eds. Raven Press, New York, 1969, p. 11.

HAMBURGH, M., L. A. MEDOZA, J. F. BURKART, and F. WEIL. "The Thyroid as a Time Clock in the Developing Nervous System." In *Cellular Aspects of Neural Growth and Differentiation.* UCLA Forum of Medical Science No. 14, D. C. Pease, ed. University of California Press, Los Angeles, 1971, p. 321.

LERNER, A. B., J. D. CASE, Y. TAKAHASHI, T. H. LEE, and W. MORI. "Isolation of Melatonin, the Pineal Gland Factor That Lightens Melanocytes." *J. Amer. Chem. Soc.* **80:** 2587 (1958).

SCHUBERT, P., and G. W. KREUTZBERG. "Parameters of Dendritic Transport." In *Advances in Neurology,* Vol. 12, G. W. Kreutzberg, ed. Raven Press, New York, 1975.

SELYE, H. General Adaptation Syndrome and Diseases of Adaptation. *J. Clin. Endocr. Metab.* **6:** 117 (1946).

STRAND, F. L. Endocrineurology. In *Humoral Control of Growth and Differentiation,* Vol. 2., A. S. Gordon and J. LoBue, eds. Academic Press, Inc., New York, 1974, p. 191.

VERNEY, E. B. "The Antidiuretic Hormone and the Factors Which Determine Its Release." *Proc. Roy. Soc. London Biol.* **135:** 25 (1947).

WURTMAN, R. J., J. AXELROD, and D. E. KELLY. *The Pineal.* Academic Press, Inc., New York, 1968.

Additional Readings

BOOKS

ARIENS-KAPPERS, J., and J. P. SCHADÉ. *Structure and Function of the Epiphysis Cerebri.* Progress in Brain Research 10. American Elsevier Publishing Co., Inc., New York, 1965.

BAJUSZ, E., and G. JASMIN. *Major Problems in Neuroendocrinology.* The Williams & Wilkins Company, Baltimore, 1964.

BRODISH, A., and E. S. REDGATE, eds. *Brain-Pituitary-Adrenal Interrelationships.* S. Karger, New York, 1973.

FORD, D. H., ed. *Influence of Hormones on the Nervous System.* S. Karger, New York, 1971.

GISPEN, W. H., TJ. B. VAN WIMERSMA GREIDANUS, and D. DE WIED, eds. *Hormones, Homeostasis, and the Brain.* Progress in Brain Research 42. American Elsevier Publishing Company, Inc., New York, 1975.

LISSAK, K., and E. ENDRÖCZI. *The Neuroendocrine Control of Adaptation.* Pergamon Press, Inc., New York, 1965.

MARTINI, L., and W. F. GANONG, eds. *Frontiers in Neuroendocrinology.* Vols. 1-4, Academic Press, Inc., New York; Vol. 4, Raven Press, New York, 1969-1976.

MARTINI, L., and W. F. GANONG, eds. *Neuroendocrinology.* Vols. 1 and 2. Academic Press, Inc., New York, 1966, 1967.

MARTINI, L., M. MOTTA, and A. PECILE, eds. *The Hypothalamus.* Academic Press, Inc., New York, 1971.

NALBANDOV, A. V., ed. *Advances in Neuroendocrinology.* University of Illinois Press, Urbana, 1963.

SCHARRER, E., and B. SCHARRER. *Neuroendocrinology.* Columbia University Press, New York, 1963.

WOLSTENHOLME, G. E. W., and J. KNIGHT, eds. *The Pineal Gland.* A Ciba Foundation Symposium. The Williams & Wilkins Company, Baltimore, 1970.

ARTICLES

Refer also to the articles on axonal transport by Grafstein and Ochs in Chapter 8.

AXELROD, J. "The Pineal Gland: A Neurochemical Transducer." *Science* **184:** 1341 (1974).

BESSER, G. M. "The Clinical Implications of the Hypothalamic Regulatory Hormones." In *Frontiers in Neurology and Neuroscience Research,* P. Seeman and G. M. Brown, eds., University of Toronto, Toronto, Canada, 1974.

COHEN, R. A., moderator. "Some Clinical, Biochemical, and Physiological Actions of the Pineal Gland." *Ann. Int. Med.* **61:** 1144 (1964).

MASON, J. W. "Specificity in the Organization of Neuroendocrine Response Profiles." In *Frontiers of Neurology and Neuroscience Research,* P. Seeman and G. M. Brown, eds. University of Toronto, Toronto, Canada, 1974.

REITER, R. J. "Pineal and Associated Neuroendocrine Rhythms." *Psychoneuroendocrinol.* **1:** 255 (1976).

SCHARRER, E. "Principles of Neuroendocrine Integration." In *Endocrines and the Central Nervous System,* R. Levine, ed. The Williams & Wilkins Company, Baltimore, 1966.

SCHUBERT, P., and G. W. KREUTZBERG, "Parameters of Dendritic Transport." In *Advances in Neurology,* Vol. 12, G. W. Kreutzberg, ed. Raven Press, New York, 1975.

STRAND, F. L. "The Influence of Hormones on the Nervous System (With Special Emphasis on Polypeptide Hormones)." *Bioscience* **25:**(9): 568 (1975).

WURTMAN, R. J. "The Effects of Light on the Human Body." *Sci. Am.,* July 1975.

Regulation of the Internal Environment

Chapter 12

Cardiovascular Regulation

Wonderful to depart!
Wonderful to be here!
The heart, to jet the all-alike and innocent blood!
To breathe the air, how delicious!
To speak—to walk—to seize something by the hand!
To prepare for sleep, for bed, to look on my
rose-color'd flesh!
To be conscious of my body, so satisfied, so large!
To be this incredible God I am!
To have gone forth among other Gods, these men
and women I love.

Walt Whitman, "Song at Sunset"

BLOOD REACHES THE tissues through a system of blood vessels that are essentially living tubes of greatly varied diameter, elasticity, and permeability. It is forced through this high resistance system by the pressure generated by the pumping of the *heart;* those vessels into which the heart first ejects the blood are subjected to the greatest pressure. These are the *arteries,* highly elastic and tough; they distribute the blood to the smaller *arterioles* and ultimately to the tiny thin-walled *capillaries,* through which the exchange of fluids and small molecules occurs. Blood returns to the heart through the *venules* and wide, thin-walled *veins.* In vertebrates, the heart and blood vessels form a *closed circulatory system,* the continuity of which results in the return of almost all of the circulating blood to the heart. The fluid and protein that do leak out into the tissue are restored to the blood through the *lymphatic circulation.*

The structure of the mammalian heart is well adapted for the complete separation of unoxygenated blood from oxygenated, permitting maximal utilization of the available oxygen. The unoxygenated blood is delivered to the lungs, where exchange of respiratory gases with the exterior air occurs, and the oxygenated blood returns to the heart. This highly important circulation through the lungs, the *pulmonary circulation,* must handle the same volume of blood in the same time as the *systemic circulation,* which pumps blood through the rest of the body.

Mammalian and avian hearts are four-chambered pumps, consisting of two thin-walled *atria,* which receive blood from the systemic and pulmonary circulations, and two heavy-walled *ventricles,* which pump the blood out.

12–1 Embryological Development of the Heart

Embryologically, the heart is a specialized part of the early blood vessels, which are among the first of the organs to form in the rapidly growing embryo. The part of these vessels destined to become the four-chambered heart thickens and twists upon its axis, while separating membranes grow up within this tube. These result in the formation of first a left and a right side, and then a division of each of these into an upper, thin-walled receiving chamber, the *atrium,* and a lower, heavy-walled pumping chamber, the *ventricle* (Fig. 12-1).

It may happen that the membranes do not completely meet so that there is a leakage of blood from one side to the other. This defect is most common in the wall separating the two atria. Sometimes almost all the wall is missing. In the fetus, there is normally a flow of blood from the right atrium to the left, so that the lungs, which are nonfunctional at this stage, are almost completely bypassed. The flow is through an oval hole, the *foramen ovale.* Immediately after birth, with the infant's first breath, the pressure on the left side of the heart rises considerably above that on the right, because of the fall in pressure in the lungs as the thorax expands. This pressure pushes a small valvelike flap over the foramen ovale, closing it. This later becomes replaced by fibrous tissue so that the "hole" is permanently closed (see also Sec. 13-6).

If the foramen ovale fails to close, so that there is a copious flow of blood into the left atrium from the right atrium, the efficiency of the circulation is dangerously reduced, and not enough oxygen reaches the tissues. The unsaturated hemoglobin gives the skin a blue color (cyanosis), which is the basis of the term *blue baby.* This foramen can be closed surgically.

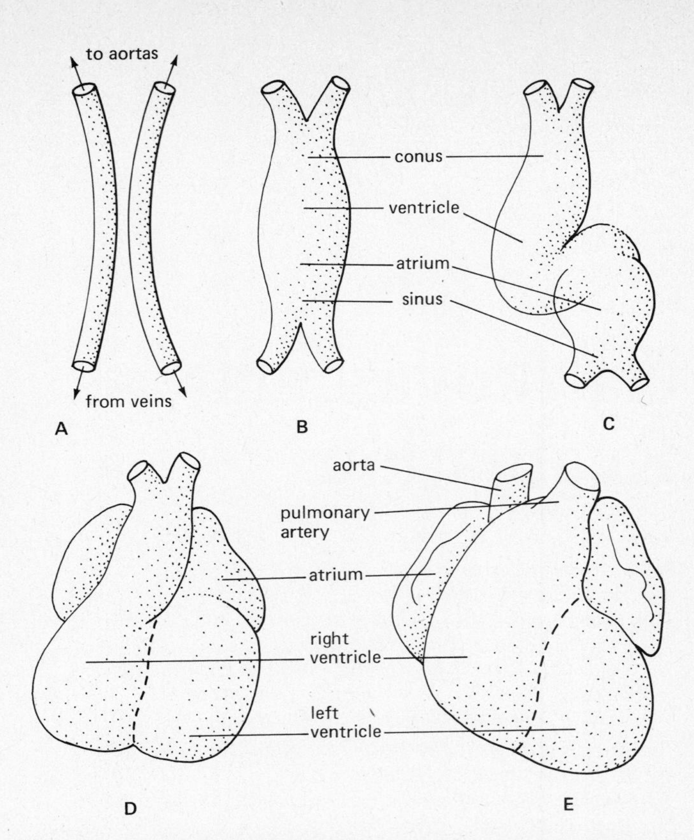

Figure 12-1. **Stages in development of the heart (ventral views). (From C. Villee, *Biology,* 7th ed., W. B. Saunders Company, Philadelphia, 1977.)**

12–2 The Adult Heart

The heart of the adult human is estimated to be roughly about the size of a closed fist. It is enclosed in a double-walled sac, the *pericardium,* which fastens it to the *mediastinum,* a thick mass of tissue separating the lungs. The rounded apex of the heart is formed entirely by the left ventricle and is located just behind the sixth rib, about 3 in. to the left of the midline of the body. The smaller right ventricle does not extend as far as the apex. The left atrium is almost vertically above the right atrium, so that the two atria form a vertical axis behind the sternum, while the two ventricles slope off slightly to the left.

12–3 Cardiac Tissues

The most important tissue of the heart is muscle, the *myocardium,* upon which the pumping action of the heart depends. However, about one half of the weight of the heart is made up of *noncontractile material,* such as the connective tissue coverings of the muscle fibers, the fibrous heart skeleton, tendons, valves, blood vessels, lymphatics, and nerves. The chambers of the heart are lined (as are the blood vessels) with *endothelium*, a thin, smooth surface that offers little resistance or damage to blood cells.

The *conduction system* of the heart is made up of cells that are modified cardiac muscle fibers, specialized for the rapid conduction of excitation but which nevertheless do contain contractile myofibrils.

Myocardium

The atrial myocardium is thin-walled, serving mainly as a passage way to collect blood, which first flows passively into the ventricle and then is propelled into the ventricle by the weak atrial contraction.

The complex ventricular myocardium is divided into the spiral muscles and the deep constrictor muscles, an arrangement that looks somewhat like a sandglass (Fig. 12-2). As a result of this complex twisting, contraction of these muscles directs the main stream of the blood toward the openings of the great arteries leaving the ventricles (the aorta and the pulmonary artery).

The myocardium is composed of muscle fibers (cells), which form a latticework of long, faintly striated cells, separated from one another by extensions and modifications of the plasma membrane, the *intercalated disks* (Fig. 12-3). Cardiac muscle fibers do not form an anatomical syncytium (network of cells lacking separate cell membranes), but they do act as a *functional syncytium* because of the very low electrical resistance through the intercalated disks. There is no significant impediment to the passage of an action potential; therefore, when any one muscle fiber of a heart chamber is excited, the excitation spreads to all other muscle fibers of that chamber.

Ultrastructure of Cardiac Muscle

The ultrastructure of cardiac muscle is very similar to that of skeletal muscle, which is discussed in detail in Sec. 23-2. Cardiac muscle differs quantitatively from skeletal muscle in having more mitochondria crowded between the contractile filaments (myofibrils), and in being richer in the red-pigmented myoglobin than most skeletal muscle.

Myoglobin is found in the sarcoplasm of cardiac muscle fibers. It is a conjugated protein containing an iron porphyrin prosthetic group closely related to hemoglobin. In addition to its somewhat limited ability to store oxygen, myoglobin facilitates the transport of oxygen from the cell membrane to the mitochondria, where tissue respiration occurs. Most myoglobin is found in those muscles characterized by relatively slow, strong repetitive contractions requiring oxidative metabolism.

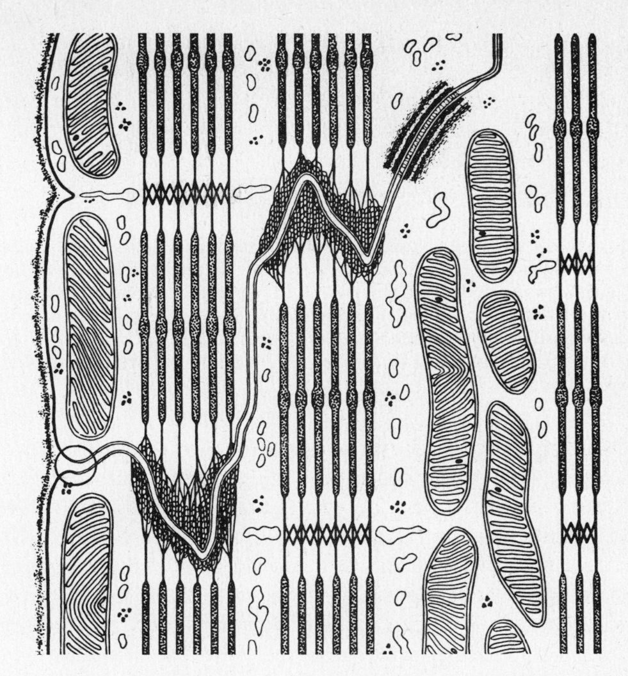

Figure 12-3. **Schematic drawing of an intercalated disk of cardiac muscle. The zigzag line from the middle left of the diagram to the top right is the disk formed as the cell membrane penetrates between the cells. The parallel rods are the contractile filaments, attached by a network of fibers to the intercalated disk. The oval bodies are mitochondria. [From F. S. Sjöstrand, S. Andersson-Cedergren, and M. Dewey, *J. Ultrastructural Res.*, Academic Press, Inc., New York, 1:271 (1958).]**

Figure 12-2. **Muscles of the ventricles of the heart.**

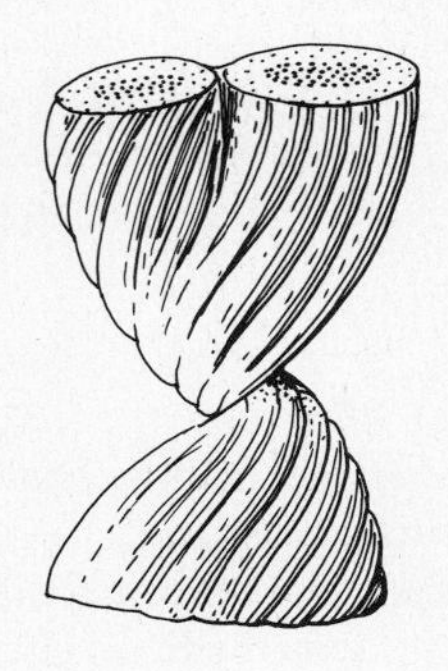

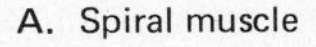

A. Spiral muscle

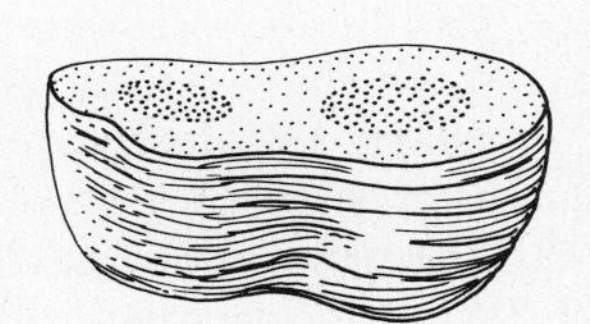

B. Deep constrictor muscle

12-4 Excitation of the Heart

Pacemaker

An isolated frog heart will beat spontaneously for hours, provided that it is in a solution of balanced ions such as Ringer's solution (Sec. 12-5). If one studies a mechanical recording of the cycle of this three-chambered heart (only one ventricle), one can recognize the systole and diastole of the two atria contracting almost simultaneously, followed by the systole of the ventricle (Fig. 12-4). The contraction of the atria is preceded by the contraction of the *sinus venosus*, which is a large vessel formed by the junction of the venae cavae (great veins) before they enter the right atrium. By

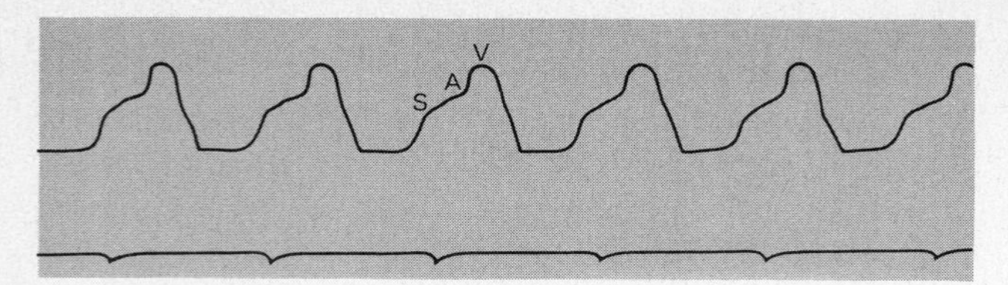

Figure 12-4. **Heartbeat of frog. The upper trace shows the three parts of each beat, the first attributed to the sinus venosus (S), or pacemaker; the second to the two atria (A) beating simultaneously; and the third to the single ventricle (V). The lower trace is a time marker indicating seconds. (Student record of R. J. Caldwell, R. P. Moshova, J. Perrine, and H. I. Winkler, Physiology Laboratory, New York University.)**

preventing the passage of the impulse from the sinus venosus to the rest of the heart, it is possible to determine that the rate of the sinus sets the pace for the rest of the heart (60 beats/min). When the atria no longer receive the sinus impulse, they beat at their own, slower rate, stimulating the ventricle to beat at this pace (40 beats/min). If the impulse to the ventricle is blocked, it will beat at its own, still slower rate (10–20 beats/min). These rates are for temperatures of about 22°C., and they are illustrated in Fig. 12-5.

In the four-chambered human heart, there is an analogous area, about 3 mm wide and 10 mm long, where the great veins enter the right atrium; although no sinus is formed, a specialized muscle fiber area, known as the sinoatrial (SA) node, originates the pace of the entire heart and is known as the pacemaker. Each part of the heart is capable of beating spontaneously, but the rate for each chamber is slower than that of the pacemaker, so that in the normal heart the atria and ventricles beat at the same rate as the pacemaker, which is approximately 72 beats/min.

The SA node is in the posterior wall of the right atrium,

Figure 12-5. **Upper trace: The atria (A) and ventricle (V) of the frog heart beat at the pace set by the sinus venosus (S). After a ligature is tied to separate the sinus from the atria, the sinus beats at the same rate as before but the atria beat more slowly, at their own inherent rate (second and third traces). Following a second ligature that separates the atria from the ventricle, the slow beat of the ventricle is seen (fourth trace). The last trace is a time marker indicating seconds. (Student record, Physiology Laboratory, New York University.)**

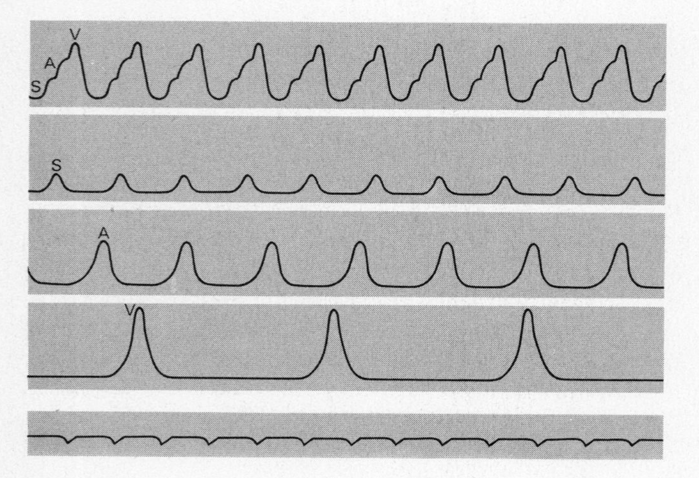

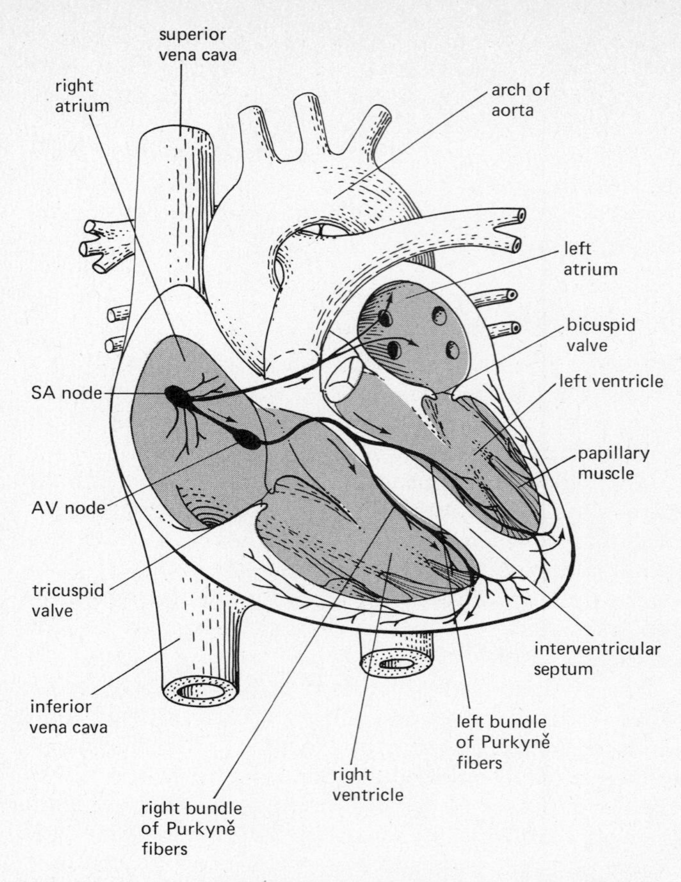

Figure 12-6. **Conduction of electrical excitement from the sinoatrial node (SA) through the rest of the heart. AV = atrioventricular node. Arrows indicate the direction of the impulse.**

below the opening of the superior vena cava (Fig. 12-6). The nodal cells are smaller than the surrounding atrial cells (3 μm in diameter as compared to 5 to 11 μm for the atrial cells). The rhythmic excitation (depolarization) originating in the SA node is responsible for the subsequent excitation, and consequently contraction, of the atria and ventricles, in that order.

Conduction System of the Heart

Excitation is conducted through the heart more rapidly than can be accounted for by conduction through typical myocardial fibers, but more slowly than would be characteristic of nerve fibers. A specialized conducting system of modified cardiac muscle fibers (Fig. 12-7) controls the excitation; the system consists of three elements:

1. *Bundles of fibers from the SA node* through the atria; these fibers activate first the right and then the left atrium. Conduction is faster in these recently discovered pathways than in the surrounding tissue.
2. *The atrioventricular (AV) node,* in the atrial septum, just above the tricuspid valve. The AV node is the chief site of delay in the conduction of excitation from the atria to the ventricles. This delay permits the atria to empty

their contents into the ventricles before the ventricles contract.

3. *The AV bundle* (*bundle of His*) is made of specialized muscle fibers, the Purkyně fibers, that originate in the AV node and form a bundle in the septum dividing the two ventricles. Purkyně fibers are shorter and broader than the contractile myocardial fibers and they have fewer myofibrils. The AV bundle divides to form a right and left bundle branch, each of which pierces the endocardium of the appropriate ventricle with a fine network of Purkyně fibers (Fig. 12-7). This is the fastest conducting tissue in the heart. Conduction through the Purkyně fibers is 1.5 to 4.0 m/sec as compared to 0.4 m/sec through contractile myocardial fibers.

Heart block

Any interference with this conducting system, through mechanical obstruction by arteriosclerotic plaques or by degeneration due to infection, will prevent the regular con-

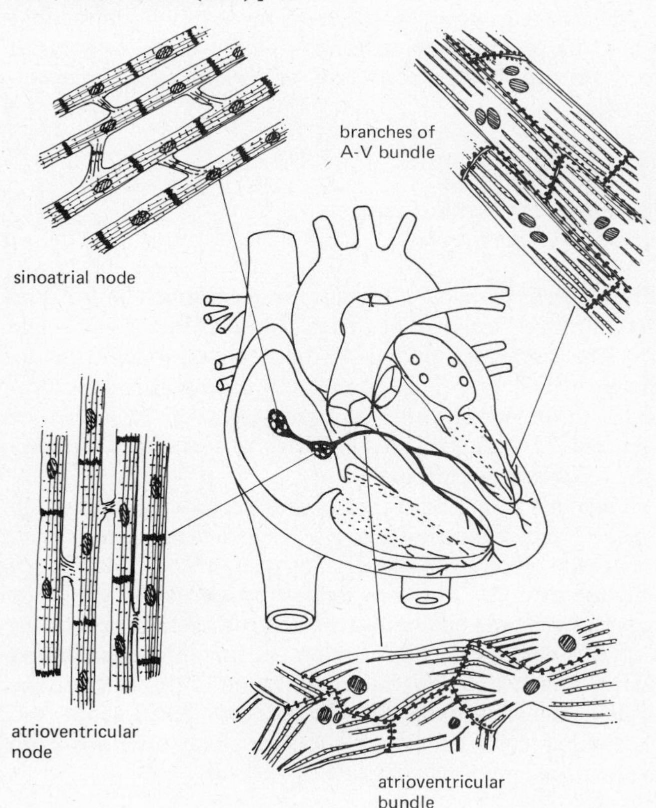

Figure 12-7. **A schematic representation of the gross anatomical, histological, and (to some extent) ultrastructural organization of the conducting system of the steer heart. The four areas represent camera lucida drawings at a magnification of about 130 times. The size of the intercalated disks and the desmosomes is somewhat exaggerated in order to facilitate a comparison of their ultrastructure. [Redrawn from J. Rhodin, P. del Missier, and L. C. Reid, *Circulation* 24:349 (1961).]**

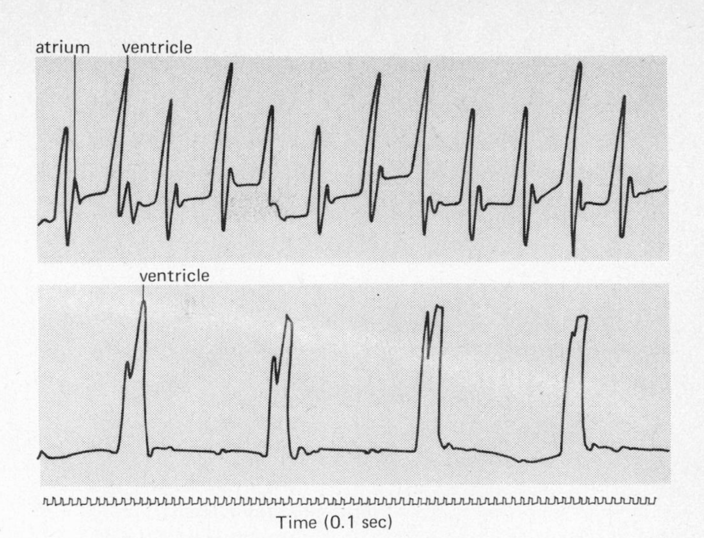

Figure 12-8. **Heart block induced in a dog by clamping the atrioventricular bundle. The ventricle now beats at its own, much slower rate. (From *The Essentials of Physiology and Pharmacodynamics,* 3rd ed., by G. Bachman and A. R. Bliss, McGraw-Hill Book Company, New York, 1940. Used with permission of McGraw-Hill Book Company. Record by G. Bachman and G. Blakiston, Philadelphia, 1940.)**

duction of impulses through the heart and cause *cardiac arrhythmia*. This disruption in the conducting system may be partial (*incomplete heart block*) or complete (Fig. 12-8). In *complete heart block,* the ventricles beat at their own inherent slow rate (about 40/min), which is insufficient to supply the oxygen needs of the body in any but the resting state.

The implantation of a small transistorized *artificial pacemaker* in the abdominal or thoracic musculature of the patient permits the electronic stimulation of the ventricles in synchrony with the SA node, so that the entire heart resumes its coordinated cycle.

Electrocardiogram

The electrical activity of the heart produces potentials at the body surface which can be recorded, following appropriate placement of the surface electrodes, as an electrocardiogram (EKG). The shape of the EKG varies with a number of factors, including electrode placement. Figure 12-9 depicts a normal EKG, consisting of three main waves, and shows their relationship to the mechanical events occurring in the heart.

1. The P wave represents *atrial depolarization,* which precedes the contraction of the atria.
2. The QRS complex represents *depolarization of the ventricles* preceding ventricular contraction. A normal QRS complex lasts about 0.06 sec.
3. The T wave represents *ventricular repolarization.* It is much smaller than the QRS depolarization complex. There is no comparative atrial repolarization visible, for it is obscured by the large QRS depolarization of the ventricles, occurring at that time.

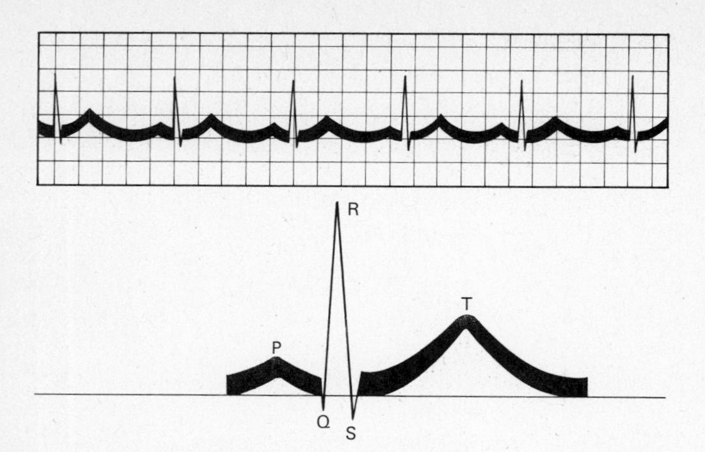

Figure 12-9. **The electrocardiogram. In the top record, the horizontal lines represent tenths of millivolts; the upright lines fifths of seconds. Below is an enlargement of one complete cardiac cycle. From *The Human Body: Its Anatomy and Physiology,* Third Edition by C. H. Best, N. B. Taylor. Copyright © 1932, 1948, 1956 by Holt, Rinehart and Winston, Inc. Reprinted by permission of Holt, Rinehart and Winston.**

An important parameter of the EKG is the *PR interval,* which indicates the length of time between the beginning of the atrial contraction and the beginning of the ventricular contraction, and thus the conduction time through the heart. A normal PR interval is 0.16 sec or less. In incomplete heart block, when conduction through the AV bundle is slowed, the PR interval may be almost doubled, and there may be two to three P waves for each QRS-T complex. In complete heart block, the ventricles "escape" from SA and atrial control, and the P waves may be completely dissociated from the QRS-T complexes.

Prolongation of the QRS complex, which normally lasts about 0.06 sec, indicates delayed conduction through the ventricles, often caused by ventricular hypertrophy. Ventricular hypertrophy may also result in an increased voltage of the QRS complex.

The *Q-T interval* coincides with the beginning and the end of ventricular contraction. It lasts about 0.30 sec.

12-5 Characteristics of Cardiac Muscle

The structure and physiological characteristics of cardiac muscle cells may be considered to be intermediate between smooth and striated muscle. Striated muscle is completely dependent on its innervation for contraction, whereas smooth muscle, especially that of the viscera, has a slow rhythmic contractility that persists even after the nerves to it are cut. The inherent contractility of cardiac muscle is much more marked. Striated muscle contracts most rapidly, then cardiac muscle, and smooth muscle contractions are long and slow.

Inherent Rhythmicity

The unceasing rhythmic activity of the heart is one of the great miracles of life. Even isolated cells from embryonic hearts grown in tissue culture show this *inherent contractility,* which originates from cardiac muscle cells and is consequently *myogenic* in nature. These heart cells in tissue culture can be divided into "leader cells" and "follower cells." The "follower cells" beat at the rate set by the "leaders" in the culture. Leader cells derive from an area of the embryonic heart that would have developed into the *pacemaker* which, in a normal heart, regulates the beating of the entire organ.

All-or-None Principle

Cardiac muscle always responds to an effective stimulus with a maximal contraction of the entire chamber of the heart, the strength of the contraction being independent of the strength of the stimulus. This is called the all-or-none principle. Once the impulse has been initiated by the internal pacemaker cells, or by an external electrical stimulus, the high conductivity of the cardiac muscle cells transmits it rapidly to other muscle fibers of that chamber.

Although the heart always responds with a maximal contraction, the maximum varies with physiological conditions. The degree to which the heart is filled with blood (stretched), hormones, changes in ionic environment, temperature changes—all modify both the rate and the strength of the heart beat.

Factors that affect the force of myocardial contractions affect the *inotropic* characteristics of the heart, in contrast to those that change heart rate, or its *chronotropic* characteristics.

Refractory Period and the Cardiac Cycle

The refractory period of cardiac muscle is much longer than that of nerve (Sec. 6-3) or skeletal muscle (Sec. 23-10). The absolute refractory period of the heart is about 0.25 sec, during which time it is completely unresponsive to additional stimulation from its own pacemaker (but can be stimulated by a high voltage electrical stimulus). The relative refractory period is 0.05 sec.

The normal cardiac cycle requires the orderly contraction (*systole*) and relaxation (*diastole*) of the heart chambers to propel the blood from the atria into the ventricles and hence into the arteries. The long refractory period of the heart prevents it from going into sustained contraction, or tetanus, and thus ensures that there is an adequate diastolic period during which the heart fills with blood. The total cardiac cycle in the adult human lasts a little less than 1 sec at rest, shortening considerably as the heart rate increases with exercise or emotional stress.

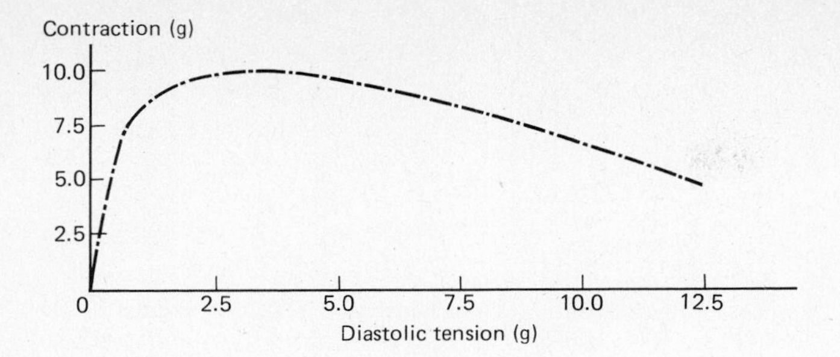

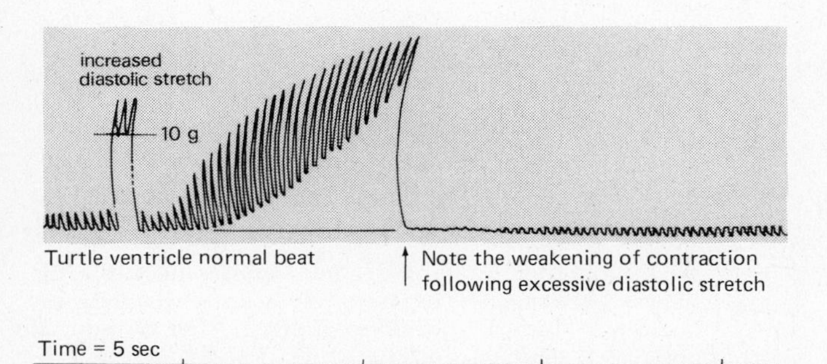

Figure 12-10. **Starling's law of the heart. The greater the diastolic stretch the greater the force of contraction. Note, however, that above a certain point, greater diastolic stretch diminishes the force of contraction. (From H. E. Hoff and L. A. Geddes, *Experimental Physiology,* Baylor Medical College, Houston, Tex., 1962.)**

Effect of Stretch on the Heart

Cardiac muscle responds to stretching of its fibers by an increase in the strength of the subsequent contractions. This is *Starling's law of the heart* (1912). Its physiological significance is that as blood is returned more rapidly to the heart during exercise, the increased volume stretches the walls of the ventricles and causes a more forceful propulsion of blood out of the heart (Fig. 12-10). This is one of the main factors balancing the amount of blood entering and leaving the heart.

Effect of Ions

Even if the heart is in an osmotically balanced solution, the rate and strength of its contractions depend on the particular ions present. The most important cations are calcium, potassium, and sodium. Ringer (1887) made a series of observations on the beating of the isolated frog heart in various salt solutions. He showed that in a solution containing only sodium chloride, the heart continues to beat rhythmically, but the strength of the beat declines. If calcium chloride is added, the beats resume their force but an excess of calcium ions causes the heart to stop beating in systole (*rigor*). This is probably caused by a direct effect of calcium on the contractile process (Sec. 23-4).

The subsequent addition of potassium chloride relaxes the heart so that normal beats are resumed, but an excess of potassium ions stops the heart in diastole. This is caused by a decrease in the resting potential of the cardiac muscle fibers, as the concentration gradient for potassium is reduced by increasing the external potassium concentration. These ionic effects are illustrated in Fig. 12-11.

On the basis of these observations (which are still being repeated by present-day physiology students) Ringer devised a solution, *Ringer's solution*, that contains the chlorides of sodium, potassium, and calcium in concentrations that maintain the beating of the isolated frog heart almost indefinitely (Table 12-1). The solution is made slightly alkaline by the addition of sodium bicarbonate.

Table 12-1. **Physiological Salt Solutions (g/liter solution)**

	Ringer's	Ringer-Locke's	Tyrode's
NaCl	6.00	9.00	8.00
KCl	0.22	0.42	0.20
$CaCl_2 \cdot 2H_2O$	0.29	0.24	0.20
$MgCl_2$	—	—	0.05
$NaHCO_3$	—	0.20	—
NaH_2PO_4	—	—	0.04
Glucose	—	1.00	1.00

Figure 12-11. **The effect of ions on the frog heart. This heart was first perfused with Ringer's solution, a balanced solution of ions that maintains the normal beat. This was replaced by isotonic solutions of individual ions, causing abnormalities in the strength and rate of the heartbeat. (Student record by R. Martocci and G. Bass, Physiology Laboratory, New York University.)**

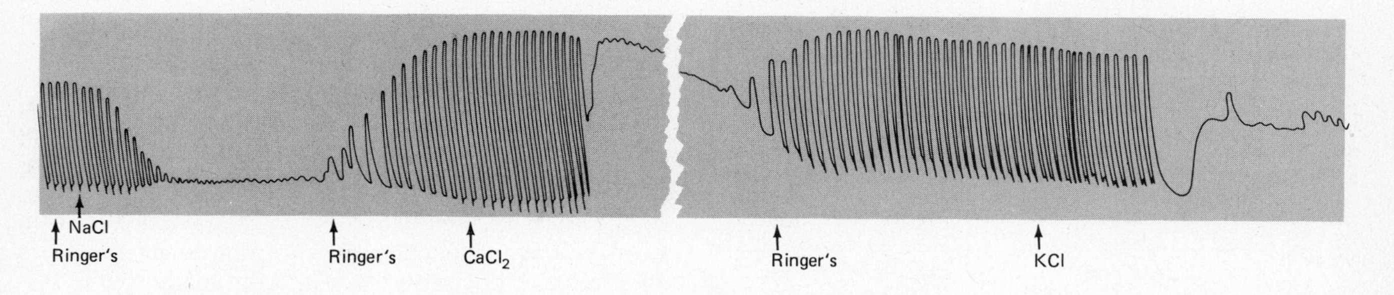

The mammalian heart responds in the same way to changes in the ionic environment. The isolated mammalian heart requires the addition of glucose to the medium (Ringer-Locke or Tyrode's solution) because its higher metabolic rate requires an additional energy source (Table 12-1).

Effect of Epinephrine and Norepinephrine

Epinephrine, the main secretion of the adrenal medulla, is released on sympathetic stimulation of this gland. At the same time, lesser amounts of norepinephrine are released. Epinephrine is a powerful cardiac stimulant, increasing both the rate and the strength of the heartbeat (Fig. 12-12). The heart, especially the denervated heart, is extremely sensitive to traces of epinephrine and can be used to detect concentrations as low as 1 part in 200,000. Norepinephrine, secreted by postganglionic sympathetic nerve fibers, has a similar though less potent effect on the heart. Both these compounds also increase the rate of conduction through the heart and decrease its refractory period. Stimulation of the sympathetic nervous system is thus a potent factor in speeding up heart rate and increasing its force.

Effect of Acetylcholine

Loewi's discovery (1921) of a chemical secreted at the endings of the vagus nerve provided the basis for our concepts of the transmission of nerve impulses to the following cell. Loewi removed two frog hearts, one with its vagi still attached to it and the other completely denervated. He perfused them with Ringer's solution in such a way that the perfusate from the first heart dripped over the denervated heart. Stimulation of the vagi of the first heart inhibited its beating, and then, after a short time lapse, the second heart stopped beating (Fig. 7-5). He named the effective chemical *"Vagusstoff"* (vagus material), and this was later determined to be *acetylcholine.* Acetylcholine is produced at the pre- and postganglionic endings of all parasympathetic nerves, not only the vagus. Acetylcholine placed directly on the heart, slows the beat, decreases the strength of contraction, increases the refractory period, and delays conduction through the heart (see Figs. 12-13 and 12-14).

The effects of epinephrine and acetylcholine are closely related to their actions on cell membrane permeability. Epinephrine effects are mediated by cyclic AMP, whereas acetylcholine acts directly on membrane receptors of cardiac muscle fibers, increasing potassium and chloride permeability and causing hyperpolarization (compare to depolarizing effects discussed in Sec. 7-5), and consequently causes inhibition.

Figure 12-12. **Epinephrine increases the strength and rate of contraction of the isolated frog heart. Time in min. (Record by R. Papke, Physiology Laboratory, New York University.)**

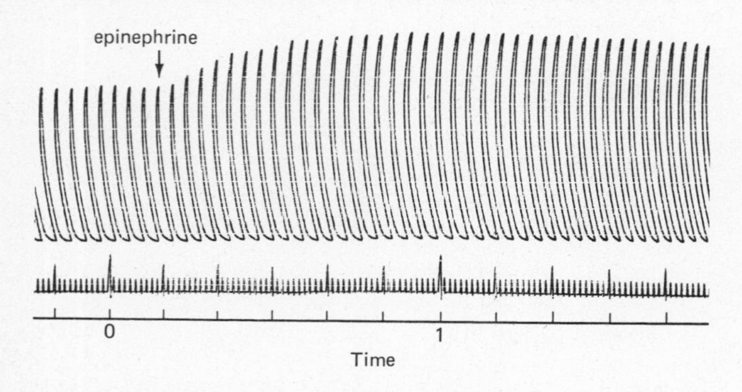

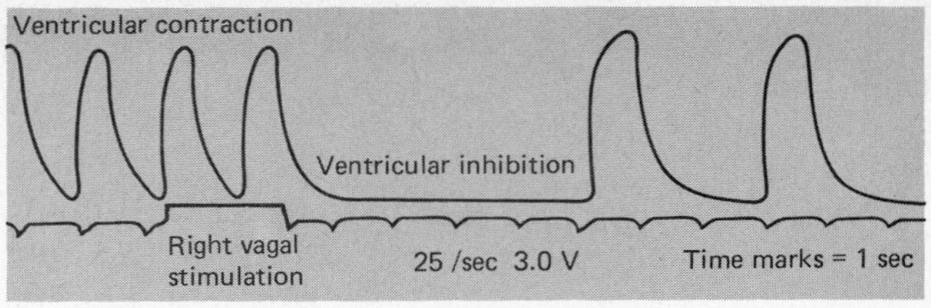

Figure 12-13. **Inhibition of the heart as a result of vagal stimulation. (From H. E. Hoff and L. A. Geddes, *Experimental Physiology,* Baylor Medical College, Houston, Tex., 1962.)**

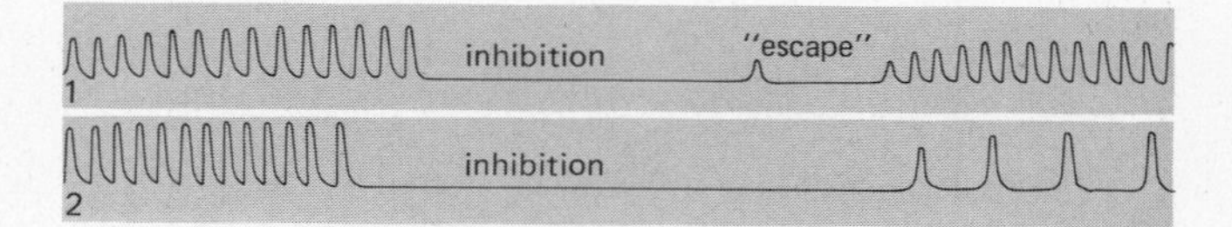

Figure 12-14. **Inhibition of the heart by acetylcholine. At 1 the Ringer's solution perfusing the heart was replaced by acetylcholine. One beat is seen to "escape" the inhibitory effect. After recovery, a more concentrated solution of acetylcholine, which had a longer-lasting effect, was added at 2. (Student record, Physiology Laboratory, New York University.)**

Effect of Temperature

The rate and strength of the heartbeat are increased by a rise in temperature up to a certain point, which coincides with the optimal temperature for enzymatic reactions for the particular animal. In warm-blooded animals, this optimum is about 40°C. If the temperature is raised considerably higher, however, enzymes and structural proteins are destroyed. Cold slows down the heart, decreasing the strength of the beat, because of the slowing of chemical reactions. This effect is reversible, and cold is sometimes used in surgical procedures on the heart when a very slow heart rate aids the surgeon.

Effect of Cardiac Glycosides

Digitalis, a drug obtained from the leaves of the lovely foxglove plant, and other similar drugs known as cardiac glycosides, such as *oubain,* exert a powerful inotropic action on the heart. The ability of these drugs to increase the force of myocardial contraction gives them great importance in the treatment of certain types of heart failure.

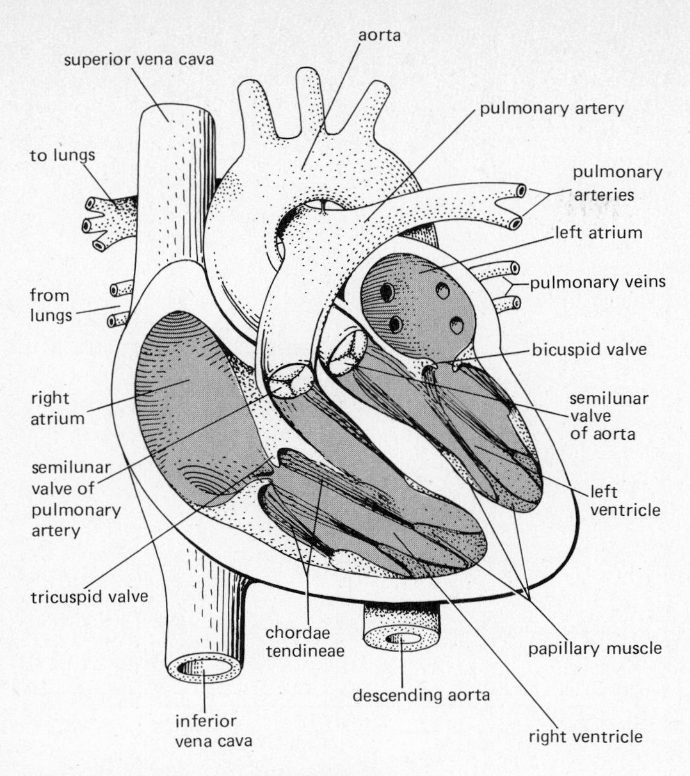

Figure 12-15. **The structure of the heart.**

Many of the effects of digitalis on heart muscle are similar to those produced by calcium, whereas potassium ions inhibit cardiac glycosidic activity. This may indicate an effect on contractile proteins but the mechanism of action of these drugs is still unclear. There is no evidence that they cause a specific increase in energy production by the heart. It has been suggested that in therapeutic concentrations, the glycosides potentiate a $Na^{+}K^{+}$-activated ATPase in heart microsomes, whereas in toxic quantities this enzyme is inhibited.

12–6 Separation of Venous and Arterial Blood in the Heart

The heart consists of a right and a left side, each side receiving blood through veins emptying into the thin-walled but muscular atrium and pumping blood out into the arteries through the powerful muscular contractions of the ventricle (Fig. 12-15). These two sides normally are completely separate, with venous blood being collected and distributed to the lungs by the right side of the heart, and arterial blood being collected and distributed by the left side of the heart.

Figure 12-16. **The fibrous skeleton of the heart is shown in relationship to the chambers of the heart and the roots of the aorta and the pulmonary artery. (Redrawn from R. F. Rushmer, *Cardiovascular Dynamics,* 2nd ed., W. B. Saunders Company, Philadelphia, 1961.)**

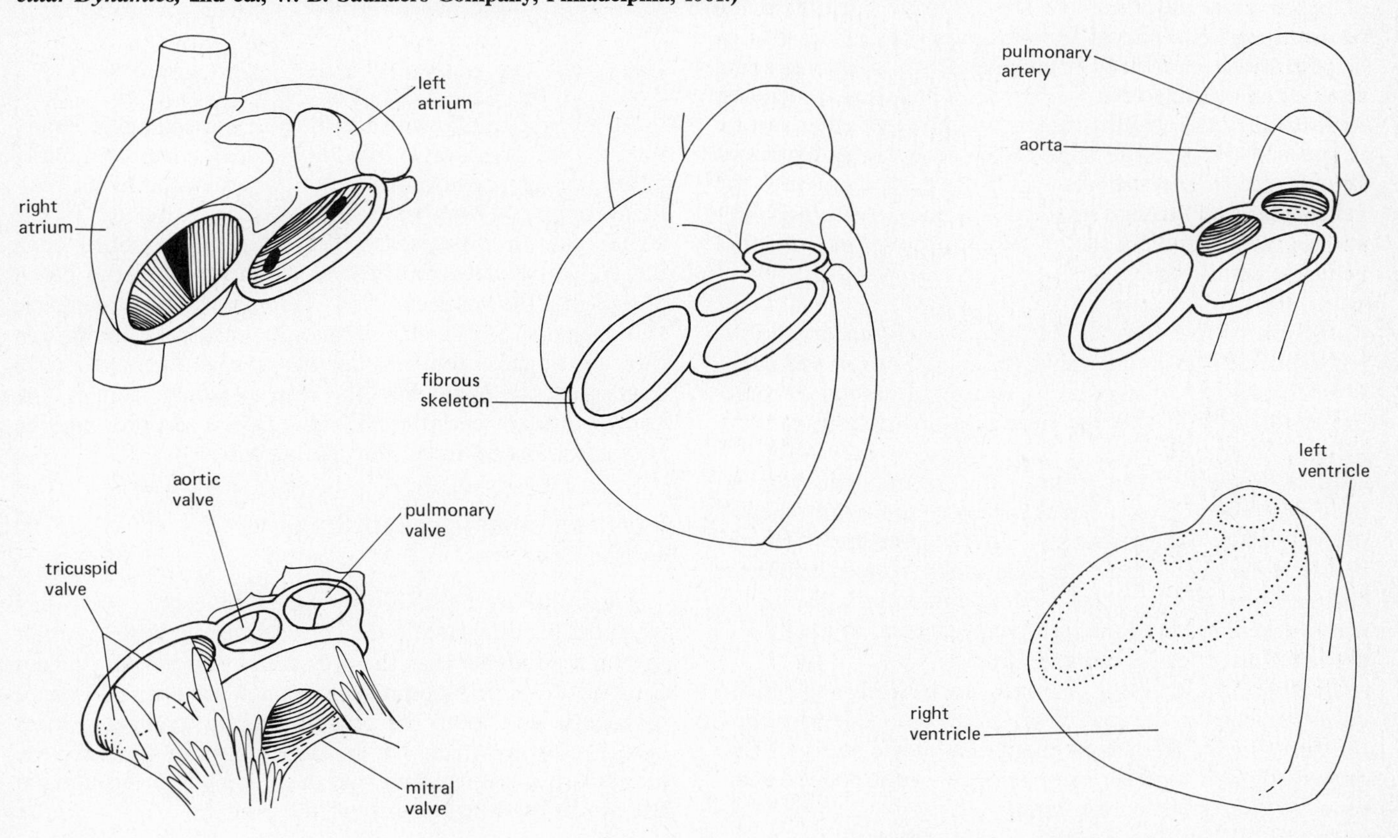

One-Way Flow in the Heart— The Valves of the Heart

The direction of flow through the heart is from the large veins into the atrium, from the atrium into the ventricle, and from the ventricle into the elastic, thick-walled artery. The entrances to the ventricle and the artery are guarded by flaps of tissue ingeniously arranged to form *valves* that are regulated by the pressure gradient on either side, permitting one-way flow in the direction described above. The excellence of the design of the heart valves can be better appreciated when it is realized that the orifices that they protect are constantly changing their shape; despite this, normal heart valves shut abruptly and permit only minimal leakage.

Anchorage of Heart Valves— The Heart Skeleton

Four interconnected fibrous rings of heavy connective tissue surround the openings of the great vessels of the heart (Fig. 12-16). These rings form the fibrous skeleton of the heart and provide a firm basis for the attachment of the heart valves, as well as help to keep the orifices open. The muscles of the ventricles are also attached to this fibrous skeleton.

ATRIOVENTRICULAR (AV) VALVES

Right Side of the Heart. Between the right atrium and the right ventricle is a specialized valve that permits the flow of blood from the atrium to the ventricle but not in the reverse direction. This is the *tricuspid valve,* which gets its name from its three flaps or cusps attached to a fibrous ring around the opening to the ventricle. The cusps are smooth on the side of the atrium, but their rough ventricular surface is fastened down to the walls of the ventricle by tendinous strands, the *chordae tendineae.* These chordae fasten to the ventricular wall through small conical muscles, the *papillary muscles,* which prevent the eversion of the valves when the ventricle contracts.

Left Side of the Heart. Between the left atrium and the left ventricle is a similar valve, the *bicuspid* or *mitral valve.* It is formed by two large cusps, instead of the three characteristic of the valve on the right side of the heart.

Arterial Valves of the Heart. Unlike the great veins of the heart that possess no true valves at the point of entry into the atria, each of the two arteries that lead from the ventricles is guarded by the beautifully formed *semilunar valves.* These strong valves are found within the *pulmonary trunk,* where it leaves the right ventricle, and at the base of the aorta, as it leaves the left ventricle.

Each semilunar valve is made up of three half-moon cusps, attached to the wall of the artery by one border, with the curved edge free inside the lumen of the artery. These curved edges fit together very closely to withstand the enormous pressures within the artery.

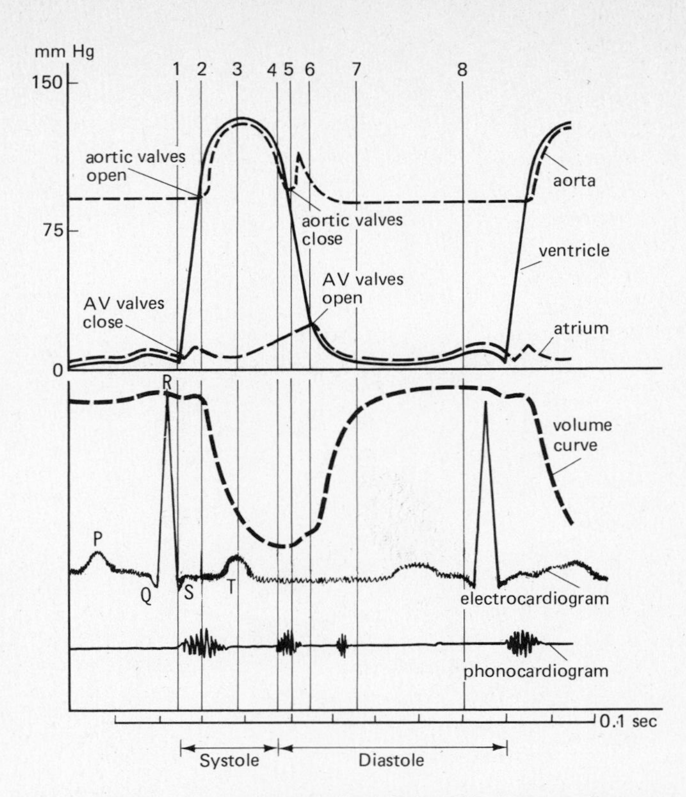

Figure 12-17. **Curves showing the relationship between the pressure changes in the heart and aorta, and the resultant opening and closing of the different valves. Note the decrease in ventricular volume as the ventricular pressure mounts. (From C. H. Best and N. Taylor, *The Physiological Basis of Medical Practice,* 9th ed., The Williams & Wilkins Company, Baltimore, 1973.)**

PRESSURE CHANGES REGULATING AV VALVES

As the atrium fills with blood from the veins that empty into it, the pressure in the thin-walled chamber rapidly mounts; because the ventricle is in diastole, and hence relatively empty at this time, the pressure is greater in the atrium than in the ventricle (Fig. 12-17). This forces open the AV valve between the two chambers, and the blood gushes into the ventricle. Now the pressure in the ventricle is higher than in the atrium, and the back pressure against the valves snaps them shut; as the powerful ventricle contracts to force blood into the artery leading from it, the rising pressure keeps the AV valve closed and prevents the backward flow of blood into the atrium.

PRESSURE CHANGES REGULATING ARTERIAL VALVES

The tightly fitted semilunar valves are forced open only when the pressure within the contracting ventricle has built up to a level higher than the pressure within the artery itself (Fig. 12-17). The blood rushes through the open valve into the artery, and when the pressure in this engorged vessel becomes higher than that in the emptying ventricle, the semilunar valves snap shut, and the blood pulses through the arteries to the lungs or rest of the body.

Heart Sounds

The sounds made by the closing of the valves of the heart and the resulting vibrations of the blood in the heart are transmitted to the chest wall. A stethoscope (a mechanical sound transmitter) can be used to amplify the sounds. The heart sounds can also be amplified by electronic means and displayed on an oscilloscope. The recording is called a *phonocardiogram,* and the heart sounds appear as vibrations (Fig. 12-17).

The *first sound,* usually heard as the "lub" of the "lub, dup, lub, dup" of the heart, is very low pitched, and it is caused by the closure of the AV valves before the beginning of the ventricular contraction. The *second sound,* "dup," is higher in pitch, with a crisp ending; it is caused by the closure of the semilunar valves.

Abnormal Heart Sounds

Any condition that causes the valves to become inflamed may result in the adhesion of the cusps of the valve. These adhesions may become scar tissue, fusing parts of the valve together. The valve now cannot prevent the backflow of blood, and this *regurgitation* of the blood is heard as a differently pitched heart sound or "murmur." These heart murmurs are often caused by rheumatic fever; the mitral valve appears to be a special site of attack by the toxins released by this streptococcus. Rheumatic fever is an autoimmune disease initiated by the streptococcal toxin. The antibodies formed in response to these toxins react with many tissues of the body to cause immunological damage (Sec. 20-4). The mitral valve appears to be especially vulnerable to these antibodies and large, hemorrhagic lesions form along the edges of the valve.

If the valve is so fused that there is only a small opening left for the blood to pass through to the next chamber or vessel, it is said to be *stenosed,* and the sound is again quite abnormal. Certain damaged heart valves can be repaired or even replaced, for modern techniques of heart surgery permit the blood to bypass the heart through the use of an artificial pumping system which circulates the blood during the operation.

Effect of Damaged Heart Valves on the Circulation

Stenosis of the aortic valve occurs when the narrowed opening to the aorta impedes the flow of blood from the left ventricle into this artery, causing a tremendous pressure to build up in the left ventricle as the blood accumulates. During systole, this pressure may reach 450 mm Hg (normal = 120 mm Hg) and is usually enough to compensate for the resistance, forcing the blood into the aorta. Only when the stenosis is extreme is the left ventricle unable to overcome the resistance in the aorta. The ventricle is said to "fail," and the resulting deprivation of arterial blood causes fatal anoxia.

Stenosis of the mitral valve causes the blood to be dammed up in the left atrium, raising the pressure in this chamber. This increases the blood pressure in the vessels of the lungs, for the blood entering the left atrium from the lungs encounters more and more resistance. Congestion and edema of the lungs result and can be fatal.

Regurgitation of the mitral valve permits the blood from the left ventricle to re-enter the atrium during ventricular systole. Although the pressure does not increase in the left atrium to the same extent as in mitral stenosis, the end effects of pulmonary congestion and edema are very similar.

12-7 The Right Side of the Heart and the Pulmonary Circulation

The right side of the heart may be considered the receiving and pumping center for the *unoxygenated blood* coming from the body and going to the lungs. The right atrium receives blood from the *inferior* and *superior venae cavae,* which drain the lower and upper parts of the body, respectively, and from the coronary vessels which drain the blood from the muscle of the heart itself.

Blood leaves the *right atrium* through the right AV valve to enter the *right ventricle.* The walls of this chamber are much more muscular than those of the right atrium, but considerably less thick than those of the left ventricle. The muscular strength developed by the left ventricle must be great enough to send the blood through the entire body, whereas the right ventricle has less resistance to work against in the short distance to the lungs.

The blood leaves the right ventricle through the *pulmonary trunk.* When the right ventricle contracts, the generated pressure forces the tricuspid valves to close; at this time the semilunar valves at the entrance to the pulmonary trunk spring open, and the blood is forced into this vessel.

The pulmonary trunk runs backward and upward, winding around the left side of the aorta, then dividing into the *right and left pulmonary arteries.* These blood vessels are carrying unoxygenated blood; but they, like the aorta, have the elastic structure of an artery that is necessary for the rhythmic adjustment to the great pressure generated by the heart. Each pulmonary artery enters its lung through the root, to divide into branches that follow the bronchioles and to end in *capillary networks* around the *alveoli* (Sec. 16-2). This close contact between the capillaries and the alveolar surfaces allows for efficient gaseous exchange, and the blood leaving the lungs through the four *pulmonary veins* is rich in oxygen. The pulmonary veins return this blood to the *left atrium,* where it is available for distribution to the entire body (Fig. 12-15).

The nonrespiratory tissues of the lung, such as the bronchial tree and the pleura, receive *arterial blood* from the *bronchial arteries,* branches of the aorta. Usually, one bronchial artery supplies the right lung, and two supply the left lung.

Changes in Pressure in the Pulmonary System

Because the two sides of the heart pump synchronously, and because the volume of blood flowing through the lungs is the same as that flowing through the systemic circulation, any backing up of blood in either system will cause changes in pressure in the other.

If the left ventricle fails or if there is an obstruction at the mitral valve, the pressure in the left atrium is increased. The pulmonary veins bringing blood to the left atrium are exposed to this increased resistance, and the pressure rises in the pulmonary veins, capillaries, and arteries, causing edema in the lungs, coughing, and difficulty in breathing. The increased load makes the right ventricle work harder, because the increased stretch on the heart increases the force of contraction (Starling's law). The *compensation,* by which the heart overcomes the additional resistance, can achieve as much as four times the normal work before the right ventricle fails.

If the cardiac output increases, the pressure in the pulmonary system shows no great rise until the output is approximately three to four times greater than normal, because the pulmonary vessels expand to accommodate the increased blood volume. This means that gaseous exchange can continue efficiently during increased cardiac output, which is, after all, the *raison d'être* of the flow of blood through the lungs.

12–8 The Left Side of the Heart and the Systemic Circulation

Like the right atrium, the *left atrium* is the receiving chamber, but the blood brought to it is *oxygenated* (Fig. 12-15). The four *pulmonary veins* empty into it, two from each lung. They have no valves, nor have the small cardiac veins that drain the heart muscle and empty some of this venous blood into the left atrium. The opening to the left ventricle, however, is guarded by the bicuspid or mitral valve (Sec. 12-6), and blood leaves the left atrium through this valve to fill the ventricle.

The *left ventricle* is an extremely powerful ejecting chamber. Its cavity is longer and narrower than that of the right ventricle, and its walls are considerably thicker. When the left ventricle contracts, the rising pressure forces the closing of the bicuspid valve and opens the semilunar valves that guard the opening to the *aorta,* thus pumping the blood into this highly elastic artery.

The aorta is the main arterial trunk of the systemic circulation. Like the other great vessels of the heart, its opening is surrounded by a fibrous ring of the heart skeleton, to which the cusps of the aortic valves are also attached. From the aorta the blood is distributed to the other arteries, arterioles, and capillaries of the systemic circulation. The detailed anatomy of the circulation of different parts of the body is considered in Chapter 13.

12–9 Return of Venous Blood to the Heart

Venous return to the heart is a vital aspect of cardiovascular physiology and is intimately involved with adaptations of the heart and circulation to different levels of activity. The heart cannot pump out more blood than it receives and therefore must be able to adjust its output to the inflow. One of the most important mechanisms by which this is achieved is through the inotropic response of cardiac muscle to increased stretch of the ventricular chambers as they are distended with blood (Starling's law of the heart).

Venous Pressure

In humans in the upright position, the long axis of the body is parallel to gravitational pull. The resulting hydrostatic pressures in the fluid system of the circulation cause the

Figure 12-18. **Effect of hydrostatic pressure on venous pressures throughout the body during quiet standing. The pressure in the abdominal cavity and in the feet is usually reduced by the action of the muscle "pumps" to less than 10 and 25 mm Hg, respectively. (Redrawn from A. Guyton, *Textbook of Medical Physiology,* 5th ed., W. B. Saunders Company, Philadelphia, 1976.)**

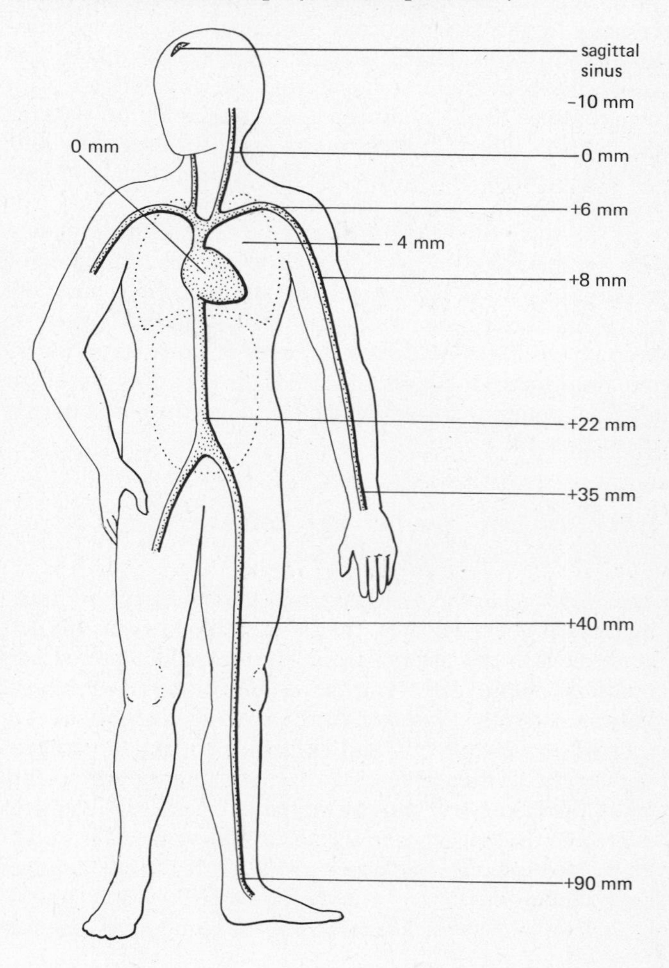

blood to pool in dependent regions, especially the feet. Because the arteries and veins function as interconnecting tubes, this does not have a direct effect on blood flow, but as the veins are highly distensible, blood tends to pool within them.

Peripheral venous pressure

In the standing individual, venous pressure varies at different points in the body (Fig. 12-18). Venous pressure is greatest in the feet, and in the standing position with the leg muscle relaxed, the pressure in the leg veins is 80 to 90 mm Hg. One can roughly estimate the peripheral venous pressure by watching the veins gradually collapse as the arm is slowly raised above the level of the heart. The peripheral venous pressure is the point at which the veins collapse, with reference to the level of the heart. Direct measurement of venous pressure is accurately obtained by inserting a needle connected to a water manometer into a vein. The venous pressure is expressed as the height in centimeters of water above the level of the tricuspid valve. This measurement can be converted to mm Hg by dividing by 1.36.

Central venous pressure

Blood from all the veins of the systemic circulation flows into the right atrium; hence *right atrial pressure* is considered to be *central venous pressure*. In the standing individual, the right atrial pressure is approximately 0 mm Hg, because the heart pumps the blood out as fast as it flows in. If right atrial pressure rises above this, it backs up the blood in the large veins, an occurrence which is normally prevented by homeostatic regulatory mechanisms. These mechanisms are discussed later in this chapter (Sec. 12-11).

Factors Involved in Venous Return

Muscle pumps and venous valves

Veins contain valves that permit the flow of blood only in the direction of the heart. When the thin-walled veins are massaged by the contractions of the limb muscles, the veins are constricted, and the blood is pushed up toward the heart (Fig. 12-19). Blood from the legs is forced toward the abdomen and thorax, where the veins are compressed by the visceral organs. Venous pressure in the abdominal region is usually 6 to 9 mm Hg higher than right atrial pressure.

The importance of the venous valves is noted best when they are defective, as in the condition of *varicose veins*. This condition often occurs after the veins have become abnormally distended from continuous prolonged standing, or from pressure changes during pregnancy. Because the valves are not stretched concomitantly, they can no longer prevent the backward flow of blood to the feet. This distends the veins still further, and they form bulbous protuberances, or *varicosites*. Associated with this distention is an increase in venous and capillary pressures and a resultant accumulation of fluid in the tissues (*edema*).

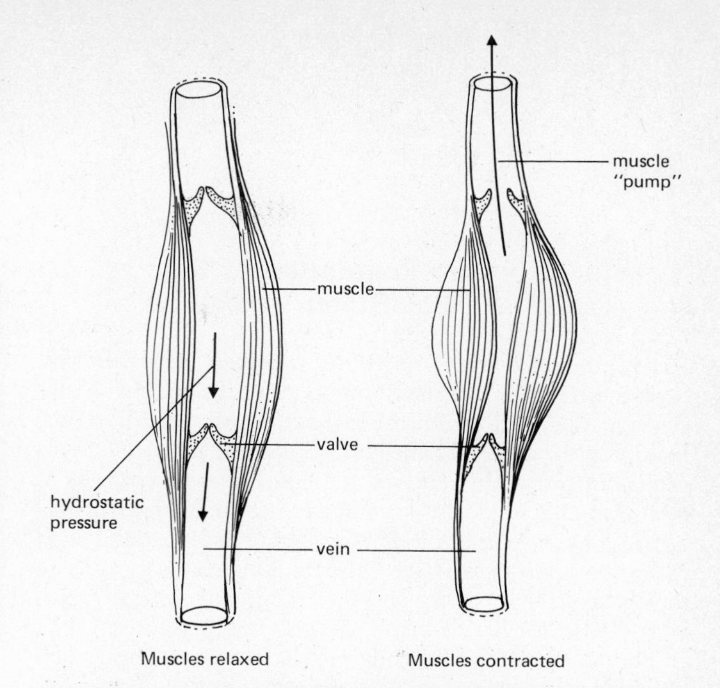

Figure 12-19. **Muscle pumps. Contraction of the leg muscles constricts the veins, forcing blood up toward the head. The one-way valves prevent most backflow. If the leg muscles are completely relaxed, a very rapid buildup of hydrostatic pressure in the lower leg occurs.**

The respiratory pump

The heart and the great veins are prevented from collapsing by the negative pressure in the thorax. Intrathoracic pressure is less than atmospheric pressure, and during inspiration it falls still lower (from −2 mm Hg to −6 mm Hg) as the thoracic cavity expands. During inspiration, contraction of the abdominal muscles and the descent of the diaphragm causes intra-abdominal to pressure rise (Secs. 16-4 and 16-5), so that blood is first pushed up and then sucked into the intrathoracic veins. During expiration the opposite changes take place, so that the respiratory movements act as a push-pull pump.

Venous reservoirs

More than 50 per cent of the total blood volume is contained within the systemic veins. Some of these venous areas hold so much blood that they are referred to as venous reservoirs. The most important reservoirs are the large veins, the spleen, and the liver.

Sympathetic stimulation of the large veins, such as occurs during exercise, causes them to constrict (vasomotion, Sec. 12-10), reducing their capacity and returning the stored blood to the heart. Blood stored in the spleen is forcefully expelled when the splenic capsule and the splenic veins contract in response to sympathetic stimulation (the importance of the spleen as a blood reservoir appears to be much less in humans than in lower animals). This return of stored blood from venous reservoirs to the heart distends it and causes it to pump blood out more strongly, adjusting its output to the inflow within a few beats. This important homeostatic control of cardiac output is discussed in more detail later in this chapter.

12–10 Nervous Regulation of the Heart and Blood Vessels

Despite the myogenic nature of the heartbeat, neural control is essential for the integrated functions of the cardiovascular system. This reflex control is exerted on several levels, including the spinal cord, the medulla and pons, the diencephalon, and certain cortical regions. Different patterns of cardiovascular response accompany different types of behavior; the response varies considerably, for example, in defensive versus aggressive fighting. This seems to indicate that psychic inputs may share common anatomical routes with cardiovascular, somatic, and visceral pathways.

The study of the nervous regulation of the heart and blood vessels is enormously complicated by the *autoregulatory ability* of many tissues: metabolites produced by tissue activity regulate the amount of blood flowing through them and consequently the resistance against which the heart has to work.

Most of our knowledge about reflex circulatory control has been obtained from experiments in which only one input has been varied. This makes for a controlled experiment but is not adequate to explain cardiovascular regulation during exercise, or hemorrhage, where many other inputs to the CNS are involved. The blood pressure control system has the ability to integrate information from many peripheral inputs and flexibly adjust the neural commands sent to the heart and vascular beds to conform to the specific needs of the organism. Modern cardiovascular experimentation is now concerned with investigating the integration of multiple information and its transformation into autonomic responses.

Innervation of the Heart

Reflex changes in heart rate are effected by *central control* of the *parasympathetic* (vagal) nerves and the *sympathetic* (accelerator) nerves to the heart. The degree of control of these two systems over the heart varies with the species. In the dog, for example, the parasympathetic nerves are dominant: slowing is induced by stimulating the vagi, while section of the vagi speeds up the heart. In the baboon, changes in heart rate are more easily obtained by stimulating or inhibiting the sympathetic nerves. The human heart appears to be intermediate in terms of the balance of autonomic control.

The *right* and *left vagi* innervate the atria, particularly the SA node and the AV node, with only a few parasympathetic fibers piercing the ventricles. The *sympathetic nerves* reach the heart by way of the inferior cervical ganglion and the sympathetic chain of ganglia. Sympathetic fibers are distributed mainly to the ventricles (Fig. 12-20).

Stimulation of the sympathetic nerves has the same effect as the application of norepinephrine to the heart, as discussed earlier in this chapter. Heart rate and contractility are increased, whereas the conduction time and refractory period are decreased. Stimulation of the parasympathetic nerves, in general, has the opposite effect, equivalent to the effects evoked by application of acetylcholine to the heart. The heart slows down, its beat is weaker, and conduction time and refractory period are longer.

Structure and Innervation of Blood Vessels

Arteries and arterioles

Arteries serve not only as a conducting system for blood pumped out of the heart under high pressure, but also as elastic buffering chambers, stretched during systole and recoiling in diastole. Their ability to stretch and recoil depends on the elastic tissue in their walls. They are prevented from "blowing out" by a firm jacket of collagen in the outer fibrous connective tissue.

All arteries have *four main layers:* an outer fibrous connective tissue layer, a layer of smooth muscle, a layer of elastic tissue, and a lining of endothelium (Fig. 12-21). In addition, arteries are supplied with their own blood vessels (vasa vasorum) and autonomic nerves.

Figure 12-20. **The innervation of the heart.**

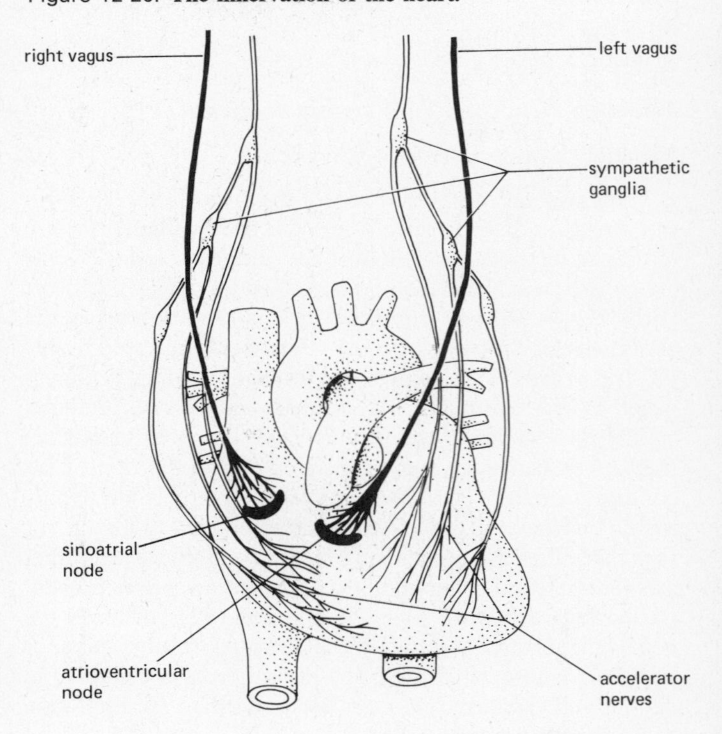

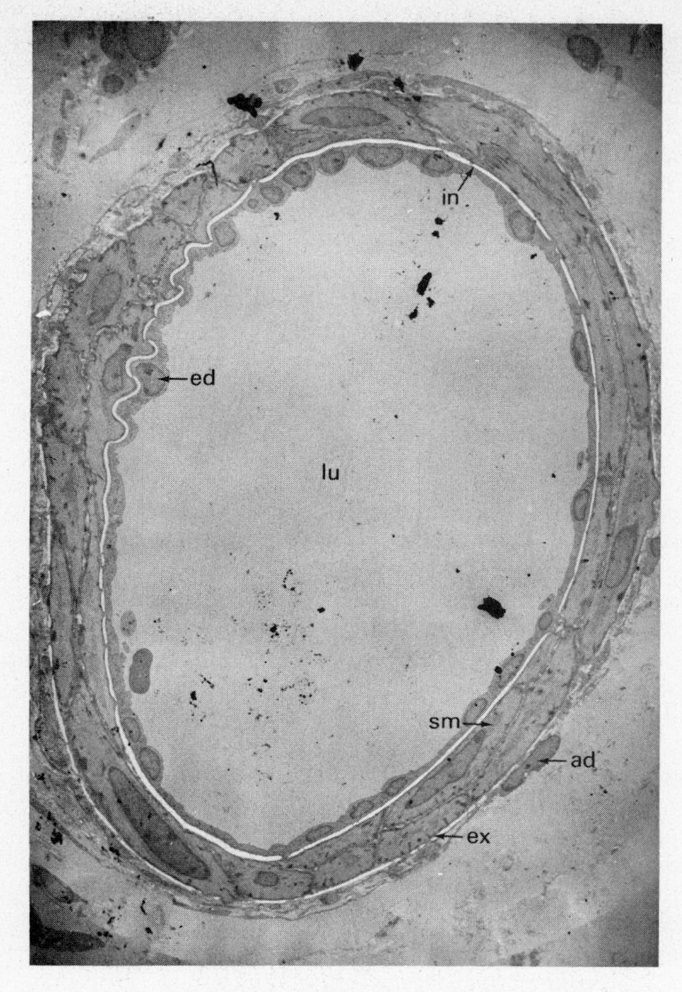

Figure 12-21. **Cross section of a small artery. The lumen (lu) is lined by a thin layer of endothelial cells (ed) on a basement membrane (the intima) that may have some elastic fibers (in). The middle layer, tunica media, contains smooth muscle cells (sm). The next layer is the elastica externa, a membrane containing a network of elastic fibers (ex). The outermost layer, the tunica adventitia (ad), is made of loose connective tissue. (From J. Rhodin, *An Atlas of Ultrastructure,* W. B. Saunders Company, Philadelphia, 1963.)**

As the arteries branch to form the small *arterioles,* the thickness of the wall decreases, but the arterioles remain relatively muscular and control the flow of blood into the capillaries through the contraction (*vasoconstriction*) or relaxation (*vasodilation*) of their smooth muscle fibers.

Microcirculation

The microcirculation is a term introduced by Zweifach (1961) to describe the closed system of branching arterioles, capillaries, and venules that can be seen only with the aid of a microscope. Blood flows from the arteriole via a *metarteriole* through a *preferential channel,* from which the *capillaries* branch off. The origin of the capillary is guarded by a ring of smooth muscle, the *precapillary sphincter,* which when opened and relaxed permits the flow of blood through the capillary (Fig. 12-22). The capillary usually makes a large loop around an area of cells before it enters a venule, so that it is customary to refer to the arterial end and the venous end of the capillary.

The capillary wall, only 1 μm thick, is made up of endothelium, flattened against a basement membrane (Fig. 12-23). These thin-walled endothelial tubes permit the diffusion of small molecules between the blood and tissues; the factors regulating this exchange are discussed in Sec. 4-2. The diameter of the capillary lumen may be considerably less than that of the red blood cell, which enters the capillary edgewise and folds like a crêpe suzette.

Other connecting chambers between the arterioles and the venules have been reported for some vascular beds, especially in the cutaneous circulation. These wide *arteriovenous (AV) anastomoses* or *shunts* are also shown in Fig. 12-22. It is assumed that they can shunt blood directly from the arteriole to the venule. This emergency mechanism precludes fluid exchange, for the walls of the thoroughfare channels and the AV shunts are thicker than those of the capillaries. This type of shunting occurs when the capillary circulation is blocked—by cold, for example—and blood is rapidly shunted from the superficial capillary beds in the extremities into the deeper AV shunts.

Figure 12-22. **Diagrammatic representation of the microcirculation. Arrows indicate the direction of blood flow.**

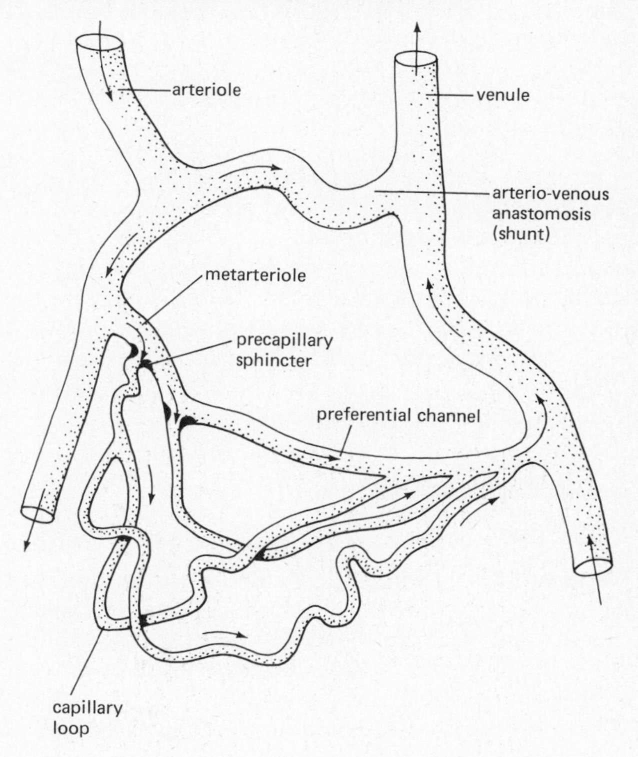

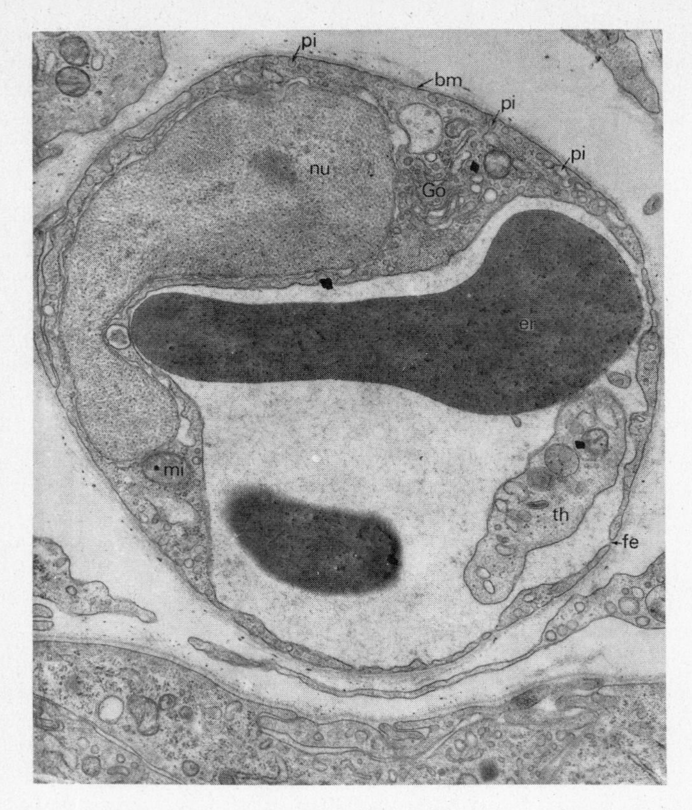

Figure 12-23. **Cross section of a thin capillary shows that the wall consists only of thin endothelium on a basement membrane (bm). Note how thin the endothelial cell is, except for the part that contains the nucleus (nu), the Golgi complex (Go), and mitochondria (mi). Many pinocytotic vesicles (pi) are seen in the endothelial membrane. The closed pores (fe) are bridged by a very thin membrane. Neither the red blood cell (er) nor the platelet (th) has a nucleus. (From J. Rhodin, *An Atlas of Ultrastructure,* W. B. Saunders Company, Philadelphia, 1963.)**

Venules and veins

The *venules* collect blood from the capillaries and deliver it into the larger *veins*. The smaller venules have no smooth muscle cells, but larger venules and veins are muscular, though thin-walled. The venous system is able to store about 60 per cent of the total blood volume, a large proportion of which can be returned to the circulation as a result of venous vasoconstriction. This means that the muscular venules play a critical role in the mobilization of blood in conditions such as hemorrhage and shock.

Vasomotion and Vasomotor Nerves

Vasomotion

Vasomotion is the intermittent vasoconstriction and vasodilation of the arterioles, metarterioles, and precapillary sphincters that regulate blood flow through the capillary beds. Vasomotion also includes the constriction of the veins in response to sympathetic stimulation.

The *autonomic nerves* regulate vasomotion, particularly in the establishment of basic arteriolar tonus, but *other factors,* such as the oxygen requirements of the tissues and their production of metabolites, are probably more important in the long-term adjustment of the local blood flow to the needs of the tissue. Oxygen lack, accumulation of carbon dioxide, and increased tissue acidity and osmolarity are all important vasodilators, causing increased blood flow to active tissues. This *autoregulation* of blood flow to the tissues is evident a few seconds after the beginning of activity, whereas neurally evoked vasomotion occurs within a fraction of a second (central control of circulatory reflexes is considered later in this section).

Vasomotor nerves

The sympathetic nervous system is of much greater importance in the regulation of vasomotion than is the parasympathetic. Sympathetic nerves innervate all blood vessels, except the capillaries, although their distribution to heart and skeletal muscle blood vessels is scanty. Parasympathetic fibers are confined to certain glands, where they cause vasodilation (and thus increased secretion) through the release of an enzyme, *bradykinin,* which acts on tissue proteins to produce a vasodilator substance.

Sympathetic control over blood vessel musculature is of two types:

1. *Sympathetic vasoconstrictor fibers.* These nerves maintain a constant vascular tone in most tissues. Stimulation of the sympathetic nervous system by stress or exercise causes the release of norepinephrine, which rapidly enhances vasoconstriction and shunts blood from the veins and visceral organs to active skeletal and heart muscle. This *splanchnic shift* is an important mechanism in the routing of blood to actively metabolizing tissues during exercise. The reverse shift occurs after a heavy meal, when the sympathetic nervous system is relatively dormant and *passive vasodilation* occurs in the gut and splanchnic area. These regions then become engorged with blood, facilitating the removal of nutrients following digestion. This process is obviously impeded if exercise follows a large meal.
2. *Sympathetic vasodilator* (*cholinergic*) *fibers.* Of lesser importance are the sympathetic fibers that cause vasodilation through the release of acetylcholine. These cholinergic sympathetic nerves are found mainly in the blood vessels of heart and skeletal muscle. The resulting vasodilation at the onset of exercise speeds up the flow of blood through these active tissues. However, sympathetic cholinergic fibers are only found in nonprimates, so their physiological significance is unclear. The local production of metabolites is probably of greater importance in the maintenance of vasodilation.

Reflex Control of Heart and Blood Vessels

Peripheral input

Those regions of the central nervous system that are concerned with cardiovascular control receive information from many peripheral inputs, including arterial baroreceptors in the systemic and pulmonary circulation, mechanoreceptors in the heart and lungs, arterial chemoreceptors, and input from skeletal muscle. The *arterial baroreceptors* are probably the most important of these and are the only ones to be discussed here, but it is the task of the central nervous system to integrate all this peripheral information and adjust its neural commands to the heart and blood vessels accordingly.

Arterial Baroreceptors. The *carotid sinus* has been the most extensively studied of the arterial baroreceptors, ever since its function was first described by Heymans in 1955 (Heymans and Neil, 1958), perhaps because it is more easily isolated than are the other baroreceptors (Fig. 12-24). The carotid sinus may be seen as a thickening in the wall of the *internal carotid artery* (just above the bifurcation of the common carotid into the internal and external carotids). The baroreceptors within the arterial walls are fine nerve endings that are stimulated by changes in the tension of the arterial wall evoked by increases in blood pressure. Figure 12-25 shows that the rate of firing of these nerves increases proportionally with increased stretch on the arterial walls, within a pressure range of 50 to 200 mm Hg. Consequently, these baroreceptors are most sensitive in the normal physiological range of arterial pressure.

Figure 12-24. **The arterial baroreceptors.**

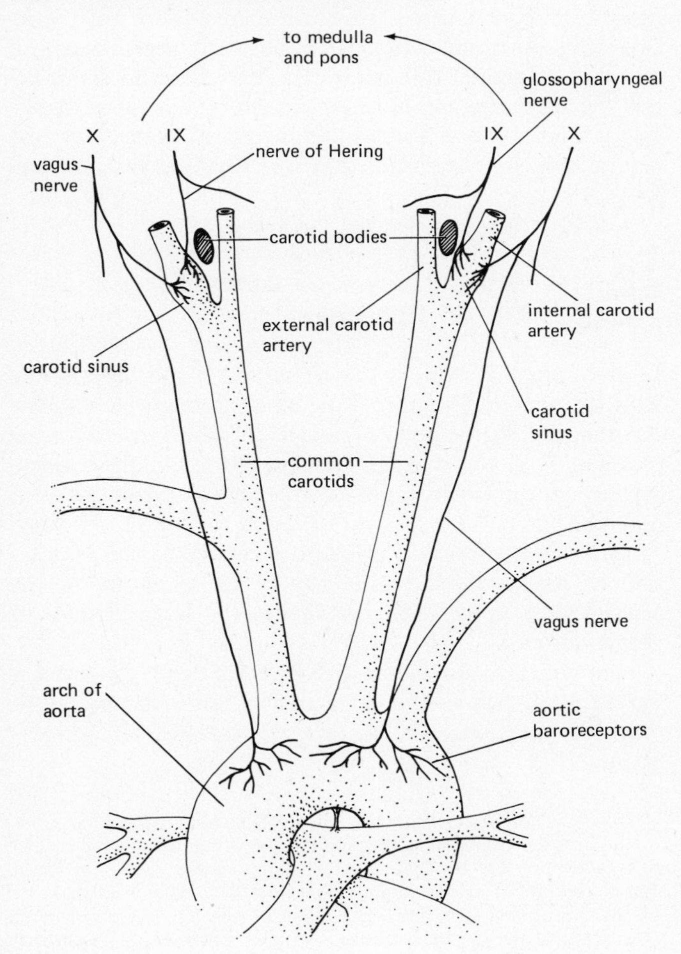

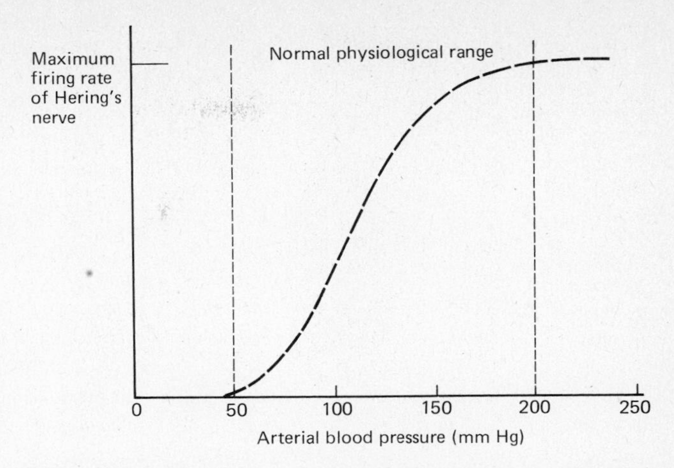

Figure 12-25. **The firing rate of the baroreceptor nerves increases with increased arterial blood pressure.**

There are similar baroreceptors in the wall of the *aortic arch* and other large arteries which appear to have a higher threshold than the sinus baroreceptors. The *pulmonary baroreceptors,* in the walls of the right and left main branches of the pulmonary arteries, signal rate of change in pressure as well as changes in mean pressure.

The carotid sinus is innervated by the fine *nerve of Hering,* which leads to the *glossopharyngeal* (IX cranial) nerve. The aortic and pulmonary baroreceptors are the terminals of branches of the *vagus* (X cranial) nerve. Impulses from these cranial nerves are conducted to the vasomotor and cardiac centers in the medulla and pons.

Central control of circulatory reflexes

Bulbar Vasomotor and Cardiac Centers. The vasomotor center is in the lower third of the pons and the upper part of the medulla (bulbar level). It consists of an *excitatory portion* that continuously fires sympathetic vasoconstrictor nerves to maintain vasomotor tone and an *inhibitory portion* that can inhibit sympathetic vasoconstriction, thereby allowing passive vasodilation (Fig. 12-26).

Also in the bulbar region is a cardiac center, which similarly consists of two functionally different regions. The *cardiac acceleratory center* causes increased heart rate and cardiac contractility through stimulation of sympathetic nerves, whereas the *cardiac inhibitory center* slows the heart and decreases its contractility through parasympathetic activity.

Baroreceptor (Buffer) Reflex. An increase in arterial blood pressure increases the rate of firing of the arterial baroreceptors, inhibiting the excitatory portion of the vaso-

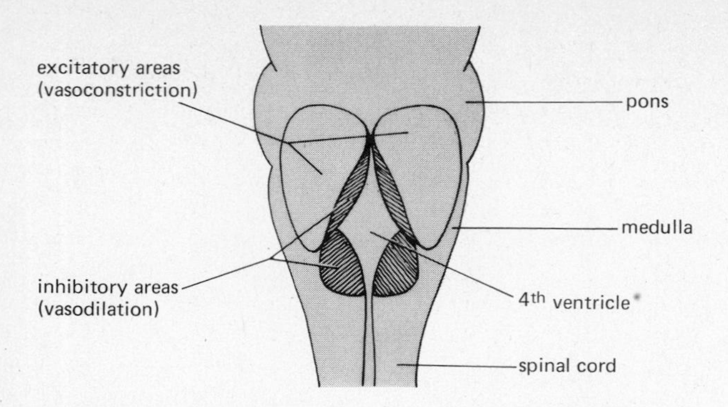

Figure 12-26. **The vasomotor areas of the medulla and pons.**

motor center and hence causing extensive vasodilation. At the same time, the cardiac inhibitory center is excited, decreasing heart rate and contractility. As a result, arterial blood pressure falls, and the firing of the baroreceptors slows (Fig. 12-27).

When the firing rate of the baroreceptor nerves decreases, sympathetic vasoconstriction is restored, parasympathetic inhibition of the heart is removed, and the blood pressure rises. Because the main function of these baroreceptors is to respond immediately to sharp changes in pressure, they are also called the *buffer nerves,* and the reflex is referred to as the *buffer reflex.* The buffer reflex is *critical in postural changes,* especially in humans, for it regulates arterial blood pressure to the brain. This subject is discussed in more detail later in the chapter. Baroreceptor reflexes, however, are probably of minor importance in adjustment to exercise, as discussed in the following section.

Suprabulbar centers

The same *diencephalic structures* that control autonomic responses to emotion (Sec. 9-4) can be shown to be of the utmost importance in the integration of reflex control of the cardiovascular system. In addition to the hypothalamus and associated limbic structures, many parts of the *cerebral cortex* (*telencephalon*) receive inputs from the glossopharyngeal and vagal nerves. These areas of the cortex may respond by stimulating or inhibiting the hypothalamus, but most cortical influences on the hypothalamus seem to be inhibitory.

There are *direct neuronal pathways* from the motor cortex to the sympathetic cholinergic vasodilator fibers of skeletal muscle. These tracts bypass the hypothalamus and account for vasodilation in muscles of nonprimates at the onset of exercise.

Figure 12-27. **The baroreceptor (buffer) reflex. Broken arrows indicate inhibition.**

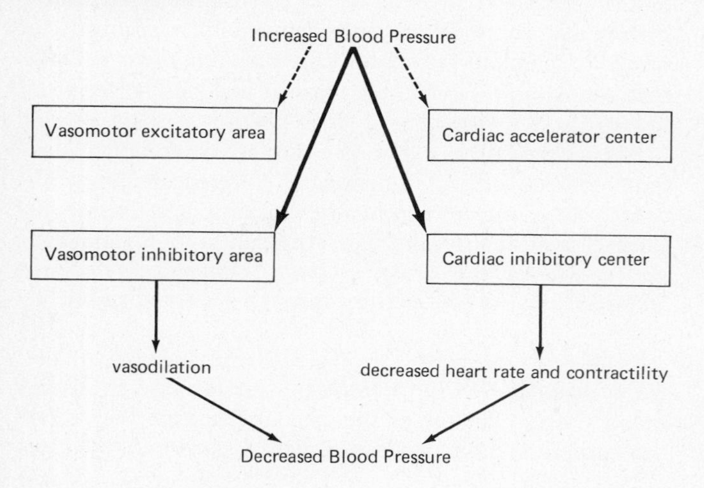

Resetting the Baroreceptor Reflex. The *suprabulbar centers* are responsible for modifying the excitability of the vasomotor and cardiac centers in the medulla and pons; in other words the suprabulbar centers reset the baroreceptor reflex in response to the barrage of peripheral signals received during increased activity. This resetting is essential for the maintenance of an adequately high blood pressure during exercise. If the baroreceptor reflex immediately lowered the elevated blood pressure, blood flow to the active tissue would decrease, and muscle activity could not be sustained. It is believed that the *hypothalamus,* which exerts central control over the pool of sympathetic and parasympathetic motoneurons, is chiefly responsible for this resetting. The role of the cortex in these control mechanisms still is not clear, apart from such common observations as increased heart rate and blushing caused by emotional stress; but it appears that the various levels of control on the central nervous system function as mutually interdependent loops.

12–11 Maintenance of Arterial Blood Pressure

In the closed system of the heart and blood vessels, the working pressure that forces the blood through the variable resistance of the vascular system is the *mean arterial blood pressure.* It is the average pressure[1] through the pulsing cardiac cycle of systole and diastole and is usually taken as 100 mm Hg.

Systolic pressure is the pressure generated by the contraction of the heart, and modified by the elasticity of the arteries. It averages 120 mm Hg in the large arteries of young adults.

Diastolic pressure is the pressure in the large arteries during relaxation of the heart and indicates the degree of recoil of the arteries. The average diastolic pressure is 80 mm Hg.

[1] Because the mean arterial pressure is closer to the diastolic rather than the systolic pressure throughout most of the cardiac cycle, the mean arterial pressure is usually slightly less than the arithmetical mean of systolic and diastolic pressures. It is about 95 mm Hg, but the value of 100 mm Hg is used for convenience.

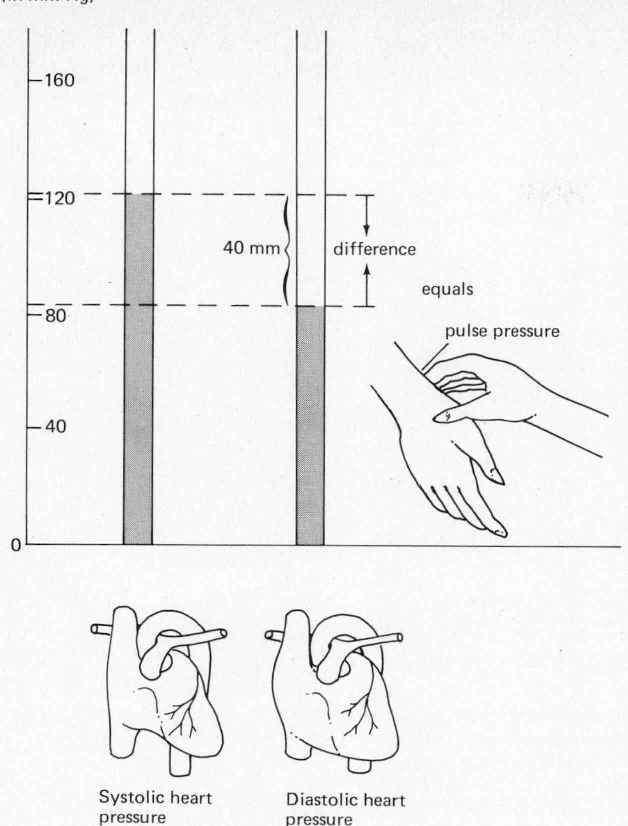

Figure 12-28. **Pulse pressure. (Redrawn from *The Physiology of Work and Play* by S. Riedman. Copyright © 1950 by the Dryden Press, A Division of Holt, Rinehart and Winston. Reprinted by permission of Holt, Rinehart and Winston.**

Pulse pressure, illustrated in Fig. 12-28, is the difference between the systolic and diastolic pressures (40 mm Hg).

The mean arterial blood pressure is dependent upon two main factors, *cardiac output* and *peripheral resistance,* which are depicted in Fig. 12-29.

Cardiac Output

The cardiac output is the amount of blood pumped by the left ventricle into the aorta per minute. Because this is the volume of blood returning to the heart from the lungs, it is also the same volume that was pumped to the lungs by the right ventricle. Consequently, in the normal heart, *the right and left sides of the heart pump the same volume of blood.*

The normal cardiac output for the young adult female, reclining, is about 5 liters/min; that for a male of the same age and size is 5.6 liters/min. Cardiac output is affected by changes in posture and by exercise. In strenuous exercise, the cardiac output of a well-trained athlete may reach levels as high as 35 liters/min. Postural and exercise effects on blood pressure are discussed in more detail later in this chapter.

Factors influencing cardiac output

Cardiac output depends on *heart rate* and *stroke volume.* If the heart rate is 72 beats/min and the volume of blood ejected by each ventricle per beat (stroke volume) is 70 ml, then the cardiac output at rest is 5040 ml/min.

Heart rate is affected by many factors, most of which have already been discussed. They include the baroreceptor reflexes on the bulbar level, the resetting by suprabulbar integrative centers, circulating epinephrine and other hormones, and changes in the temperature and ionic balance of the blood.

Stroke volume depends largely on the volume of venous blood returning to the heart (Starling's law of the heart), the size of the heart, and its contractile strength. Trained athletes usually have hearts larger than normal individuals, because of *hypertrophy* of cardiac muscle fibers as a consequence of constant work against resistance. These muscle fibers have more contractile proteins and myoglobin than is found in untrained individuals, and consequently they contract with greater strength. This increase in the inotropic force of the heart increases the stroke volume and permits the athlete to achieve the same cardiac output with a slower heart rate. This is not the case in pathological enlargement of the heart, with which the athlete's heart should not be confused. An interesting problem is the mechanism by which the rate of the heart is slowed by training; not all individuals respond to the same extent, and it must be associated with a felicitous combination between the vagal inhibitory nerves and the inherent rate of the heart.

The athlete's heart rate may be as low as 45 to 50 beats/min, but the increased stroke volume of 100 to 110 ml/min results in normal cardiac output. The advantage of this is that the athlete has to increase heart rate proportionally less during strenuous exercise than does the untrained individual. This permits better filling of the heart in diastole and contributes, together with the greater strength of the hypertrophied heart, to an enhanced cardiac output. A comparison between the stroke volume, heart rate, and cardiac output during exercise in normal and in trained individuals is illustrated in Fig. 12-30.

Figure 12-29. **Factors influencing mean arterial blood pressure.**

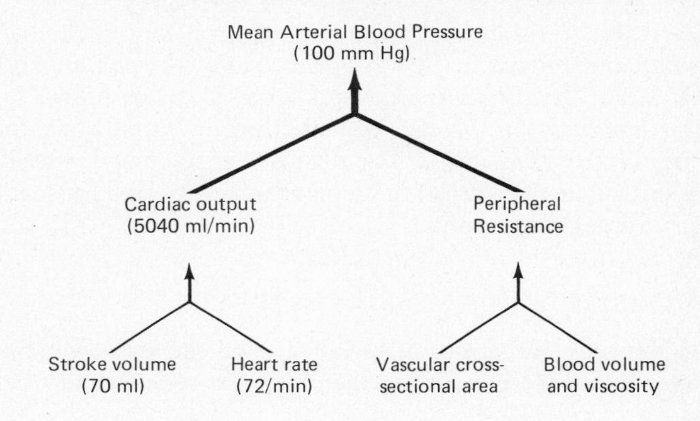

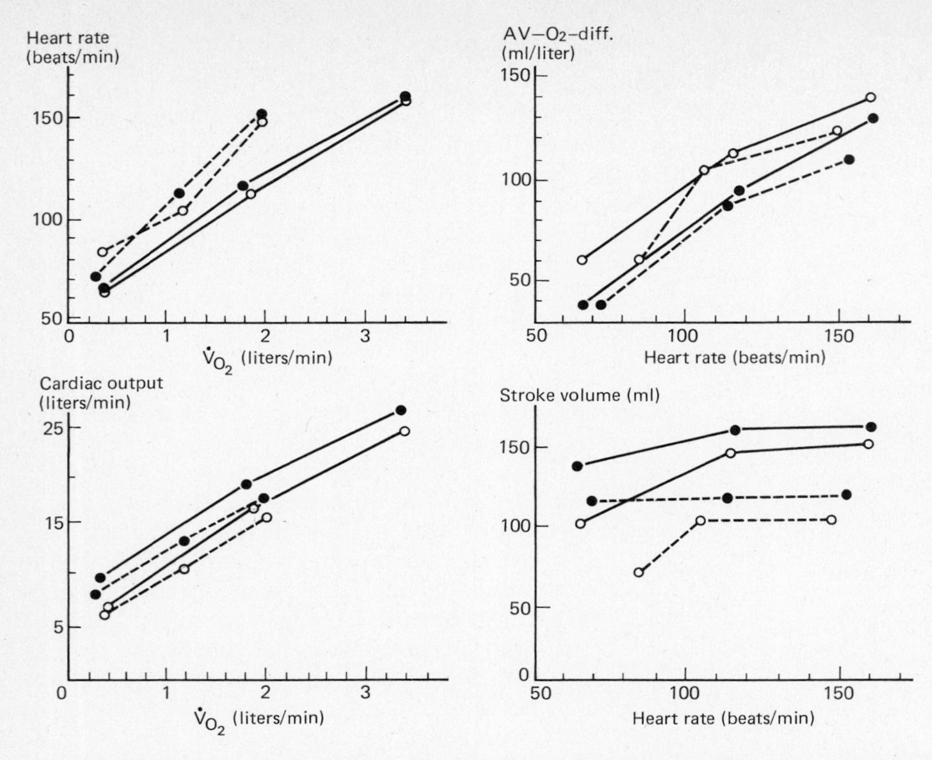

Figure 12-30. **Various cardiac parameters related to posture and exercise in normal man (broken lines) and trained athletes (solid lines). Considerable increase in stroke volume in the erect subjects (open circles) at low-intensity work is obvious. Note also that this paired study of individual subjects shows a small but significant difference of stroke volumes and cardiac output in the two positions even at high heart rates. [From S. Bevegard, A. Holmgren, and B. Jonsson, *Acta Physiol. Scand.* 57:26 (1963).]**

The contractile strength of the heart, as well as its rate, is increased by sympathetic stimulation and by circulating epinephrine, as has previously been discussed. Nervous and hormonal factors, therefore, play an important role in determining stroke volume. However, the most important regulating factor of stroke volume is the control over venous return exerted by the arterioles and other vessels of the peripheral circulation. This was discussed under vasomotion, earlier in this chapter, and will be considered further in the next section on peripheral resistance.

Peripheral Resistance

Peripheral resistance is the resistance of the entire systemic circulation offered to the force generated by the pumping of the heart. Peripheral resistance is inversely proportional to the *diameter of the blood vessels* and directly proportional to the *viscosity of the blood.* The diameter of the blood vessels will regulate the amount of blood returning to the heart and so affects cardiac output.

Diameter of the blood vessels

Not only is the *flow* of blood through the blood vessels affected by the pressure in them and the resistance of the vessel, it also requires a pressure gradient. It is important to realize that there would be no blood flow if the pressure were the same at both ends of the vessel; flow depends on the pressure's being higher at one end than at the other. Figure 12-31 shows the relationship between velocity of blood flow, blood pressure, and cross-sectional area in arteries, capillaries, and veins.

There is very little change in blood pressure as the blood flows through the large arteries and then the smaller ones. However, when the blood reaches the arterioles, the decreased bore of these vessels tremendously increases the resistance to flow, and there is a marked drop in blood pressure. When the blood enters the arterioles, the pressure drops to about 65 mm Hg, and the pulse is damped out because the arterioles have little elasticity. By the time the blood reaches the beginning of the capillary bed, the pressure falls to about 30 mm Hg. At the venous end of the capillary circulation the pressure is only about 10 mm Hg. This venous pressure decreases still further to 0 mm Hg at the right atrium. This great fall in pressure, despite the large size of the veins, is caused chiefly by the compression of the thin-walled veins by other tissues.

Neural Regulation. The role of the *sympathetic vasoconstrictor fibers* in altering the caliber of all the blood vessels

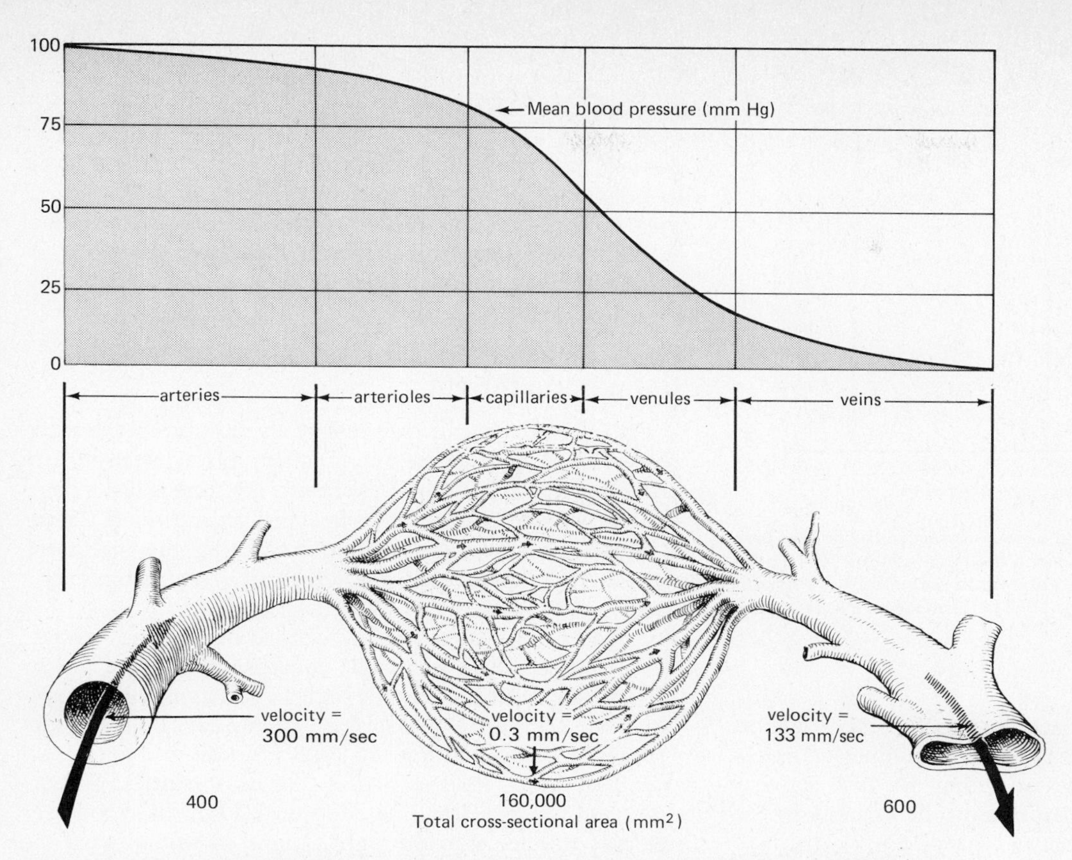

Figure 12-31. **Diagram showing the relationships between the velocity of blood flow and cross-sectional area in arteries, capillaries, and veins. Blood pressure is shown in the upper diagram in mm Hg. (From W. R. Amberson and D. C. Smith, *Outline of Physiology,* 2nd ed., The Williams & Wilkins Company, Baltimore, 1948.)**

Figure 12-32. **Cutting the cervical sympathetic nerves to the left ear results in vasodilation of the blood vessels of that ear.**

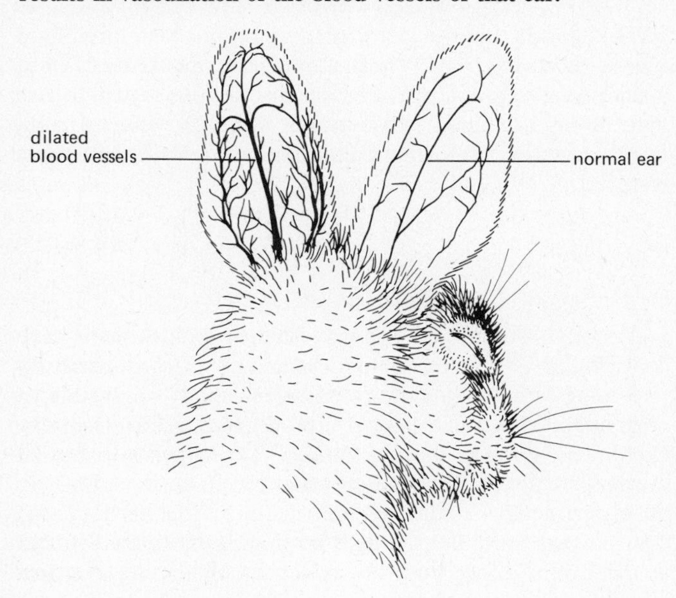

except the capillaries has already been discussed and is illustrated in Fig. 12-32. Because the circulation is a continuous circuit, blood displaced by vasoconstriction in one area is shunted to a region of vasodilation. It is the relative size of constricted and dilated areas at any one time that will determine how much blood is actively circulating and how much is accumulating in venous blood reservoirs. Consequently, sympathetic stimulation, which causes vasodilation in skeletal muscle and vasoconstriction elsewhere, will initially cause an increase in blood pressure due to the more rapid passage of blood through the dilated vessels (lowered peripheral resistance) and the increased venous return as the veins constrict and force blood back to the heart.

Autoregulation. The importance of these neural mechanisms in the maintenance of vasomotion has probably been overemphasized. Granger and Guyton (1969) have shown in a series of convincing experiments that *local control* of blood flow is of greater significance after the initial reflex vasodilation in skeletal muscle. (Remember, too, that this reflex vasodilation does not seem to occur in primates.) If an isolated limb is perfused with oxygen-poor blood, the rate at which the perfused blood flows through the tissues is almost 200 times greater than if the same limb is perfused with

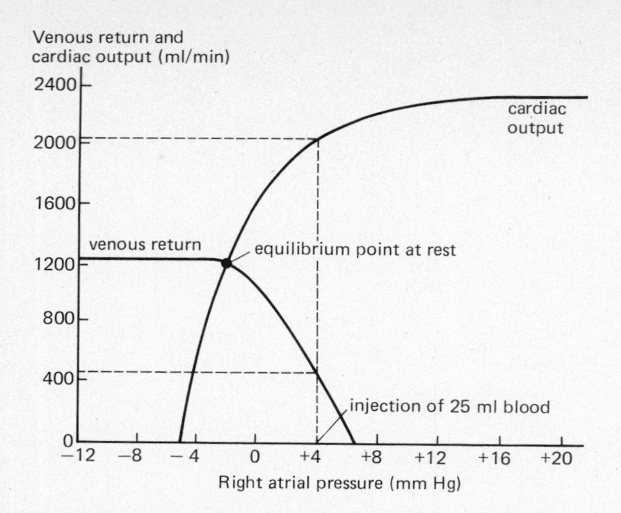

Figure 12-33. **Cardiac output and venous return in the dog, at rest, and after injection of 25 ml blood into the right atrium. Full explanation in text. (Adapted from A. Guyton, in *Handbook of Physiology—Circulation II,* Sec. 2, Vol. II, The American Physiological Society, Bethesda, Md., 1963, p. 1112.)**

oxygen-rich blood. Oxygen lack, increased carbon dioxide, and other factors already discussed permit *autoregulation of blood flow* through the tissues according to their activity. Thus venous return to the heart is controlled mainly by the tissues.

Viscosity of the blood

Viscosity is the resistance of a liquid to flow. Blood is a highly viscous fluid, composed of cells and plasma (Sec. 14-1). The viscosity of normal blood is three to four times greater than that of water, mainly because of the number of red blood cells it contains. The concentration of plasma proteins also contributes to the viscosity of the blood, but not to any significant degree.

In the normal individual, blood viscosity is fairly constant, but changes in altitude and pathological conditions that increase or decrease (polycythemia and anemia, respectively) the number of circulating red blood cells will alter the viscosity of the blood and thus the resistance in the peripheral blood vessels.

Effect of Exercise on Cardiac Output

Strenuous exercise is a very stressful situation, to which all of the regulatory systems respond. The increased cardiac output necessary to supply the exercising muscles with oxygen and nutrients is achieved through:

1. *Neural commands* from the motor cortex to the *sympathetic vasodilator fibers in skeletal muscle,* increasing blood flow through them and doubling venous return.

2. *Neural commands* from the motor cortex to the *hypothalamus,* and consequently to the *bulbar centers,* to increase sympathetic activity. This results in *increased heart rate* and *contractility* and *increased mean arterial pressure.* It also involves a resetting of the baroreceptor reflexes to prevent "buffering" of this increase in arterial blood pressure.

3. *Increased metabolism* of the active tissues causes oxygen lack and metabolite build-up and directly increases local vasodilation, blood flow, and venous return to the heart.

4. *Increased venous return* results in increased cardiac output. The relationship between these two functions is described in the following discussion.

Homeostatic control of venous return and cardiac output

The pressure changes that restore an equilibrium between venous return and cardiac output, after a brief deviation, are illustrated in Fig. 12-33. These results are from a 12-kg dog. The curves show the effect of injecting an additional 25 ml of blood into the right atrium, simulating the increased venous return that would physiologically result from exercise.

1. The solid curves indicate cardiac output and venous return at rest. There is a single equilibrium point at which the flows and the pressures for these two functions are equal. When the right atrial pressure is −2 mm Hg, both venous return and cardiac output are 1200 ml/min.

2. The dotted lines show the changes in cardiac output and venous return when an additional 25 ml of blood are suddenly injected into the right atrium, as would occur in exercise. Right atrial pressure is increased to 4 mm Hg, causing venous return to fall to approximately 500 ml/min. However, the high right atrial pressure increases cardiac output to 2000 ml/min, resulting in a 1500 ml/min discrepancy between venous return and cardiac output.

venous return	500 ml/min
cardiac output	2000 ml/min
difference	1500 ml/min

This means that the heart is pumping out far more blood than is returning to it. Within a few beats, this reduces right atrial pressure to −2 mm Hg, causing venous return to rise to 1200 ml/min and cardiac output to fall to 1200 ml/min; in other words, the original equilibrium described in (1) is restored.

Relative significance of heart rate and stroke volume in increasing cardiac output

Although studies on trained athletes and animals have contributed greatly to our understanding of cardiovascular physiology, the results are not always exactly applicable to normal, untrained individuals. *Athletes* respond to strenuous exercise by *increased stroke volume.* Ths appears to be the mechanism also for trained greyhounds. In an ingenious set of experiments, Donald, Milburn, and Shepherd (1964) showed that these dogs, which are highly motivated to run at maximum effort for five sixteenths of a mile, perform almost as well after all nerves to the heart have been cut. The increased cardiac output necessary for such maximal exertion is met mainly by an increase in stroke volume.

Rushmer's experiments (1962) show, however, that increased stroke volume is not a major factor in normal, healthy, *untrained* humans and dogs. The *normal way to increase cardiac output is through increased heart rate,* together with the *increased extraction of oxygen* from the blood by the active tissues. Only when cardiac rate is reduced by some extraneous mechanism, such as prolonged training or denervation, is increased stroke volume of major significance.

Biofeedback in Cardiovascular Regulation

In the middle 1960s Miller was able to show that the autonomic nervous system obeys the same laws of learning as the somatic nervous system, and that vasomotor responses can be learned. Blood pressure changes that are independent of heart rate changes can be learned, just as it is possible to learn to dilate the vessels of the skin of one ear and constrict those of the other ear. In humans, one of the critical features of autonomic learning is *feedback,* or knowledge of results. If the subject has feedback as to the correctness or incorrectness of the response, learning proceeds much more quickly than if no such biofeedback is provided. It is feasible to utilize this type of training therapeutically: to train patients with hypertension, for example, to lower their blood pressure.

Biofeedback probably can be used to control other systems apart from the cardiovascular system, but the degree to which it is possible varies markedly in individuals. Our understanding of the central mechanisms involved is poor, and biofeedback at the moment is being used as a tool to improve therapeutic results more than as an investigative device into the regulatory mechanisms involved in autonomic system physiology.

Effects of Posture on Blood Pressure

Traditionally, most observations on blood pressure have been made on anesthetized animals lying on their backs, or on recumbent humans. However, the erect position is really the normal, basal condition for healthy humans, and data from quadripeds, even when they are erect and awake, are not always applicable to the human. One of the most important physiological adaptations to the bipedal life is the *ability to maintain blood pressure in the standing position,* despite the fact that the head and heart are considerably above the center of gravity of the body. This adaptive ability of the circulation is reinforced by the daily training provided by a reasonably active life. Prolonged bed rest, during which time the entire body is equally affected by gravity, often results in a temporary inability to adjust to changes in posture. Suddenly sitting up or standing may induce blackout and unconsciousness.

In the *recumbent individual,* more than 50 per cent of the total blood volume is contained within the systemic veins, about 30 per cent in the intrathoracic vessels and less than 15 to 20 per cent in the systemic arteries. Consequently, shifts in blood volume most significantly involve the low-pressure venous system. *Heart size* is greatest in the recumbent position because there is little venous pooling in the legs, and most of the venous blood is in the intrathoracic compartment, available to fill the heart (Fig. 12-34).

When a human is standing, large displacements of blood volume occur, as we said earlier (Sec. 12-9). Intravascular pressures decrease above the right atrium and increase in the dependent parts of the body. The most extensive pressure changes take place in the *leg veins* (Fig. 12-34).

Much of the blood displaced to the legs during prolonged, quiet standing comes from the intrathoracic vascular compartment. This greatly diminishes the amount of blood in the heart and the pulmonary circulation. The stroke volume falls markedly and even the reflex increase in heart rate due to the baroreceptor response can only restore cardiac output to 60 to 80 per cent of its value in the recumbent position. This increased sympathetic stimulation of the heart, aided by sympathetic vasoconstriction, together result in a standing blood pressure slightly higher than that of the recum-

Figure 12-34. **Changes in posture induce great shifts in the distribution of blood in the veins. These shifts are particularly marked between the intrathoracic area and the legs. On standing, the upper parts of the pulmonary circulation and the intrathoracic veins are collapsed and venous pressure increases in the legs. In the recumbent position, the heart is large, since most of the blood is in the intrathoracic area. When the head is down and the legs up, the venous volume of the legs can easily be accommodated in the intrathoracic area. The volume of the intra-abdominal area is relatively constant. G = gravity. (Adapted from O. H. Gauer and H. L. Thron, in *Handbook of Physiology—Circulation III,* Sec. 2, Vol. III, The American Physiological Society, Bethesda, Md., 1965, p. 2415.)**

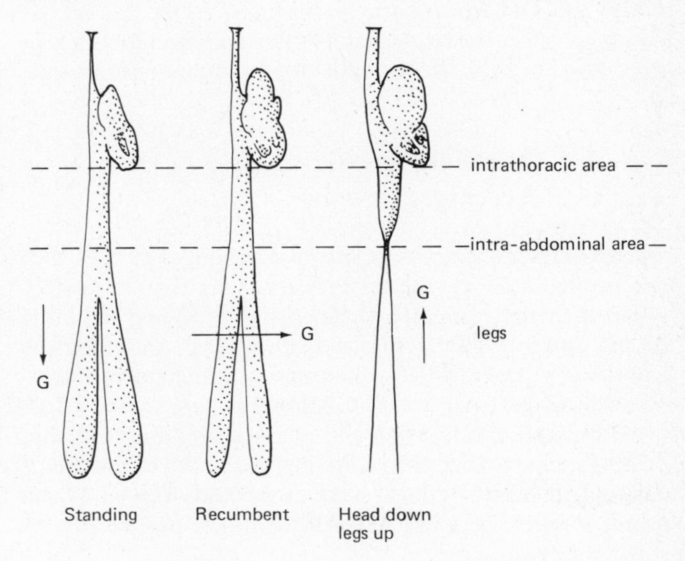

Table 12–2. **The Effect of Posture and Exercise on Blood Pressure, Pulse Pressure, and Pulse Rate. (These figures represent averages. A wide range exists between normal individuals.)**

	Reclining	Sitting	Standing	After Severe Exercise (BP measured sitting)
Arterial blood pressure (mm Hg)	112/70	120/72	125/82	160/90
Pulse pressure (mm Hg)	42	48	43	70
Pulse rate (per min)	66	72	83	140

bent or sitting position. Table 12-2 shows the effect of posture and exercise on some parameters of the cardiovascular system.

Effects of Space on Blood Pressure

When astronauts are subjected to positive gravity (2–6 G) during acceleration, the blood is centrifuged toward the lower part of the body, passively dilating the veins and permitting the blood to pool in the lower abdomen and legs. This shifts a large volume of blood into these venous reservoirs. The cardiac output drops, because the heart does not receive enough blood to maintain its output. Too little blood reaches the brain, rapidly resulting in blackout of vision, followed by unconsciousness.

Special antigravity suits are used to prevent this pooling of blood. Graduated positive pressure is applied, by means of inflatable compression bags, to the legs and lower abdomen as the G increases, forcing the blood back to the heart.

Interestingly enough, the body "interprets" the pooling of venous blood in the extremities as a loss in blood volume, because less blood reaches the *volume receptors* in the great veins and right atrium. These volume receptors, like the baroreceptors already discussed, are sensitive to stretching of the walls of the structures in which they are imbedded. With decreased venous return due to venous pooling, the volume receptors cease firing, reflexly resulting in a rapid release in ADH from the hypothalamus. ADH acts on the kidney to cause the retention of water and the reduction of urine volume (Sec. 15–4), and blood volume is increased.

12–12 Atherosclerosis and Arteriosclerosis

Atherosclerosis and arteriosclerosis are changes in the structure of the walls of the arteries, changes that ultimately impair their function. *Atherosclerosis* is considered to be the leading cause of heart attack and stroke. It is caused by the deposition of lipid, usually cholesterol, in the smooth muscle cells of the artery wall. Cholesterol deposition is followed by some calcification and is topped by a fibrous cap, forming a plaque that protrudes into the lumen of the artery and cuts off blood to those parts of the body served by the diseased vessels (Fig. 12-35). Although the precise role of

Figure 12-35. **An atherosclerotic plaque completely blocks this artery. The plaque is formed from dead cells and fat with a calcified base and a fibrous cap. (From C. Schroeder, *The Sciences,* The New York Academy of Sciences, New York, Sept. 1974, p. 12.)**

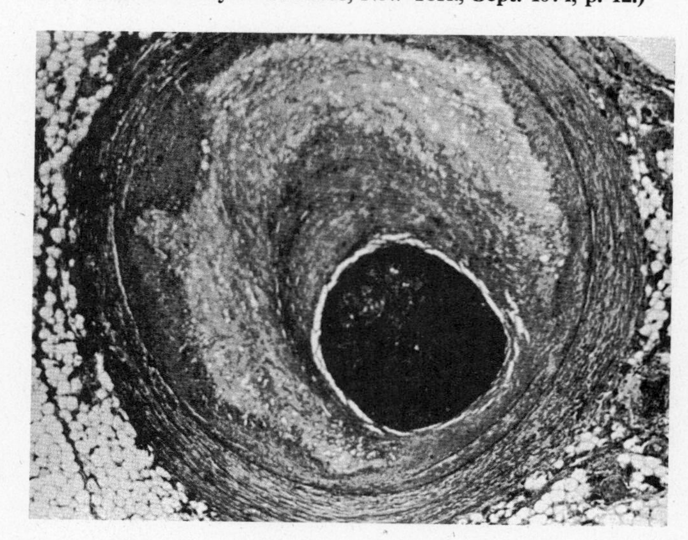

cholesterol is uncertain, the formation of atherosclerotic plaques is accelerated by abnormally high blood cholesterol and by other factors, such as cigarette smoking and high blood pressure. The cumulative effects of these factors is depicted in Fig. 12-36.

Arteriosclerosis is hardening of the arteries due to the deposition of calcium and fibrous tissue within their walls. This reduces the elasticity of the arteries and consequently causes increases in pressure within them. When blood reaches the brain under such high pressure it may rupture the small vessels, depriving the tissues of oxygen and resulting in a "stroke."

12–13 Chronic Control of Blood Pressure

While the baroreceptors and other neural mechanisms are important for the immediate and rapid adjustments of the circulation, chronic control of arterial blood pressure depends on the regulation of body fluids and electrolytes mainly by the kidney. *Autoregulation by the kidney* (dis-

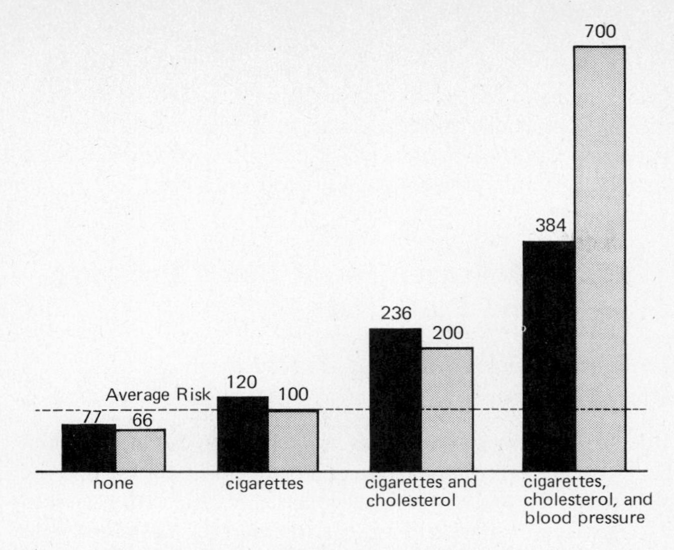

Figure 12-36. **Cigarette smoking, high cholesterol levels in the blood, and high blood pressure increase the likelihood of heart attack (black bar) and stroke (white bar) dramatically. This chart illustrates the combined effect of the three major risk factors on a 45-year-old man with an abnormal blood pressure level of 180 systolic and a cholesterol level of 130. (From C. Schroeder, *The Sciences,* The New York Academy of Sciences, New York, Sept. 1974, p. 10.)**

cussed in Sec. 15-3: juxtaglomerular apparatus) refers to the ability of the renal tissues to control the blood flow through them. When arterial blood pressure drops, the resulting fall in renal blood flow causes the retention of water and electrolytes, which in turn increases blood volume, venous return, and cardiac output, and consequently elevates arterial blood pressure.

A proposed mechanism for this negative feedback system of arterial blood pressure control is:

1. Decreased stretch of the afferent renal arterioles due to lowered arterial blood pressure causes the secretion of an enzyme, *renin,* by the juxtaglomerular cells of the kidney (Sec. 15-3).
2. Renin released into the circulation indirectly increases *aldosterone* secretion by the adrenal cortex.
3. Aldosterone acts on the kidney to increase sodium and water retention, thus increasing blood volume.
4. Renin indirectly causes peripheral vasoconstriction (see renin-angiotensin system, below).
5. Increased blood volume together with vasoconstriction increases venous return and cardiac output, thus elevating arterial blood pressure.

The Renin-Angiotensin System

Renin, a protein with a molecular weight of about 40,000, does not directly affect blood pressure. It acts enzymatically to cleave a decapeptide, *angiotensin I,* from a large plasma protein, *angiotensinogen.* Angiotensin I is converted by enzymes in the lungs into *angiotensin II,* an octapeptide, the most powerful vasoconstrictor agent known and a potent stimulus for the secretion of aldosterone into the circulation (Fig. 12-37). Angiotensin can therefore increase blood pressure by its action on the adrenal cortex, causing increased aldosterone secretion, and/or it can cause vasoconstriction of the arterioles.

Very little renin or angiotensin is found in the plasma under physiological conditions so that the significance of this mechanism in the normal individual is unclear. It is more probably that the renin-angiotensin system is involved in certain types of *hypertension,* which will now be discussed.

12–14 Hypertension

Hypertension, or chronically high blood pressure, is a prevalent and dangerous condition in humans. The high pressure generates an enormous resistance against which the heart must pump. The resulting heavy work load causes hypertrophy of the cardiac muscle fibers, but they do not receive enough blood for their size and energy requirements. The high pressure also damages the walls of the arteries, causing hemorrhage, an especially dangerous condition when it occurs in the brain or kidneys.

Figure 12-37. **The renin-angiotensin system and arterial blood pressure regulation.**

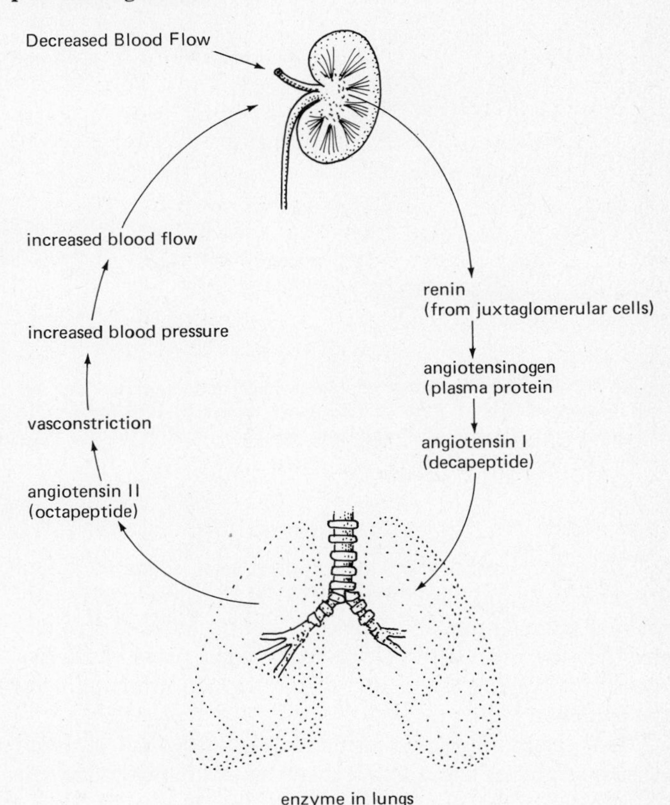

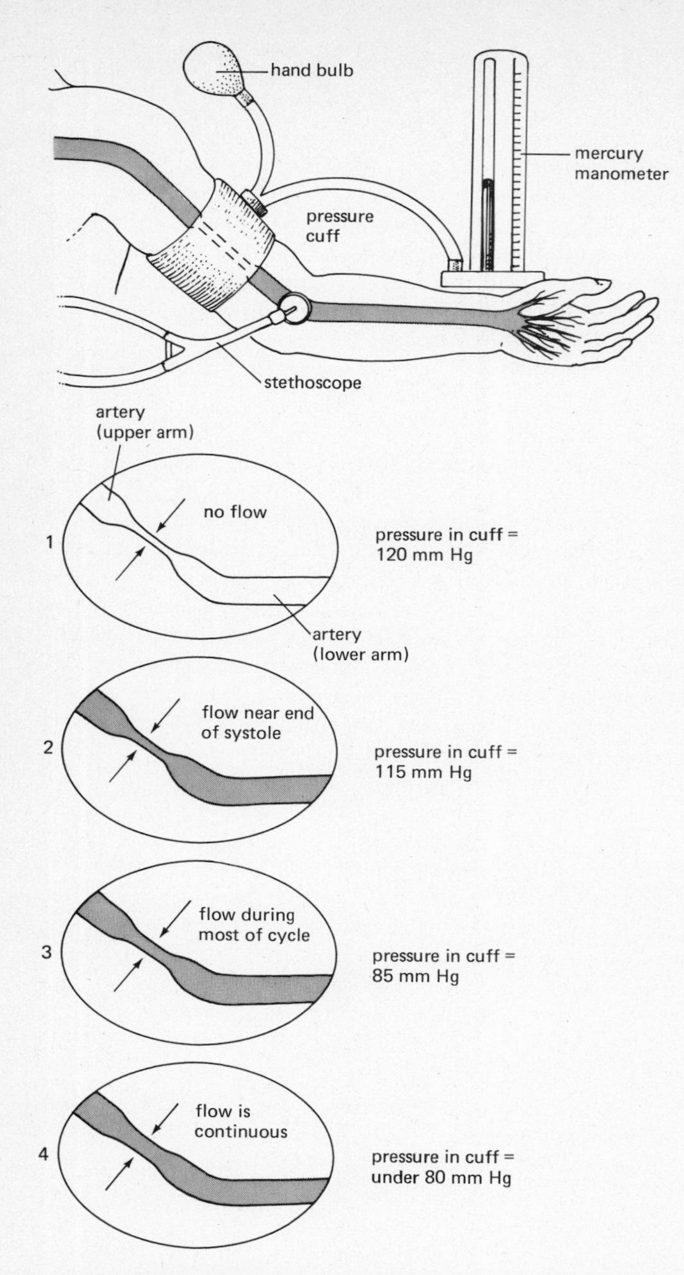

Figure 12-38. **Measurement of blood pressure. See text for an explanation of stages 1 to 4. (Redrawn from S. Riedman, *The Physiology of Work and Play,* Holt, Rinehart and Winston, New York, 1956.)**

There are several types of hypertension, some of which are renal in origin. Goldblatt (1938) demonstrated in experimental animals that hypertension can be induced by constricting the renal artery. The high blood pressure in this *Goldblatt hypertension* is due to the formation of renin by the ischemic kidney.

Many cases of hypertension do not show an elevated plasma renin but instead have disturbances in water and salt retention caused by other kidney abnormalities. It is possible that the most important role of angiotensin lies in its effectiveness in causing the secretion of aldosterone from the adrenal cortex. Aldosterone exerts a profound effect on water and salt metabolism, affecting blood volume and fluid distribution and, consequently, blood pressure.

12–15 Measurement of Blood Pressure and Pulse Rate

Direct Measurement of Arterial Blood Pressure

The first direct method was indeed very direct. Stephen Hales in 1733 connected a piece of brass tubing to the windpipe of a goose, which served as a flexible connection to a long piece of glass tubing. He inserted the brass end into the carotid artery of a horse and reported that the blood rose in the tube until it reached a height of 9 ft 6 in. (2.9 m). (This is somewhat higher than is found in humans and other animals, in whom the blood will rise to a height of about $4\frac{1}{2}$ ft. or 1.4 m.)

Today's techniques for the direct measurement of blood pressure are considerably more refined. A needle is inserted into the artery and connected by rigid tubing to an electronic or optical measuring device. The former is a pressure transducer, which changes the pressure pulses into electrical impulses that can be read accurately. The optical manometer responds to pressure changes by the movement of a beam of light. These movements are recorded on sensitized photographic paper and can measure pressure pulses lasting less than $\frac{1}{400}$ of a second.

Indirect Measurement of Arterial Blood Pressure

The arterial blood pressure may be measured with a *sphygmomanometer,* which consists of an inflatable cuff attached to a mercury manometer, and a stethoscope which amplifies the sounds within the artery. The flow of blood into the brachial artery (chosen for convenience) is temporarily cut off by inflating the cuff around the upper arm (Fig. 12-38).

1. When the stethoscope is placed above the brachial artery at the bend of the elbow, no sound can be heard after the artery is collapsed.

2. The pressure in the cuff is gradually reduced until the blood courses through with each systole of the heart. The artery snaps shut with a thump with each diastole, because the pressure in the cuff is still higher than the pressure in the artery during diastole. The beginning of the thumping sound indicates that the cuff pressure is approximately equal to the *systolic pressure in the artery,* allowing the artery to remain open during the systole of the heart.

3. The pressure is slowly reduced in the cuff, and the sound continues with each cycle.

4. As the cuff pressure is reduced further, the artery remains open even during diastole. The blood courses freely through it, and the thumping sound of the artery closing disappears. The reading on the manometer at the point of disappearance of the sound is the approximate *diastolic pressure of the artery.*

The arterial blood pressure measured in the brachial artery will not be same as that within the aorta or the smaller arterioles. It will vary also with the position of the individual; standing, sitting, and reclining all alter the efficiency of venous return to the heart and, consequently, the blood pressure (Table 12-2). Nevertheless, when the pressure in the brachial artery is measured under standard conditions, it does give a good idea of the condition of the heart and arteries.

Pulse Rate

The pulsation of the blood through the arteries in response to the contraction of the heart can be felt easily in several places in the body where the artery becomes superficial. The most convenient place is at the wrist at the side of the thumb, where the radial artery pulsates against the radius. The number of pulsations per minutes is the *pulse rate* or *heart rate.* It varies considerably between individuals, but the average is about 72 beats/min. Excitement, exercise, and changes in posture as they affect the heart will, of course, affect the pulse rate (Table 12-2).

Cited References

DONALD, D. E., S. E., MILBURN, and J. T. SHEPHERD. "Effect of Cardiac Denervation on the Maximal Capacity for Exercise in the Racing Greyhound." *J. Appl. Physiol.* **19:** 849 (1964).

GOLDBLATT, H. "Studies on Experimental Hypertension: The Production of the Malignant Phase of Hypertension." *J. Exp. Med.* **67:** 809 (1938).

GRANGER, H. J., and A. C. GUYTON. "Autoregulation of the Total Systemic Circulation Following Destruction of the Central Nervous System in the Dog." *Circ. Res.* **25:** 379 (1969).

HEYMANS, C., and E. NEIL. *Reflexogenic Areas of the Cardiovascular System.* Churchill Press, London, 1958.

LOEWI, O. "Über humorale Übertragbarkeit Herznervenwirkung." *Pflüg. Arch. ges. Physiol.,* **189:** 239 (1921).

MILLER, N. E., and A. BANAUZIZI. "Instrumental Learning in Curarized Rats of a Specific Visceral Response, Intestinal or Cardiac." *J. Comp. Physiol. Psychol.,* **65:** 1 (1968).

RINGER, S. "Further Experiments Regarding the Nature of Lime, Potassium, and Other Salts on Muscular Tissue." *J. Physiol.* **7:** 291 (1887).

RUSHMER, R. F. "Effects of Nerve Stimulation and Hormones on the Heart: The Role of the Heart in General Circulatory Regulation." In *Handbook of Physiology,* Sec. 2: *Circulation,* Vol. 1. *Am. Physiol. Soc.,* Washington, D.C., 1962, Chap. 16, p. 533.

STARLING, E. H. *Starling's Principles of Human Physiology.* Churchill Press, London, 1912.

ZWEIFACH, B. W. *Biochemical Mechanisms in Inflammation.* Charles C Thomas, Publisher, Springfield, Ill., 1961.

Additional Readings

BOOKS

BERGEL, D. H., ed. *Cardiovascular Fluid Dynamics.* Academic Press, Inc., New York, 1972.

BERNE, R. M., and M. N. LEVY. *Cardiovascular Physiology,* 2nd ed. Mosby, St. Louis, 1972.

BURCH, G. E., and T. WINSOR. *A Primer of Electrocardiography.* Lea & Febiger, Philadelphia, 1966.

BURTON, A. C. *Physiology and Biophysics of the Circulation,* 2nd ed. Year Book, Chicago, 1972.

FOLKOW, B., and E. NEIL. *Circulation.* Oxford University Press, New York, 1971.

GREGG, D. E. *Coronary Circulation in Health and Disease.* Lea & Febiger, Philadelphia, 1950.

GUYTON, A. C., ed. *MTP International Review of Science: Physiology,* Vols. 1 and 2: *Cardiovascular Physiology.* University Park Press, Baltimore, 1974, 1976.

GUYTON, A. C., C. E. JONES, and T. G. COLEMAN. *Circulatory Physiology: Cardiac Output and Its Regulation.* W. B. Saunders Company, Philadelphia, 1973.

LUISADA, A. A., ed. *Development and Structure of the Cardiovascular System.* McGraw-Hill Book Company, New York, 1961.

NOBLE, D. *The Initiation of the Heartbeat.* Oxford University Press, New York, 1975.

PICKERING, G. *Hypertension,* 2nd ed. Churchill Livingstone, Division of Longman, Inc., New York, 1974.

RUSHMER, R. F. *Structure and Function of the Cardiovascular System.* W. B. Saunders Company, Philadelphia, 1972.

ARTICLES

ADOLPH, E. "The Heart's Pacemaker." *Sci. Am.,* Mar. 1967.

BEVEGARD, B. S., and J. T. SHEPHERD. "Regulation of the Circulation During Exercise in Man." *Physiol. Rev.,* **47:** 178 (1967).

BRAUNWALD, E. "Regulation of the Circulation," Parts I and II. *N. Eng. J. Med.,* **290:** 1124 and 1420 (1974).

BURTON, A. C. "Role of Geometry, of Size and Shape, in the Microcirculation." *Fed. Proc.* **25**(6): 1753 (1966).

CARO, C. G., T. J. PEDLEY, and W. A. SEED. "Mechanics of the Circulation." In *MTP International Review of Science: Physiology,* Vol. I: *Cardiovascular Physiology,* A. C. Guyton, ed. University Park Press, Baltimore, 1974.

CHAPMAN, C. B. and J. H. MITCHELL. "The Physiology of Exercise." *Sci. Am.,* May 1965.

COLEMAN, T. G., A. W. COWLEY, JR., and A. C. GUYTON. "Experimental Hypertension and the Long-Term Control of Arterial Pressure." In *MTP International Review of Science: Physiology,* Vol. I: *Cardiovascular Physiology,* A. C. Guyton, ed. University Park Press, Baltimore, 1974.

GAUER, O. H., and H. L. THRON. "Postural Changes in Circulation." In *Handbook of Physiology,* Sec. 2: *Circulation,* Vol. 3. *Am. Physiol. Soc.,* Washington, D.C., 1965, p. 2409.

GUYTON, A. C., G. G. ARMSTRONG, and P. L. CHIPLEY. "Pressure-Volume Curves of the Entire Arterial and Venous Systems in the Living Animal." *J. Physiol.* **184:** 253 (1956).

GUYTON, A. C., T. COLEMAN, and H. GRANGER. "Circulation: Overall Regulation." *Ann. Rev. Physiol.,* **34:** 13 (1972).

HILTON, S. M. "Hypothalamic Regulation of the Cardiovascular System." *Brit. Med. Bull.* **22**(3): 243 (1966).

JOHNSON, P. C. "The Microcirculation and Local and Humoral Control of the Circulation." In *MTP International Review of Science: Physiology,* Vol. I: *Cardiovascular Physiology,* A. C. Guyton, ed. University Park Press, Baltimore, 1974.

KORNER, P. I. "Integrative Neural Cardiovascular Control." *Physiol. Rev.* **51**(2): 312 (1971).

KORNER, P. I. "Keynote Address: Present Concepts About the Myocardium." *Adv. Cardiol.* **12:** 1 (1974).

LUFT, J. H. "The Ultrastructural Basis of Capillary Permeability." In *The Inflammatory Process,* B. W. Zweifach, L. Grant, and R. T. McCluskey, eds. Academic Press, Inc., New York, 1965.

PEART, W. S. "Renin-Angiotensin System." *N. Engl. J. Med.* **292:** 302 (1975).

SCHER, A. "Excitation of the Heart." In *Handbook of Physiology,* Sec. 2: *Circulation,* Vol. 2, *Am. Physiol. Soc.,* Washington, D.C., 1962, p. 287.

SCHER, A. "The Electrocardiogram." *Sci. Am.,* Nov. 1961.

SMITH, O. "Reflex and Central Mechanisms Involved in the Control of the Heart and Circulation." *Ann. Rev. Physiol.* **36:** 93, 1974.

SPAIN, D. M. "Atherosclerosis." *Sci. Am.,* Aug. 1966.

WARREN, J. V. "The Physiology of the Giraffe." *Sci. Am.,* Nov. 1974.

WOOD, J. E. "The Venous System." *Sci. Am.,* Jan. 1968.

ZWEIFACH, B. W., "The Microcirculation of the Blood." *Sci. Am.,* Jan. 1959.

ZWEIFACH, B. W., and D. B. METZ. "Selective Distribution of Blood Through the Terminal Vascular Bed of Mesenteric Structures and Skeletal Muscle." *Angiology* **6:** 282 (1955).

Chapter 13

Special Circulatory Regulation

Surgeons must be very careful
When they take the knife!
Underneath their fine incisions
Stirs the culprit, —Life!

Emily Dickinson, "Surgeons Must Be Very Careful"

THE BASIC PLAN of the circulation is that oxygen-rich blood is brought to an organ by an artery through the driving force of the contracting heart. The artery penetrates into the substance of the organ, breaking up into numerous arterioles and capillary networks. The capillary bed is so widespread that all the cells are close enough to it to receive the diffusing dissolved gases and nutrients and to be relieved of their secretions and wastes. These are drained into venules which return the oxygen-poor blood via the veins to the heart. The return of the venous blood to the heart is due to several factors, among which is the contraction of the muscles surrounding the veins.

This basic plan is modified for several organs of the body that require supplementary circulation because of their greater metabolic activity. The *brain*, the *liver*, and the *heart* itself are prominent in this group. There is also the very special case of the pulmonary circulation: the *lungs* receive mainly arterial blood that is oxygen-poor, whereas the venous blood draining back to the heart is rich in oxygen, the blood having been aerated during its passage through the pulmonary capillaries.

The blood supply to the developing *fetus* in placental animals also represents a very special adaptation of circulation, because the venous blood from the mother is filtered through the placental blood vessels. This relatively oxygen-poor ultrafiltrate passes through the umbilical blood vessels to reach the tissues of the fetus before it is returned to the maternal lungs for oxygenation.

Finally, when the normal circulatory route to an organ is blocked, as happens with certain diseases, collateral pathways may develop that are of vital importance for the survival of the tissues affected.

13–1 The Systemic Circulation: Arterial System

The Aorta and Its Branches

The main artery of the body is the aorta, which arises from the left ventricle as the *ascending aorta*, curves backward and toward the left side of the body as the *aortic arch*, and then runs down through the thorax and abdomen as the *descending aorta*. The descending aorta ends by bifurcating into two *common iliac arteries*, which supply the legs. All the arterial blood received by the body must first pass through the aorta and one of its branches. This forms the *systemic circulation* (Fig. 13-1).

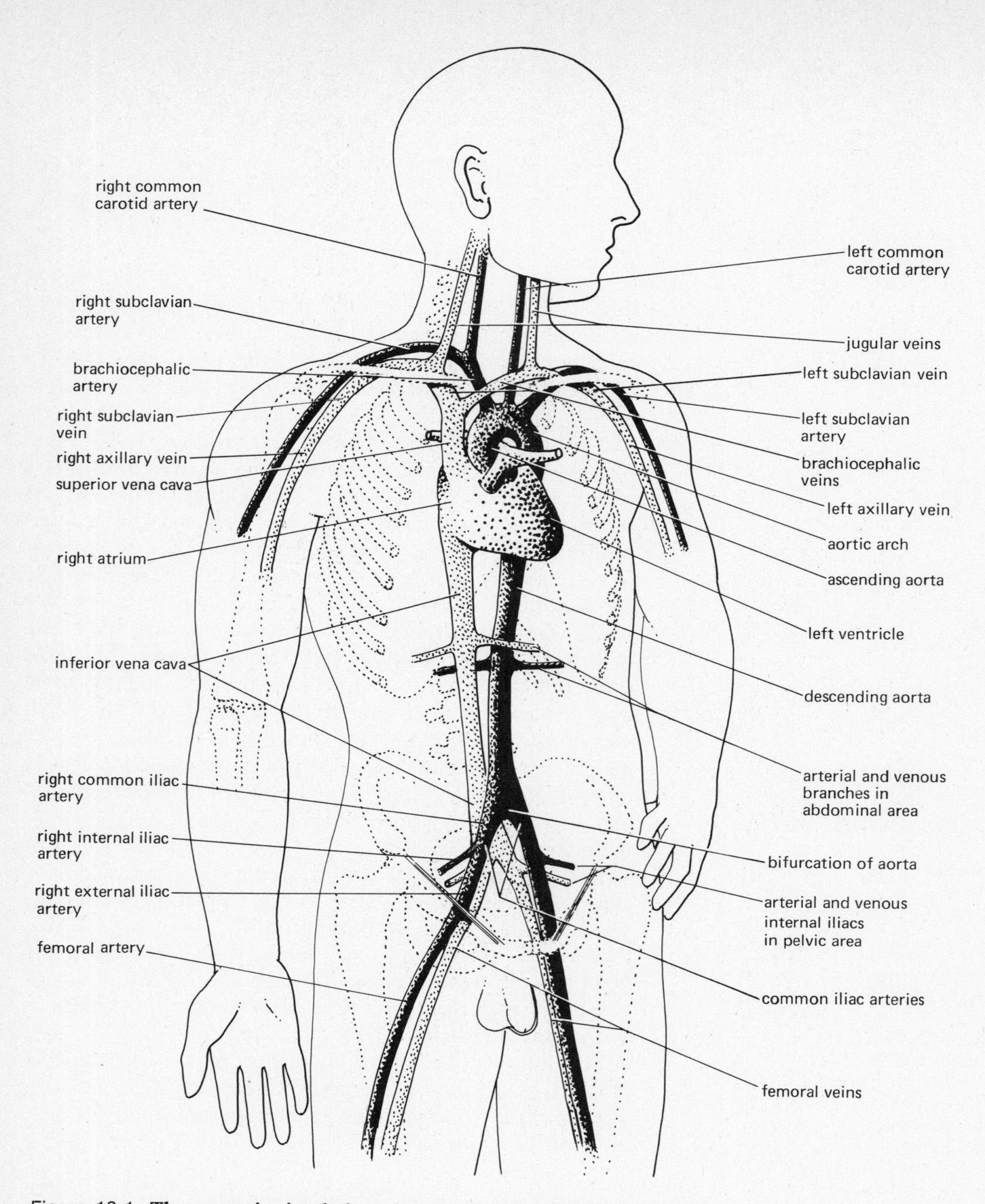

Figure 13-1. **The systemic circulation, showing the main arteries and veins.**

THE ASCENDING AORTA

This artery springs from the base of the left ventricle, and two branches from it supply the heart itself with arterial blood. These are the *right and left coronary arteries.*

THE ARCH OF THE AORTA

This begins at the level of the junction of the sternum with the second costal cartilage (second rib) and passes backward across the trachea and esophagus to end at the level of the fourth thoracic vertebra. In this short space it gives off three very large and important blood vessels that supply the head, neck, and upper arms.

1. The *brachiocephalic artery* is the first branch of the arch of the aorta. It is short and wide and divides into the *right subclavian* and *right common carotid arteries.*
2. The second branch of the aortic arch is the *left common carotid artery.*
3. The third branch is the *left subclavian artery.*

The asymmetry in arterial branches to the right and left side of the body is caused by a peculiarity in the phyloge-

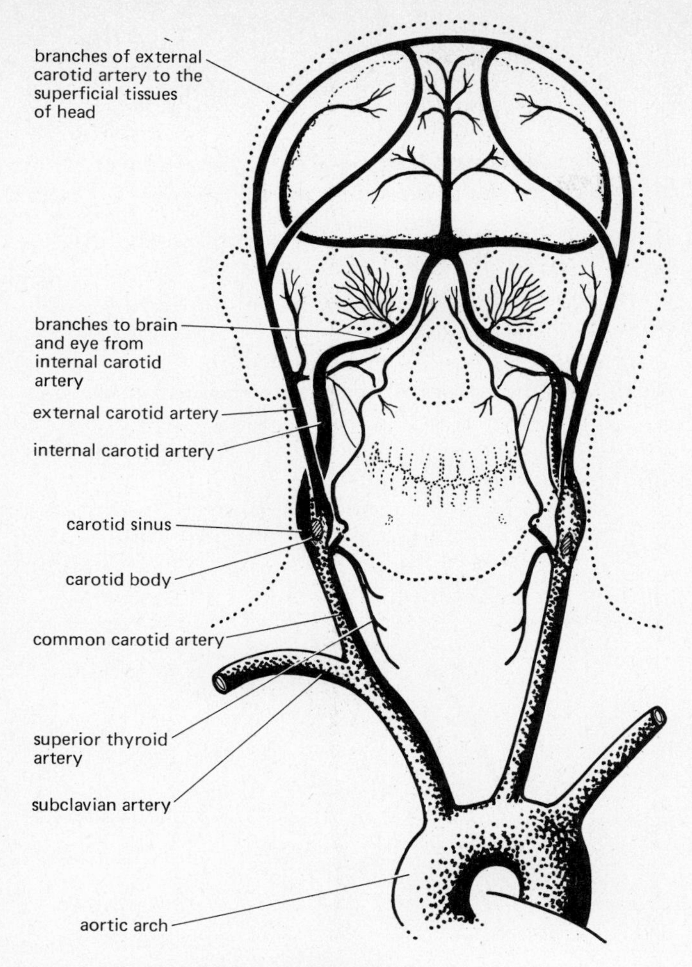

Figure 13-2. **The arteries of the head.**

netic development of the arterial system in higher animals. In lower animals, the two sides are connected by a symmetrical series of aortic arches, but many of these disappear or fuse as the circulatory system becomes adapted to the requirements of more active animals.

The descending aorta

This part continues from the arch as it runs dorsally down through the thorax and abdomen to end as the common iliac arteries at the level of the fourth lumbar vertebra. Both the thoracic and abdominal parts of the descending aorta give off numerous branches to the organs along their way.

The Arch of the Aorta: Arteries to the Head

The blood supply of the head is derived chiefly from the two common carotids, each of which divides into an *internal* and *external carotid artery* (Fig. 13-2). The internal carotid arteries bring blood to the brain and the orbit of the eye, whereas the external carotids supply the more superficial tissues of the head and neck, including the facial muscles, the muscles of the jaws, and the thyroid gland. Other arteries supplement this blood supply to the head (see Sec. 13-3).

At the junction of the internal carotid with the common carotid the walls are dilated and contain pressure receptors that are innervated by the glossopharyngeal nerve. This is the *carotid sinus,* and it is part of a reflex mechanism regulating blood pressure. Near the bifurcation of the internal and external carotids is the small neurovascular *carotid body,* innervated by both the vagus and the glossopharyngeal nerves. The carotid body is associated with reflex control of

Figure 13-3. **The blood vessels of the arm.**

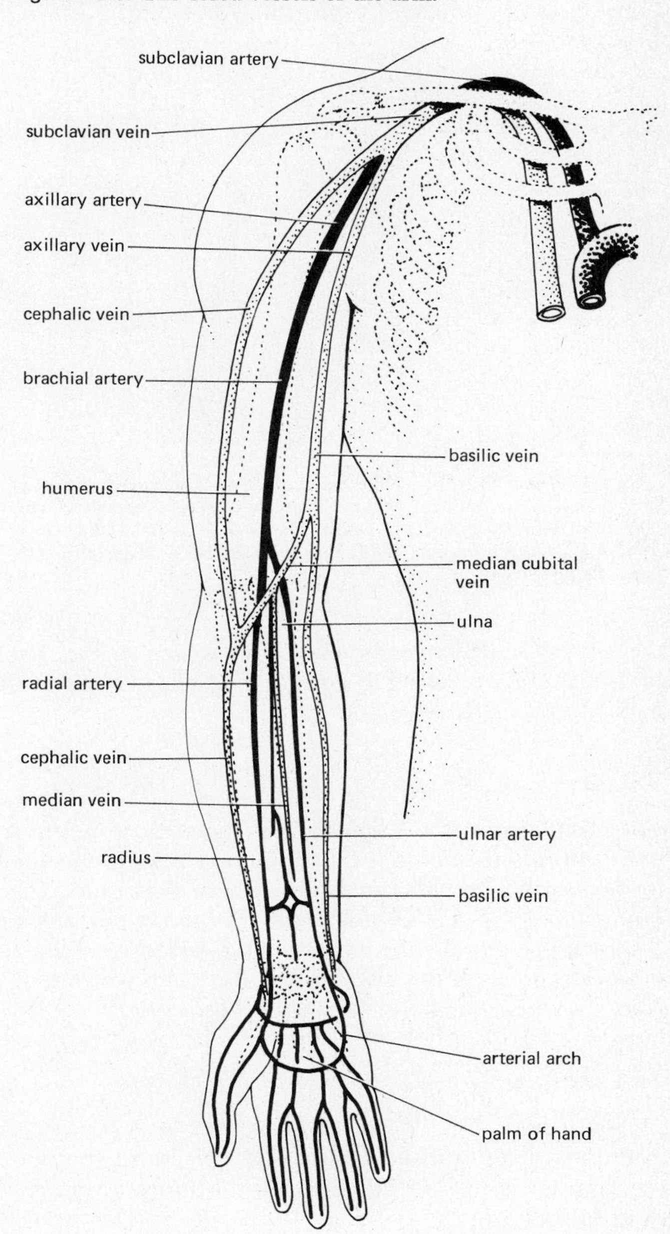

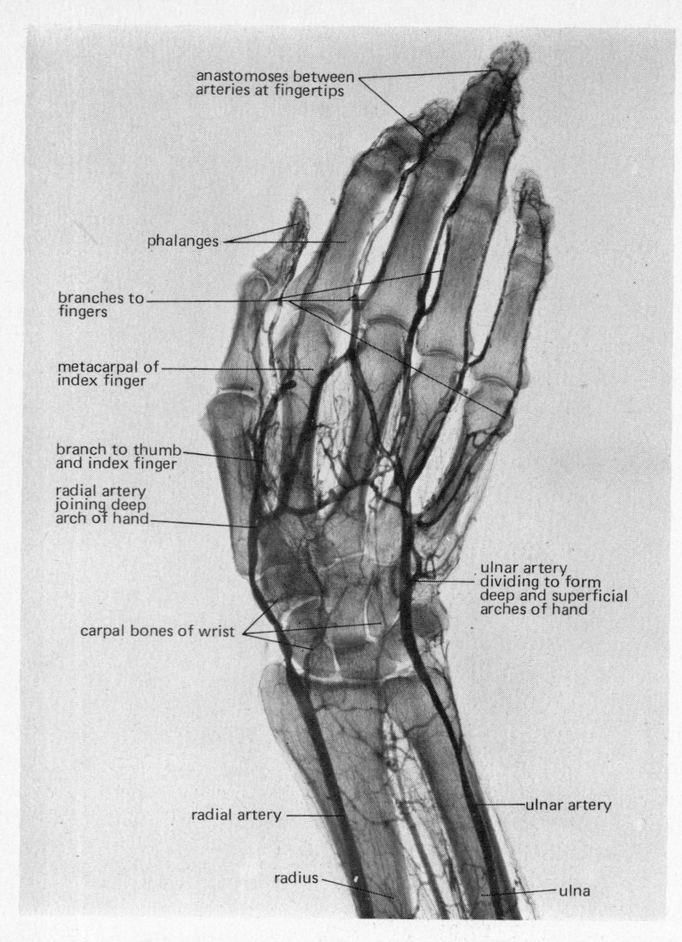

Figure 13-4. **Radiograph of the hand after radiopaque injection of the arteries. The arterial arches of the hand are especially clear. (From J. Dankmeijer, H. J. Lammers, and J. M. F. Landsmeer, *Prаčtische Ontleedkunde,* De Erven F. Bohn, N. V., Haarlem, Holland, 1955.)**

the respiration under conditions of low oxygen tension, but probably is also involved in some of the circulatory reflexes of the carotid sinus.

The Arch of the Aorta: Arteries to the Arm

The blood supply of the arm starts as the *subclavian artery* and continues through the axilla, where it is known as the *axillary artery,* to the upper arm, where it is called the *brachial artery* (Fig. 13-3). Opposite the head of the radius, the brachial artery divides into the *radial and ulnar arteries,* which run through the forearm into the hand, where they form the arterial arches of the hand (Fig. 13-4).

The Descending Aorta—Thoracic Part: Arteries to the Thorax

This runs dorsally from the level of the fourth thoracic vertebra to the level of the fourth lumbar vertebra (Fig. 13-5).

It gives off numerous paired branches to the wall of the thorax and to the organs within it, supplying the lungs, pericardium, esophagus, and mediastinum.

The Descending Aorta—Abdominal Part: Arteries to the Abdomen and Pelvis

Before the descending aorta ends as the common iliac arteries, it gives off *visceral branches,* to the abdominal and pelvic organs; and *parietal branches,* to the abdominal and pelvic walls (Fig. 13-5).

Figure 13-5. **The arteries of the thorax, abdomen, and pelvis.**

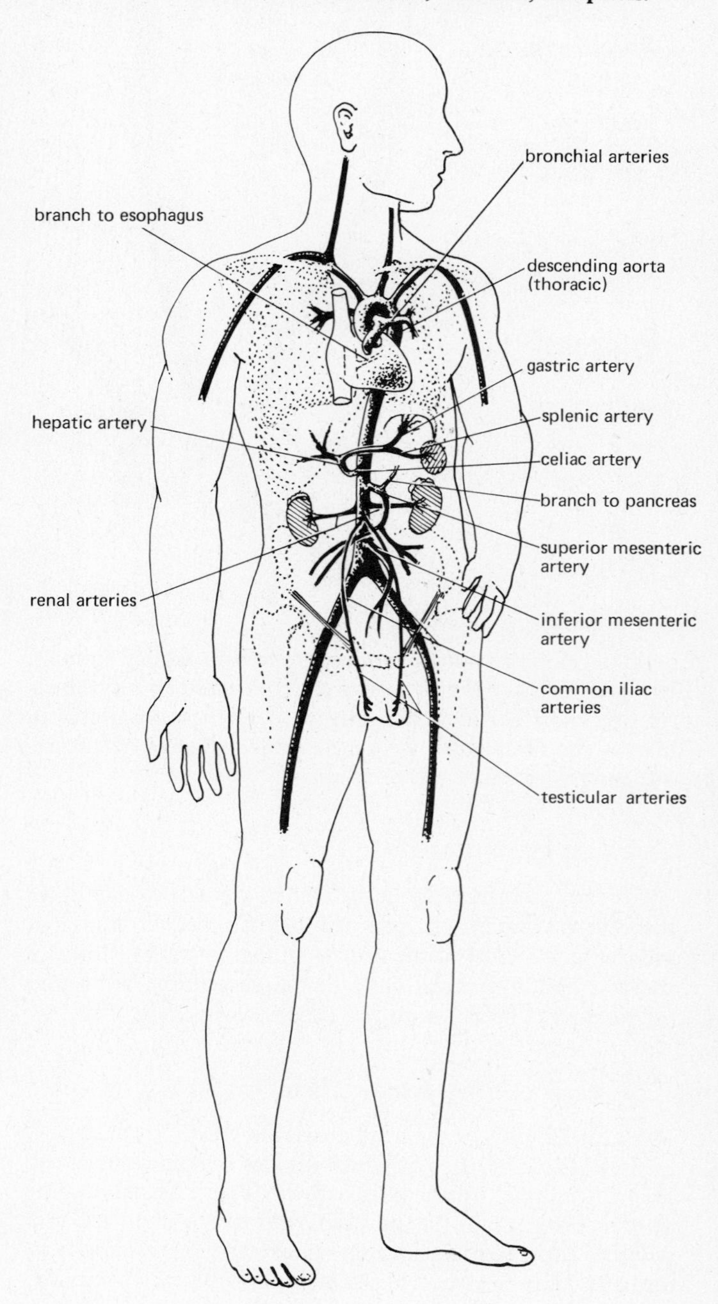

Visceral branches

The abdominal aorta gives off paired branches to the kidneys (*renal arteries*), to the adrenals (*adrenal arteries*), and to the gonads (*testicular or ovarian arteries*), and single branches to the organs of the digestive tract.

Celiac Artery. This is the first of these single visceral branches; it supplies the stomach, liver, and spleen with arterial blood. These arteries are the *gastric, hepatic,* and *splenic arteries.*

Superior Mesenteric Artery. This springs from the front of the abdominal aorta, about $\frac{1}{2}$ inch lower than the celiac artery. The superior mesenteric artery passes downward and forward to supply the viscera. Its branches bring arterial blood to all of the small intestine, and part of the duodenum, the pancreas, and the colon.

Inferior Mesenteric Artery. This artery arises from the front of the abdominal aorta about 3.8 cm above the bifurcation. It supplies the descending colon and the rectum.

Parietal branches

These branches supply the walls of the abdominal cavity. They include branches to the diaphragm (the *phrenic arteries*), five pairs of *lumbar arteries* that supply the vertebrae and muscles of this area, and a single median *sacral artery* that supplies the sacrum and coccyx.

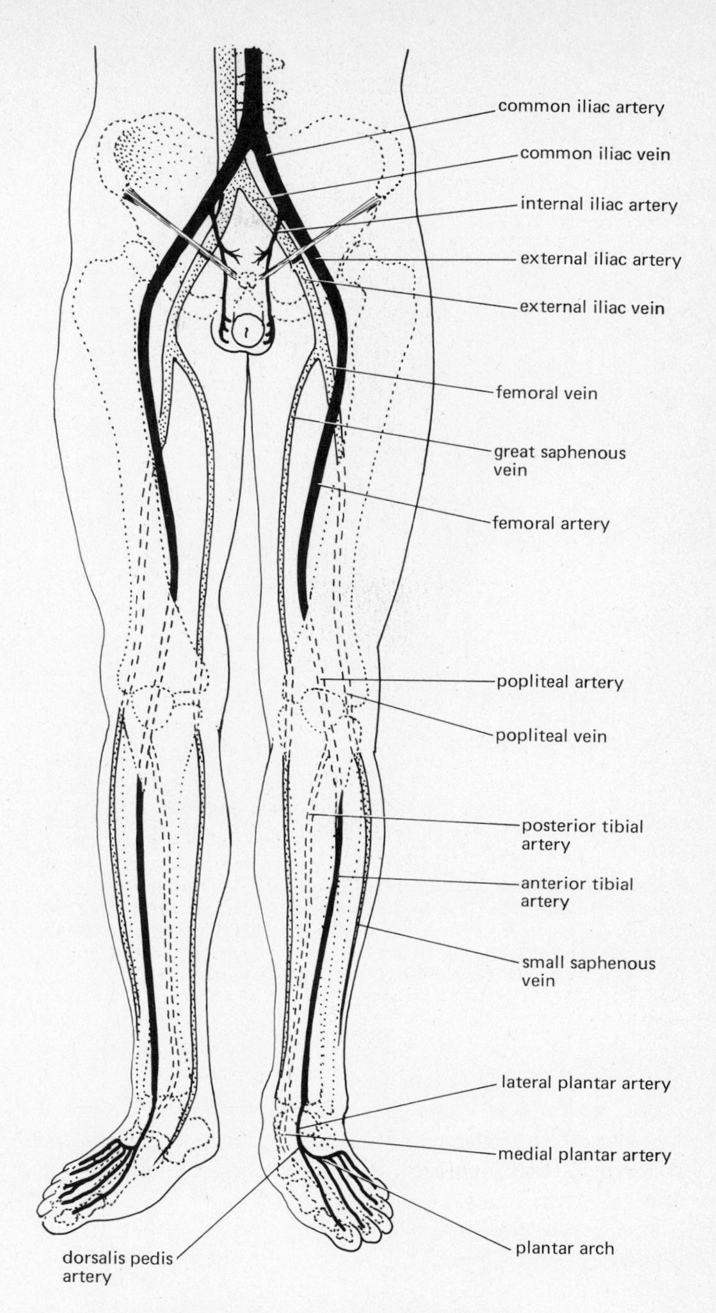

Figure 13-6. **The arteries and veins of the legs.**

Arteries to the Legs

The common iliac arteries

The common iliac arteries are the terminal branches of the aorta (Fig. 13-6). They start at the level of the fourth lumbar vertebra and end by dividing into the *external and internal iliac arteries* at the level of the lumbosacral promontory.

The internal and external iliac arteries

The internal iliac artery supplies the viscera and walls of the pelvis. Branches also supply the gluteal region and the external genitalia.

The external iliac artery continues downward through the pelvis to become the main artery of the lower extremity, the *femoral artery,* just above the pubic symphysis.

The femoral artery and its continuations

The femoral artery is the continuation of the external iliac into the thigh. It begins at the inguinal ligament and ends at the opening in the adductor magnus. It is enclosed, together with the femoral vein and nerve, in a *sheath of fascia* as it emerges from the pelvic area. The femoral artery gives off branches to the fascia and muscles of the pelvis, including the genitalia, and to the medial and posterior muscles of the thigh.

The *popliteal artery* is the continuation of the femoral behind the knee; it sends off branches to the knee before dividing into the *posterior tibial artery,* which supplies the muscles of the back of the leg, and the *anterior tibial artery* which, by passing forward from its origin at the back of the

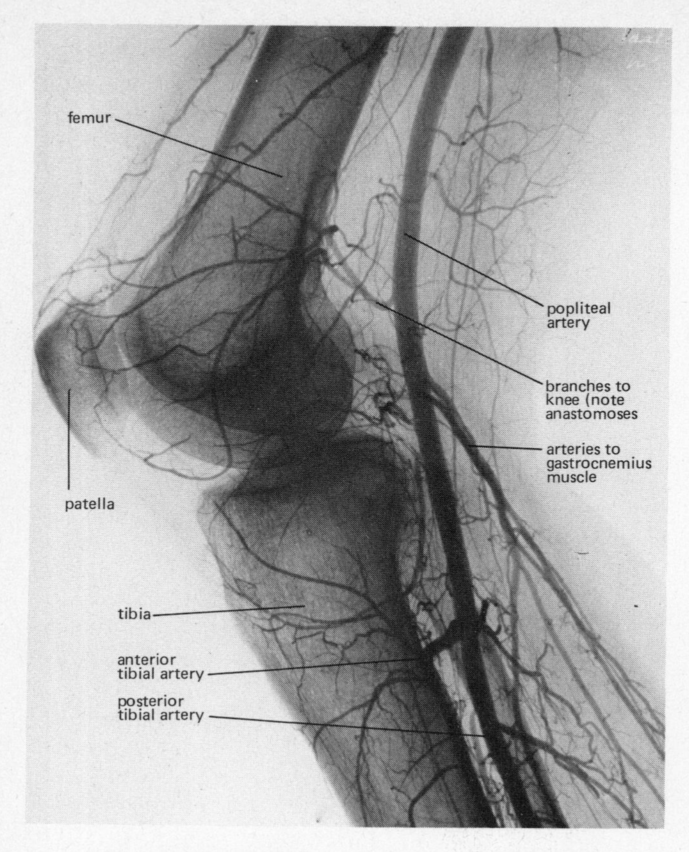

Figure 13-7. **Lateral radiograph of the knee after radiopaque injection of the arteries. (From J. Dankmeijer, H. J. Lammers, and J. M. F. Landsmeer, *Pračtische Ontleedkunde,* De Erven F. Bohn, N. V., Haarlem, Holland, 1955.)**

leg, brings blood to the anterior muscles of the lower leg, including the ankle (Figs. 13-6 and 13-7).

The *medial and lateral plantar arteries* are the terminal branches of the posterior tibial artery and so supply the plantar (or lower) surface of the foot and the medial (or inner) surface.

The *dorsalis pedis artery* is the continuation of the anterior tibial artery and supplies the front of the ankle and the toes; it finally unites with the lateral plantar artery to form the plantar arch of the foot.

13–2 The Systemic Circulation: Venous System

Between the arteries and the veins are the all-important *capillary beds* and *venous plexuses* that divide the organs into areas small enough to have their environmental needs cared for by the process of filtration. The capillaries reunite as veins, two of which usually accompany the artery. The veins often have the same name as the supplying artery, but they are usually more numerous, because they form a *double system* of drainage, a deep and a superficial system.

All the venous blood from the upper part of the body drains into the *superior vena cava,* whereas the blood from the abdomen, pelvis, and legs is ultimately emptied into the inferior vena cava. The venae cavae, in turn, empty into the *right atrium* of the heart (Fig. 13-1).

Veins of the Leg

Blood drains from the venous arches of the foot into the superficial and the deep veins of the leg (Fig. 13-6). The superficial veins are the *saphenous veins,* which often become congested and tortuous, the condition being known as varicose veins. The small saphenous vein empties into the *popliteal vein* behind the knee. The great saphenous vein flows into the *femoral vein* near the *external iliac vein.*

Veins of the Pelvis

The *internal iliac veins* receive blood from the walls and viscera of the pelvis, the external genitalia, and the gluteal and sacral areas (Fig. 13-8). In addition, the internal iliac receives some of the venous blood from the thigh, although most is returned via the external iliac to the common iliacs. The *common iliac veins* are formed by the union of the external and internal iliacs, and they extend from the brim of the pelvis to the level of the fifth lumbar vertebra. The common iliacs empty all the blood from the lower limbs into the *inferior vena cava.*

The inferior vena cava receives all the blood from the lower limbs and most of the venous blood from the abdomen and pelvis. It ascends through the abdomen, piercing the diaphragm and the pericardium to end at the back of right atrium of the heart. The most important tributaries of the inferior vena cava, apart from the two common iliac veins, are the *hepatic veins,* the *right phrenic vein,* the *right suprarenal vein,* the *right and left renal veins,* the *right testicular or ovarian vein,* and the *third and fourth lumbar veins.*

An important connection between the superior and inferior venae cavae is the *azygos vein* (Fig. 13-8). This vein leaves the back of the inferior vena cava about the level of the renal veins and passes up through the diaphragm to the thorax, where it enters the superior vena cava. On its way up it receives tributaries from veins draining the thorax. The azygos vein may become the main pathway for blood from the lower part of the body if the inferior vena cava becomes obstructed.

Veins of the Upper Part of the Body

THE SUPERIOR VENA CAVA AND ITS BRANCHES

The superior vena cava receives the blood from the head and neck, the arms, the thoracic wall, and part of the upper portion of the abdomen (as can be expected from the asymmetry in the tributaries we have cited for the inferior vena

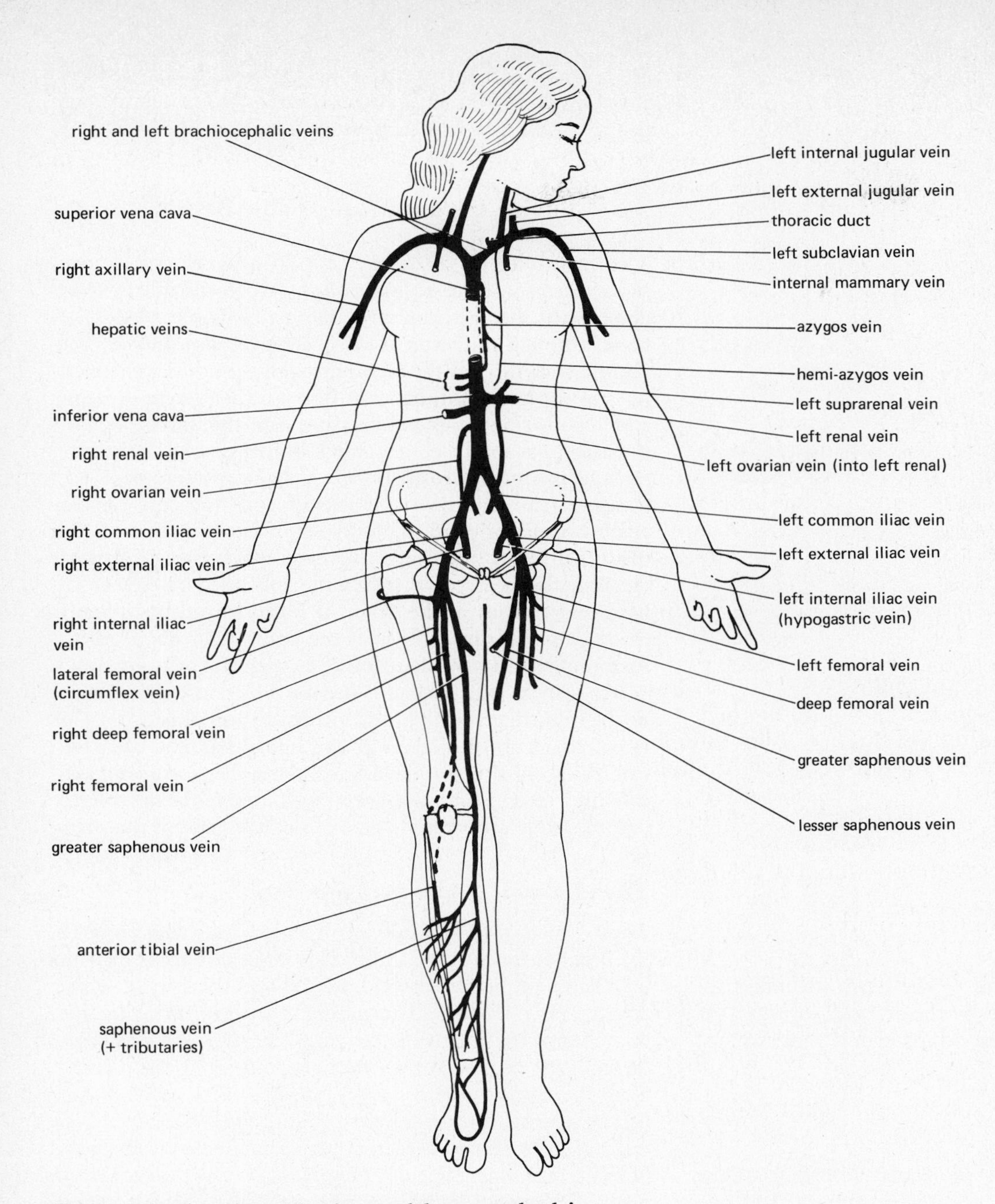

Figure 13-8. **The veins of the thorax, abdomen, and pelvis.**

cava). The upper part of the body is drained by a system of veins that ultimately empty into the *superior vena cava.* This large vein is formed by the fusion of the two *brachiocephalic veins,* which in turn receive the blood from the *jugular and subclavian veins.* The left brachiocephalic also receives the lymph from the thoracic duct, which opens into it at the junction of the internal jugular and subclavian veins. A smaller amount of lymph enters the right brachiocephalic vein through the *right lymphatic duct.* Each brachiocephalic receives blood from the veins that drain the viscera of the thoracic cavity, including some of the vertebrae.

The veins of the head and neck

The brain is supplied with a triple system of veins: two deep and one superficial (Fig. 13-9). The two deep systems lie within the cranial cavity. They are:

1. The veins of the brain that lie in the folds of the pia matter and in the subarachnoid space.
2. The *venous sinuses,* which are channels between the double layers of the dura mater. The veins empty into the venous sinuses which in turn empty into the *internal jugular vein.*

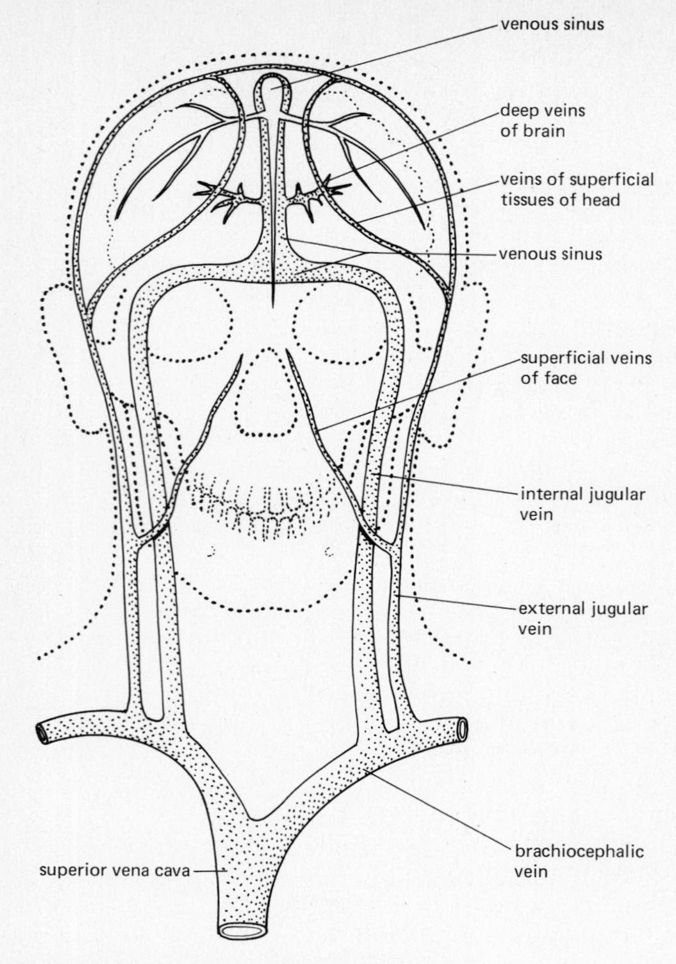

Figure 13-9. **The veins of the head and neck.**

The more superficial tissues of the head, including the face and the neck, are drained by the *external jugular vein,* which joins the *subclavian vein,* which brings venous blood from the arms.

THE VEINS OF THE ARM

The deep veins of the arm are named according to the arteries they accompany and are usually connected to one another across the artery by many *anastomoses,* direct connections (Fig. 13-3). Both the deep and the superficial veins of the arm empty into the *axillary vein,* which is continued as the *subclavian vein.*

The superficial veins begin in the veins of the hand, and although the pattern varies considerably in individuals, in general two main veins run along the forearm and upper arm:

1. The *cephalic vein* on the radial side of the arm.
2. The *basilic vein* on the ulnar side.

There is sometimes a *median vein* running along the front of the forearm to the elbow. The cephalic and the basilic veins are connected by a thick channel in the front of the elbow, the *median cubital vein.* When *venesection* (cutting of the vein) is performed, it is usually the median cubital vein or the basilic vein that is opened.

13–3 The Circulation of the Brain

The sensitivity of the brain cells to a slight decrease in oxygen tension or to an increase in carbon dioxide content emphasizes the importance of the extensive circulation of blood through the brain. This is achieved not only by the rapid distribution of the blood through the arterial channels but also by the rapid drainage through the system of veins and venous sinuses. Blood pressure is of the utmost significance, too, as might be imagined. Sufficient blood pressure must be maintained to assure adequate circulation; when the pressure falls, faintness and loss of consciousness rapidly result because of the lack of oxygen in the brain tissues. On the other hand, too high a pressure, whether generated through the heart or through the increased resistance of hardened arteries in the brain, is liable to cause a hemorrhage of one of the arteries of the brain. This deprives the surrounding areas of the required oxygen and nutrients, and death of these cells (*necrosis*) occurs. The consequence may be the immediate death of the victim, or partial paralysis, or impaired mental function. If the injury has not been too extensive, small anastomosing channels develop from surrounding arteries and supply enough blood to the neurons to permit a slow recovery of some of the functions of this area.

The Arterial Supply of the Brain

Approximately 650 to 700 ml of blood/min flow through the brain of the average adult. This large amount is distributed through the two carotid arteries and the two vertebral arteries (Fig. 13-10). The vertebral arteries are the first branches of the subclavian artery. One rises on either side through the neck, and they fuse at the level of the pons to form the *basilar artery.*

The internal carotid arteries terminate as the *anterior and middle cerebral arteries,* each of which supplies specific areas of the cerebral hemispheres (Fig. 13-11). The basilar artery gives off branches to the pons and cerebellum before ending as the *posterior cerebral artery* which, together with the anterior and middle cerebral arteries, completes the arterial supply to the cerebral hemispheres.

The Circle of Willis

A series of anastomoses between these cerebral and basilar arteries forms a circle around the pituitary gland on the ventral surface of the brain. This circle was first described by Willis in 1664 and so bears his name. The arrangement provides a copious supply of blood to the pituitary gland and ensures that blood will still reach the cerebrum if one of these major vessels is obstructed.

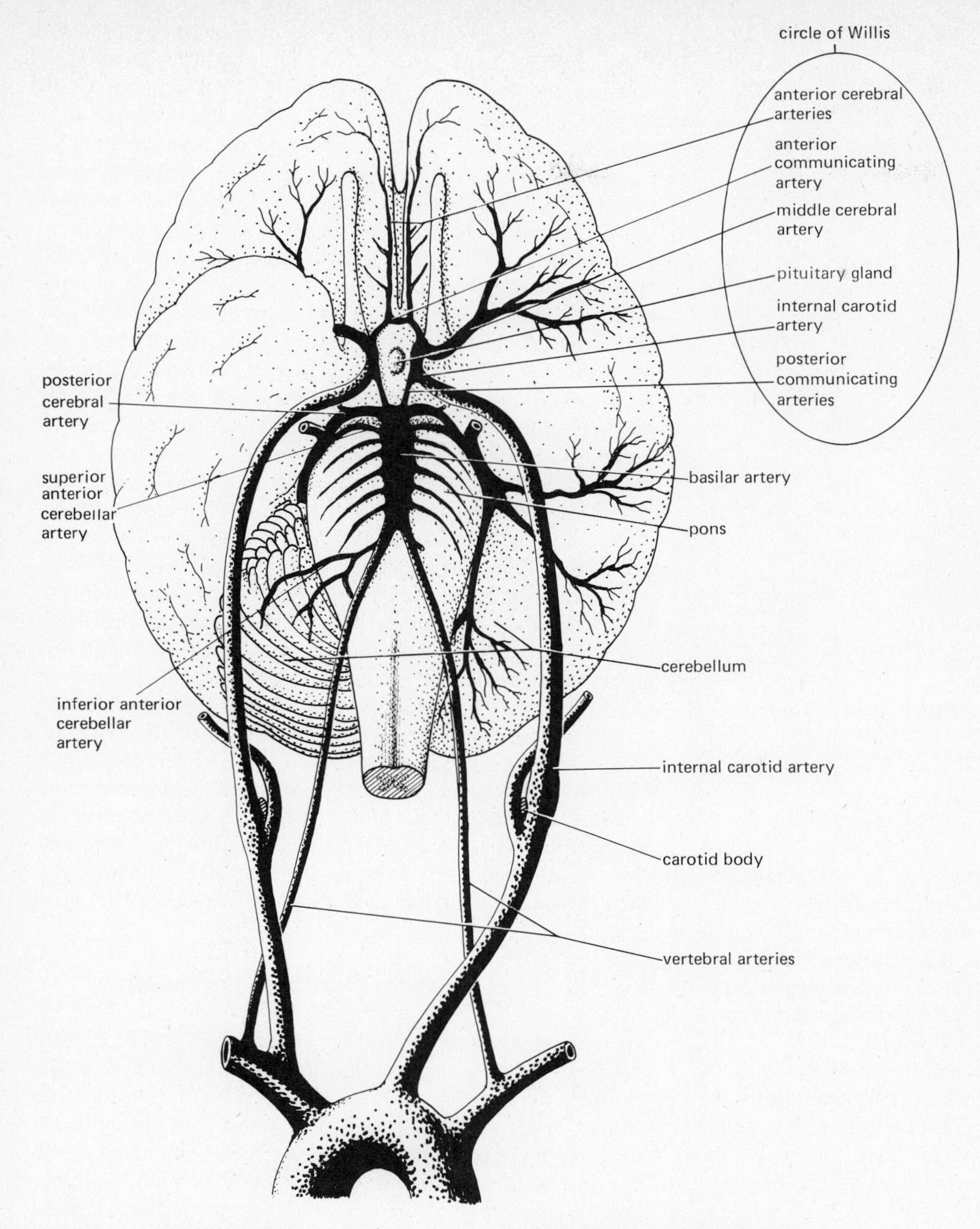

Figure 13-10. **The arterial circulation of the brain, showing the circle of Willis.**

The anastomoses are between the carotid system of arteries and the basilar system (*posterior communicating arteries*) and between the right and left anterior cerebral arteries (*anterior communicating artery*). (See Fig. 13-10.)

The Hypophyseal Portal System

The hypophysis, or pituitary gland, is regulated to a large extent by the hypothalamus, to which it is connected by the blood vessels and nerves that pass through the infundibular stalk (Fig. 10-3). Because the nerves enter the posterior lobe only, direct nervous control by the hypothalamus is exerted only on this part of the pituitary gland. The circulation of the posterior lobe is quite separate from that of the hypothalamus and anterior lobe.

The hypophyseal portal circulation closely connects the hypothalamus and the anterior lobe of the pituitary gland, making possible the chemical regulation of the anterior lobe by factors secreted in the hypothalamus.

Like the liver, the anterior lobe receives both arterial and

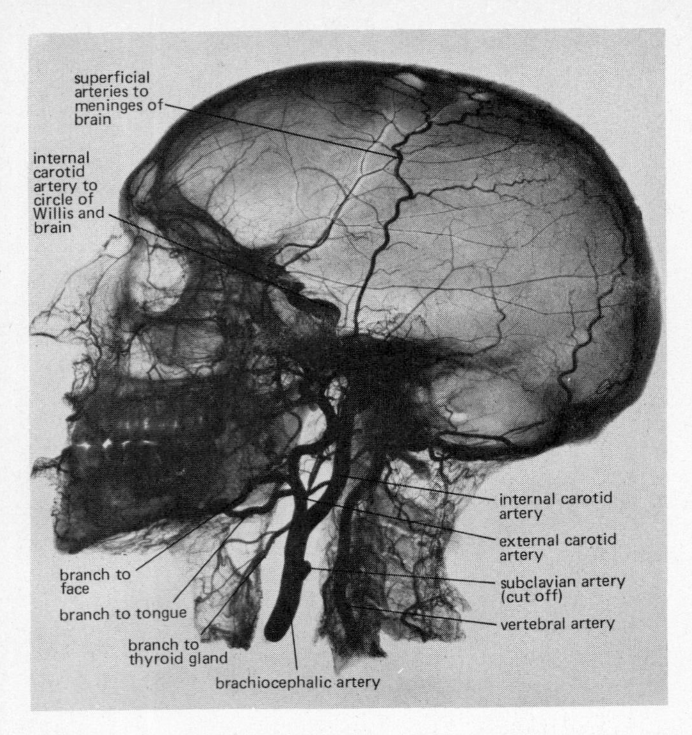

Figure 13-11. **Radiograph of the head after radiopaque injection of the arteries. (From J. Dankmeijer, H. J. Lammers, and J. M. F. Landsmeer, *Pračtische Ontleedkunde,* De Erven F. Bohn, N. V. Haarlem, Holland, 1955.)**

venous blood, the latter being delivered by the portal system. Arterial blood is brought to the anterior lobe through the *superior hypophyseal arteries,* which arise from the circle of Willis and the internal carotids. These arteries end in the sinusoids of the anterior lobe and also as an extensive capillary network in the stalk. These capillaries are continuous with the capillary bed of the hypothalamus.

Venous blood from the hypothalamus passes through the *portal venules* back to the anterior lobe, carrying with it the chemical factors that either stimulate or inhibit this gland. The anterior lobe is drained by veins that empty its hormone-laden blood into the venous sinuses of the brain.

13–4 The Coronary Circulation

The muscle of the heart itself requires a constant supply of blood delivered to the muscle fibers; the blood within the lumen of the heart passes through too rapidly to be able to nourish the heart, although there may be some flow through channels in the ventricles called the thebesian veins. Cardiac muscle receives blood from the *right and left coronary arteries,* which arise from the bulge or sinus in the aorta as it leaves the left ventricle (Fig. 13-12). The right coronary artery supplies the right side of the heart with about 20 per cent of the total coronary blood, whereas the left coronary artery delivers the main volume of the coronary blood to the left and right sides of the heart. The *veins* of the heart drain into the short, wide *coronary sinus,* which lies in the groove between the left atrium and the left ventricle, and which empties into the right atrium. There is some vestige of a valve between the coronary sinus and the right atrium.

Coronary occlusion is the plugging of the coronary blood vessels, and the degree of damage will depend on whether the occlusion is gradual, in which case a collateral blood supply has time to develop, or whether it is sudden, in which case great damage to the cardiac muscle occurs. A gradual occlusion may be due to the progressive deposition of fats containing cholesterol and cholesterol salts along the lining of the arteries (atherosclerosis). This condition is followed by the formation of fibrous tissue scars, which act as a trap for the beginning of a blood clot. As the clot grows larger, it gradually blocks off the blood vessel. Sometimes a clot formed elsewhere in the body may break loose into the circulation and become lodged in one of the coronary arteries. This is called an *embolism* and causes an immediate occlusion. The cells deprived of blood (*ischemia*) become nonfunctional, but if the patient is confined to absolute rest, these cells may recover because most of the blood flowing into the coronary vessels can flow through the normal tissue and allow anastomotic channels to develop. The degree of recovery from the coronary occlusion will depend on the size of the ischemic area and the efficiency of development of new channels in the heart muscle. The judicious use of a machine that helps pump the blood during the critical stages of recovery from a heart attack may be of great importance in permitting cardiac muscle to recover. The pumping machine is attached to an artery in the leg and is synchronized with the heart, relieving it of about 40 per cent of its task.

13–5 The Hepatic Portal Circulation

The viscera act as a reservoir for blood which can be rapidly diverted to skeletal muscle and the skin during exercise. In addition to the spleen, which can store a considerable amount of blood, the portal system draining the blood from the digestive tract into the liver has been estimated to be able to hold as much as one third of the total blood volume. This reserve blood in the spleen and portal system is called the *splanchnic reserve.*

The hepatic portal system consists of the *hepatic portal vein,* the *hepatic sinusoids,* and the *hepatic veins* (Fig. 13-13). The hepatic portal vein is a short, wide vessel that receives the blood from the veins draining the stomach and intestine, spleen and pancreas. It enters the liver, where it breaks up into many venules which empty the blood into the extensive hepatic sinusoids (see Sec. 18-2). The sinusoids are drained by the hepatic veins, which open into the *inferior vena cava* just below the diaphragm.

In addition to the nutritive-rich venous blood brought into the hepatic sinusoids via the portal system, the liver also receives arterial blood from the *hepatic artery,* a branch of the celiac artery. This means that the blood in the hepatic

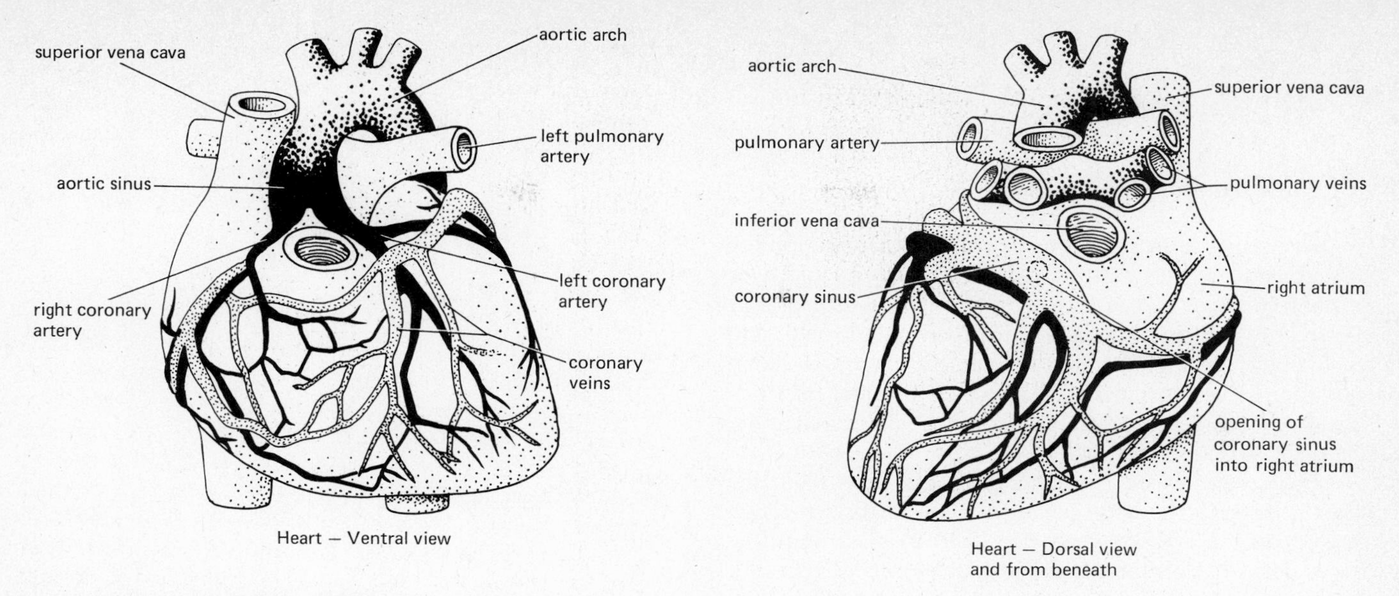

Figure 13-12. **The coronary circulation.**

Figure 13-13. **The hepatic portal circulation.**

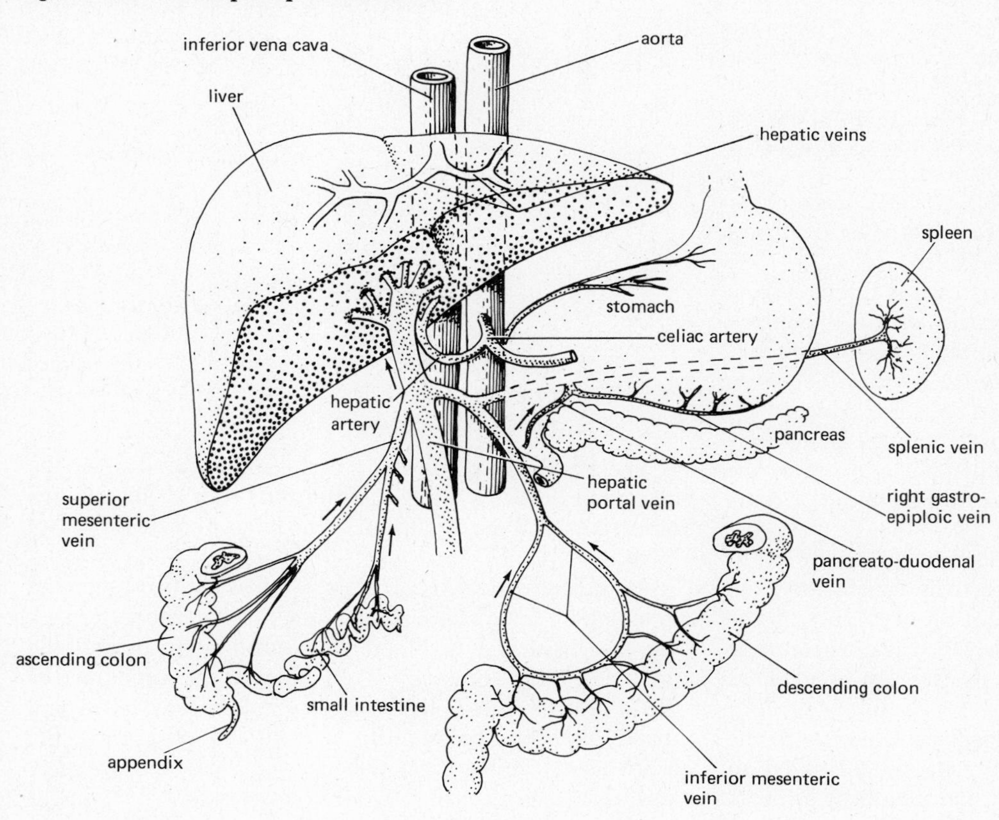

sinusoids not only contains the absorbed foods from the digestive tract but also an adequately high oxygen tension for the many metabolic reactions that occur in the liver cells.

13–6 The Fetal Circulation

The developing fetus relies on the placental circulation for both oxygen and nutrients because the fetal lungs and liver are nonfunctional before birth (Fig. 13-14). These organs receive only the blood necessary for their own growth and development during this period, but the placental blood circulation is very extensive. This means that the fetal circulation is considerably different from the adult circulation. The chief variations follow.

1. *Oxygenated blood* is delivered to the fetus from the placenta through the large *umbilical vein.* It flows through the liver to the inferior vena cava. The flow through the liver is not only through the hepatic sinusoids and veins but also through a large direct channel, the *ductus venosus.* Most of the blood flowing through the ductus venosus actually bypasses the liver tissue. From the inferior vena cava, the blood is delivered to the right atrium.

Because the blood in the maternal placenta has flowed through the sinuses of the placental tissue, it is obvious that it must have lost a fair amount of oxygen en route, so that the fetus receives only partly oxygenated blood. Nevertheless, the blood in the umbilical vein has the highest oxygen tension in the fetal circulation.

2. Circulating through the heart, the blood coming into the right atrium from the superior vena cava follows a different path from the blood entering the same chamber from the inferior vena cava. The blood from the inferior vena cava coming from the umbilical vein has more oxygen than that returning from the head and arms through the superior vena cava. Most of this *oxygen-rich blood goes from the right atrium to the left atrium through the foramen ovale,* bypassing the lungs. From there it reaches the left ventricle and is pumped directly into the aorta.

The blood from the superior vena cava is directed from the right atrium into the right ventricle, then it is pumped into the pulmonary arteries and so to the lungs. In the fetus, there is a connecting channel, the *ductus arteriosus,* between the pulmonary artery and the arch of the aorta. Most of the blood pumped from the right ventricle into the pulmonary artery therefore reaches the descending aorta to supply the rest of the body. The fetal blood is returned to the placenta through the *umbilical arteries.*

Despite the flow of blood from the right side to the left in the fetal heart, the result is unexpectedly satisfactory: the heart itself, the head, and the brain receive most of the oxygen-rich blood from the inferior vena cava, whereas the less active tissues are supplied by the oxygen-poor blood from the superior vena cava.

There are therefore three important passageways in the circulation of the fetal blood that must be closed off to accommodate the violent change to independence at birth. These three passageways are:

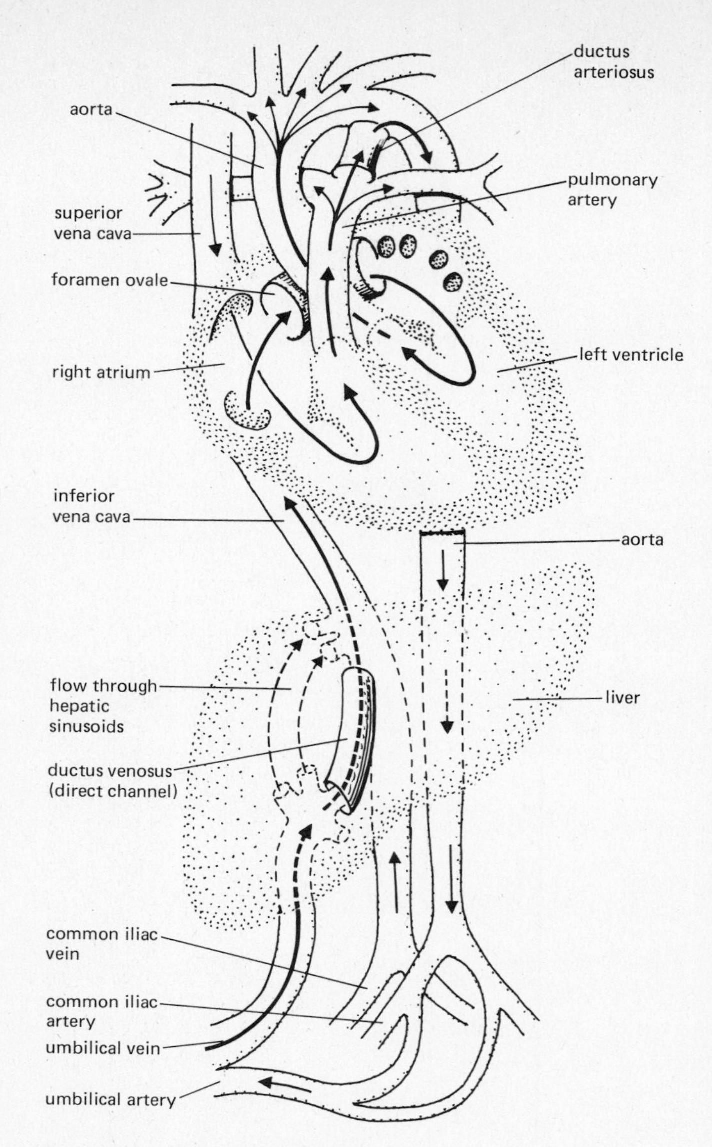

Figure 13-14. **The fetal circulation.**

1. The ductus venosus.
2. The foramen ovale.
3. The ductus arteriosus.

The chief cause for their occlusion is, of course, the change in pressure resulting from the tying off of the umbilical cord and the expansion of the lungs immediately after birth.

As is discussed in Sec. 16–7, as soon as the baby succeeds in taking its first difficult breath, the lungs expand, decreasing the resistance to blood flow through them, and the pressure in the pulmonary artery falls. At the same time, as the umbilicus is tied off, the pressure in the aorta rises, and this difference in pressure in the right and left sides of the heart closes the flap over the foramen ovale and probably also causes the closure of the ductus arteriosus. The ductus venosus, the direct channel through the liver to the inferior vena cava, degenerates slowly, since no more blood flows through it once the umbilical vessels are severed.

13–7 Patterns of Arterial Supply to the Organs of the Body

There are three main patterns of blood supply to the organs of the body, and the vulnerability of each organ to an interruption in its blood supply is dependent mainly on the type of its supply.

Double Blood Supply

Some organs, like the liver and the lungs, have a double blood supply. The liver receives blood from the hepatic artery and the portal vein; the lungs receive blood through the pulmonary arteries and the bronchial arteries. Any blockage (occlusion) of one of these systems is unlikely to lead to an *infarction.* An infarction is the death of the tissues in a localized area because of deprivation of blood supply. The second system is usually adequate to care for the needs of the tissues.

Figure 13-15. **The lymphatic circulation.**

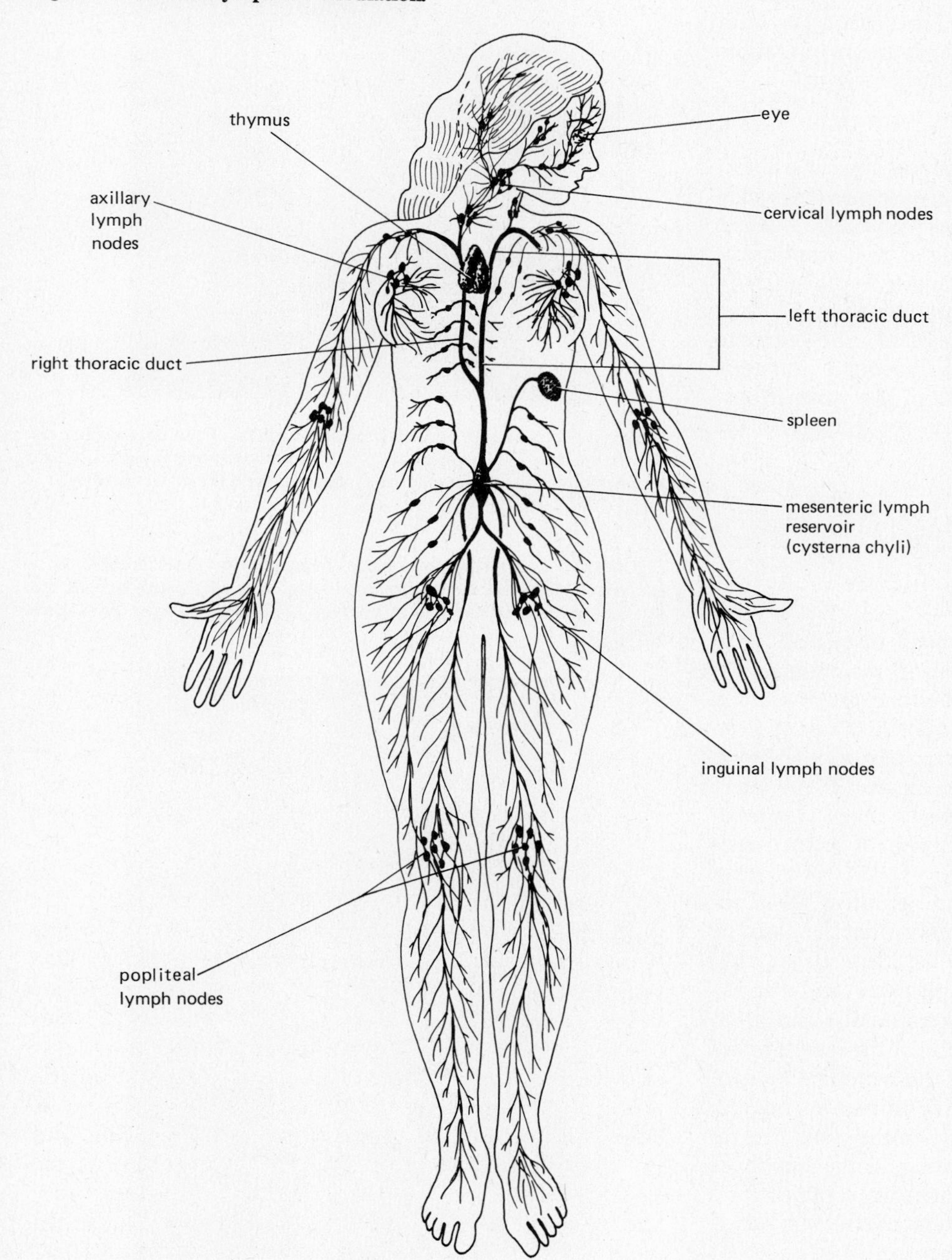

Anastomoses

Some organs receive blood through parallel arteries that are joined by many anastomoses or connecting channels. The forearm receives its blood through both the radial and ulnar arteries and any blockage of one of these arteries results in the shunting of blood through the anastomoses to the second artery. This is usually sufficient to care for the nutritive needs of the tissue. The azygos vein (Sec. 13-2) serves as a parallel venous drainage system for the inferior vena cava.

End Arteries

Most of the organs of the body receive their blood through a single artery, known as an *end artery.* The kidney, spleen, and coronary arteries are of this type, and an occlusion of one of these arteries almost always leads to an infarction.

Collateral Circulations

Even when no anastomoses between blood vessels are normally present, a complex anastomosing system may develop from the capillary circulation if an artery is *slowly occluded.* This forms a collateral circulation and is of great importance in preventing tissue destruction, particularly in the heart. Of course, a sudden and complete occlusion of an artery does not permit a collateral circulation to develop. The degree to which the heart is able to develop an adequate collateral circulation following occlusion of one of the coronary arteries will govern to a large extent the recovery from the damage.

13–8 The Lymphatic System

General Functions

The lymphatic system has already been discussed as an accessory to the blood vascular system in maintaining the *distribution of water* between the tissues and the blood and also as a specialized system in *transporting digested fats* from the lumen of the small intestine ultimately to the blood (Fig. 13-15). The lymphatic system is also part of the internal defense system of the body in terms of *phagocytic activity* and *antibody production,* a vital area of homeostasis (Sec. 20-6).

In addition, because of the "filtering" action of the reticular structure of the lymph nodes, minor infections can be localized, and the nodes respond to the bacterial or viral invasion by a proliferation of lymphocytes and macrophages. "Swollen glands" in the neck and axilla and groin are common manifestations of the action of the lymph nodes in localizing infections. The course of the lymph nodes is of great importance in the spread of malignant tumors, because the tumorous cells may break off from the main site of proliferation and either be carried along in the lymph or grow along the lymph vessels. They are then confined for a brief time within the lymph glands, the removal of which is essential to prevent further spreading.

Structure and Circulation

The lymphatic system consists of organs that produce and accumulate lymphocytes and the lymphatic vessels that transport the lymph from the tissues through these lymphatic organs to ultimately drain into the venous system (Fig. 13-16). The lymphatic organs consist of the innumerable chains of *lymph nodes,* situated both superficially and

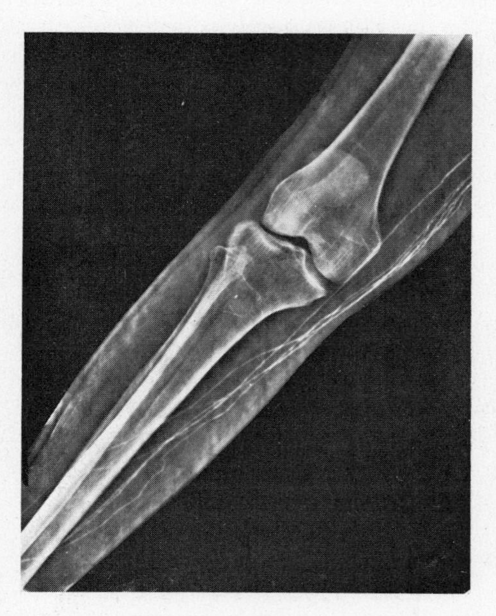

Figure 13-16. **The straight lymphatic vessels of the normal leg are shown in this X-ray photograph. The vessels were injected with a radiopaque dye. (Courtesy of C. Smith, New York University.)**

Figure 13-17. **The tortuous path of lymphatic vessels subjected to increased pressure due to improper drainage of the thigh. Note the marked edema of the leg. (Courtesy of C. Smith, New York University.)**

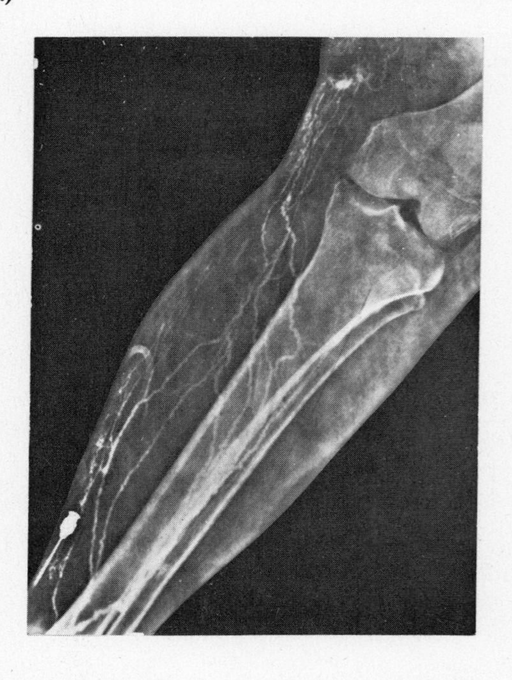

deeply in the fascia of the limbs, head, neck and trunk, the spleen, and thymus. The lymphatic organs are discussed in detail in Sec. 14-2.

The lymphatic vessels resemble the veins in structure, but their walls are thinner. They also contain numerous valves. The *lymphatic capillaries* are the smallest of these vessels, irregular in shape and ending blindly, in contrast to the continuity of the blood capillaries. Once the interstitial fluid has drained into the lymph capillaries, it is called *lymph,* but its composition is essentially unaltered except for the addition of lymphocytes as the fluid passes through the lymphatic organs. The lymph also acts as an important mechanism for returning protein from the interstitial fluid to the blood. If there is insufficient drainage of fluid and protein (see Sec. 4-2), edema of the tissue results (Fig. 13-17).

Most of the lymph entering the lymphatic organs is drained off into their veins; relatively little flows through to the efferent lymphatic vessels that ultimately unite to form the large *lymph trunks* that drain definite territories of the body (lumbar, intestinal, intercostal, mediastinal, subclavian, and jugular). These lymph trunks finally empty into the large *thoracic duct* which collects the lymph from all the body except the right side above the diaphragm. This area is drained by the smaller *right lymphatic duct.* These ducts open separately into the great veins at the base of the neck.

Additional Readings

BOOKS

ZELIS, R., ed. *Peripheral Circulations.* Grune & Stratton, Inc., New York, 1975.

Handbook of Physiology, Sec. 2: *Circulation,* Vol. 2, *Am. Physiol. Soc.,* Washington, D.C., 1962, has several excellent chapters on specialized circulations:

30. "The Physiologic Importance of Lymph," H. S. Mayerson.
31. "The Peripheral Venous System," R. S. Alexander.
39. "The Circulation Through the Skin," A. D. M. Greenfield.
40. "Circulation in Skeletal Muscle," H. Barcroft.
41. "The Hepatic Circulation," S. E. Bradley.
42. "The Flow of Blood in the Mesenteric Vessels," E. Grim.
43. "The Renal Circulation," E. E. Selkurt.
44. "Blood Supply to the Heart," D. E. Gregg and L. C. Fisher.
45. "Maternal Blood Flow in the Uterus and Placenta," S. M. Reynolds.
46. "The Fetal and Neonatal Circulation," M. Young.
47. "The Flow of Blood Through Bones and Joints," W. S. Root.
48. "Dynamics of the Pulmonary Circulation," A. P. Fishman.

ARTICLES

KORNER, P. I. "Control of Blood Flow to Special Vascular Areas: Brain, Kidney, Muscle, Skin, Liver and Intestine." In *MTP International Review of Science, Physiology,* Vol. I.: *Cardiovascular Physiology,* A. C. Guyton, ed. University Park Press, Baltimore, 1974.

MAYERSON, H. "The Lymphatic Circulation." *Sci. Am.,* June 1963.

Chapter 14

Blood as a Regulatory Organ

This was the first shape,
The shapeless hunger for being
In the black womb of the mire,
Eager for ultimate meaning
In a world beyond forseeing
On the day the sun took fire.

And I am the last shape,
The shapeless measure of pride
Caught in the mind's invention
Of a world at the senses' end
And no door leading outside
Time's one dimension.

A. M. Sullivan, "Protozoa"

THE MAINTENANCE OF the composition of the blood within a relatively narrow range is brought about by the integration of an extraordinary series of complex phenomena. Tissues through which the blood flows are continuously adding to it (secretions, metabolites, wastes, acids) and extracting from it (nutrients, oxygen, vitamins, etc.). The kidney and liver are organs that are especially adapted to homeostatically control the composition of the blood flowing through them. Changes in the composition of the blood profoundly influence the functions of many of the integrative organs, particularly nervous and endocrine centers and the liver, so that blood may be considered not only a self-regulating organ but also an important regulator of other organ systems. Finally, it will be remembered that blood contains antibodies and many of the cells capable of producing them, and so bears within its substance protective devices against viruses, bacteria, fungi, and toxins; it also spreads these agents and others, such as malignant cells, throughout the body.

Blood is pumped by the heart to reach the tissues through blood vessels that are essentially living tubes of greatly varied diameter, elasticity, and permeability. The health and integrative action of this cardiovascular system are essential to get the blood to and from the tissues rapidly and as efficiently as possible; these aspects of blood physiology are discussed in Chapter 12.

14–1 Composition and Volume of the Blood

Blood Volume

Because blood is a fluid organ, its major component is water, in which are dissolved the important biological electrolytes and gases. It also contains various larger particles ranging from the plasma proteins to the blood cells. The average *blood volume* is considered to be about 5000 ml in a 70-kg adult, although the variation among individuals, apart from sex, age, and health differences, is considerable.

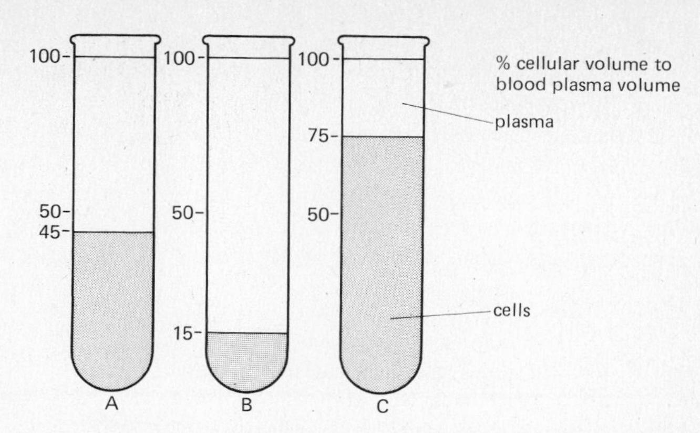

Figure 14-1. **The hematocrit. A = normal, B = anemia, and C = polycythemia.**

Hematocrit

Not only is the total blood volume significant, but the *proportion of cells,* particularly the red blood cells, to the amount of fluid within the blood is very important. When blood is centrifuged, the cells are separated out. The heavier particles form a sediment; the fluid supernate contains all other components. This fluid is called *plasma.*

Plasma makes up about 55 per cent of the total blood volume and represents part of the extracellular fluid of the body. The remaining solid mass, 45 per cent, consists of red blood cells covered by a thin layer of white blood cells. The ratio of red blood cells to blood plasma is expressed as the *hematocrit* (Fig. 14-1). The *normal hematocrit* is approximately 45 for a man and 40 for a woman; that is, 45 per cent of the blood volume in the human male is composed of red blood cells as against 40 per cent in the female. These values have a wide range, however, and can be influenced by many factors, including adaptation to high altitude (Sec. 16-20).

Figure 14-2. **The relationship between hematocrit and blood viscosity.**

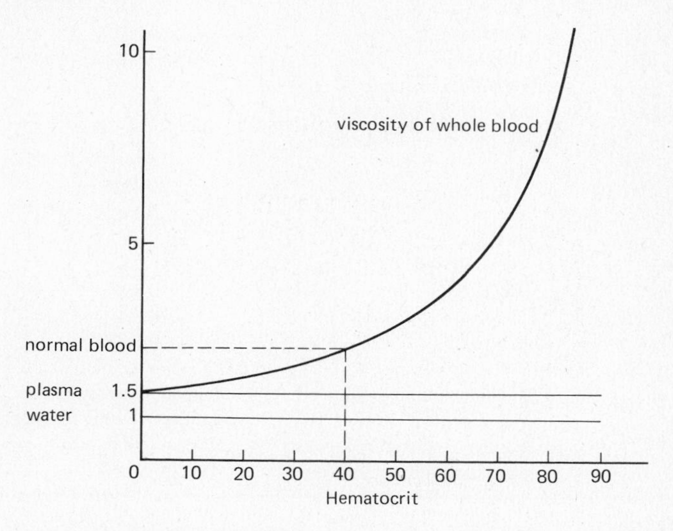

In very severe *anemia,* in which the number of erythrocytes is sharply diminished, the hematocrit may fall as low as 15 (15 per cent of the blood volume is cellular). In diseases causing excessive production of blood cells, *polycythemia,* the hematocrit may rise to 75, and the blood becomes so viscous that the blocking of the capillaries and smaller arteries causes death. The relationship between hematocrit and blood viscosity is illustrated in Fig. 14-2.

The method of determining the hematocrit by centrifugation is simple and useful, although a fair amount of plasma becomes trapped within the cellular mass, giving almost a 5 per cent overestimation of the number of erythrocytes in the blood. For purposes of clinical comparisons this method is adequate. Table 14-1 lists the physical properties, chemical composition, and cellular components of the blood.

Table 14–1. **Composition of Blood and Plasma**

	Blood	Plasma
Physical Properties		
Color	Arterial—scarlet of oxyhemoglobin Venous—dark red of hemoglobin Opaque	Transparent yellow
Specific gravity	1.059 (male) 1.056 (female)	1.027
Viscosity	4.7 (male) 4.4 (female)	
Osmotic pressure	6.7 atmospheres (isotonic to 0.9% NaCl solution—physiologic saline)	6.7 atmospheres
pH	7.4	7.4
Chemical Composition g/100 ml		
Water	78.0	90.7
Total solids	22.0	9.3
Organic substances	21.2	8.5
Glucose	0.07	0.08
Lactic acid	0.006	0.008
Fatty acids	0.36	0.37
Lecithin	0.3	0.2
Cholesterol	0.2	0.18
Ketone bodies	0.002	0.0
Total protein	19.1	7.0
Serum albumin	2.5	4.2
Serum globulin	1.38	2.6
Fibrinogen	0.25	0.3
Hemoglobin	15.0	0.0
Urea	0.02–0.035	0.026
Uric acid	0.002	0.003
Creatinine	0.001	0.003
Ammonia	0.00025	0.0
Inorganic Substances		
Na	0.19	0.33
K	0.20	0.017
Ca	0.006	0.010
Mg	0.003	0.002
Cl	0.280	0.365
NaCl	0.480	0.600
P	0.0	0.004
Fe	0.050	0.0001
Cu	0.0001	0.0001
I	0.000008	0.00001
Cellular Composition		
Erythrocytes (red blood cells)	5,400,000/mm^3 (male) 4,800,000/mm^3 (female)	
Leukocytes (white blood cells)	5,000–10,000/mm^3	
Platelets or thrombocytes	250,000/mm^3	

Plasma

Plasma consists mainly of water, in which are carried the plasma proteins, inorganic ions and salts, glucose, amino acids, fatty acids, respiratory gases, and so forth (see Table 14-1). *Plasma volume* is the result of the balance of forces resulting in the passage of fluid out of the capillaries, versus the forces that pull fluid back into the capillaries. The important role of the plasma proteins and the osmotic pressure they exert to maintain blood volume is discussed in Sec. 4-2.

FUNCTIONS OF PLASMA PROTEINS

Plasma proteins have many vital functions. They are involved in the following phenomena:

1. Osmotic pressure (Secs. 4-1 and 4-2).
2. Blood viscosity. The large, asymmetric protein molecules add greatly to the viscosity of the blood and are essential for the maintenance of blood pressure (Sec. 4-2).
3. Nutrition. Some of the plasma proteins are used in nutrition (Sec. 19-2).
4. Blood clotting. Fibrinogen, an asymmetrical, soluble plasma protein, is converted to the insoluble fibrin in the final stage of blood coagulation.
5. Acid-base equilibrium. Proteins act as buffers in the blood, maintaining its pH within an extremely narrow range.
6. Plasma immune system. The antibodies of the blood, plasma proteins collectively referred to as immunoglobulins, form part of the defense system of the body against foreign substances (Sec. 20-6).

Functions 1, 2, 3, and 6 are discussed elsewhere in this text. Functions 4 and 5 are discussed in this chapter.

SERUM

Serum is plasma from which the fibrinogen has been removed either artificially or else physiologically through the extrusion of plasma through a clot. The clear yellowish fluid that exudes from the clot is *serum,* which cannot clot because it contains no fibrinogen.

Plasma Substitutes

Plasma volume can be drastically decreased in cases of severe burns, in which a transudation of plasma through the burned surfaces occurs. Plasma volume is also markedly decreased when the water intake is too low, or following diarrhea, vomiting, or polyurea. An infusion of isotonic saline (0.9 per cent NaCl) will increase plasma volume adequately. However, if there has been a loss in *total blood volume,* such as follows hemorrhage—which involves a loss in plasma proteins, salts, and cellular elements—saline infusions serve as a temporary measure only. Because of the absence of plasma proteins, infused saline is lost rapidly from the circulation to the tissues. Adequate therapy therefore requires the transfusion of whole blood, plasma, or plasma substitutes. Blood transfusions require careful matching of blood types (Sec. 14-16); therefore, much effort has been expended on the development of substitute fluids that can be used for immediate infusion.

There are several types of blood substitutes. Some of these simply replace plasma volume: *plasma volume expanders* have been used for many years and include dextran and hydroxyethyl starch. However, dextran, although it is able to maintain osmotic pressure, has an adverse effect on blood clotting.

Blood substitutes that have *oxygen- and carbon dioxide-carrying characteristics* have been developed. The most successful of these are perfluorochemicals, *fluorocarbons* that may also carry oxygen, nitrogen, or sulfur. Of these fluorocarbons, perfluorotributylamine is the only one that can be used to replace blood completely. This compound is extremely stable and does not seem to cause adverse effects, but it is retained by the tissues for long periods of time.

The potential uses of such blood substitutes include not only the replacement of lost blood volume in emergencies, but also their use in surgery, where the perfusion of organs outside the body is required. This technique requires extremely large quantities of blood, which are not always available. Fluorocarbon substitutes would adequately substitute for many functions of the blood and thus save large amounts of natural blood.

Another use for blood substitutes would be in certain anemias, where the oxygen-carrying capacity of the blood is low, and the frequent transfusion of natural blood often induces intolerance to the blood. Individuals with sickle cell anemia are especially well suited for treatment with perfluorotributylamine. The long-lasting action of this compound means that a high oxygen concentration (P_{O_2}) can be maintained in the blood without frequent transfusions.

14-2 Blood Cells

There are three main types of blood elements, red blood cells, white blood cells, and platelets. The functions of the blood cells are largely dependent on other components of the blood (such as coagulation factors, plasma factors, and gamma globulin) and on the efficiency of the cardiovascular and respiratory systems in bringing oxygen to the blood. The study of these blood cells is *hematology.*

Blood-Cell-Producing (Hematopoietic) Organs

The chief hematopoietic organ of higher vertebrates is the *bone marrow* which has the capacity to form all the blood cell types in the adult. In the human fetus, before the bone cavities form in the fifth fetal month, hematopoeisis starts in the yolk sac and then is continued in the liver and spleen. Even before birth, however, blood cell formation is taken over by the bone marrow, which contains *multipotential cells* (*stem cells*) capable of producing all the different kinds of blood cells. This multipotentiality is clearly demonstrated by studies of *bone marrow colonies* that have been transplanted to the spleen, a highly vascularized organ which provides a favorable environment for the growth and differentiation of the bone marrow cells (Fig. 14-3).

Each colony of cells is derived from a single, multipotential bone marrow cell, yet the developed colony consists of a mixture of blood cell types. Interestingly enough, the dominant type of colony cell depends on whether it has been grown on the surface of the spleen (when most of the cells develop into red cells) or in the center of the spleen (when white blood cells dominate). The immediate environment of the developing cells obviously exerts a profound influence on the direction of hematopoeisis.

The *adult spleen* no longer is involved with blood cell formation but is an important site for selective *red blood cell destruction* and *reservoir of platelets.*

Figure 14-3. **The upper spleen is a normal control. The lower spleen contains colonies of bone-marrow cells seen as white raised plaques on the surface of the spleen. These colonies were found 7 days after total body radiation of the mouse, immediately followed by a transfusion of bone-marrow cells obtained from an isogeneic donor. Each colony is derived from a single, sequestered, multipotential stem cell. (From A. J. Erslev and T. G. Gabuzda, *Pathophysiology of Blood,* W. B. Saunders Company, Philadelphia, 1975.)**

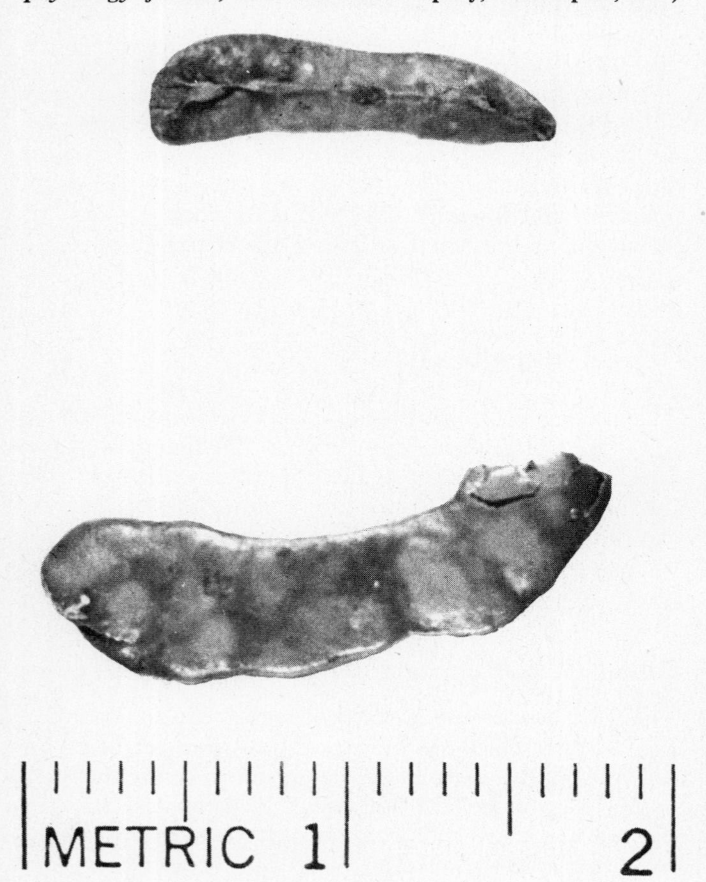

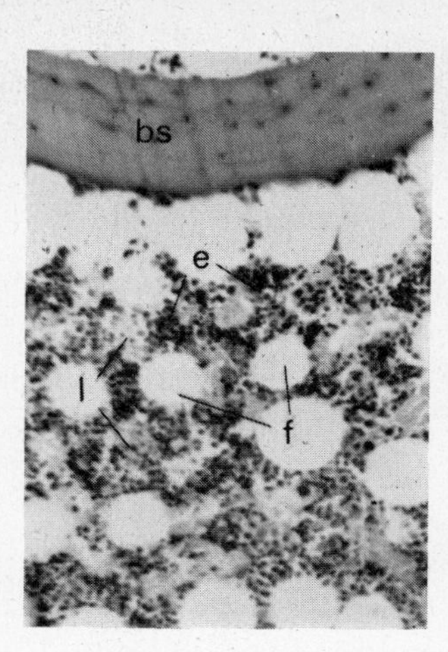

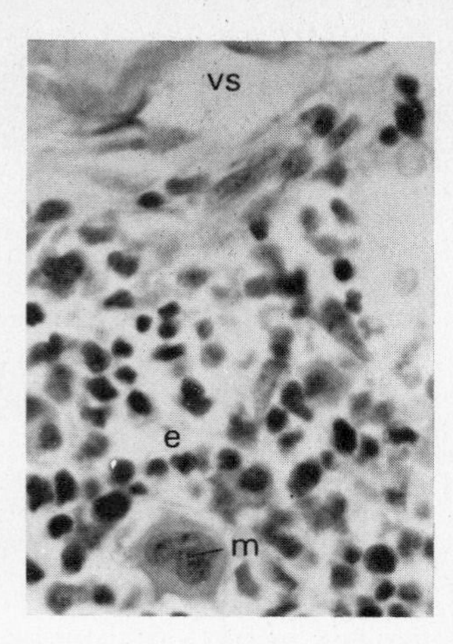

Figure 14-4. **Cellular elements of bone marrow. (Left) Section of bone marrow from the epiphysis of a femur at intermediate magnification; e = nests of erythroid, 1 = leukocytic cells, f = fat cells, and bs = bony spicule of femur. (Right) Section of marrow of a fetal vertebra at higher magnification; e = erythroid cells scattered among leukocytic cells, m = megakaryocyte, and vs = venous sinus. (Courtesy of S. Piliero, Dental School, New York University.)**

Bone marrow

The bone marrow is of two types: the *red marrow,* which is concerned with hematopoeisis, and the fatty *yellow marrow,* which gradually replaces much of the red marrow in the adult. Red marrow consists of (1) a spongelike framework of *stroma,* to which phagocytic cells are closely attached [see discussion of the reticuloendothelial system in Sec. 14-9 (phagocytes) and Sec. 20-6]; and (2) *free cells* within the meshes of the stroma. These are blood cells in all stages of development, and some fat cells, but young red and white cells predominate (Fig. 14-4). The formed blood cells reach the circulation through large vessels called *sinusoids* in the bone marrow. In the yellow marrow, most of the blood cells have been replaced by fat cells.

In the *growing infant,* all the available bone marrow space is needed for the formation of blood cells, but as the bone cavities expand in size more rapidly than the blood volume increases, much of the red marrow is replaced by fatty yellow marrow. In the *normal adult,* red marrow is found mainly in the proximal ends (*epiphyses*) of the long bones, and in the vertebrae, ribs, sternum, and skull.

Lymphocyte-producing organs

Bone Marrow. The *stem cell of the bone marrow* is the precursor of *all* the lymphocytes, which are white blood cells capable of producing antibodies. Those lymphocytes that continue their development within the bone marrow are called *B cells;* their ability to form antibody depends on their development in the specialized environment of the bone marrow.

Figure 14-5A. **Scanning electron microscope picture of a lymphocyte identified as B type immunologically, with multiple microvilli. [From A. Polliack et al., *J. Exp. Med.* 138:607 (1973).]**

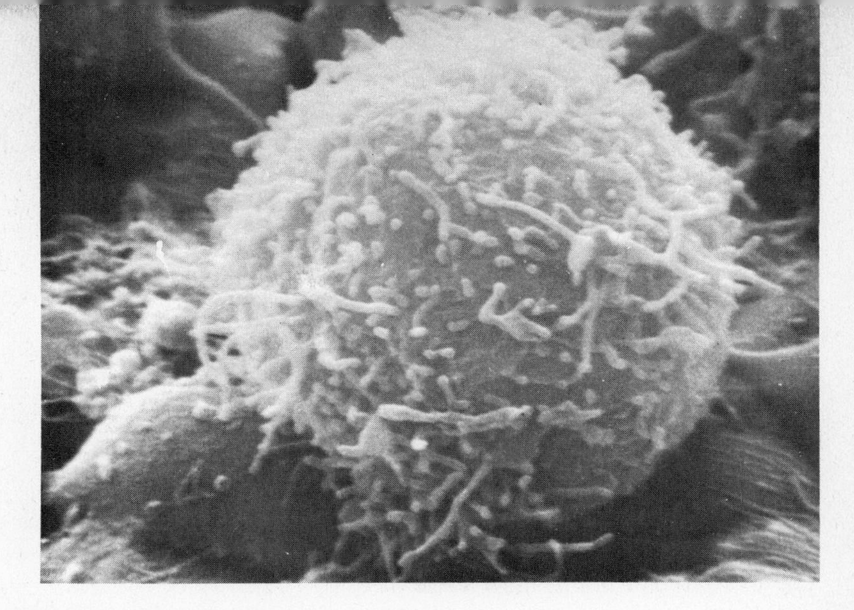

Figure 14-5B. **Scanning electron microscope picture of cultured lymphocytes, identified as T type immunologically, with smooth membrane surfaces. [From A. Polliack et al., *J. Exp. Med.* 138:607 (1973).]**

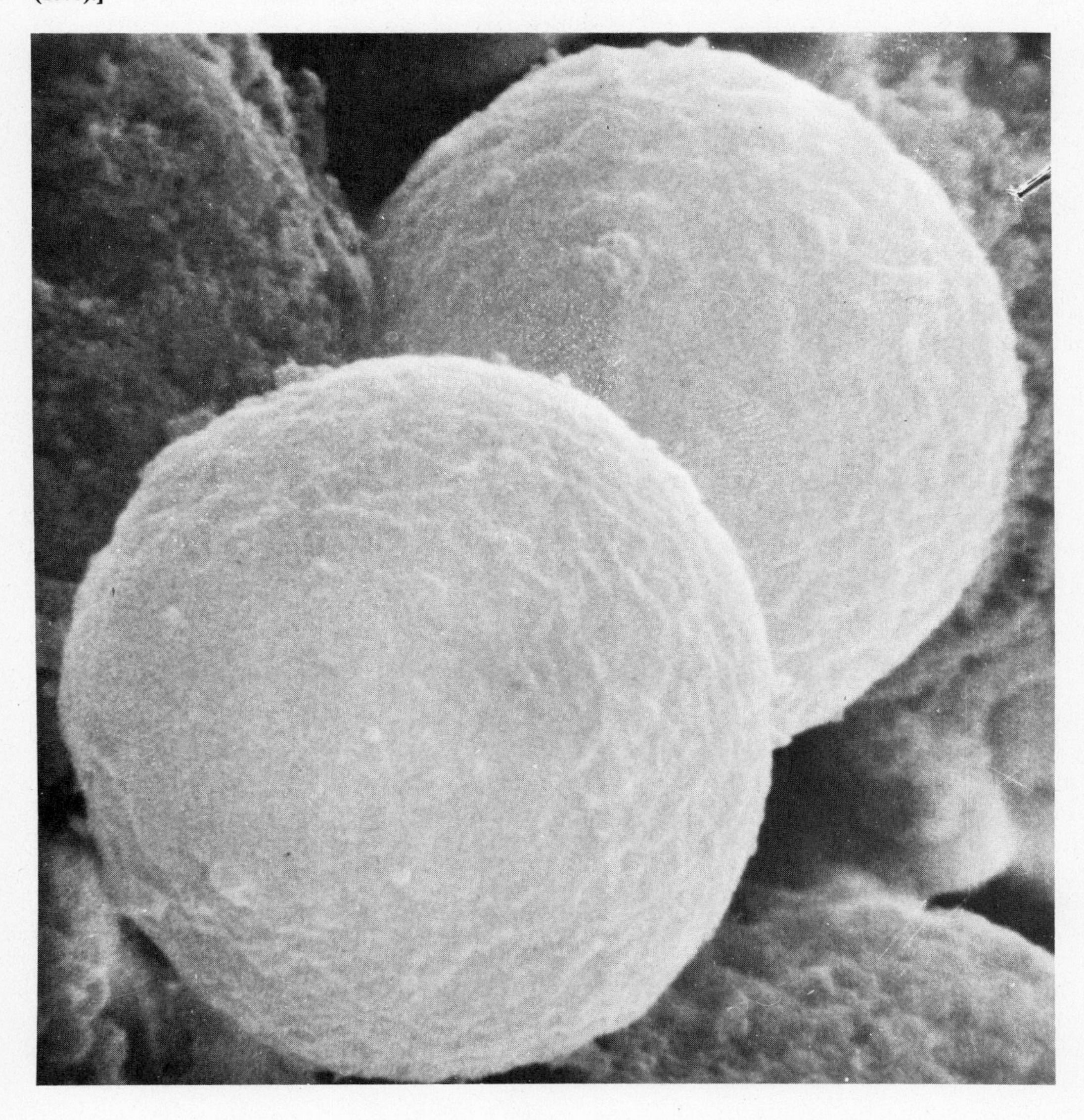

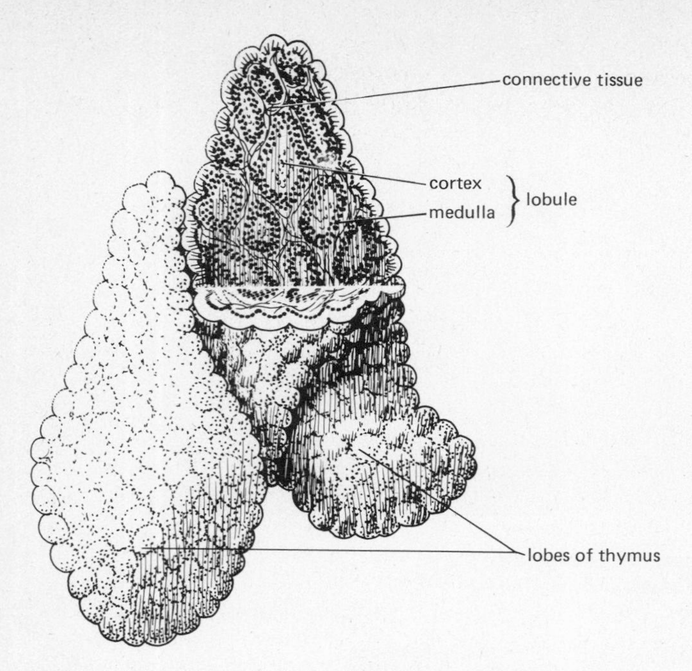

Figure 14-6. **The thymus.**

Lymphoid Organs. The thymus, lymph nodes, and spleen are lymphoid organs that contain lymphocytes in various stages of development. They originate from immature lymphocytes that have migrated from the bone marrow to the thymus, where they develop into T cells. These T cells differ from other lymphocytes in that their membranes contain specific surface antigens and consequently have different immunological characteristics from the B cells. The T cells leave the thymus to form a large proportion of the cellular elements of the spleen and lymph nodes. Figures 14-5A and 14-5B show the difference in the surface membranes of B and T lymphocytes.

The lymph nodes, spleen, and thymus store lymphocytes in various stages of development and discharge certain numbers of them into the blood, where they form part of the white blood cell complement. All these tissues are constructed of a basic reticular framework to which the macrophages and lymphocytes are attached, but the organization of the organs varies.

THYMUS. In humans, the thymus is an unpaired gland situated under the sternum, close to the pericardium and the great veins of the heart, as shown in Fig. 13-15. It consists of two lobes, each of which is composed of lobules with a peripheral cortex and central medulla (Fig. 14-6). The amount of lymphatic tissue is sharply reduced after puberty, and the thymus becomes infiltrated with connective tissue strands and fatty tissue.

Apart from its important role in the maturation and discharge of the T cells, the thymus has been associated with various endocrine functions, none of which have been satisfactorily established. A thymic factor related to growth has been suggested, but no such hormone has been identified.

The thymus is relatively largest in embryonic life and early childhood and begins to involute at puberty, a process closely associated with changes in hormonal concentrations. The thymus, like the other lymphoid organs, is responsive to the adrenal cortical hormones, but it appears to be considerably more sensitive. The thymus practically disappears in long-lasting infections and wasting diseases where the levels of adrenal cortical hormones remain high. This responsiveness to hormonal control makes the lymphocyte content of the thymus a significant *storehouse of antibodies* to be released in time of stress.

LYMPH NODES. These are dense aggregates of lymphoid tissue along the course of the lymphatic vessels (Fig. 13-15). They vary in size from the head of a pin to a pea and are usually grayish pink in color, although the lymph nodes of the lungs are usually black with carbon; those of the mesentery, creamy white after a fatty meal. The lymph nodes act as stations in which the lymph vessels are broken up into the reticular tissue. They are widely distributed in the deep and superficial fascia but absent from the entire nervous system, and possibly from bone and muscle. The node is surrounded by a connective tissue capsule that penetrates into the outer part of the organ, the *cortex,* dividing it into many sections. Within the cortex are lighter areas, the *nodules,* that produce the lymphocytes. The inner portion of the lymph node is the *medulla,* and both cortex and medulla consist of reticular fibers and dense masses of lymphocytes in various stages of development, separated by spaces known as the *lymphatic sinuses* (Fig. 14-7).

SPLEEN. The spleen is the largest of the lymphoid organs. Although the mature spleen is not involved in erythropoiesis, it is an important blood reservoir and filter, capa-

Figure 14-7. **A lymph node actively producing lymphocytes. The capsule (c) of connective tissue penetrates through the outer cortex, which contains the lymphocyte-producing nodules (n). m = medulla, a = artery, and s = sinus. The insert shows lymphocytes within a nodule at a higher magnification. (Courtesy of S. Piliero, Dental School, New York University.)**

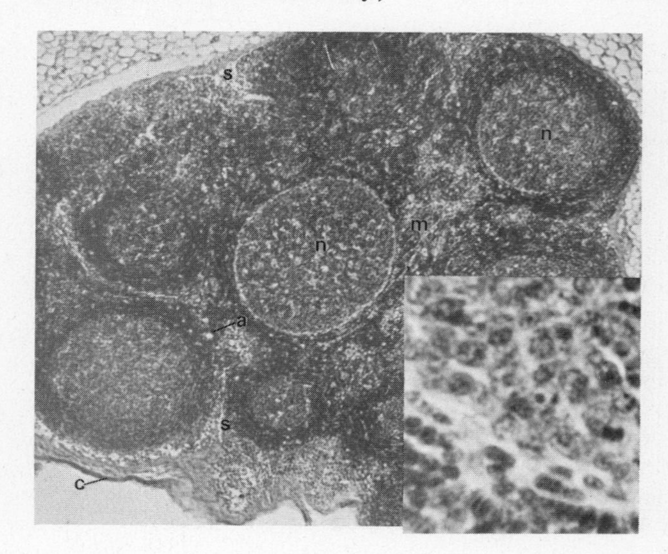

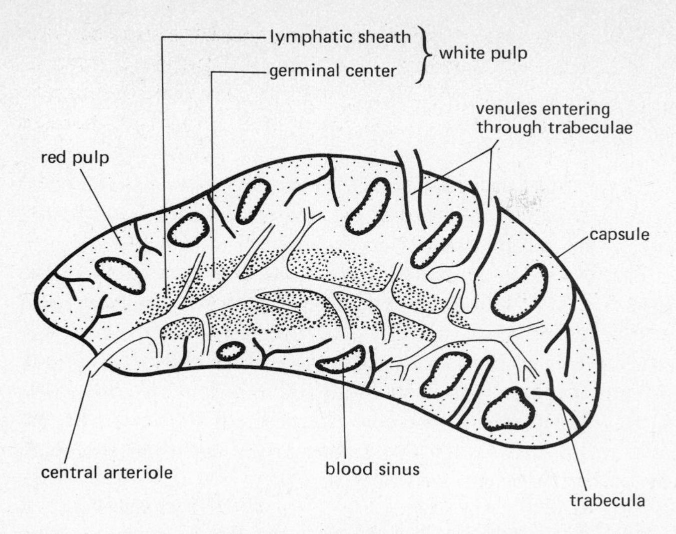

Figure 14-8. **Section through the spleen, showing the red and white pulp.**

ble of removing undesirable particles from the blood through the phagocytic activity of its many macrophages.

The spleen is a soft, pulpy, and very vascular organ, filled with blood supplied by the exceptionally wide splenic artery. It lies obliquely in the upper left part of the abdomen, under the diaphragm and behind the stomach (Fig. 17-1). The spleen is usually the size of the left lobe of the liver, but in disease it may become enormous, larger than the liver.

The spleen is covered by a connective tissue capsule which penetrates into the splenic substance to become continuous with the reticular fibers of the pulp (Fig. 14-8). The *white pulp* consists of masses of lymphocytes, fastened together by a network of reticular fibers to form a *lymphatic sheath* around a *central arteriole.* The white pulp also contains nodules or *germinal centers,* filled with lymphocytes and surrounded by macrophages.

The *red pulp* is segmented by fibrous *trabeculae* and filled with blood sinuses which give this region its red color. The spleen thus serves as a filter for both the lymphatic system and the blood. Venous blood is drained from the sinuses via the splenic veins, which leave the spleen at the hilus and empty into the portal vein.

14–3 Red Blood Cells: Erythrocytes

Function

The function of the red blood cells is to synthesize, store, and transport hemoglobin, the iron-containing protein responsible for the transport in the blood of almost all the oxygen obtained from the lungs. Although carbon dioxide is soluble enough for a fair proportion to be transported in solution from the tissues to the lungs (Sec. 16-15), the combination of the remaining carbon dioxide with hemoglobin serves as an important mechanism by which excess acidity is buffered in the blood. Hemoglobin is restricted to the stroma of the red blood cells: were it free in plasma the amount of hemoglobin required to transport oxygen in animals with high metabolic rates would seriously disturb the osmotic pressure relationships between blood and tissue fluid.

Structure

The human erythrocyte is a *biconcave disk,* which is distorted in shape as it is squeezed through the capillaries but which regains its characteristic form rapidly (Fig. 14-9). The *flexible membrane* with which it is surrounded has been the source of much information concerning cell membranes in general (Sec. 4-5). However, the red blood cell membrane is a *specialized membrane;* it is extremely permeable to anions and permits the rapid exchange of Cl^- and HCO_3^- (the *chloride shift,* a phenomenon discussed in Sec. 16-15). Within the membrane is a dense *stroma* in which the hemoglobin molecules that give the cell its characteristic reddish-yellow color are embedded. These molecules are more concentrated toward the surface of the cell; this concentration, together with the *large surface area* afforded by the unusual shape of the cell, facilitates the diffusion of gases into and out of the cell. The actual size of the cell varies according to its physiological condition. In venous blood, where CO_2 and the resultant acidity permit more water and Cl^- to enter through the cell membrane, the diameter is approximately 0.5 μm more than in arterial blood. The diameter of an average capillary varies from 5 to 20 μm, so that considerable squeezing of the red blood cells is necessary before they can pass through some of the narrower capillaries.

In lower vertebrates, the erythrocytes are nucleated, but

Figure 14-9. **Erythrocytes (1) and thrombocytes or platelets (2). In the upper right corner is a side view of an erythrocyte. (Courtesy of S. Piliero, Dental School, New York University.)**

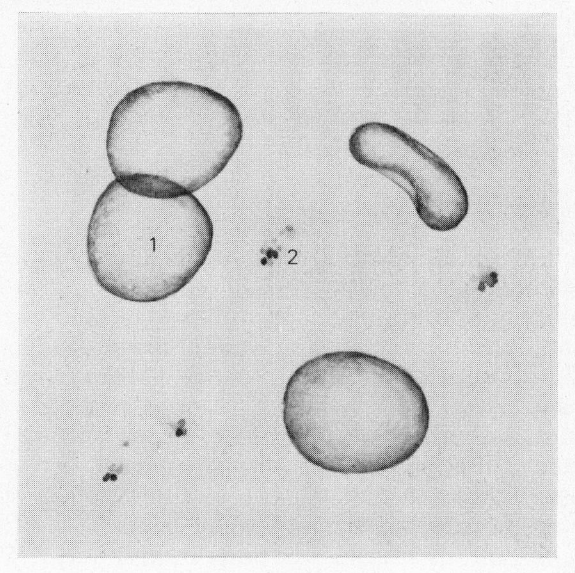

almost all mammalian red blood cells are enucleated, a feature that is usually considered advantageous in that it makes more space available for hemoglobin. However, birds and certain marsupials (e.g., the koala), warm-blooded organisms with high metabolic rates, seem to do very well with large, nucleated red blood cells.

ERYTHROCYTE NUMBER, HEMOGLOBIN CONTENT, AND LIFE SPAN

The normal number of erythrocytes in the human male is 5,400,000/mm^3 of whole blood and in the human female 4,600,000/mm^3, with a range of 600,000/mm^3 for both sexes.

The normal hemoglobin content in whole blood is 16 g/100 mm^3 in the male, 14 g/100 mm^3 in the female.

The highly specialized mature erythrocyte has an average life span of 120 days in human males and about 5 days less in females. Erythrocytes of lower mammals have a shorter life and are produced more rapidly. The life span of red blood cells is shortened by a number of diseases, including pernicious anemia and leukemia, and by lead poisoning.

DESTRUCTION OF RED BLOOD CELLS

Aging red blood cells fragment as they squeeze through narrow capillaries. The spleen is a particularly hazardous site, and many red blood cells are destroyed as they are compressed in the red pulp of the spleen. Removal of the spleen, which is not an essential organ in adults, causes a marked increase in the number of circulating erythrocytes in most animals.

The *hemoglobin* liberated by the destruction of the red blood cells is captured by phagocytes of the reticuloendothelial system. *Iron* is freed to be transported by transferrin to the hematopoietic tissues of the bone marrow. Any excess iron is stored in the liver and other tissues. The *heme* portion of the hemoglobin molecule is also retrieved and converted to the bile pigment bilirubin, which is secreted into the bile by the liver (Sec. 17-5).

14–4 Production of Red Blood Cells: Erythropoiesis

Sites of Erythropoiesis

In the fetus, the liver and spleen produce red blood cells, but during the latter part of gestation and after birth, erythropoiesis is limited to the bone marrow. The other organs act mainly as important red blood cell reservoirs, although the human spleen, unlike the spleen of other mammals, is not a significant storage site for red or white blood cells.

Until adolescence, the marrow of practically all the bones produces red blood cells, but gradually the red marrow of the long bones becomes fatty and, by the age of 20 years, only the marrow of the membranous bones, such as the sternum, ribs, and vertebrae, produces red blood cells. With age, even this marrow becomes fatty, so that a slight anemia occurs in elderly people.

Stages in Red Blood Cell Formation

Erythropoiesis consists of a series of cell stages that starts with a *pluripotential primitive stem cell* and leads up to the formation of the mature erythrocyte (Fig. 14-10).

The primitive stem cell in the bone marrow is believed to give rise to two lines of cells, one of which is the *committed erythroid cell,* a cell that can proceed only along the pathway leading to red blood cell formation. The *other line of cells* arising from the primitive stem cell can develop into white or red blood cells. These compartments of cells appear to be in communication with one another so that their interplay will affect erythropoeisis.

The primitive erythroid cell, the *early erythroblast,* is nucleated and larger than the mature erythrocyte. The stages through which the erythroblast proceeds in order to form the erythrocyte include:

1. Rapid synthesis of hemoglobin by the erythroblast. This reaches a peak at the time that both the nucleus and the total cytoplasm shrink. The resulting cell, the *late erythroblast,* is smaller than its precursor; it has a shriveled nucleus and a high concentration of hemoglobin (about 34 per cent) in the cytoplasm.
2. The *late erythroblast* gives rise to the reticulocyte, which gets its name from the remnants of endoplasmic reticulum, used to synthesize hemoglobin, that are still found in its cytoplasm. In the formation of the reticulocyte the nucleus is lost.
3. Reticulocytes are released from the bone marrow into the circulating blood, where they mature within one to two days, losing their reticulum and becoming mature *erythrocytes.* Usually, there is only a small number of reticulocytes in circulating blood, about 0.5 per cent. The appearance of large numbers of immature reticulocytes in blood indicates an abnormally rapid production of red blood cells by the bone marrow, a characteristic of certain anemias.

Factors Influencing Erythropoiesis

HORMONES

Erythropoietin (Ep). The most important factor regulating red blood cell formation is erythropoietin (Ep), or erythrocyte-stimulating factor, a glycoprotein with a molecular weight of 46,000. Erythropoietin appears to be present in all vertebrates tested, from fish to humans. The kidney tissue is most clearly associated with Ep production.

EP PRODUCTION. The main stimulus for Ep production is *hypoxia.* While it is possible that the kidney itself is able to produce this hormone, renal extracts do not generally show erythropoietic activity. Gordon and his colleagues (1973) at New York University, have suggested that the kidney synthesizes, not Ep itself, but an activator, *erythro-*

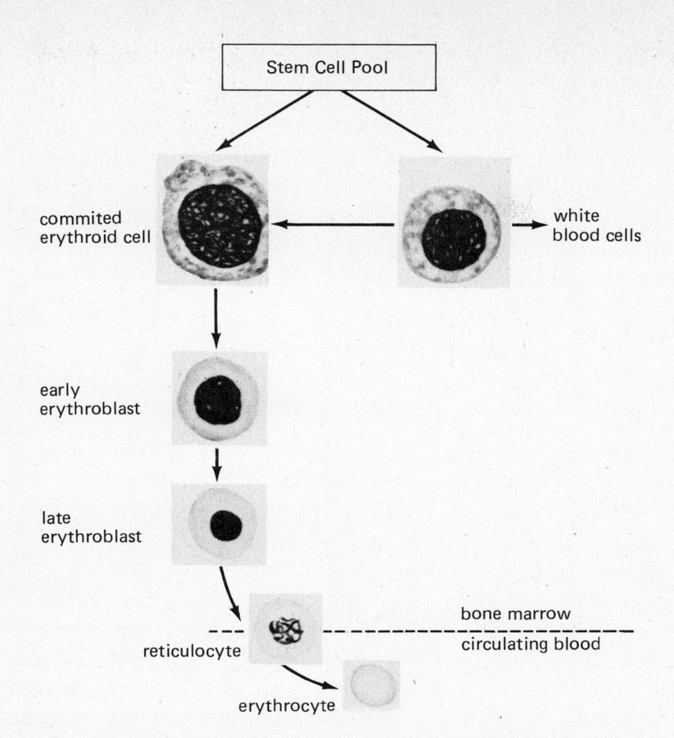

Figure 14-10. **Erythropoiesis. The two lines of cells arising from the stem cell pool in the bone marrow are shown. One line (the committed erythroid cell) produces red blood cells only. The other line can develop into either red or white blood cells. (Courtesy of S. Piliero, Dental School, New York University.)**

genin, which reacts with a circulating erythropoietin precursor, *erythropoietinogen,* to generate the active hormone. The stages of Ep production are diagrammed as follows:

Erythrogenin	(hypoxic kidney)
+	(serum)
Erythropoietinogen	
Erythropoietin	(formed in the kidney or in the circulating blood)

SITE OF ACTION OF EP. Erythropoietin acts on the committed erythroid cell, stimulating its differentiation into the early erythroblast and accelerating its production of hemoglobin. This stimulatory effect of Ep on protein synthesis appears to come from the ability of Ep to cause the sequential production of an assembly of different RNAs (messenger, transfer, and ribosomal) that are concerned with the synthesis of hemoglobin and other organic molecules in the developing erythroid cell series.

SEX STEROIDS. The sensitivity of erythropoietin production to the levels of sex steroids accounts for the difference in numbers of erythrocytes in males and females. Testosterone injections increase the production of Ep, and consequently the number of erythrocytes rises. This action can be blocked by the administration of anti-erythropoietin serum (Fig. 14-11). Estrogen tends to depress both Ep production and the number of circulating erythrocytes.

OTHER HORMONES. *Cortisone, thyroxine,* and *growth hormone,* among others, are involved in normal red blood cell production. This is demonstrated by experiments on hypophysectomized animals, in which normal erythropoiesis cannot be reestablished by the administration of testosterone alone, but requires a combination of the hormones mentioned above.

HEMORRHAGE

Hemorrhage is another potent stimulus for Ep production and release. The mechanism for this appears to be the hypoxia that results from the loss of hemoglobin.

HIGH ALTITUDE

Mammals native to high altitudes have higher red blood cell counts, a greater concentration of hemoglobin, and a more rapid rate of erythropoiesis than mammals living at sea level.

The effects of the low barometric pressure, and consequently low P_{O_2}, can be tested experimentally by placing animals in a decompression chamber. When these animals have been exposed for several days to a pressure of 322 mm Hg (as opposed to 760 mm Hg at sea level), their red blood cell counts, hematocrits, and hemoglobin values increase remarkably. These changes are closely correlated to the large amounts of circulating Ep. The relationship between

Figure 14-11. **Effects of anti-erythropoietin serum (anti-ESF) on the erythropoietic response (as judged by the rate of incorporation of ^{59}Fe into newly formed circulating erythrocytes) of normal mice to testosterone and to one of its metabolites, 11-ketopregnanolone. Vertical lines through the tops of the bars represent one standard error of the mean. These results suggest that testosterone stimulates erythropoiesis through the production of erythropoietin (since its effects are abolished by anti-ESF serum) and that the effects of the metabolite are exerted directly on the ESF responsive cells in the blood-forming tissues. Action here is not eliminated by the anti-ESF serum. (Courtesy of A. S. Gordon, Laboratory of Hematology, New York University.)**

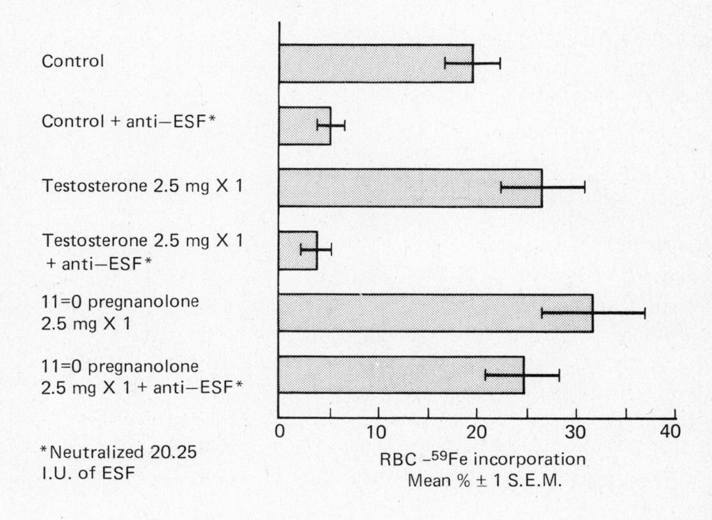

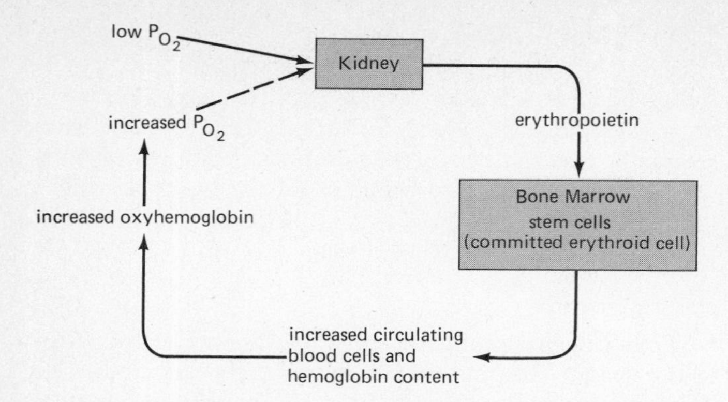

Figure 14-12. **The negative feedback mechanism by which low P_{O_2} (hypoxia) regulates the production of red blood cells. The broken arrow indicates inhibition.**

low P_{O_2}, Ep levels, and red blood cell count is shown in Fig. 14-12.

The common factor shared by these three stimuli for erythropoiesis (hormones, hemorrhage, and high altitude) is that each contributes to hypoxia of the kidney, which in turn increases the amount of Ep produced.

14-5 Hemoglobin

Structure of Hemoglobin

Hemoglobin is a molecule (molecular weight 68,000) consisting of a protein, *globin*, to which are attached four iron-containing *heme* groups (Figs. 3-12 and 14-13). The globin molecule, a *tetramer*, is composed of two α and two β polypeptide chains, each of which has its own heme group. The primary structure of globin is under genetic control, probably requiring the activity of at least five different genes. In normal *adult hemoglobin* (HbA), the α and β chains are produced in equal numbers.

Fetal Hemoglobin

Fetal hemoglobin (HbF) is made up of α_2 and γ chains, polypeptides with a different amino acid sequence from the α and β chains of HbA. Just before birth, the synthesis of β chains increases rapidly, so that HbA gradually replaces HbF in the circulating red blood cells (Fig. 14-14). HbF has a greater affinity for oxygen than does HbA, a characteristic that permits the fetal hemoglobin to load oxygen effectively at the relatively low P_{O_2} available in the placenta (Fig. 14-15).

Combination of Hemoglobin with Oxygen

The uptake and delivery of oxygen by hemoglobin is based on a structural rearrangement within the hemoglobin molecule. The extensive work of Perutz, for which he was awarded the Nobel Prize in 1962, has shown that the α and β subunits, which are similar in size and shape, complement each other's structure in a manner that permits them to form a *dimer*. When the chains are unpaired, they cannot transport oxygen and are also very unstable (Perutz, 1962, 1963, 1970).

In the oxygenation of the hemoglobin molecule, oxygen initially attaches to the heme of the α chain. This causes a spatial change in the configuration of the molecule, increasing the affinity of the β chain heme for oxygen atoms. The β chains move closer to one another by about 0.7 nm in this state of oxygenation (oxyhemoglobin, HbO_2), and they shift apart again when the oxygen is removed, forming deoxyhemoglobin (Hb).

The significance of this allosteric configurational change is that *deoxygenated hemoglobin* has a relatively *low affinity* for oxygen, so that a high P_{O_2} is needed to attach the first oxygen molecule. Once this has been accomplished, however, the molecular rearrangement exposes the other heme groups, making them more accessible and consequently increasing the oxygen affinity of the HbO_2 sharply. These changes in oxygen affinity are responsible for the sigmoid shape of the hemoglobin-oxygen dissociation curve discussed in Sec. 16-15.

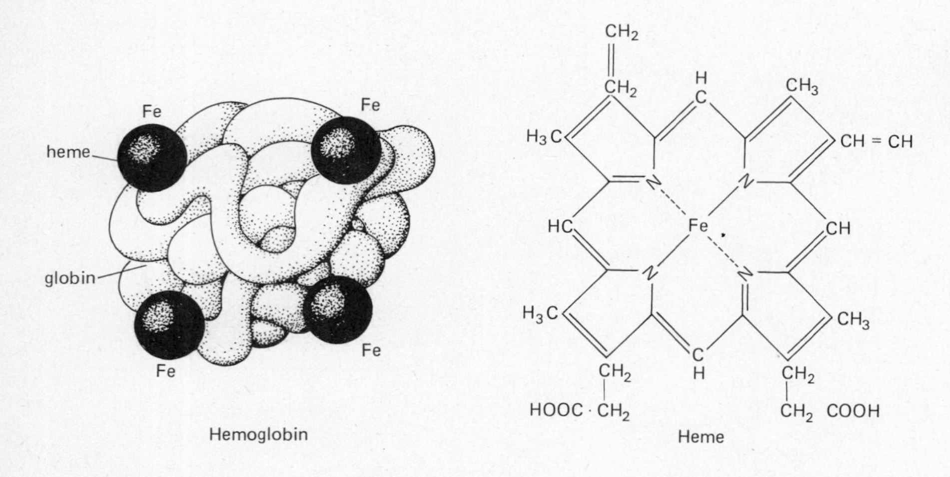

Figure 14-13. **The diagram on the left is a schematic representation of the hemoglobin molecule, with its four heme molecules attached to one molecule of globin. On the right is the structure of a molecule of heme.**

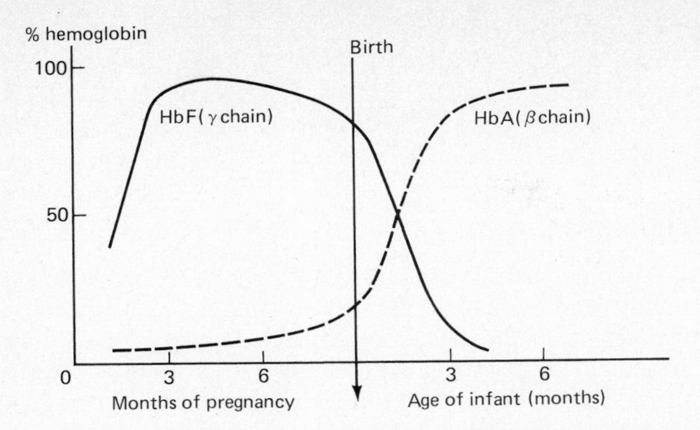

Figure 14-14. **Replacement of γ chains of fetal hemoglobin (Hb F) by the β chains of adult hemoglobin (Hb A) following birth.**

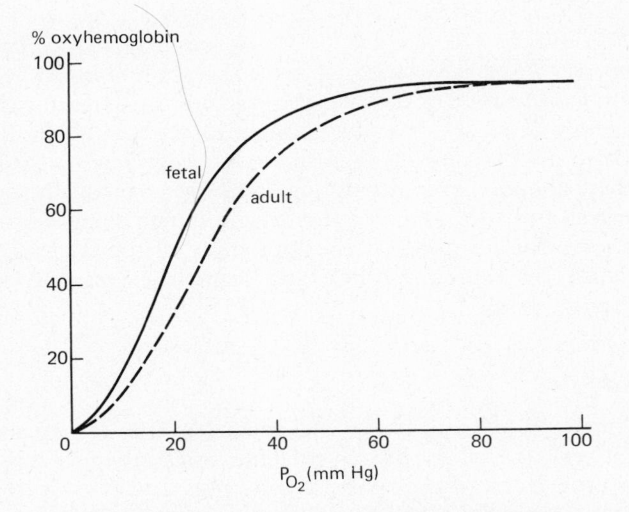

Figure 14-15. **The greater affinity of fetal hemoglobin for oxygen is shown by its ability to form more oxyhemoglobin at low P_{O_2} than can adult hemoglobin.**

Synthesis of Heme

The red bone marrow and the liver are the most important sites of production of the precursor of heme, *protoporphyrin.* Heme is synthesized from protoporphyrin and iron in a complex series of enzymatically controlled reactions. The heme that combines with globin in the erythroid precursor cells is biochemically identical to the heme found in myoglobin and in the cytochromes. The structure of heme is shown in Fig. 14-16.

Formation of Hemoglobin

The final step in the formation of hemoglobin is the assembly of a heme molecule with each of the α and β polypeptide chains. The finished hemoglobin molecule has two pairs of subunits, each linked with its own heme group into a tetrameric molecule (Fig. 14-16).

14–6 Metals and Vitamins Essential for Erythropoiesis

Iron

Iron is the most abundant heavy metal in the body; its chief function is in the synthesis of hemoglobin.

ABSORPTION

Iron is absorbed through the stomach and intestine, probably by an *active transport mechanism.* To be absorbed, however, it must be in the reduced or ferrous state (Fe^{++}), and it has been found that ascorbic acid, which is a potent reducing agent, will increase the absorption of iron by the intestine. The intestine itself seems to act *to regulate iron metabolism,* increasing absorption when needs are greatest (iron deficiency or increased red blood cell formation after a hemorrhage) and decreasing absorption when stores are adequate or when red blood cell formation is depressed. The mechanism by which the intestine functions as a homeostatic regulator is quite unknown (see also Sec. 17-6).

The amount of iron absorbed also depends on the *type of food* in which it is contained. By tagging the iron with radioactive tracers, it has been found that iron in liver, muscle, and enriched bread is much more readily absorbed than the iron in egg yolk. In the egg yolk, the iron is bound to phosphates, forming an insoluble compound. Of all foods, liver is the richest source of iron. Next to aspirin, iron pills are the most common cause of death in children, following accidental ingestion.

TRANSPORT AND STORAGE

Iron is absorbed directly into the blood stream, where it is oxidized to the ferric state and combined with a specific iron-binding protein, *transferrin.* Each molecule of this protein can combine with two atoms of iron, but there is normally more transferrin in the plasma than there is iron to be bound. Transferrin binds to specific receptors on the immature red blood cell membrane, where the iron is released to pass into the cell. Once inside the cell, iron is bound quickly by an intracellular protein that transports it to the mitochondria, where heme synthesis occurs (Fig. 14-17).

Figure 14-16. **Four polypeptide chains and four molecules of heme are distributed to two identical subunits to form the hemoglobin molecule.**

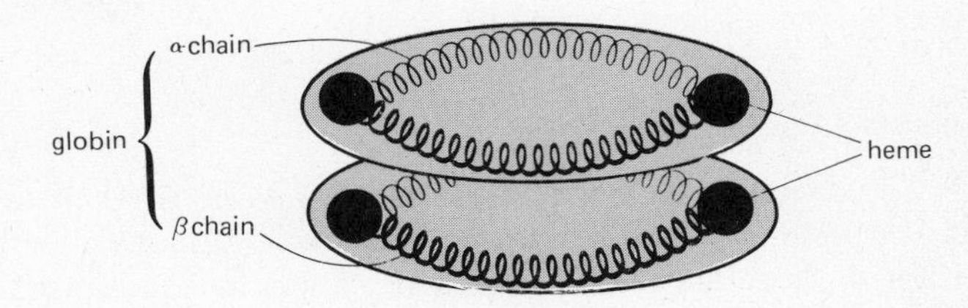

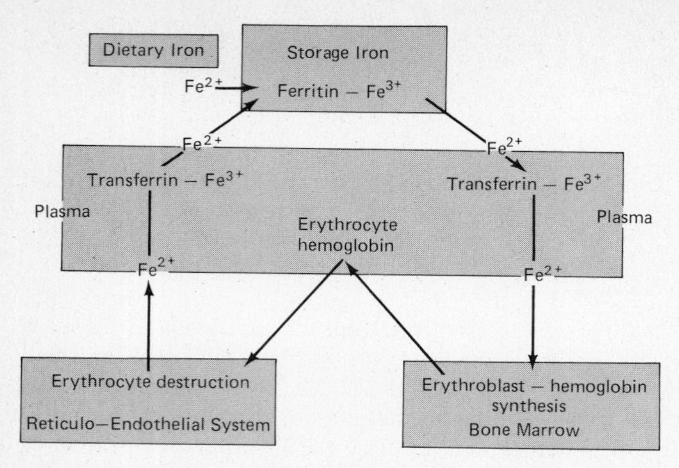

Figure 14-17. **Iron storage and transportation as related to hemoglobin synthesis and destruction. Fe^{2+} = ferrous (reduced) transport into and out of cells; Fe^{3+} = ferric (oxidized) storage and in combination with transferrin.**

Iron can enter or leave the cell only in the ferrous state, and it is possible that ascorbic acid, with the help of ATP, reduces the protein-bound ferric iron to the ferrous state, releasing it from the transferrin and permitting it to enter the cell.

The iron-free transferrin is used for shuttling iron from reticuloendothelial cells to erythroid cells. These two types of cells are in close proximity to one another in the bone marrow (Sec. 14-2).

Iron is stored in many tissues, particularly the liver, spleen, and bone marrow. It is stored in the ferric form, as *ferritin.* When more iron accumulates in the tissues, the ferritin molecules form larger aggregates, *siderotic granules,* which can be stained and seen in histologic preparations. This stored iron is readily available for hemoglobin synthesis when it is needed; conversely, the iron released by the destruction of the red blood cells through the reticuloendothelial system can be stored again in the tissues.

Loss

Iron is lost through the urine, feces, and sweat, and from cells sloughed off by the skin and mucous membranes. Although most of the iron liberated from broken-down red blood cells is incorporated into the bile pigments by the liver and secreted into the intestine, from which it may again be reabsorbed, some of the iron is lost through the gut into the feces. More iron is lost in the urine and in sweat and also from cells desquamated from the skin and mucosal surfaces. The loss of blood during menstruation must also be considered a form of iron excretion.

Balance

The average diet permits the *absorption of from 0.6 to 1.5 mg of iron/day.* The normal adult male requires about 1 mg of iron daily and so normally stays in equilibrium without difficulty, but in women the menstrual loss of blood averages an additional depletion of 2.0 mg/day, when calculated over a month. The growing child requires an excess of about 0.6 mg absorbed per day, over and above the amount lost or excreted, for normal growth (Table 14-2).

Deficiency

Although iron is conserved avidly by the body, which reutilizes more than 90 per cent, *anemia* is one of the most widespread of the deficiency diseases. It may be caused by an actual lack of iron in the diet, poor absorption, or greatly increased iron needs. It is sometimes due to chronic internal hemorrhage, as may occur with gastric ulcers. It is not surprising that growing children and young women during the years of menstruation and childbearing should be the most susceptible to this form of anemia. The changes in the daily iron requirement during pregnancy are depicted in Fig. 14-18. Anemia due to lack of iron is *hypochromic microcytic anemia,* which indicates that the red blood cells are *pale,* because of the lack of hemoglobin, and *small.* (For a more detailed discussion of anemias, see Sec. 14-7.) Because of the consequent inefficiency of oxygen transport, there are unrelieved exhaustion, headache, and usually gastric disturbances. One of the most common symptoms of severe iron deficiency is the appearance of the fingernails, which are thin and brittle and seem to be concave or "spoon-shaped."

Copper

Copper is essential for normal erythropoiesis. It forms an integral part of the plasma protein, *ceruloplasmin.* Ceruloplasmin catalyzes the oxidation of ferrous iron to ferric iron. Iron enters the blood from the liver storage cells in the ferrous form and must be oxidized before it can combine with transferrin, the only means of transport for iron through the blood to the bone marrow cells.

The copper-containing, blue ceruloplasmin has many

Table 14–2. **Recommended Daily Dietary Allowance of Iron**

	Iron (mg)
Man (154 lb, 70 kg)	16
Woman (128 lb, 58 kg)	18
Pregnancy (latter half)	25
Lactation	22
Children	
0–1 year	1 mg/kg body weight
1–3 years	8
3–6 years	8
6–9 years	8
9–12 years	15
Over 12 years	15

Source: Food and Nutrition Board, National Academy of Sciences, National Research Council, Washington, D.C., revised 1974.

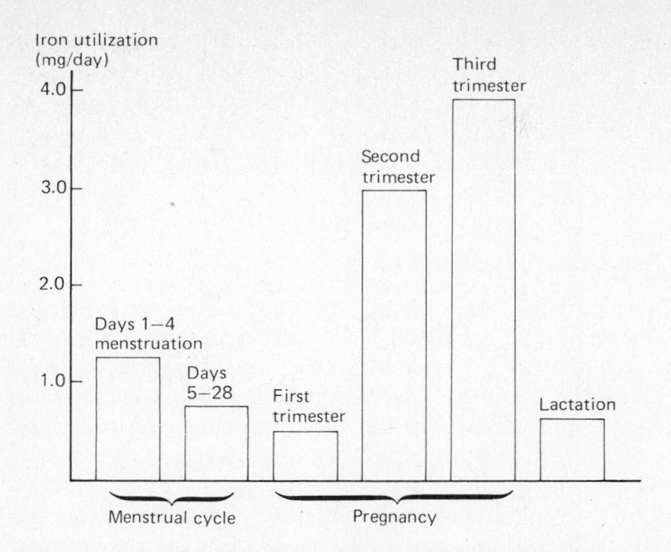

Figure 14-18. **Changes in the daily iron requirement of women during pregnancy and lactation.**

other functions. It is required for the catalytic activity of several important enzymes:

1. *Cytochrome oxidase,* the terminal oxidase in the electron-transport chain of mitochondria (Sec. 5-12).
2. *Monoamine oxidase,* which inactivates the monoamine neurotransmitters epinephrine, norepinephrine, and serotonin (Sec. 9-5).
3. *Tyrosinase,* an enzyme essential for the synthesis of melanin pigment from tyrosine (Sec. 9-5).

Copper is plentifully distributed in nature, and a *copper deficiency* has never been reported in humans; the copper requirement of 2 mg/day is present in most diets. In some areas of the world, however, livestock suffer from a copper deficiency, which results in a severe anemia. This anemia cannot be cured by iron administration, but the addition of copper to the soil on which the animals are grazing prevents or cures the disease.

Cobalamin (Vitamin B_{12}) and Folic Acid

Cobalamin and folic acid are essential for the normal proliferation of the erythroid cells. In the absence of these vitamins, the cells enlarge to form giant cells, *megalocytes,* instead of dividing and maturing into the normal red blood cell. This is characteristic of *pernicious anemia.* Both cobalamin and folic acid act as coenzymes in many biochemical reactions, particularly those affecting DNA synthesis. Thus their effect on the multiplication processes of red blood cells is not surprising.

Cobalamin is built around *cobalt,* a *trace metal,* in much the same way that heme is built around iron. The clinical investigations by Castle in the 1920s and 1930s showed that the absorption of cobalamin (the *extrinsic factor*) depends uniquely upon its binding to the *intrinsic factor,* a glycoprotein produced by the parietal cells of the stomach. The complex formed by the intrinsic factor–cobalamin attaches to specific receptors on the surface of the microvilli of the ileum where absorption occurs. Less than 1 per cent of cobalamin can be absorbed independently of the intrinsic factor.

Cobalamin is also needed for the proper functioning of the nervous system. In its absence, demyelination of the large nerve axons of the spinal cord occurs. Consequently, the anemia resulting from cobalamin deficiency may be accompanied by uncoordinated gait, changes in reflexes, and even impaired mental processes. Treatment with synthetic cobalamin and folic acid is usually effective in reversing these anemic and nervous system symptoms.

Folic acid (pteroylglutamic acid) as it is found in natural food contains 1 molecule of pteroic acid bound to several glutamic acid residues. This pteroyl-polypeptide must be shortened by hydrolysis in the cell to its active form.

Both folic acid and cobalamin are potent growth promotors as well as important factors for the maturation of red blood cells. These vitamins are discussed in more detail in Sec. 21-7.

Ascorbic Acid (Vitamin C)

Ascorbic acid is a strong reducing agent, sensitive to reversible oxidation by many tissues. Plants and most animals can synthesize this vitamin, but the guinea pig and primates, including humans, cannot. Although ascorbic acid can be shown experimentally to influence the activity of many enzyme systems, its exact physiological function is uncertain. However, it is known to be important in hemoglobin formation, probably through its facilitation of iron transport into the cell (see earlier discussion) and for the maturation of red blood cells. In the latter case, ascorbic acid may be acting through its potentiation of folic acid effect. Other functions of ascorbic acid are discussed in Sec. 21-7.

Ascorbic acid is also necessary for the development of *intercellular substance.* This is especially evident in bone and blood vessels: in the absence of ascorbic acid, bone does not ossify properly and the blood vessels are so fragile that they rupture easily, leading to frequent hemorrhages.

In the absence of adequate amounts of ascorbic acid, the deficiency disease *scurvy* results. It was the correlation of scurvy with the lack of fresh fruits and vegetables that first showed that a disease could be caused by the omission of a specific type of food, and it was Lind, a British naval surgeon, who demonstrated that the disease could be cured by food alone. He wrote that citrus fruits "contain something that neither medicine, surgery, or physic could supply," and as a result of his efforts, it became part of the regime of the British navy that all members of the fleet receive one fluid ounce of lemon juice daily (and also resulted in the term "limey" for an Englishman). Scurvy disappeared from the British navy after this.

Bottle-fed infants may develop scurvy unless their milk and cereal diets are supplemented by ascorbic acid in the form of vegetables or fruit juices. Chronic alcoholics may develop scurvy because the highly caloric alcohol satiates the appetite without providing vitamins. Treatment with ascorbic acid results in very rapid improvement within 24 hr, but without treatment, death may occur within a few weeks.

14–7 Anemia

Anemia, or lack of hemoglobin, may be due to:

1. Decreased number of red blood cells with the normal Hb concentration in each.
2. Normal number of red blood cells containing, however, a subnormal amount of Hb.

In addition to anemia resulting from dietary deficiencies of iron or copper (discussed in Sec. 14-6), other types of anemias include acute blood loss, chronic blood loss, bone marrow aplasia, pernicious anemia, and inherited anemias.

Acute Blood Loss

Acute blood loss (hemorrhage) decreases the number of red blood cells because both cells and plasma are lost, but plasma volume is replaced from tissue fluid within about 24 hr. This obviously dilutes the blood and decreases the hematocrit, but the cells are normal with the normal amount of hemoglobin per cell. Gradually over the following weeks, the marrow produces new red blood cells until the normal condition is reestablished.

Chronic Blood Loss

When the blood loss is chronic, such as in internal bleeding, there is a depletion of the body stores of iron, so that, although the bone marrow is producing large numbers of red blood cells, the lack of iron results in a lowered concentration of hemoglobin. The red blood cells are small (microcytic) and pale (hypochromic). Oral administration of iron is quite effective in correcting this type of anemia.

Bone Marrow Aplasia

If the bone marrow is relatively inactive, it produces fewer numbers of cells, even though they are normal in size and hemoglobin concentration. Severe bone marrow aplasia is caused by excessive exposure to X rays or gamma radiation (resulting from fallout from an atomic explosion). Cancerous invasion of the bone marrow or poisoning with chemicals such as benzene will also cause aplastic anemia.

Pernicious Anemia

In the absence of vitamin B_{12} or folic acid, proliferation of red blood cells is impaired, and large megalocytes are produced. These immature cells are exceedingly fragile and are rapidly destroyed in the circulation. Although the megalocytes are larger than normal erythrocytes, they contain the normal amount of hemoglobin, but anemia results from the very low red blood cell count, which may plunge to 1 million/mm^3.

Inherited Anemias

There are several types of inherited anemias in which the fragility of the red blood cell is increased, so that the circulating number is low because of the rapid disintegration of the cells. One of these anemias is *sickle cell anemia*, found in Mediterranean and African races, in which the red blood cells, instead of being biconcave disks, have a sickle shape or other bizarre and distorted forms (Fig. 14-19). The hemoglobin in these abnormal cells (HbS) differs from normal Hb as a result of a one-gene change that replaces glutamic acid by valine.

Sickling appears to be due to a gelation of HbS, which results in an irreversible deformation of the red cell membrane. This may cause hemolytic anemia as the deformed blood cells block the capillaries and sometimes even major arteries and veins, causing death.

Newborn infants do not suffer from sickle cell disease because HbF does not gelate. However, as HbF gradually is replaced by HbS, symptoms of the disease become apparent.

Sickle cell disease appears only in individuals homozygous for HbS. Heterozygous carriers of the sickle cell trait (HbA·HbS) show no abnormalities except under conditions of great stress, such as severe exercise. Interestingly enough, associated with the inheritance of sickle cell disease is an increased resistance to malaria.

Another inherited anemia is *familial hemolytic anemia*, in which the red blood cells are extremely small and fragile; as they disintegrate, large amounts of bilirubin (Sec. 14–3) accumulate, causing yellowing or jaundice of the skin and the whites of the eyes.

Thalassemia is an inherited disease in which the synthesis of globin chains is abnormal. The presence of a defective gene results in the continued production of HbF instead of HbA. In Cooley's anemia (β thalassemia) very few β chains are produced, and severe hemolysis and anemia result from the inbalance of α and β chains. Frequent transfusions are needed beginning in early childhood, and the spleen and liver become greatly enlarged as they work strenuously to destroy the abnormal red cells.

Erythroblastosis Fetalis

This disease is not inherited but is caused by incompatibility of blood types of the mother and the fetus, as a result of which immune bodies from the mother pass through the placenta to destroy the fetal erythrocytes. Usually, the immune bodies in the mother do not reach high enough concentrations in the first pregnancy to do too much damage to the baby, but if the placenta is not properly formed or, as the

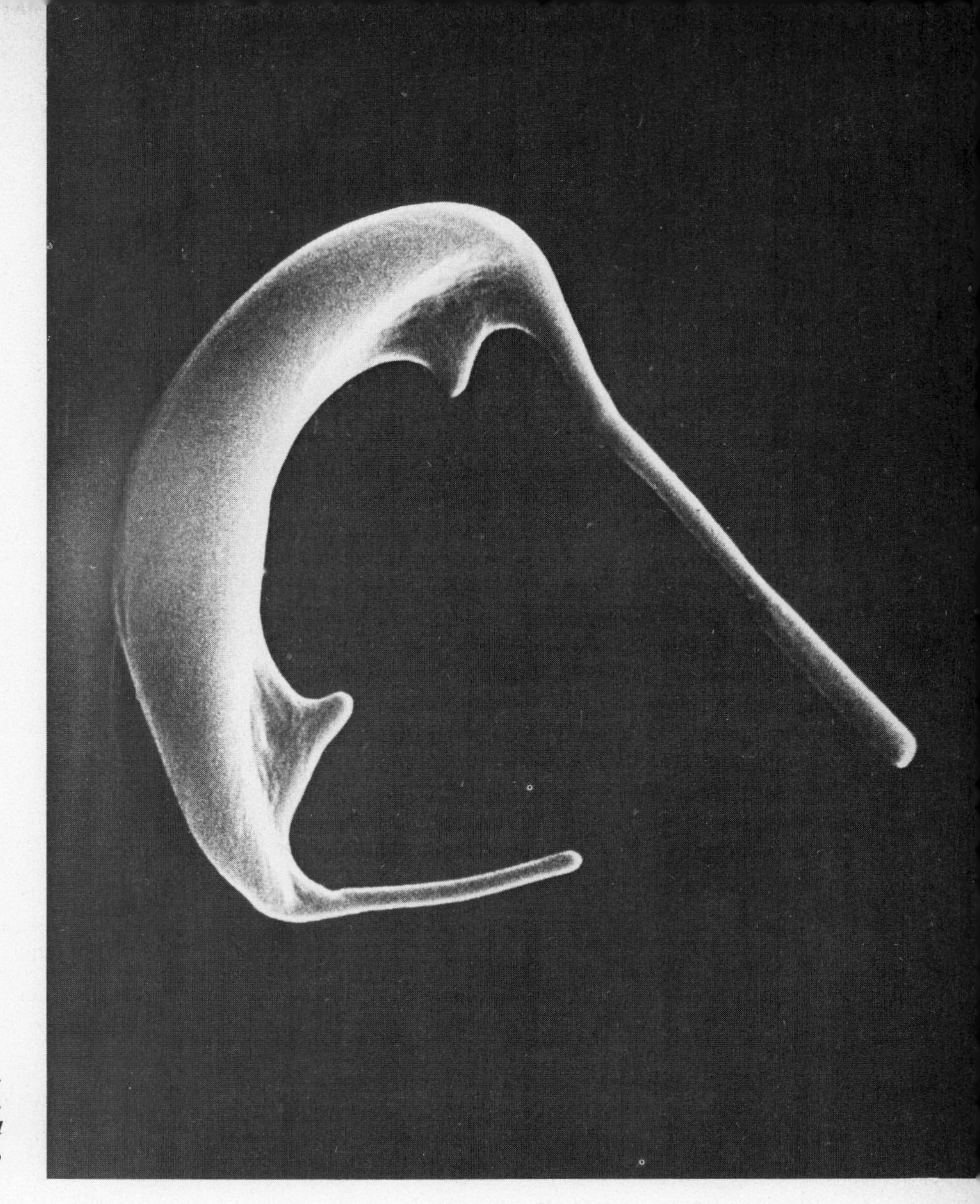

Figure 14-19. **Sickled erythrocytes take many strange shapes when exposed to low oxygen pressures. (From M. Bessis, in *Living Blood Cells and Their Ultrastructure,* Springer-Verlag New York, Inc., New York, 1973.)**

maternal antibodies increase with subsequent pregnancies, more and more immune bodies reach the fetus, and the baby develops severe anemia (Sec. 14-16).

14-8 Polycythemia

Polycythemia is an increase in the number of red blood cells in the circulation. The adverse effects are caused by the resulting increase in blood viscosity and blood volume rather than by changes in oxygen delivery. Blood flow through the vessels is very slow, and the circulation time may be almost twice normal (120 sec instead of about 60 sec).

An increase in red blood cells may be *physiological,* as seen in people living permanently in high altitudes. Residents of areas at 5000 m may have blood cell counts averaging around 7 million per mm^3, as compared to the normal 4.6 to 5.4 million.

Polycythemia may be *pathological,* as in *polycythemia vera.* This disease results from a tumor of the bone marrow, and the number of red blood cells may reach 11 million per mm^3. As the sluggish passage of the blood causes the reduced hemoglobin to remain longer in the venous plexuses of the skin, the skin develops a bluish (cyanotic) color rather than the normal pink tinge.

14-9 White Blood Cells: Leukocytes

Functions

The main function of the leukocytes is to protect the organism against foreign invaders, including bacteria and viruses. Regions of the body that are particularly susceptible to such invasion, such as the respiratory tract, the mouth, and the gastrointestinal tract, have reservoirs of leukocytes to ward off infection. Leukocytes can be functionally and morphologically divided into two main groups:

1. *Phagocytes,* which destroy and remove foreign material and also the aged and damaged cells of the body itself.
2. *Immunocytes,* which give specificity to the action of the phagocytes, selecting the object of their attack. These immune responses are essential for immunity to certain diseases, rejection of foreign tissue transplants, and possibly even for the defense of the body against the growth of malignant cells.

Because most leukocytic activity occurs in the tissues, blood serves mainly as a transport vehicle for the leukocytes. Consequently, this chapter will restrict itself to the identification of the various leukocytes found in blood; the phagocytic and immunological functions of these cells will be discussed in Chapters 19 and 20.

Structure and Differentiation

The white blood cells in general receive their name because, unlike the red erythrocytes, they possess no hemoglobin. In the usual blood smear, which is stained with different dyes, the leukocytes show varying coloration according to their affinity for acid or basic dyes. This makes differentiation of the several types much easier. The leukocytes are distinguished in terms of the form of their nucleus, general size and shape, and color, and the presence or absence of granules in the cytoplasm.

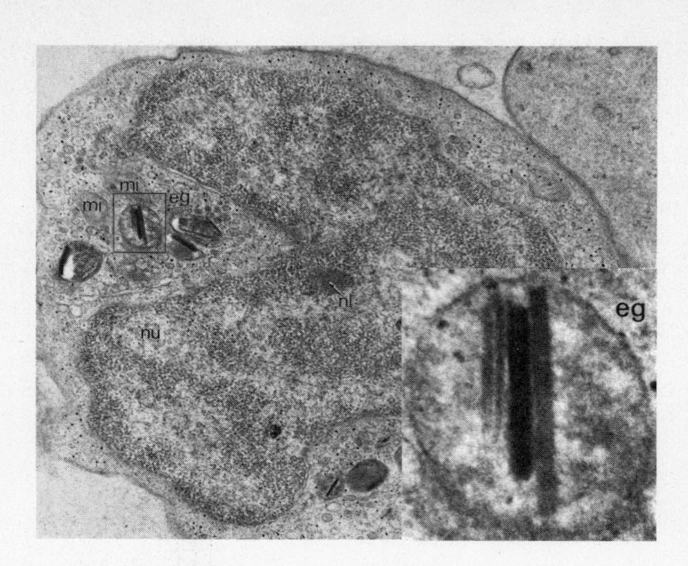

Figure 14-21. **Eosinophil from goldfish blood. The small inset shows the typical crystals found within the eosinophil granules. Magnification of the cell is ×34,500, that of the crystals is ×74,800. nu = nucleus, mi = mitochondria, eg = eosinophil granules, and nl = nucleolus. [From E. L. Weinreb, *Anat. Rec.,* 147:237 (1963).]**

Figure 14-20. **Leukocytes of peripheral blood. The cells in the right column are granulocytes. (Courtesy of S. Piliero, Dental School, New York University.)**

Lymphocyte

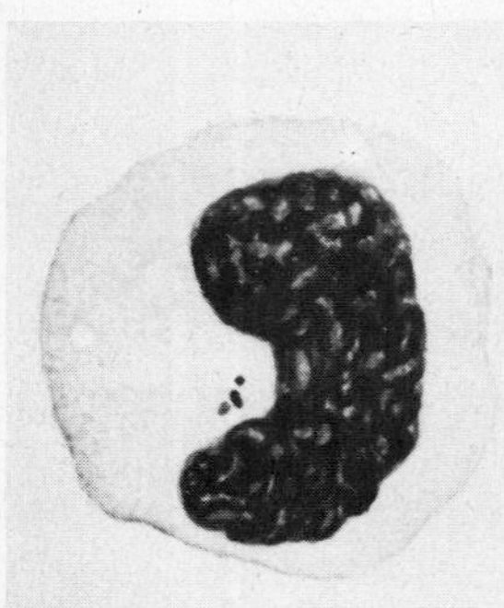

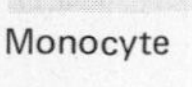

Monocyte

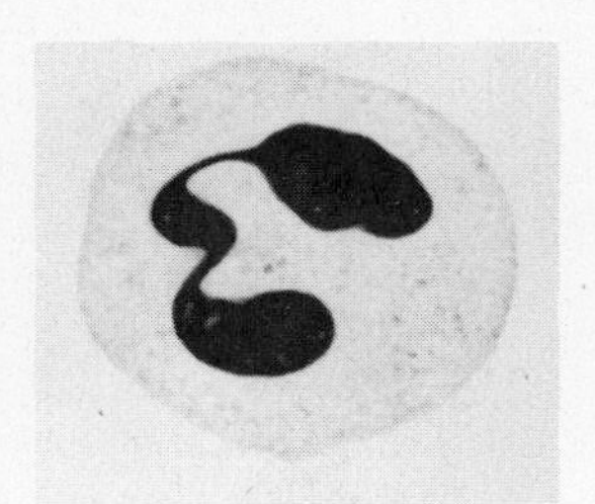

Segmented neutrophil

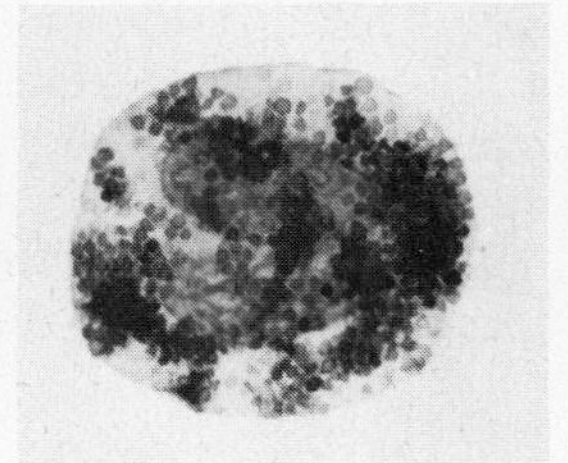

Basophil

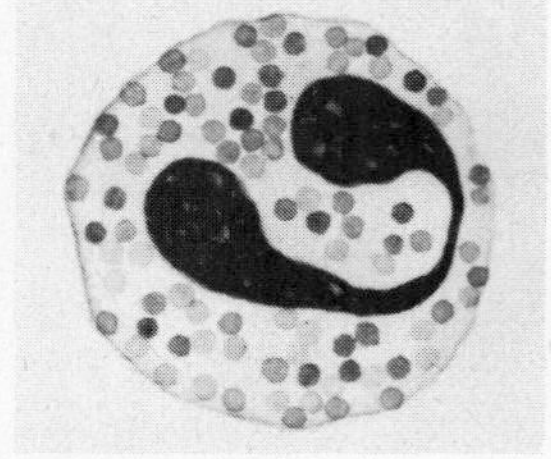

Eosinophil

Phagocytes

This group of leukocytes can in turn be divided into two groups, the *granulocytes* and the *monocyte-macrophages.* Both types originate from the bone marrow. They are illustrated in Fig. 14-20.

Granulocytes. These cells get their name from the presence of granules in the cytoplasm. As the granulocytes also are distinguished by the possession of a large lobed nucleus, they are often called *polymorphonuclear* cells, or *"polys."* Granulocytes may be divided into three main types according to the characteristics of the cytoplasmic granules. When the cells are stained with Wright's stain, which is a combination of methylene blue and several other dyes, they can be identified chiefly by the color of their granules, but with the electron microscope differences in the structure of the granules can be seen (Fig. 14-21):

1. *Neutrophils,* small pink granules. These cells are very active and, in the living state, send out pseudopodia.
2. *Eosinophils,* coarse reddish-orange granules, fairly uniform in size.
3. *Basophils.* The cytoplasmic granules are much more irregular in size and number and stain reddish purple to blue black. The granules often partly obscure the nucleus.

Granulocytes spend less than a day in the circulation before they migrate through the endothelial wall of the capillary and are stored or destroyed in various tissues.

Monocyte-Macrophages. These cells have a large nucleus and scanty cytoplasm without any conspicuous granulation. The mature monocyte is a large cell with a diameter between 20 and 30 μm. The cytoplasm is grayish blue, and Wright's stain also shows many pinkish lysosomes. In the living state, the monocyte is very active and constantly forms pseudopodia. The monocyte escapes from the blood by amoeboid movement through the capillary wall. Once in the tissues, the monocyte is transformed into a large macrophage, 50 μm or more in diameter (Fig. 14-22). These macrophages are filled with lysosomes, ready to digest any engulfed foreign material.

Some of the macrophages are *mobile* and are found in the alveoli of the lungs, the peritoneal cavity, and the inflammatory exudate (Sec. 19-4). Other macrophages become *attached* to tissues, where they are able to proliferate and wall off invading particles. Many of these tissue macrophages, or *histiocytes,* are found lining the sinuses of the lymph nodes, forming part of the walls of the alveoli of the lungs, and lining the liver sinuses, where they are called *Kupffer cells.*

A further line of defense is formed when these macrophages become attached to the reticular meshwork of the spleen and bone marrow, forming part of the *reticuloendothelial system.*

IMMUNOCYTES

These leukocytes can be subclassified into *lymphocytes* and *plasma cells* (Fig. 14-23). As was previously described, all lymphocytes originate from stem cells in the bone marrow, but some continue their differentiation in the bone marrow (B lymphocytes), whereas others differentiate in the thymus (T lymphocytes).

Figure 14-22. **Transformation of a monocyte into a macrophage that is ingesting foreign particles by phagocytosis.**

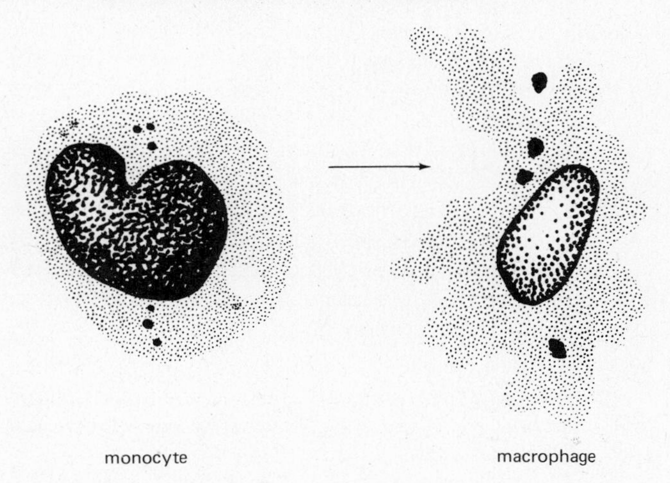

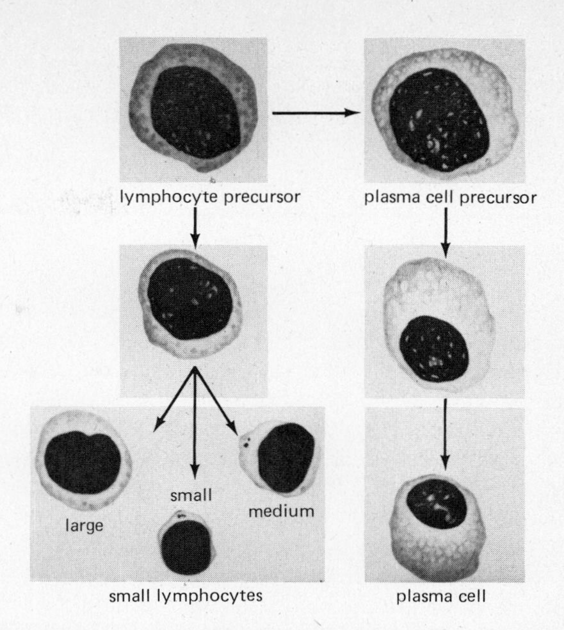

Figure 14-23. **Lymphocytes and plasma cells. (Courtesy of S. Piliero, Dental School, New York University.)**

B Lymphocytes. B lymphocytes depart from the bone marrow to settle and proliferate in the spleen and lymph nodes, where they form a relatively stable pool and do not circulate. These large lymphocytes are characterized by a large nucleus and bright blue cytoplasm after staining with Wright's stain. The cytoplasm may contain a few reddish-violet granules. The lymphocytes mature into the nondividing *plasma cell,* which can be recognized by its eccentrically placed nucleus and deep blue-green cytoplasm. Plasma cells are usually loaded with secretory vacuoles or granules, for they are highly specialized to synthesize a wide variety of antibodies.

T Lymphocytes. Immunocytes that have come from the thymus, T lymphocytes, are constantly circulating through the blood to the lymph nodes and spleen and thence back to the blood. Their movements can be followed by tracing the fate of radioactively labelled lymphocytes over a period of several months. The T lymphocytes mediate immunity but do not secrete circulating antibodies. Most of the circulating lymphocytes in the blood and lymph are T lymphocytes.

Differential White Blood Cell Count

The *number of circulating phagocytes* varies considerably with the physiological condition (age, activity, sex, etc.); it also varies in pathological states, ranging from a mild respiratory infection to acute *leukocytosis* (severe increase in leukocyte numbers), and *leukopenia* (severe drop in leukocyte numbers). In cases of increased leukocyte production, as might be expected, many younger immature forms are found liberated into the blood from the bone marrow, and these cells

Table 14–3. **Differential White Blood Cell Count (per cent)**

PHAGOCYTES (granulocytes and monocyte macrophages)				IMMUNOCYTES
Neutrophils	Eosinophils	Basophils	Monocytes	Lymphocytes
59	2.7	0.5	4	34

do not yet have the typically lobed nucleus. Gordon and his co-workers, at New York University (1973), have shown that there is a specific factor in the plasma that causes the release of white blood cells from the bone marrow. This *leukocytosis-inducing factor* (LIF) can be demonstrated in animals that have been depleted of large numbers of leukocytes: the injection of LIF rapidly increases the number of circulating white blood cells.

The *number of immunocytes* produced and liberated from the bone marrow and thymus may bear little relationship to the number found in the blood, for many are trapped in the tissues or lymph. In the blood, however, they form about 30 per cent of the leukocyte count (Table 14–3).

The normal leukocyte count is 5000 to 10,000/mm^3 blood.

The final circulating number of white blood cells is regulated in inflammation by stimulating and inhibiting factors apparently produced by the inflamed tissue itself (Sec. 19–4).

Differential Leukocyte Numbers in Disease

Apart from the important role of leukocytes in protecting the organism from infection, there is a relationship between specific changes in numbers of leukocyte types in many diseases, but the causal relationship is unclear.

Plasma cells

Plasma cells increase markedly in number in German measles (rubella) and infectious hepatitis. They produce large quantities of antibody.

Eosinophils

Eosinophils increase considerably in numbers in allergies and in parasitic diseases. The decrease in eosinophils during stress has been associated with the increased amounts of ACTH and cortisone secreted.

Basophils

Basophils also disappear during stress or cortisone treatment. They may be associated with the production of the anticoagulant *heparin*.

Lymphocytes

Lymphocytes may increase in number in some acute childhood diseases (whooping cough and acute infectious lymphocytosis). Lymphocyte numbers may rise as high as 50,000 to 100,000/mm^3.

Infectious mononucleosis, a disease prevalent in young adults, is believed to be caused by a virus. This infection causes only a modest increase (20,000/mm^3) in lymphocyte numbers, but there is a disproportionate number of young and atypical cells.

An increase in lymphocyte numbers that is not associated with infection or endocrine disorders may indicate a pathological proliferation of these cells, as in lymphatic leukemia.

14–10 Platelets: Thrombocytes

Repair of Ruptured Blood Vessels

Although platelets have been described in the circulation since 1849, most of our knowledge of their structure and function has been obtained in the last 10 years. Platelets are small, complex, and fragile cells found in large numbers in the blood (250,000/mm^3). Platelets are immediately mobilized to prevent blood loss when a blood vessel is damaged. They seal off the leak in the blood vessel in several ways:

1. Platelets form a *temporary platelet plug* in the wall of the damaged vessel by their ability to *adhere* physically to cut surfaces where collagen is exposed. The development of an adhesive surface appears to be associated with a very rapid change in shape from a disk to a spiny form (Fig. 14-24).
2. The turbulence and movement at the damaged area causes the *aggregation* of more platelets and the release from the platelets of *ADP*, a most powerful inducer of platelet aggregation. This positive feedback mechanism causes the aggregation of still more platelets and the strengthening of the platelet plug. Simultaneously, platelets release *calcium*, an essential factor for aggregation.
3. Platelets contain most of the *serotonin* of the body, and this powerful vasoconstrictor is released when the platelets fragment. *Vasoconstriction* aids in plugging the vascular leak.
4. A *permanent seal* is obtained following the release of an essential coagulation factor, *thromboplastin*, from the platelets. Thromboplastin initiates a series of reactions that results in the formation of a permanent clot, in which the platelets themselves, as well as fibrin, are incorporated. Blood coagulation is discussed in more detail in the next section.

All the mechanisms described above contribute to *hemostasis*, which is the arrest of bleeding after the blood vessel

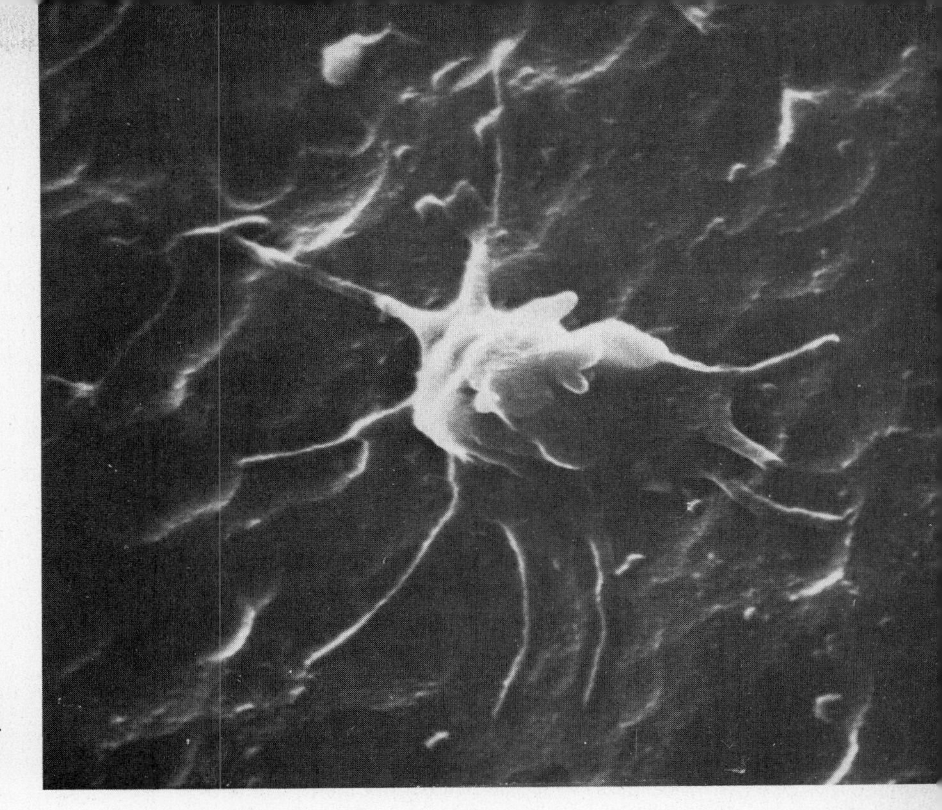

Figure 14-24. **A scanning electronmicrograph of a spiny platelet. The spines are pseudopodia spreading on the surface to which they are attached. (Courtesy of T. Hovig, Oslo University Institute of Pathological Anatomy, Oslo, Norway.)**

wall is ruptured. It is of obvious importance that the triggering of the change in adhesiveness of the platelet plasma membrane be under control, because the formation of clots within the intact circulation is extremely serious, blocking off vital circulation to the tissues. Apparently, the presence of collagen in the deeper layers of the torn vessel is one of these controlling factors; in its absence, adhesion and aggregation do not normally occur.

Maintenance of Normal Blood Vessels

Platelets are also needed to maintain the normal health of the blood vessels, and individuals lacking the proper number of platelets (*thrombocytopenia*) have a decreased capillary resistance. An increase in internal blood pressure or a slight external pressure will cause leakage of blood into the tissues (*petechiae*) in these individuals.

Figure 14-25. **Growth of megakaryocytes (giant cells with excess DNA) and the formation of platelets. (Courtesy of S. Piliero, Dental School, New York University.)**

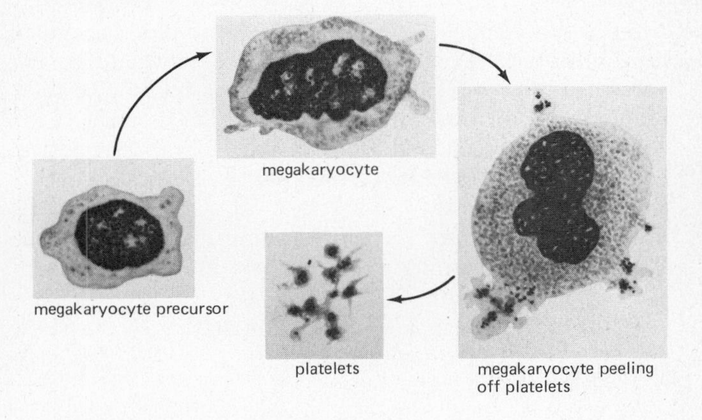

Formation of Platelets

Platelets, like all other blood cells, originate from a multipotential stem cell in the bone marrow. One of these stem cell compartments is committed to form *megakaryocytes,* giant cells that are the precursors of the platelets. Megakaryocytes may be as large as 30 μm in diameter, and in the process of differentiation the nucleus divides several times without any accompanying division of the cytoplasm. The resulting giant cell has many times the normal diploid compliment of DNA. At this stage, the cytoplasm becomes channeled, and the platelets peel off from the mother cell (Fig. 14-25). Although these fragments of the megakaryocyte are nonnucleated, they are packed with granules, vesicles, microtubules, and mitochondria.

The newly formed thrombocytes leave the bone marrow to circulate for about 8 to 10 days before they are destroyed by the reticuloendothelial tissue of the liver and spleen. The spleen holds about one third of the total number of intact platelets, and sympathetic stimulation, resulting from the stress of hemorrhage, will cause the spleen to contract and release the stored platelets into the circulation.

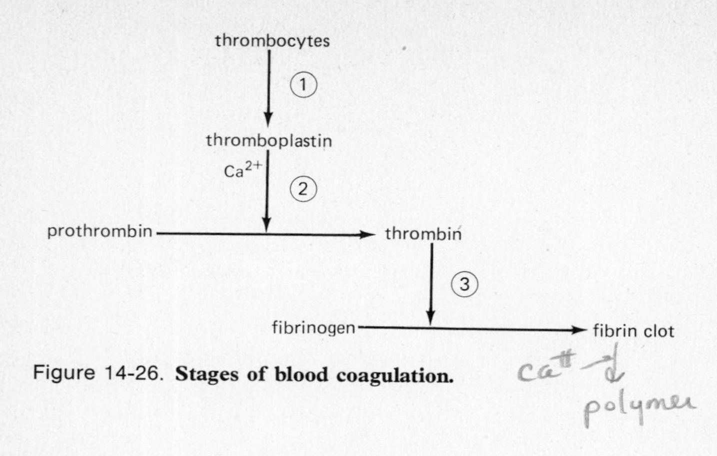

Figure 14-26. **Stages of blood coagulation.**

14–11 Blood Coagulation

The basic theory of blood coagulation was developed in 1904 by Morowitz, Fuld, and Spiro and is still fundamentally correct. It may be divided into three stages, as illustrated in Fig. 14-26:

1. The release of thromboplastin from the thrombocytes in blood and from damaged tissue cells.
2. The activation of prothrombin to thrombin by thromboplastin in the presence of calcium ions.
3. The conversion of soluble fibrinogen to insoluble fibrin by thrombin. The fibrin strands form the clot that blocks blood flow through the damaged vessel.

Since this original description of *in vitro* blood coagulation, many additional factors have been shown to be involved in the coagulation of blood in the body. The total number of known coagulation factors at the moment is 13: 12 different proteins and ionic calcium. These coagulation factors are designated by Roman numerals in the order in which they were discovered (Factor VI is no longer used). Table 14-4 lists these factors by number and description. We shall be concerning ourselves here only with Factors I-IV and XIII.

In the modern "cascade" concept of clotting, as proposed by MacFarlane (1970), the simple series of three reactions depicted in the table is extended to include the many other factors known to be involved in coagulation. Each of the protein coagulation factors is seen as existing in an inactive (procoagulant) state. The activation of one factor specifically activates the next factor in the series, with the final cascade of reactions achieving the formation of the fibrin clot.

Table 14–4. **Plasma Coagulation Factors and Their Synonyms**

Factor I	Fibrinogen
Factor II	Prothrombin
Factor III	Tissue thromboplastin
Factor IV	Calcium
Factor V	Proaccelerin
Factor VII	Proconvertin; SPCA
Factor VIII	Antihemophilic globulin (AHG)
Factor IX	Plasma thromboplastin component (PTC), Christmas factor
Factor X	Stuart-Prower factor
Factor XI	Plasma thromboplastin antecedent (PTA)
Factor XII	Hageman factor
Factor XIII	Fibrin stabilizing factor

Release of Thromboplastin

FROM THE TISSUES (EXTRINSIC)

Practically all tissues contain thromboplastin (Factor III). When the tissues are damaged, thromboplastin is released; in the presence of calcium ions (Factor IV), the thromboplastin initiates coagulation outside the blood vessels. This seems to be a less complex reaction than the release of thromboplastin from the platelets.

FROM BLOOD PLATELETS (INTRINSIC)

Contact with a damaged surface, and with exposed collagen in the walls of the broken vessel, causes the platelets to adhere and aggregate at the site of the injury. This is the stimulus for the release of thromboplastin and calcium, as was described previously. Several other factors, activated in a cascade manner, are necessary for the generation of thromboplastin from the thrombocytes.

Activation of Prothrombin to Thrombin

Prothrombin (Factor II), a medium-sized protein with a molecular weight of 62,700, is produced by the liver. Its formation depends on adequate amounts of Vitamin K (naphthoquinone), which is widely distributed in natural foods, especially in alfalfa and other green plants. The dietary supply is also supplemented by the synthetic activity of intestinal bacteria.

Thromboplastin, helped by three other coagulation factors, rapidly converts prothrombin to thrombin. The thrombin formed not only acts "forward" to precipitate the next series of reactions in clot formation, but also acts "back" on the platelets to increase platelet aggregation, another example of a positive feedback mechanism.

Fibrinogen Polymerization to Fibrin

produced by liver

Fibrinogen (Factor I) is a plasma protein found in much greater concentration in the blood than are the other coagulation factors. The normal concentration of fibrinogen is 200 to 400 mg/mm³ of blood. Fibrinogen, with a molecular weight of 340,000 is a large molecule made of two identical subunits, each consisting of three polypeptide chains, α, β, and γ. The chains are bound together by disulfide bonds at each end of the molecule (Fig. 14-27).

When thrombin attacks fibrinogen, it splits off small pieces, fibrinopeptides, from the parent monomer, which then rapidly undergoes intermolecular association to form a polymer. This is the elongated, insoluble fibrin molecule, held together by hydrogen bonds.

The next step stabilizes the clot by linking the fibrin threads to one another, side by side, through cross-linking peptide bonds. This stabilization requires the presence of the active form of Factor XIII.

14–12 The Blood Clot and Clot Retraction

The blood clot is made up of the fibrin network of tangled threads in which blood cells, platelets, and plasma are trapped. A few minutes after the clot has formed it begins to shrink, or *retract,* and most of the fluid it contains oozes out in the form of *serum.* Serum contains no fibrinogen and very few of the other coagulating factors, most of which are trapped in the meshes of the clot.

The retraction of the clot enables it to pull the surfaces of small wounds together. This shrinking is probably caused by a contractile protein that is released from the platelets caught in the clot. This protein uses ATP as an energy source just as the contractile proteins of muscle do, a very interesting example of the repeated use of the same efficient energy source for apparently very different physiological processes.

Dissolution of the Blood Clot

Another system takes care of the removal or dissolution of the blood clot. The inactive plasma protein component of this system is a *plasminogen.* The presence of a fibrin clot causes the rapid release of a *plasminogen activator* into the clot, where it converts plasminogen to *plasmin,* a proteolytic enzyme that digests the fibrin threads, causing lysis of the clot.

A plasminogen activator (urokinase) is also found in the urinary tract and presumably prevents clots from forming and blocking the flow of urine.

14–13 Mechanisms That Prevent Coagulation in Normal Blood Vessels

Physical Factors

Normally, blood does not clot as it flows through the vascular system because the smooth endothelial lining of the blood vessels minimizes abrasive surface contact with the platelets. In addition, the negative charge of the monomolecular protein layer that is adsorbed on the endothelial cell surface repels the negatively charged protein coagulation factors and the platelets.

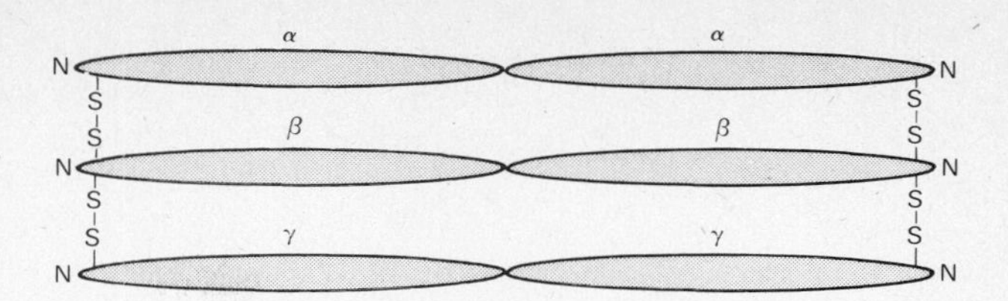

Figure 14-27. **Each of the two subunits of fibrinogen is made up of an α, a β, and a γ polypeptide chain. The chains are linked by disulfide bonds (—S—S—). The N-terminal amino acids are indicated by N.**

When the blood vessel is damaged, both its smooth lining and its negative charge are lost; these changes initiate the reactions of the intrinsic pathway of clotting.

Anticoagulants

Blood contains several important anticoagulants.

Fibrin

Fibrin, which absorbs 90 per cent of the thrombin formed, prevents the clot from spreading too far.

Antithrombin III

An alpha globulin that binds excess thrombin formed, antithrombin III also inactivates the thrombin that has been trapped and adsorbed on the fibrin threads of the clot.

Heparin

A powerful anticoagulant, heparin is a polysaccharide that is produced by many different types of cells, especially the mast cells of connective tissue and the basophils of circulating blood. The liver and lungs are especially well supplied with heparin-producing mast cells.

Dicoumarol

Dicoumarol is a toxin found in fermented sweet clover and acts on the liver, preventing the formation of prothrombin. Cattle that have eaten this clover suffer from hemorrhagic disease caused by the failure of the blood to clot.

14–14 Prevention of Blood Coagulation (In Vitro)

Blood removed from the body with a hypodermic syringe coagulates rapidly because of the disintegration of the platelets under this harsh treatment. If the needle and syringe are coated with *silicone,* the smooth surfaces do much less damage to these fragile cells, and coagulation does not occur. The coating of the syringe and needle with *heparin* is even more effective.

Another mechanism that inhibits blood clotting is the removal of calcium ions from the blood, since calcium is required for several stages in blood coagulation. The addition of anions such as *oxalate* and *citrate,* which combine with Ca^{2+} to form insoluble salts, effectively removes the Ca^{2+} from the blood and inhibits coagulation. Citrate anticoagulants are clinically preferable to oxalate anticoagulants, for the oxalates are toxic. Moderate amounts of blood treated with citrate can be safely transfused, for the liver is able to metabolize citrate through the usual glucose pathways. Large amounts of citrated blood, however, rapidly transfused, deplete the calcium levels in the blood and can result in convulsions and death.

14–15 Disorders of Blood Coagulation

Inadequate Numbers of Platelets (Thrombocytopenia)

Lack of platelets may be caused by inadequate hematopoiesis resulting from a folic acid or cobalamin (vitamin B_{12}) deficiency, or else the thrombocytopenia may be the result of an increased rate of platelet destruction. Infection, antibodies, or drug sensitivity can all enhance platelet destruction.

When there are too few circulating platelets, bleeding time is prolonged and clot retraction impaired. Small red patches (petechiae) are seen on the skin and mucous membranes, and more severe hemorrhages may also occur.

Deficiencies in Components of the Coagulation System

DIETARY

Deficiencies due to dietary inadequacies are rare. A deficiency of naphthoquinone (vitamin K) is not usually due to too little in the diet but is more likely the consequence of antibiotic medication interfering with the normal synthesis of this vitamin by the intestinal bacteria. Another common cause for naphthoquinone deficiency is inadequate absorption of this fat-soluble vitamin. If the liver is diseased, or if the common bile duct is obstructed, the bile salts necessary for absorption are absent, and naphthoquinone deficiency results.

Because the synthesis of prothrombin and other protein-coagulating factors depends upon adequate supplies of naphthoquinone the concentration of these plasma proteins falls, and hemorrhages occur. Naphthoquinone therapy is usually able to correct the bleeding tendencies within a few hours.

Table 14–5. **Agglutinogens and Agglutinins in Different Blood Groups**

	Protein in rbc (agglutinogen)	Protein in Plasma (agglutinin)
Type A blood	A	AntiB
Type B blood	B	AntiA
Type AB blood	A and B	None
Type O blood	None	AntiA and antiB

INHERITED

Hemophilia is an inherited deficiency of certain components of the coagulating system. There are several types of hemophilia: the classical hemophilia is a sex-linked disorder transmitted by female carriers to half their sons. In this type of hemophilia there is deficiency of the antihemophilic factor (Factor VIII).

Severe hemophilia manifests itself by uncontrolled bleeding on injury, and by arthritis and destruction of the joints because of chronic bleeding into the joints. The transfusion of normal fresh plasma or of the antihemophilic factor will correct the bleeding tendency for a few days.

14–16 Blood Groups

Evolutionary Significance of the Proteins of the Blood

The proteins of the blood owe their specific structure to the genetic code built into the DNA in the nucleus. It is not surprising to find that animal groups that are most closely related have the greatest similarities in their types of proteins. Although there are great differences in blood proteins between the primates and the birds or amphibia, for example, among primates the differences are less marked. Newer research techniques have been able to analyze the immune reactions of the proteins of primates, and this fascinating work has helped greatly in determining the closeness of the relationship between humans and various other primates. It has been shown that humans are more closely related to the African apes (chimpanzee and gorilla) than they are to the Asiatic apes (gibbon and orangutan). There is probably as much evolutionary time between the gibbon and the chimpanzee as there is between humans and the chimpanzee.

Blood proteins also vary considerably among individual members of the human species. There are *four main types and many subtypes* of blood proteins in the human being. these, too, are genetically determined (Sec. 27-3) and demonstrate again the closeness of the relationship between individuals. Identical twins have identical blood groups. Because these groups are created by the presence or absence of specific proteins, and because the introduction of a foreign protein into the circulation induces the formation of an antibody (Sec. 20-6), any exchange of blood between individuals, as in a blood transfusion, must be strictly limited to the use of the same type of blood. Today we have become much more wary in giving blood transfusions unless they are absolutely essential, because even a slight interaction of the bloods can result in *agglutination* or sticking together of the red blood cells, which can block some small vessels and lead to damage of the kidneys or other vital organs.

Nature of the Agglutinin-Agglutinogen Reaction

The *four main blood groups* in humans are A, B, AB, and, most common in Caucasian races, O. We can consider them as representing the presence of a protein A, or protein B, or both A and B (AB), or neither (O), on the membrane of the red blood cells of the individual. These proteins are the *agglutinogens,* antigens that cause agglutination.

Present in the plasma are other proteins, the *agglutinins,* which act as antibodies to a foreign agglutinogen. They attach themselves to the surface of the red blood cells and cause them to clump together and agglutinate. Obviously, an antiA agglutinin could not be present normally in the blood of an individual with the A-type protein in his red blood cells, but antiB will be present. Similarly, a B-group individual will have antiA in his plasma. Because the AB individual has both types of agglutinogen, he can have neither agglutinin in the serum, whereas the O individual has both antiA and antiB. This is represented in Table 14–5.

Blood Transfusions

When a blood transfusion is given, the plasma of the donor is rapidly diluted by the plasma of the recipient, so that the concentration (titer) of the agglutinins of the donor is so low that it cannot cause agglutination. However, if the protein of the donor's blood cells is of a different type from that of the recipient, the agglutinin of the recipient's plasma will agglutinate the foreign cells.

Because type O blood has no agglutinogen in the red blood cells to be agglutinated, this type can be given theoretically to any recipient. Type O is therefore called the *universal donor.*

Type AB blood, on the other hand, has no agglutinins in the plasma to react with foreign agglutinogens in the donated red blood cells; therefore, regardless of the type of cells introduced, there should be no agglutination. Type AB is the *universal recipient.*

Because there are many other blood groups—M, N, P, S, and several others, as well as the important Rh (Rhesus) factor in human blood—it is considered inadvisable to use the universal donor or universal recipient as the basis for blood transfusions, unless the emergency is so great that there is no time to get the matching blood group. In all cases of blood transfusions, however, it is essential that the bloods of the donor and recipient be *individually matched* before the transfusion is given. That means that samples are taken of the blood of recipient and would-be donor, and the cells of one suspended in the plasma of the other. If no clumping of the cells of either donor or recipient occurs, it is deemed safe to make the transfusion.

The Rh Factor

This is a type of blood protein that was first shown to be present in the blood of the rhesus monkey. If it is present in the blood, the blood is termed *Rh positive;* if it is absent, *Rh negative.* The Rh factor differs from the four blood groups described above in that there is normally no antiRh protein present in the plasma of Rh-negative individuals until that person has been exposed to relatively large amounts of Rh proteins. This occurs if an Rh-negative individual is given Rh-positive blood, or if an Rh-negative mother bears the children of an Rh-positive father. The Rh-negative individual then builds up increasingly large titers of AntiRh agglutinins, and if any further Rh-positive blood is introduced, agglutination of these cells will occur.

Erythroblastosis fetalis

This disease of the newborn is caused by the formation of antiRh agglutinins in the blood of the mother during pregnancy. It occurs in cases in which the mother is Rh negative, the father Rh positive, and the baby will have inherited the Rh-positive characteristic from the father. Probably all the cells of the baby's body have this Rh protein, and when some cells degenerate, the protein slowly diffuses through the placenta to the mother, starting the process of *sensitization* to the Rh protein. Even intact red blood cells appear to be able to pass through the placenta. In response to this foreign protein in her circulation, the mother produces the antiRh agglutinins. These agglutinins diffuse through the placenta to the developing embryo, but usually it is not until the second or third pregnancy that the mother builds up sufficiently great titers of the agglutinin to damage the child.

The agglutinins of the mother that enter the circulation of the fetus cause the baby's red blood cells to clump together, and these clumps block small blood vessels. As the clumps disintegrate, the cells break up, releasing hemoglobin, which is converted by the reticuloendothelial system into bilirubin, a yellow pigment. The newly born child appears yellow, or *jaundiced,* as a result of the accumulation of this pigment, but far more serious is the continued effect of the mother's agglutinins within the baby's circulation. More and more red blood cells are destroyed, the hemoglobin level falls, and severe and even fatal anemia can ensue. The liver and spleen, the blood-producing tissues of the embryo and newborn, desperately attempt to replace these cells with immature red blood cells, the erythroblasts. The presence of large numbers of erythroblasts in the circulation has given this disease its name.

The treatment is to replace the entire blood volume of the baby with Rh-negative blood over a period of about one and one-half hours. Now there will be very little destruction of the red blood cells. Slowly, over a period of weeks, the baby will replace this with its own Rh-positive blood, but by this time the agglutinins it received from its mother will have been destroyed, and there will be no more destruction of the Rh-positive cells.

A recent, fairly effective treatment is *passive immunization of the mother* with antiRh agglutinins, shortly after delivery of the first baby. This serves as prophylaxis for subsequent children, who will not be exposed to maternal agglutinins as a result. Since the introduction of this new drug, RhoGAM, the incidence of this disease has dropped from 45/1000 births in 1970 to 23/1000 births in 1974, the last year for which figures were available. This topic is discussed further in Sec. 20-8.

14–17 Regulation of the pH of the Blood

The pH *of the Blood and Body Fluids*

In the various chapters throughout this book, we have been considering the maintenance of the internal environment in its many facets as the most essential function of the various organ systems. It has been obvious that the regulation of the tissue fluids is dependent on the constancy of the composition of the blood; it must be evident also that alterations in tissue activities with the resulting production of acids and alkalis will temporarily affect the acid-base equilibrium of the blood.

Not only are acids produced by the metabolic processes of the body, but most of the products of the breakdown of foods are acidic, and the chief problem of the body is buffering against acids rather than against bases.

The surprising aspect is that despite the many chemical reactions that result in the production of strong acids and large volumes of carbon dioxide, the pH of blood and the other body fluids remains within a very narrow range. The pH of venous blood is slightly lower (7.36) than that of arterial blood (7.4), and the interior of the cell is slightly more acid (about 7.1), and this range is delicately controlled by three important systems. These are:

1. The buffer systems of the blood.
2. The respiratory system.
3. The regulatory action of the kidneys.

Disturbances of the pH *of the Blood*

Acidosis

The vital role of these systems is seen most clearly when, because of some pathologic condition, they do not function properly. An upset of the acid-base balance is seen frequently in infants suffering from severe diarrhea, which results in loss through the feces of much of the buffer system (bicarbonate) of the blood. The pH of the blood falls, and the resulting *acidosis* may affect the respiration and the central nervous system so markedly that the baby dies after being in a state of coma.

In *diabetes mellitus,* when the normal pathway for the metabolism of glucose is disturbed, so much strong acid is formed that the blood is not able to neutralize it nor the kidney to excrete it. Similarly, if the kidneys are so inflamed that they cannot function properly (*nephritis*), even normal levels of acid cannot be excreted, and the accumulation of acid in the blood results in acidosis.

Alkalosis

The accumulation of alkaline substances in the blood does not occur very often because they are formed by very few of the metabolic processes of the body. Alkalosis may occur after taking large quantities of alkaline drugs, including sodium bicarbonate. Or it may occur from excessive vomiting, when large amounts of hydrochloric acid are lost from the stomach, depleting the body of acidic ions.

Mechanism of a Buffer System

A buffer system is one that contains enough of a weak acid to neutralize additional base and enough of a weak base to neutralize any acid that is added to it.

A weak acid dissociates in water to yield hydrogen ions (H^+) and a base:

$$\underbrace{\boxed{\text{H}\cdot\text{base}}}_{\text{weak acid}} \rightleftharpoons \text{H}^+ + \text{base}^-$$

The dissociated ions exist in equilibrium with the undissociated acid, and the particular equilibrium point is expressed for each acid as its *dissociation constant* or *K*. Carbonic acid (H_2CO_3), the weak acid that forms a very important part of the buffer system of the blood and body fluids, dissociates to yield hydrogen ions (H^+) and bicarbonate ions (HCO_3^-) as the base.

The equilibrium between the undissociated carbonic acid and the H^+ and HCO_3^- formed by dissociation may be shown as follows:

$$\frac{\text{H}^+ \times \text{HCO}_3^-}{\boxed{\text{H}_2\text{CO}_3}} = K$$

Because K is a constant for carbonic acid, any change in the concentration of H^+ will change the proportion of HCO_3^- to H_2CO_3. The normal ratio is about 20 : 1 in blood plasma, which represents a pH of 7.4.

If acid is added to this buffer system, the increase in the number of hydrogen ions will decrease the ratio of bicarbonate ions to carbonic acid, because the hydrogen ions combine with the bicarbonate ions (thus decreasing their concentration) to produce carbonic acid. This is a particularly efficient way to neutralize acid in the body, because the carbonic acid quickly decomposes into carbon dioxide, which is removed through the lungs.

$$\boxed{H^+} + HCO_3^- \rightleftharpoons \boxed{H_2CO_3} \rightleftharpoons HOH + \boxed{CO_2} \uparrow \text{exhaled}$$

increase; Carbonic acid; Water; Carbon dioxide

If *base is added to this buffer system,* there is not much of a decrease in the hydrogen ion concentration. The strong base, sodium hydroxide, dissociates to form sodium ions (Na^+) and hydroxyl ions (OH^-). These ions then combine with the buffer as follows:

1. $NaOH \rightleftharpoons Na^+ + OH^-$

2. $Na^+ + \boxed{OH^-} + H^+ + HCO_3^- \rightleftharpoons \boxed{NaHCO_3} + \boxed{HOH}$

increase; Sodium bicarbonate (a salt)

The efficiency of the bicarbonate buffer system is greater, however, in its ability to buffer acids.

Buffers of the Blood: Hemoglobin, Plasma Proteins, and Plasma Alkali Reserve

The most important buffers of the blood are the *proteins,* especially the protein *hemoglobin,* contained within the red blood cells. Proteins are very effective buffers because they can combine either with acid or with base (see the section on amino acids as buffers in Sec. 3-3.1). The proteins in the plasma are also important buffers, but hemoglobin has almost five times the buffering power of any of the other buffer systems and is by far the most important factor in buffering the blood against the large volumes of carbon dioxide that it transports.

The carbon dioxide that enters the blood from the tissues may form carbonic acid with the water of the plasma or of the red blood cell, but almost all the carbon dioxide enters the red blood cell (see Sec. 16-15) where the enzyme, *carbonic anhydrase,* speeds up the formation of carbonic acid.

$$CO_2 \longrightarrow CO_2 + H_2O \rightleftharpoons H_2CO_3$$

The carbonic acid in the red blood cell dissociates to yield hydrogen ions and bicarbonate ions. The hydrogen ions readily combine with the negatively charged hemoglobin (Hb^-) or oxyhemoglobin. This leaves a large amount of HCO_3^- free in the cell, and much of it diffuses back into the plasma, where it forms a very large part of the buffer system of the plasma, called the *alkali reserve.*

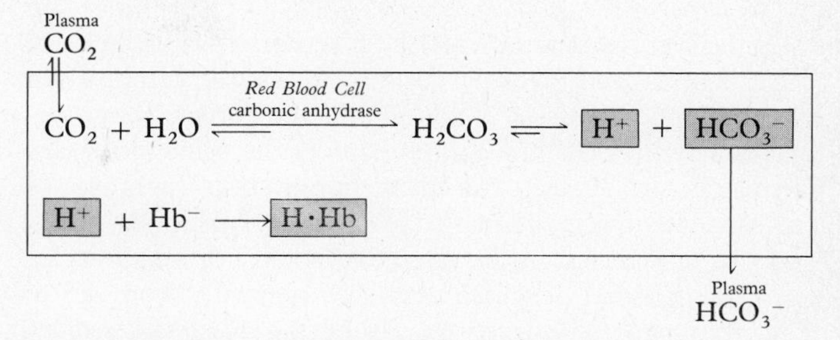

ALKALI RESERVE OF PLASMA AND BODY FLUIDS

Bicarbonate and Phosphate Buffer Systems. The alkali reserve forms the second type of buffer system in the blood and of the lymph and tissue fluids. The bicarbonate system, as we have already described it, is the chief component of the alkali reserve, but a small amount of a similar phosphate buffer system is also present. The kidney is extremely efficient, however, in excreting phosphates so that the amount of phosphate buffer is very low in blood but is the most important buffer of normal urine.

Effect of Respiration on the pH of the Blood

The respiratory center in the medulla is *stimulated* by two mechanisms mediated through the blood circulating to it (refer also to Sec. 16-17):

1. An increase in P_{CO_2}.
2. An increase in H^+ concentration (acidosis).

This stimulation causes breathing to become faster and deeper, shifting the following equation to the right and getting rid of the excess acid or the excess carbon dioxide.

$$\boxed{H^+} + HCO_3^- \rightleftharpoons H_2CO_3 \rightleftharpoons H_2O + \boxed{CO_2} \nearrow \text{exhaled}$$

Increase in H^+ or P_{CO_2}

The respiratory center in the medulla is inhibited by two mechanisms mediated through the blood circulating to it:

1. A fall in P_{CO_2}.
2. A decrease in H^+ concentration (alkalosis).

Breathing becomes slower and shallower, permitting carbon dioxide to accumulate and resulting in the formation of more carbonic acid. This restores the pH of the blood to normal because the carbonic acid dissociates, releasing more H^+, shunting the reaction to the left.

The Kidneys as Blood pH *Regulators*

Conservation of the alkali reserve of the blood

In the section on the homeostatic activities of the kidneys, it was emphasized that in order to maintain the composition of the blood and tissue fluids within the narrow range demanded by the cells, the composition of the urine must vary tremendously. Among the many homeostatic functions of the kidney is the maintenance of the pH of the body fluids, and this is reflected in the degree of acidity or alkalinity of the urine, the pH of which may vary normally from 4.5 to 7.8. The urine is buffered mainly by the *phosphate buffers,* removed so efficiently from the blood by the kidneys.

In response to an increase in acid in the blood, even when it is buffered, the kidneys excrete an *acid urine.* This saves the bicarbonate of the alkali reserve, preventing a drop in the pH of the blood. If the kidney excreted these acids in their buffered form, the blood would soon be depleted of its alkali reserve.

Production of an acid urine

The cells of the distal part of the kidney tubule can secrete hydrogen ions into the urine when the blood is too acid, producing an acid urine. At the same time, the kidney cells reabsorb more bicarbonate ions, returning them to the blood. Both secretion and reabsorption are active processes requiring energy.

	Blood	Kidney Cell	Fluid in Kidney Tubule
Excess	H^+ H^+ H^+	H^+ — secretion →	(acid urine) H^+ H^+ H^+
	HCO_3^- ←	HCO_3^- (reabsorption)	HCO_3^-

These processes are reversed when the blood becomes too alkaline, so that very little bicarbonate ion is reabsorbed, and practically no hydrogen ions are secreted into the tubules.

Conservation of the sodium and potassium of the blood

The kidney uses an ingenious device that results in the exchange of an ammonium ion (NH_4^+) for the biologically more valuable sodium and potassium ions. The cells of the kidney tubule synthesize ammonia (NH_3), which reacts with carbonic acid to form ammonium bicarbonate. This dissociates to form the ammonium ion (NH_4^+) and the bicarbonate ion.

$$NH_3 + H_2CO_3 \rightleftharpoons NH_4HCO_3 \rightleftharpoons NH_4^+ + HCO_3^-$$

(NH_4HCO_3: Ammonium bicarbonate)

The ammonium ion now will replace a sodium or potassium ion from a salt, returning sodium bicarbonate (or potassium bicarbonate) to the plasma and freeing an ammonium salt to be excreted in the urine.

$$\text{Na salt} + NH_4^+ + \underset{HCO_3^-}{\rightleftharpoons} NH_4\ \text{salt} + NaHCO_3$$

(NH_4 salt: Excreted in urine; $NaHCO_3$: Returned to plasma)

Cited References

Gordon, A. S., E. D. Zanjani, A. S. Gidari, and R. Kuna. "Erythropoietin: The Humoral Regulator of Erythropoiesis." In *Humoral Control of Growth and Differentiation,* Vol. 1: *Vertebrate Regulatory Factors,* J. LoBue and A. S. Gordon, eds. Academic Press, Inc., New York, 1973.

MacFarlane, R., ed. *The Hemostatic Mechanism in Man and Other Animals.* Academic Press, Ltd., London, 1970.

Perutz, M. F. "Relation Between Structure and Sequence of Haemoglobin." *Nature* **194:** 194 (1962).

Perutz, M. F. "Stereochemistry of Co-operative Effects in Hemoglobin." *Nature* **228:** 726 (1970).

Perutz, M. F. "X-Ray Analysis of Hemoglobin." *Science* **140:** 863 (1963).

Additional Readings

BOOKS

Davenport, H. W. *The ABC of Acid-Base Chemistry,* 6th ed. University of Chicago Press, Chicago, 1973.

Erslev, A. J., and T. G. Gabuzda. *Pathophysiology of Blood.* W. B. Saunders Company, Philadelphia, 1975.

Gordon, A. S. *Blood Cell Physiology.* BSCS Pamphlet No. 8. American Institute of Biological Sciences, Biological Sciences Curriculum Study. D. C. Heath & Company, Lexington, Mass., 1963.

Krantz, S. B., and L. O. Jacobson. *Erythropoietin and the Regulation of Erythropoiesis.* University of Chicago Press, Chicago, 1970.

LoBue, J., and A. S. Gordon, eds. *Humoral Control of Growth and Differentiation,* Vol 1. Academic Press, Inc., New York, 1973.

Matoth, Y., ed. *Erythropoiesis: Regulatory Mechanisms and Developmental Aspects.* Academic Press, Inc., New York, 1970.

Murayama, M., and R. M. Nalbandian, *Sickle Cell Hemoglobin.* Little, Brown and Company, Boston, 1973.

Platt, W. R. *Color Atlas and Textbook of Hematology.* J. B. Lippincott Company, 1975.

Rapaport, S. I. *Introduction to Hematology.* Harper & Row, Publishers, New York, 1971.

Surgenor, D. M. *The Red Blood Cell,* 2nd ed. Academic Press, Inc., New York, 1975.

ARTICLES

See also Chapter 20 for references on immune reactions.

ADAMSON, J. W., and C. A. FINCH. "Hemoglobin Function, Oxygen Affinity, and Erythropoietin." *Ann. Rev. Physiol.* **37:** 351 (1975).

BOGGS, D. R. "Homeostatic Regulatory Mechanisms of Hematopoiesis." *Ann. Rev. Physiol.* **28:** 39 (1968).

CLARKE, C. A. "The Prevention of "Rhesus" Babies." *Sci. Am.,* Nov. 1968.

COHN, Z. A. "The Metabolism and Physiology of the Mononuclear Phagocytes." In *The Inflammatory Process,* B. W. Zweifach, L. Grant, and R. T. McClusky, eds, Academic Press, Inc., New York, 1965.

DAVIE, E. W., and O. D. RATNOFF. "Waterfall Sequence for Intrinsic Blood Clotting." *Science* **145:** 1310 (1964).

DEMPSEY, H. "Hemostasis." In *Biologic Basis of Wound Healing,* L. Menaker, ed. Harper & Row, Publishers, New York, 1975.

FORTH, W., and W. RUMMEL. "Iron Absorption." *Physiol. Rev.* **53:** 724 (1973).

FROMMEYER, W. B. "Blood Coagulation." In *Biologic Basis of Wound Healing,* L. Menaker, ed. Harper & Row, Publishers, New York, 1975.

GESNER, B. M. "The 'Life History' and Functions of Lymphocytes." In *The Inflammatory Process,* B. W. Zweifach, L. Grant, and R. T. McClusky, eds. Academic Press, Inc., New York, 1965.

GEYER, R. P. "'Bloodless' Rats Through the Use of Artificial Blood Substitutes." *Fed. Proc.* **34**(6): 1499 (1975).

GEYER, R. P. "Potential Uses of Artificial Blood Substitutes." *Fed. Proc.* **34**(6): 1525 (1975).

GORDON, A. S., and E. D. ZANJANI. "Humoral Control of Erythropoiesis." *Adv. Intern. Med.* **18:** 39 (1972).

HIRSCH, J. G. "Neutrophil and Eosinophil Leucocytes." In *The Inflammatory Process,* B. W. Zweifach, L. Grant, and R. T. McClusky, eds. Academic Press, Inc., New York, 1965.

JERNE, N. K. "The Immune System." *Sci. Am.,* July 1973.

LAKI, K. "The Clotting of Fibrinogen." *Sci. Am.,* Mar. 1962.

LERNER, R. A., and F. J. DIXON. "The Human Lymphocyte as an Experimental Animal." *Sci. Am.,* June 1973.

MCMANUS, T. J. "Comparative Biology of Red Blood Cells. *Fed. Proc.* **26**(6): 1821 (1967).

MORENO, H. "Platelet Function." In *Biological Basis of Wound Healing,* L. Menaker, ed. Harper & Row, Publishers, New York, 1975.

PITTS, R. F. "Renal Regulation of Acid-Base Balance." In *Physiology of the Kidney and Body Fluids,* 3rd ed., R. F. Pitts, ed. Yearbook Medical Publishers, Inc., 1974.

SEEGERS, W. H. "Blood Clotting Mechanisms: Three Basis Reactions." *Ann. Rev. Physiol.* **31:** 269 (1969).

ZUCKER, M. B. "Blood Platelets." *Sci. Am.,* Feb. 1961.

Chapter 15

Regulation of Water Balance

What am I, Life? A thing of watery salt
Held in cohesion by unresting cells,
Which work they know not why, which never halt
Myself unwitting where their Master dwells,
I do not bid them, yet they toil, they spin. . . .

John Masefield, "What Am I, Life?"

WATER BALANCE IS dependent on the equilibrium between the amount of water ingested and that lost from the body. The volume of water obtained from food and drink and metabolic reactions is fairly constant in each individual, although it varies over short periods of time. Hunger, appetite, and thirst are complex neural and psychological problems that will be discussed in Chapter 21.

Water may be *lost* from the body via the *kidneys* (1000-1500 ml daily), the *skin* (450-1050 ml daily), the *lungs* and *air passages* (evaporation of 250-300 ml daily), and the *intestine* (50-200 ml daily in the feces). On occasion, additional water may be lost in the form of milk secretion in the lactating woman, tears, or in disturbances of the digestive tract, such as vomiting and diarrhea (Table 15-1).

The relatively large amount of water that is lost through evaporation and secretion from the skin is attuned to temperature and humidity changes and acts as a homeostatic mechanism for temperature regulation rather than water balance. Nevertheless, it plays an important role in loss of water from the body. This loss is compensated for by the activity of the kidney, which through delicate adjustments of reabsorbing mechanisms is able to conserve water by excreting a scanty, concentrated urine when the water levels of the body are low (*antidiuretic mechanisms*) and to remove water when the blood is diluted by forming a copious, watery urine (*diuresis*).

Table 15–1. **Loss of Water Each Day (in milliliters)**

	Normal Temperature	Hot Weather	Prolonged Heavy Exercise
Insensible loss:			
Skin	350	350	350
Lungs	350	250	650
Urine	1400	1200	500
Sweat	100	1400	5000
Feces	200	200	200
Total	2400	3750	6700

Source: A. C. Guyton, *Medical Physiology,* 5th ed., Table 33-1, Saunders, Philadelphia, 1976.

In earlier chapters it was emphasized that the optimal functioning of cells depends on the characteristics of the internal environment, requiring controlled concentrations of nutrients and electrolytes, a constant fluid volume, and the elimination of metabolic end products. In addition, the pH of body fluids must be maintained within a very narrow range.

The distribution of body water into its various compartments was discussed in Chapter 4. This distribution is dependent on the electrolyte and protein concentration of the body fluids. Water and salt balance are inextricably interdependent, and both are regulated largely by the homeostatic activity of the kidney. The relative constancy of the internal environment is obtained at the expense of a highly variable urine produced by the kidney. The kidney is attuned to minute changes in the concentration of blood constituents and constantly adjusts the amounts of these substances returned to the blood or removed in the urine.

The final balance between water intake and water loss is the responsibility of the hypothalamus. Its neuroendocrine role is displayed by its regulation of diuresis through the amount of antidiuretic hormone it secretes, and its neural control is demonstrated by its regulation of behavioral patterns of controlling water intake.

15–1 The Skin

The skin forms an airtight, fairly waterproof covering for the delicate tissues beneath, protecting them from changes in the external environment and maintaining the integrity of the *milieu interne.* Yet the skin itself is made of living cells, also sensitive to environmental alterations; indeed, the skin contains many special sensory endings that enable us to perceive and react to the external environment. Touch, pain, temperature, and pressure receptors are located at various depths through the skin, and these in turn are protected by the uppermost epidermal layers of flattened *dead and dying cells.* These dead cells have been gradually pushed up from the germinative lower layers, which are in close contact with the blood supply, toward the surface where they are exposed to continuous pressure and literal "wear and tear." This, together with the decreasing blood supply, flattens and kills them, and as their cytoplasm hardens, it forms a substance *keratin,* which makes this layer of cornified cells both hard and waterproof.

The *surface area* covered by the skin is surprisingly large; in a young man 5′11″ (1.8 m) tall, weighing 154 lb (70 kg), it would comprise 1.9 m^2. The surface area of a young woman, 5′6″ (1.7 m) tall, weighing 130 lb (59 kg), is approximately 1.65 m^2. Destruction of even 20 per cent of this total skin area is almost always fatal, unless there is constant replacement of fluid lost through the denuded areas. Burns involving less of the skin area are nevertheless very serious and have to be treated with great care. Exposure of the underlying tissues results in the loss of large quantities of body water through evaporation and seepage; this loss is aggravated by the toxins formed in the damaged tissues, because they increase capillary permeability and expedite the further passage of water out of the circulation. This, in turn, can cause an alarming fall in circulating blood volume and blood pressure.

The loss of water through the intact skin takes two forms. One is the continuous, scarcely perceptible evaporation of water from even the resting individual. This is *insensible perspiration* and accounts for 300 to 800 ml of water daily, depending on the external temperature and humidity. This perspiration is most marked on the palms of the hands and soles of the feet, less marked on the back of the hands, the neck, and the face, and least on the rest of the body. The second form of water loss through the skin is an active secretion of the sweat glands and can result in the loss of as much as 5000 ml of water in a day. *Sweat* can be produced as a response to a rise in internal or external temperature or intense excitement or fear and varies in composition, according to the evoking stimulus, from a very dilute, watery secretion with an alkaline pH to a more acid secretion with a pungent odor. There is really a difference in chemical composition between a "normal" sweat caused by increased temperature and a "cold" sweat caused by fear or excitement.

Structure of the Skin

The skin is an organ made up of several different types of tissues, as would be expected from its dual origin from both ectodermal and mesodermal germ layers (Fig. 15-1). The ectoderm gives rise to the outer skin layers, or *epidermis,* which forms four layers of varying thickness as the cells are pushed up from the lowest, germinative layer, the *stratum germinativum,* toward the tough *stratum corneum,* with its flattened, cornified cells. The mesoderm gives rise to the *dermis or corium,* lying beneath the epidermis; the dermis is the highly sensitive and vascular layer of the skin containing the specialized nerve endings and the rich plexuses of blood vessels. The glands of the skin and the hair follicles also originate in this lower layer.

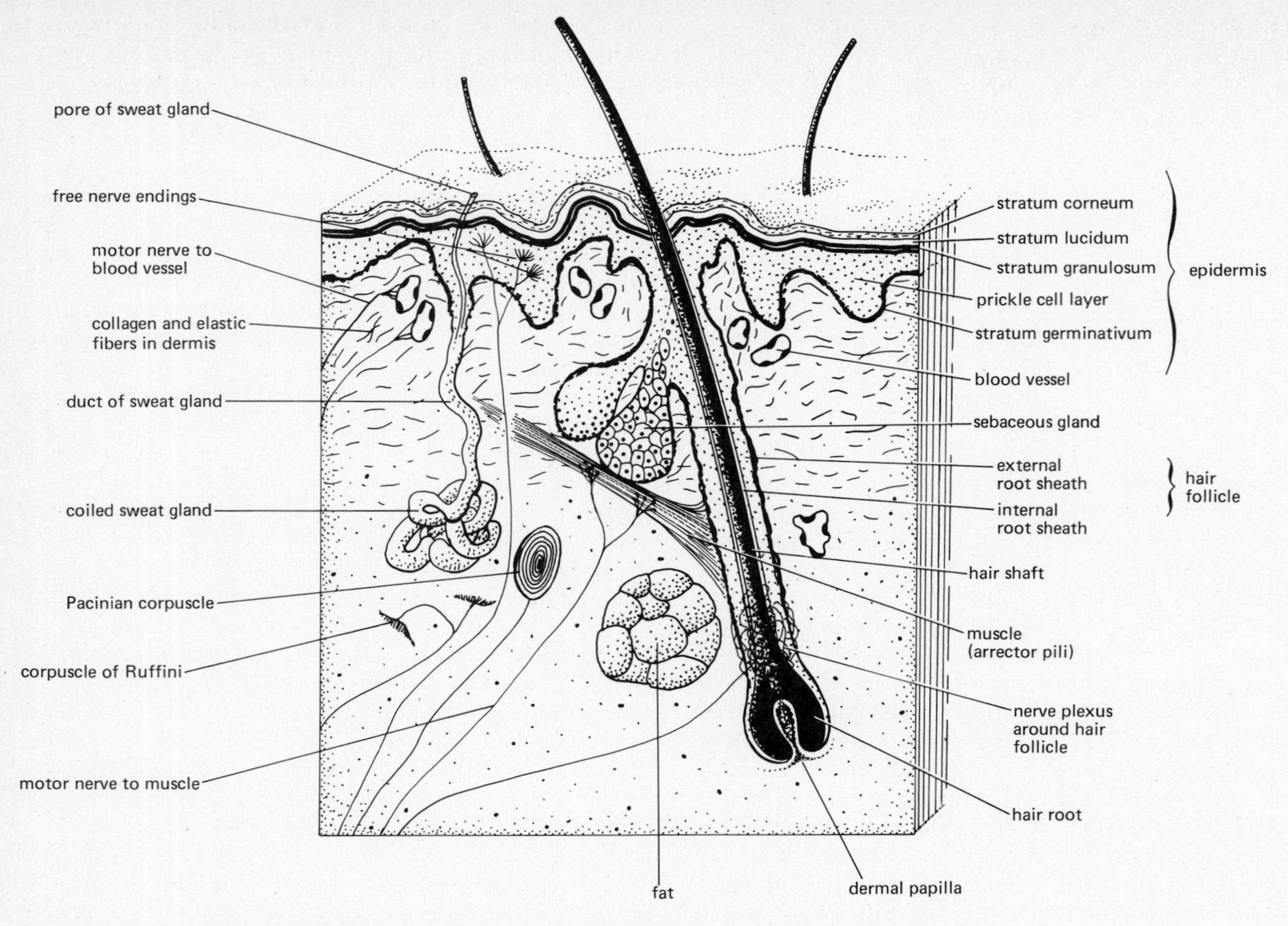

Figure 15-1. **Section through skin showing a hair follicle, a sweat gland, and different types of nerve endings.**

Epidermis

The epidermis consists of the following four layers (Fig. 15-2):

1. *The stratum germinativum* is made up of spiny, columnar cells, connecting to form intercellular bridges. These cells are very active mitotically and produce the epidermal cells that work their way up to the skin surface to replace those cells that are being gradually sloughed off. The cells of the lowest layer of the stratum germinativum contain fine granules of melanin, the pigment responsible for varying shades of brown in the skin. The pigmentation of the skin of the black races is due to the presence of greater amounts of this pigment, distributed through all the layers of the epidermis. This pigment is affected by light, which results in tanning of the skin in the sun, and by hormones, which are responsible for the darkening of the nipples and abdominal striae in pregnancy. The lower border of this germinative layer is pushed in at irregular intervals by dermal protrusions or *papillae,* many of which contain the tactile corpuscles, especially in areas where the sense of touch is particularly developed.
2. *The stratum granulosum* is a thin layer of flattened, granular cells between the stratum germinativum and the upper horny layers.
3. *The stratum lucidum* is a clear layer of flattened cells, lacking granules and nuclei.
4. *The stratum corneum,* the uppermost layer, is made up of the dead, cornified cells that are constantly being replaced by the cells pushing up from the stratum germinativum. The material forming these cells is keratin.

The four layers of the epidermis form a corrugated covering of *stratified squamous epithelium,* which varies in thickness and smoothness over the various parts of the body. It is thickest in areas subjected to pressure, such as the sole of the foot, and thinnest in delicate areas, such as the eyelid. Pressure is not the only factor involved, however, for even in the fetus the epidermis of the soles is considerably thickened. Very early in fetal development, the epidermis on the tips of the fingers and toes is thrown into characteristic whorls and ridges, which remain constant throughout life and therefore enable the fingerprints to be used as a source of personal identification.

Dermis (or corium)

The dermis is the sensitive, highly vascularized part of the skin immediately beneath the stratum germinativum of the epidermis. Not only the specialized nerve endings are to be found here in profuse variation, but also the sweat and sebaceous glands and the hair follicles. These structures, together with the network of blood vessels and lymphatics, are held in a connective tissue matrix containing many elastic fibers and collagen bundles. The elasticity of the skin is dependent on the amount and condition of these fibers, which gradually lose their elastic qualities as the organism ages. The lower layers of the dermis are held to the underlying tissues, such as muscle, by a subcutaneous layer of areolar connective tissue called *fascia.* This is an extremely important structure which binds the skin down, penetrates through the muscles and visceral organs in very well-defined sheets, and provides channels for many of the blood vessels. It is of great significance in localizing infections by acting as a barrier to the spread of bacteria.

Figure 15-2. **Thick skin from the palm. ep = epidermis, de = dermal connective tissue, sc = subcutaneous connective tissue, c = stratum corneum, sl = stratum lucidum, gr = stratum granulosum, g = stratum germinativum, p = papillary layer of dermis, r = reticular layer of dermis, d = duct of sweat gland, s = sweat glands, and f = fat cells. (Courtesy S. Piliero, Dental School, New York University.)**

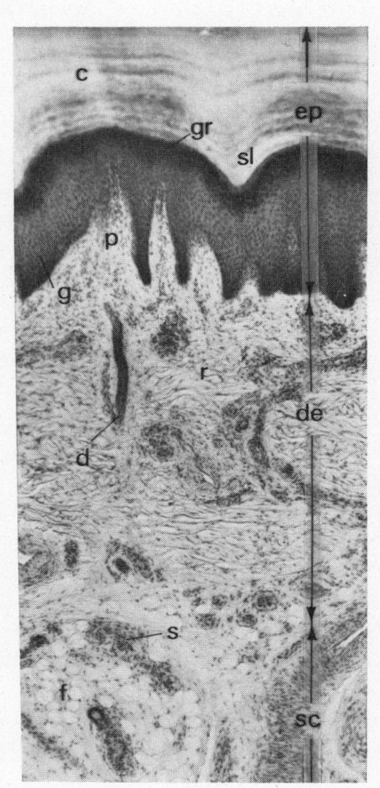

Specialized Structures of the Dermis

1. *Blood Vessels.* Branches of the arteries supplying the skin form rich networks of arterioles and capillaries that branch to all the dermal structures and provide nourishment also to the stratum germinativum of the epidermis, which receives no direct blood flow. The skin of the white races, which has very little pigment in the epidermis, derives its pink color mainly from these blood vessels in the dermis, because the epidermal layers are fairly transparent, especially around the nails. Large amounts of blood may be shunted to these capillary beds in the dermis by dilation of the arterioles. This will increase the temperature of the skin and the amount of heat lost by radiation and sweating. The skin will be pink and warm in contrast to the cold, bluish-tinged skin that is seen when the arterioles and capillaries are constricted by cold.
2. *Lymphatics.* The skin is well supplied with lymph vessels that drain it and transmit the fluid to the deeper lymphatic system.
3. *Nerves.* The extreme sensitivity of the skin is caused by the wealth of nerve endings found in the dermis. They vary in structure from the simplest free nerve endings to the complex capsulated corpuscles (Figs. 15-1 and 22-53).

Sensory Fibers

1. *Touch.* Meissner's corpuscles and the free nerve endings around the hairs.
2. *Pressure.* Pressure receptors called the Pacinian corpuscles.
3. *Temperature.* Cold receptors (end organs of Krause); heat receptors (end organs of Ruffini).
4. *Pain.* Free nerve endings.
5. *Other sensations,* such as tickling, itching, and so on, are supposed to be due to weak stimulation of touch and pain receptors and perhaps also to a mixture of weak sensations of several different types of receptors.

Motor Fibers. Motor fibers are classified as follows:

1. *Vasomotor fibers,* which control the vasodilation and vasoconstriction of the arterioles within the dermis and thus regulate the amount of blood flowing through the skin, or various parts of the skin.
2. *Motor nerve fibers,* which innervate the hair muscles (arrector pili) and cause contraction of these muscles, resulting in the hair standing "on end" and the so-called "gooseflesh."

Sweat Glands. The coiled sweat glands are found in the skin of the entire body, except the lips, glans penis, and nail bed. The secretory portion forms a coiled ball in the dermis, and its duct extends up through the epidermis to open on the surface of the skin (Fig. 15-1). The composition of sweat varies in different parts of the body, reflecting differences in the structure of these glands. The sweat glands of the external genitalia and the axilla produce a more concentrated and stronger smelling sweat than those of the rest of the body.

The composition of the sweat also varies according to the cause that has produced its secretion: sweat produced by heat is more acid than that evoked by exercise, and there are also differences in the concentration of salts and ions.

Hair. The biological function of hair is to increase the sensitivity of the skin surface for tactile stimuli as well as to preserve heat. Lower mammals and most primates, but not man, also have special sensory hairs or whiskers that have a particularly well-developed nerve supply. The shaft of the hair acts as a lever, magnifying and transmitting any minute mechanical change to the network of nerve endings surrounding the hair follicle. Neither of these functions, sensitivity nor heat conservation, is of much importance to humans.

Each hair arises in a deep invagination of the skin, the hair *follicle,* which is made up of *epidermis* and a connective tissue *sheath.* This sheath is a continuation of the *dermal papilla,* a projection of the dermis into the base of the follicle. Both the dermis and the papilla are richly supplied with blood vessels. Within the connective tissue sheath is a noncellular glassy membrane, which is closely applied to the outermost layer of the hair.

The hair itself consists of a *shaft,* covered by an external and an internal sheath and a cuticle. The shaft is made up of keratinized cells, pigment, and air vacuoles, the proportions of which determine the color of the hair. The keratin is formed as microfibrils in the cytoplasm, which then become imbedded in a cement or matrix, together with the shriveled remnants of the nucleus. The bottom of the sheath forms the bulbous *root,* which is penetrated by the dermal papilla (Fig. 15-1).

In the development of the hair, a bud grows down from the epidermis and forms the follicle. Cells near the base of the bud then proliferate to form a hair. The continued proliferation of these cells pushes the hair up through the epidermis. Hair growth stops, or only tiny hairs are formed, when the efficiency of the circulation to the active cells of the dermis and papilla is impaired. However, changes in the structure of the glassy membrane and the connective tissue sheath seem to precede the degeneration of the blood vessels that accompanies *baldness.* In typical male baldness, the number of hairs per square centimeter is about the same in bald and nonbald areas, but the hairs are extremely small and fine in the bald spots. Genetic and hormonal factors, as well as disease, are involved in baldness, but local factors are also to be considered: hairbearing skin transplanted to areas that normally do not grow hair continue to form normal hair in their new site. If the transplant has been made from an area that later becomes bald, the transplanted skin also loses its hair at this time. It has been suggested that each individual follicle is genetically predisposed to respond or not to respond to the inhibitory influence of the male sex hormone (see also Sec. 26-3).

STRUCTURES ATTACHED TO THE HAIR FOLLICLE. Bands of smooth muscle form the *muscles* of the hairs (arrector pili). When these muscles contract in response, for example, to cold, they pull the sloping hair into a vertical position, pushing up the skin around the hair. Also attached to the follicle is one or more *sebaceous glands,* which empty their fatty secretion into the upper part of the follicle. In nonhairy parts of the skin, such as the lips, mammary papillae, and penis, the sebaceous glands oil the skin by emptying directly onto its surface. The largest of these glands are found at the sides of the nose. Only the palms and soles are devoid of sebaceous glands.

15-2 The Kidney

Man's problem, as a land animal, is chiefly water conservation and elimination of nitrogenous wastes in a nontoxic, concentrated form. The truly independent land animals are the higher arthropods (spiders, insects, etc.) and the higher vertebrates (reptiles, birds, and mammals), all of which have solved the dual problem of protection against loss of water from the body surfaces through the development of *horny coverings, feathers,* or *hair,* and of excessive water excretion through the production of a *hypertonic urine.*

Evolutionary Development of Structure and Function

A comparative study of the anatomy of the kidney shows in a most interesting way the adaptation of structure to function.

WATER BALANCE, REGULATION OF COMPOSITION OF THE BLOOD

Fresh-water fish, living in a *hypotonic* environment, must rid themselves of the excess water that passes into the blood through the gills and oral membranes. This elimination is attained by the development of millions of small filtering structures (glomeruli) within the kidneys which filter tremendous amounts of blood daily, permitting most of the water to be excreted as a copious, dilute urine. The glomerular filtrate is modified by the long renal tubule, through which it passes before being eliminated from the kidney. The distal end of this tubule reabsorbs some water and much salt, thus conserving salt against the constant tendency to dilution of the blood. Many fresh-water fish have special salt-secreting cells, in addition, to maintain their blood salt concentration.

In fish that live in salt water, the problem is akin to that of land animals, that is, water conservation, with the additional complication of a constant inflow of salt into the

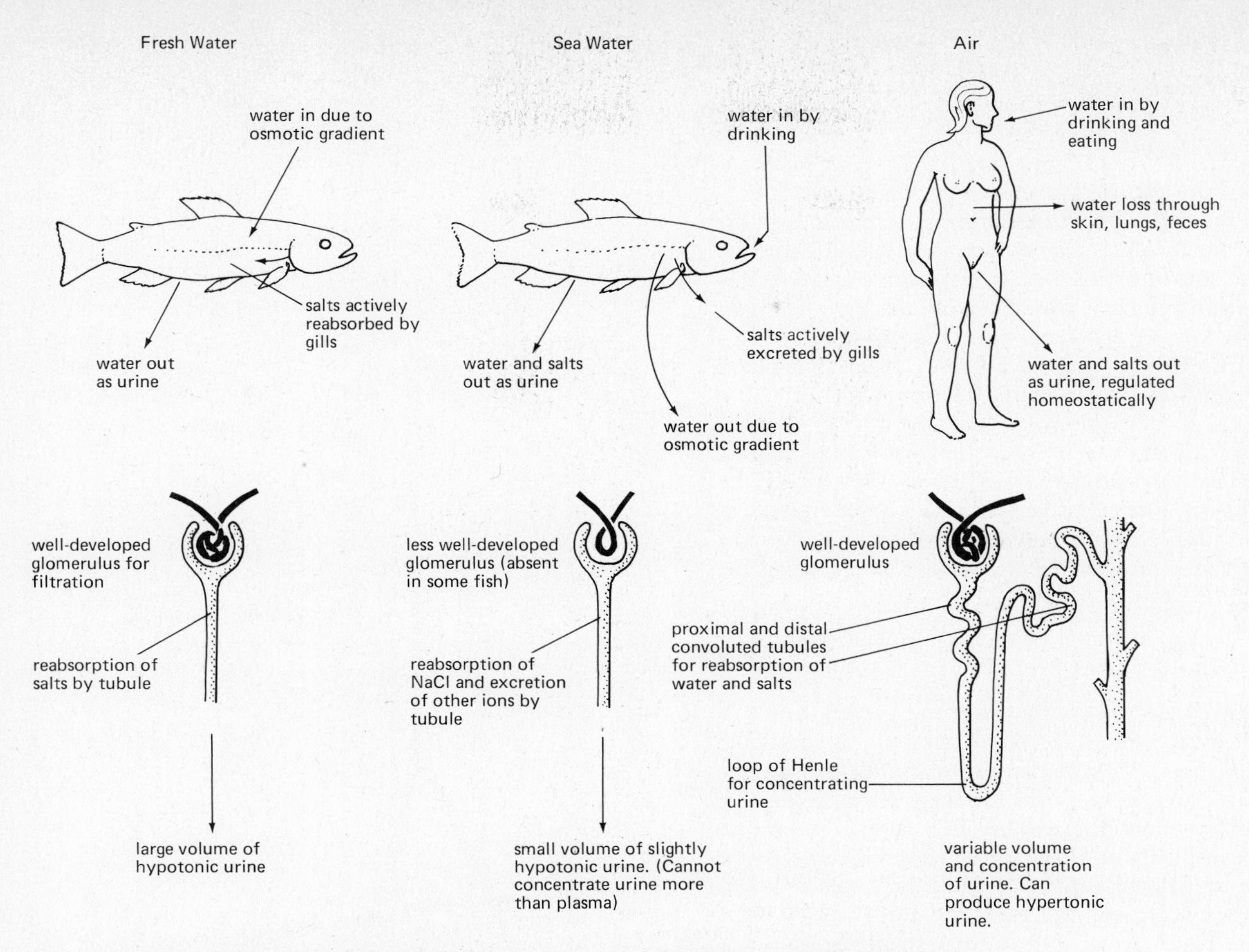

Figure 15-3. **Comparison of salt and water balance in fresh-water fish, salt-water fish, and humans.**

blood via the digestive tract. These salt-water fish, with a body fluid concentration of salts much less than that of the surrounding sea water, are surviving in a *hypertonic* medium. Their water-excreting kidneys are now a liability, and in some forms the glomeruli and distal portions of the renal tubules have degenerated or are altogether absent. To obtain water, the fish drink great amounts of sea water and eliminate the excess salt through excretion by the gills (Fig. 15-3).

In humans, the problem of water conservation is fortunately not complicated by a constant salt inflow. *Humans can excrete hypertonic urine* because the activity of the glomerulus, which is basically a device for removing excess water, is modified by cells of the renal tubule which reabsorbs water. These specialized tubule cells also selectively return certain substances to the blood and excrete others, so that the composition of the blood and tissue fluids remains constant, within physiological limits. In addition, the renal tubules are delicately regulated by the activity of various *endocrine glands,* which in turn are affected by changes in water and ion concentration.

Work Done by the Kidney

To maintain fluid balance, the normal human adult filters about 160 liters of plasma a day but excretes in the final urine only about 1 per cent of the salt and water of the filtrate. This feat requires specialized ion and water transport mechanisms and the expenditure of considerable amounts of metabolic energy. Table 15-2 shows a comparison of the composition of plasma and urine.

Table 15–2. **Comparison of Plasma and Urine Composition**

Constituent	% Plasma	% Urine
Water	90–93	95
Glucose	97	0
Protein	7	0
Sodium	0.3	0.35
Ammonia	0.004	0.05
Phosphate	0.009	0.5
Urea	0.03	2.0
Sulfate	0.002	0.18

15–3 Organization of the Urinary System

Kidney and Ureters

The kidneys are two bean-shaped structures, located dorsally about the level of the upper lumbar vertebrae, with the left kidney usually a little higher than the right, reaching the level of the eleventh rib. The indentation that gives this organ its beanlike form is the *hilum,* and through the hilum pass the *renal artery* and *renal vein* (Fig. 15-4).

Passing from the hilum of each kidney, behind the blood vessels, is the *ureter,* which drains the urine from the inner portion of the kidney (the *pelvis*) into the urinary bladder. The kidneys are surrounded by an adipose tissue capsule, whose fat acts as a protective layer and also helps keep the kidney in place. When a fat person loses a great amount of weight, the kidney may slip out of position to such an extent that it may block the passage of urine by compressing the ureter.

Urinary Bladder and Urethra

The *urinary bladder* lies in the lower and posterior part of the pelvic cavity, behind and above the pubic bones. The three openings in the bladder form a triangle on its posterior wall. Two of these openings are the inlets from the *ureters;* the third is the outlet to the *urethra* (Fig. 15-4). The urethra is the final canal through which the urine is discharged from the bladder to the exterior. It is an inch and a half in length in the female, whereas in the male, the urethra is approximately 8 in. long, having to pass through the penis.

The urinary bladder, ureters, and urethra are all muscular structures, lined with a mucous membrane of transitional epithelium, impermeable to the normal soluble substances of the urine, and of connective tissue containing blood vessels and elastic fibers. The terminal portion of the urethra is lined with stratified squamous epithelium. The muscular coat of the bladder is particularly strong, and at the point where the urethra exits, the smooth muscle fibers form two rings, the *urinary sphincters.* These sphincters are under nervous control; increased tension brought about by filling causes reflex relaxation of one set of these muscle fibers, the coordinated contraction of the wall of the bladder, and the emptying of its contents through the urethra. This reflex (*micturition*) can be voluntarily controlled in the child at about 1 to 2 years of age, when the cerebral cortex is developed sufficiently to permit conscious inhibition of a basically involuntary act.

The Nephron

The structural and functional unit of the kidney is the *nephron.* There are approximately 1 million nephrons in each human kidney, and painstaking microdissection has shown their individual length to be about 30 to 38 mm. The nephrons are arranged in an orderly manner through the two urine-forming zones of the kidney, the outer *cortex* and the central *medulla.* The innermost zone of the kidney is the *pelvis,* which connects the medulla with the transport facilities of the ureter (Fig. 15-5). Toward the medulla the pelvis forms one or two large outpocketings, the major *calyces,* which in turn are subdivided into several minor calyces.

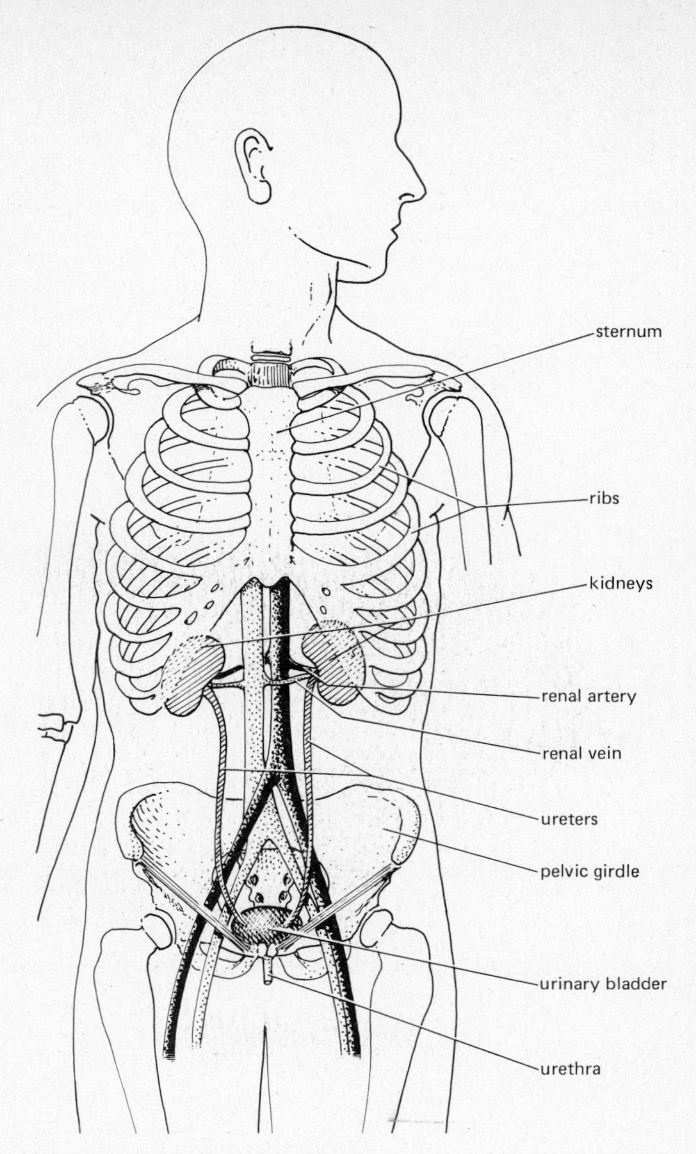

Figure 15-4. **The urinary system.**

There are two morphologically distinct kinds of nephrons, the superficial *cortical nephron,* which makes up 80 to 90 per cent of the nephron population in the human kidney (somewhat less in the kidneys of other species), and the *juxtamedullary nephron* (Fig. 15-5). The cortical nephrons originate in the cortex of the kidney, they are more superficial and so more accessible to study. They extend only a short distance into the outer medulla. The juxtamedullary nephrons originate at the junction between the cortex and the medulla and extend deep into the medulla, where the formation of hypertonic urine takes place. There is a definite correlation between the number of juxtamedullary loops, the length of the loops, and the ability of an animal to concentrate its urine. The desert rat, a small rodent living in a very arid environment, has a relative medullary thickness almost three times that of man and can concentrate its urine practically six times as much.

The nephron is composed of the *glomerulus* and the *tubules*. The tubules form a continuous system that modifies the filtrate formed by the glomerulus and transports this modified fluid towards the kidney pelvis. The parts of the tubules have different structural and functional characteristics and are known as the proximal convoluted tubules, the loop of Henle, the distal convoluted tubule, and the collecting tubule (Fig. 15-6).

The glomerulus

The kidneys receive the greatest blood flow in proportion to their weight (about 20–25 per cent of the cardiac output at rest) of any organ of the body. The blood is brought to the kidney by the renal arteries, each of which forms branches that ultimately divide into the interlobular artery, which then divides to supply blood to 80 or more glomeruli through the *afferent arterioles* [Figs. 15-6 and 15-8(A)].

The *glomerulus* is formed by the branching of the afferent arteriole into an interconnecting tuft of capillaries. This tuft is cupped by a double-layered membrane formed by the indentation of the blind end of the proximal convoluted tubule. The cup is known as the *renal,* or *Bowman's, capsule.*

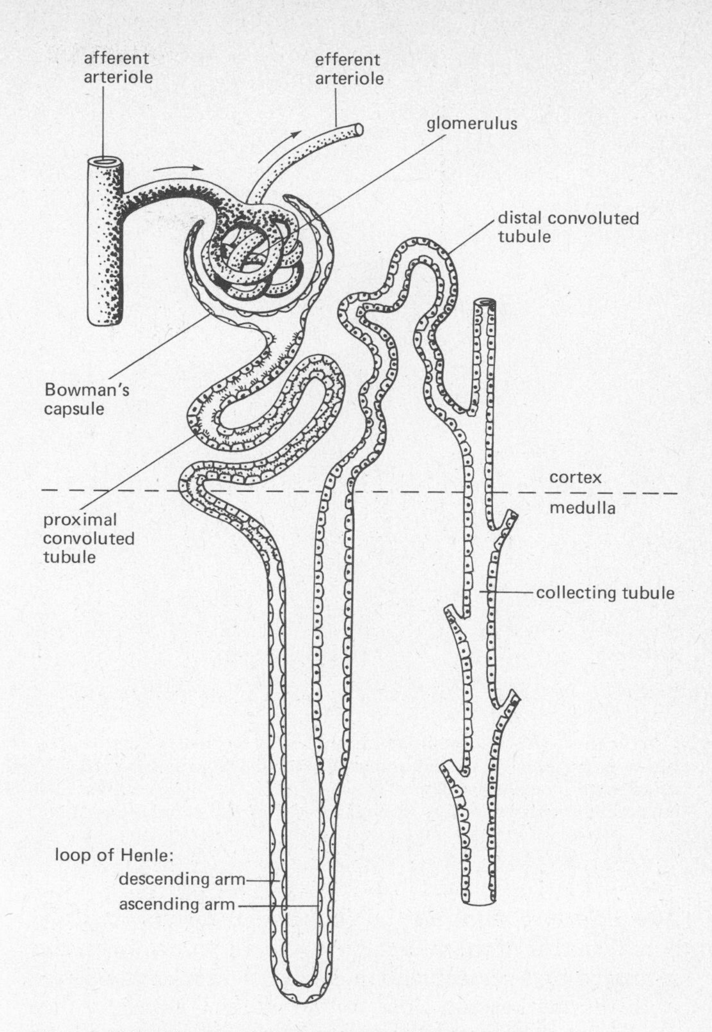

Figure 15-6. **The functional regions of the nephron.**

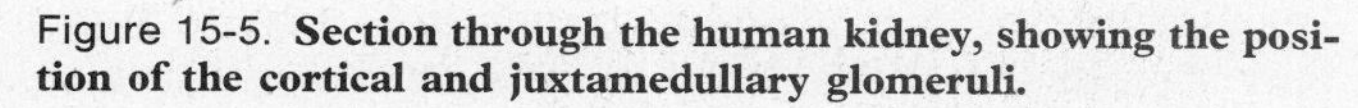
Figure 15-5. **Section through the human kidney, showing the position of the cortical and juxtamedullary glomeruli.**

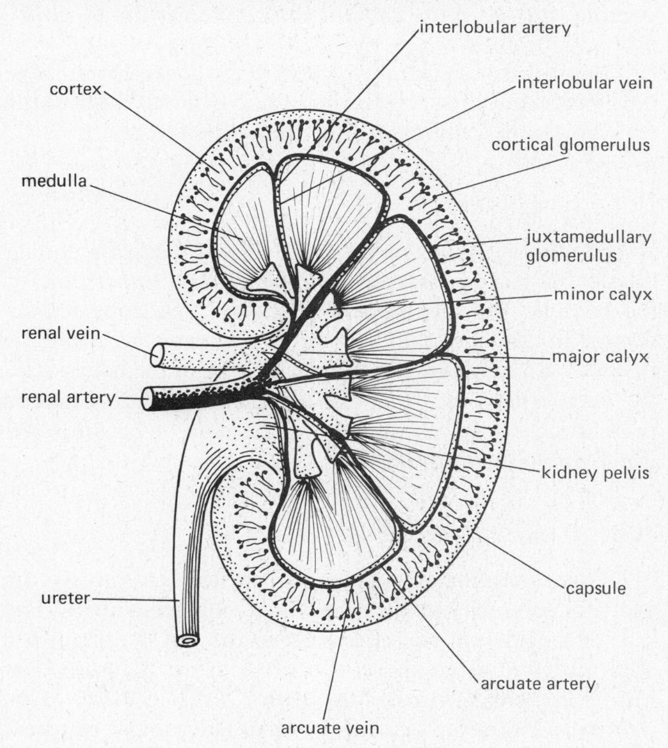

The Glomerular Membrane. The structure of the glomerular capillary wall and of its closely adhering capsule determines the characteristics of the filtrate that pass into the tubule (Fig. 15-7). The membrane consists of three layers: the *capillary endothelium,* thin flattened cells with fenestrations between them; an acellular *basement membrane;* and a third layer of *epithelium.* This epithelium possesses some peculiar-looking cells called *podocytes,* the foot processes of which project into the basement membrane. It is believed that these podocytes form split pores, which restrict the passage of colloids but permit free passage of water and small molecules. As a result of the permeability characteristics of the three layers, cellular elements, plasma proteins, and other macromolecules are prevented from passing into the tubule.

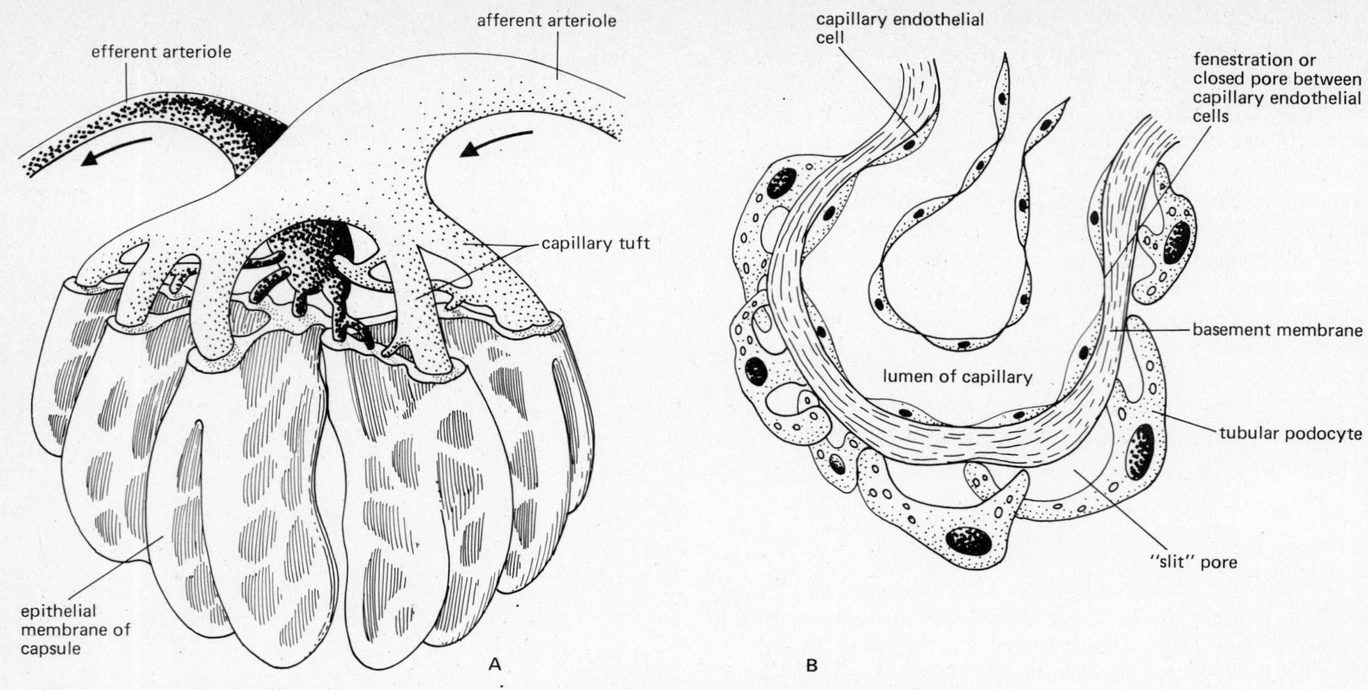

Figure 15-7. **(A) The capillary glomerulus is closely invested by the membrane of the capsule, which is formed by the invaginated end of the proximal convoluted tubule. (B) The glomerular membrane consists of three main layers: the thin capillary endothelium; the thick, acellular basement membrane; and the podocytes of the tubular epithelium.**

The Efferent Arteriole. Possessing a slightly narrower bore than that of the afferent arteriole, the efferent arteriole is formed by a recombination of the glomerular capillaries. It leaves the capsule close to the afferent arteriole, then subdivides to form a second set of capillaries around the tubules, *the peritubular capillaries,* before joining the venous system (Fig. 15-8). It is essential to realize the significance of this secondary capillary circulation. It permits the tubular fluid to be modified by additions from the blood (secretion) and by removal of some of its constituents into the blood (reabsorption).

The Efferent Arteriole of the Juxtamedullary Nephron. This arteriole is unique. It is wider than the afferent arteriole; after supplying a network to the tubule in the outer medulla, it plunges down into the inner medulla as two or three straight branches, the *vasa recta.* Each of the vasa recta makes a U turn at the innermost part of the medulla and returns to the venous circulation near the junction of the medulla and cortex [Fig. 15-8(A)].

Proximal convoluted tubule

The beginning of the proximal convoluted tubule (pct) is continuous with Bowman's capsule, but the epithelium changes from flattened to high cuboidal (Figs. 15-6 and 15-8). These cells are complex. Beneath the basement epithelium are many invaginated slits that open to the extracellular space beneath the basement membrane. These invaginations, together with microvilli that form a brush border on the luminal side, create an extensive cell surface, essential for the remarkable reabsorptive capability of the pct.

The pct forms an irregular spiral in the general vicinity of the glomerulus, then projects straight down into the medulla to form the descending arm of the loop of Henle.

Loop of Henle

The loop of Henle consists of a thin-walled, descending tubule that makes a sharp hairpin bend in the upper third of the medulla for the cortical nephrons, and considerably deeper in the medulla for the juxtamedullary nephrons (Figs. 15-6 and 15-8). The ascending limb immediately after the bend is thin, but near the cortex it becomes wide and thick. At the level of its own glomerulus, it becomes the distal convoluted tubule.

Juxtaglomerular apparatus

As the ascending arm of the loop of Henle returns to its own glomerulus, it comes into close contact with it. The cells of the afferent and efferent arteriole and the tubule are quite different from those of related areas; together they form the juxtaglomerular apparatus [Fig. 15-8(B)]. This intimate relationship between vascular and tubular elements reflects a functional relationship: the enzyme *renin*, produced by the juxtaglomerular cells, can cause a systemic increase in blood pressure (see renin-angiotensin system, Sec. 12-13), and in addition can regulate the blood flow through the glomerulus. It appears that *increased sodium concentration* in the tubule in this area activates the enzyme and causes constriction of the afferent arteriole. This constriction reduces glomerular filtration and conserves water, decreasing sodium blood levels. This feedback mechanism, operating at the level of the nephron, is one way in which *autoregulation of glomerular filtration rate* occurs.

Distal convoluted tubule

The distal convoluted tubule (dct) commences as a short, highly convoluted coil in the cortex or juxtamedullary region (Figs. 15-6 and 15-8). Like the glomeruli, these tubules have been extensively studied using micropipettes because their surface loops are relatively easily accessible. The water permeability of the cuboidal cells of the dct, and the size of the intercellular spaces, are dependent to a large extent on the amount of circulating antidiuretic hormone (ADH).

Collecting tubules

In the cortex, each collecting tubule (ct) drains an increasing number of ducts of dct, descending to the tip of the medulla to form the collecting ducts of Bellini, which ultimately empty into the kidney pelvis via the calyces (Figs. 15-6 and 15-8). The cells of the ct are high-columnar, and like the dct cells, their permeability and the extent of the intercellular spaces are controlled by ADH.

Figure 15-8. **(A) The circulation of the cortical and juxtamedullary nephrons. Note the specialized vasa recta that surround the long loops of Henle of the juxtamedullary nephron. The vasa recta form the secondary arteriolar circulation of these nephrons, while the peritubular capillaries serve the cortical nephrons and the other parts of the juxtamedullary nephrons. (B) The juxtaglomerular apparatus consists of specialized cells in the walls of the afferent arterioles near the glomerulus, and the macula densa of the neighboring distant convoluted tubule.**

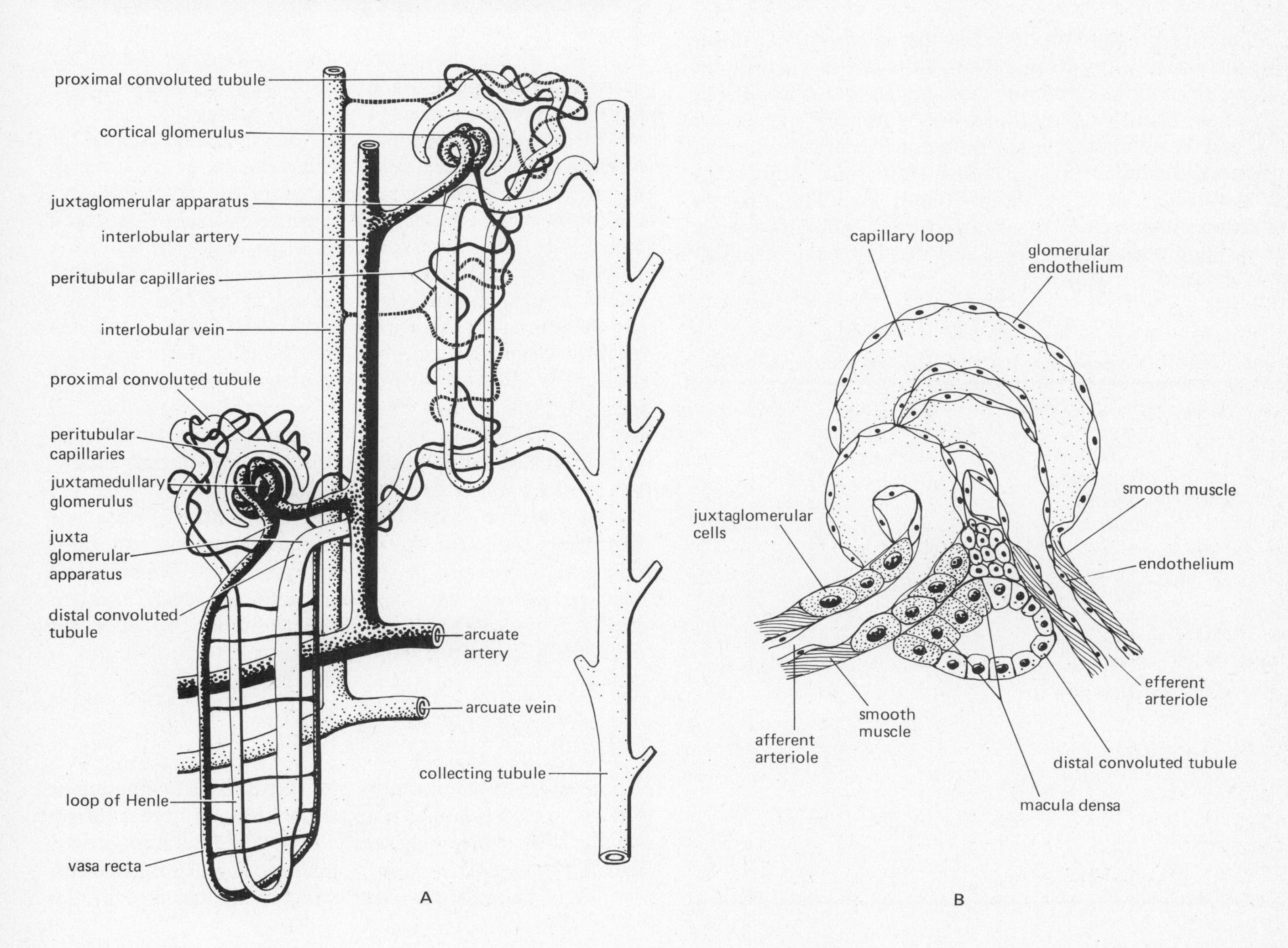

15–4 Urine Formation

There are three main processes by which the kidney produces the concentrated solution of metabolic wastes that is the urine and regulates the concentration of the constituents of the body fluids. These processes are *filtration, reabsorption,* and *secretion.* The *average composition* of the final product is shown in Table 15–3. As has previously been emphasized, the volume and composition of urine are extremely varied, changing with diet, exercise, metabolism, and so forth.

Glomerular Filtration

The structure of the glomerulus permits it to act as an ultrafiltration device. For a fluid to be considered as an ultrafiltrate of plasma, it must be protein-free and contain all crystalloids in the same concentration as in the plasma. The hydrostatic pressure of the blood within the glomerular capillaries must be sufficient to produce this filtrate at the rate at which it is known to be formed.

FILTRATION PRESSURE

The back pressure produced by the narrower bore of the efferent arteriole results in a hydrostatic pressure within the glomerular tuft that is considerably higher than in capillaries elsewhere in the body. Although it is difficult to measure this exactly in humans, it has been done accurately in other species. Calculations for the human glomerulus estimate the glomerular pressure as about 70 mm Hg. The effective filtration pressure can be calculated in the same way as for tissue capillaries (Sec. 4–2). The forces opposing capillary pressure are the *intracapsular pressure* exerted by the fluid in Bowman's capsule and the colloid *osmotic pressure* generated by the plasma proteins.

$$\underset{70 \text{ mm Hg}}{\text{glomerular capillary pressure}} \xrightleftharpoons{\text{opposed by}} \underset{30 \text{ mm Hg}}{\text{osmotic pressure}} + \underset{15 \text{ mm Hg}}{\text{intracapsular pressure}}$$

$$\text{effective filtration pressure} = 70 - (30 + 15) \text{ mm Hg} = 25 \text{ mm Hg}$$

Table 15–3. **Composition of Urine (units as indicated/24 hr)**

Total volume	1,200	ml
Water	1,140	ml
Total solids	50	g
Glucose	0	
Protein	0	
Ketones	0	
Urea	30	g
Creatinine	1.6	g
Creatine	0.1	g
Hippuric acid	0.7	g
Urobilinogen	0–4	mg
Porphyrins	50–300	μg
Uric acid	0.7	g
NaCl	15.0	g
K	3.3	g
Ca	0.3	g
Mg	0.1	g
Fe	0.005	g
SO_4	2.5	g
PO_4	2.5	g
NH_3	0.7	g

COMPOSITION OF THE ULTRAFILTRATE

The meticulous studies of Richards in 1924 were the first to demonstrate the composition of the fluid in Bowman's capsule. Using the micromanipulator developed by Robert Chambers in 1921, Richards, working with Wearn, was able to puncture the capsule in the kidneys of living amphibia, where the glomeruli are readily accessible to micropuncture (Fig. 15-9). Analyzing the capsular fluid extracted, he found its composition to be essentially identical to that of plasma with respect to water and low molecular weight solutes (glucose, chloride, sodium, potassium, phosphate, urea, creatinine, and uric acid). The recent discovery of a unique strain of rats with glomeruli on the surface of the kidney has permitted micropuncture studies that have confirmed glomerular ultrafiltration in mammalian kidneys.

The integrity of the glomerular membrane is essential for ultrafiltration, and it is believed that the size of the pores in the membrane is chiefly responsible for the molecular sieving that restrains the passage of particles larger than about 5 nm. Pore size is not the only factor involved, for many pores are 7.5 nm, but even though a particle may be smaller than a pore, as its size relative to that of the pore increases, the chances of its bouncing off instead of entering increase (*steric hindrance*). In addition, the asymmetry of many large molecules will affect their passage through pores: a few long, narrow molecules will randomly enter when their thin end hits the pore (Fig. 15-10). Plasma albumin, a relatively small protein, is filtered in small amounts by the normal kidney, but most of it is reabsorbed in the pct.

The presence of red blood cells or protein in the urine indicates a defect in the glomerular membrane, or excessively high pressure in the glomerular tuft. White blood cells sometimes find their way into the urine through ameboid movement across capillary walls: this does not necessarily indicate a fault in glomerular structure but is usually indicative of bacterial infection.

GLOMERULAR FILTRATION RATE

The glomerular filtration rate (GFR) is the amount of filtrate formed per minute in all nephrons of both kidneys. In the adult male, this rate is about 125 ml/min; in the female about 110 ml/min. Under normal circumstances, variations in urine flow are effected by changes in the

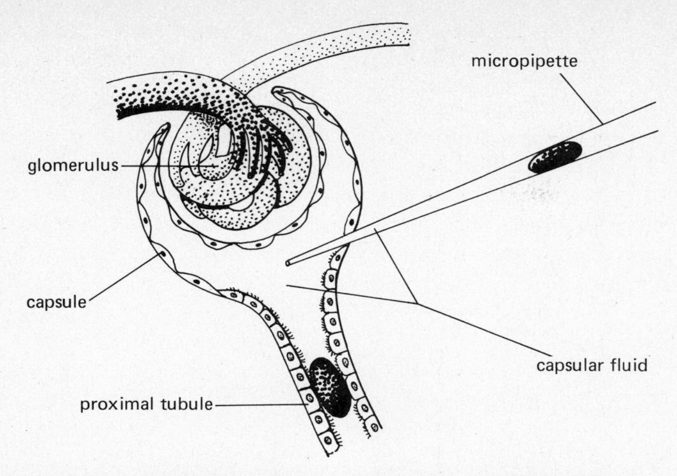

Figure 15-9. **Richard's technique for collection of the glomerular ultrafiltrate in the frog kidney. Oil droplets block the proximal tubule and the micropipette.**

amount of water reabsorbed, not by changes in GFR. In extreme dehydration or severe hemorrhage, when the circulating blood volume is considerably reduced, GFR drops because of the fall in pressure in the afferent arteriole. Normal physiological alterations in blood pressure do not affect GFR because the pressure relationships in the glomerulus are maintained by the poorly understood phenomenon of *autoregulation,* one aspect of which was discussed in the section on the juxtaglomerular apparatus.

Measurements of GFR are indirect because micropuncture techniques can sample only a few nephrons at a time. GFR calculations are based on the important concept of plasma clearance, which is described later in this chapter.

Figure 15-10. **Relationship between molecular weight, molecular dimensions, and glomerular sieving of solutes. (From Pitts, R. F.: *Physiology of the Kidney and Body Fluids,* 3d edition. Copyright © 1974 by Year Book Medical Publishers, Inc., Chicago. Used by permission.)**

Substance	Molecular weight	Dimensions in Angstrom Units: Radius from diffusion coefficient	Dimensions in Angstrom Units: Dimensions from X-ray diffraction	Filtrate / (Filtrand)
Water	18	1.0		1.0
Urea	60	1.6		1.0
Glucose	180	3.6		1.0
Sucrose	342	4.4		1.0
Insulin	5,500	14.8		0.98
Myoglobin	17,000	19.5	54, 8	0.75
Egg albumin	43,000	28.5	88, 22	0.22
Hemoglobin	68,000	32.5	54, 32	0.03
Serum albumin	69,000	35.5	150, 36	<0.01

Tubular Secretion

The observation that certain substances are found in the urine in greater concentration than can be accounted for by filtration alone led to the inference that these substances may be *added to the filtrate* by secretion. In secretion, the transported substances are moved from the peritubular capillaries across the tubular cells and into the tubular fluid. In reabsorption, the movement is from the tubular fluid back into the circulation. Active secretion and active reabsorption both imply transport against an electrochemical gradient and require metabolic energy; the difference is the *direction of the transport mechanism.* Most of the metabolic energy of the kidney is required for active transport processes.

The most convincing evidence for tubular secretion comes from studies in vitro. The tubules of the kidney of the chick embryo when cultured in vitro round up into hollow cysts; if they are placed in a dilute solution of phenol red, they secrete the dye in high concentration in their lumina. This concentration of dye is blocked by metabolic inhibitors such as cyanide, indicating an active transport process. Only cysts formed from pct will secrete phenol red. Similar results have been obtained with kidney tubules of marine fish (e.g., flounder), which have only a pct, and with mammalian and human kidneys.

Characteristics of active secretion

Transport Maximum Limited Secretory Mechanisms. Active transport processes have a *transport maximum* (Tm), which represents the saturation of the carrier mechanism involved. When concentrations are above the Tm, the excess remains in the blood, if the mechanism is secretion, or is excreted in the urine if the mechanism is reabsorption.

There are two secretory mechanisms that have a Tm, and both are limited to the pct. One mechanism is responsible for the secretion of a heterogeneous group of substances, including phenol red, creatinine, penicillin and para-aminohippurate (PAH) (Fig. 15-11). Many of these substances are *carboxylic or sulfonic acids.* Urological contrast media such as the complex iodine compound Diodrast are also secreted. It is difficult to explain the development of mechanisms that secrete foreign substances, but the ability to secrete drugs, poisons, and other foreign agents into the urine, as well as secreting the metabolic toxins normally produced by the body, conveys a considerable advantage over a kidney that could only filter and reabsorb.

The other Tm-limited secretory mechanism transports a group of *strong organic bases,* including choline, guanidine, histamine, and thiamine, and a variety of foreign compounds such as tetraethylammonium (TEA), a drug that selectively blocks potassium conductance.

Competitive studies have proven that these are two separate secretory mechanisms. None of the members of the organic acid group will depress the secretion of the strong bases, nor will the converse occur. However, within each

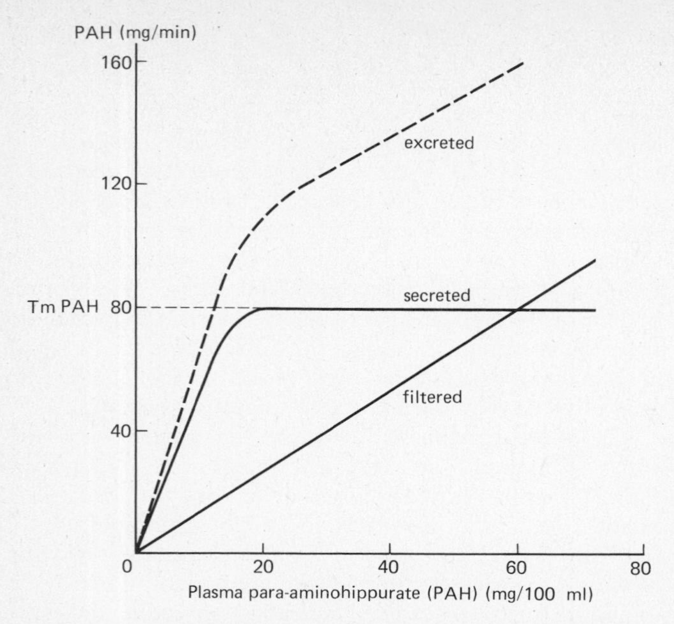

Figure 15-11. **Rates of secretion and excretion of para-aminohippurate (PAH) with increasing concentrations of plasma PAH. At very low plasma PAH concentrations, most of the PAH in the urine is due to secretion. When plasma PAH concentrations rise above 60 mg/100 ml, most of the urine PAH is due to filtration. (Adapted from Pitts, R. F.: *Physiology of the Kidney and Body Fluids,* 3d edition. Copyright © 1974 by Year Book Medical Publishers, Inc., Chicago. Used by permission.)**

group, competition for the carrier does occur, as is shown by the inhibition of TEA secretion in the presence of guanidine.

Concentration Gradient–Limited Secretory Mechanisms. This type of secretion apparently has no Tm but is regulated by concentration gradients. The *secretion of potassium and hydrogen ions* is concentration gradient–limited. These ions are secreted by a common transport mechanism confined to the dct. They compete for the carrier: if the intracellular concentration of K^+ is high and that of H^+ is low, K^+ is transported into the tubular fluid to the exclusion of H^+. If the intracellular H^+ concentration is high, then large quantities of H^+ are secreted. Sodium ions are reabsorbed from tubular fluid in exchange for secreted H^+ and K^+ ions. This renal regulation of acid-base balance is discussed in Sec. 14–17.

Passive secretion

Passive secretion occurs by diffusion down a concentration gradient. Although energy is not required for the actual transport of the substance, it is essential for the maintenance of the gradient. Passive secretion accounts for the movement of *weak bases,* such as ammonia, from the peritubular fluid into the kidney tubules. In the rat, both the pct and dct secrete ammonia; it is uncertain whether this mechanism is confined to the dct in humans.

Tubular Reabsorption

The kidney tubules reabsorb most of the ultrafiltrate formed by the glomeruli, including many of its valuable constituents. This is accomplished by a variety of passive and active transport processes; as a consequence, many filtrable components of plasma are absent from the urine. Table 15–4 illustrates the amount of water and ions filtered, reabsorbed, and excreted by the human kidney. Experimental evidence for reabsorption has come from micropuncture experiments and sophisticated adaptations of this technique.

Proximal convoluted tubule

Water and Salts. Almost 80 per cent of the glomerular filtrate is reabsorbed in the pct. This is referred to as *obligatory reabsorption* and occurs regardless of the degree of hydration of the organism. This means that homeostatic regulation of urine volume is confined to 20 per cent of the filtrate, still a considerable amount.

Analysis of tubular fluid withdrawn by micropuncture has shown that it remains isotonic to plasma, indicating that electrolytes and water are reabsorbed at osmotically identical rates. The rate of water transport is always proportional to the rate of solute transport; that is, *water reabsorption is a passive process* responding osmotically to the *active transport of sodium salts,* the most important of which are NaCl and $NaHCO_3$. While it can be shown that Na^+, Cl^-, and HCO_3^- are actively reabsorbed in the pct, the detailed mechanism of these transport systems is largely unknown. Almost all the K^+ filtered is reabsorbed in the pct, and it is probable that the K^+ found in the urine is secreted by the dct.

Table 15–4. **Work Done by the Kidney: Water and Electrolytes Reabsorbed**

	Concentration in Serum (mg/100 ml)	Amount Filtered (g)	Amount Excreted (g)	Amount Reabsorbed (%)
Sodium	320	428	4.6	99
Potassium	17	23	2.7	88
Chloride	405	555	7.4	99
Phosphate	4.3	6	1.1	80
Water	94 ml/100 ml	180 liters	1.5 liters	99

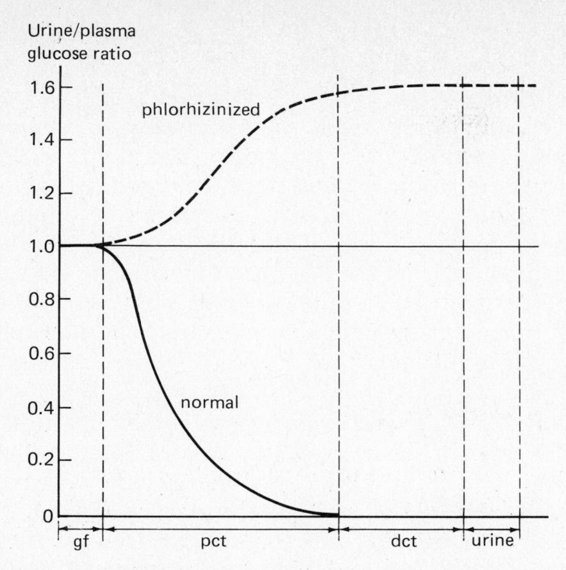

Figure 15-12. **Reabsorption of glucose from the nephron of the normal and the phlorhizinized frog. The drug prevents reabsorption of glucose by proximal convoluted tubule (pct); as water is reabsorbed, the glucose becomes more concentrated (gf = glomerular filtrate, dct = distal convoluted tubule).**

Urea. Urea is the most important of the solutes present in the glomerular filtrate to be concentrated as a result of fluid reabsorption. As urea is very freely diffusible, about 40 per cent of the filtered urea is passively reabsorbed through the pct epithelium. The amount reabsorbed varies widely with the rate of urine flow, falling sharply when urine flow is reduced.

Glucose. Richards' micropuncture experiments in frogs showed that the glucose concentration becomes progressively less as the filtrate passes down the pct, indicating that glucose is absorbed more rapidly than water (Fig. 15-12). Administration of the drug *phlorhizin,* which specifically blocks pct reabsorption of glucose in all glomerular vertebrates, increases pct glucose concentrations, and glucose appears in the urine (*glucosuria*).

GLUCOSE TRANSPORT MAXIMUM. The highly efficient reabsorption mechanism for glucose, which also transports other monosaccharides, is an active process with a transport maximum. As the plasma concentration of glucose is increased above a critical level (the *renal plasma threshold*), the Tm is exceeded, and glucose appears in the urine (Fig. 15-13). Tm for adult males is about 365 mg/min. There is a significant sex difference in this parameter: it is considerably lower in adult women, approximately 303 mg/min.

Renal glucosuria may occur in some individuals with normal blood glucose. This is caused by a low Tm for glucose resulting from a kidney defect, rather than by a lack of the pancreatic secretion insulin, as in *diabetic glucosuria.*

Figure 15-13. **Reabsorption of glucose as a function of plasma glucose concentration. Once the renal threshold for glucose has been reached, increasing amounts of glucose are excreted in the urine. The transfer maximum (Tm) for glucose is between 300 and 370 mg/min.**

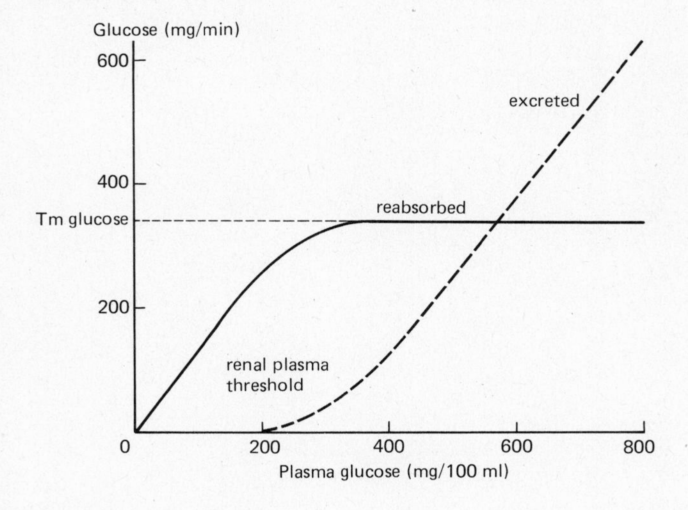

Amino Acids, Acetoacetate Ion, Vitamins, and Protein. Amino acids, acetoacetate ion, vitamins, and the small amount of protein that does get into the ultrafiltrate, all are reabsorbed in the pct by active processes.

LOOP OF HENLE

The fluid entering the loop of Henle is *isotonic* to plasma, with essentially the same concentration of Na^+, Cl^-, and HCO_3^- as plasma, but it has otherwise been considerably modified from the original glomerular filtrate. Its volume represents 20 per cent of the filtrate, all glucose and other nutritionally valuable molecules have been removed, and urea has been concentrated.

The tubular fluid leaving the ascending arm of the loop near or in the cortex and entering the dct is *hypotonic* to plasma, but is otherwise unchanged after its passage through the loop of Henle. This important observation caused considerable confusion; it implied that the function of the loop of Henle is to dilute the tubular fluid. Where then does urine concentration occur?

Accurate determinations have shown that the *tip of the medulla* is considerably *hypertonic* to plasma; in fact there is a sodium concentration gradient that increases from the cortex to the innermost medulla. It is the function of the loop of Henle to establish this concentration gradient, the function of the vasa recta surrounding the loop to maintain it, and the function of the collecting tubules to utilize the gradient to produce a concentrated urine.

FORMATION OF A HYPERTONIC URINE

There are three mechanisms involved in the concentration of urine:

1. Countercurrent multiplication (loop of Henle).
2. Countercurrent exchange (vasa recta).
3. Osmotic exchange (dct and ct).

Countercurrent Multiplication (Loop of Henle). To understand these processes, refer to Fig. 15-14, which shows the hairpin turn of the loop of Henle bringing the thin descending arm in close apposition to the ascending arm. The tubular fluid is flowing in opposite directions (countercurrent flow) in the two arms. A small osmolar concentration difference of 200 mOsmol/liter can be established between the fluid contents of the two limbs at each level in the medulla. This is achieved by the transport of sodium chloride out of the ascending arm into the surrounding interstitium of the medulla. A new osmotic equilibrium is established in a stepwise manner between the interstitium at each level and the fluid in the arm. Consequently, the fluid in the ascending arm becomes progressively *more dilute* as it approaches the cortex. The fluid in the descending arm and the interstitium become simultaneously and dependently *more concentrated* toward the tip of the medulla, with an ultimate concentration gradient of 300 to 1200 mOsmol/liter developed from the cortex to the medullary tip.

It has recently been shown that it is the *chloride ions* that are *actively transported* out of the thick ascending arm and into the interstitium, establishing a potential difference across the tubular cells of about 5 to 8 mV, positive on the luminal side. This potential is the driving force for the associated movement of sodium ions out of the tubule. The high sodium permeability of the tubular cells, together with the low electrical resistance, make the *passive transport of sodium* a highly efficient process.

This countercurrent multiplication of concentration utilizes the expenditure of metabolic energy for the active pumping out of chloride by the ascending arm. This process is also dependent upon the *difference in water permeability* between the two arms of the loop of Henle. The ascending arm is impermeable to water and as it is constantly pumping out sodium chloride, its fluid becomes more dilute as it passes toward the cortex. The descending arm, however, is freely permeable to water so that as the osmotic gradient of the interstitium increases, water diffuses out and the tubular fluid becomes more and more concentrated as it approaches the bend in the loop.

Countercurrent Exchange (Vasa Recta). In its hairpin course through the medulla, the loop of Henle has established a *concentration gradient* from the cortex (isotonic to plasma) to the medullary tip (hypertonic). The blood in the vasa recta follows the same course as the loops of Henle of the juxtamedullary nephrons: it is exposed to the same concentration gradient in the interstitium, and because the walls of the vasa recta are permeable to both water and sodium chloride, the blood concentration reflects that of the interstitium at all levels through the medulla (Fig. 15-15).

This process is a *passive osmotic exchange* between the interstitium and two closely apposed vessels, with fluid flowing in opposite directions. This *countercurrent exchange* maintains the osmotic gradient established by the countercurrent multiplier of the loop of Henle. Consider the situa-

Figure 15-14. **Countercurrent multiplication in the loop of Henle. The shaded columns indicate the concentration gradient in the medulla. The progressive increase in concentration in the descending arm of the loop of Henle is due to the active pumping of NaCl from the ascending loop. The subsequent equilibrium between the descending arm and the medulla is dependent on the permeability of the walls of the descending arm to NaCl and water. The effectiveness of the stepwise countercurrent multiplier requires that the walls of the ascending tubule be impermeable to water. Concentrations in mOsmol/liter.**

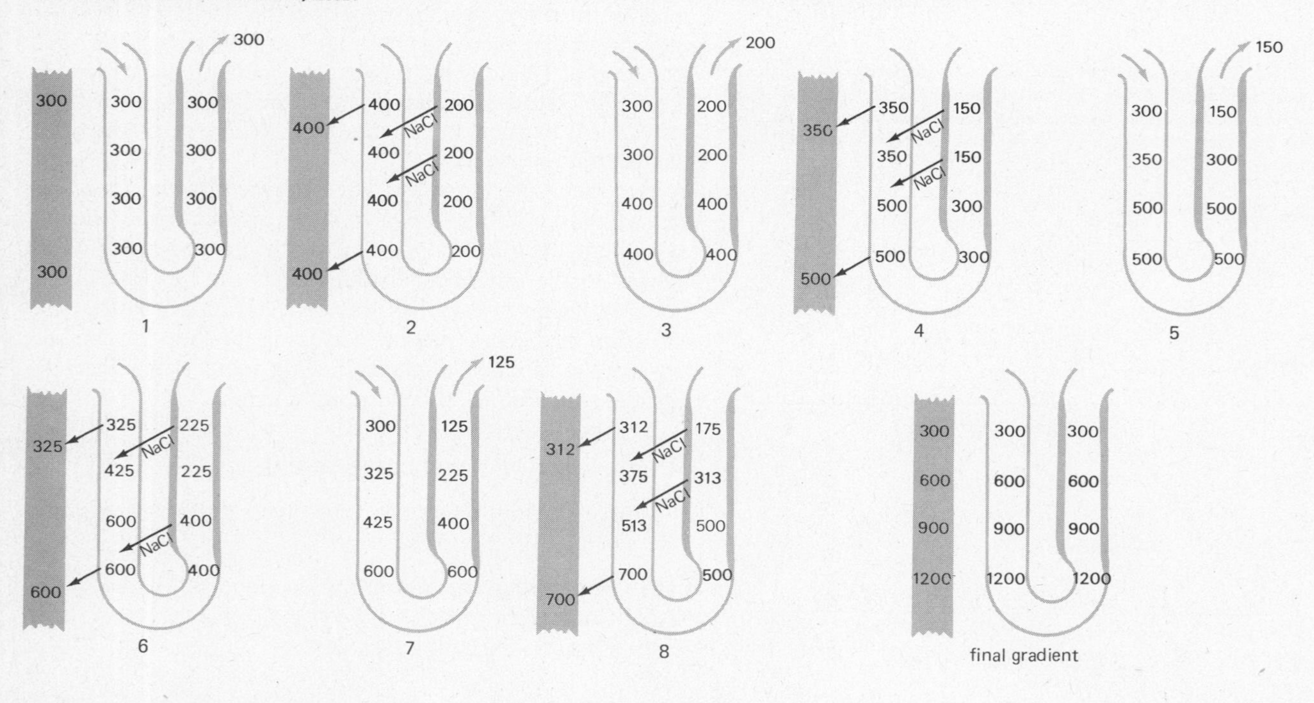

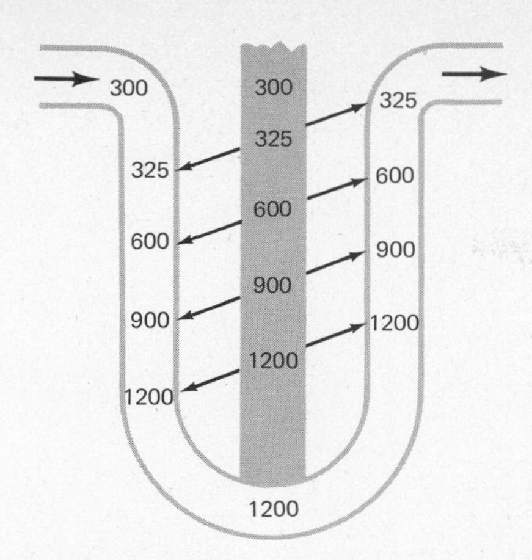

Figure 15-15. **Countercurrent exchangers. The vasa recta do not actively transport solutes or water, but they establish a concentration equilibrium between the blood and the medulla. Because of the countercurrent flow, the blood leaving the medulla is only slightly hypertonic to arterial plasma. The shaded column represents the concentration gradient in the medulla. Concentrations in mOsmol/liter. This countercurrent loop shows blood flow through the vasa recta.**

tion if the blood were to flow through to the tip of the medulla and then directly into the venous circulation: all of the salt would be rapidly flushed out of the interstitium and into the circulation. By looping back through the decreasing concentration gradient toward the cortex, sodium is returned to the medulla, and the blood enters the veins only slightly hypertonic to arterial plasma.

Osmotic Exchange (*Distal Convoluted and Collecting Tubules*). As the hypotonic fluid from the ascending arm passes through the *dct,* it gradually becomes isotonic to the cortical interstitium. The volume of fluid entering the dct is now only about 15 per cent of that of the filtrate. This volume varies, however, with the amount of circulating *ADH,* which affects the water permeability of the dct epithelial cells, and with the level of *aldosterone,* which affects their reabsorption of sodium. As water passively follows sodium reabsorption, the tonicity of the fluid in the dct reflects that of the cortical interstitium and thus enters the ct essentially isotonic to plasma.

The main function of the *ct* is to produce a hypertonic urine from the isotonic fluid that enters them. Micropuncture studies of various levels from the cortex to the medulla have shown that an equilibrium exists between the tubular fluid and the medulla at each level tested, just as was seen in the loop of Henle. In the ct, however, this is not a countercurrent mechanism but a *passive establishment of osmotic equilibrium;* no active pumping takes place, and the fluid that leaves the ct enters the kidney pelvis at approximately the same concentration as in the medullary tip.

The degree of urine concentration, which in mammals varies from very dilute to extremely hypertonic, is ultimately determined by the permeability of the ct epithelium to water (Fig. 15-16). This in turn is dependent upon the amount of circulating ADH, as we shall see.

Hormonal regulation of tubular reabsorption

Antidiuretic Hormone (*ADH*). Urine volume normally reflects the state of hydration of the organism and is chiefly regulated by ADH, a hypothalamic octapeptide (Sec. 11-3). *Diuresis* is the rapid excretion of large volumes of urine and is prevented by the action of ADH on the water permeability of the cells of the dct and ct.

In *well-hydrated states,* following the ingestion of a large amount of water, the titers of ADH are very low, and the epithelium of the dct and ct is impermeable to water. The spaces between the cells disappear and the hypotonicity of the fluid entering the dct is maintained, despite its subsequent passage through the hypertonic medulla. The final urine volume is large, and the urine is very dilute (*water diuresis*).

Destruction of the integrity of the supraoptic nuclei of the hypothalamus, the infundibular stalk, and the posterior pituitary gland, all of which are involved in the synthesis and storage of ADH (Secs. 10-6 and 11-3) causes the condition known as *diabetes insipidus,* which is characterized by the excretion of a large volume of pale dilute urine. This is a distinctly different condition from *diabetes mellitus* (*mellitus* = "sweet"), which is caused by a pancreatic malfunction resulting in increased blood glucose and consequently glucosuria accompanied by an increase in urine volume.

CELLULAR MECHANISM OF ACTION OF ADH. It has not been possible to measure the intracellular content of collecting tubule cyclic AMP, but studies on broken cell preparations of kidney cortex and medulla show that ADH increases both adenyl cyclase activity and the concentration of

Figure 15-16. **Osmotic exchangers. The collecting tubules can establish an equilibrium between their fluid and the medullary interstitium (shaded column) if the walls of the collecting tubule are freely permeable to water (A). In the presence of antidiuretic hormone (B), the walls become water-impermeable and a dilute urine is excreted. Concentrations in mOsmol/liter.**

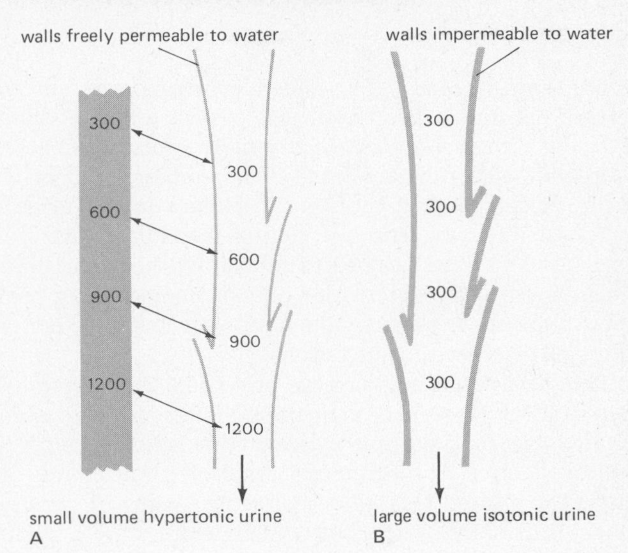

cyclic AMP. The addition of cyclic AMP to isolated collecting tubules markedly increases their permeability. It seems justified to infer from these experiments that ADH, like other polypeptide hormones, exerts its physiological effects on target hormones through cyclic AMP. It is possible that the resulting change in permeability is brought about by a kinase-activated phosphorylation of proteins in the cell membrane, converting a hydrophobic region into a hydrophilic one and thus increasing the number of pores in the membrane.

HOMEOSTATIC CONTROL OF ADH SECRETION. There are several mechanisms that control ADH secretion. The classical experiments of Verney in 1947 demonstrated that if the blood flow to the hypothalamus were made hypertonic, through injections of hypertonic saline into the carotid artery, a previously induced state of diuresis is changed dramatically to a state where urine flow is inhibited (Fig. 11-5). Verney induced that there were cells in the hypothalamus that responded to changes in tonicity of the blood by alterations in ADH secretion; he called these cells *osmoreceptors*. It has subsequently been shown that the osmoreceptors are located in the supraoptic nuclei of the hypothalamus and that they respond to other stimuli as well as changes in plasma osmotic pressure.

ADH secretion is strongly influenced by stimulation of various *stretch receptors* in the circulatory system, particularly in the left atrium. These stretch receptors are really *volume receptors* and reflexly stimulate ADH secretion when the volume of blood passing through the atrium drops. Consequently, the increase in water reabsorption induced by the secretion of ADH restores blood volume to normal. *Pain* or *surgical stress* are other potent stimulators of ADH release. *Alcohol* inhibits the release of ADH, resulting in a marked diuresis.

A redistribution of blood from central to peripheral regions, such as occurs during *acceleration in space* when gravitational forces are suddenly increased, is recognized by these vascular volume receptors as a drop in blood volume and reflexly results in increased ADH secretions.

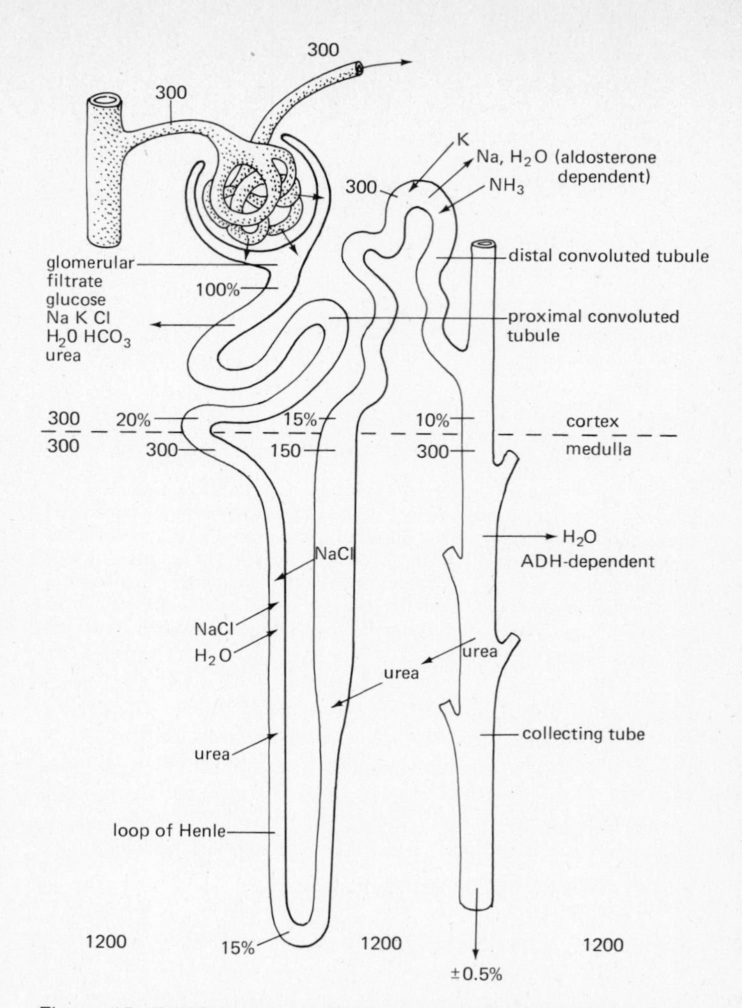

Figure 15-17. **Diagrammatic summary of modifications in urine volume (expressed as percentage glomerular filtrate) and tonicity (figures represent concentrations in mOsmol/liter) as the tubular fluid flows through the nephron.**

Mineralocorticoids. The mineralocorticoids are adrenocortical hormones (Sec. 10-6) that possess a high sodium-retaining ability. *Aldosterone* is the most important mineralocorticoid, and in its absence *sodium reabsorption* does not occur. Accompanying the loss of sodium in the urine is increased water excretion, depleting the extracellular volume of the body and leading to circulatory shock and death. There is also a related retention of potassium, so that plasma potassium rises as plasma sodium falls, reversing the normal ratios of these two major cations.

This devastating impairment of sodium reabsorption is seen in adrenalectomized animals and in persons with *Addison's disease,* a condition in which the adrenal cortex is destroyed, often by tuberculosis. Adrenalectomized rats can survive if sodium balance is maintained with 0.9 per cent saline as drinking water, but humans require the administration of aldosterone. However, for experimental animals or Addisonian patients to survive any stressful situation, glucocorticoid replacement therapy is essential.

SITE AND MECHANISM OF ACTION. Micropuncture studies have shown that aldosterone has a direct effect on the kidney tubule, profoundly affecting the ability of the dct to *reabsorb sodium.* The collecting tubules may be affected to a lesser extent. The regulation of sodium transport by aldosterone action appears to be initiated by the binding of the steroid to specific hormone receptors in the cytoplasm of the kidney cells. As discussed in Sec. 10-8, this induces the synthesis of specific protein kinases, which presumably results in permeability changes in the dct cells and greater availability of ATP at the site of the sodium pump.

REGULATION OF ALDOSTERONE SECRETION. Aldosterone secretion is increased by a fall in extracellular sodium or a rise in extracellular potassium; it may be that it is the ratio of sodium to potassium that is the decisive stimulus.

Other postulated mechanisms for the regulation of aldosterone secretion include ACTH and the renin-angiotensin system. Infusion of ACTH or of angiotensin into the circulation causes a marked increase in aldosterone secretion by the adrenal cortex.

The main modifications in urine volume and tonicity are diagrammatically represented in Fig. 15-17.

15–5 pH Regulation

The kidney regulates the acid-base balance of body fluids through a complex series of reactions that involve the reabsorption of sodium and potassium, bicarbonate ion excretion into the urine, ammonia secretion into the tubules, and hydrogen ion secretion.

The blood contains buffers that neutralize metabolically produced acid by combining with them to form salts; the kidney "saves" the buffer by selectively reabsorbing it as bicarbonate or phosphate and excreting the acid. This mechanism is discussed in Sec. 14-17. As a result of the kidney's ability to excrete an acid urine, the pH of the urine may drop to 4.8 instead of its usual 6.0, but the pH of the blood remains constant at 7.4. Occasionally, if the amount of alkali in the blood increases markedly, an alkaline urine with a pH of as high as 8.2 may be excreted.

15–6 Diuretics

A diuretic increases the rate of urine output, either by acting on the kidney itself or by inhibiting the secretion of ADH. The *glomerular filtration rate* can be increased by agents that increase systemic blood pressure, such as norepinephrine, but this effect is offset by a concomitant constriction of the afferent arterioles. The end result is only a moderate diuresis. The most useful diuretics are the organic mercurial compounds and the thiazides, which act on the *kidney tubules* to increase the excretion of water by depressing electrolyte reabsorption. Regulation of ADH secretion has been discussed in Sec. 15-4.

15–7 The Artificial Kidney

Although the kidney is probably one of the most complex organs of the body, an artificial replacement for it has been devised and has become part of the standard equipment of most modern hospitals. The artificial kidney does not resemble the natural kidney in either form or function, although the final result of regulating the concentration of individual substances in the blood is the same. The basic process used by the artificial kidney in removing substances from the blood, or adding substances to the blood, is *dialysis,* or movement of particles across a porous membrane which separates two solutions of unequal concentrations. The particles must be small enough to penetrate the membrane, and although they will move back and forth across it, the net result will be more movement from the solution of high concentration to the solution with the lower concentration.

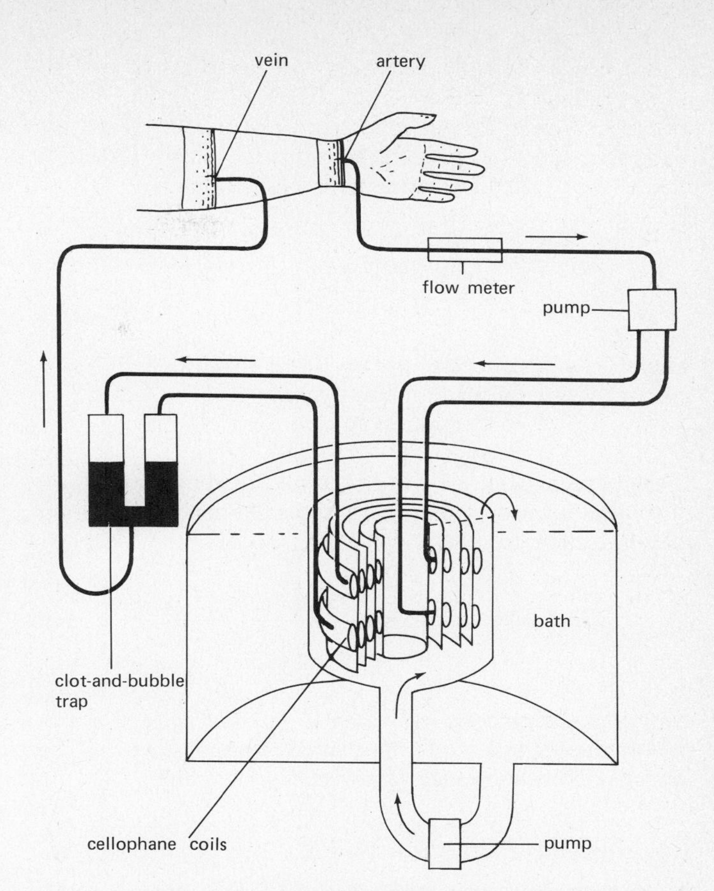

Figure 15-18. **The artificial kidney. The coils of cellophane tubing through which the blood is pumped act as a semipermeable membrane. Water, crystalloids, glucose, amino acids, gases, and urea are freely filtered into the dialyzing solution surrounding the tubing. This solution contains most of these substances in the same concentration as blood, with the exception of urea, which it lacks. Consequently, the dialysate must be constantly replaced to maintain the urea concentration gradient in the right direction. The dialyzed blood, cleared of urea, is returned to a vein.**

In the artificial kidney, the patient's blood is pumped through coils of cellophane, while being maintained at the correct temperature and oxygen pressure (Fig. 15-18). The cellophane acts as the dialyzing membrane across which particles can be moved into or out of the blood, depending on the relative concentration maintained in the dialyzing solution on the other side of the membrane. The ability of the artificial kidney to add or remove practically any component selectively is particularly valuable in cases of poisoning, where it can speed the removal of toxins by supplementing the work of the normal kidney.

15–8 Determination of Clearance, GFR, Reabsorption, and Secretion

Plasma Clearance

Plasma clearance is the *volume of plasma that is cleared* of a substance per minute. It represents the end result of filtration, reabsorption, and secretion—all renal processes that affect the final concentration of a substance in the plasma. Clearance was first determined for urea as a clinical reference for kidney function. The normal value for plasma clearance of urea, with adequate urine flow, is 75 ml/min.

Clearance can be calculated from the following formula:

$$\frac{Ux \cdot V}{Px}$$

where Ux = concentration of x in 1 ml urine
V = ml urine formed/min
Px = concentration of x in 1 ml plasma

If the concentration of x in the urine is 125 mg/ml, 1.0 ml of urine is formed per minute, and the plasma concentration of x is 1.0 mg/ml, then

$$\text{plasma clearance of } x = \frac{125 \text{ mg/ml} \times 1.0 \text{ ml/min}}{1.0 \text{ mg/ml}} = 125 \text{ ml/min}$$

Note that clearance is expressed in ml/min because it expresses the volume of plasma cleared of a substance, *not* the amount of the substance removed.

Glomerular Filtration (GFR)

If a substance is freely filtered, neither secreted nor reabsorbed, and is biologically inert (not metabolized), then its clearance demonstrates the rate of glomerular filtration. It should also be nontoxic and accurately measurable in plasma and urine to be a useful tool. The substance that fulfills all these criteria is *inulin,* a large polysaccharide with a radius of 1.5 nm. It is obtained from dahlia roots and Jerusalem artichokes. Unlike most other carbohydrates, inulin is not metabolized; neither is it reabsorbed or secreted. Consequently, inulin clearance represents only filtration and can be used as an indirect measure of GFR. In the example of clearance given above, x represents inulin, and consequently the GFR is equivalent to the inulin clearance, which is 125 ml/min.

It is important for the validity of this concept that the rate of filtration of inulin not change with the amount of inulin in the plasma. Figure 15-19(A) shows that the amount of inulin found in the urine is directly proportional to the concentration in the plasma, and Fig. 15-19(B) shows that inulin clearance remains constant, regardless of plasma inulin concentration.

Figure 15-19. **(A) The amount of inulin excreted in the urine is directly proportional to the plasma inulin concentration over a wide range. (B) Inulin clearance remains constant over a wide range of plasma inulin concentrations. As the amount of inulin filtered does not change with plasma inulin concentration, and as inulin is neither filtered nor reabsorbed, inulin clearance may be taken as a measure of the glomerular filtration rate.**

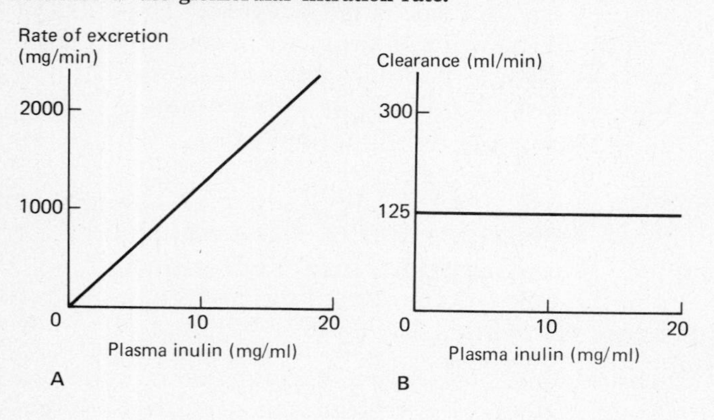

Reabsorption

The rate of reabsorption is the difference between the *rate of filtration* and the *rate of excretion* in the urine. This assumes that there is no secretion involved and that the substance is freely filtered and not bound to plasma proteins. Glucose is a good example of such a substance. Its rate of reabsorption can be calculated by inducing hyperglycemia (high blood glucose) with glucose infusions, so that the Tm for glucose is surpassed and the amount excreted can be determined. If inulin clearance is simultaneously measured, then the GFR is known.

clearance of inulin = 125 ml/min
plasma glucose = 300 mg% = 3.0 mg/ml
urine glucose = 15 mg/ml
urine flow = 5 ml/min

glucose reabsorbed = glucose filtered—glucose excreted
= (125 ml/min × 3 mg/ml) − (15 mg/ml × 5 ml/min)
= 375 mg/min − 75 mg/min
= 300 mg/min

Secretion

To determine secretion, one has to use a substance that is freely filtered and not reabsorbed. Diodrast and PAH are often used for this purpose. Simultaneous determination of inulin clearance is necessary for the calculation of GFR.

clearance of inulin = 125 ml/min
plasma Diodrast = 100 mg% = 1 mg/ml
urine Diodrast = 50 mg/ml
urine flow ⊆ 3.5 ml/min

Diodrast secreted = Diodrast excreted − Diodrast filtered
= (50 mg/ml × 3.5 ml/min) − (125 ml/min × 1 mg/ml)
= 165 mg/min − 125 mg/min
= 40 mg/min

These calculations become more complicated and less accurate when substances are both reabsorbed and secreted.

Cited References

CHAMBERS, R., and R. T. KEMPTON. "Indications of Function of the Chick Mesonephros in Tissue Culture with Phenol Red." *J. Cell. Comp. Physiol.* **3:** 131 (1933).

VERNEY, E. B. "The Antidiuretic Hormone and the Factors Which Determine Its Release." *Proc. Roy. Soc. Biol.* **135:** 25 (1947).

WEARN, J. T., and A. N. RICHARDS. "Observations on the Composition of Glomerular Urine with Particular Reference to the Problem of Reabsorption in the Renal Tubules." *Amer. J. Physiol.* **71:** 209 (1924).

Additional Readings

BOOKS

BRENNER, B. M., and F. C. RECTOR, eds. *The Kidney.* W. B. Saunders Company, Philadelphia, 1976.

DEETJEN, P., J. W. BOYLAN, and K. KRAMER. *Physiology of the Kidney and of Water Balance.* Springer-Verlag New York, Inc., New York, 1975.

GUYTON, A. C., and K. THUREAU, eds. *MTP International Review of Science, Physiology,* Vols. I and II. *Kidney and Urinary Tract Physiology.* University Park Press, Baltimore, 1974 and 1976.

JARRETT, A., ed. *The Physiology and Pathophysiology of the Skin,* Vols. 1 and 2. Academic Press, Inc., New York, 1973.

KOUSHANPOUR, E. *Renal Physiology: Principles and Functions.* W. B. Saunders Company, Philadelphia, 1976.

MASSRY, S. G., ed. *Symposium on Kidney and Hormones. Nephron* **15**(3–5), 1975. (Entire volume.)

PITTS, R. *Physiology of the Kidney and Body Fluids,* 3rd ed. Year Book Medical Publishers, Inc., Chicago, 1974.

ROBINSON, J. R. *Fundamentals of Acid-Base Regulation,* 4th ed. J. B. Lippincott Company, Philadelphia, 1972.

SMITH, H. W. *From Fish to Philosopher.* Anchor Books (Doubleday and Co.), Garden City, New York, 1961. (Paperback.)

SMITH, H. W. *Lectures on the Kidney.* Oxford University Press, New York, 1943.

SMITH, H. W. *The Kidney: Structure and Function in Health and Disease.* Oxford University Press, New York, 1951.

VANDER, A. J. *Renal Physiology.* McGraw-Hill Book Company, New York, 1975.

WESSON, L. G., Jr. *Physiology of the Human Kidney.* Grune & Stratton, Inc., New York, 1969.

WILLIAMS, P. C., ed. *Hormones and the Kidney.* Academic Press, Inc., New York, 1963.

ARTICLES

BENTLEY, P. J. "Adaptations of Amphibia to Arid Environments." *Science* **152:** 619 (1966).

BURG, M. B., and N. GREEN. "Function of the Thick Ascending Limb of Henle's Loop." *Amer. J. Physiol.* **224:** 659 (1973).

KOKKO, J. P., and F. C. RECTOR, Jr. "Countercurrent Multiplication System Without Active Transport in Inner Medulla." *Kidney International* **2:** 214 (1972).

MALI, J. W. "Transport of Water Through the Human Epidermis." *J. Invest. Dermat.* **27:** 451 (1956).

MOSES, A. M., and M. MILLER. "Osmotic Influences on the Release of Vasopressin." In *Handbook of Physiology,* Sec. 7, Vol. 4, Part 1., R. O. Greep and E. B. Astwood, eds. The Williams & Wilkins Company, Baltimore, 1974.

NASH, F. D. "Control of Antidiuretic Hormone Secretion—Introductory Remarks." *Fed. Proc.* **30**(4): 1376.

SAWYER, W. H. "The Mammalian Antidiuretic Response." In *Handbook of Physiology,* Sec. 7, Vol. 4, Part 1, R. O. Greep and E. B. Astwood, eds. The Williams & Wilkins Company, Baltimore, 1974, p. 443.

SCHMIDT-NIELSEN, K., and B. SCHMIDT-NIELSEN. "The Desert Rat." *Sci. Am.,* July 1953.

SCHMIDT-NIELSEN, K. "Water and Osmotic Regulation." In K. Schmidt-Nielsen, *Animal Physiology, Adaptation and Environment.* Cambridge University Press, New York, 1975.

SHARE, L., and J. R. CLAYBOUGH. "Regulation of Body Fluids." *Ann. Rev. Physiol.* **34:** 235 (1972).

SOLOMON, A. K. "Pumps in the Living Cell." *Sci. Am.,* Aug. 1962.

WALKER, A. M., and J. OLIVER. "Methods for the Collection of Fluid from Single Glomeruli and Tubules of the Mammalian Kidney." *Amer. J. Physiol.* **134:** 562 (1941).

Chapter 16

Regulation of Respiration

The smoke of my own breath,
Echoes, ripples, buzz'd whispers, love-root, silk-thread, crotch and vine,
My respiration and inspiration, the beating of my heart, the passing of blood and air through my lungs,
The sniff of green leaves and dry leaves, and of the shore and dark-color'd sea-rocks, and of hay in the barn,
The sound of the belch'd words of my voice loos'd to the eddies of the wind,
A few light kisses, a few embraces, a reaching round of arms,
The play of shine and shade on the trees as the supple boughs wag,
The delight alone or in the rush of the streets, or along the fields and hill-sides,
The feeling of health, the full-noon trill, the song of me rising from bed and meeting the sun.

Walt Whitman, "Song of Myself"

TO MAINTAIN LIFE, oxygen must reach the cells in amounts adequate for the many metabolic reactions that are ultimately dependent on molecular oxygen. Deprivation of oxygen for only a few minutes results in cell death, brain cells being particularly sensitive to anoxia. Under normal circumstances, however, the respiratory system is capable of adjusting, through a variety of regulatory controls, to changes in the oxygen requirements of the tissues.

As a result of cellular metabolism, large amounts of carbon dioxide are constantly being produced by the tissues. Small amounts of carbon dioxide are essential for the synthesis of various organic compounds and for the efficient functioning of many cells and organs. The accumulation of carbon dioxide is toxic, however, causing cell death. Most of the carbon dioxide is removed from the body by the lungs, although fair amounts are excreted by the kidney as urea. Adjustments to variations in the internal concentration of carbon dioxide are mainly through the homeostatic adjustments of the respiratory system.

16–1 Respiration

Respiration is the transport of oxygen to the cells from the external atmosphere and the return transport of carbon dioxide from the cells to the atmosphere. Oxygen and carbon dioxide are generally referred to as the respiratory gases, although other gases are present in the air.

Respiration may be divided into two stages, external and internal.

External Respiration

External respiration includes (1) passage of air through the respiratory passages and lungs (ventilation), (2) diffusion of the respiratory gases between the alveoli of the lungs and the pulmonary capillaries, (3) transport of oxygen and carbon dioxide through the blood, and (4) diffusion of the respiratory gases between the blood and the tissues.

Internal Respiration

Internal respiration, or cellular respiration, involves the utilization of oxygen and the production of carbon dioxide by the tissues, essential metabolic reactions in the production of energy from food.

Changes in the composition of the inspired air and changes in the metabolic activities of the body affect the rate and depth of breathing through nervous and chemical feedback mechanisms, maintaining optimal levels of the respiratory gases in the internal environment.

16–2 The Respiratory System

The nose, mouth and pharynx, trachea, bronchi, and lung make up the respiratory system (Fig. 16–1). This system is lined with specialized ciliated cells and terminates in one of the most delicate of all living structures, the *respiratory membrane* that separates the external environment (the inhaled air) from the internal environment (the blood). In fish, this membrane is represented by the gills, and exchange takes place readily between the gases of the water in which the fish lives and the gases of its blood. In the transition to a terrestrial environment, the principle of gaseous exchange across a living membrane in an aqueous environment has remained the same. Oxygen is much more plentiful in air than in even well-aerated water, but there is constant danger of the drying out of the respiratory membrane. The development of the respiratory system in higher animals serves a triple purpose: the protection of the respiratory membrane by the inpocketing of the lungs; the passage of air to these hidden structures, in the course of which the air is

Figure 16-1. **The respiratory system.**

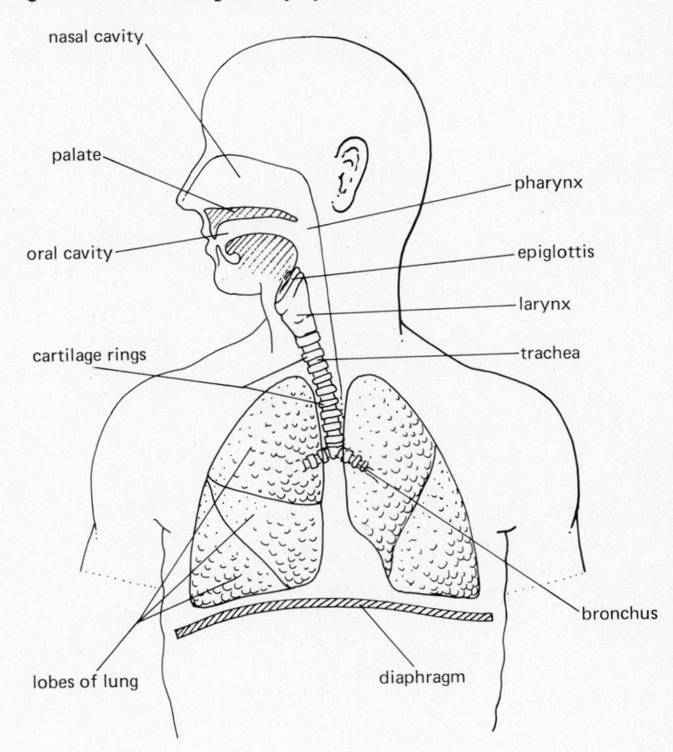

warmed, humidified, and filtered; and the increase of the available area for gaseous diffusion. Thus the air reaching the living barrier between the external and internal environments is modified in several important ways.

Nose and Nasal Passages

The external nose consists of a movable lower portion, made up chiefly of cartilage and skin, and a rigid upper portion with a bony skeleton. The part of the nasal cavity immediately above each nostril—the *vestibule* of the nose—is lined with skin from which hairs grow to protect the entrance. The *nasal cavity* is divided by a *septum* into a left and right half, the posterior apertures of which open into the pharynx.

Bones of the nose

The *roof* or *bridge* of the nose is a narrow shelf made up of the *nasal bones* and portions of other bones of the skull, including the nasal part of the frontal bone and the frontal processes of the maxillae (bones of the upper jaw). In the middle is a portion contributed by the *perpendicular plate of the ethmoid bone,* which like the frontal bones and the maxillae is pierced with air spaces or *sinuses* that communicate with the cavity of the nose (Fig. 16-2). During a severe cold, the inflammation often spreads from the mucous lining of the nose to the lining of the sinuses, causing headache and pain between the eyes. The *floor of the nose* is formed by the *bones of the hard palate* (the palatine process of the maxillae and the palatine bones).

Figure 16-2. **Structure of the nose and the sinuses of the skull. The tear duct is shown as a dotted line leading from the eye to the inferior (lowest) nasal meatus.**

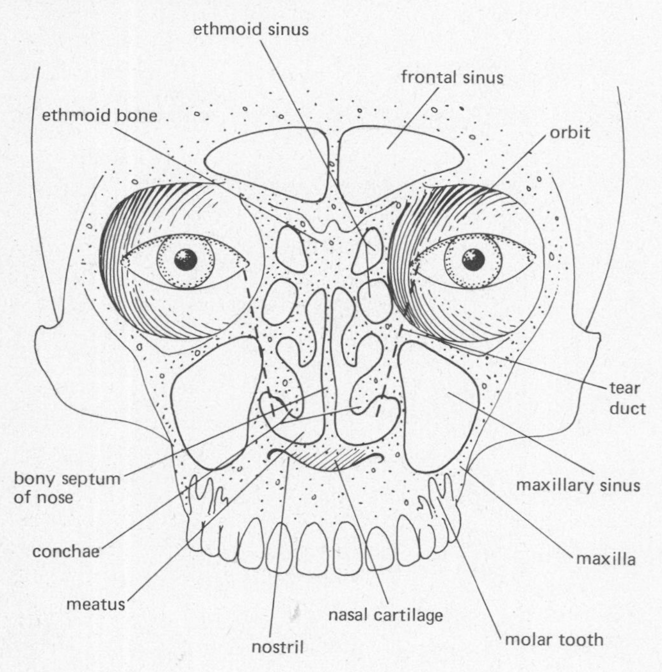

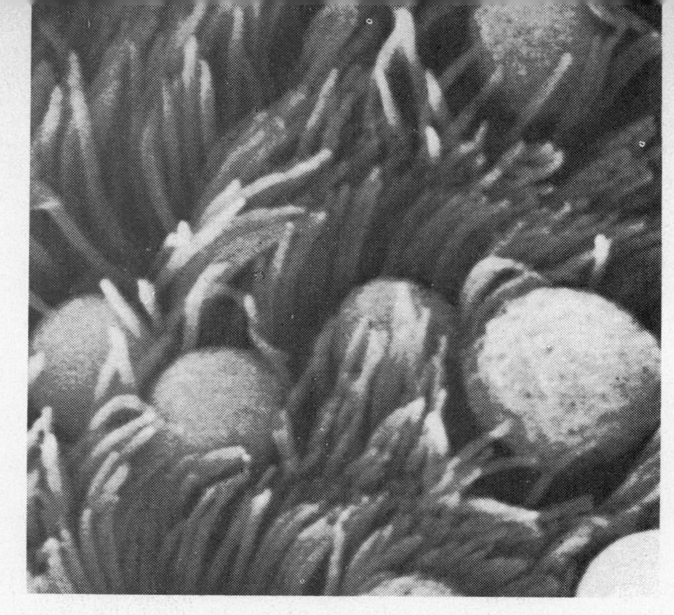

Figure 16-3. **Ciliated surface of tracheal epithelium. ×15,000. (Courtesy of du Pont/Sorval Instruments.)**

The *lateral wall* of the nasal cavity is very uneven because it is formed by three shell-like bones that curve down into the cavity. These bones are called the *nasal conchae,* and the spaces they separate are the *meatuses.* They are of importance, for they open into the *sinuses,* or air spaces, of the skull. The nasal portion of the tear duct opens into the lowest (inferior) meatus.

Mucous lining of the nose

Although the vestibule of each nasal cavity is lined with skin, the rest of the lateral wall is lined with a delicate mucous membrane, which is highly vascular and contains many mucous glands. These structures, together with the increased surface area and turbulent effect afforded by the convoluted conchae, facilitate the respiratory function of the mucous membrane, which is the warming and moistening of the inspired air. The epithelium is also ciliated, and the cilia beat in the direction of the pharynx, causing inhaled particles to become trapped in the mucus of the nose.

The uppermost portion of the nasal mucosa is specialized for the olfactory function of the nose. It possesses the many olfactory cells, fine processes of which form the olfactory nerves, which then pass through the small openings in the cribriform plate of the ethmoid bone at the roof of the nose to pierce the protective sheaths of the brain and end in the olfactory bulb of the brain (see also Sec. 22-8).

Pharynx, Epiglottis, Larynx, and Trachea

Both the mouth and the nose open into the pharynx, from which proceed the esophagus and the trachea as well as the eustachian tubes to the middle ear (Sec. 22-6). Section 17-2 describes the way in which *swallowing* raises the larynx against the epiglottis, closing off the trachea briefly. In addition, coughing and sneezing reflexes help keep the respiratory passages clear of foreign matter, forcing air outward and expelling with it the irritating foreign particles.

The *trachea,* or windpipe, is a large elastic tube kept open by the curved cartilaginous bands within its walls. These bands are open posteriorly, so that the posterior surface of the trachea which rests on the esophagus, is always flat. (The thyroid gland lies on top of it in the upper part of the neck.) The trachea is lined with ciliated epithelium (Fig. 16-3). The one-way movement of the cilia helps clear the respiratory tract of particles.

The trachea leads into the *right and left principal bronchi* (singular: *bronchus*), which enter the lungs where they divide into smaller bronchi, which divide the lungs into rather specific areas. This is important clinically for the localization of diseased areas or foreign bodies. The right principal bronchus is wider and more vertical than the left so that inhaled foreign bodies pass more frequently into it than into the left bronchus. The bronchi divide into successively smaller *bronchioles,* the smallest of which have no cartilage in their walls. Finally, the bronchioles end in the thin-walled air sacs, the *alveoli.* The entire respiratory passage, with the exception of the alveoli, is lined with ciliated mucous membrane.

The *larynx,* or *voice box* (Adam's apple), forms the upper expanded portion of the trachea, and through its construction of cartilages, ligaments, membranes, and muscles it is elaborately modified for the production of the voice. It lies below the hyoid bone and the tongue and in front of the third to sixth cervical vertebrae and moves in its position according to the movements of the head, speech, and swallowing. The most essential parts of the larynx for the production of the voice are the *vocal folds* or *cords,* ridges of mucous membrane that extend from the front to the back of the laryngeal cavity. These folds enclose the *vocal ligaments,* which are firmly attached also to part of the cartilaginous structure of the larynx. Between the vocal folds is an elongated fissure, the *glottis,* and it is through changes in the width of this fissure that changes in the pitch of sound can be controlled (Fig. 16-4). When the vocal ligaments are widely separated, through the action of the muscles of the larynx, the glottis is open, and the pressure of exhaled air and rotating cartilages causes the emission of a low-pitched sound. When the glottis is reduced by the tension on the vocal folds to a mere slit, forced passage of exhaled air causes a high-pitched note to be produced. The pitch of a sound depends on the number of cycles per second of the vibrations: in the case of a musical instrument, the strings vibrate; in the human being it is the vocal folds. (Above and parallel to the true vocal folds are two mucous folds, known as the false vocal cords, which are not involved in voice production.) The larger, heavier larynx of the male results in a deeper-pitched voice. This structural change in the larynx occurs at puberty, causing the voice to change suddenly—one octave deeper for boys, only two notes lower for girls.

Removal or destruction of the vocal folds prevents speech. Patients who have had their vocal folds removed surgically, usually following malignant destruction of these organs, can learn to speak again by forcing swallowed air into the esophagus and then back into the pharynx. Words can be formed, but they have no tone; nor is it possible to learn how to sing.

Figure 16-4. **The vocal folds or cords open (right) and closed (left).**

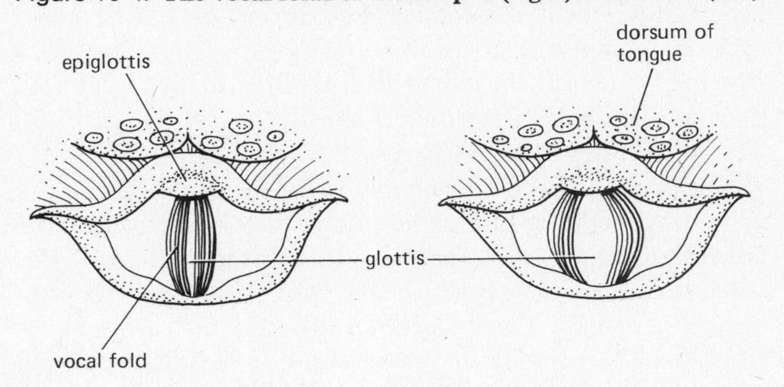

Lungs

General structure and function

The lungs are a pair of light, spongy organs that are peculiarly resilient because of the elastic fibers in their walls. The lungs lie free in the thoracic cavity, attached only by their roots, through which the bronchus, the pulmonary artery and veins, the nerves, and the lymph vessels enter or leave the lung. Each lung is surrounded by a delicate membrane, the *pleura,* which is reflected back over the root of the lung to become continuous with the pulmonary ligament. These two pleural layers, the *visceral pleura* immediately adhering to the surface of the lungs and the *parietal pleura* that lines the chest wall, are separated by an infinitesimal space, the *intrapleural cavity.* The normal pressure of the fluid in the intrapleural space is -10 to -15 mm Hg. This unusually negative pressure in the intrathoracic space is caused by the continuous reabsorption of fluid into the pleural capillaries by a special lymph-pump system.

The right lung is subdivided into three lobes, the left into two, but the further subdivisions made as a result of the branching of the bronchi are of more importance. The single pulmonary artery that enters each lung gives off branches that correspond to those of the bronchi, as do the tributaries of the two pulmonary veins from each lung (Fig. 16-5).

Respiratory membrane

The respiratory membrane, through which gaseous exchange takes place, consists of the *thin lining of the alveoli,* the *endothelium of the capillaries,* and a *delicate interstitial connective tissue layer* (Fig. 16-6). It is formed as follows:

The *bronchioles* branch into fine terminal branches that have smooth muscle in their walls, capable of contracting and regulating the amount of air passing into the alveoli. (In

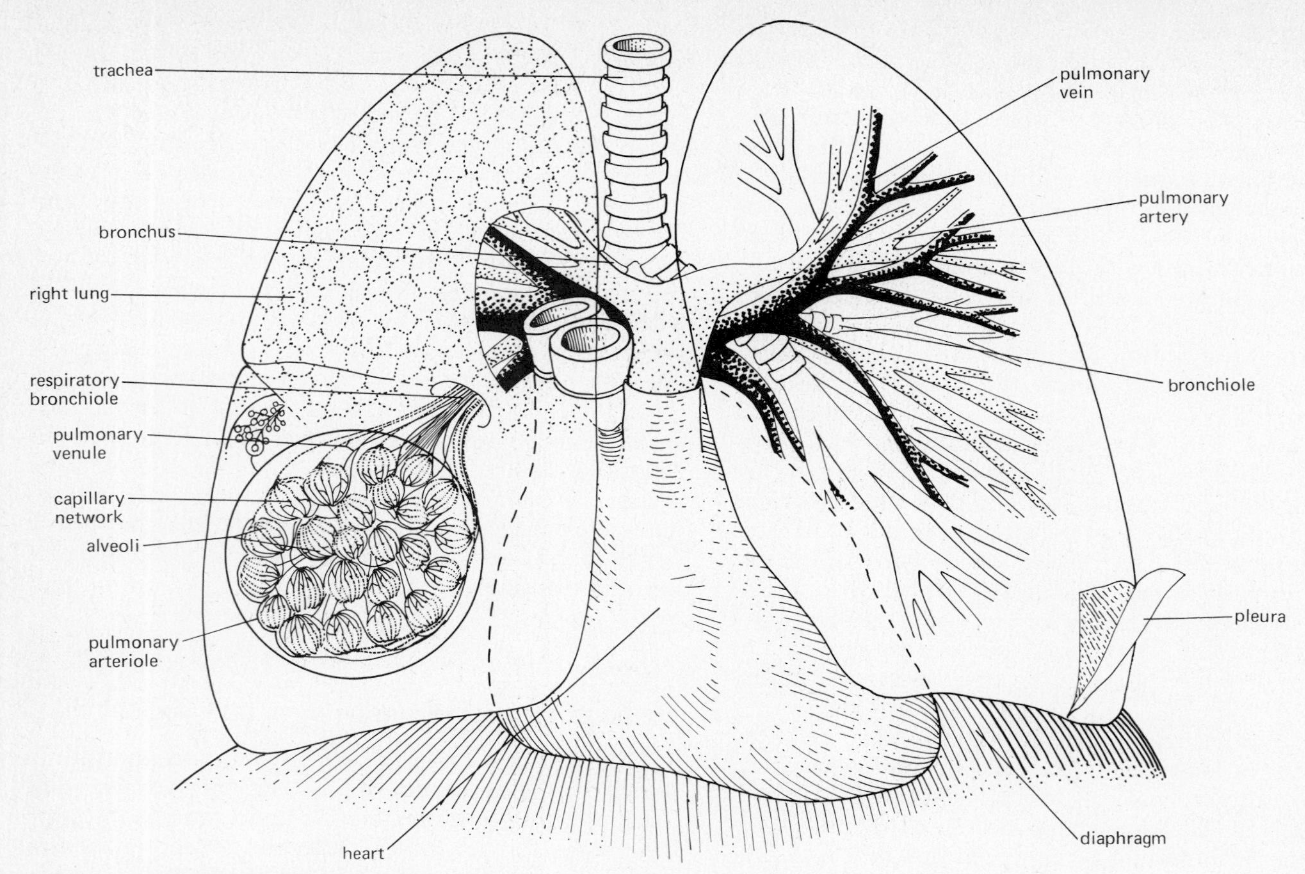

Figure 16-5. **The blood vessels of the lungs. The structures shown in the right lung are drawn at much greater magnification than those in the left lung. The arteries and veins follow the ramifications of the bronchioles and form intricate capillary networks around the alveoli, which are the terminal air sacs of the bronchioles.**

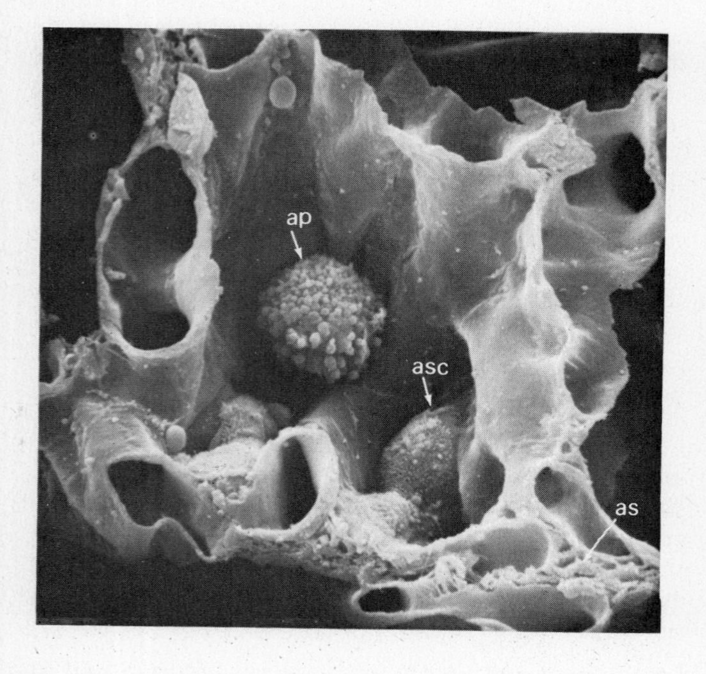

Figure 16-6. **Scanning electronmicrograph of an alveolus of rat lung. The flattened epithelial cells line the alveolar septa (as) that separate the alveoli. Bulging alveolar secretory cells (asc) are also part of the alveolar epithelium and may secrete the surfactant that lowers the surface tension in the lungs. Alveolar phagocytes (ap) are seen in the alveolar spaces. They remove dust and carbon particles. ×2420. (Courtesy of R. G. Kessel and R. Kardon, University of Iowa.)**

asthma, these muscles are contracted, and relief can usually be produced by the injection of a broncho-dilating agent such as epinephrine or atropine.) These branches then fork into the true *respiratory bronchioles,* which in turn open into the *alveoli* (via their atria and alveolar ducts) (Fig. 16-5). The alveoli are lined with very thin epithelium.

The branchings of the *pulmonary artery,* which carries the venous blood from the right side of the heart, follow the branchings of the bronchioles, so that the arterial capillaries come to lie in very close contact with the alveolar epithelium (Fig. 16-5). The *endothelial lining of the capillaries* is separated from the *alveolar epithelium* by a thin interstitial layer of connective tissue. The capillaries gradually lead into larger branches, which then form the two pulmonary veins from each lung, returning the aerated blood to the left side of the heart.

Nerves

Fine branches of the *vagi* (parasympathetic) and branches of thoracic *sympathetic* ganglia enter the lung at its root. Specialized *pressure receptors* are located in the walls of the alveoli, and impulses from these receptors travel along the afferent branches of the vagus to the hindbrain and midbrain.

Ventilation of the lungs

The lungs are ventilated through the combined action of the muscles of respiration, which include not only the diaphragm and the intercostal muscles of the thorax, but also abdominal and accessory chest and neck muscles. The spinal motoneurons that innervate these muscles receive impulses from higher levels of the brain, from the medulla and pons, from the reticular formation, and also from the cerebral cortex.

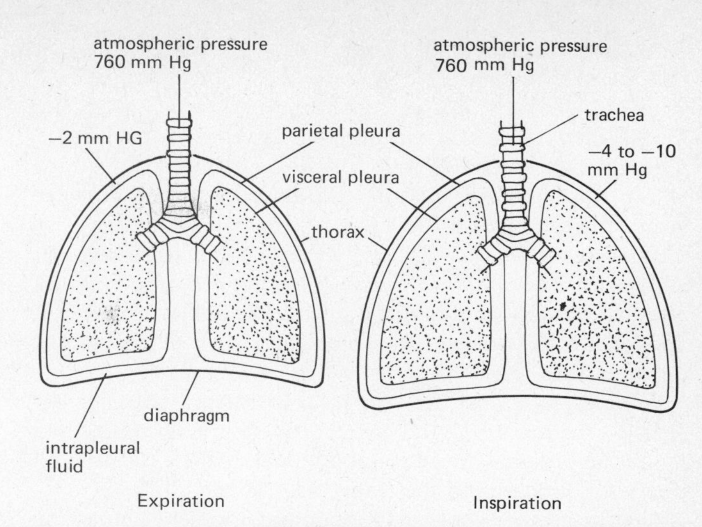

Figure 16-7. **Changes in intrapleural pressure are responsible for air entering and leaving the lungs. In inspiration, expansion of the thorax, aided by descent of the diaphragm, decreases intrathoracic pressure from −4 to −10 mm Hg, and air rushes into the lungs (the lungs are open to the atmospheric pressure of 760 mm Hg). In expiration, the size of the thorax is decreased, the intrathoracic pressure is raised to −2 mm Hg, and air is forced out of the lungs. The visceral and parietal pleurae are actually very close together, separated only by a thin layer of intrapleural fluid.**

16–3 Pressure Changes in the Thorax

The thorax acts as a closed cavity, which by increasing or decreasing its volume alters the pressure upon the elastic lungs enclosed within it (Fig. 16-7). When the thorax enlarges, the *negative pressure* in the thorax increases from −4 to −10 mm Hg, decreasing the pressure on the lungs. Atmospheric pressure remains the same, so that air rushes in through the respiratory passages. Thus in *inspiration* the lungs are extending passively in response to the various mechanisms that result in an increase in thoracic volume. In *expiration* the thoracic volume is sharply decreased, increasing the pressure on the lungs to −2 mm Hg and forcing air out. This process is aided by the elasticity of the lungs, which enables them to recoil to their previous form and size.

Forces Balancing Expansion and Collapse of the Lungs

Intrathoracic fluid pressure

While the very negative *intrathoracic fluid pressure* (about −10 mm Hg) holds the lungs against the wall of the thorax in an expanded position, the elasticity of the lung walls tends to make them shrink back and collapse. Another force tending to cause collapse of the lungs is the surface tension of the outermost layer of fluid that lines the alveoli and that accounts for more than half of the lung's elasticity. This surface tension is regulated by the surface active properties of the *surfactant,* a secretion produced by certain cells in the walls of the alveolus.

Surfactant is a complex lipid-protein substance. The important lipid component is lecithin. Surfactant lowers the surface tension of the alveoli when the surface area of the lung decreases during exhalation and thus prevents their collapse.

Intrathoracic pressure

This is the pressure needed to prevent the elasticity and surface tension of the lungs from causing them to collapse. It is about −4 mm Hg, the difference between the negative intrathoracic fluid pressure and the opposing lung elasticity and surface tension. In the absence of surfactant, the intrathoracic pressure would be less negative, making the work of expanding the lungs much greater.

16–4 Movements of the Thorax in Respiration

The *bony structure* is formed by the thoracic vertebrae, the ribs with their cartilages, and the sternum. The ribs are joined to the thoracic vertebrae on a somewhat higher level than they are joined by their cartilages to the sternum in front, resulting in a sloping of the bony cage from posterior down to the anterior surface. In addition, the ribs increase in length from the first to the seventh, after which they decrease again. The last two ribs, not attached to the sternum (the floating ribs), do not take part in respiration.

The *only movement possible for the ribs* is a slight rotation around the joints with the vertebrae. The sternum moves with the ribs, and because of the greater motility of the middle ribs, it moves *upward and forward* in inspiration. The ribs now extend directly forward, instead of downward and forward, and this increases the anteroposterior diameter considerably (Fig. 16-8).

Thus movement of the ribs in inspiration increases the diameter of the thorax in the anteroposterior diameter. Muscles that elevate the thoracic cage are muscles of inspiration and include the *external intercostals* and the neck and shoulder muscles.[1]

Expiration returns the ribs to their sloping position, decreasing the size of the thoracic cavity. This is accomplished by the abdominal muscles (abdominal recti, external and internal obliques, and the transversi). When all these muscles contract, they squeeze the abdominal viscera against the liver, which pushes the relaxed diaphragm up and pulls the ribs and sternum down.

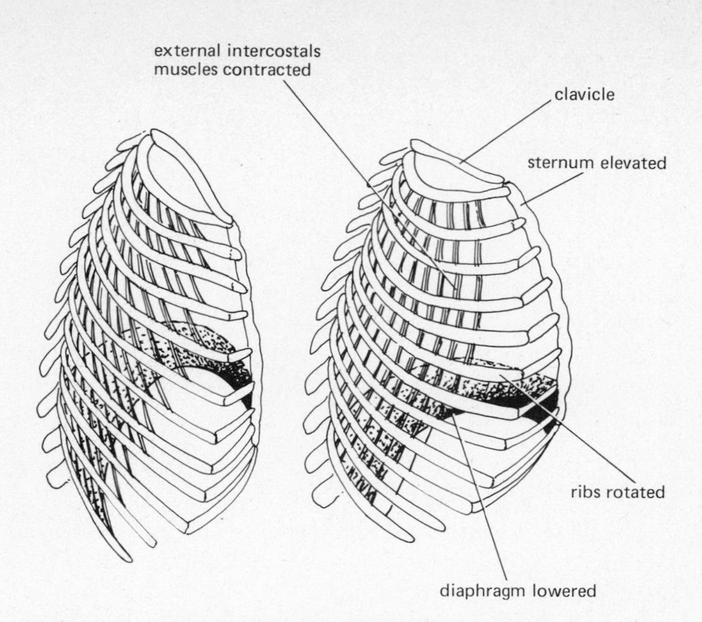

Figure 16-8. **Increase in size of the thorax in inspiration.**

16–5 Muscles of Respiration

Diaphragm

The diaphragm is the most important of the respiratory muscles. It is a thin, movable partition between the thorax and the abdomen, forming a vaulted roof to the abdomen. The diaphragm is higher on the right than on the left. The central part is a very strong tendon that is attached to the margins of the outlet from the thorax. The muscular part forms a dome-shaped structure originating on the ribs and sternum or on the upper lumbar vertebrae posteriorly. The diaphragm has *three large apertures* and several small ones, through which pass the *aorta,* the *esophagus,* and the *inferior vena cava.* In addition, tl.. thoracic duct and azygos vein pass along with the aorta.

MOVEMENTS OF THE DIAPHRAGM

Contraction of the diaphragm increases the size of the thoracic cavity in three diameters, but chiefly by lengthening its vertical diameter as the diaphragm descends (Figs. 16-7 and 16-8). The cross-sectional area of the diaphragm dome is approximately 250 cm^2. If the diaphragm descends 10 mm, it will increase the volume of the thoracic cavity by 250 cc. The diaphragm is the main muscle of inspiration. When it relaxes, passive expiration results, which is sufficient for quiet breathing.

[1] The sternomastoids, the scapular elevators plus anterior serrati, and the scaleni. (Sternomastoids lift upward on the sternum, the anterior serrati lift many of the ribs, and the scaleni lift the upper ribs. The anterior serrati lift the ribs only if the scapulae are simultaneously adducted so that the shoulder muscles that adduct the scapulae are also muscles of inspiration.) If the diaphragm is entirely paralyzed, as a result of poliomyelitis, such patients have been known to breathe with their chest and neck muscles (scaleni) lifting the thoracic cage, and with their abdominal muscles expelling the air.

External and Internal Intercostals

The *external intercostals* are the muscles mainly responsible for the elevation of the ribs in normal, quiet breathing. They are inserted between two neighboring ribs, sloping forward and downward (Fig. 16-9). Relaxation of these muscles brings about *passive expiration* (Table 16-1).

The *internal intercostals* form a deeper layer of muscle between the ribs with the fibers running in the opposite direction, from above downward and backward. On contraction, these muscles depress the ribs, aiding in expiration during forced or very deep breathing. This is *active expiration.*

16–6 Innervation of Respiratory Muscles

Phrenic Nerves

The most important nerves in the continuous rhythmic contraction and relaxation of the diaphragm are the *phrenic nerves,* which arise in the neck from the third, fourth, and fifth cervical nerves. Each nerve passes down through the thorax a little in front of the root of the lung to enter the diaphragm.

Intercostal Nerves

The *intercostal nerves* are also part of the peripheral nervous system, arising from the upper eleven thoracic nerves and connected with the sympathetic trunk. They give off branches to the intercostal muscles.

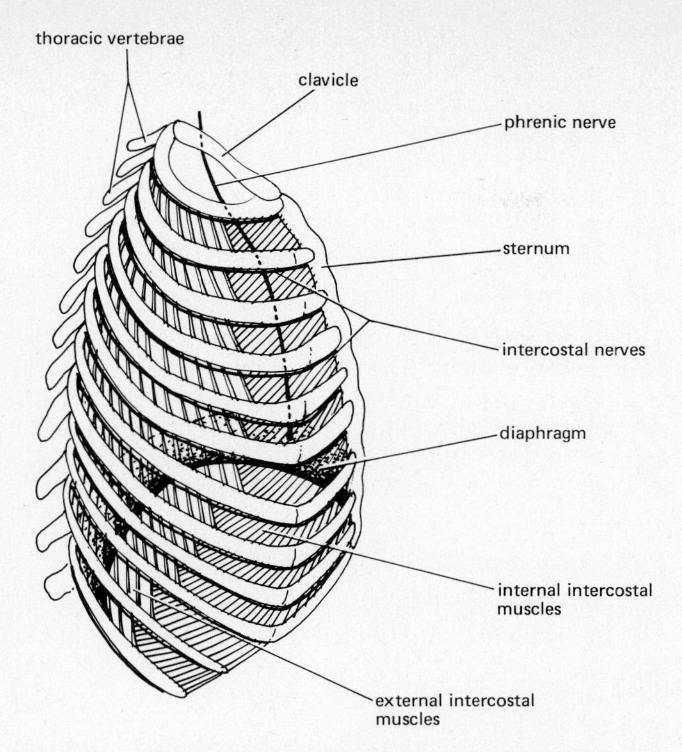

Figure 16-9. **The external and internal intercostal muscles.**

16–7 Respiration at Birth

The newborn infant has to change rapidly from one oxygen source (the placenta) to another (oxygen obtained from the lungs). In addition, the initial resistance of the lungs has to be overcome in the first few breaths. This requires pressures 10 to 15 times greater than the pressure needed for breathing once the lungs have been filled with air. Adequate amounts of *surfactant* (Sec. 16-3) are essential to keep the surface tension of the alveoli low enough so that the newborn can overcome the lung resistance.

Initiation of Respiration

Any of the traumatic experiences that a baby undergoes at birth could be the essential stimulus that initiates rhythmic respiration. These experiences include a change from the warm, moist uterine environment to the cooler, dry, nonsupporting environment of the hospital. The infant has also endured a physical pummelling at birth and a slight hypoxia as the umbilical cord is squeezed, then cut. Although it is generally believed that these changes are responsible for the initiation of respiration, there is little experimental evidence to support these ideas.

It is difficult, of course, to control these birth conditions in humans, but experiments in fetal lambs (Dawes, 1965) that have been delivered by caesarian section show that normal quiet breathing can be induced without any change in blood oxygen or carbon dioxide, or in blood pH. Contrary to the idea that a sharp spank will initiate breathing, this only elicits a short gasp from the unanesthetized lambs with an intact cord. Consequently, it is possible that cutting the cord sets up some generalized sympathetic discharge that is responsible for breathing. The mechanism involved in the initiation of breathing is still problematic: fortunately, babies do not wait to breathe until physiologists can explain how they do it.

Respiratory Distress Syndrome

Respiratory distress syndrome, or *hyaline membrane disease* (so-called due to the glassy pinky tinge of the plasma that leaks out and covers the lungs), occurs in very small or premature infants. The lungs of these babies are too immature for sufficient amounts of surfactant to be produced. As a result, the alveoli collapse after each expiration and have to be newly expanded for each inspiration, as though the baby had to blow up a new balloon with each breath. The tremendous effort required rapidly exhausts the infant, and death ensues from physical exhaustion as well as oxygen lack.

Current therapy uses *positive pressure to force air* into the baby's lungs. This procedure has to be maintained until the alveolar cells *mature enough to produce surfactant.* During this time, the oxygen pressure in the air must be carefully controlled. Too much oxygen may damage the retina, causing blindness. Too little oxygen may result in brain damage.

16–8 Pneumothorax and Atalectasis

If air enters the pleural cavity, the condition is known as pneumothorax. The increased pressure causes the lung to collapse (*atalectasis*). This results in serious respiratory difficulties: the diaphragm is no longer drawn into the thorax by the elastic force of the lung, so that it loses its

Table 16–1

Muscles of Inspiration (and passive expiration)		
Main	Accessory	Muscles of Expiration (active)
External intercostals Diaphragm	Muscles of neck and shoulder Abdominal muscles	Abdominal muscles Internal intercostals

domelike shape, and the normal lung (the pleural cavities are separate) also suffers a decrease in respiratory efficiency. Under normal, quiet conditions, the one lung can maintain life, and a pneumothorax is often induced in cases of severe tuberculosis to permit the seriously infected lung to heal. It can be inflated subsequently by the proper pressure adjustments.

16–9 Artificial Respiration

The physiological mechanism of respiration in higher vertebrates is *negative pressure breathing,* with air filling the lungs due to the negative pressures in the thorax; but *positive pressure breathing,* induced by artificial means, can adequately compensate when the normal respiratory processes fail. Artificial respiration administered in this manner depends on rhythmically increasing the pressure on the thorax and/or abdomen to force air out, or on forcing air into the lungs by blowing it into the mouth.

Schafer Prone Method

The person applying the artificial respiration places the subject on his stomach and applies pressure once every 5 sec to the lower ribs and posterior abdominal wall. This compresses the abdomen against the ground and forces the ribs upward into the chest cavity, forcing about 500 cc of air out of the lungs with each cycle. This is usually not enough to maintain life, unless the patient can help a little with the breathing.

Back-Pressure, Arm-Lift Method

This is recommended as being more efficient, although it is more difficult to learn. The subject is placed prone on his stomach with his elbows bent. The operator kneels at the head of the patient and places his hands firmly on the subject's back, rocking himself slowly forward until his arms are almost vertical. This places almost all his body pressure on the back of the subject, forcing air out. Then the operator slowly rocks back, sliding his hands along the patient's arms and pulling them forward. This increases the thoracic dimensions and permits inspiration. This technique, properly applied, permits about 1000 cc of air to enter the lungs per cycle (about 12 times/min).

Mouth-to-Mouth Breathing

This is frequently used with babies, although it can be satisfactory for adults. The operator rapidly inspires a deep breath then exhales it into the mouth of the subject. If the operator breathes in rapidly, the amount of oxygen expelled into the patient will be almost atmospheric, and the small amount of carbon dioxide will not be harmful.

Pulmotor and Resuscitator

The pulmotor is basically a machine that delivers blasts of oxygen rhythmically to the patient by way of a mask. It has been modified in the resuscitator, which has safety mechanisms that prevent blowing the lungs up to too great a pressure and permits automatic feedback of the lung's gases to the atmosphere.

Tank Respirator (The Iron Lung)

In this type respirator, the patient is placed inside the tank with the head protruding. The pressure in the tank is rhythmically increased and decreased, mechanically causing the thorax to expand and contract and to simulate normal respiration.

Electrophrenic Respirator

This is an electrical stimulator that rhythmically stimulates one or both of the phrenic nerves, causing the diaphragm alternately to contract and relax. The difficulty with this

Table 16–2. **Subdivisions of Lung Volume (in milliliters)**

		Male	Female
Functional residual capacity	Volume of air the lungs can hold when the muscles of inspiration and of expiration are relaxed	2300	1820
Tidal volume	Volume of air breathed in and out in one respiratory cycle	500	380
Inspiratory capacity	Volume of air that can be taken in with a maximum inspiration	3800	2420
Expiratory reserve volume	Volume of air that can be forced out of the lungs following a normal expiration	1000	730
Residual volume	Volume of air that cannot be forced out of the lungs despite all effort	1200	1100

Note: The maximum amount of air that can be taken in per minute is the maximum respiratory minute rate. It equals the inspiratory capacity × the maximal respiratory rate. (3800 ml × 35 = 133 L/min for males; 2420 ml × 35 = 84.7 L/min for females.)

type of respirator is that the phrenic nerve must be exposed, or the electrode must be held directly over it in the neck. This is uncomfortable for the patient and can be maintained only for short periods of time. In addition, it is obviously unsuitable for patients who have paralysis of the diaphragmatic muscles.

16–10 Inhalation Anesthesia

Inhalation anesthetics, including nitrous oxide, cyclopropane, ether, ethylene, and chloroform, are gases that are absorbed via the lungs into the blood, where they pass to the brain and prevent the conduction across the synapses. The higher centers of the brain are affected first, so that the patient becomes unconscious and unaware of sensation. Fairly large amounts of the anesthetic are needed to saturate the blood, but after this level is reached, it is important that no more gas be given, for the lower centers that control respiration and circulation of the blood would then be affected. The anesthetic machine is so constructed that once the desired concentration of anesthetic gas has been delivered, the patient rebreathes the expired anesthetic gas, thus maintaining the same concentration in his blood. The expired carbon dioxide is removed, however, by soda lime, and additional quantities of oxygen are supplied because only the oxygen is used up.

16–11 Distribution and Measurement of Air Within the Respiratory System

Terms, Definitions, and Measurements

In order to be able to determine the efficiency of individual respiration in many diseases, as well as in physiological conditions (including athletic training), it is important to know the normal amount of air present within the respiratory system and the rate at which it is exchanged during the *respiratory cycle* (inspiration and expiration). This is simply measured with an instrument called the *spirometer,* into which the subject breathes through a tube placed in his mouth. Expiration forces the floating drum of the spirometer up; in inspiration the drum descends. The floating drum is connected to a recording device so that its upward and downward movements can be measured.

The maximum amount of air that a person can expel from his lungs after taking the deepest breath possible is known as the *vital capacity.* This dimension is important clinically because it gives three kinds of information:

1. The strength of the respiratory muscles.
2. The distensibility of the lungs and the thorax.
3. The physical size of the thoracic cage. The average woman has a vital capacity of about 4000 ml. That of the average man is about 5000 ml.

Some useful subdivisions of lung volume are listed in Table 16-2 and depicted in Fig. 16-10.

Dead Space

The amount of air in the nose, trachea, and bronchi is known as the *dead space* because they contain expired gases that are immediately inspired into the alveoli on the next breath. The dead space amounts to about 150 ml, so that when the tidal air volume is 500 ml only 350 ml of fresh air reaches the alveoli per breath. This is sometimes referred to as the *ventilatory air.* The dead space is of importance when the tidal volume is low, as proportionately less air is available for gaseous exchange with the blood.

Vital Capacity in Disease

Paralysis of the respiratory muscles will decrease the vital capacity most markedly by affecting the expansive ability of the thorax. In severe *tuberculosis,* where fibrous tissue has replaced most of the normal elements of the lungs, the

Figure 16-10. **The subdivisions of lung volume. [From J. H. Comroe, *Fed. Proc.* 9:602 (1950).]**

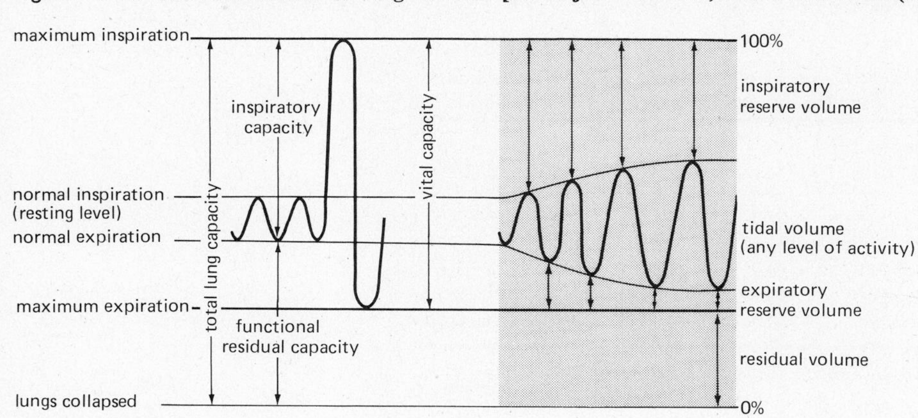

elasticity is greatly diminished, and the vital capacity is correspondingly decreased. Similarly, fibrous changes of the pleura will decrease the distensibility of the lungs. In diseases of the left side of the heart, in which blood from the lungs cannot flow normally into the left atrium (see Sec. 12-7), a back-pressure may be built up in the lungs, forcing fluid into the alveoli and the pleural spaces, greatly decreasing the vital capacity.

Another widely occurring disease of the lungs is *emphysema.* As a result of chronic respiratory illnesses, such as asthma or persistent coughs and colds, the lungs sometimes remain overinflated as the elastic tissue is destroyed by the continuous stretching. Not only is the elasticity of the lungs decreased but the septa between the alveolar sacs degenerate, reducing the surface area of the respiratory membrane considerably. This means that the vital capacity is decreased (due to loss of elasticity), and respiratory exchange is far less efficient, because the air that does reach the alveoli has a smaller surface for gaseous exchange. The resulting fall in P_{O_2} in the blood causes a severe restriction on physical activity, and even slight exertions leave such patients short of breath and cyanotic.

Terms Used in Describing Respiration

Eupnea. Normal respiration.
Apnea. Cessation of respiration.
Dyspnea. Irregularities of respiration.
Hyperpnea. Increase in respiratory rate and depth.

16–12 External Respiration

Composition of the Atmosphere

The external atmosphere, at 760 mm Hg and on a warm day, consists of 78 per cent nitrogen and other inert gases, 21 per cent oxygen, 0.04 per cent carbon dioxide, and 0.5 per cent water vapor. It is believed that life as we know it can exist only under these conditions—that considerably less oxygen or more carbon dioxide or other noxious gases would prevent the development of, or destroy, existing life.

INERT GASES

The large volume of inert gases, mainly nitrogen, that is inspired normally passes into the blood with oxygen and carbon dioxide, but the chemical inactivity of these gases prevents them from participating in the metabolic reactions of the body.

ACTIVE GASES—OXYGEN, CARBON DIOXIDE, AND WATER VAPOR

On warm humid days, the amount of water vapor in the air may rise from 0.5 to as much as 6.2 per cent. The atmospheric pressure remains the same, but because each gas exerts a pressure in terms of its own relative volume, the pressure exerted by each, or its partial pressure, will be different from its partial pressure on a dry day (Table 16-3).

The partial pressure exerted by each gas determines the rate with which it diffuses through cellular membranes (Sec. 16-14). Gas molecules diffuse according to their concentration gradient, regardless of the concentration of the other gases in the solution. In the lungs, the inspired air enters the thin-walled alveolus or air sac; it is separated from the blood, with its high concentrations of carbon dioxide, by a delicate membrane, the *respiratory membrane* (Sec. 16-2). It is across this membrane that a concentration gradient for the active respiratory gases is established.

In alveolar air, the partial pressure of oxygen is 100 mm Hg, and that of carbon dioxide is 40 mm Hg. On the other side of the alveolar membrane, oxygen partial pressure in the blood returning from the tissues is 40 mm Hg, whereas that of the carbon dioxide is 45 mm Hg. Because each gas diffuses according to its own concentration gradient, the net result is the diffusion of oxygen into the lung capillary and the diffusion of carbon dioxide from the capillaries to the alveolus.

The exchange of gases in the lung according to their partial pressures (concentration gradient) is

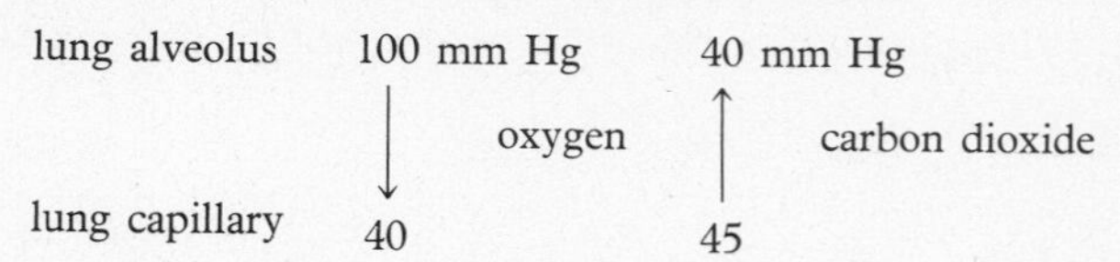

Table 16–3. **The Partial Pressure of Respiratory Gases as They Enter and Leave the Lungs (Sea Level)* (Partial pressure in mm Hg. Figures in parentheses are volume %.)**

Gas	Atmospheric Air (cool, clear day)		Atmospheric Air (warm, humid day)		Alveolar Air*		Expired Air*	
N_2	597.0	(78.62%)	563.5	(74.19%)	569.0	(74.9%)	566.0	(74.5%)
O_2	159.0	(20.84%)	149.35	(19.57%)	104.0	(13.6%)	120.0	(15.7%)
CO_2	0.15	(0.04%)	0.15	(0.04%)	40.0	(5.3%)	27.0	(3.6%)
H_2O	3.85	(0.5%)	47.0	(6.2%)	47.0	(6.2%)	47.0	(6.2%)
Total	760.0	(100.0%)	760.0	(100.0%)	760.0	(100.0%)	760.0	(100.0%)

Source: A. C. Guyton, *Textbook of Medical Physiology,* 5th ed., Table 40-2. Saunders, Philadelphia, 1976.
* Calculated for an average, cool, clear day.

16–13 Exchange of Gases Across the Respiratory Membrane

Surface Area

The total area of the respiratory membrane in the normal adult has been estimated to be about 50 to 70 m^2, and the total amount of blood in the capillaries is about 60 to 100 ml; this large amount of surface for relatively little blood permits the rapid diffusion of gases across the membrane.

Partial Pressure of Gases Within the Lungs

Table 16-3 shows that on a humid day the amount of water vapor in the air increases greatly, lowering the partial pressures of the other gases. In the passage of air from the outside through the respiratory passages to the lungs, the air becomes very humid, and the partial pressures of carbon dioxide and oxygen within the alveoli are equivalent to that of the humidified air cited in this table. When the air is unusually dry and burning hot, the respiratory passages cannot humidify it adequately, and destruction of the delicate membrane results.

Carbon Dioxide and the Minute Volume

The normal *rate of respiration* in the adult is 14 breaths/min, but it is more rapid in children (up to 30/min), and during strenuous exercise it may increase to 60 or more breaths/min. Each inspiration admits approximately 350 ml of new air to mix with the 2500 ml of old air present in the lungs. The quantity of new air that enters the lungs per minute is known as the *minute volume*. In the average adult it is approximately 4900 ml.

$$\underset{4900\text{ ml}}{\text{minute volume}} = \underset{350\text{ ml}}{\text{amount of new air}} \times \underset{14}{\text{respiratory rate/min}}$$

The rate of carbon dioxide exchange across the respiratory membrane is the result of its *concentration gradient* across the membrane. This in turn represents a balance between the ventilatory rate and the rate at which the tissues are forming carbon dioxide and liberating it into the blood. An increase in ventilatory rate will remove carbon dioxide from the alveoli rapidly enough to lower the alveolar CO_2 partial pressure in the lungs and increase the rate at which carbon dioxide diffuses from the blood into the alveolus. If, at the same time, the tissues are producing increased quantities of carbon dioxide, as occurs in strenuous exercise, the concentration gradient will be further exaggerated with more efficient removal of carbon dioxide from the blood. Since an increase in CO_2 concentration in the blood acts as a stimulus to the respiratory centers, increasing the ventilatory rate, this serves as an excellent example of homeostatic control between blood and respiration.

Table 16–4. **Changes in the Alveolar Air, Volume of Breathing, and Hemoglobin of the Blood During Acclimation to Altitude and After Return to Sea Level**

Altitude (ft)	Days After Arrival	Alveolar Carbon Dioxide (mm Hg)	Volume of Breathing (percentage of sea-level volume)	Hemoglobin (percentage of normal)
Sea level	100+	40	100	100
6,000	1	37	108	100
	2	36	111	100
	3	35	115	101
14,000	1	32	125	103
	2	31	129	106
	3	30	133	108
	5	29	138	109
	10	28	143	113
	20	27	148	116
	35	26	154	120
6,000	1	29	138	112
	2	31	129	110
	3	33	121	108
Sea level	1	36	111	105
	2	39	102	104
	5	40	100	102
	30	40	100	100

Source: Y. Henderson: *Adventures in Respiration.* Williams & Wilkins, Baltimore, 1938.

Oxygen and the Minute Volume

The concentration gradient for oxygen is similarly the result of the rate at which *oxygen enters the alveoli* and the rate at which the *blood absorbs oxygen.* During strenuous exercise, the tissues utilize oxygen very rapidly, increasing the amount absorbed from the blood. This increases the rate of oxygen diffusion from the alveolus into the blood and would rapidly lower the partial pressure of oxygen (P_{O_2}) in the alveolus, unless it were accompanied by an increase in the minute volume, replenishing the oxygen content of the alveoli.

16–14 Changes in Barometric Pressure

If the barometric pressure is sharply increased or decreased, the total pressure and consequently the partial pressures of the inhaled gases are affected, with resultant changes in the concentration gradient of oxygen across the respiratory membrane (Table 16-4).

Physical laws concerning the volume, temperature, and pressure of gases may be summarized as follows:

1. *Relationship of Pressure to Volume* (*Boyle's Law*). If the pressure on a given amount of gas is increased and the temperature remains constant, the volume of the gas will be decreased, i.e., the gas molecules which were free to move are now compressed. Conversely, with constant temperature but decreased pressure, the molecules can move more freely, and the volume of the gas is increased. Thus the *volume varies inversely with the pressure, $V = 1/P$.*

2. *Relationship of Temperature to Volume* (*Gay-Lussac's Law*). If the pressure on a given amount of gas remains constant but the temperature is increased, the kinetic energy of the gas molecules is increased, they move more rapidly, and the volume occupied by the gas is increased. Similarly, a decrease in temperature slows down the molecules, decreasing the volume. Thus *volume varies directly with the temperature, $V = KT$.*

3. *The Gas Law* (*1 + 2*). The first two laws may be combined to express the relationship between the volume, pressure, and temperature of a gas:

$$\text{volume} = \frac{nR\ \text{temperature}}{\text{pressure}}$$

where n = the quantity of the gas and R is a constant depending on the units of measure used for the other factors.

Physiologically, this is significant when the *atmospheric pressure decreases* as at high altitudes, because the same mass of air occupies a larger volume, so that it is "thinner" or contains fewer molecules per milliliter. This means that the total and partial pressures of the gas are decreased, and the P_{O_2} in the alveoli decreases.

4. *Partial Pressures.* In a mixture of gases, each gas exerts a pressure that depends on the percentage of that gas in the mixture. The pressure exerted by each gas is the partial pressure of that gas.

$$\text{partial pressure} = \frac{\text{total pressure} \times \%\ \text{concentration of gas}}{100}$$

If a gas mixture of 20 per cent oxygen and 80 per cent nitrogen has a total pressure of 760 mm Hg (the total pressure will be determined by the factors discussed above, including volume and temperature), then

$$\underset{(P_{O_2})}{\text{partial pressure of oxygen}} = \frac{760 \times 20}{100} = 152 \text{ mm Hg}$$

and

$$\underset{(P_{N_2})}{\text{partial pressure of nitrogen}} = \frac{760 \times 80}{100} = 608 \text{ mm Hg}$$

When the partial pressure of a gas is calculated, all the gases including water vapor must be accounted for.

When gases are dissolved in a liquid, each gas goes into solution independently of the other, in accord with its own solubility and partial pressure.

16-15 Transport of Oxygen and Carbon Dioxide by the Blood

Oxygen Transport

The protein *hemoglobin,* contained within the red blood cells, forms a reversible combination with either oxygen or carbon dioxide, depending on the concentrations of these gases in the blood. Some of the oxygen that enters the blood from the alveoli, and the carbon dioxide entering from the tissues, go into simple solution in the plasma. This accounts for only a very small amount and would never be adequate to sustain the high metabolic activity of the cells of higher animals. The combination of these gases with hemoglobin, on the other hand, allows for the transportation of 50 to 100 times as much gas as could be transported in simple solution. In addition, the retention of hemoglobin within the red blood cells adds immensely to the efficiency of this transportation system, because a noncellular solution of hemoglobin rapidly escapes into the tissues and is destroyed or eliminated (see Fig. 16-11).

Structure of Hemoglobin

Hemoglobin is a conjugated protein, made up of *heme* (a prosthetic group containing iron) and *globin* (the protein portion). The heme is the same in all species, but the globin may vary somewhat in its characteristics.

Hemoglobin is composed of two *alpha chains* (141 amino acids) and two *beta chains* (146 amino acids). These polypeptide chains are folded and twisted to form the globular hemoglobin molecule, with the four heme groups symmetrically incorporated. Hemoglobin (molecular weight 68,000) can fragment into four subunits, each with either an alpha or a beta chain, and a molecular weight of about 17,000 (Sec. 14-5 and Fig. 14-16).

Figure 16-11. **The quantity of oxygen dissolved in the water of the blood and in combination with hemoglobin when exposed to very high oxygen pressures. (From A. C. Guyton, *Textbook of Medical Physiology,* 5th ed., W. B. Saunders Company, Philadelphia, 1976.)**

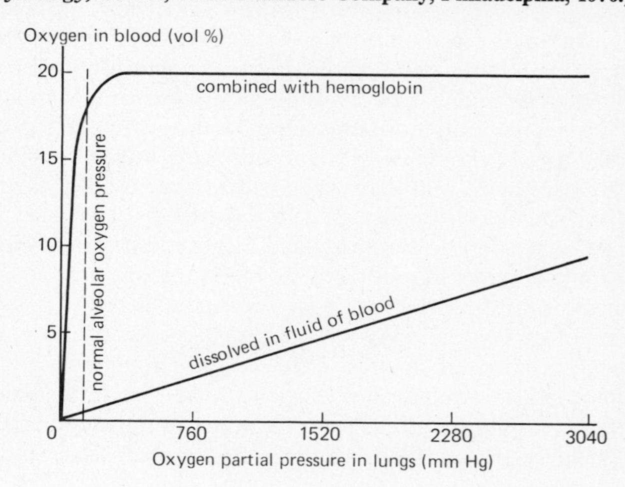

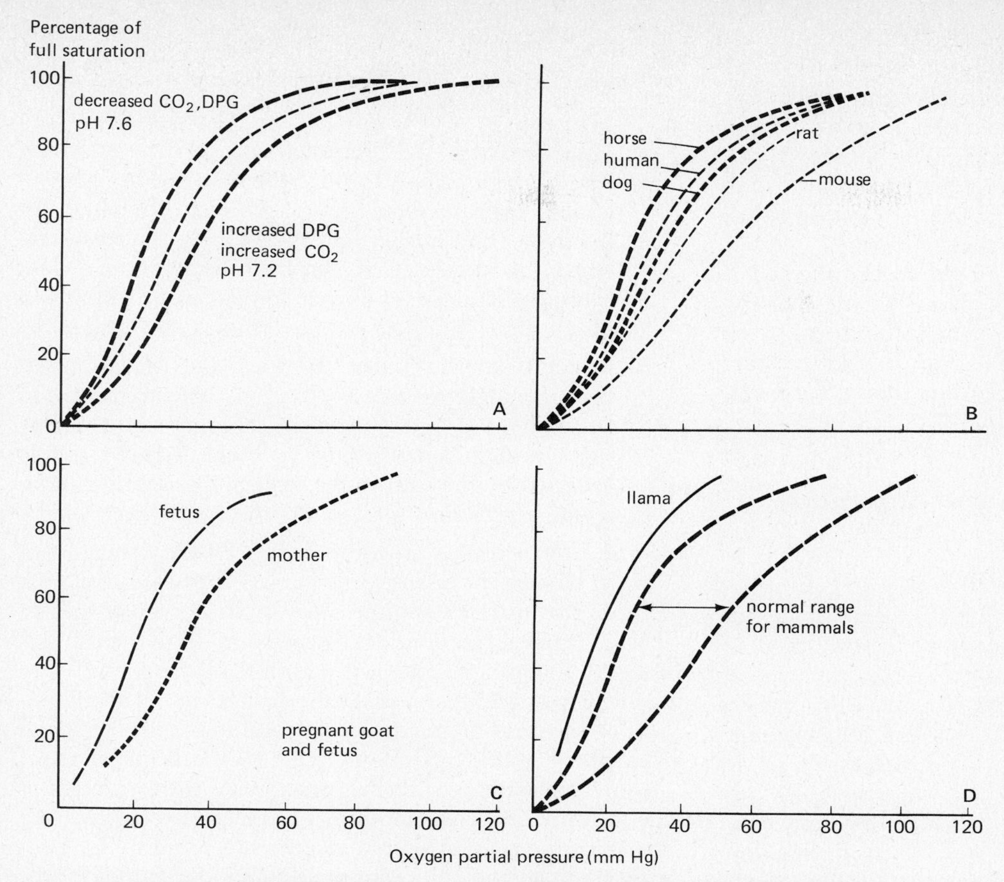

Figure 16-12. **The oxygen dissociation curves for different animals at different pressures. A curve located to the right signifies that oxygen is more readily given up by hemoglobin. (A) Increased CO_2, decreased pH, or increased 2,3 DPG (see Chap. 14) all shift the curve to the right. The curve is shifted to the left by a decrease in CO_2, an increase in pH, or a fall in 2,3 DPG (the Bohr effect). (B) Small animals that need more oxygen/gram of tissue have blood that gives up oxygen readily. (C) Fetal blood binds oxygen more readily than maternal blood. (D) The llama, which lives at the low oxygen pressure of the high Andes, has blood that binds oxygen more readily than does the blood of other mammals. (Adapted from Knut Schmidt-Nielsen, *Animal Physiology,* 2nd Edition, © 1964, p. 23. Reprinted by permission of Prentice-Hall, Inc., Englewood Cliffs, New Jersey.)**

Small changes in the position or type of individual amino acids in any of these chains results in a *mutant* form of hemoglobin. Some 200 mutants of hemoglobin are known, most of which do not appear to affect the function of the hemoglobin molecule; but some mutants alter the affinity of the hemoglobin for oxygen sufficiently to cause anemia. Sickle cell anemia, common in black populations, and thalassemia (also called Cooley's anemia or Mediterranean anemia), which is found in several Mediterranean races, are examples of hereditary abnormalities of hemoglobin structure. These are discussed in Sec. 14-7.

Combination of Hemoglobin with Oxygen

Each atom of iron in the heme portion of hemoglobin combines with a molecule of oxygen to form *oxyhemoglobin,* which carries oxygen in the molecular rather than in the ionic form. This loose combination with oxygen is termed *oxygenation* rather than *oxidation.* Oxygen is given off to the tissues in the form of molecular oxygen (O_2) and not the ionic form (O^{2-}). Because each hemoglobin molecule possesses four atoms of iron, it can transport four molecules of oxygen.

Oxygenation of hemoglobin begins with the alpha chains, or subunits, and progresses through the other chains, weakening the bonds. This process greatly enhances the affinity of heme iron for oxygen and accounts for the sigmoid shape of the oxygen-hemoglobin dissociation curve depicted in Fig. 16-12.

Oxygen-Hemoglobin Dissociation Curve

If one takes a series of tubes containing the same concentration of hemoglobin solution and exposes them to different oxygen pressures, it is found that there is a considerable difference in the amount of hemoglobin that has become oxygenated in each tube. The English physiologist Barcroft (1909) showed that when hemoglobin is exposed to 100 mm Hg P_{O_2} (the normal P_{O_2} for blood leaving the lungs) approximately 97 per cent of the hemoglobin is combined with oxygen to form oxyhemoglobin. There is a very gradual decrease in the amount of bound oxygen until the P_{O_2} drops to 40 mm Hg. At this point, only 70 per cent of the hemoglobin is in the form of oxyhemoglobin. Below this P_{O_2} most of the oxygen is released. The relationship between the amount of oxyhemoglobin formed and the oxygen pressure to which the blood is exposed is shown in the *oxygen-hemoglobin dissociation curves* in Fig. 16-12.

Oxygen Capacity, Content, and Saturation of Blood

The normal blood of men has approximately 16 g of hemoglobin/100 ml blood, that of women 14 g/100 ml. One gram of hemoglobin can combine with a maximum of 1.3 ml of oxygen. This means that the *oxygen capacity,* or maximum amount of oxygen that blood with these amounts of hemoglobin can carry, is 21 ml of oxygen for males and 18 ml for females. This is often expressed as 21 or 18 *volumes per cent.*

Although the oxygen capacity indicates the maximum amount of oxygen that can be bound by the hemoglobin contained in 100 ml blood, the *oxygen content* is the amount that actually is bound. The relationship between the oxygen capacity and the oxygen content (excluding the small amount dissolved in plasma) is the *oxygen saturation* of the blood.

$$\text{oxygen saturation} = \frac{\text{oxygen content} - \text{dissolved oxygen}}{\text{oxygen capacity}}$$

If, for example, the oxygen capacity is 10 volumes per cent, and the blood only contains 5 volumes per cent, then its saturation is 50 per cent. The low oxygen capacity may be caused by anemia, resulting from either too few red blood cells or from a low hemoglobin content of a normal number of red blood cells.

Other Factors Affecting the Dissociation of Oxygen from Hemoglobin

The most important function of hemoglobin is its ability to form a loose and reversible combination with oxygen, binding oxygen in the pulmonary capillaries, where P_{O_2} is high, and releasing oxygen to the tissues, where P_{O_2} is low. The amount of oxygen released is also influenced by tissue activity and the velocity with which blood flows through that tissue. As tissue activity increases—e.g., in exercise—the additional oxygen required for the higher metabolic rate is supplied by the following regulatory mechanisms.

CARBON DIOXIDE

The increased carbon dioxide produced by metabolism results in:

1. *Vasodilation* of the arterioles and capillaries, which speeds up blood flow through the tissues. This permits a more rapid *extraction of oxygen* from the blood by the tissues and further lowers blood P_{O_2}, which accelerates the release of oxygen from hemoglobin.
2. More rapid release of oxygen to the tissues. There is a direct effect of P_{CO_2} on the binding of oxygen to hemoglobin: the higher the P_{CO_2}, the more rapidly does hemoglobin release oxygen. This relationship between the P_{CO_2} and the dissociation of oxyhemoglobin was first demonstrated by Bohr in 1904 and is reflected by the shift in the oxygen dissociation curve to the *right.* This shift is known as the *Bohr effect.*

The Bohr effect is probably caused by the concomitant drop in blood pH caused by the increased amount of carbon dioxide that diffuses into the blood from the active tissues. These chemical reactions are discussed in detail later in this chapter. Figure 16-12 depicts the shift in the curve.

In Chapter 14, we discussed the changes in the configuration of the hemoglobin molecule resulting from the initial attachment of the oxygen molecule to the heme of the α chain. This is the process that precipitates the separation of the β chains from one another. The resulting change, from the *low affinity for oxygen* of deoxyhemoglobin to the *high-affinity* oxyhemoglobin, is responsible for the sigmoid shape of the oxygen dissociation curve. A decrease in pH decreases the oxygen affinity of oxyhemoglobin considerably, unloading much more oxygen to the anoxic tissues.

Consequently, as a result of the increased carbon dioxide produced by active tissues, oxyhemoglobin readily releases its oxygen, and the local vasodilation lets the tissues extract the oxygen from the blood more efficiently. These interactions make carbon dioxide the most important local regulator of oxygen delivery to the tissues.

pH AND TEMPERATURE

Other acid metabolites produced during activity lower the pH and enhance the dissociation of oxygen from hemoglobin. The increased temperature resulting from increased metabolic activity also moves the dissociation curve to the right.

2,3-DIPHOSPHOGLYCERATE

A phosphorylated compound 2,3-diphosphoglycerate (DPG) plays an important role in the release of oxygen to the tissues. An increase in DPG concentration is associated with decreased oxygen levels such as are found in anemia or cardiac insufficiency. Diphosphoglycerate reacts with hemoglobin to reduce the affinity of hemoglobin for oxygen and so makes more oxygen available to the anoxic tissues. In some genetic abnormalities, DPG is reduced and the red blood cells have a shortened survival time.

Significance of the Oxygen-Hemoglobin Dissociation Curve

In the lungs, the blood in the pulmonary capillaries is exposed to the P_{O_2} of the alveoli, which is about 100 mm Hg. At this P_{O_2}, the blood is 97 per cent saturated, and there is only a slight decrease in oxygen saturation, despite a fall in P_{O_2} to about 60 to 70 mm Hg as the blood circulates through the arterial system. This is indicated by the relatively flat part of the curve (Fig. 16-12).

When the blood reaches the tissues, it is suddenly exposed to P_{O_2} levels of from 30 to 40 mm Hg. Tissue P_{O_2} may be as low as 20 mm Hg during strenuous exercise. Now hemoglobin readily releases oxygen, which diffuses into the tissues.

$$Hb + O_2 \xrightarrow{\text{lungs } (P_{O_2} = 100 \text{ mm Hg})} HbO_2 \text{ (oxyhemoglobin)}$$

$$HbO_2 \xrightarrow{\text{tissues } (P_{O_2} = 40 \text{ mm Hg})} Hb + O_2 \text{ (deoxyhemoglobin and oxygen)}$$

During exercise, there is a fall in tissue P_{O_2}, an increase in P_{CO_2}, and an increase in hydrogen ion concentration, local temperature, and DPG concentration. All of these factors promote the release of oxygen from hemoglobin (moving the oxygen-hemoglobin dissociation curve to the right) and increasing the efficiency of oxygen delivery to the active tissues.

Carbon Dioxide Transport

Carbon dixoide entering the blood may be transported either in solution in the plasma or in the red blood cells. Because carbon dioxide enters rapidly into chemical combination with many substances, it is usually found in the combined form, rather than in solution (Fig. 16-13).

In the plasma

Carbon dioxide combines with water to form *carbonic acid,* which in turn dissociates into bicarbonate and hydrogen ions:

$$CO_2 + HOH \rightleftharpoons \underset{\text{carbonic acid}}{H_2CO_3} \rightleftharpoons \underset{\text{bicarbonate ion}}{HCO_3^- + H^+}$$

These ions then combine with the buffers of the blood (see Sec. 14-17). The rate of formation of carbonic acid in the plasma is relatively slow, however, so that only a small proportion (perhaps 5 per cent) of the total carbon dioxide is transported in this manner.

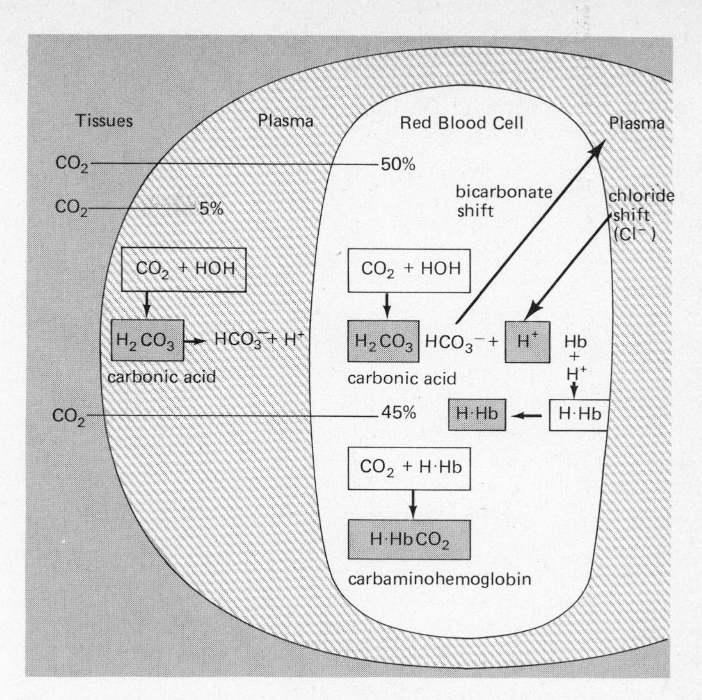

Figure 16-13. **Transport of carbon dioxide in the blood and the chloride shift. See text for explanation.**

In the red blood cell

About 95 per cent of the carbon dioxide enters the red blood cell, where it may either form carbonic acid with the water in the cell or enter into a loose and reversible combination with hemoglobin (*carbaminohemoglobin*).

Formation of Carbonic Acid. This reaction proceeds extremely rapidly in the red blood cell because of the presence of the enzyme *carbonic anhydrase,* which accelerates the reaction 200 to 300 times. Thus despite the low concentration gradient (tissue P_{CO_2} is 45 mm Hg, that of arteriolar capillaries is 40 mm Hg) carbon dioxide diffuses rapidly from the tissues into the blood. This is a good example of facilitated diffusion.

The Chloride Shift. About 50 per cent of the carbon dioxide is transported to the lungs as carbonic acid and as its ion, the bicarbonate ion. The hydrogen ion released as a product of carbonic acid dissociation is buffered by the hemoglobin within the red blood cell. The bicarbonate ion diffuses out of the red blood cell, disturbing the electrical equilibrium on either side of the membrane sufficiently to cause the passage of other negatively charged ions, chiefly chloride ions, into the red blood cell. This increase in the concentration of chloride ions within the red blood cells of venous blood is called the *chloride shift* and is the direct consequence of the passage of bicarbonate ions out of these cells in the *bicarbonate shift* (Fig. 16-13).

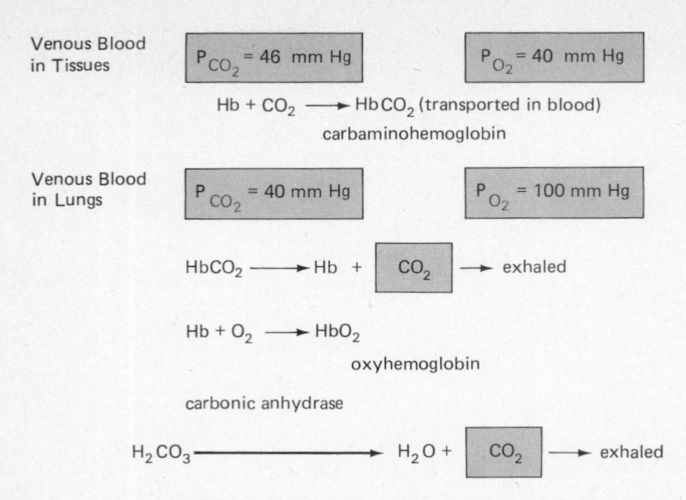

Figure 16-14. **Exchange of carbon dioxide in the lungs and in the tissues.**

Formation of Carbaminohemoglobin. The remaining 45 per cent of the carbon dioxide forms carbaminohemoglobin. The carbon dioxide rapidly forms a loose complex with the hemoglobin molecule, the stability of which is easily disturbed by an increase in the P_{O_2}. Although both oxygen and carbon dioxide theoretically can combine with the same molecule of hemoglobin, the affinity of the oxyhemoglobin for carbon dioxide is so slight as to preclude this as an important physiologic phenomenon. This means that when the carbaminohemoglobin reaches the lungs, where the P_{O_2} is high, the affinity of the hemoglobin is much higher for oxygen than for carbon dioxide; simultaneously the release of the carbon dioxide is facilitated by the slightly lower P_{CO_2} of the blood in the lungs brought about by the loss of carbon dioxide to the alveoli (Fig. 16-14).

Combination of Hemoglobin with Carbon Monoxide

The carbon monoxide molecule is a rapid respiratory toxin, combining with hemoglobin at the same active site of the molecule as does oxygen to form carboxyhemoglobin. Carbon monoxide bonds with hemoglobin so firmly that it requires 230 times as much oxygen as carbon monoxide to combine with the same quantity of hemoglobin. Displacement of carbon monoxide from the blood requires large amounts of pure oxygen. The therapy is still more efficient if some carbon dioxide is simultaneously administered, thus increasing the ventilatory rate and so enhancing the passage of carbon monoxide from the pulmonary capillaries into the alveoli.

Carbon monoxide accumulates in large amounts in the fumes of fires, from car exhausts, and in the lungs of tobacco smokers, in whom the levels of carboxyhemoglobin may reach levels as high as 10 per cent of the total hemoglobin.

16–16 Internal Respiration: Energy Release and Coupling

Internal respiration may be considered the sum of all the metabolic reactions occurring within the cell that ultimately utilize oxygen and yield carbon dioxide. Green plants in the light are able to reverse this process and utilize the carbon dioxide yielded by respiration to synthesize carbohydrates and other organic molecules, but animal cells can utilize carbon dioxide to a very limited degree, as a basis for synthetic reactions. The formation of urea is a synthetic reaction (see Sec. 19-2), but urea is chiefly a waste product and serves as a soluble means of excretion for amino radicals and carbon dioxide. The accumulation of carbon dioxide to any marked degree is toxic to the cell because the continuously forming carbon dioxide cannot be incorporated to any marked degree within organic molecules. The capacity of the base of the cell to combine and form bicarbonates and other neutral salts is also limited; therefore, the rapid transportation of this gas from the site of formation is essential to life.

Similarly, a constant flow of oxygen to the tissues and cells is essential; an interruption of more than a few minutes can cause cellular death. Some of the energy-requiring reactions of the body can continue briefly in an anaerobic environment, but when the immediately available stores of energy-rich chemicals such as ATP and CP have been used up, no resynthesis of these chemicals occurs in the absence of oxygen. To list even a few of the physiological reactions dependent on a constant replenishment of ATP through oxidative reactions, we must include the *electrical changes* that maintain the selective permeability of all cells, as well as those that are involved in the passage of the action potential along the membrane of nerve and muscle cells; the *harnessing* of the *energy derived* from the breakdown of glucose to the *contraction of muscle fibers;* and the synthetic reactions, particularly of the mitochondria, involving the enzymes. These reactions have been discussed in Chapter 5.

16–17 Nervous and Chemical Control of Respiration

Both the rate and the depth of respiration can be regulated by neural and chemical factors. There is, of course, a limit to the efficiency of increasing both these factors simultaneously, because an increase in respiration above a certain rate decreases the time available for the filling of the lungs.

Nervous Control

RESPIRATORY CENTERS IN THE HINDBRAIN

Motor Centers in the Medulla and Reticular Formation. There are two basic types of respiratory neurons, inspiratory and expiratory. It is convenient to think of these neurons as being grouped into centers, an *inspiratory center* in the medulla oblongata and the ventral part of the reticular formation, and an *expiratory center* in the medulla oblongata, caudad to the inspiratory center (Fig. 16-15). Recent experiments indicate, however, that this is an oversimplification derived from studies in anesthetized animals. In awake animals, the respiratory neurons appear to be widely dispersed in both the medulla and the reticular formation.

The *functional concept* of respiratory motor centers in the hindbrain is still valid. Electrical stimulation of inspiratory neurons causes inspiration, and if this electrical stimulation is continuously maintained a prolonged inspiration without expiration ensues. This condition is *apneusis.*

Electrical stimulation of the expiratory neurons results in expiration, but prolonged stimulation does not cause a sustained expiration because inspiration is dominant over expiration. As might be expected, simultaneous stimulation of both inspiratory and expiratory neurons causes inspiration.

The respiratory cycle consists of a rhythmic alternation of inspiration and expiration. This rhythm appears to be controlled by *oscillating circuits* between the inspiratory and expiratory centers themselves, and between these centers and coordinating centers in the pons. The inspiratory center probably originates the respiratory impulses, which are modified by higher centers and also by impulses from spinal afferent nerves, especially those from the lungs.

Motor nerves from the inspiratory and expiratory centers innervate the muscles of inspiration and expiration, respectively.

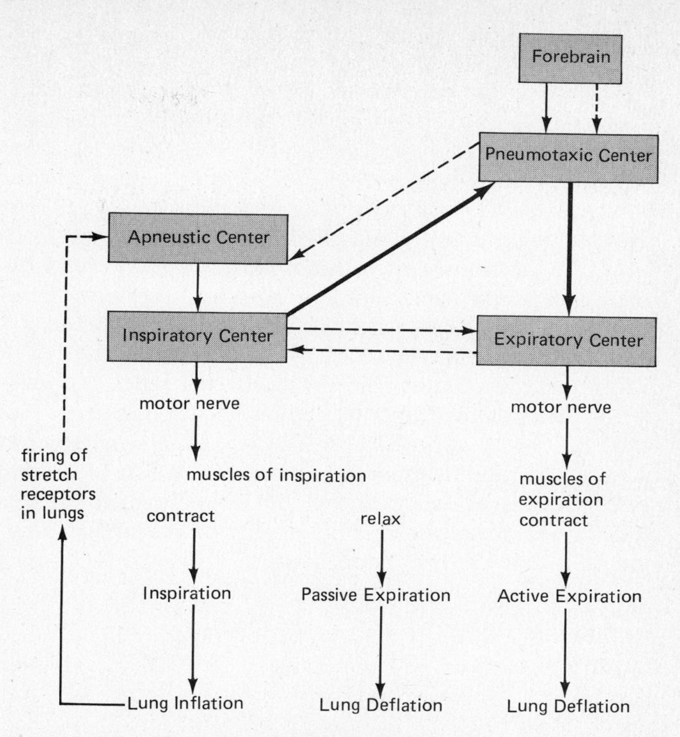

Figure 16-16. **Normal respiration involves reverberating neuronal circuits between the rhythmic centers in the medulla and probably between the medullary and pontine respiratory areas. The Hering–Breuer reflex is initiated by inflation of the lungs and requires the integrity of the vagal nerves coming from stretch receptors in the lungs. Firing of these nerves inhibits the apneustic center and results in expiration. Solid arrows represent stimulation; broken arrows indicate inhibition.**

Figure 16-15. **The respiratory centers of the brain stem. Although there is much neuronal overlapping, the areas indicated by horizontal striping and by stippling are active on inspiration. The shaded areas are active on expiration.**

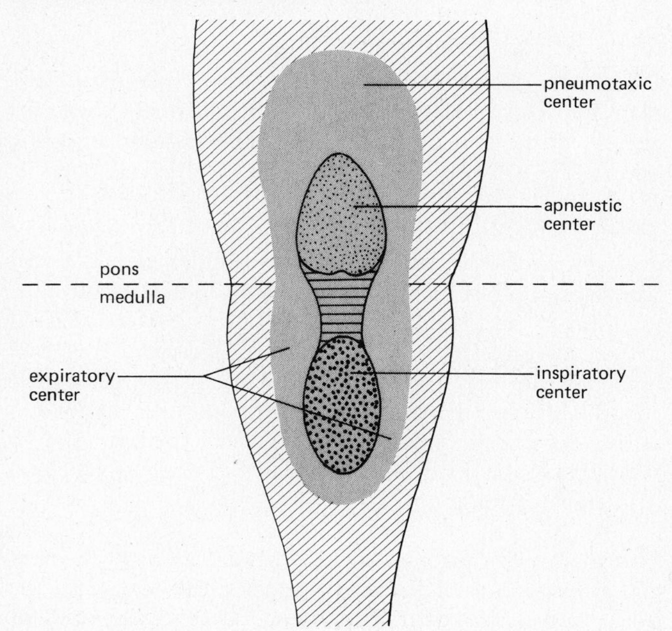

Coordinating Centers in the Pons. The function of many respiratory neurons in the pons is to smooth or coordinate the transition from inspiration to expiration and vice versa, thus bringing about the regularity of the respiratory rhythm (Fig. 16-16). Destruction of these pontine centers, or their connections to the medullary centers, causes irregular respiration that is mostly inspiratory in nature.

PNEUMOTAXIC CENTER. The pneumotaxic center, in the upper pons, is a component of an oscillating circuit between the inspiratory and expiratory centers and itself. The pneumotaxic center does not originate impulses but is stimulated by the activity of the inspiratory center. When the pneumotaxic center is active, it stimulates the expiratory center in the medulla and inhibits the other pontine center, the apneustic center. This facilitates expiration.

APNEUSTIC CENTER. The apneustic center, in the middle and lower pons, stimulates the inspiratory center in the medulla and causes inspiration. Prolonged electrical stimulation of the apneustic center causes prolonged inspiration or apneusis. In normal respiration, the apneustic center is rhythmically inhibited by the pneumotaxic center, thus permitting expiration to occur.

Hering-Breuer reflex

This reflex, first described by Hering and Breuer in 1868, involves both the medullary and the pontine respiratory centers. It is *initiated by inflation of the lungs,* which stimulates *stretch receptors* in the walls of the lungs and results in an increase in the firing rate of these neurons (Fig. 16-16). These impulses travel along afferent branches of the vagus nerve into the brain stem, causing *inhibition of the apneustic center* and permitting expiration to occur. Section of these vagal branches, or destruction of the apneustic center, abolishes the Hering-Breuer reflex.

In normal respiration, the Hering-Breuer reflex prevents overinflation of the lungs by inhibiting inspiration as the lungs fill. There is probably also a reverse Hering-Breuer reflex which inhibits expiration when the stretch receptors cease firing as air is forced out in expiration. This *reflex damping* of inspiration and expiration facilitates the smoothness of the respiratory rhythm.

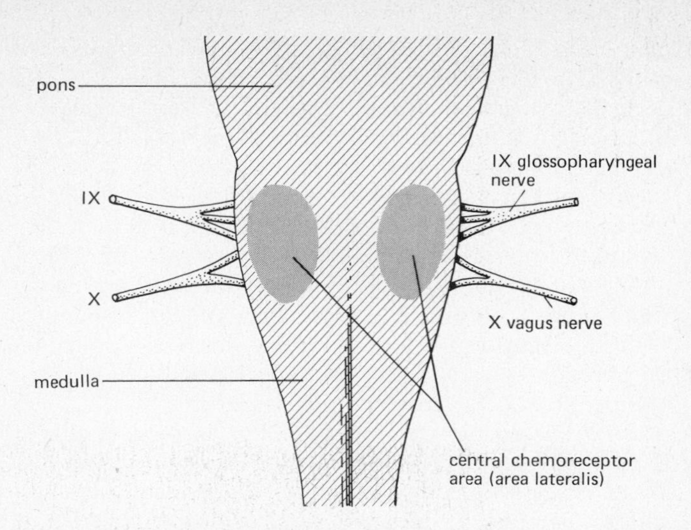

Figure 16-17. **The central chemoreceptor areas of the medulla.**

Forebrain mechanisms in the control of respiration

Neural structures above the pons play an important role in the respiratory adjustments involved in speech and behavior. Breathing is influenced by voluntary controls, by swallowing, talking, laughing, coughing, sneezing, and yawning. Postural changes, emotional stimuli, and mental concentration all influence the pattern of respiration. Although little is known about the specific pathways involved, it is certain that respiration may be affected by structures at all levels of the central nervous system, including the cerebral cortex.

The interrelationship between these neural control mechanisms is indicated in Fig. 16-16.

Chemical Control

The concentrations of carbon dioxide, oxygen, and hydrogen ions in arterial blood and the cerebrospinal fluid are important regulatory agents of the respiratory rate and depth. These substances exert this homeostatic effect through changes they evoke in the respiratory chemoreceptors of the body. *Chemoreceptors* are nerve cells or nerve endings that respond to specific chemical changes in their internal environment, the tissue fluid. There are two types of chemoreceptors, the *central chemoreceptors,* which are sensitive to changes in P_{CO_2} and pH, and the *peripheral chemoreceptors,* which respond to a lack of O_2.

Central chemoreceptors

Central chemoreceptors are found on either side of the medulla, near the entry of the IXth and Xth nerves (Fig. 16-17). This region is called the *area lateralis,* and the cells in this area are activated by an *increase in* P_{CO_2} of the blood and cerebrospinal fluid. Carbon dioxide diffuses freely between these fluids and causes a tremendous increase in respiratory rate and depth. This quickly removes the excess carbon dioxide from the body through the lungs.

The response of the central chemoreceptors to changes in pH is less direct because the blood-cerebrospinal barrier is practically impermeable to hydrogen ions. However, as carbon dioxide freely enters the cerebrospinal fluid, increased amounts of carbonic acid are formed. Consequently, as this acid dissociates, the hydrogen ion concentration of the cerebrospinal fluid quickly rises and this is a potent stimulus for increased activity of the central chemoreceptors.

Increased P_{O_2} has little if any effect on the central chemoreceptors, but it is possible that they are depressed by a marked drop in P_{O_2}.

Peripheral chemoreceptors

In 1929, the father-and-son team of physiologists, Heymans and Heymans, demonstrated that there are chemoreceptors in the two carotid bodies and the group of aortic bodies (Fig. 16-18).

The Carotid Body. The carotid body (bilateral) is a tiny organ found at the bifurcation of the common carotid artery (compare, contrast but do not confuse, with the carotid sinus involved in the modulation of blood pressure, as discussed in Sec. 12-10). Each carotid body is innervated by a fine branch of the nerve of Hering, which joins the glossopharyngeal (IX) nerve and then enters the medulla. The blood flow to the carotid body is 15 times that to the heart and 30 times that to the brain, when calculated in terms of ml blood delivered per minute to 100 g of tissue.

The Aortic Bodies. Located along the arch of the aorta and around some of the major arteries near the heart, the aortic bodies are innervated by branches of the afferent

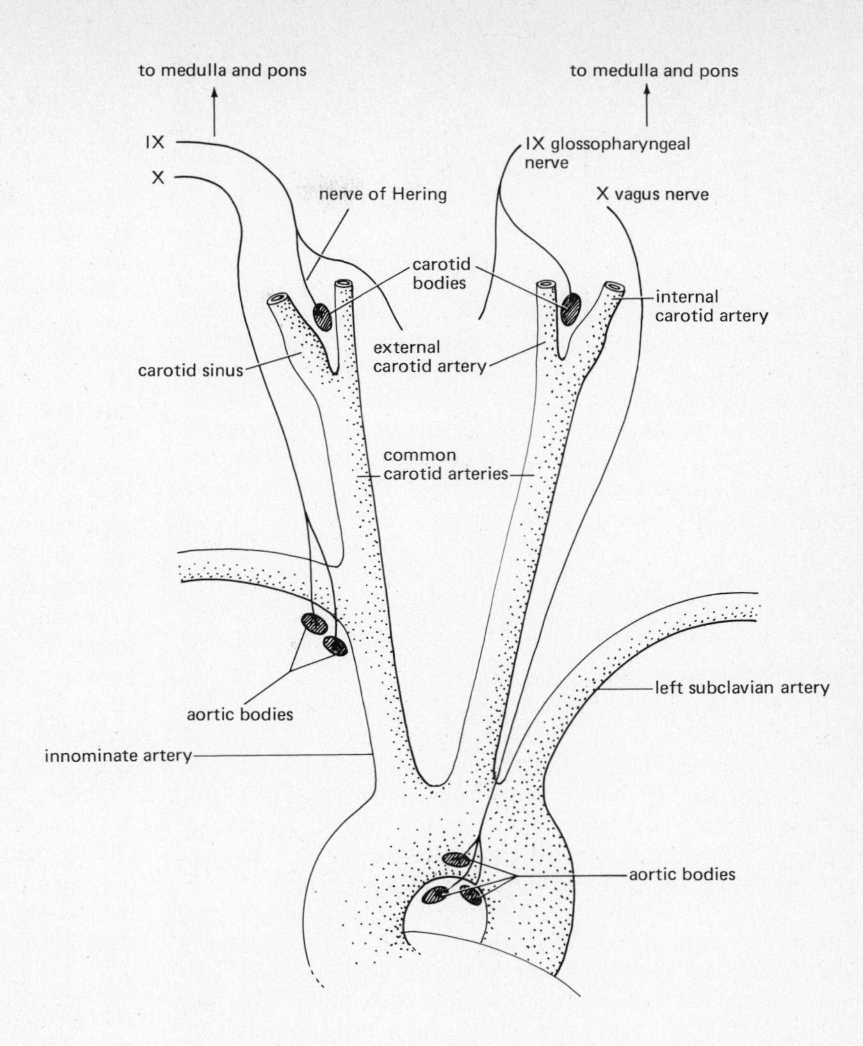

Figure 16-18. **The arterial chemoreceptors: the carotid and aortic bodies.**

vagus, and each receives arterial blood through a small, individual branch of an artery.

Response of Peripheral Chemoreceptors to Changes in P_{O_2}

The carotid and aortic bodies are extremely sensitive to a decrease in arterial P_{O_2}, especially in the physiological range of between 70 and 30 mm Hg. A fall in P_{O_2} increases the frequency and amplitude of nerve impulses from the carotid body nerves to the respiratory centers in the medulla. The resulting increase in alveolar respiration is a feedback mechanism that increases arterial P_{O_2}.

However, this response is seen most clearly in experimental situations where P_{CO_2} and pH are kept constant. Under normal circumstances, P_{O_2} is not a very important regulator of respiration because until P_{O_2} falls to 60 mm Hg or less, oxygen remains adequately bound to hemoglobin. A drop in P_{O_2} from 100 to 80 mm Hg hardly affects the oxygen saturation of the blood. During exercise, when the tissue utilization of oxygen rises enormously, the carbon dioxide concentration increases accordingly. This *increase in P_{CO_2}*, and the accompanying *rise* in *hydrogen ion concentration*, are the main factors increasing ventilatory rate and maintaining P_{O_2} in the alveoli at 100 mm Hg.

16–18 Respiration During Exercise

Humoral Factors

Exercise is one of the most potent ways to increase ventilation. In moderate exercise, ventilation increases in direct proportion to the amount of carbon dioxide formed. When the exercise becomes more strenuous, lactate is also formed, and the resulting increase in arterial hydrogen ion concentration acts as an additional stimulus to pulmonary ventilation. These changes in P_{O_2}, P_{CO_2}, and pH (humoral factors), however, can account for only a small amount of the increased respiration during exercise.

Neural Factors

PROPRIOCEPTORS

Even when the P_{CO_2}, P_{O_2}, and pH are artificially maintained at resting levels, by ingenious cross-circulation techniques, exercise results in a marked increase in respiration. This increase is caused by impulses reaching the respiratory center from proprioceptors in the contracting skeletal muscles. Passive "pumping" of these limbs to excite these proprioceptors will also increase pulmonary ventilation.

Motor cortex

An additional neural source of respiratory stimulation comes from the motor cortex. The motor cortex responds to the stress of exercise by exciting the respiratory center by way of the reticular formation. This is similar to the role of the cortex in increasing cardiac output during exercise (Sec. 12-11). Consequently, the *respiratory and circulatory adjustments* to exercise *are well integrated* through these higher levels of the nervous system.

16–19 Voluntary Control of Respiration

It is possible to increase the rate of respiration or to briefly inhibit respiration voluntarily. However, voluntary control of respiration is very short because the resulting chemical changes in the blood directly and reflexly stimulate the respiratory centers, which are fortunately far more powerful in controlling breathing than are the higher centers. A child's temper tantrum can no more result in voluntary asphyxiation than we can forget to breathe.

Holding the breath is possible for short periods of time, until the accumulated carbon dioxide of the blood reaches a level at which the stimulation of the inspiratory center is so strong that the higher centers can no longer inhibit it. Rapid, deep breathing (*hyperventilation*), voluntarily undertaken, can be more dangerous, however, for it may deplete the blood carbon dioxide to such an extent that dizziness and unconsciousness set in, thus inactivating the higher centers as well as the respiratory centers, and resulting in cessation of breathing (*apnea*). When the blood carbon dioxide has built up again to sufficient concentration, stimulation of the inspiratory center results in the resumption of respiration (Figs. 16-19 and 16-20).

16–20 Respiration at High Altitudes

Lowered Atmospheric Pressure

In an atmosphere depleted of oxygen, or at high altitudes where the oxygen pressure may be 60 instead of 160 mm Hg, severe symptoms of *hypoxia* (lack of oxygen) appear; after a few minutes, they may become fatal. Even if the subject recovers, permanent injury, especially to the nervous system, may remain. These symptoms include dizziness, nausea, loss of memory, cyanosis (blueness of the lips and nails), and changes in heart rate and respiration. Ventilatory rate increases, removing carbon dioxide from the blood in such amounts that the alkali reserve of the blood falls. These symptoms are seen in varying degrees of severity in "mountain sickness," depending on the altitude and the extent to which the individual has become acclimatized. These symptoms are exacerbated by exercise, which increases the tissue utilization of oxygen. It is usually necessary to begin breathing oxygen at altitudes of 20,000 feet (6600 m) or even lower. In order to maintain the homeostatic mechanisms between respiration and blood, a mixture of oxygen (95 per cent) and carbon dioxide (5 per cent) is usually administered. Pure oxygen at normal atmospheric pressures is toxic after a few hours.

Under these conditions of very low alveolar P_{O_2}, the *peripheral chemoreceptors* play a larger role in reflexly stimulating the ventilatory rate, but the maximum increase is only from the normal 5 liters/min to 8 or 9 liters. This increase is relatively insignificant when compared to the increase due to carbon dioxide of 50 to 60 liters/min.

Acclimatization to High Altitude

The span of physiological adjustments to high altitude is best seen in humans living as high as 18,000 ft (6000 m), where the atmospheric P_{O_2} is only 80 mm Hg. These adaptations include the enlargement of the thorax with a corresponding increase in vital capacity, an increase in red blood cell number and hemoglobin content, and a hypervascularization of the tissues. These processes make available more hemoglobin to bind oxygen (i.e., the oxygen capacity

Figure 16-19. **Record of breathing taken by means of a simple pneumograph: downward strokes correspond to inspiration. The tracing from left to right shows: (1) normal breathing; (2) a 2-min period of forced breathing (voluntary hyperventilation); (3) apnea of 130-sec duration; and (4) transitory periodic breathing, which gradually gives rise to normal regular breathing. The time tracing at the bottom shows intervals of 10 sec. (From C. G. Douglas and J. G. Priestley, *Human Physiology*, 3rd ed., The Clarendon Press, Oxford, 1948.)**

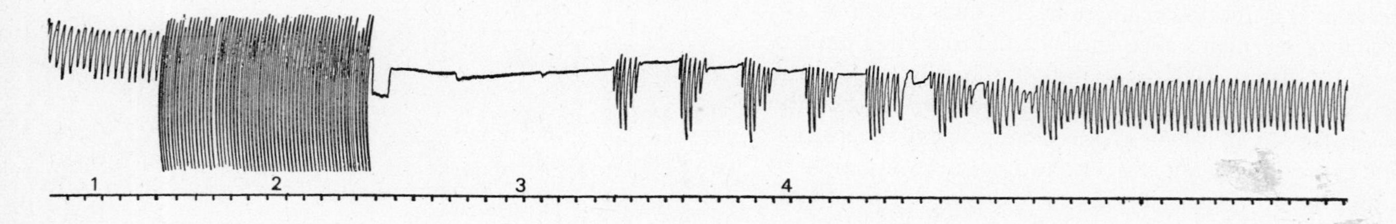

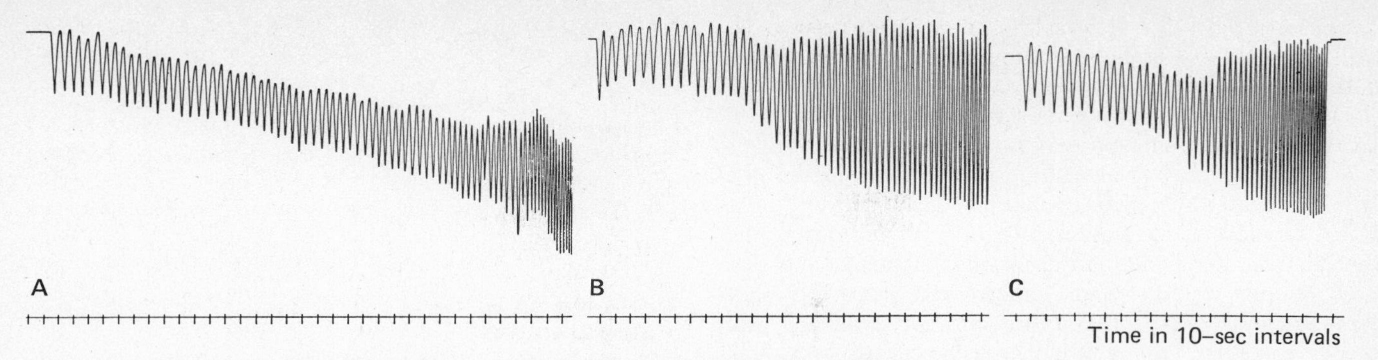

Figure 16-20. **Effect of a deficiency of oxygen or an excess of carbon dioxide on respiration. (A) Effect of increasing the oxygen deficiency, (B) effect of increasing carbon dioxide excess (no oxygen deficiency), and (C) effect of increasing carbon dioxide excess together with an oxygen deficiency. (From C. G. Douglas and J. G. Priestley, *Human Physiology,* 3rd ed., The Clarendon Press, Oxford, 1948.)**

of the blood is increased) and more capillaries to transport the blood and provide a surface for gas exchange. Some of these changes are depicted in Fig. 16-21.

In addition, people living at high altitudes show important differences in their respiratory responses to changes in P_{O_2} and P_{CO_2} as compared to sea-level dwellers. The highlanders show an insensitivity to increased P_{CO_2}. This indicates that they rely more on the physiological adjustments mentioned in the preceding paragraph than on increased respiratory ventilation to get oxygen to the tissues.

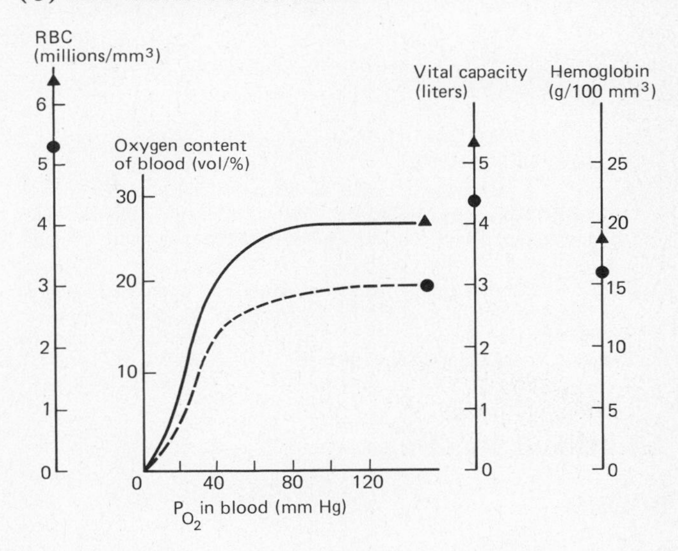

Figure 16-21. **Comparisons of the oxygen dissociation curves, red-blood-cell counts, and hemoglobin content of the blood of residents at high altitude (4500 m), and of residents at sea level. Vital capacities are also compared. Values for the mountain dwellers (▲) were measured at 4500 m; values for sea-level residents (●) were measured at sea level.**

16-21 Respiration Under the Sea or in Tunnels

Increased Atmospheric Pressure

Increased atmospheric pressure does not seem to cause serious physiologic discomfort, but the *decompression,* or return of the individual to normal pressure, may be accompanied by serious disturbances known as "caisson disease." This may be seen in deep-sea divers returning from depths where the atmospheric pressure may have been increased as much as 10 atmospheres (1 atm = 760 mm Hg), in tunnel workers to ground level, and in aviators ascending rapidly to high altitudes. As the *atmospheric pressure increases* at great depths, the gases go into solution in the blood, and on sudden decrease in pressure they are released in the form of small bubbles. Oxygen is rapidly absorbed by the cells, but nitrogen, which is the least soluble of the gases in the blood, remains and may block off some of the smaller blood vessels and even the large veins supplying the heart and lungs (air embolism). Rapid decompression is accompanied by the excruciating symptoms of "the bends": pains in the bones and joints, asphyxia, and paralysis due to embolisms in the spinal cord. The only treatment, if symptoms of rapid decompression appear, is equally rapid compression in a special pressure chamber until the bubbles have gone into solution and then a very slow decompression lasting over several hours. Miners' and divers' compression suits and the space suits of the present-day astronauts are all designed to control the range of atmospheric pressure to which the body is exposed.

Animal Divers

Animal divers—dolphins, seals, whales, and so forth—do not suffer from the bends because no gas at high pressure has been forced into the lungs. Many animals breathe out before diving and allow their lungs to collapse, thus preventing even a small quantity of nitrogen from dissolving in the plasma. Whales and dolphins have a specific motor center for the control of respiratory muscles. This center can achieve the emptying and refilling of the lungs in less than half a second.

During the dive, oxygen must be supplied to the brain and heart; there is a reflex shunting of blood to these organs and corresponding vasoconstriction to other tissues, the latter so marked that muscles cut open during a dive do not bleed. Obviously, respiration must be inhibited during the dive, and the immersion of the nostrils initiates a reflex that both inhibits respiratory movements and slows the heart (bradycardia). These mechanisms prevent drowning and conserve oxygen. Diving reflexes are present but very weak in humans.

16–22 Respiration in Outer Space

Very little is known of the respiratory adjustments of humans in space because an artificial atmosphere is always provided. In American spaceships, the atmosphere selected is almost pure oxygen, but the total pressure is maintained at 200 to 250 mm Hg. This provides enough oxygen for the physiological requirements yet minimizes the danger of explosive decompression of the spaceship.

Cited References

BARCROFT, J., and M. CAMIS. The Dissociation Curve of Blood. *J. Physiol. (London)* **39:** 118 (1909).

BOHR, C., K. A. HASSELBALCH, and A. KROGH. "Ueber, einen in biologischer Bezichung wichtigen Einfluss, den die Kohlensaurespannung der Blutes auf dessen Sauerstoffbinding uebt." *Skand. Arch. Physiol.,* **16:** 402 (1904).

DAWES, G. S. "Oxygen Supply and Consumption in Late Fetal Life and the Onset of Breathing at Birth." In *Handbook of Physiology,* Sec. 2: *Respiration,* W. O. Fenn and H. Rahn, eds. American Physiological Society, Washington, D.C., 1965, pp. 1313–1328.

Additional Readings

BOOKS

CIBA FOUNDATION SYMPOSIA. *High Altitude Physiology.* Churchill Livingstone, Division of Longman Inc., New York, 1971.

DAWES, G. S. *Foetal and Neonatal Physiology.* Year Book Medical Publishers, Inc., Chicago, 1968.

FINK, B. R. *The Human Larynx: A Functional Study.* Raven Press, New York, 1975.

GUYTON, A. C., and J. G. WIDDICOMBE, eds. *MTP International Review of Science, Physiology,* Vol. 2: *Respiratory Physiology.* University Park Press, Baltimore, 1974.

KAO, F. F. *An Introduction to Respiratory Physiology.* American Elsevier Publishing Co., Inc., New York, 1972.

KROGH, A. *Comparative Physiology of Respiratory Mechanisms.* University of Pennsylvania Press, Philadelphia, 1941.

LEE, D. H. K., ed. *Physiology, Environment and Man.* Academic Press, Inc., New York, 1970.

MURRAY, J. F. *The Normal Lung.* W. B. Saunders Company, Philadelphia, 1976.

WEST, J. B. *Respiratory Physiology.* The Williams & Wilkins Company, Baltimore, 1974.

ARTICLES

ADEY, W. R. "The Physiology of Weightlessness." *Physiologist* **16:** 178 (1974).

ANDERSON, H. T. "Physiological Adaptations in Diving Vertebrates." *Physiol. Rev.* **46:** 212 (1966).

AVERY, M. E., N. WANG, and H. W. TAEUSCH, JR. "The Lung of the Newborn Infant." *Sci. Am.,* Apr. 1973.

BAKER, P. T. "Human Adaptation to High Altitude." *Science* **163:** 1149 (1969).

BATES, D. V. "Physiological Effects on Man of Air Pollutants." *Fed. Proc.* **33**(10): 2133 (1974).

BEHNKE, A. R., JR., and E. H. LAMPHIER. "Underwater Physiology." In *Handbook of Physiology,* Sec. 3, Vol. 2. The Williams & Wilkins Company, Baltimore, 1965.

BISCOE, T. J. "Carotid Body: Structure and Function." *Physiol. Rev.* **51:** 427 (1971).

BULLARD, R. W. "Physiological Problems of Space Travel." *Ann. Rev. Physiol.* **34:** 205 (1972).

BURNS, B. D. "The Central Control of Respiratory Movements." *Brit. Med. Bull.* **19:** 7 (1963).

COHEN, A. B., and W. M. GOLD. "Defense Mechanisms of the Lungs." *Ann. Rev. Physiol.* **37:** 325 (1975).

COMROE, J. H., JR. "The Lung." *Sci. Am.,* Feb. 1966.

COMROE, J. H., JR. "The Peripheral Chemoreceptors." In *Handbook of Physiology,* Sec. 3: *Respiration,* Vol. 1. The Williams & Wilkins Company, Baltimore, 1964, p. 557.

HOCK, R. J. "The Physiology of High Altitude." *Sci. Am.* Feb. 1970.

HONG, S. K., and H. RAHN. "The Diving Women of Korea and Japan." *Sci. Am.,* May 1967.

MCCUTCHEON, F. H. "Organ Systems in Adaptation: The Respiratory System." In *Handbook of Physiology,* Sec. 3: *Respiration.* Vol. 1. The Williams & Wilkins Company, Baltimore, 1964, p. 167.

PACE, N. "Respiration at High Altitude." *Fed. Proc.* **33:** 2126 (1974).

SMITH, C. A. The First Breath. *Sci. Am.,* Oct. 1963.

VAIL, E. G. "Hyperbaric Respiratory Mechanics." *Aerosp. Med.* **42:** 536 (1971).

WANG, S. C., and S. H. NGAI. "Respiration Coordinating Mechanisms of the Brain Stem—A Few Controversial Points." *Ann. N.Y. Acad. Sci.* **109:** 560.

Chapter 17

Regulation of Digestion and Absorption

Once I was fool enough to think
That brains and sweetbreads were the same,
Till I was caught and put to shame,
First by a butcher, then a cook,
Then by a scientific book.
But 'twas by making sweetbreads do
I passed with such a high I.Q.

Robert Frost, "Quandary"

FOOD, WATER, AND electrolytes are ingested through the mouth and pass in assembly-line fashion through the digestive tube. This assembly line breaks down products rather than building up products. Before the nutrients (carbohydrates, fats, and proteins) can be utilized by the body, they must be broken down by enzymatic hydrolysis (*digestion*) into compounds small enough to pass into the digestive tube cell (*absorption*), and thence into the circulation. Digestion occurs as the foods are passed along the digestive tube (the *gastrointestinal* or G.I. *tract*) and subjected to the action of the secretions produced by the digestive glands. The digestive tube and its glands make up the digestive system.

Absorption in the digestive tract involves many factors, including simple diffusion, facilitated diffusion, active transport, and in some special cases, pinocytosis. The basic principles of these mechanisms have been discussed in Chapter 4.

17–1 The Digestive System

Development

The digestive system develops very early in the embryo from the endoderm, forming an incomplete tube from the head end to the tail end, with the middle portion open to connect with the yolk sac in lower animals and with the placenta in higher mammals. A placental embryo, such as a human embryo, depends on the food and oxygen that diffuse through the placenta to the embryonic blood vessels, so that the digestive system, although it develops enormously in size and complexity, does not become functional until after birth. After the placental connection is severed by the cutting of the umbilical cord, the digestive system is responsible for the hydrolytic reactions necessary to transform the ingested food into molecules small enough to be absorbed.

Figure 17-1. **Development of the adult digestive tract from the embryonic digestive tube.**

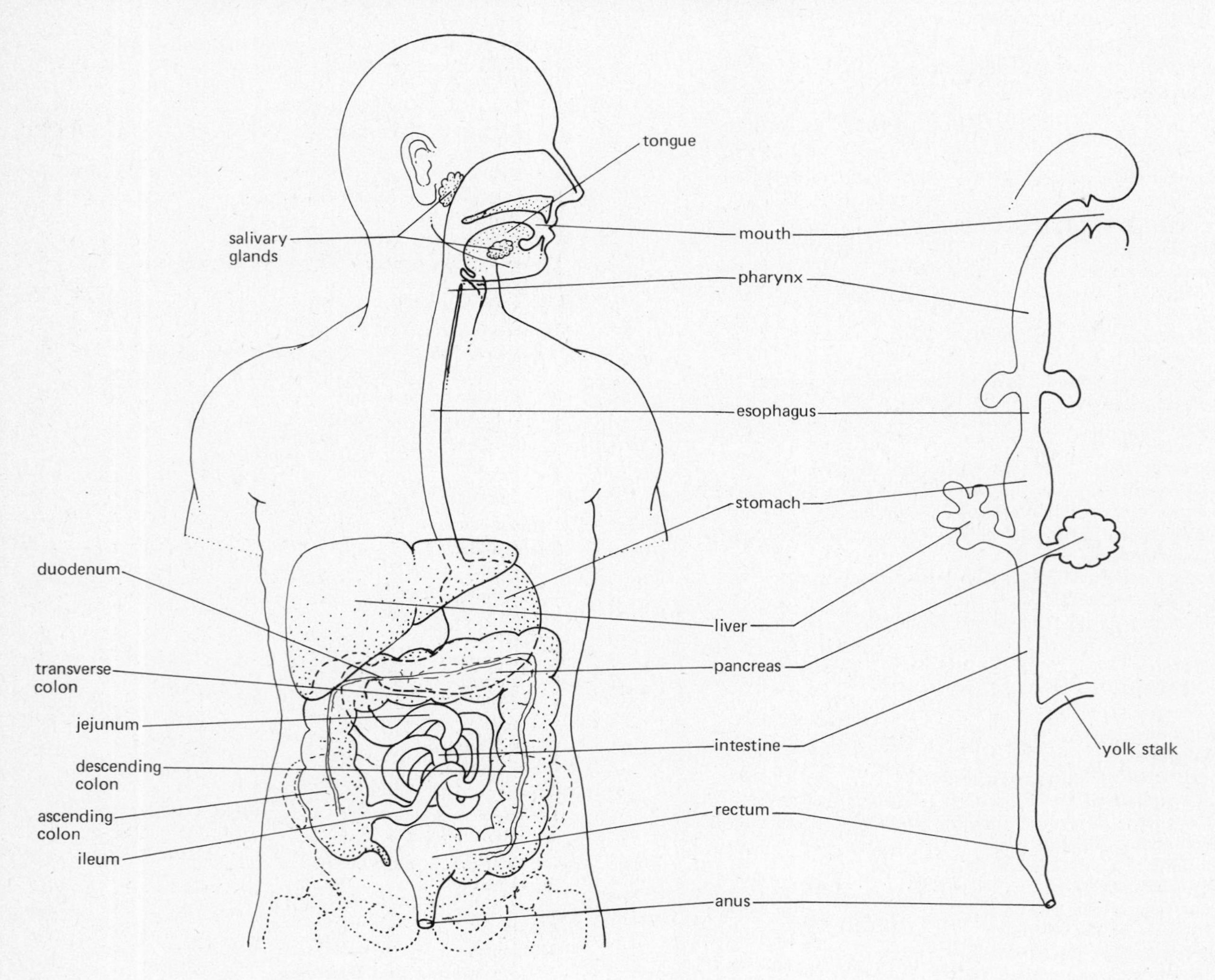

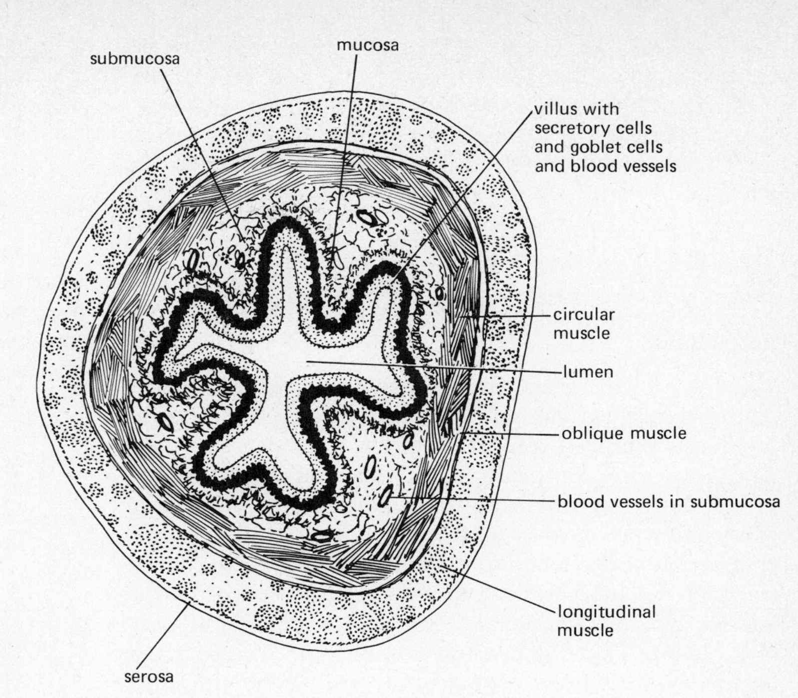

Figure 17-2. **Structure of the wall of the small intestine, showing the various layers that are common to the digestive tube.**

Prior to birth, and for a few days thereafter, the digestive system secretes mainly a type of protective mucus, but the secretion of digestive enzymes rapidly increases.

The aforementioned development in complexity includes the growth and organization of many glandular organs both *within the walls of the digestive tube* and as *separate organs with ducts* leading back into the lumen of the digestive tube (Fig. 17-1). These glands are referred to as exocrine, in contrast to the ductless endocrine glands.

Basic Structure

The digestive tube includes the *mouth, pharynx, esophagus, stomach, small intestine, large intestine, rectum,* and *anus.* Each of these areas is specialized in some way; most are closely associated with the digestive glands external to the tube.

Despite the area specializations, the basic organization of the tube consists of the following layers (Fig. 17-2): the mucosa, submucosa, muscularis externa, and serosa.

Mucosa

This is the inner lining of the tube, formed from endoderm that has become modified for secretion. The mucosa produces the digestive juice which consists of digestive enzymes in large volumes of fluid. The pH of the digestive fluid varies considerably in the different regions of the tube, as a result of the acid or alkali secreted.

The mucosa consists of three layers:

1. A superficial layer of *epithelium,* invaginated to form the secretory glands.
2. The *lamina propria,* a delicate layer of connective tissue.
3. The *muscularis mucosa,* a thin layer of muscle fibers.

Submucosa

This layer of connective tissue is between the mucosa and the muscle layers. It contains the intricate network of blood vessels and nerves that supply the digestive tract. Very often, the glands from the epithelial layer become large enough to extend into the submucosa. In certain areas, the submucosa is filled with lymphatic nodules.

Muscularis externa

Two muscular tubes encircle the mucosa and submucosa. The inner tube consists of circularly arranged smooth muscle fibers, which on contraction constrict the lumen of the digestive tube. The outer tube consists of longitudinally arranged smooth muscle fibers, which shorten the tube when they contract. Between the *inner circular* and the *outer longitudinal* muscle layers is sometimes a third layer of *oblique smooth muscle. Nerve plexuses* associated with visceral reflexes are abundant between these various muscle layers.

Serosa

This is a thin layer of connective tissue covering the digestive tube and the organs associated with it. It is continuous with the serous membrane lining the walls of the abdominal cavity and so acts to attach the viscera to the abdominal wall. That part of the membrane that covers the viscera is the *visceral peritoneum;* that part lining the abdominal wall is the *parietal peritoneum.* That portion of the membrane that is reflected back from the viscera to the somatic peritoneum is the double-layered *mesentery,* which carries the blood vessels, nerves, and lymphatics to the intestine.

Movement of Food Through the Digestive System

PERISTALSIS

Because of the arrangement of the muscle layers of the digestive tube, an unusual form of propulsion is seen as the food mass is propelled from the pharynx to the anus. Alternate *progressive* contraction and relaxation of the muscle layers along the length of the tube result in an area of constriction preceded by an area of dilation. In addition, the contraction of the longitudinal muscle results in the shortening of the tube, so that the food mass is pushed from behind, by constriction, into the wider area ahead of it (Fig. 17-3). The distance traveled may be merely a few centimeters at a time, or it may involve a *peristaltic rush,* propelling the mass all the way from the small intestine to the anus. This type of peristalsis occurs rarely, only under conditions of extreme gastrointestinal irritation. Peristalsis is a local reflex of the smooth muscle layers and their intrinsic nerve plexuses.

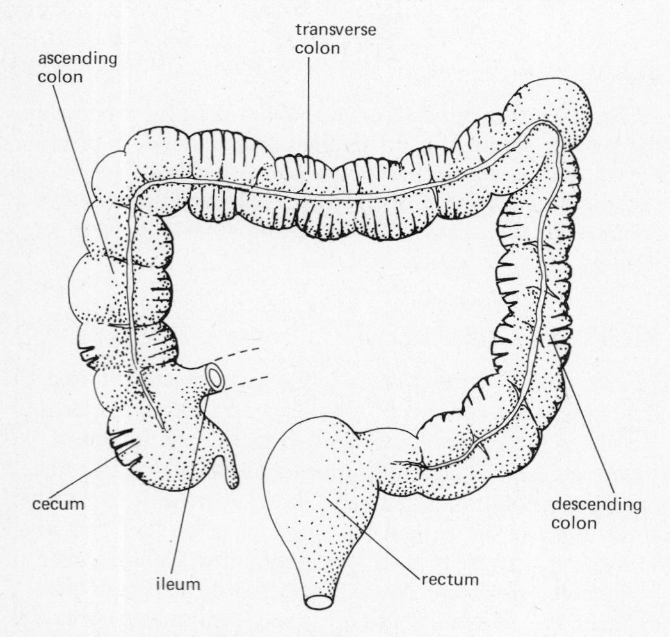

Figure 17-3. **Peristalsis in the large intestine.**

Figure 17-4. **Different forms of segmentation in the intestine.**

MIXING

The smooth muscle of the digestive tube exhibits another type of movement that is responsible for the mixing of the intestinal contents, rather than their propulsion. This movement is a *rhythmic segmentation* of a loop of the intestine, dividing it up to look like a string of bloated beads on a thread (Fig. 17-4). A constriction then appears in the middle of the beads while the threadlike portion dilates. This is repeated 6 to 10 times a minute in humans.

The mixing movements vary somewhat in the stomach, small intestine, and large intestine, but the end effect is that the contents of the lumen are rolled around and mixed with the digestive juices, exposing as much of the food mass as possible to the hydrolytic action of the enzymes.

TONUS

Normal peristaltic activity is superimposed upon a background of constant low-grade muscle contraction, known as tonus. This is common to all types of muscle contraction—the efficiency of the contraction depends on a previous base of tension. Tonus is especially important in those parts of the digestive tube where contents are stored for various periods of time, in the stomach and large intestine, for instance. Without the contractile influence of tonic activity, these areas tend to become permanently distended. Poor gastrointestinal tonus is sometimes associated with chronic constipation.

Nervous Control of Gastrointestinal Motility

The smooth muscle layers and the glands of the digestive tube are supplied by a network of sensory and motor nerves integrated by groups of nerve cells, or *ganglia.* This extensive system forms the *myenteric plexus,* controlled through the autonomic nervous system. The plexus does have its own *inherent pacemaker,* however, for isolated strips of intestine show peristaltic activity. The frequency and force are much greater near the duodenal end, diminishing toward the large intestine. The metabolic activity of the intestine also follows this gradient, so that although myogenic contractions are poorly understood, they are probably dependent on the metabolic activity of the tissues.

SYMPATHETIC NERVOUS SYSTEM

The *sympathetic nervous system* acts to *inhibit* intestinal motility and activity. However, it causes the contraction of some of the sphincter muscles which guard the opening of one part of the tube into the next. The anal sphincter is constricted by sympathetic activity so that stimulation of the sympathetic nervous system can almost completely block the passage of food through the gut and inhibit defecation. But because cutting the sympathetic nerves has little effect on the activity of the gut, this part of the autonomic nervous system does not appear to exert a tonic effect under normal circumstances.

Parasympathetic nervous system

The *parasympathetic* nervous system, on the other hand, is of paramount importance for the normal function of the gastrointestinal tract. It exerts a *constant stimulating effect* on gut muscle and so is responsible for the tonus of the muscle. Without adequate muscle tonus, peristalsis is weak and ineffective. As the parasympathetic nerves also cause relaxation of most of the sphincter muscles, the propulsion of the tubal contents is facilitated. This is especially important for defecation reflexes, which are discussed later in this chapter. Cutting the parasympathetic nerves profoundly affects gastrointestinal function, impairing tonus, peristaltic movement, and defecation.

Localized gut reflexes: the intramural plexus of the digestive tract

There are two extensive networks of neurons extending from the esophagus down to the anus. The *motor network* lies between the longitudinal and circular layers of muscle: this is the *myenteric plexus*. Stimulation of these myenteric neurons increases the tone and contractility of the muscles.

The second network is formed by *sensory cells* in the submucosa: this is the *submucosal* or *Meissner's plexus,* and is far less extensive than the motor plexus. Many of the reflexes that involve the gastrointestinal tract are local circuits between the sensory and motor plexuses of the gut itself. Other reflexes involve the afferent branches of the vagal nerves, the brain stem, and the efferent pathways of either the parasympathetic or sympathetic nerves.

Emotion, pain, and *stresses* of many types influence gastrointestinal motility, but the particular effect is unpredictable. These stresses may induce a reversal of the direction of peristalsis, one of the factors involved in the vomiting reflex, which is discussed later.

Other Factors Affecting Gastrointestinal Motility

The presence of food in the gastrointestinal tract distends the gut and is a potent stimulus for peristalsis. Irritation of the mucosa, changes in the levels of blood glucose, and other metabolites also affect gut motility. Hormones such as serotonin, which is probably the neurotransmitter in the myentric plexus, and gastrin, a polypeptide hormone secreted by the gastric mucosa, both stimulate intestinal contractions.

17–2 The Mouth

Structures of the Mouth and Pharynx

The mouth forms the first part of the digestive tube, and it is lined throughout with mucous membrane, which is protected in the region near the lips by a stratified squamous epithelium very similar to that of the skin (Sec. 15–1). The many glands in the mucosa secrete mucus and enzymes, but the main source of the secretion of the mouth, *saliva,* is the *salivary glands.* The muscles of the mouth are under voluntary control. In front of the actual oral cavity are the muscular *lips,* which help to get the food into the mouth and to keep it there while it is being *masticated.* This is the process by which the large lumps of food are broken down into smaller pieces that can be swallowed by the biting and grinding of the *teeth,* the rolling action of the *tongue,* and the lubrication provided by the saliva.

The oral cavity itself is bounded in front by the *gums* and teeth; posteriorly it opens into the *pharynx;* the floor is formed by the tongue. The vaulted roof of the mouth is built of the *hard and soft palate.* The posterior margin of the soft palate is free, and its apex hangs down as the *uvula,* to rest on the tongue.

Gums and teeth

The gums are composed of dense fibrous tissue, covered by mucous membrane. They are attached to the bony structure of the jaws, in which the *roots* of the teeth are embed-

Figure 17-5. **Structure of a tooth.**

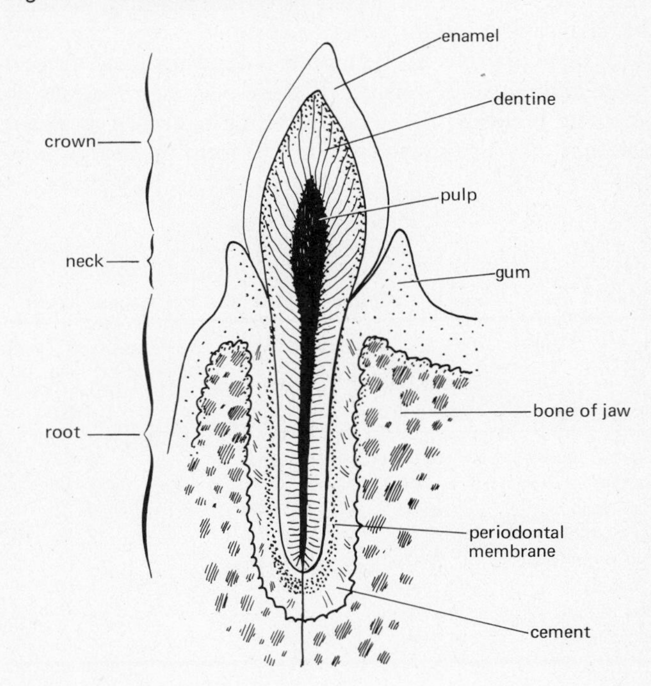

ded. That part of the tooth which projects above the gum is known as the *crown;* the junction between root and crown is the *neck.* The body of the entire tooth is composed of *dentine,* which is slightly harder than bone. The dentine surrounds a soft central core, the *dental pulp,* which contains the blood vessels and nerves of the tooth. In the region of the crown, the dentine is surrounded by *enamel,* the hardest structure of the body, but the nonexposed areas of dentine are covered by a bonelike substance, the *cement.* Surrounding the root of the tooth and transmitting the blood vessels and nerves to it is the periodontal membrane, which also plays an important role in the formation of the tooth.

The teeth are actually the products of the skin, the enamel being elaborated from the ectoderm (epidermis) and the dentine, pulp, and cement by the mesoderm (dermis) (Fig. 17-5). In the eruption, or cutting, of the teeth the crown is pushed through the gum. The first set of teeth, the milk teeth, is gradually pushed and pressed out of place by the growth of the secondary, permanent teeth (Table 17-1). During childhood and adolescence, the jaws grow sufficiently to accommodate 12 molar teeth that have no counterpart in the milk dentition. There are 32 teeth in the adult, 16 in each jaw.

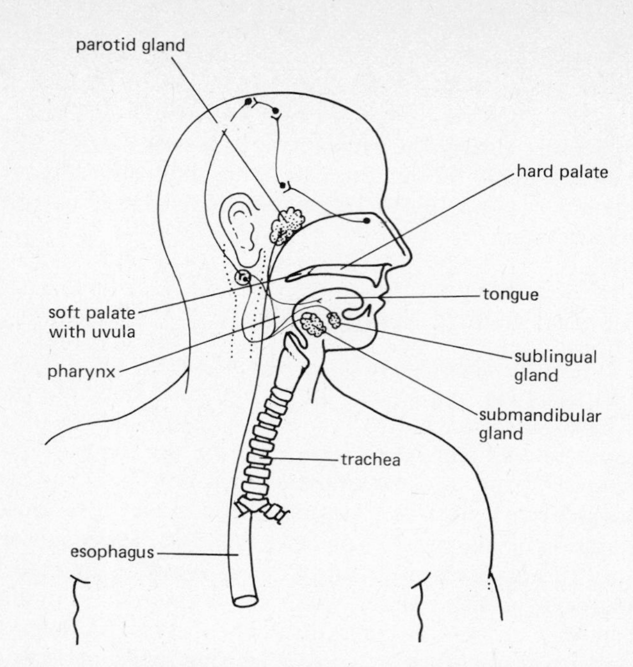

Figure 17-6. **The salivary glands and the reflex secretion of saliva.**

PALATE

Mammals are the only vertebrates that chew their food, and the development of the palate and salivary glands is a decidedly mammalian characteristic. The palate separates the mouth from the nasal respiratory passages and so enables the young to breathe and suck (and later to chew) at the same time.

The palate is formed by the growth of two projections from the jaw; these fuse in the midline of the oral cavity. Anteriorly, they are connected also to the septum, dividing the nasal cavity, and this anterior portion becomes the bony, hard palate. Farther back, no bone is deposited, and the soft palate and uvula are formed. If these palatine processes do not unite properly, the defect resulting is known as a *cleft palate;* it may also involve the lips to form the *cleft or hare lip.*

Table 17–1. **Time of Eruption of Milk and Permanent Teeth**

	Milk Teeth		Permanent Teeth
	Tooth Erupts (months)	Tooth Sheds (years)	Tooth Erupts (years)
Central incisors	6–7½	6–7	6–8
Lateral incisors	7–9	7–8	7–9
Canines	16–18	9–12	9–12
1st premolars	12–14	10	12–13
2nd premolars	20–24	11	12–13
1st molars	—	—	9–10
2nd molars	—	—	14–16
3rd molars	—	—	18–25

PHARYNX

This is a wide, muscular tube through which both food and air must pass. It is also lined with mucous membrane. The pharynx lies behind the nasal cavities and the mouth and is continuous with the larynx and esophagus. The auditory, or eustachian, tubes from the middle ear open into the muscle layers of the pharynx, which thus regulates the pressure in the middle ear. In the side walls of the pharynx, behind the tongue, are a pair of lymphoid organs, the *tonsils,* which serve as filters and traps for bacteria and toxins. The *adenoids* are lymph nodules in the nasal part of the pharynx.

TONGUE

The tongue is composed mainly of striated muscle, covered by a mucous membrane that is studded with large projections, the *papillae.* The papillae give the tongue its characteristic roughness. They vary somewhat in shape but consist basically of a connective tissue core covered by epithelium. In certain of these papillae are found the *taste buds,* or receptors for the nerves of taste. Taste buds are found also on the pharynx and soft palate and in the walls of the cheeks. They are discussed in more detail in Sec. 22-7.

Salivary Glands

Three pairs of large glands empty their secretions into the mouth. The smallest of these, the *sublingual,* lie in the floor of the mouth, and their ducts lead into the mouth under the tongue. The *submaxillary* glands are partly covered by the mandible on each side of the mouth, and their ducts also open under the tongue. The largest of the salivary glands are the *parotids,* which are wedged into a hollow behind the mandible, extending also into the region in front of the ear. The parotid duct opens into the mouth opposite the second upper molar tooth, and the little papilla where it opens is quite easily seen. (The facial nerve runs through the parotid gland, and any infection of the gland is likely to cause irritation of the nerve.) These glands are shown in Fig. 17-6.

There are two types of cells in these glands, the *serous cells* that secrete the watery enzyme-containing juice and the *mucous cells* that produce the viscous mucus. The glands are richly supplied by blood vessels, and they are innervated by both sympathetic and parasympathetic nerves.

Table 17–2. **Reflex Pathway for Salivation**

Stimulus	Presence of food or other object in mouth
Receptor	Nerve endings in the mouth
Afferent neurons	Trigeminal (fifth cranial) nerve Facial (seventh cranial) nerve Glossopharyngeal (ninth cranial) nerve Vagus (tenth cranial) nerve
Synapse	Salivary centers in medulla
Efferent neurons	Craniosacral nerves (parasympathetic) within the seventh and ninth cranial nerves
Effector	Salivary glands
Response	Secretion of saliva
Effect	Enzyme-containing lubricant poured into mouth

MUMPS

Mumps is a virus disease affecting the salivary glands, chiefly the parotids. One or both of these glands may swell up as the tissue responds to the virus with an accumulation of fluid, leukocytes, and macrophages. Other glands may become involved, especially the pancreas and the testes. If the inflammation is particularly intense in the testes, adult males may become sterile.

Regulation of Salivary Secretion

The amount and composition of saliva vary with the nature of the stimulus. Parasympathetic stimulation causes vasodilation and the secretion of a large volume of serous fluid by the parotid glands, and a large volume of mucous juice from the submaxillary glands. Sympathetic stimulation causes vasoconstriction and a scanty secretion containing mostly mucus.

The total amount of saliva produced per day varies enormously but may reach 1 liter in a normal individual. There are many physical stimuli for the reflex evocation of salivary secretion. Any foreign body in the mouth will induce salivation, but the amount and composition will vary with the object. Sucking a lemon results in a copious secretion, whereas sand almost completely inhibits it. Meat in the mouth induces the secretion of a juice very rich in enzymes and comparatively poor in mucus.

The actual presence of food in the mouth may be unnecessary for salivation; the effect of the physical object may be induced by any stimulus associated with food, such as the appetizing sight, smell, or even thought of a favored dish. The replacement of the original physiological stimulus by an associated one results in the *conditioned reflex,* and the end result is the same, or very nearly the same as that of the unconditioned reflex. The conditioned reflex was first demonstrated by the Russian physiologist Pavlov, who induced salivation in dogs by the ringing of a bell alone, after the bell had been rung consistently at the time of presentation of food to the animal. Similarly, stimuli associated unpleasantly with food, such as a dirty plate or a putrefying odor can completely inhibit salivation. The pathways involved in the salivary reflex are shown in Fig. 17-6 and Table 17-2.

COMPOSITION AND FUNCTIONS OF SALIVA

As has been discussed above, the composition of saliva varies considerably, but it consists mainly of a *watery base* in which ions are generally present in the same amounts as in extracellular fluid; *salivary amylase* (ptyalin), the enzyme that hydrolyzes starch to maltose; *mucus,* a glucose-protein compound that lubricates the passage of the chewed, moistened food ball (the bolus) into the esophagus. The pH of saliva is almost neutral or slightly alkaline and when secreted in large amounts will decrease considerably the acidity of the stomach.

Other functions of saliva, apart from digestive and lubricative, include the protection of the delicate mucous membrane of the mouth by keeping it moist, the cleansing and bactericidal effect on the teeth and other structures of the mouth, and the moistening of the tongue and cheeks that is essential for their movement in speech. Saliva is also necessary to dissolve substances to be tasted. Salivary secretion may cease almost entirely following water loss, and the subsequent drying of the mouth gives rise to the sensation of *thirst.*

DIGESTIVE ACTION OF SALIVARY AMYLASE

The human diet may contain carbohydrates in the form of the polysaccharides glycogen and starch, the disaccharides sucrose (cane sugar) and lactose (milk sugar), and the naturally occurring monosaccharides glucose, fructose, and galactose. In addition, the product of anaerobic fermentation of glucose, alcohol, may be ingested. Our diets also contain a certain amount of the polysaccharide cellulose, but as the digestive system of man, unlike that of bovines, possesses no enzymes able to digest cellulose, it cannot be considered a food. Most natural starches are protected by a thin cellulose coat that must be destroyed by cooking before the starch can be digested.

The monosaccharides and disaccharides are unaffected by salivary amylase. Alcohol passes undisturbed to the stomach, where it is rapidly absorbed. *Starch and glycogen,* the polysaccharides, are the substrates upon which salivary amylase acts.

Salivary amylase, the only enzyme present in saliva, acts upon starch to hydrolyze it to *maltose*, a disaccharide. Because food is swallowed quite rapidly, not much hydrolysis occurs in the mouth itself, but the amylase activity continues with the bolus until the low pH of the gastric juices inactivates it. Despite the continuation of amylase activity within the stomach, only about 40 per cent of the carbohydrate is hydrolyzed before the enzyme is inactivated. All further carbohydrate digestion takes place in the small intestine.

Swallowing (Deglutition)

After the food has been formed into a bolus by the muscles of the mouth and tongue, it is propelled to the back of the tongue, which is depressed while the anterior portion of the tongue is raised, closing off the front of the mouth. The tongue moves upward to push the bolus sharply into the pharynx. This first stage of deglutition is under voluntary control. (Try swallowing without moving the tongue.)

Once the food is in the pharynx, the muscles of the tongue prevent the bolus from returning to the mouth; the bolus is prevented from entering the nasal areas by contraction of the muscles of the palate and the uvula, which raises the soft palate. (When these muscles are paralyzed, as they sometimes are by the diphtheria toxin or the polio virus, liquid food may come out through the nose, rather than going down the esophagus.) The third possibility, entrance into the larynx, is prevented by the elevation of the larynx against a cartilaginous flap, the epiglottis. This closes off the respiratory tract, and the food is propelled into the esophagus (Fig. 17-7). The opening to the esophagus is guarded by the *upper esophageal sphincter,* whose muscles relax during the first part of swallowing, opening the sphincter to permit the passage of the bolus from the pharynx into the esophagus. This stage is purely reflex, as is the next—the passage of the bolus through the esophagus to the stomach.

This third phase, the esophageal phase, is helped by peristaltic waves along the esophagus. Gravity normally assists the passage of the food, although humans can swallow when standing on their heads. Liquids pass rapidly down the esophagus, so that corrosive fluids that are swallowed damage mainly the upper end of the esophagus and the cardia, the entrance to the stomach.

Respiration is completely inhibited during swallowing, and this reflex inhibition is closely associated with the reflex control over the pharynx and esophagus (Table 17-3).

Table 17-3. **Reflex Pathway for Swallowing**

Stimulus	Presence of food bolus in pharynx
Receptor	Nerve endings in pharynx and soft palate
Afferent neurons	Glossopharyngeal (ninth cranial) nerve Trigeminal (fifth cranial) nerve
Synapse	Swallowing center in medulla
Efferent neurons	Glossopharyngeal nerve Vagus nerve
Effector	Muscles of pharynx and larynx Muscles of soft palate, pharynx, and upper esophagus
Response	Soft palate and uvula pulled up to close off nasal passages Closure of the trachea through elevation of larynx, pressure of epiglottis against the glottis, and closing of vocal folds Relaxation of upper part of esophagus
Effect	Propulsion of food from the pharynx into the esophagus

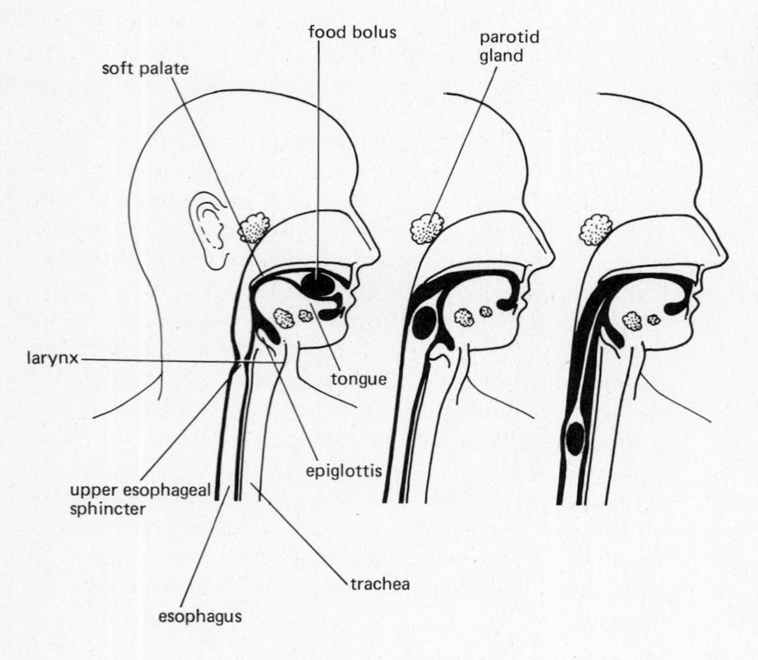

Figure 17-7. **Passage of the food bolus from the mouth into the esophagus. Note the changes in the position of the tongue, the movements of the larynx, and the peristaltic movement in the esophagus.**

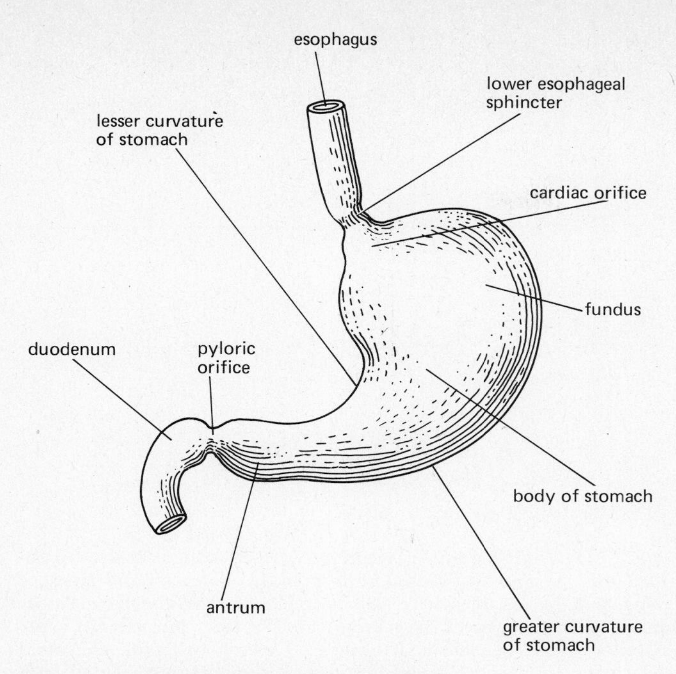

Figure 17-8. **The stomach and its openings.**

17–3 The Esophagus

The esophagus is a tube that connects the pharynx to the stomach; it is specialized only for the propulsion of the bolus. The upper esophagus consists of striated muscle, partly under voluntary control. The rest of the esophageal muscle is smooth and reflexly controlled.

The mucosa of the esophagus produces large amounts of mucus, important for lubrication and the protection of the esophagus. No digestive enzymes are secreted by the esophagus.

Between the esophagus and the stomach is the *lower esophageal sphincter* (Fig. 17-8), which is normally tightly closed. But when peristalsis commences in the esophagus, the peristaltic wave is preceded by a wave of relaxation, which relaxes the sphincter, the stomach, and even the upper part of the small intestine. Consequently, there is no impediment to the passage of the bolus into the stomach. The lower esophageal sphincter acts as a one-way valve, permitting food to enter the stomach but preventing reflux of the stomach contents back into the esophagus.

17–4 The Stomach

Functions of the Stomach

The stomach is not an indispensable organ: it can be removed, and, provided that the individual eats frequent, small meals, a normal life may be pursued. However, the normal functions of the stomach include *storage* of food and the transformation of the bolus into a soft, homogeneous, semifluid mass, the *chyme,* by a combination of mechanical churning and chemical activity. The chyme is released to the small intestine in small spurts, the small volume of which permits proper digestion and absorption in the small intestine.

The third and very important function of the stomach is the initiation of *protein digestion.* The glands of the stomach produce the proteolytic enzyme *pepsin* and the *acid* required for its optimal activity. In addition, other gastric glands produce *mucus,* which protects the stomach from the digestive action of its own juices.

Structure of the Stomach

The stomach is a hollow, muscular organ, which in humans has the shape of a J or a reversed L. The concave border of the J is the *lesser curvature;* the convex forms the *greater curvature* (Fig. 17-8). The stomach lies in the upper left part of the abdominal cavity. It is joined by the esophagus on its right side through the *cardiac orifice*. Above this junction is the rounded portion, the *fundus,* which is usually full of gas and is separated from the heart only by the diaphragm. The main part of the stomach is the *body* (*corpus*). The part of the stomach near its opening into the first part of the small

Figure 17-9. **The gastric glands of the body and fundus of the stomach. The three types of cells—chief, oxyntic, and mucous—are shown.**

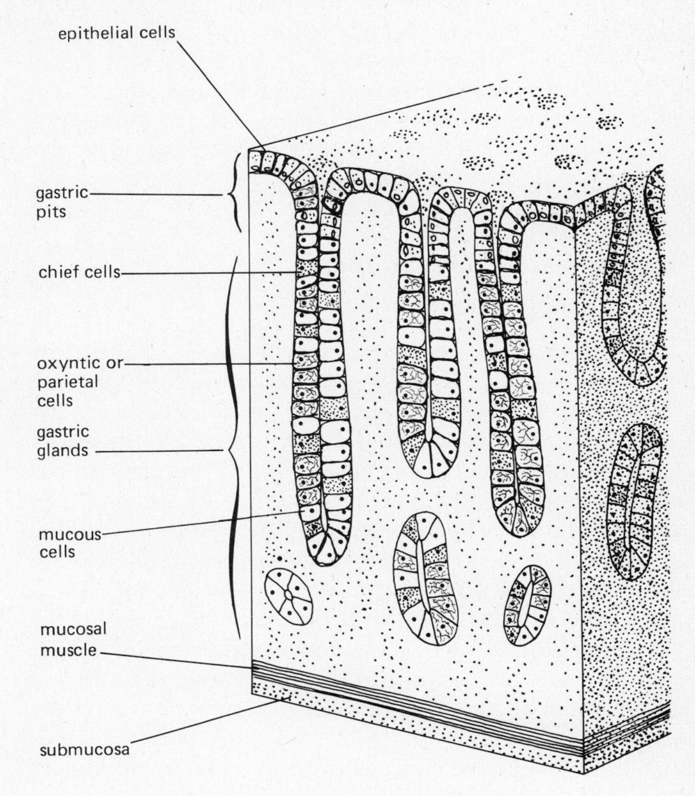

intestine, the duodenum, is the *antrum.* The antrum opens into the duodenum via the *pyloric orifice* guarded by the *pyloric sphincter.* (*Pyloros* means "gatekeeper.")

The stomach is especially well supplied with arteries derived from all three branches of the celiac artery. Parasympathetic fibers from the vagus nerve as well as sympathetic fibers from the celiac plexus innervate it.

Modifications of the basic structural layers in the digestive tube are seen chiefly in the thick *gastric mucosa.* This membrane is thrown into many folds or *pits* when the stomach is empty, and is stretched smooth when the stomach is distended. The gastric mucosa contains two quite different types of glands, the *pyloric glands* in the antrum, which secrete only *mucus,* and the *gastric glands* of the body and fundus, which produce both the *digestive juices* and the protective *mucus.*

The gastric glands consist of three kinds of cells (Fig. 17-9):

1. *Chief cells.* Large pyramidal cells, filled with secretory granules that contain pepsinogen, the precursor of the enzyme pepsin.
2. *Oxyntic cells* (parietal cells). Pale oval cells, most numerous in the neck region of the gland. The oxyntic cells secrete *hydrochloric acid.* The exact mechanism for the production of this acid is not known, but the presence of carbon dioxide is essential for the process. It probably consists of the formation of carbonic acid from water and carbon dioxide, catalyzed by carbonic anhydrase, and the resulting dissociation of the carbonic acid into bicarbonate and hydrogen ions. This is similar to the process for the transport of carbon dioxide by the red blood cell, already discussed in Sec. 16-15. In the oxyntic cells, hydrogen ions are actively transported through special regions of the endoplasmic reticulum into the lumen of the digestive tract.
3. *Mucous cells.* These cells are filled with pale transparent globules of mucus.

Movements of the Stomach

Hunger contractions

Contractions of the gastric antrum can be recorded by having the subject swallow a small balloon or pressure transducer, attached to a recording device. In the empty stomach, the muscles of the walls are contracted, practically obliterating the lumen, and waves of vigorous contractions occur at intervals which are multiples of 21 seconds. The arrival of food causes the reflex relaxation of these muscles, as we have seen. This relaxation ensures that the volume of the stomach cavity increases to adapt to the amount of food entering it, so that there is little pressure change despite the distention. Figure 17-10 shows the disappearance of hunger contractions when recorded from the antrum 1 hr after a normal meal, and the reappearance of contractions after the second hour.

The correlation between gastric contractions and hunger

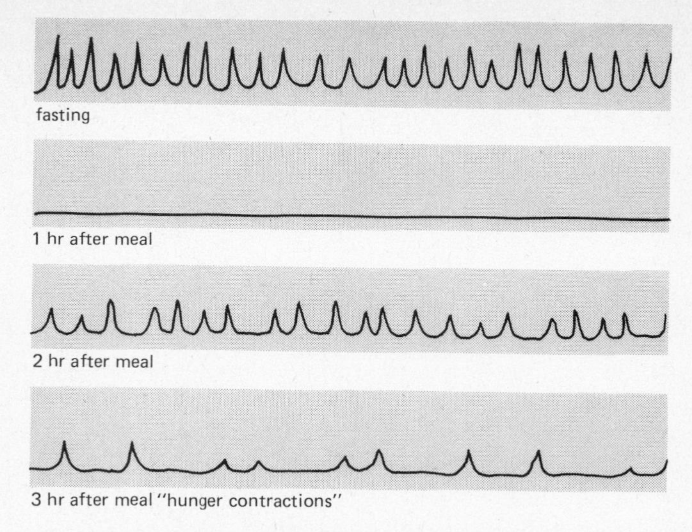

Figure 17-10. **Relationship between gastric contractions, as recorded from a small inflated balloon in the stomach, and "hunger pangs." Although the severe gastric contractions of the fasting state disappear after a meal, they reappear 2 hr later but are not associated with the sensation of hunger. Three hours after the meal the slower, and more irregular, contractions are correlated with hunger.**

pangs is very poor. The sensation of hunger is a complex phenomenon, involving psychological and physiological parameters, some of which are discussed in the chapter on appetite and satiety (Chap. 21).

Contractions of the full stomach

Mixing Waves. Once the food enters the relaxed stomach, mixing waves churn and mix the newly arrived food with older food deposited around the walls of the stomach. Food and gastric juices are mixed and the semifluid chyme is propelled toward the antrum.

Strong Peristaltic Waves. These contractions, probably initiated by the mixing waves, spread from the antrum to the pyloris and even the duodenum, emptying the stomach. These movements are called the *pyloric pump.*

Factors that inhibit the pyloric pump

The rate at which the pyloric pump empties chyme into the duodenum is determined by the fluidity, acidity, and type of food the chyme contains, as well as the amount of chyme already present in the duodenum. The pyloric pump is reflexly influenced by these factors (*the enterogastric reflex*) and hormonally influenced through release of the hormone *enterogasterone.*

The enterogastric reflex, which inhibits gastric peristalsis, is elicited when there is too much chyme in the duodenum, when the chyme is too acid, or when there are irritants or obstructions in the duodenum. This means that gastric emptying is slowed down, giving the duodenum time to neutralize the acid and digest the contents.

The secretion of enterogastrone by the mucosa of the small intestine is evoked by large amounts of fatty acids in the chyme. Enterogastrone is rapidly absorbed by the blood and returned to the stomach, where it slows the pyloric pump, thus allowing more time for digestion of fats in the small intestine.

Factors that speed up the pyloric pump

Distention of the stomach and intestine after a meal initiates *gastrocolic* and *duodenocolic reflexes* that speed up the movement of food through the stomach and the intestine. These reflexes involve the local circuits through the intramural plexus and also the pathways leading from the afferent vagi to the brain stem to stimulate the parasympathetic nerves to the gut. As a result, the presence of a large amount of food in the stomach causes the food to leave the stomach more rapidly and to be propelled through the intestines faster. Eating a large meal, consequently, often is accompanied by a desire to defecate, a response that is particularly marked in infants, who cannot control this reflex voluntarily.

Composition and Functions of Gastric Juice

The fine experimental work of an American surgeon, W. Beaumont, laid the foundation of our modern concepts of gastric secretion. Beaumont had as his patient, and later retained in his service, a hunter, Alexis St. Martin, who had shot himself accidentally in the stomach (in 1822). The wound did not heal properly, and a passage between the stomach and the external wall of the abdomen remained. This type of abnormal opening is known as a *fistula*. Beaumont utilized this fistula to remove gastric juice under controlled conditions and so was able to study its composition and regulation. Pavlov's experiments on dogs with surgical fistulae have extended these findings to detailed descriptions of the nervous mechanisms underlying gastric secretion.

Composition of gastric juice

The fasting stomach of humans secretes about 8 to 15 ml of gastric juice in 24 hr. It is a clear, colorless fluid, isotonic to blood and strongly acid because of its *hydrochloric acid* content. In addition to acid, and to the *mucus* that covers the mucosa with a tenacious, slimy protective coat, gastric juice contains several *enzymes*. The most important of these enzymes is *pepsinogen,* the inactive precursor of *pepsin*.

Pepsinogen, with a molecular weight of 42,500, is split to the active pepsin, which has a molecular weight of 35,000, when it comes in contact with hydrochloric acid. The newly formed pepsin then facilitates the activation of more pepsinogen molecules in the acid medium. Pepsin is maximally active at a pH between 1 and 2, the pH of gastric juice, and loses its enzymatic activity in neutral or alkaline solutions.

Other gastric enzymes include a *gastric lipase,* which is weakly active at the beginning of gastric digestion before much acid is secreted, and *rennin,* an enzyme found in the stomach of calves. Rennin clots milk and speeds up its digestion. Its presence in the stomach of human infants is debatable, and the function of rennin in the human may be taken over by pepsin. The pH of gastric juice in infants is slightly higher than in adults, around 3.5.

Gastric juice also contains an *intrinsic factor,* secreted by the mucosa, which combines with an *extrinsic factor* from digested meat in the stomach to form the antianemic factor (Sec. 14-7).

Protein digestion in the stomach

Both pepsin and hydrochloric acid are essential for protein digestion in the stomach.

Hydrochloric Acid. The acid content of gastric juice not only activates pepsinogen to pepsin, but also denatures the protein substrate (protein in food), thus exposing more bonds to pepsin action. Its extreme acidity also kills ingested bacteria, serving as a bactericidal agent to prevent putrefaction.

Pepsins. There are at least two pepsins active in gastric juice. Their proteolytic action on intercellular collagen fibers and on cell membranes permits the progressive penetration of digestive enzymes into cells. This liberates not only proteins, but also carbohydrates and fats, exposing them to the action of digestive enzymes further in the small intestine. Very little fat digestion occurs in the stomach, and carbohydrate digestion is only a continuation of salivary amylase activity in the bolus, before the pH becomes too acid. The main digestive function of the stomach is the proteolytic breakdown of cells and the digestion of proteins into very large polypeptides. Pepsin does not effectively split proteins into their component amino acids (Sec. 3-3.1) for only about 15 per cent of the digested protein is broken down in the stomach to amino acids. The rest of protein digestion occurs in the small intestine.

Control of gastric secretion

The control of gastric secretion is both nervous and chemical.

Nervous. The *first stage* of gastric secretion is reflexly elicited by the presence of food in the mouth. This physiological stimulus may be replaced by the conditioned stimulus, just as salivation may be stimulated by the sight, smell, or thought of food, or by any stimulus associated pleasantly with eating. Similarly, unpleasant associations will inhibit both salivary and gastric secretion. These conditioned reflexes are not seen in the infant, where association has not yet taken place. A certain degree of development of the cerebral cortex is probably necessary for this association.

Chemical. The *second stage* of gastric secretion is both chemical and nervous. The amount of gastric juice secreted is considerably reduced after the nerve supply (the vagi) to the stomach has been cut, but the presence of food in the stomach, especially meat, will cause the production of a dilute secretion of gastric juice.

In addition, the distention of the stomach and the presence of partly digested protein act as a stimulus on the gastric mucosa to cause the release of a digestive hormone, *gastrin,* which stimulates the secretion of hydrochloric acid by the gastric glands. When gastric pH falls below 2, gastrin release is inhibited, an efficient negative feedback control.

Histamine, released by almost all tissues when they are damaged, is another potent stimulus for the secretion of hydrochloric acid and consequently may play a role in ulcer formation, as discussed later.

Table 17–4. **Reflex Pathway for Vomiting**

Stimulus	Irritants in stomach or intestine
Receptors	Nerve endings in mucosa of gastrointestinal tract
Afferent neuron	Vagal and sympathetic fibers from digestive tract
Synapse	Vomiting center in medulla
Efferent neuron	Vagus nerve Phrenic nerves Motor nerves
Effector	Smooth muscle of stomach Diaphragm Skeletal muscle of abdominal wall
Response	Relaxation of stomach, especially the cardia Contraction of diaphragm Contraction of abdominal wall
Effect	Sharp pressure on relaxed stomach forces contents out through mouth

Summary of Digestion in the Stomach

1. *Protein.* The activation of pepsinogen by hydrochloric acid to pepsin results in the hydrolysis of protein to polypeptides.
2. *Carbohydrates.* No gastric digestion of carbohydrate occurs. There is instead an inhibition of the salivary amylase activity when the acid of the stomach penetrates to the interior of the bolus.
3. *Fat.* Gastric lipase may hydrolyze some emulsified fats to fatty acids and glycerol.

Absorption in the Stomach

Very little absorption occurs in the stomach. Only a few highly lipid-soluble substances, like alcohol and certain drugs, are absorbed.

Vomiting

Vomiting is the coordinated action of the muscles of the stomach, esophagus, and abdominal wall to eject the contents of the stomach through the mouth (Table 17–4). At the same time respiration is inhibited, the glottis is closed, and the soft palate elevated so that the stomach contents do not enter the nose or trachea. It is vigorous pressure exerted by the *contracted abdominal musculature upon the relaxed stomach* that results in the evacuation of its contents through the cardia of the stomach; if the cardia itself does not relax, vomiting is impossible. The cardia does not relax during coughing or defecation, and so the stomach contents remain in place.

The stimulus for vomiting is usually an irritant in the duodenum or stomach, but it can also be due to some drug (tartar emetic, mustard, ipecac) or to the mechanical stimulation of the back of the pharynx. These stimuli result in a sensation of nausea, which precedes the vomiting. Direct stimulation of the vomiting center of the medulla will also initiate the vomiting reflex. This local irritation may be due to excitement of neighboring centers such as occurs in seasickness (during which the vestibular nucleus receives barrages of impulses) or accompanying encephalitis, brain tumors, or meningitis. It is thought that the nausea and vomiting of early pregnancy may be due to the effect of the increased hormone production (especially the chorionic gonadotropin) on the vomiting center; other theories contend that changes in carbohydrate metabolism are responsible, since low blood sugar is often associated with the vomiting. Drinking a sweet beverage before arising often prevents both nausea and vomiting.

Ulcers

Gastric (peptic) ulcers are inflammations of the mucosa and may occur at the gastric end of the esophagus, in the stomach itself, or in the duodenum. The ulcers are associated with excessive secretion of pepsin and hydrochloric acid, acting on an existing area of damaged mucosa. This damage may be a result of nervous stimulation or inadequate circulation. Histamine released by the damaged mucosa may exacerbate the peptic ulcer by further increasing the secretion of hydrochloric acid.

17–5 The Small Intestine, Pancreas, and Liver

One function of the small intestine is to continue the *digestion* of carbohydrate and protein that began in the mouth and stomach, and to digest fats, most of which reach the small intestine completely undigested. These digestive reactions are aided by the exocrine secretions of the pancreas, bile produced by the liver, and the secretions of the intestine itself. The secretions of the liver and pancreas reach the duodenum by way of the common bile duct (Fig. 17-11).

In addition, the small intestine is the chief site of *absorption* of large quantities of water, salts, and the products of digestion. The structure of the small intestine is modified to provide an enormous surface area for these transport processes.

Structure and Glands of the Small Intestine

The small intestine is divided functionally into three parts, but the basic structure of these regions is very similar. The *duodenum* is the first and widest part of the 20-ft-long small intestine. It receives the chyme from the pyloric end of the stomach, and most digestive processes are completed in the duodenum. The remaining two thirds of the small intestine consists of the *jejunum* and the *ileum,* where most of the absorption of the digested food occurs. The coils of the intestine are suspended from the body wall by a double-layered, slippery membrane, the *mesentery,* which holds them in place while it permits peristalsis. Through the mesentery travel the many blood vessels and nerves supplying the intestine. Large numbers of lymph nodes and variable amounts of fat are also found in the mesentery (Fig. 17-11).

SURFACE AREA

The very large surface area of the intestinal mucosa is created by the many surface folds into which it is thrown. These folds are covered with countless fingerlike projections called *villi* (singular *villus*) (Fig. 17-12). A *villus,* which is 0.5 to 1.5 mm long, consists of a central connective tissue core containing blood vessels and a special lymph vessel, the *lacteal.* This arrangement permits rapid passage of absorbed material into the circulation (Fig. 17-13).

Figure 17-11. **Lipid absorption in the duodenum of the rabbit. The bile duct can be seen entering the duodenum in the upper left; the pancreatic duct, unlike that of the human, enters the duodenum in the lower right. The lymphatics beyond the point of the entry of the pancreatic duct are white and filled with absorbed fats. The many lymph nodes can be seen in the mesentery, which also carries blood vessels and nerves.**

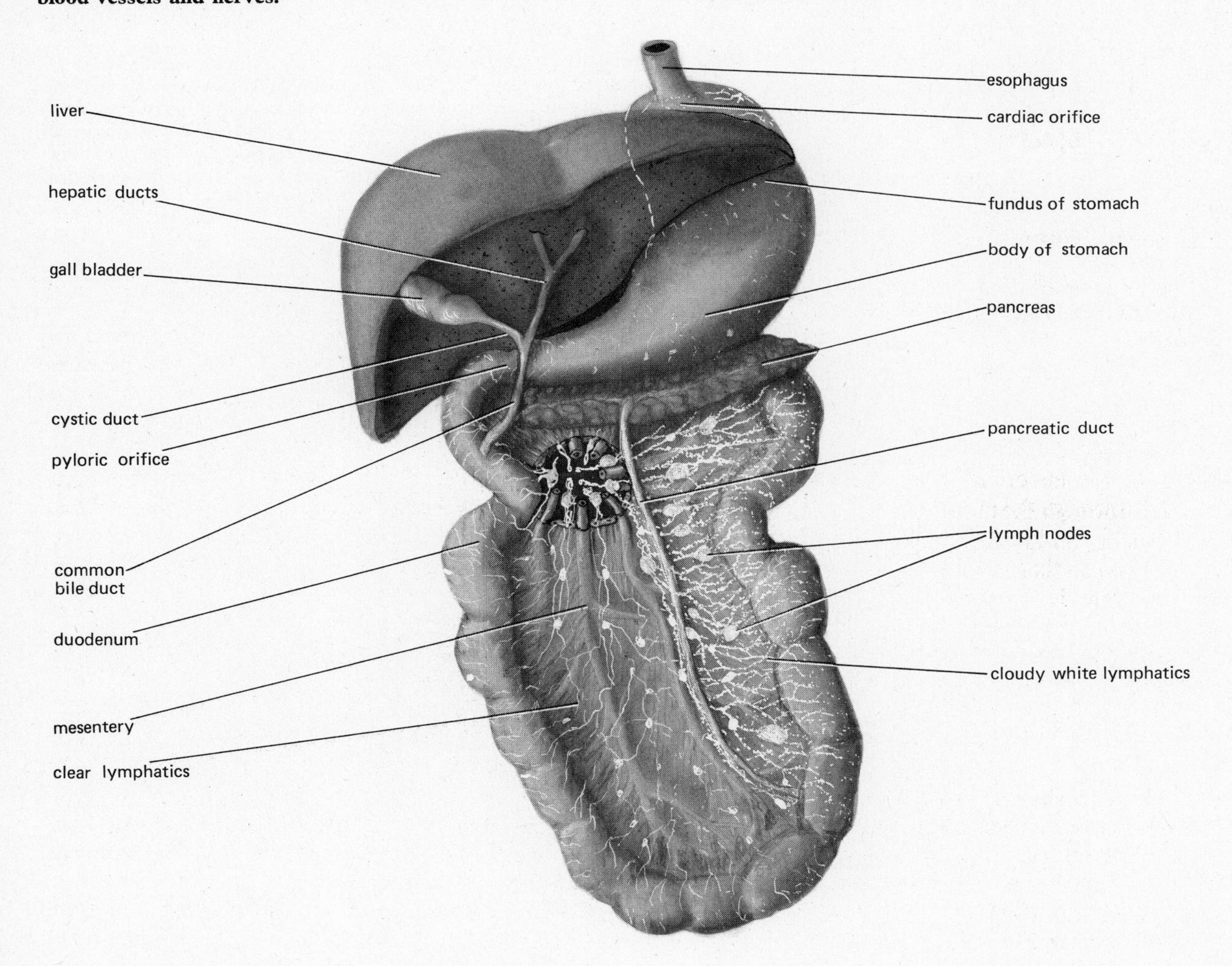

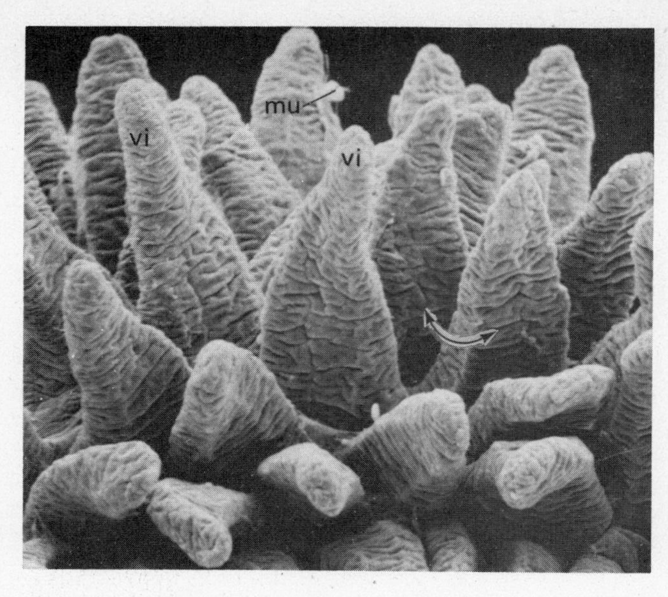

Figure 17-13. **Villi of mouse ileum. vi = villi, mu = mucus. Curved double-headed arrow shows fold on villus. ×120. (From R. G. Kessel and C. Y. Shih, in *Scanning Electron Microscopy in Biology*, Springer-Verlag New York, Inc., New York, 1974.)**

Figure 17-12. **The structure of intestinal villi and a crypt, showing the functional cell types.**

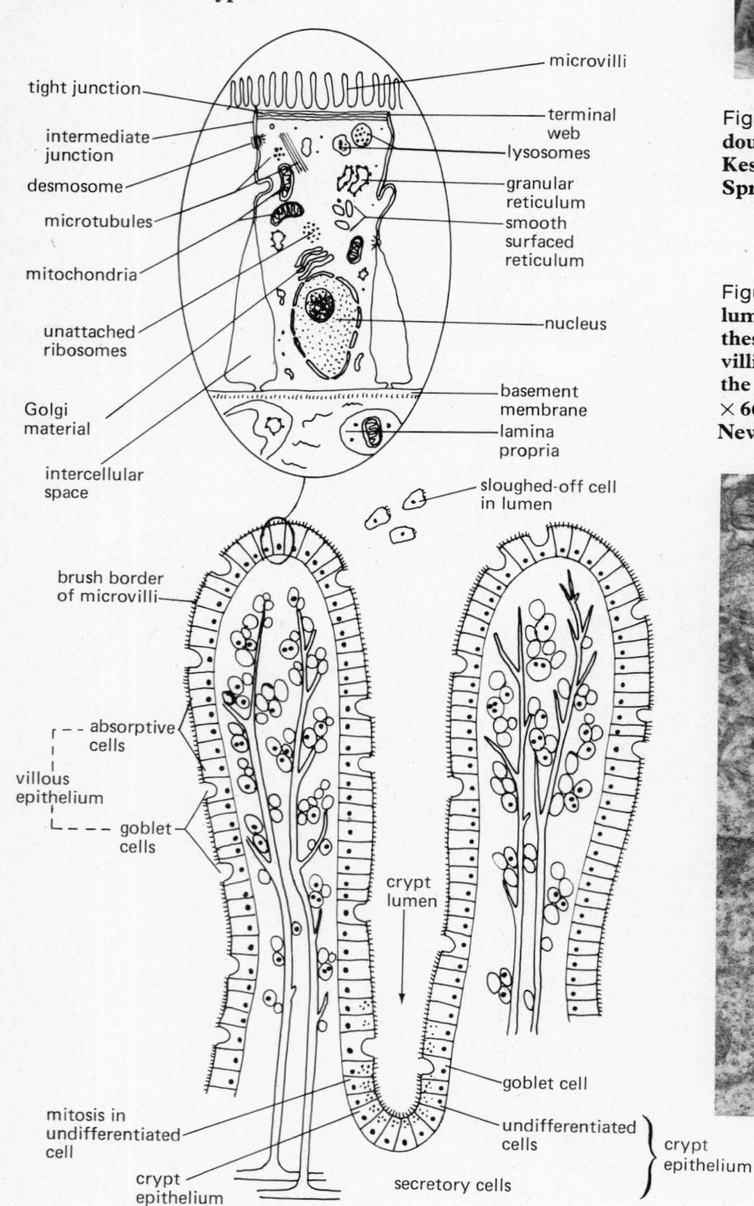

Figure 17-14. **Intestinal cells with microvilli projecting into the lumen of the digestive tract. Note the tight junctions (t) between these cells at their apical ends. Dark areas at the tips of the microvilli probably represent attachment of contractile filaments within the microvilli. mv = microvilli, c = cells and m = mitochondria. × 6600. (Courtesy of A. Moorthy, Laboratory of Cytogenetics, New York University.)**

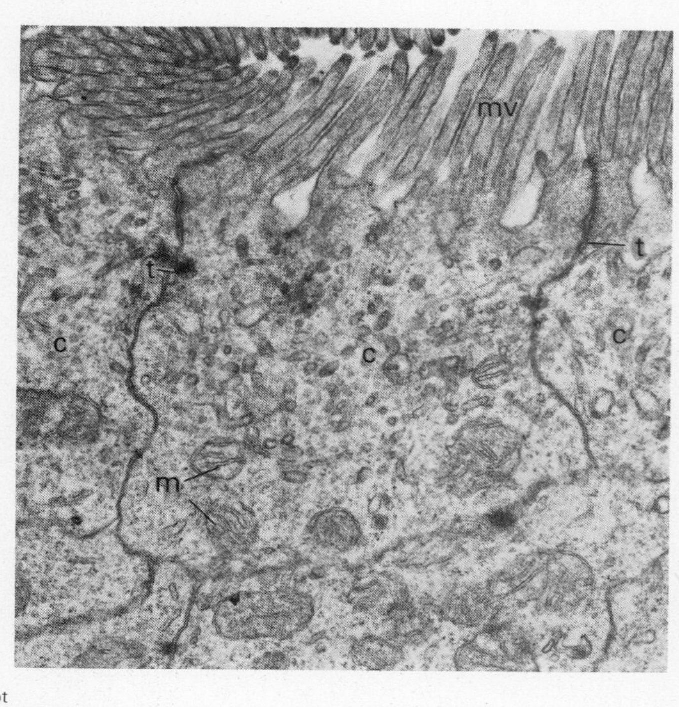

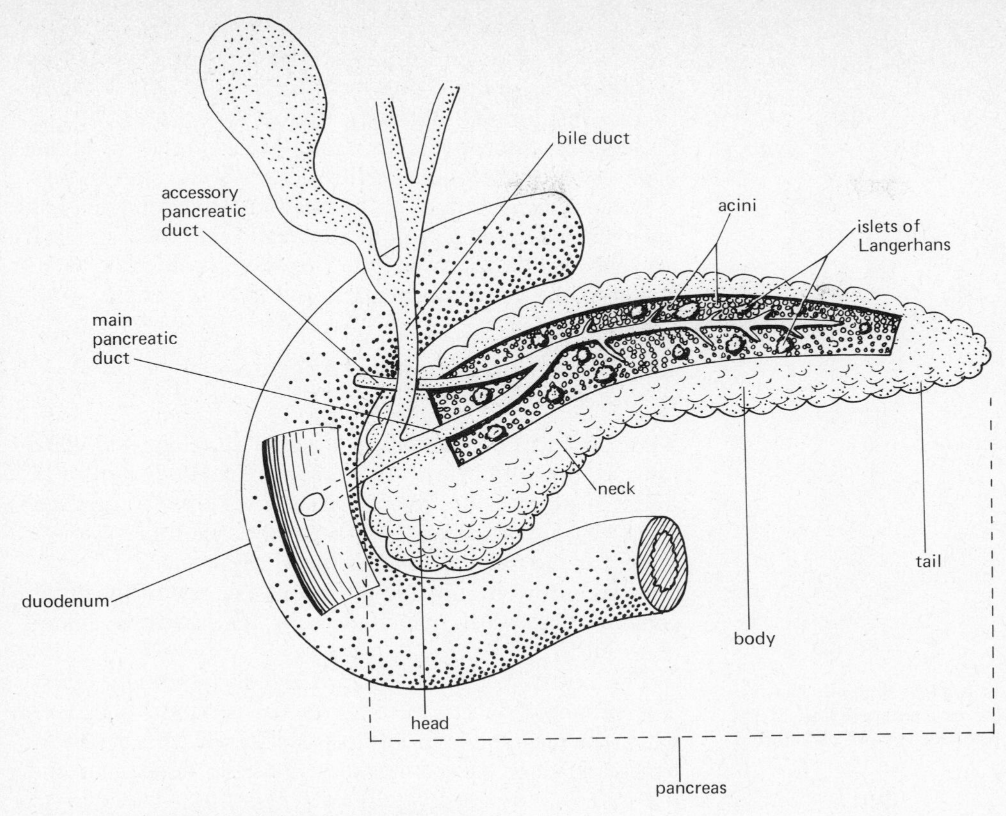

Figure 17-15. **The structure of the pancreas.**

Each villus, the surface area of which is covered with columnar epithelium, has its surface area greatly increased by minute extensions—*microvilli*—of the free border of the columnar cells. These microvilli form a brush border (Fig. 17-14), under which are many mitochondria that supply energy for the active transport processes involved in the absorption. The combination of surface folds, villi, and microvilli results in an increase of the surface area of the small intestine of approximately 600-fold.

Cell types

Epithelium of the Villus. The epithelium of the villus consists mainly of *absorptive cells,* and of *goblet cells* that secrete mucus. Mucus-secreting cells, arranged in glands, are particularly abundant in the upper part of the duodenum, where the mucus protects the duodenal mucosa from the digestive activity of the acidic chyme. These mucous glands are inhibited by sympathetic stimulation; hence this area of the intestine is particularly susceptible to peptic ulcers of nervous origin.

Specialization of the Absorptive Cells

1. The *permeability* of the apical and basilar membranes of the absorptive cells varies considerably; this variation, coupled with active transport mechanisms, is responsible for the preferential transport of many substances in the direction of the serosa rather than toward the mucosa.
2. The apical membrane is *thicker* than most cell membranes, being 10 to 11 nm wide as compared to the usual 7 to 8 nm. The significance of this is unknown.
3. Another specialization of these cells is the fusion of their lateral cell membranes with those of the adjacent absorptive cells to form *tight junctions* near the apical end (Fig. 17-14). Below these apical tight junctions, the intercellular space may be quite wide, so that absorbed substances must pass to adjacent walls *below* the tight junctional area.
4. Isolation of the apical brush border has shown that it contains disaccharidases and probably other *enzymes.* Thus the microvilli not only increase the absorptive surface of the mucosa but are involved in the digestion of certain substances.
5. The microvilli are covered by a *fuzzy surface coat* of mucopolysaccharide, in addition to the mucous coat secreted by the goblet cells. The "fuzz" on the surface appears to contain specific binding sites for certain substances, a fact that may explain the selective absorption of substances in different regions of the gastrointestinal tract.

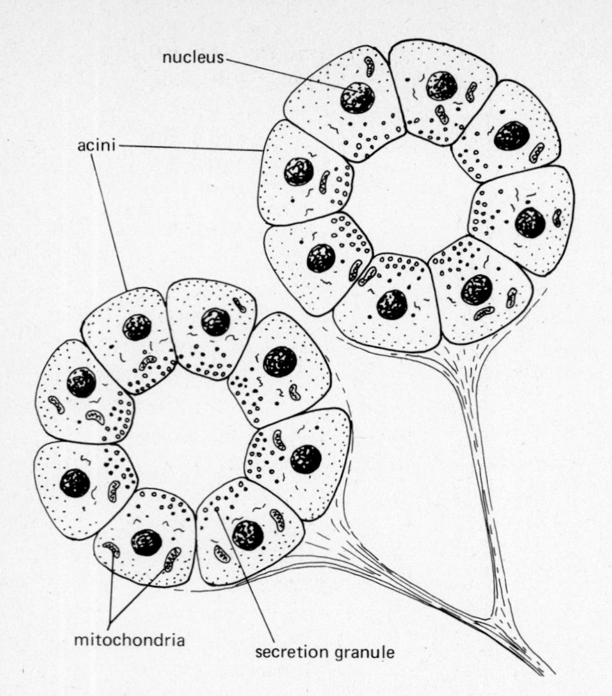

Figure 17-16. **The acini of the pancreas contain the cells that secrete the digestive enzymes. This is the exocrine portion of the pancreas. The enzymes are liberated into the venous circulation.**

Crypts Between the Villi. The crypts between the villi contain *secretory cells* that produce the intestinal juice, *goblet cells* that secrete mucus, and *undifferentiated cells* that proliferate rapidly and migrate to the surface of the villi, replacing the absorptive cells. Complete replacement of the epithelium of the human villus occurs every 5 to 6 days.

Because the secretions of the small intestine act mainly on the hydrolytic products of the pancreatic enzymes, intestinal enzymes will be discussed following the consideration of the action of the various pancreatic and liver secretions.

Structure and Exocrine Secretions of the Pancreas

The pancreas (sweetbread) is a yellowish, elongated gland lying behind and below the stomach (Fig. 17-1). It can be divided superficially into a head, neck, body, and tail (Fig. 17-15). In the tail region are the cell concentrations known as the *islets of Langerhans,* which produce the endocrine secretions insulin and glucagon. The rest of the organ contains the *acini,* which secrete the digestive juices (Fig. 17-16).

The digestive juice is drained by *two pancreatic ducts.* The principal duct is fused with the common bile duct, which pierces the wall of the duodenum; the accessory pancreatic duct opens separately into the duodenum and

Table 17–5. **Action of Digestive Enzymes**

Enzyme	Site of Secretion (fluid volume secreted)	Activity Requirements	Reaction Catalyzed
Salivary amylase (ptyalin)	Salivary glands (1200 ml/day)	pH 6–8 Inactivated by gastric HCl	Starch (cooked) → Maltose
Pepsin	Stomach (2000 ml/day)	pH 1–2 Pepsinogen activated by gastric HCl	Protein → Polypeptides
Trypsin	Pancreas (1200 ml/day)	pH 7–9 Trypsinogen activated by enterokinase	Chymotrypsinogen → Chymotrypsin
Chymotrypsin		pH 7–9 Chymotrypsinogen activated by trypsin	
Lipase		pH 7–9 Bile salts	Fats → Fatty acids + glycerol
Amylase		pH 6–7	Starch → Maltose
Enterokinase	Duodenal mucosa (3050 ml/day)	pH 6–9	Trypsinogen → Trypsin
Disaccharidases			
Maltase Lactase		pH 7–9	Maltose → Glucose Lactose → Glucose + Galactose
Sucrase		pH 7–9	Sucrose → Glucose + fructose
Lipase		pH 7–9 Bile salts	Fats → Fatty acids + glycerol
Peptidases		pH 7–9	Polypeptides → Dipeptides + Dipeptides → amino acids

may become of importance in cases of blockage of the main pathway by a gallstone or other impediment.

Pancreatic juice

This secretion contains such a wide variety of enzymes important in carbohydrate, fat, and protein digestion that it alone could probably support sufficient digestion to maintain life. It also contains a high concentration of bicarbonate ion, which accounts for its alkalinity; this is of great importance in neutralizing the acid chyme entering from the stomach and in providing an optimum pH for enzyme action in the small intestine. Table 17-5 shows the pH optima for the most digestive enzymes.

Pancreatic Protease. Trypsin and chymotrypsin are both produced as the inactive precursors trypsinogen and chymotrypsinogen. On activation, they hydrolyze proteins and very large polypeptides to small polypeptides and dipeptides.

Pancreatic Amylase is responsible for almost 60 per cent of the digestion of carbohydrate, ingested as starch. The remaining fraction is hydrolyzed by salivary amylase. As a result of the activity of these amylases, all polysaccharides are present in the duodenum in the form of *disaccharides.*

Pancreatic Lipase is the most important digestive enzyme for fats. Removal of the pancreas still permits a fair amount of protein and carbohydrate digestion, but almost all ingested fat is eliminated in the feces. Pancreatic lipase hydrolyzes fats to *fatty acids and glycerol,* after the fats have been emulsified through the action of the *bile salts.*

Summary of Action of Pancreatic Digestive Juice

As a result of the hydrolytic action of pancreatic enzymes, following salivary and gastric digestion, all foods have been broken down to polypeptides or dipeptides, disaccharides, and fatty acids and glycerol. These processes are summarized in Table 17-5.

Digestive Secretions of the Liver

The structure and metabolic functions of the liver are discussed in Sec. 19-2. Only the digestive role of the liver is considered here.

Bile

The liver cells continually produce bile, which is both an excretion and a secretion. It is considered an excretion because it contains the *bile pigments* that are excreted in the feces and urine, and a secretion because it contains the secreted *bile salts,* which are essential to fat digestion.

Functions of Bile Salts. Bile salts are formed from cholesterol, combined with certain amino acids in the liver. They act as detergents to lower the surface tension of ingested fat globules and thereby emulsify the fats, increasing the surface area exposed to the hydrolytic action of the digestive lipases.

Bile salts also help in the absorption of fat soluble substances such as cholesterol, lipids, and fatty acids, presumably by adsorbing these molecules and thus changing the electrical charge. This facilitates their passage through cell membranes.

Fate of Bile Pigments. The bile pigments *biliverdin* and *bilirubin,* are breakdown products of hemoglobin and give the bile its characteristic color. In humans, bilirubin is the predominant pigment, so that human bile is golden yellow, and an obstruction of the common bile duct leads to a yellowish color in the tissues, as seen in jaundice.

Normally, bilirubin is converted in the intestine into the more soluble *urobilinogen,* and this pigment, together with other breakdown products, colors the feces brown. Some of the urobilinogen may be reabsorbed by the intestine and recycled to the liver through the circulation, and small amounts may be excreted by way of the kidneys into the urine.

Storage of Bile. Bile is drained from the liver by the hepatic ducts, which then enter into the common bile duct.

Figure 17-17. **The gall bladder, the biliary ducts, and the pancreatic duct.**

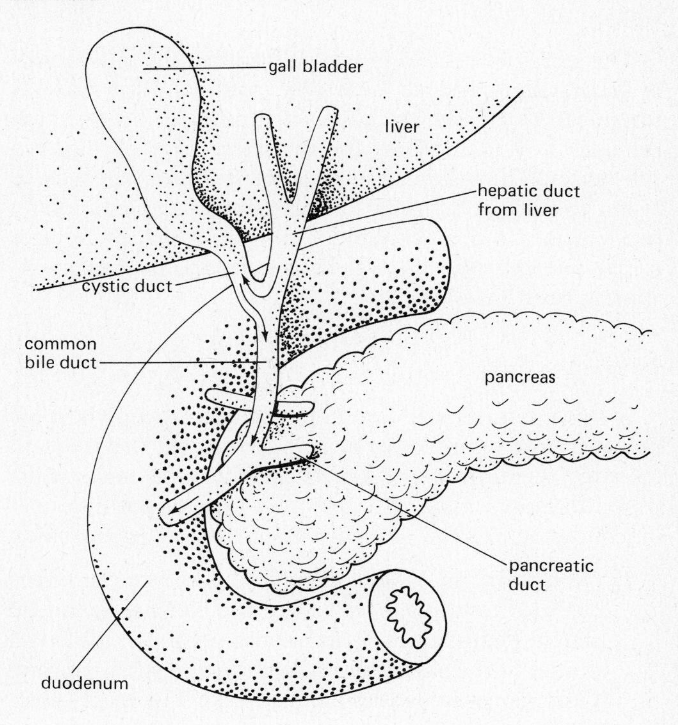

Most of the bile does not go directly into the duodenum but backs up through the cystic duct into the gall bladder, which is a small sac fastened by connective tissue to the liver. Figure 17-17 shows the relationship between these biliary structures, the liver, pancreas, and duodenum.

Bile is stored and concentrated in the gall bladder. The concentrating ability of the gall bladder permits radiological visualization of its position in the living body because an opaque dye, injected into the organism, will be concentrated with the bile and consequently visible on X-ray examination.

Release of Bile. Bile is released spasmodically during meals as a result of contraction of the muscles of the gall bladder. The bile passes through the cystic duct again and through the common bile duct to the opening to the duodenum.

Two digestive hormones, both produced by the duodenal mucosa, are responsible for the release of bile: *secretin* (which also stimulates the pancreas) and *cholecystokinin,* a hormone secreted specifically in response to the presence of fat in the duodenum. Consequently, the amount of bile released is proportional to the amount of fat in the meal.

Secretions of the Small Intestine

The secretions of the small intestine complete the digestive process. Polypeptides and dipeptides are hydrolyzed to amino acids, and disaccharides to monosaccharides; any remaining fat is digested to fatty acids and glycerol. The digestive enzymes of the small intestine are of two main types, extracellular and intracellular.

Extracellular

The secretory cells in the crypts between the villi (crypts of Lieberkühn) produce the watery *intestinal juice,* which is isotonic to blood and which has a neutral pH. This secretion provides a fluid medium for the absorption of digested substances. The intestinal juice, when collected carefully so as not to damage the cells, has been shown to contain only two enzymes, a small amount of *amylase,* which hydrolyzes starch, and *enterokinase,* which activates the pancreatic enzyme trypsin.

Intracellular

If the intestinal cells are not ruptured during their removal from the mucosa, it can be shown that they contain most of the enzymes previously thought to be constituents of the intestinal juice. The secretory cells of the intestinal mucosa contain large amounts of the following enzymes:

1. *Disaccharidases.* These enzymes hydrolyze the disaccharides resulting from pancreatic enzyme action on starch. Digestion of disaccharides occurs on the brush border of the absorptive cells of the intestinal mucosa. Only monosaccharides are absorbed. The main disaccharides are: (a) maltase (maltose hydrolyzed to two molecules of glucose); (b) lactase (lactose hydrolyzed to glucose and galactose); and (c) sucrase (sucrose hydrolyzed to glucose and fructose).
2. *Peptidases.* There are several intestinal peptidases which selectively attack the final peptide linkages between specific amino acids of small polypeptides and dipeptides, releasing free amino acids.
3. *Enteric Lipase.* Lipase hydrolyzes fats to fatty acids and glycerol; these enzymes, like pancreatic lipases, require the action of bile salts.

Table 17-5 summarizes the action and activity requirements of all of the digestive enzymes from the mouth to the small intestine. The shaded portion of the last column of this table shows the final products of digestion, molecules small enough to be absorbed by the intestinal mucosa.

Control of Secretion of Digestive Enzymes into the Duodenum

The main stimulus for the release of digestive juices from both the small intestine and the pancreas is the presence of chyme in the duodenum. As the *acid* chyme from the stomach passes into the duodenum, at rates dependent on the composition of the chyme, it stimulates the release of the polypeptide hormone *secretin* from the duodenal mucosa. (Some secretin may also be released through reflex stimulation.) This hormone is absorbed through the mucosa into the blood and circulated to the pancreas, where it causes the release of a copious flow of *alkaline* fluid. At the same time, the *food contents* of the chyme provoke the release of the

Figure 17-18. **Factors that regulate the secretion of digestive enzymes into the duodenum.**

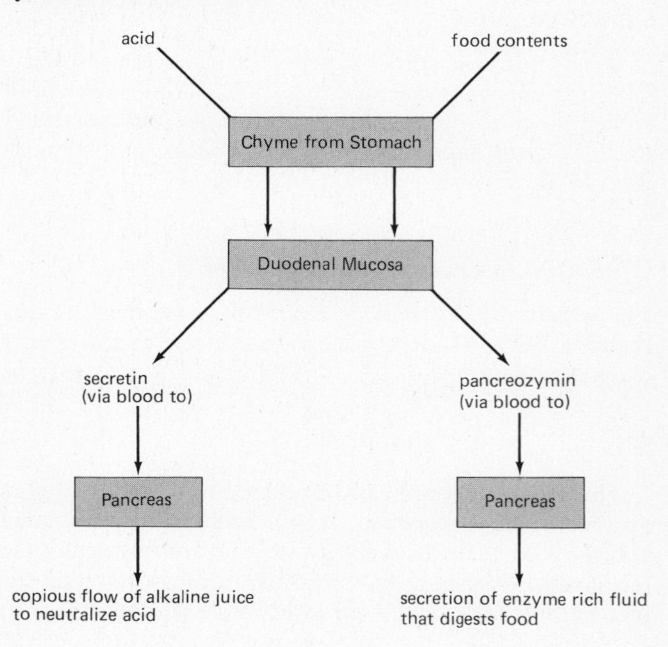

hormone *pancreozymin,* also from the duodenal mucosa. This hormone also circulates through the blood ultimately to reach and stimulate the pancreas to produce an *enzyme-rich* secretion.

Note that the nature of the substances within the duodenum determines the type of duodenal hormones secreted and the composition of the pancreatic juice released. Acid released from the stomach between meals evokes merely a copious flow of alkaline juice from the pancreas to neutralize the acid; following a meal, an enzyme-rich alkaline fluid is secreted into the duodenum (Fig. 17-18).

The control of the secretion of intestinal juice is less clearly understood. Apparently, the presence of food in the intestine acts as a local stimulus for the secretion of intestinal juice—perhaps through the mediation of a hormone, *enterocrinin,* perhaps through local reflexes set up in the myenteric plexus. It is possible also that the composition of the digestive juice varies to some extent according to the type of food in the intestine.

Table 17–6. **Relative Rates of Absorption of Sugars from Small Intestine***

		Normal	Poisoned
Galactose	(C_6)	115	53
Glucose	(C_6)	100	33
Fructose	(C_6)	44	37
Mannose	(C_6)	33	25
Xylose	(C_5)	30	31
Arabinose	(C_5)	29	29

* Selected data taken with permission from Wilbrandt and Laszt, *Biochem. Z.* **259:**398, Springer, Berlin, 1933.

17–6 Absorption in the Small Intestine

Nonelectrolytes

SUGARS, SMALL PEPTIDES, AND AMINO ACIDS

Active transport is the major mechanism for the absorption of nutrients in the small intestine. The cell membrane acts as an effective barrier for nonlipid soluble particles with molecular weights over 80. The absorptive cells of the small intestine transport nutrients from the lumen into the blood even if the nutrient concentration is greater in the blood than in the lumen. This one-way transport can be shown in vitro, by a technique devised by Wilson (1954) in which everted intestinal sacs are used. As a result of eversion, the serosal side of the intestine is turned to the inside of the sac; consequently, if substances are transported in the normal direction (mucosa to serosa), they will become concentrated within the sac (Fig. 17-19).

The everted sacs are filled with a test solution, tied off, and incubated under aerobic conditions to test the passage of small particles into the sac. By using this technique, it can be shown that glucose, galactose, and most other monosaccharides are actively transported into the sac. Recently, it has been demonstrated that not only free amino acids but also dipeptides and tripeptides are absorbed by the mucosa by active transport mechanisms. Some interesting characteristics of the intestinal active transport processes have been demonstrated by testing the sacs in the presence of amino acids, small peptides, and sugars.

Characteristics of Intestinal Active Transport. CARRIER MECHANISM. *Selective transport.* Sugars, dipeptides, and tripeptides, and amino acids are transported through the membrane by a carrier in the membrane. Experiments show that there is a selective transport of certain sugars, with galactose and glucose being transported most readily. Similarly, there is a selective transport of particular small peptides and amino acids. Table 17-6 shows the relative rates of absorption of sugars of different sizes from the small intestine. If absorption were purely a matter of diffusion, the small C_5 sugars (xylose and arabinose) should pass through faster. However, galactose and glucose obviously are absorbed much more quickly. The rate of absorption of these two sugars is drastically lowered when the gut is poisoned; but some absorption still occurs, indicating that the normal passage of glucose and galactose involves both diffusion and active transport.

Competition for the carrier. Glucose transport is markedly slowed if large amounts of other sugars are added to the system. Sugars, especially galactose, also inhibit the transport of amino acids. Amino acids, in turn, inhibit sugar

Figure 17-19. **The tied, everted, mucosal intestinal sac. When this sac is placed in an oxygenated, buffered solution at 37°C, substances in the solution that normally are transported actively from the inner mucosal surface to the outer serosal surface are now moved in the opposite direction. This permits easier measurement of the concentration of substances within the sac.**

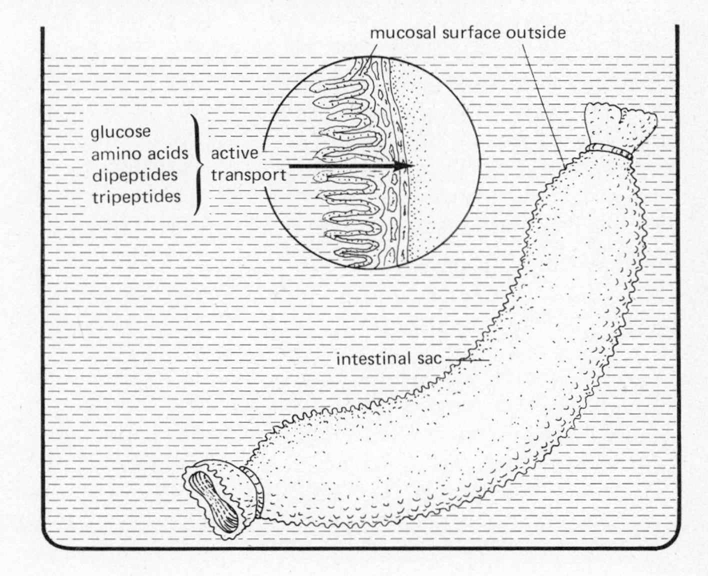

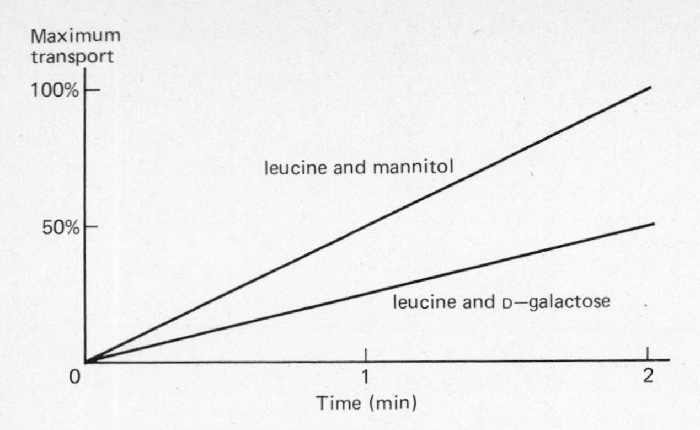

Figure 17-20. **Competition of an amino acid (leucine) with a sugar (D-galactose) for a common carrier for active transport. Mannitol is an inert carbohydrate that is not transported by the gut.**

transport. These observations indicate a competition between individual sugars and amino acids for a particular carrier (Fig. 17-20).

Because it is difficult to imagine that such structurally different substances as a sugar and an amino acid share a common binding site, the attractive suggestion has been made that the binding sites are interrelated in the membrane surface, like pieces of a mosaic, to form a *polyfunctional unit.* An excess of any substance capable of binding this complex will disrupt the stereogeometry of the polyfunctional unit and prevent the efficient binding of other substances present in lesser amounts.

Independent uptake of small polypeptides and amino acids. The absorptive capacity of the small intestine is greater for mixtures of amino acids and peptides than for either amino acids or peptides alone. This indicates that these substances are taken up quite independently by the intestinal mucosa and do not share a common carrier.

SODIUM-DEPENDENT TRANSPORT. As active transport is a translocation of a solute from a low energy state to a higher energy level, energy must be supplied to the system. The most important energy source for active transport processes is ATP, and lack of oxygen or the presence of metabolic inhibitors, such as cyanide or dinitrophenol, will effectively block sugar and amino acid transport. This transport is also dependent on sodium: if sodium ions are eliminated from the system, sugar and amino acid transport is abolished.

FATS

Unlike sugars and amino acids, which cannot diffuse through the cell membrane of the epithelial cells of the intestinal mucosa, fatty acids and glycerol, which are highly lipid soluble, become dissolved in the membrane of the brush border of these cells and diffuse into the cell. These fatty acids and glycerol molecules are then adsorbed on to the endoplasmic reticulum membranes and resynthesized into triglycerides (see Sec. 18-4). The triglyceride molecules are packaged by these cells into small fat droplets about 0.5 μm in diameter, together with phospholipid and cholesterol, and coated with a thin layer of protein. The complete protein-covered droplet is a *chylomicron,* and it is relatively stable in aqueous solution because of its hydrophilic protein covering.

Chylomicrons are released by the epithelial cells through their lateral cell membranes, below the level of the apical tight junctions discussed earlier (Fig. 17-21). This demonstrates clearly the large differences in permeability in cell membrane in different regions of the same cell. Fats have to be hydrolyzed to fatty acids and glycerol before they can enter the brush border of the epithelial cells, yet the much

Figure 17-21. **Fat globules are digested outside the intestinal villi into fatty acids and glycerol, which then penetrate into the absorptive intestinal cell. In the cell they are resynthesized to triglycerides and covered with a layer of protein, to form chylomicrons. The chylomicrons pass through the lateral spaces into the lacteals. Amino acids and monosaccharides pass directly into the venous capillaries of the villus.**

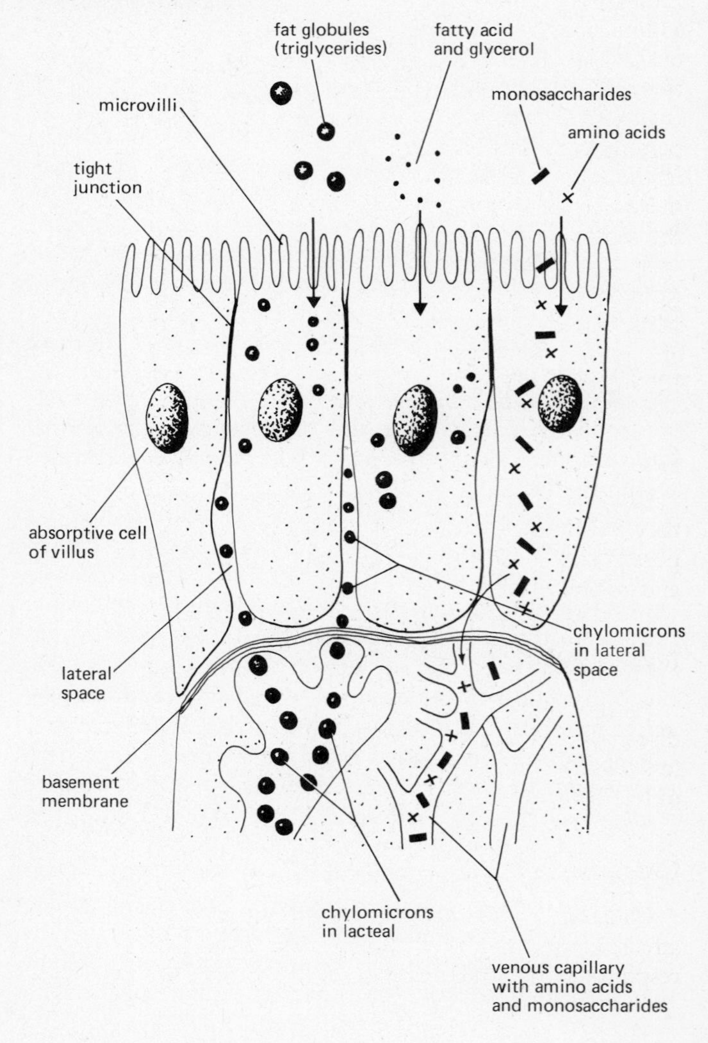

larger chylomicrons escape easily through the lateral cell membranes. From there they pass into the central lacteal of the villus, to be transported through the lymphatic system into the circulation (see Sec. 18-4). After a fatty meal, the lacteals are filled with a milky suspension of chylomicrons; it is from this milky appearance that the lacteals get their name.

Electrolytes

Sodium

Absorption of sodium occurs by *passive diffusion* when the electrochemical gradient is in the direction from lumen to blood, and by *active transport* when the electrochemical gradient is from blood to lumen, or "uphill." In the human ileum, active absorption of sodium ions can be measured by determining the rate of movement of sodium across pieces of isolated intestine. This movement must be measured in both directions to determine the *net flux*.

flux (lumen to blood) − flux (blood to lumen) = net flux

Passive Diffusion. Figure 17-22 shows the movement of sodium across the intestinal mucosa when there is no previously established electrochemical gradient. Loops of intestine are filled with isotonic saline to which a small amount of radioactive sodium (^{24}Na) has been added. Over a 30-min period, there is no change in the total amount of sodium ions in the loop, but the amount of ^{24}Na decreases. This must mean that under these conditions, sodium enters the loop in the same amounts as it leaves.

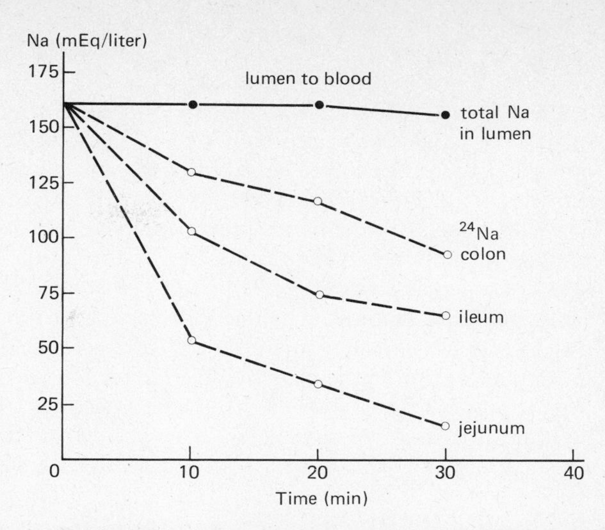

Figure 17-22. **Movement of sodium across the intestinal mucosa occurs in both directions. Intestinal loops were filled with isotonic saline, containing trace amounts of radioactive sodium (^{24}Na). Although the total Na content of the loops remains constant, the ^{24}Na rapidly falls, showing that the amount of sodium that enters the loop must be the same as the amount that leaves. [From M. B. Visscher, et al., *Amer. J. Physiol.* 141: 488 (1944).]**

Active Transport. If the intestinal loops are filled with saline of increasing sodium concentration, active absorption of sodium occurs up to a concentration gradient of 100 mEq/liter and an electrical gradient of 5 to 15 mV.

Chloride

Chloride ions are absorbed both by passive diffusion as they follow the electrical gradient established by the transport of sodium ions, and also to a certain extent (in the dog and human, but not in many other species) by active transport.

Potassium

Potassium moves across the intestinal mucosa in both directions as does sodium, but the potassium flux is much less than that of sodium, and potassium absorption does not differ in the ileum and jejunum.

Calcium

Calcium absorption is most rapid in the duodenum and is affected by many factors that regulate calcium absorption in response to the metabolic status of the organism. Calcium absorption varies with the *dietary intake* of calcium and the plasma levels of calcium. *Age* also affects calcium absorption: sacs of intestine taken from 1-month-old rats can absorb calcium many times faster than can intestinal sacs from older rats. Similar experiments show that *pregnancy* increases the rate of calcium absorption.

Parathyroid hormone is another factor that regulates calcium absorption from the intestine, doubling the concentration gradient against which calcium can be moved. Because it is the calcium level of the plasma that regulates the secretion of parathyroid hormone, this negative feedback mechanism delicately stabilizes the calcium level of the blood.

Vitamin D also enhances calcium absorption, and this vitamin is essential for the active transport process specific for ionized calcium. This transport system does not move other divalent ions, like Mg^{2+}.

Iron

Iron is absorbed in the duodenum and jejunum and stored briefly in the mucosal cells before being actively transferred to the plasma. This storage appears to be a homeostatic device for delivering the right amount of iron to the plasma in response to metabolic requirements. If the need for iron is high, for example during pregnancy or menstruation, or following other hemorrhages, much of the iron moves from the mucosal cells into the plasma. When the need is low, iron remains in the cells. This stored iron is soon lost, because the mucosal cells have a very short life before they are sloughed off into the feces.

Iron is absorbed much more efficiently in the ferrous

form than in the ferric. *Ascorbic acid* reduces iron to the ferrous form and therefore increases the rate of iron absorption. Once inside the cell, iron may be stored diffusely through the cytoplasm or in a combined form with a protein to form the complex *transferritin,* in which form iron circulates through the blood. Transferritin is stored in many other tissues, especially the liver, spleen, and bone marrow. Accumulation of transferritin results in the formation of aggregates known as *hemosiderin,* which can be stained and seen in histological sections. The finely dispersed ferritin is not visible. Release of iron from the stored form is regulated by the amount of iron in the plasma, which in turn regulates the rate at which iron is absorbed from the intestine.

The rate of iron absorption is really very slow, so that only a small amount is absorbed from a meal rich in iron.

COPPER

The upper part of the small intestine absorbs most of the ingested copper, and the amount absorbed increases with increasing copper concentration, up to a maximum of about 40 per cent of the daily intake. The rest is excreted in the feces. In a genetically determined disease, copper may accumulate in many tissues, especially the liver, the lens of the eye, and the central nervous system because of an inability to excrete excess copper.

BICARBONATE

The net flux of bicarbonate depends upon the concentration gradient between the lumen and plasma. When bicarbonate in the lumen is high, it is absorbed by the mucosal cells; when blood bicarbonate is high, bicarbonate is secreted into the lumen. Other factors, such as a coupling of bicarbonate secretion with sodium absorption to maintain electrical neutrality, are also involved.

PHOSPHATE

Phosphate is absorbed both actively and passively by all parts of the small intestine.

Absorption of Water

Of the 5 to 10 liters of fluid that enter the small intestine, only about half a liter reaches the large intestine. Water is passively absorbed by the cells of the small intestine, following the absorption of the osmotically active constituents of digestion. This keeps the intestinal fluid approximately isotonic to blood. Most of the water is absorbed by the duodenum.

Changes in Absorption Characteristics of the Intestine with Maturation

Young animals cannot absorb glucose as rapidly as older animals do when the concentration of glucose rises markedly in the intestine. This inability may be caused by differences in the concentration of certain enzymes (carrier molecules) within the mucosal cells. Another interesting characteristic of the very young intestine is illustrated by Fig. 17-23. Large molecules, such as proteins and polypeptides, normally do not pass through the adult intestinal wall. When insulin is administered orally to 10-day-old rats, it effectively lowers blood sugar, indicating that it has passed through the intestinal mucosa. It is without effect on the blood sugar of adult rats. A rapid maturation of the young intestine occurs when cortisone is administered; it becomes impermeable to insulin and to antibodies, and the rate of glucose absorption increases considerably.

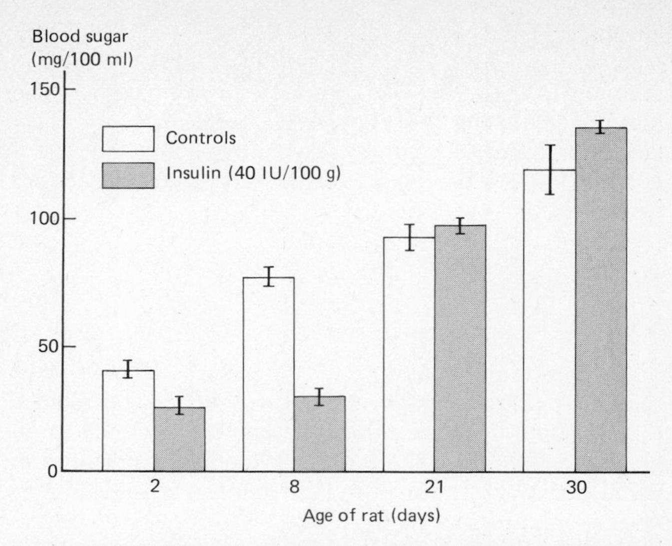

Figure 17-23. **Effect of oral administration of insulin on the blood sugar of rats of different ages. Note that oral administration is effective only in the very young animal. (From O. Kodolvsky, E. Faltova, P. Hahn, and E. Vacek, in *The Development of Homeostasis,* Academic Press, New York, 1960.)**

Similarly, the absorption of antibodies from the mother's milk by suckling animals also depends on the ability of the immature intestine to absorb protein, probably by pinocytosis.

17–7 Large Intestine

Structure

The large intestine is a loose sac with thin walls, similar in basic structure to the rest of the digestive tube but wider in diameter and more distensible with a large capacity for retention of its contents (Fig. 17-24). This is aided by the special arrangement of the *taenia coli,* long flat muscular bands that draw the colon into saccules or pockets. The large intestine, though wide, is much shorter than the small intestine, being about 1.5 to 1.7 m long, as contrasted to the 6 m of the former. The large intestine consists of three parts; the *first part, the cecum,* starts at the junction with the small intestine. It is a large, somewhat flaccid pouch from which a narrow, blind tube, the *vermiform appendix,* hangs. In humans, the appendix is functionless but occasionally reminds one of its existence when it becomes inflamed or congested. It has no fixed position but moves with the cecum, in the general area of the lower right abdomen. The *second part, the colon,* ascends toward the liver on the right side, where it is identified as the *ascending colon,* to continue as the arched *transverse colon* from in front in the right kidney to the lateral border of the left kidney. Here it flexes sharply downward to become the *descending colon.* The *third part* of the large intestine is the *rectum and anal canal.* The descending colon empties its fairly solid, bacteria-rich contents, now called the *feces,* into the rectum. The rectum is about 13 cm long, and its diameter varies according to the bulk of its contents. Although its name implies "straight" (for it is straight in lower animals), the human rectum has three bends or flexures. The anal canal is the end portion of the large intestine, and it opens to the exterior as the *anus,* which is surrounded by strong fascia and muscles, the *internal and external anal sphincters* and the *levator ani,* which hold the walls of the anus together except at *defecation,* the discharge of the feces from the body.

Functions

In both the large and the small intestine, propulsion, secretion, and absorption occur, but because the small intestine is so much more efficient in these processes, the colon proceeds with its duties in a rather lazy and haphazard manner.

Its functions are:

1. To receive and store material from the small intestine.
2. To secrete and excrete.
3. To absorb water and salts.
4. To provide a locus for the action of bacteria that release food substances and produce vitamins and gases. This bacterial synthetic action is destroyed by large doses of antibiotics, resulting in quite marked symptoms of vitamin deficiency.

Figure 17-24. **The large intestine.**

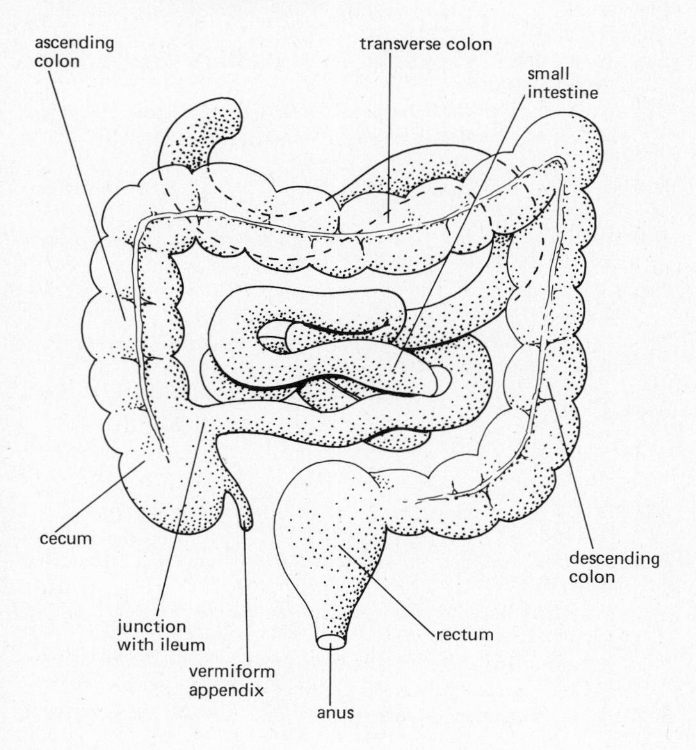

Feces

Normal feces have a water content of from 65 to 80 per cent, the rest of the water having been removed by the first part of the colon. The solid matter is almost one third dead bacteria, and the rest is 2 to 3 per cent nitrogen, 10 to 20 per cent fat, and 10 to 20 per cent inorganic material. The normal feces are brown, because of the presence of modified bile pigments. Fat in the stool imparts a pale color, blood a red or dark brown or black coloration. Even during fasting, a small amount of fecal material is produced.

Defecation

When the rectum has become sufficiently distended, the pressure of its contents against its walls initiates the stimulation of the internal sphincter muscles of the anus, causing them to relax. If at the same time the external sphincter muscles, which are under voluntary control, are relaxed, reflex increase in peristalsis along the lower colon propels the fecal matter toward and through the relaxed anus. By voluntarily contracting the external sphincter muscles, it is possible to inhibit the act of defecation for hours or even days. This voluntary inhibition can be learned by the child from about one to one and one-half years of age, when the pathways involved have reached the appropriate stage of development. Defecation can be initiated by increasing the abdominal pressure, which not only causes compression of the rectum but seems to initiate the defecation reflexes mentioned above. (Abdominal pressure is increased voluntarily by contracting all the abdominal muscles, contracting the diaphragm, and closing the glottis.)

Defecation is controlled by the ratio of the pressure in the rectum to the degree of contraction of the internal and external sphincters of the anus. The *internal sphincter is reflexly relaxed* by the parasympathetic nerves, but the *external sphincter is under voluntary control* (at least, after the age of about 2 years). When the pressure in the rectum increases because of the accumulation of the feces, reflex defecation will occur, unless voluntarily inhibited by the closing of the external sphincter. When defecation is voluntarily initiated, the sphincter is relaxed and the abdominal wall muscles are vigorously contracted. This, together with the contraction of the diaphragm and closure of the glottis, so increases the

Table 17-7. **Reflex Pathway for Defecation**

Stimulus	Increased pressure in the rectum
Receptor	Nerve endings in mucosa of rectum
Afferent neuron	Visceral afferent fibers
Synapse	Autonomic centers in sacral part of spinal cord
Efferent neuron	Parasympathetic fibers
Effector	Smooth muscle of colon (stimulated)
	Sphincter muscles of anus (relaxed)
Response	Propulsion of fecal material through the colon and into the rectum, from the rectum out through the anus
Effect	Abrupt fall in rectal pressure

intra-abdominal pressure that the pressure in the rectum pushes the feces out through the relaxed anal sphincter. Apparently, this stimulates reflex defecation once more, permitting the emptying of higher parts of the colon as well (Fig. 17-25 and Table 17-7).

Constipation and diarrhea

Constipation, irregular or inadequate defecation, may be caused by many factors, including mechanical obstruction of the colon, decreased excitability of the musculature, and psychological problems. Each of these conditions requires different treatment, but the era has passed when it was considered that irregular defecation was of paramount importance in undermining health and that it could cause autointoxication. Chronic distention of the rectum does produce symptoms of discomfort and headache on occasion, but no toxic effects have been demonstrated.

Diarrhea, excessive motility of the bowel resulting in the rapid and uncontrollable propulsion of the contents of the colon, may be caused by irritation of the bowel by bacterial toxins or by cathartics, such as castor oil or epsom salts. It may also be caused by excessive nervous stimulation. Although usually the stimulation of the sympathetic nervous system results in inhibition of intestinal motility and the parasympathetic activity stimulates motility, strong emotions involving the sympathetic nervous system are often accompanied by diarrhea. Diarrhea may result in dehydration of the tissues, because the intestinal contents pass through the colon too rapidly to permit the proper resorption of water by the mucosa of the colon, and the total loss of water and salts per day in severe diarrhea is often large enough to be extremely dangerous.

Figure 17-25. **Reflex and voluntary control of defecation.**

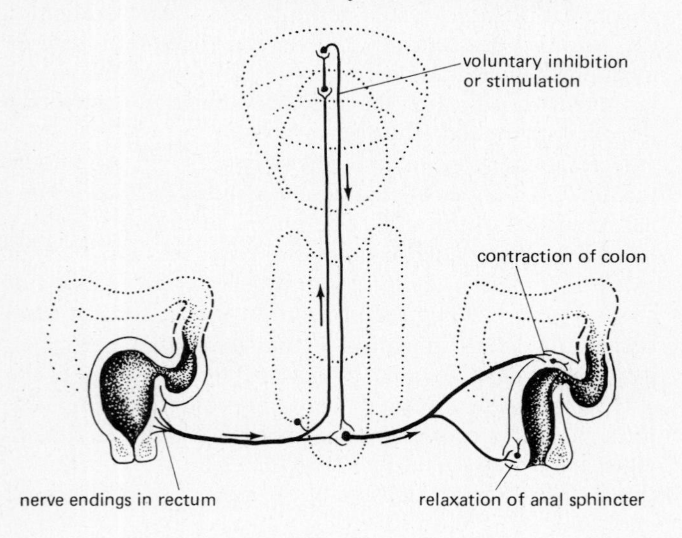

Cited References

Beaumont, W. *Experiments and Observations on the Gastric Juice and the Physiology of Digestion*. F. P. Allen, Plattsburgh, N.Y., 1833.

Pavlov, I. P. *The Work of the Digestive Glands.* Charles Griffin & Co. Ltd., London, 1902.

Wilson, T. H. "A Modified Method for the Study of Intestinal Absorption in Vitro." *J. Appl. Physiol.* **9:** 137–140 (1954).

Additional Readings

BOOKS

Brooks, F. P. *Control of Gastrointestinal Function: An Introduction to the Physiology of the Gastrointestinal Tract.* Macmillan Publishing Co., Inc., New York, 1970.

Csaky, T. Z., ed. *Intestinal Absorption and Malabsorption.* Raven Press, New York, 1975.

Davenport, H. W. *Physiology of the Digestive Tract,* 2nd ed. Year Book Medical Publishers, Inc., Chicago, 1971.

Dworkin, H. J. *The Alimentary Tract.* W. B. Saunders Company, Philadelphia, 1974.

Gall, E. A., and F. K. Mostofi, *The Liver.* The Williams & Wilkins Company, Baltimore, 1973.

Guyton, A. C., ed. *MTP International Review of Science, Physiology,* Vol 4: *Gastrointestinal Physiology.* University Park Press, Baltimore, 1974.

Jenkins, G. N. *The Physiology of the Mouth,* 3rd ed. F. A. Davis, Co., Philadelphia, 1965.

Maclagan, N. F., ed. *The Liver: Some Physiological and Clinical Aspects. Brit. Med. Bull.* **12**(3) (1957). (Entire volume.)

Payne, W. S., and A. M. Olsen. *The Esophagus.* Lea and Febiger, Philadelphia, 1974.

Smyth, D. H., ed. *Intestinal Absorption,* Vols. 4A and 4B. Plenum Publishing Corporation, New York, 1974.

Wilson, T. H. *Intestinal Absorption.* W. B. Saunders Company, Philadelphia, 1962.

ARTICLES

Anderson, S. "Secretion of Gastrointestinal Hormones." *Ann. Rev. Physiol.* **35:** 431 (1973).

Beck, I. T. "The Role of Pancreatic Enzymes in Digestion." *Am. J. Clin. Nutr.* **26:** 311 (1973).

Borgström, B. "Fat Digestion and Absorption." *Biomembranes* **4B:** 555 (1974).

Bortoff, A. "Digestion: Motility." *Ann. Rev. Physiol.* **34:** 261 (1972).

BRAUER, R. W. "Liver Circulation and Function." *Physiol. Rev.* **43:** 115 (1963).

CODE, C. F., and H. C. CARLSON. "Motor Activity of the Stomach." In *Handbook of Physiology,* Sec. 6, Vol. 4. The Williams & Wilkins Company, Baltimore, 1968.

DAVENPORT, H. W. "Why the Stomach Does Not Digest Itself." *Sci. Am.* Jan. 1972.

GROSSMAN, M. I. "The Digestive System." *Ann. Rev. Physiol.* **25:**165 (1963).

KAPPAS, A., and A. P. ALVARES. "How the Liver Metabolizes Foreign Substances." *Sci. Am.,* June 1975.

KRETCHMER, N. "Lactose and Lactase." *Sci. Am.,* Oct. 1972.

MATTHEWS, D. M. "Intestinal Absorption of Peptides." *Physiol. Rev.* **55:** 537 (1975).

NEURATH, H. "Protein Digesting Enzymes." *Sci. Am.,* Dec. 1974.

TRUELOVE, S. C. "Movements of the Large Intestine." *Physiol. Rev.* **46:** 457 (1966).

TURNBERG, L. A. "Absorption and Secretion of Salt and Water by the Small Intestine." *Digestion* **9:** 357 (1973).

VAN CAMPEN, D. "Regulation of Iron Absorption." *Fed. Proc.* **33:** 100 (1974).

WALSH, J. H. "Circulating Gastrin." *Ann. Rev. Physiol.* **37:** 81 (1975).

WOOD, J. D. "Neurophysiology of Auerbach's Plexus and Control of Intestinal Motility." *Physiol. Rev.* **55:**307 (1975).

Chapter 18

Regulation of Carbohydrate and Fat Metabolism

A man said to the universe:
"Sir, I exist!"
"However," replied the universe,
"The fact has not created in me
A sense of obligation."

Stephen Crane, "A Man Said to the Universe"

CARBOHYDRATES AND FATS are the main sources for the energy required by the physiological processes of the cell. These processes include muscle contraction, temperature maintenance, nerve impulse generation and propagation, glandular secretion, synthetic reactions of all kinds, and active transport processes. The chemical reactions through which this energy is captured, stored in high energy phosphate compounds such as ATP, and then released in a controlled and stepwise manner, have been discussed in Secs. 5-8 through 5-10.

Reserves of carbohydrate are found stored as glycogen, chiefly in the liver and in skeletal muscle. Fat is stored chiefly in adipose tissue. These energy stores are mobilized and used according to the food intake and metabolic needs of the body. The complex control of food intake is discussed in Sec. 21-2. The present chapter concerns itself with the maintenance of blood glucose, which is of primary importance among the homeostatic mechanisms regulating the environment of the cell.

Both carbohydrate and fat reserves (and in extreme cases, such as starvation, even protein) can be drawn upon to increase blood glucose. Similarly, excess carbohydrate can be stored as fat as well as glycogen. The balance between glucose utilization and glucose storage is crucial and is delicately controlled by several hormones. The level of blood glucose itself, through a negative feedback process, regulates the secretion of these controlling hormones, thus influencing its own level.

18-1 Carbohydrate Metabolism

Among the many organs contributing to the regulation of blood glucose levels are the following.

Liver

Absorbed monosaccharides pass through the hepatic portal vein to reach the liver. Most of the simple sugars are converted by isomerases to glucose in the liver. By variations in the amount of glucose converted to glycogen (*glycogenesis*) and stored in hepatic cells, and in the amount of glucose that is released into the blood (*glycogenolysis*), the liver acts as an important regulator or buffer for blood glucose. Hepatic cells can store up to 8 per cent of their weight as glycogen.

Muscle

Muscle acts both as a site of glycogen storage and as the main consumer of energy derived from the breakdown of glucose. Muscle may make up as much as 50 per cent of the body weight.

Pancreas

The level of circulating glucose, acting as a metabolic feedback mechanism, controls the amount of *insulin* and *glucagon* secreted by the endocrine portion of the pancreas.

Adrenal Gland

The secretion of *epinephrine* by the *adrenal medulla* is partially controlled by the level of blood glucose and partially regulated through its sympathetic innervation.

The *glucocorticoids* secreted by the *adrenal cortex* play an important role in diverting protein and fat metabolism to glycolytic pathways when blood glucose is low (*gluconeogenesis*).

Anterior Pituitary Gland

Somatotropin (growth hormone) is secreted by the anterior pituitary gland in response to many different types of stimuli, one of which is a fall in blood glucose. Somatotropin appears to inhibit glucose utilization by muscle and so raises blood glucose.

Thyroid

Thyroxine increases tissue metabolism in general and so affects the rate of glucose utilization in a nonspecific manner.

Hypothalamus

Somatostatin, secreted by the hypothalamus, directly inhibits both insulin and glucagon secretion. Because somatostatin appears to be a more effective inhibitor of glucagon than insulin, it may be useful for those forms of diabetes mellitus that are caused by an excess of glucagon.

18–2 Liver

Structure, Position, and Blood Supply

The liver is the largest gland in the body. In addition to its vital role in protein, fat, and carbohydrate metabolism, it acts as an important blood reservoir, blood filter, and detoxification center. It is a very vascular organ and receives arterial blood from the *hepatic artery*, a branch of the celiac artery (Fig. 18-1). In addition, the liver receives venous blood, rich in nutrients, from the *hepatic portal vein*, which drains the spleen, pancreas, and gall bladder as well as the stomach, small intestine, and the upper part of the large intestine. The liver is drained by the *hepatic veins*, which empty into the inferior vena cava.

The liver is reddish brown and covered by the glistening peritoneum. It is partly divided into the small left lobe and the larger right lobe, on the inferior surface of which is attached the *gall bladder* (see also Sec. 17-5). The liver is in very close contact with the diaphragm and is sheltered by the ribs, a very necessary protection because the soft tissue is easily ruptured, with very serious aftereffects. The bulging of an infant's abdomen is caused by the large size of the liver and the relatively small pelvis, which only later is large enough to room the viscera that subside into it.

The short, wide hepatic portal vein and the hepatic artery enter the liver together at its hilum. The delicate connective tissue capsule of the gland surrounds the blood vessels, following them into the substance of the liver and dividing it into innumerable small *lobules*. From these branches of the blood vessels arise capillaries that penetrate into the lobules themselves; these capillaries form large irregular spaces into which the afferent blood is poured. These spaces are known as *sinusoids*, and as a result of this type of blood supply, the hepatic cells that make up the lobule are richly supplied both with arterial blood and blood from the digestive tract. In the center of the lobule is the central vein that drains the lobule and ultimately joins the hepatic veins (Fig. 18-2; see also Fig. 18-1).

Accompanying the branches of the hepatic artery and the portal vein through the liver are the *interlobular bile ducts*, which are formed from delicate *bile capillaries* that run between the hepatic cells. The larger bile ducts finally form the common hepatic duct, which unites with the cystic duct to form the *common bile duct* (Sec. 17-5).

The cells that receive such specialized blood supply are astonishingly efficient in performing a vast variety of chemical reactions. As far as is known, there is no difference between the individual cells that make up the *hepatic cords*, so that separation of the chemical reactions must occur on an intracellular level, probably based on the structural compartmentalization provided by the cytoplasmic reticulum. The *hepatic* cells themselves are polyhedral in shape and contain large numbers of mitochondria, as would be expected from their great metabolic activity. They also contain substantial numbers of *glycogen droplets*, for the liver contains as much as 5 per cent of its weight as stored glycogen.

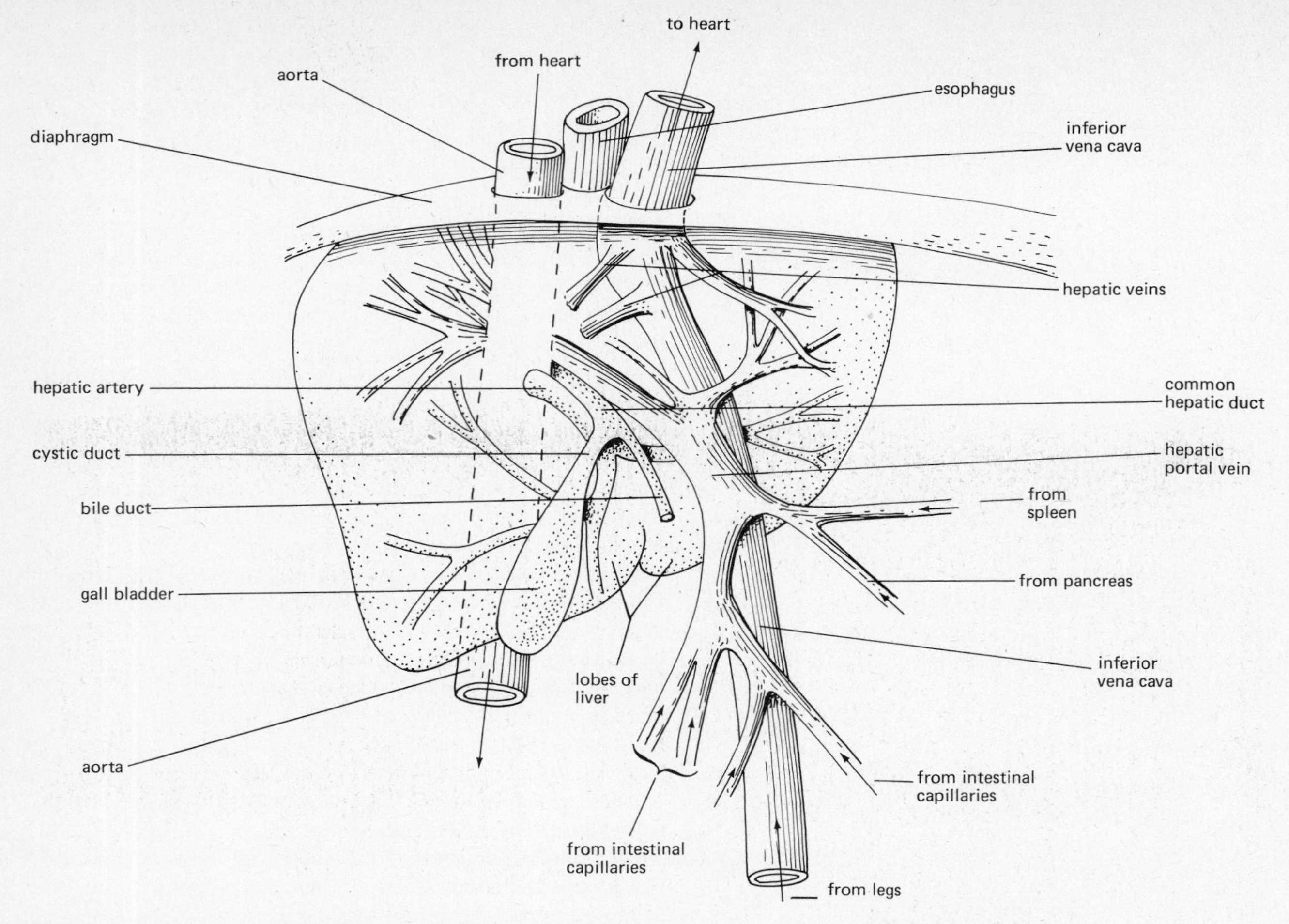

Figure 18-1. **The circulation of the liver. The main blood supply of the liver comes from the hepatic portal vein. The liver receives arterial blood from the hepatic artery, a branch of the aorta. Blood is drained from the liver by the hepatic veins, which empty into the inferior vena cava. This is a view of the liver from below.**

Often *fat droplets* can be seen in the cytoplasm of these cells (Figs. 2-10 and 2-11).

Reticuloendothelial portion of the liver

Apart from the hepatic cells, specialized cells of quite different structure and function line the sinusoids. These are stellate cells, the *Kupffer cells,* which are part of the *reticuloendothelial system.* This system consists of a widespread series of cells capable of actively engulfing foreign particles by *phagocytosis* (*phagein,* Greek "to eat"). (Sec. 14-9.)

Functions of the Liver

1. The liver is the chief homeostatic organ in the maintenance of the blood sugar and carbohydrate metabolism.
2. The liver plays an important role in protein synthesis and degradation and in the formation of urea from nitrogenous wastes (Sec. 19-2).
3. The liver is vitally involved in the synthesis, storage, and utilization of fats.
4. The formation of bile by the liver is essential for proper fat digestion.
5. The liver inactivates many chemicals, including hormones, and detoxifies many poisons. The reticuloendothelial cells within its sinusoids engulf foreign particles and are involved in antibody formation.
6. The liver absorbs and stores the antianemic factor necessary for the normal maturation of erythrocytes.

Role of the Liver in Carbohydrate Metabolism

Activation of glucose

When the portal blood enters the liver, it flows slowly through the blood sinuses in close contact with the hepatic cells. Glucose and fructose enter the liver cells very rapidly and may either diffuse back in the same form into the blood or be prepared by the enzyme systems of the hepatic cells for synthesis to glycogen. This preparation consists of an activation process, the addition of inorganic phosphate, very much in the way that amino acids are activated for synthesis to protein. The energy for activation is again donated by the ready pool available in the ATP molecule:

$$\text{Glucose} \xrightarrow[\text{Glucokinase}]{\text{ATP} \to \text{ADP} + \text{inorganic phosphate} + \text{energy}} \text{Glucose 6-phosphate}$$

The *phosphorylated glucose* may now enter into the glycolytic pathway for the release of energy or it may be stored as glycogen. The particular pathway it takes is directed by a series of hormonal controls, ultimately determined by the level of blood glucose. These reactions are discussed a little in this chapter.

The most important *storage sites* for glycogen are liver and muscle, but only the liver has the necessary enzyme systems to reverse the phosphorylation of glucose and *release free glucose* into the blood. Phosphorylated glucose does not pass through cell membranes easily, so that only the liver can release glucose in a form that can be used by all tissues.

Carbohydrate metabolism in the normal, fasting, and feasting states

The *normal level* of blood glucose, in between adequate meals, is approximately 90 mg glucose/100 ml blood (90 mg/100 ml). After a *meal rich in carbohydrate,* blood glucose may rise to 130 to 150 mg/100 ml, and the concentration of glucose in the hepatic portal vein may be even higher. This indicates that the liver has stored some of the excess glucose as glycogen, acting as a buffer and preventing the flooding of the tissues with large amounts of glucose.

Habitual excessive ingestion of carbohydrate results in the conversion of glucose to fat rather than to glycogen. This fat is stored in the liver (fatty degeneration of the liver also follows excessive alcohol intake) and also in other tissues, particularly in the large peritoneal folds in front of the intestines. Commercially, the cornfed pig is a good example of the metabolic efficiency by which inexpensive carbohydrate (corn) is transformed into the high-priced delicacy bacon (chiefly fat).

Figure 18-2. **A liver lobule, showing the hepatic cords in the upper-right-hand section. The liver cells that make up the cords are surrounded by sinusoids—wide, blood-filled spaces formed by the capillaries of the portal vein. The arterial capillaries also empty into the sinusoids. The lobule is drained by the central vein, which empties into one of the hepatic veins. The bile capillaries, which also penetrate between the hepatic cells, empty into the bile duct via the common hepatic duct.**

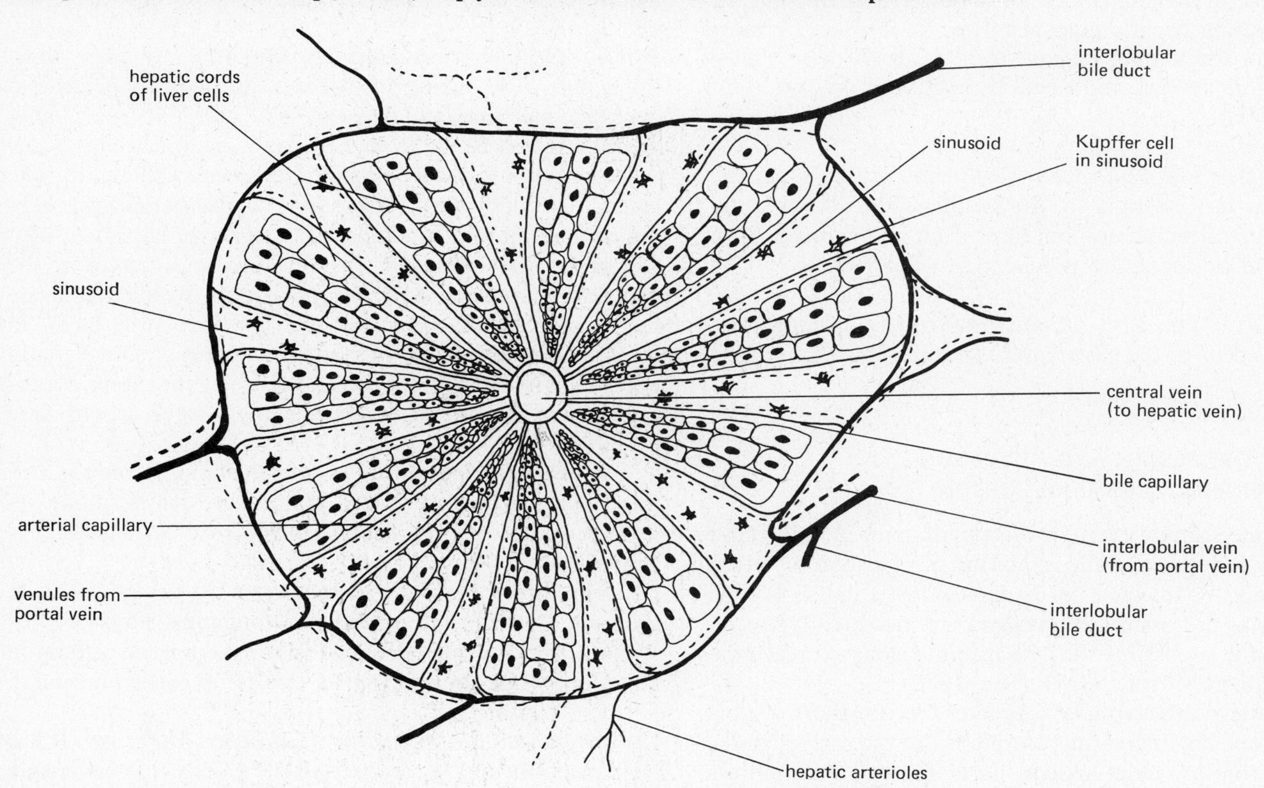

Following a *fast,* or prolonged *strenuous exercise,* blood glucose may fall to 60 mg/100 ml for a brief time. The liver rapidly compensates for this *hypoglycemia* by the release of stored glycogen as free glucose. The balanced glycogenetic and glycogenolytic reactions in the liver maintain blood glucose at a rather regular level after a short-term elevation or depression.

In the absence of sufficient carbohydrate, which is the preferred fuel for most tissues, the liver is able to utilize both fat and protein degradation products, routing them through the common intermediary, acetyl coenzyme A (Sec. 5-12), into the glycolytic pathway. The conversion of fatty acids and de-aminated amino acids to glucose is *gluconeogenesis,* and this process is stimulated by the glucocorticoids and by glucagon.

18–3 Endocrine Glands Controlling Carbohydrate Metabolism

Endocrine Portion of the Pancreas—Islets of Langerhans

Interspersed among the digestive glands of the pancreas, and particularly concentrated in the tail region, are small circles of lightly staining cells, the *islets of Langerhans* (Fig. 17-15). These islets are composed of three types of cells, alpha, beta and delta cells (Fig. 18-3).

Alpha Cells. The *alpha cells* secrete *glucagon,* a *hyperglycemic* factor which is much more powerful than epinephrine in causing hepatic glycogenolysis. It is also extremely effective in stimulating gluconeogenesis. Both these effects of glucagon are accomplished through the activation of cyclic AMP.

Beta Cells. The *beta cells* secrete *insulin,* a *hypoglycemic* factor, which is essential for the synthesis and conservation not only of carbohydrate, but also of fat and protein. Both insulin and glucagon are polypeptides (Sec. 10-7).

Delta Cells. The *delta cells* secrete *gastrin,* a hormone that stimulates cells in the gastric mucosa to produce the gastric juice.

Release of insulin and glucagon and their mechanisms of action

The same stimulus, a high-carbohydrate meal, will cause a release of insulin and the inhibition of glucagon secretion in the normal subject. This is depicted in Fig. 18-4. The responses of the tissues to glucagon are mediated by cyclic AMP, and it is possible that the insulin-regulated reactions are mediated by cyclic GMP (Sec. 10–8).

It is believed that *insulin* facilitates the transport of glucose into cells by increasing the affinity of the carrier molecule for glucose. Once glucose is in the cell it is almost immediately phosphorylated and thus able to enter into the glycolytic pathway, or to be synthesized to glycogen. In either case, blood glucose levels are lowered (Fig. 18-5). In the absence of insulin, blood glucose levels may rise to as much as between 300 to 1000 mg/100 ml. If too much insulin is administered, blood glucose levels can drop precipitously to 20 mg/100 ml, causing diabetic coma.

Diabetes mellitus

Diabetes mellitus (sweet urine) is a disease characterized by a high blood sugar and a concomitant excretion of glucose in the urine. It is usually caused by an *insufficient insulin secretion* by the pancreas. The role of pancreatic extract in alleviating the symptoms of diabetes was discovered by Banting, a Canadian surgeon, and a graduate student, Charles Best, in 1922. The insulin deficiency may be due to an initial defect in the beta cells, or it may be the result of an overstimulation, followed by exhaustion, of the beta cells. The stimulating factor may be the *diabetogenic factor* found in crude extracts of the pituitary gland, or it may be *somatotropin.* In addition, abnormally *high blood sugar* levels maintained for a long period of time may exhaust the beta cells and cause diabetes. Or it is possible that there may be a genetically determined sensitivity of the beta cells to the diabetogenic factor or to a high-carbohydrate diet. This would explain the *inherited susceptibility* to this disease.

Recent work has implicated the liver, which normally reactivates the insulin circulating through it. In certain types of diabetes, this mechanism fails, and large amounts of inactivated insulin are found in the blood of these diabetics.

Role of Glucagon in Diabetes Mellitus. Whereas insulin is normally lacking in diabetes, glucagon is present in excessive amounts, and this excess of glucagon is probably as important a factor in diabetes as is insulin lack.

The study of glucagon action has been facilitated by the recent discovery that *somatostatin,* the hypothalamic hormone that inhibits the release of growth hormone, also suppresses the release of insulin and glucagon from the pancreas. Experiments with somatostatin strongly indicate that it is an *excess of glucagon* that is the essential factor in the development of diabetes: administration of somatostatin together with insulin is very effective in preventing hyperglycemia in diabetic patients following a high-carbohydrate meal.

The practical application of somatostatin therapy is still uncertain, for because it also inhibits growth hormone release it cannot be used in young diabetics. Somatostatin is active for only a very short time and must be given by infusion, directly into the veins. Roger Guillemin (1972) at the Salk Institute is currently investigating ways of prolonging and enhancing somatostatin action on glucagon secretion, while minimizing its effects on other hormones.

Carbohydrate Metabolism in Diabetes Mellitus. It has been mentioned in the chapter on the kidney that pancreatic

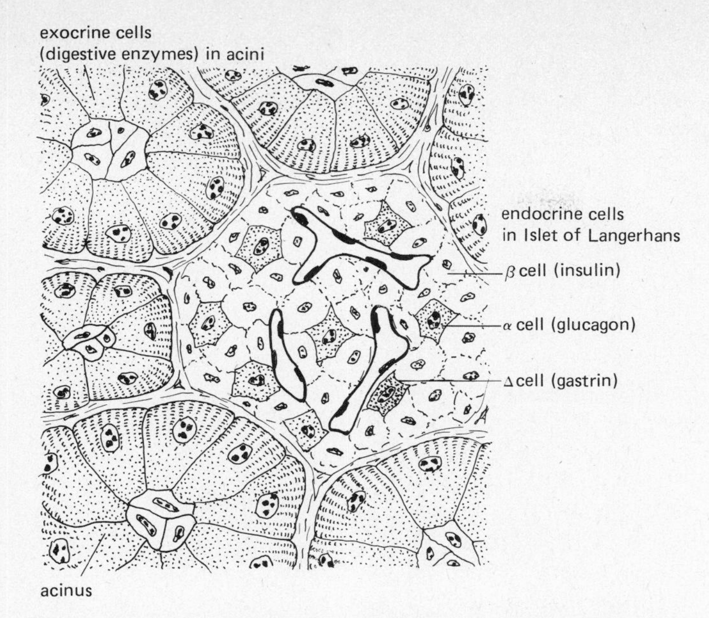

Figure 18-3. **The exocrine and endocrine cells of the pancreas.**

diabetes is accompanied by the loss of large amounts of glucose through the urine because the cells of the kidney tubule are not able to handle the vast amounts of glucose entering in the filtrate. The *liver* and *muscle* of the diabetic are *low in glycogen,* which means that the high blood glucose of the diabetic is maintained through the depletion of the tissue glycogen stores. Nor is this high blood glucose available to the tissues for energy: the patient is weak and apathetic. In other words, the circulating glucose cannot enter the cell to become part of the glycogen synthesis or glucose oxidation pathway. This lends credence to the view that insulin acts to speed up the passage of glucose into the cell. The diabetic can no more utilize the rich energy source of his high blood sugar than can the shipwrecked sailor utilize the water of the sea around him. Instead, the diabetic shifts from carbohydrate metabolism to fat metabolism for energy, which results in the formation of keto-acids; in severe cases, acidosis develops.

ADMINISTRATION OF INSULIN TO DIABETICS. Proper treatment of diabetes mellitus requires the administration of the right amount of insulin so that carbohydrate metabolism is balanced. This obviously must be coordinated with the *carbohydrate intake,* which preferentially is kept very low, and with *energy output*. Because insulin is a polypeptide and is not absorbed through the intestinal mucosa, it must be administered by intramuscular injection.

ORAL HYPOGLYCEMIC AGENTS. Patients who have some beta cell function respond to an orally effective drug, *tolbutamide,* which causes the release of insulin into the blood. Another type of oral hypoglycemic agent, the *biguanides,* lower blood glucose by acting on the peripheral tissues. There is some indication that patients taking these drugs suffer from a higher incidence of cardiovascular complications than insulin-treated patients; therefore the use of these oral hypoglycemic drugs is controversial at the moment.

Tests for Diabetes Mellitus: Glucose and Insulin Tolerance Tests. In the discussion on the kidney, it was mentioned that the presence of glucose in the urine does not necessarily indicate lack of insulin secretion. It could be caused by a disturbance in the function of the proximal convoluted tubule, resulting in failure of the tubule cells to absorb the filtered glucose back into the blood. Even a routine blood glucose determination is not always diagnostic of the pancreatic origin of the high blood glucose (there are liver diseases that can simulate this), and the following two tests are routinely done to determine the origin of the meta-

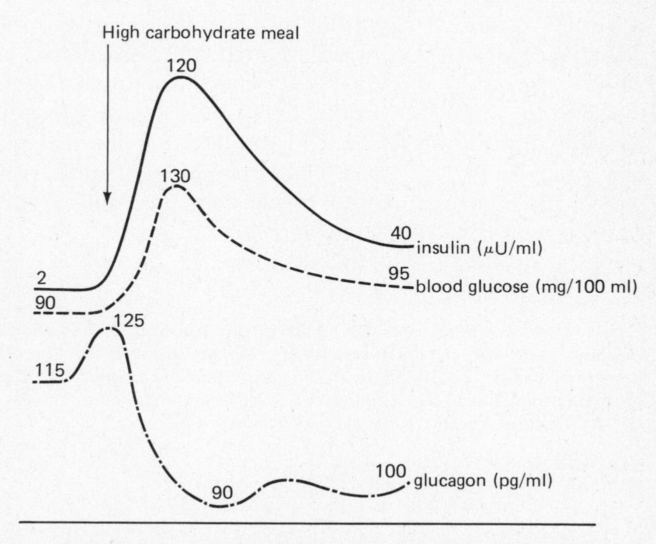

Figure 18-4. **A high-carbohydrate meal raises blood glucose rapidly, causing an inhibition of glucagon secretion and stimulating insulin release. Only selected values are shown for each curve.**

Figure 18-5. **The effect of an intravenous injection of one unit of insulin upon blood glucose.**

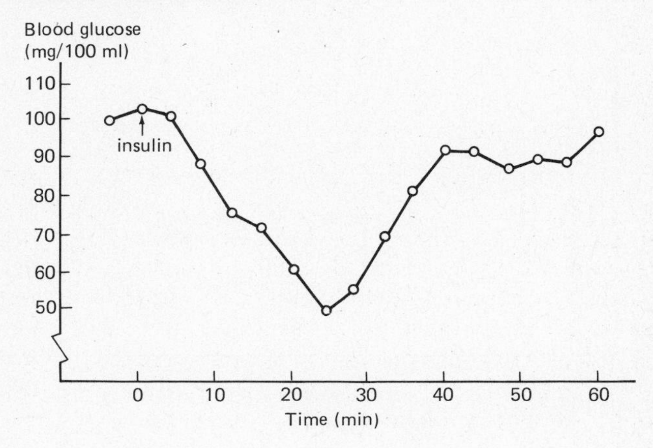

bolic error; they dramatically underline the physiological principles.

The *glucose tolerance test* indicates graphically the response of the body to the ingestion of glucose. In a normal, fasted individual, given 1 g glucose per kilogram body weight, the blood glucose level rises from about 90 mg/100 ml to 140 mg/100 ml and then falls to normal within about 3 hr. This fall comes from the release of insulin following the elevation of blood glucose. In the diabetic, the fall is very much slower and usually fails to drop below the control level, which may be as high as 300 mg/100 ml (Fig. 18-6).

The *insulin tolerance test* indicates the response of the body to the administration of insulin. It is usually called the *insulin sensitivity test,* because the diabetic is extraordinarily sensitive to minute amounts of insulin. In a normal individual, insulin causes a slight fall in the blood sugar, but the diabetic response is greatly exaggerated, and a sharp drop of blood sugar results (Fig. 18-6). Too much insulin may cause the diabetic to go into insulin shock, due to the *hypoglycemia;* extreme *hyperglycemia* in the uncontrolled diabetic may result in *diabetic coma.* These pathological states emphasize the importance of the homeostatic mechanisms that normally maintain the blood sugar within a relatively narrow range.

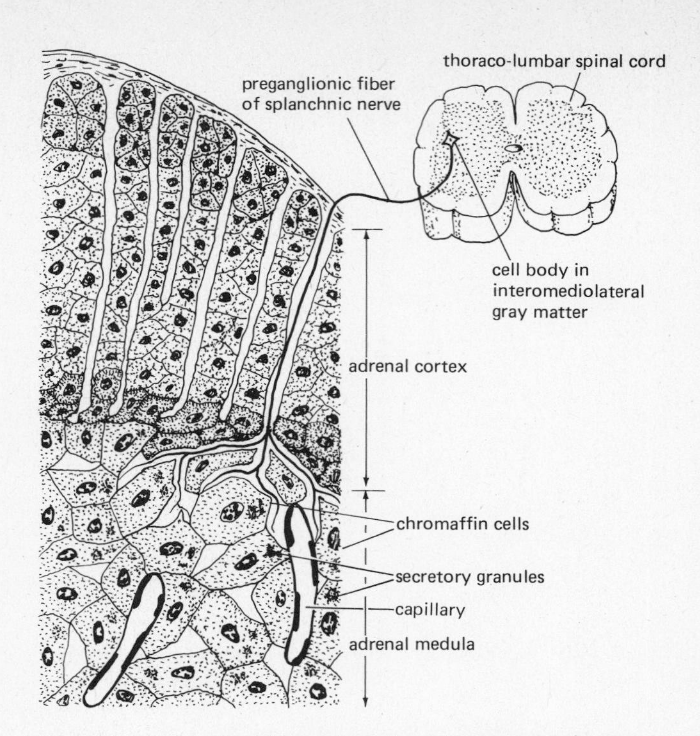

Figure 18-7. **The cells of the adrenal medulla are neurosecretory cells, innervated by preganglionic sympathetic nerve fibers. The chromaffin cells of the medulla secrete mostly epinephrine, but small amounts of norepinephrine are also produced.**

Figure 18-6. **Glucose tolerance tests in normal and diabetic individuals. Note the marked sensitivity of the diabetic to glucose ingestion. (From F. Bertram, in *Thannhauser's Textbook of Metabolism and Metabolic Disorders,* N. Zöllner and S. Estren, eds., Georg Thieme Verlag, Stuttgart, Germany, 1962.)**

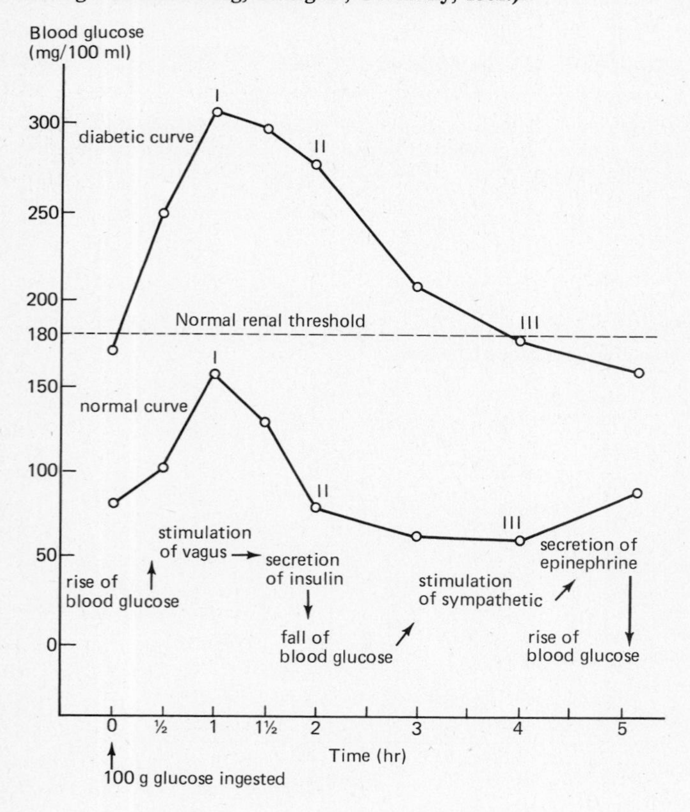

Adrenal Medulla

In an earlier discussion of the adrenal gland in Sec. 10-6, the structural and functional differences between the adrenal cortex and the medulla have been mentioned, but the emphasis has been on the adrenal cortex and its hormones.

The *adrenal medulla* consists of cells that are derived embryologically from neural crest tissue. The medulla can be considered a modified sympathetic ganglion, with the ganglionic cells specialized for secretion instead of conduction. These neurosecretory cells, like other sympathetic postganglionic cells, are stimulated by preganglionic cholinergic fibers of the sympathetic nervous system (Fig. 18-7). In the case of the medulla, these preganglionic fibers form the splanchnic nerve. The adrenal medulla is more richly innervated than any other organ.

The secretory cells of the medulla are called *chromaffin cells,* because of the presence of darkly staining granules, containing epinephrine. The development of the technique of fluorescent microscopy by Falck in 1961, to a degree where it has become possible to distinguish between the different catecholamines present in cells, has shown that there are some medullary cells that secrete norepinephrine. However, only small amounts of norepinephrine are present in the medulla, and the main secretion is epinephrine. Norepinephrine functions mainly as a neurotransmitter in the sympathetic and central nervous system.

Von Euler, who won the Nobel Prize in 1970 for his work on norepinephrine, has shown that both norepinephrine and epinephrine appear in the urine after stress. The relative proportions of these two amines vary with the type of stress and also with its severity. In *emotional* stress, associated with pain and anxiety, more epinephrine is released. In *cold stress,* on the other hand, there is a greatly increased secretion of norepinephrine.

Biosynthesis of epinephrine

The synthesis of the catecholamine neurotransmitters nornephrine and dopamine from the amino acid tyrosine was discussed in Sec. 9-5. The adrenal medulla is the only mammalian organ that has the enzyme *N*-methyltransferase (NMT), which can convert norepinephrine to epinephrine by the addition of a methyl group. The activity of NMT appears to be under the influence of adrenal cortical hormones and ACTH.

HO–(ring, OH)–CH(OH)–CH$_2$–NH$_2$

Norepinephrine

N-methyltransferase →

HO–(ring, OH)–CH(OH)–CH$_2$–NH–CH$_3$

Epinephrine

Release of epinephrine and its metabolic effects

Epinephrine is released into the circulation in response to sympathetic stimulation by way of the splanchnic nerve and in response to a fall in blood glucose levels. The central importance of epinephrine in preparing the animal for "flight" or "fight" can be seen from the wide variety of short- and long-term physiological effects (Fig. 18-8).

Carbohydrate Metabolism. Epinephrine increases blood glucose and blood lactate by stimulating glycogenolysis, first in the liver and then in muscle. This *hyperglycemic* effect of epinephrine is mediated through cyclic AMP and activation of the phosphorylase system (Sec. 10-8). Epinephrine may also add to its hyperglycemic effect by inhibiting the release of insulin and by increasing glucocorticoid release as a result of epinephrine-induced ACTH secretion.

Fat Metabolism. Epinephrine stimulates lipolysis in adipose tissue and raises the level of circulating free fatty acids in the blood as much as 100 per cent.

Oxygen Consumption. Epinephrine causes vasodilation in skeletal muscle and vasoconstriction in the splanchnic and skin circulations (splanchnic shift; see Sec. 12-10). This shunts the glucose- and fatty-acid-rich blood to the muscles, greatly increasing the amount of oxidizable substrate available to the muscles. As the strength of heart contraction is also increased by epinephrine, blood flows more rapidly through the muscles, and the total increase in oxygen consumption may be as high as 30 per cent.

Carbohydrate Metabolism During Exercise

The physiological importance of epinephrine is seen best during exercise, when the sympathetic nervous system is highly excited. If we consider an athlete in a long-distance race, the immediate effects of epinephrine are on the heart and circulation. It increases heart rate and blood pressure and, through alterations in the diameter of the arterioles, shifts the blood from the viscera to the muscles and skin. All these changes are discussed in detail in the section on circulation, but they are of great significance here. The large amounts of blood passing through the muscles bring glucose and epinephrine to the hard-working tissue; as the muscles continue to contract and utilize the energy resulting from glucose breakdown, the blood glucose falls. This sets into action a series of regulatory mechanisms:

1. The liver releases glucose from its glycogen stores.
2. Pancreatic secretion of insulin is inhibited.
3. The low blood sugar stimulates the sympathetic nerve

Figure 18-8. **Effect of epinephrine (0.2 mg/kg body weight) on the blood glucose and tissue glycogen levels of the normal rat, as compared with quick response of the cardiovascular system. (From E. Frieden and H. Lipner, *Biochemical Endocrinology of the Vertebrates,* Fig. 7.6. Foundations of Modern Biochemistry Series. Prentice-Hall, 1971. Reprinted by permission of Prentice-Hall, Inc., Englewood Cliffs, New Jersey.)**

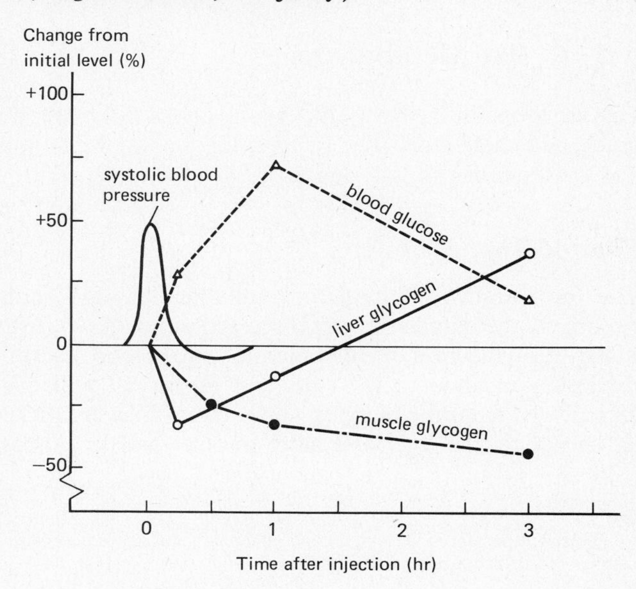

centers and causes stimulation of the adrenal medulla by way of the sympathetic nerves, resulting in the increased secretion of epinephrine.

4. The low blood sugar directly stimulates the adrenal medulla, increasing epinephrine secretion.
5. The circulating epinephrine enhances glycogenolysis in the liver and muscle, raising blood sugar and making glucose-6-phosphate available to the muscle for energy. This rise in blood sugar may be as great as 10 to 20 mg/100 ml per minute.
6. The breakdown products of glucose metabolism may be liberated into the circulation, and depending on whether the activity is very strenuous and thus partly anaerobic, or milder and thus mainly aerobic, glucose may be *incompletely metabolized to lactic acid* or *completely and efficiently oxidized to carbon dioxide and water* (Sec. 5-12). The energy yield from the oxidative reactions is much greater than from the anaerobic, but the body, with its customary economy, utilizes even the lactic acid produced. It is transported to the liver, where it is converted back to glucose.
7. If the level of exercise is not maintained at a rate that utilizes the supplementary glucose as fast as it is provided, a temporary hyperglycemia results, reversing the above phenomena, causing glycogenesis, insulin secretion, and epinephrine inhibition.

Homeostatic Control of Blood Glucose Concentration

Homeostasis is achieved through *temporary swings above or below the "base line"* of the normal blood sugar. An initial hypoglycemia evokes compensatory reactions that tend to overcompensate, causing a brief hyperglycemia, which slowly subsides to the normal level. This type of physiological response is encountered in a great many instances of homeostatic regulation.

18–4 Fat Metabolism

Fats are found in the body chiefly as triglycerides, phospholipids, and cholesterol. The relationship among these lipids was shown in Figs. 3-18 and 3-19.

Fate of Absorbed Fats

The formation of chylomicrons and their passage in the plasma via the lymphatics was discussed in Sec. 17-6. After a fatty meal, the level of chylomicrons in the blood reaches a maximum in about 2 to 4 hr, and almost all will have disappeared from the blood in about 8 hr. They are picked up chiefly by the liver and heart muscle[1] and by adipose tissue. These tissues may break down the chylomicrons, liberating into the blood the various components, that is, triglyceride, cholesterol, phospholipid and protein. In addition, these tissues may hydrolyze the triglyceride, releasing fatty acids, or they may store it as fat. The individual fat molecules now in the blood are coated lightly with a film of protein, which prevents their sticking together to form large globules that might block the blood vessels. These protein-coated fat molecules are known as *lipoproteins* and are much smaller than the chylomicrons formed by the small intestine.

Some hydrolysis of these lipoproteins takes place in the blood, under the influence of the enzyme *lipoprotein lipase.* This enzyme frees fatty acids, which immediately combine with plasma albumin to form what is called the *free fatty acids* of the blood. This appears to be an important lipid transport mechanism. The remaining lipids are transported as cholesterol, phospholipids, triglycerides and lipoproteins. The total concentration of these lipids in the plasma is approximately 700 mg/100 ml plasma. Of this amount about 180 mg/100 ml is cholesterol.

[1] This is important in view of the fact that heart muscle, unlike skeletal muscle, can utilize lipids for fuel.

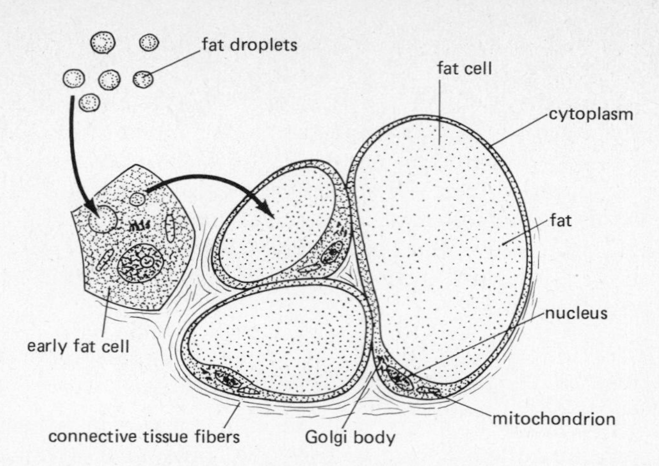

Figure 18-9. **Progressive deposition of fat into cells to form adipose tissue. The cytoplasm of the fat-filled cell is compressed into a narrow rim that contains all the cell organelles.**

Fat Deposition

Lipids are deposited in the *liver,* which stores them as triglycerides or degrades them into small compounds (acetyl coenzyme A) that can be used for energy or for the synthesis of other lipids, especially cholesterol.

Lipids can be deposited in *adipose tissue,* where they are stored as triglycerides that can be mobilized for energy or retained as heat insulation.

Structure of Adipose Tissue and Fat Cells

Fat cells are large and swollen by the lipid accumulated within them. Isolated fat cells are spherical, but when they are crowded together in adipose tissue, they are distorted into polygonal shapes. The cytoplasm is compressed into a thin rim around the cell, with the nucleus flattened and pushed to one side of the cell (Fig. 18-9). Like all actively metabolizing cells, the fat cell has mitochondria and a Golgi complex, all compressed into the narrow rim of cytoplasm. Around each individual fat cell is a delicate network of fibers, forming a coat for the plasma membrane. These fat cells then form aggregates in connective tissue, building up the fat pads or *adipose tissue* which is distributed in specific regions of the body as superficial or deep fat.

Fat Distribution

Fatty tissue appears to develop from special primitive connective tissue cells, but superficial fat is deposited only in certain regions of the body such as the abdominal wall, thighs, buttocks, and shoulders (Fig. 18-10). Even in the very obese individual, certain parts of the body never develop fat (e.g., the eyelid, external ear, and penis).

The distribution of fat is also regulated in part by the *endocrine glands,* for although the pattern of fat distribution is the same in both sexes before puberty, in normal women the average thickness of subcutaneous fat is almost twice that of normal men, and a greater proportion of the fat is on the breasts, legs, and hips, causing the softly rounded contours of the normal woman. In the male, fat is deposited primarily on the trunk, especially on the abdomen, the back, and the nape of the neck. *Deep fat* is concentrated on the mesenteries and around the kidneys.

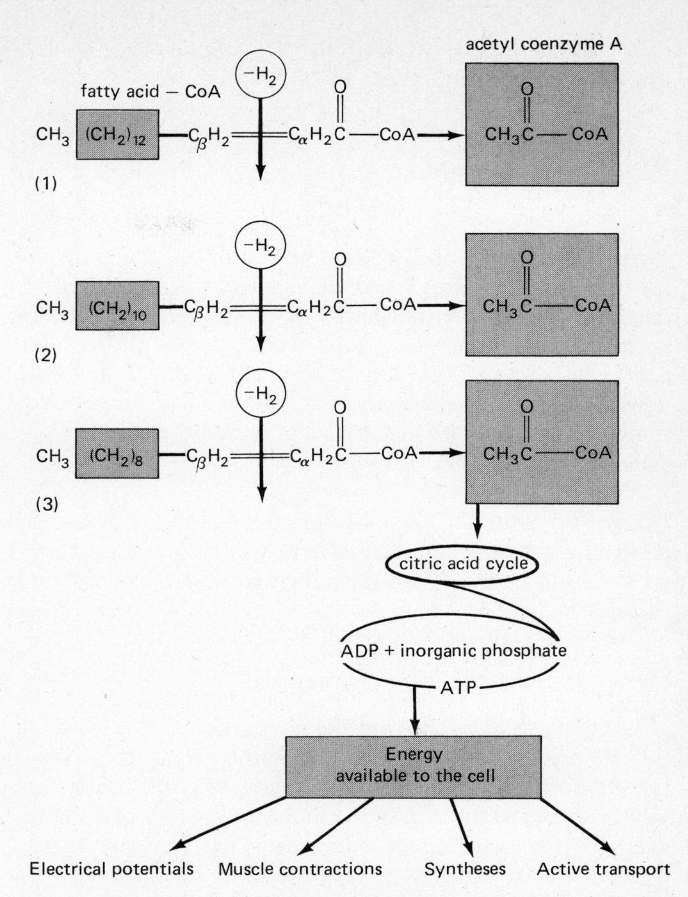

Figure 18-11. **Beta oxidation of fatty acids to yield acetyl coenzyme A and energy. In reaction 1, the long $(CH_2)_{12}$ fatty acid first combines with coenzyme A (CoA) to form fatty acid—CoA. The loss of two hydrogen atoms (oxidation) from the alpha and beta carbons results in the formation of a double bond between these carbons. This double bond is then split to yield acetyl coenzyme A and a $(CH_2)_{10}$ fatty acid. In reaction 2, the $(CH_2)_{10}$ fatty acid combines with another molecule of coenzyme A and the oxidation process is repeated to yield acetyl coenzyme A and a $(CH_2)_8$ fatty acid. Reaction 3 shows the continuation of the stepwise degradation of the fatty acid. Ultimately, the entire fatty acid is split into acetyl coenzyme A, which enters the Krebs tricarboxylic (citric acid) cycle.**

Figure 18-10. **Fat distribution in the normal man and woman, and in an obese woman.**

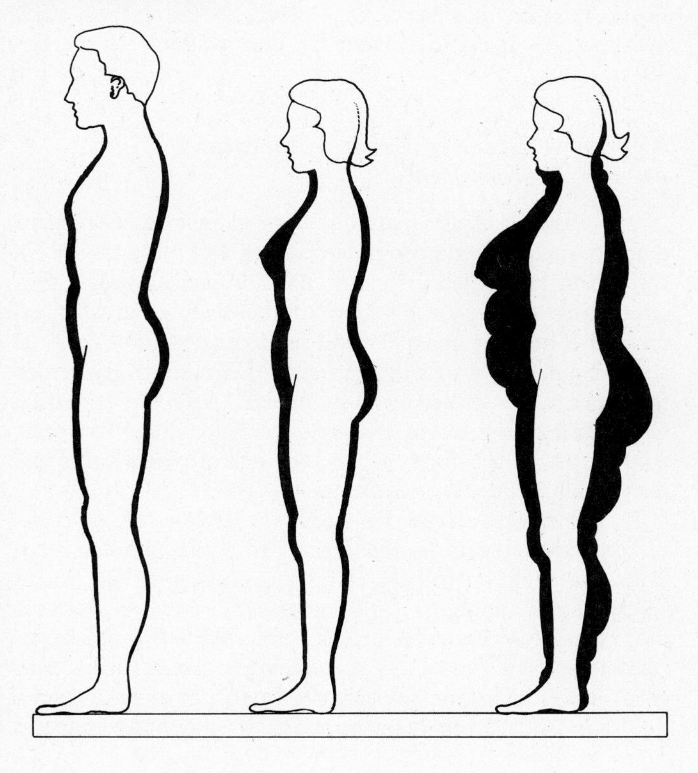

Oxidation of Fats to Yield Energy—Beta Oxidation

Beta oxidation occurs in the liver and in adipose tissue. The long-chain fatty acids are oxidized in steps. Oxidation and splitting of the molecule usually occur at the second carbon atom from the end carboxyl group. This is the beta carbon (Fig. 18-11). At each step the fatty acid chain is shortened by 2 carbon atoms, through the release of a 2-carbon fragment combined with coenzyme A, to form acetyl coenzyme A. In the complete oxidation of fatty acids, acetyl-coenzyme A enters the same metabolic pathway as the acetyl-coenzyme A derived from pyruvic acid in glucose metabolism. The fatty acids are oxidized ultimately to CO_2 and H_2O and energy. The energy yield is high, for a net gain of 146 molecules of ATP is achieved by the complete oxidation of one molecule of fatty acid.

Functions of Fats

Stored fat—triglycerides

Energy Reserve. Most fat is stored in fat depots in the form of triglycerides. Even in a normal person, there is enough fat reserve stored in the tissues to allow for normal energy consumption for 3 to 7 weeks. Fat is utilized for energy, however, only when the level of carbohydrate is low. This is sometimes called the *fat-sparing effect* of carbohydrate. Because fats are metabolized before proteins, they are called *protein sparers.*

Insulating Effect. The fat layer covering the body acts as an insulator to prevent excessive heat loss. Emaciated persons tend to have a low body temperature and to chill very easily.

Cholesterol and phospholipids

1. Cholesterol is used by the adrenal cortex to produce the adrenal steroid hormones and by the gonads to produce the sex hormones, although other pathways of synthesis are also available to these tissues. Cholesterol is essential, however, for the synthesis of bile salts. The liver uses cholic acid, which it forms from cholesterol to produce the bile salts (Fig. 18-12), which are of great importance in the digestion and absorption of fats (Secs. 17-5 and 17-6).

2. Both *cholesterol* and *phospholipids* are essential for the normal structure and permeability of the cell, contributing mainly to the *structure of the membranes* surrounding and penetrating the cell. This means that not only the cell membrane itself, but the endoplasmic reticulum, the mitochondrial membranes, the Golgi apparatus, and the nuclear membrane are dependent for their integrity on these water-insoluble substances, together with certain insoluble proteins (Sec. 4-5).

3. The *water impermeability* of the skin is maintained by cholesterol and phospholipid, preventing excessive evaporation of water and making it resistant to the absorption of water-soluble substances.

Figure 18-12. **Cholic acid is formed from cholesterol in the liver. The liver then synthesizes bile salts from the cholic acid.**

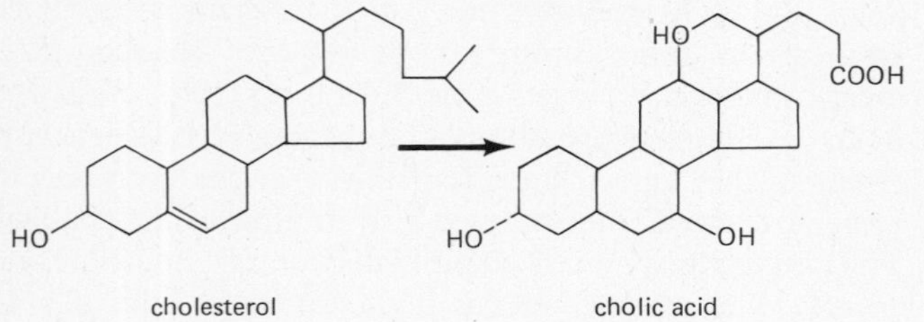

cholesterol → cholic acid

4. Blood from which lipids have been removed does not *coagulate.* This is probably caused by the absence of a lipoprotein found in the platelets. (See Secs. 14-11 and 14-15 for a discussion of the importance of the platelets to blood coagulation.)

5. High concentrations of circulating cholesterol and other lipoproteins are associated with lipid deposition in the walls of the arteries (atherosclerosis). This has been discussed in Sec. 12-12.

Hormonal Regulation of Fat Metabolism

Hormones control the balance between lipolysis and fat deposition. Epinephrine, glucagon, growth hormone, and ACTH stimulate lipid breakdown and the release of fatty acids from lipid stores. These hormones activate lipases, probably through the intermediate action of cyclic AMP.

Insulin is the most important regulator of triglyceride synthesis and deposition. It is an indirect mechanism, for it is achieved by promoting glucose passage into the cell, so that glucose is preferentially used for oxidation, thus permitting fatty acids to be used for the synthesis of triglycerides and cholesterol.

The fatty acid content of the blood is also an important regulator of fat metabolism, controlling the rate of lipid oxidation by the tissues. Storage of fats will not occur in the starved animal given insulin, but under normal physiological conditions, the homeostatic mechanisms that regulate insulin and epinephrine secretion are closely associated with the circulating glucose levels; a *high blood sugar* stimulates the release of insulin, which then facilitates fat synthesis and storage; a *fall in blood glucose* causes the secretion of *epinephrine,* which has the reverse effect, liberating fatty acids into the blood and increasing the fuel available to the tissues.

Ketosis—Effect of Lack of Insulin on Fat Metabolism

In the absence of adequate amounts of insulin, as seen in diabetes mellitus, adipose tissue is unable to build triglycerides from fatty acids, so that the amount of fatty acids circulating through the blood to the liver is greatly increased. The liver normally will oxidize these fatty acids to acetyl coenzyme A or will synthesize them into triglycerides for storage. *In the absence of insulin, only the oxidative mechanism can function properly,* so that neither liver nor adipose tissue can store fats, and the rate of oxidation of the fatty acids is greatly speeded up.

In the normal individual, acetyl-coenzyme A is completely oxidized as it passes through the ATP-coupled reactions that yield energy for cellular processes. When excessively large amounts are produced, however, acetyl coenzyme A accumulates and is converted to form *ketones,* such as acetone (CH_3COCH_3). Ketones can be utilized in small amounts as fuel by other tissues, but they are produced in such large concentrations in untreated diabetic patients that they accumulate in the blood. The normal ketone level in the blood is rarely more than 1 mg/100 ml, the normal urinary excretion less than 500 mg in 24 hr. In uncontrolled diabetic patients, the blood ketone level may reach 368 mg/100 ml (*ketosis*), and urinary excretion 75 g in 24 hr (*ketonuria*). These ketones combine with base in the tissue fluids to form salts that deplete the alkali reserve of the body (Sec. 14-17). This loss of base causes an acidosis, which if severe enough can lead to coma.

It is important for the diabetic patient to remember that because carbohydrate exerts a fat-sparing effect, small amounts of carbohydrate, taken at frequent intervals, will prevent the formation of ketones.

Neural Regulation of Fat Metabolism

INNERVATION OF ADIPOSE TISSUE

In addition to its sensitivity to hormonal concentrations, fat is regulated by nervous activity. Most adipose tissue is well supplied with nerve fibers—sympathetic, parasympathetic, and sensory. Stimulation of the *parasympathetic* nerves increases fat deposition, whereas *sympathetic* stimulation, resulting in the release of norepinephrine, accelerates the release of fatty acids from the stored fat.

Destruction of Fatty Acids—Oxidation to Hydroperoxides

Highly unsaturated fatty acids are readily oxidized in the presence of oxygen. This *rapid oxidation,* unlike stepwise beta oxidation, results in the formation of hydroperoxides. Hydroperoxides, which are also produced by ionizing radiations, are strong oxidizing agents and are extremely toxic. The formation of hydroperoxides does not appear to occur in humans to the same extent as in experimental animals.

Protective Action of Alpha Tocopherol (Vitamin E): Alpha Tocopherol and Aging

Alpha tocopherol, found in most vegetable seed oils except safflower seeds, is an antioxidant and appears to prevent the oxidation of unsaturated fatty acids (Sec. 21-7). It is important that unsaturated fats be protected from oxidation during storage in foods, digestion in the gut, and in the body. In the absence of alpha tocopherol, symptoms similar to those of muscular dystrophy appear, perhaps because of the toxic effects of the hydroperoxides on muscle tissue.

It has been suggested that alpha tocopherol may slow the aging process, in humans as well as in experimental animals, through its antioxidant effects. Antioxidants act to absorb or "quench" the reactive energy of *free radicals,* which are extremely reactive chemical entities containing an unpaired electron. In the reduction of molecular oxygen, important reactive intermediates are formed, such as $\cdot OH$ and $HO_2\cdot$ radicals.

One of the *theories of aging* is that the accumulation of free radicals damages vital molecules such as nuclear DNA, lipids in cell membranes, and collagen molecules in connective tissue. Consequently, experiments have been carried out in which antioxidants have been fed to animals. In some of these experiments, the average life span of the animals (mice) has been lengthened from 725 to 900 days, but Comfort et al. (1971) have reported that untreated mice of the same strain also live to 900 days.

Cited References

BANTING, F. G., and C. H. BEST. "The Internal Secretion of the Pancreas." *J. Lab. Clin. Med.* **7:** 251 (1922).

COMFORT, A., I. YOUHOTSKY-GORE, and K. PATHMANATHAN. "Effect of Ethoxyquinone on the Longevity of C_3H Mice." *Nature (London)* **229:** 254 (1971).

FALCK, B., J. HAGGENDAL, and C. H. OWMAN. "The Localization of Adrenaline in Adrenergic Nerves in the Frog." *Quart. J. Expl. Physiol.* **48:** 253 (1963).

GUILLEMIN, R., and R. BURGESS. "The Hormones of the Hypothalamus." *Sci. Am.,* Nov. 1972.

VON EULER, U. S. *Noradrenaline: Chemistry, Physiology, Pharmacology and Clinical Aspects.* Charles C Thomas, Publisher, Springfield, Ill, 1956.

Additional Readings

Refer also to readings for Chapter 17.

BOOKS

FALLS, H. B., ed. *Exercise Physiology.* Academic Press, Inc., New York, 1968.

NEWSHOLME, E. A., and C. START. *Regulation in Metabolism.* John Wiley & Sons, New York, 1973.

RODAHL, K., and B. ISSIBUTZ eds. *Fat as a Tissue.* McGraw-Hill Book Company, New York, 1964.

WHELAN, W. J., ed. *Control of Glycogen Metabolism.* Academic Press, Inc., New York, 1968.

ZÖLLNER, N., and S. ESTREN, eds. *Thannhauser's Textbook of Metabolism and Metabolic Disorders,* 2nd ed., Vols. 1 and 2. Grune & Stratton, Inc., New York, 1962.

ARTICLES

BARRINGTON, E. J. W. "Hormones and Digestion; Hormones and Metabolism I and II." In *An Introduction to General and Comparative Endocrinology,* 2nd ed. Clarendon Press, Oxford, 1963, Chaps. 2 to 4.

BENTLEY, P. J. "Hormones and Nutrition." In *Comparative Vertebrate Endocrinology.* Cambridge University Press, New York, 1976, p. 173.

CHAPMAN, C. B., and J. H. MITCHELL. "The Physiology of Exercise." *Sci. Am.,* May 1965.

FREINKEL, N. "Aspects of the Endocrine Regulation of Lipid Metabolism." In *Metabolism and Physiological Significance of Lipids,* R. M. C. Dawson and D. N. Rhodes, eds. John Wiley & Sons, New York, 1964.

FRIEDEN, E., and H. LIPNER. "Insulin and Glucagon; The Adrenal Hormones; Other Vertebrate Hormones." In *Biochemical Endocrinology of the Vertebrates.* Prentice-Hall, Inc., Englewood Cliffs, N.J., 1971, Chaps. 4, 7, and 9.

HOCH, F. L. "Metabolic Effects of Thyroid Hormones." In *Handbook of Physiology,* Sec. 7, Vol. 3. The Williams & Wilkins Company, Baltimore, 1974.

HUIJING, F. "Glycogen Metabolism and Glycogen Storage Diseases." *Physiol. Rev.* **55:** 609 (1975).

SHIMAZU, T., H. MATSUSHITA, and K. ISHIKAWA. "Cholinergic Stimulation of the Rat Hypothalamus: Effects on Liver Glycogen Synthesis." *Science* **194:** 535 (1976).

STEINBERG, D. "Fatty Acid Mobilization—Mechanisms of Regulation and Metabolic Consequences." In *The Control of Lipid Metabolism,* Biochemical Soc. Symp. No. 24, J. K. Grant, ed. Academic Press., Inc., New York, 1963.

TEPPERMAN, J. "The Adrenal Medulla"; "Endocrine Function of the Pancreas." In *Metabolic and Endocrine Physiology,* 3rd ed. Year Book Medical Publishers, Inc., Chicago, 1974. Chaps. 9 and 10. (Paperback.)

TOPOREK, M., and P. H. MAURER. "Metabolic Biochemistry." In *Pathological Physiology: Mechanisms of Disease,* W. A. Sodeman, Jr., and W. A. Sodeman, eds. W. B. Saunders Company, Philadelphia, 1974.

UNGER, R. H., and L. ORCI. "Physiology and Pathophysiology of Glucagon." *Physiol. Rev.* **56:** 778 (1976).

WEBER, G., R. L. SINGHAI, N. B. STAMM, and S. K. SRIVASTAVA. "Hormonal Induction and Suppression of Liver Enzyme Biosynthesis." *Fed. Proc.* **24:** 745 (1965).

YOUNG, V. R., and N. S. SCRIMSHAW. "The Physiology of Starvation." *Sci. Am.,* Oct. 1971.

Chapter 19

Protein Regulation and Tissue Growth, Healing, and Regeneration

To see a World in a grain of sand,
And a Heaven in a wild flower,
Hold Infinity in the palm of your hand,
And Eternity in an hour.

William Blake, "Auguries of Innocence"

THE GROWTH OF tissues, their inflammatory response to injury, and their subsequent healing or regeneration, are all dependent upon protein metabolism. Protein metabolism in turn is controlled by hormones that regulate protein synthesis (*anabolism*) and utilization (*catabolism*).

19-1 Nitrogen Balance

Although protein is not stored in the tissues in the sense of being an inactive accessory inclusion within the cell, it is present as about 15 per cent of the dry weight of the tissues in the form of the structure of the cells and as metabolically active molecules. In order to increase the number and size of the cells, protein in the requisite amounts must be available, and this protein must include the essential amino acids. During periods of growth or repair of injured portions of the body, during pregnancy and lactation, where the synthesis of complex molecules requires the availability of large amounts of protein, the amount of nitrogen ingested in the form of protein is far greater than the amount eliminated as nitrogenous waste in the urine and feces. The body is said to be in *positive nitrogen balance*.

During periods of starvation, high fevers, and wasting diseases, where the ingested protein is far less than the amount continuously eliminated through the breakdown of the tissues, the body is in *negative nitrogen balance*. This protein deficiency prevents the proper repair of tissues, synthesis of plasma proteins and hemoglobin is interrupted, the resistance to infection is sharply diminished due to the decrease in antibodies, and the very structure of the cells breaks down. If carbohydrate and fat are ingested, despite a brief lack of ingested protein, most structural protein is spared. *Nitrogen equilibrium* occurs when the amount of nitrogen taken in as food is equal to the amount eliminated as waste; the *minimum protein requirement* can be calculated as the smallest amount of protein needed by the individual to maintain nitrogen equilibrium; this amount will obviously differ during the various physiological and pathological states mentioned. Although the figure given for the minimum protein requirement varies widely, the recommended protein intake for the average adult male is 56 g/day, and that for the adult female is 46 g/day. A pregnant woman requires an additional daily intake of 30 g of protein; during lactation the additional amount of protein needed is 20 g/day.

19–2 Protein Metabolism and Tissue Growth

Protein Metabolism

REGULATION BY THE LIVER

Amino acids absorbed from the small intestine pass directly from the blood vessels that drain the digestive tract into the large hepatic portal vein that empties into the liver sinuses. The metabolic fate of the amino acids is decided in the liver. The specific pathway entered is regulated by several hormones and other feedback mechanisms.

Synthesis of Plasma Proteins. More than 85 per cent of the plasma proteins (fibrinogen, albumin, and some of the globulins) are made by hepatic cells. The remaining 15 per cent are the gamma globulins synthesized by the plasma cells (Sec. 14-9). The amount of circulating plasma protein appears to regulate the rate of synthesis of plasma proteins: a marked loss of these proteins, which occurs in some severe diseases of the kidney (nephrosis), causes a tremendous compensatory hypertrophy of the liver and a rapid synthesis of plasma proteins in an attempt to replace the lost protein.

Synthesis of Amino Acids—Transamination. The liver can synthesize the nonessential amino acids (Sec. 3-3.1) by transferring an amino group from one amino acid to a keto-acid with a structure otherwise the same as that of the amino acid to be formed. This is transamination.

$$\underset{\text{amino acid } R_1}{NH_2\overset{R_1}{\overset{|}{C}}HCOOH} + \underset{\text{keto acid } R_2}{O{=}\overset{R_2}{\overset{|}{C}}COOH} \rightleftharpoons \underset{\text{keto acid } R_1}{O{=}\overset{R_1}{\overset{|}{C}}COOH} + \underset{\text{amino acid } R_2}{NH_2\overset{R_2}{\overset{|}{C}}HCOOH}$$

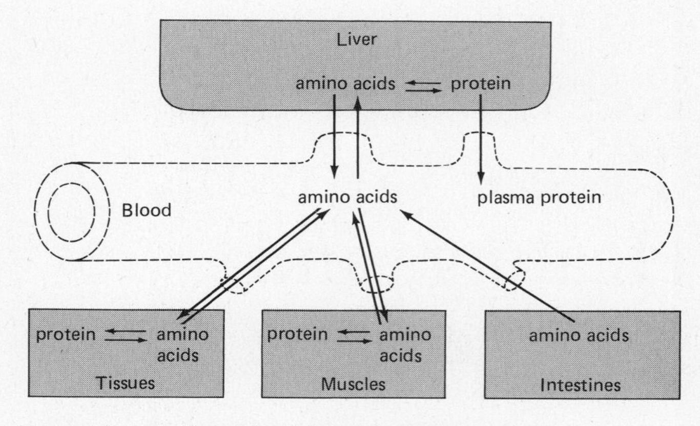

Figure 19-1. **Amino acid and protein homeostasis.**

Figure 19-2. **The ornithine cycle.**

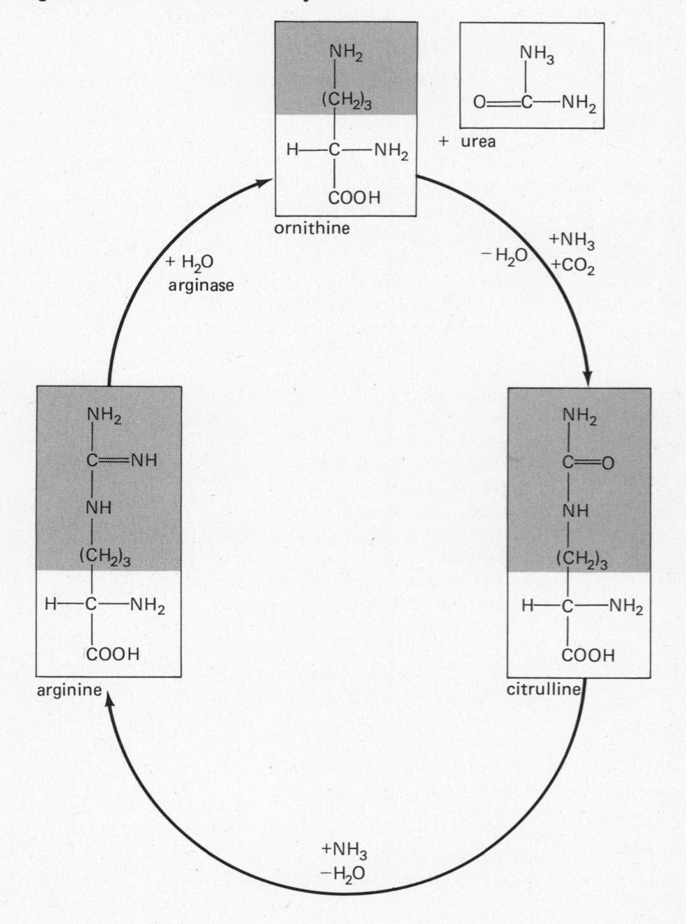

Entry into Other Metabolic Pathways—Deamination. The liver is the main site of deamination of amino acids, an essential reaction before they can enter into other metabolic pathways. The possible pathways include the entry of deaminated compounds into the citric acid cycle to yield energy, or their conversion into carbohydrate or fat. These metabolic pathways were discussed in Sec. 5-12.

Storage of Amino Acids. The liver acts as a *homeostatic regulator* of amino acid concentration, storing the excess and releasing amino acids into the circulation for their use by other tissues when the level in the blood is low (Fig. 19-1). The average amino acid content of circulating blood is 30 mg/100 ml, a level that may rise immediately after a meal but returns to normal within half an hour.

Removal of Ammonia Via Urea Formation. The liver is essential for the formation of urea from ammonia. Most of the ammonia for urea formation is derived from deamination and transamination in the kidney, and the kidney in turn is responsible for the excretion of the urea synthesized by the liver. Ammonia accumulates in the blood when the liver is absent or seriously diseased and acts as a toxin that may induce coma.

A subject on a mixed diet excretes about 9 to 13 g of nitrogen daily, most of which is in the form of urea. The normal range of urea in the blood is 18 to 35 mg/100 ml. Higher levels are abnormal and indicate a severe disorder of the kidney.

THE ORNITHINE CYCLE. The formation of urea is a complex procedure, involving many enzymatic reactions. Basically, it consists of the *stepwise addition of NH_2 or NH_3 groups,* which have been split off from many different amino acids, to one specific amino acid, *ornithine* (Fig. 19-2). Ornithine has one nitrogen-containing group in its residue. The addition of ammonia and carbon dioxide transforms it to *citrulline.* The addition of another ammonia molecule to citrulline results in the formation of *arginine,* which has three nitrogenous groups in its residue. Ultimately, the *enzyme arginase splits off urea* from arginine, leaving ornithine, free once more to pick up nitrogenous fragments. This is the Krebs' "ornithine cycle."

The liver also produces another nitrogenous product, *hippuric acid,* but in small quantities (0.7–1.0 mg/day). Its usefulness is chiefly in its diagnostic value for normal liver function.

REGULATION BY THE KIDNEY

Ammonia Formation. The kidney normally *retains plasma proteins* through the integrity of the walls of the glomerular tuft, so that these molecules do not appear in the glomerular filtrate. The kidney *synthesizes ammonia* (NH_3) about 0.4 to 1 g/day on a mixed diet, the amount of which varies according to the ingestion of acids or bases and according to the pH of the urine. This mechanism is important for saving sodium or potassium ions and regulating the pH of the blood (Sec. 14-17).

REGULATION BY MUSCLE

Muscle also *stores amino acids,* the amount sometimes being 10 to 15 times more than the amount found in blood. As a result of muscle activity, an energy-rich nitrogenous compound, *creatine phosphate,* is broken down to yield creatine, phosphate, and energy. This is the energy that is coupled to ATP synthesis and muscle contraction. In this series of reactions, *creatinine* is formed from creatine and excreted in the urine. After urea, creatinine is the nitrogenous compound found in greatest concentration in the urine. Its excretion amounts to approximately 1 to 1.5 g/day.

19–3 Hormonal Regulation of Protein Metabolism

Provided that sufficient amounts of protein are available in the diet (or in the case of the fetus, from the maternal circulation), the rate of growth is controlled by hormones. In the human there are regular growth spurts: the growth rates of the embryo and fetus are considerably greater than that of the child, and the rate of growth continues to slow until puberty, at which time growth accelerates markedly.

Anabolic processes leading to tissue growth are dependent upon the secretions of the endocrine glands, which in turn are controlled by the hormones of the hypothalamus. The long-term cyclic mechanisms responsible for the time-related growth surges are not understood. However, the resulting hormonal fluctuations can be correlated to patterns of growth.

Anabolic Hormones and Tissue Growth

SOMATOTROPIN (GROWTH HORMONE)

Regulation of Somatotropin Release. There is much physiological evidence to show that the hypothalamus regulates the release of somatotropin through hypothalamic secretion of somatotropin-releasing hormone (SRH). The release of somatotropin is prevented by hypothalamic *somatostatin* (Sec. 11-3).

Apart from the long-term cycles of growth, there is a fine, day-to-day and hour-to-hour control of growth hormone release that is correlated to the immediate *metabolic state* of the organism. A drop in blood sugar due to prolonged fasting, strenuous exercise, or insulin administration rapidly raises somatotropin levels. Feeding causes a drop in plasma levels of somatotropin (Fig. 19-3). These changes in carbohydrate metabolism may affect the hypothalamus through its glucose-sensitive cells (Sec. 21-2).

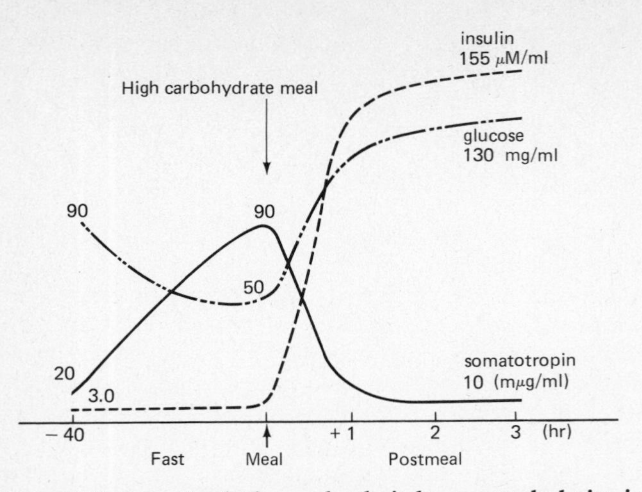

Figure 19-3. **Low blood glucose levels induce a marked rise in somatotropin titers, whereas insulin levels are very much depressed. A meal rich in carbohydrates raises both blood glucose and insulin levels, and somatotropin concentration falls. [Adapted from R. Roth, S. M. Glick, R. S. Yalow, and S. A. Berson, *Science* 140:987 (1963). Copyright 1963 by the American Association for the Advancement of Science.]**

Metabolic Effects of Somatotropin. Somatotropin stimulates protein synthesis by increasing the *transport of amino acids* across cell membranes and by affecting transcription or translation steps leading to protein synthesis. Plasma amino acid levels are raised. This hormone also inhibits glucose utilization by muscle (*diabetogenic action*), thereby elevating blood glucose. Somatotropin mobilizes fatty acids from fat depots: this is its *ketogenic effect.* The increase in available fat may play a role in providing extracellular energy for growth.

Somatotropin and Growth Patterns. FETAL AND EARLY POSTNATAL GROWTH. In the human up to about 3 years of age fetal and early postnatal growth do not appear to be dependent on growth hormone, but its absence after this time slows or completely stops growth. Although one would expect increased levels of somatotropin during *puberty,* the values in adolescents do not seem to be markedly different from the adult range. This may be explained by a more rapid turnover of somatotropin by the growing tissues, or it may be caused by the wide range of fluctuations of hormone levels during the day, depending upon the metabolic state.

DURING PREGNANCY. Somatotropin levels increase during pregnancy, probably because of the production of this hormone by the placenta, as well as an increased synthesis by the maternal pituitary gland.

DWARFISM. Dwarfism of pituitary origin seems to be related to very low growth hormone levels. Although research in this field is comparatively new, it is extremely active; it is hoped that SRH, if available in sufficient amounts, may be useful in inducing normal growth in children. Somatotropin, unlike the androgenic anabolic hormones, has no virilizing effects.

GIGANTISM AND ACROMEGALY. Gigantism and acromegaly are the result of abnormally rapid growth of the tissues. Hormones exert their effects against the individual genetic background and nutritional environment of the cells, and the results depend largely upon the stage of growth already achieved. Secretion of somatotropin in excessive amounts at an early age will cause the skeletal system and the soft tissues to grow at an abnormally rapid rate, and *gigantism* will result. If, however, growth has been normal during the first two decades of life, but subsequently a tumor of the pituitary gland causes secretion of somatotropin after the longitudinal growth of the bones has been

Figure 19-4. **Normal variations in body size: (1) child, (2) pregnant woman, and (3) adult male. Abnormal variations in body size: (4) extreme emaciation due to starvation, (5) giant, (6) acromegaly, (7) pituitary dwarf, and (8) cretin.**

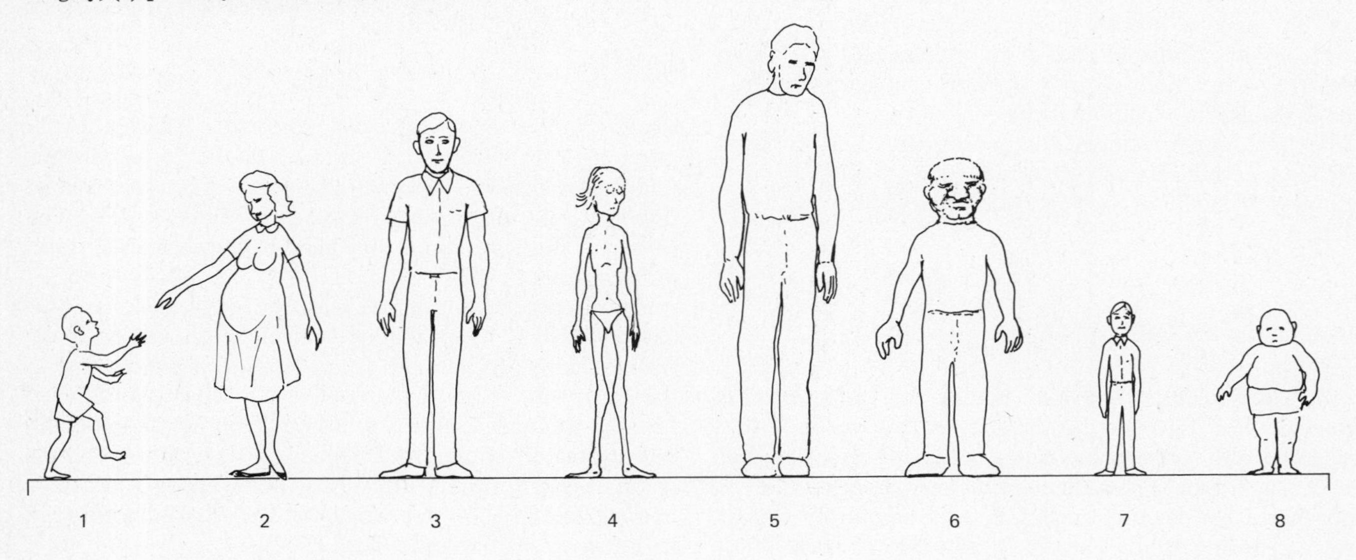

completed, the bones will develop in girth. This thickening of the bones is particularly noticeable in the jaws, the cheek bones, and the skull, resulting in an apelike appearance. This type of abnormal growth is called *acromegaly,* or literally "large extremities." Acromegaly is accompanied also by an increase in size of the viscera and a thickening of the skin and subcutaneous tissues. Figure 19-4 illustrates some normal and abnormal growth patterns.

Prolactin

Prolactin is a pituitary hormone associated with lactation, but human prolactin is almost indistinguishable in most of its properties from human growth hormone. This is not true for other species, however, in which prolactin is quite distinct from growth hormone and has only lactogenic properties.

Somatomedins

Serum contains a number of substances that promote growth. Some of these factors are dependent upon growth hormone and in turn mediate the action of growth hormone on protein synthesis and tissue growth. This interdependence of the serum factors and growth hormone action was first demonstrated experimentally by Salmon and Daughaday (1958).

In their experiments, these investigators showed that there is a very low uptake of labeled sulfate into the cartilage of hypophysectomized rats. Hypophysectomized animals do not grow. The administration of growth hormone to these animals increases the uptake of labeled sulfate into cartilage in vivo, but has no effect in vitro. However, serum taken from normal rats will enhance sulfate uptake in vitro by cartilage from hypophysectomized rats. This means that normal serum has a growth-promoting factor that is dependent upon growth hormone, for this factor is not found in the serum of hypophysectomized animals. These experiments also indicate that the serum factor is necessary for growth hormone to cause skeletal growth.

There are several such growth-promoting polypeptides in serum, and those that are dependent on growth hormone and that exert an anabolic effect on tissues are called *somatomedins* (they *mediate* the effect of somatotropin on tissue growth).

In humans, high somatomedin levels are associated with acromegaly, and somatomedin levels are extremely low in pituitary dwarfism.

Insulin

Insulin is an extremely effective anabolic hormone that acts synergistically to increase the effects of somatotropin on growth. A hypophysectomized rat, with its pancreas also removed, will not grow at all. Growth cannot be adequately restored by somatotropin or insulin alone, but a combination of these two hormones enhances growth remarkably (Fig. 19-5).

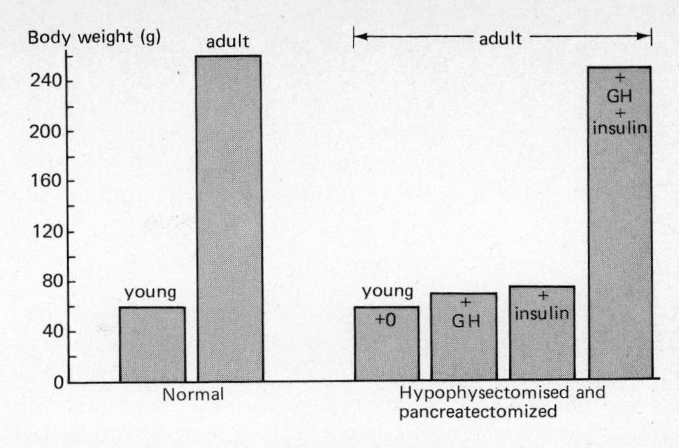

Figure 19-5. **The synergistic effect on growth of growth hormone (GH) and insulin. Young rats that have been hypophysectomized and pancreatectomized do not grow at all without hormonal replacement.**

The anabolic effects of insulin can be attributed to:

1. *Increased active transport of amino acids* into cells, where they can be used for protein synthesis.
2. *Increased DNA transcription* in the nucleus, so that more messenger RNA is produced.
3. *Acceleration of the translation* of messenger RNA by the ribosomes, which speeds up protein synthesis.

   These effects of insulin on protein synthesis are linked to its *depressant action on cyclic AMP levels.* Insulin binds with specific insulin membrane receptors and may inhibit adenyl cyclase. This is in direct contrast to the action of epinephrine on this enzyme (Sec. 10-8).
4. Insulin may also enhance protein synthesis indirectly by *facilitating glucose transport* into the cell, thus making more amino acids available for protein synthesis while the glucose molecules can be used for energy.

In the absence of insulin, protein continues to be broken down as usual, but there is no replacement. The wasting of the tissues and their consequent weakness is a very serious result of diabetes mellitus.

Androgens

The development of muscle is dependent on the testicular hormones. Castrate males usually have smaller, weaker muscles than normal males. Females with adrenal tumors that produce androgens develop strong masculine-type muscles.

It is interesting that the anabolic effects of androgens are most marked in those muscles intimately associated with sexual activity, such as the muscles of the perineal region (the area between the posterior part of the external genitals and the anus). Androgens have a somewhat lesser effect on those muscles involved in the general mating activity of the male (shoulders, chest, and upper back in man; head, neck, and forequarters in the guinea pig), and only a slight effect on muscles in general.

Like insulin, androgens act synergistically with somatotropin. Unlike the continuous effects of growth hormone, however, the effects of testosterone on protein synthesis cease after a few months of prolonged administration.

Specific growth factors

Other growth-promoting polypeptides have been identified. These include nerve growth factor, epidermal growth factor and fibroblast growth factor. These differ from the generalized anabolic effects of growth hormone in that the growth factors selectively stimulate growth of certain tissues or cells.

Nerve Growth Factor. First isolated by Levi-Montalcini (1968), the nerve growth factor is widely distributed in nature and most abundant in the salivary glands of rodents and in snake venom. The importance of this polypeptide in the development and maintenance of sensory and sympathetic neurons was discussed in Sec. 8-2.

Recent investigations have demonstrated that nerve growth factor receptors are present at synapses in the central nervous system. It is possible that this factor may also be needed for the normal development of central neurons, as well as peripheral ones. Nerve growth factor appears to be very similar in structure and metabolic function to insulin.

Epidermal Growth Factor. A polypeptide originally isolated from the submaxillary glands of mice, the epidermal growth factor causes proliferation and keratinization of epidermal tissue. Minute quantities injected daily will cause the precocious opening of the eyelids and toothbud eruption in newborn mice, because of the enhanced epidermal growth and keratinization of these tissues. This growth factor stimulates the synthesis and accumulation of protein and RNA. A similar substance isolated from human urine will stimulate epithelial cell growth (Fig. 19-6).

Fibroblast Growth Factor. A polypeptide that has been isolated from the brain and from the pituitary gland, the fibroblast growth factor provokes cell division in fibroblasts and also in several other cell types. This growth factor seems to be able to induce limb regeneration in frogs.

Figure 19-6. **Epithelial cells of chick embryo cornea grown in organ culture. (A) Control cornea, in culture medium only, is one to two cells thick. (B) When mouse epithelial growth factor (EGF) is added (20 ng/ml), the cornea thickens to six to eight cells deep. (C) If both antiserum to mouse EGF and mouse EGF are added, the effect seen in (B) is prevented. (D) Human urine extract also contains an EGF that thickens chick cornea. [From R. H. Starkey, S. Cohen, and D. N. Orth, *Science* 189:800 (1975).]**

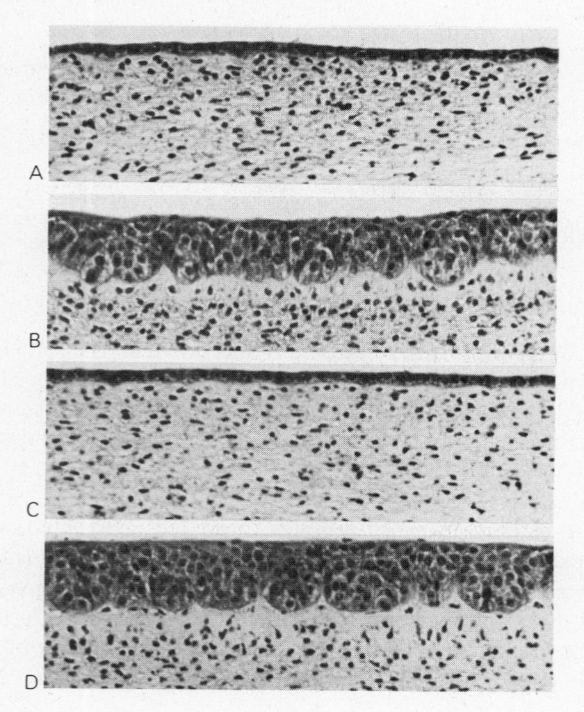

Thyroxine and Tissue Growth

The principal effect of thyroxin is to increase metabolic activity. Oxygen consumption and the utilization of foods for energy rise precipitously. The increased rate of protein metabolism includes an increase in both protein synthesis and protein catabolism. Administration of thyroxine to growing children accelerates growth markedly.

The ways by which these metabolic activities are influenced are very poorly understood. It is possible that the increased protein synthesis seen almost immediately after the administration of thyroxine is caused by an increase in protein formation by the ribosomes, whereas the later, more prolonged effects may be caused by an increase in RNA transcription (the process by which RNA is synthesized by the genes).

Growth cannot occur in the absence of thyroid hormone. Thyroxine is essential for the proper chronological development of synaptic contacts in the *developing brain,* a topic discussed in more detail in Sec. 8-2. These effects on neuronal growth may be mediated through stimulation of protein synthesis, whereas thyroidectomy depresses protein synthesis.

In *children,* hypothyroidism severely limits growth, and the resulting mental and physical retardation is known as *cretinism.* These effects are reversible only if the retardation is slight and treatment is begun early. If the cretinism is severe, even early treatment with thyroxine is ineffective.

The *adult brain,* unlike most other tissues of the body, is unresponsive to the metabolic effects of thyroxine, including enhanced protein synthesis (Fig. 19-7). The mental disturbances seen in hypothyroid humans must be caused by general metabolic disturbances such as alterations in the intracellular concentration of sodium.

A very characteristic effect of lack of thyroid hormone is the deposition of a mucoprotein substance in the skin and

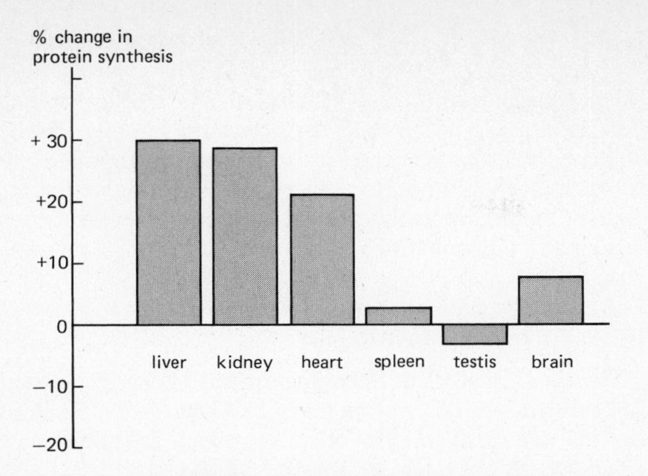

Figure 19-7. **The administration of thyroxine increases protein synthesis in most organs except the brain, spleen, and testis.**

other tissues (*myxedema*). This is probably the same protein that, in small amounts, forms the normal intercellular cement between tissue cells. The administration of thyroxine diminishes the excessive accumulation of this mucoprotein.

Antianabolic Hormones

Glucocorticoids

The glucocorticoids (cortisol and cortisone) of the adrenal cortex reduce the protein stores of all cells except the liver cells. In vitro studies of muscle show that this is accomplished both by decreased protein synthesis and by an increase in catabolism of previously deposited protein.

These antianabolic effects may be attributed to a depression of RNA formation in extrahepatic tissues and to a decreased transport of amino acids into these tissues, under the influence of the glucocorticoids. This lowers the intracellular concentration of amino acids. At the same time, the catabolic activities of these extrahepatic cells release amino acids into the circulation, increasing the plasma amino acid level. Thus glucocorticoids *mobilize amino acids* from the tissues.

Figure 19-8. **Regulatory effects of glucocorticoids on protein metabolism.**

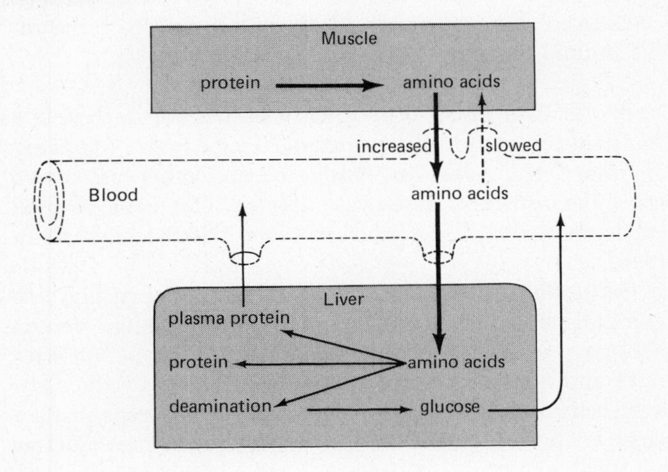

The effects of glucocorticoids on *liver* are quite different because protein stores are increased in this organ. Amino acid transport into hepatic cells is accelerated, and this movement is helped by the increase in available plasma amino acids, mobilized from extrahepatic tissues. Figure 19-8 shows the net flow of amino acids from muscle into the liver cells. As the concentration of amino acids rises inside hepatic cells, the amino acids become available for entry into several metabolic pathways:

1. Deamination.
2. Protein synthesis.
3. Formation of plasma proteins.
4. Conversion to glucose (gluconeogenesis).

Prostaglandins

Prostaglandins act antagonistically to growth hormone, at least insofar as the lipolytic effects of growth hormone are concerned. They also stimulate adenyl cyclase activity and cyclic AMP formation, activities that result in a depression of growth.

19–4 Tissue Healing and Regeneration

Inflammation

Tissue inflammation is the result of a complex series of reactions that are coordinated to isolate and destroy injurious agents and to prepare the tissue for subsequent healing or regeneration. A surprisingly uniform response is evoked by widely varied stimuli, including mechanical cuts and bruises, chemicals, irradiations, bacterial toxins, and antigen-antibody complexes.

An important response common to all conditions leading to inflammation is the release of *vasoactive substances* from the tissues. These substances profoundly affect the fluid exchange in the tissues.

Regulation of fluid exchange in the tissues

Pressure Mechanisms. The regulation of fluid exchange between the blood and the tissues involves many factors including the selective shunting of blood through capillary beds, variations in blood pressure in the capillaries, variations in osmotic pressure, and variations in the volume of blood flow. These regulatory mechanisms were discussed in Sec. 4–2.

Vasoactive Substances. Many vasoactive substances are released by the tissues themselves, causing local changes in vasomotion. Potent vasodilators include nonspecific metabolites like carbon dioxide and changes in ionic levels, especially increases in potassium concentration. The amines *histamine* and *serotonin* (5-hydroxytryptamine) and the polypeptide *bradykinin* are powerful vasodilators. *Prostaglandins* are also implicated in the inflammatory response.

HISTAMINE. Histamine is present in almost all tissues. It is found in mast cells, in basophils, and in platelets, and may even be present in the smooth muscle and endothelial cells of the blood vessels. Histamine is involved in certain types of allergic responses in which the antigen-antibody reaction causes the cells to release histamine. Antihistaminic compounds reduce the intensity of these hypersensitivity phenomena (see also Sec. 20-7).

The effect of histamine on vascular permeability is illustrated in Fig. 19-9. If large dye particles are injected into the circulation, they act as markers of areas of increased permeability. After histamine administration, the dye readily escapes from the microcirculation and can be seen as accumulations in the tissue, especially around the venules. The same selective increase in venule permeability is seen in inflammation.

SEROTONIN. Serotonin (5 hydroxy-tryptamine) is contained within blood platelets as well as in most other tissues, especially those of the gastrointestinal tract. It affects the caliber of the blood vessels, especially the small veins, but its physiological role is difficult to evaluate because it may cause constriction or dilation of the vascular bed depending upon the particular tissue and its physiological state. Similarly, serotonin increases capillary permeability more effectively than histamine in the rat but has no effect on capillary permeability in other species, including humans.

BRADYKININ. The peptide bradykinin is one of a group of vasodilator polypeptides (*kinins*), which also include the kinins of wasps and hornets. These insects mix a little histamine, serotonin, and acetylcholine to add to the pain of the sting.

PROSTAGLANDINS. These hormonelike substances were discussed in Sec. 10-9. PGE_1 induces both vasodilation and increased vascular permeability and may also stimulate leukocyte migration through capillary walls. Prostaglandins are thought to be responsible for the pain and fever that often accompany inflammation.

Figure 19-9. **The administration of histamine increases the permeability of the microcirculation so that large dye particles, injected as markers, escape from the small blood vessels, especially from the venules.**

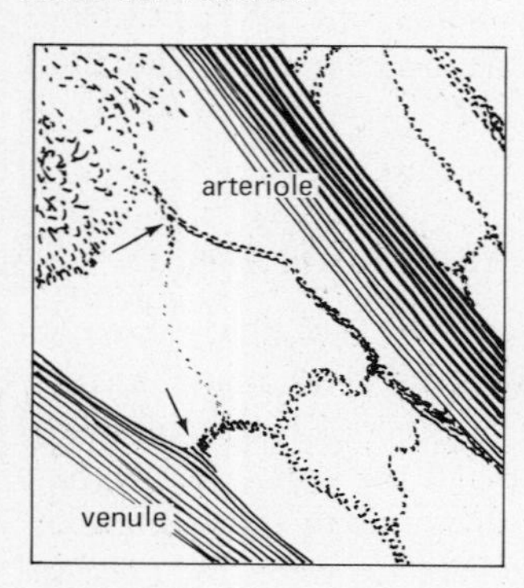

A. Control

B. After histamine administration

SIGNS OF INFLAMMATION

The classical signs of inflammation have been recognized for centuries; the description given by Galen (A.D. 130–200) lists the five *cardinal signs* of inflammation as follows.

Redness (rubor). Redness (rubor) is caused by the dilation of small blood vessels following the release of vasodilator substances from the injured tissues.

Heat (calor). The increased diameter of the blood vessels increases the flow of warm blood to the injured area, raising the temperature of the tissue.

Edema or Swelling (tumor). The vasodilator substances not only affect the caliber of the small blood vessels but also increase their permeability. A fluid *exudate,* which also includes proteins and blood cells, accumulates in the tissues.

Pain (dolor). Pain resulting from the pressure of the exudate on the local sensory nerve endings, is characteristic of inflammation. The vasodilator substances *histamine, serotonin,* and *bradykinin* may also irritate these nerve endings, causing pain.

Loss of function (functio laesa). The loss of function is probably due to the accompanying pain caused by muscle movements.

STAGES OF INFLAMMATION

The stages of inflammation are progressive and should be considered as a continuum leading ultimately to repair of the injured tissues.

Vascular Changes. VASODILATION. Although there is a transient contraction of the arterioles immediately following an injury, the release of histamine and other vasoactive substances quickly causes vasodilation. This is aggravated by neural axon reflexes, and the blood flow becomes more rapid.

INCREASED VASCULAR PERMEABILITY. Normally, the adjoining endothelial cells of the microcirculation vessels are firmly sealed by desmosome junctions (Sec. 2-2). Histamine and other vasoactive substances released by the damaged cells cause these cells to separate, leaving gaps that are large enough to permit the escape of protein molecules as

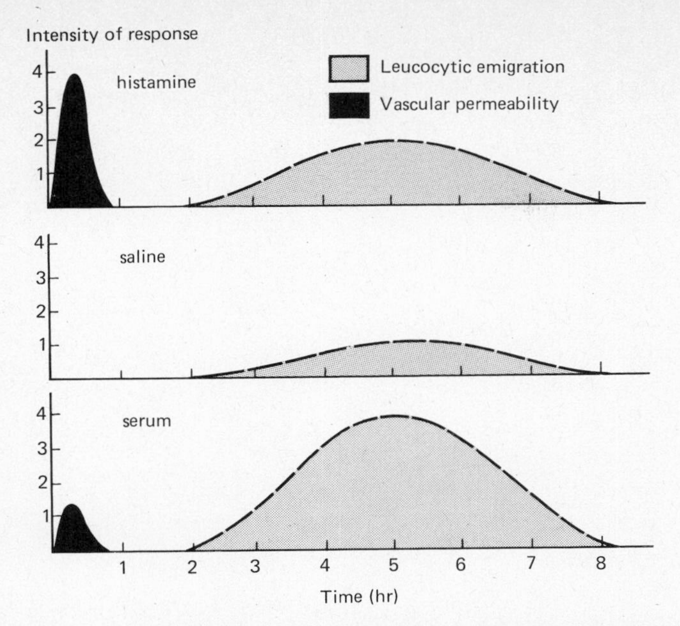

Figure 19-10. **Diagrammatic representation of the time course and intensity of leukocytic emigration and of increased vascular permeability following a single intradermal injection of isotonic saline, histamine, and serum, respectively. [From J. V. Hurley, *Ann. N. Y. Acad. Sci.* 116:918 (1964).]**

well as fluid into the tissues. This increased permeability is especially evident in the region of the venules (Fig. 19-9).

STASIS. Due to the loss of fluid and solutes, the blood becomes more concentrated and the cells clump together, forming large aggregates. This slows down the blood flow and may even stop it.

Leukocyte Responses. MARGINATION OR ADHESION. Leukocytes in particular, and other blood types to a lesser extent, adhere to the damaged endothelial surface. Although the mechanism responsible for this margination is not well understood, the adhesion of the cells may be caused by the roughness of the injured endothelium, similar to the adhesion of the platelets in hemostasis, discussed in Sec. 14–10.

EMIGRATION. The marginated leukocytes develop pseudopodia, which force the endothelial cells apart and permit the motile leukocytes to squeeze through the basement membrane and emigrate into the surrounding connective tissues. The endothelial cells close ranks after this, and no visible gaps remain.

Sometimes, red blood cells are passively forced through the endothelial gaps before they close. This passage of red blood cells into the tissues is *diapedesis.*

CHEMOTAXIS. A multitude of chemical mediators attracts the migrating leukocytes to the site of injury. This movement of the cells toward a chemical attractant is chemotaxis. Chemotaxic factors include exudates from bacteria and from injured cells. Serum proteins, especially those of the serum complement system, are potent chemotaxic agents (Fig. 19-10). The importance of the complement system in immune responses is discussed in Sec. 20–5.

PHAGOCYTOSIS. The ingestion of bacteria or other foreign particles by a cell is called phagocytosis. The main phagocytes of the blood are the *polymorphonuclear leukocytes* and the *monocytes*. When the leukocyte comes in contact with the foreign particle, the cytoplasm of the leukocyte flows out in two embracing pseudopodia, engulfing the particle. This incorporates the particle into a vesicle formed by the plasma membrane of the phagocyte (Fig. 19-11). The invaginated vesicle attracts the adhesion of the lysosomes of the leukocytes. The *lysosomes,* filled with digestive hydrolases, fuse with the membrane of the vesicle (Sec. 2–2) to form a digestive vacuole within which the foreign particle is destroyed.

After the phagocytes have destroyed the bacteria and engulfed the tissue debris, they themselves become degranulated and die. In the tissue cavity left after the battle, a central mass of fluid is formed. This fluid, *pus,* contains the necrotic tissues and the dead phagocytes. Pus may be extruded through extension of the cavity to an external or internal surface of the body, or it may be absorbed by the surrounding tissues.

Certain serum proteins, including those of the complement system, enhance the process of phagocytosis. These proteins, known as *opsonins,* coat the foreign particle, making it more susceptible to phagocytic action. Many impor-

Figure 19-11. **Phagocytosis of foreign particle (t = thorium) by a blood cell. The particles enter the cell by pinocytosis. The arrows show small vesicles forming in the cell membrane. They transport the particles into the center of the cell, where they form large granules. n = nucleus, p = pseudopodia, m = mitochondria, Go = Golgi apparatus, and re = rough endoplasmic reticulum showing ribosomes. (Courtesy of Dr. Eva Lurie Weinreb.)**

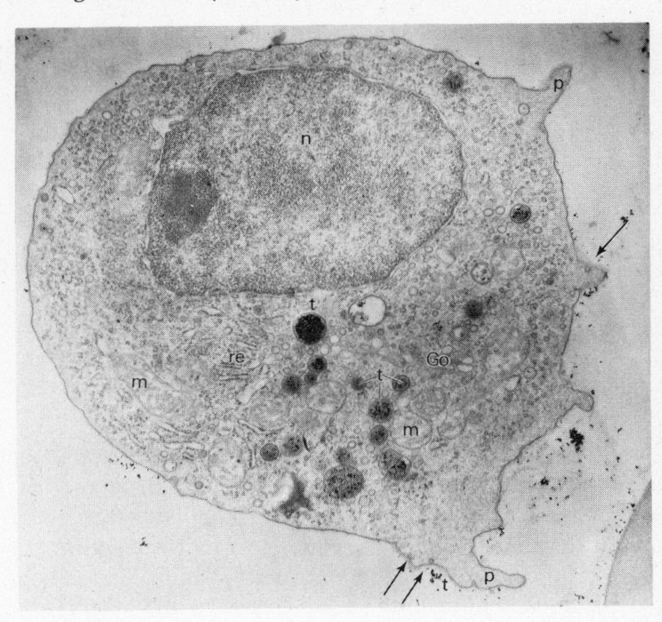

tant opsonins are antibodies, which accounts for the specificity of bacterial destruction by phagocytes. This process is discussed in more detail in Sec. 20-3.

Anti-Inflammatory Agents

Although many drugs are capable of reducing the inflammatory responses, the salicylates and the adrenal glucocorticoids are the most effective anti-inflammatory agents.

SALICYLATES

Acetylsalicylic acid (aspirin) is a weak organic acid that is readily absorbed through the stomach. It reduces the inflammatory response by decreasing histamine production and by stabilizing the membrane of the lysosome, which hinders the release of the digestive hydrolases. It is also possible that the salicylates act to suppress the production of prostaglandins and so prevent the development of pain and fever as well as the inflammatory response.

GLUCOCORTICOIDS

Glucocorticoids are extremely effective anti-inflammatory agents. They suppress all of the stages of inflammation, including lysosomal digestion. Obviously, it is unwise to administer glucocorticoids to suppress inflammation resulting from bacterial infection, because inflammation serves the extremely useful purpose of localizing infection. In the absence of the inflammatory response, the infection may spread throughout the body.

Wound Healing

Both the age and the physiological condition of the tissue affect the rate and extent of repair possible. Usually, younger individuals show greater capacity for healing, and this is associated with a more efficient blood supply to the affected part. Impaired blood supply has been shown to hinder repair markedly—wounds in immobilized parts or paralyzed limbs heal extremely slowly, if at all. Other factors certainly are involved, among them the hormonal environment, which is discussed in more detail later in this chapter. Apart from these factors, basic differences in cellular types are responsible for the degree of healing possible.

CELLULAR REGULATION OF WOUND HEALING

Cells may be divided into three main types: labile cells, stable cells, and permanent cells.

Labile Cells. These cells continue to multiply throughout life and replace those cells that are constantly being desquamated or sloughed off. Epithelial surfaces are made up of labile cells and include the stratified squamous epithelium of the skin; the surfaces of the oral cavity, vagina, and cervix; the columnar epithelium of the digestive and respiratory tracts; the lining of the excretory glands; and the transitional epithelium of the urinary bladder. The most dramatic example of regeneration is evinced by the monthly regrowth of the endometrial lining of the uterus following each menstrual period.

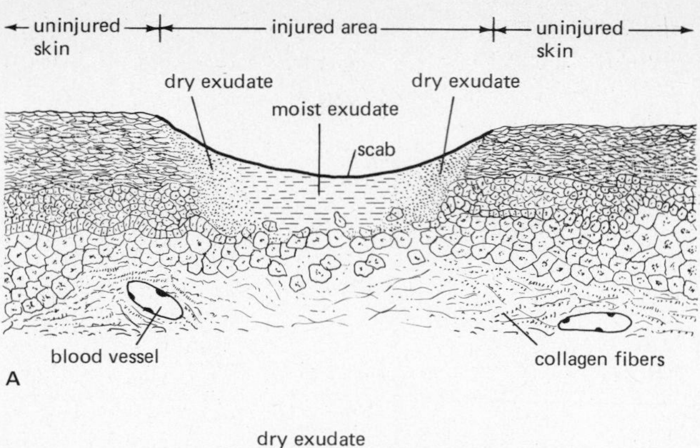

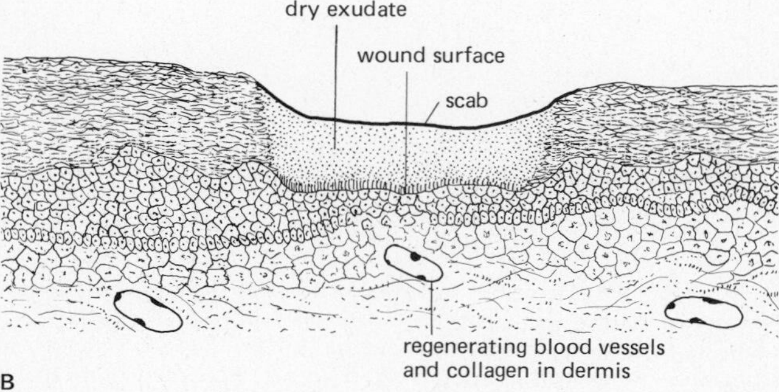

Figure 19-12. **The healing of superficial wounds.**

Stable Cells. These cells do not normally multiply during adult life but retain a latent ability to undergo mitotic division under appropriate stimulation. Perfect repair may occur in the cells of almost all the glands of the body, including the liver, spleen, pancreas, endocrine glands, kidney tubular cells, and so forth, *provided that the basic framework of the tissue is present.* Similarly, the cells that produce connective tissue, the fibroblasts, and the blood-forming elements achieve perfect repair, if the tissue stroma is intact. Without this underlying tissue to serve as support and guide, the regenerating cells proliferate haphazardly, and a completely disorganized mass of cells replaces the original highly organized tissue. Deep injuries to the skin, affecting the underlying tissue, may eventually be repaired by overgrowth of neighboring epithelial cells, but the complex structures of the dermis such as the hair follicles and sweat glands are permanently lost, resulting in *scars.*

Permanent Cells. These cells cannot reproduce mitotically in adult life and are completely unable to replace destroyed elements. In tissues such as muscle and nerve, which are made up of permanent cells, repair consists only of the proliferation of the fibrous connective tissue, resulting in the formaton of fibrous scar tissue. *The destruction of a neuron* in the central nervous system is a permanent loss, and the area becomes filled in by the supporting cells of the central nervous system, the glial cells. If the damage is to the cell body of a peripheral nerve, the entire structure, cell body and axon, degenerates; if the damage is confined to the axon, the portion peripheral to the wound degenerates, and the remaining axon may grow out again along the original trail to link up with the organ originally innervated. Unless the trail is followed, the proliferation results only in a mass of tangled fibers. Reparative surgery is sometimes able to provide an artificial trail for the axon tip to ensure the proper innervation later. *The destruction of muscle cells* is also irreplaceable, and the destroyed area becomes infiltrated with scar tissue. This is of special importance in cardiac muscle, where repeated hemorrhages in the muscle may impair the blood supply to the muscle of the heart itself, causing death of the cells (necrosis), and ultimately interfering critically with the physiological functioning of the heart.

Processes involved in wound healing

The skin, bone, and liver are organs that have excellent repair and regenerative capabilities. The skin, in particular, has been extensively studied, using animal and human models of wound healing.

The following sequence of events has been observed in the healing of superficial skin wounds (Fig. 19-12):

1. About 18 hr after injury, the acute phase of inflammation dies down, and the exudate and some of the dermis dry out.
2. The epidermis moves out from under this dry layer and migrates from the wound edge to meet other sheets of epidermal cells moving from opposite directions. Desmosomes are formed and contact inhibition may play a role in stopping these moving cells when they reach each other.
3. New dermis forms below the regenerating epidermis. By the ninth day, the newly formed collagen fibers and blood vessels are visible.
4. Around the eleventh day, the new dermis contracts, pulling the edges of the wound together.
5. Healing is usually completed by the fortieth day.

Open Versus Covered Wounds. The epithelial cells usually migrate under the dry scab, which acts as a mechanical obstruction. If the wound is covered with a thin polyethylene film which prevents drying and the formation of a thick crust, then healing is speeded up. The entire process may be shortened from 40 to 20 days. If there is no deep scab, the epithelial cells can migrate at the level of the undamaged epidermis and not below it, so that no scars are left (Fig. 19-13). Although these covered wounds heal more rapidly than open wounds, the danger of bacterial infection is significantly greater.

Hormones regulating wound healing

The appropriate amounts and the proper sequence of the hormones that enhance healing are essential elements in the healing process.

The trauma of a wound is a typical stress stimulus, and the endocrine glands participate in the complex defense mechanisms activated. Consequently the whole array of *stress-related hormones* may be involved directly or indirectly in the homeostatic response of the organism. For example, vasopressin (anti-diuretic hormone) will help regulate water balance to restore fluid loss, and the catecholamines will

Figure 19-13. **Covered wounds form thin scabs and permit faster migration of epithelial cells to repair the damaged tissue. Healing is complete in this human incision wound in 24 hr. Uncovered wounds exposed to air form thick scabs that slow epithelialization so that 3 days are required for healing. [From Rovee, D. T., et al.: "Effect of Local Wound Environment on Epidermal Healing," in Maibach, H. L., and Rovee, D. T. (eds.): *Epidermal Wound Healing.* Copyright © 1972 by Year Book Medical Publishers, Inc., Chicago. Used by permission.]**

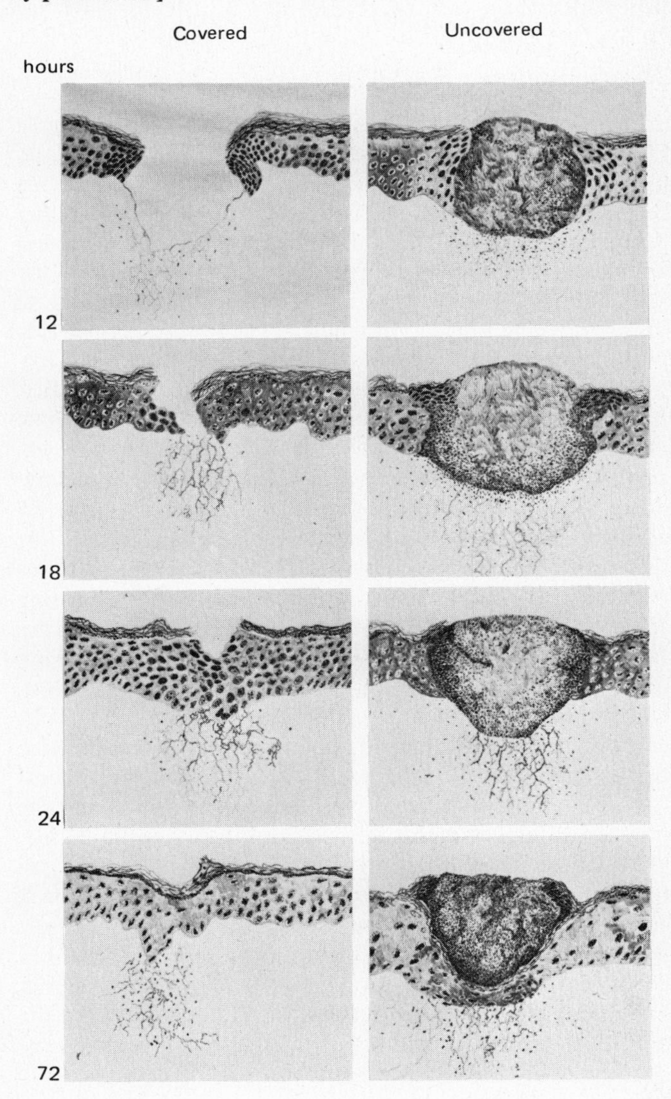

activate the circulatory and muscle responses to adjust to the oncoming flight or fight. Water balance and circulation are obviously of prime importance to adequate wound healing, but here we shall be concerned only with those hormones that directly affect growth and replacement of cells in the injured area.

Hormones That Enhance Wound Healing. *Growth hormone,* the *somatomedins, insulin,* and *thyroxine* have important anabolic effects that promote wound healing in all damaged tissues. In addition, the *specific growth factors* (nerve, epidermis, and fibroblast) accelerate the repair and growth of these cells. *Androgens* assist in the repair, but not the regeneration, of muscle cells.

Hormones That Inhibit Wound Healing. The *glucocorticoids* and the *prostaglandins* delay tissue healing. In addition to its antianabolic effects, cortisone inhibits the synthesis of collagen and mucopolysaccharides. These essential elements in the formation of new connective tissue are necessary for the healing process.

Regeneration

In humans the ability to regenerate missing parts is almost completely absent. Once the underlying tissue is absent, no re-formation of the tissue is possible; therefore any true regeneration is precluded. The liver still possesses greater regenerative powers than any other organ, but a certain amount of tissue must be present; no regeneration of the liver is possible after surgical removal of the entire organ. "Still" refers to the fact that most lower animals, in particular the amphibia and reptiles, possess remarkable regenerative abilities, regenerating new limbs or a tail quite readily. Regeneration assumes the initial dedifferentiation of some remaining tissue to a nonspecific cell mass, where it can produce all the various specialized cells of a new organ and do so in an organized way so that the tissues are properly arranged and integrated. Have cells of higher animals lost this dedifferentiating ability? Is there a specific stimulus necessary to initiate this dedifferentiation? A fascinating study by Singer (1958) has been made on frogs, based on the fact that tadpoles retain the ability to regenerate missing limbs, but frogs do not. What has happened to these regenerative powers following metamorphosis to the frog? Singer's work has shown that surgical increase of the number of nerves supplying the amputated area will induce regeneration of the limb in the frog; in other words, regrowth may be induced in higher forms if the appropriate conditions are provided to permit the latent regenerative powers of the tissue to develop. The possibilities inherent in this approach are magnificent, for if one can control the growth of cells, it is feasible that regeneration of missing or damaged organs in the human may one day be induced. It is also possible that knowledge about this subject may provide a solution to the unbridled growth of malignant cells, most of which have dedifferentiated in the process.

Cited References

Levi-Montalcini, R., and P. U. Angeletti. "Nerve Growth Factor." *Physiol. Rev.* **48:** 534 (1963).

Salmon, W. D., Jr., and W. H. Daughaday. "A Hormonally Controlled Serum Factor Which Stimulates Sulfate Incorporation by Cartilage in Vitro." *J. Lab. Clin. Med.* **49:** 825 (1958).

Singer, M. "The Regeneration of Body Parts." *Sci. Am.* Oct. 1958.

Additional Readings

BOOKS

Brown, H., ed. *Protein Nutrition.* Charles C Thomas, Publishers, Springfield, Ill., 1974.

Denvenyi, T., and J. Gergely. *Amino Acids, Peptides, and Proteins.* American Elsevier Publishing Co., Inc. New York, 1974.

Gardner, L. I., and P. Amacher, eds. *Endocrine Aspects of Malnutrition: Marasmus, Kwashiorkor, and Psychosocial Deprivation.* Raven Press, New York, 1973.

Goss, R. J. *Principles of Regeneration.* Academic Press, Inc., New York, 1969.

Goss, R. J., ed. *Regulation of Organ and Tissue Growth.* Academic Press, Inc., New York, 1973.

LoBue, J., and A. S. Gordon, eds. *Humoral Control of Growth and Differentiation,* Vol. 1. Academic Press, Inc., New York, 1973.

Luft, R., and K. Kall, eds. *Advances in Metabolic Disorders,* Vol. 8, *Somatomedins and Some Other Growth Factors,* R. Luft and K. Hall, eds. Academic Press, Inc., New York, 1975.

McGilvery, R. W. *Biochemical Concepts.* W. B. Saunders Company, Philadelphia, 1975.

Menaker, L., ed. *Biologic Basis of Wound Healing.* Harper & Row, Publishers, New York, 1975.

Munro, H., and J. B. Allison. *Mammalian Protein Metabolism,* Vols. 1 and 2. Academic Press, Inc., New York, 1967.

Newsholme, E. A., and C. Start. *Regulation in Metabolism.* John Wiley & Sons, Inc., New York, 1973.

Polezhaev, L. V. *Organ Regeneration in Animals.* Charles C Thomas, Springfield, Ill., 1972.

Zweifach, B. W., L. Grant, and R. T. McCluskey, eds. *The Inflammatory Process.* Academic Press, Inc., New York, 1965.

ARTICLES

Bergen, W. G. "Protein Synthesis in Animal Models." *J. Anim. Sci.* **38:** 1079 (1974).

Bourne, H. R., L. M. Lichenstein, K. L. Melmon, C. S. Henney, Y. Weinstein, and G. M. Shearer. "Modulation of Inflammation and Immunity by Cyclic AMP." *Science* **184:** 19 (1974).

Daughaday, W. H., A. C. Herrington, and L. S. Phillips. "The Regulation of Growth by Endocrines." *Ann. Rev. Physiol.* **37:** 211 (1975).

Fernstrom, J. D., and R. J. Wurtman. "Nutrition and the Brain." *Sci. Am.,* Feb. 1974.

Frieden, E., and H. Lipner. "Thyroxine and Triiodothyronine." In *Biochemical Endocrinology of the Vertebrates,* Foundations of Modern Biology Series. Prentice-Hall, Inc., Englewood Cliffs, N.J., 1971, p. 48.

Gardner, L. I. "Deprivation Dwarfism." *Sci. Am.,* July 1972.

Hogue-Angeletti, R., R. D. Bradshaw, and W. A. Frazier. "Nerve Growth Factor (NGF): Structure and Mechanism of Action." In *Advances in Metabolic Disorders,* Vol. 8, *Somatomedins and Some Other Growth Factors,* R. Luft and K. Hall, eds. Academic Press, Inc., New York. 1976, p. 285.

JACOBSON, MARCUS. "Effects of Nutrition, Hormones, and Metabolic Factors on the Development of the Nervous System." In *Developmental Neurobiology.* M. Jacobson, ed. Holt, Rinehart and Winston, New York, 1970, p. 196.

LATHAM, M. C. "Protein-Calorie Malnutrition in Children and Its Relation to Psychological Development and Behavior." *Physiol. Rev.* **54:** 541 (1974).

LIPMANN, F. "What Do We Know About Protein Synthesis?" *Basic Life Sci.* **1:** 1 (1973).

PECILE, A., and E. E. MULLER. "Control of Growth Hormone Secretion." In *Neuroendocrinology,* Vol. 1, L. Martini and W. F. Ganong, eds. Academic Press, Inc., New York, 1966, p. 445.

PITTS, R. F. "The Renal Metabolism of Ammonia." *Physiologist* **9:** 97 (1966).

REICHLIN, S. "Control of Thyrotropic Hormone Secretion." In *Neuroendocrinology,* Vol. 1, L. Martini and W. F. Ganong, eds. Academic Press, New York, 1966, p. 445.

ROSS, R. "Wound Healing." *Sci. Am.,* June 1969.

STARKEY, R. H., S. COHEN, and D. N. ORTH. "Epidermal Growth Factor: Identification of a New Hormone in Human Urine." *Science* **189:** 800 (1975).

STRAND, F. L. "Endocrineurology." In *Humoral Control of Growth and Differentiation,* Vol. 2, J. LoBue and A. S. Gordon, eds. Academic Press, Inc., New York, 1974, p. 191.

STRAND, F. L. "The Influence of Hormones on the Nervous System, with Special Emphasis on Polypeptide Hormones." *Bioscience* **25:** 568 (1975).

TEPPERMAN, J. "The Thyroid." In *Metabolic and Endocrine Physiology,* 3rd ed. Year Book Medical Publishers, Inc., Chicago, 1973, p. 102.

THORNTON, C. S. "Amphibian Limb Regeneration." In *Advances in Morphogenesis,* M. Abercrombie and G. Brachet, eds. Academic Press, Inc., New York, 1968.

TOPOREK, M., and P. H. MAURER. "Metabolic Biochemistry." In *Pathologic Physiology: Mechanisms of Disease.* 5th ed., W. A. Sodeman, Jr., and W. A. Sodeman, eds., W. B. Saunders Company, Philadelphia, 1974.

WILLIAMS-ASHMAN, H. G. "Metabolic Effects of Testicular Androgens." In *Handbook of Physiology,* Sec. 7, Vol. 5, R. O. Greep and E. B. Astwood, eds., The Williams and Wilkins Company, Baltimore, 1975.

Chapter 20

Regulation of Immune Processes

Vinay Likhite

Assistant Professor
of Biology
New York University

A symbol from the first, of mastery,
experiments such as Hippocrates made
and substituted for vague
speculation, stayed
the ravages of a plague.

A "going on": yes, anastasis *is the word*
for research a virus has defied,
and the virologist
with variables still untried—
too impassioned to desist.

Suppose that research has hit on the right one
and a killed vaccine is effective
say temporarily—
for even a year—though a live
one could give lifelong immunity,

knowledge has been gained for another attack.
Selective injury to cancer
cells without injury to
normal ones—another
gain—looks like prophecy come true.

Now, after lung resection, the surgeon fills space.
To sponge implanted, cells following
fluid, adhere and what
was inert becomes living—
that was framework. Is it not

like the master-physician's Sumerian rod?
staff and effigy of the animal
which by shedding its skin
is a sign of renewal—
the symbol of medicine.

Marianne Moore, "The Staff of Aesculapius"

20–1 Introduction

Immunology began as, and still is, the art and science of making vaccines in order to protect ourselves against fatal diseases. The promise of such immediate practical applications for human (and animal) health continues to stimulate our interest in this field and assures its continued development. Dreaded diseases such as smallpox, yellow fever, and poliomyelitis have been brought under control through immunological expertise, and vaccines for tooth decay and cancer are under development. In Table 20-1 the diseases that have fallen to (or will fall to) immunology are listed.

Our ancestors were aware of the phenomenon of immunity but did not understand it well enough to manipulate it to their advantage. One of the earliest evidences of this awareness is the written account by Thucydides of the plague that struck Athens in the year 430 B.C.: "All speculation as to its origins and causes, if causes can be found adequate to produce so great a disturbance, I leave to other writers. . . . For myself, I shall simply set down its nature, and explain the symptoms by which it may be recognized by the student, if it should ever break out again. This I can the better do, as I had the disease myself, and watched its operation in the case of others. . . . Yet it was with those who had recovered from the disease that the sick and the dying found most compassion. These knew what it was from experience, and had now no fear for themselves; *for the same*

Table 20–1. **Human Diseases Preventable by Vaccination**

Disease	Vaccine
Smallpox	Attenuated Variola vaccinae (cowpox) virus
Rabies	Killed (inactivated) virus
Diphtheria	Inactivated corynebacterium diphtherae bacterial toxin
Yellow fever	Attenuated virus
Cholera	Extract of cholera vibrios bacteria
Pertusis (whooping cough)	Killed Bordetella pertusis bacteria
Tetanus (lockjaw)	Inactivated clostridium tetanii bacterial toxin
Tuberculosis	BCG (bacille Calmette Guérin): attenuated Mycobacteria tuberculosis
Plague	Extract of plague bacillii
Meningitis	Polysaccharide extract from Neisseria meningitidis
Pneumonia	Polysaccharide extract from diplococcus pneumoniae
Poliomyelitis	Live virus (Sabin) or killed virus (Salk)
Measles (Rubella)	Attenuated virus
Mumps	Attenuated virus
Influenza	Killed virus
Yellow fever	Killed virus

man was never attacked twice—never at least fatally. And such persons not only received the congratulations of others, but themselves also, in the elation of the moment, *half entertained the vain hope that they were for the future safe from any disease whatsoever. . . .*"

These writings clearly indicate that Thucydides, 2500 years ago, recognized the most fundamental principles of immunology:

1. Prior exposure to the disease-causing agent (*antigen*) was a necessary condition for the induction of protection (*immunity*).
2. After recovery the body remembers the event by not succumbing to the same disease following a second exposure (*immunological memory*).
3. This protection afforded by the immune system is specific only for that disease (*immunological specificity*).

Such first-rate observations were seized upon by some of the physicians of those days, who began to explore ways to mimic this naturally acquired immunity. The successful induction of acquired immunity to the dreaded smallpox (variola) by deliberately coming into physical contact with

Table 20–2. **Highlights in the Development of Modern Immunology**

430 B.C.	Thucydides of Athens—his writings indicate an awareness of the most fundamental principles of immunology: (1) prior exposure to the disease causing agent is obligatory to the induction of protective immunity; (2) the body remembers this encounter by not succumbing (fatally) to a re-exposure to the same disease; and (3) the immunity induced is operative specifically only against that disease.
A.D. 1000	Chinese, Hindus, and African tribes develop the technique of deliberate exposure to prevent smallpox.
1798	Edward Jenner demonstrates that infection with cowpox protects against smallpox.
1880	Louis Pasteur, using Jenner's approach, develops "attenuated" vaccines against chicken cholera, anthrax in sheep and cows, and rabies in man.
	Robert Koch discovers delayed hypersensitivity, the forerunner of cell-mediated immunity. Nobel Prize, 1905.
1890	Emil von Behring discovers antibodies as a result of curing diphtheria with blood serum from immunized donors. Winner of the first Nobel Prize, 1901.
1896	Paul Ehrlich develops theoretical concepts of humoral immunity. Nobel Prize, 1908.
1897	Rudolf Kraus discovers the precipitin reaction.
1901	Karl Landsteiner discovers human blood groups, thereby making blood transfusions safe and practical. Nobel Prize, 1930.
1902	Charles Richet and Paul Portier discover that immunization can result in allergies. Nobel Prize, 1919.
1958	Rodney Porter and Gerald Edelman determine the complete chemical structure of the antibody molecule. Nobel Prize, 1972.
1959	F. Macfarlane Burnet and Peter Medawar demonstrate that rejection of transplants is an immunological phenomenon. Nobel Prize, 1960.

one with the active disease marks the beginning of modern immunology. In the seventeenth century smallpox was endemic everywhere in Europe and probably throughout the world. Probably the most infectious disease known to man, smallpox lasted for two weeks or longer. Its first signs were fever, backache, and vomiting. The fever then subsided, and many small bumps appeared on the skin, particularly on the face, the chest, and the arms. Over several days the bumps enlarged as they filled with fluid. The fever then returned and the bumps became inflamed and swollen with pus that broke and formed a soft yellow crust with an offensive odor. Finally, if the person survived (and in many epidemics 40 per cent of the sick died), the crusts fell off, revealing the characteristic pox, or depressed scars, that gave the disease its name. It was not unusual for the person also to be left blind.

Thus, although incurable, smallpox could be prevented! By the early seventeenth century, the technique of direct physical contact with a smallpox victim had evolved into the method of *variolation:* immunity was induced by inserting pus from smallpox bumps directly into slight cuts in the skin or by inhalation of the dried pus and powdered scabs. Then in 1798 Edward Jenner announced that immunization with the pus from the boils of cows with cowpox protected humans against smallpox. This astonishing discovery was not taken seriously, and almost a century elapsed before Jenner's approach was applied to induce protection against other diseases. In the 1880s Louis Pasteur, the first great experimental immunologist (among his many other talents), extended Jenner's observations and developed vaccines for protection against various diseases, such as cholera in chickens, anthrax in sheep and cows, and rabies in humans. Pasteur's dramatic and celebrated public demonstrations, aided by his strong and influential personality, aroused great interest among physicians and provided a potent stimulus to the development of immunology. In Table 20-2 the highlights in the development of modern immunology are depicted.

20–2 Antigens

Early beliefs held that only material from a sick individual (e.g., smallpox pus, dried crust powder, etc.) would induce immunity. Although the precise nature or composition of such preparations was unknown, they nevertheless sufficed to induce immunity. As our understanding of the nature of infectious disease progressed, it was recognized that disease-causing agents such as bacteria, viruses, and fungi and their toxic products could induce immunity. Even harmless materials such as milk, eggs, serum, plants, and tissue like meat also induced immunity. We now know that the fundamental basis for the induction of immunity is the molecules (both large and small) of which the more complex organisms are made. Included among these molecules are proteins, carbohydrates, nucleic acids, and lipids, as well as a variety of synthetic molecules of known chemical structure (Table 20-3). Even the atoms of chromium, nickel, and beryllium have been shown to induce an immune response. Substances that elicit a specific immune response when introduced into the tissues of an animal are called *antigens* (Ag or Ags). In addition, antigens are capable of reacting in some detectable manner with the immunized animal.

Table 20–3. **Some Commonly Studied Antigens**

Antigen
Sheep red blood cells (SRBC)
Tobacco mosaic virus (TMV)
Ovalbumin (OA)
Bovine serum albumin (BSA)
Human serum albumin (HSA)
Bovine gamma globulin (BGG)
Ragweed pollen
Myoglobin
Ribonuclease
Lysozyme
Keyhole limpet hemocyanin (KLH)
Bacterial polysaccharides
Flagellin
Polylysine

It is known that the entire Ag molecule does not participate in the induction of immunity. Rather, certain restricted portions within the parent Ag, called *antigenic determinants* or *epitopes,* determine the specificity and direction of the immune response. These determinants vary in length, being about 10 amino acids long for proteins, and 6 monosaccharides for carbohydrates. The number of determinants increases directly in proportion to the size of the Ag molecule. This number is also called the *valence* of the Ag, and Ags are usually multivalent. Several empirical criteria determine the *immunogenicity* of an Ag, that is, its ability to provoke a vigorous immune response. Good immune responses are usually produced by Ags of a high molecular weight, and whose determinants are easily accessible to and are recognized as being *"foreign"* or *"alien"* by the immunological machinery of the responding animal. Important also is the amount of Ag used, the route of its introduction into the animal, and the use of *adjuvants* (substances that increase the immunogenicity of an Ag).

Haptens, usually low-molecular weight substances, are not immunogenic by themselves but become so when attached to other Ags that act as "carriers." However, haptens do not require the presence of the carrier in reacting with the immunized animal either directly or in vitro with its antibodies and lymphocytes. It should be noted that immunogenicity is not an inherent property of a molecule, as for example is its molecular weight or its absorption spectrum. Rather, immunogenicity is operationally dependent upon the biological system and the conditions employed for immunization.

20–3 Antibodies

Antibodies (Ab or Abs) were first recognized by Emil von Behring (1890) as proteins present in the blood serum of diphtheria-immunized animals and capable of neutralizing the fatal toxicity of the diphtheria toxin. But it was not until 50 years later that Tiselius and Kabat (1938) demonstrated through the use of the electrophoretic technique that Ab activity is localized in the γ-globulin fraction of serum. We now know that Abs are an extremely heterogeneous group of protein molecules, all being capable of reacting with the specific Ag and separable into classes and subclasses on the basis of their chemical properties. In humans, to date five major classes and several subclasses of proteins with Ab activity are recognized. They are collectively termed *immunoglobulins* or Igs (to distinguish them from γ-globulins without Ab activity): IgG, IgM, IgA, IgD, and IgE. The important characteristics and subclasses of Abs are listed in Table 20-4.

The heterogeneity of the Abs in normal serum made structural studies of the molecules extremely difficult. However, the discovery of homogeneous Igs (structurally similar to Abs) in the blood and urine of humans with certain forms of bone cancer (myeloma and macroglobulinemia) provided a unique source of large amounts of material for the determination of Ig (and Ab) structure.

Structure of Antibodies

The most common type of Ab present in serum, and in high concentration, is IgG. Furthermore, IgG is also relatively easy to isolate and purify. For these reasons it has been the most frequently studied. The high molecular weight (150,000) of IgG clearly indicated that its chemical structure was quite complex. Until the late 1950s it was generally believed that the IgG molecule (and other Abs) consisted of a single, long polypeptide chain made up of various amino acids. However, subsequent discoveries have shown that this concept was erroneous.

The Four-Chain Model for IgG Antibody

The treatment of IgG with the enzyme papain results in the formation of three major fragments termed I, II, and III; these fragments can be separated by chromatography. Porter (1959) showed that fragments I and II are identical to each other and react with the Ag. These fragments are also termed *Fab* (*fragment-Ag-binding*). Fragment III did not bind Ag but could be crystallized from solution (now termed *Fc* for *fragment-crystallizable*).

Further studies by Porter (1959), and by Edelman (1959), showed that treatment of IgG with reducing agents (such as mercaptoethanol), which break disulfide (S—S) bonds, resulted in the formation of subunits. These subunits then could be separated by chromatography into a high molecular weight (50,000) or H (for "heavy") fraction and a low molecular weight (25,000), or L (for "light") fraction. The actual data showed that one molecule of IgG yielded two molecules of H-fraction and two molecules of the L-fraction. Porter interpreted these results to mean that the IgG molecule was made up of two heavy subunits (or H chains) and two light subunits (or L chains). The four polypeptide chain model for the IgG molecule as proposed by Porter is shown in Fig. 20-1.

Table 20-4. **Characteristics of Human Antibodies**

Characteristic	IgG	IgM	IgA	IgD	IgE
Electrophoretic mobility	γ	Between γ and β	Slow β	Between γ and β	Slow β
Sedimentation coefficient	7S	19S	7S, 11S, 13S	7S	8S
Molecular weight	160,000	900,000	170,000–350,000	180,000	185,000
Subclasses	IgG_1, IgG_2, IgG_3, IgG_4		IgA1, IgA2		
Heavy chains	γ_1, γ_2, γ_3, γ_4	μ	α_1, α_2	δ	ϵ
Light chains	κ, λ	κ, λ	κ, λ	κ, λ	κ, λ
Molecular formula	$\gamma_2\kappa_2$, $\gamma_2\lambda_2$	$(\mu_2\kappa_2)_5J$	$\alpha_2\kappa_2$, $\alpha_2\lambda_2$ $(\alpha_2\kappa_2)_2J$, $(\alpha_2\lambda_2)_2J$ $(\alpha_2\kappa_2)_2SJ$, $(\alpha_2\lambda_2)_2SJ$	$\delta_2\kappa_2$, $\delta_2\lambda_2$	$\epsilon_2\kappa_2$, $\epsilon_2\lambda_2$
Valence	2	10	2	?	2
Concentration in serum (mg/ml)	13.0	1.2	4	0.4	0.00030
Carbohydrate, %	3	12	8	13	12
Half-life in serum (days)	23	5	6	3	25
Occurrence	Serum	Serum	Serum, secretions, Colostrum, saliva, tears	Serum	Serum

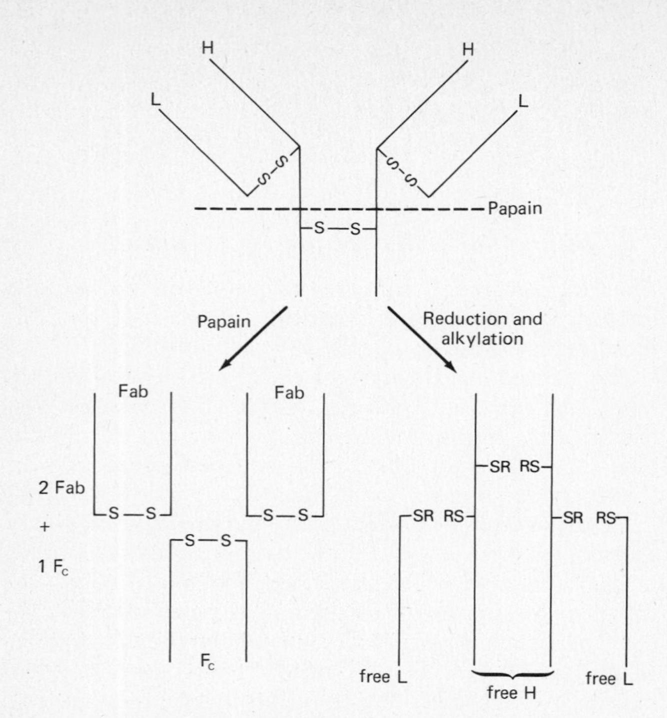

Figure 20-1. **The four-chain structure of the IgG antibody molecule.**

A molecule of IgG consists of two identical H chains and two identical L chains, linked together by covalent S—S bonds and by noncovalent bonding forces (hydrogen bonds, van der Waals forces, etc.). The H chains (of human IgG) contain 450 amino acids and are about twice as long as the L chains with 220 amino acids. This general structure for IgG has been confirmed by electron microscopy and by X-ray diffraction studies.

We now know that the four-chain structure is basic for all Ig classes, and in each case the H chains are specific for each class. Thus the γ H chains are specific for IgG, α for IgA, μ for IgM, δ for IgD, and ϵ for IgE. However, in each case the L chains are similar. The L chains have been separated into kappa (κ) and lambda (λ) types. In humans, 65 per cent of Igs contain κL chains, the rest containing λL chains. However, in any given Ig molecule the two L chains are always of the same type, and both types may be associated with any H chain class.

IgM and IgA molecules are made of several basic four-chain units joined by J (for joining) chains, and secretory IgA contains an additional subunit. This S (secretory) or T (transport) "piece" is thought to facilitate its transport into exocrine secretions (saliva, colostrum, etc.) across the cell membrane. The physiochemical characteristics of human antibody Igs are shown in Table 20-4, and their biological activities will be discussed in Sec. 20-5. Note the heterogeneity of Igs with regard to molecular weight. This results from the presence of polymeric forms (IgM and IgG) and also from differences in H-chain structure.

20-4 Antigen-Antibody Interaction

When Ags and Abs are mixed together in solution (in vitro or in vivo) they unite spontaneously and reversibly to form the Ag-Ab complex:

$$\text{Ag} + \text{Ab} \rightleftharpoons \text{Ag-Ab}$$

The formation of the Ag-Ab complex is one of the most fundamental reactions in immunology. It manifests itself in a variety of readily detectable forms: formation of *precipitates, clumping* of bacteria and red blood cells, *neutralization* of various toxins, *protection* against disease, and certain *allergic* phenomena (see Sec. 20-7).

Precipitation Reactions

The reaction of Abs with univalent Ags (e.g., haptens) results in Ag-Ab complexes that remain in solution. However, with multivalent Ags the complexes tend to aggregate, become insoluble, and separate from solution as precipitates. This is the basis for the *precipitin* reaction, and the Abs involved are also called precipitins. The precipitin reaction was discovered by Krause (1897) and studied extensively by Heidelberger and his "school" in the 1920s. The quantitative precipitin reaction is shown in Fig. 20-2.

As increasing amounts of Ag are added to a fixed amount of Ab, the quantity of Ab precipitated increases. After the addition of a small amount of Ag, some precipitate is formed, and free Ab is detectable in the supernatant; this is the *Ab excess zone.* Here the ratio of Ab to Ag in the Ag-Ab complex depends on the valency of the Ag. The addition of larger amounts of Ag results in increased precipitation until a point is reached where no free Ab or Ag is detectable in the supernatant. This is the *equivalence zone.* It is theorized that at this point optimal proportions of Ab and Ag form a

Figure 20-2. **The quantitative precipitin reaction.**

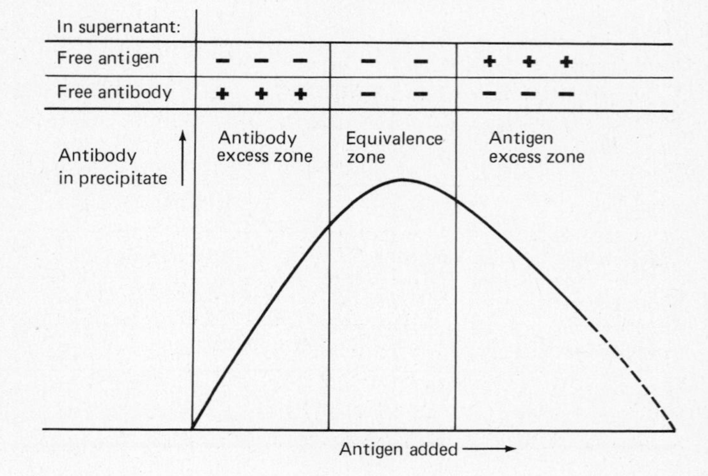

Table 20–5. **Biological Activities of Human Antibodies**

Activity	IgG-1	IgG-2	IgG-3	IgG-4	IgM	IgA-1	IgA-2	IgD	IgE
Crosses placenta	+	±	+	+	–	–	–	?	–
Fixes complement	+	±	+	–	+	–	–	?	–
Fixes to human mast cells	–	–	–	–	–	–	–	–	+
Present in exocrine secretions	–	–	–	–	–	+	+	–	+
Binds to macrophages	+	–	+	–	–	–	–	–	–

continuous, stable Ag-Ab "lattice," which precipitates. At high levels of Ag, free Ag appears in the supernatant; at the same time, precipitate formation is still maximal. This is the first stage of the *Ag excess zone.* In extreme Ag excess, the amount of precipitate is greatly reduced, because of the formation of soluble complexes. This results from the excess free Ag competing for the Ab sites in the precipitate with subsequent formation of soluble complexes of the Ab_1Ag_2 type. Both the quantity and the quality of Ab are important in determining whether a precipitate will be formed. The Ag valence is also critical, since the formation of a lattice is impossible if the Ag is univalent, that is, if it has a single binding site.

Agglutination Reactions

Lattice formation with Ab and multivalent Ag results, as we have said, in precipitation. However, when Ab reacts with antigenic sites located on the surface of particles such as those on bacteria or red blood cells, clumping or *agglutination* occurs. Even inert particles such as latex, to which Ag is chemically attached, are also agglutinated by Ab. The lattice theory is also applicable to the formation of clumps. Agglutination reactions have been extremely useful in detecting certain Rh antibodies.

Strength of the Ag-Ab Bond: Antibody Affinity

The Ab *affinity* is a quantity that is a measure of the strength of the interaction between the antigenic determinant and its homologous Ab. Detailed physicochemical and mathematical considerations in calculating affinities can be obtained from any immunology textbook, and will not be discussed here. In addition to the heterogeneity with respect to class and subclass, a given Ab population is also heterogeneous with regard to the affinity of the Ab for the Ag: some Ab molecules are of a higher affinity than others regardless of their class or subclass. Thus a high-affinity Ab is one that forms a strong bond with the antigenic determinant to give an Ag-Ab complex with a low tendency to dissociate. On the other hand, a low affinity Ab forms a complex with Ag that is less stable and has a tendency to dissociate more easily. It follows that the higher the affinity of the Ab, the greater will be the amount of Ag bound to Ab at equilibrium.

20–5 Biological Activities of Antibodies

Apart from reacting with Ags, Abs have many other biological activities. These are associated with the Fc portion of the H chains of certain Ig molecules and become manifest only after the Ab has reacted with the Ag. These properties differ within the various Ig classes and subclasses, as shown in Table 20–5.

The antibody *IgG,* the most important serum Ig in humans, functions to neutralize viruses and bacterial toxins and binds to bacteria, facilitating their phagocytosis and elimination by phagocytes. It is the only Ig able to cross the placenta in humans and is thus a major defense mechanism against infection in the early part of an infant's life. The four subclasses of IgG differ in their biological activities, especially in their ability to fix complement (see Section 20–6). Complement is fixed most efficiently by IgG3, followed by IgG1, then IgG2; however, IgG4 does not fix complement at all.

Although serum *IgA* has Ab activity, it is in colostrum, saliva, tears, gastrointestinal juice, and in respiratory tract secretions that this Ig is most important. In such secretions, IgA is thought to provide an immunological barrier against microorganisms, that are located on the mucosal surfaces exposed to the environment.

The antibodies *IgM* are polyvalent, so that they have high functional affinity for multivalent Ags. This property, together with their effectiveness in agglutination, complement fixation, cytolysis, and predominant intravascular localization, suggests that IgM Abs are particularly important in dealing with multivalent Ags such as bacteria and viruses that invade the blood stream.

Although *IgD* Abs have been demonstrated to bind certain Ags, their general biological function remains unknown at present.

The antibody *IgE* is present in the serum in extremely low levels, and higher amounts are characteristically found in the serum of people with hay fever and asthma. The IgE Abs bind firmly to *mast cells* independent of any reaction with Ag. If an appropriate Ag (now called an *allergen*) subsequently enters the host and reacts with the tissue fixed IgE Ab, a chain of events is triggered. The mast cells degranulate, vasoactive amines (e.g., histamine) are released, and subsequent clinical symptoms of immediate hypersensitivity or "allergy" such as sneezing and urticaria are produced. Current ideas on the mechanisms involved in immediate hypersensitivity are discussed in Sec. 20-7.

Complement Fixation

Complement (C) is a complex biological system consisting of 11 protein components the first of which, when activated by Ag-Ab complexes, is able to activate the next component in the sequence, which itself activates the next component and so on, thus producing a "cascade" effect. The activation of C has profound effects on cell membranes and can cause cell death by producing holes through the cell membrane on which C is fixed.

In addition to cell lysis, C activation also promotes *chemotaxis*—the attraction of polymorphonuclear phagocytes to the site of the Ag-Ab reaction—and other aspects of inflammation such as increased vascular permeability (Sec. 19-4). These properties of C activation are possessed by breakdown or fusion products of the activated C proteins. The chemistry of this process has been studied in great detail, and the individual C proteins have been separated and purified. In Figure 20-3 is depicted the C-mediated "cascade" reaction. Genetically inherited deficiencies of the C system are known in humans. One of these is called *angioneurotic edema,* which is usually fatal in the afflicted individual.

Figure 20-3. **Complement mediated cascade reaction.**

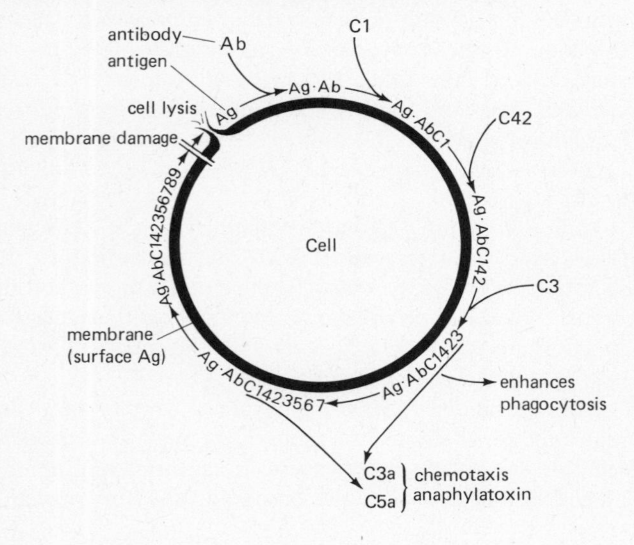

20-6 Biosynthesis of Antibodies

The entrance of an Ag into the body of an animal of an appropriate genetic makeup activates a complex series of interactions with the *lymphoid system* (organs and cells); these interactions ultimately result in the synthesis and release of free Ab into the blood and other body fluids. In addition, certain "sensitized" cells (*lymphocytes*) appear; these act as effectors of *cell-mediated immunity* expressed in such immunological phenomena as the *rejection of skin transplants* and *delayed-type allergies* (Sec. 20-7).

The Primary Immune Response

When an animal is exposed to an Ag for the first time, a *primary response* will occur. The following sequence of events accompanies the primary response:

1. An initial induction period that varies from 2 to 4 days, during which no Ab is detectable in the serum. Thereafter (depending on the sensitivity of detection method employed), Abs begins to appear in the serum.
2. Usually by the fourth or fifth day Abs are readily detected; initially they consist entirely of IgM. By the seventh day, IgG Abs begin to appear; these reach much higher concentration than the IgM Abs. The IgM concentration usually begins to wane before the IgG Ab has reached its peak (usually around the fourteenth or fifteenth day). After the IgG Ab peaks, the total Ab level begins to decline, and may become virtually undetectable within 30 to 40 days after the first antigenic exposure. This sequence will vary according to the animal and Ag used. The events of a primary response are outlined in Fig. 20-4.

The Secondary Response

A second (or booster) exposure to the same Ag, following the primary response, results in the development of a *secondary response.* Ab production is accelerated this time, the rate of synthesis is greater, the induction period is shorter, Ab levels peak higher, and the Abs disappear less rapidly from the blood, at times persisting for several years. The decreased rate of disappearance can be accounted for on the basis of the observations that IgG has a longer serum half-life than IgM, and that in the secondary response, the amount of IgG synthesized is greater than the amount of IgM. Because the animal seems to "remember" its first contact with the Ag, the accelerated and intensified second-

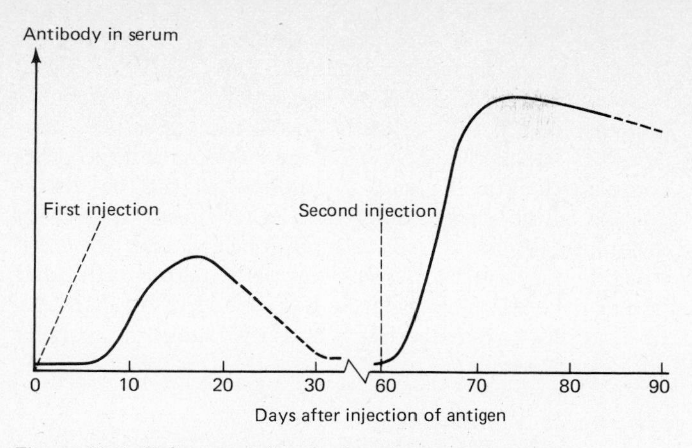

Figure 20-4. **Primary and secondary immune response.**

ary response is also called the *memory* or *anamnestic response.*

Secondary (or anamnestic) responses demonstrate a most important feature of immune responses—*immunological memory.* Even when the first exposure to Ag may have occurred in some remote past, a second exposure to that Ag can be expected to produce an anamnestic response. Anamnestic responses provide the basis for using multiple doses of vaccines (Ag) in some immunization programs (e.g., the Salk poliomyelitis vaccine, diphtheria and tetanus toxoid, etc.).

Organs Involved in Antibody Formation

The immunological machinery of the body involved in the production of immunity is located in the *lymphatic system.* It is a complex system whose components are dispersed throughout the tissues of the body. In humans the immune system weighs about 4 to 5 lb. It consists of a series of discrete lymphoid organs such as the *bone marrow, thymus, spleen, lymph nodes, tonsils,* and *appendix,* which are connected with each other via the *lymphatic vessels.* The "tree" of lymphatic vessels collects *lymphoid cells* and their products (Abs), along with other cells and molecules. The interstitial fluid that bathes all the body's tissues pours its contents via the lymphatic ducts back into the blood stream by joining the subclavian veins behind the collarbone. The human lymphatic system is shown in Fig. 13-15.

Cells Involved in Antibody Formation

As we have indicated, the organized tissues of the lymphatic system contain cells of the lymphoid series: *lymphocytes, macrophages,* and *plasma cells,* collectively referred to also as *leukocytes* or white blood cells (see Sec. 14-9). These cells are also distributed diffusely through the entire body—the reticular connective tissues, mesentery, intestinal wall, liver, and so forth. Immunologists have convincingly demonstrated that lymphoid cells are responsible for immunological reactions, such as Ab formation and cell-mediated immunity. The source of the lymphoid cells is the *bone marrow.* The principal cells involved are the *lymphocytes.* Although they are morphologically indistinguishable under the microscope, these cells can be classified by functional differences and surface markers into two distinct types: one type that requires the thymus gland for development—the *thymus-dependent lymphocytes* or *T-cells*—and a second that requires the *bursa of Fabricius* as found in birds (Fig. 20-5) or its mammalian equivalent (thought to be the bone marrow by some) for development; the latter are *B-lymphocytes* or *B-cells.* (It should be noted that all lymphoid cells develop from a stem cell precursor that originates from the bone marrow.)

The T-cells are responsible for the reactions of cell-mediated immunity, and the B-cells are precursors of cells that synthesize and secrete Ab—the plasma cells. It is not certain if T-cells themselves produce Ab, but they do cooperate with B-cells in Ab production, which has led to their description in this context as *"helper"* cells. A third cell type—the *macrophage*—has also been implicated in the immune response (Fig. 20-6). Although macrophages do not themselves synthesize Ab, their major role appears to be that of Ag processing, but there have also been reports that macrophages are able to transfer specific information to lymphocytes (Figs. 20-7A and B).

The classification of lymphocytes into two distinct immunologically competent types is based on experimental evidence in birds and rodents, and on observations from certain diseases in humans. Removal of the thymus in rodents virtually abolishes the cell-mediated immune responses but has almost no effect on Ab production. In birds, the removal of the bursa of Fabricius results in a total loss of Ab production, but cell-mediated immunity remains rela-

Figure 20-5. **Organs of the immunological system in the bird. The bursa of Fabricius is not present in mammals.**

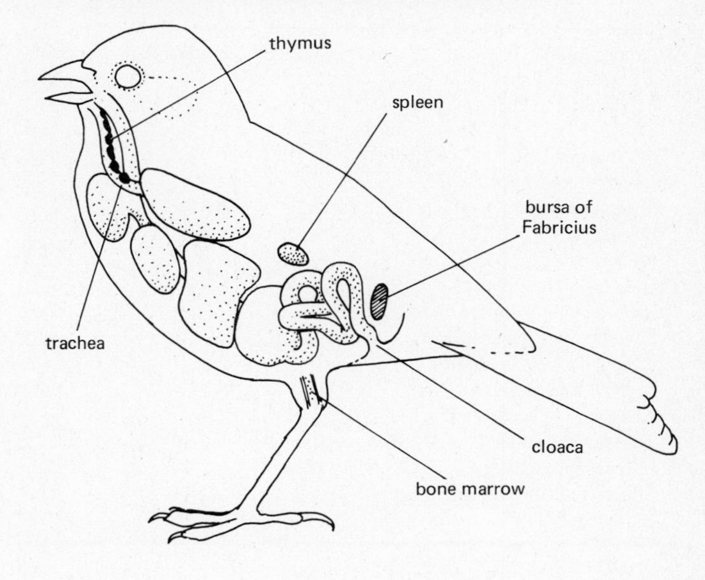

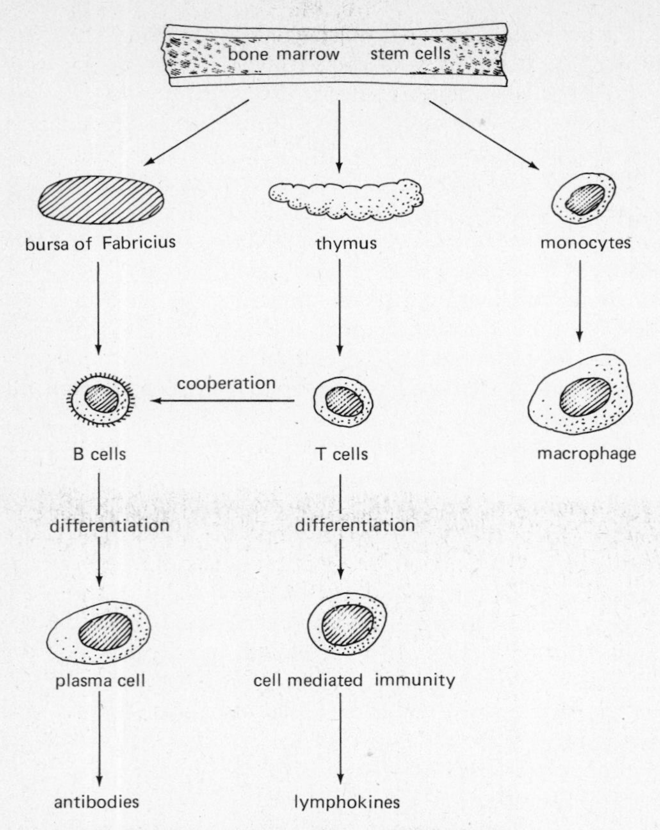

Figure 20-6. **Cells involved in the immune response.**

Table 20–6. **Characteristics of T and B Lymphocytes**

Properties	T-cells	B-cells
Differentiation	Thymus	Bursa of Fabricius
Cell surface antigens:		
θ (Thy 1)	+	−
Ly	+	−
PC (plasma cell)	−	+
Surface receptors for:		
Sheep red cells	+	+
Fc of IgG	−	+
C3 receptor	−	+
Surface immunoglobulin	−	+
Responsiveness to mitogens:		
Phytohemagglutinin (PHA)	+	−
E. coli lipopolysaccharide (LPS)	−	+
Polymerized flagellin	−	+
Functions:		
Secretion of antibody molecules	No	Yes
Helper function	Yes	No
Effector cell of cell mediated Immunity	Yes	No

tively unaffected. The bursa equivalent in mammals, including humans, remains unknown. The gut-associated lymphoid tissues such as appendix, tonsils, and mesenteric lymph nodes, as well as the bone marrow, have been suggested as bursa-equivalents, but the evidence is largely circumstantial. The T- and B-lymphocytes can be distinguished on the basis of differences in certain *surface Ags.* In the mouse, for example, T-cells have the characteristic θ Ag, which is not found on Ab-forming cells (plasma cells), and B-cells have the mouse-specific B-lymphocyte Ag (MBLA). In Table 20-6 are listed the various distinguishing markers on T- and B-cells.

Antibody Synthesis

A widely accepted working hypothesis for Ab synthesis is the *clonal selection theory* of Burnet (1957). This theory suggests that individual lymphocytes have the genetic capacity to make one, or possibly a few, particular Abs. The lymphocytes have Ig or Ig-like receptor molecules on their surface. These receptors have the same specificity for Ag as the Ab that the cell can make when differentiated. Thus, when an Ag enters the body of an animal, those lymphocytes with receptors that can react with the Ag will be stimulated to differentiate and will ultimately produce a clone of cells producing Ab with specificity toward the inducing Ag. Not all of the Ag-stimulated cells differentiate into Ab-forming B-cells (plasma cells); some become transformed into long-lived Ag-reactive *"memory cells."*

It is now clear that both B- and T-cells can bind Ag by means of surface receptors, and that clonal selection, as postulated by Burnet, does occur with both types. What is not clear is the detailed chemical nature of those receptors, particularly the T-cell receptors. The B-lymphocyte receptors are Ig, as shown by techniques using specific antisera to various Ig determinants. Each cell has approximately 10^4–10^5 Ig molecules on its surface. The IgG, IgM, IgA, and IgD receptors have been demonstrated; and in humans, subclasses of IgG have been shown on B-cell surfaces. The Ig receptors are thought to be oriented with the Fc regions toward the B-cell surface, leaving the Ab binding sites free to react with the Ag. The nature of the T-cell receptors is controversial, and convincing proof of their Ig nature has yet to be provided. Indeed, it has been suggested that nonIg receptors, such as the products of *immune response* (I_r) *genes,* are involved in T-cell recognition of Ag. The role of T- and B-cell in immune reactions has been reviewed recently (Greaves et al., 1975).

20–7 Allergy

Although the immune response is of major importance in protection and recovery from disease, it can also be responsible for many harmful reactions in humans and animals. A group of such harmful effects is the *allergies* responsible for common diseases like hay fever, asthma, and poison ivy sensitivity.

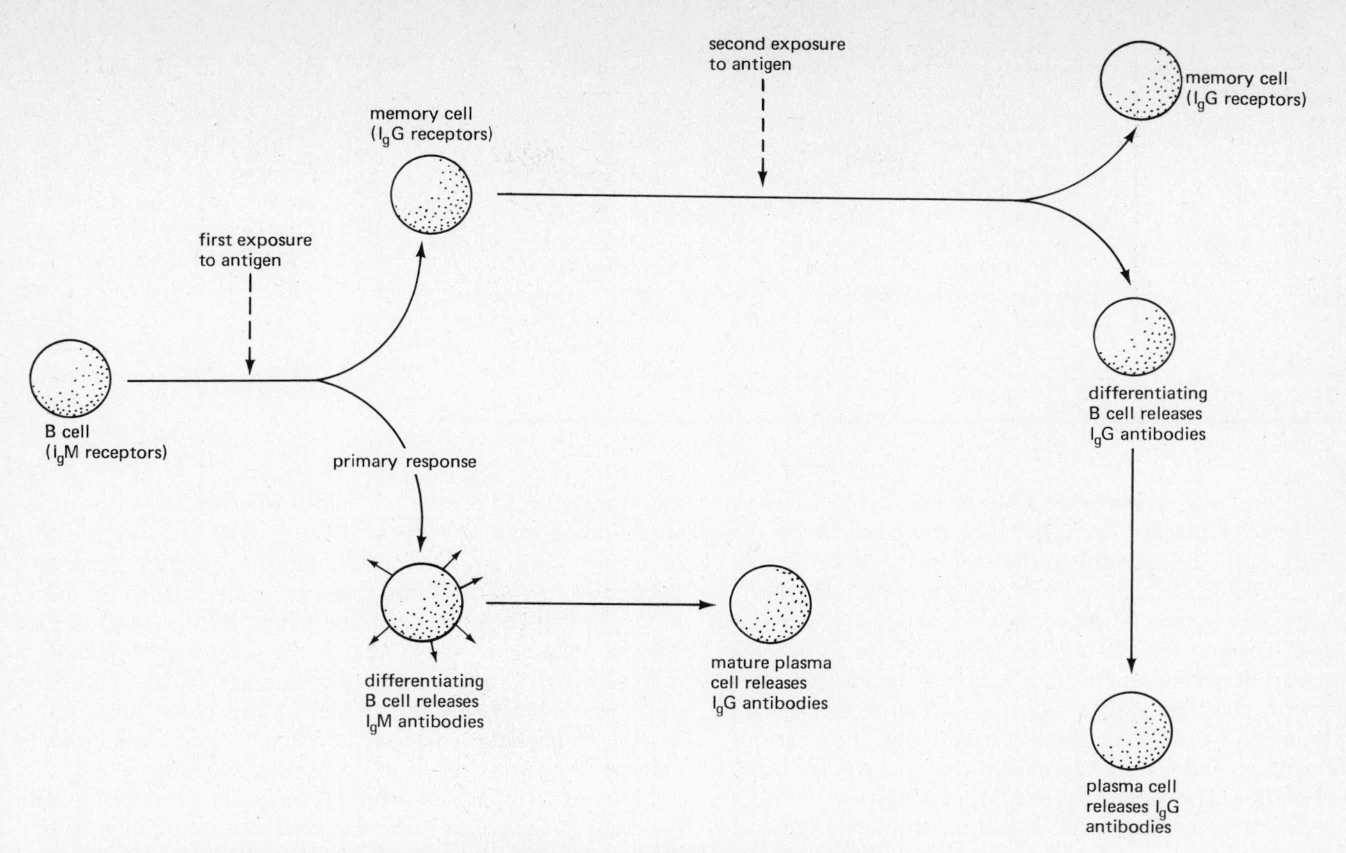

Figure 20-7A. **Cellular interaction in the immune response: in the absence of a T cell.**

Figure 20-7B. **Cellular interaction in the immune response: in the presence of a T cell.**

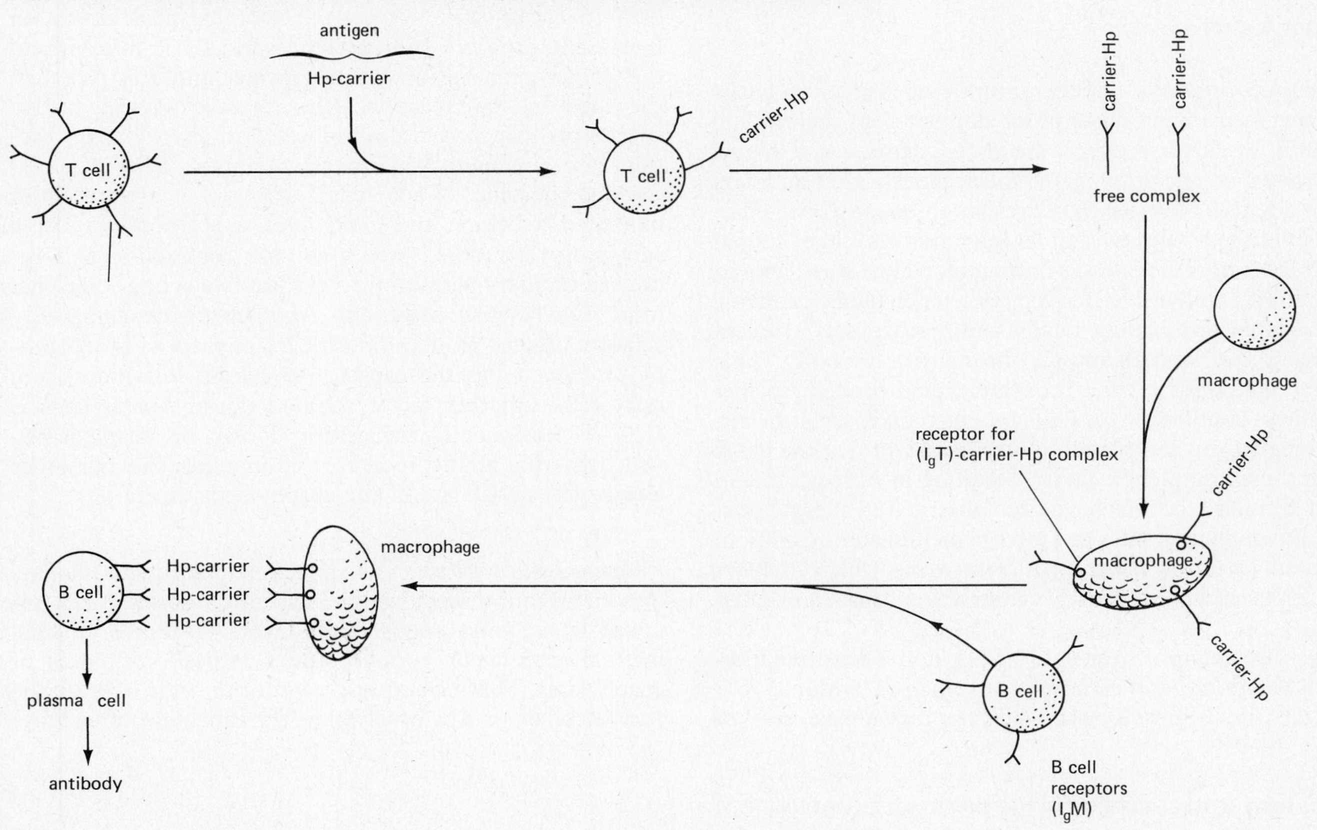

Table 20–7. **Major Differences Between Immediate- and Delayed-Type Allergies**

Property	Immediate	Delayed
Synonym	Humoral or antibody mediated hypersensitivity; Type I, II, III hypersensitivity	Cell-mediated hypersensitivity; Type IV hypersensitivity
Time course	Minutes to hours	24–48 hr or more
Visual skin reaction	Wheal or edema and flare or erythema	Erythema and induration (skin thickening)
Histology	Infiltration by polymorphonuclear leukocytes	Infiltration by mononuclear cells, mostly macrophages and lymphocytes
Passive transfer	With serum antibodies called reagins or IgE	Lymphoid cells (T-cells)

Allergy (or *hypersensitivity*) denotes an altered immune reactivity to a usually harmless substance, by which the substance becomes harmful to the recipient host. Allergies can be further classified into two large groups: *immediate-type* and *delayed-type*. These terms were originally based on the time required for the two kinds of reaction to become apparent, but the distinguishing criterion now is whether they can be transferred passively to a normal recipient with the serum of a sensitized (immunized) donor. The various characteristics and manifestations of these allergies are listed in Table 20–7. It is now clear that immediate-type allergies are mediated by a special kind of serum Abs, while those of the delayed-type are mediated by specific T-cells.

Antibody-Mediated Hypersensitivity (Immediate-Type Allergy)

Anaphylaxis

Active Anaphylaxis. When a guinea pig is given an initial injection (sensitizing dose) of an allergen (Ag) such as egg albumin, no adverse effects are noted. However, a *second injection* (shocking dose), given intravenously after an interval of about 10 days or more, results in *anaphylactic shock* almost immediately (within a few minutes): the animal shows signs of restlessness, a drop in blood pressure, labored respiration, convulsive movements, and finally convulsions—often terminating fatally. Autopsy reveals a severe contraction of smooth muscle throughout the body. This contraction constricts the bronchioles and bronchi so that the animal continues to breathe in, but cannot expel the air. The lung alveoli are dilated to such an extent that the lungs fill up the entire pleural cavity, resulting in suffocation and death. Similar reactions have been observed in allergic people following insect bites and after penicillin administration. Often only a timely injection of *epinephrine,* which counters bronchiolar smooth muscle contraction, has prevented death.

The intracutaneous injection of Ag into a sensitive individual results in *local cutaneous anaphylaxis:* within 2 to 3 min the site begins to itch; then appears a pale area of swelling—due to increased fluid—called a *wheal,* with a surrounding area of redness called a *flare* (also erythema, hives, or urticaria). The *wheal and flare* reaction becomes maximum in about 10 min, persists another 10 to 20 min, then slowly disappears. Approximately 10 per cent of the U.S. population is *atopic,* that is, has acquired "spontaneous" sensitivity to various environmental substances (pollen of ragweed, grasses, fungi, animal dander, house dust, etc.), leading to hay fever, asthma, and hives. Atopy is inherited and has been recognized in dogs, cats, and mice.

Many of the effects of anaphylaxis can be produced by an injection of *histamine;* histamine can be liberated in vitro by pieces of lung, intestine, uterine muscle, or other tissues from sensitized guinea pigs on contact with Ag. Pieces of intestine or of uterine muscle from sensitized animals, placed in a bath containing a balanced electrolyte solution and oxygenated, contract upon the addition of Ag to the bath (*Schultze-Dale reaction*). Histamine also causes an increase in *capillary permeability,* especially in humans and the guinea pig, and an intracutaneous injection produces the wheal and flare reaction. Other pharmacologically active substances that have been implicated in anaphylactic reactions are *serotonin* (5-hydroxytryptamine); *SRS-A* (slow-reacting substance of anaphylaxis), a lipid or lipoprotein of unknown structure; and *bradykinin,* a polypeptide. These substances all cause smooth muscle contraction and increased capillary permeability, the last two acting somewhat more slowly than serotonin. Another pharmacologically active substance produced during anaphylaxis is called *anaphylatoxin;* it has the capacity to release histamine from mast cells. Another factor released during anaphylaxis is *ECF-A* (eosinophil chemotactic factor of anaphylaxis), which has the ability to attract eosinophils, but not other leukocytes, to the site of the anaphylactic reaction.

Passive Anaphylaxis. The injection of adequate amounts of serum from a sensitized animal into a normal one, followed by an intravenous injection of Ag after a suitable interval (the latent period), causes typical symptoms of anaphylaxis. This technique allows one to study in detail the various kinds of Abs involved in the induction of anaphylactic shock and to compare Abs from the same or different species in producing anaphylaxis. A widely used and simpler method is that of *passive cutaneous anaphylaxis* (PCA), which allows the comparison of different Abs, or of varying quantities of Ab, in a single animal. The PCA method is performed by injecting intracutaneously various amounts of Ab, followed after a latent period by an intravenous injection of Ag mixed with Evans blue dye. In a few minutes blue spots appear at the sites of Ag-Ab reaction and reach a maximum in about 30 min.

The human analogue of the PCA is the *Prausnitz-Kustner* (PK) reaction. Serum from the allergic person is injected intracutaneously into a normal person, and after a latent period a small amount of the Ag injected into the same site causes a wheal and flare reaction. The "wheal and flare" Ab, present in extremely small amounts, is also called *reagin.*

Reagin Ab (at least that involved in the PK reaction to ragweed pollen) has been identified as IgE; it interacts via the Fc region with mast cells by virtue of a specific receptor on their surface, and contact with Ag triggers degranulation with the release of histamine and other mediators (Table 20-8).

Other Immediate-Type Allergies

Arthus reaction

The Arthus reaction, discovered by Maurice Arthus (1903), is an allergic reaction involving severe tissue damage caused by the formation of Ag-Ab complexes and their deposition in tissues. The Ag-Ab complexes fix C and attract polymorphonuclear leukocytes; the subsequent release of lysosomal enzymes causes tissue damage, with *vasculitis*—the inflammatory destruction of small blood vessels. As with anaphylaxis, repeated injections of an Ag into the skin cause inflammation, hemorrhage, and necrosis at the sites of later injections, although the earlier injections are completely innocuous.

Table 20–8. **Types of Antibody-Mediated (Immediate-Type) Allergic Reactions**

Type	Examples	Cells Involved	Pharmacological Mediators
Anaphylaxis	Hay fever Asthma Hives Wheal and flare	Mast cells Basophils	Histamine Serotonin Kinins SRS-A
Arthus reactions	Serum sickness Immune complex diseases, e.g., glomerulonephritis and rheumatoid arthritis	Neutrophils	Lysosomal enzymes
Reactions to transfused blood	ABO blood group incompatibilities (does not include Rh disease)	None	None

Serum sickness

Serum sickness is a man-made disease, which originated with the use of horse Abs for passive immunotherapy. Seven to 12 days after the injection of a large volume of serum, recipients may show swelling of the lymph nodes, urticarial wheals, and erythematous (reddish) areas, with itching and often edema of the eyelids, face, and ankles followed by joint pains and fever. These symptoms appear at a time when an Ab response or active sensitization to Ag takes place. The symptoms are a consequence of the interaction of newly formed Ab with excess free Ag in the tissue fluids and blood.

Drug allergy

A small number of people develop sensitivity to drugs and to certain chemicals to which they are repeatedly exposed. These substances, usually of low molecular weight, are not antigenic by themselves, but they or their metabolic products may react with tissue proteins of the recipient to form complete Ags. The sensitivity induced is most often of the *delayed type* but may also be of the *immediate type.* Perhaps the most important example of drug allergy is allergy to penicillin, which results from its widespread use in human and veterinary medicine.

Cell-Mediated Hypersensitivity (Delayed-Type Allergy)

The development of delayed-type allergy, also called tuberculin-type allergy or hypersensitivity, is essentially no different than the formation of the usual types of Ab and immediate-type allergy. In both cases initial contact with Ag is required to sensitize, the sensitivity is specific for the Ag, and a reaction can be obtained in the sensitized animals after contact with Ag. Both delayed- and immediate-type reactions are produced after contact with Ag, and it is often difficult to distinguish which manifestations are due to delayed hypersensitivity (DH) and which are associated with immediate hypersensitivity (IH).

Delayed hypersensitivity, which occurs during or after various infections (bacterial, fungal, and viral), after contact with certain chemicals (penicillin, poison ivy, and dinitrochlorobenzene), and after the rejection of organ transplants and tumors (Table 20-9), is recognized by the intracutaneous injection of Ag or by directly applying Ag to the skin. A positive reaction appears a few hours after contact with Ag and reaches a maximum intensity after *24 to 48 hr.* It differs from the wheal and flare and Arthus reactions and is usually a pinkish hardened lump (*induration*) that may vary from a few millimeters to several centimeters in diameter. Histologically, the reaction site is heavily infiltrated by masses of large *mononuclear cells,* macrophages and lymphocytes,

Table 20–9. **Types of Cell-Mediated (Delayed-Type) Allergic Reactions**

Type	Examples	Cells Involved	Mediators
Tuberculin sensitivity	Bacterial allergies; resistance to infectious agents (including parasites)	Macrophages, monocytes, and T-cells	Transfer factor; "immune" RNA
Contact dermatitis	Poison ivy allergy; nickel dermatitis; allergy to cosmetics; some drug allergies	Macrophages, monocytes, and T-cells	?
Rejection of organ transplants	Skin graft rejection; rejection of kidney transplants; graft vs. host reactions; rejection of malignant tumors	Macrophages, monocytes, and T-cells	Transfer factor; "immune" RNA
Autoallergic diseases	Experimental allergic encephalomyelitis (EAE); Hashimoto's thyroiditis; allergic aspermatogenesis	Macrophages, monocytes, and T-cells	?
In vitro reactions	Inhibition of cell migration	Macrophages and lymphocytes	Migration inhibition factor (MIF)*
	Blast transformation	Lymphocytes	Blastogenic factor (BF)*

*MIF and BF belong to a large and diverse "family" of various mediators collectively called "products of activated lymphocytes" (PALs) or "lymphokines." The precise identity of most of these mediators remains unknown.

whereas the wheal and flare reaction is essentially a deposition of fluid at the site.

Delayed hypersensitivity can be differentiated from IH by its inability to be transferred passively to normal recipients with serum of animals or persons with delayed sensitivity. However, passive transfer can be effected by lymphoid cells of sensitive donors. Extracts of lymphoid cells have been found to transfer DH in man and animals. The active principle in such extracts of human lymphoid cells, which is called the *"transfer factor,"* is resistant to DNase, RNase, and trypsin; is dialyzable; and has a molecular weight of less than 10,000. The sensitivity transferred in humans by cells or cell extracts persists for several months and even as long as 2 years. The "transfer factor" is released into solution upon exposure of sensitized cells to Ag in tissue culture. Several tests for detecting DH through the use of cell suspensions have been developed. One of the most interesting is the *inhibition of macrophage migration.* Peritoneal exudate cells (macrophages) from tuberculin-sensitive guinea pigs placed in a capillary tube in small tissue culture chambers migrate out of the capillary in the form of a "fan" in 24 to 48 hr. Such migration is strikingly inhibited when the specific Ag is also present in the chamber; cells from normal animals are unaffected (Fig. 20-8). The active cell responsible for migration inhibition has been shown to be the lymphocyte with the macrophage being merely an indicator cell. Upon contact with Ag the sensitive lymphocyte releases a soluble *migration inhibitory factor* (MIF), which by an as yet unknown mechanism inhibits the migration of the macrophage. Several theories have been proposed to account for the differences between DH and IH, including:

1. DH is entirely cellular, being an intrinsic property of the sensitized cell.
2. DH is mediated by an Ab with a high affinity for tissue cells, so that it does not usually exist free in the circulation.
3. DH is associated with minute amounts of circulating Ab with a very high binding affinity for Ag.
4. Some sort of transfer factor is involved.

Figure 20-8. **Cell-mediated hypersensitivity: inhibition of macrophage migration. Capillary tubes filled with spleen cells from tumor-immunized mice are placed on a culture medium and the migration of the cells photographed. In (A), the control, no antigen is added and the extent of the migration (m) is seen as the large, pale halo at the top of each tube. (B) In the presence of antigen, the extent of the migration (m) is markedly diminished. The white blobs at the bottom of each photograph are spots of silicone grease that hold the tubes in place. (From Ph.D. thesis of V. Likhite, 1971.)**

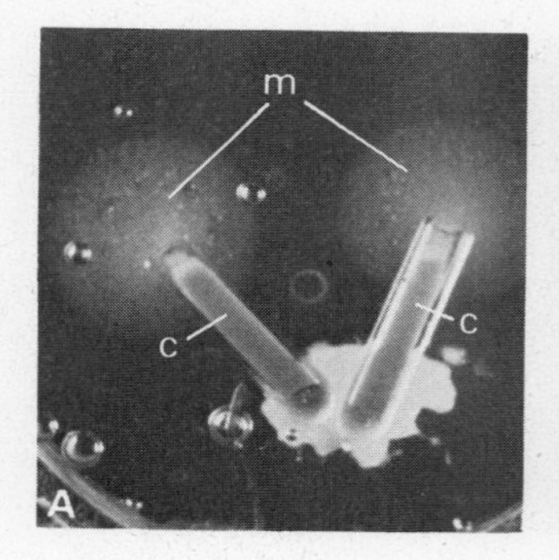

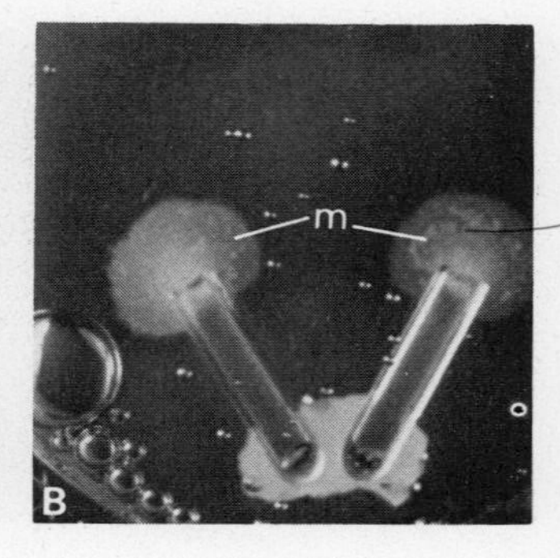

5. DH is an early stage in the formation of Ab of the usual types.

These theories are viewed with varying degrees of enthusiasm by different workers in the field.

20–8 Blood Groups, Transplantation, and Tumor Immunology

Blood Groups (Immunohematology)

Serous fluids and tissues contain substances that are immunogenic. Of interest in this section are the *isoantigens* (iso-Ags or *alloantigens*), which play a significant role in blood transfusion, transplantation, and certain immunological disorders such as hemolytic disease of the newborn or *erythroblastosis fetalis*. The iso-Ags are expressions of different alleles that occur at one genetic locus within different members of one interbreeding population (genetic polymorphism). They can be considered as the factors that make a person's tissues and blood cells immunologically unique (i.e., unlike those found in another individual).

ABO AND RH BLOOD GROUPS

The red blood cells of some people are clumped (agglutinated) in the presence of sera of other people. In humans, the ABO and Rh red cell iso-Ags are medically important in transfusions and, to some extent, in tissue transplantation. They are found on the surface of red blood cell membranes and on various cells scattered throughout the body. The Rh iso-Ags, on the other hand, can be demonstrated only in the erythrocytes of humans and monkey (Rh stands for rhesus monkey). The characteristics of the ABO and Rh iso-Ags are summarized in Table 20–10. The ABO iso-Ags have reciprocal Abs (*isohemagglutinins*) present in the blood of individuals that lack them. These isohemagglutinins appear soon after birth and are primarily IgM Abs. The Rh iso-Ags have no reciprocal "natural" Abs, but are so highly immunogenic that improper transfusion from an Rh-positive (Rh^+) to an Rh-negative (Rh^-) individual results in anti-Rh^+ Ab production in the Rh^- recipient. Similarly, the transmission of red cells during birth from an Rh^+ infant to an Rh^- mother (e.g., due to rupture of the placental barrier) can stimulate anti-Rh^+ Abs in the mother. An Rh^+ fetus conceived in later pregnancies can be seriously affected by the anti-Rh^+ IgG Abs carried by the mother because these antibodies are able to cross the placenta and enter the fetal circulation. Massive destruction of the infant's cells results in abortion, stillbirth at term, or death after birth. This disorder is known as *erythroblastosis fetalis* or *hemolytic disease* of the newborn. Recently, it has become possible to prevent this disease. Because the sensitization of the mother mainly occurs just at the time of birth, injection of anti-Rh^+ Abs (RhoGam®) into the Rh^- mother just after delivery of an Rh^+ infant will coat any fetal (Rh^+) cells that may have entered the circulation of the mother. The Ab-coated cells are rapidly removed from the maternal circulation, and the mother is prevented from developing her Abs to fetal cells.

Table 20–10. **Human Blood Groups**

a. Major Characteristics of the ABO Blood Group System

Serum of Group	Agglutination of Red Cells from Group: A	B	O	AB	"Natural" Abs in Serum	Distribution in U.S. Population (per cent)	Ag on Red Cells
A	–	+	–	+	anti-B	43	A
B	+	–	–	+	anti-A	10	B
O[1]	+	+	–	+	anti-A and anti-B	44	H
AB[2]	–	–	–	–	–	3	A and B

[1] "Universal donor."
[2] "Universal recipient."

b. Some of the Known Human Blood Groups

Blood Group	Blood Group Antigens	Year of Discovery
ABO	A(A1, A2), B, H	1900
MNSs	M, N, S, rs	1927
P	P1, P2, Pk	1927
Rh	C, C^w, c, D, D^u, d*, E, e	1940
Lutheran	Lu^a, Lu^b	1946
Kell	K, k, k_p^a, K_p^b, J_s^a, J_s^b	1946
Lewis	Le^a, Le^b	1946
Duffy	F_y^a, F_y^b	1950
Kidd	J_k^a, J_k^b	1951

The ABO blood group system is genetically controlled by three alleles at one genetic locus (three allelic genes). These are the A, B, and O alleles, where O is an amorph (i.e., it does not specify any red cell iso-Ag). The O allele is expressed as the *absence* of the A and B alleles and is involved in the expression of the *heterogenetic* (*H*)-iso-Ag (i.e., the Ag is formed by a variety of unrelated species). Subgroups (A1 and A2) are now known in the A group. The *secretory* (Se) genes, distinct from the ABO loci, regulate the secretion of the A, B, and H iso-Ags in body fluids (especially saliva). The ABO iso-Ags have not been completely characterized structurally, but the terminal residues are known.

There are approximately 12 other blood group iso-Ags of the human red blood cells. They have been primarily useful in medico-legal and anthropological problems. With the exception of rare transfusion difficulties, the contribution of these groups to diseases has not been determined.

Transplantation Immunology

When skin from one individual is grafted onto a genetically distinct individual, it initially appears to be accepted and becomes infiltrated with blood vessels. However, within 10 to 14 days it is invaded by white blood cells and necrosis begins, leading quickly to *graft rejection.* The immunological basis of rejection can be demonstrated by comparing the speed at which the first graft is destroyed to the rate at which a second graft is rejected. Second grafts using tissue from the same donor are rejected within 3 to 4 days (*accelerated rejection*), whereas a second graft using an unrelated donor again takes 10 to 14 days to be sloughed off. These experiments demonstrate that graft rejection involves immunological memory and specificity.

Immunologists first suspected that circulating Abs mediated graft rejection; but then it was found that intact small lymphocytes are the active agents. Thus accelerated rejection can be transferred to normal recipients with living lymphoid cells of immunized donors while the serum of these animals is completely ineffective. These lymphocytes can recognize the surface of grafted cells as either native or foreign and specifically destroy the cells bearing foreign *"transplantation"* Ags. Genetic analysis of highly inbred mouse strains reveals the existence of a set of contiguous genes (the *H-2 complex*) that predominate in the graft rejection process (Table 20–11). They determine the specificity of the major mouse transplantation (*histocompatibility*) Ags, the H-2 proteins. In humans, a similar linked group of major histocompatibility genes is called the HL-A complex. Within each H-2 (HL-A) complex, there is evidence for the coding of two distinct antigenic determinants, and so each diploid cell expresses four different specificities. *The HL-A (H-2) antigen is an immunoglobulin-like structure.*

It is now known that the major histocompatibility Ags (mw 125,000) have a structure very similar to that of an Ig. They are built up from two L-chains (mw 12,000) and two H-chains (mw 50,000) held together by S-S bonds. The L-chain appears to be identical to *β_2 microglobulin,* a serum protein whose amino acid sequence has been worked out. This protein contains an S-S region strikingly homologous to the S-S linked CH3 region of the IgG1 H-chain. This suggests that the various HL-A (H-2) specificities must all reside within the H-chain. Recent analysis of the H-2 Ag suggests that each chain consists of three S-S bonded regions, each of which also resembles a C-region of an H Ig chain. The close homology between the histocompatibility Ags and the Igs suggests that, early in the evolution of vertebrates, Igs evolved by the duplication of genes coding for histocompatibility-like surface Ags. Attempts to minimize or prevent graft rejection, for replacement of diseased organs with healthy ones, involve:

1. *Matching* donor and recipient at the HL-A (H-2) locus. Indeed, in the case of human kidney transplant, it is clear that the closer the match, the better the survival of the graft.
2. General *immunosuppression* by use of substances that nonspecifically interfere with the induction or expression of the immune response. Included among such agents are whole body irradiation, immunosuppressive drugs (azathioprine, methotrexate, cyclophosphamide), and antilymphocyte serum (ALS), all of which have been used to prolong the survival of transplants. Because these agents act nonspecifically, people on immunosuppressive therapy may be susceptible to infection and also prone to develop cancer.
3. The use of antigenic specific depression of transplantation immunity through the induction of *immunological tolerance,* which is the specific lack of immunological responsiveness to a given Ag (in this case the transplantation Ags). An important characteristic of the immune response is its ability to recognize a molecule as foreign. Thus the body does not make Abs against its own proteins and nucleic acids. This lack of responsiveness has nothing to do with an inherent lack of antigenicity of these molecules. For example, almost any human protein will induce the formation of specific Abs if injected into a rabbit. Similarly, rabbit proteins are antigenic in humans. However, if a foreign Ag is injected into a newly born

Table 20–11. **The Histocompatibility (H-2) Locus in the Mouse**

Chromosome 17 ← To centromere	← H-2 Complex →																			
	H-2K					Ir					Ss				H-2D					
H-2 Alleles	Antigenic Specificities																			Mouse Strains in Which Alleles Are Present
	1	2	3	4	5	6	8	10	11	13	14	22	25	27	28	29	31	32	33	
a	+	−	+	+	+	+	+	+	+	+	+	−	+	+	+	+	−	−	−	A/J, AKR · K
b	−	+	−	−	+	+	−	−	−	−	+	+	−	+	+	+	−	−	+	C57B1/6, 129
d	−	−	+	+	−	+	+	+	−	+	+	−	−	+	+	+	+	+	−	BALB/c, DBA/2
k	+	+	−	−	+	−	+	−	+	−	−	−	+	−	−	−	−	+	−	CBA, C3H, AKR

Note: Different H-2 alleles (a, b, d, and k) correspond to different patterns of antigenic specificity (1,2,....., 33), which are found in the inbred mouse strains (A/J, C57B1/6, etc.). Each of the alleles, representing a combination of linked alleles of two genes (K and D) in the locus, corresponds to a haplotype in the human HL-A locus. Gene Ss specifies certain serum proteins associated with the complement system; the Ir (immune response) region is a cluster of genes that control the ability to make immune responses to a variety of antigens. Although only 4 alleles are shown, more than 20 alleles, each controlling a different set of antigens, have been identified. Thus far 15 histocompatibility loci have actually been discovered in the mouse. Of these, 5 autosomal loci (H-1, H-2, H-3, H-4, and H-13), a Y-linked locus, and an X-linked locus, have now been assigned to specific linkage groups. Only a portion of the H-2 locus is depicted in this table. The human histocompatibility locus is equally complex.

animal before its immunological machinery has matured, then, in adult life, the animal is unable to form Abs against the early injected foreign Ag. The foreign Ag is recognized as though it were its own protein.

Tumor Immunology

The possibility of inducing protective immunity against malignant tumors has intrigued immunologists for many years. It is known that although malignant tumors are fatal to some individuals they are nevertheless rejected as transplants by other members of the same species. Indeed, there exists an active immune response, both Ab- and cell-mediated, against tumor Ags, not only in individuals that have rejected tumors but also in those in which the tumors continue to thrive.

The ability to reject transplants is thought to confer survival advantage on the host. It is possible that the immune system "polices" the body for antigenically altered somatic cells that might become malignant. Effective *"immunological surveillance"* depends on the presence of tumor-specific Ags on the surface of malignant cells, which enable them to be recognized as "foreign" to be destroyed by immunological reactions. These tumor Ags may be virally controlled, embryonic, or idiotypic. The failure of this "surveillance mechanism" is thought to enable malignant tumors to survive in an otherwise immunologically competent host. Tumors are thought to overcome immune surveillance by: (1) producing enhancing or blocking factors (Abs?) that can abrogate the destructive effect of lymphocytes (killer cells) on the cells of the tumor; (2) inducing specific immunological tolerance to tumor-specific Ags; and (3) releasing excess amounts of tumor-specific Ags, which result in an inhibition of the effector limb of the immune response.

Acknowledgments

The author wishes to thank his many graduate students, assistants and research fellows, and Dr. Param Chawla for their excellent help in preparing this review article.

Cited References

BURNET, F. M. *Self and Not-Self.* Cambridge University Press, New York, 1969.

EDELMAN, G. M. "The Structure and Function of Antibodies." In *Immunology.* W. H. Freeman and Company, Publishers, San Francisco, 1976.

HELDELBERGER, M., and F. E. KENDALL. *J. Exp. Med.* **50:** 809 (1929).

LANGER, W. L. "The Prevention of Smallpox Before Jenner." *Sci. Am.,* Jan. 1976.

PORTER, J. R. "Louis Pasteur Sesquicentennial (1822–1972)." *Science* **87:** 416 (1972).

PORTER, R. R. "The Structure of Antibodies." In *Immunology.* W. H. Freeman and Company, Publishers, San Francisco, 1976.

TISELIUS, A., and E. A. KABAT. *Science* **87:** 416 (1938).

Additional Readings

BOOKS

BILLINGHAM, R., and W. SILVERS *The Immunobiology of Transplantation.* Prentice-Hall, Inc., Englewood Cliffs, N.J., 1971.

EISEN, H. N. *Immunology,* 2nd ed. Harper & Row, Publishers, New York, 1975.

GREAVES, M. F., J. J. T. OWEN, and M. C. RAFF. *T and B Lymphocytes: Origins, Properties and Roles in Immune Responses.* Excerpta Medica, American Elsevier Publishing Co., Inc., New York, 1975.

KABAT, E. A. *Structural Concepts in Immunology and Immunochemistry,* 2nd ed. Holt, Rinehart and Winston, New York, 1976.

ROITT, I. R. *Essential Immunology.* 2nd ed. Blackwell, Oxford, 1975.

STANWORTH, D. R. *Immediate Hypersensitivity: The Molecular Basis of the Allergic Response. Frontiers of Biology.* Vol. 28, A. Neuberger and E. L. Tatum, eds. North Holland/American Elsevier, New York, 1973.

TURK, J. L. *Delayed Hypersensitivity.* 2nd rev. ed, *Frontiers of Biology,* Vol. 4, A. Neuberger and E. L. Tatum, eds. North Holland/American Elsevier, New York, 1975.

ZMIJEWSKI, C. M., and J. L. FLETCHER, *Immunohematology.* 2nd ed. Appleton-Century-Crofts, New York, 1972.

Chapter 21

Energy Balance: Regulation of Body Temperature, Food and Water Intake, and Metabolism

Have you listened for the things I have left out?
I am nowhere near the end yet and already hear
the hum of omissions

A. R. Ammons, "Unsaid"

THE ENERGY BALANCE of the body is determined by the inflow of energy, in the form of ingested food, and energy outflow. Energy output involves the energy required to maintain the body temperature within a relatively narrow range and to sustain body reserves of energy, chiefly stored fat. Because the food intake and energy output vary widely in an individual over short periods of time while the body weight remains relatively constant, the error in energy balance over long periods is extremely small.

This delicate control over energy balance involves many physiological regulatory systems that are integrated in the hypothalamus. The outstanding difference between the cells of the hypothalamus and the other cells of the brain lies in the *sensitivity of hypothalamic cells* to direct thermal and chemical changes in their environment. In contrast, other brain cells, such as cells in the thalamus, respond only to information delivered in the form of nerve impulses that has been processed by peripheral receptors.

Not only is the hypothalamus responsive to different modes of stimulation; simultaneously, it is uniquely able to regulate the different physiological systems involved. To compare it again with the thalamus, the thalamus serves only as a relay station, sending the information it receives from sensory receptors on to the cortex which has to execute the response. The hypothalamus not only can *transduce* the various types of stimuli it receives, it also *organizes* the extremely complex responses required for homeostasis. Temperature regulation, for example, is achieved through control of cardiovascular responses, metabolic rate, and sweating. These responses in turn involve "lower-priority" systems such as osmotic pressure, blood glucose, and food intake. The hypothalamus has to determine the level of priority for these regulatory mechanisms. We are still far from understanding how these multiple inputs and outputs are coordinated.

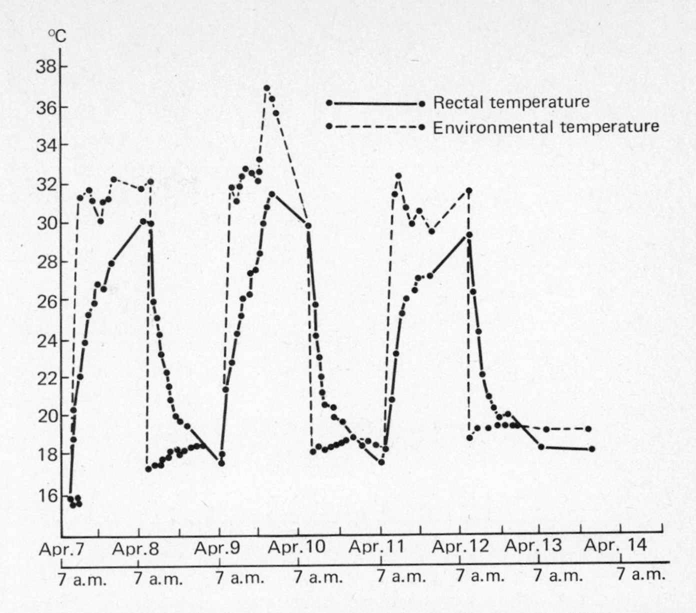

Figure 21-1. **The body temperature of the python snake follows that of the environment. (From F. Benedict, *The Physiology of Large Reptiles,* The Carnegie Institution, Publ. 425, 1932. Courtesy of the Carnegie Institution of Washington.)**

21–1 Regulation of Body Temperature

Mammalian cells are particularly sensitive to extremes in temperature: a marked increase in the environmental temperature rapidly kills off most cells, whereas severe cold can slow down metabolic processes so much that tissue necrosis and gangrene result.

Normally, body temperature, and hence the temperature of the cellular environment, are maintained within a rather narrow range by the thermostatic activities of the *hypothalamus.* This regulation requires that the hypothalamic thermostat both receive information about body temperature and have the physiological mechanisms that can respond to adjust the temperature. Through feedback mechanisms between the rate of firing of hypothalamic neurons and the temperature of the blood circulating through it, the hypothalamus balances the amount of heat produced and the amount of heat lost by the body. These mechanisms are discussed in detail later in this chapter.

Animals that maintain a relatively constant body temperature are *homeotherms.* Most mammals and birds are homeotherms: they are also called *endotherms* to indicate that they produce and control their own sources of heat. Other vertebrates, such as fish, amphibia, and reptiles, are *ectotherms* (*poikilotherms*), so called because their body temperatures vary with fluctuations in the external environment. Figure 21-1 shows how the body temperature of a snake follows that of the external environment. Ectotherms can gain heat almost exclusively from the environment.

Hibernating mammals differ from poikilotherms in that they still regulate their body temperatures but at a very much lower level. A hibernating ground squirrel, for example, will maintain its body temperature at about 5°C; if the temperature drops still lower, the squirrel will spontaneously increase its metabolic rate or wake up. Figure 21-2 shows some of the physiological events that accompany entering hibernation and awakening from it. Hibernators also lower the melting point of their body fats, just prior to entering hibernation, as though they were changing from summer to winter oil. They accomplish this feat metabolically by increasing the proportion of desaturated fatty acids in the fat.

Behavioral and Physiological Regulation of Body Temperature

Behavior may be viewed as a physiological response. Behavioral and physiological thermoregulatory adaptations appear to share common neural systems, involving the hypothalamus and limbic system.

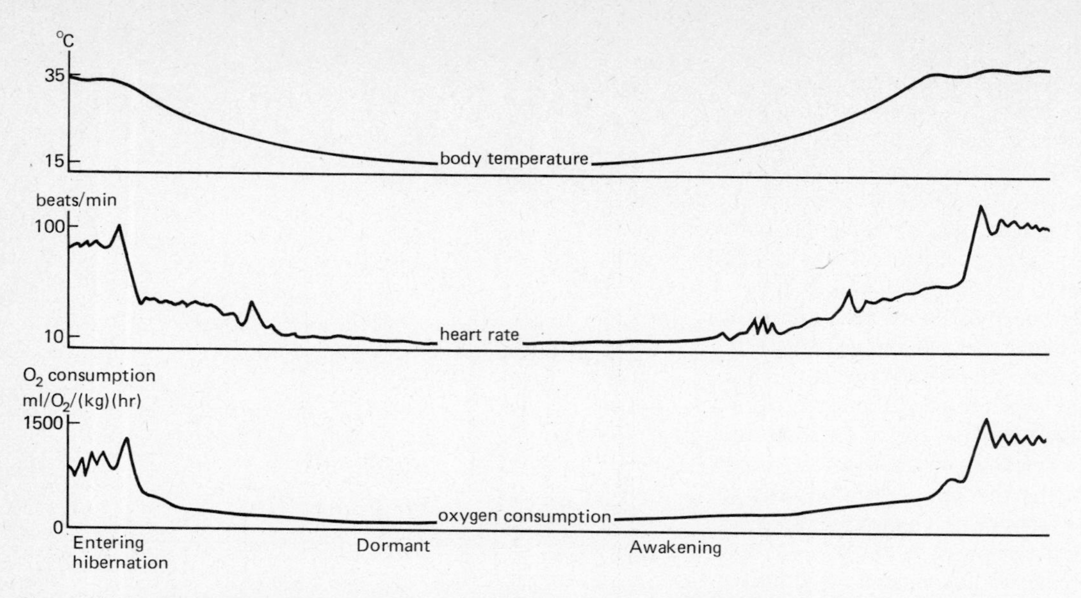

Figure 21-2. **Body temperature, heart rate, and oxygen consumption during entry into, and awakening from, hibernation. [Adapted from E. Folk, Jr., *Amer. J. Physiol.* 194:83 (1958).]**

Behavioral adaptations are used by both endotherms and ectotherms to extend the range of temperature environments in which they can survive. Snakes and lizards bask on warm rocks in the sun; desert rats stay in cool burrows underground until the evening cool descends. Humans have learned to come in out of the cold, to wear layers of warm cloth to maintain as much heat as possible, and to regulate the temperature and humidity of the rooms in which they work and live.

Normal Temperature Range

Homeothermia appeared some 200 million years ago when mammals and birds first evolved from their reptilian ancestors. The deep body (core) temperature for mammals averages about 38°C, whereas that for birds is set about 1°C higher.

In humans, the normal resting temperature when measured orally ranges between 97 and 99°F. The average normal temperature is considered to be 98.6°F (37°C) *orally*. When the temperature is measured *rectally*, it is about one degree higher (99.6°F or 37.6°C). Neither of these measurements is an absolute indication of the internal body temperature and, experimentally, more accurate determinations can be made by taking the temperature at the ear drum membrane (tympanic temperature). This is of little value clinically, however, where minute changes in temperature are less important than ease and comfort of measurement.

Any variations in temperature must be compared with the normal resting temperatures for that particular individual. The ease with which the temperature rises varies considerably from person to person, and it varies particularly with age. The thermostat in the hypothalamus is easily set off balance in young children, who may extremely rapidly develop a high fever that returns to normal in a matter of a few hours. Adults usually have more stable temperatures (see Appendix II for variations in body temperature with age and sex).

Thermography

Thermography is a new technique for measuring *skin temperature* by photographing the skin with an infrared camera, which reproduces a thermal map of the skin. Where the circulation is rapid and close to the surface, the film appears light; where the metabolism is low or circulation impeded, the film is dark. Abnormal growths with high metabolic activity show up on the thermogram as areas of higher temperature.

Cyclic Temperature Changes

Daily rhythm

A daily temperature rhythm has been noted for most people, but the high and low points over a 24-hr period vary for each individual. There is some correlation between the high point and the time of greatest efficiency and activity. Early risers appear to have their high point in the morning but reach their lowest temperature later in the day. Those who are sluggish and dazed in the mornings record their lowest temperature at this time, and there is an elevation later in the day, or even at night, when they feel most active and enterprising.

Attempts to change the cycle in individuals have met with dubious success, but the correlation between temperature and activity remains. Subjects who were able to reverse their habits of work and sleep also changed their temperature cycles. Those who were unable to make the change successfully did not make the physiological temperature adjustment (Fig. 21-3).

Monthly rhythm

A monthly temperature rhythm has been established conclusively for women during the reproductive part of their lives. This temperature pattern is dependent on the cyclic secretion of the female sex hormones. These cyclic changes in hormone production are also regulated by the hypothalamus, a topic which is discussed in more detail in Chapters 10 and 26. At the time of ovulation, approximately halfway through the menstrual cycle, there is a brief fall in temperature, followed by a sharp rise (Fig. 21-4). This elevated temperature is maintained for the second half of the menstrual cycle, falling to its lower level just before menstruation, remaining low until the brief fall and sharp rise that occur at ovulation. This rhythmic change in body temperature, if measured routinely for several months, can be used to indicate the approximate time at which a woman is fertile. She can conceive only if mating occurs within a period of 24 hr before and after ovulation. The temperature remains high if pregnancy occurs.

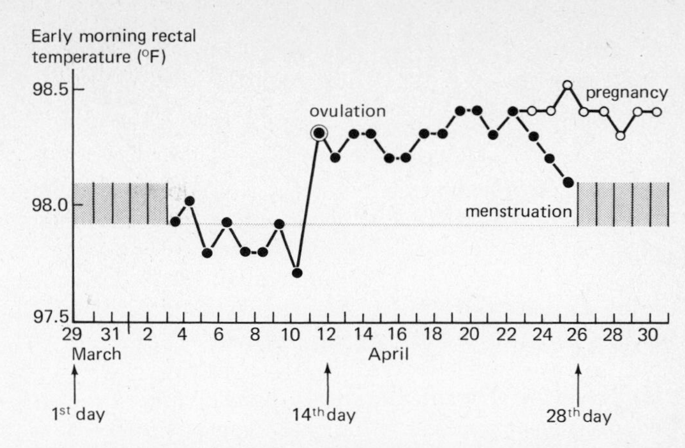

Figure 21-4. **Variations in rectal temperature of women during the menstrual cycle and pregnancy. Note that there is a fall in temperature just prior to ovulation, after which temperature remains high if pregnancy occurs. If pregnancy does not occur, the temperature drops a few days before the beginning of the next menstruation.**

Production, Retention, and Loss of Heat

Heat production

Heat is gained by the body both as a result of the metabolic activity of the tissues and from the external environment if the latter is hotter than body temperature. Heat generated by the body increases as the external temperature decreases. *High ambient temperatures* also decrease the amount of heat generated by the body. Metabolic activities are slowed, work output is decreased, and muscle tonus slacks as lethargy sets in. The heat-generating mechanisms involved are increased metabolic activity of the tissues, increased muscle activity, and nonshivering thermogenesis.

Increased Metabolic Activity of the Tissues. By increasing the amount of fuel oxidized by the tissues, the body

Figure 21-3. **Correlation of body temperature with differences in sleep cycles. (A) This individual maintained the 24-hr temperature cycle despite alterations in the sleep cycle from a 20- to a 28-hr day. (B) Temperature cycles in this subject followed the changes in the length of the day, probably indicating an ability to perform more efficiently under these varying circumstances than subject A.**

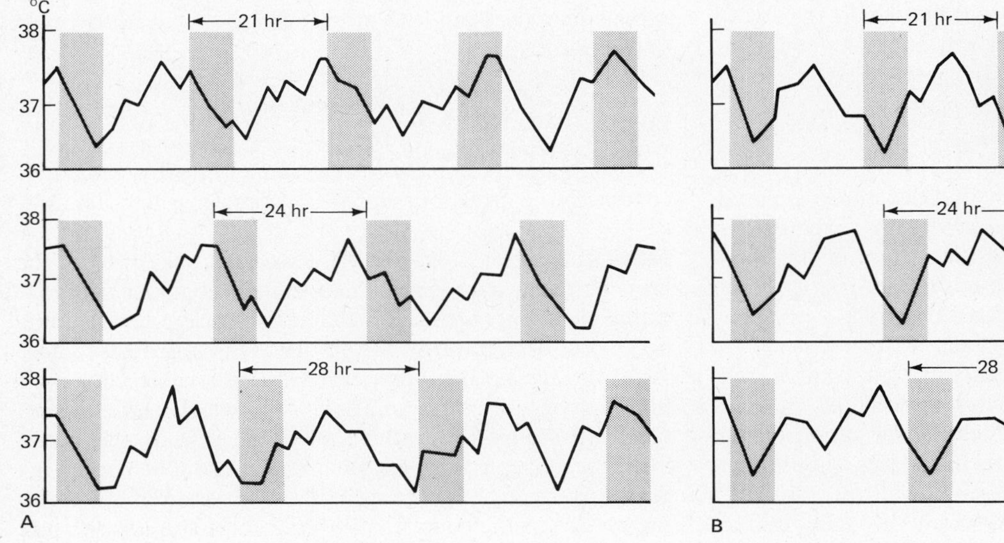

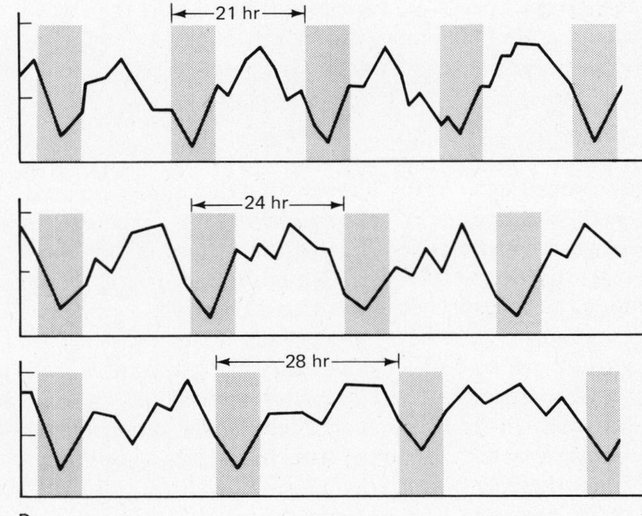

liberates heat. Both epinephrine and thyroxine increase tissue oxidation.

Increased Muscle Activity. The muscles make up almost 50 per cent of the body by weight, and the heat liberated by their *tonic contraction* (the maintenance of a certain amount of constant tension) can be increased by the involuntary tensing of the muscles. This activity can be considerably augmented by *shivering,* a rhythmic involuntary contraction of the muscles. Of course, active *voluntary exercise* will increase the heat production still further.

Nonshivering Thermogenesis. Nonshivering heat production (thermogenesis) is the chief way in which animals and humans acclimatize to cold. An increase in the rate of lipid metabolism (and to a lesser extent, of carbohydrate metabolism) produces heat, quite independent of muscle contraction. This has been demonstrated in several ways, one of which is to paralyze the muscles of cold-acclimatized rats by the use of the drug curare. These animals cannot shiver or move, but they are able to double their heat production when exposed to 5°C, and so maintain their body temperature.

Another indication of the *chemical nature* of this type of thermogenesis comes from young mammals. Many newborn mammals, including humans, cannot shiver, yet they can generate heat in a cool environment. These young mammals have a special type of adipose tissue, *brown fat,* which is found chiefly between the shoulder blades, along the spine, behind the breastbone, and around the neck (Fig. 21-5). Brown fat has a unique metabolic device for oxidizing fatty acids and turning the resulting chemical energy mainly into heat. The rate of heat production is controlled by the release of *norepinephrine* from sympathetic nerves, which directly innervates the fat. Nonshivering thermogenesis cannot occur in sympathectomized, cold-acclimatized rats exposed to cold.

Other hormones, such as the adrenal glucocorticoids and thyroxine, are released in response to cold stress and may potentiate the effects of norepinephrine. Again the hypothalamus, as the integrating center for autonomic and neuroendocrine responses, is involved.

Heat retention

Vasoconstriction. Heat is retained in the animal exposed to cold through local and reflex *vasoconstriction,* a regulatory mechanism that routes blood from the surface of the body down to the deeper tissues. This minimizes heat loss by radiation, a topic discussed in the next section.

Insulation. A coat of fat under the skin and a furry coat over it serve to prevent heat loss from the body to the cold environment. In hairy animals, a reflex contraction of the small muscles at the base of the hairs in the skin causes these hairs to stand erect (*piloerection*), forming a warm insulating blanket to trap the heat radiating from the skin. In the relatively hairless human being, this is seen merely as "gooseflesh," which serves no practical purpose. Humans supply their own insulating layers by the addition of clothing.

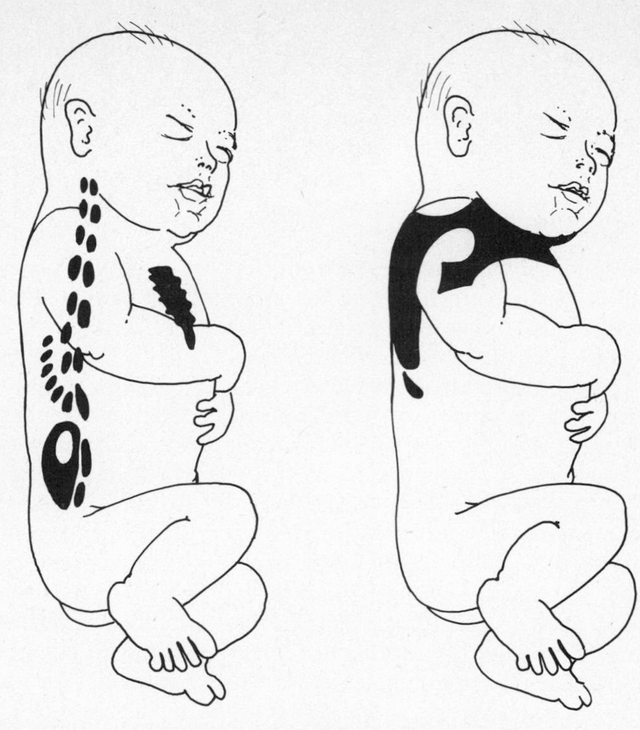

Figure 21-5. **Brown fat is deposited in young infants between the shoulder blades, along the spine, behind the breastbone, and around the neck.**

Rete Mirabile (Wonderful Net). The extremities of animals exposed to severe cold are protected from heat loss by a network of small arteries and veins arranged so that the warm blood leaving the trunk arteries heats the cool blood returning in the veins by a mechanism of *countercurrent flow* (Fig. 21-6).

Heat loss

Heat is lost from the mammalian body by radiation, conduction, convection, and the evaporation of water.

Radiation and Convection. Radiation of heat from the body occurs when the body temperature is higher than that of the atmosphere (the ambient temperature). As the warm body loses heat, it warms the air that is in immediate contact with it. This hot air rises, being lighter, while cooler air moves in to take its place. These air movements, or *convection currents,* assist in the loss of heat from the body. Stronger movements of air, such as wind, will be considerably more effective in permitting the dissipation of heat from the surface of the body. Loss of heat through radiation will be most efficient when the air is cooled, dehumidified, and kept moving—the basis of air conditioning.

Conduction. Heat is conducted from the deeper organs to the surface of the skin through the blood vessels. The skin acts as a large radiating surface for the loss of heat to the environment. *Vasodilation,* a local and reflex response to temperature, routes more blood through the capillary beds in the skin, permitting greater heat loss by radiation to the atmosphere.

Evaporation of Water from the Skin

1. The body constantly loses heat because of the continuous evaporation of water from the *skin* and *lungs.* This is elegantly termed *insensible perspiration* and reminds one of the genteel saying that "horses sweat, men perspire, but ladies glow." Water molecules are constantly diffusing through the skin and also through the alveoli of the lungs to form water vapor with a consequent heat loss. This mechanism is not significantly altered with changes in body temperature and so is not a homeostatic phenomenon.

2. Heat loss through the evaporation of water from the skin can be heightened greatly by the increased production of *sweat,* a watery secretion of the sweat glands in the skin that has basically the same composition as the other body fluids. If sweating is profuse over a few hours, however, the cells of these glands produce a more dilute secretion by retaining sodium and chloride. This is a peculiar phenomenon, for it is much more marked in people who have become acclimatized to tropical environments. Their sweat glands save considerably more of these electrolytes than when they first were exposed to the heat. Nevertheless, even after acclimatization, extreme sweating will deplete the body's supply of electrolytes, and these should be replaced by salt tablets.

3. Furred animals do not have functional sweat glands. Those that *pant,* like the dog, lose heat by the rapid movement of air over the hot, moist tongue and air passages, cooling these structures as water evaporates from them. The rat, which does not pant, increases its secretion of saliva, which it spreads over its body.

THE ENVIRONMENT AND HEAT LOSS

The discomfort experienced in a hot, moist environment is due to the deprivation of two important mechanisms for heat loss: radiation and evaporation. In a hot, dry climate, the evaporation of sweat permits the loss of more than 1.5 liters of water/hr, which is the equivalent of a heat loss of 870 Calories. When the humidity of the air is too high to permit the evaporation of the sweat, however, sweat drops are formed which simply roll off the body, without having contributed to heat loss through evaporation. Because the difference in temperature between the body and the air is very slight, or even nonexistent, no heat can be lost by radiation.

Regulation of Body Temperature by the Hypothalamus

TEMPERATURE SENSORS

Nerve structures sensitive to temperature or temperature changes are called temperature sensors. Central temperature sensors are found chiefly in the hypothalamus; there are also some temperature sensors in the spinal cord. Peripheral temperature sensors are distributed in the skin.

Central Temperature Sensors. The most important temperature sensors involved in the regulation of body

Figure 21-6. **When atmospheric temperature is low, warm blood leaving the arm arteries loses heat to the chilled blood returning from the hand through the veins. Thus warmed venous blood is returned to the body. Heat loss is further minimized by vasoconstriction of surface blood vessels.**

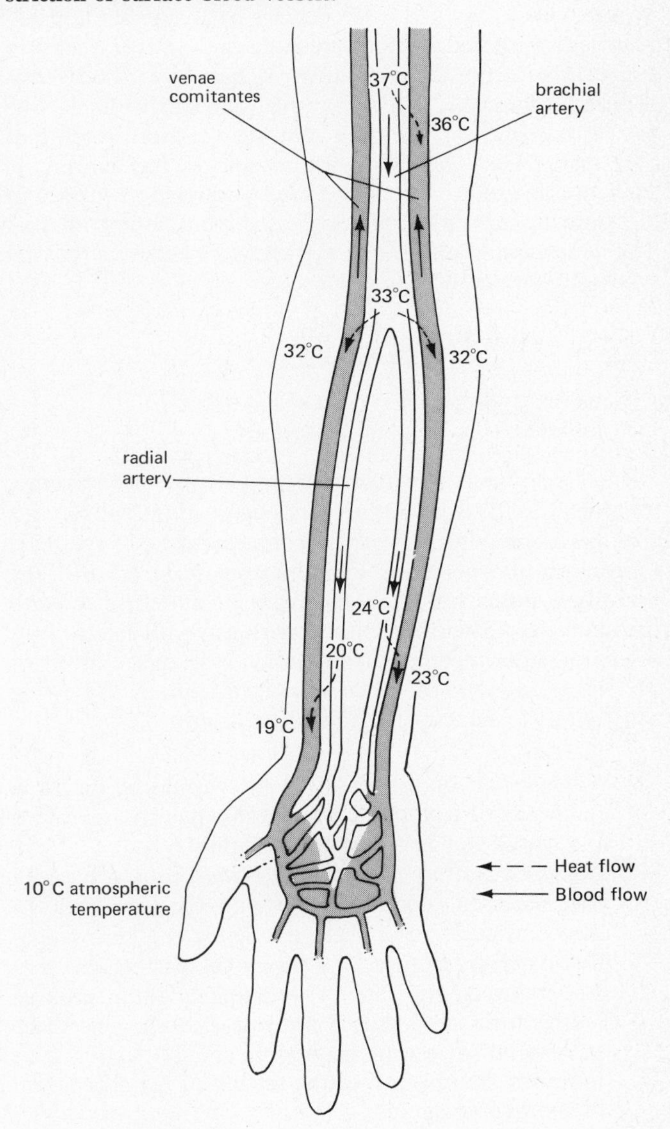

temperature are heat-sensitive neurons in the *anterior (preoptic) region of the hypothalamus.* These neurons respond to changes in the temperature of the blood circulating through this organ. A change of only 0.01°C in the temperature of the blood circulating through the hypothalamus has been shown to stimulate the thermostatic mechanism and cause an adaptive response. The degree of the response is so finely matched to the change in temperature that precisely enough heat is generated or lost to restore the temperature of the blood to normal. There are some cold-sensitive hypothalamic neurons, but their specific contribution to thermoregulation is uncertain. The existence of these temperature-sensitive neurons has been established electrophysiologically by inserting hot or cold probes into specific areas of the hypothalamus and recording the changes in firing rate of the neurons affected (Fig. 21-7).

Peripheral Temperature Sensors. Temperature sensors in the skin, for both heat and cold, relay information to the hypothalamus. Marked discrepancies between skin temperature and tympanic temperature indicate, however, that skin temperature is not the regulating mechanism for body temperature in humans, although in many small mammals, like the rat and the cat, the skin may play a more important regulatory role. The difference in control mechanisms is probably due to the larger body mass relative to surface in big animals; this ratio could lead to the buildup of internal heat deficits or surpluses unless there were sensitive internal sensors to control temperature.

Anterior (preoptic) hypothalamus and the prevention of overheating

The preoptic region of the hypothalamus is the center for reflex mechanisms that prevent overheating. Cells in this area make inhibitory synaptic contacts with sympathetic neurons in the posterior hypothalamus (Fig. 21-8). The preoptic hypothalamic cells also make synaptic contact with parasympathetic neurons in the anterior hypothalamus. As a result of parasympathetic stimulation and the removal of sympathetic tone to blood vessels in the skin, the following adaptations to increased temperature occur:

1. *Vasodilation* of the small blood vessels in the skin, increasing the blood flow and consequently increasing the amount of heat lost by radiation.
2. *Increased activity of the sweat glands,* caused by both the increased blood flow through the skin and the direct stimulation of these glands by the parasympathetic nerves. Figure 21-9 shows the perfect relationship between the rate of sweating and the internal temperature as measured at the ear drum. The sharp increase in sweating beginning at 36.9°C (98.4°F) indicates the position of the setting of the thermostat of the hypothalamus.

Posterior hypothalamus and the conservation of heat

The posterior hypothalamus responds mildly to local cooling but gets a much stronger input from the peripheral skin receptors when the body is exposed to cold. The posterior hypothalamus becomes extremely active as the cold-temperature skin sensors increase their firing rate. Activity in the posterior hypothalamus stimulates the sympathetic nervous system and probably inhibits the parasympathetic to a certain extent. As a result of this shift to sympathetic activity, several important mechanisms for the conservation of heat are initiated:

1. *Vasoconstriction* of the small vessels supplying the skin reduces the amount of blood flowing through the superficial tissues of the body and consequently decreases heat loss by evaporation.
2. *Inhibition of the activity of the sweat glands* is due both to the decreased blood supply and to direct inhibition through the sympathetic nerves.
3. *Increased metabolic rate of the tissues occurs.* The activity of the sympathetic nervous system involves also the stimulation of the adrenal medulla, which responds by secreting more *epinephrine* into the circulation. Epinephrine increases the rate of tissue oxidation, as does thyroxine. These are also the hormones involved in chemical (nonshivering) thermogenesis following cold acclimatization, as was previously discussed.
4. *Shivering* is caused by the increase in activity of the *primary motor centers for shivering,* which is also in the posterior hypothalamus. Neurons from this center stimulate the reticular formation in the brain stem, resulting in facilitation of the motor neurons in the spinal cord. This increases muscle tonus and is followed by rhythmic shivering, which can increase heat production four to five times above normal levels.

The Effect of Exercise on Body Temperature

With strenuous muscular exercise, a rise in internal temperature of 1 to 2°C may occur, but this rise in turn sets off the thermostatic activity of the hypothalamus, so that the temperature of the skin may fall by 1 to 2°C. This fall is due, of course, to the compensatory mechanisms of vasodilation in the skin and increased sweating. The movements of the body, too, help lose some heat through convection currents.

Resetting the Hypothalamic Thermostat

Ionic changes

The set point of the mammalian hypothalamus, usually at 37°C, appears to be determined by the ratio of *sodium* and *calcium* ions in the hypothalamus. Artificially increasing the concentration of sodium by perfusing sodium through implanted cannulae directly into the hypothalamus causes a sharp rise in body temperature (*hyperthermia*). A similar

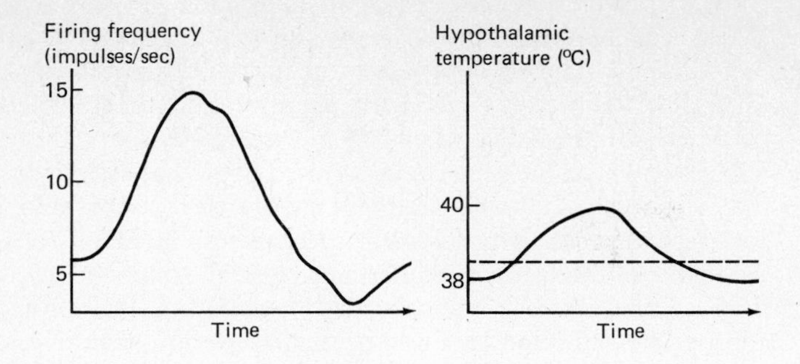

Figure 21-7. **Increase in firing frequency of a single hypothalamic cell as the temperature of the hypothalamus is raised locally. The broken line indicates midbrain temperature.**

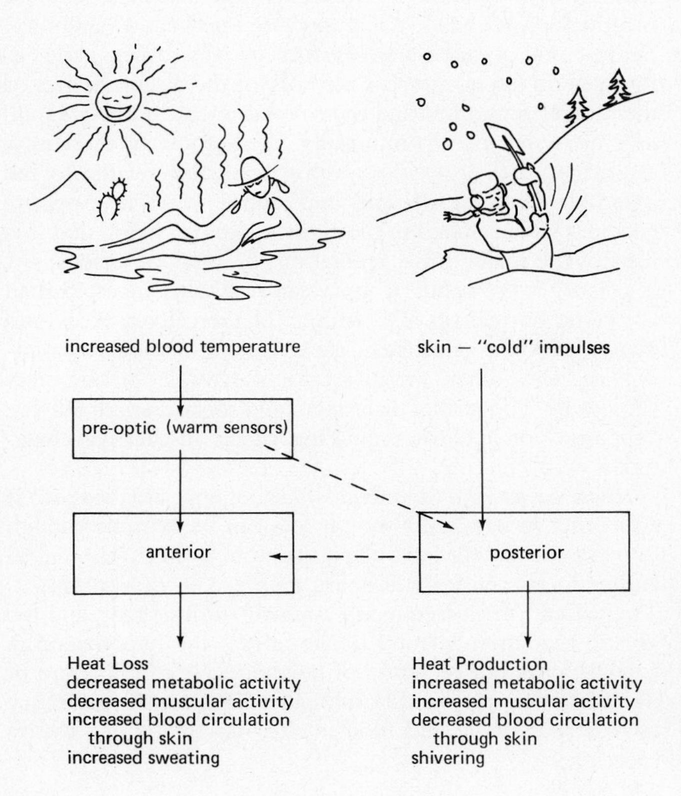

Figure 21-8. **Hypothalamic control of body temperature. The pre-optic area responds to changes in blood temperature by initiating reflexes that prevent overheating. This involves stimulation of the anterior hypothalamus and inhibition of the posterior hypothalamus. The posterior hypothalamus responds only to nerve impulses coming from skin receptors. Cold stimulates these nerves, causing activation of the posterior hypothalamus and reflex mechanisms of heat production. There is also simultaneous inhibition of heat-loss mechanisms of the anterior hypothalamus.**

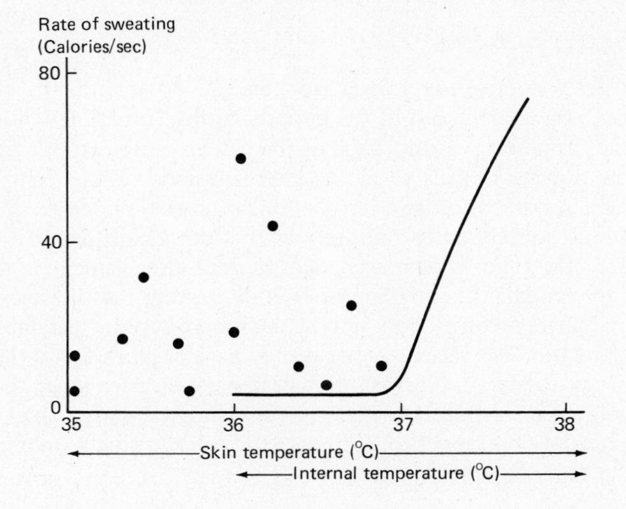

Figure 21-9. **There is a close correlation between the rate of sweating and the internal temperature, measured near the hypothalamus (solid line). However, when the skin temperature is measured (scattered dots), no correlation with sweating is seen.**

increase in hypothalamic calcium concentration produces just the reverse effect, an intense *hypothermia.* Experimental animals in which the set point has been altered now thermoregulate around this new temperature. It has been suggested that a tremendous saving in metabolic expenditure could be achieved by permanently setting the hypothalamus one or two degrees lower—this device not only would save energy but might permit the organism to live longer.

Pyrogens

Although there is much controversy over a possible resetting of the thermostat in physiological conditions such as exercise, there are many known chemical agents that do seem to raise the temperature set point. Certain *bacterial toxins* are potent pyrogenic (fever-inducing) agents and can raise the body temperature up to 106°F, an effect exerted by the toxins upon the hypothalamus. The increased temperature accelerates metabolic activities of all cells beyond the point at which adequate oxygen can be supplied, and anoxia results. Brain cells are particularly sensitive to hypoxia and respond first with increased firing, so that convulsions may occur. If the hyperthermia is maintained, the cells degenerate and die. Once they have been destroyed they cannot be replaced.

Cooling of the body usually can be speedily accomplished by sponging with either cool water or alcohol because these liquids evaporate rapidly from the hot skin and the heat of evaporation that is lost cools the body considerably. The administration of antipyretics, drugs that have a special affinity for the thermostat regions of the hypothalamus, has a longer-lasting effect.

Prostaglandins are naturally occurring pyrogens. The anterior hypothalamus is especially sensitive to these hormonelike substances. If a purified preparation of prostaglandin E_1 is injected into this region, fever develops immediately. This reaction is not seen when prostaglandins are injected into other brain regions.

Antipyretics

Many drugs, known as antipyretics, counteract the effects of pyrogens on the hypothalamus and rapidly bring the temperature down to normal levels. Aspirin (acetylsalicylic acid) is an excellent antipyretic in febrile subjects but has practically no effect on normal body temperature. The mechanism of action of antipyretic drugs is not known except for the fact that they bind specifically to hypothalamic cells.

21–2 Regulation of Food Intake

Hunger and Appetite

The average adult maintains a fairly constant weight over long periods of time, despite daily variations in diet and energy expenditure. This represents a rather sensitive homeostatic mechanism, controlling the nutrient balance. Any excess food absorbed beyond the energy requirements will be stored as fat, whereas of course an inadequate caloric diet will result in the fat deposits being used up. What regulates the food intake? Hunger and appetite are the two psychic correlatives of the physiological mechanisms involved for the intake of food. It is bewildering to read the many definitions of these two commonly used terms, but the most appropriate definition may well be that *hunger* is a primitive, unconditioned mechanism inducing the individual to ingest food. It is associated with *disagreeable sensations* (hunger contractions of the stomach, increased nervous excitability, nausea, and weakness). *Appetite,* on the other hand, is founded on the learning or memory of the disappearance of the hunger sensations and their replacement by the *pleasurable sensations* of a comfortably filled stomach, of satiety, relaxation, and drowsiness. Appetite is also related to the agreeable taste, smell, and appearance of food. Appetite becomes so developed in the civilized human being that it is usually the major force regulating the intake of food.

Obesity is the result of more calories being ingested than are being burned up. The causes for the caloric excess are many, involving psychiatric disturbances in which overeating replaces other gratifications, metabolic disturbances caused by endocrine disorders, and disturbances of the "appestat" or appetite-regulating center in the hypothalamus.

Anorexia nervosa (*anorexia* = lack of appetite for food) is a disorder in which the weight loss can be extreme enough to make the patient look like a skeleton. Both psychological and endocrine disturbances are involved in this condition. The patient (most frequently a female) will not eat, and the drastic reduction in food intake affects the hypothalamic-anterior pituitary secretion of hormones. A combination of psychiatric therapy and hormonal replacement is necessary to restore appetite and food intake, but this is not always successful.

Hypothalamic Regulation of Food Intake

Satiety center

The controlling center for energy balance is in the *ventromedial nucleus* of the hypothalamus, sometimes called the "appestat." Stimulation of the center causes satiety, and the animal refuses to eat. Bilateral lesions of the ventromedial nucleus cause a tremendous increase in body weight, which subsequently stabilizes at this new, high level. Figure 21-10 shows weight changes in mice after damage to the ventromedial hypothalamus. It is interesting that these animals will become obese only if the food offered is palatable: if the food is unattractive or if they have to work for it, they prefer to starve. Normal animals are not so particular; they will eat in response to internal cues (hunger) rather than to external cues (appetite). The satiety center normally inhibits the feeding center.

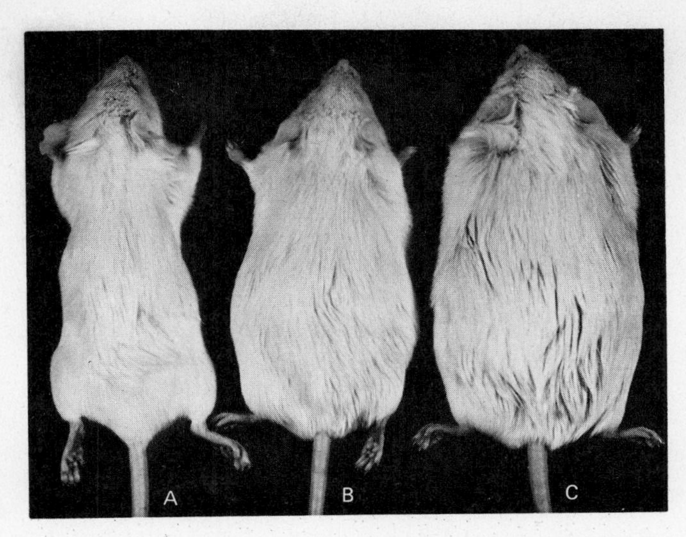

Figure 21-10. **Weight changes in mice after damage to the ventromedial hypothalamus. A is the control mouse; B and C show increasing obesity proportional to the extent of the destruction of the satiety center. [From R. A. Liebelt, *Ann. N.Y. Acad. Sci.* 110:723 (1963).]**

FEEDING CENTER

The *lateral hypothalamus* is considered to control feeding behavior because stimulation of this area causes the animal to eat voraciously. Bilateral lesions of the lateral hypothalamus will prevent eating, and the animal may starve itself to death.

Although it is convenient to define specific hypothalamic areas as feeding or satiety centers, this definition is probably too simplistic. The complex interactions between the different hypothalamic centers concerned with feeding, drinking, and thermoregulation are discussed a little later in this chapter (Sec. 21-4).

Feedback Mechanisms to Regulate the Hypothalamic "Appestat"

Feeding behavior, and the consequent control of body weight, are regulated on a long-term basis. Over a 24-hr period, there is little correlation between food intake and energy expenditure, but the correlation between intake and output is excellent over several weeks. Physiologists are still looking for a blood-borne constituent to account for the feedback signal to the hypothalamus. Among the suggestions are the following.

GLUCOSTATIC HYPOTHESIS

A glucostatic hypothesis for the control of food intake, proposed by Mayer (1955) and harshly criticized by others, attempts to explain short-term, daily feeding. The signal is the *utilization of glucose* by the cells. Glucose utilization can be measured as the difference between the arterial and the venous blood glucose levels. When arterial blood glucose is low, the amount available for utilization by the cells is low; this activates the feeding center. Conversely, when blood glucose levels are high, the satiety center becomes more active and inhibits the feeding center.

A relationship between *gastric contractions* and feelings of *hunger* was suggested by Cannon in 1912. Mayer has extended this to the glucostatic concept: glucose infusions, which increase the arterial-venous difference, inhibit gastric contractions, and still the feeling of hunger (Fig. 21-11). This effect of high glucose levels is not seen in the diabetic, presumably because the lack of insulin prevents glucose utilization and maintains a low arterial-venous difference, so that the gastric contractions and hunger persist.

Glucoreceptors. CENTRAL GLUCORECEPTORS. The cells of the *ventromedial nucleus* have been described as glucoreceptors because they selectively concentrate glucose and increase their firing rate in response to increased blood glucose. It is difficult to explain their role in the hyperphagia of diabetics, however. Brain cells, unlike other cells, do not need insulin to utilize glucose; thus the cells of the satiety center should respond to the high blood glucose of the diabetic by increasing their activity and inhibiting the feeding center.

PERIPHERAL GLUCORECEPTORS. A more recent suggestion by Russek is that there are glucoreceptors in the

Figure 21-11. **Hunger pangs and hunger gastric contractions are both abolished by the intravenous infusion of 25 g glucose. This does not occur in the diabetic.**

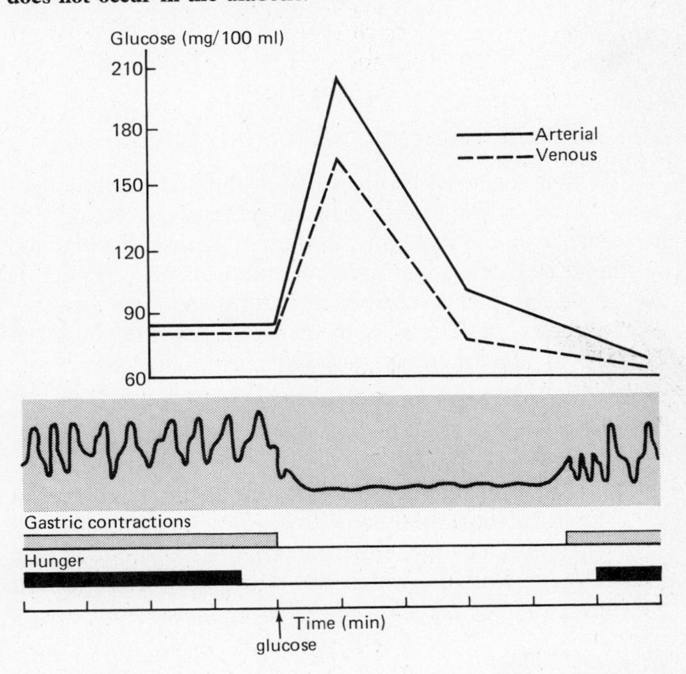

liver, glucose-sensitive fibers of the vagus nerve that fire when hepatic glucose is low and decrease their "hunger discharge" when intracellular glucose rises.

It is quite possible that both central and peripheral glucoreceptors exist and respond to a decreased glucose level by initiating a state of hunger that is followed by eating.

Lipostatic hypothesis

Short-term control of food intake may be glucostatic, but *long-term regulation* is probably lipostatic. An increase in body weight of a nongrowing adult is chiefly caused by the accumulation of fat; thus it seems logical to look at fat metabolism for a possible source of feedback to the central nervous system. Unfortunately, no correlation at all has been found between circulating fatty acids, or triglycerides, and food intake.

An alternative proposal implicates the *number of fat cells* in the body. Obese individuals have more fat cells than normal, and it is believed that this cellular proliferation occurs early in life, perhaps in babies overfed with high-calorie commercial foods. The number of fat cells after this critical period is then set for life, and the adult has to constantly replenish them with fat in order to reach the set point of the hypothalamus.

How does the information about the amount of fat in adipose tissue reach the hypothalamus? An intriguing suggestion is that this information is conveyed by two types of *prostaglandins*—one stimulates feeding, whereas the other type causes anorexia. It is possible that overloaded fat cells produce the prostaglandin that inhibits eating, whereas depleted adipose tissue sends out the prostaglandin that excites the feeding center. We really have no satisfactory explanation as yet to account for long-term control of feeding behavior.

Central cholinergic control of feeding

The hypothalamus receives neural input from the olfactory portion of the brain, from the reticular activating system, and from the thalamus. Because the sense of smell and a state of alertness are essential components of feeding, it is not surprising that these brain structures are able to modify feeding behavior. The chief mediator for this stimulation of feeding is *acetylcholine;* norepinephrine, an adrenergic transmitter, suppresses feeding. Evidence in support of this concept is:

1. Local application of acetylcholine to the ventromedial hypothalamus stimulates feeding.
2. Atropine, an anticholinergic drug, reduces food intake.
3. D-amphetamine, which has adrenergic activity, suppresses appetite, a fact familiar to dieters.

21–3 Regulation of Water Intake

Thirst

Water balance requires that the water intake be matched to the water need. The sensation of thirst is a basic one; if water is lacking, thirst dominates all other sensations and activities. Thirst is normally finely adjusted to meter the amount of water to be ingested under a wide variety of different conditions. Thirst is affected not only by a lack of body water, which induces *primary* drinking, but also a host of factors that are not directly related to the need for water. These include feeding behavior, the nature of the diet, and the climatic conditions. The drinking induced by these factors is *secondary* drinking.

Secondary drinking is regulatory in that it permits an animal to anticipate its future requirements accurately. Under normal circumstances, there is a *thirst threshold* that ensures that small losses of fluid do not arouse thirst. The animal does not have to constantly divert its attention to replacing minimal changes in water content. The insatiable thirst of the patient who has *diabetes insipidus,* and who lacks antidiuretic hormone, necessitates copious and frequent drinking to the exclusion of most other activities. This maintains proper water balance, but it disrupts the normal pattern of life.

Regulation of Drinking

There are three different explanations to account for drinking: peripheral stimulation, cellular dehydration, and extracellular dehydration.

Peripheral stimulation

Nerve endings in the mouth and pharynx are stimulated by dryness. Dryness of the mouth is associated with lack of saliva, and, although salivary secretion is correlated to body water content, the dry-mouth theory involves other factors as well. Dryness of the mouth occurs with nervousness, excessive smoking, and administration of certain drugs; none of these conditions involves the overall state of body hydration. Also, removal of the salivary glands or denervation of the mouth and pharynx does not impair water intake.

Cellular dehydration

The sensitivity of osmoreceptors to changes in blood tonicity was discussed in Secs. 11-3 and 15-4. Verney's experiments in 1947 demonstrated that the injection of hypertonic NaCl into the internal carotid artery caused the release of antidiuretic hormone. Subsequent work by Andersson (1960) has shown that the hypothalamus also has *thirst receptors;* in the goat these are in the middle hypothalamic region, but in most other species the *drinking center* is in the *lateral hypothalamus.* A single injection of hypertonic

NaCl into the middle hypothalamus can cause a well-hydrated goat to drink more than 9 liters of water. This *polydipsia* (excessive drinking) cannot be elicited by similar injections into other hypothalamic regions. Injection of hypertonic solutions into the supraoptic nuclei causes the release of antidiuretic hormone but does not increase drinking; the hypothalamic centers involved in antidiuretic hormone secretion and in drinking are separate.

EXTRACELLULAR DEHYDRATION

Extracellular dehydration is another regulatory mechanism involved in thirst and drinking. It may be differentiated experimentally from cellular dehydration by lowering the extracellular sodium (e.g., a low-sodium diet and sweating in humans). This sodium depletion results in a loss of water from the extracellular space; some of this water is excreted as urine but some moves into the cells, which become overhydrated. Despite the cellular overhydration, drinking increases.

Extracellular dehydration is also the result of hemorrhage. Thirst often accompanies hemorrhage, for reasons that probably involve the *volume receptors* in the left atrium (Sec. 15-4). Stimulation of these receptors by the fall in blood volume not only causes antidiuretic hormone release from the supraoptic and paraventricular hypothalamic nuclei, but also reflexly stimulates the drinking center in the hypothalamus.

21–4 Interrelationship of Drinking, Feeding, and Thermoregulatory Centers

Feeding and Drinking

In many animals there is a close relationship between eating and drinking. At least 70 per cent of the total water intake is closely associated with meals. The type of food in the diet also affects drinking: a high-protein diet induces a larger turnover of water than do diets rich in carbohydrate or fat. There seems to be an ability to regulate body water by anticipating future water requirements on a meal-to-meal basis. This anticipatory behavior is called *feed-forward* control.

When the feeding center is destroyed, the animals not only do not eat but they also refuse to drink. If rats with lesions in the lateral hypothalamus are kept alive by tube feeding, and offered wet, palatable food in addition to the routine dry food pellets, they eventually recover both feeding and drinking behavior. This recovery is only partial, however, and it is *nonregulatory.* A normal animal will eat more in response to low blood sugar, but the lesioned animals are unable to adjust their food intake to changes in blood sugar.

Thermoregulation, Drinking, and Feeding

Humans and animals drink more in a warm environment than in a cold one. This fact can be shown to be a direct effect of heat on drinking and not a secondary response to water loss. Local warming of the preoptic hypothalamus causes a well-hydrated animal to drink, whereas cooling this area of the brain will prevent a dehydrated animal from drinking.

Food intake is similarly affected by temperature. The amount of food eaten is reduced when the ambient temperature rises. Local warming of the preoptic hypothalamus also reduces feeding.

Although laterally lesioned animals that have resumed eating are unable to regulate their food intake according to changes in blood sugar, they nevertheless eat more in the cold and less when it is warm. Thermoregulation of food intake is still intact. The hypothalamic drinking, feeding, and thermoregulatory centers seem to be as interrelated functionally as they are close anatomically.

21–5 Hypothalamic Integration of Sensory and Motor Pathways

Hypothalamic-obese animals (with ventromedial lesions) are also *hypoactive.* It must be remembered that eating and drinking are behaviors involving alertness and locomotor activity: almost 25 per cent of the daily water intake of the rat is associated with its grooming activities rather than with the state of hydration of its body. It is very likely that the regulation of feeding and drinking incorporates overlapping stimulation of other neurons in the brain, especially the reticular formation.

The hypothalamus, particularly the lateral hypothalamus, is closely associated with *stereotyped feeding and drinking behavior,* such as chewing, licking, and swallowing. This is accomplished through important feedback mechanisms that are integrated in the hypothalamus. This structure receives input from receptors for taste, smell, touch, vision, hearing, and muscle sense. All *sensory information,* together with the responses of the thermoregulatory, osmoregulatory, and glucoregulatory cells, is passed to *motor systems* in the brain-stem reticular formation. This sensory input is essential to maintain those movements associated with getting and ingesting food and water.

It is clear that the pleasures of eating and drinking are not mere gifts to enhance the joy of life but are fundamental regulatory components needed for survival.

21–6 The Well-Balanced Diet

The well-balanced diet is composed of protein, carbohydrate, and fat (to supply the basic needs of the structure of the body and its energy-requiring reactions) together with

Table 21–1 **Food and Nutrition Board, National Academy of Sciences—National Research Council Recommended Daily Dietary Allowances,[a] Revised 1974 (Designed for the maintenance of good nutrition of practically all healthy people in the U.S.A.)**

	Age	Weight		Height		Energy	Protein	Fat-Soluble Vitamins				Water-Soluble Vitamins							Minerals					
								Vitamin A Activity		Vitamin D	Vitamin E Activity[e]	Ascorbic Acid	Folacin[f]	Niacin[g]	Riboflavin	Thiamin	Vitamin B_6	Vitamin B_{12}	Calcium	Phosphorus	Iodine	Iron	Magnesium	Zinc
	(years)	(kg)	(lb)	(cm)	(in.)	(kcal)[b]	(g)	(RE)[c]	(IU)	(IU)	(IU)	(mg)	(μg)	(mg)	(mg)	(mg)	(mg)	(μg)	(mg)	(mg)	(μg)	(mg)	(mg)	(mg)
Infants	0.0–0.5	6	14	60	24	kg × 117	kg × 2.2	420[d]	1400	400	4	35	50	5	0.4	0.3	0.3	0.3	360	240	35	10	60	3
	0.5–1.0	9	20	71	28	kg × 108	kg × 2.0	400	2000	400	5	35	50	8	0.6	0.5	0.4	0.3	540	400	45	15	70	5
Children	1–3	13	28	86	34	1300	23	400	2000	400	7	10	100	9	0.8	0.7	0.6	1.0	800	800	60	15	150	10
	4–6	20	44	110	44	1800	30	500	2500	400	9	10	200	12	1.1	0.9	0.9	1.5	800	800	80	10	200	10
	7–10	30	66	135	54	2400	36	700	3300	400	10	40	300	16	1.2	1.2	1.2	2.0	800	800	110	10	250	10
Males	11–14	44	97	158	63	2800	44	1000	5000	400	12	15	400	18	1.5	1.4	1.6	3.0	1200	1200	130	18	350	15
	15–18	61	134	172	69	3000	54	1000	5000	400	15	45	400	20	1.8	1.5	2.0	3.0	1200	1200	150	18	400	15
	19–22	67	147	172	69	3000	54	1000	5000	400	15	45	400	20	1.8	1.5	2.0	3.0	800	800	140	10	350	15
	23–50	70	154	172	69	2700	56	1000	5000		15	45	400	18	1.6	1.4	2.0	3.0	800	800	130	10	350	15
	51+	70	154	172	69	2400	56	1000	5000		15	45	400	16	1.5	1.2	2.0	3.0	800	800	110	10	350	15
Females	11–14	44	97	155	62	2400	44	800	4000	400	12	45	400	16	1.3	1.2	1.0	3.0	1200	1200	115	18	300	15
	15–18	54	119	162	65	2100	48	800	4000	400	12	45	400	14	1.4	1.1	2.0	3.0	1200	1200	115	18	300	15
	19–22	58	128	162	65	2100	46	800	4000	400	12	45	400	14	1.4	1.1	2.0	3.0	800	800	100	18	300	15
	23–50	58	128	162	65	2000	46	800	4000		12	45	400	13	1.2	1.0	2.0	3.0	800	800	100	18	300	15
	51+	58	128	162	65	1800	46	800	4000		12	45	400	12	1.1	1.0	2.0	3.0	800	800	80	10	300	15
Pregnant						+300	+30	1000	5000	400	15	60	800	+2	+0.3	+0.3	2.5	4.0	1200	1200	125	18+[h]	450	20
Lactating						+500	+20	1200	6000	400	15	80	600	+4	+0.5	+0.3	2.5	4.0	1200	1200	150	18	450	25

[a] The allowances are intended to provide for individual variations among most normal persons as they live in the United States under usual environmental stresses. Diets should be based on a variety of common foods in order to provide other nutrients for which human requirements have been less well defined.

[b] Kilojoules (kJ) = 4.2 × kcal.

[c] Retinol equivalents.

[d] Assumed to be all as retinol in milk during the first 6 months of life. All subsequent intakes are assumed to be half as retinol and half as β-carotene when calculated from international units. As retinol equivalents, three fourths are as retinol and one fourth as β-carotene.

[e] Total vitamin E activity, estimated to be 80 per cent as α-tocopherol and 20 per cent other tocopherols.

[f] The folacin allowances refer to dietary sources as determined by *Lactobacillus casei* assay. Pure forms of folacin may be effective in doses less than one fourth of the recommended dietary allowance.

[g] Although allowances are expressed as niacin, it is recognized that on the average 1 mg of niacin is derived from each 60 mg of dietary tryptophan.

[h] This increased requirement cannot be met by ordinary diets; therefore, the use of supplemental iron is recommended.

the vitamins, minerals, and trace elements that are discussed in this chapter.

The distinction between vitamins, minerals, and trace elements is becoming less and less significant as the structure of the vitamins has been determined and their role in the metabolism of the cell demonstrated. In general, the vitamins act as *active organic groups,* coupled with the protein portion of an enzyme, to catalyze reactions. This is equally true of many of the minerals and trace elements such as iron or copper, which act as *active inorganic groups* in enzyme reactions. The distinction between minerals and trace elements is merely a matter of quantity; sodium, potassium, calcium, and chloride, minerals basically necessary for the distribution of fluid in the body and the maintenance of electrical potentials, are found in relatively large amounts in the body, whereas copper, manganese, cobalt, and zinc are present in minute concentrations as trace elements.

Table 21-1 shows the recommended daily dietary allowance of the Food and Nutrition Board, National Research Council, in terms of calories from carbohydrate and fat, protein, vitamins, and some minerals. The percentage of fat recommended to complete the daily caloric supply varies from 31 to 0.3 per cent, obviously too wide a range to be meaningful. If one takes into consideration the recommendations of the Department of Health, City of New York (Diet and Coronary Heart Disease Study Project), the ratio of polyunsaturated fatty acids to saturated fatty acids should be 1.5 to 2.0.

If adequate amounts of these foods are in the diet, there will normally be enough vitamins, minerals, and trace elements for the average adult, provided the milk is fortified with vitamin D. Administration of additional vitamins, particularly the water-soluble ones, is wasteful, because the body simply excretes the excess. The fat-soluble vitamins (A, D, E, and K) can be stored, and although cases of vitamin poisoning have been reported, especially with overdosages of vitamins A and D, these cases are relatively rare.

Deficiency Diseases

Vitamin deficiencies rarely occur alone; there is usually a *combination of deficiencies* leading to a confusing medley of symptoms, and they may all be one aspect of undernutrition. Undernutrition is often due to protein deficiency, because a protein-rich diet is one only the economically secure can afford, but severe food shortage may restrict caloric intake to the starvation level, so that deficiencies of all nutrients occur.

In humans, dietary *mineral deficiencies* are rare, except for *iron* and *iodine.* In other animals, with a less varied diet, many symptoms of mineral deficiencies arise.

Protein deficiencies lead to a very widespread disease in poverty-stricken countries. This is *kwashiorkor,* characterized by a wasting away of fat and muscle and degeneration of many of the internal organs. One of the most interesting signs clinically is the "flag" sign, in which a depigmented zone of the hair marks or "flags" a period of severe protein deficiency, which has been corrected. In children, the swollen pot-belly is characteristic of the edema caused by changes in fluid distribution. The children are lethargic and miserable (Fig. 21-12A).

Carbohydrate and fat deficiencies are usually secondary to protein deficiency, because the caloric requirement is filled first, when at all possible, and carbohydrates are the cheapest foods. However, carbohydrate diets that previously contained enough vitamins are made deficient by the refining and polishing of grain. The ideal diet of the West has emphasized dairy products, but now the interest in the relationship between dietary fat and heart disease indicates that our ideal diet may be creating problems as well as solving them.

Carbohydrates, fats, and proteins have been discussed in detail in previous chapters. The functions of minerals and trace elements are included throughout this text where they are most appropriate. This section will therefore only consider *vitamins,* substances that have already been alluded to

Figure 21-12A. **The potbelly of this undernourished child is a sign of malnutrition. This child is receiving milk in a hospital in Tanzania. (Photograph courtesy of World Health Organization.)**

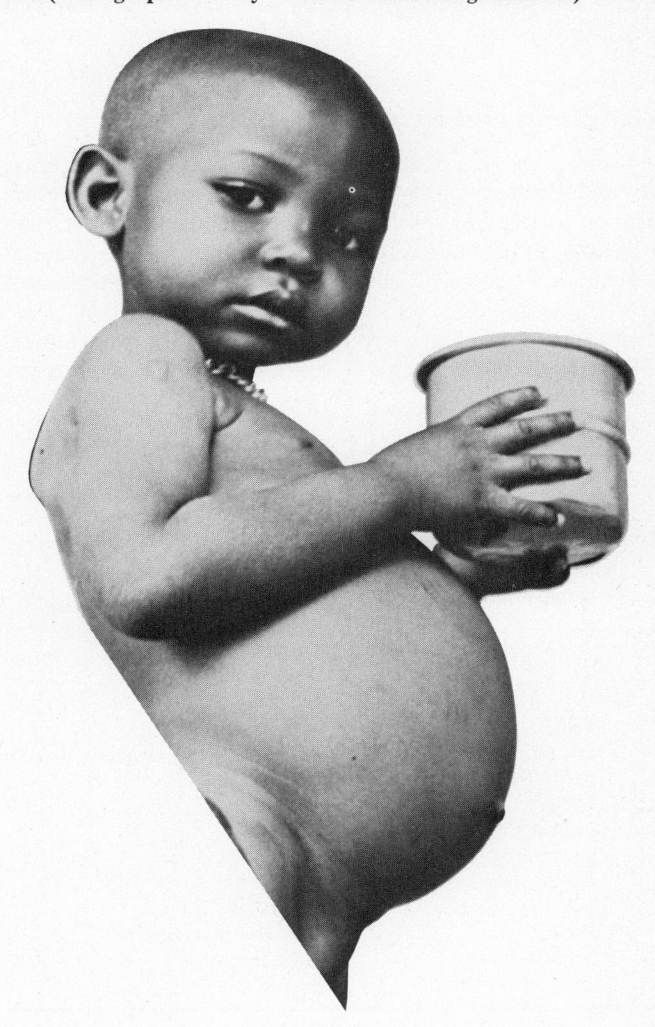

in Sec. 5-11 with reference to their biochemical role as *coenzymes.*

21-7 Vitamins

Nomenclature

The original word "vitamin" is a tribute to the vital properties of a nitrogen-containing substance that was shown to be effective against beriberi, a deficiency disease still prevalent in the Far East, where rice is the chief constituent of the diet, and where polishing removes the hull with its vitamin content. This nitrogen-containing compound was called vitamin B, but it was later discovered that several different substances were actually present, the antiberiberi factor being *vitamin* B_1, or *thiamine.* Many of the so-called vitamins do not have this nitrogen-containing base (amine), but the name remains. The tendency today is to replace the letter name with the specific name of the substance as it becomes known.

Table 21-2 lists the old and new names for the vitamins, with the new name in parentheses if it has not been accepted yet into common usage.

Definition and Requirements

A vitamin may be defined as an organic substance required in small amounts by the body, which usually cannot synthesize it, so that it must be supplied by the diet. Although vitamins do not themselves yield energy, they are essential to complete the metabolism of energy-yielding substances (carbohydrates, fats, and proteins).

A vitamin deficiency will result in a complex of metabolic disorders that give rise to characteristic clinical symptoms.

Table 21-2. **Old and New Nomenclature for Vitamins**

Old Names	New Names
Vitamin A	Vitamin A (retinol)
Vitamin B complex	
B_1	Thiamine
B_2 (British)	Riboflavin
B_6	Pyridoxine
B_{12}	Cobalamin
niacin	Niacin
folic acid	Folic acid (pteroylglutamic acid)
pantothenic acid	Pantothenic acid
biotin	Biotin
inositol	Inositol
choline	Choline
G (American)	Riboflavin
Vitamin C	Ascorbic acid
Vitamin D	Vitamin D (cholecalciferol)
Vitamin E	α tocopherol
Vitamin K	Naphthoquinone

The deficiency need not be due to a lack of this vitamin in the diet, for disorders in the absorption or metabolism of the vitamin can waste the vitamin supply.

Absorption

Even in the presence of adequate vitamins within the lumen of the digestive tract, *chronic irritation* of the mucosal lining will result in very rapid peristalsis and diarrhea, with the food mass being passed along the gut too rapidly for proper absorption of the vitamins. Fat-soluble vitamins cannot be absorbed properly in the absence of the *bile salts,* so that an obstruction of the bile duct may cause a vitamin deficiency.

Metabolism

Once the vitamins have been absorbed into the blood stream, they are usually incorporated into more complex molecules. The liver, one of the chief sites of such syntheses, also stores large amounts of the fat-soluble vitamins (vitamins A and D, tocopherol, and naphthoquinone). Similarly, the intestine synthesizes some vitamins from their precursors. Diseases of the liver and intestine interfere drastically with normal vitamin metabolism.

Special physiological stresses

Vitamin requirements are usually met by the well-balanced diet, provided that absorption and metabolism are normal. However, special physiological conditions such as *rapid growth, pregnancy,* and *lactation* impose new demands on the metabolism, increasing the vitamin requirement. Any marked *increase in the metabolic rate* of the body, such as a high fever, prolonged strenuous activity, or recovery from a wasting disease, means that more vitamins will be required to fulfill their role in the speeded-up chemical reactions of the tissues.

Antivitamins

Certain chemicals that are very similar in structure to the vitamins compete with them for a place on the enzyme to which the vitamin normally attaches itself. The antivitamin replaces the vitamin from its position and so changes the function of the enzyme. The administration of such antivitamins causes the appearance of all the symptoms of vitamin deficiency.

Antibiotics

Antibiotic drugs, when administered over a long period of time, destroy not only the pathogenic bacteria but also the bacteria of the intestine that normally produce significant quantities of some of the vitamins (pantothenic acid and naphthoquinone).

Retinol (Vitamin A)

The precursor (carotene) of the fat-soluble *vitamin A* is obtained from the yellow carotinoid pigments of plants and metabolized by the intestinal mucosa to the colorless vitamin known as *retinol.* Vegetable sources rich in carotene are carrots, fruits, and green leaves. Vitamin A appears in animals in two main forms: vitamin A_1 is found in mammals and salt-water fish, whereas vitamin A_2 is found in fresh-water fish. An interesting side line on evolutionary development is the discovery that the tadpole possesses the A_2 form, common to fresh-water denizens, but upon metamorphosis to the amphibious frog there is a change to A_1.

The typical signs of vitamin A deficiency in human beings are night blindness, thickening and drying of the conjuctiva and the cornea of the eye, skin lesions, and interference with growth. Vitamin A is essential for resynthesis of the visual pigments of the eye (Sec. 22-5), but its site of action in normal skin development is not known. The administration of the vitamin will clear up all the deficiency symptoms in a few weeks.

Vitamin A deficiency is uncommon in the United States. It has been seen, however, in groups of war prisoners, kept on carbohydrate diets with little meat and no green vegetables.

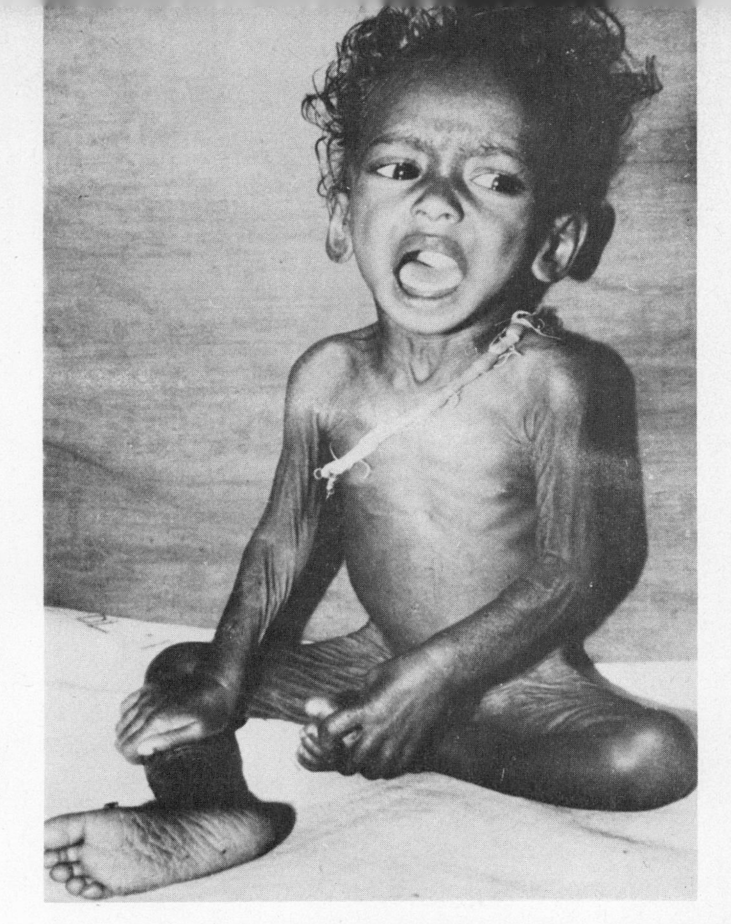

Figure 21-12B. **This extremely malnourished child has the rough skin, dark in patches, characteristic of the niacin-deficiency disease pellagra. Irritation and inflammation of the mouth and gastrointestinal tract are also characteristic of pellagra. (Photograph courtesy of World Health Organization.)**

Thiamine (Vitamin B_1)

Thiamine is the specific *antiberiberi factor,* and it was first isolated from rice polishings. It also contains an important growth factor. It has since been synthesized and shown to be the active group, or coenzyme, of carboxylase. Through its action on pyruvic acid, splitting off the carbon dioxide to permit its transformation to acetyl coA, it exerts a basic influence on carbohydrate metabolism. Animals deprived of thiamine accumulate pyruvic acid, and consequently any deprivation of thiamine will interfere drastically with general metabolism.

The deficiency disease resulting from inadequate thiamine intake over a long period of time is *beriberi.* It received its name from the Singhalese word *beri,* which means "weakness." This is caused by changes in the *peripheral nervous system,* involving sensory disturbances of the legs ranging from hypersensitivity to anesthesia. Ultimately, the arms and legs become paralyzed. Unlike pellagra, which comes from a deficiency of niacin, beriberi does not usually affect the central nervous system, but disorders of the *heart and circulation* are common.

The administration of thiamine improves the utilization of carbohydrate, and after a short time the deficiency symptoms of beriberi disappear. Thiamine is not stored in the body, and signs of deficiency can be seen when there is any interference with normal absorption from the intestine.

Unpolished rice and wheat germ are rich in thiamine. Most meat, especially pork, has adequate amounts of thiamine.

Niacin (Nicotinic Acid)

Niacin is found in the body as the coenzyme nicotinamide adenine dinucleotide (NAD) and also as the phosphate NADP. These coenzymes act as hydrogen acceptors in the electron transport chain (Sec. 5-12) of all tissues; consequently, deficiencies in niacin have widespread effects, the chief of which are seen clinically as disturbances of the gastrointestinal tract, neurological symptoms, and dermatitis (Fig. 21-12B).

The widespread deficiency disease resulting from the lack of niacin is *pellagra.* The term means "rough skin," from the Italian *pelle agra,* and the disease was first noted in the peasants of Spain and Italy. It is also endemic in the southern United States. Pellagra is characterized by lesions of the skin; a red, sore tongue; disturbances of the intestinal tract; and changes in the nervous system. It can be prevented by adequate amounts of a well-balanced diet, but niacin is the specific *pellagra-preventive factor,* dramatically evoking an improvement in the symptoms in the first 24 hr. Foods rich in niacin are lean meat, fish, eggs, milk, fruit, and vegetables (all the foods, that is, that are usually beyond the means of the poverty-stricken).

Riboflavin (Vitamin B_2)

This fluorescent yellow pigment is water-soluble. It forms an essential part of the coenzymes flavin mononucleotide (FMN) and flavin adenine dinucleotide (FAD), both of which usually accept hydrogen atoms from NAD and pass them on to the cytochrome system (Sec. 5-12). Riboflavin is an essential *growth factor* for many species, including humans.

Lack of riboflavin may not be noted for several months, or it may appear in pregnancy or lactation, when the requirements are suddenly increased. Clinical signs of riboflavin deficiency are difficult to distinguish at first from those of other deficiency states affecting the *skin, eyes,* and *intestine.* In advanced cases, however, lesions of the mucosa of the lips and fissures in the lips and at the corners of the mouth are characteristic of the disease. Particularly, riboflavin deficiency appears to be associated with changes in the sclera of the eye. Foods rich in riboflavin are milk, liver, kidney, and heart, and many vegetables contain adequate amounts of it. The bacterial flora also synthesize riboflavin to a certain extent, but whether it is absorbed to act as a supplement to the diet in this manner is unknown.

Pyridoxine (Vitamin B_6)

Pyridoxine has been shown to be a coenzyme in many enzyme reactions, especially those involved in the metabolism of amino acids. Its action is not confined to carboxylases; decarboxylation reactions in which it is involved have been shown to be of great importance. For example, the removal of CO_2 from glutamic acid converts the acid to a substance essential for the metabolism of the brain. Convulsions may occur when this reaction is blocked. Pyridoxine is also involved in the formation by a decarboxylation reaction of serotonin, a chemical transmitter in the central nervous system (Sec. 9-5). Pyridoxine is also a coenzyme in the metabolism of linoleic acid, one of the essential fatty acids.

Pyridoxine is found in both vegetable and animal foods, although in slightly different forms. It is also synthesized in the intestine as the result of bacterial activity, but it is probably not absorbed. A mixed diet of natural foods will contain enough pyridoxine to prevent a deficiency of this vitamin in adults, although in pregnant women, metabolic abnormalities sometimes develop that have been traced to inadequate pyridoxine intake for the increased needs of pregnancy.

A pure pyridoxine deficiency is found only under experimental conditions, for in human beings it would normally be associated with all the other deficiency symptoms of an inadequate diet. However, the treatment of a mixed vitamin B-complex deficiency with niacin, riboflavin, and thiamine does not relieve the dermatitis (itching, scaly skin), the irritability, and sleeplessness. These are specifically treated with pyridoxine.

Cobalamin (Vitamin B_{12})

Cobalamin is a coenzyme acting as a hydrogen acceptor. It is unique in that it contains cobalt as an integral part of the molecule. It can be extracted from liver as a red, crystalline compound, or it can be produced by bacterial fermentation in the colon, but very little of it is absorbed. Cobalamin is found in extremely small amounts in foods of animal origin only. The human body can store this vitamin for considerable periods of time, so that deficiency symptoms appear only after a long period of deprivation.

Cobalamin designates all of the fairly large group of substances that have vitamin B_{12} activity. This activity includes the following functions:

1. Cobalamin plays a vital role in the reduction of ribonucleotides to deoxyribonucleotides in the *formation of the gene.*
2. Probably because of its influence on gene formation, cobalamin is essential for the maturation of the red blood cell (Sec. 14-6). *Pernicious anemia* results from the lack of this vitamin.
3. Cobalamin is required for the *myelination* of the large nerve fibers of the spinal cord, especially the dorsal columns. Deficiencies of vitamin B_{12} sometimes result in loss of sensation and paralysis.
4. Again because of the effect of cobalamin on gene formation, it is required for *growth* of all tissues.

Folic Acid (Pteroylglutamic Acid)

Folic acid is a water-soluble vitamin present in a great many foods, especially the dark green, leafy vegetables, liver, kidney, and dried brewer's yeast. It is composed of two fragments (pteroic acid and glutamic acid) which, even if administered simultaneously, cannot be combined in the body to form the vitamin pteroylglutamic acid.

Folic acid is essential for the *formation of genes* because it is involved in the synthesis of purines and thymine required for the production of deoxyribonucleic acid. Consequently, like cobalamin, it is a vital *growth* factor and also essential for the proper maturation of the red blood cells (Fig. 21-13). Cobalamin and folic acid are both needed for *normal erythropoiesis* because they act on different aspects of this process. Folic acid is of benefit in the anemia seen in *sprue,* a disease in which various vitamins are not properly absorbed from the intestine.

Ascorbic Acid (Vitamin C)

Ascorbic acid, a strong reducing agent, is involved in the metabolism of tyrosine and phenylalanine, the amino acid precursors of *epinephrine* and *insulin* and of the dark-brown pigment *melanin.* In infants lacking ascorbic acid, the urine becomes dark, showing the abnormal oxidation of these substances. Despite many experiments in vitro that show the importance of ascorbic acid in oxidation-reduction reactions, there is no definite proof yet of its specific place in the

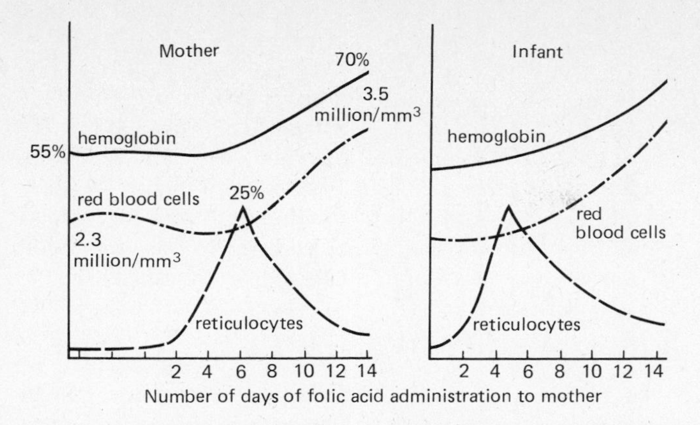

Figure 21-13. **The administration of folic acid to a nursing mother increases the red-blood-cell count and hemoglobin concentration in both mother and child. The brief increase in reticulocytes indicates the release of immature red blood cells into the circulation. (Redrawn from T. Spies, in *Clinical Nutrition,* 2nd ed., N. Jolliffe, ed., Harper & Row, Publishers, New York, 1962.)**

metabolism of the body, apart from the definite symptoms that appear in its absence.

Ascorbic acid is associated with the synthesis of *adrenal cortical steroids,* particularly the glucocorticoids released during stress. Ascorbic acid is necessary for the development of *intercellular substances.* This is especially evident in the walls of the blood vessels and in the development of bone. In the absence of ascorbic acid, bone does not ossify properly and the fragile blood vessels rupture easily, leading to frequent hemorrhages.

Ascorbic acid is a vital factor in the formation of *hemoglobin* and the *maturation of red blood cells,* as we saw in Sec. 14-6.

The deficiency disease *scurvy* develops in the absence of ascorbic acid. This disease is seen only in humans and other primates, and in the guinea pig, all of which are unable to synthesize this vitamin. Plants and most other animals can synthesize ascorbic acid.

It was the correlation of scurvy with the lack of fresh fruit and vegetables that first showed that a disease could be caused by the omission of a specific type of food, and it was Lind, a British naval surgeon, who demonstrated that the disease could be cured by food alone. He wrote that citrus fruits "contain something that neither medicine, surgery, or physic could supply," and as a result of his efforts, it became part of the regime of the British navy that all members of the fleet receive one fluid ounce of lemon juice daily (and also resulted in the term "limey" for an Englishman). Scurvy disappeared from the British navy after that.

Scurvy may develop in bottle-fed infants unless their milk and cereal diets are supplemented by ascorbic acid in vegetables or fruit juice. Characteristically in scurvy, the limbs are extremely tender and movement is painful. The gums are swollen and bleed easily, and the skin hemorrhages frequently. Heart and respiratory rates are increased.

The symptoms of scurvy are essentially the same in the adult. Chronic alcoholics may develop scurvy because the high caloric alcohol satiates the appetite without providing vitamins. Treatment with ascorbic acid results in very rapid improvement within 24 hr, but without treatment death may occur within a few weeks.

Cholecalciferon (Vitamin D)

Cholecalciferon, one of two main types of vitamin D, is produced in the skin of animals, including humans, when skin cholesterol is exposed to ultraviolet rays from the sun. This produces the *natural* vitamin D, cholecalciferon. Vitamin D can also be obtained from plant sources as ergosterol, which can be irradiated to form the *synthetic* vitamin D.

Vitamin D increases *calcium absorption* from the intestine by enhancing active transport of calcium through the intestinal cells. Even in the presence of large amounts of ingested calcium, practically no absorption can occur in the absence of vitamin D.

The *absorption of phosphorus* appears to be dependent upon calcium absorption. Because vitamin D also prevents phosphorus excretion by the kidneys, the net effect of this vitamin is to increase blood levels of both calcium and phosphorus.

The relationship between vitamin D and overall calcium metabolism, and especially with regard to calcium deposition in bone, is discussed in Sec. 24-7.

Deficiency diseases due to a lack of vitamin D are diseases of *civilization,* for the short wavelengths of the sun are completely filtered out by window glass and partially absorbed by the smoke and dust in the atmosphere of large northern cities. Dark-skinned races require even more sunlight to form vitamin D, because the pigment in their skin filters out much of the ultraviolet light. Whether vitamin D is taken by mouth or formed in the skin, the physiological effect is the same: it is absorbed into the blood stream and distributed to all parts of the body.

The deficiency disease arising from lack of vitamin D in early life is *rickets,* which was a very prevalent disease until about 30 years ago, because babies were fed a diet almost exclusively of cow's milk. The calcium in milk can be absorbed only in the presence of adequate amounts of vitamin D, and this vitamin is almost completely absent in boiled milk. In the winter especially, the vitamin D content of the milk was low; the infants were kept tightly bundled up on the rare occasions they were taken into the open; and until they were almost a year and a half, their diet was mostly easily digestible carbohydrates. Breast-fed children suffered less, because of course the mother's milk was not sterilized and, if the woman were on a mixed diet, usually had more vitamin D.

As a result of impaired calcium (and phosphorus) metabolism, bone is formed in the usual way (Sec. 24-7), but *no calcium salts are available for deposition.* The cartilage is not properly ossified, and the soft bones easily become deformed. The situation is aggravated by the *parathyroid gland*, which is stimulated by the low calcium levels in the blood to increase its hormone secretion, thereby causing the dissolution of bone by certain bone cells. This liberates calcium into the blood. Because the bone matrix is continuously laid down without being hardened, the child suffering from rickets has soft, deformed bones, with characteristic thickenings on them. These thickenings are particularly marked on the ribs, forming a chain down both sides of the thorax (the rachitic rosary). The legs may be bowed or knock-kneed and the pelvis deformed if the rickets persists (Fig. 21-14).

A deficiency of vitamin D in the adult leads to analogous softening of the bones due to reabsorption of the matrix. This is sometimes called "adult rickets," but *osteomalacia* is a much better term, implying poor bone formation. The soft bones cannot support the strains imposed on them; they fracture easily and become markedly deformed.

Rickets and osteomalacia can be prevented by the addition of vitamin D to the milk or by the administration of the vitamin in the form of irradiated ergosterol, irradiated cholesterol, or cod liver oil. Because vitamin D is stored in the body for as long as 3 months, rickets is sometimes treated with a massive injection of vitamin D.

Figure 21-14. **The classical signs of rickets are seen in this child from West Benghal: big abdomen caused by weak abdominal muscles, big forehead, soft leg bones and vertebrae, the rachitic rosary on the ribs, and the swelling of the wrists. (Photograph courtesy of Pan American Health Organization, World Health Organization.)**

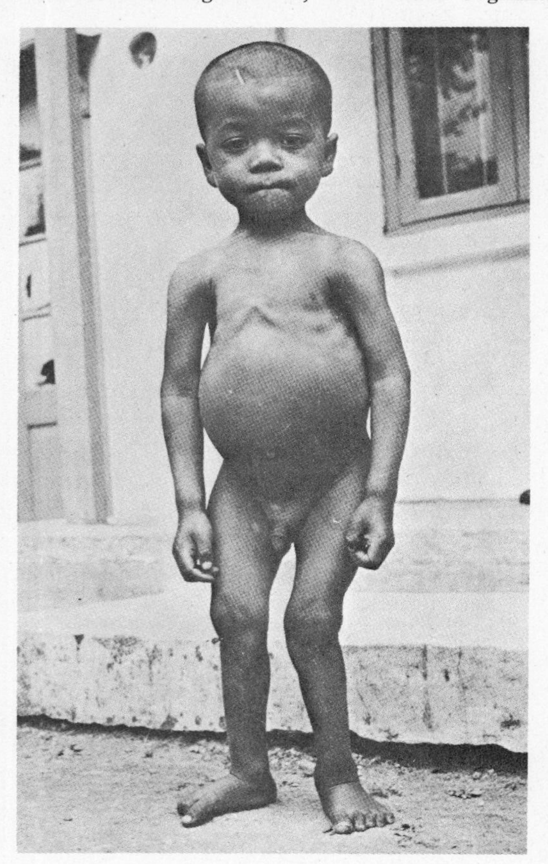

Vitamin D is not as widely distributed in foods as are the other vitamins. Fish oils, fish liver, and eggs of all kinds are the only foods with significant amounts of vitamin D. Milk contains small quantities but not enough to prevent rickets, unless it has been fortified by the addition of this vitamin or the cow has been fed large quantities of vitamin D.

Alpha Tocopherol (Vitamin E)

Although this fat-soluble vitamin is essential for human nutrition, no supplements are necessary because adequate amounts are present in the diet. It has been suggested that alpha tocopherol plays an important role in preventing the rapid oxidation of fatty acids to hydroperoxides (Sec. 18-4). No specific deficiency disease in humans has been attributed to a lack of alpha tocopherol, but in lower animals very severe diseases result. In many animals, a lack of alpha tocopherol causes abnormal development of the placenta and death of the fetus. In the male, tocopherol deficiency causes sterility.

Again in lower animals only, a deficiency of alpha tocopherol may cause muscular dystrophy. This is of great interest to investigators working on this disease, but although tocopherol is an effective cure for this disease in rodents, it is ineffective in humans.

The vegetable seed oils are particularly rich in tocopherol.

Naphthoquinone (Vitamin K)

Naphthoquinine is distributed widely in nature, and the dietary supply is also supplemented by the activity of the intestinal bacteria. Signs of naphthoquinone deficiency are more likely to be associated with *interference* with the synthetic activity of these bacteria (because of antibiotics) or to inadequate absorption of this fat-soluble vitamin in the absence of the bile salts. If the common bile duct is obstructed, as in obstructive jaundice, the bile does not reach the intestine, and naphthoquinone is absorbed very poorly if at all.

The only sign in human beings of naphthoquinone deficiency is a hemorrhagic tendency, which is due to a deficiency of *prothrombin* in the blood. Prothrombin, an essential component of the blood-coagulation mechanism (Sec. 14-15), is produced in the liver, provided minute amounts of naphthoquinone are present. Foods rich in naphthoquinone are widely distributed in nature. Alfalfa and similar green plants contain particularly abundant supplies of it, and many species of bacteria synthesize it.

Pantothenic Acid

Pantothenic acid is a sulfur-containing substance that is widely distributed in nature and an important part of the coenzyme A molecule. Coenzyme A functions in a large number of metabolic reactions, through its ability to activate acetic acid into acetyl coenzyme A. The metabolism of lipids and the synthesis of steroids, the *detoxication* of many substances in the liver, as well as the final *energy-yielding reactions* of cellular respiration, are all dependent on the formation of acetyl coenzyme A.

Pantothenic acid, like so many of the vitamins, exerts an important effect on *growth*, but in humans no well-defined clinical entity is seen as a result of pantothenic acid deprivation, perhaps because the intestinal bacteria can synthesize all that is normally required. There is some association between leg cramps experienced by pregnant women and the sensations of "burning feet" in elderly people and pantothenic acid deficiency.

In lower animals, a pantothenic acid deficiency retards growth, impairs reproduction, causes graying of the hair, and damages the adrenal cortex. Practically all foods contain some pantothenic acid, but green vegetables, liver, and the yolk of egg have a particularly high pantothenic acid content.

Biotin

Biotin may be considered one of the most potent of the vitamins because minute quantities will satisfy metabolic requirements. Because it is widely distributed in food, a biotin deficiency does not normally occur in humans. Experimentally, a biotin deficiency results in dermatitis, muscle pains, and disturbances of the gastrointestinal tract. Both biotin and *para-aminobenzoic acid* (PABA) are vital growth factors for microorganisms, which is important therapeutically, because the sulfonamide drugs destroy bacteria by interfering with PABA metabolism.

Biotin is a coenzyme for carboxylase, probably involved in the reversal of decarboxylation; that is, it is necessary in the body for the *addition of carbon dioxide to pyruvic acid* to form a 4-carbon compound (oxalic acid) essential for the further reactions of cellular respiration. Interference with this step through a lack of biotin could obviously cause disturbances in the metabolic efficiency of the cell.

Inositol

Because inositol is one of the most widely distributed of the vitamins, a deficiency of this substance is not normally encountered in man. It is essential as a growth factor for microorganisms and for some animals, such as rats, in which an inositol-deficient diet stunts growth, causes loss of hair, and leads to the development of a fatty liver. This observation excited interest as to whether inositol deficiency might be a cause of fatty liver in humans, but there is no evidence for this as yet.

The foods in which most inositol is found are citrus fruits and grains. Inositol is a carbohydratelike substance that is treated by the body in a way similar to glucose. It is excreted in large quantities in diabetes mellitus, and this is corrected by the administration of insulin.

Choline

Choline is a substance containing three methyl groups (CH_3), one of which it can donate for the synthesis of lipids from fatty acids. In the absence of choline, the fatty acids accumulate in the liver, causing the development of fatty liver. Choline is also a precursor of acetylcholine, an important chemical transmitter at nerve endings.

A choline deficiency leads to fatty degeneration of the liver, and this condition, as well as cirrhosis of the liver caused by excessive alcohol, can be greatly improved by a well-balanced diet with additional choline. The richest sources of choline are egg yolk, liver, kidney, and heart.

21–8 The Rate of Metabolism

The Respiratory Quotient

The metabolic activity of the tissues is reflected in the volume of oxygen they utilize and carbon dioxide they produce in the chemical reactions involved in cellular respiration. The ratio of carbon dioxide produced to oxygen consumed is known as the respiratory quotient or RQ.

$$\text{respiratory quotient} = \frac{\text{volume of carbon dioxide produced}}{\text{volume of oxygen consumed}}$$

In these reactions, as the fuel is used up, energy is released, much of it in the form of heat. The heat liberated is measured in terms of Calories. A *Calorie* (spelled with a capital *C*) is the amount of heat required to raise the temperature of 1 kg of water from 15 to 16°C. A small calorie (spelled with a lower-case *c*) is $\frac{1}{1000}$ of a large Calorie and is too minute a measure to be useful in discussing the energy production of the body.

Types of Fuel and Heat Liberated

The amount of heat that is liberated depends on the particular fuel being oxidized.

Carbohydrates liberate 4 Calories per gram.
Fats liberate 9 Calories per gram.
Proteins liberate 4 Calories per gram.

The tissues utilize these foods in the order named here so that proteins, the structural part of the cell, will be oxidized only when there are no carbohydrate or fat reserves.

The basic reaction may be represented as follows:

$$\text{fuel} + \text{oxygen} \longrightarrow \text{carbon dioxide} + \text{water} + \text{energy} \quad \text{(heat)}$$

One may gauge roughly the metabolic rate by determining either the amount of heat liberated or the amount of oxygen consumed.

Measurement of the Amount of Heat Liberated

The amount of heat produced by the body can be measured directly by placing the subject in an insulated room, in which the temperature of the outer walls is maintained constant. Any change in the temperature within the room must be result of the heat produced by the subject. One way of measuring the heat change is through the change in temperature of water circulating through coils in the ceiling. Another, more sensitive, technique is the use of a thin lining of foil on the inner surface of the chamber. The foil is interlaced with thousands of thermoelectric junctions that record local changes in temperature all over the room. These changes are then integrated as a single measurement of heat output.

Many factors must be taken into consideration, including the heat lost through the evaporation of insensible perspiration. This technique is better adapted for research purposes than for clinical use.

Measurement of Oxygen Consumed

It is much simpler to measure the volume of oxygen consumed under standard conditions than the amount of heat generated. There is a direct relationship between the volume of oxygen utilized and the heat (in Calories) produced. This relationship is dependent, however, on the type of food being used for fuel, because not only do different foods yield different amounts of calories, but they require different volumes of oxygen for their complete oxidation.

If the fuel is chiefly carbohydrate, 0.812 liter of oxygen is required to burn each gram of carbohydrate, and the physiologically available energy released is 4 Calories. At the same time, 0.812 liter of carbon dioxide is produced. This is understandable in terms of the familiar equation:

molecules

$$6\ C_6H_{12}O_6 + 6\ O_2 \longrightarrow 6\ CO_2 + 6\ H_2O + \text{energy}$$

grams

$$1\text{ g sugar} + \underset{\text{oxygen}}{0.812\text{ l}} \longrightarrow \underset{\text{carbon dioxide}}{0.812\text{ l}} + H_2O + 4\text{ Calories}$$

The respiratory quotient (carbon dioxide produced/oxygen consumed) is 1.

If the fuel is fat, more oxygen is required to burn each fat molecule. The proportion of oxygen to carbon dioxide is consequently different, and the RQ is 0.71. The energy produced is 9.0 Calories/g of fat.

If the fuel is protein, the RQ is 0.81, and the caloric yield 4.0 Calories/g.

A measurement of the RQ indicates the dominant type of metabolism for the individual. A diabetic person with faulty carbohydrate metabolism caused by lack of insulin will burn chiefly fat, and the RQ would reflect this by being close to 0.71.

Technique for Determining the Volume of Oxygen Consumed

The metabolator used for the indirect measurement of metabolic rate consists of a chamber of oxygen that rests on a floating drum. The drum in turn is connected to an ink-

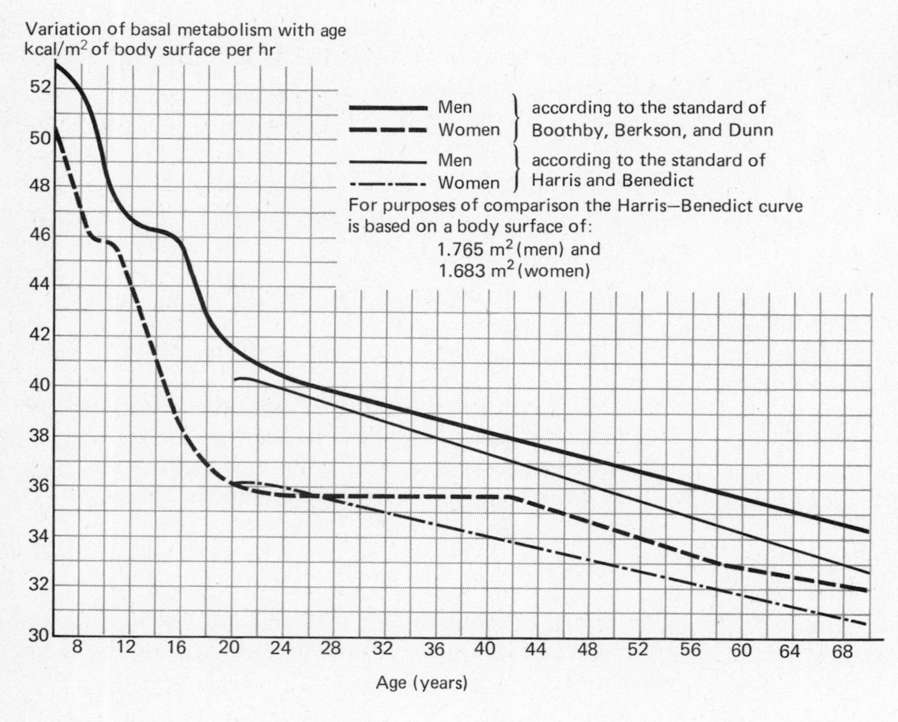

Figure 21-15. **Variations in basal metabolic rate with age. (From *Documenta Geigy, Scientific Tables,* 6th Edition 1962. Courtesy of Ciba-Geigy Limited, Basle, Switzerland, 1962.)**

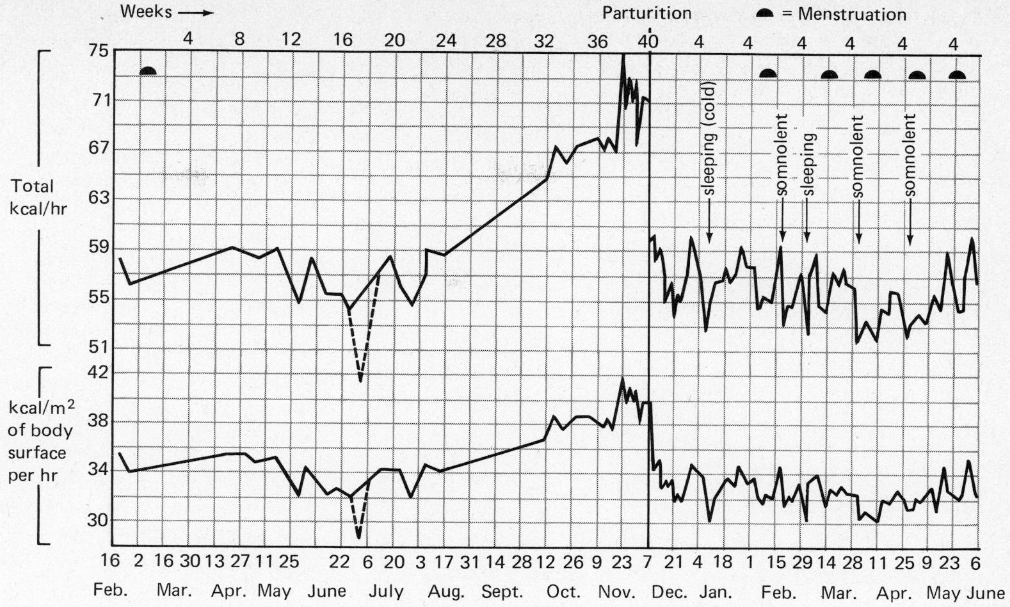

Figure 21-16. **Basal metabolism before, during, and after pregnancy. The upper curve gives the total calories, the lower curve the calories per square meter of body surface calculated from the Du Bois and Du Bois formula. [From Sandiford and Wheeler, *J. Biol. Chem.* 62:329 (1924). Adapted from *Documenta Geigy, Scientific Tables,* 6th Edition 1962. Courtesy of Ciba-Geigy Limited, Basle, Switzerland.]**

writing kymograph in such a way that it records the descent of the drum into the water on which it floats as the oxygen is used up. The slope of the line on the kymograph is calibrated to the volume of oxygen.

The subject is attached to the oxygen chamber through two rubber tubes inserted into a mouthpiece placed in the mouth. The nose is pegged closed. Valves in the two tubes direct the exhaled air through a container of soda lime that absorbs the carbon dioxide, while the subject inhales from the oxygen chamber. The volume of oxygen consumed can then be used to calculate the total quantity of energy liberated within the body. The subject is assumed to be on a mixed diet, and the RQ is taken as 0.82. On this basis, approximately 4.825 Calories of energy will be liberated per liter of oxygen consumed.

If 15 liters of oxygen (under standard conditions) are utilized in 1 hr, the total amount of energy liberated by the body in this time must be 15×4.825 Calories $= 72.4$ Calories.

21–9 Factors Affecting Metabolic Rate

The metabolism of the tissues is increased considerably by exercise and decreased to its lowest level in deep sleep. It varies with age and sex, and, as we have seen, the energy produced will also vary with the diet. Cold will lower the metabolic rate, and heat will increase it. If one is measuring the overall metabolism of the body, obviously a larger individual will utilize more energy than a smaller one, although the metabolic rate of the tissues may be identical. It has been found that this variation can be correlated best with *body surface,* and detailed tables have been made to facilitate the determination of body surface from height and weight.

In addition, it is a poorly understood but well-documented fact that for several hours after a meal, the metabolic rate may be increased anywhere from 20 to 50 per cent above normal. The extent of the increase depends again on the type of food ingested, with the greatest effect occurring after a protein-rich meal. This increase in metabolism after intake of food is called the *specific dynamic action of food.* It is important, naturally, that measurement of the metabolic rate be made some time after the last meal has been digested and metabolized.

Basal Metabolic Rate (BMR)

If all these variants affect the rate of tissue metabolism, then in order to get some norm for metabolic rate, it is essential to control as many of these factors as possible. The basal metabolic rate, which is supposed to represent the *activity of the tissues at rest,* is therefore measured under the following conditions:

1. The subject should not eat any food for 24 hr prior to the test.
2. The subject should have had a restful night and have come relaxed to the test.
3. The temperature of the air should be maintained between 62 and 87°F, and be fairly dry.
4. The values obtained must be corrected for surface area, age, and sex (Fig. 21-15), and for physiological condition, such as pregnancy (Fig. 21-16).

Normal Range of the Basal Metabolic Rate

The basal metabolic rate is usually expressed as a percentage above or below normal, as determined from a chart such as is shown in Fig. 21-15. From this chart the mean BMR for a 20-year-old girl is seen to be 35.0 Calories/(meter^2)(hr). If a 20-year-old girl were found to have a BMR of 45.0 Calories/(m^2)(hr), this would be expressed as a 28.6 per cent increase above normal, or *plus 28.6*. The normal range should fall between plus 10 per cent and minus 10 per cent of the mean, and plus 28.6 is an indication of abnormal metabolism, probably caused by overactivity of the thyroid gland. *Thyroxine is one of the most important factors in regulating metabolic rate,* and excessive secretion of this hormone can raise the BMR to as much as 100 per cent above normal. An underactive thyroid can result in the BMR falling to 50 per cent of normal. Thus by controlling all other variables, the BMR gives some reference to the state of activity of the thyroid gland.

Cited References

ANDERSSON, B., S. LARSSON, and N. PERSSON. "Some Characteristics of the Hypothalamic 'Drinking Centre' in the Goat as Shown by the Use of Permanent Electrodes." *Acta Physiol. Scand.* **50:** 140 (1960).

CANNON, W. B., and A. L. WASHBURN. "An Explanation of Hunger." *Amer. J. Physiol.* **29:** 441 (1912).

MAYER, J. "Regulation of Energy Intake and the Body Weight: The Glucostatic Theory and the Lipostatic Hypothesis." *Ann. N.Y. Acad. Sci.* **63:** 15 (1955).

VERNEY, E. B. "The Antidiuretic Hormone and the Factors Which Determine Its Release." *Proc. Roy. Soc.* (*London*) *Biol.* **135:** 25 (1947).

Additional Readings

BOOKS

BLIGH, J. *Temperature Regulation in Mammals and Other Vertebrates.* American Elsevier Publishing Co., Inc., New York, 1973.

FOLK, G. E., JR. *Textbook of Environmental Physiology,* 2nd ed. Lea & Febiger, Philadelphia, 1974.

GOODHART, R. S., and M. E. SHILS, eds. *Modern Nutrition in Health and Disease,* 5th ed. Lea & Febiger, Philadelphia, 1973.

GUYTON, A. C., ed. *MTP International Review of Science: Physiology,* Vol. 7: *Environmental Physiology.* University Park Press, Baltimore, 1974.

JENNINGS, I. W. *Vitamins in Endocrine Metabolism.* Charles C Thomas, Publisher, Springfield, Ill., 1970.

LEBLANC, J. *Man in the Cold.* Charles C Thomas, Publisher, Springfield, Ill., 1976.

LEE-MINARD, D. K., ed. *Physiology, Environment and Man.* Academic Press, Inc., New York, 1970.

MOGENSON, G. J., and F. R. CALARESU, eds. *Neural Integration of Physiological Mechanisms and Behavior.* University of Toronto Press, Toronto, Canada, 1975.

NOVIN, D., W. WYRWICKA, and G. A. BRAY, eds. *Hunger: Basic Mechanisms and Clinical Implications.* Raven Press, New York, 1975.

SCHMIDT-NIELSEN, K. *"Animal Physiology, Adaptation and Environment."* Cambridge University Press, New York, 1975.

WHITTOW, G. C. *Comparative Physiology of Temperature Regulation,* Vols. 1 and 2. Academic Press, Inc., New York, 1970 and 1971.

WILLIAMS, R. J. *Physician's Handbook of Nutrition.* Charles C Thomas, Publisher, Springfield, Ill., 1974.

ARTICLES

ADOLPH, E. F. "Regulation of Water Intake in Relation to Body Water Content." In *Handbook of Physiology,* Sec. 6, Vol. 1. The Williams & Wilkins Company, Baltimore, 1967, p. 163.

ANAND, B. K. "Central Chemosensitive Mechanisms Related to Feeding." In *Handbook of Physiology,* Sec. 6, Vol. 1. The Williams & Wilkins Company, Baltimore, p. 249.

ANDERSSON, B. "The Effect of Injections of Hypertonic NaCl Solutions into Different Parts of the Hypothalamus of Goats." *Acta Physiol. Scand.* **28:** 188 (1953).

BENZINGER, T. H. "The Human Thermostat." *Sci. Am.,* Jan. 1961.

BENZINGER, T. H. "The Thermal Homeostasis of Man." In *Homeostasis and Feedback Mechanisms, Symp. Soc. Exp. Biol. 18.* Academic Press, Inc., New York, 1964, p. 49.

BRAY, G. A. "Nutritional Factors in Disease." In *Pathologic Physiology: Mechanisms of Disease,* 5th ed. W. B. Saunders Company, Philadelphia, 1974, p 839.

CABANAC, M. "Temperature Regulation." *Ann. Rev. Physiol.* **37:** 415 (1975).

CROSS, B. A. "The Hypothalamus in Mammalian Homeostasis." In *Homeostasis and Feedback Mechanisms, Symp. Soc. Exp. Biol. 18.* Academic Press, Inc., New York, 1964, p. 157.

DAWKINS, M. J. R., and D. HULL. "The Production of Heat by Fat." *Sci. Am.,* Aug. 1965.

FITZSIMONS, J. T. "Thirst." *Physiol. Rev.* **52:** 468 (1972).

FRIEDEN, E. "The Chemical Elements of Life." *Sci. Am.,* July 1972.

GALE, G. C. "Neuroendocrine Aspects of Thermoregulation." *Ann. Rev. Physiol.* **35:** 391 (1973).

GARCIA, J., W. G. HANKINS, and K. W. RUSINIAK. "Behavioral Regulation of the Milieu Interne in Man and Rat." *Science* **185:** 824 (1974).

GELINEO, S. "Organ Systems in Adaptation: The Temperature Regulating System." In *Handbook of Physiology,* Sec. 4. The Williams & Wilkins Company, Baltimore, 1964, p. 259.

HARDY, J. D. "Posterior Hypothalamus and the Regulation of Body Temperature." *Fed. Proc.* **32:** 1564 (1973).

HART, J. S. "Insulative and Metabolic Adaptations to Cold in Vertebrates." In *Homeostasis and Feedback Mechanisms, Symp. Soc. Exp. Biol. 18.* Academic Press, Inc., New York, 1964, p. 31.

HENSEL, H. "Thermoreceptors." *Ann. Rev. Physiol.* **36:** 233 (1974).

HOEBEL, B. G. "Feeding: Neural Control of Intake." *Ann. Rev. Physiol.* **33:** 533 (1971).

IRVING, L. 1966. "Adaptations to Cold." *Sci. Am.,* Jan. 1966.

LADELL, W. S. S. "Terrestrial Animals in Humid Heat: Man." In *Handbook of Physiology,* Sec. 4. The Williams & Wilkins Company, Baltimore, 1964, p. 625.

LEE, D. H. K. "Terrestrial Animals in Dry Heat: Man in the Desert." In *Handbook of Physiology,* Sec. 4. The Williams & Wilkins Company, Baltimore, 1964, p. 551.

LEPKOVSKY, S. "Newer Concepts in the Regulation of Food Intake." *Amer. J. Clin. Nutr.* **26:** 271 (1973).

MAYER, J., and E. A. AREES. "Ventromedial Glucoreceptor System." *Fed. Proc.* **27:** 1345 (1968).

MAYER, J., and D. W. THOMAS. "Regulation of Food Intake and Obesity." *Science* **156:** 328 (1967).

MORGANE, P. J., ed. "Neural Regulation of Food and Water Intake." *Ann. N.Y. Acad. Sci.* **157**(2) (1969).

NISBETT, R. E. "Hunger, Obesity and the Ventromedial Hypothalamus." *Psychol. Rev.* **79:** 433 (1972).

SMITH, R. E., and B. A. HOROWITZ. "Brown Fat and Thermogenesis." *Physiol. Rev.* **49:** 330 (1969).

Regulatory Responses to the External Environment

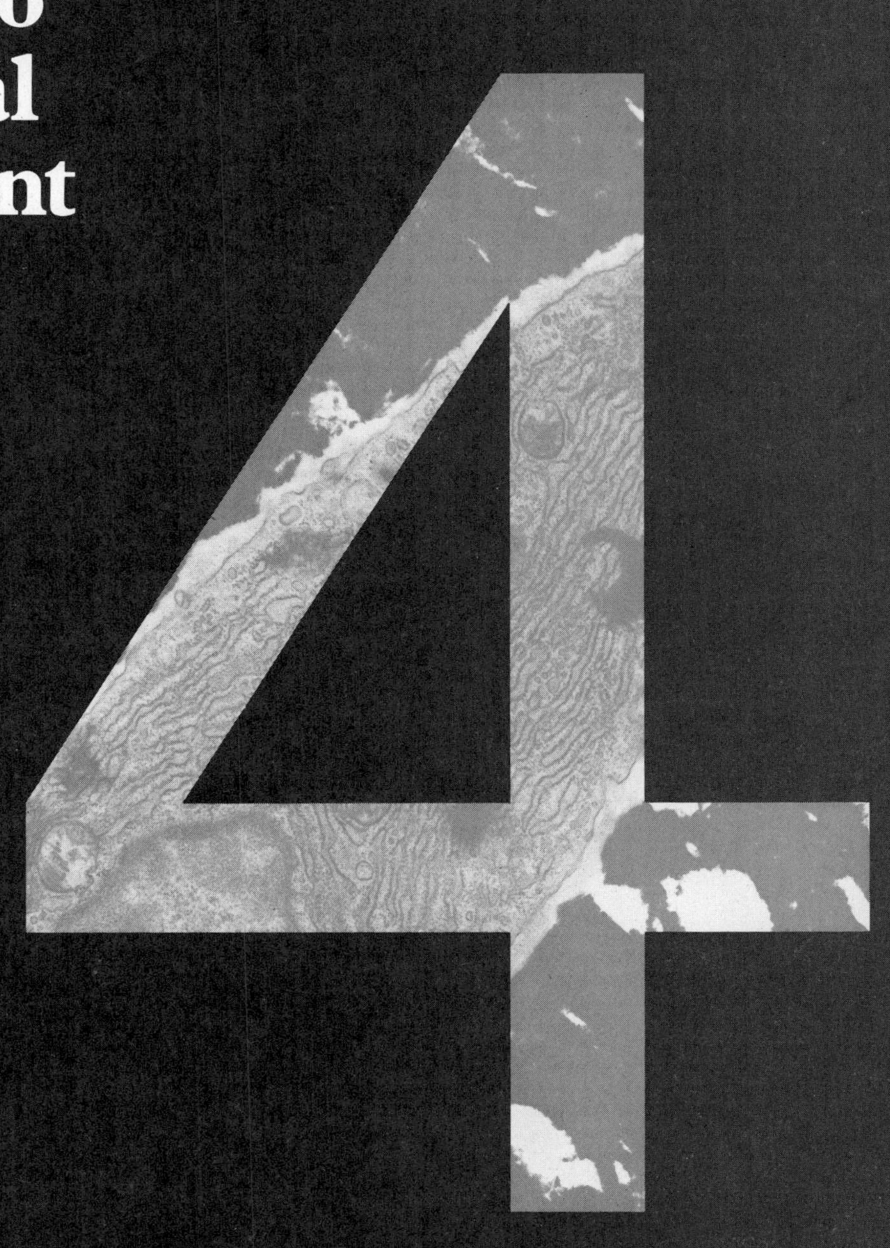

Chapter 

Regulatory Responses to the External Environment: Sensory Systems

Round, angular, soft, brittle, dry, cold, warm,
Things are their qualities: things are their form—
And these in combinations, even as bees,
Not singly but combined, make up the swarm:

And when qualities like bees on wing,
Having a moment clustered, cease to cling,
As the thing dies without its qualities,
So die the qualities without the thing.

Where is the coolness when no cool winds blow?
Where is the music when the lute lies low?
Are not the redness and the red rose one,
And the snow's whiteness one thing with the snow?

Even so, now mark me, here we reach the goal
Of Science, and in little have the whole—
Even as the redness and the rose are one,
So with the body one thing is the soul.

Titus Lucretius Carus, "No Single Thing Abides"

THE BEHAVIOR PATTERNS necessary for animals to reproduce and survive are regulated by information received from their sensory systems. The peripheral receptors in the skin, muscles, and joints, the taste and olfactory receptors, the eyes and the ears, all contribute information about the external environment. This information is both received and processed by the *peripheral receptors* before it is sent on to the *central nervous system* for further processing and evaluation. Much of the processing is accomplished through *selective inhibition* at progressively higher levels in the brain, the result of which is to eliminate extraneous signals and thus to sharpen and concentrate on the important cues.

22-1 Comparative Importance of Sensory Systems

Usually, the different sense organs function together to provide an integrated picture of the environment, but in many instances one sensory system may be poorly developed, and the animal must depend on the other systems for its survival. Some nocturnal animals, such as owls and bats, have poor vision, but their superb sound localization permits them to pounce on their prey with awe-inspiring accuracy, even in complete darkness.

Although the *auditory system* is the most sensitive to minute changes in stimulus energy, the vertebrate *visual system* is the most highly developed in terms of the richness of the information it can process. Unlike those animals that depend chiefly upon their hearing and their sense of smell, humans rely much more on vision. Yet it seems that adjustment to blindness is easier than adjustment to complete deafness. Blind people may be able to use certain auditory cues for the localization and recognition of obstacles. On the other hand, although we are not usually aware of the importance to our lives of the contribution by the *proprioceptive* and *cutaneous systems,* complete loss of these systems would be fatal. The adaptive value to survival of each sensory system is of greater significance than the amount of sensory information the system can handle.

22–2 Transduction and Perception

In order to respond to environmental stimuli, cells must be able to transduce extremely small amounts of light, sound, chemical, mechanical, or thermal energy into electrical energy. This electrical energy, in the form of nerve action potentials, is sent over *specific pathways* to the brain to excite *defined areas of the cerebral cortex.* Despite this localization of sensory input into the brain, several sensory areas are usually involved, so that a *patterned overlapping of sensations* is usual. One must remember that electrophysiological studies done on single cells cannot take this complexity into account.

Stimulation of sensory cortical areas is interpreted as *vision, hearing, taste, smell, touch,* and *position in space.* Although perception is known to occur in the cerebral cortex, it is not understood how it occurs, nor do we know why some people are more sensitive than others to certain sensations and more creative in responding to them. We cannot explain the talent of a composer on neurological or electrophysiological grounds. A musical genius and an individual indifferent to music both have adequate receptors in the internal ear; the auditory nerve and the auditory cortex are physiologically the same, as far as we know. Perhaps the difference lies in minute variations in sensitivity, synaptic connections, and membrane characteristics that cannot as yet be detected.

This chapter will deal with the more factual aspects of sensation, starting with those characteristics that are common to all receptors, and then discussing the specializations of each type of receptor that permit it to respond with a heightened sensitivity to a particular form of stimulus energy. Muscle spindle receptors are not included in this chapter; they are described in detail in Chapter 23.

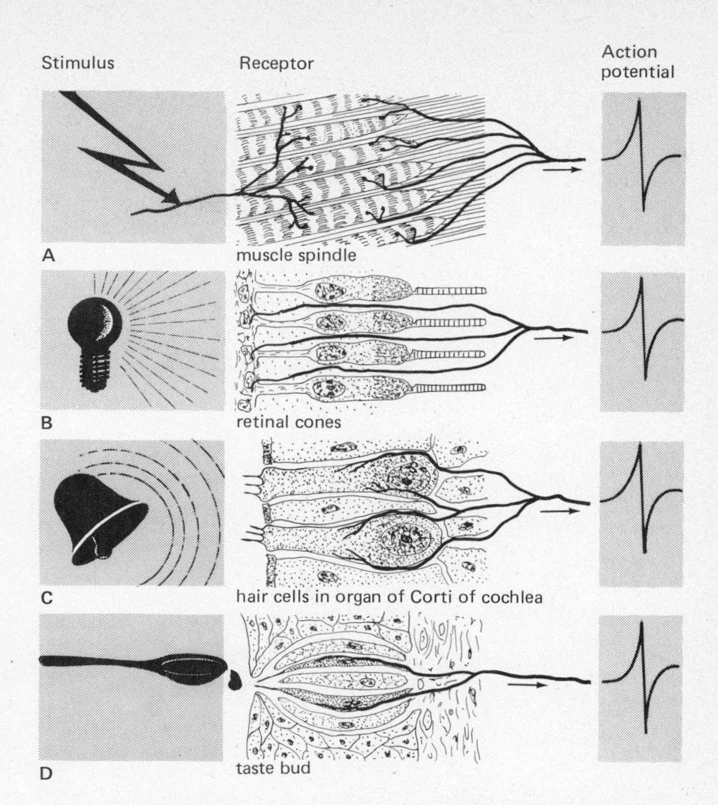

Figure 22-1. **Receptors as transducers of (A) electrical energy, (B) light energy, (C) mechanical vibratory energy, and (D) chemical energy.**

22–3 Characteristics Common to All Receptors

Sensory Transduction

Sensory transduction is a combination of processes by which the physical energy of a stimulus is converted to a *rhythmic discharge of impulses* by the receptor (Fig. 22-1). The different receptors use different energy-conversion mechanisms, each of which consists of a series of distinct processes that will be discussed under the appropriate specialized receptor later in this chapter.

Amplification of Signal

Not only do receptors transduce different forms of stimulus energy into rhythmic electrical firing of the nerves, but the signal from the stimulus is amplified enormously, so that a very small amount of stimulus energy triggers a large response. The amplification mechanism of the inner ear in hearing is the most remarkable of all the sense organs: it responds to a mechanical displacement of 0.1 of the diameter of a hydrogen atom.

Receptor and Generator Potentials

Receptor cells may produce either receptor or generator potentials, both of which are graded, nonpropagated potentials that can be summated. The difference between a receptor potential and a generator potential is based upon the anatomical difference between the two receptor cell types. *A cell that produces a receptor potential does not have its own axon;* instead, it synapses almost immediately upon another neuron, so that chemical mediation is invoked at the synapse. The postsynaptic cell is excited, and an action potential is produced which conducts the information to the central nervous system. *A cell that produces a generator potential has its own axon,* along which an action potential can be conducted without the immediate intervention of a synapse. These relationships are diagrammed in Figs. 22-2 and 22-3.

All-or-None Law

The nerve impulse follows the all-or-none law. Once the generator potential has reached the critical threshold, the action potential that is produced reaches its maximal ampli-

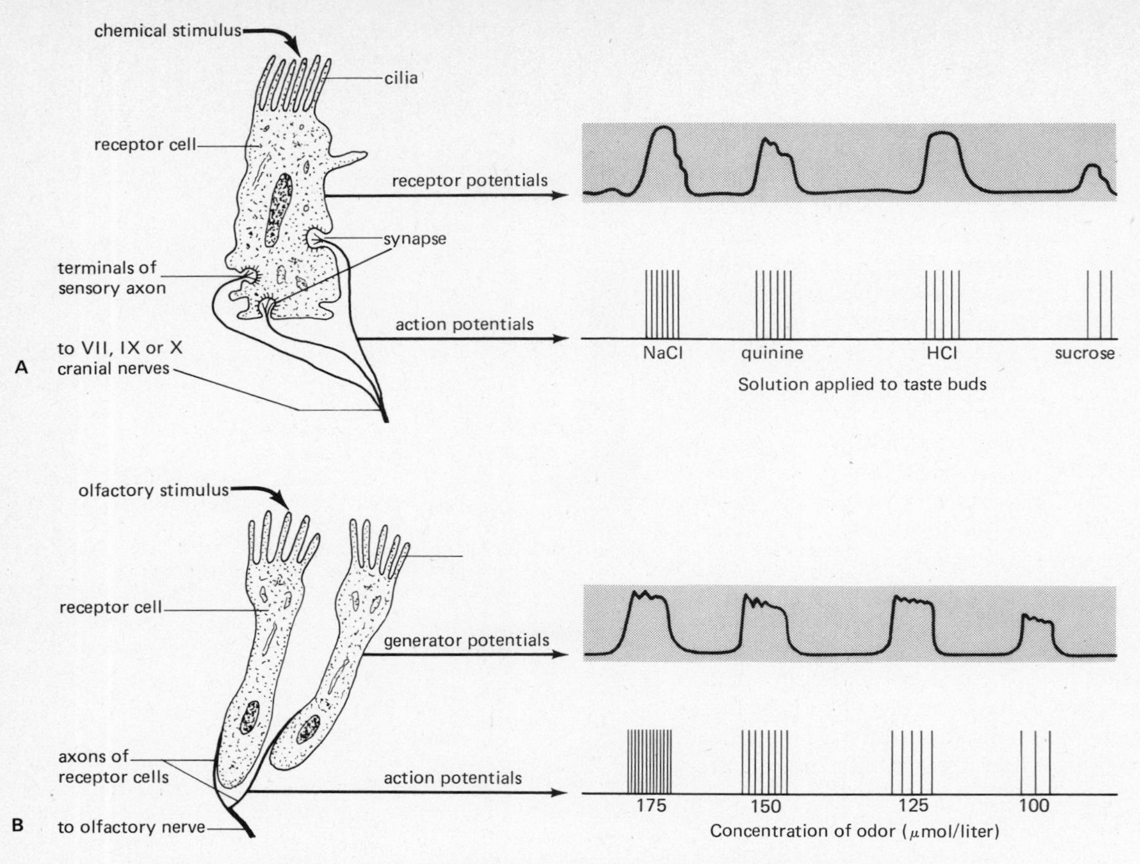

Figure 22-2. **Anatomical comparison of a cell that produces a receptor potential [receptor cell of a taste bud in (A)] and a cell that produces a generator potential [olfactory receptor cell in (B)]. The taste receptor has no axon of its own, but many small sensory axons synapse into specialized invaginations of the receptor cell. The olfactory receptor has its own axon so that an action potential is generated without an intervening synapse. On the right of both (A) and (B) are shown the graded receptor and generator potentials evoked by different stimuli. The all-or-none action potential shows corresponding changes in the frequency of impulses.**

tude for the particular physiological conditions. Any variation in action potential height reflects changes in the metabolic state of the axon and is not related to any characteristic of the stimulus. Consequently, information about the stimulus must be conveyed to the central nervous system in a code that does not incorporate the magnitude of the nerve impulse conducted along an individual nerve fiber.

Information Coding

There are many ways by which the neural response is known to code sensory information, and other possible coding mechanisms are constantly being tested. It is important to realize that the *perception* of the dimensions of a stimulus varies quite remarkably when *other parameters are simultaneously changed.* For example, the *loudness* of a sound varies not only with the frequency of the wavelength but also with the sound pressure level. On the other hand, increasing the intensity of a stimulus beyond a certain range changes the *quality* of the perceived sensation, altering the hue or the pitch, or even evoking pain.

Some of the coding mechanisms may be considered to be concerned with *temporal* aspects of the stimulus, others with its *spatial* and *magnitude* parameters, but most are probably coordinated into coding patterns. These responses regulate sensory perception on a neural basis: psychological perception involves complex phenomena beyond the scope of this text.

Temporal patterns of coding

Shape of Response of the Generator Potential. Graded potentials vary in their *shape* according to the rate of rise of the wave, the duration, and the slope with which it falls away. The amplitude of the generator potential increases with the strength of the stimulus [Fig. 22-2(B)]. These parameters may be codes sending specific information to the brain.

The great significance of the information contained in graduated and nonpropagated signals is gaining increasing appreciation as the many different levels of neural coding are becoming apparent.

Frequency of Firing. The frequency of the action potentials along a sensory nerve increases as the stimulus becomes

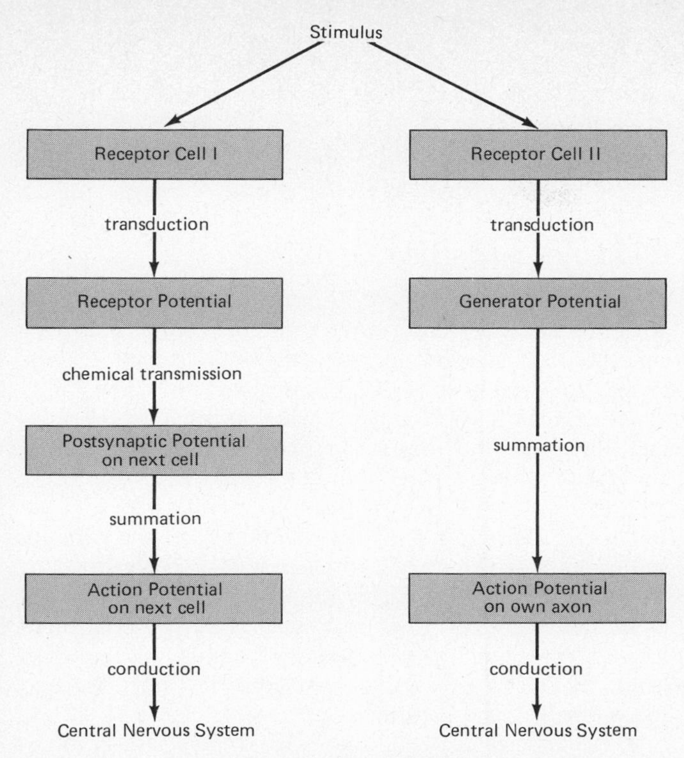

Figure 22-3. **Comparison of the electrophysiological events characterizing cells that produce receptor potentials and those that produce generator potentials.**

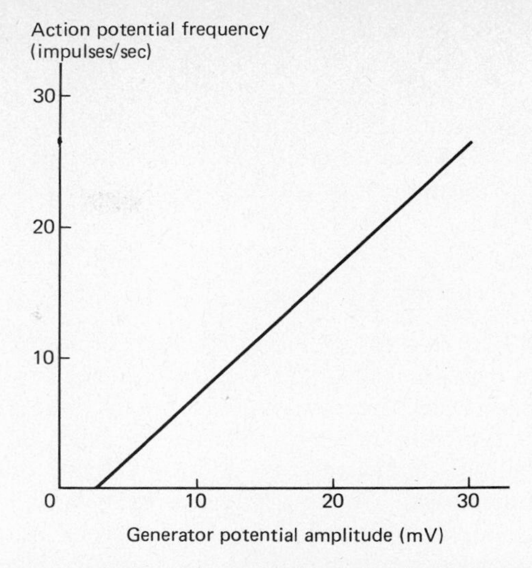

Figure 22-4. **The linear relationship between the strength of the generator potential and the frequency of the nerve firing.**

more intense (Figs. 22-2 and 22-4). We interpret changes in the quantity of the stimulus, such as increased brightness of light or loudness of sound, by the increase in the number of action potentials reaching the brain within a given time.

Fluctuations in Frequency Firing Pattern. A pattern of nerve impulses that changes rapidly from a very low frequency, and then slowly diminishes, is perceived differently from a firing pattern in which the frequency slowly increases and then rapidly diminishes. Even an additional impulse or a missed impulse may carry information about the stimulus. This type of frequency fluctuation response is seen very often in electrophysiological recordings from single neurons.

"On" and "Off" Responses. Painstaking dissection of optic nerve fibers in the frog has permitted us to demonstrate the existence of two antagonistic neuron systems, one of which is *excited* by light (the "on" neurons) and the other of which is *inhibited* by light (the "off" neurons). These two systems act antagonistically upon ganglion cells and so regulate the discharge of the ganglion cell. Figure 22-5 shows the spontaneous firing of three different fibers of the optic nerve and their responses to illumination. Note that some of the fibers respond to both the onset and the cessation of the light stimulus. This is shown in another form in Fig. 22-6, where the organization of "on" and "off" regions in the cat's retina is illustrated.

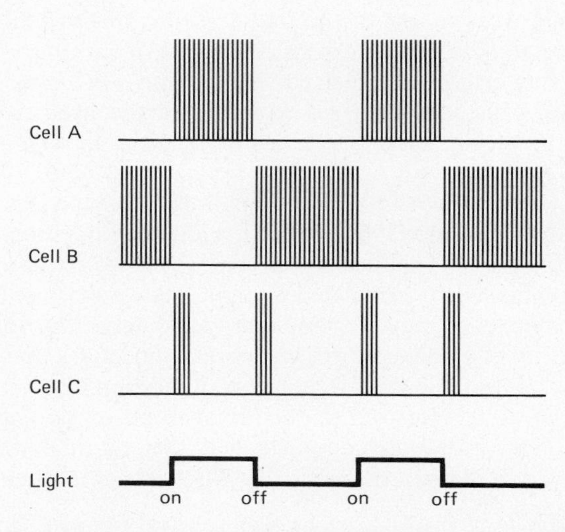

Figure 22-5. **"On" and "off" fibers of the optic nerve. Cell A fires when the light is on, cell B fires when the light is turned off, and cell C fires at both the onset and the cessation of the light stimulus.**

Figure 22-6. **Organization of "on" and "off" regions in the retina of the cat. The middle region is an area of "on-off" sensitivity.**

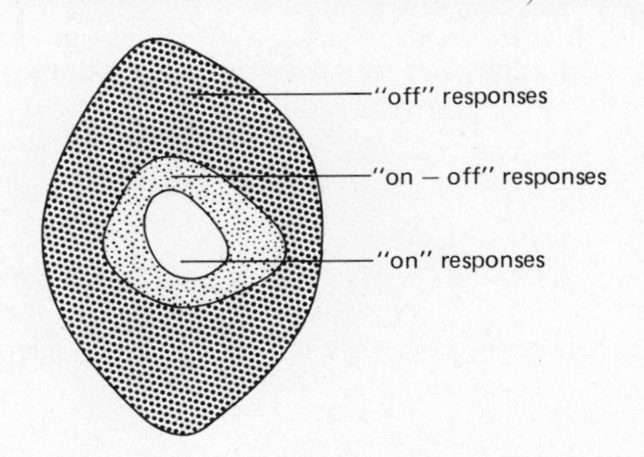

Temporal Comparisons. Important information about the stimulus is derived from a comparison of information arriving at slightly different times. The ear is a particularly good example of this *comparison device*, permitting accurate localization of sound on the basis of the difference in times at which the auditory receptors in the cochlea are stimulated.

Spatial patterns of coding

Numbers of Receptors Responding. The number of receptors that are excited and conduct impulses to the central nervous system increases as the strength of the stimulus increases. Different cells vary in their thresholds but when the stimulus is sufficiently strong, many more cells will respond than when a weak stimulus is applied.

An increase in the number of excited neurons means also that the frequency with which the brain sensory centers are stimulated is increased. *Coding by place* and *coding by frequency* must be closely associated.

Filtering. Much more information is received by some receptors than is sent to the brain. In the frog retina, approximately 1 million receptor cells synapse with 3 million connecting cells, which then converge on only half a million ganglion cells, the axons of which form the optic nerve. Thus the retina acts more as a filter than a photographic film.

Lateral inhibition is a phenomenon that was first demonstrated by Hartline (1956) in the horseshoe crab (*Limulus*), an invertebrate. Hartline showed that when two retinal cells are simultaneously stimulated by light, the response of each is less than when they are stimulated separately. This inhibitory effect is mediated through lateral connectives between the cells, and so is known as lateral inhibition.

In the vertebrate eye, the anatomical and physiological situation is much more complicated, but lateral inhibition is employed to sharpen the image by the elimination of excess information (Fig. 22-7).

Müller's Law of Specific Nerve Energies. This concept is true in a gross way and represents a means of spatial coding. The law states that when specific nerves are excited, only a specific sensation is elicited. Stimulation of the optic nerve always evokes a visual sensation; stimulation of the auditory nerve produces a sensation of sound. However, the validity of this specificity of nerve energies is challenged by the finding that *afferent fibers can be excited by more than one mode of stimulation.* Skin receptors, for example, will respond to pain, touch, temperature, and some intermediate sensations that are also produced.

Figure 22-7. **A stimulus that excites a wide range of receptors is sharpened (striped region) by lateral inhibition (stippled area).**

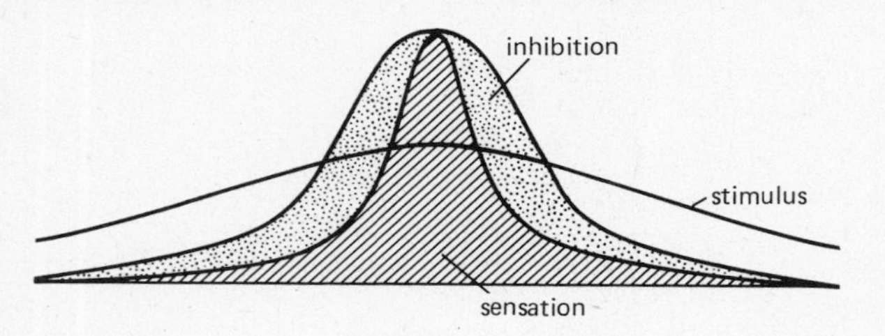

Coding of magnitude

Threshold and the Weber-Fechner Law. An important aspect of signal detection is the threshold of the receptor; a concept of an absolute sensory threshold cannot be sustained, however, even for each type of receptor. The ability to detect a stimulus does not depend as much on the absolute energy of the stimulus as it does upon the relative amount of energy in the stimulus compared to the background.

This relationship between the amplitude of the stimulus and changes in threshold and perception is represented by the *Weber-Fechner law.* It is important to realize that this logarithmic law describes only one aspect of the stimulus—its intensity; whereas sensory coding, on the other hand, includes many other characteristics such as time, space, quality, and pattern.

Weber reported in 1832 that it is the *proportional difference* between stimuli that we perceive. His original experiments used weights placed in the palm of the hand. Weber found that although a subject could distinguish between a weight of 30 g and a second weight of 31 g, the addition of 1 g to a 300-g weight would not be perceived. Only when the second weight is brought to 310 g is differentiation possible; that is, we can distinguish between 30 and 31 g and between 300 and 310 g. The difference in stimulus intensity must be $\frac{1}{30}$. The *smallest discernible difference* is a constant fraction of the weights themselves, the so-called *Weber fraction.*

The mathematician Fechner (1860) expressed this concept in the formula

$$\frac{\Delta I}{I} = C$$

in which ΔI means a just noticeable difference (jnd) in the intensity of the stimulus.

The percentage of change perceived depends on the individual sense organ. For the eye, the difference in intensity of light must be approximately $\frac{1}{167}$ to be perceived. (How pale the moon is, when the sun is still out; how bright the stars on a moonless night; how luminous a candle in a dark room, and how insignificant its light when it burns in a room lit by a 60-watt electric bulb.) The ear requires a difference in sound intensity of between $\frac{1}{9}$ to $\frac{1}{20}$.

Fechner (1860) also described the Weber fraction in more imaginative terms: "A dollar has, in this connection, much less value to a rich man than to a poor man. It can make a beggar happy for a whole day, but it is not even noticed when added to the fortune of a millionaire."

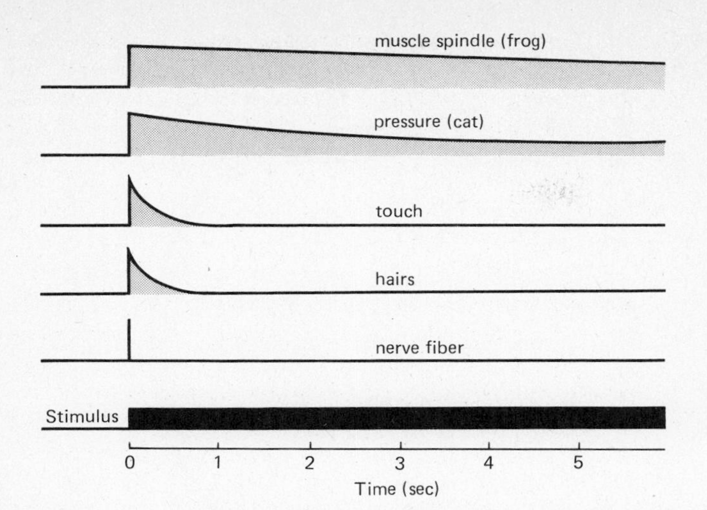

Figure 22-8. **Adaptation of various sense organs in response to a constant stimulation. The height of the curve indicates the relative rate of firing. (Reprinted from *The Basis of Sensation* by E. D. Adrian. By permission of W. W. Norton & Company, Inc. Copyright 1928 by W. W. Norton & Company, Inc.)**

Adaptation as a coding device

The *rate of discharge* of various receptors changes when they are exposed to constant stimulation (Fig. 22-8). The changing response to long-lasting stimulation is known as *adaptation.* This is another way of coding in the nervous system, for it accentuates the significance of a *changing* stimulus as compared to a *constant* stimulus.

All the sense organs, except the pain receptors, *adapt or accommodate* to continuous stimulation. In other words, if a stimulus is maintained over a long enough period of time, it no longer excites the receptor. We are familiar with this phenomenon in everyday occurrences. We often become aware that we have been surrounded by continuous noise only when there is a sudden silence. This is an "on-and-off" response, with accommodation occurring when the noise background as a stimulus is maintained constant. The receptors also use this mechanism to signal the *rate of change* of a stimulus.

Adaptation is particularly marked in the *eye.* Adaptation to light can increase the threshold of the rods of the retina more than a million times. This remarkable decrease in sensitivity occurs extremely rapidly and accounts for the total inability of the light-adapted eye to see in the dark until the reverse phenomenon, dark adaptation, occurs. The chemical basis for these adaptive responses is discussed later in this chapter.

"Behavioral" patterns of coding

The visual stimuli to which an animal responds are closely associated with the behavioral patterns of that animal. In 1959, Lettvin and his colleagues published a paper entitled "What the Frog's Eye Tells the Frog's Brain." Their experiments emphasized that the *retina actively processes the visual stimulus* and does not merely act as a passive mosaic transmitter.

Planning their experiments around the concept of a frog's world containing small moving prey (flies) and large moving predators (hawks), they recorded from optic nerve fibers when the eye was exposed to small and large, moving and static, metal disks. They found that one group of nerve fibers responded only to small, moving spots (a fly?). Another group of optic fibers, very rapidly conducting ones, responded to any overall dimming of the visual field (a hawk overhead?). This indicates a remarkable ability of the ganglion cells, the fibers of which form the optic nerve, to respond to specific spatial and temporal patterns related to the behavior of the animal.

22-4 Topographical Representation of Sensation in the Brain

There are four main sensory areas in the cerebral cortex (Sec. 8-6). Immediately behind the motor area and separated from it by the central sulcus is the *somesthetic sensory area* for reception of information from the cutaneous sense organs and the deeper ones of pressure and movement. The receptive area for *vision,* the striate cortex, is at the occipital pole of the brain. The *auditory* area extends through the temporal lobe, and *olfaction* is on the undersurface of the temporal lobe, closely connected to the limbic system. *Taste* is probably located at the base of the somesthetic area. In addition to these primary receiving areas, there are *association areas* that are closely connected to the primary areas and are concerned with the *interpretation of the sensation.* The primary visual area, for example, responds to information from the eyes according to the pattern of firing neurons excited in the retina. This pattern is interpreted and recognized by the association areas. For example, if the visual association areas are destroyed, the individual can still see an object but cannot give it a meaning or a name. It is believed that the difficulty some children have in understanding the meaning of words they can see perfectly well (*dyslexia*) is due to some abnormality in the *visual association areas.*

The sensory areas of the brain can be mapped by stimulating different parts of the body and recording the arrival of the sensory impulses. This response occurs even when the animal is anesthetized. In the monkey and in humans, a large part of the somesthetic cortex is devoted to the face and hand and very little to the trunk. This means that information coming from the face and hand, particularly the fingers, can be interpreted in a great deal more detail than information from the trunk, where we are conscious only of somewhat gross differentiation of stimuli. There appears to be some sort of protective mechanism against complete loss of sensation through destruction of these primary areas in the brain in the presence of a *second receptive area* for the sensations.

All the pathways connecting the receptors with the brain *cross over* from one side to the other, so that stimuli received by the left side of the body cause impulses to go to the right cerebral hemisphere. These pathways, and the cortical representation of each of the special senses, are discussed in the appropriate sections of this chapter.

22–5 Vision

General Structure of the Eye

The eye contains not only the *receptor cells of the retina,* the fibers of which send coded impulses to the visual cortex, but a *complex optical system* by which light from the exterior is refracted through various structures to be focused on the sensitive retina. Only a small portion of the globular eye is exposed in front; the rest is hidden in the bony socket of the orbit, on a cushion of fat and connective tissue. The eye can turn freely within a certain range vertically, horizontally, or obliquely through the delicately controlled actions of *three pairs of muscles* (two pairs of rectus muscles and one pair of oblique muscles), which permit the eye to move quickly first in one direction, then another.

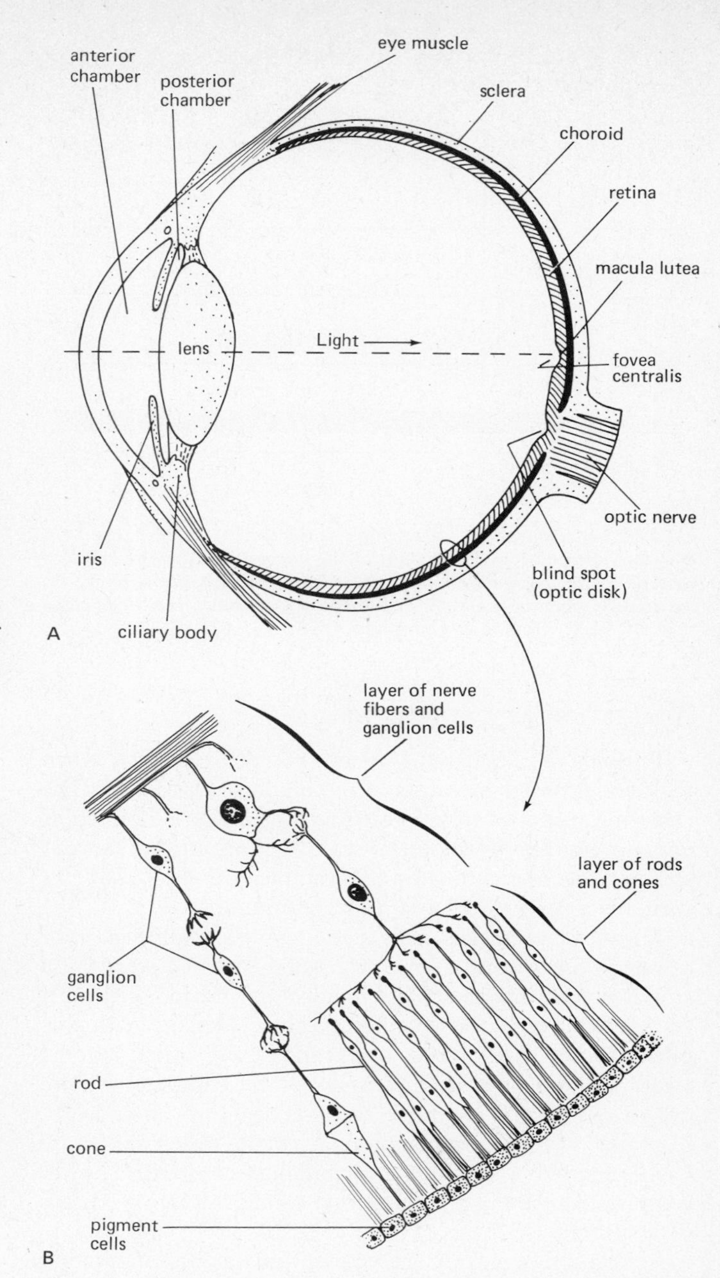

Figure 22-9. **(A) The structure of the eye. (B) The wedge of tissue shows the cell layers of the retina.**

Eyelids and tears

The *eyelids* are two movable curtains in front of the eyeball that close reflexly to protect the eye when the light is too bright or when any object approaches the eye rapidly. The three or four rows of eyelashes also serve to trap foreign particles as well as to beautify the eye. The eyelids are lined with the same delicate membrane, the *conjunctiva,* that covers the surface of the eyeball. The eye is lubricated by the secretions of the *lacrimal gland,* about the size of an almond, which is situated in the upper lateral corner of each orbit. The secretions, the *tears,* flow across the eye and drain through the tiny canals that begin in the small holes that are seen in the little elevation on each eye (lacrimal papilla) near the nose. The tears then run down the *nasolacrimal duct* (tear duct), to empty into the inferior meatus of the nose, behind the nostril (Fig. 16-2).

The three coats of the eye and muscles of the eye

The eyeballs are surrounded by three coats, the outermost of which is the *sclera* (meaning "hard"), a middle vascular layer, the *choroid,* and finally the delicate and intricate *retina,* which contains the receptor cells and the other nerve cells that connect with them (Fig. 22-9).

The Opaque Sclera. In front the opaque sclera is known as the "white of the eye." In the center it merges with the transparent round window, the *cornea,* and is separated from it by a small canal, called Schlemm's canal. The front of the cornea is highly curved and convex, acting as a lens, and forming the first of the refracting surfaces of the eye.

The Dark Choroid Coat. The blood vessels that supply nutrients to the eye are found in the dark choroid coat. Its inner surface (close to the retina) contains pigment cells, which absorb scattered light and so increase the sharpness of the focused image. In the front part of the eyeball, the choroid becomes thicker and forms the round and ruffled *ciliary body,* which in turn is hidden by the muscular ring of colored membrane, the *iris.* The ciliary body contains many blood vessels and the *muscles of accommodation,* by which the eye can change its focus from objects in the distance to those within a few inches of the eye. The iris (from the Greek word for rainbow) contains pigment, which varies in amounts in different people and causes the eye to appear blue (very little pigment) or brown (considerable pigment). Babies' eyes shortly after birth are nearly always blue because the pigment develops somewhat later. In the center of the iris is a hole, the *pupil,* through which the light penetrates before being focused on the retina. The diameter of the pupil is reflexly controlled by the *muscles of the iris,* which contract the pupil when the light is too bright, dilate it when there is less light. Pupillary reflexes are discussed in more detail later in this chapter.

The Light-Sensitive Retina. The innermost of the three coats of the eye, the retina, is exceedingly thin and transparent. The living retina is purplish red, because of the visual purple of the rods, but it becomes white and opaque soon after death because the visual pigment is irreversibly bleached by light. Despite its thinness, the retina is made up of at least 10 separate layers of cells.

Figure 22-10. **Synaptic contacts between cells of the various layers of the retina. The horizontal and amacrine cells make lateral connections with the cells in the retina above and below them, and with one another. Areas of synaptic contact are darkened.**

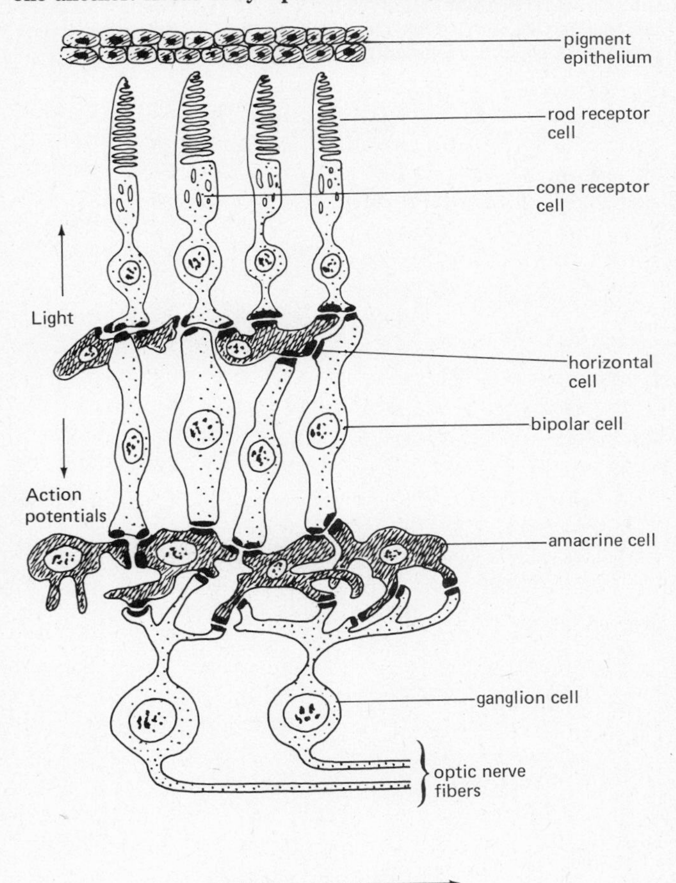

ORGANIZATION OF THE RETINA. The photoreceptor cells, the rods and cones, are in the external layers of the retina, backed by a layer of pigment cells. Before light can reach the rods and cones, it must first penetrate between the many nerve cell bodies and processes that form the remaining layers of the retina.

THE BLIND SPOT. These nerve fibers all come together in a thick bundle near the back of the eye (slightly to the nasal side of center) to form the *optic nerve,* which completely interrupts the retina. Because there are no receptors in this area, no sensation of vision is elicited when light falls upon it, and it is called the *blind spot* (Fig. 22-9).

The existence of the blind spot is quite easily demonstrated by putting a row of five coins on the table about three inches apart, and looking fixedly at the middle coin with one eye closed. If the right eye is closed, the coins to the left of the middle one will become invisible, because their image will be thrown on the blind spot of the left eye. We are normally quite unaware of the rather large gap in our visual field, because the movements of the eyes continuously place the image on that part of the retina where the sharpest vision occurs, the *yellow spot,* or *macula lutea.*

The macula lutea lies a little to the temporal side of center, about 3.5 mm lateral to the blind spot, and it forms a disk, the outer rim of which contains both rods and cones. Toward the center the rods disappear, so that the central area or *fovea* contains only the cones, closely packed together, with perhaps as many as 150,000 cones/mm^2. At the fovea, the layers of connecting nerve cells almost disappear, and each *cone* is connected to only one *ganglion cell* (Fig. 22-10). This means that the image transmitted from the fovea to the brain is much sharper than that from the rest of the retina. Very few cones are in the peripheral part of the retina, the photosensitive portion of which is composed almost exclusively of rods.

CELLULAR CONNECTIONS WITHIN THE RETINA. The retina in all vertebrates is constructed from *five basic types of cells.* Three of these cell types—the *receptor cells* (the rods and cones), the *bipolar cells,* and the *ganglion cells*—form a connecting sequence that carries information from the region of light transduction in the external retinal layer to the optic nerve fibers leading to the brain. The remaining cell types, the *horizontal and amacrine cells,* form lateral synapses that relate activity across different visual fields. These cellular relationships are depicted in Fig. 22-10. Figure 22-11 indicates the functional significance of these multiform synapses in determining retinal sensitivity.

PIGMENT LAYER OF THE RETINA. The outermost layer of the retina, behind the rods and cones, is the pigment layer, the cells of which contain the blackish-brown pigment *melanin.* As in the choroid, this pigment absorbs excess light, preventing the reflection of light through the eyeball and so permitting precise image formation.

Albinos, who genetically lack melanin, show this deficit also in the pigment layer of the retina and in the iris. The resulting diffusion of light through the eye makes sharpness of vision impossible.

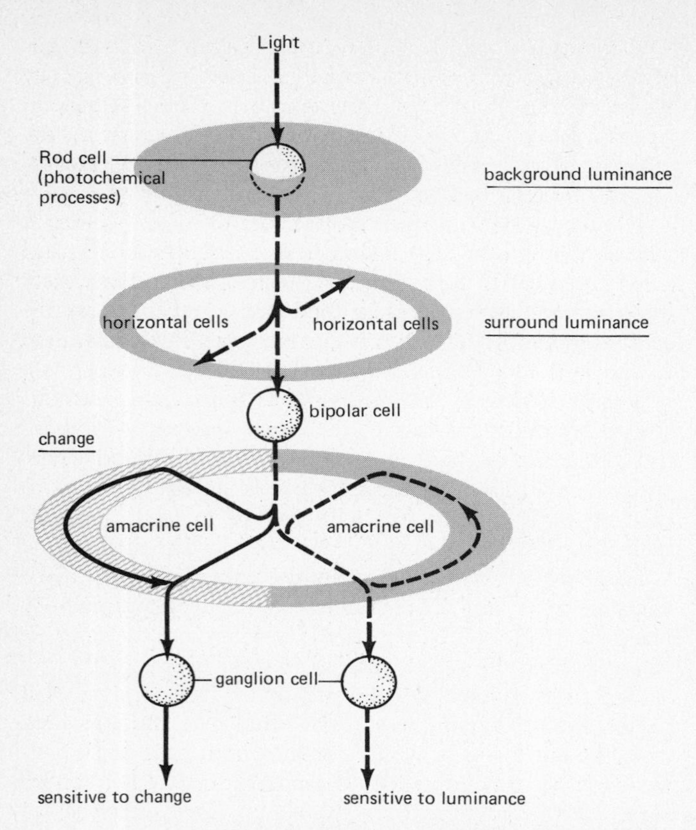

Figure 22-11. **The sensitivity of the retina is determined on at least three levels. (1) The response of the rod cell to light will vary with the background luminance to which this cell is exposed. (2) Surround luminance at the next level of the horizontal cells. (3) Changes in the interactions between amacrine and bipolar terminals will interact with the responding rod and horizontal cells. The resulting information leaving the ganglion cells will be a composite of all these interactions.**

The pigment layer also stores large amounts of *vitamin A,* an important precursor of the photosensitive pigments. The role of vitamin A in the photochemical reactions of the eye is discussed later in this chapter.

Receptors of the Eye: Rods and Cones

Development of receptor cells from ciliated epithelium

Despite the diversity of structure and function in various sense organs, many receptor cells appear to be modifications of simple, ciliated cells. In lower animals—the unicellular paramecium, for example—ciliated cells have a dual sensory and motor function and are characterized by their nine pairs of peripheral filaments and two central filaments (Sec. 2-2). During the processes that result in the specialization of receptor cells, the motor role is lost. In the retinal rod, a specialized ciliated cell, this loss is reflected in the absence of the two central filaments.

Vitamin A (retinol) appears to be an important influence in the development of receptor cells; cells with cilia or flagella always contain vitamin A, which may be essential for the differentiation of the ciliary apparatus. In the absence of vitamin A, receptor cells degenerate, a phenomenon that has been studied in great detail in the eye.

The rods and cones are so named because of their shapes (Fig. 22-12). Although there are overlapping shapes in different animal groups, the two groups of photoreceptors differ considerably in their structure and function.

The beautiful drawings of Young (1975) (Figs. 22-12 and 22-13) show some of the details of rod and cone structure that have been elucidated through the electron microscope. Both rods and cones consist of an *outer segment* (considered to be the expanded top of the cilium) and an *inner segment,* with the remains of a cilium connecting the two.

The outer segment of the rod

The outer segment of the rod is considerably longer than the pyramidal outer segment of the cone, but both types of cells contain invaginated membranes, or *disks,* in their outer segments. It is on these layered disks that the photosensitive pigments are organized. The pigment in the rods is *rhodop-*

Figure 22-12. **A rod and a cone cell from frog retina. The connecting cilium could be identified only through use of the electron microscope. (From "Visual Cells," by R. W. Young, *Scientific American,* Oct. 1975, p. 83. Copyright © 1975 by Scientific American, Inc. All rights reserved.)**

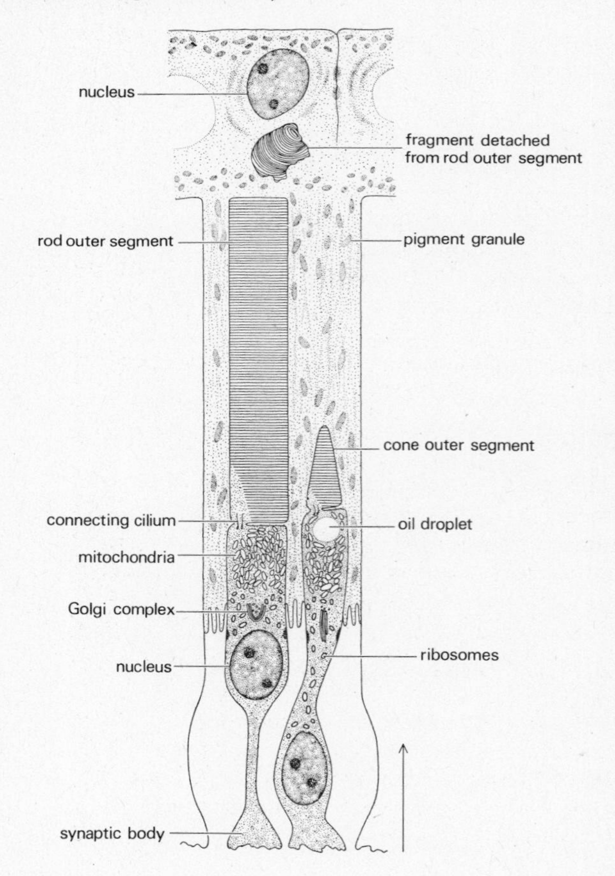

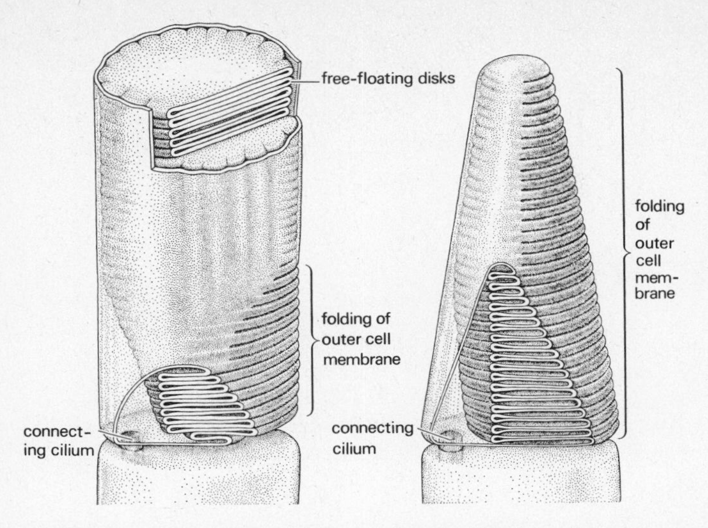

Figure 22-13. **Drawing of a greater magnification of the outer segments of a rod and a cone, showing the differences in the method of disk formation. See text for details. (From "Visual Cells," by R. W. Young, *Scientific American*, Oct. 1975, p. 84. Copyright © 1975 by Scientific American, Inc. All rights reserved.)**

sin; that in the cones is *iodopsin.* Both of these visual pigments consist of *retininal* (formerly retinine), a carotenoid pigment related to vitamin A (discussed in Sec. 21-7), combined with a protein, *opsin.* The opsins differ slightly in the rods and cones, but their photochemical reactions may be considered to be essentially the same. The rod opsin is *scotopsin,* and the opsin of the cones is *photopsin.*

An important difference between the rods and cones appears to be the way in which the disks are formed. Using radioactive tracers, Young (1975) has shown that in the rods, the disks originate near the bottom of the outer segment as invaginations of the cell membrane, and then migrate toward the tip. The *rod disks* appear to be *constantly regenerated.* This does not seem to be the case for the cones: once the disks have been formed they remain fused to the outer cell membrane, and there is no further disk production (Fig. 22-13). The outer segment of the cones also contains an oil droplet that may be involved in the optics of color vision.

The inner segment of both rods and cones

The inner segment of both rods and cones is rich in mitochondria, which provide energy-yielding oxidative reactions needed for the synthetic reactions occurring in the outer segment. The footlike process extending from the inner segment contacts the bipolar cells of the retina.

Effect of light on the visual pigments

Decomposition of Rhodopsin. George Wald (1968), whose brilliant studies on the visual pigments have led the research in that field, has shown that light alters the shape of the pigment part of the rhodopsin molecule so that it no longer fits the protein portion, and the chemical bonds between the parts can be hydrolyzed. This means that the *rhodopsin* molecule is split into *opsin* and *retininal.* The change in shape of the retininal molecule is called *isomerization* and represents the straightening of a side chain in the molecule. The form of retininal with the *twisted side chain* is called *11-cis-retininal;* the *straightened* one is *all-trans-retinal.* The *cis-trans* isomerization is basic to every visual system known. These changes in retinal structure are shown in Fig. 22-14, and the alterations in fit between retinal and opsin are diagrammed in Fig. 22-15.

Several other intermediate pigments are obtained during the bleaching of retinal to retinol. These include lumirhodopsin and metarhodopsin, which are extremely unstable compounds.

Retinal-Retinol (Vitamin A) Cycle. Also shown in Fig. 22-14 is the *reduction of retinal to retinol* (vitamin A). Although this is a reversible reaction, as we shall see, the direction of the reaction is in favor of vitamin A, permitting excess vitamin A to be stored in the pigment cells of the retina where it is readily available to the rods.

Resynthesis of Rhodopsin. OXIDATION OF RETINOL TO ALL-*TRANS*-RETINAL. In the dark, retinol is oxidized to all-*trans*-retinal through the catalytic action of the enzyme *alcohol dehydrogenase,* a reaction that requires NAD (Fig.

Figure 22-14. **Light splits rhodopsin into opsin and retinal, which is subsequently reduced to retinol (vitamin A).**

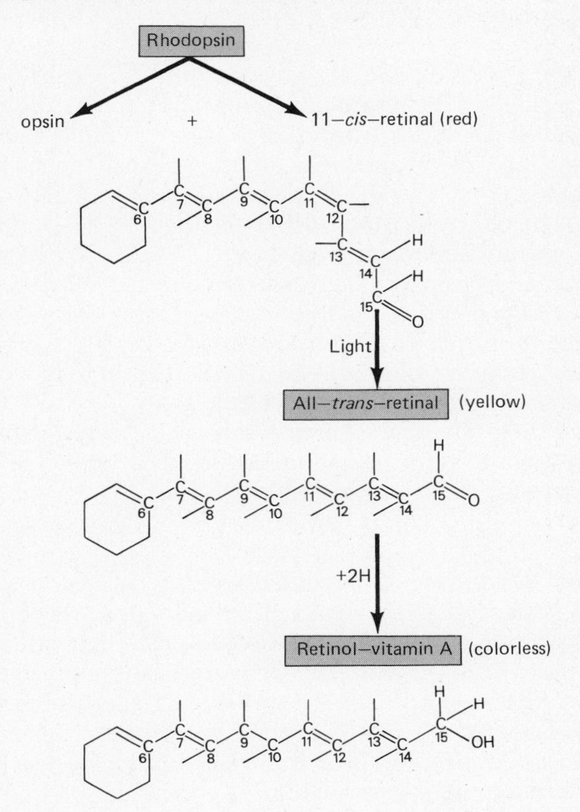

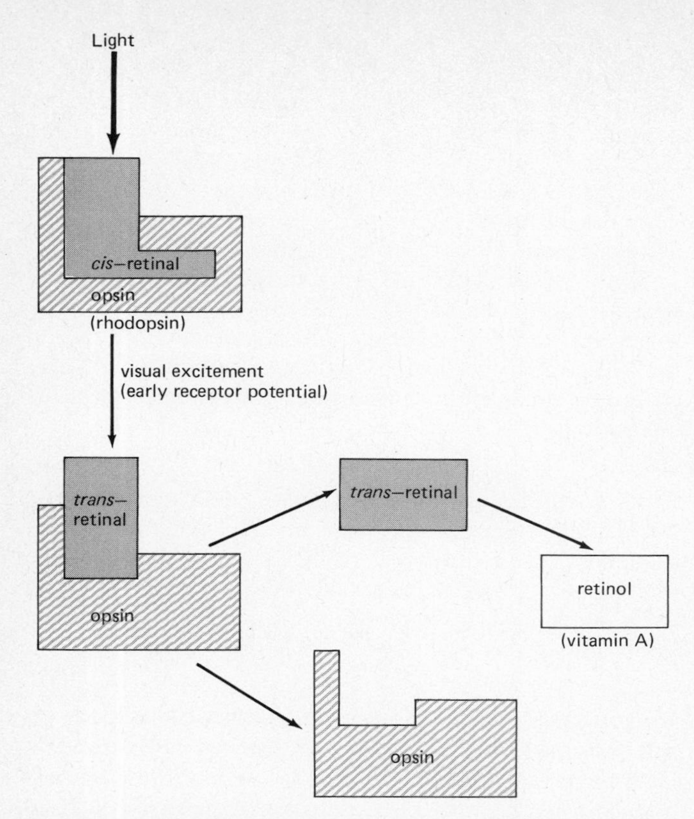

Figure 22-15. **The bleaching of retinal to retinol by light involves the isomerization of *cis*-retinal to *trans*-retinal. This frees the pigment part of the rhodopsin molecule from the protein opsin. Subsequent reactions include the hydrolysis and reduction of colored intermediary pigments to the colorless retinol (vitamin A).**

22-16). This is a relatively slow reaction in comparison to the rapidity of the bleaching reactions that decompose rhodopsin. If the retina is exposed to light, bleaching rapidly occurs, and the re-formation of *trans*-retinal cannot take place.

ISOMERIZATION OF ALL-*TRANS*-RETINAL TO 11-*CIS*-RETINAL AND COMBINATION WITH OPSIN. The combination of retinal and opsin is a stereospecific one and requires the 11-*cis* form of retinal. As the bleaching of rhodopsin results in the formation of all-*trans*-retinal, the reverse reactions involve isomerization to 11-*cis*-retinal. The enzyme *retinal isomerase* is needed for this to take place.

Once 11-*cis*-retinal is formed, it spontaneously combines with opsin to form rhodopsin. A simplified version of the rhodopsin cycle is shown in Fig. 22-16.

Night Blindness. If the diet is severely deficient in vitamin A, the available amounts of retinol, retinal, rhodopsin, and iodopsin are all decreased. Consequently, light sensitivity drops considerably. This is particularly noticeable at night, when too little light is available for adequate vision. This inability to see well at night because of a vitamin A deficiency is *night blindness.* It can be cured in less than 1 hr by an injection of vitamin A.

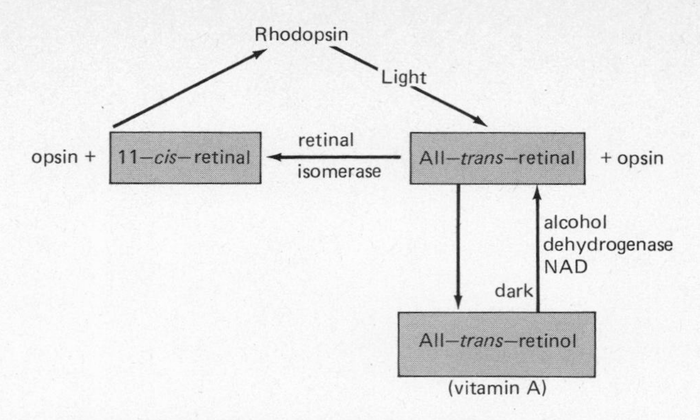

Figure 22-16. **The chemical reactions involved in the exposure to light of rhodopsin (opsin-retinal) to retinol (vitamin A) and its subsequent resynthesis in the dark.**

Transduction of Light Energy by Photoreceptors

VISUAL EXCITEMENT

Visual excitement occurs at the stage when sufficient light energy is absorbed by rhodopsin to isomerize 11-*cis*-retinal to all-*trans*-retinal. In this process, the retinal is displaced from its opsin site, and reactive sulfhydryl groups of the opsin molecule are exposed (Fig. 22-15).

THEORIES OF TRANSDUCTION

The precise mechanism by which the photochemical reactions excite the rods is not known. One suggestion explains this excitation by the exposure of the sulfhydryl groups of the opsin as the all-*trans*-retinal splits away from the "nest." This changes the charge of the membranous disks, and consequently their permeability, producing the receptor potential.

Another hypothesis relates the visual excitement to the parallel alignment of the rhodopsin molecules in the disks of the outer segment, for the intact segment is needed for the electrophysiological response. It has been suggested that the photochemical reactions may alter the *position* of the rho-

Figure 22-17. **The electroretinogram. The small negative α wave is followed by the large positive β wave. Alcohol diminishes the α wave but increases the size of the β wave.**

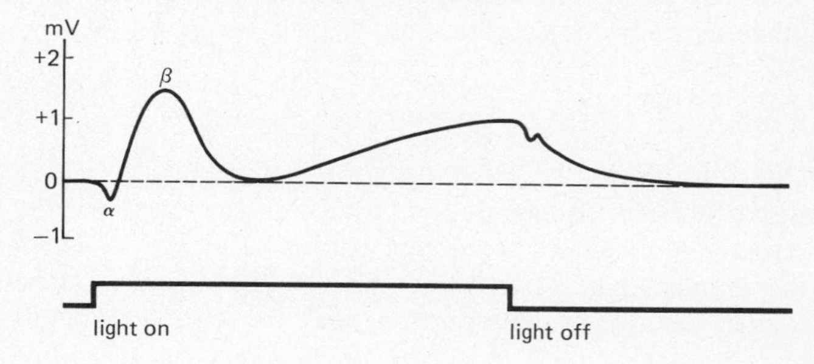

dopsin molecules in the membrane, initiating a very fast response even before the photochemical disruption of rhodopsin occurs. This very fast visual response is known as the *early receptor potential* (ERP).

Evidence for this concept comes from studies of the *electroretinogram* (ERG), which is measured by placing an active electrode on the cornea and an indifferent electrode elsewhere on the body. When a dark-adapted eye is exposed to a flash of light, there is a short latent period followed by a biphasic ERG (Fig. 22-17).

If the eye is exposed to a much stronger light, then the very fast ERP appears before the ERG. The latency of the ERP compares to the time it needs to push the rhodopsin molecules out of alignment in the disks: it is too fast to account for the decomposition of the rhodopsin.

THE RECEPTOR POTENTIAL

Once the rod is excited, the sodium conductance of the membrane of the outer segment is *decreased, hyperpolarizing* the membrane. This is different from most other receptors, in which depolarization occurs.

The receptor potential has two phases, the first of which is *directly* proportional to the amount of light and is probably associated with the photochemical reactions. The second phase of the receptor potential is proportional to the *logarithm* of the light energy and is associated with the decreased sodium conductance.

It is believed that this biphasic response contributes to the remarkable ability of the eye to discriminate a very wide range of light intensities.

Duplicity Theory of Vision

The duplicity theory of vision postulates that the rods are concerned with night vision and the cones with day vision.

NIGHT VISION (SCOTOPIC VISION)

The rods are not sensitive to color, responding to light wavelengths at the blue end of the spectrum only. This type of sensitivity can be quantitatively related to the absorption characteristics of rhodopsin.

ABSORPTION AND ACTION SPECTRA

For a quantum of light to be effective, it must be absorbed. An *absorption spectrum* is a plot of the number of quanta absorbed by a molecule at various wavelengths of light (Fig. 22-18).

An *action spectrum* plots the number of quanta of the different wavelengths required for a photochemical response. To relate this to the *efficiency of the different wavelengths* in evoking a visual response, the reciprocal of the number of quanta is plotted. Figure 22-18 shows that the most effective wavelength for human rhodopsin is blue

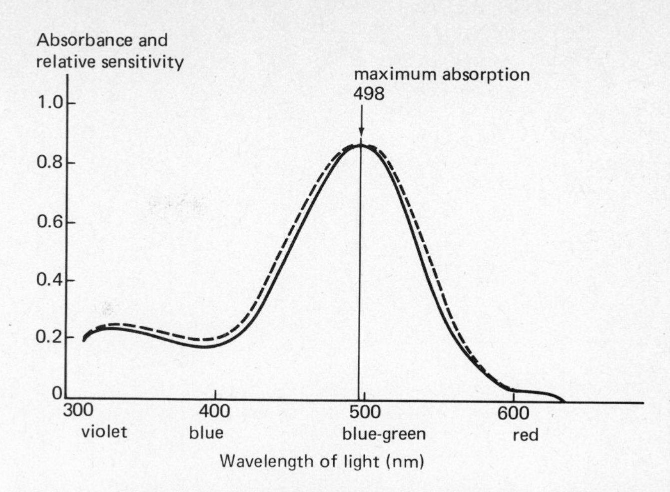

Figure 22-18. **The absorption spectrum of rhodopsin (solid curve) shows that the maximum absorption of light coincides with the greatest sensitivity (action spectrum) of the retina to scotopic luminosity (blue-green). The broken-line curve represents the action spectrum.**

green (500 nm). Red light and violet light are much less effective. The absorption spectrum corresponds to the action spectrum for rhodopsin, showing that scotopic vision is based upon light absorption in this wavelength.

CHARACTERISTICS OF NIGHT VISION

The inability to see anything in a dark room for the first 20 to 30 min corresponds to the time needed for the resynthesis of rhodopsin in the light-bleached (*light-adapted*) retina. Slowly, the shape and position of objects become visible, and the eye is *dark-adapted* (Fig. 22-19). There are several differences in the image seen at night from that of daylight vision.

1. It has no color but is black or gray.
2. The edges are not as sharp.
3. Fixing one's eye on the object, which is so helpful in the light, is a hindrance at night.

These differences are caused by the insensitivity of the rods to wavelengths other than blue green and the fact that the rods are not concentrated in one area of the retina, as are the cones. The cones are concentrated in the macula lutea, whereas the rods are spread over the periphery of the retina. Consequently, there is no one area where a sharp image can be focused. The vagueness of the image at night, and the inability to fixate upon it, may be the basis for the ghosts and similar hallucinations seen by the imaginative.

DAY VISION (PHOTOPIC VISION)

In bright light, only the cones of the retina respond to the visual stimulation. Absorption spectra studies on cone pigment have shown that there are at least three visual pigments, each contained within a different type of cone. The

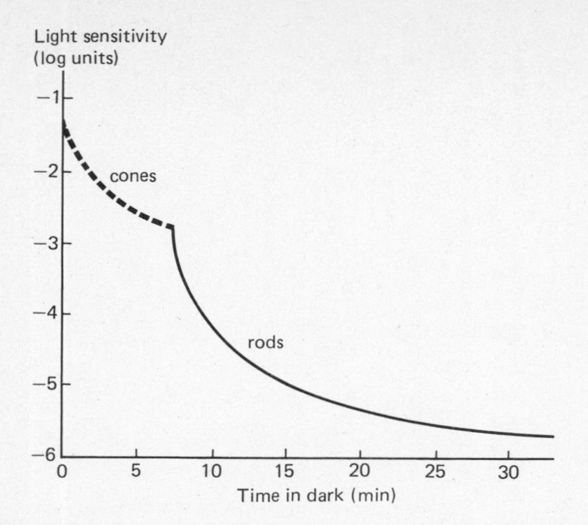

Figure 22-19. **The lower part of the curve represents the dark adaptation of the rods, showing their increased sensitivity to dim light with time. The upper part of the curve shows the more rapid, but very limited, adaptation of the cones.**

peak absorbancies are at 430, 535, and 575 nm, respectively, making the cones sensitive to *blue, green,* and *red light.*

Color Vision. Our perception of the three primary colors of light (red, green, and blue) and their various combinations is a century-old physiological conundrum that is still not solved. The theory proposed by Helmholtz (1866), and modified recently by Granit (1945) and Wald (1964), is the simplest and most likely one. There are *three main types of color receptors* in the *cones,* containing pigments that have their main spectral sensitivity to red, green, or blue wavelengths. Various combinations of these three main types can give rise to a sensation of almost any color. Simultaneous stimulation of all types of cones by equal amounts of colored light will give rise to the sensation of *white light* (Fig. 22-20).

Afterimages. The study of afterimages has supported some of the theories of color vision. If one looks at a bright light intently for a few minutes and then closes one's eyes, the *positive afterimage* of the light can still be seen. This is because the chemical products of the splitting of the visual pigments are still stimulating the retinal cells for a second or two after the light stimulus has disappeared. The positive afterimage appears with the original bright area as bright, dark areas as dark.

If one looks at a brightly colored pattern intently for about 10 sec and then at a sheet of white paper, one sees a *negative afterimage,* in which the bright parts seem dark, the dark parts white, and the *colors of the negative afterimage are complementary* to those in the original. Yellow appears blue, red appears green, black appears white. This is caused by the *fatigue* of the highly stimulated color receptors in the retina. The pigments in the receptor are chemically altered by the light and diffuse out presumably to stimulate the previously unstimulated receptors for the complementary colors. This can be interpreted as meaning that when a receptor for red light is stimulated, the pigment within that red-sensitive cone is altered in some way, diffuses out of the cell, and affects the green-sensitive cone. When you evoke a negative afterimage, notice that there is an evident time lapse after you focus your eyes on the white paper before the negative afterimage appears.

Color Blindness. When one type of color-receptive cone is missing from the retina, that person will not be able to distinguish this color from some others. If the red-sensitive cones are lacking, the wide range of wavelengths that are normally interpreted as red will fall into the sensitivity range of the green-sensitive cones. Because it is a *ratio* of the number of cones stimulated that results in our interpretation of color, and because only green cones will be stimulated, only green will be perceived. The person is said to be red-green color-blind when he is missing either the red- or the green-sensitive cones. Blue-yellow color blindness is less often encountered, and there is no really satisfactory explanation for it that fits in with the other theories of color vision.

Sometimes complete color blindness occurs, in which presumably all the color-sensitive cones are absent or not functioning. This is very rare. Color blindness in its various forms is inherited as a sex-linked characteristic, showing up in males, therefore, far more often than in females (Sec. 27-4).

Characteristics of Day Vision

1. The cones are stimulated by daylight.
2. Their stimulation gives rise to the sensation of color.
3. Because of their concentration and one-to-one relationship to nerve cells in the fovea, the image is sharp when the eye is fixed on the object, so that the light rays focus on the fovea.

Figure 22-20. **The absorption curves of the three color-sensitive pigments of the eye. Simultaneous stimulation with equal amounts of these wavelengths results in the sensation of white light. The open circles indicate the sensitivity of the blue photopigment.**

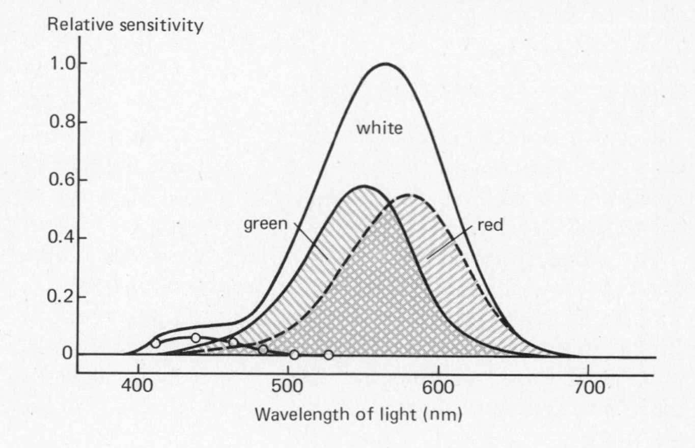

Stimulation of the cones is involved, therefore, in daylight (photopic) vision, and it is only in daylight that we can perceive color. (We can see neon lights in color at night because the intensity of the color is so high that it stimulates the cones.)

Pathway for Nerve Impulses from the Retina to the Brain

To the striate (visual) cortex

All the nerve fibers from the retina pass backward toward the brain as the *optic nerve.* Just in front of the pituitary gland, at the base of the brain, the optic nerves form the *optic chiasma,* through a partial crossing over of their fibers. Those nerve fibers that originated in the *temporal halves* of the two retinae remain uncrossed, whereas those coming from the *nasal halves* cross over. Thus the continuations of the optic nerves, after the chiasma, are composed of the temporal fibers of the one retina, together with the nasal fibers of the other. These two bundles of mixed nerves are called the *optic tracts.* Each optic tract synapses with cells in the *lower centers for vision* in the *lateral geniculate bodies* under the surface of the thalamus. The fibers then spread out to end in the visual cortex (striate cortex) in each occipital lobe. These are the *primary visual centers* (Fig. 22-21).

The peculiar anatomic distribution of the nerve fibers from the two retinae means that damage to one optic tract will affect the nasal part of one retina and the temporal part of the other. It is a good exercise in optics to determine which part of the visual field will then be missing for each eye. Light waves from the temporal field fall upon the nasal retina; light waves from the nasal field fall upon the temporal retina (Figs. 9-10 and 22-21).

To the reticular system

The reticular system is associated with the activation of many other parts of the cortex that are associated with vision and visual impressions, but not the primary visual image.

To the superior colliculus in the midbrain

The superior colliculus receives nerve impulses directly from the optic nerve (Fig. 22-21). It also receives sensory information from the *auditory* and *somatic* systems and impulses from the *visual* cortex. This information is organized topographically in the various layers of the colliculus. Collicular cells have large receptive fields and are relatively insensitive to the size, shape, and orientation of stimuli. They are very responsive, however, to the *direction of a moving stimulus.* In some way this information about the movement of a stimulus is coordinated with head and eye movements to keep the stimulus in the visual field. Animals with collicular lesions cannot orient themselves to visual, auditory, or somatic stimuli.

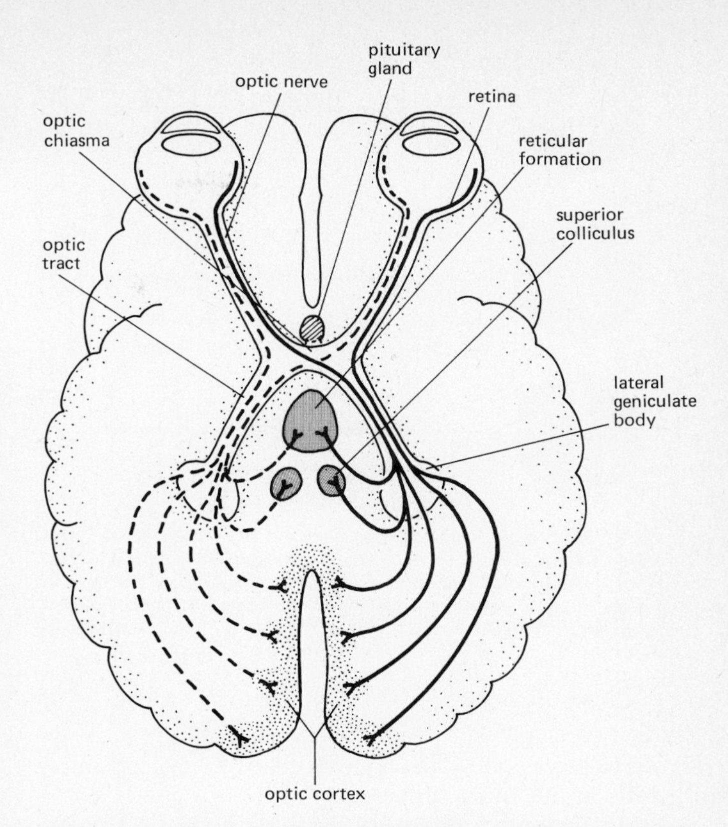

Figure 22-21. **Nerve pathways from the retina to the visual cortex of the brain.**

Projection of Retinal Cells to the Visual Cortex

The direct visual pathway from the retina, through the lateral geniculate body to the visual cortex, is a relatively recent phylogenetic development in higher mammals. At each stage in the pathway, cells receive excitatory and inhibitory connections from the previous stage, so that information is processed from the periphery as well as in the visual cortex.

The visual cortex is the first part of the pathway to receive impulses from *both* eyes. The receptive field in one eye seems to be mimicked exactly in the other eye. Because each neuron in the striate cortex can simultaneously "look" in two directions (through each eye), the visual cortex has been called the "cyclopean eye" of the binocular animal. This dual information makes it possible for the cyclopian eye to resolve the different directions of a stimulus as seen by each eye and to use *binocular parallax* (see later in this chapter) to judge the distance of an object.

Specialization of Cells in the Visual Cortex

Unlike cells in the retina, cortical neurons require precisely defined stimuli if they are to fire, and this precision has been

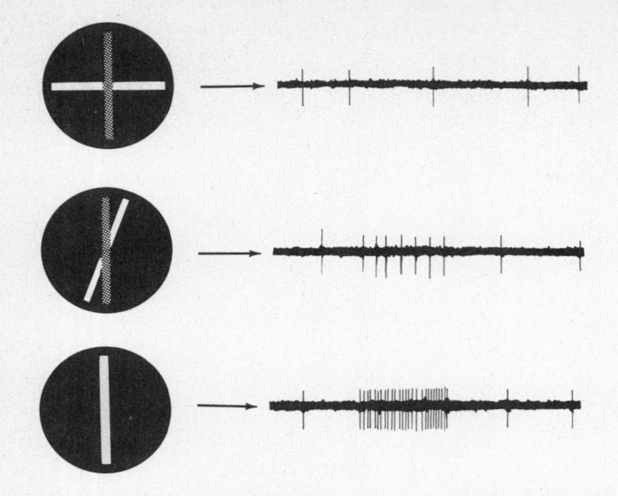

Figure 22-22. **Simple cortical cells respond to changes in the orientation of a slit of light to which they are exposed. (Top): The cell does not respond to a horizontal orientation. (Middle): If the light is at a slight angle, it evokes slow firing of the cell. (Bottom): Vertical orientation of the light is the most exciting to the cell, resulting in rapid firing.**

shown to be caused by the high degree of specialization of the cells in the visual cortex.

A very important discovery concerning this increasing cellular specialization was made by Hubel and Wiesel of Harvard University in 1959. They recorded the activity of single cells in the visual cortex, while presenting a cat with a simple visual shape, such as a bright bar of light. They found three hierarchies of cells: *simple, complex,* and *hypercomplex.*

Simple cells

The simple cell fires when the bar is presented at a certain angle; some fire when it is horizontal, some when it is vertical, and so on (Fig. 22-22). These cells also require that the bar be in a precise position in the visual field—higher, lower, or in the middle, depending upon the cell. The simple cells receive a fairly direct input of fibers from the lateral geniculate body and an array of inhibitory connections from surrounding neurons. This causes the simple cells to respond only to stimuli on narrowly defined areas of the retina.

Complex cells

The complex cells respond to the bright bar when it is presented at a certain angle, regardless of its position in the field; that is, if the cell responds to a horizontal bar, it is unimportant whether the bar is in the middle of the field, above, or below. It is thought that the complex cells receive input from many simple cells and so are able to respond over a larger area of the retina.

Hypercomplex cells

The hypercomplex cell adds another dimension to that of the complex cell. The bar must be in the *preferred position,* regardless of its precise position, *but it must not project beyond a certain length.* For example, a curved line, or a line with a break or corner, will stimulate a hypercomplex cell, but a long simple line is ineffective.

This electrophysiological analysis indicates that information about the world enters the brain as highly abstract information, with *abstraction occurring at each step in sensory transformation.* The remarkable specialization of cells required for all the possible combinations of stimulus form, orientation, position, and length depends to a great degree on the *selective destruction of sensory information.*

The Optical System of the Eye

Refractive structures

The many refractive structures in the eye through which the light must pass before it can be focused on the retina account for the *inversion of the image on the retina.*

The main refraction occurs at the *cornea* as the light passes through this transparent coat into the *aqueous humor,* which lies between the cornea and the iris. The space in which this watery fluid is found is the *anterior chamber.*

Behind the iris is the highly elastic crystalline *lens,* made up of intricately interlacing fibers and covered by a capsule to which fine ligaments are attached and which attach it in turn to the *ciliary body.* This permits changes in tension on the lens, allowing the anterior surface to bulge forward or flatten out. These changes are of great importance in *altering the refractive strength of the lens* and are essential for the phenomenon of accommodation, which will be discussed.

The space between the lens and the retina is the *posterior chamber,* and it is filled with a gelatinous substance, the *vitreous humor,* which is composed of fine fibrils in a jelly-like medium. This jelly also contains protein and hyaluronic acid. There are no cells in the vitreous humor, and it has been suggested that the hyaluronic acid is responsible for this lack.

Figure 22-23. **The passage of light rays through (A) flat glass, (B) convex lens, and (C) concave lens.**

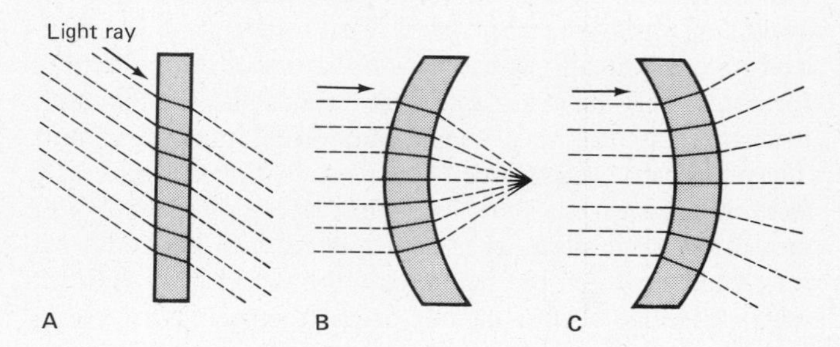

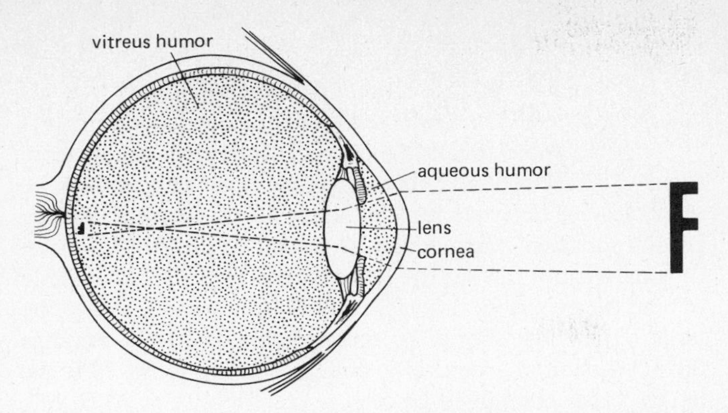

Figure 22-24. **The combined refractive power of the structures of the eye.**

For simplicity, we shall consider the total refractive power of the eye to be the sum of the refractive powers of the individual structures.

Passage of light through refractive structures

The speed of light through air or a vacuum is 300,000 km per second. When light passes through a denser medium, its speed is slowed down. The denser the medium, the more it will decrease the speed of light, and the greater the *refractive index* of that medium as compared to that of air, which is taken arbitrarily to be 1.0.

When light waves pass from one medium to another with a different refractive index, the waves are simply slowed down, unless they hit the new medium at an angle. If they do enter it at an angle, the light waves are *bent* or *refracted.* Those light waves that enter perpendicular to the new medium pass through without refraction (Fig. 22-23).

Parallel light rays entering a *convex* lens, which has a curved surface, will be refracted toward the center of the lens, except for those that enter through the middle of the lens (perpendicular to it). These will not be refracted at all. As the rays leave the lens, they are refracted once more, to converge at a single point focus some distance from the lens. The strength of a convex lens is measured in diopters; a lens has an arbitrary value of +1 diopter when it can focus parallel beams of light at a point 1 m beyond the lens. The stronger the lens, the closer to the lens will be the focal point, and a lens with a refractive power of 10 diopters will *converge* parallel light rays to a focal point only 10 cm beyond the lens (Fig. 22-23).

Parallel light rays entering a *concave* lens are refracted away from the center of the lens, except for those entering the middle of the lens, which pass through unchanged in direction. Thus the concave lens *diverges* light rays, and no focal point is formed, as depicted in Fig. 22-23.

Light rays entering the eye pass through the refractive media of the cornea, aqueous humor, lens, and vitreous humor and are focused on the retina in the fovea (Fig. 22-24). If we consider all these as additive, the refractive index of the eye varies from 59 diopters, when the lens is at rest (and the anterior convex surface relatively flat), to about 71 diopters, when the lens is bulging in maximum accommodation.

This change in the refractive power of the lens occurs when it is necessary to *change the focus from a distant object to a near object.* Points of light from a distant object come into the lens almost parallel and so are converged to a focus fairly close to the lens. When the eye now fixes on a nearby object, the points of light come into the lens at diverging angles so that when this divergence is compensated for, the light rays are focused farther away from the lens. For the object to be sharply focused on the retina in both cases, either the distance between the retina and the lens must be variable, or the lens must be able to change its refractive power and thus its focal distance. It is the *change in strength of the lens* that is the physiological mechanism of accommodation (Fig. 22-25).

The lens will bulge forward when the tension exerted on it by the pull of the ligaments on its capsule is relaxed. This is accomplished by the contraction of the radial and circular muscle fibers of the ciliary muscle. This muscle is at the insertion of the ligaments into the ciliary body, and when it contracts, the insertions of the ligaments are moved forward, shortening the distance between ciliary body and lens and thus releasing the tension on the lens capsule.

Figure 22-25. **Changes in the shape of the lens and the size of the pupil in accommodation.**

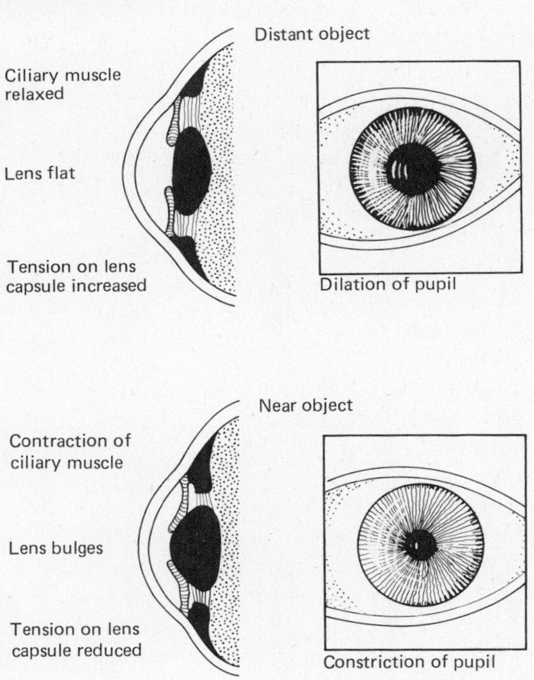

Accommodation reflex

The accommodation reflex occurs when the eye changes its focus from a distant object to a nearby one. The reflex contraction of the ciliary muscle that increases the strength of the lens is under *parasympathetic control.* At the same time as the *ciliary muscle contracts, the pupil constricts,* which increases the sharpness of the image on the retina (Fig. 22-25).

With increasing age, the lens fibers become more rigid, so that even when the tension on the capsule is relaxed, the lens does not bulge anteriorly but remains flat. This means that the eye cannot accommodate to near objects, and we have all seen people who hold the newspaper at arms' length in order to read it. Corrective lenses that increase the refractive power of the eye are necessary to permit focusing on nearby objects. This loss of accommodation with age (due to the loss of the elasticity of the lens) is called *presbyopia* and is considered a reliable index of physiological age, more trustworthy than the patient's version sometimes.

Figure 22-26. **The effect of lens strength on vision. The upper diagram shows the position of the image in the normal eye. The next two diagrams show that the image which is "in front" of the retina in the myopic eye is corrected by a concave lens. In the lower two diagrams, the image which in the hyperopic eye is "behind" the retina is focused on the retina by a convex lens.**

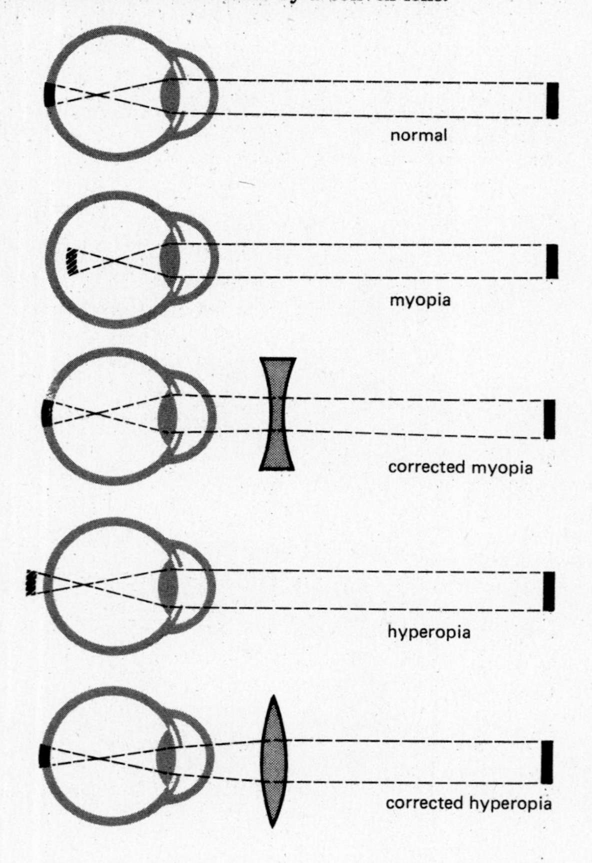

Effect of lens strength on vision

The eye is considered *normal or emmetropic* if, when the ciliary muscle is completely relaxed (i.e., when the lens is at its flattest and weakest), parallel light rays are sharply focused on the retina. This means that distant objects can be seen clearly, but that for the sharp focus of near objects, accommodation must occur. A person with 20/20 vision (normal vision) can see the letters one is supposed to be able to read on a chart 20 ft away. If at 20 ft, one can see only the large letters that should be recognizable at 100 ft, one has 20/100 vision or $\frac{1}{5}$ normal vision. If one can see at 20 ft the letters small enough to be seen normally only at 15 ft, one has 20/15 vision or $1\frac{1}{3}$ normal vision (Fig. 22-26).

If the distance between the lens and the retina is too short or the power of the lens too weak, light rays coming into the eye from a near object will be focused on an imaginary spot behind the retina, so that the retina receives only a blurred impression. To sharpen the image, the lens must increase its strength through the accommodation reflex. This type of eye is referred to as *hyperopic or far-sighted,* because the person can see distant objects (with light rays coming in almost parallel) much better than close-by ones.

When the distance between the lens and the retina is too great or the lens is too strong, the image will be focused in front of the retina, again causing a blurring of the retinal stimulation. This is seen in the *myopic* or *near-sighted eye.* The point sources of light from nearby objects that require strong refraction can be focused on the retina, but the parallel rays from distant objects are too sharply refracted, and accommodation must occur to weaken the power of the lens.

When the power of accommodation is constantly used to correct the faults in the optical system of the eye, corrective lenses should be worn to permit the proper image formation for far and near objects and to lessen eye strain.

Corrective lenses

The hyperopic eye must constantly accommodate to focus on objects that are not in the distance. This causes a chronic strain on the eye, and it can be relieved by the use of lenses that increase the power of the biological lens (Fig. 22-26). This consists simply of a convex lens, which will add to the refractive power of the eye and permit the image to be focused on the retina instead of behind it. Usually, these lenses are chosen in a somewhat hit-and-miss fashion, lenses of different strengths being tried until the sharpest image is obtained.

The myopic eye requires a lens that will decrease the refractive power of the eye and so permit the image to be focused on the retina instead of in front of it. A *concave lens* will do this.

The presbyopic eye, which cannot accommodate to near vision, requires a convex lens, much like the hyperopic eye. If the individual also has difficulty in adjusting to far vision,

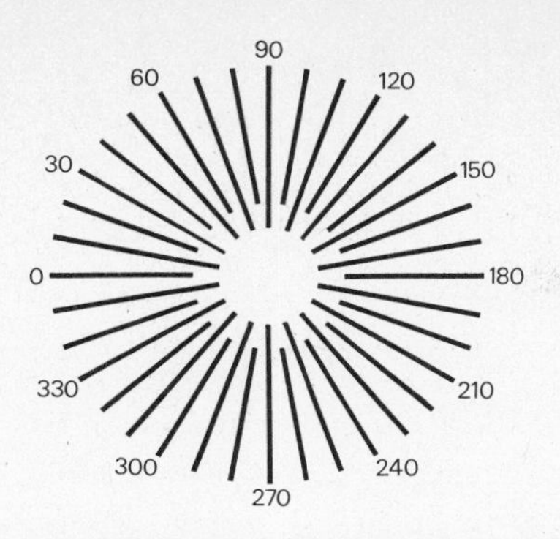

Figure 22-27. **Astigmatism chart. The normal eye sees all the lines clearly. In astigmatism, some of the meridians are blurred. (From *Human Physiology* by B. Houssay. Copyright 1955, McGraw-Hill Book Company, New York. Used with permission of McGraw-Hill Book Company.)**

another pair of spectacles with concave lenses is needed to permit proper vision at a distance. It is a considerable nuisance to be constantly changing spectacles to look at one's watch and to rush off down the street. *Bifocal spectacles,* in which an upper concave and a lower convex lens have been ingeniously combined in one piece of glass, are very helpful for such people. It does require a little time to become adjusted to them and learn to look through the right segment for distant vision or for reading.

Other variations in the optical system of the eye that will affect vision follow.

Spherical Aberration. The biological lens is not a perfectly formed optical lens, and light rays passing through the peripheral portion are not always completely refracted so as to come to a perfect focus with all the other rays passing through the lens. This causes a blurring of the edges of the image.

Chromatic Aberration. If the lens refracts the different wavelengths of light differently, the white light entering will be split up into the various wavelengths or colors as it emerges from the lens. This means that the far point for red will be farther, for example, than the far point for blue.

Astigmatism. If the curvature of the lens varies from one plane to another, light rays passing through the different planes will be unequally refracted and cannot come to a common focal point. This means that the image will be blurred in one plane (Fig. 22-27). Special lenses must be worn to compensate for this astigmatism.

Cataracts. Cataracts result when the lens becomes opaque through calcification and may be associated with age or certain diseases. The cataracts form at the beginning in the center of the lens, so that blurred images are obtained from light rays passing through the periphery of the lens. As the vision becomes worse, it is necessary to remove the lens and replace it with a strong convex glass lens in front of the eye.

Contact lenses are the only correction possible when the cornea is abnormal, so that the image cannot be focused on the retina. These fine lenses, which are held to the eyeball by fluid adhesion, can be ground to fit the surface of the cornea and correct the refractive faults.

Pupillary Reflexes

Constriction and Dilation of the Pupil

The size of the pupil of the eye can vary from 1.5 mm in bright light to almost 10 mm in complete darkness. This response of the pupil to light is, of course, really the response of the muscular iris that surrounds it. There is a *circular band of smooth muscle* surrounding the pupil, which on contraction *constricts* the pupil. This circular muscle is innervated by cranial fibers of the *parasympathetic nervous system,* originating in a nucleus (the Edinger-Westphal nucleus) in the upper brain stem.

The other layer of smooth muscle of the iris is arranged *radially,* originating from the ciliary muscle and converging toward the margin of the pupil. When these muscles contract, they pull on the margin and dilate the pupil (Fig. 22-25). The muscles are innervated by fibers of the sympathetic nervous system, which also innervate the *upper eyelid.* During the day, when the sympathetic nervous system is active, the eyelid is partly elevated, keeping the eyes open. At night, when fatigue depresses the sympathetic nervous system, the eyelid droops to close the eye. (The muscles of the eyelid are also under voluntary control.)

The stimulus for constriction of the pupils is bright light, which acts through the retinal cells, the optic nerves, and optic tracts to an intermediary nucleus in the brain stem and finally to the Edinger-Westphal nucleus, which is also in the brain stem. Sudden exposure to bright light causes the immediate constriction of the pupil, protecting the retina from too much light (Fig. 22-25). Because the pupil can constrict to varying degrees, it regulates the amount of light entering the eye and so maintains good contrast of dark and light areas in vision.

Accommodation reflex to near vision is really a reflex associated with the change in the power of the lens in accommodation for near vision. In order for the lens to bulge anteriorly, the ciliary muscle contracts. The contraction is regulated by parasympathetic fibers which also originate in the Edinger-Westphal nucleus. The electrical excitement overflows to stimulate the parasympathetic fibers that cause constriction of the pupil.

The stimulus for dilation of the pupil is emotion, an excitement arising from fear, pain, or a sudden, sharp sound. Stimulation of the skin at the back of the neck also causes dilation of the pupil. Dilation of the pupil is mediated

Table 22–1. **The Reflex Pathways for Constriction and Dilation of the Pupil**

	Constriction	
Stimulus	Bright light impinging on the retina	
Receptors	Retinal nerve cells	
Afferent neurons	Fibers in the optic nerves (second cranial) and optic tracts	
Synapses	Intermediary nucleus and the Edinger-Westphal nucleus in the brain stem	
Efferent neuron	Parasympathetic fibers of the oculomotor (third cranial) nerve	
Effector	Circular muscles of the iris	
Response	Constriction of the pupil in proportion to the amount of light	
Effect	Reduction of amount of light entering eye	
	Dilation	
Stimulus	Fear, pain, excitement, dim light	Pinching the skin at the back of the neck
Receptor	Relayed through cerebral cortex to lower centers?	Ending of sensory nerves in skin
Afferent neuron	?	Sensory nerve to spinal cord at cervical level
Synapse	Hypothalamus	Sympathetic nuclei
Efferent neuron	Sympathetic nerves	Sympathetic nerves
Effector	Radial muscles of iris	Radial muscles of iris
Response	Dilation of pupil	Dilation of pupil
Effect	Increased amount of light entering eye	Increased amount of light entering eye

through the sympathetic nerves but involves the inhibition of the dominant parasympathetics as well.

Reflex pathways for pupillary constriction or dilation are shown in Table 22–1.

Projection of the Image to the Brain

The image is projected upside down on the visual cortex, and we use many cues to interpret it right side up (Fig. 22-28). A remarkable series of experiments has shown that when an inversion lens is worn, so that the image is actually projected the right side up, the brain continues for some time to invert the image. This results in a grotesque clash of the information coming from the eyes, which shows the room and its contents to be upside down, while the information from the skin, muscles, and joints shows the body to be standing upright. After about 10 days of the lens being worn, the brain learns to reinterpret the visual cues and adapts to the new reality.

We therefore *interpret sensation* on the basis of our *experience* and can quite easily be misled (see some examples of optical illusions in Figs. 22-29A and B). Our eyes do not deceive us, but our brain does.

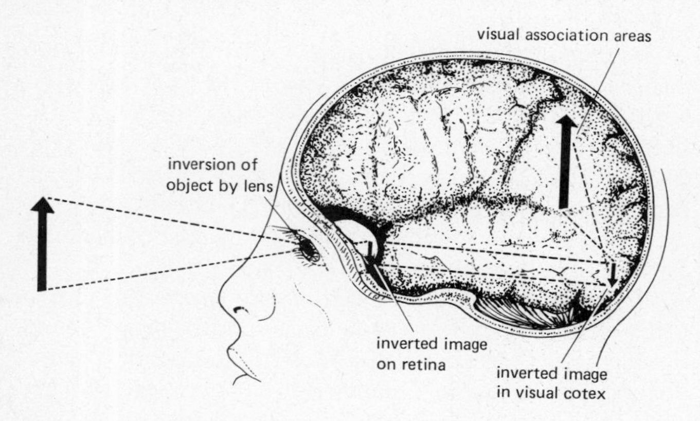

Figure 22-28. **Projection of the image on the visual cortex (upside down) and the reinterpretation of the image by other areas of the brain.**

Binocular Vision

In humans, most primates, cats, and predatory birds, the two eyes are placed in the front of the head, and the nerve fibers from each eye pass after partial decussation to both the right and the left visual cortex (Figs. 9-10 and 22-30). This placement requires a very delicate harmony between the motor mechanisms of the eyes to permit them to act in unison and produce a single visual image in the brain, an accomplishment that has become so perfected that we are usually quite unaware of the two images being formed, although we know that the visual field of one eye is slightly different from that of the other (Fig. 22-30). In many animals, the eyes are opposite each other on the two sides of the head, and there is little, if any, common field of vision. This has its advantages: a squirrel, for example, can see an object coming toward it from behind without turning its head. The disadvantage of such monocular vision is that there is little perception of depth or distance.

Figure 22-29A (right). **Waterfall, an imaginative version of the impossible triangle. Falling water keeps a millwheel in motion and subsequently flows along a sloping channel between two towers, zigzagging down to the point where the waterfall begins again. If one follows the various parts of the construction, no mistake can be detected. Yet it is an impossible whole because changes suddenly occur in the interpretation of distance between our eye and the object. The impossible triangle is fitted three times into the picture. (M. C. Escher—"Waterfall 1961," litho 378 × 300. Escher Foundation, Haags Gemeentemuseum, The Hague, Holland.)**

Figure 22-29B (below). **Optical illusions. (1) The different segments of the vertical lines are all of equal length. (2) Both figures are perfect squares of the same size but appear to be rectangles. (3) The distance from A to B is equal to that from B to C. (4) The white and black disks have the same diameter, but the white one appears larger. (5) The long horizontal lines are straight parallel lines. (Redrawn from *Human Physiology* by B. Houssay. Copyright 1955, McGraw-Hill Book Company. Used with permission of McGraw-Hill Book Company.)**

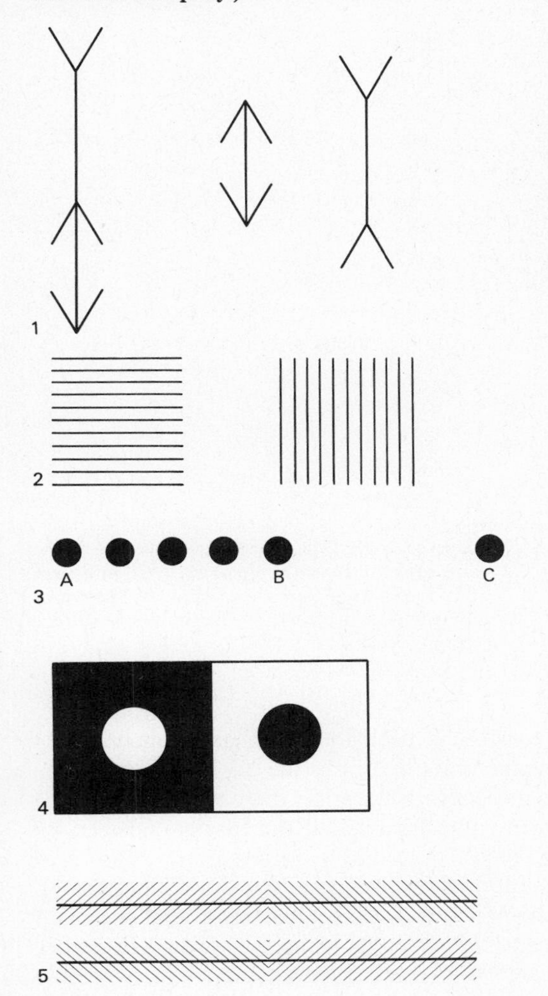

Figure 22-30. **Binocular vision. The slightly different images received by the left and right retinae are merged into a single, three-dimensional image in the brain.**

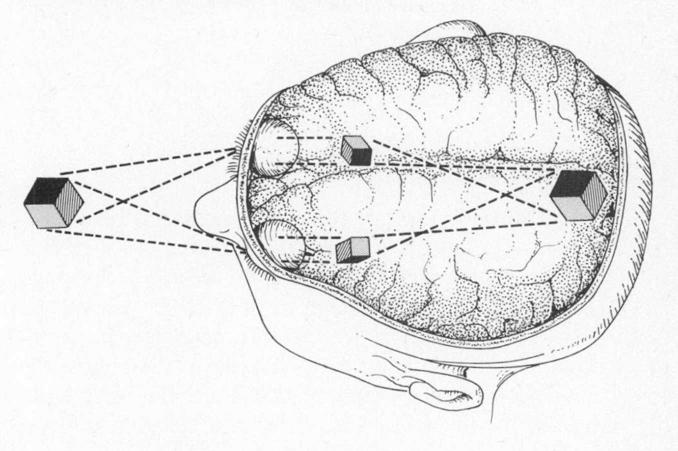

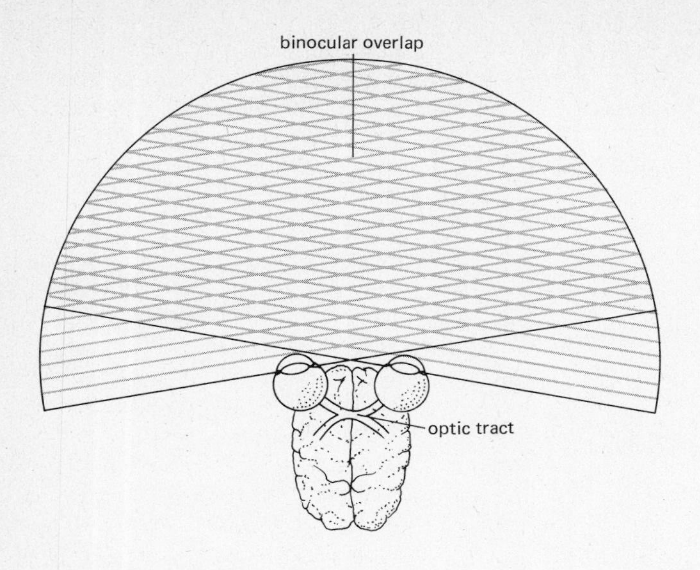

Figure 22-31. **Binocular overlap in humans is almost complete; the number of crossed and uncrossed fibers from the retina to the optic cortex is almost equal.**

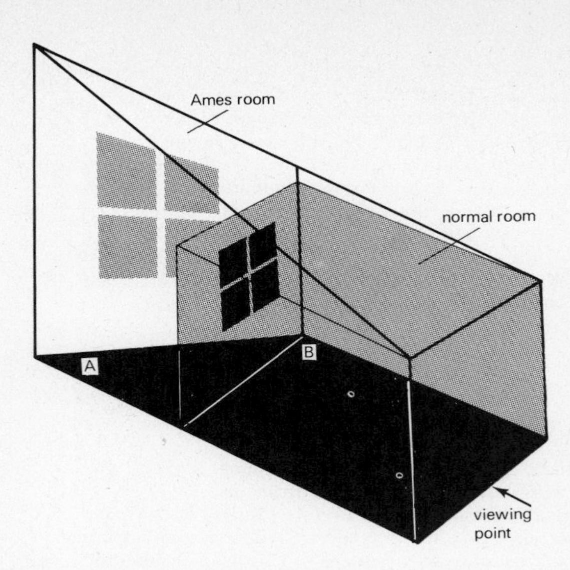

Figure 22-32. **The geometry of the Ames distorted room. The further wall recedes from the observer to the left. The walls and windows are arranged to give the same retinal image as a normal rectangular room, so that objects at A and at B appear to be at the same distance but of different size, although it is the distances that are different.**

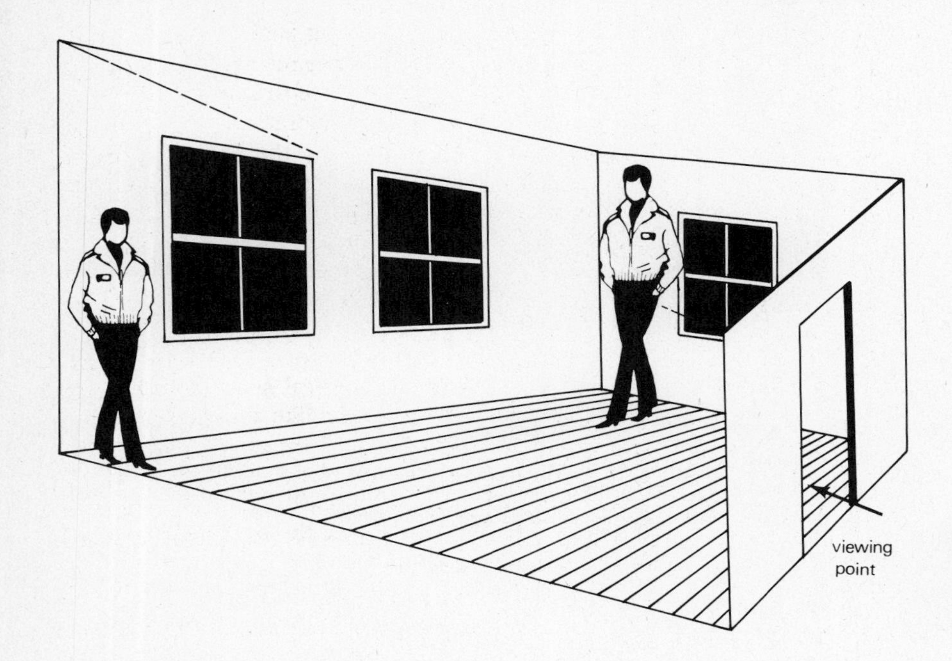

Figure 22-33. **The two figures are the same size and the room is abnormal. We interpret this in terms of our experience and perceive the room as normal and the figures as different in size.**

Depth and distance perception

The great number (50 per cent) of uncrossed fibers from the retina to the visual cortex gives humans an almost complete binocular overlap of the two visual fields (Fig. 22-31). In the rabbit, by contrast, most fibers cross over, so that there is little binocular overlap and practically no depth perception. Depth perception depends upon the slight difference in information about an object sent to the brain from the right and the left eyes. The difference between the two views is called *binocular parallax.*

Binocular parallax is used for depth and distance estimates, because the brain interprets the *disparity between the two images* as a specific distance on the basis of experience and comparison with other cues. If the cues are falsified, as in the Ames Distorted Room, a normal adult appears to shrink to the size of a small child (Figs. 22-32 and 22-33). We are so accustomed to a rectangular room that the brain prefers to accept the person as being the wrong size rather than the unlikely (but true) fact that the room is a most peculiar shape.

22–6 The Ear: Hearing and Equilibrium

A discussion of the ear may be conveniently divided into three parts, describing the *external,* the *middle,* and the *inner* ear. The *air-filled* external ear and middle ear act as conducting and pressure-amplifying channels for sound waves before the waves reach the inner ear.

The inner ear possesses two quite different sensory systems: the *cochlea,* which contains the *receptors for sound,* and the *vestibular apparatus,* which is associated with the *sense of balance.* It is in the *fluid-filled* inner ear that energy transduction by the mechanoreceptors takes place. Despite the complete functional separation of the auditory and vestibular regions of the inner ear, these two sensory systems are anatomically connected, and both are filled with the same fluid (Fig. 22-34).

Hearing

General structure and function of the ear

The highly efficient acoustical ear possessed by humans and other higher vertebrates has developed from the simple fluid-filled sac that was chiefly involved with the determination of body position. It responded, as do our proprioceptors in the inner ear, to sudden movements and slow vibrations. It gradually evolved into an *aquatic ear,* responding to vibrations in the water. Sound vibrations do not travel well from one medium to another; hence a fisherman is at liberty to shout and whistle all he pleases, although it usually pleases him to be quiet.

The evolution of the *aerial ear* required some mechanism for *transforming* and *amplifying* the sound waves from the air as they enter the fluid of the cochlea, because the transmission loss would otherwise be so great that hearing would be extremely poor. The amplification mechanism of the cochlea is extraordinarily efficient, as shall be described later.

External Ear. The external ear consists of the conspicuous flap, the *pinna,* which collects sound waves from a wide area, and like a hearing horn funnels them into the external ear passage, the *external auditory meatus.*

The pinna gets its shape and irregularities from a supporting sheet of cartilage, covered with thin skin. Only the lobe of the ear has no cartilage, a fact taken advantage of by those who pierce their ears. In humans, the pinna is essentially vestigial: we have lost control of the muscles that permit lower animals to cock their ears in the direction of the sound. Nevertheless, even in humans, the pinna probably plays an important role in sound localization. If the pinna is distorted by bending or strapping, localization of sound is lost.

The entrance of the external ear is the opening of the external auditory meatus, the channel that curves up toward the middle ear. Many *glands* within this portion of the ear secrete the wax that tends to accumulate in the ear. Sound waves are usually conducted undistorted through the external meatus to the ear drum or *tympanic membrane,* the outer boundary of the middle ear.

Middle Ear. The middle ear lies in a cavity in the outer part of the *temporal bone* (the petrosal portion). It is filled with air conveyed to it by the *eustachian tube,* which connects the middle ear to the *pharynx* (Fig. 22-34). The eustachian tube is usually closed but can be opened by yawning, chewing, and swallowing. Its function is to equalize the pressure on the two sides of the eardrum. Any imbalance of pressure reduces the conduction of sound through the middle ear, a familiar phenomenon encountered when the pressure in a plane abruptly increases on landing. The indignant yelling of small children is an effective way of opening the eustachian tube and equalizing the pressure. This pressure imbalance may be aggravated by respiratory infections which cause inflammation of the eustachian tube and prevent efficient restoration of pressure balance. The eardrum may be painfully deformed and even ruptured by the built-up pressure.

Conducting mechanism of the middle ear. Sound is conducted in the middle ear by the mechanical distortion of the *tympanic membrane,* which separates the outer ear from the middle ear, and by the vibrations of a chain of three tiny intricately formed bones, the *auditory ossicles.* These bones convey the vibrations from the tympanic membrane to the membrane that separates the middle and inner ears, the *oval window.*

The tympanic membrane is stretched across the meatus and held in place in a little groove by a bony ring. The first ossicle, the *malleus* (hammer), is attached to the tympanic membrane in such a way that it pulls the center of the membrane inward, giving it a conical surface. This form adds *rigidity* to the membrane and, like a loudspeaker diaphragm, adds to the *effective area,* because it permits the central portion to work as a unit and amplify the sound energy received over its entire surface. This becomes concentrated in the central area where the malleus is attached and so is transmitted to the ossicles.

The malleus articulates with the second ossicle, the *incus* (anvil), and except at very high frequencies, the malleus and the incus move as one. The long handle of the incus connects to the *stapes* (stirrup). The foot of the stapes is firmly implanted in the oval window, and it is held there by a ligament. *Ligaments* also hold the ossicles together and suspend them within the middle ear.

Sound vibrations travel well through bone; although there is very little mechanical advantage of the lever system of the ossicles, the *total force is concentrated in a very much smaller area* at the oval window than it was at the tympanic membrane. The pressure on the fluid within the cochlea is thus considerably amplified. The vibration pressure then sets the fluid in the cochlea of the inner ear in motion.

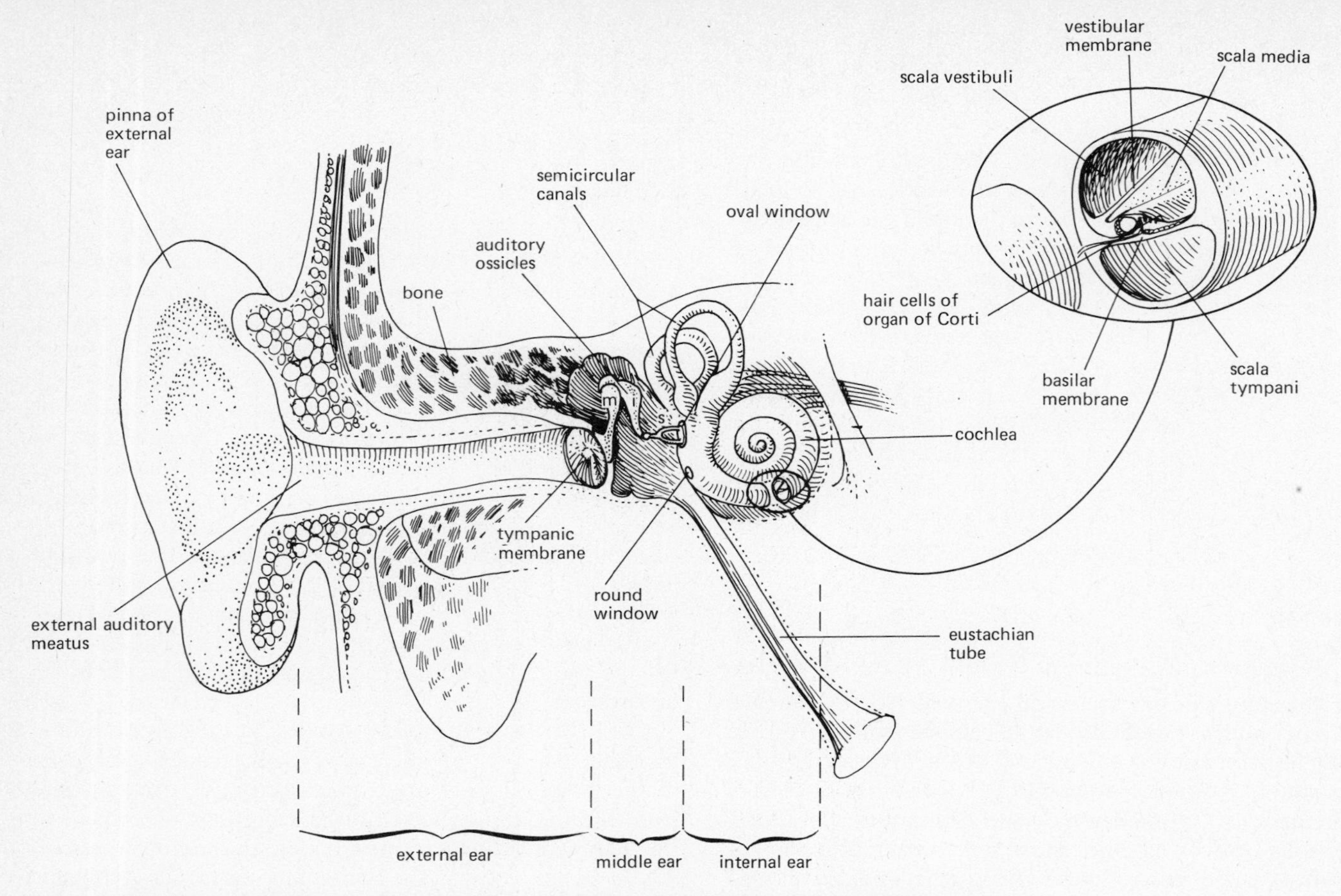

Figure 22-34. **The passage of sound waves through the ear involves the transformation and amplification of the waves as they pass from the air in the external meatus, to the auditory ossicles of the middle ear, and finally to the fluid-filled inner ear. m = malleus, i = incus, and s = stapes. A section of the cochlea has been removed from the diagram and shown enlarged in the upper right.**

TYMPANIC REFLEX. There are two small muscles in the inner ear, the *tympanic muscles,* attached respectively to the malleus and stapes. These appear to have a *protective* function, *damping* the sound rather than amplifying it. They contract reflexly in response to sounds, but they cannot give much protection against sudden sounds like explosions, which are effective before the reflex can occur.

Inner Ear: The Cochlea and the Receptors for Sound. The *cochlea,* so named because of its snail-like spiral form (*cochlea* is the Latin word for "snail"), is a membranous tube coiled within a bony labyrinth hollowed out of the densest part of the temporal bone (Fig. 22-35). The membranous tube is separated from the bone by a fluid called *perilymph,* which also fills two of the three tubes into which the spiral cochlea is divided. This division of the membranous cochlea into *three longitudinally spiralling tubes* is accomplished by two membranes, *Reissner's membrane* and the *basilar membrane.*

The middle tube, the *scala media,* is the portion containing the auditory receptor cells that form part of the *organ of Corti.* The scala media is separated from the upper *scala vestibuli* by the very thin Reissner's membrane, and from the lower *scala tympani* by the basilar membrane and its attachments to the bony labyrinth on each side.

At the basilar end of the cochlea, the vestibular tube communicates with a chamber, the *vestibule,* by an *oval window;* the tympanic tube is closed where it connects to the vestibule by a thin membrane, the *round window.* Sound waves are transmitted from the external and middle ears to the vestibule through air and bone and then to the fluid of the cochlear tubes. The detailed structure of the cochlea, including the position of the organ of Corti, is shown in Fig. 22-35. Figure 22-36 is a photograph of a human dissected cochlea.

PERILYMPH AND ENDOLYMPH. Both the *scala vestibuli* and the *scala tympani* are filled with *perilymph;* the two tubes are connected to one another through a small hole, the

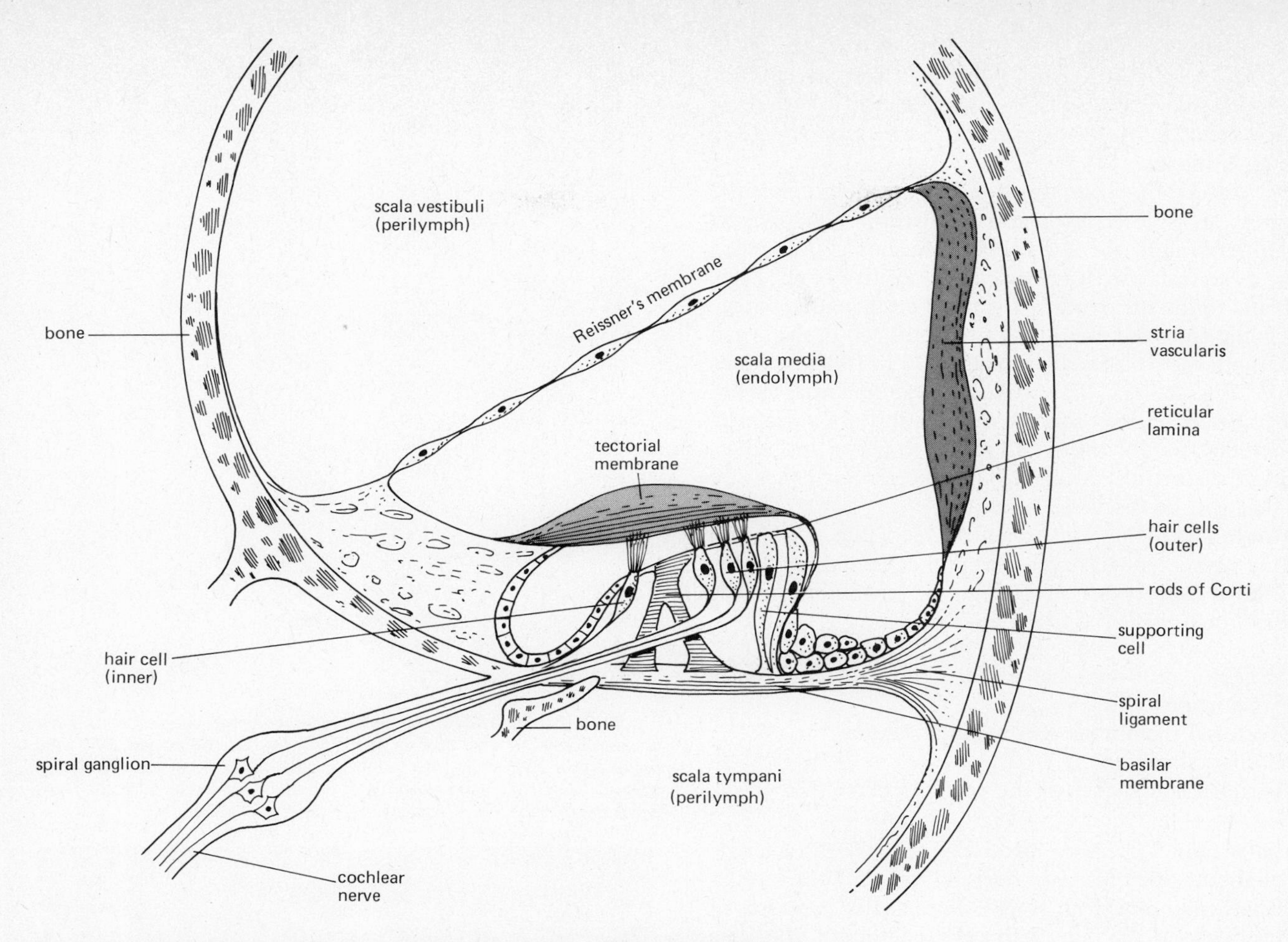

Figure 22-35. **A diagram of the structure of the cochlea and the organ of Corti.**

Figure 22-36. **Photograph of a dissected human cochlea showing its spiral structure, the oval window (ow), and the round window (rw). [From J. G. Johnsson and J. E. Hawkins, *Arch. Otolaryng.* 85:599 (1967).]**

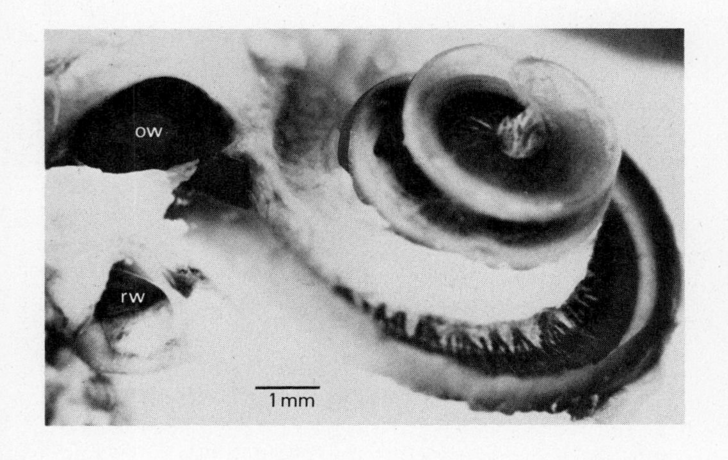

Table 22–2. **Composition of Spinal Fluid, Perilymph, and Endolymph**

		Spinal Fluid	Perilymph	Endolymph
Potassium	mEq/1	4.2	4.8	144.1
Sodium	mEq/1	152.0	150.3	15.8
Chloride	mEq/1	122.4	121.5	107.1
Protein	mg/100 ml	21.0	50.0	15.0

Source: Adapted from C. A. Smith, O. H. Lowry, and M. L. Wu, *Laryngoscope* **64:** 141 (1954).

helicotrema, at the apical end of the cochlea. The *scala media* is filled with *endolymph.* The composition of these two fluids differs from one another, and from that of spinal fluid, as is shown in Table 22-2.

The cells in the organ of Corti are not supplied with blood vessels; the sensitivity of the receptor cells to movement would probably result in "hearing" the blood flow, which would interfere intolerably with their sensitivity to sound vibrations. These cells receive nourishment through the endolymph, which is produced by a highly vascularized strip of tissue, the *stria vascularis,* which runs along one side of the scala media.

An extremely important function of the endolymph is the production of an electrical potential between the endolymph and the perilymph, the *endocochlear potential.* The endolymph in the scala media is electrically positive to the perilymph and to the surrounding tissues. The change in potential in going from the interior of a receptor cell in the organ of Corti into the scala media is from −70 to +80, or about 150 mV (Fig. 22-37). This endocochlear potential is believed to be the energy source responsible for the tremendous amplification of the sound signal to permit transduction into the electrical energy of the *cochlear microphonic,* a topic that will be discussed later.

ORGAN OF CORTI. The organ of Corti (Figs. 22-35 and 22-37) is a complex sensory structure that extends along the three and a half twists of the basilar membrane in the cochlea. The organ of Corti consists of:

1. Receptor cells—the hair cells.
2. Supporting cells.
3. Tectorial membrane and reticular lamina.
4. Basilar membrane.
5. Peripheral branches of the cochlear nerve.

Receptor cells. There are two types of hair cells, a single layer of about 3500 *inner hair cells* and three to four rows of 20,000 *outer hair cells* (Fig. 22-38). The hairlike processes of the apical ends of the outer hair cells extend into the scala media and are firmly imbedded in the rooflike tectorial membrane. The hairs of the inner cells, however, do not contact the tectorial membrane but are relatively free at that end. This anatomical difference is believed to account for a difference in function between these two types of sensory cells. Recent evidence indicates that the outer hair cells are primarily concerned with the *displacement* of the basilar membrane by the sound waves, whereas the inner hair cells respond to the *velocity of movement* of the basilar membrane.

The hair cells, like the cilia from which they are derived, have a *basal body* just under the hairs. This basal body probably plays an important role in the transduction of the mechanical signal to a neural signal, but the mechanisms involved are by no means clear. According to some studies, vesicles containing neurotransmitter are concentrated at the base of the cell.

Supporting cells. The lower ends of the hair cells are cupped by the invaginated tips of the supporting cells, which rest firmly on the basilar membrane. The several types of supporting cells probably lend nutritive as well as physical support to the hair cells, but they are not involved in sound transduction.

Tectorial membrane. The tectorial membrane is a flap of fibrous and gelatinous tissue that grows out from tissue attached to the wall of the scala media. The outer edge of the tectorial membrane is a strong rigid plate, the *reticular*

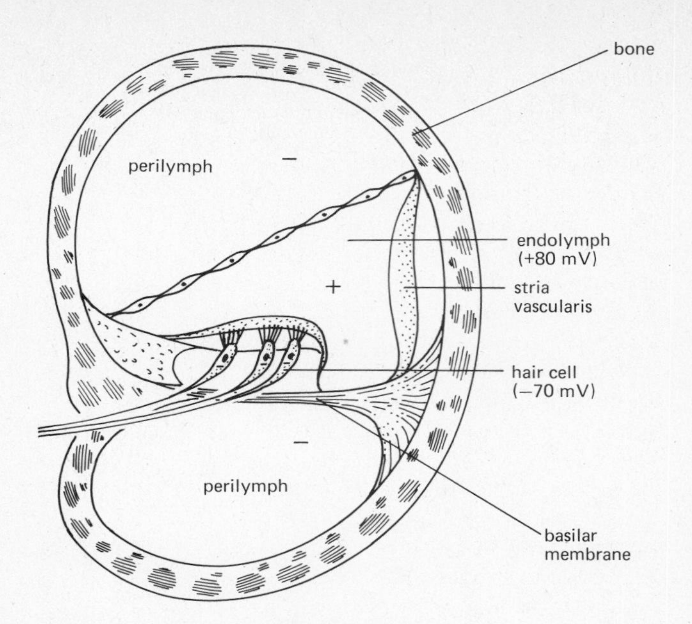

Figure 22-37. **The endocochlear potential.**

Figure 22-38. **The scanning electronmicrograph shows the uniform arrangement of the supporting cells (s) in the organ of Corti. ihc = inner hair cells, ohc = outer hair cells. (From R. G. Kessel and C. Y. Shih, in *Scanning Electron Microscopy in Biology,* Springer-Verlag New York, Inc., New York, 1974.)**

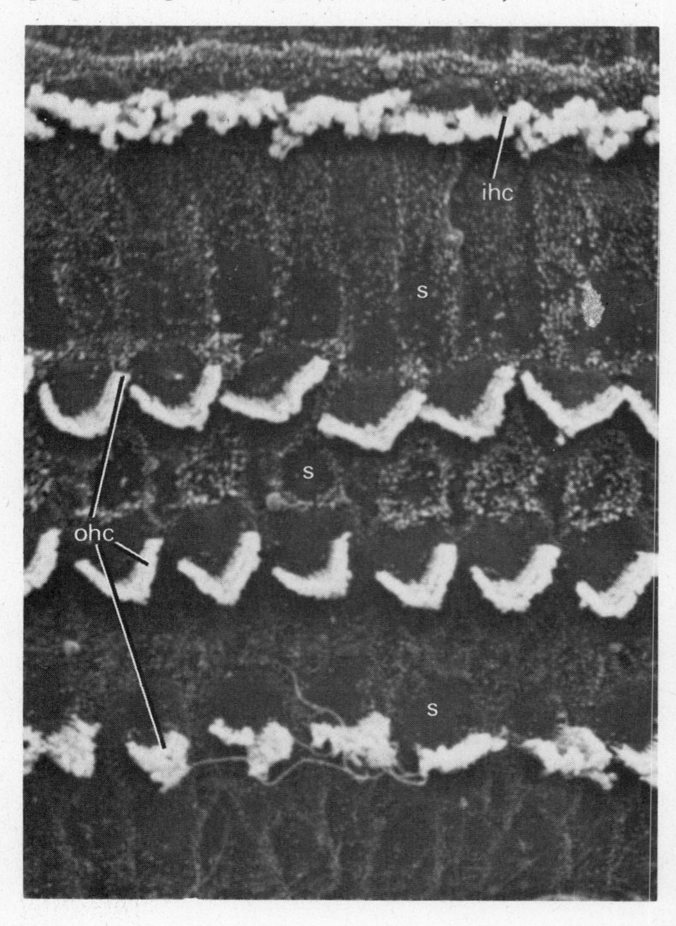

lamina, which is supported by two firm pillars, the *rods of Corti,* that are anchored to the basilar membrane. Some of the supporting cells also cushion this reticular lamina.

Basilar membrane. The basilar membrane is of limited elasticity; it widens gradually from 0.04 mm at the stapes to 0.5 mm at the helicotrema, with a variation in stiffness of about 100:1 from one end to the other (Fig. 22-39). The width of the *apical* end permits the basilar membrane in this region to move only in response to *low frequencies,* whereas the *narrow basal end* moves in response to *all frequencies* within the audible range.

Peripheral branches of the cochlear nerve. Like the rods of the retina, the hair cells of the organ of Corti have no axons but synapse immediately with endings of the cochlear nerve. These nerve terminals lie between the hair cells and the supporting cups (Fig. 22-35).

The neurons of the cochlear nerve, like those of the sensory nerves in the spinal cord, are bipolar cells with their cell bodies in ganglia. The bodies of these cochlear neurons are in the *spiral ganglion,* deep in the bone of the skull near the cochlea.

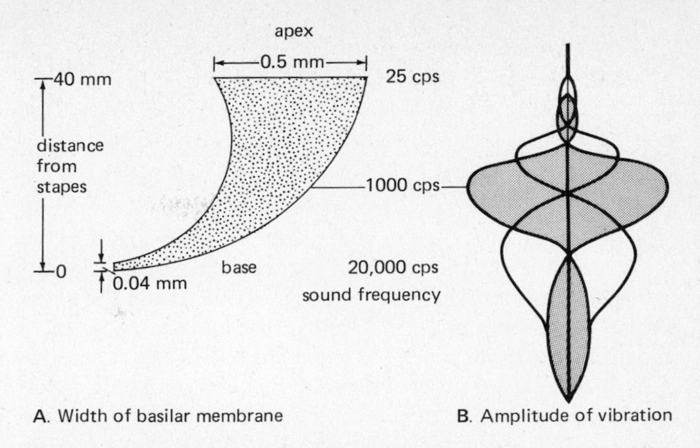

Figure 22-39. **The wide apical end of the basilar membrane responds only to low frequencies of sound. (A) As the membrane narrows toward the base at the stapes, the sensitivity to high frequencies increases. (B) The amplitude pattern of vibration of the basilar membrane for a sound frequency of 1000 cps.**

Passage of sound waves through the cochlea

The guinea pig cochlea has been extensively studied, as it lies comparatively superficially in a bulge of the temporal bone. This makes it possible to examine it under the microscope in the living state, and also to insert fine microelectrodes into the various scalae, and even into the hair cells. Von Békésy, who received the Nobel Prize in 1961 for his work on the ear, has described in detail the movements of the basilar membrane when sound is introduced into the inner ear (1960).

Sound waves that have travelled through the middle ear reach the inner ear at the oval window, pushing it inward. This puts pressure on the confined fluid within the vestibular canal (scala vestibuli), and the pressure is transmitted to the fluid in the tympanic canal (scala tympani), with which it is in contact (Fig. 22-34). This causes the round window to bulge outward.

As a result of these *fluid-conducted vibrations,* the basilar membrane bulges upward or downward, with the extent of the movement varying with the width of the membrane, and the frequency of the vibrations as described and as depicted in Fig. 22-39. The sound range in humans is from 20 to 20,000 cps, with the greatest sensitivity around 1000 cps. This range varies in different animals—dogs can hear the "silent whistle" because they are sensitive to higher frequencies than are human beings.

Figure 22-40. **Bending of the hair cells of the organ of Corti in one direction depolarizes the membrane, resulting in the receptor potential. Bending in the other direction hyperpolarizes the membrane.**

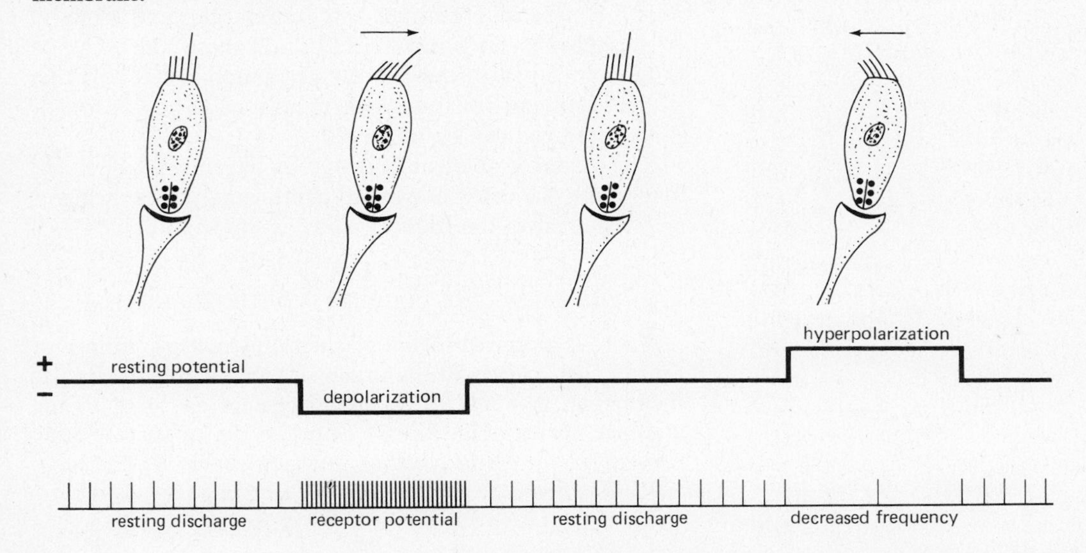

Transduction of mechanical energy by the hair cells

Bending of the Hair Cells. The *hairs* of the outer hair cells are imbedded in the reticular lamina and the tectorial membrane (Fig. 22-35). The *lower ends* of the hair cells are firmly attached to the basilar membrane by the supporting cells. Consequently, when the sound vibrations bulge the basilar membrane, the rods of Corti rock and pivot the stiff tectorial membrane. The shearing action caused by the basilar membrane and the tectorial membrane moving in different directions bends the imbedded hairs of the sensory cells. It is believed that bending of the hairs in one direction causes *depolarization* of the cell membrane and the buildup of a receptor potential, whereas bending in the opposite direction *hyperpolarizes* the receptor cell (Fig. 22-40).

Cochlear microphonics and the receptor potential

It has been known since 1930 that there is an electrical counterpart of the mechanical movement of the basilar membrane. This electrical potential is the *cochlear microphonic,* an ac potential that reflects the pattern and the amplitude of the mechanical movement (Fig. 22-41). The cochlear microphonic is symmetrical; it has no threshold, no all-or-none characteristics, and no refractory period. It appears across the basilar membrane and is associated with the upper end of the hair cells.

One of the outstanding investigators in this field, Davis (1965), has presented convincing evidence that the *cochlear microphonic* is the *receptor potential* and is generated by the change in resistance across the tops of the hair cells. This receptor potential then is directly responsible for the stimulation of the nerve endings of the fibers of the cochlear nerve. Thus the cochlear microphonic causes the firing of these nerves.

Figure 22-41. **Cochlear microphonics recorded at the base and at the apex of the cochlea in response to different sound frequencies. Human sensitivity to sound is greatest around 1000 cps. There is decreased sensitivity to high frequencies near the apex.**

Frequency (cps) | Sound | Base | Apex

500

1,000

2,000

5,000

10,000

Table 22–3. **Sequence of Mechanical and Electrical Events in Hearing**

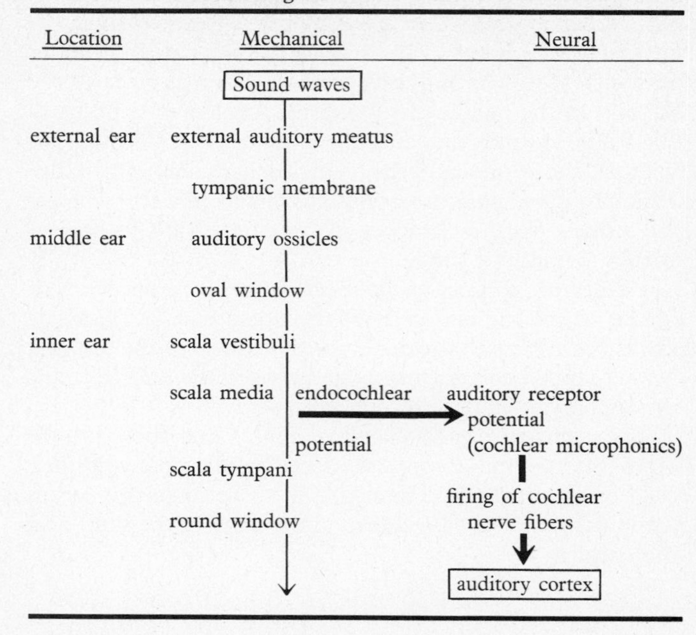

Location	Mechanical		Neural
	Sound waves		
external ear	external auditory meatus		
	tympanic membrane		
middle ear	auditory ossicles		
	oval window		
inner ear	scala vestibuli		
	scala media	endocochlear potential →	auditory receptor potential (cochlear microphonics)
	scala tympani		firing of cochlear nerve fibers
	round window		auditory cortex

Amplifier role of the endocochlear potential

The calculated threshold movement for hearing at the basilar membrane is less than one tenth of the diameter of a hydrogen atom. The energy needed to amplify this ultrasmall signal, according to Davis, comes from the very large endocochlear potential of about 150 mV. This extraordinarily high potential is found in the human as well as in the guinea pig cochlea.

Although it would seem logical to attribute this potential to differences in potassium and sodium concentrations in perilymph and endolymph (Table 22–2), these concentration differences are also found in the perilymph and endolymph of the vestibular apparatus, where no such large transmembrane potential can be measured.

The sequence of events initiated by the entrance of sound waves into the external ear and culminating in the firing of selected fibers of the auditory nerve is shown in Table 22–3.

Pathway from the cochlea to the brain

There is a precise spatial orientation from the endings of the cochlear nerves in the organ of Corti along the basilar membrane, through all the relay centers en route to the auditory cortex (Fig. 22-42). Some of the important relay centers are the *spiral ganglion,* the *cochlear nuclei,* the *superior olivary nuclei,* and the *inferior colliculi.*

Peripheral fibers of the cochlear nerve enter the *spiral ganglion,* in which their cell bodies are located. These bipolar neurons send fibers centrally to the *cochlear nuclei* in the medulla. All the cochlear nerve fibers synapse in these nuclei. Most of the neurons originating in the cochlear nuclei send their axons to the *superior olivary nucleus* in the same side of the medulla: some fibers, however, pass to the superior olivary nucleus on the other side of the medulla. This means that these olivary centers receive signals from *both* ears.

Most fibers from the cells in the olivary nuclei pass to the inferior colliculi in the midbrain, where they synapse again with neurons leading to the auditory cortex, deep in the lateral fissures of the cerebral hemispheres.

Along their course, many of the nerves give off *collaterals,* especially to the cerebellum and the reticular formation, both of which are immediately activated by a sudden, loud noise.

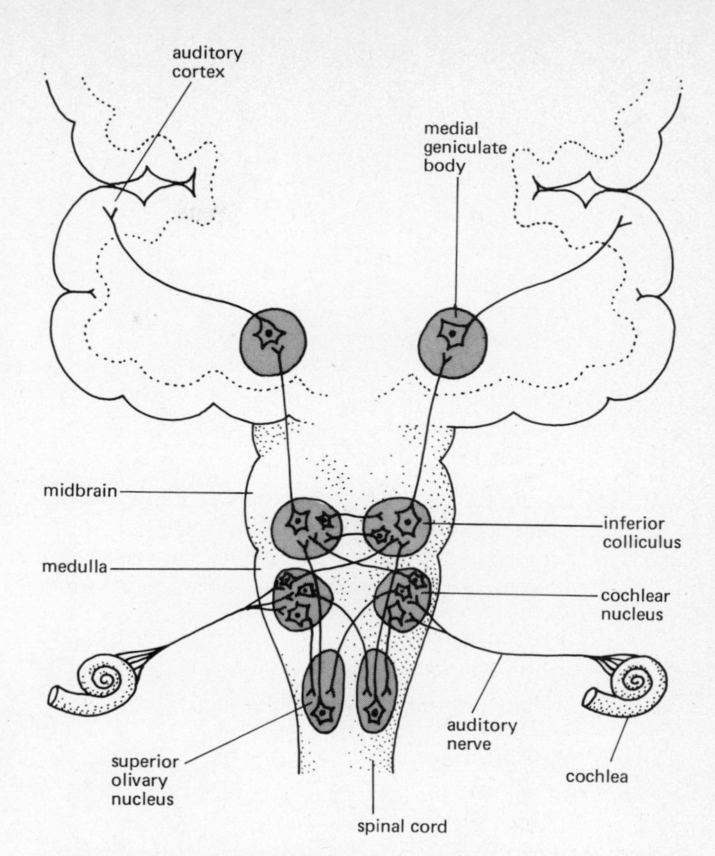

Figure 22-42. **The ascending pathway of the auditory system. The auditory nerves synapse in the ipsilateral cochlear nuclei. Neurons in the cochlear nuclei send axons to both olivary nuclei and to the contralateral inferior colliculus. From the inferior colliculus, fibers pass via the medial geniculate body to the primary auditory cortex.**

Central and peripheral discrimination of sound

Qualities of Sound. PITCH. The most important quality of sound is its pitch. High pitches are associated with high-frequency stimulation and low pitches with low-frequency stimulation. This simple description omits many other dimensions, such as the perception of a mixture of sounds. Most sounds to which we are normally exposed are a complex mixture of different pitches that the ear is able to separate. We can identify a particular voice or instrument in a concert, and both the pitch and the overtones, which give the pitch its *timbre,* produced by the larynx or instrument, are involved in our recognition.

LOUDNESS. Loudness is the intensity of the sound. The ear is able to discriminate between sounds varying in intensity from a soft whisper to an extremely loud explosion. A change in sound intensity is expressed in *decibels.* A tenfold increase in sound energy is one *bel;* one tenth of a bel is a *decibel.*

The threshold for hearing different frequencies of sound will vary with the intensity of the sound. The complete range of frequencies from 30 to 20,000 cpm can be heard by the human ear only with intense sounds.

Peripheral Discrimination of Sound. PITCH. The most commonly accepted theory of pitch discrimination is the *place theory,* proposed by von Békésy (1960). This theory is based upon his direct observations, under the microscope, of the movement of the basilar membrane as a function of the frequency or amplitude of sound.

Von Békésy observed that the sound initiated a *pattern of traveling waves* that moved from the oval window toward the helicotrema at the apex of the cochlea. *High frequencies* caused the greatest movement at the base near the stapes, whereas *low frequencies* caused greater movement near the apex. Depending upon the frequency of the travelling wave, it would show a maximum amplitude at one point along the basilar membrane, with lesser amplitudes before and after this region (Fig. 22-39). This rather *broad spread of the wave,* over 20 to 40 cpm, means that the ability to discriminate differences as small as 2 or 3 cpm must be caused by some other fine sharpening mechanism. It is possible that this sharpening is the responsibility of *central inhibitory mechanisms,* similar to those discussed for vision.

This concept of localization of sound frequencies in the cochlea is supported by experiments on rhesus monkeys in which selective regions of the basilar membrane were destroyed by antibiotics. Kanamycin, streptomycin, and neomycin are antibiotics that specifically destroy the hair cells of the cochlea, beginning at the basilar end. By varying the duration of the drug administration, it was possible to correlate the destruction of regions of the basilar membrane with loss of hearing. High-frequency loss occurred first and was followed by a progressive loss of the middle and low frequencies, as the cellular destruction proceeded toward the apex.

LOUDNESS. Discrimination of loudness is achieved by both temporal and spatial summation. As the intensity of the sound increases, the movement of the basilar membrane also

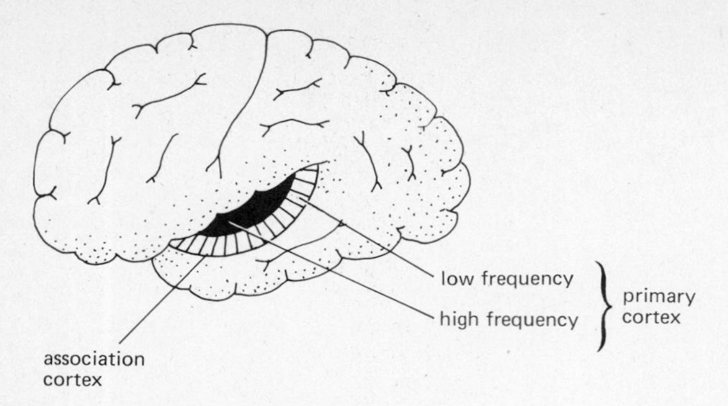

Figure 22-43. **The auditory cortex.**

increases causing the hair cells to fire more rapidly (*temporal summation*). In addition, neighboring hair cells are stimulated (*spatial summation*). It is also possible that certain hair cells, with a very high threshold, will fire only when a sound is very loud.

Constant exposure to intense noise, equivalent to listening to rock music (2000 cpm octave band at 120 decibels), destroys the hair cells of the organ of Corti.

Central Discrimination of Sound. There is a *spatial organization of neurons,* according to their sensitivity to different frequencies, starting from the cochlear nerve as it emerges from the basilar membrane, through the levels of the ascending pathways. This spatial organization is also seen in the auditory cortex, where certain areas of the *primary auditory cortex* respond to high frequencies, whereas other areas respond to low frequencies (Fig. 22-43).

The *auditory association cortex* does not respond directly to sound stimuli. It apparently integrates information from the primary auditory cortex with information from other sensory areas. If the auditory association areas are destroyed, sounds are still heard, but they lose their meaning.

It would appear, then, that most sound discrimination occurs at the level of cochlear receptors, but that fine discrimination is achieved through the various relay sites in the ascending pathway and in the auditory cortex itself.

Figure 22-44. **The sonic shadow cast by the head prevents the far ear from receiving as much sound as the near ear. The far ear also receives the sound later.**

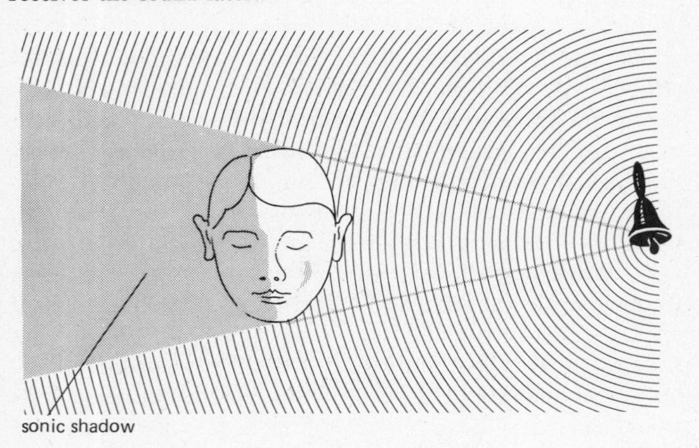

Sound localization

Why the Ears Are Placed on Both Sides of the Head. Binocular vision requires both eyes to be in the front of the head, but binaural cues depend upon the ears being placed on either side of the head. The head casts a *sonic shadow* between the sound source and the farther ear, so that one ear receives less sound than the other, and there is also a time difference (Fig. 22-44). One usually attempts to compensate for this by turning the head in the direction of the sound, which changes the intensity and the timbre of the sound. This is all information that can be received, integrated, and interpreted by the central nervous system, in terms of both distance and direction, to permit sound localization.

The interaural time difference is accentuated in some owls, who depend on consummate accuracy of sound localization to capture their prey in the dark. These birds have their ears asymmetrically placed and far apart on either side of the head, so that the time difference of sound arrival at the two ears is emphasized.

Hearing abnormalities

Because hearing is dependent on the proper *conduction and pressure amplification* of sound waves through the middle ear and on the response of the *receptor cells and the transmission of impulses to the brain,* it is obvious that interference with either of these systems will result in deafness. If the cochlea or the cochlear nerve is completely destroyed, there is no way to alleviate this *nerve deafness.* If the deafness is *conduction deafness* only, it is possible to use other means of conducting sound to the receptor cells.

The type of deafness can be determined through the use of an *audiometer,* which tests the threshold of hearing for pure tones from the very low to the very high and also has an electronic vibrator for testing the conduction of these tones through bone.

Types of Conduction Deafness. If the *external meatus* is blocked, as by a plug of wax which may be large enough to almost completely close the passage and which lies against the tympanic membrane, hearing may be seriously affected.

If the *tympanic membrane* is perforated, as sometimes happens when children stick a sharp object into their ears, or as a result of an infection in the middle ear, there is impairment of hearing. The hearing loss seems to be proportional to the size of the area damaged.

The *auditory ossicles* may become fixed to one another as a result of infection in the middle ear. (This fixation occurs quite infrequently at the present time, because it can be controlled by antibiotics before fibrous tissue has a chance to form.) More serious is the fixation that occurs in *otosclerosis,* a disease of the bone resulting in fixation of the foot of the stapes in the oval window. One of the most outstanding surgical accomplishments has been the development of the *fenestration* (*window*) *operation,* which basically provides a new functional window in the place of the one that has become rigid.

The electronic hearing aid is often of great help to those suffering from conduction deafness. It consists of a sensitive microphone that through the use of a battery or transistor converts sound waves into electrical impulses which are then amplified, converted back to louder sound waves at auditory threshold, and transmitted to the inner ear.

Equilibrium: The Vestibular Apparatus

Within the membranous labyrinth of the internal ear are the cochlea, which is concerned only with hearing, and the *vestibular apparatus,* which is concerned with *equilibrium.* The vestibular apparatus consists of two large chambers, the *utricle* and the *saccule,* and the *three semicircular canals.* Each of these structures is filled with endolymph.

The utricle and saccule are essential for vertical static equilibrium; in addition, the utricle responds to changes in *linear acceleration.* The semicircular canals are stimulated by *rotational* (*angular*) *acceleration* in three planes, horizontal, sagittal, or coronal, corresponding to the position of each of the pairs of semicircular canals in the inner ear.

EQUILIBRIUM RECEPTORS: THE HAIR CELLS OF THE VESTIBULAR APPARATUS

Like the receptors in the organ of Corti, the receptor cells of the vestibular apparatus are *hair cells.* Bending of the hairs in one direction is believed to cause depolarization and the *receptor potential,* whereas bending in the other direction causes hyperpolarization and inhibition. These sensory cells synapse with naked endings of the *vestibular nerve,* which is part of the sixth cranial nerve (the cochlear nerve forms the other part).

PATHWAYS FROM THE VESTIBULAR RECEPTORS TO THE BRAIN

The vestibular nerve sends impulses to the vestibular nuclei in the medulla. From these nuclei, impulses are relayed to the *cerebellum,* the *reticular formation,* and the *several nuclei* that control the movements of the *muscles* of the eyes, head, and neck. In addition, there are several direct *descending pathways* to *skeletal muscles.* These intricate connections to subcortical motor centers and to the skeletal muscles permit the extremely rapid coordination seen to perfection in ballet dancers and acrobats.

There are also pathways that lead to the cerebral cortex from some of the motor centers mentioned above. These cortical pathways are not adequately traced as yet, but they do not seem to be necessary for the rapid vestibular control of movement.

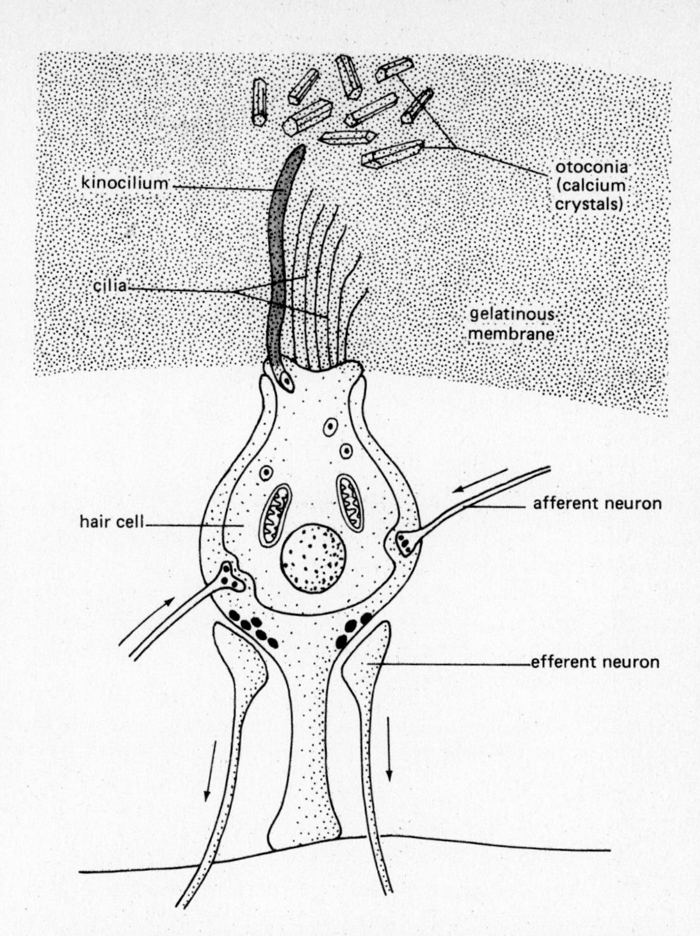

Figure 22-45. **The receptor area of the utricle, showing the hair cells projecting into the gelatinous layer. These cells have both afferent and efferent innervation.**

UTRICLE AND SACCULE

Both the utricle and the saccule have a small specialized area called the *macula.* Within the macula is a thick gelatinous layer in which are imbedded tiny crystals of calcium carbonate. These crystals are called *otoconia.* Projecting into this jelly are the hairs of the hair cells (Fig. 22-45).

Vertical Static Equilibrium. The utricle and saccule are extremely sensitive to changes in the position of the head with respect to the force of gravity. The receptors are most sensitive when the head is upright, and they can detect a change as small as half a degree in this position. This information is vital to permit the maintenance of the upright position.

Linear Acceleration. When the body is thrust forward by a sudden linear acceleration (or when there is a rapid ascent or descent, as in a rapidly moving elevator), the otoconia, being heavier than the endolymph in the utricle, fall against the hair cells, distorting them with their weight (Fig. 22-46). This impact causes a volley of impulses from the nerve endings.

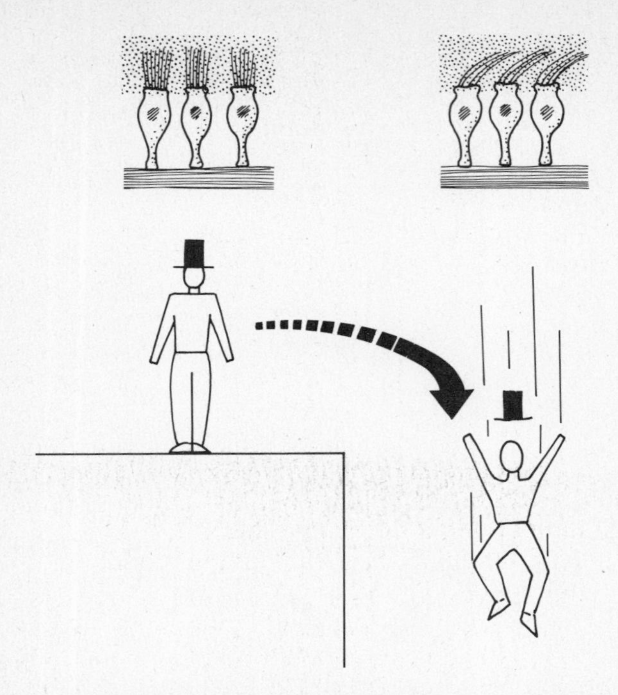

Figure 22-46. **Linear acceleration stimulates the receptors in the utricle, as the otoconia press on the hair cells and distort them.**

Semicircular canals

The semicircular canals are three extremely delicate, hollow tubes, one in each plane (superior, posterior, and lateral). They are filled with endolymph. Each one widens out into an *ampulla,* after which they connect with one another. The ampulla contains the receptors for *rotational acceleration.* Within the ampulla is a gelatinous layer called the *cupola,* into which project the hairs of the hair cells. Like the cells of the macula, they are surrounded by free nerve endings of the vestibular nerve. The endolymph within the canals, like all fluids, has a certain inertia, and when the head is suddenly rotated in one plane, the endolymph within the canal in the same plane tends to remain still for a moment, while the canals turn with the head. This passively causes movement of endolymph in the opposite direction to that of the actual rotation of the head. The flow of endolymph bends the cupola, distorting the hair cells and stimulating the nerve endings around them.

If the movement continues, the endolymph catches up and flows at the same speed as the canals are moving, so that there is no further stimulation of the nerves.

When the movement stops, the reverse occurs: the endolymph continues to flow in the direction of the rotation a little longer, pushing against the hair cells in the opposite direction. This produces a rather curious phenomenon, *past-pointing,* which is a conscious attempt to compensate for changes in acceleration, except that the clues given the brain are false. If the eyes are closed during and immediately after rotation, the subject feels he is falling because of the delayed excitation of the hair cells by the endolymph. In an attempt to compensate for this, he leans over in the opposite direction, sometimes toppling right over (Fig. 22-47). If the eyes are kept open, this past-pointing of the body does not occur, because the subject can see that he is upright. Past-pointing can be demonstrated also by having the subject try to put his finger on his nose with his eyes closed, immediately after the rotation. He will deviate, missing his nose, past-pointing to the side as he mistakenly compensates for the sensations initiated by the delayed excitation.

Not only do the impulses that are generated in the vestibular apparatus excite the centers in the hindbrain that are directly involved with posture and equilibrium, but their activity radiates to stimulate other closely located centers, so that a wide variety of responses occur that are associated with "motion sickness." The feelings of nausea, headache, and eye strain all can be traced to a secondary stimulation of such centers.

Nystagmus

Closely associated with the vestibular reflexes during the rotation of the head are reflex movements of the eyes, which permit the eye to fix on an object while the head is turning. To do this requires *very rapid movement in the direction of the rotation,* to fix on something still ahead, and then a *slow movement of the eyes backward,* keeping the fixed object in view. This permits us to keep an object in sight even during rapid movement, a very necessary ability for our tree-living ancestors as they swung from branch to branch, and a highly perfected mechanism in ballet dancers as they "spot" during a swift *fouetté.* The forward movement and the slow backward movement of the eyes is *nystagmus,* and it can occur physiologically in *any of the three planes,* depending on which of the semicircular canals is stimulated. We have all experienced the nystagmus induced by linear movement while watching the scenery from a rapid car or train. To a lesser extent, these movements of the eyes are part of all our movements—running, walking, talking, and rotating the head—and probably are essential for a continuous clear formation of the visual image.

The reflex pathways that are involved in equilibrium are summarized in Table 22-4.

Figure 22-47. **Movement of the endolymph in the semicircular canals with horizontal rotation. When the movement is abruptly stopped, the endolymph continues its movement.**

Table 22–4. **Reflex Pathways Involved in Equilibrium**

	Reflexes of the Utricle
Stimulus	Stopping or starting in a linear direction (forward, up, or down)
Receptor	Nerve endings in the hair cells of the utricle
Afferent neuron	Vestibular nerve (part of eighth cranial nerve)
Synapse	Vestibular nucleus in the brain stem
Integrating center	Facilitative and inhibitory centers in reticular formation and cerebellum
Efferent neuron and second synapse	Vestibulospinal and reticulospinal fibers to ventral horn motor cells and small cells
Efferent neuron	Alpha motor fibers Gamma motor fibers
Effector	Postural muscles and muscle spindles
Response	Change in tension in muscle and spindle, causing change in position of limbs and body
Effect	Maintenance of equilibrium and notification of position to higher centers by subsequent impulses to cortex
	Reflexes of the Semicircular Canals
Stimulus	Stopping or starting rotation of the head
Receptor	Nerve endings of the hair cells in the canal in the plane of rotation
Afferent neuron	Vestibular nerve
Synapse	Vestibular nucleus in brain stem
Integrating center	The reticular formation and cerebellum Ascending fibers to cortex
Efferent neuron and second synapse	Vestibulospinal and reticulospinal fibers to ventral horn motor cells and small cells
Efferent neuron	Alpha motor fibers Gamma motor fibers
Effector	Postural muscles and muscle spindles
Response	Contraction of extensor muscles to adjust position of the body Change in tension in muscle spindle
Effect	Maintenance of equilibrium and notification of higher centers of change in position by subsequent impulses to cortex
	Nystagmus
Stimulus	Rotation or linear movement
Receptor	Nerve endings in the hair cells in the vestibular apparatus
Afferent neuron	Vestibular nerve
Synapse	Vestibular nucleus in the brain stem
Integrating center	Ascending fibers to the nucleus of the oculomotor nerve (third cranial nerve) just above the pons
Efferent neuron	Oculomotor nerve
Effector	External muscles of the eye
Response	Rapid forward movement of eyes, slow backward movement
Effect	Brief fixation on an object during rapid movement

22–7 Taste

The senses of taste and smell regulate to a large degree an animal's selection of food and mates. Taste, unlike vision and hearing, appears to have undergone very little evolutionary development, and it is poorly represented in the cerebral cortex. There does seem to be a common *phylogenetic* basis for taste: practically all vertebrates show a preference for sweet substances and reject bitter or sour substances.

These taste preferences, however, can be modified by the *physiological state* of the animal, contributing to the general regulatory mechanisms of the body. An adrenalectomized animal will select a 1 per cent sodium chloride solution rather than tap water, thus restoring some of the sodium that is continuously being lost through the kidneys because of the lack of aldosterone. When the sensory nerves to the tongue are cut, the animals no longer can select the salt solution and the amount of circulating blood sodium drops rapidly.

There is considerable evidence, too, that taste is *modified by previous substances tasted:* water may taste bitter or sour following adaptation of the tongue to sodium chloride. A taste-modifying berry known as miracle fruit, found in southwest Africa, modifies taste to make even extremely bitter substances taste sweet. Artichokes have been reported to have a similar effect, to a much lesser extent. Other substances have the opposite effect, suppressing sweetness.

Taste Receptors: The Taste Buds

Primary taste sensations

Classically, the taste sensations have been presented as consisting of four types—bitter, sweet, sour, and salt—and the attempt has been made to correlate the kind and distribution of the taste buds to these sensations. This has not been very satisfactory, because there are specific sensations for many isolated compounds, chemically very similar. A chick will accept a rather thick, viscous solution of glucose but will reject all xylose (a closely related sugar) solutions even at very weak concentrations. There are perhaps *overlapping fields* for taste, with different combinations giving different sensations, as there are overlapping visual fields and overlapping somesthetic fields.

Distribution of taste buds

These intricately formed receptors are found on the tongue, inside the lips and cheeks, and also on the back of the pharynx. Taste buds whose chief sensitivity is to sweetness are localized predominantly on the anterior tip of the tongue. Salty and sour substances excite mainly taste buds on the sides of the tongue, whereas bitter substances excite taste buds on the back of the tongue (Fig. 22-48). As most taste buds respond to two or three different types of tastes, the ability to perceive a specific taste may be the *ratio* of those taste buds that respond more strongly to the substance (fire rapidly) to those that respond weakly (slow or sporadic firing).

Figure 22-48. **Distribution of tongue areas showing major sensitivity to different tastes. The striped areas are most sensitive to salt taste. All regions, however, show considerable overlap.**

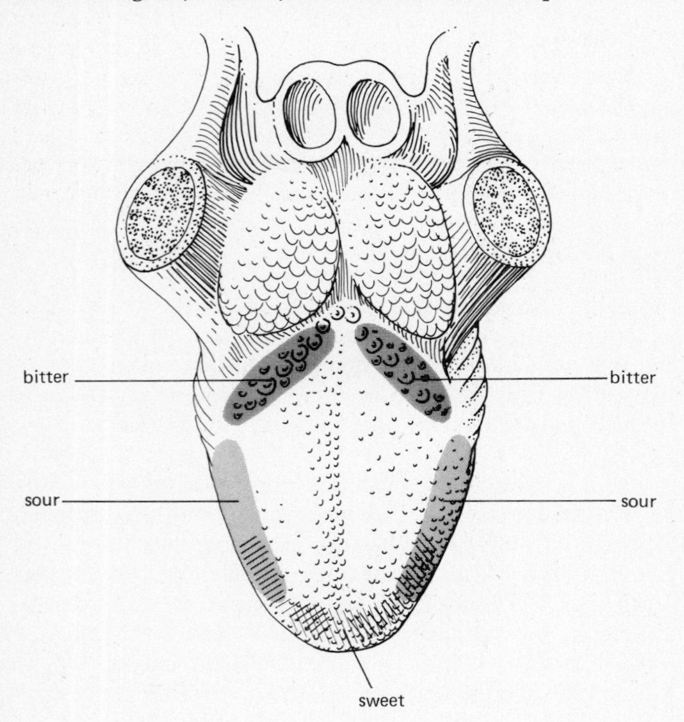

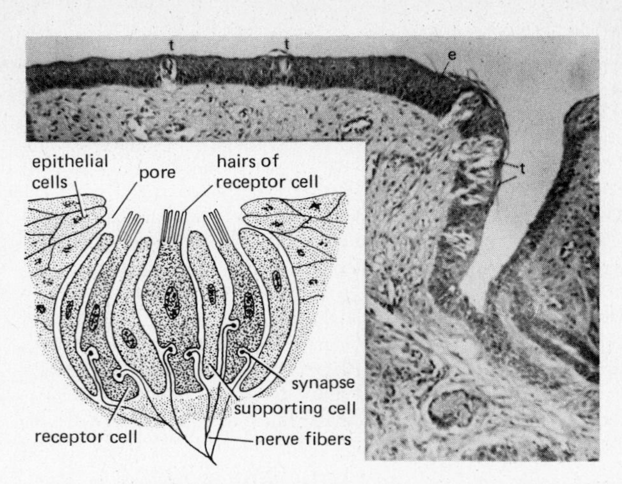

Figure 22-49. **(A) Taste buds (t) in the epithelium (e) of the papillary ridges of the tongue. (B) The detailed structure of a taste bud. (Photograph courtesy of S. Piliero, Dental School, New York University.)**

Taste is probably a continuum of separate but interacting sensations, some of which are pleasant and some unpleasant. In the gourmet, taste discrimination may become an extremely critical sense, leading into fields of sensation apparently unknown to lesser mortals.

Structure and innervation of taste buds

The taste bud is a cluster of *epithelial cells* around a *central pore* (Fig. 22-49). From the tip of each cell protrude several *hairs,* which are believed to be the receptor surface. Nerve endings of the *glossopharyngeal* (ninth cranial nerve) and the *chorda tympani* (seventh cranial nerve) form networks around these receptor cells, invaginating deeply into the cluster. The taste buds degenerate completely if these nerves are cut.

Unlike most other receptors, which are carefully protected from the environment, taste buds are exposed to many corrosive and destructive juices. Their life span is only about 10 days, and they are constantly being renewed from the surrounding epithelial cells.

Transduction of chemical energy and the pathway for impulses to the brain

It appears that the taste-provoking substance combines with a receptor on the hair surface of the taste bud, resulting in a depolarization of the cell and a *receptor potential.* There must be more than one type of receptor on a taste bud, because they respond to different tastes, but the receptor potentials evoked vary considerably in strength and duration.

All the taste fibers synapse in the *tractus solitarius* in the brain stem. Neurons in the tractus solitarius send their axons to the *thalamus,* where a second synapse is made with cells

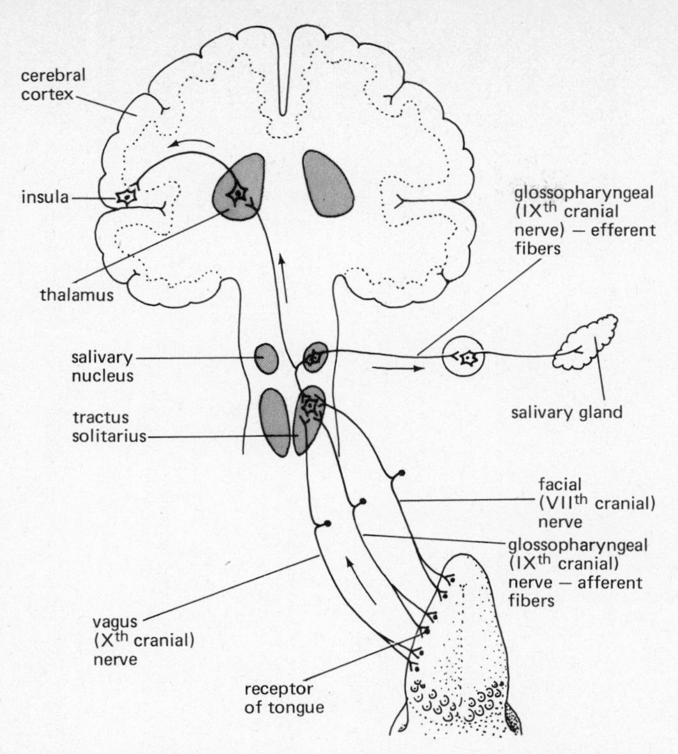

Figure 22-50. **A simplified representation of the taste pathway and the salivation reflex.**

that terminate in the *insula* of the cerebral cortex (Fig. 22-50). This is probably a great oversimplification of the taste pathway, which certainly involves other projections, as yet undetermined, in the central nervous system.

Taste Reflexes

Taste stimuli, especially sour tastes, induce *salivation* through taste reflexes. Impulses from the tractus solitarius stimulate the salivary nuclei in the brain stem, causing the salivary glands to increase their secretions (Fig. 22-50). Appetizing foods also cause salivation, but this response involves mediation of the hypothalamus.

Nontasters (Taste Blindness)

The chance discovery that phenylthiocarbamide (PTC) tastes bitter to some people but is tasteless to others led to investigations that showed that some types of taste sensitivity are genetically influenced. In European and North American populations, about one third are nontasters. In many other races, the proportion of nontasters is much smaller. PTC inheritance is discussed in Sec. 27-3.

22-8 Olfaction

In most animals, olfaction is an important sense primarily involved in the regulation of behavior such as mating, the detection of territorial boundaries, and the selection of food. It is believed that the sense of smell is poorly developed in humans, yet the sensitivity to certain odors, such as that of methyl mercaptan, permits the detection of one part in a billion. Methyl mercaptan is the substance that is smelled in a gas leak; it is added in minute quantities as a protective device to natural gas, which has no odor.

Primary Odors

It is clear that color vision is dependent upon the three primary colors; it is somewhat less clear that taste is composed of four primary tastes. But the finite number, if any, of the primary odors is quite unclear. At the last count, there appear to be more than 30 primary odors.

Much of the research in humans has come from a study of *odor-blind* (*anosmic*) individuals, who cannot detect one particular odor. In the general population studied, 10 per cent could not smell the poisonous hydrogen cyanide, 2 per cent could not detect the musky odor of isovaleric acid (found in axillary sweat and vaginal secretions), and 0.1 per cent could not smell the mercaptanlike odor of the skunk.

Table 22-5. **Possible Pheromonal Significance of Human Primary Odors**

Probable Primary Odor	Established Primary Odorant	Presence in Human Secretions	Possible Pheromonal Significance
Sweaty	Isovaleric acid	Peri-anal secretion[b]	Alarm pheromone
		Vaginal secretion	Ovulation indicator
Spermous	1-Pyrroline	Male pubic sweat[b]	Aphrodisiac
		Semen[b]	Mating marker
Fishy	Trimethylamine[a]	Menstrual sweat[b]	Oestrus indicator[a]
		Menstrual blood	Menstruation indicator
Urinous	Δ^{16}-Androsten-3-one[a]	Male axillary sweat[b]	Dominance pheromone
		Male urine	Territorial marker

Source: J. E. Amoore, in *Olfaction and Taste V,* D. A. Denton and J. P. Coghlan, eds., Academic, New York, 1975. Modified by Amoore, 1977.

[a] For some species.

[b] Suspected.

Specific anosmias have been demonstrated for more than 70 distinct compounds. If this failing is considered to be an inborn molecular defect in the odor receptors, it implies that there must be more than 70 different receptors.

Amoore (1967) has listed four primary odors which he believes are possible pheromones in humans. They are sweaty, spermous, fishy, and urinous, and they are all produced by the human body (Table 22-5). Their role as pheromones may be vestigial, but recent experiments have shown that both men and women accurately classify sweat as coming from either a male or female (they had to smell unidentified sweat shirts). Alex Comfort (1972) has stated that human pheromones are not really vestigial; they are only thinly veiled by the taboos of civilization and the promotion of deodorants.

Olfactory Epithelium

The olfactory epithelium lies in a yellowish patch in the upper part of the inside of the nose (Fig. 22-51). It is composed of the *olfactory cells,* which are the receptors, the *supporting cells,* and the *mucus-producing cells.*

OLFACTORY RECEPTORS

The olfactory cells are bipolar sensory cells derived from the brain. The thick peripheral portion has a crown of small cilia, the *olfactory hairs,* which project into the mucous lining of the nose. These hairs are presumed to be the receptor surface for odors. The slim central end of these sensory cells is an axon that leads back to the central nervous system (Fig. 22-52).

Figure 22-51. **Position of the taste and olfactory receptors.**

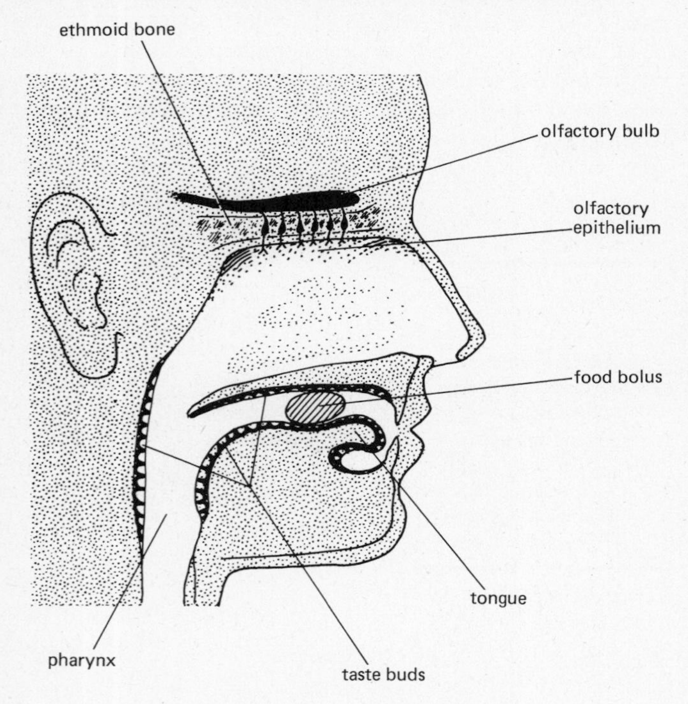

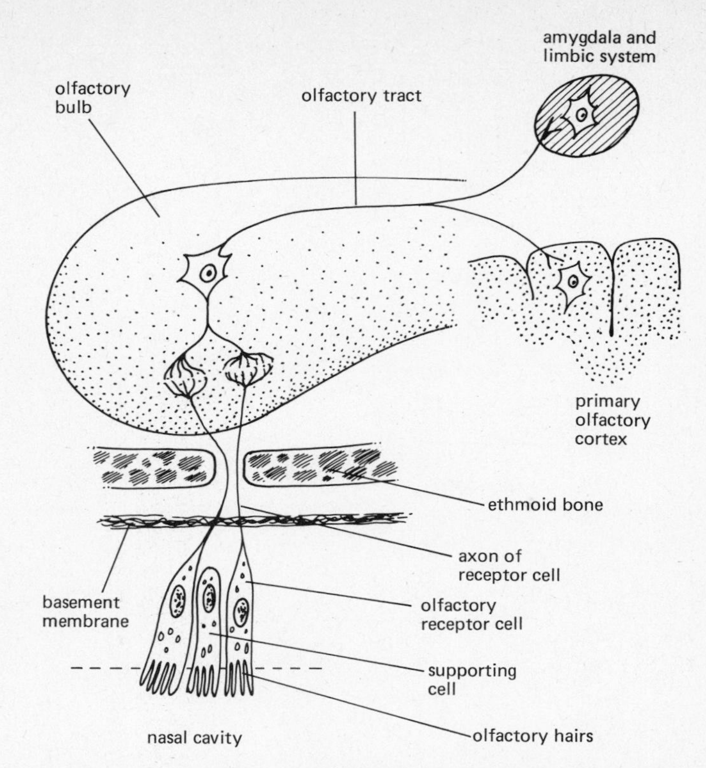

Figure 22-52. **The olfactory epithelium and the olfactory pathway to the brain.**

Stimulation of the Olfactory Receptors. Despite much ingenious research, little is known about the way in which a specific odor excites an olfactory cell. One theory holds that it is the *chemical nature* of the odorant that determines whether or not it will combine with the receptor. However, many substances with quite different chemical characteristics have very similar shapes and odors. This lends credence to the view that it is the *physical characteristics* of the odorant that permit it to adsorb to specific receptor sites.

It has been suggested that an effective olfactory stimulant must have a degree of water solubility (to penetrate the mucous lining) and be fairly lipid soluble (to penetrate the lipid membrane of the hair cells). However, if the receptors are on the surface membrane of the cells, the necessity for much lipid solubility is difficult to perceive. Perhaps Amoore's suggestion (1967) of a combined *stereochemical configuration* is the most plausible at the moment.

Transduction of Chemical Energy by the Olfactory Receptors

Regardless of the mechanism by which the odorant excites the hair cells, it is presumed that a depolarization occurs, setting off the *generator potential* (no synapse at the bottom of the olfactory hair cell). The generator potential results in the action potential that is conducted along the axon toward the olfactory bulb in the brain.

Adaptation

We are all familiar with the extremely rapid adaptation to strong odors: our strong reaction in coming into a smoke-filled room in which the smokers are completely unaware of the smell, or the inability to smell perfume on oneself after the first minute. Yet the olfactory receptors, which adapt to about 50 per cent of their initial activity within the first second or two, subsequently retain this level of activity. Adaptation may therefore be a central phenomenon.

The importance of *centrally controlled adaptation* has also been stressed by Cairncross and his associates in Australia (1972). They devised a cannula that could be chronically implanted in the nasal cavity of a rat, thus permitting the delivery of specific concentrations of odorant for precise periods of time. Their results indicated that sensitivity to odors could be increased or decreased by different types of *training*, and that the specific physicochemical characteristics of the odorant were not of fundamental importance.

Pathway from the Olfactory Receptors to the Brain

The olfactory receptor cells send their axons, without synapsing, up through perforations in the ethmoid bone at the base of the skull. These slender axons make up the *olfactory nerves* that terminate in the *olfactory bulbs,* a part of the rhinencephalon (inelegantly termed by some "the smell brain"). The next neurons in the chain send their axons by way of the *olfactory tracts* to the *olfactory areas* in the brain (Fig. 22-52).

The olfactory tracts, like much of olfaction, are not well defined but connect the olfactory bulbs with the amygdala, with areas in the frontal cortex near the hippocampus, and to the hippocampus itself.

Other important pathways connect the olfactory bulbs with the hypothalamus and the limbic system, permitting the very close coordination between the sense of smell and feeding and emotional behavior. This integrated activity is still clearly seen in humans: a smell associated with food that has once caused a gastric upset will evoke feelings of nausea years later, and a pleasant smell can recall fond memories and emotions long forgotten.

22–9 Somatosensation

Somatosensory Receptors: Touch, Temperature, Pressure, Muscle Sense, and Pain

The skin, muscles, and joints are richly supplied with nerves and organs that are specialized to transduce physical energies. However, there is considerable dispute about the degree of specialization and the ability of different somesthetic end organs to respond to different sensory modalities. Many of the classical theories concerning the somesthetic receptors have not been supported by modern research, but that research has no completely satisfactory explanation to substitute.

STRUCTURAL CLASSIFICATION OF SOMATOSENSORY RECEPTORS

The simplest classification, that on the basis of structure, divides these receptors into those that are *encapsulated* and those that are *free nerve endings.* Some of these receptors are illustrated in Fig. 22-53.

Encapsulated Nerve Endings. Encapsulated receptors include *Meissner's corpuscles,* which are extremely sensitive to tactile stimulation, and the *Pacinian corpuscles,* which respond best to vibrations or rapid mechanical changes. The *muscle spindles* are encapsulated sensory receptors, responding to change in length, and rate of change in length, of the muscles. The spindles are discussed in Sec. 23–13.

Free Nerve Endings. Free nerve endings include those responsive to pain, but a great many free nerve endings also respond to touch and to temperature changes. *Ruffini's end*

Figure 22-53. **Some types of nerve endings in the skin.**

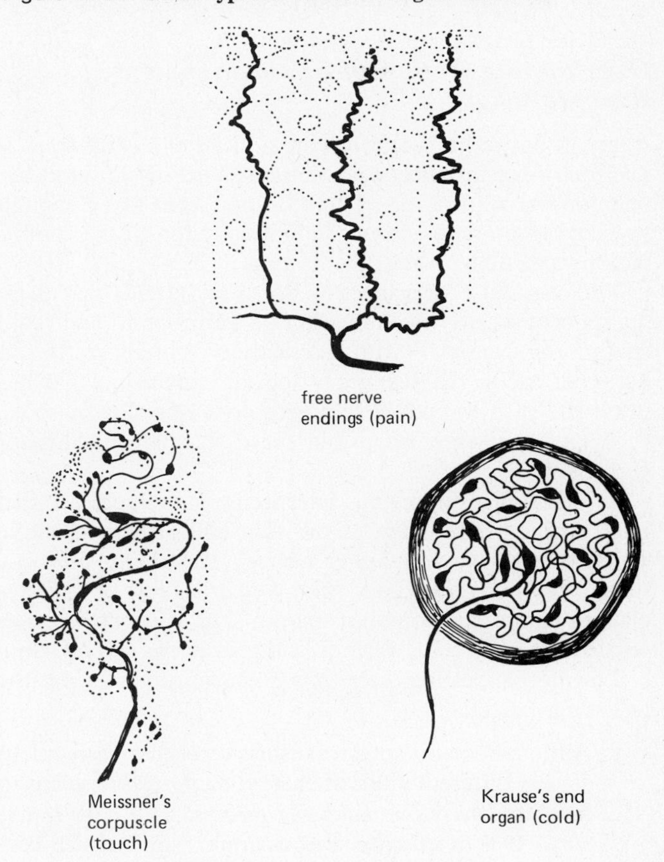

organs (warmth) and the *bulbs of Krause* (cold) are more complex forms of free nerve endings and are also shown in Fig. 23-53, as are the *nerve plexuses around the hair follicles.* These plexuses respond when the hair is moved or pulled. The *Golgi tendon organs* are leaflike free endings; their functions are discussed in Sec. 23-13.

It is probable that an important function of the capsule is to protect the nerve endings from the trauma to which they are exposed. Loewenstein's beautiful experiments on the Pacinian corpuscle (1958), in which he peeled off successive layers of the onionlike capsule, demonstrated clearly that the capsule has only a nonneural role in the formation of the generator potential. This is discussed further in the section on transduction.

FUNCTIONAL CLASSIFICATION OF SOMATOSENSORY RECEPTORS

The somatosensory receptors have been grouped according to their functions:

1. *Contact receptors.* Touch, pressure, vibration, tickle, and itch.
2. *Thermal.* Cold and warmth.
3. *Kinesthetic.* Sense of position in space and sense of the position of the limbs relative to the body.
4. *Pain.* Burning, prickling, and aching pain, which sometimes is the result of extreme stimulation of any of the other somatosensory receptors.

Overlapping Action of Somatosensory Receptors

Many, if not most, sensations are a *blend of several types of sensation*—for instance, cold wetness, hard or soft smoothness, or warm fuzziness. There do not seem to be enough specific receptors to account individually for all the combinations and dimensions of sensation.

The classical observation by von Frey (1895) that there are spots of sensitivity on the skin—"warm spots" and "cold spots," for example—with insensitivity to temperature in between, led to the seemingly logical conclusion that because the distribution of skin receptors was also in spots, each receptor was responsible for a specific modality of sensation.

Evidence that refutes this interpretation includes detailed microscopic examination of the skin under these sensitive spots; there is *no correlation between these areas and any specific skin receptor.* Hairy skin, which has few of the specialized end organs found in smooth skin, is nevertheless as sensitive to pain, cold, warmth, and pressure as smooth skin.

The most appealing *theory of somatosensation* includes the following postulates:

1. Although many somatosensory receptors respond to several different kinds of energy, *each receptor* seems to have its *lowest threshold and greatest sensitivity to one single type of stimulus.* For example, the Pacinian corpuscle responds preferentially to deformation, but this response may be considerably influenced by temperature. Probably the transduction activities of most receptors are modified by changes in temperature, a not surprising observation.
2. The information received by a single receptor may be *coded differently* according to the type of stimulus. Figure 22-54 shows how the pattern of firing of a spinal neuron changes in frequency and duration depending on whether it is responding to touch, heat, pain, itch, or vibration.

Transduction by a Somatosensory Receptor: The Pacinian Corpuscle

There is not much detailed information concerning transduction by somatosensory receptors because the overlapping of their sensitivities makes experimentation extremely difficult. There is one notable exception, however: the relatively large Pacinian corpuscle. This receptor responds to mechanical change, and thus, like the organ of Corti in the cochlea, is a *mechanoreceptor.* The experiments by Loewenstein (1958) on the Pacinian corpuscles, found in the mesentery of the cat, provide the most specific information on somatosensory transduction.

Figure 22-55 shows that the Pacinian corpuscle has a nonmyelinated nerve fiber tip in its central core. The naked tip is surrounded by the concentric layers that make up the capsule. Using micromanipulators, Loewenstein meticulously stripped off layer after layer of the capsule. As each

Figure 22-54. **The frequency of firing of sensory neurons depends on the modality of the stimulus.**

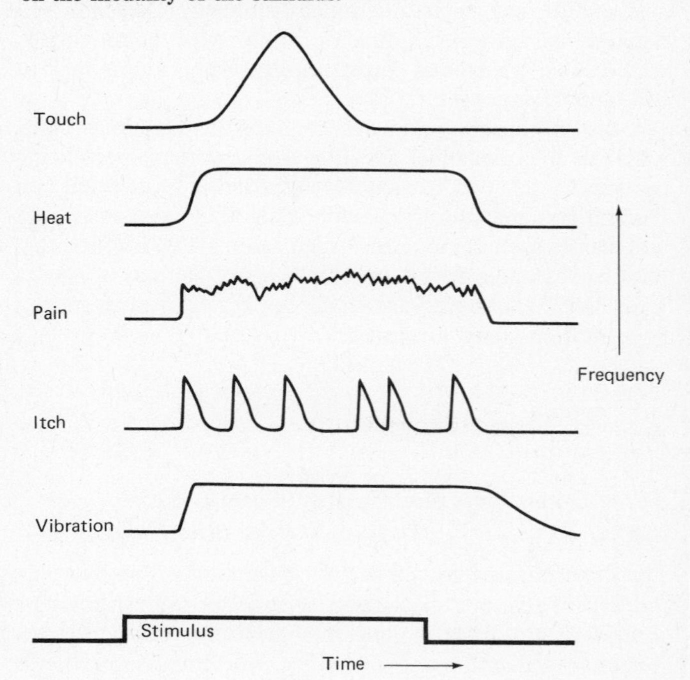

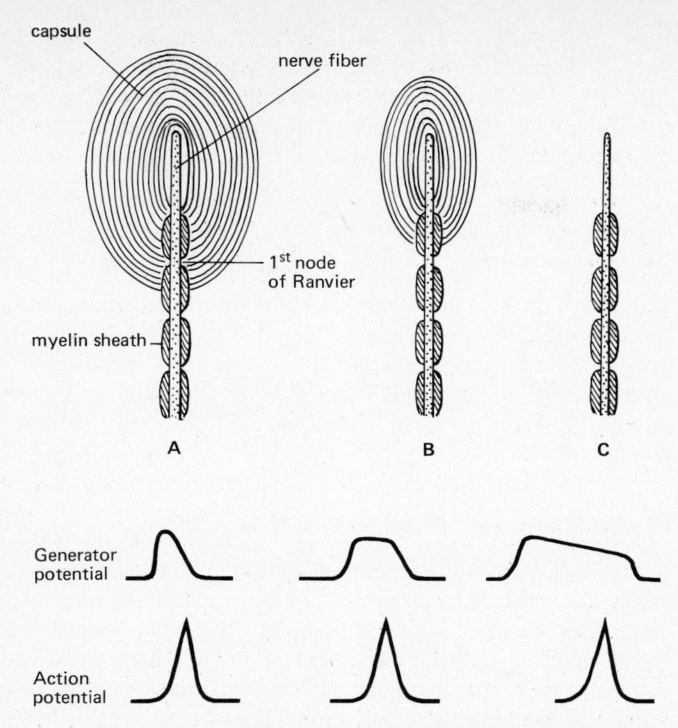

Figure 22-55. **The Pacinian corpuscle. The naked nerve fiber tip within the corpuscle responds to mechanical distortion with a generator potential (A), the size of which is independent of the progressive stripping off of the surrounding capsule (B and C). The form of the generator potential is changed by this procedure, however. The action potential arises at the first node of Ranvier and is unaffected by removal of the capsule. [Adapted from W. R. Lowenstein and M. Mendelson, *J. Physiol.* 177:377 (1965).]**

layer was stripped away, he stimulated the remaining capsule and found that the *generator potential* evoked had the same magnitude regardless of the presence or absence of the capsule.

What is the *function,* then, *of the capsule?* Figure 22-55 shows that its presence *changes the wave form* of the generator potential. When the capsule is present, the potential is evoked only at the beginning and the end of the stimulation, adapting very rapidly. When the capsule is removed, there is very little adaptation to the stimulus, so that the capsule appears to peak the generator potential at the beginning and end of the stimulus.

It is supposed that the mechanical deformation of the intact capsule will compress, elongate, or deform in some way the central nerve core of the fiber. This physical deformation may alter the conformation of the nerve membrane to cause a change in permeability, resulting in depolarization and the generator potential. Like any other *generator potential,* the change is *graded* and *nonpropagated;* if it reaches a critical level, it will set off the *conducted action potential.* This occurs at the *first node of Ranvier,* which is still within the capsule of the Pacinian corpuscle. This node, of course, is part of the series of nodes involved in saltatory conduction of the action potential along a myelinated axon (Sec. 6-3).

Somatosensory Pathways to the Brain

CUTANEOUS PATHWAYS

There are two separate pathways for sensations from the skin, a fast and a slow pathway.

Fast Pathway: The Dorsal Column and the Medial Lemniscus. The fast pathway is the route taken for the rapid, highly specific senses of *fast touch* and *pressure.* These impulses travel along a pathway known as the *medial lemniscus,* which travels through the brain stem to the thalamus (Fig. 22-56).

To reach the medial lemniscus, impulses generated at the nerve endings pass along the peripheral nerves which have their cell bodies in the dorsal ganglia. These fibers then ascend, uncrossed and without synapsing, through the dorsal columns of the spinal cord to the medulla. In the medulla the fibers cross over to form the *medial lemniscus,* which ends in the *thalamus.* The final neurons in this pathway originate in the thalamus and terminate in the *somatosensory* (somesthetic) *areas* of the cortex.

SPATIAL ORGANIZATION OF THE FAST CUTANEOUS PATHWAY. The dorsal columns and the thalamus have separate areas for the *cutaneous nerves* from different parts of the body. Those nerves entering from the legs and trunk are directed to the *central regions* of the columns. Nerves from the arms, neck, and head are distributed in layers toward the *sides of the dorsal columns.*

This same distribution of fibers, depending on their origin, is seen in the *thalamus,* but as the fibers cross over in the medulla, the right side of the body is represented in the left side of the thalamus. This spatial orientation of the fibers is

Figure 22-56. **Fast pathway for touch and pressure sensations from the skin through the dorsal column of the spinal cord and the medial lemniscus, which is between the medulla and the thalamus. From the thalamus, axons ascend to the sensory cortex.**

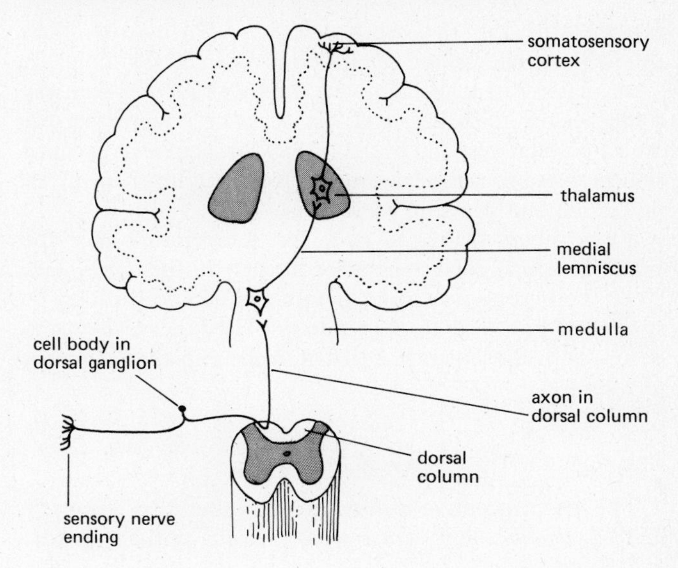

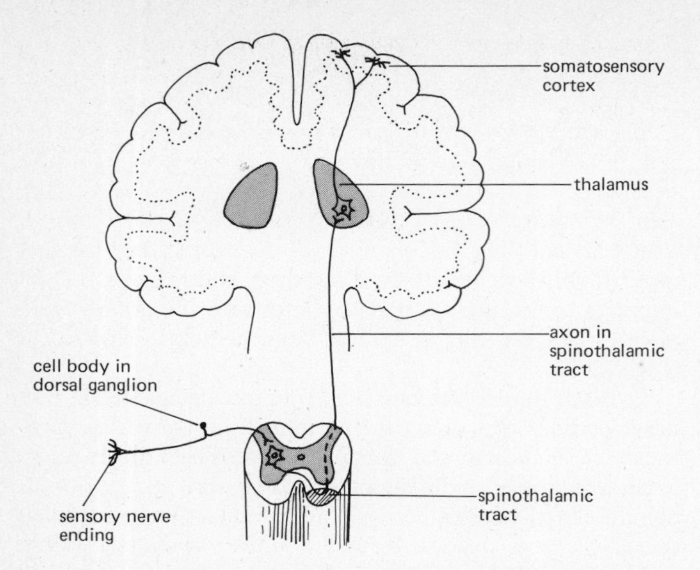

Figure 22-57. **The spinothalamic pathway. This slow pathway for pain, temperature, and sexual sensations involves sensory fibers that synapse directly with dorsal horn cells in the spinal cord, then cross over and travel up the spinothalamic tracts to the thalamus. From the thalamus, axons ascend to synapse in different regions of the sensory cortex.**

continued to the *somatosensory cortex,* so that the specific areas of the body are represented in specific cortical regions, similar to the representation of body areas in the motor cortex, as depicted in Fig. 8-30.

DIVERGENCE OF STIMULI REACHING THE BRAIN. Each of these relay stations plays a role in increasing the number of cortical cells that will ultimately be excited by the initial stimulus. Marked divergence occurs in the medulla and in the thalamus, permitting a weak stimulus to affect only the central cortical neurons in a sensory field, whereas a stronger stimulus may excite many more peripheral neurons. However, the neurons in the central field always fire at a higher frequency.

Slow Pathway: The Spinothalamic Pathway. *Pain, temperature,* and *sexual sensations* are conducted along the other major cutaneous pathway, the *spinothalamic* pathway, a much slower, cruder, and less accurate system than the dorsal column system (Fig. 22-57). Consequently, localization of these sensations is much less accurate than for the fast touch and pressure sensations.

Nerve fibers from the pain and temperature receptors enter the spinal cord to synapse directly in the *dorsal horn.* They then cross over immediately and ascend through the spinothalamic tracts to the *thalamus.* Here the final synapse occurs with those neurons that terminate in the *somatosensory cortex.*

PROPRIOCEPTIVE PATHWAYS

The proprioceptive pathways carry impulses from the *muscles, tendons,* and *joint receptors,* utilizing the same medial lemniscus pathway to the thalamus and somatosensory cortex as do the cutaneous fibers. A significant difference, however, is that the proprioceptive fibers also project to the *cerebellum.* This enables the proprioceptors to regulate muscle tonus, posture, and the integration of movement.

Effects of Destruction of the Somatosensory Cortex

If much of the somatosensory cortex is destroyed, sensations can no longer be specifically localized. They are then interpreted as coming from relatively large body areas. The awareness of the position of the body in space and the relative position of the different parts of the body to one another is also lost.

Pain Pathways: Fast and Slow Fibers

There is a *fast* pathway for *pain* along small A fibers, conducting from 3 to 20 m/sec. These reach the *somatosensory cortex* to give rise to a pricking pain. *Slow* impulses travelling along the thinner C fibers at 0.5 to 2 m/sec end in the *reticular activating system* and the *thalamus,* resulting in a generalized slow, burning pain. These pain signals are poorly localized and additive, so that they become increasingly disturbing if ignored. This mechanism ensures the activation of the reticular activating system itself, which

Figure 22-58. **One explanation of referred pain is on the basis of spread of excitation from neurons from the viscera to neurons that receive pain impulses from the skin. Consequently, both visceral and skin sensations travel along spinothalamic pathways for skin. Subsequent interpretation by the cortex is that the pain originates from the skin.**

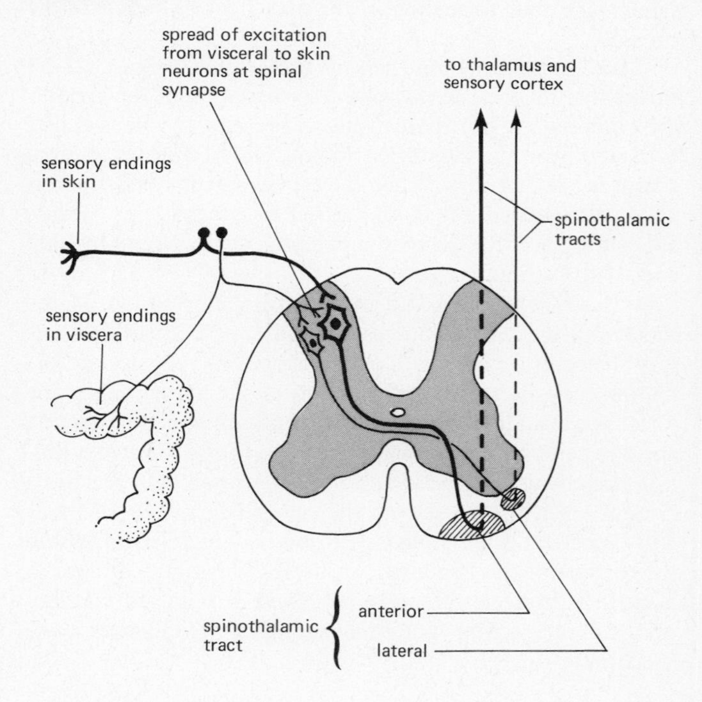

regulates the arousal and protective responses of the organism.

Response to pain

The individual response to pain varies widely, and the differences are more likely central, rather than being differences in receptor threshold. The response to pain is affected by the mores of the society as well as by emotional differences between people, a fact that demonstrates the important role of both the cerebral cortex and the limbic system in pain perception.

Referred pain

Referred pain is pain initiated in one of the visceral organs and felt elsewhere in the body, commonly in the skin. It is believed that the referral of pain occurs when visceral pain fibers are so severely stimulated that they activate synapses in the spinal cord and thalamus that normally only receive impulses from the skin (Fig. 22-58). This is interpreted by the somatosensory cortex as pain coming from the skin.

Cited References

AMOORE, J. E., and D. VENSTROM. "Correlations Between Stereochemical Assessments and Organoleptic Analysis of Odorous Compounds." In *Olfaction and Taste II*, Proc. Second International Symp., T. Hayashi, ed. Pergamon Press, Inc., Elmsford, N.Y., 1967, p. 3.

BÉKÉSY, G. VON. *Experiments in Hearing.* McGraw-Hill Book Company, New York, 1960.

COMFORT, A. *The Joy of Sex.* Simon and Schuster, New York, 1972.

DAVIS, H. "A Model for Transducer Action in the Cochlea." *Cold Spring Harbor Symp. on Quant. Biol.* **30:** 181 (1965).

FECHNER, G. *Elements of Psychophysics.* H. E. Adler, trans. Holt, Rinehart, and Winston, New York, 1966.

FREY, M. VON. "Beitrage der Sinnesphysiologie der Haut (Dritte Mitteilung)." *Akad. Wiss. Leipzig, Mathnaturw. Kl. Berichte.* **47:** 166 (1895).

GRANIT, R. "The Colour Receptors of the Mammalian Retina." *J. Neurophysiol.* **8:** 195 (1945).

HARTLINE, H., H. WAGNER, and F. RATCLIFF. "Inhibition in the Eye of *Limulus*." *J. Gen. Physiol.* **39:** 651 (1956).

HUBEL, D. H., and T. N. WIESEL. "Receptive Fields of Single Neurons in the Cat's Striate Cortex." *J. Physiol. (London)* **148:** 574 (1959).

LAING, D. G., K. D. CAIRNCROSS, M. G. KING, and K. E. MURRAY. "The Use of Behavioural Methods in Studying the Physiology of Olfaction." *Proc. Aust. Physiol. Pharmacol. Soc.* **3**(2): 203 (1972).

LETTVIN, J. Y., H. R. MATURANA, W. S. MCCULLOCH, and W. H. PITTS. "What the Frog's Eye Tells the Frog's Brain." *Proc. Inst. Radio Engr.* **47:** 1940 (1959).

LOEWENSTEIN, W. R., and R. RATHKAMP. "The Sites for Mechano-Electric Conversion in a Pacinian Corpuscle." *J. Gen. Physiol.* **41:** 1245 (1958).

VON HELMHOLTZ, H. *Handbuch der physiologischen Optik,* 1st ed., Hamburg and Leipzig, 1866. English ed., J. P. C. Southall, trans., *Treatise on Physiological Optics,* Vols. 1–3. Optical Society of America, Rochester, N.Y., 1924–1925.

WALD, G. "Molecular Basis of Visual Excitation." *Science* **162:** 230 (1968).

WALD, G. "The Receptors of Human Color Vision." *Science* **145:** 1007 (1964).

YOUNG, R. W. "Visual Cells." *Sci. Am.,* Oct. 1975.

Additional Readings

BOOKS

ARMINGTON, J. C. *The Electroretinogram.* Academic Press, Inc., New York, 1974.

BONICA, J. J., P. PROCACCI, and C. A. PAGNI. *Recent Advances in Pain.* Charles C Thomas, Publisher, Springfield, Ill., 1974.

BRINDLEY, G. S. *Physiology of the Retina and Visual Pathway,* 2nd ed. The Williams & Wilkins Company, Baltimore, 1970.

DAVSON, H., ed. *The Eye,* 2nd ed. Academic Press, Inc., New York, 1969.

DAVSON, H. *The Physiology of the Eye,* 3rd ed. Academic Press, Inc., New York, 1972.

DENTON, D. A., and J. P. COGHLAN, eds. *Olfaction and Taste V.* Academic Press, Inc., New York, 1975.

FIELD, J., H. W. MAGOUN, and V. E. HALL, eds. *Handbook of Physiology,* Sec. 1, Vol. 1. Williams & Wilkins Company, Baltimore, 1959, Chaps. 15–31.

GRANIT, R. *Receptors and Sensory Perception.* Yale University Press, New Haven, Conn., 1955. (Paperback.)

GREGORY, R. L. *Eye and Brain: The Psychology of Seeing.* World University Library, McGraw-Hill Book Company, New York, 1973. (Paperback.)

MOLLER, A. R., ed. *Basic Mechanisms in Hearing.* Academic Press, Inc., New York, 1973.

PADGHAM, C. A., and J. E. SAUNDERS. *The Perception of Light and Color.* Academic Press, Inc., New York, 1975.

UTTAL, W. R. *The Psychobiology of Sensory Coding.* Harper & Row, Publishers, New York, 1973.

VON BÉKÉSY, G. *Sensory Inhibition.* Princeton University Press, Princeton, N.J., 1967.

ZOTTERMAN, Y., ed. *Olfaction and Taste: A Symposium.* Pergamon Press, Inc., Elmsford, N.Y., 1963.

ARTICLES

BRADLEY, R. M., and C. M. MISTRETTA. "Fetal Sensory Receptors." *Physiol. Rev.* **55:** 325 (1975).

BROWN, J. L. "Visual Sensitivity." *Ann. Rev. Psychol.* **24:** 151 (1973).

CASEY, K. L. "Pain: A Current View of Neural Mechanisms." *Am. Sci.* **61:** 194 (1973).

DAVIS, H. "Biophysics and Physiology of the Inner Ear." *Physiol. Rev.* **37:** 1 (1957).

DAW, N. W. "Neurophysiology of Color Vision." *Physiol. Rev.* **53:** 571 (1973).

DEREGOWSKI, J. B. "Pictorial Perception and Culture." *Sci. Am.,* Nov. 1972.

ELDRIDGE, D. H., and J. D. MILLER. "Physiology of Hearing." *Ann. Rev. Physiol.* **33:** 281 (1972).

ERULKAR, S. D. "Comparative Aspects of Spatial Localization of Sound." *Physiol. Rev.* **52:** 237 (1972).

GOMBRICH, E. H. "The Visual Image." *Sci. Am.,* Dec. 1972.

GORDON, B. "The Superior Colliculus of the Brain." *Sci. Am.,* Dec. 1972.

GRANIT, R. "The Development of Retinal Neurophysiology." Nobel Prize Lecture. *Science* **160:** 1192 (1968).

HARTLINE, H. K. "Visual Receptors and Retinal Interactions." Nobel Prize Lecture. *Science* **164:** 270 (1969).

HESS, E. H. "The Role of Pupil Size in Communication." *Sci. Am.,* Nov. 1975.

HODGSON, E. S. "Taste Receptors." *Sci. Am.,* May 1961.

HUBBARD, R., and A. KROPF. "Molecular Isomers in Vision." *Sci. Am.,* June 1967.

LI, C. L. "Neurological Basis of Pain and Its Possible Relationship to Acupuncture-Analgesia." *Amer. J. Clin. Med.* **1:** 61 (1973).

MCILWAIN, J. T. "Central Vision: Visual Cortex and Superior Colliculus." *Ann. Rev. Physiol.* **34:** 291 (1972).

MICHAEL, C. R. "Retinal Processing of Visual Images." *Sci. Am.,* May 1969.

MILLER, W. H., F. RATCLIFF, and H. K. HARTLINE. "How Cells Receive Stimuli." *Sci. Am.,* Sept. 1961.

MOULTON, D. G., and L. M. BEIDLER. "Structure and Function in the Peripheral Olfactory System." *Physiol. Rev.* **47:** 1 (1967).

NASHOLD, B. S., JR. "Central Pain: Its Origins and Treatment." *Clin. Neurosurg.* **21:** 311 (1974).

NOTON, D., and L. STRAK. "Eye Movements and Visual Perception." *Sci. Am.,* June 1971.

OAKLEY, B., and R. M. BENJAMIN. "Neural Mechanisms of Taste." *Physiol. Rev.* **46:** 173 (1966).

OSTER, G. "Auditory Beats in the Brain." *Sci. Am.,* Oct. 1973.

PETTIGREW, J. D. "The Neurophysiology of Binocular Vision." *Sci. Am.,* Aug. 1972.

RATCLIFF, F. "Contour and Color." *Sci. Am.,* June 1972.

ROCK, I. "The Perception of Disoriented Figures." *Sci. Am.,* Jan. 1974.

ROSS, J. "The Resources of Binocular Perception." *Sci. Am.,* Mar. 1976.

TOATES, F. M. "Accommodation Function of the Human Eye." *Physiol. Rev.* **52:** 828 (1972).

WALD, G. "Molecular Basis of Visual Excitation." Nobel Prize Lecture. *Science* **162:** 230 (1968).

WARREN, R. M., and R. P. WARREN. "Auditory Illusions and Confusions." *Sci. Am.,* Dec. 1970.

WERBLIN, F. S. "The Control of Sensitivity in the Retina." *Sci. Am.,* Sept. 1972.

YOUNG, R. W. "Proceedings: Biogenesis and Renewal of Visual Cell Outer Segment Membranes." *Exp. Eye Res.* **18:** 215 (1974).

Chapter 23

Regulation of Movement. I. Muscle and Nerve as a Functional Unit

From dark the striped muscles sprang,
The lion mane, the spotted foot.
Across the dark was gently put
The lengthening and hungry fang.

Onto the open firmament,
With each black star a stepping stone,
Fierce, supple, silent and alone,
Light, the first creature, softly went.

Winifred Welles, "God's First Creature Was Light"

MUSCLES ARE SPECIALIZED tissues which through their unique structure have been able to capture the universal property of protoplasmic contractility and harness it for the production of controlled, forceful, and extremely rapid mechanical work. They are regulated through increasingly complex controls by the nervous system. *Skeletal muscle,* in vertebrates, is responsible for the movements of the body, in cooperation with the passive system of levers and supports afforded by the skeletal system.

In the course of the evolution of humans, with the change from the quadripedal to the bipedal position, many changes have occurred in the weight-bearing components of the *jointed skeleton,* especially those of the pelvic girdle and the vertebral column. The S-shaped vertebral column has adjusted through these changes in shape to the different burdens upon it, as first the head and then the trunk and finally the entire body have had to be supported in the erect position. The pelvic girdle has become shorter and broader, tilting forward to permit the muscles to keep the thigh and leg extended while the trunk is erect. The muscles that must stabilize the joints in this new erect position have become heavier and more powerful; these are chiefly the muscles of the back, gluteal region, and legs. With the lower part of the body assuming the responsibility for the support of the body and for locomotion, the arms are free for the manipulation of tools. This, of course, involves the development of the appropriate *neuromuscular mechanisms of control,* so that one must assume that there is constant interaction and dependency between the use of the muscles and the development of the nervous system; the more the arms and hands are used in fine movement, the greater the influx of impulses to the central nervous system, the finer its control over the muscles, and the more delicate the resulting movement.

The characteristics of *muscle as a tissue* will be considered in this chapter, as well as the *maintenance of posture* and the *coordination of movement* achieved through regulation of muscle by different levels of the central nervous system.

The application of these basic principles to muscle action in stabilizing or moving the joints of the skeleton is discussed in the following chapters.

23–1 Cellular Contractility

Nonspecialized

CYCLOSIS

Cyclosis is the *streaming* of protoplasm seen in all living cells, plant and animal. This streaming is especially dramatic in the slime mold, a primitive noncellular mass in which a rhythmic flow in alternating directions can be seen and even recorded. Cyclosis is caused by changes in the viscosity of the cytoplasm, as it changes regionally from the sol to the gel state. However, although the pulsations can be correlated to electrical changes, the source of the rhythm is quite unknown.

AMEBOID MOVEMENT

A more coordinated protoplasmic streaming is seen in the ameba, in which pseudopod formation is utilized both for movement and for capturing food. *Ameboid movement,* with its continuous extension and withdrawal of streaming fingers of protoplasm, as seen under a microscope becomes almost hypnotic to the observer. One sees the rapid movement of the granules in the liquefying cytoplasm of a newly forming pseudopod stopped by the production of a clear, thick cap of gelled cytoplasm at the tip of the pseudopod. As pseudopodia are produced at one side, they are usually retracted elsewhere, and if a pseudopod attaches to a surface, it can then contract and pull the entire animal along.

We still are far from an adequate understanding of amoeboid movement. According to one of several theories, when the anterior end of the ameba begins to flow forward to form new pseudopodia, the rear of the ameba is contracting, for the cytoplasm in the rear becomes more gel-like (Fig. 23-1). This pushes the interior sol-like cytoplasm still further forward, extending the pseudopod. The stimulus for the initiation of pseudopod formation may be the proximity of a juicy prey, which emits some chemical that affects the membrane of the ameba (chemotaxis). *Integration of the solation occurring in the interior and at the anterior end,*[1]

[1] It may be necessary to define anterior and posterior here, for the ameba certainly has no permanent front or rear. Biologists use a surprisingly pragmatic term of definition: anterior is that portion of the organism that first enters the environment. Consequently, because the ameba is moving into its environment through the formation of pseudopodia, it is sensible to refer to this as its anterior end.

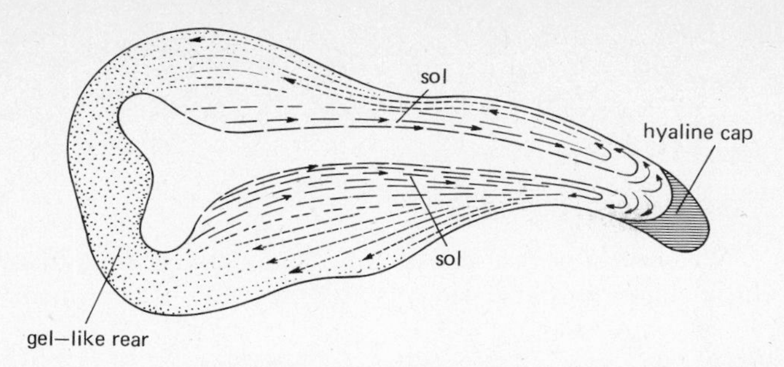

Figure 23-1. **Pseudopod formation in the amoeba may occur through chemotaxis, which stimulates the anterior end to become more sol-like, whereas the rear contracts and becomes gel-like.**

with the *gelation that occurs at the posterior end and laterally,* presumably comes from a series of metabolic and electrophysiological cues. The chemical stimulation results in bioelectrical potentials, which are conducted over the entire surface of the organism. In some unknown manner, this potential change at the site of stimulation must cause solation: gelation occurs further from the stimulus. It is known that these changes in cytoplasmic viscosity can be elicited by alterations in such factors as pH, temperature, and pressure. We can only speculate that the change in membrane permeability must initiate some metabolic change that alters viscosity. It may be an alteration in carbon dioxide production, amino acid metabolism, or rate of metabolism that affects temperature locally. *Chemical regulation* predominates.

This type of protoplasmic movement is dependent on the characteristics of the *protein molecules* that form the ultrastructure of cytoplasm. However, despite the ability of the proteins to form long chains, which not only affect the *viscosity* of the cytoplasm but also are *contractile,* this form of movement is relatively nonspecialized. There is no definitive organization of the proteins into regions utilized mainly for contraction and locomotion, as opposed to those regions concerned with the synthetic activities of the cell.

Specialized Contractile Structures

Cilia and flagella

Cilia and flagella are the whiplike organs of locomotion in many free-swimming unicellular organisms, including the spermatozoan. Cilia and flagella from different types of organisms show the same *basic plan* of nine pairs of peripheral filaments, surrounding an inner core of two filaments (Figs. 2-12 and 2-13). The filaments are *microtubules,* made of the protein tubulin and are attached to the main body of the cell through a *basal plate* which is anchored by rootlets (Fig. 2-14). The entire bundle of filaments is surrounded by the same cell membrane that envelops the rest of the cell. This is obviously specialization of a definite region within a cell, and the filaments are made up of highly organized protein molecules. There is still further specialization of the filaments: the central ones are contractile and so responsible for the movement of the cilium or flagellum, whereas the outer filaments are conductile and, together with the cell membrane, are responsible for transmitting electrical impulses to the central filaments.

The strong beat of cilia is coordinated to beat more strongly in one direction than the other. Cilia of the respiratory mucosa are quite capable of moving particles weighing as much as 5 to 10 g. The cilia will move the particles in one direction only, and this direction is predetermined. If a patch of ciliated cells is removed and grafted with reverse polarity, the cells will continue to beat in their original direction, working against the beating of the neighboring cilia (Fig. 23-2).

Very little is known about the processes that regulate ciliary movement, but both *neural and mechanical* integration have been suggested. It is believed that the cilia whip back and forth as a result of the microtubules sliding over one another. Like other energy-requiring processes, this movement requires ATP and calcium.

Muscle cells

Muscle cells are yet more highly specialized for contraction, and the protein filaments within the cell are even more complex in their organization. In addition, skeletal muscle cells are bound together in parallel bundles by fibrous connective tissue, which ultimately fastens the entire muscle firmly to the fibrous covering of bone at one end and, by way of a tendon, to another bone or rigid structure at the other end. This arrangement efficiently channels the total contractile power of the many muscle fibers that contract together.

Vertebrate Muscle Types. The three main types of vertebrate muscle, *striated, cardiac, and smooth,* were discussed briefly in Sec. 2-5. The physiology of cardiac muscle was described in some detail in Secs. 12-3 and 12-5. This chapter is concerned only with *striated* muscle. Far more is known about the contractile properties of striated muscle than of other muscle types, but it is believed that all share a common molecular mechanism for contraction. It must be realized, however, that there are differences in the details of the contractile responses.

Figure 23-2. **The direction of ciliary beat is predetermined as indicated by the central cilium, which has been transplanted from an area in which beat direction is opposite to that of its new site.**

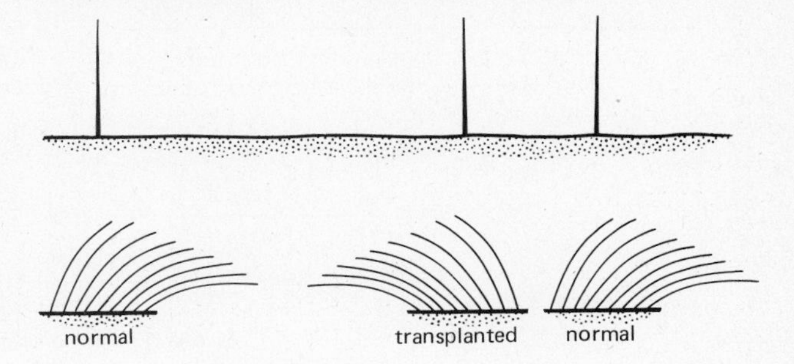

23–2 Ultrastructure of Muscle

The relationships among a muscle, the many muscle fibers (cells) that it contains, and the intracellular contractile elements are depicted in Fig. 23-3. In addition to the usual biochemical machinery of highly active cells, muscle fibers contain specialized *contractile filaments* and a *dual system of membranes* that is intimately involved with the excitation of these filaments and the regulation of their contraction and relaxation.

Muscle Fiber Structure

STRIATED MUSCLE CELL

The *striated muscle* cell is long and slender, containing many nuclei asymmetrically placed near the plasma mem-

Figure 23-3. **The dimensions and arrangement of the contractile components in a muscle. The whole muscle (A) is made up of fibers (B) that contain cross-striated myofibrils (C). The two types of myofibrils are shown in (D). (Adapted from H. E. Huxley and J. Hanson, in *Structure and Function of Muscle,* 2nd ed., Vol. I, G. H. Bourne, ed., Academic Press, Inc., New York, 1973.)**

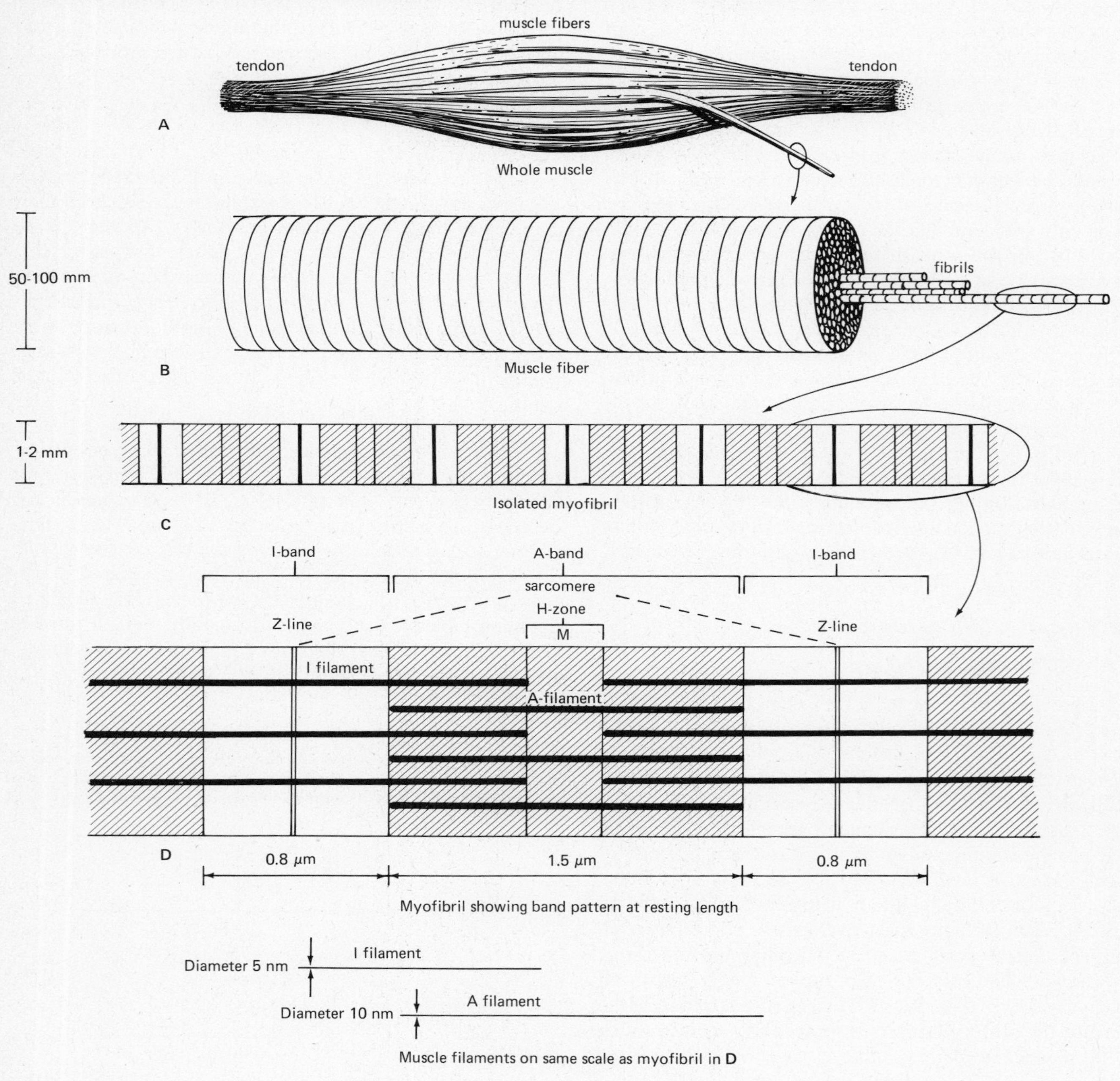

brane, or *sarcolemma.* The cytoplasm of a muscle fiber is the *sarcoplasm* (*sarco* = "flesh"). The sarcoplasm contains many *mitochondria,* as is to be expected in such an active tissue, and *glycogen granules* that can be utilized to yield energy. *Fat droplets* may also be present in varying amounts. The muscle fibers of salmon are heavy with fat droplets just prior to the annual migration upstream, providing a ready source of energy for their exhausting journey.

Myoglobin

Muscle fibers also contain variable amounts of the red, iron-containing protein *myoglobin.* Myoglobin is similar in structure to hemoglobin, but it has a higher affinity for oxygen than hemoglobin does. Consequently, myoglobin acts as an oxygen-storage device, yielding oxygen to the muscle fiber only when cellular P_{O_2} is relatively low.

Muscle enzyme systems

The muscle enzyme systems that are involved in glycolysis, the Krebs citric acid cycle, and oxidative phosphorylation have been thoroughly studied, as have the enzymatic characteristics of the contractile protein *myosin,* which acts as an *ATPase.* Differences in the activities of these enzymes distinguish the three different types of striated muscle fibers, which are discussed a little later in this chapter.

Contractile elements: The myofibrils

The contractile myofibrils of the muscle fiber are precisely oriented longitudinally in the muscle fiber. The structure of the myofibril has been analyzed by several techniques, including X-ray diffraction, electron microscopy, chemical extraction, and immunological reactions, and these different approaches have yielded a coordinated and satisfactory picture. Myofibrils are composed of two types of filaments, the *thick* and the *thin filaments.* It is the repeat pattern of thick and thin filaments, and the areas of their overlapping, that give striated muscle its striped appearance under the microscope (Figs. 23-3 and 23-4A and B).

Thick and Thin Filaments. The two types of myofilaments, the thick and thin filaments, are arranged parallel to one another, overlapping in certain regions. The thick filaments are 11 nm in diameter and 1.5 μm long and form the dark, anisotropic band (*A band*); the thin filaments are 5 nm in diameter and are only 1.0 μm long and form the light, isotropic band (*I band*). (Because of a less ordered arrangement of its structural components, the I band appears light under polarized light. The more orderly arrangement of the A band makes it appear dark under the polarized light microscope.)

In the center of the A band is a narrow, light area called the *H zone.* Because the I bands do not reach this far into the center of the relaxed fiber, this area is lighter than the rest. In the center of the H zone is a dark thin line, the *M line* (Fig. 23-3).

Sarcomeres. One end of each thin filament is attached to the Z line, a flat protein structure which appears as a thin, dark line, perpendicular to the filaments. The distance between two Z lines is a *sarcomere* (Figs. 23-3 and 23-4). Sarcomeres vary considerably in length from the usual 2 to 3 μm of most vertebrate muscles to very long ones (33 μm) found in some invertebrates. When vertebrate muscles increase in length, it is the *number* of sarcomeres that increases, not the length of each sarcomere. The lengths of the thick and the thin filaments of sarcomeres are therefore identical in the neonate and in the adult muscle.

Proteins of the Myofilaments. The *thick filaments* are composed of several hundred long, asymmetrical molecules of the protein *myosin.* The *thin filaments* are made up of three different types of proteins: the globular proteins *actin* and *troponin,* and a long, thin protein, *tropomyosin.* One thin filament probably contains between 300 and 400 actin molecules and about 50 tropomyosins and 50 troponins. The role of these proteins in the contraction and relaxation of muscle is discussed later in this chapter.

Cross-Bridges. Under the electron microscope, the two sets of myofilaments appear to be connected by *cross-bridges* except in the central H zone, which is devoid of such connections (Figs. 23-24A and B). These cross-bridges represent extensions of the *myosin* molecules of the thick filaments, which contact the actin molecules of the thin filaments. The opening and the locking of the cross-bridges are essential factors in the contraction and relaxation of muscle, and the regulation of these processes is discussed in detail in subsequent sections of this chapter.

The dual membranous system

In 1902, long before the advent of the electron microscope, Veratti produced beautiful drawings of the dual membranous system of muscle. Some of the membranes were parallel to the contractile filaments, and others were at right angles, surrounding the filaments like a series of bracelets. These membranes occupy about 15 per cent of the total cell volume. Electron microscopic studies, particularly by Franzini-Armstrong (1964), have permitted the visualization of much more detailed structure of these two systems.

Transverse Tubules. The sarcolemma penetrates by invagination into the substance of the muscle, forming transverse tubules (TT). The mouth of each tubule is open to the extracellular space, but its other end is closed. Hugh Huxley (1969) showed that when a muscle is soaked in a solution of ferritin (large iron particles), these particles travel into the TT as far as the blind end, where they then accumulate (Fig. 23-5).

Figure 23-4A. **Electronmicrograph of several myofibrils, showing the dark and light bands. The thick filaments of the A band and the thin filaments of the I band can be seen. ×53,000 (From H. E. Huxley and J. Hanson, in *Structure and Function of Muscle,* 2nd ed., Vol I, G. H. Bourne, ed., Academic Press, Inc., New York, 1973.)**

Figure 23-4B. **Two adjacent myofibrils are shown at higher magnification than in Fig. 23-4A. Note that the thin filaments extend into the A band, up to the H-zone (clear area in the center). Cross-bridges between the thick and thin filaments are visible. The vertical lines at the top and the bottom of the micrograph are the Z lines that delineate a sarcomere. ×147,000. (From H. E. Huxley and J. Hanson, in *Structure and Function of Muscle,* 2nd ed., Vol. I, G. H. Bourne, ed., Academic Press, Inc., New York, 1973.)**

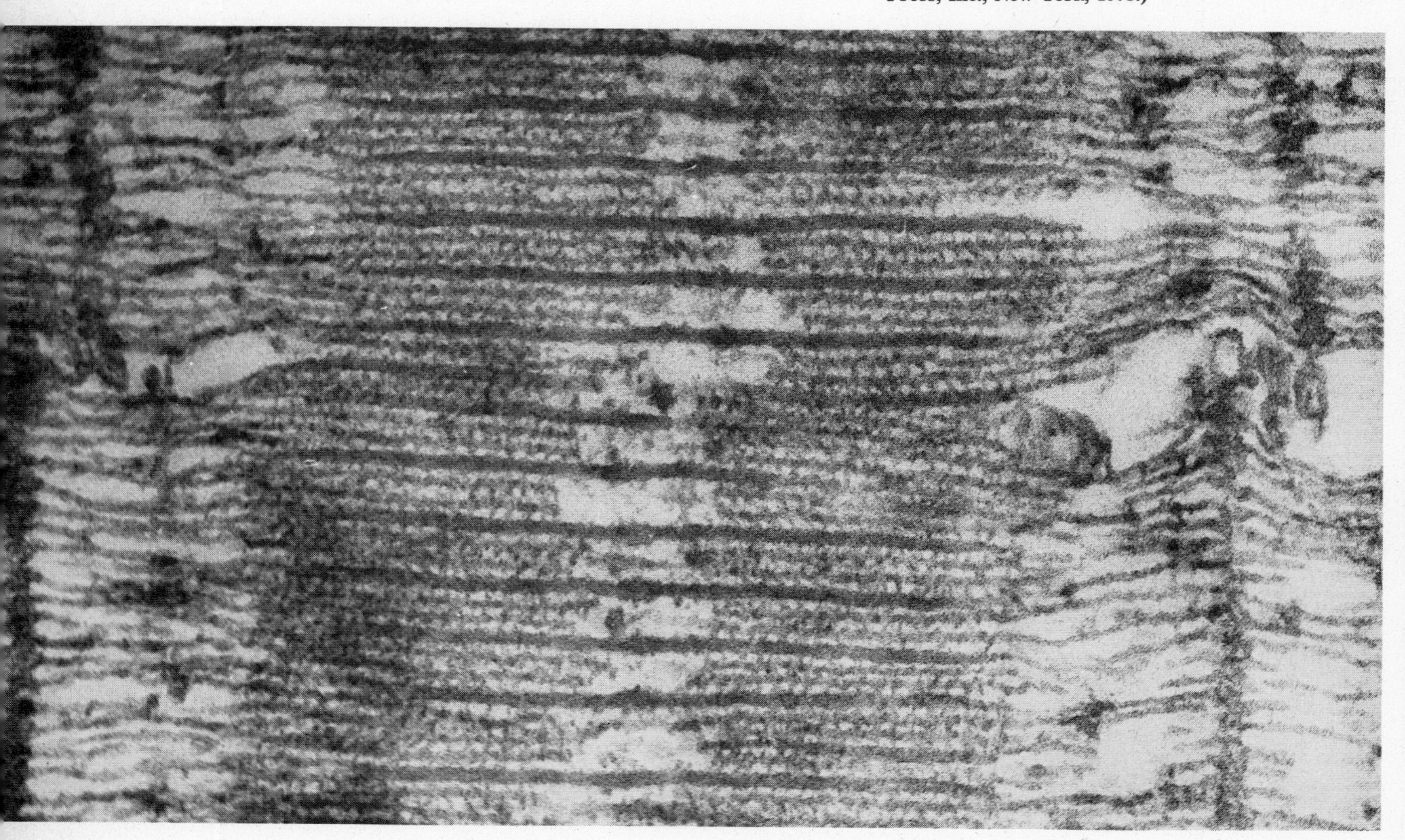

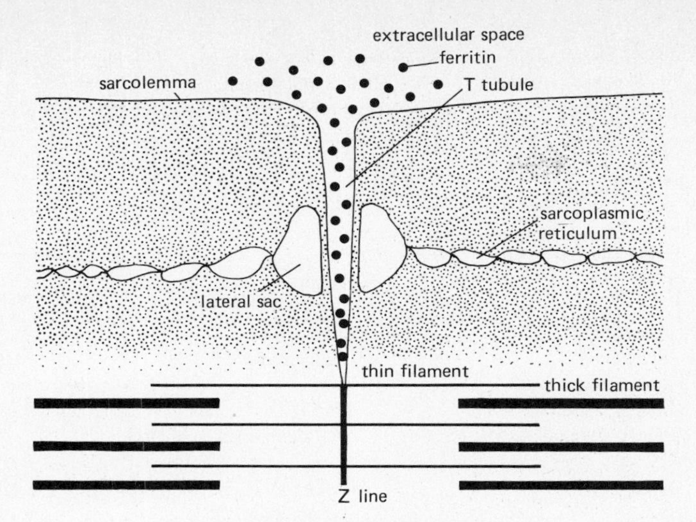

Figure 23-5. **The transverse tubular (TT) system of muscle is open to extracellular space at one end but closed at the other, as shown by the ferritin particles that have penetrated to the end of the tubule. The sarcoplasmic reticulum is a longitudinally arranged series of vesicles, linked together. The enlarged vesicle on either side of the TT is a lateral sac: two lateral sacs and a TT form a triad.**

Figure 23-6. **Electronmicrograph of the body muscle of the black Mollie fish. At every Z line the sarcolemma penetrates into the muscle fiber (arrows) to form the walls of the T tubule. The sarcoplasmic reticulum (SR) has no connection with the sarcolemma and does not open to the outside. At the level of the A band the SR sacs are parallel and oriented longitudinally (asterisk). At the level of the I bands these sacs fuse together to form the dilated lateral sacs (1s) of triad. ×28,000. [From C. Franzini-Armstrong and K. R. Porter, *Nature* 202:355 (1964).]**

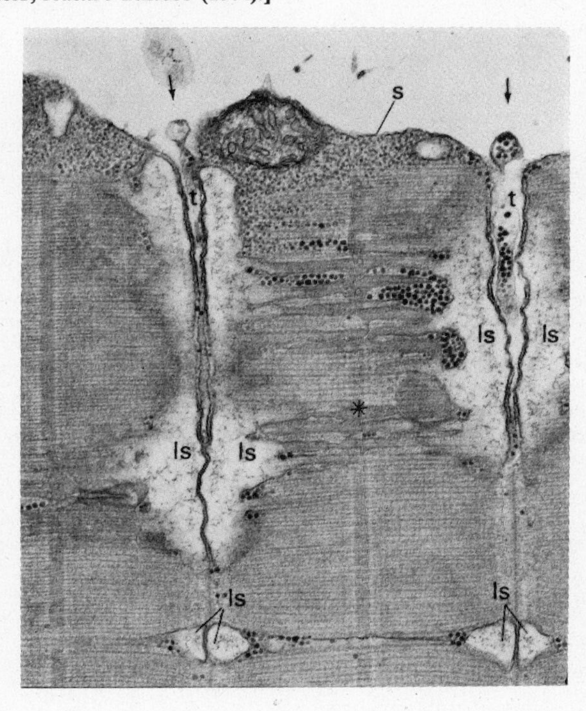

Sarcoplasmic Reticulum. The other membrane system is completely closed and consists of longitudinally oriented, connecting cisterns that run parallel to the contractile filaments. This longitudinal system is the *sarcoplasmic reticulum* (*SR*). Unlike the TT system, the SR is not continuous with the sarcolemma. The SR makes important membranous connections with the TT system, forming *triads* (Fig. 23-6).

Triads. On either side of a transversely oriented tubule is an enlarged cistern, a *lateral sac,* which is continuous with the sarcoplasmic reticulum. The lateral sacs come in close apposition to the TT but are not in direct contact with it. The sacs cover almost 30 per cent of the surface of the TT. Thus a *triad* consists of *two lateral sacs* of the SR and *one TT.* These elements of a triad are close to each other but not continuous (Figs. 23-5 and 23-6). The functional significance of this arrangement will become apparent later in this chapter.

23–3 Types of Striated Muscle Fibers

Mammalian skeletal muscles differ in color, depending upon the amount of myoglobin they contain, varying from deep red to almost white. Because early observations indicated that a *slow* muscle, like the *soleus,* is red, whereas the *rapidly contracting gastrocnemius* muscle is much paler,[2] it was believed that all slow muscles were red and all fast muscles white.

More recently, however, histological and histochemical studies have classified striated muscle into three different fiber types, *slow, fast, and intermediate.* These fibers differ in the type of neuromuscular junctions they contain, in the activities of their glycolytic and oxidative enzymes, and in certain contractile characteristics.

Physiologically, this classification is not satisfactory, for not all red fibers are slow. A better functional classification of the three fiber types is into *fast white, fast red, and slow intermediate* twitch fibers. Evidence for this concept comes mainly from the soleus muscle, which is composed almost entirely of slow twitch fibers, all of which are of the intermediate type. Most other muscles are heterogeneous and contain all three histologically distinct types. However, each motor unit (Fig. 7-6) contains only one type of twitch fiber. Table 23-1 indicates the distribution and characteristics of fiber types in some muscles.

Functions of Fiber Types

Close (1972) has pointed out a satisfactory correlation between the overall functions of these three muscle types and their *physiological properties:*

1. The *fast red fibers,* with a short contraction time, high oxidative enzyme activity, and many mitochondria, are

[2] The functions of these muscles of the lower leg are discussed in more detail in Chap. 25.

Table 23–1. **Distribution and Characteristics of Muscle Fiber Types**

	Red	Intermediate	White
% Fibers			
Diaphragm muscle	60	20	20
Soleus muscle		90–100	
Gastrocnemius muscle	30	20	50
Physiological Properties			
Speed of contraction	Fast	Slow	Fast
Fatigue	Moderate	Little	Rapid
Histological Characteristics			
Fiber diameter	Small	Intermediate	Large
Neuromuscular junction	Small, simple	Intermediate	Large, complex
Number of mitochondria	Many	Intermediate	Few
Z line	Broad	Intermediate	Thin
Sarcoplasmic reticulum	Well developed, with triads	Poorly developed, with dyads*	Well developed, with triads
Histochemical Characteristics			
Myoglobin	High	High	Low
Glycogen content	High	Low	Intermediate
Myosin ATPase activity	High	Low	High
Glycolytic enzyme activity	Intermediate	Low	High
Innervation			
Motoneurons	Phasic, fire at 30–60/sec	Tonic, fire at 10–20/sec	Phasic, fire at 30–60/sec

* One lateral vesicle forms a junction with a transverse tubule.

quite resistant to fatigue. These characteristics make the fast red fibers ideal for *prolonged phasic contractions,* typical of a muscle like the diaphragm.

2. The *fast white fibers,* although they possess a well-developed glycolytic system (Sec. 5–12) for immediate bursts of energy, have relatively few mitochondria and rather low oxidative enzyme activity. These fibers fatigue rapidly and so are most suited for *short-term, powerful phasic contractions,* characteristic of muscles used in running, like the gastrocnemius.
3. The *slow intermediate fibers* are low-speed, economical units. These fibers have many mitochondria, a highly active oxidative enzyme system, and a rather poorly developed glycolytic system. These slow fibers show little or no fatigue and so are well adapted for the sustained tonic activity necessary for the maintenance of posture. The soleus is a good example of this type of muscle.

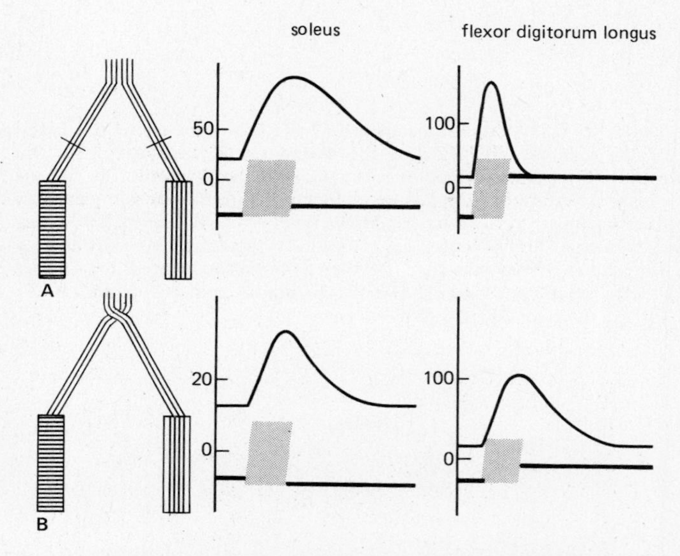

Figure 23-7. **Cross-union of motor nerves between slow (soleus) and fast (flexor digitorum longus) changes the contractile characteristics of these muscles. (A) shows the contractions of the soleus and flexor muscles after the motor nerves to these muscles had been severed and rejoined as a control operation. (B) shows the increased speed of the soleus and slowing of the flexor muscle after the cross-suturing of their motor nerves (diagram on left). The decade-counter device (lower trace) gives the contraction time in msec. [Redrawn from A. J. Buller, J. C. Eccles, and R. M. Eccles, *J. Physiol.* 150:417 (1960).]**

Regulation of Fast and Slow Fibers by Nerves (Trophic Action)

Cross-union of motor nerves

In the early embryological development of muscle, all striated muscle fibers are slow, but as the fibers differentiate, the fast fibers shorten their time of contraction. Eccles and his co-workers began an ingenious set of experiments in 1960 designed to investigate the influence of peripheral nerves on developing muscles in very young kittens (Eccles, 1967). They united a fast (or tonic) motor nerve to a slow muscle, and a slow (or phasic) motor nerve to a fast muscle (Fig. 23-7). These experiments, and subsequent work by other scientists on the biochemical and physiological changes in developing and adult muscles, yielded important conclusions. As a result of cross-union of motor nerves, the following changes occur:

1. The contraction of fast muscles is slowed.
2. The contraction of slow muscle is accelerated.
3. The enzymatic characteristics of fast and slow muscles

are reversed. That is, the activity of myosin ATPase in fast muscle decreases after innervation by the slow soleus nerve, and the activity of this enzyme is increased in slow muscles after innervation by the fast gastrocnemius nerve.
4. The maximal capacity of the SR to take up calcium is reversed in slow and fast fibers.

Denervation sensitivity

An important role for nerve in the regulation of the sensitivity of muscle to acetylcholine has been shown by experiments in which the peripheral motor nerve to the muscle has been cut. Not only does the muscle *atrophy* following denervation, but the entire surface of the muscle becomes highly sensitive to acetylcholine. This *supersensitivity* appears to be related to a change in the number and distribution of *acetylcholine receptors* on the sarcolemma. The acetylcholine receptors of a fully innervated muscle are restricted mainly to the tops of the junctional folds in the region of the neuromuscular junction. Acetylcholine sensitivity is confined accordingly to this area. In the denervated muscle, however, acetylcholine receptors are distributed over the entire sarcolemma (Fig. 23-8). It is not known whether these receptors are entirely new membrane proteins synthesized by the denervated cell, or whether previously existing "blind" or "covered" receptors become exposed and available to bind acetylcholine, once the influence of the nerve has been removed.

These observations have led to the concept of motor nerve regulation of the morphological, biochemical, and contractile properties of skeletal muscle. It is assumed that there is a *trophic* (nourishing) *substance* in the nerve that travels down the axon by axonal transport, and that this trophic substance (tentatively called mysterine) is essential for the healthy development and maintenance of muscle.

Figure 23-8. **(A) In normal muscle, the tops of the junctional folds of the sarcolemma contain almost all the acetylcholine receptors. However, in denervated muscle (B), the acetylcholine receptors are spread over the entire sarcolemma.**

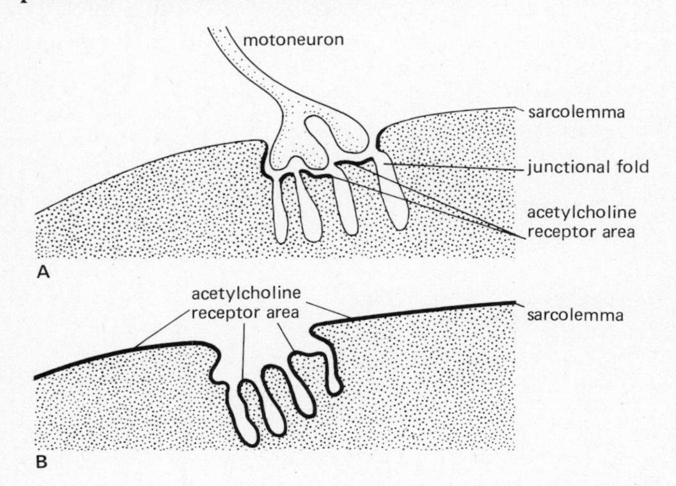

It is interesting that the time needed for denervation sensitivity to set in is directly proportional to the length of the remaining nerve stump still attached to the muscle. One may infer that the longer the stump, the more of the trophic nerve substance remains available to regulate the muscle.

23–4 Sliding Filament Theory of Contraction

Maintenance of the Length of the Filaments During Contraction

Although the muscle cell itself shortens when it contracts, careful analysis of the bands and filaments within the fiber shows that there is no change in length of either the thick or the thin filaments. The H zone in the middle of the A band disappears, however, and the Z line becomes considerably darker. The sliding filament theory proposed by Hanson and Huxley (1955) neatly includes all these factors (Fig. 23-9).

The theory convincingly explains the *disappearance of the H zone* on the basis of the *sliding-in of the I filaments* toward the center. This means that the central area now contains both thick and thin filaments and therefore not only is as dark as the rest of the A band, but may even be darker if the I bands from the one side overlap those from the other. As the thin filaments slide into the central region, they pull the Z lines with them, and the ends of the thick filaments appear to become entangled in the Z lines, making them darker and heavier.

Molecular Mechanisms Regulating the Sliding of the Filaments

The four proteins of the contractile filaments are *myosin, actin, tropomyosin,* and *troponin.* These proteins can be extracted and purified and many of their properties studied in vitro. In the 1940s, Szent-Györgyi showed that if actin is added to myosin, a complex, *actomyosin,* is formed. In the presence of ATP, actomyosin *superprecipitates,* an in vitro reaction comparable to contraction. Since that time, the mechanisms involved in the regulation of the rapid cycle of contraction and relaxation in the living muscle have been under extensive study.

Organization of the proteins in the myofilaments

The *thick filaments* are composed of *myosin,* an elongated protein with a tail and two globular heads. The heads consist of *heavy meromyosin,* a protein with a molecular weight of 490,000; the tail consists of the smaller, *light meromyosin,* with a molecular weight of 135,000.

The myosin molecules are oriented on either side of the H zone with their tails, which form the core of the thick filament, pointing toward this central bare region devoid of

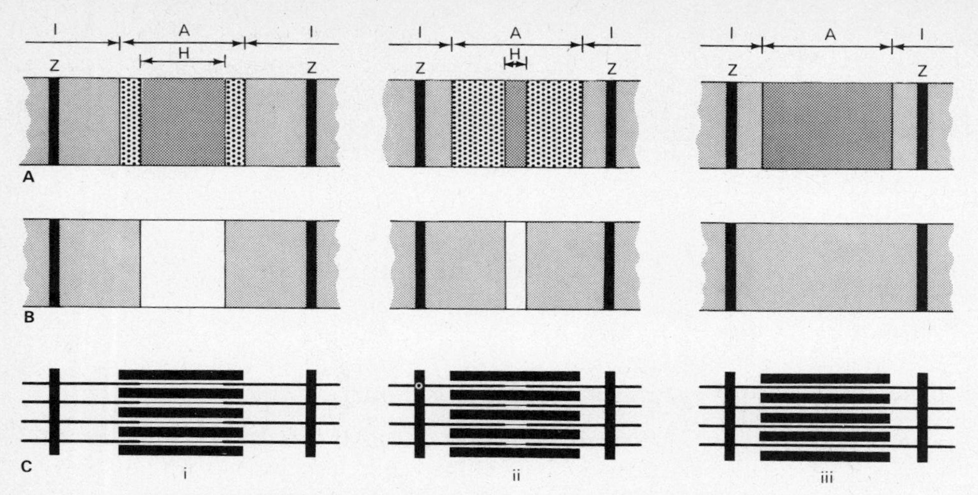

Figure 23-9. **Structural changes in muscle with contraction and extension. The three diagrams in the central column (ii) represent the sarcomere at resting length. The top row (A) shows the band pattern of the intact fibrils. The next row (B) shows the pattern after extraction of myosin: the A band disappears. The bottom row (C) shows the position of the thick and thin filaments at resting length.**

In the column on the right (iii) are shown the changes on contraction. The H zone has disappeared and the I filaments touch in the center of the sarcomere.

In the column on the left (i) are shown the changes on extension. The H zone is much wider because the I filaments have slid farther apart, leaving a free area in the center. (From H. E. Huxley and J. Hanson, in *Structure and Function of Muscle,* 2nd ed., Vol. I, G. H. Bourne, ed., Academic Press, Inc., New York, 1973.)

Figure 23-10. **It is the heads of the myosin molecules that form the cross-bridges with actin molecules. Because there are no myosin heads in the central H-zone, there are no cross-bridges either in this region.**

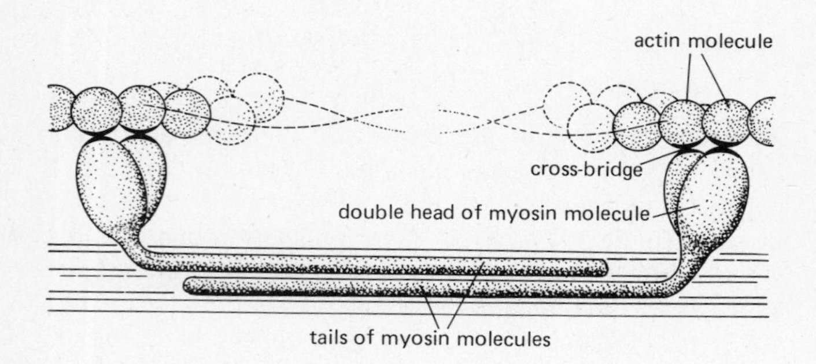

myosin heads (Fig. 23-10). The heads of the myosin molecules project out to contact the actin molecules of the thin filaments. It is the *myosin heads,* therefore, that form the *cross-bridges* linking the thick and thin filaments.

In addition to their ability to bind to actin, the myosin heads *bind and hydrolyze ATP,* an energy-yielding reaction essential for the breaking of the cross-bridges prior to contraction.

The *thin filaments* contain the other three proteins *actin, troponin,* and *tropomyosin.* The small, globular actin molecules form most of the thin filament. The actin molecules, which have a "front-to-back" polarity, are arranged like a double strand of beads, with a groove between them, and this strand is twisted into a helix. In this configuration, all the "fronts" of the actin molecules face in one direction.

Figure 23-11. **The long, thin tropomyosin molecules run in the groove between the double twisted strand of actin molecules. The globular troponin molecules mark the end of each tropomyosin molecule.**

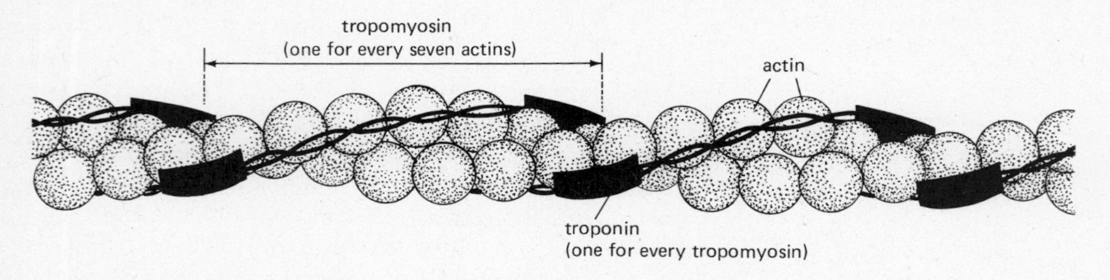

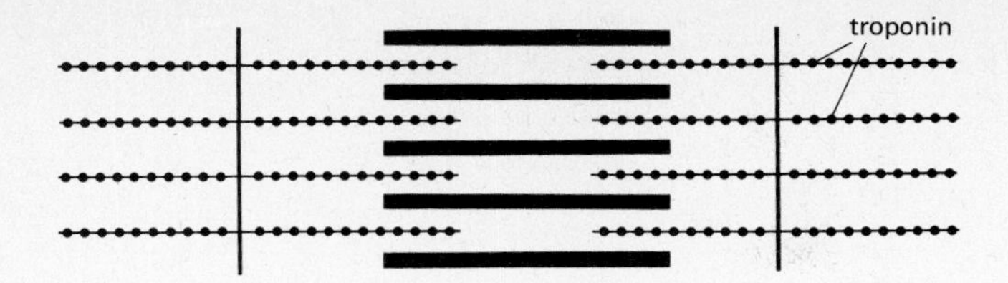

Figure 23-12. **When muscle fibers are treated with antibody to troponin, the reaction sites show the troponin molecules to be regularly distributed every 40 nm along the thin filament.**

Running along the groove of each actin strand is a band of tropomyosin, a long thin molecule that covers the length of seven actin molecules. The globular troponins are spaced about 40 nm apart, approximately marking the end of each tropomyosin molecule (Fig. 23-11). When muscle is treated with an antibody against troponin, the antibody-troponin complexes can be located every 40 nm along the thin filaments (Fig. 23-12).

REGULATION OF MUSCLE CONTRACTION

When a muscle contracts, the myosin heads on either side of the H zone contact the actin molecules, and by swivelling in opposite directions, pull the opposed sets of thin filaments in toward each other (Fig. 23-13). Each sarcomere is shortened as the thin filaments slide in towards the center, with the thick and thin filaments retaining their original length. The cross-bridges on the opposite sides of the bare H zone swivel in opposite directions to shorten the sarcomere.

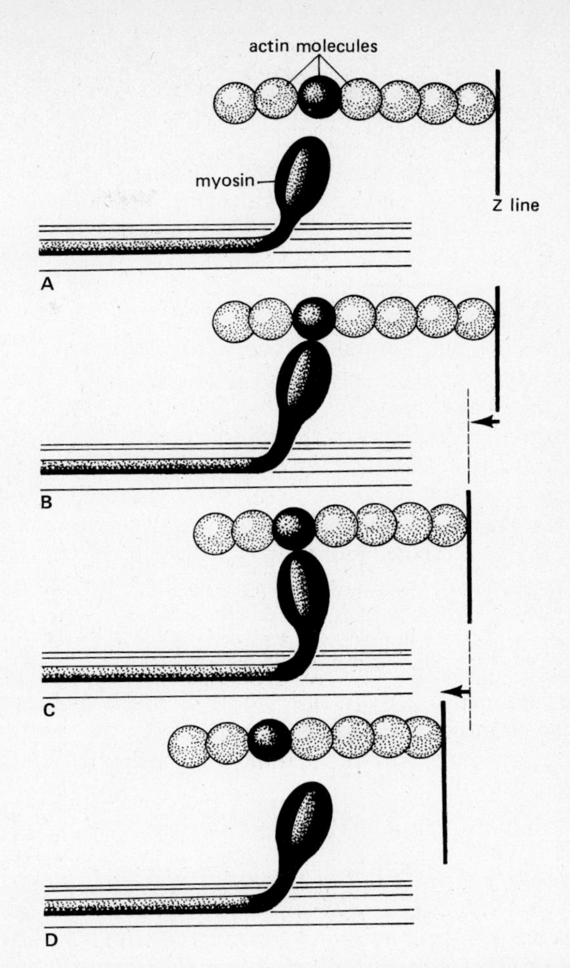

Figure 23-13. **Muscle contracts when the action potential initiates events that cause myosin heads to attach to actin [change from (A) to (B)], swivel (C), then break contact (D), propelling the thin filaments past the thick filaments and shortening the sarcomere.**

Regulatory Action of Calcium. Myosin is not only involved in the mechanical aspects of the contractile process; it also acts as an enzyme. Each myosin head has two binding sites, one that binds to actin and a second site that has a very high affinity for ATP. Myosin, acting as an ATPase, hydrolyzes ATP to release the energy required for contraction.

It has long been known that the hydrolysis of ATP by myosin is absolutely dependent upon calcium ions. Artificially prepared actomyosin filaments to which ATP has been added will contract in the presence of Ca^{2+} but will relax if Ca^{2+} is absent.

$$\text{actomyosin} + \text{ATP} + Ca^{2+} \longrightarrow \text{contraction}$$
$$\text{actomyosin} + \text{ATP} - Ca^{2+} \longrightarrow \text{relaxation}$$

Note that ATP is also needed for relaxation; however, in the absence of calcium, ATP is not hydrolyzed. The significance of this will be discussed later.

REGULATION OF CALCIUM CONCENTRATION BY THE SR AND TT. *Calcium Uptake by the SR.* Our understanding of the regulatory action of calcium on muscle contraction began with the discovery by Marsh (1952) that extracts of homogenized muscle caused actomyosin filaments to which ATP has been added to relax. This extract, named the *relaxing factor,* was unable to induce relaxation if calcium ions were added to the system.

$$\text{actomyosin} + \text{ATP} + \text{relaxing factor} \longrightarrow \text{relaxation}$$
$$\text{actomyosin} + \text{ATP} + \text{relaxing factor} + Ca^{2+} \longrightarrow \text{contraction}$$

Next came the suggestion by Annemarie Weber in 1959 that the sarcoplasmic reticulum might be the relaxing factor. The SR actively takes up calcium from the sarcoplasm, thus permitting relaxation. Muscles treated with oxalate show the deposition of insoluble calcium oxalate salts exclusively in the SR.

Calcium Release and Excitation-Contraction Coupling. Calcium is stored in the cisterns of the SR, to be released when the muscle is stimulated. The stimulus travels as an action potential along the sarcolemma and down the transverse tubules. In some unknown manner, this signal is "sensed" by the SR. There is no direct contact between these two membrane systems, so that it is unlikely that they communicate by direct electrical coupling. However, as a result of the close proximity of the lateral sacs and the TT that forms the triad, there may be a change in the membrane potential of the SR and a resulting change in permeability of the cisterns. *Depolarization of the TT is followed by release of calcium ions from the SR.* The increased Ca^{2+} concentration in the sarcoplasm permits the hydrolysis of ATP, and the muscle contracts.

In this manner, through the cooperative action of the two membrane systems of muscle in regulating the cycling of Ca^{2+}, the excitation evoked by the stimulus is coupled to the contractile mechanisms of the muscle fiber. This is *excitation-contraction coupling* (Sandow, 1965). Both calcium release and calcium uptake are extremely fast processes that take place within a fraction of a second. Calcium release is enhanced by caffeine, another way in which this substance may act to delay fatigue, apart from its inhibitory effect on the enzyme that activates cAMP (Sec. 10-8).

Further proof for the importance of the TT in excitation coupling is provided by experiments in which the TT have been disrupted by hypertonic glycerol. These tubules still remain connected to the sarcolemma, and the excitation mechanism remains, but the contractile response is eliminated.

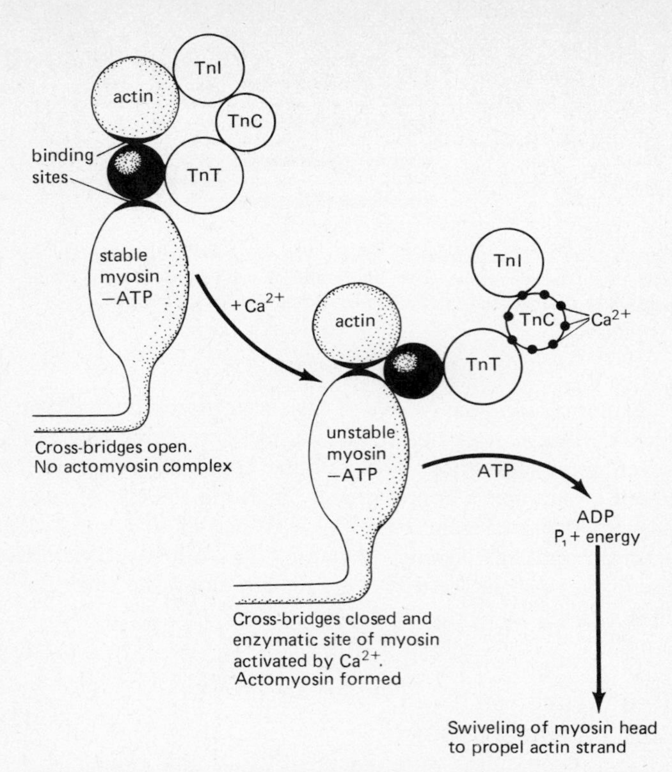

Figure 23-14. **Calcium plays a dual role in muscle contraction. In the absence of Ca^{2+}, tropomyosin (black ball) blocks the binding sites of actin and myosin. The three subunits of troponin (TnT, TnC, and TnI) stabilize this blocking position of tropomyosin. When Ca^{2+} is released in the active state, TnC binds the Ca^{2+}, strengthening the bonds between the Tn subunits and pulling them, together with tropomyosin, away from the binding sites of actin and myosin. The actomyosin complex is formed, and the cross-bridges close. The second role of Ca^{2+} is to charge the stable myosin-ATP complex, making it unstable with a great tendency to bind to actin; at the same time the enzymatic site of myosin is activated, and ATP is hydrolyzed to ADP, Pi (inorganic phosphate), and energy.**

Regulatory Role of the Troponin (TN)-Tropomyosin (TM) System. The calcium sensitivity of actomyosin is lost if the actin and myosin components are highly purified. Under these conditions, actomyosin will superprecipitate in the presence of ATP, whether or not calcium is in the system. Ebashi (1974) concluded that the missing ingredients needed for normal calcium sensitivity were tropomyosin and/or troponin. By adding these proteins separately and together to actomyosin and ATP, he found that *both* tropomyosin and troponin were needed to restore calcium sensitivity. Thus, the complete system in the highly purified state requires:

$$\text{actomyosin} + \text{tropomyosin} + \text{troponin} + \text{ATP} + Ca^{2+} \longrightarrow \text{superprecipitation}$$

In order to explain the role of the Tn-Tm system in the regulation of calcium sensitivity of actomyosin, a model based on the differing characteristics of the *three subunits of troponin* has been formulated as follows:

1. The largest subunit of troponin binds to tropomyosin and is appropriately called *TnT* (molecular weight 24,000).
2. A small inhibitory unit that binds to actin is *TnI* (molecular weight 24,000).
3. A very small unit that binds Ca^{2+} is *TnC* (molecular weight 18,000).

These subunits affect the formation of the actomyosin complex through their ability to *cover or expose* the necessary *binding sites* for actin to bind to myosin. The relationship between Ca^{2+} and the subunits is described below and illustrated in Fig. 23-14. Much of the evidence for the changes in position of the tropomyosin molecules comes from X-ray diffraction studies of living frog muscle.

In the resting state, tropomyosin effectively blocks the binding sites of actin and myosin, preventing the formation of the actomyosin complex; consequently no cross-bridges can be made. Tropomyosin is held in this blocking position by the TnT subunit to which it is bound. In addition, this active state is stabilized by the position of the TnI subunit, which also binds to actin.

In the active state, following the release of Ca^{2+} by the SR, TnC binds the calcium ions, thus strengthening the links between the Tn subunits and pulling them, together with the attached tropomyosin, away from the binding sites of actin. This permits myosin to bind actin and form the *actomyosin complex.* The cross-bridges are consequently closed.

Regulatory Role of ATP. Although calcium is essential for the actomyosin complex to be formed, the sliding of the filaments requires the cross-bridges to undergo a rhythmic process of attachment, swivelling, detachment, and reattachment further along the thin filament. For this to occur, the formation and breaking of the cross-bridges must be under rigorous control; the controlling molecule is ATP.

1. Because of the very high affinity of myosin for ATP, there is normally one ATP molecule attached to the special ATP binding site on each myosin head. *This myosin-ATP^1 complex is stable.*
2. Probably because of the release of Ca^{2+} from the SR, this stable myosin-ATP complex becomes a charged intermediate of very high instability and rapidly binds to actin (the binding sites have been exposed by the removal of the troponin-tropomyosin complex by Ca^{2+}).

$$\underset{\text{(charged)}}{\text{myosin-ATP}^1} + \text{actin} \rightarrow \underset{\text{(active complex)}}{\text{actomyosin-ATP}}$$

3. In the active complex of actomyosin-ATP, the enzymatic site of myosin becomes activated, and ATP is rapidly hydrolyzed to ADP, inorganic phosphate, and energy. The energy is utilized for the swivelling of the myosin heads. The thin filaments are pulled in a notch, equivalent to about 0.5 nm for each side, making the total movement 1 nm for every two molecules of ATP hydrolyzed. *Magnesium* is essential for the hydrolysis of ATP.

$$\text{actomyosin-ATP}^1 \rightarrow \text{actomyosin-ADP} + \text{Pi} + \text{energy}$$

4. *Actomyosin-ADP* is a very low-energy, stable complex and requires the addition of a new molecule of ATP (ATP^2) to free the actin and myosin molecules. This opens the cross-bridges and permits sliding of the filaments. (ATP^1 and ATP^2 indicate additional molecules of ATP, not differences in molecular structure).
   (a) *If ATP^2 is available,* the cross-bridges open, and myosin-ATP^2 is formed. *If Ca^{2+} is present,* the recharged myosin-ATP^2 can reattach to another actin molecule further along the filament, pulling the thin filament further in and *increasing the amount of contraction.*

$$\underset{\text{(stable)}}{\text{myosin-ATP}^2} + Ca^{2+} \rightarrow \underset{\text{(charged)}}{\text{myosin-ATP}^2} + \text{actin} \rightarrow \underset{\text{(active complex)}}{\text{actomyosin-ATP}^2}$$

   (b) *If ATP is available but Ca^{2+} has been removed* by the SR, then the cross-bridges open, myosin-ATP^2 (uncharged) is formed, but there is no recharging of this stable complex. Consequently, no new cross-bridges develop, and the freed filaments slide apart in relaxation.
   (c) *If there is no ATP* available, the stable actomyosin-ADP complex persists, the cross-bridges are locked, and rigor (irreversible contracture) occurs. This is the *rigor mortis* of death, for no ATP is being synthesized, and the muscle becomes completely rigid and inextensible.

From this discussion, it is apparent that ATP is required for both contraction and relaxation. For *contraction* to occur, the ATP must be hydrolyzed and the energy used for the swivelling of the myosin heads and the pulling in of the thin filaments. For *relaxation* to occur, the ATP is required to free the myosin from the actin, opening the bridges and permitting the sliding apart of the filaments. The regulatory processes involved in the sliding of the myofilaments are summarized in Fig. 23-15.

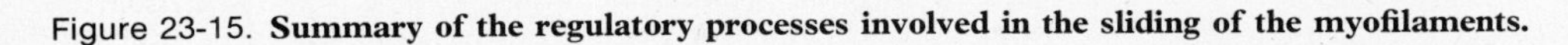
Figure 23-15. **Summary of the regulatory processes involved in the sliding of the myofilaments.**

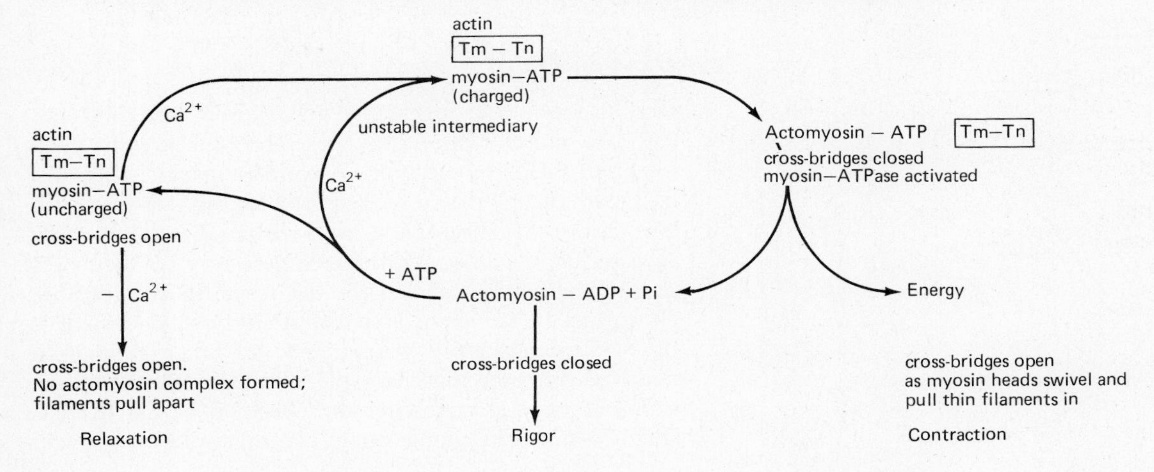

23-5 Sources of Energy for Muscle Contraction

Anaerobic Reactions

From the preceding discussion, it is obvious that the direct source of energy for the contraction of the muscle comes from the breakdown of ATP. There are limited stores of this high-energy compound in the tissues, and for its resynthesis the tissues must draw on other reserves of energy. The immediate source of energy for ATP resynthesis comes from the breakdown of another high-energy compound known as *creatine phosphate*, a reaction catalyzed by *creatine kinase*. Creatine phosphate (CP), in turn, must be resynthesized because it, too, is present only in small quantities in the tissues. The ultimate source of energy, then, comes from the constantly replenished supply of ingested food, especially *carbohydrate* and *fat* (Secs. 18-3, 18-4: Fig. 23-16).

Muscle is able to contract for a short time in the *absence of oxygen* because it can use the energy coming first from the *breakdown of ATP* present in the tissues, and secondly from the ATP that can be resynthesized while the supply of CP holds out. These chemical reactions do not require oxygen. A third energy yielding series of reactions that does not require oxygen is the *glycolytic pathway*. Through this pathway, glucose is anerobically metabolized to pyruvic acid. If no oxygen is available, pyruvic acid is rapidly reduced to lactic acid.

Figure 23-16. **Chemical reactions supplying energy for muscle contraction. The energy-yielding breakdown of ATP and CP are anaerobic reactions, as are the glycolytic reactions that yield pyruvic acid. The chemical reactions that yield the greatest amount of energy, however, are the aerobic reactions following decarboxylation of pyruvic acid to acetyl coenzyme A, which enters the citric (tricarboxylic) acid cycle. In the absence of oxygen, lactic acid is formed and when oxygen once more is available, four fifths of the lactic acid enters pathways that result in the resynthesis of glycogen. The energy required for these synthetic reactions comes from the remaining one fifth of lactic acid passing through the citric acid cycle.**

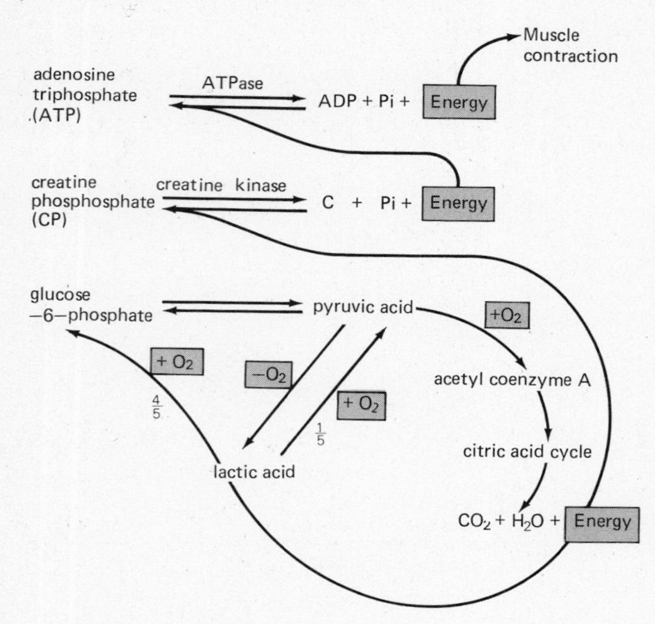

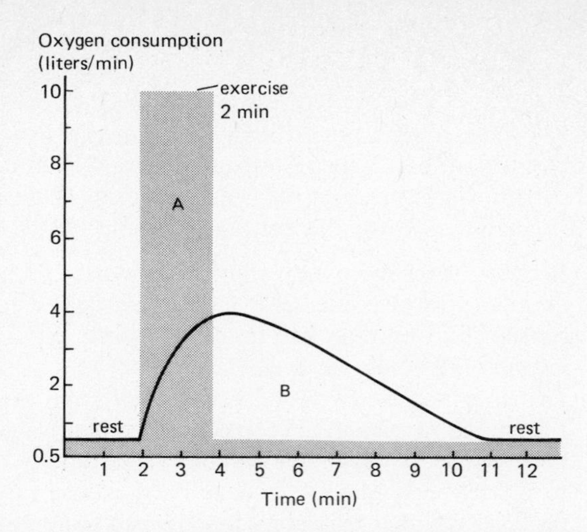

Figure 23-17. **Oxygen need (shaded area) and oxygen consumption (curve) during strenuous exercise. Area (A) represents the amount of oxygen needed above the actual amount consumed (oxygen debt). Area (B) represents the oxygen consumption above the actual need after exercise (payment of the oxygen debt). From *Human Physiology* by B. Houssay. Copyright 1955, McGraw-Hill Book Company, New York. Used with permission of McGraw-Hill Book Company.)**

Aerobic Reactions

In moderate activity, the supply of oxygen to the muscles is usually adequate to prevent the accumulation of lactic acid, and pyruvic acid is metabolized to acetyl coenzyme A, which enters the citric (tricarboxylic) acid cycle (Sec. 5-12 and Fig. 23-16). The energy yield from each turn of this wheel, together with the energy derived from the oxidation of hydrogen atoms through the electron transport chain, is very large. Each molecule of glucose that is completely oxidized yields 36 molecules of ATP, but these reactions depend entirely on adequate supplies of oxygen reaching the muscles.

Oxygen Debt

Despite the loss in long-range efficiency in anaerobic metabolism, muscles can utilize these anaerobic energy stores for brief, intensive exertions. A sprinter may actually hold his breath during the 10 sec required for the 100-yard dash. The cost of this anaerobic work is measured in terms of the *oxygen debt*, the amount of oxygen necessary for the oxidation of the lactic acid and for the resynthesis of creatine phosphate. In strenuous activity, the lactic acid content of the blood rises sharply; when it is so high that it cannot be removed from the muscles rapidly enough, muscle fatigue and soreness occur, forcing the cessation of the exercise (Fig. 23-17). This homeostatic control limits the oxygen debt to about 15 liters, an amount that will vary individually with the efficiency of the mechanisms that remove lactic acid from the circulation.

Repayment of the oxygen debt

Repayment of the oxygen debt takes place after the exercise has ceased. The oxygen intake remains above normal until the lactic acid has been removed by:

1. Oxidation to pyruvic acid.
2. Resynthesis in the liver to glycogen.
3. Neutralization by the bicarbonate of the blood and its excretion as sodium lactate by the kidneys.

23-6 Work and Heat Production by Muscle

In muscular contraction, a chemical reaction is coupled with a mechanism capable of generating tension and liberating *heat,* or capable of shortening and producing both *work* and *heat.*

Isometric Contractions

When muscle is prevented from shortening—by a weight that it cannot move, for example—it generates tension when it is stimulated, but its *length remains constant.* This is isometric contraction. The myofilaments do not slide in isometric contraction. Most of the energy derived from the chemical reactions that proceed during isometric contraction is transduced into *heat* and *tension.*

In the strictly physical sense, *no work* is done during isometric contractions because the muscle does not *move* any load. However, in biological terms, much energy is expended to maintain the load, as anyone who has tried to hold a weight above the head for any length of time is aware. This type of contraction is analogous to the tension developed by the postural muscles acting against gravity. The muscles of the back and the extensors of the legs and thighs contract against the weight of the body, the tension that they generate being adjusted by the proprioreceptors to balance exactly against the force of gravity and keep the body upright (refer also to Sec. 25-2).

Isotonic Contractions

If the muscle is permitted to shorten, the tension generated remains relatively constant, and this type of contraction is referred to as *isotonic contraction.* As the muscle shortens, it moves the load a certain distance and work is done:

$$\text{work} = \text{load} \times \text{distance moved}$$

Using the same example of a weight being lifted over the head, isotonic contractions of the arm muscles result in the *movement* of the arm to the desired position; the work done will be the weight lifted through this distance. Once the arm is above the head, however, isometric contractions of the antagonistic muscles keep the elbow joint rigid and *hold the weight in place.* Most of the responses of the muscles of the body are a similar combination of isotonic and isometric contractions. This is discussed further later in this chapter.

Although muscle is a highly efficient transducer of chemical energy into mechanical energy, even in isotonic contractions much of the energy is liberated as *heat.* This is not a biologically wasteful by-product, for this muscle-produced heat is used to maintain body temperature, as was discussed in Sec. 21-1.

Heat Production by Muscle

In the studies of frog metabolism for which they received the Nobel Prize in 1922, Hill and Meyerhof (Hill, 1931) showed that the contraction of muscle is always accompanied by the liberation of heat, which under aerobic conditions may be differentiated into the *initial heat* and the *delayed heat* (Fig. 23-18).

Initial heat

The initial heat is liberated in two bursts during the time that corresponds to the mechanical contraction and relaxation of the muscle, and is over in less than one second. The first component of the initial heat is known as the *activation heat* and accounts for 40 per cent of the total initial heat. The second component is the *shortening heat,* which is responsible for the remaining 60 per cent. Neither of these two bursts of initial heat is affected by an *anaerobic environ-*

Figure 23-18. **Correlation of the heat output of muscle with the contractile events of the twitch and with oxygen dependency. The shaded areas of the graph are unaffected by anaerobic conditions. The initial heat may be divided into activation heat (A) and shortening heat (B). The first part (C) of the delayed heat is independent of oxygen, but the large burst of delayed heat requires oxygen. Note the break in the time scale.**

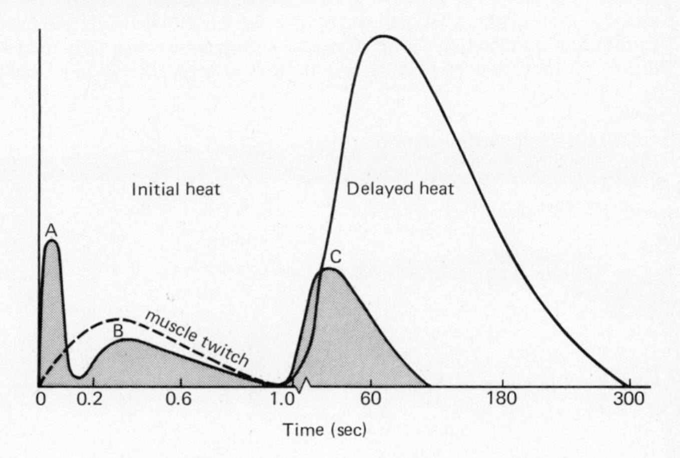

ment (an atmosphere in which the oxygen content is replaced by nitrogen).

DELAYED HEAT (RECOVERY HEAT)

The delayed heat follows the initial heat and may take up to 300 sec. It corresponds to the *recovery* processes in muscle, starting only when the muscle has completed its mechanical relaxation. There is a small, short burst of recovery heat that is independent of oxygen, followed by a much greater, longer-lasting surge of oxygen-dependent recovery heat (Fig. 23-18).

23-7 Correlation of Chemical and Thermal Events of Muscle Contraction

Attempts to correlate the chemical and thermal events occurring during and after the mechanical response of muscle have not been completely successful.

Anaerobic Reactions

INITIAL HEAT

Activation Heat. The hydrolysis of ATP and CP account for part of the heat liberated during this precontraction period, and it is possible that the rest of the heat is generated by the liberation of Ca^{2+} from the SR.

Shortening Heat. The source of shortening heat is not known. It is possible that some of it may be derived from the heat liberated by the formation of new cross-bridges between actin and myosin.

DELAYED HEAT

The first, small portion of the delayed heat corresponds to the breakdown of glucose, through the glycolytic pathway, to pyruvic acid. A relatively small amount of energy is liberated by this anaerobic pathway.

Aerobic Reactions

The large burst of delayed heat, which requires oxygen, corresponds to the aerobic recovery reactions, during which pyruvic acid is oxidized through the citric acid cycle and the electron transport chain, to yield CO_2 and H_2O and large amounts of energy.

23–8 Effect of Metabolic Inhibitors on Muscle Contraction

Metabolic inhibitors will interfere with the supply of energy needed for muscle contraction and drastically shorten the length of time a muscle can continue to contract. An inhibitor such as *iodo-acetic acid* (*IAA*), which inhibits the enzyme glyceraldehyde phosphate dehydrogenase (Sec. 5–12), stops glycolysis at the stage of 3-phosphoglyceraldehyde. The muscle is able to contract only while the supplies of ATP, CP, and some of the glycolytic substrates remain available: because no oxidative reactions can occur, no resynthesis of these materials results. The effect of IAA on the fatigue curve is shown in Fig. 23-19.

23–9 The Initiation of Contraction

Motor Units

Skeletal muscle fibers are functionally arranged in *motor units,* which consist of several muscle fibers innervated by

Figure 23-19. **The effect of iodoacetic acid (IAA) on the fatigue curve. The muscle soaked in IAA fatigues much more rapidly than the normal muscle because IAA interferes with oxidative reactions required for energy. Treppe is the increase in height of the first few contractions. The primary contracture is a failure of the repeatedly stimulated muscle to relax properly at the beginning of contractions. The secondary contracture is a failure of relaxation that is indicative of fatigue. (Student record of A. Sinesi and N. Falanga, Physiology Laboratory, New York University.)**

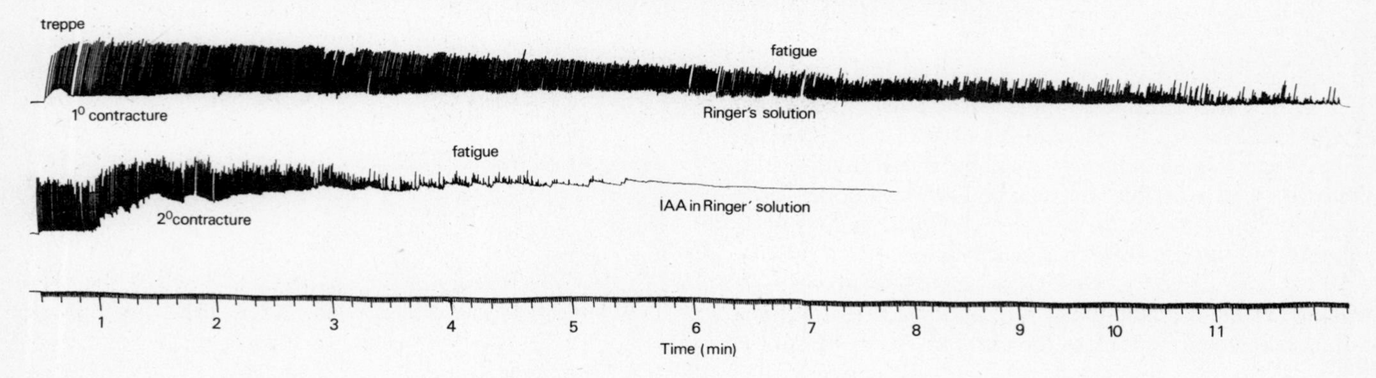

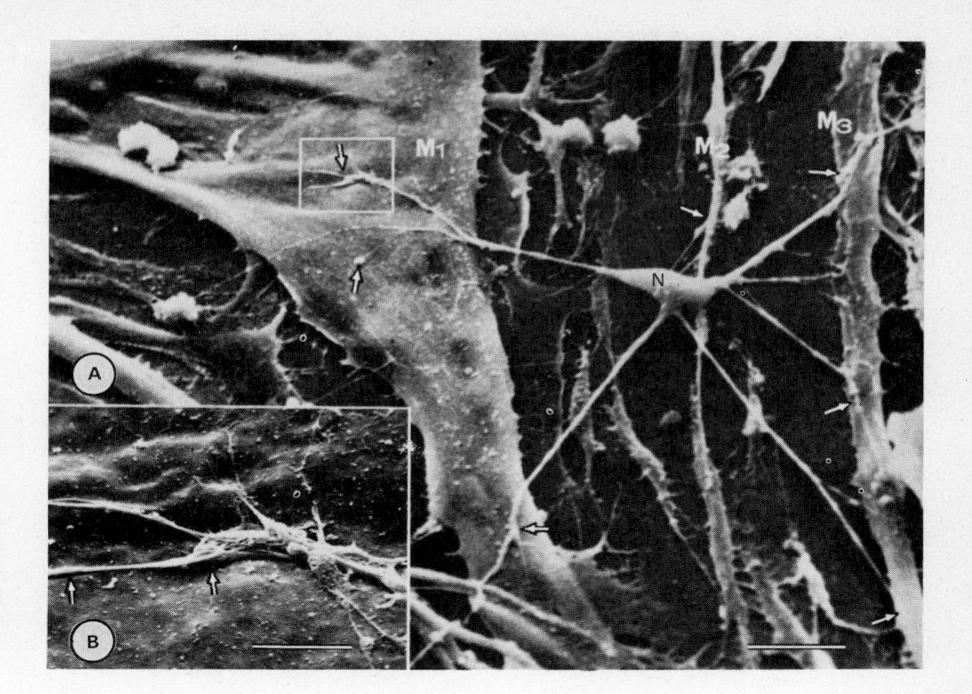

Figure 23-20. **(A) Nerve-muscle cells grown in culture show the development of neuromuscular junctions. A single neuron (N) may innervate several muscle fibers (M_2 and M_3) and also send two branches to a large muscle fiber (M_1). Scanning electronmicrograph, ×880. Bar = 20 μm. (B) The square in the upper left of (A) is shown at higher magnification (×3,800). The nerve fiber branches extensively at the arcolemma, and many of the fine terminals exhibit oval swellings (arrows). [From Y. Shimada and D. A. Fischman, *Developmental Biology* 31:200 (1973).]**

one alpha motoneuron, that is, by a ventral horn motor cell and its axon (Fig. 7-6). Tissue cultures containing nerve and muscle cells show the nerve fibers growing out to make contact with, and presumably innervate, several muscle fibers (Fig. 23-20). In the large muscles of the calf, one motoneuron may innervate as many as 1900 muscle fibers: each time the motoneuron is effectively stimulated, all 1900 of the muscle fibers will contract.

Fine control of muscle contraction is obtained when the motor unit is reduced to five or six muscle fibers, as in the small muscles that move the eyes. When the motor unit is small, its action is rapid and a delicate and precise movement is obtained, but it lacks power. The individual fibers of the motor unit are not necessarily next to one another, but may be distributed throughout the muscle, so that a balanced contraction of the entire muscle can be obtained even when relatively few motor units are involved.

For a *powerful contraction* to be generated in a muscle, many motor units act at the same time, and the larger each unit is, the stronger the contraction will be (Fig. 23-21). Usually, the motor units of a muscle act in relays so that not all the fibers fatigue at once.

Excitation of motor units

The concepts of threshold and excitation have been discussed in detail in Sec. 6-6, with respect to nerve. In the normal muscle, the fibers that make up the motor unit appear to have the same *threshold* to stimulation at any given time, so that if the stimulus is strong enough to excite the motor unit, all its constituent fibers will respond with a maximum contraction. This *all-or-none law* is a useful concept, provided that one be aware of the conditions under which it operates. A muscle unit that has been fatigued, or affected by changes in its chemical or thermal environment, will still respond with a maximum contraction—maximum, that is, *for these particular circumstances.* An entire muscle is capable of graded responses until all the motor units are responding, at which time the muscle is really giving its all (Fig. 23-22). This additive effect of increasing stimulus strength to increase the number of responding units is *spatial summation,* which was also discussed for nerve in Sec. 7-6.

In the intact organism, an efficient system of *alternation of motor unit activity* is utilized to increase the strength of contraction of the entire muscle and to delay muscle fatigue.

Central inhibition of motor units

In vertebrate skeletal muscle, *all motor nerves stimulate* the muscle fibers to contract; consequently, relaxation requires inhibition of the firing of these nerves. This is accomplished by central inhibition of the ventral horn motor cell, resulting from an excess of IPSPs over EPSPs impinging on that motoneuron (Sec. 7-6 and Fig. 7-30).

Desynchronization of motor units: fasciculation

In *denervated* muscle, or in conditions in which the ventral horn motor cells are diseased, the intricate mechanisms that coordinate the individual muscle fibers of the motor

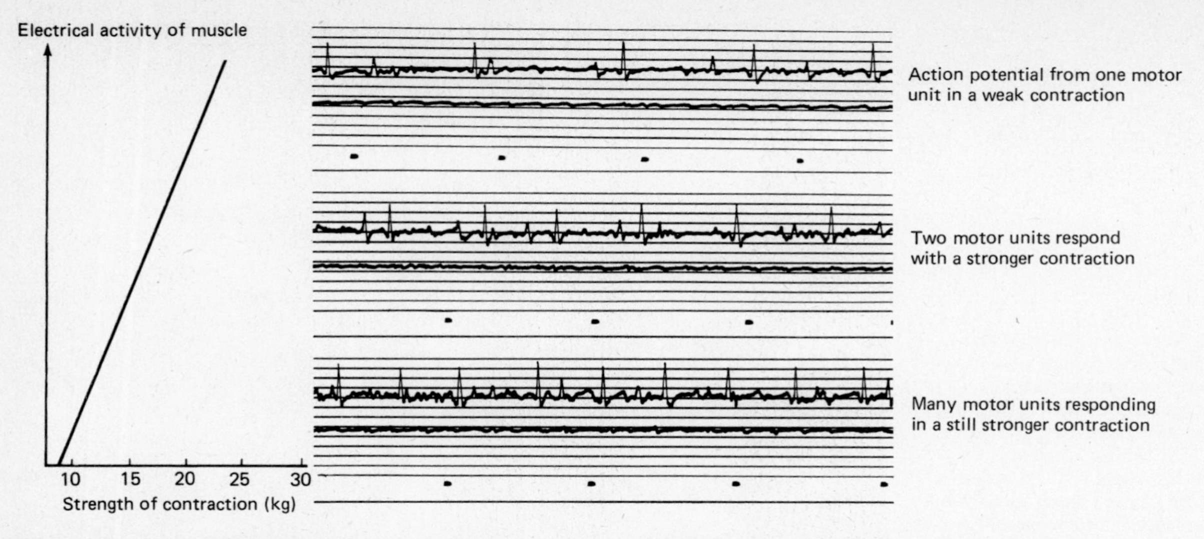

Figure 23-21. **Increase in the number of active motor units with increasing strength of muscle contraction. The top part of the figure shows the individual motor units responding, as recorded on an oscilloscope. The graph beneath shows the increase in total electrical activity. (From G. Friedebold, *Deutschen Orthopaedischen Gesellschaft,* Ferdinand Enke, Stuttgart, Germany, Sept. 1956.)**

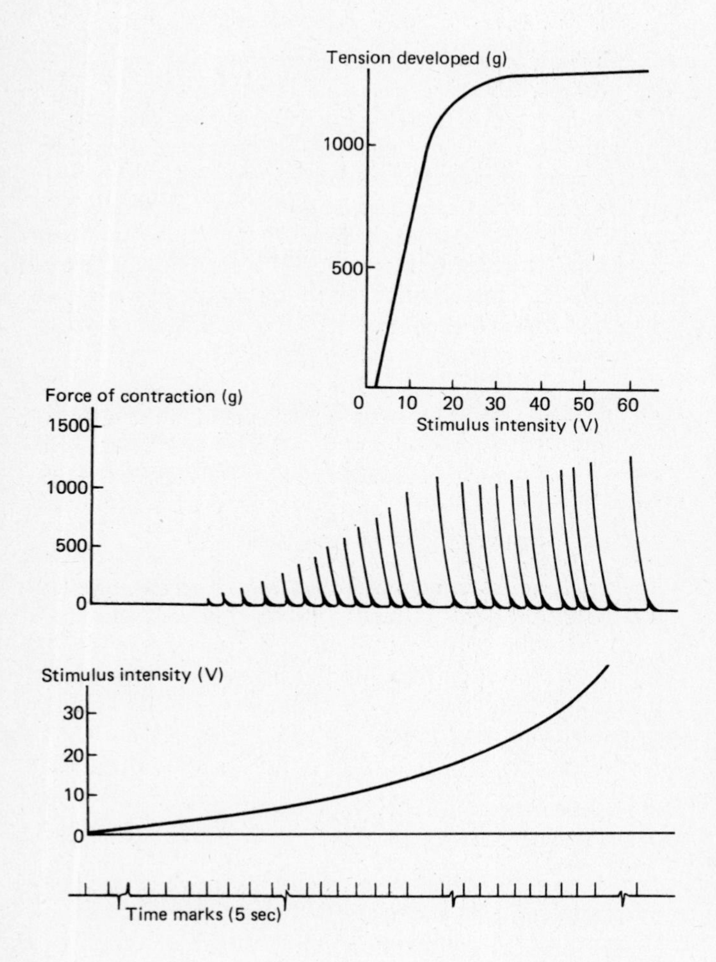

Figure 23-22. **Graded responses of skeletal muscle with increasing strength of stimulation until all muscle fibers are activated. Once this occurs, a further increase in stimulus strength does not cause a greater response. Notice that a certain intensity of stimulus must be applied (threshold) before any contraction occurs. (From H. E. Hoff and L. A. Geddes, *Experimental Physiology,* Baylor Medical College, Houston, Tex., 1962.)**

unit are disturbed, and these fibers no longer contract synchronously. This gives rise to small, irregular spontaneous contractions known as *fasciculations.*

The Neuromuscular Junction

The neuromuscular junction is a specialized type of synapse in which the axon of the nerve terminates as a *motor endplate* on the muscle fiber. The ultrastructure and electrophysiology of the neuromuscular junction were discussed in detail in Sec. 7-4.

The action potential travels along the motoneuron to the nerve terminals, where *acetylcholine* is released to diffuse across the synaptic cleft and act on the membrane of the muscle fiber itself. If sufficient acetylcholine is present to depolarize the membrane, an action potential will be generated and conducted along the muscle fiber. This occurs simultaneously on all the muscle fibers of a motor unit so that all fibers contract. The rapid inactivation of acetylcholine by the enzyme *acetylcholinesterase* permits the repolarization of the end plate so that the muscle can respond to the next stimulus.

23–10 Characteristics of Muscle Contraction

Recording from the Isolated Muscle

Muscle, acting as a transducer, responds to an electrical change with a mechanical response. The extent and the duration of the mechanical force generated depend on several factors, which can best be studied in the *isolated muscle.* Much of our information about the physiology of muscle has been derived from isolated muscle fibers, isolated muscles, and the nerve-muscle preparation that generally comes from the frog, an inexpensive and easily maintained laboratory animal.

The *nerve-muscle preparation* usually consists of the sciatic nerve and the gastrocnemius muscle that it innervates. The muscle is fixed firmly at one end through its femur, which is placed in a bone clamp. The other end of the muscle is attached to the arm of an electronic *transducer* by way of the tendon of Achilles (Fig. 23-23). When the muscle contracts, it pulls the arm up and the *transducer* converts this mechanical energy into electrical energy. The electrical signal is led into an electronic *recording module,* which has a chart drive capable of moving recording (calibrated) paper at one of several selected speeds. The amplified signal from the transducer causes deflections of a pen across the moving paper. Consequently, mechanical contractions of the muscle are recorded permanently on calibrated paper, moving at a known speed, permitting accurate measurement of the amplitude and duration of the contractions.

Figure 23-23. **Recording of the mechanical responses of a frog gastrocnemius muscle preparation. The stimulating electrodes (e) deliver the shock from the stimulator to the muscle (m), which is attached through its tendon to the arm of an electronic transducer (t). The mechanical contraction is transduced to an electrical signal that, when amplified, causes deflections of the pen (p) across the moving chartpaper (c). The second pen is a timer. The insert shows the muscle attachment to the arm of the transducer more clearly. The small beaker on the stimulator is filled with Ringer's solution, used to keep the muscle moist. (Physiology Laboratory, New York University.)**

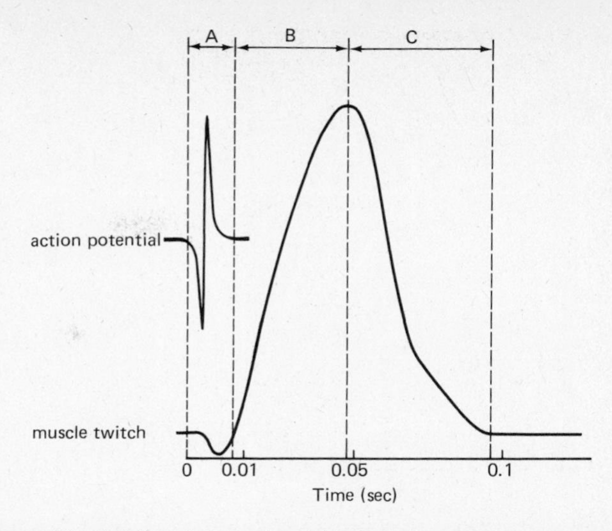

Figure 23-24. **Electrical and mechanical changes in the simple muscle twitch: A = latent period, B = contraction period, and C = relaxation period. The slight relaxation in (A) is the latency relaxation, usually seen only with very sensitive recording apparatus, as depicted in Fig. 23-25.**

Although a muscle will respond by contracting to any adequate energy change (stimulus), be it chemical, thermal, mechanical, or electrical, electrical stimulation is preferred because it can be quantitatively measured and does the least harm to the tissues. An *electrical stimulator* that delivers square wave shocks (see Sec. 6-6) of desired frequency, duration, and strength is also shown in Fig. 23-23.

The Simple Twitch

LATENT PERIOD

When a muscle is stimulated to contract, there is a short time interval (in a frog muscle, about 7.5 msec) between the muscle action potential and the beginning of the mechanical contraction (Fig. 23-24). During this time, the *latent period,* many important changes are occurring in the muscle. Some of these changes are reflected in the *latency relaxation,* a brief relaxation (1.5 msec) occurring in the latent period before the development of tension. The latency relaxation may be associated with the coupling of the electrical impulse and the contractile process, since latency relaxation occurs just after the first phase of the action potential (Figs. 23-24 and 23-25).

Sandow (1965) has suggested that part of the resting tension of a muscle is borne by the SR and that this tension is decreased as Ca^{2+} is released from the SR in the excitation-contraction coupling process. The amount of Ca^{2+} that is released then determines the magnitude of the latency relaxation.

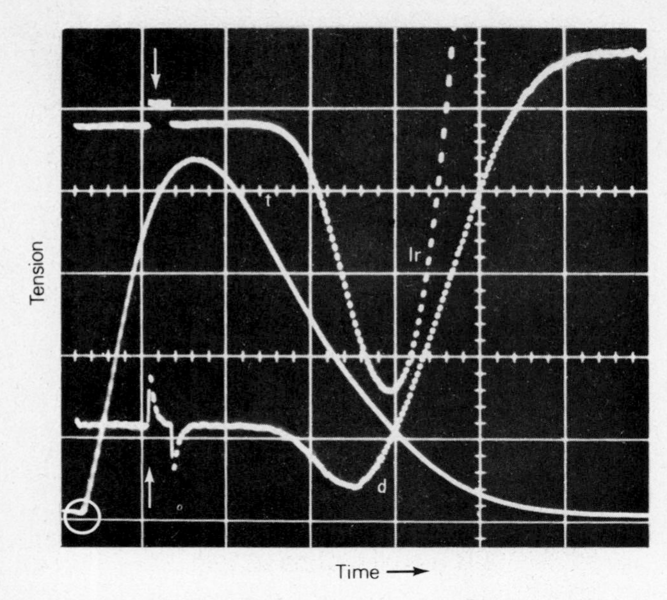

Figure 23-25. **Latency relaxation of muscle precedes the building up of tension characteristic of muscle contraction. Latency relaxation (lr) is seen in the upper trace (↓). It is so small (represented by the circled area) in comparison to the whole twitch (t) that it can be shown only with a different calibration. Twitch time calibration is 20 msec/square; tension is 5g/square; and lr is 1 msec/square and 50 mg/square. It is during this time of lr that the coupling of the electrical excitement with the contractile process occurs. The bottom trace is the derivative (d), which indicates the rate at which tension is generated. Time for derivative d is 1 msec/square, and the derivative is 0.25g/msec/square. The arrows represent the shock artifact (instant of stimulation). (Record made by Jan P. Koniarek in the laboratory of A. Sandow, New York University.)**

This, of course, implies that the entire latent period represents the time it takes for Ca^{2+} to diffuse from the lateral cisternae to the overlap region of the A band. The latent period continues until the tension has risen to its original resting value and begins to increase, as contraction is initiated.

Contraction period

Almost immediately following the latent period, the contractile filaments develop their maximum power, and the muscle shortens abruptly. It reaches its peak of contraction in about 0.04 sec in the frog.

Relaxation period

The muscle fibers relax and lengthen, a process that takes slightly longer than contraction. For the frog, the relaxation period is about 0.05 sec, making the *entire twitch time* approximately 0.1 sec. The rapid insect muscles complete a single twitch in 0.003 sec, whereas in humans it requires 0.03 sec.

Effect of Temperature on Muscle Contraction

All these values are altered considerably by *changes in temperature.* This finding is not at all unexpected, for we know that the chemical reactions involved are enzymatic and that enzyme activity is affected by temperature. An increase in temperature will increase the rate and height of contraction up to an optimum, beyond which muscular efficiency decreases, irreversibly if the temperature has risen high enough to denature the proteins. Cold slows the speed of the response and decreases the height of contraction (see Fig. 23-26). The optimum temperature for mammalian muscle is between 37 and 40°C, the body temperature range for most warm-blooded animals. For the frog, and other cold-blooded animals, the temperature optimum is between 25 and 30°C.

The Refractory Period

Like nerve, skeletal muscle has a period lasting for a fraction of a millisecond following stimulation, during which it is *absolutely refractory,* and then a brief *relative refractory period,* during which it will respond only to a stimulus stronger than the normal one needed to elicit a response. This time must represent the time needed for the membrane to become repolarized and for the reticulum to regain its calcium. The relative refractory period of skeletal muscle, although longer than that of nerve, is considerably less than that of cardiac muscle. This fact is important; it means that skeletal muscle can respond to a series of rapid stimuli.

Summation and Tetanus

When a muscle is stimulated repeatedly during the period of contraction, the tension generated by the stimuli is added or summated, so that the ultimate tension may be nearly 10 times as much as that of a single twitch. This is an example of temporal summation, which was also discussed for central

Figure 23-26. **Effect of temperature on the amplitude and duration of the mammalian muscle twitch. The effect of the high temperature is irreversible because of the denaturation of the muscle proteins.**

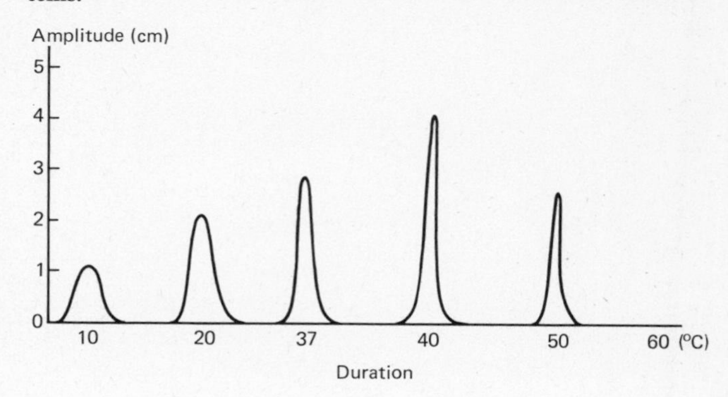

neurons (Sec. 7-6). Not only is the height of the contraction increased by the repetitive stimuli, but because the muscle is allowed no time to relax, it maintains a *smooth contraction.* This smooth contraction, with many times the amplitude of a simple twitch, is called *tetanus.* If the stimuli are less frequent, an *incomplete tetanus* is evoked, in which a partial relaxation is seen (Fig. 23-27).

Tetanic, or sustained, smooth contractions, are the *normal* physiological responses of muscle to the barrage of nerve impulses to which they are subjected. The simple twitch previously described is of value chiefly in the laboratory for the analysis of muscle contraction.

Post-Tetanic Potentiation

If one compares the height of a single muscle twitch immediately before and after a short tetanus, the post-tetanic twitch is found to be larger than the pre-tetanic twitch. This phenomenon is called *post-tetanic potentiation* (Fig. 23-28). Instead of fatiguing the muscle, the prolonged stimulation potentiates contraction, presumably through the release of large amounts of Ca^{2+} from the SR.

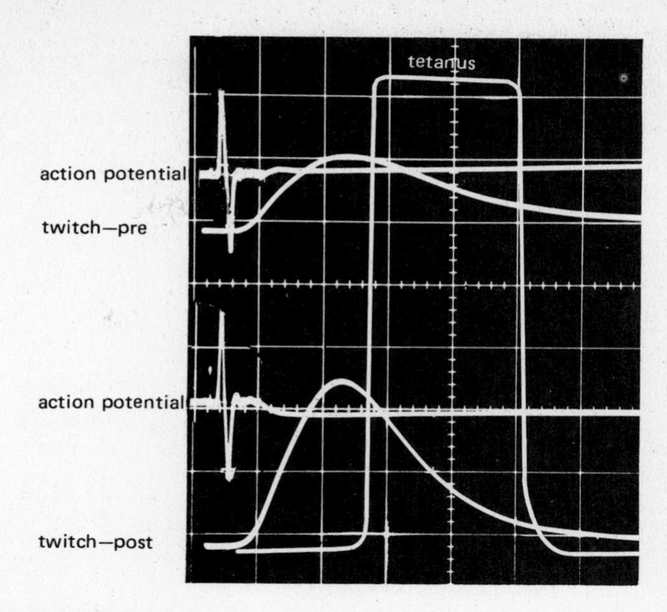

Figure 23-28. **Posttetanic potentiation is seen in the increased tension of the twitch following the tetanus (twitch-post) compared to the height of the twitch immediately before the tetanus (twitch-pre). Each vertical grid represents 20-g tension, each horizontal grid is equal to msec. Note the fivefold increase in tension of the tetanus over the pretetanic twitch. Rat muscle. (Courtesy of E. Gonzalez and B. Brady, Physiology Laboratory, New York University.)**

Treppe

An isolated muscle responding to a series of supermaximal stimuli (stimuli stronger than needed to excite all the muscle fibers) will show a progressive increase in the amplitude of the first few contractions. This is not caused by the recruitment of additional muscle fibers, because each stimulus is more than adequate to excite them all. Figure 23-19 shows this stepwise rise in amplitude, which is called *treppe*—German for "staircase."

Figure 23-27. **Increasing the frequency of stimulation causes a summation of the height of the additional contractions. When the stimuli are frequent enough, there is no time for the muscle to relax; a smooth, sustained contraction, or tetanus, is maintained. Under these circumstances, summation occurs without any increase in the strength of the stimulus.**

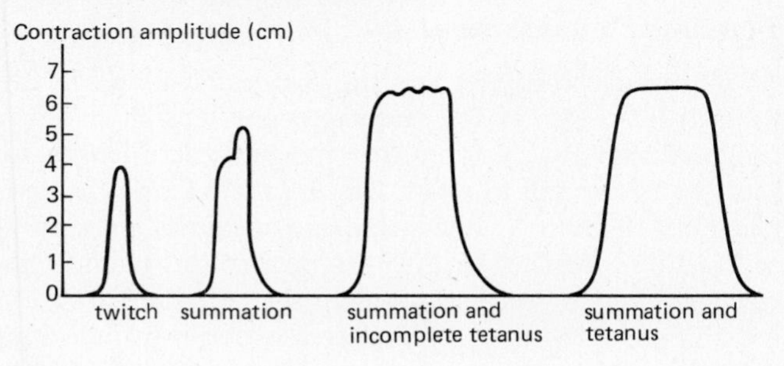

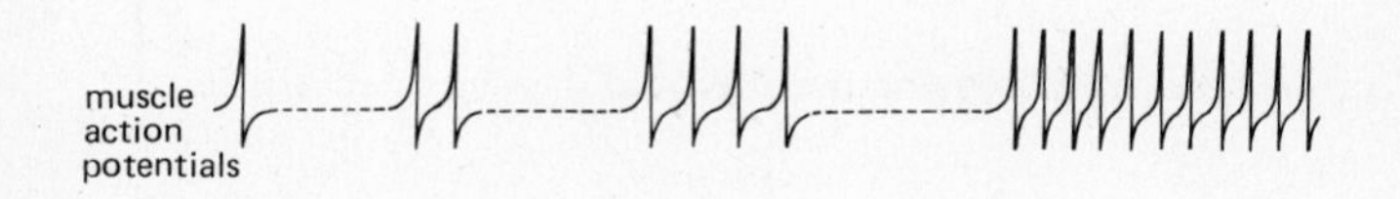

23–11 Regulatory Adjustments of the Contractile System to Load

The force and speed with which a muscle can lift a load are related to the *number* of cross-bridges formed and the *rate* at which these cross-bridges are cycled. The cycling of these cross-bridges, in turn, depends upon the rate of hydrolysis of ATP.

Our understanding of the mechanisms that link these physical and chemical reactions in muscle is by no means clear, but some relationships can be demonstrated experimentally.

The Active State

The active state is a term used to describe some of the changes that occur as the inactive muscle becomes active. Following the slight drop in tension seen in the latency relaxation, there is an increase in the resistance of the muscle to stretch. Hill (1960) defined the active state as the load that a muscle can just bear without lengthening. The active state

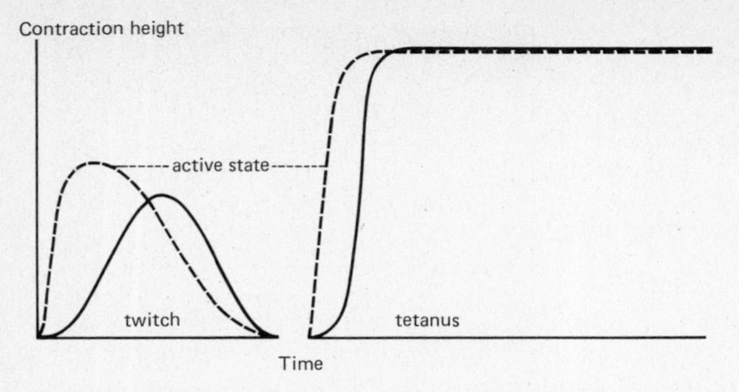

Figure 23-29. **In the short-lived twitch, the active state declines before the muscle begins to shorten. In the summated and sustained tetanus, the muscle maintains the active state for the duration of the contraction.**

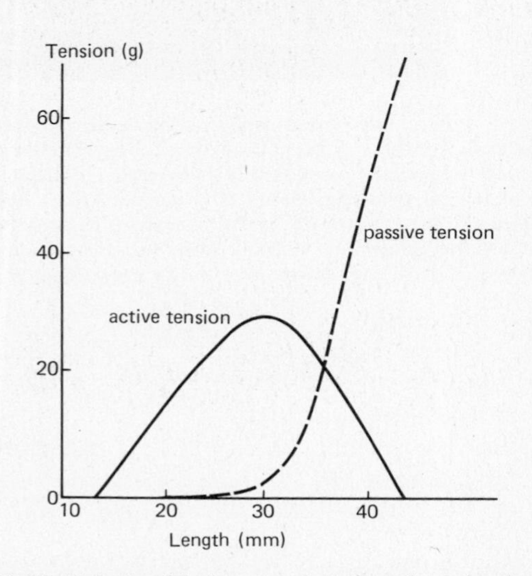

Figure 23-30. **Relationship between length and tension in a mammalian muscle.**

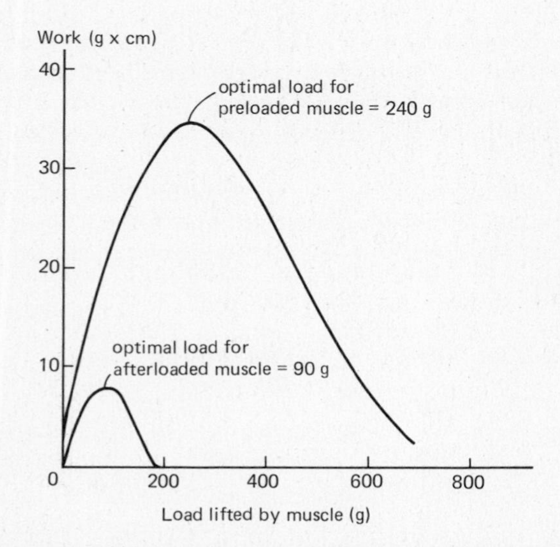

Figure 23-31. **Comparison of the work done by the preloaded and afterloaded muscle.**

is accompanied by the sudden burst of activation heat that has been described.

NONCONTRACTILE ELEMENTS OF THE MUSCLE

Sarcoplasm, the sarcolemma, the SR, and muscle connective tissue (including tendons) are all viscous, elastic structures. The contractile elements of muscle have to pull and stretch these noncontractile parts to overcome their resistance. This effort delays the achievement of actual muscle shortening, so that the active state may be declining before the muscle begins to shorten. In other words, if the active state is short (as it is in a twitch), the contractile elements may be relaxing by the time the muscle reaches its maximal shortening. Figure 23-29 shows the relationship between the *active state* and *tension development* in a twitch and tetanus.

Load and Oxygen Consumption: The Fenn Effect

The amount of *oxygen consumed and heat liberated* by muscle is much greater when the muscle is doing work (contracting isotonically) than when it is contracting isometrically. This is known as the *Fenn effect* and appears to be a reflection of the way in which a muscle "senses" the load put upon it and reacts metabolically to adjust its level of activation. It is possible that the troponin-tropomyosin-Ca^{2+} system is involved in this sensing mechanism, through its ability to regulate the number of reactive sites exposed, and the cross-bridges activated.

Velocity of Contraction and Load

The speed with which a muscle contracts becomes progressively less as the load on the muscle is increased, even though the tension increases with the increasing load (Fig. 23-30). When the load becomes too heavy for the muscle to move, no contraction is possible (velocity of contraction is accordingly zero), and isometric contraction results.

Tension Development and Muscle Length

As the muscle contracts and the thin filaments are pulled toward the center of the sarcomere, the tension increases progressively. By the time that the sarcomere length has shortened from 2.2 to about 2.0 μm, the ends of the actin filaments begin to overlap, and the maximal strength of contraction is reached (Fig. 23-9). Beyond this point, as the sarcomere length decreases still further to about 1.65 μm, the strength of the contraction decreases abruptly, indicating that *the greater the number of cross-bridges* that can be formed, *the greater the strength of contraction.*

Muscle Efficiency and Load

A muscle contracts more efficiently if it is *slightly stretched* beforehand, for the tension generated is dependent to a certain degree on muscle length. In the intact body, the elaborate system of proprioceptors acting through the reticular formation and the gamma loop keep most muscles continuously tensed, through the maintenance of tonus.

In the laboratory, one can artifically arrange the muscle so that it is stretched by a series of increasing weights only while it is lifting them, rather than prior to contraction. This type of loading is called *afterloading* and results in less work being done than when the muscle is stretched first by the load and then stimulated to contract (*preloaded*). Figure 23-31 shows that the preloaded muscle can perform far more work than the afterloaded muscle.

23–12 Fatigue

If the *isolated muscle* is made to contract repeatedly by direct stimulation to the muscle, a decrease in the height of the contractions becomes apparent. The tetanically responding muscle will fatigue much more rapidly than will a muscle responding to a series of single stimuli. Relaxation is also affected; the muscle fails to return to the base line after contracting, and goes into a state of sustained contraction known as *contracture* (Fig. 23-19). Ultimately, the muscle can no longer contract: it has exhausted its energy reserves. This is similar to *rigor mortis,* for all the cross-bridges are irreversibly locked because of depletion of ATP in the muscle fibers.

Causes of Fatigue

In the *isolated muscle,* it is probably the excitation-contraction coupling mechanism that fatigues first, and other elements of the contraction-relaxation process (such as the active transport of Ca^{2+} into the SR) are also seriously affected. Metabolic wastes accumulate, and energy stores are depleted in the isolated muscle. Because oxygenation is poor under these circumstances, lactic acid accumulates, the pH of the cells drops, and permeability changes that decrease muscle efficiency occur.

The problem is quite different and far more complex in the *intact organism,* where the circulation plays such an important role. In the first place, *glucose* is constantly being brought to the muscle, not just to the surface layers but through the extensive capillaries that penetrate to each muscle cell. *Oxygen* supply is adequate except for prolonged and extreme exercise, and wastes are efficiently removed. The activity of the *motor units* is staggered by control through the central nervous system, so that not all the units are actively engaged at one time. *Hormone* levels markedly affect susceptibility to fatigue. This is not surprising; hormones regulate metabolic rate, glycogen stores, blood pressure, and the excitability of the nervous system. Figure 23-32 shows the rapid rate of fatigue of a hypophysectomized animal as compared to an intact animal and the reversal of this effect by the administration of ACTH.

One must also consider the overall *physiological and psychological influences* that can cause fatigue in the individual. One can explain fatigue after exhausting physical work on the basis of lack of fuel or accumulation of waste products or an imbalance of electrolytes, but fatigue following mental activity, during which very little energy is expended, must be due to other causes. Boredom and staleness, which give rise to fatigue, have been shown to delay the transmission of the impulse across the synapses in the central nervous system, slowing responses and decreasing alertness. This type of fatigue is difficult to analyze and yet is of great importance, particularly when it becomes chronic.

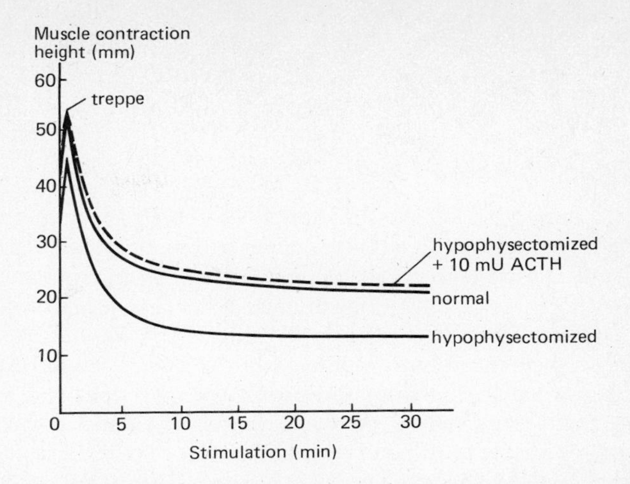

Figure 23-32. **Hypophysectomy increases the rate of fatigue of a continuously stimulated rat muscle. The response of the muscle can be restored to normal by the administration of ACTH. (Data from the laboratory of F. Strand, New York University.)**

23–13 Regulation of Muscle by the Nervous system

Normal control of movement is executed by means of various levels of the central nervous system, involving not only *spinal segments,* but also the *cerebellum, motor nuclei in the midbrain,* the *basal ganglia,* and *motor areas in the cerebral cortex.* These motor centers are integrated with one another and coordinated through information received from many parts of the body, especially from the *muscles* themselves, the *skin,* the *eyes,* and the *organs of equilibrium in the inner ear.*

One of the most urgent problems that the first terrestrial vertebrates had to overcome was the constant struggle with the *force of gravity.* The development of *special tonic reflexes* utilized gravity as a stimulus for muscle contraction, thus opposing the direction of the pull of gravity. Refinement of this mechanism permitted the addition of central nervous system regulation of the muscle spindles and the development of a finely tuned, *automatic compensation for changes in load* during muscle contraction. In Sec. 7-6, the idea of peripheral and central control of muscle spindles was introduced. Here the regulatory devices that integrate these reflexes into the general pattern of movement will be discussed.

Final Common Path

All contraction of skeletal muscle is routed through the alpha ventral horn motoneuron, regardless of whether the impulses originate in the motor cortex (as in voluntary movement) or as reflexes involving any level or levels of the CNS. Realizing the significance of this, Sherrington (1906) called the ventral horn motoneuron the *final common path* for movement. If this motoneuron is destroyed, as it is by the polio virus, muscle paralysis results. The affected muscles are unable to respond to voluntary or to reflex stimulation.

Under normal circumstances, all the muscle fibers innervated by a motoneuron contract when the cell fires, so that the *motor unit* is actually the functional final common path.

Figure 23-33. **The gamma loop. The twitch fibers of the muscle are innervated by the axons of the alpha motoneurons. The intrafusal fibers of the muscle spindle are innervated by the gamma motoneurons. The muscle spindle also is innervated by 1a sensory fibers that synapse directly on the alpha motoneurons.**

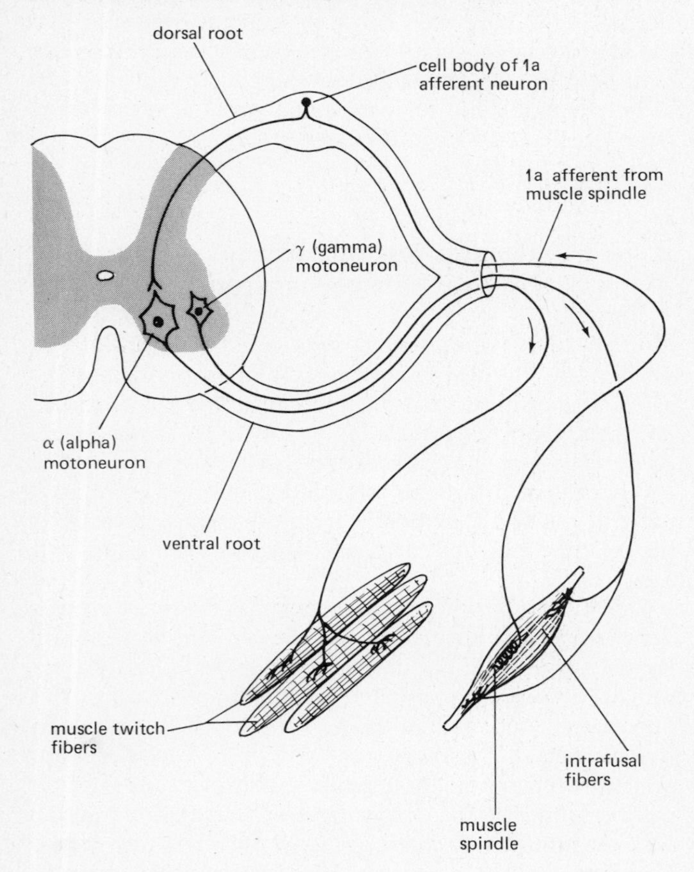

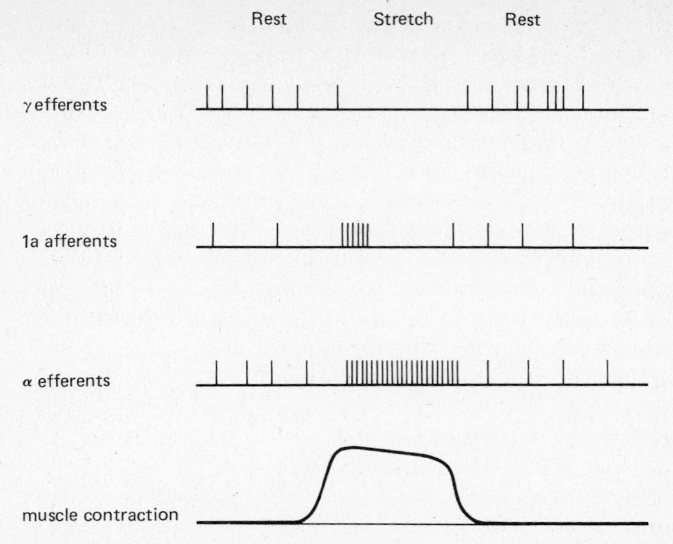

Figure 23-34. **Electrophysiological responses of gamma loop and muscle fibers to stretch. At rest, there is an irregular discharge of the gamma fibers because of central stimulation via the reticular formation. Stretching the muscle stimulates the 1a afferents from the muscle spindles, which cause the alpha motoneurons to fire. This results in muscle contraction. Muscle contraction releases the stretch on the spindles, which stops 1a afferent firing.**

Converging Controls on Spinal Motoneurons

REGULATION AT THE SPINAL LEVEL

The Gamma Loop. Muscles normally receive a constant barrage of impulses through their *dual motor innervation,* the *alpha and gamma fibers* (Sec. 7-6). The interdependence of these two motor systems and their integration through the sensory systems of the muscle spindle are important and complicated, and deserve a more detailed explanation than could be given in Chapter 7.

The *gamma loop* is the pathway that involves the alpha and the gamma neurons and the 1a sensory fibers that connect them (Fig. 23-33). The gamma loop is essential:

1. To set the general level of *muscle tone* precisely, and so to control posture.
2. To *preset* the muscle to contract to lift a load and to *make adjustments* in contraction rate and force *during* the contraction.

MUSCLE TONE (TONUS). Muscle tone, or tonus, is a sustained, low-grade reflex contraction of muscle. Adequate muscle tonus is essential as a background for subsequent contraction: a flabby, atonic muscle does not respond well to any stimulation. Muscle tonus is maintained by the constant pull of gravity on the *spindles* of the antigravity muscles. This stretch on the intrafusal fibers of the muscles pulls apart the primary endings of the *1a afferents* that are wrapped around these fibers and causes the nerves to fire. This results in *monosynaptic stimulation of the ventral horn alpha motoneuron.* Consequently, the twitch fibers are stimulated to contract, the muscle shortens, and the stretch on the intrafusal fibers is eliminated, as is the firing of the 1a afferents. Figure 23-33 shows the anatomical pathways involved, whereas Fig. 23-34 depicts the electrophysiological changes.

PRESETTING AND ADJUSTING: THE MUSCLE SPINDLE AS A SERVOMECHANISM. The spindle afferents are sensitive to *changes in length between the spindle and the twitch fibers,* which are oriented parallel to one another. Because the spindles have their own separate motor innervation through the gamma system, the spindles can be activated while the twitch fibers remain passive, and vice versa. This independence of firing gives the spindle a means to regulate the rate of contraction of the main muscle, increasing it if the load proves unexpectedly heavy or damping it if the load is lighter than anticipated. Experiments on humans have shown quite clearly that the alpha motoneurons always initiate the muscle contraction and that the gamma adjustments follow.

Central and Peripheral Regulation of the Spindle Firing. If the spindle is set to contract at the required rate *centrally* through the nervous system, and the muscle is contracting *less rapidly,* the spindles will be stretched and consequently additionally excited (*peripherally*). This increased spindle excitation will cause alpha motoneuron firing by way of the stretch reflex, increasing muscle contraction, decreasing the stretch on the spindles, and consequently slowing their rate of firing.

Merton (1972) has likened the spindles to a servomechanism similar to that of power steering in a car. Spindle contraction is analogous to turning the steering wheel, whereas contraction of the main muscle is comparable to turning the road wheels. The sensory endings of the spindle act as *misalignment detectors,* small adjustments of which, to the right or left, keep the wheels on a straight path.

Figure 23-35 shows the results of some classical experiments by Hunt and Kuffler (1951), and Granit (1955) demonstrating central and peripheral control of the spindles. Recordings were made from 1a afferents of muscle spindles:

1. When the spindles were stretched by increasing weights.
2. When stretch was accompanied by stimulation of the gamma fibers.
3. During stretch-induced muscle contraction.
4. During stretch-induced muscle contraction and gamma fiber stimulation.

These results show clearly that stretching the spindle causes an increased rate of firing of the 1a afferents, and that concomitant gamma stimulation can enhance this increase. It is also seen that spindle afferent firing is inhibited during muscle contraction (peripheral stretch removed), but that simultaneous gamma stimulation (central stimulation) is capable of exciting the spindle afferents even when the spindle is not under stretch.

Figure 23-35. **Effect of increased stretch on the muscle spindles and muscle contraction. Stretch is indicated by the weight in grams placed on the muscle. The greater the stretch, the more frequently do the 1a afferents fire, and the resulting stimulation of the alpha fibers causes muscle contraction. Both 1a and gamma fibers are inhibited by muscle contraction. If the gamma fibers are centrally stimulated (last column on the right) at the same time that the muscle is contracting, then they will cause firing of the spindle afferents despite the absence of stretch.**

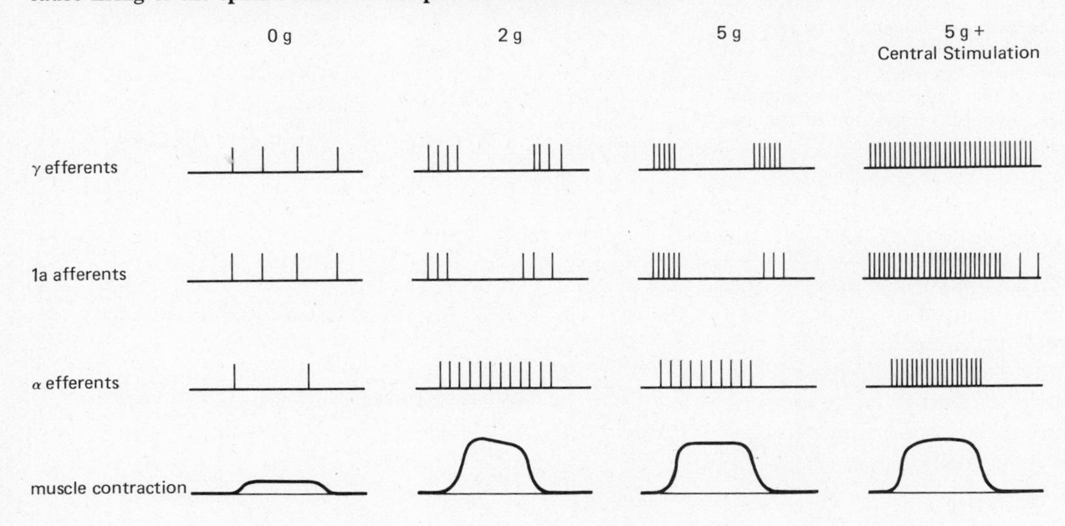

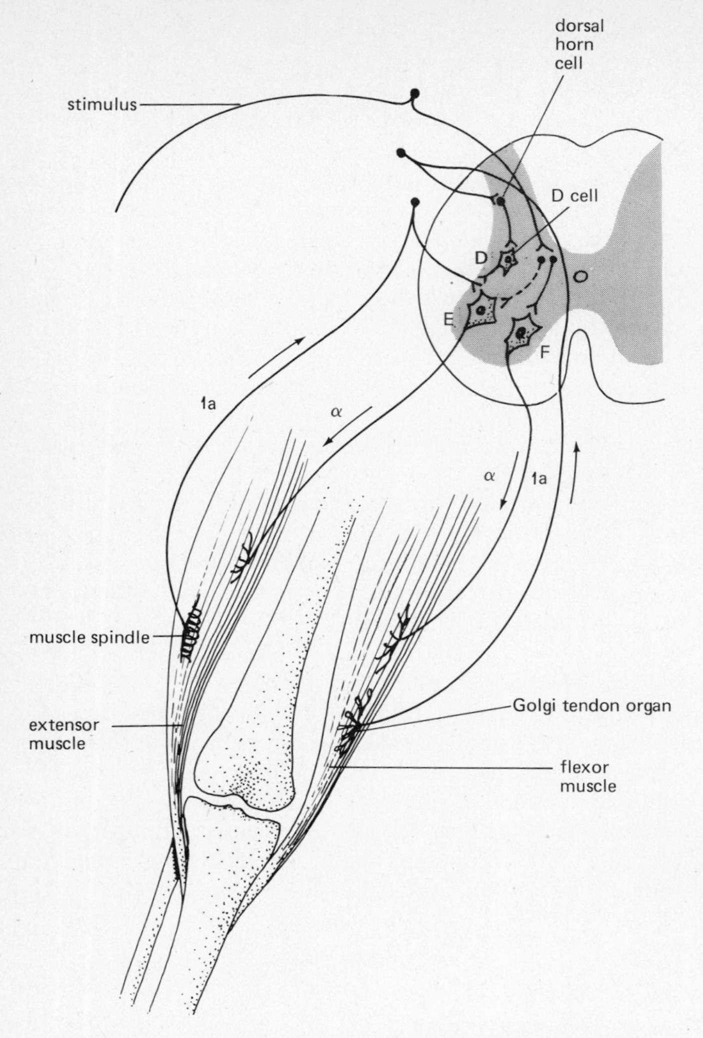

Figure 23-26. **Cooperative action of muscle antagonists via central nervous system pathways. Stimulation of the flexor ventral horn motor cell (f) causes contraction of the flexor muscle. Contraction of the flexor fires the Golgi tendon organ in the flexor muscle; the fibers of these receptors ultimately synapse on D cells in the spinal cord, which presynaptically inhibit the terminals of the 1a afferents coming from muscle spindles in the extensor muscle. This presynaptic inhibition prevents the stretch reflex of the antagonistic extensor muscle, which would otherwise occur when the flexor contracts and stretches the extensor. The inhibitory interneuron (broken line) is responsible for postsynaptic inhibition of the extensor motoneuron (E).**

Presynaptic Inhibition of the Gamma Loop. The phenomenon of presynaptic inhibition was introduced in Sec. 7-6, but its significance in the integration of spinal reflexes is more appropriately discussed here. Eccles and his colleagues (1962) have beautifully demonstrated the role of interneurons in the coordination of flexors and extensors: these "antagonistic" muscles are really cooperating muscles. Some of the pathways for the *cooperative action of antagonistic muscles* are shown in Fig. 23-36.

1. The flexor muscle is stimulated to contract reflexly through a sensory neuron, which synapses through an *excitatory interneuron* on the flexor alpha motoneuron, F.
2. The same afferent neuron synapses through an *inhibitory interneuron* with the extensor alpha motoneuron E, causing the extensor muscle to relax (central inhibition).
3. Relaxation of the extensor muscle *stretches the spindles* in this muscle, causing the 1a spindle afferents to fire. This spindle discharge would excite E by way of the gamma loop, *but*
4. The contraction of the flexor muscle causes receptors in the flexor tendons (the *Golgi tendon organs*) to fire. This discharge is conducted along the Golgi afferents to result in the stimulation of a *two-interneuron pathway.* The first of these interneurons is in the dorsal horn; when it is fired, it excites a second interneuron, a D cell. The D cell synapses on the terminals of the 1a spindle afferent, depolarizing these terminals and preventing the release of the excitatory transmitter acetylcholine. The *D cell thus presynaptically inhibits E,* permitting the extensor muscle to remain relaxed.

Figure 23-37. **A simplified version of the two types of muscle spindle fibers (intrafusal fibers) and their sensory and motor innervation.**

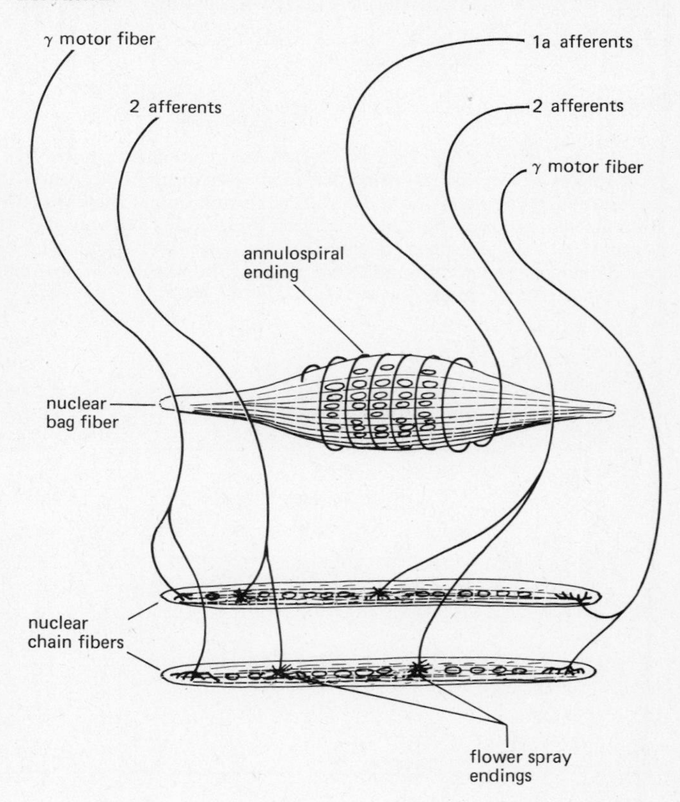

Without this prolonged presynaptic inhibition of the stretch reflex, it would be impossible to keep the extensor muscle relaxed long enough to permit smooth contraction of the flexor.

Complete removal of presynaptic inhibition—which occurs as a result of tetanus toxin poisoning, for example—results in simultaneous contraction of the extensors and flexors, locking the joint and preventing movement around it.

The Muscle Spindle as an Informant. Muscle spindles are capable of relaying far more information than merely changes in length. Because of the presence of another set of sensory nerve endings within the spindle, the *secondary (2) fibers,* the spindles can also convey information concerning the rate at which a muscle is changing its length. The 2 fibers form "flower-spray" endings around the ends of long, thin, intrafusal fibers, known as the *nuclear chain fibers*. The annulospiral endings of the 1a fibers wrap around the central region of a different type of intrafusal fiber, the *nuclear bag fiber*. There may be a total of 5 to 10 intrafusal fibers in a spindle, and several 1a and 2 nerves (in addition, of course, to the gamma motor fibers). A simplified version of muscle spindle structure is shown in Fig. 23-37.

Although the structure of the muscle spindle is second in complexity only to the receptors of the eye and ear, specific correlation of muscle spindle structure to function is unsatisfactory. It is believed that the elasticity and low viscosity of the nuclear bag region, as compared to the high viscosity of the nuclear chain region, may account for the different information conveyed by the 1 and 2 fibers.

Destination of Spindle Afferent Information

1. The shortest, simplest route for the impulses from the spindle afferents is along the gamma loop to the *spinal motoneurons.*
2. It has been well established for many years that information concerning the mechanical state of the muscle is conveyed from the spindles to the *cerebellum,* which plays a major role in the subconscious control of movement, as is discussed a little further on in this chapter.
3. There is now some evidence that the spindle afferents may also contribute to the perception of limb position, with the impulses travelling along recently discovered neural pathways to the *cerebral cortex.* This perception may be more of a subtle recognition of a misalignment ("something is out of place") than a specific awareness of limb position, to which the proprioceptors of the joints and tendons contribute the major input.

Spinal Afferents from Other Receptors. Receptors in the skin and tendons send impulses along 1a afferents to the spinal cord. These sensory neurons may synapse directly onto the alpha motoneuron in a *monosynaptic pathway* characteristic of the stretch reflex, or they may affect the motoneurons through *excitatory or inhibitory interneurons* (*polysynaptic pathways*).

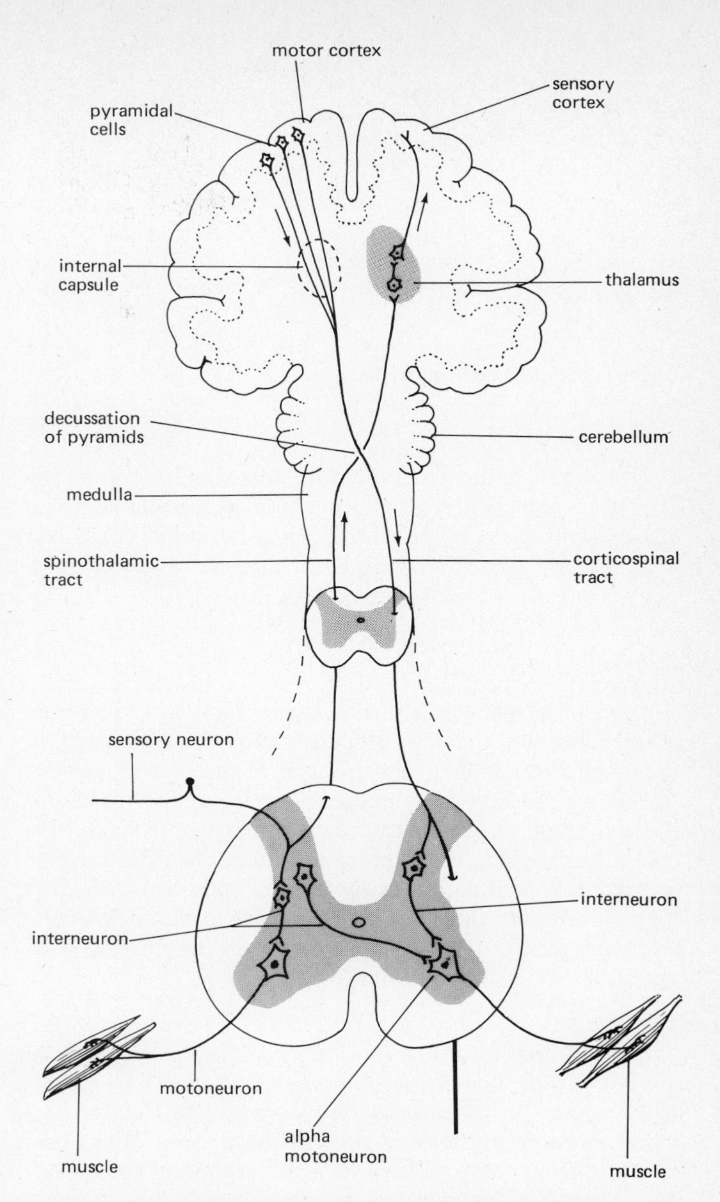

Figure 23-38. **Sensory impulses from receptors and motor impulses from higher cortical centers (corticospinal tract) impinge upon the ventral horn motor cells, either directly or through interneurons. Information from the 1a afferents (sensory neurons) is also conveyed to higher centers in the brain through branches that synapse in the thalamus (spinothalamic tract).**

The involvement of these interneurons in the flexor and crossed extensor reflexes was discussed in Sec. 7-6; it is illustrated in conjunction with voluntary input through the pyramidal (corticospinal) tracts in Fig. 23-38.

In addition to their convergence upon ventral horn motoneurons, these 1a afferents convey sensory information by

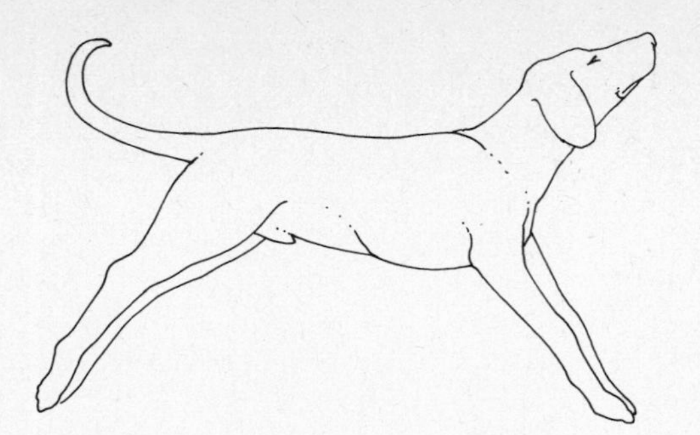

Figure 23-39. **Decerebrate rigidity.**

way of the thalamus to the sensory cortex. This communication is accomplished through synapses of the same sensory neurons with cells in the dorsal horn of the spinal cord. The axons of the dorsal horn cells form the *spinothalamic tracts* that terminate in the thalamus (Fig. 23-38).

REGULATION BY HIGHER CENTERS

Supraspinal Regulation of Muscle Tonus.

The tonus of the muscles of the body is continuously being adjusted by the integrated action of higher levels of the central nervous system on the muscle spindles. The *reticular formation* in the midbrain plays a particularly important role in this central regulation. The dominant influence of the reticular formation is stimulation of the gamma motoneurons. This highly excitatory influence is normally held in check by the *inhibitory* action of the *cerebral cortex* on the reticular formation.

DECEREBRATE RIGIDITY. The relationship between the *cerebral cortex*, the *reticular formation*, and the *gamma motoneurons* can be demonstrated by separating the cerebral hemispheres from the brain stem (*decerebration*). This operation destroys the pathways for inhibition, and the *decerebrate* animal shows an exaggerated *rigidity* of all the joints of the body (Fig. 23-39). The extensor muscles are particularly affected by the unimpeded excitatory output of the reticular formation acting on the gamma motoneurons. The continuous firing of these motoneurons results in the tetanic contractions of the muscles by way of the gamma loop.

Voluntary Movement: Pyramidal System.

The motor areas of the cerebral cortex have been described in Sec. 8-6. The primary motor cortex, which lies just anterior to the central fissure, contains many large *pyramidal cells*, the axons of which form the *pyramidal tracts* (corticospinal tracts) depicted in Fig. 8-18.

As the pyramidal tract axons pass through the brain, they give off numerous branches to the *cortex* itself, to the *basal ganglia* and *brain stem*, and to the *cerebellum*. These brain areas respond to stimulation by way of feedback pathways that affect the output of the highly excitable pyramidal cells. The complex interactions that coordinate and smooth voluntary movements are discussed a little later.

In primates, most of the axons of the pyramidal cells cross over in the medulla to travel down the other side of the spinal cord and make monosynaptic connections with motoneurons at various levels. This direct pathway is responsible for the rapid, efficient movements initiated by the pyramidal system.

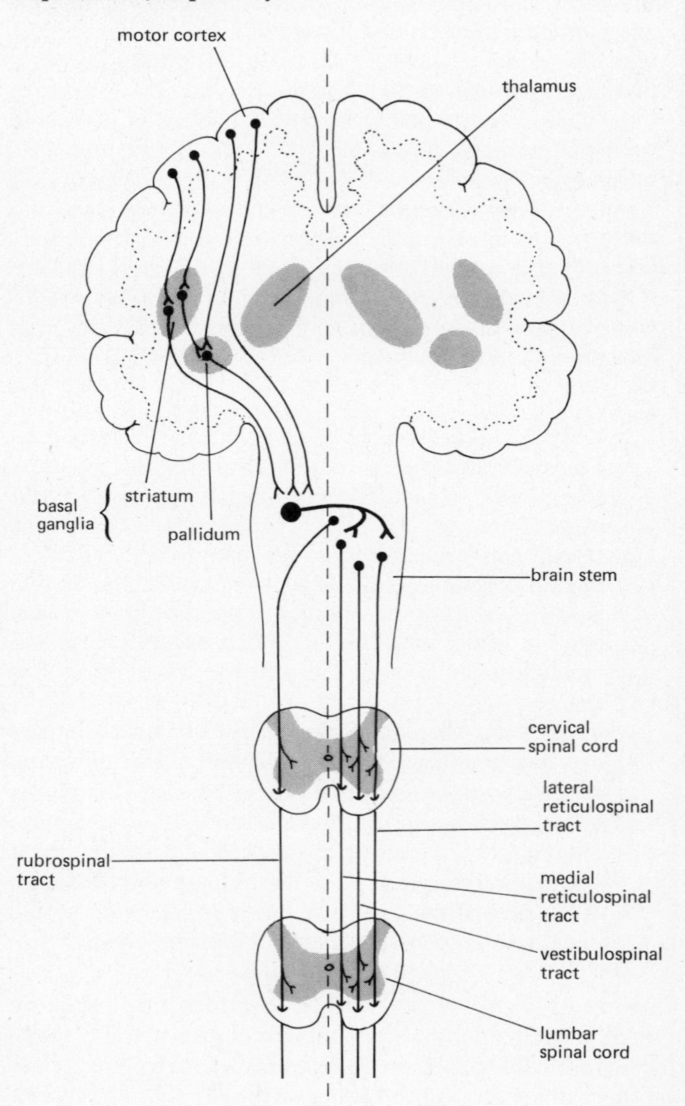

Figure 23-40. **Extrapyramidal tracts. Most extrapyramidal tracts cross in the brain stem to synapse with brain nuclei in this region. On each side of the brain stem, the red nucleus, the reticular formation, and the vestibular nucleus give rise to the rubrospinal tract, the lateral and medial reticulospinal tracts, and the vestibulospinal tract, respectively.**

Involuntary Movement: Extrapyramidal System. The extrapyramidal tracts include all the pathways for motor control from the brain to the spinal cord, except the pyramidal tract. The most important of the brain areas involved in the extrapyramidal system are the *basal ganglia,* the *reticular formation* in the brain stem, and the *cerebellum.* As has just been mentioned, these areas also receive branches from the pyramidal fibers, so that the extrapyramidal centers are kept informed of all pyramidal activity and in turn feed back to the pyramidal cells.

The output of the extrapyramidal system reaches the spinal motoneurons chiefly through the *reticulospinal* and *vestibulospinal tracts.* None of the extrapyramidal tracts connects the cerebral cortex directly to the spinal cord; one or more synapses interrupt the pathways. These extrapyramidal pathways are depicted in Fig. 23-40.

Interaction Between Pyramidal and Extrapyramidal Systems. More than 100 years ago, the British neurologist Jackson discovered that electrical stimulation of what is now known as the *primary motor cortex* produced specific movements of the muscles on the opposite side of the body (Sec. 8-6). This discovery led to the belief that the motor cortex acted as the head of the neural chain of command over the muscles.

Recent work has indicated, however, that the primary motor cortex is actually controlled by *lower cortical motor areas,* by the *basal ganglia,* and by the *cerebellum.* Evidence for this conclusion comes from many sources, including the effects of lesions in these areas on movement in animals and humans, and even more convincingly from the recording of the activity of individual nerve cells in animals performing learned movements.

EVIDENCE FROM LESIONS. Lesions to the *motor cortex* will result in *paralysis,* but damage to the *basal ganglia or cerebellum* is accompanied by disturbances in *posture* and by *abnormal movements.* The type of abnormal movements is quite opposite in conditions involving the basal ganglia from those resulting from cerebellar lesions.

Disorders of the *basal ganglia* are characterized by tremor, muscular rigidity, and a marked difficulty in initiating slow movements. There may be no problem when using the same muscles for fast movements. This syndrome is characteristic of patients with Parkinson's disease, a condition associated with a deficiency of the neurotransmitter *dopamine* in the pathway connecting the basal ganglia to the black nucleus in the brain stem (*nigrostriatal pathway*). Such a patient will reach out an arm, slowly and with much difficulty, to lift a glass of water; holding the glass steady is almost impossible because of the tremor. Yet, under stress (such as a fire alarm) the same patient may be able to leap up out of the wheelchair and run to safety, only to collapse once more into the previous state.

In disorders of the *cerebellum,* the tremor almost completely disappears when muscles are inactive but is most conspicuous during voluntary movement (*action tremor*). The smooth control of movement is lost. The resulting lack of coordination of voluntary movements is called *ataxia.* This condition is made worse by the accompanying loss of ability to estimate the distance of movement (*dysmetria*): for example, the patient overshoots the desired distance in moving the arm to touch the nose.

Clinical studies indicate that the commands given by the motor cortex are normally greatly modified by the basal ganglia and the cerebellum, insuring that inappropriate movements (tremor) are eliminated, rigidity does not develop, and muscle contractions are precisely estimated and damped to prevent overshoot.

The *rigidity* resulting from damage to extrapyramidal system differs from the decerebrate rigidity discussed earlier. In decerebrate rigidity, it is the antigravity extensor muscles that become *spastic* (go into spasms): lesions involving the extrapyramidal system cause spasms of the flexors as well, stiffening the limbs.

EVIDENCE FROM MICROELECTRODAL RECORDINGS. Microelectrodes implanted in different areas of the brains of monkeys trained to depress a telegraph key when a light appears show that cells in the motor cortex become active *before* muscle contraction. Cells in the nearby sensory cortex fire just *after* the muscle contraction. Therefore, although the sensory cells influence subsequent motor discharge, they are not initially involved in the original command.

An unexpected finding of great interest was that cells of the cerebellar cortex and the basal ganglia, as well as the motor cortex, all became active *before* muscle contraction. Other cortical regions may be involved, as the *readiness potential* (Sec. 8-6, motor areas of the forebrain) is observed almost 0.8 sec prior to the movement, over a wide area of the cerebral cortex.

Concepts of Cerebellar Function in Learning Movements. From these and other experiments and observations, the concept has arisen that although the phylogenetically older parts of the cerebellum (Sec. 8-6) are involved in the regulation of slower muscle activity required for equilibrium and posture, the newer cerebellar cortex is an active participant in the *learning and execution of skilled movements.*

The major role of the cerebellum is to preprogram and supervise all motor output from the cerebral pyramidal tract. This return circuit, from the branches of the pyramidal tract into the cerebellum, through its computerlike memory bank, and back to the cerebral cortex, takes about one fiftieth of a second in the human brain.

This type of cerebro-cerebellar circuit provides a continuous series of corrections and adjustments of voluntary movements. This system is of particular importance in the performance of highly skilled movements like golf and skiing, and is aided by feedback from the muscles themselves. These concepts are discussed in detail by Eccles (1977).

23–14 Reflex Regulation of Posture and Locomotion

The alternating contraction and relaxation of tonically active muscle is essential for the maintenance of posture, the upright position, and locomotion. Through the integrative action of subcortical centers and proprioceptive information from the muscles, joints, pressure receptors of the skin, and receptors for equilibrium in the inner ear, the tension of the different extensor muscles of the body is constantly being adjusted (Fig. 23-41). The rhythmic shifting of weight from one side of the body to the other permits the body to remain in equilibrium despite changes in the position of the limbs. Although lower animals can retain their "righting reflexes" and still know which side is up after removal of the cortex, this equilibrium is not possible in humans, who are too dependent on accessory information from the eyes and other sources.

The important regulatory effects of the information coming to the central nervous system from the equilibrium receptors in the vestibular apparatus of the inner ear were discussed in Sec. 22-6.

Figure 23-41. **Shifting of weight from one side to the other in walking is more evident in the human female than in the male because the anatomy of the female pelvis causes more rotation than in the male.**

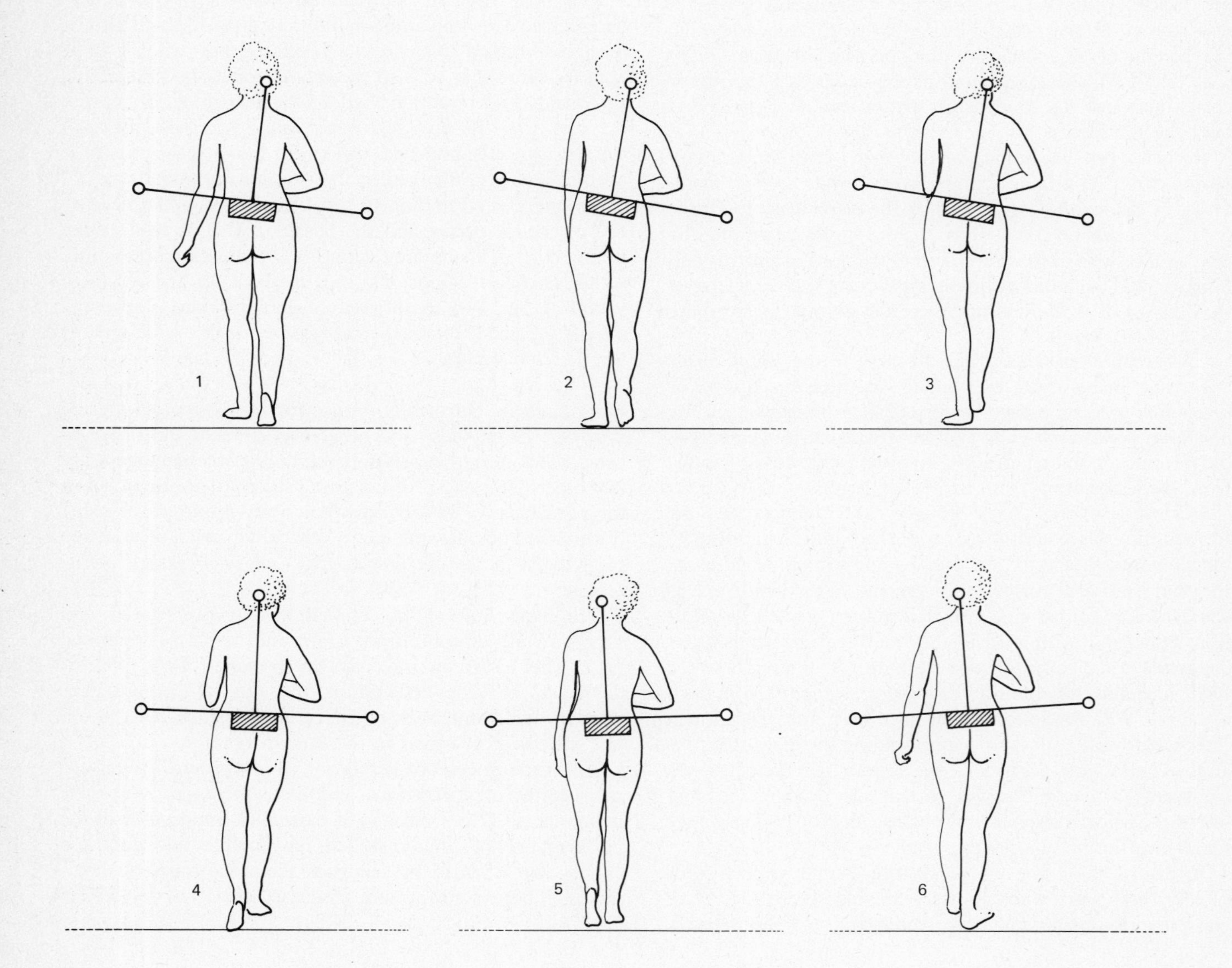

Table 23–2. **Reflex Pathways Involved in Simple Rhythmic Reflexes**

Flexor Reflex and Reciprocal Innervation	
Stimulus	Painful stimulus (heat, cold, electricity, etc.) to limb
Receptor	Nerve endings or specialized receptors in the skin
Afferent neuron	Sensory fibers in spinal nerve through dorsal root
Synapse	(a) Connecting neuron(s) in gray matter of cord; (b) flexor ventral horn motoneurons stimulated, extensor ventral horn motoneurons inhibited
Efferent neuron	Alpha motor fibers of spinal nerve through ventral root
Effector	Flexor muscle stimulated; antagonistic extensor muscle centrally inhibited
Response	Flexor muscle contracts; extensor relaxes
Effect	Withdrawal of stimulated limb from the painful stimulus
Crossed Extensor Reflex (This reflex incorporates all the above to evoke flexion on the stimulated side and, in addition, the following pathway.)	
Stimulus	Painful stimulus to limb (stronger than the one that elicits only the flexor reflex)
Receptor	Nerve endings or specialized receptors in the skin on the stimulated side
Afferent neuron	Sensory fibers in spinal nerve through dorsal root
Synapse	(a) Connecting neurons in gray matter of cord crossing over to other side; (b) extensor ventral horn motoneurons on other side of cord stimulated, flexor ventral horn motoneurons inhibited
Efferent neuron	Alpha motor fibers of spinal nerve through ventral root
Effector	Extensor muscle stimulated—contracts; antagonistic flexor muscle centrally inhibited—relaxes
Response	Extension of limb on side opposite the stimulation
Effect	Withdrawal of body from the painful stimulus

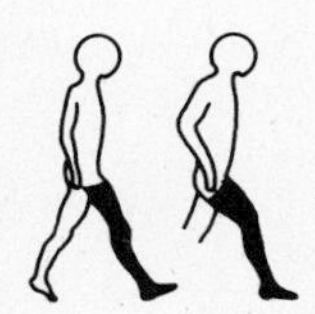

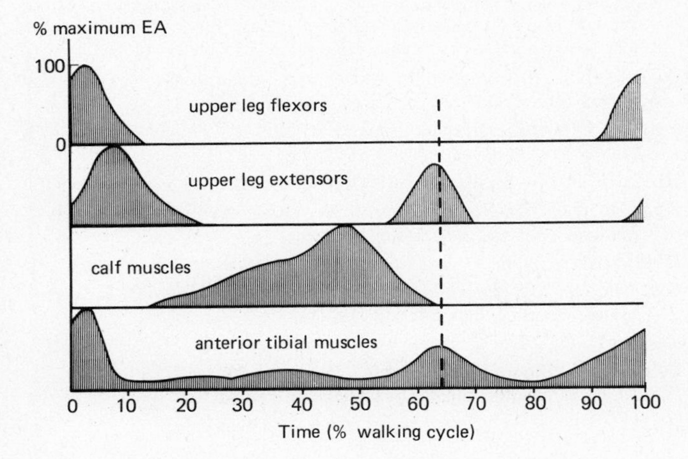

Figure 23-42. **The electrical activity (EA) of muscles used in normal, level walking. [Adapted by H. Stoboy from C. W. Radcliffe, *Artificial Limbs* 6:16 (1962).]**

Rhythmic Reflexes and Walking

Rhythmic reflexes of alternate flexion of one limb, extension of the other, can be built up on the *crossed extensor reflex* (Table 23-2 and Fig. 23-42). These rhythmic movements in a body with sufficient *extensor tone* to support itself against gravity, and with *proprioceptive regulation* controlling equilibrium, could theoretically be the basis for walking. The *initiation of impulses* must come from the cortex and the *coordinating* ones from the basal ganglia, cerebellum, and reticular formation, but the *end result* is possible only through the rhythmic reflexes.

Cited References

CLOSE, R. I. "Dynamic Properties of Mammalian Skeletal Muscles." *Physiol. Rev.* **52:** 129 (1972).

EBASHI, S. "Regulatory Mechanism of Muscle Contraction with Special Reference to the Ca-Troponin-Tropomyosin System." *Essays Biochem.* **10:** 1 (1974).

ECCLES, J. C. "The Effect of Nerve Cross-Union on Muscle Contraction." In A. T. Milhorat, ed. *Exploratory Concepts in Muscular Dystrophy and Related Disorders, Excerpta Medical Foundation,* 1967, p. 151.

ECCLES, J. C. *The Understanding of the Brain,* 2nd ed. McGraw-Hill Book Company, New York, 1977.

ECCLES, J. C., P. G. KOSTYUK, and R. F. SCHMIDT. "Central Pathways Responsible for Depolarization of Primary Afferent Fibers." *J. Physiol.* **161:**237 (1962).

Franzini-Armstrong, C. "Fine Structure of Sarcoplasmic Reticulum and Transverse Tubular System in Muscle Fibers." *Fed. Proc.* **23**(5): 887 (1964).

Granit, R. *Receptors and Sensory Perception.* Yale University Press, New Haven, 1955.

Hanson, J., and H. E. Huxley. "The Structural Basis of Contraction in Striated Muscle." In *Fibrous Proteins and Their Biological Significance. Symp. Soc. Exper. Biol.* **9:** 228 (1955).

Hill, A. V. "Production and Absorption of Work by Muscle." *Science* **131:** 897 (1960).

Hill, A. V. "The Heat of Shortening and the Dynamic Constants of Muscle." *Proc. Roy. Soc. (London) Biol.* **126:** 136 (1931).

Hunt, C. C., and S. W. Kuffler, "Further Study of Efferent Small Nerve Fibres to Mammalian Muscle Spindles." *J. Physiol (London)* **113:** 283 (1951).

Huxley, H. E. "The Mechanism of Muscular Contraction." *Science* **164:** 1356 (1969).

Jackson, J. H. "Unilateral Epileptic Seizures, Attended by Temporary Defect of Sight." *Med. Times Gaz.* **1:** 588 (1863).

Marsh, B. B. "The Effect of Adenosine Triphosphate on the Fiber Volume of Muscle Homogenates." *Biochim. Biophys. Acta* **9:** 247 (1952).

Merton, P. A. "How We Control the Contraction of Our Muscles." *Sci. Am.,* May 1972.

Sandow, A. "Excitation-Contraction Coupling in Skeletal Muscle." *Pharmacol. Rev.* **17:** 265 (1965).

Sherrington, C. S. *The Integrative Action of the Nervous System.* Cambridge University Press, New York, 1947.

Szent-Györgyi, A. G., D. Mazia, and A. Szent-Györgyi. "On the Nature of the Cross-Striation of Body Muscle." *Biochim. Biophys. Acta* **16:** 339 (1955).

Veratti, E. "Investigations on the Fine Structure of Striated Muscle Fiber." *J. Biophys. Biochem. Cytol.* **10**(suppl): 3 (1902; republished 1961).

Weber, A. "On the Role of Calcium in the Activity of Adenosine 5′-Triphosphate Hydrolysis by Actomyosin." *J. Biol. Chem.* **234:** 2764 (1959).

Additional Readings

BOOKS

Andersen, P., and J. K. S. Jansen, eds. *Excitatory Synaptic Mechanisms.* Universitetsforlaget, Oslo, Norway, 1970.

Basmajian, J. V. *Muscles Alive,* 3rd ed. The Williams & Wilkins Company, Baltimore, 1974.

Bourne, G. E., ed. *The Structure and Function of Muscle,* 2nd ed., Vols. 1–4. Academic Press, Inc., New York, 1973.

Carlson, F. D., and D. R. Wilkie. *Muscle Physiology,* Biological Science Series. Prentice-Hall, Inc., Englewood Cliffs, N.J., 1974.

Denny-Brown, D. *The Basal Ganglia: Their Relation to Disorders of Movement.* Oxford University Press, New York, 1962.

Falls, H. B. *Exercise Physiology.* Academic Press, Inc., New York, 1968.

Granit, R. *The Basis of Motor Control.* Academic Press, Inc., New York, 1970.

Gutmann, E., ed. *The Denervated Muscle.* Plenum Publishing Corporation, New York, 1962.

Guyton, A. C., ed. *MTP International Review of Science: Physiology,* Vol. 3. University Park Press, Baltimore, 1974.

Hubbard, J. I., ed. *The Peripheral Nervous System.* Plenum Publishing Corporation, New York, 1974.

Kuffler, S. W., and J. G. Nicholls. *From Neuron to Brain.* Sinauer Associates, Inc., Sunderland, Mass., 1976.

Matthews, P. B. C. *Mammalian Muscle Receptors and Their Central Actions.* The Williams & Wilkins Company, Baltimore, 1972.

Physiology of Voluntary Muscle. Brit. Med. Bull. **12**(3) (1956). (Entire volume.)

Roberts, T. D. M. *Neurophysiology of Postural Mechanisms.* Plenum Publishing Corporation, New York, 1967.

The Contractile Process. Symposium of the American Heart Association. Little, Brown and Company, Boston, 1967.

Walton, J. N. *Disorders of Voluntary Muscle.* Churchill Livingstone, a division of Longmans, Inc., New York, 1974.

Waser, P. G., ed. *Cholinergic Mechanisms.* Raven Press, New York, 1975.

Yahr, M. D., and D. P. Purpura, eds. *Neurophysiological Basis of Normal and Abnormal Motor Activities.* Raven Press, New York, 1967.

Zachs, S. I. *The Motor Endplate.* W. B. Saunders Company, Philadelphia, 1964.

ARTICLES

Allen, G. I., and N. Tsukahara. "Cerebrocerebellar Communication Systems." *Physiol. Rev.* **54:** 957 (1974).

Brooks, V. B., and S. D. Stoney, Jr. "Motor Mechanisms: The Role of the Pyramidal System in Motor Control." *Ann. Rev. Physiol.* **33:** 337 (1971).

Bucy, P. C. "Neural Mechanism Controlling Skeletal Muscular Activity and Its Unsolved Problems." In *Neurosciences Research,* Vol. I, S. Ehrenpreis and O. C. Solnitzky, eds. Academic Press, Inc., New York, 1968, p. 251.

Cohen, C. "The Protein Switch of Muscle Contraction." *Sci. Am.,* Nov. 1975.

Eccles, J. C. "Circuits in the Cerebellar Control of Movement." *Proc. Nat. Acad. Sci.* **58:** 336 (1967).

Evarts, E. V. "Brain Mechanisms in Movement." *Sci. Am.,* July 1973.

Fambrough, D. M., H. C. Hartzell, J. A. Powell, J. E. Rash, and N. Joseph. "On the Differentiation and Organization of the Surface Membrane of a Postsynaptic Cell—The Skeletal Muscle Fiber. In *Synaptic Transmission and Neuronal Interaction,* M. V. L. Bennett, ed. Raven Press, New York, 1974.

Fischbach, G. D., M. P. Henkart, S. A. Cohen, A. C. Breuer, J. Whysner, and F. M. Neal. "Studies on the Development of Neuromuscular Junctions in Cell Culture." In *Synaptic Transmission and Neuronal Interaction,* M. V. L. Bennett, ed. Raven Press, New York, 1974.

Glickstein, M., and A. R. Gibson. "Visual Cells in the Pons of the Brain." *Sci. Am.,* Nov. 1976.

Goodwin, G. M., D. I. McCloskey, and P. B. C. Matthews. "Proprioceptive Illusions Produced by Muscle Vibration: Contribution by Muscle Spindles to Perception." *Science* **145:** 1382 (1972).

Granit, R., and J. P. van der Meulen. "The Pause During Contraction in the Discharge of the Spindle Afferents from Primary End Organs in Cat Extensor Muscles." *Acta Physiol. Scand.* **55:** 231 (1962).

Grillner, S. "Locomotion in Vertebrates: Central Mechanisms and Reflex Interaction." *Physiol. Rev.* **55:** 247 (1975).

Guth, L. "Trophic Influences of Nerve on Muscle." *Physiol. Rev.* **44:** 645 (1968).

Hoyle, G. "How Is Muscle Turned On and Off?" *Sci. Am.,* Apr. 1970.

Hubbard, J. I. "Microphysiology of Vertebrate Neuromuscular Transmission." *Physiol. Rev.* **53:** 674 (1973).

Huxley, H. E. "The Contraction of Muscle." *Sci. Am.,* Nov. 1968.

Krnjevic, K. "Chemical Nature of Synaptic Transmission in Vertebrates." *Physiol. Rev.* **54:** 418 (1974).

Lester, H. A. "The Response to Acetylcholine." *Sci. Am.,* Feb. 1977.

Lippold, O. "Physiological Tremor." *Sci. Am.,* Mar. 1971.

Llinás, R. R. "Motor Aspects of Cerebellar Control." *Physiologist* **17:** 19 (1974).

LLINÁS, R. R. "The Cortex of the Cerebellum." *Sci. Am.*, Jan. 1975.

MARGARIA, R. "The Sources of Muscular Energy." *Sci. Am.*, Mar. 1972.

METTLER, F. A. "Muscular Tone and Movement: Their Cerebral Control in Primates." In *Neurosciences Research*, Vol. 1, S. Ehrenpreis and O. C. Solnitzky, eds. Academic Press, Inc., New York, 1968, p. 175.

MOMMAERTS, W. F. H. M. "Energetics of Muscle Contraction." *Physiol. Rev.* **49:** 427 (1969).

MURRAY, J. M., and A. WEBER. "The Cooperative Action of Muscle Proteins." *Sci. Am.*, Feb. 1974.

PORTER, K. R., and C. FRANZINI-ARMSTRONG. "The Sarcoplasmic Reticulum." *Sci. Am.*, Mar. 1965.

SANDOW, A. "Skeletal Muscle." *Physiol. Rev.* **32:** 87 (1970).

SATIR, P. "Cilia." *Sci. Am.*, Feb. 1961.

SHIK, M. L., and G. N. ORLOVSKY. "Neurophysiology of Locomotor Automatism." *Physiol. Rev.* **56:**465 (1976).

STEIN, R. B. "Peripheral Control of Movement." *Physiol. Rev.* **54:** 215 (1974).

TAYLOR, E. W. "Chemistry of Muscle Contraction." *Ann. Rev. Biochem.* **41:** 577 (1972).

WEBER, A., and J. M. MURRAY. "Molecular Control Mechanisms in Muscle Contraction." *Physiol. Rev.* **53:** 612 (1973).

WHITTAKER, V. P., and H. ZIMMERMANN. "Biochemical Studies on Cholinergic Synaptic Vesicles." In *Synaptic Transmission and Neuronal Interaction*, M. V. L. Bennett, ed. Raven Press, New York, 1974, p. 217.

Chapter 24

Regulation of Movement. II. Formation and Organization of the Skeleton

A science—so the savants say,
"Comparative Anatomy,"
By which a single bone
Is made a secret to unfold
Of some rare tenant of the mold
Else perished in the stone.
So to the eye prospective led
This meekest flower of the mead,
Upon a winter's day,
Stands representative in gold
Of rose and lily, marigold
And countless butterfly

Emily Dickinson, "With a Daisy"

BONE IS AN ancient skeletal material in the history of vertebrates, and modern ideas of evolution emphasize a reduction in the amount of bone, especially bones that were mostly of a protective nature. In the development of the embryo, a tiny model is formed first in *cartilage,* which is then gradually removed and replaced by *bone,* retaining essentially its original form. The use of cartilage as a model for those bones that are intricately associated with blood vessels, nerves, and muscles permits an enormous increase in growth without a change in the architectural relationships. Although cartilage is far inferior to bone as a supporting material, it can grow by internal expansion because its cells can divide and new material can be deposited between them. Bone, on the other hand, imprisons its cells within a rigid matrix, and only those cells on its surface can form new bone. Some bones, like those of the skull, are platelike and simple in form and have no complex relations with other organs. Their development is fairly simple, and no cartilage model is required. They develop within membranes, and new layers are simply added to the surface.

24-1 Functions of Bone

Bone has a triple function: it not only *supports* the body and acts as a system of levers for *movement,* but it is vitally involved in the *regulation of mineral homeostasis.* The third function of bone is the formation of blood cells by the bone marrow, a topic that was discussed in Sec. 14–2.

Support and Movement

Humans and all other vertebrates have an endoskeleton, which acts as a firm *support* for the soft body and its movements and also protects certain internal organs by forming a bony cage around them. Because the skeleton is jointed, *movement* between the bones is permitted to varying degrees, and the bones are held together by specialized connective tissue fibers, *tendons* and *ligaments. Muscles* are attached to the jointed skeleton in such a way as to move the bones as though they were a series of levers; the type of lever and the efficiency of movement depend on the angle and site of attachment of the muscle to the bone.

Regulation of Mineral Homeostasis

Bone is remarkably active metabolically despite its apparent inanimate nature. Radioactive tracer studies have shown that the matrix of bone functions as a reservoir of essential ions such as calcium, phosphate, magnesium, and sodium. In an adult, 3 to 5 per cent of the total skeleton is being actively remodelled to adjust to the changing environmental and physiological demands, and as much as one fourth of bone phosphate may be replaced by new phosphate within two months. This turnover rate is even more rapid, of course, in actively growing children. Yet despite these changes in bone form and size, the ionic concentrations in cells and in blood must remain constant for optimal cellular function. The regulation of these ionic balances is largely dependent upon hormones, as will be shown later in this chapter.

24–2 Structure of Bone

Bone is a connective tissue derived from mesoderm, in which relatively few cells are embedded in an *intercellular matrix.* The matrix is synthesized by bone cells, *osteoblasts,* provided that the necessary proportion of calcium and phosphate salts is available.

Compact and Spongy Bone

The density of different bones, and parts of bones, varies. The denser portions are referred to as *compact bone,* the more porous parts as *spongy bone.*

Compact Bone. The *shafts of the long bones* of the limbs—the femur and tibia, for example, in the leg and the humerus in the arm—are mostly made up of compact bone. The shaft is a thick hollow tube of compact bone surrounding the marrow cavities. The ends of the long bones are mostly spongy bone, covered with a *thin shell of compact bone* (Fig. 24-1).

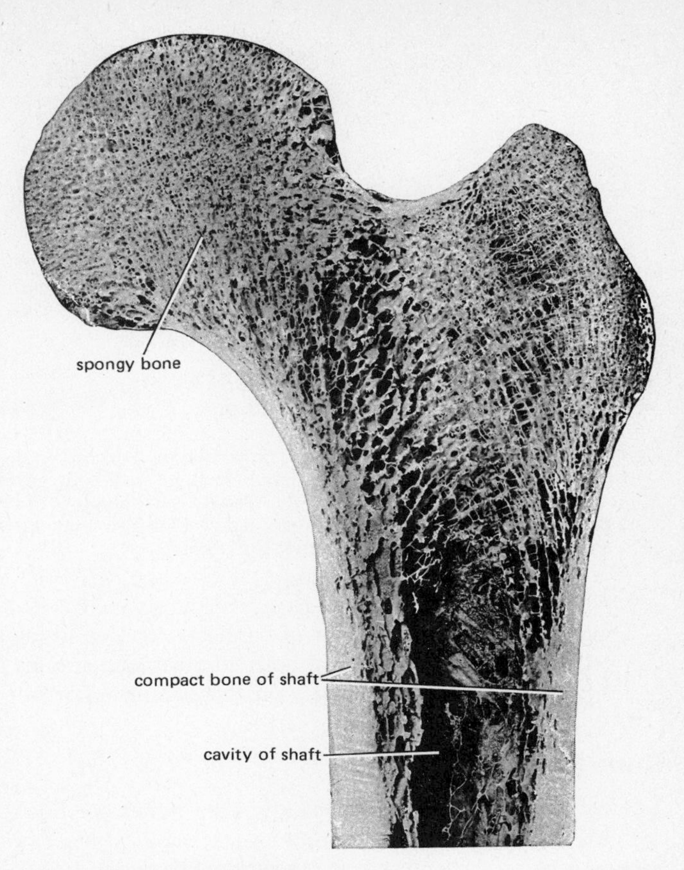

Figure 24-1. **Compact and spongy bone in the upper end of the femur (thigh bone). (From J. Sobotta and J. P. McMurrich, *An Atlas of Human Anatomy,* Vol. I, G. E. Steichert and Co., New York, 1936.)**

Spongy Bone. The vertebrae, most of the flat bones (the bones of the skull, for example), and the ends of the long bones are composed of spongy bone, made up of a network of fine partitions, the *trabeculae,* which enclose cavities containing either red or yellow marrow.

Periosteum and Endosteum

Except at its articular ends, bone is covered with a fibrous, vascular layer, the *periosteum.* The inner side of the periosteum is lined with osteoblasts, or with precursors of osteoblasts that mature into the osteoblast as a result of certain irritations. Among these irritants can be included a fracture of the bone. This damage seems to initiate the release of large numbers of active osteoblasts, which immediately start to secrete the material that will be *deposited* as new bone on the *periosteal surface* (Fig. 24-2).

Bone may be laid down so rapidly at the break that an excess, or callus, is formed at the injured site. At that time the level of alkaline phosphatase in the blood is considerably elevated, because the osteoblasts enzymatically split off phosphate from organic compounds for deposition in the new bone.

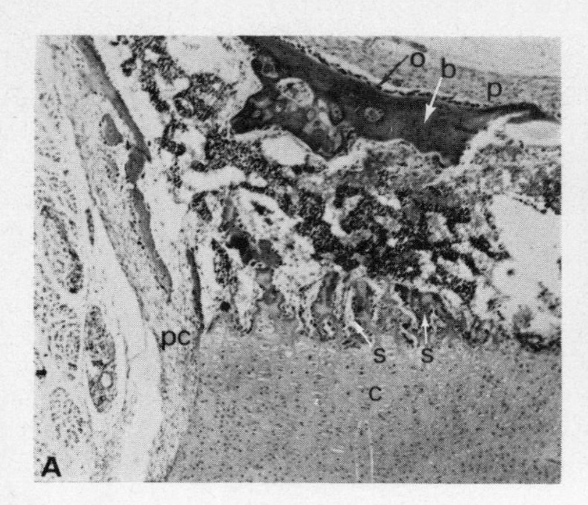

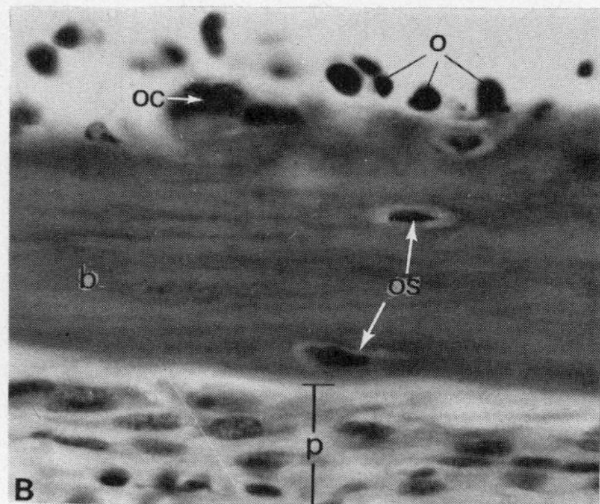

Figure 24-2. **(A) Deposition of bone (b) in cartilage (c) occurs on the surface of the periosteum (p) by osteoblasts (o). s = spicules of calcified cartilage with bony covering. pc = perichondrium, the connective tissue covering of cartilage. ×50. (B) Osteoblasts (o) imprisoned in the lacunae of new bone (b) become osteocytes (os). An osteoclast (oc) is also seen. ×400. (Both photographs courtesy of S. Piliero, Dental School, New York University.)**

The *endosteum* is a thin layer of connective tissue lining the walls of the bone cavities in both compact and spongy bone. Bone *resorption* takes place on the *endosteal surface.*

Bone Marrow

The cavities within the shafts of the long bones are filled with the fatty, yellow marrow that replaces the red marrow found there in the embryo and newborn mammal. In the adult, red marrow is found only in the cavities of the *flat bones,* such as the ribs, sternum, and skull, and in the epiphyses of two *long bones,* the femur and humerus. The hematopoietic functions of the bone marrow were discussed in Sec. 14-2.

24-3 Bone Matrix and Bone Strength

Composition of Bone Matrix

Inorganic crystals

About 70 per cent of the matrix of bone is made of inorganic salts, principally salts of calcium and phosphate. These salts are deposited as flat *crystal plates of hydroxyapatites,* which have a variable ratio of calcium and phosphorus. Many other ions, particularly fluoride, magnesium, sodium, potassium, and carbonate adsorb on to the hydroxyapatite crystals but do not themselves seem to be incorporated into bone as crystals. The enormous *compressional strength* of bone is derived from these hydroxyapatite crystals.

We have also learned to our cost that bones avidly take up *radioactive elements* such as strontium, uranium, and plutonium, produced as a result of atomic fallout. Large amounts of these radioactive substances adsorb on to the hydroxyapatite crystals and cause prolonged irradiation of the bone tissue. This not only can cause malignant growth of the bone (sarcoma), but can destroy the blood-forming elements of the bone marrow.

Organic matrix

Inorganic crystals give bone its compressional strength, but the *organic matrix,* which is almost entirely made up of *collagen fibers,* provides *tensile strength.* The collagen fibers are arranged along the lines of tensional force for each bone and are surrounded by a homogeneous *ground substance* of mucopolysaccharides and glycoproteins, containing chondroitin sulfate, hyaluronic acid, and sialic acid.

The combined tensile strength of the organic matrix and the compressional strength provided by the inorganic hydroxyapatite crystals give bone a strength that is comparable to that of reinforced concrete.

Regulation of Bone Strength by Load Stress

Bones that carry heavy compressional loads are *thicker* than those subjected to light loads; athletes, for example, have heavier bones than untrained individuals. The *loss of bone* in response to the removal of mechanical stress is dramatically seen after bed confinement, and also in astronauts, who are subjected both to weightlessness and recumbency. Therapeutic exercises during long space flights now prevent this decalcification of the bones.

If a fractured limb is immobilized, and no mechanical stress is placed upon it, the bone becomes thin and decalcified. If, however, the limb is set in a cast, which fixes the fracture but still permits the bone to be used and stress to be placed upon it, then osteoblast activity increases markedly and healing is accelerated.

It is believed that the deposition of bone in regions subjected to deformational compressional stress is caused by a transduction of mechanical energy into an electrical change (*piezoelectric effect*). The resulting electric currents in the bone may stimulate osteoblast activity at the negative end of the current flow. The mechanism by which piezoelectricity is generated is not clear but it requires the presence of both hydroxyapatite and collagen in organized bone.

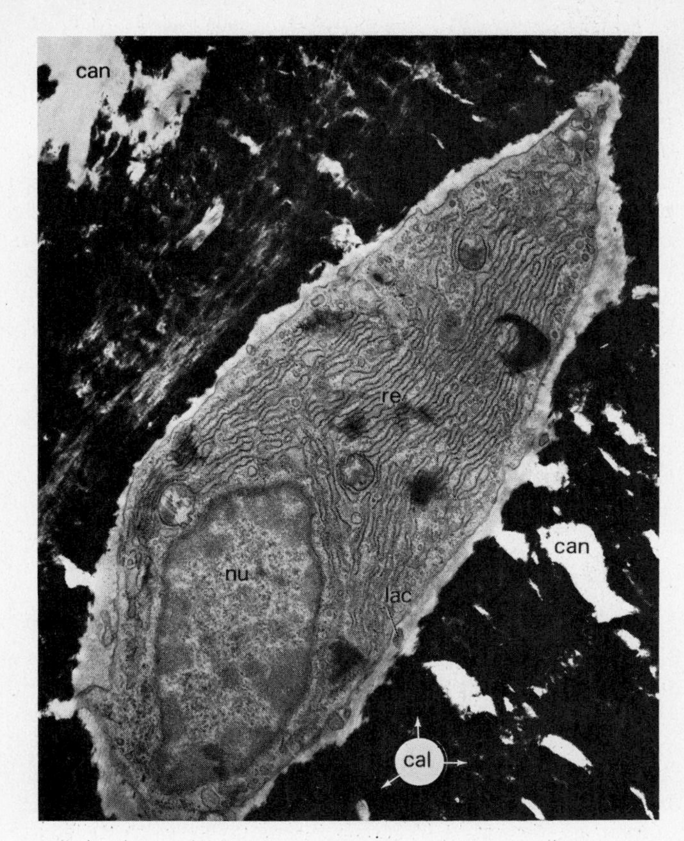

Figure 24-3. **An osteocyte. This mature bone cell is surrounded by a very narrow space, the lacuna (lac) and separated from the other osteocytes by the calcified matrix (cal). Small channels, the canaliculi (can), join the lacunae; nu = nucleus and re = endoplasmic reticulum. ×8000. (From J. Rhodin, *An Atlas of Ultrastructure*, W. B. Saunders Company, Philadelphia, 1963.)**

24-4 Cellular Elements of Bone

The cells of bone are present on bone surfaces and throughout bone, and they are responsible for both the formation and the removal of the tissue that houses them. The three main types of bone cells are the *osteoblasts*, the *osteocytes*, and the *osteoclasts*.

Osteoblasts

Osteoblasts *synthesize the organic matrix*, both collagen and ground substance. They are probably also involved in the regulation of calcium and phosphorus exchange and deposition. The osteoblast mitochondria may act as reservoirs for these two ions. Osteoblast activity is regulated by hormones of the parathyroid gland and by estrogen, as well as by vitamins A, C, and D.

Osteocytes

Osteocytes are imprisoned osteoblasts, cells that have become trapped by the newly formed bone matrix and lie in small spaces, called *lacunae*, in the bone. The osteocytes may still contribute to bone growth, for they have an elaborate cytoplasmic structure (Fig. 24-3) and appear to be metabolically active, but their precise function is not known.

Osteoclasts

Osteoclasts are the cells that *resorb bone*. They are very large, multinucleated cells and are highly mobile. They are filled with lysosomes, which contain the *proteolytic enzymes* that digest the organic matrix of bone. It is presumed that the secretion of *citric* and *lactic acid* by the osteoclasts is responsible for the solution of the bone crystals.

Under the electron microscope, long fine processes can be seen to extend from the osteoclasts to the bone surface through which they are tunnelling. The resulting debris of destroyed collagen and mineral salts is collected within these cytoplasmic channels. Osteoclast activity is affected by *calcitonin*, a thyroid hormone (Sec. 24-7).

24-5 Formation of Bone

Bones Modelled in Cartilage

Most bones are modelled in cartilage; the cartilage is gradually replaced by bone, first by *calcification* and then by *ossification*. In the process, the *osteoblasts* become imprisoned in the lacunae and are now the *osteocytes*, cells that have lost the ability to form new bone.

Calcification means simply the deposition of calcium salts in the connective tissue matrix, but ossification means the organization of the tissue into the characteristic structure of bone, composed of units called *osteons* (Fig. 24-4).

Figure 24-4. **The osteon consists of a central Haversian canal (H) surrounded by concentric lamellae (l) in which the osteocyte-containing lacunae (os) are clearly seen. Canaliculi (c) radiate between adjacent lacunae. ×400. (Courtesy of S. Piliero, Dental School, New York University.)**

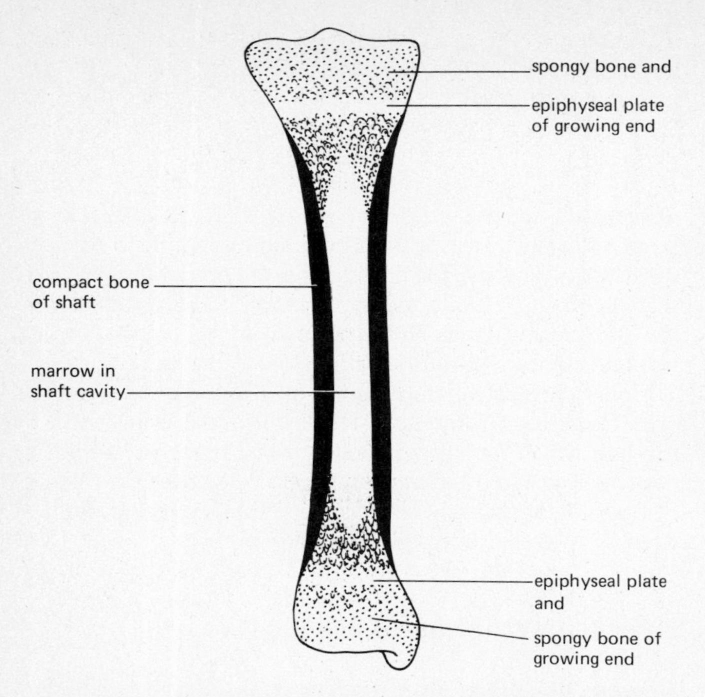

Figure 24-5A. **The structure of a long bone showing the epiphyses.**

Figure 24-5B. **Formation of new bone at the epiphysis of a long bone. ep = epiphyseal ossification center, rc = reserve cartilage, dc = dividing cartilage, mc = mature cartilage, epl = epiphyseal plate, and s = shaft of bone. ×50. (Courtesy of S. Piliero, Dental School, New York University.)**

Osteon formation

Osteon formation begins at different times for different bones, but centers for ossification usually appear at about the second month of intrauterine life. Processes of cartilage destruction and bone building occur simultaneously; large numbers of cartilage-destroying and bone-destroying osteoclasts move in through the cartilage, making space for new blood vessels and enlarging the central marrow cavity. Meanwhile, the osteoblasts are busy forming new bone and becoming confined as osteocytes in the process (Bourne, 1973).

The osteocytes in their lacunae are concentrically arranged around a central *Haversian canal,* through which a blood vessel runs. The Haversian canal and its surrounding rings of lacunae form an osteon.

Bone growth

Spongy bone, with spikes of bone separated by spaces, is formed by the destruction of the first layers of bone by the osteoclasts and the deposition of new bone by the osteoblasts. The bone grows in thickness and in strength, with the osteocytes left in concentric layers around the central cavity. Secondary ossification centers, *epiphyses,* develop at the ends of the long bones of the limbs. New bone is deposited at the epiphyses, but a plate of cartilage remains between the shaft of the bone and the growing end (Figs. 24-5A and B). When this cartilage in turn becomes ossified, it indicates the end of the lengthwise growth of the bone. Any further growth can be only in the girth (see Secs. 19-3 and 24-7 for the effect of hormones upon growth and the rate of closure of the epiphyses).

Figure 24-6. **The ossification of membranous bone. In the infant skull, ossification is incomplete, and the softer, membranous areas are called fontanelles.**

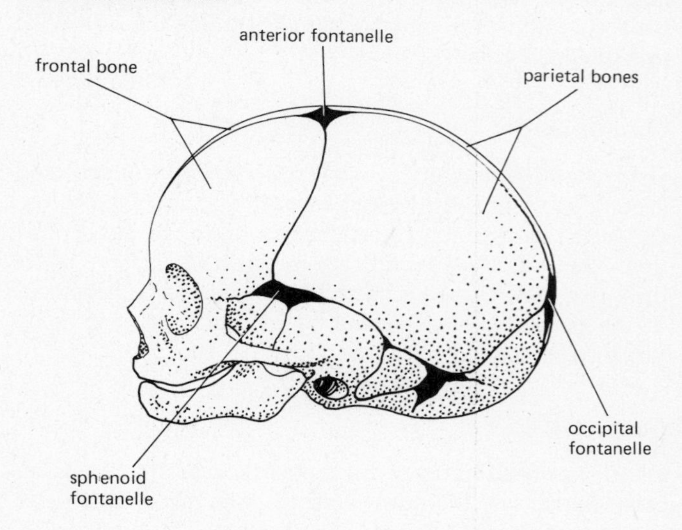

Bones formed in membrane

Some bones are formed directly by the ossification of membranes, without an intervening cartilaginous stage. These bones include the bones of the skull and the mandible and parts of the clavicle. The incomplete ossification of these membranes can be clearly seen in young infants as the "soft" part of the skull or the *fontanelles.* The largest of these fontanelles is the anterior one, between the two halves of the frontal bones and the parietal bones. It usually is closed during the second year (Fig. 24-6).

24–6 Regulation of Calcification

Calcification is the deposition of calcium salts in the collagen and ground substance secreted by the osteoblasts. Although blood is supersaturated with calcium and phosphate ions (compared to the concentration of these ions in the bone), the precipitation of calcium phosphate, which precedes the formation of the large hydroxyapatite crystals, is extremely slow. Some active intervention is needed to initiate the process; once started, this site of precipitation then acts as a *nucleus* or *seed* for further, rapid precipitation.

Theories of Nucleation

It used to be thought that it was the formation of *collagen fibers* from the collagen molecules that initiated the precipitation of calcium salts. Next the unique *glycoproteins* of bone were implicated. Now it is considered more likely that the *osteoblasts concentrate* and *secrete calcium and phosphate,* which they then liberate into the matrix. The resulting small areas of highly concentrated calcium phosphate salts then act as nucleating sites for further salt deposition (Irving, 1973).

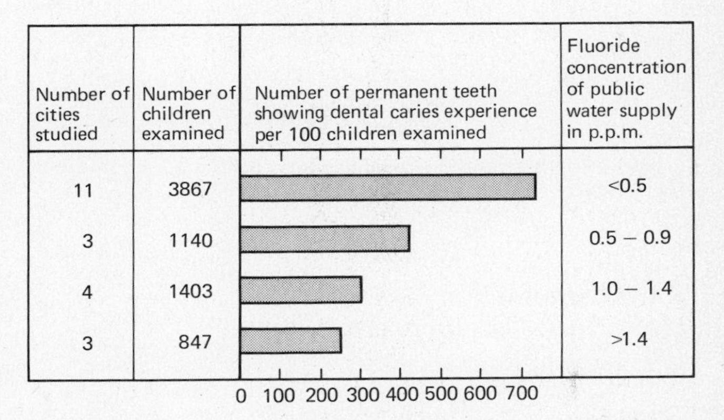

Figure 24-7. **Amount of dental caries (in permanent teeth) observed in 7257, selected, 12- to 14-year-old white school children of 21 cities of four states, classified according to the fluoride concentration of the public water supply. [H. T. Dean, P. Jay, F. Arnold, and E. Elvove, *Pub. Health Rept.,* 56:365 (1941). Copyright 1941 by the American Association for the Advancement of Science.]**

Figure 24-8. **Geographic distribution of mottled enamel in the United States, 1941. (From H. T. Dean, *Fluorine and Dental Health,* Publication No. 19 of the American Association for the Advancement of Science. Copyright 1941 by the American Association for the Advancement of Science.)**

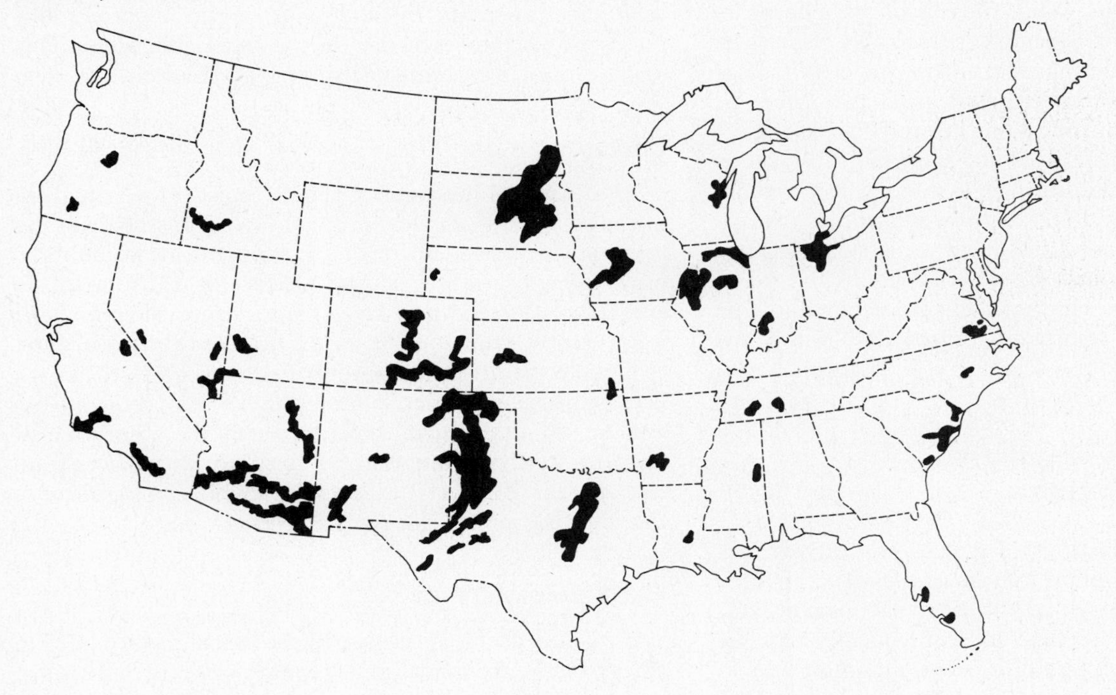

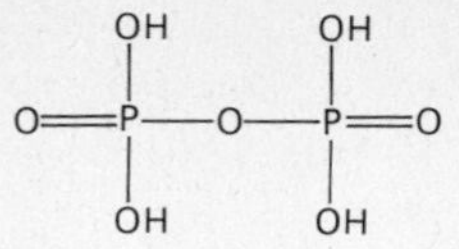

OH OH
O=P—O—P=O
OH OH

inorganic pyrophosphoric acid

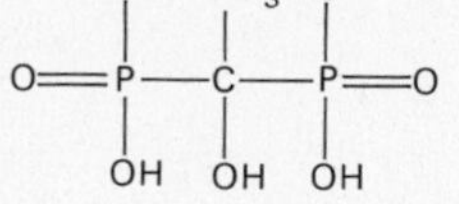

OH CH_3 OH
O=P—C—P=O
OH OH OH

ethane–1–hydroxy–1,
1–diphosphonic acid (EHDP)

Figure 24-9. **Structure of pyrophosphoric acid and a diphosphonic acid, EHDP.**

Nonhormonal Factors Regulating Calcification

Fluorine

Calcification may be enhanced by fluorine, which accelerates the growth of hydroxyapatite crystals. Fluorine has been used therapeutically in the prevention of dental caries and for the proper development of teeth in children (Fig. 24-7). An excess of fluorine, however, causes mottling of the teeth (Fig. 24-8) and enlargement of the bones.

Pyrophosphates

Inorganic pyrophosphates, substances that are present in plasma and in most tissues including bone, inhibit the precipitation of calcium salts. While the role of pyrophosphate in normal calcification is debatable, presumably it must be destroyed by alkaline phosphatase, an enzyme closely associated with calcification, before calcification can occur.

In diseases in which calcification is disturbed, the administration of pyrophosphate does not help; the substance is not absorbed by the gut, and if it is injected it is rapidly inactivated by the body. Recently, however, a new investigative tool has been found in the form of stable *diphosphonates.* These compounds have P-C-P bonds that resist enzymatic action and, in experimental animals, inhibit bone calcification and bone resorption. Figure 24-9 shows the structure of pyrophosphoric acid and a diphosphonic acid.

24–7 Hormonal Regulation of Calcification

The hormone of the parathyroid glands, *parathyroid hormone,* and a thyroid hormone known as *calcitonin,* are intimately involved in many aspects of the metabolism of vitamin D, calcium, and phosphate and consequently regulate the deposition and absorption of calcium phosphate salts in bone. Plasma levels of calcium and phosphate act as feedback regulators of the secretion of parathyroid hormone and calcitonin (Copp, 1970).

Calcium, Vitamin D, and Parathyroid Hormone

Calcium absorption and excretion

Calcium Absorption. The absorption of calcium was discussed in Sec. 17-6. Because calcium compounds are relatively insoluble, adequate calcium absorption is completely dependent upon the presence of *natural vitamin D,* (*cholecalciferon*). The active form of cholecalciferon is produced in the kidney under the influence of parathyroid hormone, so that calcium absorption is ultimately dependent upon the hormone of the parathyroid glands.

Calcium Excretion. Calcium is mostly excreted through the feces, with lesser amounts being excreted in the urine. The kidney regulates calcium excretion, however, by delicately balanced homeostatic mechanisms, increasing calcium excretion when calcium levels in blood are high and decreasing calcium excretion when blood levels of that ion are low. This renal control, in turn, is regulated by parathyroid hormone.

Calcium deposition

Ninety-nine per cent of the calcium in the organism is in the skeleton and teeth. It is deposited in certain areas of the bone as tricalcium phosphate crystals and gives the skeleton its rigidity. This calcium is not a permanent deposit, for bone is constantly being dissolved and redeposited, and the wear and tear results in the loss of several hundred milligrams of calcium each day in the feces and urine. This amount must then be replaced through the diet.

Calcium intake

The National Research Council considers that a daily intake of 800 mg of calcium is advisable for adults, but the recommended calcium intake by European nutritionists is only 450 mg, whereas in the Orient the calcium intake averages 250 to 350 mg/day. This very low calcium intake is apparently without effect on the health of the population.

Despite the advice of such folk heroes as Popeye, spinach eaten in large quantities is not healthy: it may contain large amounts of oxalic acid, which prevents the absorption of calcium from the intestine by forming *insoluble calcium oxalate salts.* Similarly, excessive amounts of phosphate in the intestine inhibit calcium absorption.

Calcium deficiency

A calcium deficiency caused by the lack of dietary calcium is very rare. Calcium deficiency usually arises because of the lack of adequate vitamin D, or to excessive amounts of other minerals that prevent calcium absorption. This is not true, however, in pregnancy and lactation, at which time the demands for calcium rise immensely. An allowance of 500 mg calcium in addition to that of the usual diet is recommended for the last 6 months of pregnancy and during lactation. Of course, the vitamin D intake must be increased also to permit the absorption of the increased dietary calcium. If the calcium supply of the mother is not adequate for the maintenance of her normal needs, plus those of the growing fetus, calcium still will be supplied in sufficient amounts to the fetus through the placenta against a concentration gradient. This means, of course, that the additional calcium supplied to the fetus is at the expense of the stored calcium in the skeleton of the mother. This situation becomes even more severe during lactation, when 300 to 400 mg of calcium are lost daily in the milk. Unless this amount is replaced, the maternal skeleton will be damaged. The diseases resulting from insufficient calcium for the development and maintenance of bone (rickets and osteomalacia) have been discussed under vitamin D deficiency in Sec. 21-7.

PLASMA CALCIUM

The small amount of calcium in the blood is very important for such basic phenomena as cell permeability, excitability, release of neurotransmitters, muscle contraction, and blood coagulation, all of which are discussed in appropriate places in the text.

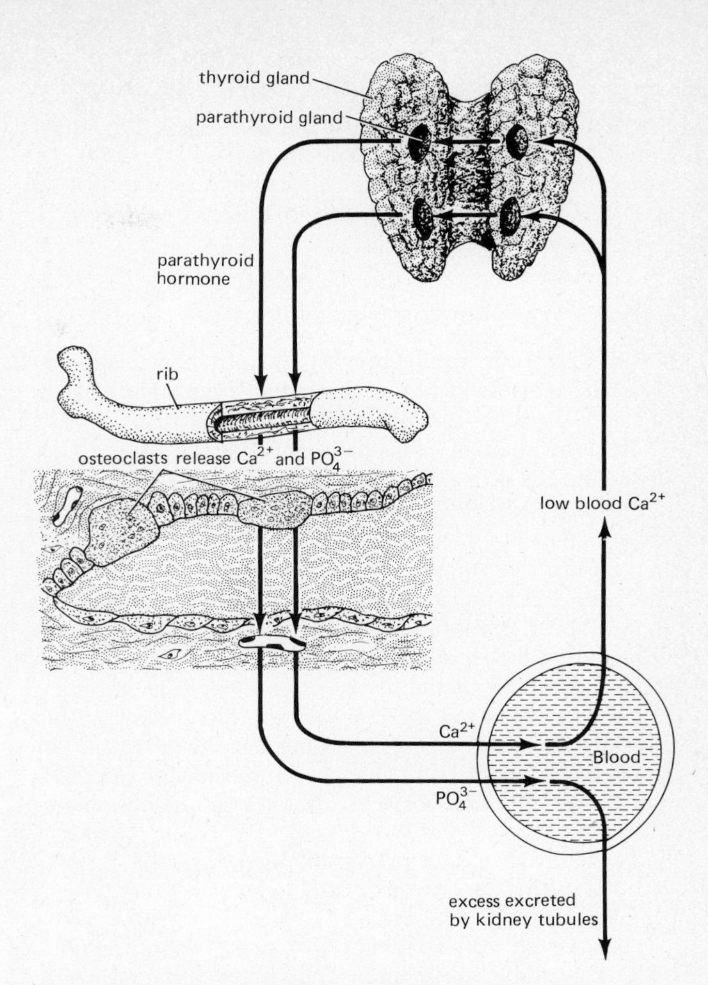

Figure 24-10. **Regulation of the level of blood calcium by the parathyroid glands.**

Interdependence of Plasma Calcium and Parathyroid Hormone

The four parathyroid glands in the human are small organs placed immediately behind the thyroid gland. They are quite difficult to distinguish, and their relationship to calcium metabolism was inadvertently discovered when the parathyroids were accidentally removed during thyroidectomy.

The parathyroid glands are extremely sensitive to changes in the level of circulating calcium. If the amount of calcium in plasma is low, parathyroid hormone is released, raising blood calcium by stimulating osteoclast absorption of bone. This process releases calcium into the blood. At the same time phosphates are liberated into the blood, but because the parathyroid hormone also inhibits the reabsorption of phosphate from the kidney tubules, there is ultimately a fall in the circulating phosphate (Fig. 24-10).

A decrease in the amount of calcium in the blood (as a result of removal of the parathyroid glands, for example) causes extreme excitability of the nervous system with spontaneous discharge of nerve fibers, resulting in tetany and convulsions.

Excessive amounts of calcium in the blood (caused by a tumor of the parathyroids, for example) can cause muscular weakness, depress the central nervous system, act as a mental depressant, and cause the heart to stop in systole. Actually, calcium concentrations do not need to vary much from normal to cause either hyperirritability or hypoirritability. The role of the parathyroids, therefore, is extremely important.

Phosphate and Parathyroid Hormone

INORGANIC AND ORGANIC PHOSPHATE

Inorganic phosphate is the major cation of the *intracellular fluids.* The ion is of enormous importance in the formation of the energy-rich phosphate bonds of such compounds as ATP. Phosphate also is able to combine reversibly with a multitude of enzyme systems, and so operates in the majority of metabolic reactions within the cell.

Inorganic phosphate is present in the extracellular fluids as well, in both the plasma and in the interstitial fluid. This inorganic phosphate exists in the form of phosphate ions (both HPO_4^{2-} and $H_2PO_4^-$). These phosphate ions are closely involved in the maintenance of the acid-base stability of the blood, as discussed in Sec. 14-0. A very small amount of phosphate ion is found in the plasma as PO_4, but this is undoubtedly of great importance in determining the rate of deposition and reabsorption of bone.

Organic phosphate is present in the blood in the form of *phospholipid,* and the phosphate from this compound can be liberated for bone formation.

Phosphate absorption and excretion

Milk and bone meal are very rich in phosphate, but phosphate is also widely found in many animal and vegetable foods.

The absorption of phosphate from the intestine is dependent on the proper absorption of calcium, which in turn is dependent on the vitamin D concentration. This means that a deficiency of vitamin D will indirectly affect phosphate absorption. Apparently, if calcium is not absorbed at a rapid rate from the intestine, the calcium and phosphate tend to form insoluble calcium-phosphate salts that are eliminated in the feces. Excess phosphate is *excreted* mainly through the kidneys; the *parathyroid hormone* is extremely effective in inhibiting the reabsorption of phosphate, thus causing a rapid drop in plasma phosphate. The tissues are not nearly as sensitive to changes in phosphate concentration, however, as they are to changes in calcium levels.

Deposition in bone: interrelationship with calcium

It has been mentioned that calcium phosphate, $Ca_3(PO_4)_2$, is a very insoluble substance, and it will precipitate when the concentration of either the calcium or the phosphate reaches a certain level. This is obviously an advantage in bone formation, in which the deposition of calcium phosphate is necessary for the formation of the rigid matrix, but it would be highly undesirable in blood or interstitial fluids. The relation between the ionized calcium and the phosphate in the blood is also affected by the pH of the blood and the carbon dioxide of the blood, according to the following formula:

$$\frac{Ca^{2+} \times HCO_3^- \times HPO_4^{2-}}{\text{pH}} = K \text{ (a constant)}$$

If the carbon dioxide and the pH of the blood are kept constant, the product of the calcium and phosphate of the blood will be between 30 and 40, because the normal calcium content is 10 to 11 mg/100 ml and that of phosphate, 3.0 to 3.5 mg per cent. That is,

$$10 \times 3.0 = 30 \text{ or } 11 \times 3.5 = 38.5 \text{ mg/100 ml}$$

This product is just below the level needed for precipitation, which therefore does not occur in the circulating fluids. In bone, however, the activity of bone-forming cells, the osteoblasts, produces the enzyme phosphatase, which splits off inorganic phosphate from the phospholipid and other organic phosphate constituents of blood, thereby increasing the concentration of phosphate at the surface of the newly forming bone and causing precipitation of calcium phosphate salts as crystals.

Alkaline phosphatase

The secretory osteoblasts and other cells of the body produce *alkaline phosphatase,* which is active at a pH of 9. This enzyme *liberates inorganic phosphate* from organic phosphate compounds, and it is present in the blood in increased concentrations whenever normal or pathologic bone formation is occurring. Alkaline phosphatase levels in the blood would be high in growing children, for example, and following major fracture of bones, as well as in a disease in which bone is being destroyed and then replaced.

Calcitonin

Calcitonin, a peptide hormone produced by the *thyroid gland* in mammals, is also present in high concentrations in certain fish, amphibia, reptiles, and birds. The cells that secrete calcitonin are the parafollicular or C cells, which lie between the follicles of the thyroid gland, and it is these C cells that are derived from the branchial glands of the lower vertebrates.

Calcitonin consists of a chain of 32 amino acids, with a molecular weight of about 3000.

Interdependence of calcitonin, plasma calcium, and phosphate

Calcitonin rapidly lowers plasma calcium and phosphate, especially in young animals and children, because sensitivity to calcitonin is closely related to the rate of bone turnover. Because this response to calcitonin is also seen in parathyroidectomized animals, it cannot be acting by simply depressing parathyroid hormone activity. In fact, most experiments indicate that calcitonin lowers plasma calcium and phosphate levels primarily by a direct effect on bone, preventing or decreasing osteoclast activity. This results in more deposition of bone than bone absorption (Rasmussen and Pechet, 1970).

Comparison of calcitonin and parathyroid hormone effects

It is evident that calcitonin and parathyroid hormone have opposite effects on blood calcium and phosphate, and that the feedback mechanisms controlling the release of these two hormones are also opposite in direction. Figure 24-11 shows the rapid increase in calcitonin secretion and the abrupt fall in parathyroid hormone secretion induced by an increase in plasma calcium.

Parathyroid hormone is undoubtedly of much greater significance in maintaining the *long-term equilibrium* of cal-

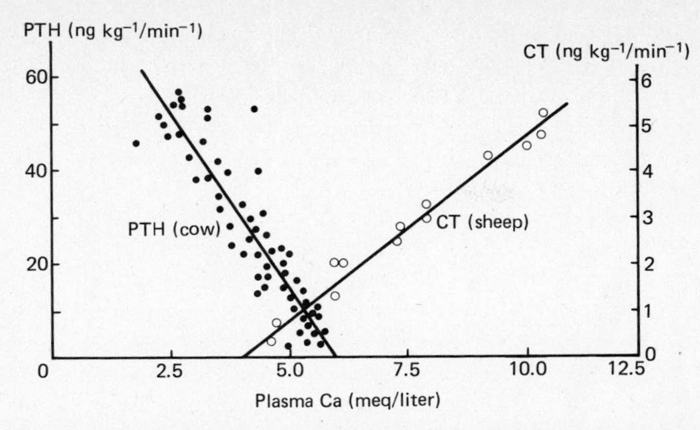

Figure 24-11. **Effect of plasma calcium concentration on the rate of secretion of parathyroid hormone (PTH) and calcitonin (CT). [From D. H. Copp, *J. Endocrinol.* 43:137 (1969).]**

cium between blood and bone in adults. Calcitonin has its peak effect within 1 hr, whereas parathyroid hormone has a slower and much longer-lasting action. Calcitonin is most effective in young people in whom bone modeling is active, for calcitonin checks osteoclast resorption. Its significance in adults, however, is unclear.

THERAPEUTIC USE OF CALCITONIN

Calcitonin has been used with considerable success in some bone disorders such as hypercalcemia and Paget's disease. The latter disease is characterized by resorption and softening of the bone and replacement of the normal bone matrix by a soft, poorly mineralized tissue. Calcitonin appears to be able to return bone turnover and structure to normal.

Calcitonin does not seem to be as effective in *osteoporosis,* which is the most common of the bone diseases. In osteoporosis, osteoblast activity is low and consequently bone deposition is poor. This demonstrates again the complexity of the regulatory mechanisms involved in the maintenance of calcium equilibrium between bone and plasma; disturbances in this equilibrium cannot easily be corrected by a single factor.

Testosterone, Estrogen, and Growth Hormone

Despite the spurt in growth that is seen in both males and females at puberty, testosterone and estrogen have been shown to accelerate the ossification or closing of the epiphyses and thus bring bone growth to a halt. Males castrated prior to puberty grow somewhat taller than normal individuals. It is uncertain whether the inhibitory action of the sex hormones is exerted directly on bone or whether these hormones act on the hypothalamus and/or the anterior pituitary gland to decrease the secretion of growth hormone.

In children, excessive secretion of growth hormone before the epiphyses have become ossified, causes gigantism, whereas increased growth hormone secretion in adults causes acromegaly, or growth in the girth of the bones. Some types of dwarfism are associated with an abnormally low level of growth hormone secretion (see Sec. 19-3 for a more detailed discussion of the effects of hormones on growth).

24-8 The Basic Organization of the Skeleton

The axial skeleton, which in the human is with difficulty maintained in the upright position, consists of the long, curved pillar of the *dorsal, segmented vertebral column;* the sword-shaped ventral *sternum* (breastbone); 12 pairs of *ribs;* the *skull,* including the lower jaw or mandible; and also the *hyoid* bone that is in the front of the neck, below the mandible and the tongue (Fig. 24-12). A strict listing of the bones of the axial skeleton would also include the *tympanic ossicles* of the middle ear (Sec. 22-6).

The appendicular skeleton consists of the *upper and lower limbs* and their attachments to the axial skeleton. The basic plan of the *pectoral* (shoulder) and *pelvic* (*hip*) *girdles* is similar, but has been greatly modified in humans and arboreal apes to permit the pelvic girdle and the lower limbs to support the weight of the body, whereas the upper limbs and pectoral girdle have a far greater freedom of movement for grasping and manipulation.

Bones of the Axial Skeleton

THE VERTEBRAL COLUMN

In the embryo and young child, 33 vertebrae are placed in a series one above the other to form the long, curved vertebral column. The vertebrae are held together by cartilaginous disks between their bodies and by the muscles and ligaments that connect them with their neighbors. In the adult, because several of the vertebrae have fused to form the sacrum and the coccyx, there are only 26 individual vertebrae. The adult vertebral column also has two additional curvatures, the first of which, the cervical curvature, begins to appear when the child holds its head up and then sits erect (Fig. 24-13). The marked lumbar curvature (more marked in women than in men) develops with the ability to stand erect and run and jump.

The basic plan of a vertebra consists of a heavy anterior *body,* an *arch,* and various bony *processes* attached to the arch (Fig. 24-14). The bodies of the vertebral column give it most of its supporting strength, and they get larger as they get closer to the sacral end. The vertebral arches enclose the *vertebral canal,* a long, jointed, bony passage for the spinal cord. Where one vertebra rests on the next, two holes are formed. A spinal nerve passes through each of them, left and right. Smooth articulating surfaces on the vertebrae mark the points of contact with adjacent vertebrae. The arches also have *dorsal spines* and *transverse processes* for the at-

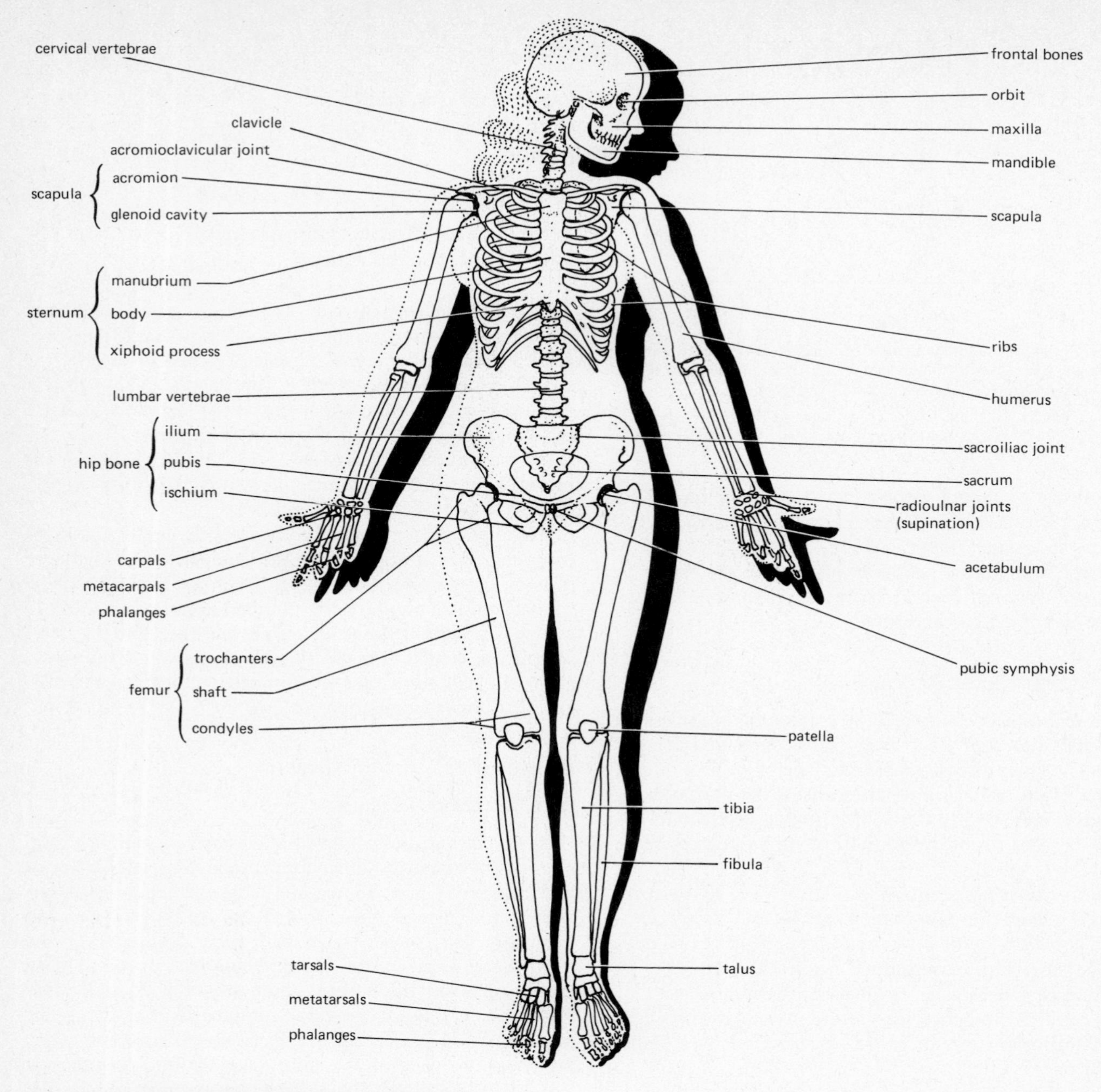

Figure 24-12. **The basic organization of the skeleton.**

tachment of muscles. The modifications in the vertebrae as they affect the range of movement possible are discussed in Sec. 25-1.

The sternum and the ribs

The *sternum* is a long, dagger-shaped bone about 7 in. in length. The widest part is the *manubrium* (handle), and the clavicles and the first pair of ribs articulate with it. The second to the seventh pairs of ribs articulate with the narrower *body* of the sternum. The sternum ends in a small, pointed piece of cartilage, which becomes ossified with age. This is the *xiphoid process.*

There are 12 pairs of ribs (in both sexes), and all of them are attached to the thoracic vertebrae dorsally, but only the first 7 pairs are attached to the sternum. These are called the *true ribs,* whereas the lower 5 are the *false ribs.* Although the anterior ends of the eighth, ninth, and tenth ribs are joined to each other by cartilage, the cartilages of the eleventh and the twelfth are free; these are the *floating ribs.*

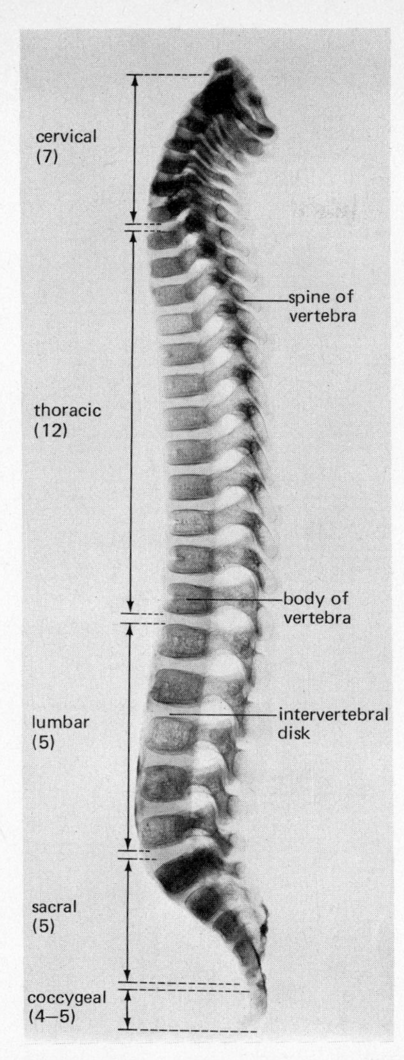

Figure 24-13. **Radiograph of the vertebral column of a child. (From J. Dankmeijer, H. J. Lammers, and J. M. F. Landsmeer, *Practische Ontleedkunde,* De Erven F. Bohn N. V., Haarlem, Holland, 1955.)**

Figure 24-14. **The basic plan of the vertebrae. Three thoracic vertebrae are shown.**

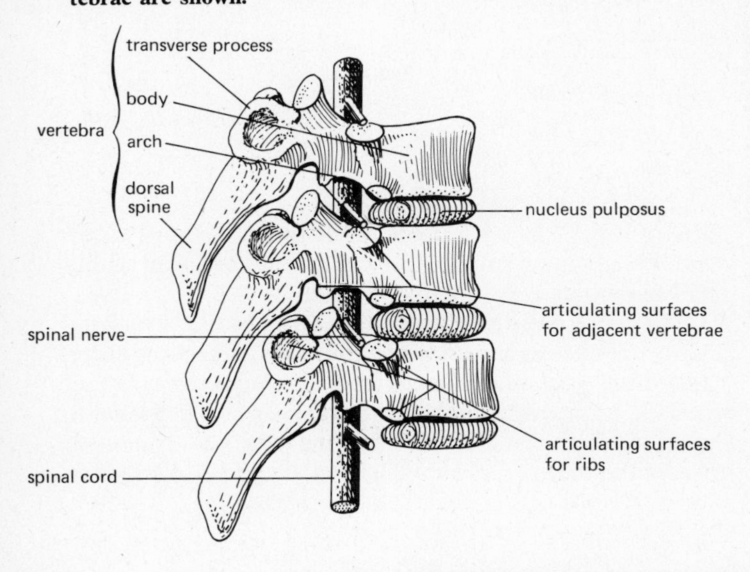

The skull

The skull really acts as an exoskeleton for the brain and inner ear, and an endoskeleton for the face. It consists of many bones that become inseparably joined by jagged *sutures,* where the membranes between them have become ossified (Fig. 24-15). The part of the skull that covers the brain consists of four fused bones, with names similar to those of the underlying lobes of the brain. The *frontal bone* is a large shell-like bone that forms the forehead and the top of the skull; the two *parietal bones* meet in the midline at the top of the skull and extend down to form the sides; the *occipital bone* forms most of the back of the skull, and the large hole in it, the *foramen magnum,* is the place where the wide hindbrain narrows before entering the vertebral column as the spinal cord.

The *temporal bones* form the lower part of the side of the skull, and their mastoid processes house the inner ear.

Bones of the face

The many bones that make up the orbit (eye socket), external nose, and upper and lower jaws vary greatly in form with age, race, and sex, as well as resulting in the individuality of the features. Some of these bones can be seen in Figure 24-15. It is beyond the scope of this text to discuss them in any detail. Cunningham's *Textbook of Anatomy* gives an excellent description of these bones.

Bones of the Appendicular Skeleton

The pectoral girdle and the upper limb

The pectoral girdle consists of the *scapula* (shoulder blade) and the *clavicle* (collar bone). These bones *articulate* (are united by a joint) with one another at the *acromioclavicular joint,* but although the clavicle also articulates with the upper end of the sternum, the scapula is held in place only by muscles (Figs. 24-12 and 24-15).

The bone of the *upper arm* is the *humerus,* extending from the shoulder to the elbow and articulating with a shallow depression, the *glenoid cavity,* of the scapula to form the shoulder joint. At the elbow, the humerus articulates with the two bones of the *forearm,* the radius and the ulna, to form the elbow joint. The radius and the ulna articulate with one another at both their proximal and distal ends to form the radioulnar joints.

The *ulna* is the longer bone of the forearm, and when the arm is in the *supine* position with the thumb away from the side of the body, the ulna and radius are parallel to one another, with the ulna on the medial side. The upper end of the ulna has a deep notch for articulation with the humerus and a shallower notch for the radius. It has a very large protuberance, the *olecranon,* which forms the back of the elbow and prevents the forward dislocation of the arm.

The *radius* is the lateral and shorter bone of the forearm, and its head is a flat disk that articulates with the humerus and the ulna. The annular ligament encircles most of the head of the radius, which rotates within this ring during

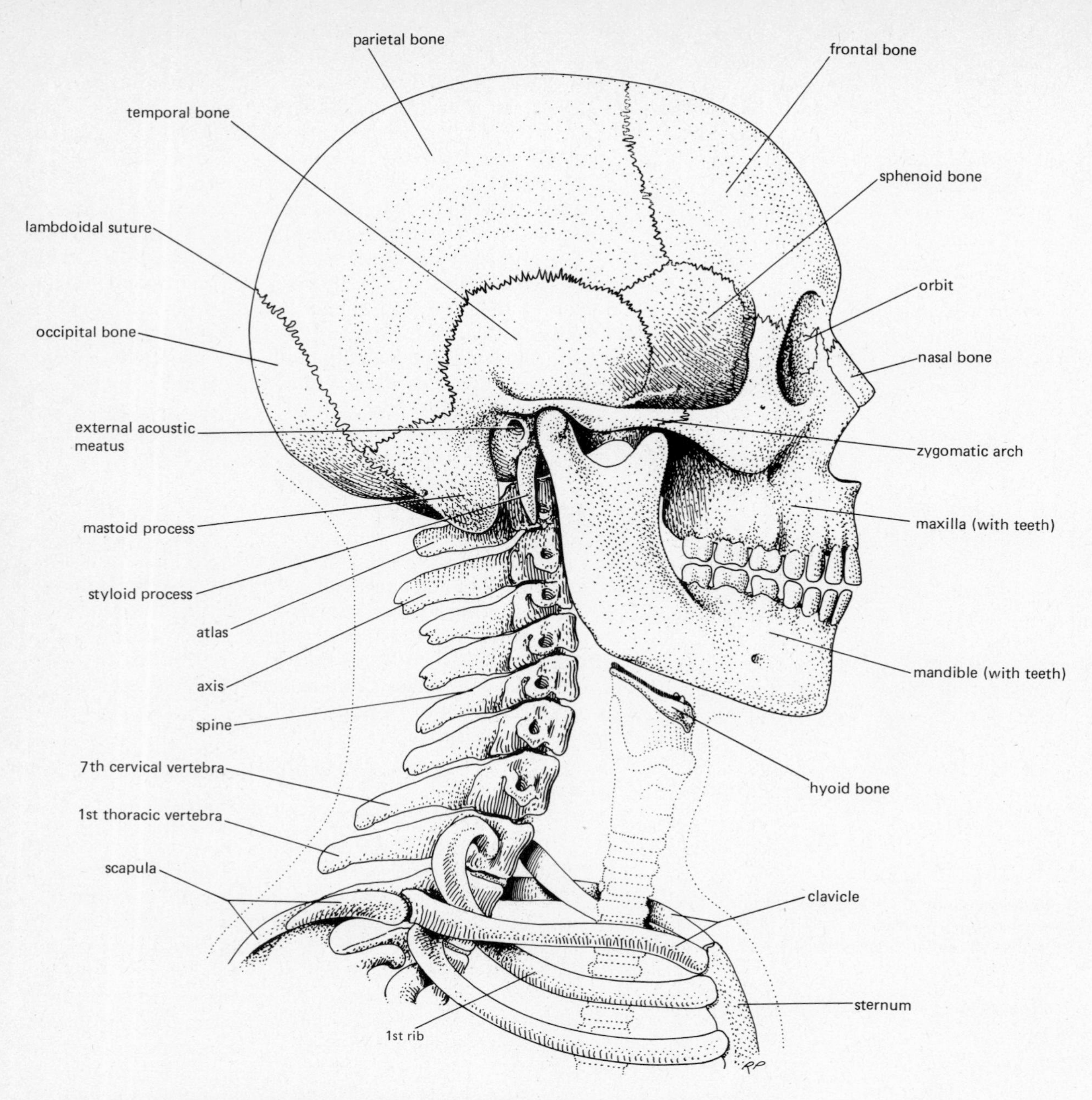

Figure 24-15. **The bones of the skull and the lower jaw, the neck, and the pectoral girdle.**

supination and pronation. To bring the arm into the *prone* position, the shaft of the radius is brought across the stationary ulna. The crossing of the bones is easily felt in the living arm. In full pronation, the thumb is next to the side of the body. The most comfortable natural position appears to be midway between pronation and supination.

The *wrist* is formed by eight small bones, the *carpals,* arranged in two rows. The radius articulates with three of the carpals in the proximal row and also with an articular disk, forming the wrist joint. Each carpal also articulates with its adjoining carpals to form a series of joints within the wrist.

The *hand* is given its elongate form by the five long, narrow *metacarpals,* with which the distal carpals articulate. The distal ends of the metacarpals form the first row of knuckles. The thumb and the fingers are known as the *digits,* each of which has three articulating bones or *phalanges,* except the thumb (pollex) which has only two. The proximal phalanx articulates also with the distal end of its metacarpal bone.

The large number of small joints within the wrist and hand permits the great variety of delicate movements of which the hand is capable. Both the thumb and the little finger are opposable, which makes grasping possible. When any of these joints is immobilized by disease (e.g., arthritis) the range of movement is drastically reduced.

The pelvic girdle and the lower limb

The ring-shaped pelvic girdle is made up of the two strong hip bones, which articulate ventrally with one another to form the pubic symphysis. Dorsally, each hip bone articulates with the sacrum (the lower part of the vertebral column) to form the sacroiliac joint. The hip bone itself is made up of the fusion of three bones, the ilium, ischium, and pubis (Figs. 24-12 and 24-16).

The *ilium* is the largest and broadest of these three fused bones. Its upper end is expanded and curved into the iliac crest, which lies below the waist in the upper part of the hip. The ilium transmits the weight of the body to the sacrum and increases considerably in strength and size after the third year of life, when the child's activities start to include running and jumping.

The *ischium* is a thick, three-sided bone that forms most of the lower, dorsal part of the hip bone. Part of the ischium surrounds an aperture, the *obturator foramen,* and then extends forward to fuse with the pubis. In the sitting position, the body rests on the two ischia (from which "sciatic" is derived).

Figure 24-16. **The fused bones of the hip bone. The drawing indicates the smooth articulating surface of the acetabulum and the roughness of those surfaces to which muscles are attached by fibrous tissue and tendons.**

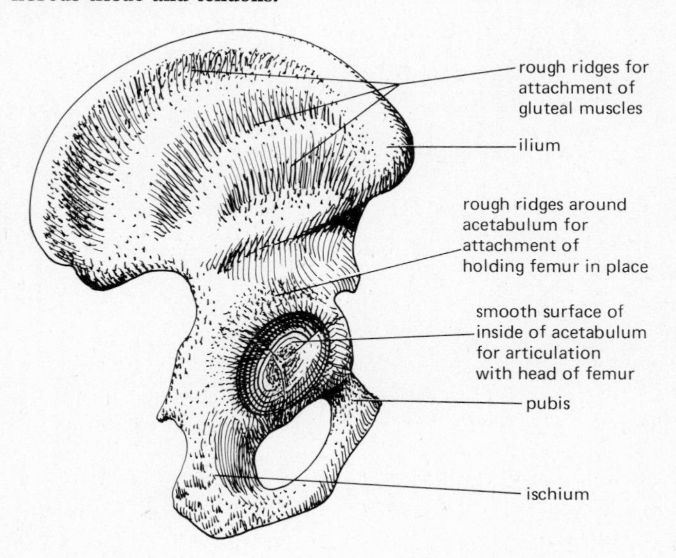

The *pubis* can be felt at the lowest part of the front of the abdomen, where its body is joined to the body of the other pubis by a plate of fibrocartilage forming the *pubic symphysis.* The body of the pubis is extended upward and backward in two thick branches, one of which fuses with the ilium, the other with the ischium.

All three of these bones contribute to the structure of the large cup-shaped cavity, the *acetabulum,* on the lateral side of the hip bone. The acetabulum is slanted forward and downward and articulates with the head of the femur.

The pelvic girdle, together with the sacrum and coccyx of the axial skeleton, is referred to as the *pelvis.* The entire pelvis is normally tilted forward, and its cavity is continuous with that of the true abdominal cavity. The female pelvis, being modified for child-bearing, is roomier and shorter than the male pelvis, and the angle below the pubic arch is much wider than that of the male pelvis.

The *thigh,* like the upper arm, has only one bone, but this bone, the *femur,* is the longest in the body. It articulates with the acetabulum of the hip bone, and at the knee it articulates with the tibia (shin bone) and the small bone of the kneecap, the *patella.* The upper end of the femur is easily distinguished by its smooth, round head on a thick neck of bone. This neck joins the head to the shaft at an angle, which enables the femur to move freely in the acetabulum but also renders it susceptible to fracture. There are two rough projections for the attachment of muscles at the base of the neck. These are the greater and lesser *trochanters.*

The lower end of the femur is made up of two large, smooth condyles which form the protuberances at the side of the knee. Their anterior surface articulates with the patella. When the knee is extended, the tibia articulates with the lower surfaces of the condyles; when the knee is flexed, it articulates with their posterior surfaces (Fig. 25-6).

The *lower leg,* like the forearm, has two bones that articulate with one another at both ends. These bones are the heavy *tibia* and the slender *fibula.* Only the tibia articulates with the femur at the knee joint, but both tibia and fibula articulate with one of the bones of the foot, the talus, to form the ankle joint. A strong sheet of fibrous tissue is stretched between the tibia and fibula.

The *foot* is greatly modified in form from the structure of the hand. Because the weight-bearing foot is at right angles to the leg, its upper surface is *dorsal* and its lower surface, the sole, is *plantar.* There are seven small bones, the *tarsals,* which form the back of the foot and correspond to the eight carpals of the wrist. The largest tarsal bone is called the calcaneum (heel bone), and the talus is placed above it. The remaining five tarsals form an arch in front of them. The tarsals articulate with one another and with the five *metatarsals* that extend to the toes or digits. Like the thumb, the big toe (*hallux*) has two phalanges, whereas the other digits have three apiece.

In order to support the body, the tarsals and metatarsals are formed into strong arches, maintained by ligaments, tendons, and muscles, but the human foot has lost its ability to grasp objects: the digits are very short and none of them is opposable.

24–9 The Surfaces of Bones

The surfaces of bones bear the marks of their various functions. The *articular surfaces* that move freely on other bones within a joint are covered with smooth, shining articular cartilage. Even the bone beneath this cartilage is smooth. When movement is limited, fibrous tissue may connect the bones, and the articular surfaces are rough. Some bones are completely fused, so that no movement is possible (Fig. 24-16).

Because movement at a joint is controlled by muscles, a firm attachment of the muscle to the bone is essential. Muscle fibers do not connect directly to bone but are attached to the *rough protuberances* of the bones by fibrous tissue and tendons. The point of attachment of a muscle at one end is called the *origin* and is a functional term for that end of the muscle that is fixed or stationary. The attachment at the other, movable end is the *insertion.* The origin and insertion may be reversed, depending on the direction of the movement. If the arm moves a weight, the insertion of the biceps is below the elbow as the forearm is flexed, but when the arms are used to support the weight of the body when hanging from a branch of a tree, the insertion is at the shoulder joint, and the elbow is held rigid.

Cited References

BOURNE, G. H. *The Biochemistry and Physiology of Bone,* 2nd ed., Vols. 1, 2, and 3. Academic Press, Inc., New York, 1973.

BRASH, J. C., ed. *Cunningham's Textbook of Anatomy.* Oxford University Press, New York, 1964.

COPP, D. H. "Endocrine Regulation of Calcium Metabolism." *Ann. Rev. Physiol.* **32:** 61 (1970).

IRVING, J. T. "Theories of Mineralization of Bone." *Clin. Orthop.* **97:** 225 (1973).

RASMUSSEN, H., and M. M. PECHET. "Calcitonin." *Sci. Am.,* Oct. 1970.

VAUGHAN, D. J. M. *The Physiology of Bone.* Oxford University Press, New York, 1970.

Additional Readings

BOOKS

FROST, H. M. *The Laws of Bone Structure.* Charles C Thomas, Publisher, Springfield, Ill., 1964.

FROST, H. M. *The Physiology of Cartilaginous, Fibrous, and Bony Tissue.* Charles C Thomas, Publisher, Springfield, Ill., 1972.

HANCOX, N. M. *Biology of Bone.* Cambridge University Press, New York, 1972.

IRVING, J. T. *Calcium and Phosphorus Metabolism.* Academic Press, Inc., New York, 1973.

LITTLE, K. *Bone Behavior.* Academic Press, Inc., New York, 1973.

MENCZEL, C., and A. HARELL, eds. *Calcified Tissue—Structural, Functional and Metabolic Aspects.* Academic Press, Inc., New York, 1971.

RASMUSSEN, H. *The Physiological and Cellular Basis of Metabolic Bone Disease.* The Williams & Wilkins Company, Baltimore, 1974.

ARTICLES

ARNAUD, C. D., JR., A. M. TENENHOUSE, and H. RASMUSSEN. "Parathyroid Hormone." *Ann Rev. Physiol.* **29:** 349 (1967).

BENTLEY, P. J. "Hormones and Calcium Metabolism." In *Comparative Vertebrate Endocrinology.* Cambridge University Press, New York, 1976.

BORLE, A. B. "Calcium and Phosphate Metabolism." *Ann. Rev. Physiol.* **36:** 361 (1974).

DAUGHADAY, W. H., A. C. HERINGTON, and L. S. PHILLIPS. "The Regulation of Growth by Endocrines." *Ann. Rev. Physiol.* **37:** 211 (1975).

DELUCA, H. F. "Regulation of Vitamin D Metabolism: A New Dimension in Calcium Homeostasis." *Biochem. Soc. Symp.* **35:** 271 (1972).

GRAY, T. K., C. W. COOPER, and P. L. MUNSON. "Parathyroid Hormone, Thyrocalcitonin, and the Control of Mineral Metabolism." In *MTP International Review of Science: Physiology,* Vol. 5, A. C. Guyton, ed. University Park Press, Baltimore, 1974, p. 239.

HARRISON, H. E., and H. C. HARRISON. "Calcium." *Biomembranes* **4B:** 793 (1974).

HIRSCH, P. F., and P. L. MUNSON. "Thyrocalcitonin." *Physiol. Rev.* **49:** 548 (1969).

JENKINS, G. N. "Current Concepts Concerning the Development of Dental Caries." *Int. Dent. J.* **22:** 350 (1972).

LOOMIS, W. F. "Rickets." *Sci. Am.,* Dec. 1970.

RASMUSSEN, H. "The Parathyroid Hormone." *Sci. Am.,* Apr. 1961.

RASMUSSEN, H., P. BORDIER, K. KUROKAWA, N. NAGATA, and E. OGATA. "Hormonal Control of Skeletal and Mineral Homeostasis." *Am. J. Med.* **56:** 751 (1974).

TEPPERMAN, J. "Hormonal Regulation of Calcium Homeostasis." In J. Tepperman, ed. *Metabolic and Endocrine Physiology,* 3rd ed., Year Book Medical Publishers, Chicago, 1973, p. 225.

Chapter 25

Regulation of Movement. III. Muscle and Bone as a Functional Unit

Professor Günter Friedebold

Director
Oskar-Helene-Heim
Orthopedic Clinic and Polyclinic
Free University
Berlin, Germany

Translated by Fleur L. Strand

Illustrious every one!
Illustrious what we name space, sphere of unnumber'd spirits
Illustrious the mystery of motion in all beings, even the tiniest insect,
Illustrious the attribute of speech, the senses, the body,
Illustrious the passing light—illustrious the pale reflection on the new moon in the western sky,
Illustrious whatever I see or hear or touch, to the last.

Walt Whitman, "Song at Sunset"

Skeletal muscle and bone cooperate so closely in the mechanical support of the body and in its movements that they can almost be considered as a single functional organ. They contribute greatly to the form of the body, in both its length and width, and with the individual pattern of fat distribution determine the type of body constitution. Although bone plays mainly a passive role, providing support and protection and forming a series of jointed levers with different ranges of movement upon which the muscles act, bone itself responds by changes in its internal structure and form to stresses that are placed upon it. It is a living tissue, capable of growth, repair, and adaptation, and subject to disease and atrophy.

25–1 Bones and Their Movements

Characteristics of Bone

Rigidity and elasticity

Bone derives its characteristics of firmness, rigidity, and elasticity from the composition and organization of its matrix. The basic structural unit of bone is the *osteon,* which consists of organized lamellae of collagen (protein) fibers that have become impregnated with crystals of inorganic salts, especially calcium salts. It is these inorganic salts that give bone its hardness. Despite this hardness, however, the position of the osteons is constantly being shifted by pressures of growth and external stress (Fig. 25-1). This results in a certain amount of elasticity of bone, especially notable in young bones, in which ossification is not yet complete, and the periosteum is strong and tough. With age, reabsorption of the collagenous part of the matrix occurs as protein synthesis slows down, making the bones considerably lighter and more fragile. Because of these physiological variations in the composition of bone during the lifespan of the individual, the type of fracture and the rapidity of its repair vary with age.

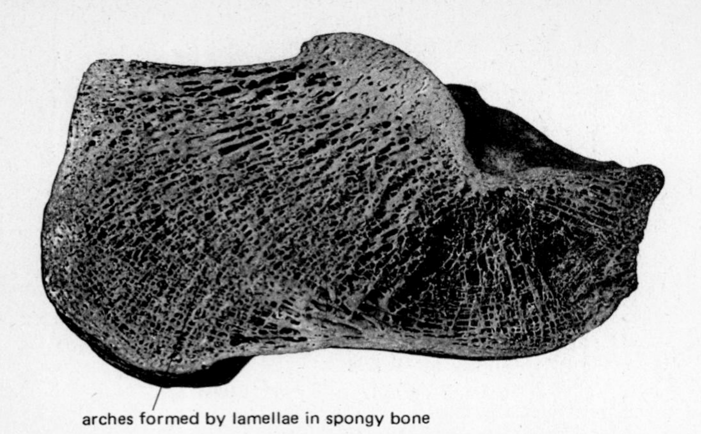

Figure 25-1. **The osteon and its structural adaptation to the support of body weight through the formation of weight-bearing arches.**

Adaptation to stress

Because the structure of bone is constantly influenced by its physiological environment, it adapts very readily to change. A most important external influence on the growth and structure of bone is the degree of compression or extension to which it is subject. The long bones of the body, like the femur and the humerus, clearly show the result of the constant strain along their longitudinal axis. The osteons of the *shafts* of these bones are tightly organized in parallel arrays, forming *compact bone;* the strength of the bone is thus increased, but its flexibility is reduced. The *spongy ends* of these bones develop a series of arched lamellae that curve and cross, giving greater elasticity to this portion of the bone.

Any stress that is placed upon the skeletal muscles, tendons, or ligaments will be transmitted to the bones. In many places the attachment of the muscles and the ligaments is fan-shaped, so that the stress is applied over a larger surface of the bone, but even so, when bone is subjected to *chronic strain* in a particular area, it responds by the local accumulation of osseous tissue, becoming thicker and stronger in proportion to the degree of strain. A strain on the bone along its *longitudinal axis,* if applied over a long period of time, will cause it to increase its thickness; the bones of athletes are considerably thicker and heavier than those of nonathletes. The hypertrophy of the normally slender fibula, in the absence of the tibia, is an excellent example of the response of bone to an increased functional load. If pressure is applied *across* the bone, however, a gradual reabsorption of bone occurs, and the bone decreases in girth. (Of course, an *acute strain* on the bone, beyond the resistance of the tissue, will cause the bone to break. Heavy pressure on the bone, beyond the range of its elasticity, will compress it, and a break may occur because of the bending of the bone.)

This characteristic ability of bone to adapt to external forces is by no means pathologic. The rigidity of a bone is not affected by the amount of activity or inactivity, but its adaptability to stress is due perhaps to reserve strength that is normally not needed, but which may be marshaled to withstand excessive forces applied to it. Muscles, tendons, and ligaments have narrower regions of tolerance.

Joints Between Bones

Although the bones, especially those of the lower trunk and legs, bear a major part of the weight of the body, they are assisted considerably by the cartilaginous covering of their joints. The type of joint by which the single bones of the skeleton are bound to one another is determined by the function of those bones (Fig. 25-2).

No movement—bones joined by bone

In this type of joint, *synostosis* (*syn,* with; *os,* bone), osseous tissue forms between the individual bones, immobilizing them. The five vertebrae that form the sacrum are completely fused together by bone, forming an inseparable, rigid block of great stability and strength.

Slight movement—bones joined by fibrous membrane

In this type of joint, a *syndesmosis,* a tough layer of fibrous connective tissue binds the bones firmly together, but despite firmness of the bond, a certain amount of movement is permitted by the flexibility of the membrane. The interosseous membrane between the tibia and the fibula joins these bones in a syndesmosis at their distal ends where they provide a socket for the upper part of the talus of the foot. These three bones form the ankle joint, and the interosseous membrane between the tibia and fibula plays an essential part in maintaining the springiness and tautness of this joint.

In *dorsiflexion* (bending the foot upward), the wider front part of the asymmetric talus is pushed back into the narrow, posterior portion of the tibiofibular socket, forcing these two bones slightly apart. In *plantarflexion* (bending the foot downward), the narrower part of the talus pushes forward into the broader region of the socket, so that the tibia and fibula spring together again, under the tension exerted by the interosseous membrane. The bones of the socket keep a firm grip on the talus in this manner.

Little movement but some elasticity—bones joined by cartilage

When bones are exposed to a constant pressure that must be withstood by a certain amount of elasticity and springiness, they are bound together by a layer of cartilage, forming a *synchondrosis.* The joints between the ribs and the sternum are of this type.

Free movement—the true joint

This is by far the most important type of union between bones. Through its specialized construction, the true joint

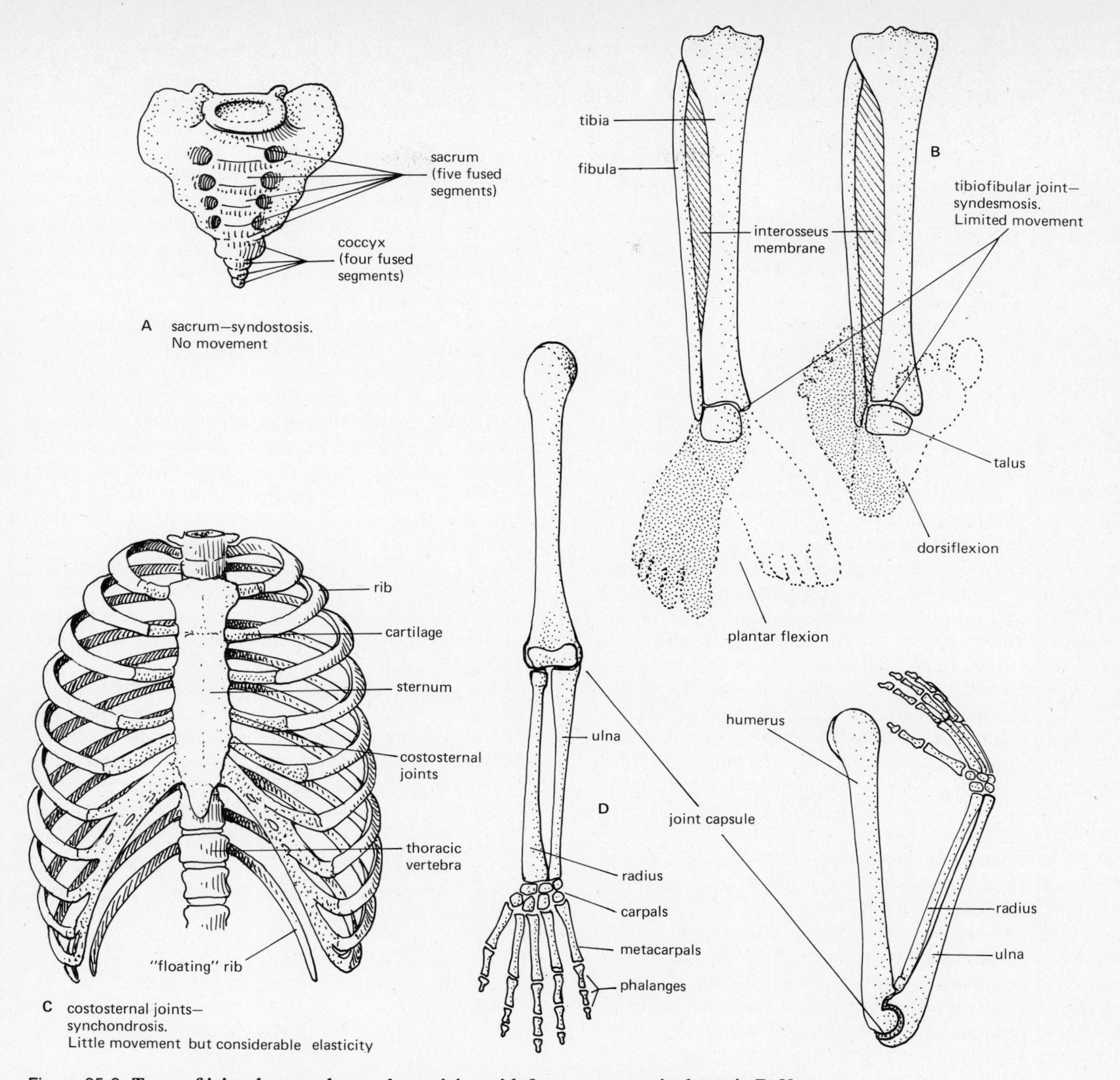

Figure 25-2. **Types of joints between bones. A true joint, with free movement, is shown in D. Note the joint capsule.**

permits a varying range of movement, coupled with a dependable resistance to the load placed upon it. Although the structures of different joints vary according to their specific functions, a basic structural plan is common to all.

Hyaline Cartilage: Pressure Resistance and Fluid Exchange. The most important element of the true joint is the firm, smooth *hyaline cartilage,* which covers the articulating ends of the bones. The convex part of the joint, the ball or head, is covered with cartilage that is considerably harder than that covering the concave socket with which it articulates. This hardness of the cartilage increases the mechanical efficiency of the head of the joint.

Cartilage is so important in the structure of the joint because its unique composition gives it an *elastic resistance to pressure,* regardless of the axis along which the pressure is exerted. The unit of cartilage is the *chondron,* a cartilage cell surrounded by a shell of hyaline ground substance. The chondrons may be considered as small balls of cartilage fastened by fibrous threads, which run parallel to the surface

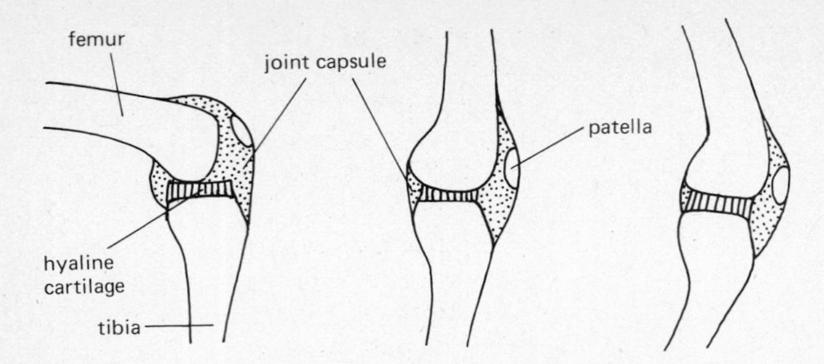

Figure 25-3. **Changes in shape of the joint capsule with movement at the joint. The stippled areas are lined with synovial membrane. Note how the movement of the bones squeezes and releases the hyaline cartilage.**

of the cartilage, then curve in arcs down to its deeper levels. The fibers then radiate out in the layer of basal cartilage that separates the bone from the articulating cartilage.

As a result of this structure, pressure-resistant balls of cartilage are embedded in tension-resistant bundles of fibers. When pressure is exerted upon the cartilage, it is transferred from the chondrons to the fibrous shell, which is then placed under tension. This allows for a limited amount of distortion of the form of the cartilage and so gives this part of the joint some elasticity despite the absence of elastic elements.

This plasticity of the cartilaginous covering of the joint is important every time a load is placed on the joint and results in the fitting together of both surfaces of the members of the joint. During this process, the cartilaginous tissue is squeezed between the bony members of the joint much like a sponge, and like a sponge it will slowly fill up with the fluid contained in the joint cavity and return to its original form when the pressure is removed (Fig. 25-3). This movement of fluid nourishes the nonvascular cartilage, and a disturbance of this mechanism can result in regressive changes in the tissues of the joint.

The efficiency of this *fluid exchange* is dependent on the equal distribution of pressure over the entire surface of the joint; a serious deformation of the skeleton, particularly of the axial skeleton, will result in very uneven application of pressure on the joints of the lower extremities. Although this interferes with the normal distribution of nutrients to the hyaline cartilage, this tissue appears to be so modest in its requirements that only a severe, long-lasting fluid depletion leads to serious changes in its structure. Normally, this cartilage is capable of withstanding a wide range of fluid disturbances.

When the pressure on the joint is continuously greater than the resistance of the cartilage, degenerative changes in the tissue will occur. In cases of knock-knees or bowlegs, the intra-articular pressure is placed upon the outer edge (knock-knees) or the inner edge (bowlegs) of the joint surfaces, and this pressure ultimately results in pathological changes in the joint (Fig. 25-4). It is also a very clear example of the dependence of form and function on statics. The importance of such skeletal malformations in causing deformation of the joints has been emphasized by calling them "prearthritic deformities."

Joint Capsule: Strength and Flexibility. The ends of the bones forming a joint are connected together by the joint capsule. The true capsule consists of an outer *fibrous layer,* the function of which is to protect the joint against excessive stretch, and an inner lining, the *synovial membrane.*

The outer fibrous layer is strengthened by several strong *ligaments,* made resistant to stretch by their parallel bundles of strong, nonelastic collagen bundles. These bundles are oriented in the direction of the major forces of stretch exerted upon them, and their strength is a reflection of the amount of stretch they must withstand.

The *strongest ligaments are found in the hip joint,* the strongest joint of the body. They are particularly well developed in the front of the joint, where they run spirally and prevent an overextension of the joint, thus increasing the passive stability of the hip joint so necessary to the support of the body (Fig. 25-5). The *side ligaments of the knee joint* act as protection against the snapping-off of the lower leg against the thigh. The knee joint has special stretch-resistant structures to enable it to withstand the tremendous weight it must bear in all positions. Two remarkably strong crossed ligaments, the *cruciate ligaments,* inside the knee joint prevent any forward or backward movement of the femur over the flat top of the fixed tibia (Fig. 25-6).

The tautly stretched *ligaments of the vertebral column,* attached from spine to spine of the vertebrae, passively protect the axial skeleton by limiting the amount of bending possible. Similarly, the fan-shaped *ligaments on either side of the ankle joint* guide and limit the movement at this joint.

The capsule must adapt to the degree of movement per-

Figure 25-4. **The effect of uneven pressure upon the joint cartilage and underlying bone.**

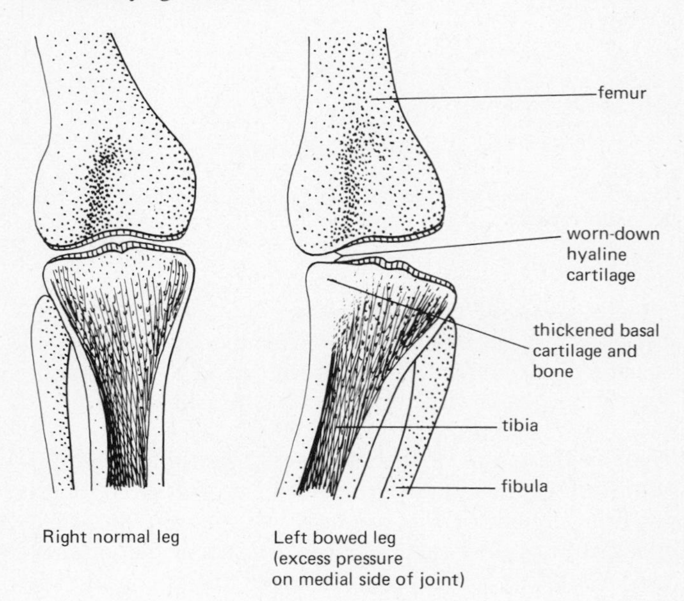

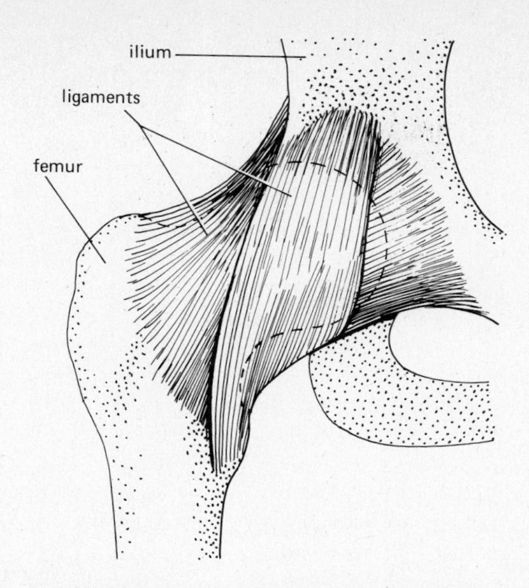

Figure 25-5. **Ligaments that strengthen the joint capsule. The hip joint is drawn.**

mitted at each particular joint: it must be able to expand sufficiently on the stretched side of the joint to permit flexion on the other side of the joint. The many peculiarly shaped recesses and pockets of the capsule give to it this protean ability to change its shape.

The *synovial membrane* is a highly vascular, delicate membrane that lines the capsule of the joint. It has many folds and villi that increase its surface area markedly and

Figure 25-6. **The strong, crossed cruciate ligaments inside the knee joint prevent forward or backward slippage of the femur.**

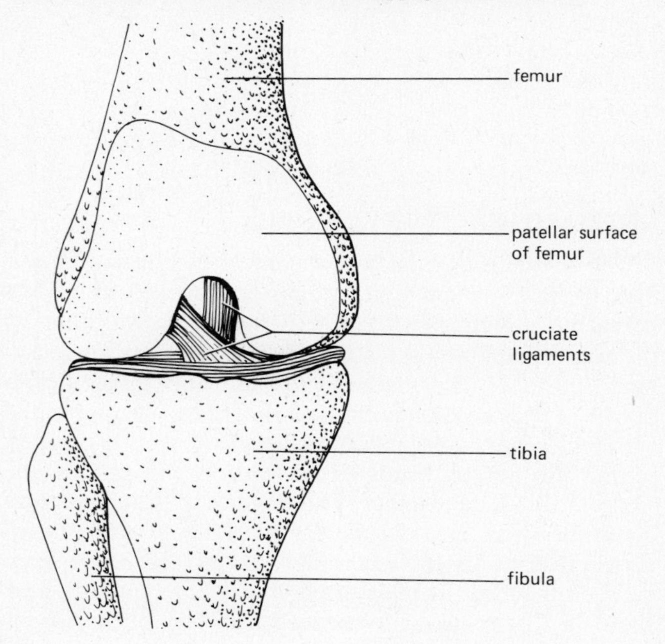

permit the very rapid exchange of fluids within the joint. This membrane both secretes and reabsorbs the fluid within the joint cavity, the *synovial fluid.* The synovial fluid is not only a filtrate of plasma, but it is also the product of the active secretion of the cells of the synovial membrane. It is rich in phagocytes, which remove microorganisms and debris, and in mucin, which has a lubricating function. The secretion and reabsorption of this fluid by the synovial membrane, together with the spongelike action of the hyaline cartilage, ensure a continuous movement of fluid within the joint and into the tissues.

Synovial fluid has a double role, biological and mechanical. The *biological* function is, of course, the supply of nutrients and removal of wastes to maintain the health of the tissues of the joint, especially of the nonvascular hyaline cartilage. The *mechanical* function is that of acting as a lubricant between the opposing surfaces of the joint; because it forms only a thin layer of viscous fluid on these surfaces, movements of the joint do not result in any hydrostatic pressure changes.

Degree of movement at a true joint

The form of the head and socket of the joint determines the degree of movement at a joint, which in turn affects the form and extent of the capsule. Joints are divided into three main categories, in terms of the *range of movement* permitted:

1. Single-axis or hinge joints.
2. Double-axis or ellipsoidal joints.
3. Triple-axis or ball-and-socket joints.

The classic example of a *large hinge joint* is the knee joint, which permits movement on a horizontal axis only, through flexion and extension. In extreme flexion the sharply curved dorsal surface of the condyle of the head of the femur is brought into contact with the flat socket of the tibia. This inequality is compensated for by the shifts in position of two wedge-shaped *semilunar cartilages* that are incompletely fastened to the surfaces of the knee joint. In this position a certain rotation is possible and the knee joint acquires an additional dimension of movement.

Classical hinge joints of a *smaller* type are the proximal and distal interphalangeal joints in the fingers and toes.

The wrist joint is a good model for an *ellipsoidal joint.* The articulation between the carpals and the distal end of the surface of the radius permits not only flexion and extension, but also *abduction* (movement toward the body midline) and *adduction* (movement away from the body midline) (Fig. 25-7).

The shoulder and hip joints are *ball-and-socket joints* with different degrees of freedom of movement. A comparison of the structure of these two joints explains clearly how they differ in function (see Figs. 24-12, 25-7, and 25-10). The great range of movement involved in the gripping and handling characteristic of higher primates is reflected in the extremely shallow socket provided by the scapula, with

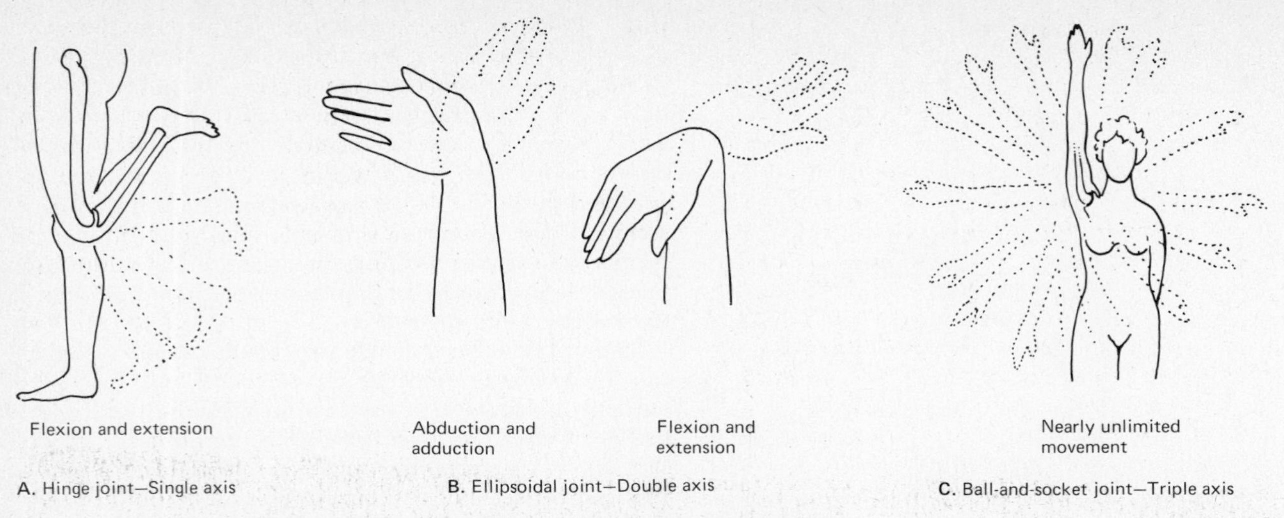

Figure 25-7. **The range of movement varies with the structure of the true joint.**

which the articulating surface of the head of the humerus barely makes contact. Although the joint at the hip provides a lesser, but still wide, range of movement, it must, in addition, support the weight of the body, which it in turn transmits to the joints of the lower extremities. The strength and mobility are made possible by the form of the head of the femur; it is practically a sphere supported by a short neck, and half of the articulating surface of this sphere is surrounded by the bony socket, the *acetabulum,* of the hip bone. The socket is made into a deep, hollow cup by a compact, sturdy sheet of connective tissue, the transverse ligament. This firm enclosure of the head of the femur results in great mechanical stability, enhanced by the strong ligaments of the capsule that connect the head of the femur to the hip bone.

Joints of the Vertebral Column

Thirty-three vertebrae are intricately fitted together to form the *vertebral column,* which has a flattened-out S shape when viewed from the side. The articulation of so many single structures upon one another, in short or long segments, results in an enormous variety of possible movements. These movements are basically forward and backward in the flexion and extension of the trunk, sideward bending, and rotational movements.

The *static function* of the vertebral column is to act as a support for the body and head. This is achieved through the combined strength of the entire series of the vertebral bodies (i.e., the anterior part of the vertebral column), which are elastically but tautly held together by the ligamentous disks that lie between them. The *dynamic functions* of the vertebral column are derived from the attachment of innumerable muscles to the spines of each vertebra (vertebral arch) and to their transverse processes (i.e., the posterior part of the vertebral column) (Figs. 24-14 and 25-8).

Figure 25-8. **The intervertebral disks and protective ligaments of the vertebral column.**

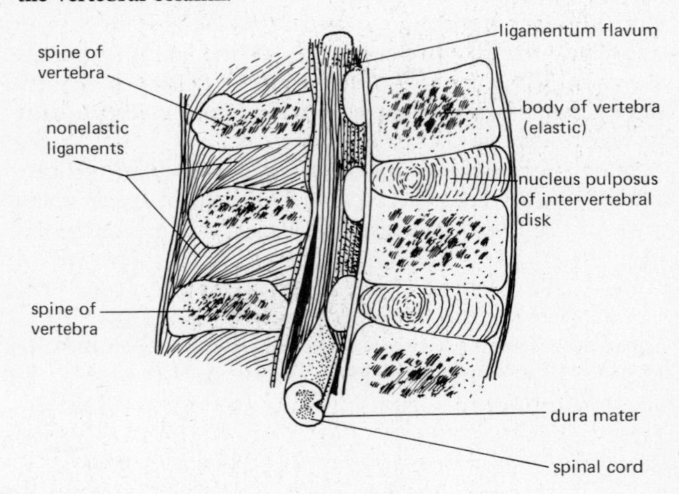

Intervertebral disks

The unique construction of the intervertebral disks gives them the role of a semijoint. Spirally running fibrous threads fasten the adjoining surfaces of two vertebrae, forming at the same time a ringlike structure around a central gelatinous substance, the *nucleus pulposus.* This gives a high degree of resistance to mechanical shocks and stresses to the vertebral column: the nucleus pulposus acts as a damper, setting the fibrous bands under tension and pulling the adjacent vertebrae together. The strength of these ligaments diminishes with age, and the damping power of the intervertebral disks lessens accordingly. A marked discrepancy between the load the intervertebral disk is required to bear and its ability to adjust to the load will damage the fibrous ligaments or displace the nucleus pulposus; malformations of the vertebral column increase the chances of such damage.

Protective ligaments

Protective ligaments between the vertebral bodies and connecting their spiny processes prevent damage to the spinal cord during movement of the vertebral column (Fig. 25-8). The highly elastic *ligamentum flavum* (yellow ligament) surrounds the entire length of the vertebral canal, filling the spaces between the adjacent vertebral arches. Because the ligamentum flavum is so elastic, it permits the separation of the vertebrae in flexion, yet on extension protects the dura mater of the spinal cord from being compressed by the bodies of the vertebrae and arches. Strong, *nonelastic ligaments* attach the spiny process of adjacent vertebrae to one another and so prevent the overextension of the column.

Range of movement of the vertebral column

The seven *cervical vertebrae,* with their specially constructed articulating surfaces, permit a *wide range* of movement of the neck and head. The modifications of the first two cervical vertebrae especially allow for the rotation of the head on the neck, as the weight-bearing atlas (first cervical vertebra) swivels on the ring of the axis (second cervical vertebra) (Fig. 25-9).

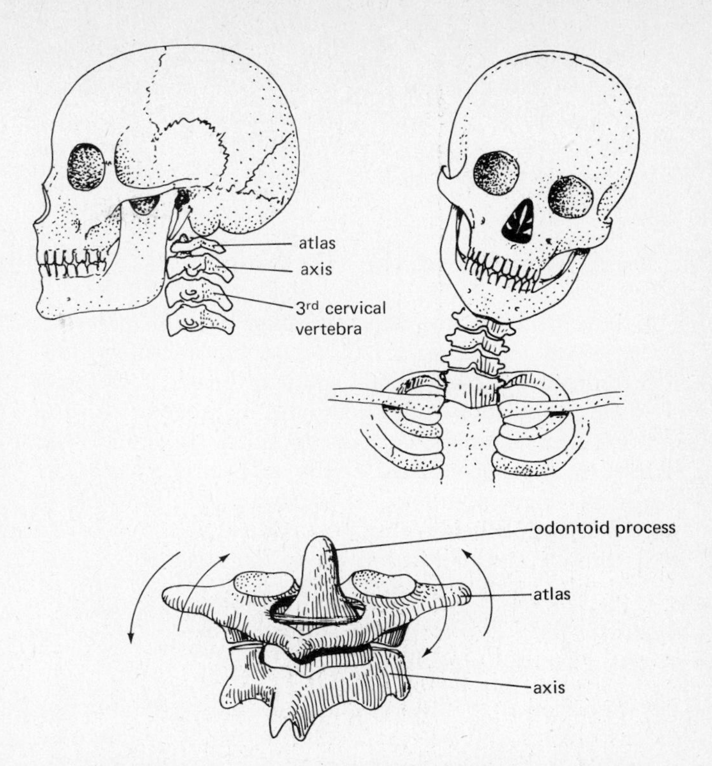

Figure 25-9. **Rotation of the atlas on the axis permits rotation of the head. The atlas pivots around the toothlike (odontoid) process of the axis when the head is moved from side to side. Strong ligaments limit the range of the movement.**

The 12 *thoracic vertebrae,* being firmly attached to the 12 pairs of ribs, are far more *limited* in their range of movement. The increase in the size of the thoracic cavity during inspiration is caused by the longitudinal rotation of the ribs at the *costal joints,* the joint formed by the rib and the transverse processes of adjacent thoracic vertebrae (Fig. 24-14). Because the ribs slope down and out, a lengthwise rotation at this joint pulls them up and forward, increasing the diameter of the thoracic cavity (Sec. 16-4).

The five *lumbar vertebrae* are mainly responsible for the movements of the trunk, but rotation is very restricted (Fig. 25-10). The five fused vertebrae that form the *sacrum* obviously are unable to move as individual vertebrae and act as an anchor for the rest of the vertebral column. The bones of the pelvic girdle are fastened securely to the sacrum by

Figure 25-10. **Movements of the vertebral column. Note the wide range that is possible in the lumbar and cervical regions and the relative immobility of the thoracic region.**

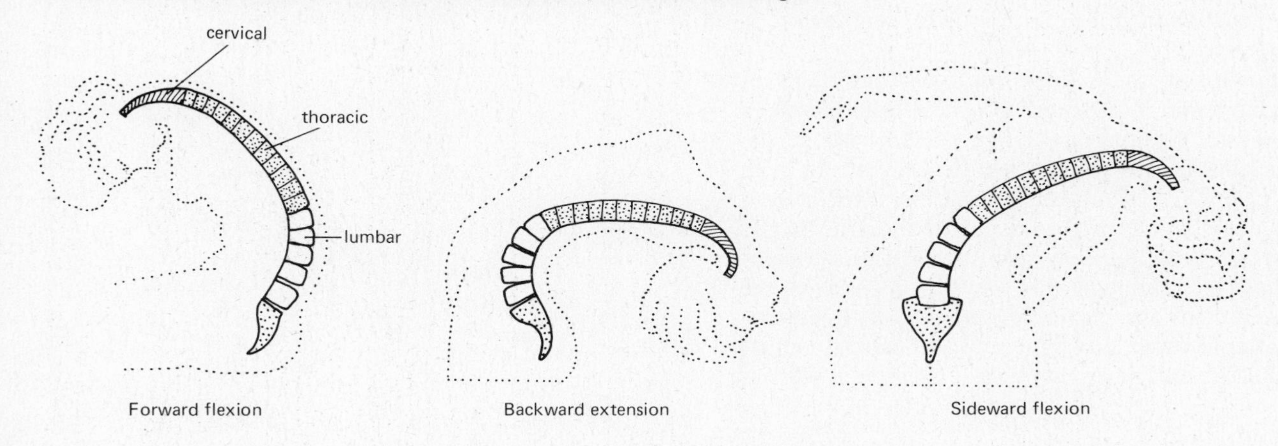

the extremely strong and taut ligaments that bind the ilium and sacrum into the sacroiliac joint. This provides a firm support for the weight of the trunk and the internal organs. The rudimentary *coccyx,* formed by the fusion of four small bones, has no function in humans.

25-2 Muscles of the Skeleton

The movements of the joints and their fixation in certain positions are *active processes* that must be carefully graduated to achieve a harmonious and coordinated reciprocity between the posture of the body and its movements. The working elements in these processes are the skeletal muscles because of their specialized ability to alternately contract and relax. The importance of the contractile elements in muscle cells is enhanced by the anatomic structure of the entire muscle and its relationship to the skeleton.

Isotonic and Isometric Contraction in the Intact Organism

The importance of the principles of muscle contraction under isometric and isotonic conditions that were discussed in Sec. 23-6 can be demonstrated in the movements of the muscles within the body. In the body, instead of a single, isolated stimulus, the muscles receive pulses of stimuli. Whereas the isotonically contracting muscle increases the *amount of its shortening,* the isometrically contracting muscle increases its *tension* as the frequency of stimulation increases. This tetanic contraction is the basis for the activity of skeletal musculature. Except for certain reflexes, there are no single muscle twitches under true physiological conditions.

The normal voluntary contraction of skeletal muscle is always a combination of processes; the lifting of a weight from the ground induces an *increased tension* in the isometrically contracting muscle until the tension is equivalent to the load to be lifted and the weight of the lower arm. Only when this correspondence has been reached is there an increase in the stimulation reaching the biceps brachii to result in *isotonic contraction;* as this muscle shortens, the elbow is bent, and the resistance or load lifted from the ground. This combined form of muscle contraction is termed *auxotonic.*

The usefulness of the concept of "work" in terms of isometric contraction is questionable. The deltoid muscle (a triangular muscle forming the bulge of the back of the shoulder) is isometrically active when the arm is held in an horizontal position. Because no distance has been moved during the maintenance of the weight of the arm in this position, no true work in the physical sense has been accomplished. However, although this is what appears in the macroscopic dimensions, *microscopically* there is a continuous, alternating contraction of the muscle fibers, accompanied by increased energy expenditure. This type of work is comparable to that of a helicopter, which remains stationary in the air by means of the expenditure of energy.

The structure of a muscle is remarkably well adjusted to its function, in terms of power, speed, and range of movement. Although each muscle is capable of contracting isotonically or isometrically, one may distinguish between those muscles that are chiefly involved in the execution of movement and those that regulate static processes such as the maintenance of posture.

Muscles Chiefly Concerned with Movement

These muscles are found in those parts of the skeleton where a magnification of the movement due to leverage is found.

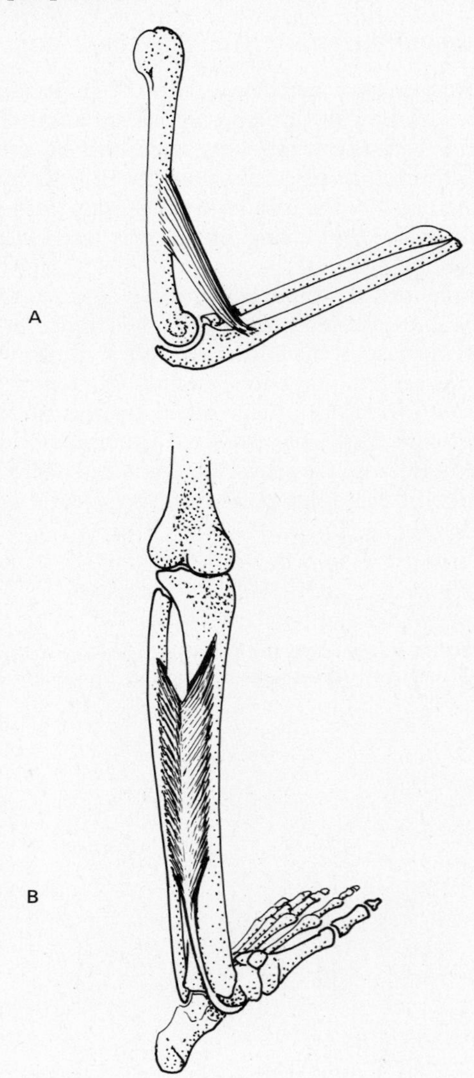

Figure 25-11. **The arrangement of muscle fiber attachment to bone. (A) Muscles specialized for isotonic contractions, for example, brachialis. (B) Muscles specialized for isometric contractions, for example, peroneus.**

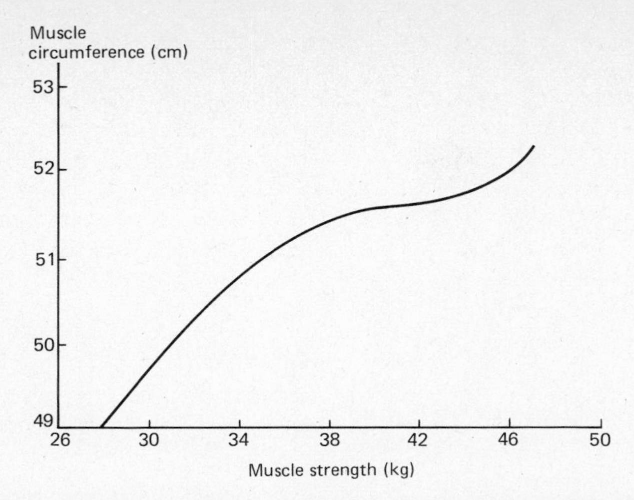

Figure 25-12. **Increase in the strength of muscle with hypertrophy. The increase in total muscle circumference is due to an increase in individual muscle fiber size as a result of isometric training.**

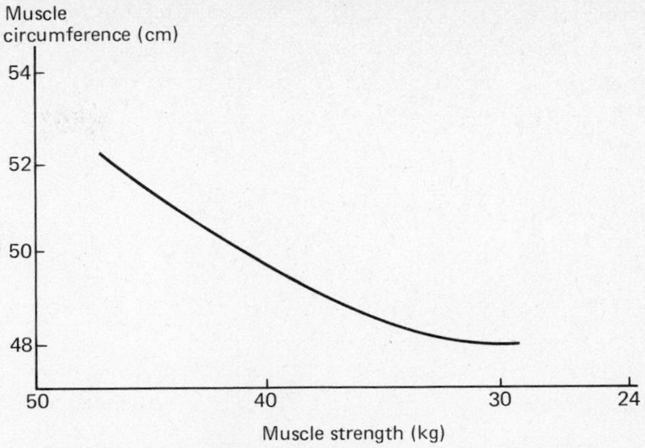

Figure 25-13. **Decrease in the strength of muscle with atrophy. The decrease in the total muscle circumference is due to a decrease in individual muscle fiber size following prolonged immobilization in a plaster cast.**

This is particularly characteristic of the movements of the upper extremities. The variety of movements of which the human arm is capable is caused by the fact that most of the muscle fibers extend from the origin to the insertion of the muscle, running parallel to one another (Fig. 25-11). These long fibers permit the passage of many successive waves of contraction during tetanic contraction. This, together with relatively small motor units, results in a delicate gradation of the possible strength of contraction.

Muscles Chiefly Concerned with Posture

The muscles concerned with posture, through the active stabilization of joints, usually have a more featherlike construction (Fig. 25-11). The muscle fibers are comparatively short and converge obliquely to end at a common tendon. In addition, the motor units of postural muscle are made up of large numbers of muscle fibers, so that even with a minimum level of stimulation these muscles exert a comparatively large force upon the noncontractile tendon, and all the force of the muscle fibers is brought to bear on the tendon.

Adaptability of Muscle

Like all living tissue, muscle possesses the capacity to *adapt.* The adaptability of muscle to the type and extent of work to which it is exposed is far greater than that of bone or connective tissue. Muscles that are systematically subjected to increased use, as in training, very rapidly show an increase in strength; unused muscles quickly weaken. This change in strength can be accurately correlated to the change in size of the muscle itself: there is an almost linear relationship between the increase in muscle size (*hypertrophy*) due to training and its increased strength. There is a similar relationship between muscle *atrophy* due to inactivity and its decreased strength (Figs. 25-12 and 25-13).

It is *isometric activity* of muscle that results in its hypertrophy, as is seen clearly in those people whose physical work consists mainly of static activity and in certain types of athletes, for example, wrestlers and weight-lifters. This increase in strength evoked by isometric training is achieved without an increase in the number of motor units utilized for contraction, but endurance is decreased. If the training is chiefly *isotonic* (e.g., sprinting or skiing), there is a considerable increase in the endurance of the individual, but muscle size is only slightly affected.

The most marked muscle atrophy is found in patients immobilized by plaster casts. Confinement to bed rapidly results in an atrophy of the extensor muscle in the thigh, the quadriceps femoris.

Regulation of Posture and Movement

Muscle tonus

The posture and poise of an individual reflect his personality to others. To a certain extent they are dependent on his basic skeletal structure and the form and development of his muscles. It is the muscles that confer upon the human form its grace. However, subjective influences, such as depression and elation, fatigue and vigor, are depicted by the musculature, which is influenced by impulses reaching it from the nervous system, chiefly from levels below that of consciousness (Tinbergen, 1974). The nervous control of muscle tonus and the importance of the stretch reflex, the gamma loop system, and the muscle proprioceptors for the maintenance of tonus have already been discussed (Secs. 7-6 and 23-13).

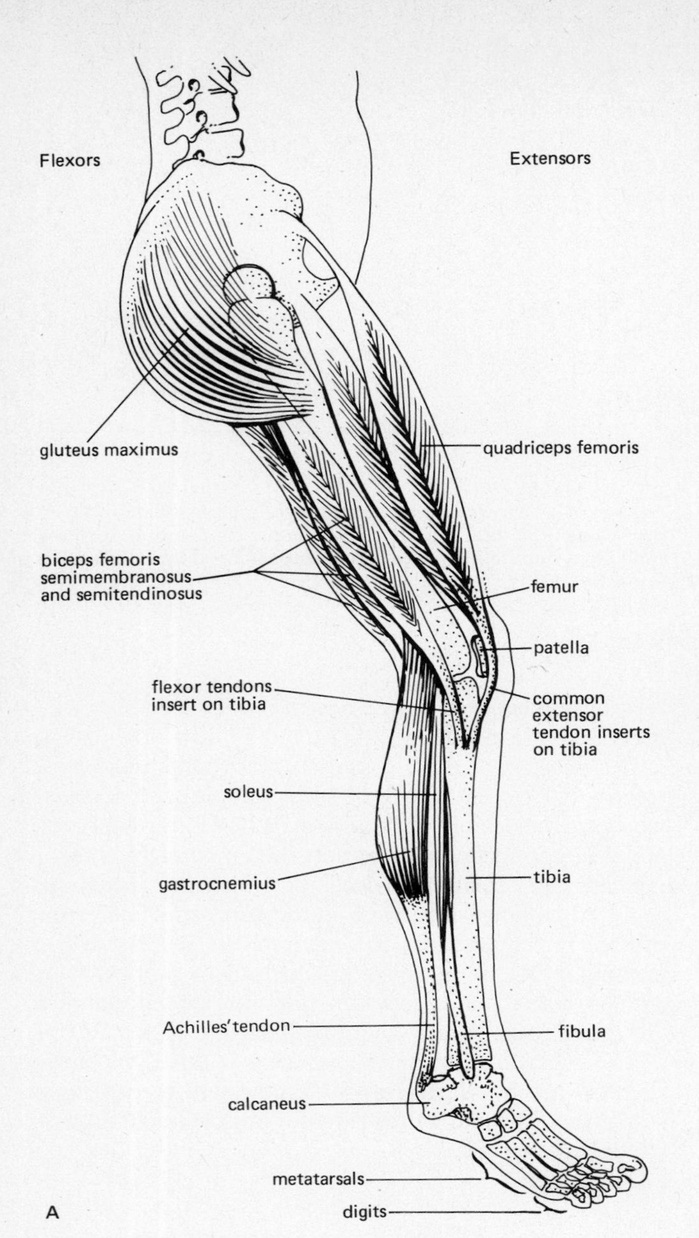

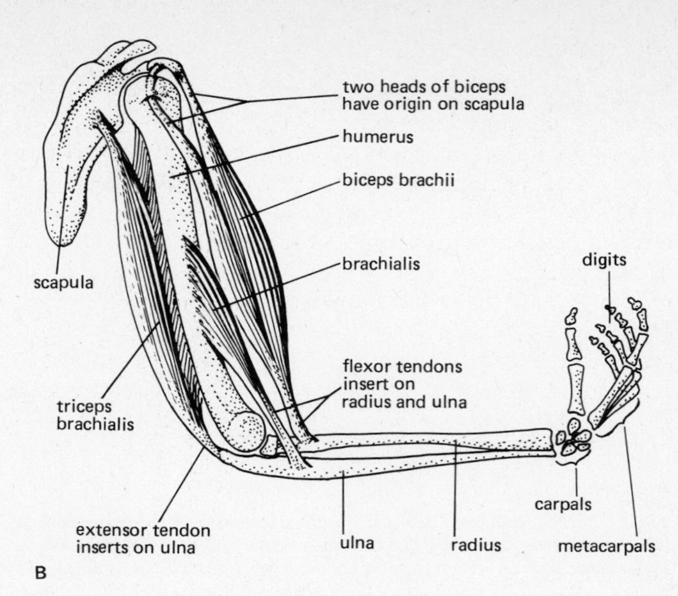

Figure 25-14A. **Antagonistic and synergistic muscles of the leg (A) and arm (B). When the extensor muscles contract synergistically, their antagonists, the flexors, relax. This permits extension of the limb. The synergistic contraction of the flexors and the coordinated relaxation of the antagonistic extensors result in flexion of the limb.**

Figure 25-14B. **Electrical activity of the antagonistic muscles of the underarm during rhythmic squeezing of the fist in untrained (A) and trained (B) individuals. 1 = electromyograph (EMG) of the extensor (M. anconeus), 2 = EMG of the flexor (M. flexor carpi ulnaris), 3 = mechanogram, and 4 = time and event marker. (From H. Stoboy, in *Handbuch für Orthopädie,* Vol. 1, Georg Thieme Verlag, Stuttgart, Germany, 1957.)**

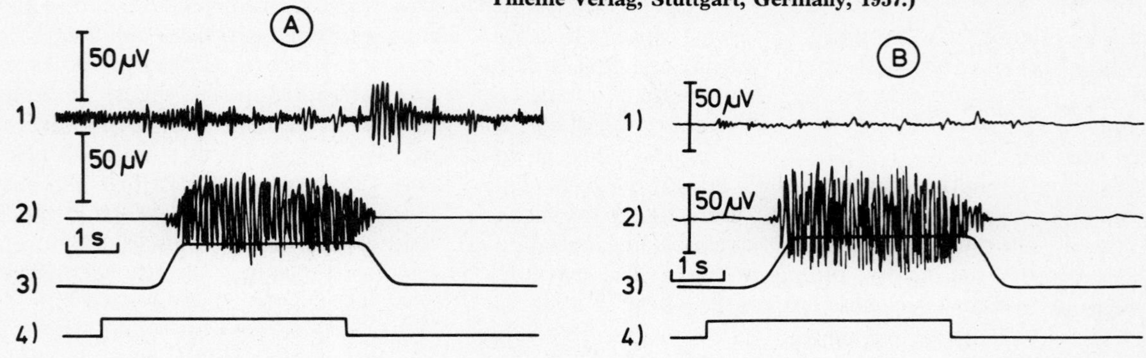

Antagonistic muscles

Superimposed on this background of tonus is the coordination of antagonistic muscles through the nervous system. The active mover contracts while its antagonist relaxes to bring about the movement at the joint. Despite their name, antagonistic muscles really assist one another, for the one must relax in order to permit the full range of operation of its antagonist.

Antagonistic muscles are involved in many different types of movement at a joint: flexion and extension, abduction and adduction, pronation and supination, and internal and external rotation. These types of movement are illustrated in Fig. 25-7.

Synergistic muscles

Synergistic muscles contract together to accomplish the same movement, whereas their antagonists must relax together. During the flexion of the lower arm at the elbow, the biceps brachii and the brachialis (flexor muscles) act together synergistically as the active movers, while the triceps brachii (extensor muscle) relaxes. Similarly, the biceps femoris, semimembranosus, and semitendinosus are all active in the flexion with some rotation of the knee joint, while the antagonistic extensor (the quadriceps femoris) relaxes (Figs. 25-14A and B).

This antagonistic action of muscles makes a fixation of the joint in different positions possible, but the degree of contraction and relaxation of the opposing muscles must be minutely controlled to permit such graduations. Because of the activity of the proprioceptors, the relaxation of the extensor during flexion at the elbow, for example, is not a passive collapse, for the extensor builds up sufficient tension so that the elbow joint is stabilized in each successive position.

Antigravity muscles

The antigravity muscles are particularly involved in the stabilization of the joints. These are the muscles that continuously oppose the effect of gravity on the body. In the upright human being, the main antigravity muscles are the extensors of the leg (the gluteus maximus, the quadriceps femoris, and the soleus) and the muscles of the back (Fig. 25-15).

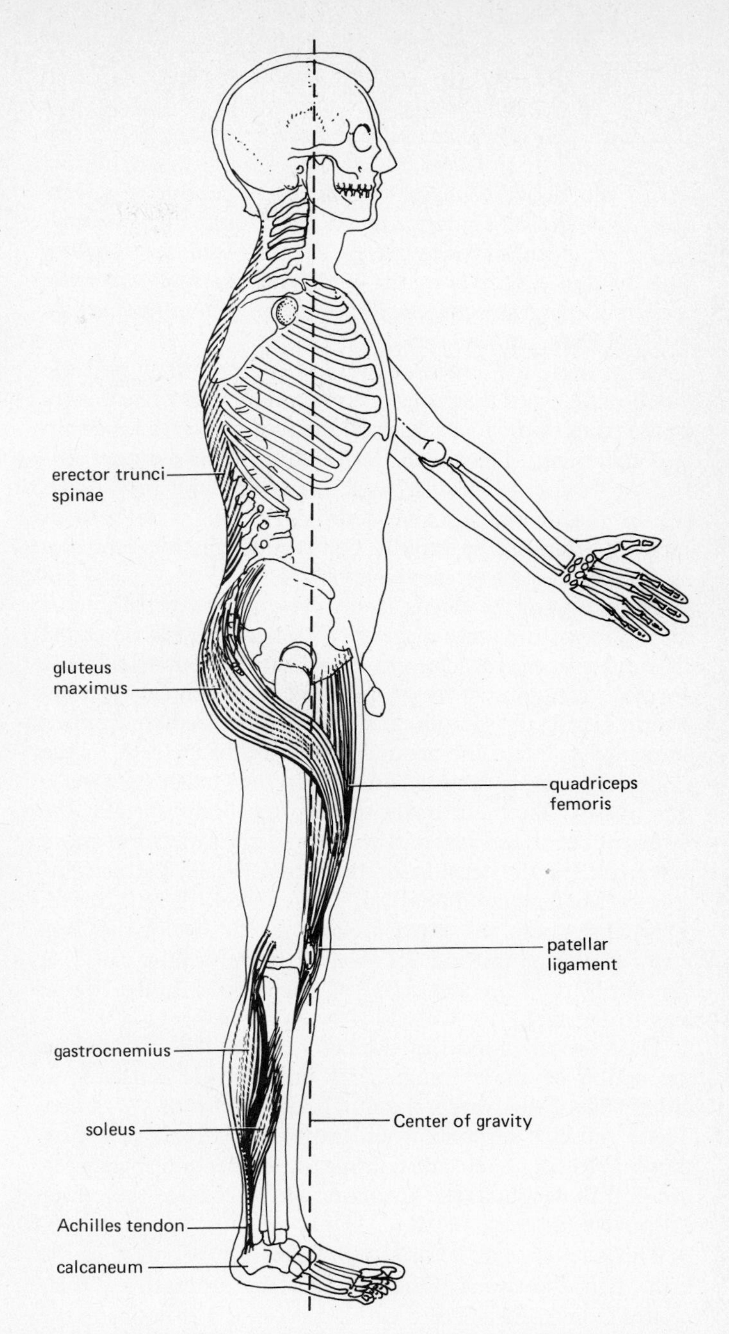

Figure 25-15. **The antigravity muscles are the strong extensor muscles of the back and legs. The center of gravity (broken line) of the upright body is in front of the ankle.**

The *gluteus maximus,* which forms the folds of the buttocks, is the heaviest muscle of the body, its development being associated with the assumption of the upright position. Its origin gives it a firm hold on parts of the ileum, sacrum, and coccyx. It inserts on the proximal end of the femur and, also, through the strong fibrous covering of the thigh that ends on the tibia, it practically inserts on the tibia. Therefore the gluteus controls the thigh firmly as an extensor and also prevents the body from being pulled too far forward and down by the force of gravity.

The *quadriceps femoris,* which forms the main part of the front of the thigh, fixes the knee joint. Its four heads originate from the ileum and the femur and run down to converge on the patellar ligament, which then inserts on the tibia. When the quadriceps contracts, it pulls the lower leg forward, extending the knee.

The *soleus* is a thick, flat muscle on the back of the lower leg, hidden by the two curved bellies of the *gastrocnemius.* Together these two muscles make up most of the bulk of the calf, the size and strength of which is associated with the ability to walk and run in the upright position. The two bellies of the gastrocnemius arise from the femur and the back of the capsule of the knee joint, whereas the soleus has its origin on the tibia and fibula and from the strong fibrous membrane between these two bones. The soleus inserts upon a powerful tendon, almost 6 in. long, the *tendocalcaneus* or *Achilles' tendon*, as it is more familiarly called. This tendon descends to the heel where it fuses with the tendon of the gastrocnemius before it inserts on the back of the heel bone (*calcaneum*).

When these two extensor muscles contract, they pull the foot down toward the ground in *plantar flexion*. (The flexors on the front of the lower leg pull the foot up toward the shin in *dorsiflexion*.) The slowly contracting soleus is more important in the antigravity action, pushing the foot down against the ground to give a firm grip and wide base of support, whereas the rapidly contracting gastrocnemius is used more for active movements.

The action of the soleus at the *ankle joint* is just sufficient to counteract the force of gravity, so that there is no actual shortening of the muscle, and the activity is isometric. In the normal "standing at rest" position, the center of body weight is near the sacrum, and a perpendicular line dropped from this point to the ground runs slightly in front of the ankle. The farther forward one bends, the farther forward is this line displaced. In order to keep the body upright, the isometric contractions of the soleus must become more powerful. As the tension of the extensors of the calf increases, the flexors on the front of the lower leg relax more. This tenses both the extensors and the flexors of the *knee joint*, so that a shifting forward of body weight causes a fixation of the knee and stabilizes the support given by the legs to the trunk.

The vertical support of the body by the legs is aided by the action of the symmetrically antagonistic muscles on either side of the vertebral column, the *erector trunci spinae*. These muscles respond when the trunk is bent sideward, those on the opposite side contracting and counterbalancing the shift in weight as the viscera are displaced and the center of gravity changed.

The *extensor muscles of the back* also act as brakes as the body bends forward. One can feel the strength of their contraction while walking with a heavy load of books in the arms. Their activity is calibrated to just counteract the forward pull and to keep the body upright. Motor unit activity stops only when the trunk is bent forward as far as possible. Now the position of extreme flexion is held passively, with the expenditure of minimum energy, through the passive pull of the ligaments of the back. The extensor muscles become active again when they pull the body back to the upright position, an action requiring much power, as is emphasized by the fact that the extensor muscles of the back in humans are twice as massive as the flexors.

This coordinated activity of different muscle groups, whether in the maintenance of posture or in movement, is enhanced by the sensitivity to stretch of the proprioceptors of the muscles and tendons, especially those in the capsules of the joints. The resulting barrage of nerve impulses is channeled through the central nervous system which integrates the action of the muscles.

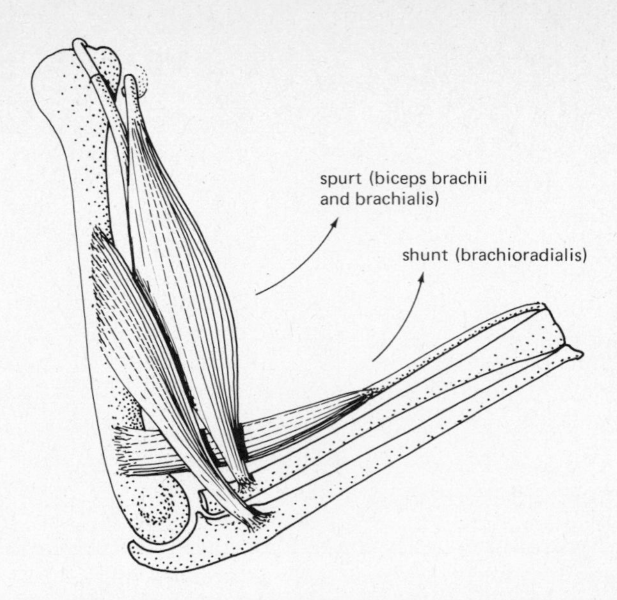

Figure 25-16. **The shunt and spurt muscles of the arm. All the muscles illustrated are flexors, but by means of a partially antagonistic action the brachioradialis slows down the action of the spurt muscles.**

"Shunt" and "spurt" muscles

An interesting concept of "shunt" and "spurt" muscles widens the role of antagonistic muscles in the regulation of posture. A delicate degree of control of certain movements is afforded by a partial *antagonistic action of the synergistic muscles*. This is most clearly demonstrated by movements of the elbow. The *biceps brachii* and the *brachialis* insert on the radius and ulna close to the axis of the elbow joint. They are called "spurt" muscles because the angle at which they insert on the forearm makes it possible for a short contraction to cause a large and rapid displacement of the hand (Fig. 25-16).

The long *brachioradialis* extends the length of the forearm from its origin on the humerus to the fibers that insert almost parallel on the distal end of the radius. This muscle is an important flexor of the elbow, but due to the angle of its insertion, it acts along the long axis of the moving bone to provide centripetal force. The rapid flexion brought about by the "spurt" muscles causes the intervention of the brachioradialis, the "shunt" muscle, the contraction of which slows down the fast movement to yield a smoother, graduated flexion.

The brachioradialis remains inactive during a slow, weak flexion of the elbow or one that takes place against a weak resistance. However, all three flexors are activated when the elbow is extended, acting as brakes against too rapid extension. During slow extension of the elbow, they act as the antigravity muscles of the upper extremity, maintaining the normally slightly flexed position of the arms. This important role is often ignored.

It is obvious from some of the examples that have been discussed that human posture is a most complex problem. A unified concept of posture must incorporate the functional and structural regulating mechanisms common to all muscles involved in the maintenance of posture, but must be modified by many other factors according to the position of the muscles and the demands placed upon them.

Development of muscular control

The higher centers of the nervous system, especially the cerebellum, are deeply involved with the regulation of peripheral functions. We can see from the development of a child that only a fraction of the necessary elements for the control of posture and movement is present at birth. Lifting the head, sitting up, seizing objects, and taking the first step are all stages of development that are constantly expanded by the widening of the child's experience. With each process learned, however, there is an accompanying increase in the control of *involuntary* posture and movement. The regulation becomes more precise. The multiplicity of response now becomes the distinguishing feature of the conscious brain, the great development of which has placed human beings in their unique and individual position in the realm of primates.

Cited Reference

Tinbergen, N. "Ethology and Stress Diseases." *Science* **185:** 20 (1974).

Additional Readings

BOOKS

Astrand, P., and K. Rodahl. *Textbook of Work Physiology.* McGraw-Hill Book Company, New York, 1970.

Carlsoö, S. *How Man Moves.* Heineman, London, 1972.

Jensen, V. R., and G. W. Schultz. *Applied Kinesiology.* McGraw-Hill Book Company, New York, 1970.

Jokl, E., ed. *Biomechanics I.* University Park Press, Baltimore, 1968.

Knuttgen, H. G., ed. *Neuromuscular Mechanisms for Therapeutic and Conditioning Exercise.* University Park Press, Baltimore, 1976.

Komi, P. V., ed. *Biomechanics V,* Vols. 1 and 2. University Park Press, Baltimore, 1976.

MacConaill, M. A., and J. V. Basmajian. *Muscles and Movements.* The Williams & Wilkins Company, Baltimore, 1969.

Morton, D. J., and D. D. Fuller. *Human Locomotion and Body Form.* The Williams & Wilkins Company, Baltimore, 1952.

Rasch, P. J., and R. K. Burke. *Kinesiology and Applied Anatomy,* 5th ed. Lea & Febiger, Philadelphia, 1971.

Williams, M., and H. R. Lissner. *Biomechanics of Human Motion.* W. B. Saunders Company, Philadelphia, 1962.

ARTICLE

Napier, J. "The Antiquity of Human Walking." *Sci. Am.,* Apr. 1967.

Regulation of Reproduction and Inheritance

Chapter 26

Regulatory Processes in Reproduction

Instead of needing lots of children,
we need high quality children.

Margaret Mead

REPRODUCTION IN MAMMALS, and especially in humans, consists of a series of interwoven regulatory processes that exemplify feedback mechanisms of several types. Neuroendocrine regulatory feedback dominates the development of the male and female gonads and reproductive tracts and the production of sperm and eggs. The sex hormones produced by the gonads, in turn, regulate not only the neuroendocrine functions of the hypothalamus, but also the behavioral pathways essential for mating. Changes in the female reproductive tract with pregnancy affect hypothalamic secretions that control the environment of the growing fetus and the duration of its gestation time in the uterus. Even after birth, neuroendocrine feedback regulates the processes involved in lactation and maternal behavior. Suckling the infant, when included in maternal behavior, triggers a nervous feedback mechanism to the hypothalamus, delaying the normal reproductive cycles and diminishing the chance of an immediate pregnancy.

The social, economic, and emotional welfare of the family, and ultimately of the world population, depends upon the proper regulation of these reproductive processes. Although human fertility was a most important survival factor for millions of years, the startling increase in the population of the world brought about by the decrease in mortality rates has made reproduction a *negative survival feature.* The regulation of conception and contraception, which can be achieved only by a combination of physiological and social measures, has become an urgent, worldwide problem.

26-1 Mating and Reproduction

The essential feature of the survival of a species is the production of the ovum, bearing the genetic characteristics of the female, and its fertilization by the sperm, with its complement of genes from the male. Both the fertilization of the ovum and the subsequent development of the zygote involve many complicated behavioral patterns, for in almost all species not only is mating preceded by courting, but after mating and the fertilization of the ovum, the developing embryo requires lengthy protection. In primates, this need involves not only the development of the placenta for the intrauterine nourishment and protection of the embryo, but also postnatal care and education of the young for increasingly long periods of time.

The reproductive tract of the female is adapted to produce ova capable of being fertilized and then to nourish and contain the rapidly growing zygote. After the birth of the young, the mammalian mother continues to feed her young through the production of milk by the mammary glands.

The reproductive tract of the male is adapted to produce large numbers of active, motile sperm to fertilize the ova. The biological necessity for the production of such vast numbers of sperm is questionable, and it has been suggested that it is a holdover from the lower forms in which fertilization was external, and the chance of a sperm encountering an ovum much less than in internal fertilization. In internal fertilization, the male inserts the sperm into the body of the female; the reproductive tracts of both sexes are modified to adapt to this process.

Fertilization of the ovum obviously requires sexual attraction between male and female. Both the desire of the male to mate and the willingness of the female to accept him are clearly regulated by the sex hormones in lower animals, but in human beings this attraction is immensely complicated by social patterns and traditions and by individual variations in the interpretation of these mores. The human female is the only animal that will accept the male at any time in her reproductive cycle or at any season of the year. With other species, "the mating season" is very often closed.

26–2 Hormonal Regulation of the Development of Sex

The cells that are destined to develop into either the male or female gonad share a common origin very early in embryonic development. Genetically, the sex of the embryo has been determined at the instant of fertilization (Sec. 27–4), but these cells appear to be so labile that they can be misdirected from their normal development into a testis or ovary by changes in the hormonal environment. In cattle, it has been clearly shown that when twin embryos share a single placenta, the hormones produced by a female twin will prevent her male twin from developing proper testes, and he becomes an intersex. In human beings the situation is not as clear, but it may be that some human intersexes are produced as a result of excessive maternal hormones inhibiting the gonadal development of the male embryo. However, female-to-male sex reversal is much more common than male to female; in fact this difference is enough to account for a significant part of the differential sex ratio at birth. It has been stated recently that completely masculinized genetic females may form 1 in 400 of the masculine population. This obviously cannot be caused by the effect of maternal hormones. Much more specific information on human intersexes has come from studies of abnormalities in the number of sex chromosomes (Sec. 27–4).

If the embryonic testes are removed early in development, the reproductive tract develops as an incomplete female; removal of the embryonic ovaries also results in the development of a female pattern of the reproductive tract, indicating that the male pattern is superimposed upon a more "neutral" female-type development, and that the hormones produced by the embryonic testes are necessary for the proper development of the male reproductive tract. The effect of hormones on the development of the hypothalamus in the male or female direction was discussed in Sec. 11-2.

Puberty: The Beginning of Sexual Maturation

Puberty starts at the age of 12 to 14 years, varying with the genetic background of the individual. In response to the increased amounts of gonadotropin-releasing hormones from the hypothalamus and consequently increased secretion of gonadotropins from the pituitary gland, the gonads begin the production of the sex hormones and gametes. Together with the *maturation of the internal reproductive tract* caused by the sex hormones are the more obvious changes in the *external genitalia* and the *secondary sexual characteristics;* in both sexes, there is a marked development of hair in the pubic region and under the axilla, and in males the beginning of a beard. Boys' voices deepen, too, at this time, because of changes in the bony structure of the larynx, and differences in the distribution of fat in the two sexes now appear. Curves develop in young girls, the breasts and hips rounding out as fat is deposited there. There is a sharp spurt of *skeletal growth* at this time, too, due to an increased production of another anterior pituitary secretion, *somatotropin or growth hormone.*

There is no evidence that the other pituitary hormones, such as adrenocorticotropic hormone (ACTH) or thyroid-stimulating hormone (TSH) are increased at this time, which means that there is a specific control over those cells in the anterior pituitary gland that produce the gonadotropins and growth hormone.

26–3 Male Reproductive Tract

The essential functions of the male reproductive tract are the production of spermatozoa and of the male sex hormones and the deposition of the spermatozoa in the female. Spermatozoa are produced in the male gonads, the *testes,* two oval bodies which in the mature male hang suspended in the scrotal sacs outside the abdominal cavity. The testes produce the male gametes, the *spermatozoa,* and the male sex hormones, the *androgens.* The testes also secrete estrogens. (Both sexes secrete both types of hormones, with the balance in favor of estrogens in the normal female.)

Androgens are vital for the *development* and *functioning of the male reproductive tract,* which is especially adapted to transport spermatozoa from the testes through a tortuous pathway of tubes, and, by way of the male penetrating organ, the *penis,* into the body of the female. Figure 26-1 illustrates the reproductive organs and the external genitalia of the mature male. The structural and functional relationship of the tubes and glands that make up the *accessory sex organs* (*epididymis, vas deferens, seminal vesicles,* and *prostrate gland*) are discussed later in this section and illustrated in Fig. 26-7.

Testes

Position and Structure

Descent into the Scrotal Sacs. The testes originate in the retroperitoneal region of the abdominal cavity from an area called the *genital ridge* of the embryo. In the last months of fetal life, they descend from the abdominal area through the inguinal canal into the *scrotal pouches.* These canals normally close in the human male, preventing the return of the testes into the abdominal cavity, but occasionally the closure is incomplete, and these organs tend to protrude back through the canal in a condition known as *inguinal hernia,* which can be corrected surgically by closing the canals. The significance of the descent of the testes is the *change in temperature* from the warmth of the abdominal cavity to the relatively cool scrotal pouch; the evidence seems to be convincing that spermatogenesis can occur only

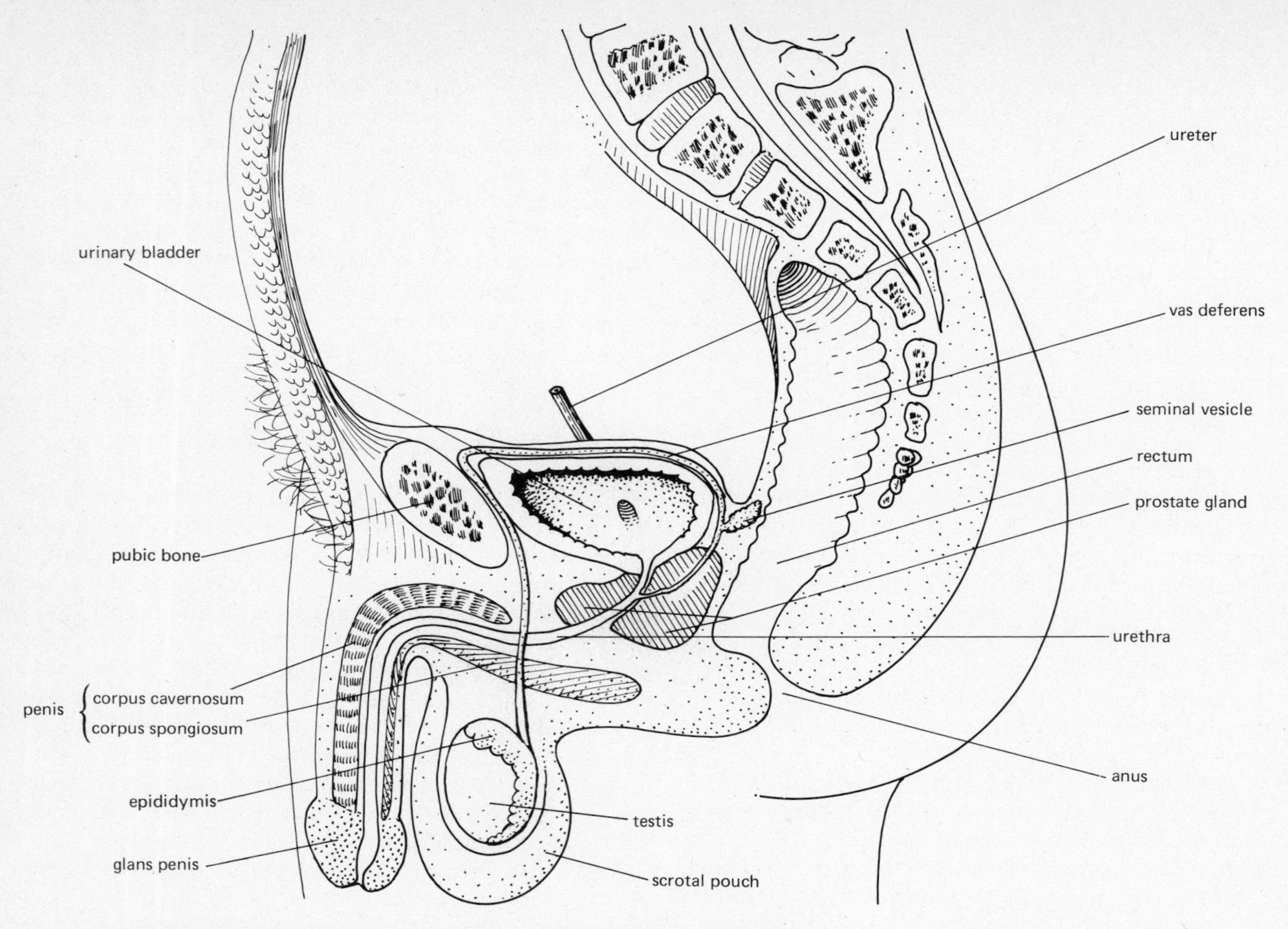

Figure 26-1. **The reproductive tract of the mature male, shown in sagittal section.**

at the lower temperature. Certainly, testes that are retained in the abdominal cavity do not produce sperm. This condition is called *cryptorchism,* from the Greek *cryptos,* "hidden," and *orchis,* "testis." If there is no mechanical obstruction of the inguinal canals, the injection of testosterone will usually cause the testes to descend.

Seminiferous Tubules. The outer wall of the testis is a tough layer of connective tissue, covered on the outside by the scrotal sac and lined on the inside with a vascular layer. Most of the testis is composed of the highly convoluted *seminiferous tubules,* which contain the *spermatogonia,* the cells from which the spermatozoa are produced, and the various stages of the differentiating spermatozoa. The spermatogonia (single: *spermatogonium*) are organized into two or three layers along the outer edges of the seminiferous tubules; the process of their differentiation occurs toward the lumen of the tubule (Fig. 26-2).

Scattered irregularly between the spermatogonia are the large supporting *Sertoli cells,* which produce nutrients, hormones, and enzymes necessary for the maturation of the spermatozoa.

Interstitial Cells. Between the seminiferous tubules of the testes are clumps of *interstitial cells* (*cells of Leydig*), the secretory cells that produce the androgens, the most potent of which is *testosterone.*

Hypothalamic and anterior pituitary regulation of the testes

Both the structure and the function of the testes are dependent upon hormonal stimulation by the pituitary gonadotropins, which in turn are regulated by hypothalamic secretions.

The hypothalamus of the male secretes at least one, and perhaps two, gonadotropin-releasing hormones, *luteinizing hormone-releasing hormone* (*LRH*) and *follicle-stimulating hormone-releasing hormone* (*FRH*). These hypothalamic peptide hormones cause the release of LH and FSH from the anterior pituitary (Secs. 10-6 and 11-3). LRH has been isolated from the hypothalamus and has also been synthesized. It is a polypeptide with 10 amino acids, and either the extract or the synthetic decapeptide will cause the release of both LH and FSH from the anterior pituitary. It is therefore uncertain whether there is an additional FRH in the hypothalamus. Because of its dual function LRH is sometimes referred to as *LRH/FRH.*

Both LH and FSH are identical in males and females, their effects being entirely dependent upon the nature of their gonadal target organ. In the male, there is still some uncertainty about the relative roles of FSH and LH.

Regulation of Testosterone Secretion by LH. The most important function of LH in the male is to stimulate the interstitial cells of Leydig to secrete testosterone. When FSH is present, it has a synergistic action on the LH effect, greatly increasing the amount of testosterone produced.

Regulation of Spermatogenesis by FSH. The follicle-stimulating hormone has been shown to bind specifically to the Sertoli cells, indicating the importance of FSH to spermatogenesis. Also, FSH is known to stimulate spermatogenesis but cannot complete the maturation process into spermatozoa unless testosterone is present. Therefore, FSH and LH must be secreted simultaneously by the anterior pituitary for proper spermatogenesis to occur.

Figure 26-2. **Stages in the differentiation of spermatozoa in a seminiferous tubule of the testis. The androgen-producing interstitial tissue is also shown.**

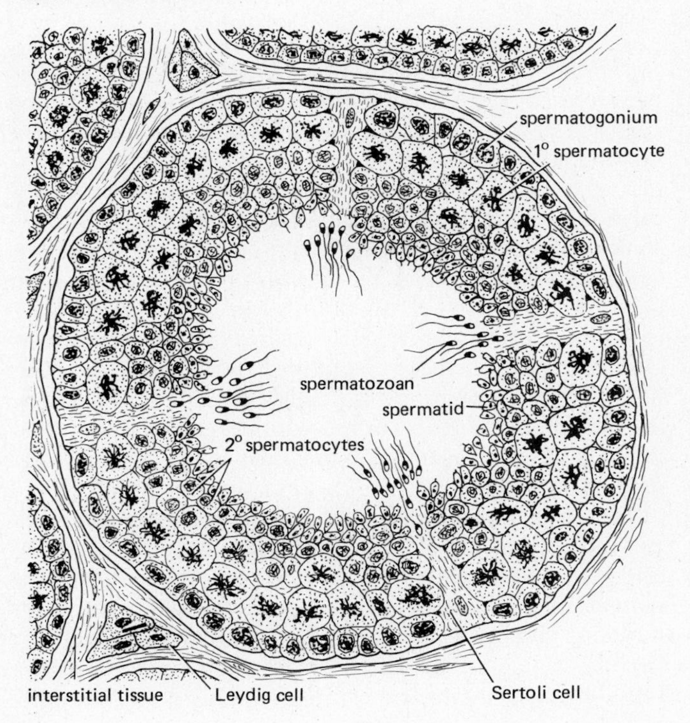

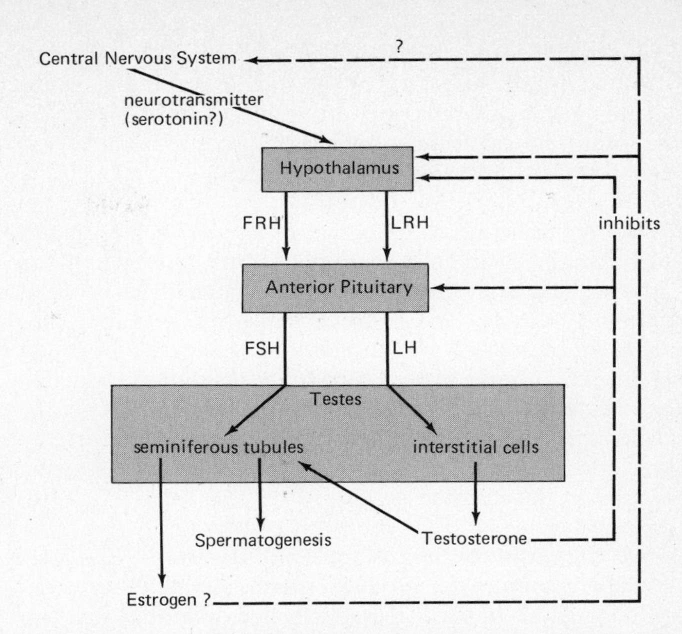

Figure 26-3. **Feedback relations between the hypothalamus, the anterior pituitary, and the testis. Broken arrows indicate inhibition.**

Negative feedback regulation of the hypothalamic secretion

It is considered that the main feedback control of the hypothalamic FRH and LRH is exerted through the level of circulating testosterone. However, these hypothalamic neurosecretory cells are also affected by neurotransmitters such as serotonin, so that neural pathways as well as hormonal may be involved. In addition, estrogen levels inhibit FRH and LRH, a mechanism that may be of more importance in the bull and stallion, which produce large amounts of estrogen, than in human males, in whom the estrogen secretion is relatively low. These feedback relations among the hypothalamus, the anterior pituitary, and the testes are illustrated in Fig. 26-3.

Testosterone secretion in the fetus and child, and during puberty and aging

Fetal Development. During fetal development, there is an early secretion of testosterone, which is vital for the proper development of the undifferentiated embryonic tissues into the male reproductive structures. This testosterone secretion is caused by stimulation of the interstitial cells by chorionic gonadotropin from the placenta. The interstitial cells at this time are already numerous and responsive to this stimulation.

The small amounts of testosterone secreted by the fetal testes are essential for the development of the typical male type of acyclic hypothalamus, as discussed in Sec. 11-2.

Childhood. During childhood, the hypothalamus appears to be extremely sensitive to the inhibitory effects of any small amount of circulating testosterone. Consequently, no gonadotropin-releasing hormones are produced, and no testosterone is secreted by the testes.

Puberty. The mechanisms that release the hypothalamus from childhood inhibition are not known, but in humans, around the age of 11 to 13 years, the hypothalamus becomes extremely active, and large amounts of gonadotropin-releasing hormones, growth hormone, and other hormones are secreted. In the male, a surge of testosterone production by the testes results, stimulating the development of the male reproductive tract and the secondary sex characteristics, effects of testosterone that are discussed in a following section in this chapter.

Male Climacteric: Sexual Aging in the Male. The male usually continues to produce sperm and sex hormones throughout his life, with a gradually decreasing efficiency no different from the normal aging processes of all cells. Testosterone secretion reaches a peak at about 20 years of age and then slowly diminishes between the ages of 40 and 80 years (Fig. 26-4). Well-authenticated cases of men in their eighties siring children are frequent.

Sometimes, however, there does seem to be a rather abrupt cessation of sexual functions in men in their fifties, and this is referred to as the *male climacteric.* It is accompanied by hot flashes and behavioral disturbances, similar to those experienced by women during the menopause.

The male climacteric appears to be caused by a failure in testosterone production by the interstitial cells, because testosterone administration alleviates the symptoms. It is not caused by hypothalamic or pituitary failure; very large amounts of FSH are secreted by the anterior pituitary gland because of the absence of the inhibitory effects of endogenous testosterone.

Reproductive cycles in the male

Many of the lower vertebrates show a definite reproductive periodicity, particularly marked in certain fish that migrate for hundreds of miles to spawn. Birds, too, migrate as far as a thousand miles in order to breed, and in most of these species it has been shown that the stimulus is change in the amount of light due to the lengthening of the day as summer approaches. The light stimulus reaches the pituitary gland by way of the hypothalamus; in response there is an increase in gonadotropin production and correspondingly an increased testosterone secretion and production of mature spermatozoa. This response to light is found in many rodents as well. The human male, however, apparently is independent of such environmental variables, apart from extremes of temperature that make him too uncomfortable to be interested in sex. Spermatogenesis in the human being is a continuous process from the age of puberty until senescence.

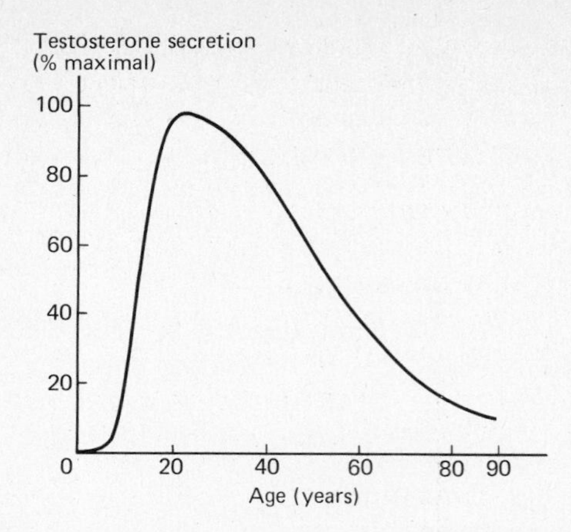

Figure 26-4. **Secretion of testosterone throughout the life of the male.**

Functions of the Testes

Spermatogenesis: formation of the male gamete

All the characteristics that must be transmitted from the complex, multicellular adult organism have to be concentrated within the single cell of the *gamete,* be it ovum or sperm. Only certain cells, known as the germ cells or germ plasm, are capable of forming the gametes, and these cells are found within the gonads. In the male, the *spermatogonia* are the cells destined to form the gametes. Of all the millions of cells in the body, only these germ cells will transmit their chromosomes and consequently the instructions for development to the zygote. Only alterations to the structure or the order of the genes of these germ cells will affect the subsequent generations.

Significance of Meiosis. Because the adult organism has been formed from the union of a sperm and an ovum, all the cells that have developed from the resulting zygote contain the double or *diploid number* of chromosomes, one set inherited from the mother and one set from the father. When this adult in turn produces gametes, the number of chromosomes must be reduced to one half, or the *haploid number.* Without this mechanism, each succeeding generation would have twice the number of chromosomes as its parents.

Although each gamete must receive one of each pair of chromosomes, whether it receives the maternally or paternally derived chromosome is a matter of chance. This is true for each of the *23 pairs* of chromosomes in *humans,* so that the possible number of recombinations of the chromosomal pairs is enormous.

Meiosis is the collective term for the processes by which the diploid cell divides to form four haploid cells. The importance of this phenomenon is not only that reducing the number of chromosomes in the gametes permits the chromosomal number of the zygote to remain constant, but also that the biological door is open to evolution through the introduction of *controlled variability.* This is in sharp distinction to *mitosis,* in which the daughter cells are qualitatively identical, and variability is at a minimum (Sec. 2-4).

Stages in Spermatogenesis

1. The spermatogonia enlarge to form the primary spermatocytes. No division is involved in this process, so that the primary spermatocyte has the same diploid number of chromosomes (23 pairs in humans) as the spermatogonium (Fig. 26-5).
2. The primary spermatocyte undergoes *meiotic division* to form two *secondary spermatocytes,* each with half the number of chromosomes (haploid number) representing one chromosome from each of the 23 pairs.
3. The secondary spermatocyte undergoes *mitotic* division to form two *spermatids,* each with 23 unpaired chromosomes.
4. The final stage of spermatogenesis is the *maturation* of the haploid spermatids into the haploid *spermatozoa.* Each diploid spermatogonium has produced four haploid spermatozoa.

Stages of Meiosis. THE FIRST STAGE OF MEIOSIS: SYNTHESIS, SYNAPSIS, AND DISTRIBUTION. The *primary spermatocyte* is extremely active, synthesizing DNA and protein *preparatory to entering the prophase* of the first meiotic division. At this time of synthesis, the individual chromosomes cannot be seen, although the nucleoli are visible. It is during this period that the DNA content is doubled so that when the chromosomes become visible in prophase, each one has already produced its mirror image (see Fig. 2-23 for a diagrammatic representation of this process).

Figure 26-5. **Chromosomal changes during spermatogenesis and the maturation of spermatozoa.**

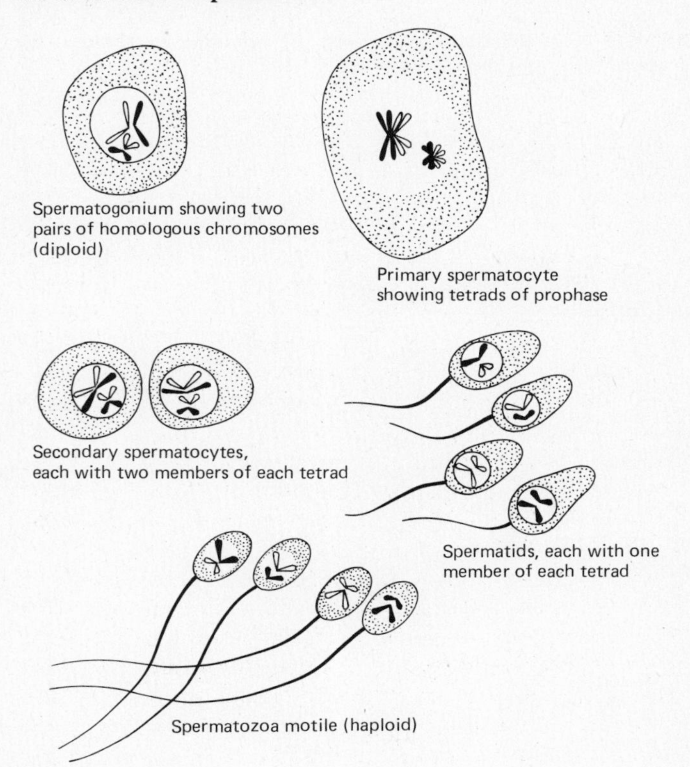

Now begins the really interesting aspect of meiosis from the genetic point of view.

Prophase shows each doubled chromosome attracted by some unknown mechanism to its homologous mate, so that *tetrads* are formed. Each tetrad thus consists of four members, or two pairs of mirror-image chromosomes. They are connected to one another by their *centromeres,* the constriction somewhere along the length of each chromosome, which not only gives the chromosome its recognizable shape but is probably the organ of movement of the chromosome. The members of the tetrad come to lie close to one another, or *synapse,* like the sides of a zipper. Sometimes the zipper gets tangled, and when the members are separated, a piece of one chromosome may have been exchanged for a piece of one of its mates. This exchange of genetic material in the tetrad is called *crossing over.* Crossing over is responsible for a large part of the qualitative differences between the resulting gametes, because it permits some genes that had been linked together along one chromosome to be exchanged with others from its mate (see also Sec. 27-3).

By this stage in prophase, the *spindle* has been formed between the two centrioles, and the nuclear membrane has disappeared. *The subsequent stages in meiosis distribute the members of the tetrads to the daughter cells.*

Metaphase is most easily recognized under the microscope, for the tetrads are lined up along the equator of the spindle so that two members of each tetrad face each pole. They are attached to the fibers of the spindle by their centromeres.

Anaphase starts by the separation of the centromeres, which permits each pair of tetrad members to travel to its pole, tugged along by its centromere.

Telophase sees the nuclear membrane re-forming around the chromosomes at each pole, separating the chromosomes from the cytoplasm. The cytoplasm is cleaved in two, and the *two secondary spermatocytes* have been formed. *These cells are still diploid but are no longer genetically identical,* since both members of some of the chromosome pairs may be from the male, some pairs may derive purely from the female, and many may have exchanged parts during synapsis. These stages of distribution of the chromosomes to the daughter cells are very similar to those illustrated in Fig. 2-23, showing mitosis in an animal cell.

THE SECOND STAGE OF MEIOSIS: REDISTRIBUTION ONLY. The *secondary spermatocyte* enters the next cycle of division very rapidly and, apart from the fact that there is no duplication of the chromosomes, it is just like any mitotic division: the diploid number of chromosomes is being distributed to two daughter cells, each of which will receive half the number of chromosomes. By the end of the second division, therefore, the *haploid number* has been achieved, and the chromosomes in each daughter cell are *qualitatively*

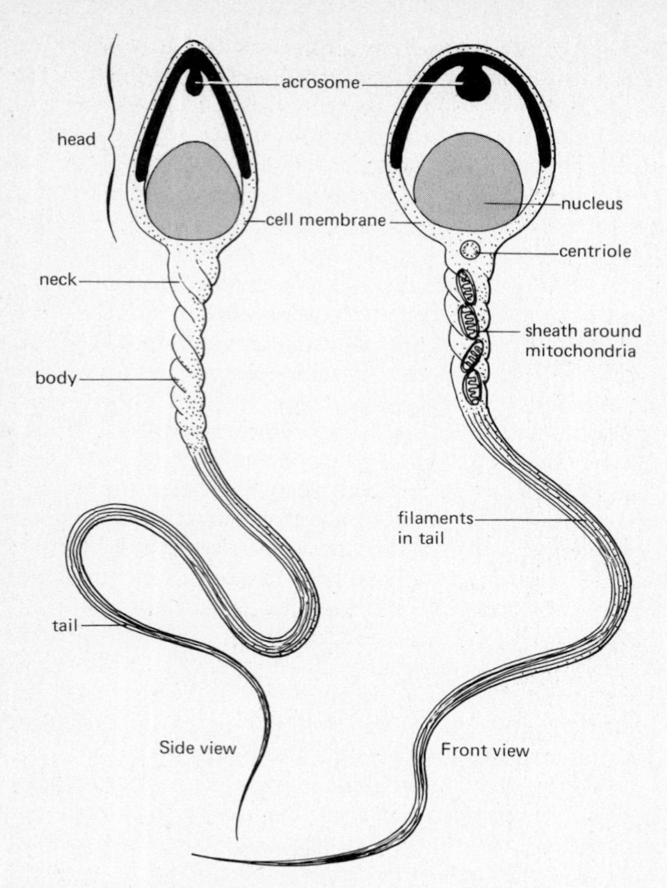

Figure 26-6. **A normal human spermatozoan seen from the front and from the side.**

different. These cells are called *spermatids,* and they are the immature gametes, still incapable of movement.

Maturation of the Gametes. The nonmotile spermatids now undergo a complex maturation process that will turn them into the specialized motile *spermatozoa* (Fig. 26-6). The nucleus becomes reorganized, and the Golgi complex, together with a part of the nuclear membrane, forms a special penetrating point, the *acrosome,* at the tip of the spermatozoan, which probably helps it to pierce the membrane of the ovum. The centrioles organize the formation of the long, whiplike *tail.* It is fascinating to realize, as electron microscope studies show, that cilia, flagella, and the spermatozoan tail all have the same internal structure: an outer ring of nine paired filaments surrounding two central contractile filaments (Fig. 2-13). Regardless of whether the flagellum is a part of a single, independent cell, or whether it represents the reproductive gamete of a highly evolved multicellular organism such as man, the basic structural arrangement for movement is the same.

In the maturation process, the cytoplasm of the spermatid is discarded, leaving a highly motile, short-lived cell, loaded with precious genetic material and having a long and somewhat precarious trip ahead of it before it reaches the ovum in the upper part of the female reproductive tract.

Passage of Sperm from the Testes. The sperm formed in the convoluted seminiferous tubules pass into small, *straight tubules,* which in turn lead into about 20 little ducts that pierce the fibrous covering of the testis to enter the tortuous duct known as the *epididymis* (Fig. 26-7). The epididymis has the shape of a comma, loosely attached to the surface of the testis; when straightened out, it is about 20 ft long. After traversing this twisted tube, the sperm reach the much wider *vas deferens,* which runs back through the inguinal canal to the neck of the bladder. The vas deferens, the blood vessels, and the nerves of the testis are bound together in a loose bundle, the *spermatic cord.*

The sperm in the vas deferens are still quite dormant, and because their metabolic activity is so low they can be stored in this tube for many days. They utilize glucose and fructose present in the tubular secretions for their energy requirements.

Semen. The sperm secretions of semen are considerably increased in volume and nutritive value by the contributions of the various glands that empty into the right and left *ejaculatory ducts.* The most important of these are the paired, saclike *seminal vesicles* and the single, large *prostate gland* that surrounds the junction of the urethra and the urinary bladder (Fig. 26-1). The ejaculatory ducts pass through the prostate and receive a thin alkaline fluid from it.

Figure 26-7. **Relationship between the seminiferous tubules, the epididymis, and the vas deferens.**

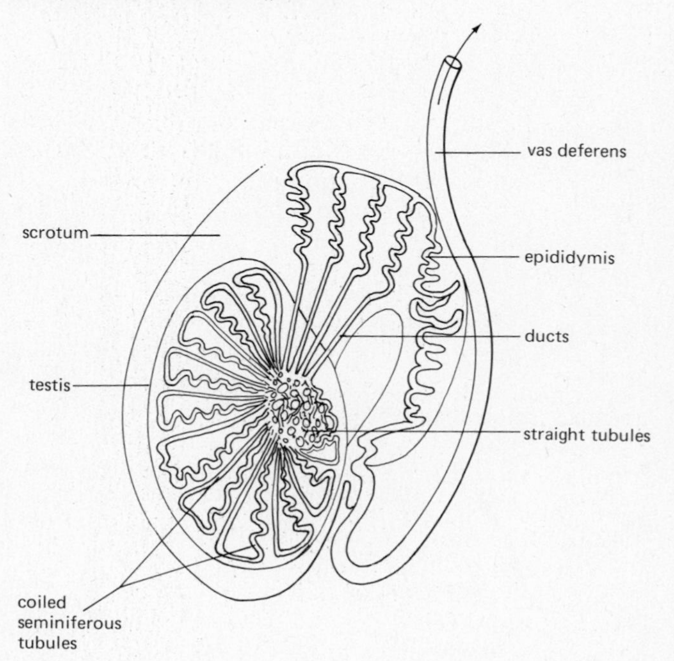

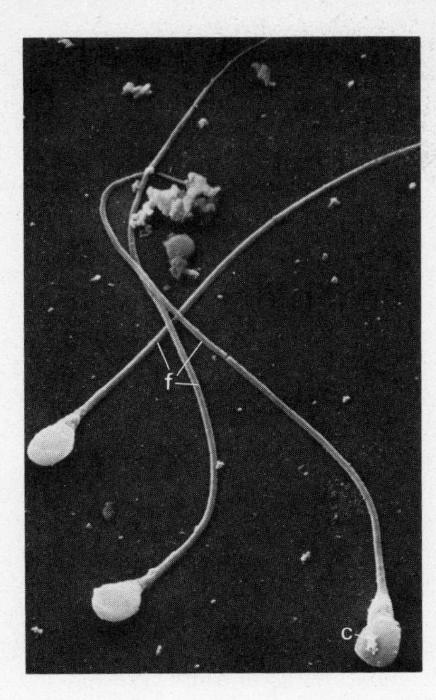

Figure 26-8. **Human spermatozoa seen with a scanning electron microscope. The sperm head is covered with the acrosomal cap (c) that contains the enzymes necessary to penetrate the ovum; f = flagella. ×2000. (Courtesy of D. M. Phillips, The Population Council, New York.)**

This mixture of fluids is called the *semen.* It consists of the sperm from the seminiferous tubules, the fluid from the vas deferens, seminal vesicles, and prostate, and the mucus from a small pair of bulbourethral glands that empty into the urethra. Human semen is the richest natural source of prostaglandins.

All these secretions are added to the contents of the vas deferens during the process of ejaculation. *Ejaculation* is the final stage in the male sexual act when, as a result of intense sexual stimulation, the reflex contraction of the musculature of these glands forces their contents into the ejaculatory ducts; simultaneously the peroneal muscles contract and force the fluid through and out the urethra.

The semen is quite alkaline (pH about 6.5), mainly because of the alkaline nature of the prostatic secretion. This alkalinity neutralizes the naturally acid secretions of the vagina and so enhances the motility and fertility of the sperm. The sperm do not live longer than 24 to 48 hr after ejaculation, but they can be frozen and stored and will become active on being brought back to body temperature.

Capacitation. The fertility of the spermatozoa is increased during their passage through the epididymis and vas deferens, and considerably more during the time they are in the female reproductive tract. This increase in fertility, essential for the penetration of the spermatozoan into the ovum, is *capacitation.*

It is not known exactly what causes capacitation, but it probably involves the sequential release or activation of a series of hydrolytic enzymes that digests the outer protective layers of the ovum. The most important of these enzymes include *hyaluronidase* and *proteinases,* which are stored in the acrosomes of the sperm. Hyaluronidase digests the intercellular cement between the cells of the protective layers. The proteinases probably digest the proteins of the cells themselves.

Sperm Count, Morphology, and Fertility. An average human ejaculate contains 400 million sperm, a couple of which are shown in Fig. 26-8. Although only one sperm is needed to fertilize the ovum, large numbers of accompanying sperm are needed to produce the digestive enzymes to open a passage for the victor. If a man's sperm count falls below 20 million, therefore, he will probably be sterile.

Even with a normal sperm count, if there are many abnormally formed sperm—such as sperm with two heads or distorted tails—the male is likely to be infertile. Figure 26-9 shows some types of abnormal human sperm.

SECRETION OF ANDROGENS: ORIGIN, METABOLISM, AND EFFECTS OF ANDROGENS

Origin. The testes are certainly the most important site of production of the androgens, although the adrenal cortex may produce small amounts. In the female, the ovaries are capable of secreting androgenic compounds. Most of the experimental evidence favors the interstitial cells of the testes as the site of androgen production, particularly of testosterone. Other naturally occurring androgens may be produced, or they may be simply degradation products of testosterone.

Figure 26-9. **Some types of abnormal human spermatozoa.**

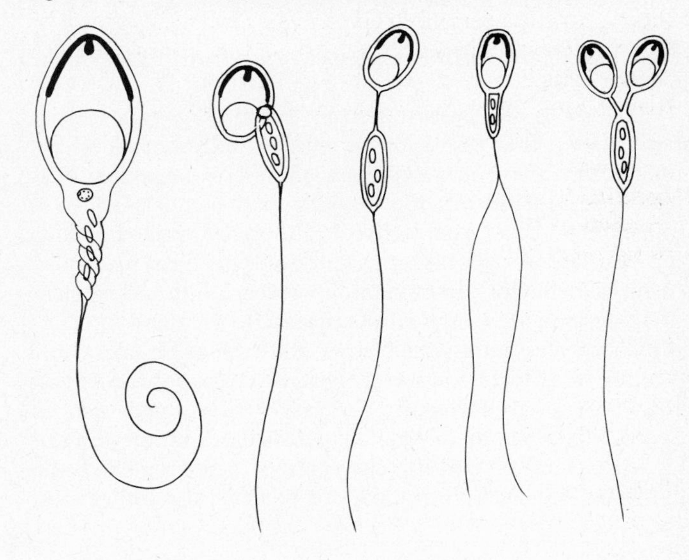

Metabolism. Testosterone is a steroid hormone that can be synthesized in the body or in vitro, from cholesterol (Sec. 10-7). Less potent androgens are formed by the metabolism of testosterone in the liver, and these substances (androsterone, among others) can be recovered in the urine.

Effects of Androgens. EMBRYONIC DEVELOPMENT. In early embryonic development, the secretion of testosterone by the embryonic testes is essential for the undeveloped reproductive tract to *differentiate* along male lines. Later in development, testosterone is necessary for the descent of the testes into the scrotum. Testosterone also organizes the female-type hypothalamus into the male type of hypothalamus (Sec. 11-2).

SPERMATOGENESIS AND THE GROWTH, DEVELOPMENT, AND MAINTENANCE OF THE REPRODUCTIVE SYSTEM. Testosterone is required for the completion of *spermatogenesis* in the seminiferous tubules. It is needed for the growth and development of the *sex-specific tissues:* the testes, scrotum, penis, prostate gland, and seminal vesicles.

SECONDARY MALE SEXUAL CHARACTERISTICS. Testosterone is responsible for the typical pattern of *hair growth* (on the face, chest, back, and pubic area), and for *hair loss* too, because baldness is the result of androgens acting upon the appropriate genetic background. Although eunuchs (castrated males) keep a fine head of hair, baldness does not always indicate excessive virility. Women with the right (wrong?) type of genetic inheritance and an androgenic tumor will become bald.

Fat distribution is also affected by testosterone. Males tend to accumulate fat in the abdominal area, rather than on the hips and thighs, as is characteristic of females.

PROTEIN SYNTHESIS. Testosterone induces *protein synthesis* and therefore nitrogen retention, especially in muscle (Sec. 19-3). The body configuration of males thus develops increased muscle mass, large bones, skin thickness, thickening of the laryngeal mucosa and enlargement of the larynx, which causes the typical bass voice of the male. As a result of this increased rate of protein synthesis, the entire metabolic rate of the male is somewhat higher than that of the female.

SEX DRIVE AND AGGRESSION. From early times it has been known that castration of animals and humans affects their sexual drive and decreases aggressive behavior. Testosterone implants into the medial hypothalamus of rats will induce male copulatory activity (although similar implants into the lateral hypothalamus cause the male rat to display maternal behavior). In humans, however, libido and patterns of aggression are complex manifestations of many interacting social and emotional factors and cannot be correlated directly to testosterone levels. Despite the experiments that Brown-Séquard performed in 1889 (mentioned also in Sec. 10-5), when he injected testicular extracts into himself and then claimed startling rejuvenation, there seems to be no evidence that even large doses of testosterone can reverse the effects of age.

26–4 The Male Sexual Act

In order for the male to deposit sperm within the female, the *penis,* which is the organ of intromission, must first become erect. The penis is composed mainly of erectile tissue, which is made up of a spongework of spaces that can be filled with blood. This erectile tissue is divided into three longitudinal columns, two of which form the *corpora cavernosum* that occupies the dorsum and sides of the penis. The urethra passes through the third column, the *corpus spongiosum.* This portion is much smaller than the rest of the penile erectile tissue (Fig. 26-10).

The penis is capped by the *glans penis* (partially covered by a fold of skin, the *prepuce*), which contains a highly sensitive end-organ system, stimulation of which gives rise to specific sexual sensations. These nerve endings are the receptors for the initiation of the male sexual act. They can be stimulated by direct touch, as occurs during sexual intercourse, or by psychic suggestion. Thinking erotic thoughts, seeing sexual pictures, or reading sexually stimulating books may be sufficient stimulus to cause erection and ejaculation. Nocturnal emissions or *wet dreams* are the result of erotic dreams, which occur most frequently during the teens.

Table 26–1. **Sexual Reflexes in the Male**

	Erection	
Stimulus	Tactile stimulation	Psychic stimulation
Receptors	Nerve ending in the glans penis and adjacent areas of the body	↓
Afferent neuron	Pudendal nerve to sacral plexus	↓
Synapse	Sacral part of the spinal cord	↓
Efferent neuron	Parasympathetic fibers from sacral cord ⟵	Cortical activity
Effector	Smooth muscle of arterioles	
Response	Vasodilation and engorgement of venous sinuses with blood	
Effect	Erection of the penis	
	Ejaculation	
Stimulus	Intense sexual stimulation, usually of erect penis, in coitus	
Receptors	Nerve endings in glans penis	
Afferent neuron	Pudendal nerve via sacral plexus	
Efferent neuron	Sympathetic fibers	
Effector	Smooth muscle of reproductive tract	
Response	Rhythmic contractions	
Effect	Propulsion of semen through urethra followed by flaccidity of penis	

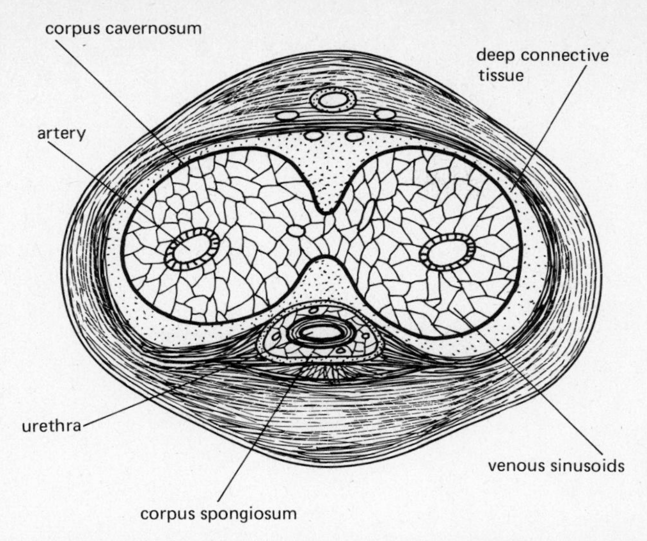

Figure 26-10. **Cross section of the penis showing the three columns of erectile tissue, (the two corpora cavernosa and the corpus spongiosum) that fill with blood to cause the penis to become erect.**

Erection, Lubrication, Ejaculation, and Orgasm in the Male

During sexual excitement, the erectile tissues of the penis become engorged with blood, and mucus from the bulbourethral and other glands is poured into the urethra and helps lubricate the glans. Most of the necessary mucus is provided by the secretions of the female if she is sexually excited. This permits the entrance of the erect penis into the vagina. The tactile sensations further excite the reflex, rhythmic muscular contractions that involve the entire reproductive tract, propelling the semen out of the urethra in a series of short spurts. This is *ejaculation* (Table 26–1). The violent muscular movements of the body, the changes in respiration and heart rate and blood pressure, together with the sensations aroused by these overall changes, form the *orgasm.* These sensations are often followed by a deep relaxation of the muscles and sometimes a feeling of fatigue.

26–5 Female Reproductive Tract

The two essential functions of the female reproductive tract are the *production of ova* and the *nurture and protection of the fertilized ovum* until it has developed into an individual capable of living a fairly independent life outside the maternal environment. From *conception,* the instant of fertilization, until *parturition,* the birth process, is the time of *gestation.* The gestation time varies considerably from species to species, but in the human being it is about 280 days, or nine and one-half months. Nor are the physiological duties completed by the birth of the child: the female continues her exertions by suckling the baby with milk produced by the mammary glands or breasts. All these processes, including lactation, are under hormonal control.

The entire female reproductive tract is shown in sagittal section in Fig. 26-11. Note that the urinary bladder is anterior to the uterus and vagina. The rectum lies behind the uterus.

Ovaries and Oviducts

The ova are produced in the ovaries, a pair of flattened organs about the size of an almond. Like the testes, they are derived from the germinal ridges at the back of the abdominal cavity near the kidneys. The ovaries descend into the pelvis during the early part of fetal life, but because they remain in the pelvis, they travel a much shorter distance than do the testes. The ovaries are kept loosely in place by a fold of peritoneum that holds them to the broad ligament of the uterus (Figs. 26-11 and 26-12).

Each ovary is in close connection with the fringed end (*fimbriae*) of a tube, the *oviduct,* which leads to the uterus. There is no direct connection between the ovary and the opening of the oviduct, however, and it is possible for an ovum to be freed from the ovary and disappear into the abdominal cavity rather than pass down the oviduct. This happens very rarely, because the fringed opening of the oviduct probably contracts rhythmically, causing the ovum to be sucked into the oviduct.

The wall of the oviduct is made up of muscle layers and a mucous lining. It produces a fluid that increases the fertilizing capacity of the sperm (capacitation), and it is also lined with cilia. Whether the cilia help to move the ovum down the tube, the sperm up the tube, or neither, is still a matter of considerable debate.

The ovary is richly supplied with blood vessels and nerves that enter it through a hilum.

Uterus, Vagina, and External Female Genitalia

UTERUS

The uterus is an extremely muscular, thick-walled, hollow organ. The strong muscular layer is the *myometrium;* the inner mucosa is called the *endometrium* (Fig. 26-12). Both these layers are extremely sensitive to the hormonal secretions of the ovary. The cyclic buildup and sloughing off of the endometrium when pregnancy does not occur form the flow of *menstruation;* its continued development and great increase in thickness at the site of implantation of the blastocyst when pregnancy does occur are responsible for the proper formation of the *placenta.* The myometrium is quiescent during this time, but its violent rhythmic contractions at the end of gestation expel the fetus through the vagina into the outside world.

The uterus is held in place by the bony pelvis and the

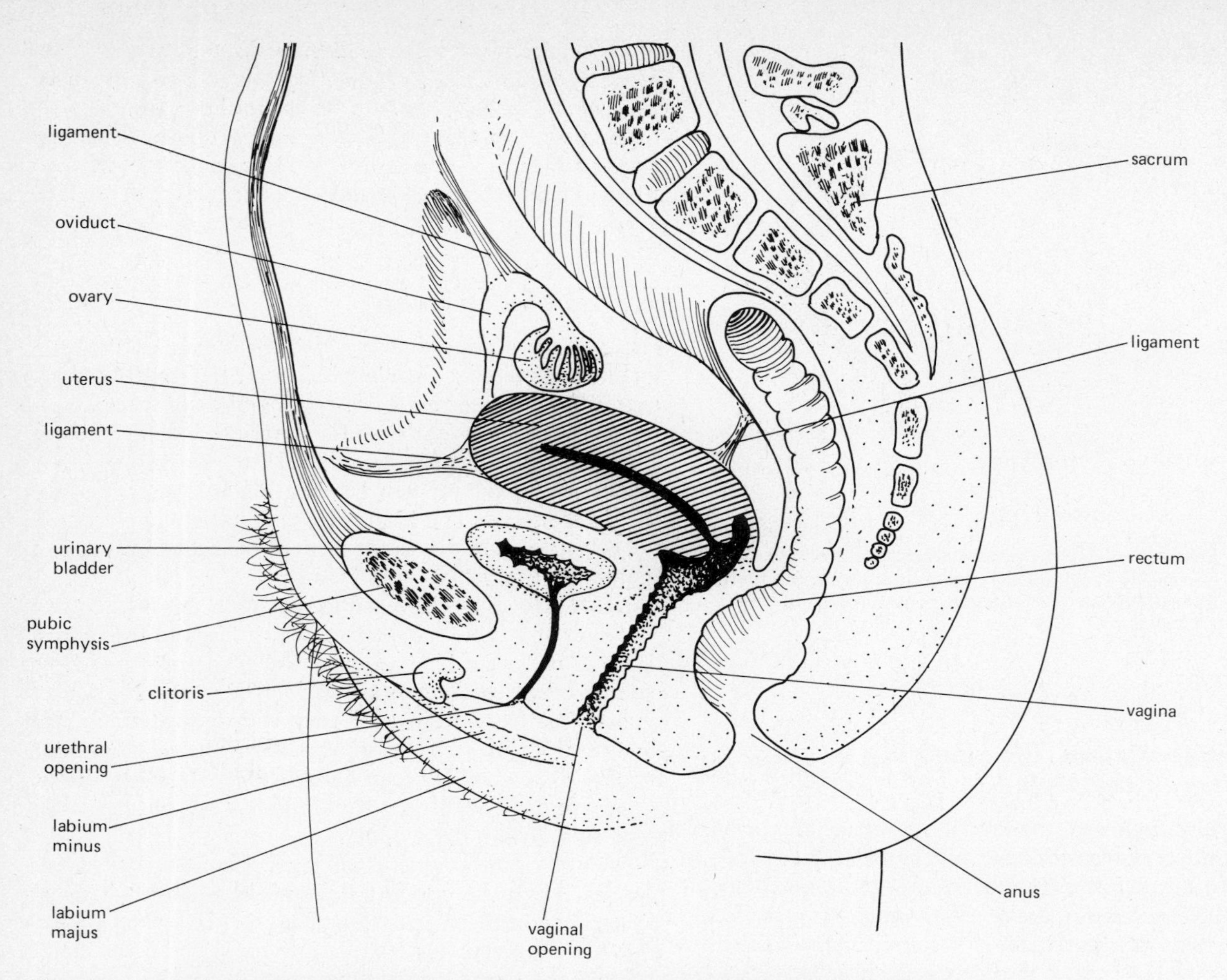

Figure 26-11. **Sagittal section through the female pelvis, showing the reproductive tract and its opening, the vagina. Also shown are the urinary bladder, which opens to the exterior via the urethra; and the rectum, the external opening of which is the anus. Note the strong ligaments holding the uterus in place.**

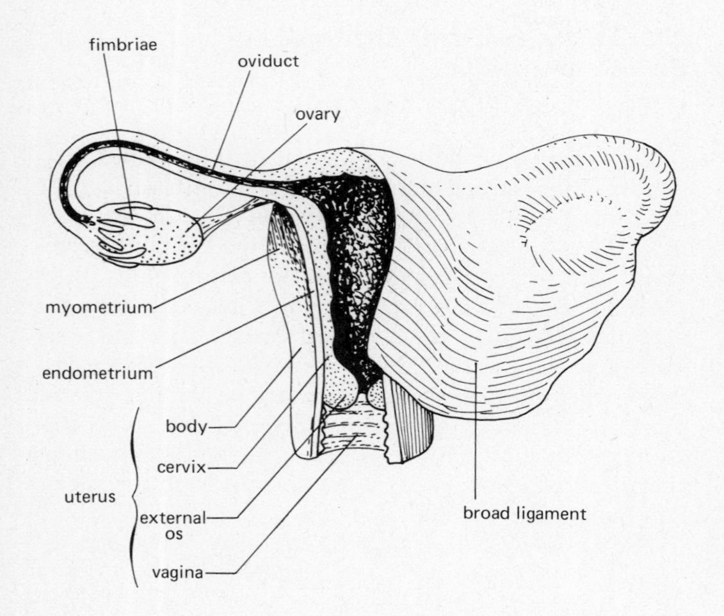

Figure 26-12. **The female reproductive organs. The broad ligament has been removed from one side to show the underlying structures.**

strong muscles and fibers between the anal canal and the urogenital area. This area is often called the *perineum* (Fig. 26-13) and is especially susceptible to tearing during childbirth. The ligaments of the uterus also help to keep it in place (Fig. 26-12).

There are *three openings to the uterus:* the two lateral ones where the *oviducts* (uterine tubes) enter, and a single small opening in the narrow neck (*cervix*) that projects into the vagina. This opening is the *external os,* and sperm must pass through the os to traverse the uterus and penetrate into the oviducts. The debris of menstruation also must pass down through this opening to reach the vagina and the exterior.

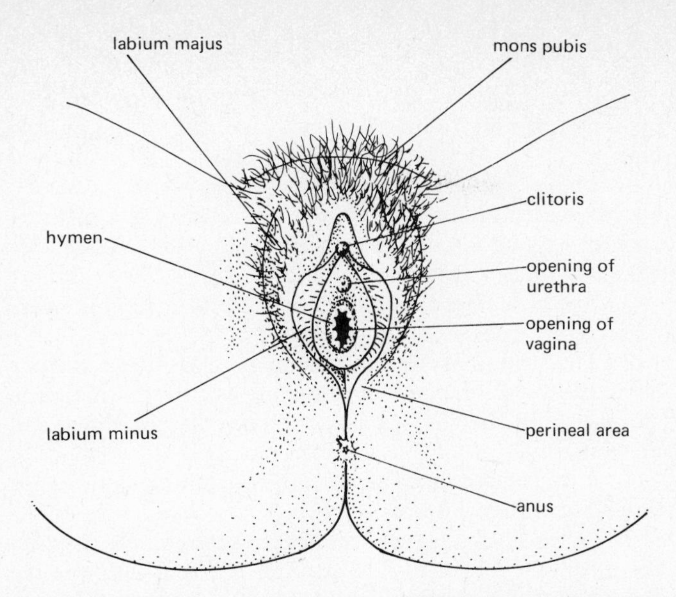

Figure 26-13. **The female external genitalia and the vaginal, urethral, and anal openings.**

Functions of the Uterus. The uterus has three related functions:

1. It provides the site and the conditions necessary for the implantation of the fertilized ovum and then responds to implantation by developing the maternal part of the placenta necessary for the nourishment of the embryo.
2. It grows and changes in shape to adapt to the rapidly growing fetus.
3. Its contractions at the end of the gestation period deliver the baby to the external world.

Vagina

The *vagina* is a slightly curved muscular canal about 3 in. long. The cervix of the uterus enters its upper end; at its lower end the vagina opens into a cleft between the labia minora, part of the external genitalia. This cleft is partly closed in virgins by a thin membrane, the *hymen* (Fig. 26-13).

The vagina is lined by a thick layer of stratified squamous epithelium, which in lower animals undergoes cyclic changes that can be correlated to the various hormonal changes during the estrous cycle. In primates, partly because of the high bacterial count in the vagina, the conditions of the vagina vary too greatly to be of much use in diagnosing ovarian conditions.

External female genitalia

The *external female genital organs* are formed by the two small folds, the *labia minora,* that guard the entrance to the vagina and the larger *labia majora,* which correspond embryonically to the scrotal sacs of the male (Fig. 26-13). They are made of fatty tissue, covered with skin and hair. Under the labia minora are two small glands, the glands of Bartholin, which secrete the mucus essential for the proper penetration of the penis into the vagina. Between the anterior ends of the labia minora is a small organ of erectile tissue, the *clitoris.* Like the penis of the male, to which it corresponds in embryologic derivation, it is made up of spongy tissue which distends during sexual excitement as it fills with blood.

The cushionlike bulge over the pubic symphysis is the *mons pubis* (or mons veneris), which at puberty is formed by the deposition of fat under the influence of estrogen. The mons pubis is covered with hair, which extends beyond its immediate region to form the straight border of the triangle of pubic hair characteristic of the female.

Ovarian Follicles

Primary follicles

The cortex of the ovary is made up of the *germinal epithelium* that in the fetus contains the 300,000 to 500,000 *primary follicles* of the two ovaries (Fig. 26-14). Each primary follicle consists of an ovum (which, at this stage, is a *primary oocyte*) surrounded by a single layer of *granulosa*

Figure 26-14. **Stages in development of the ovarian follicle and corpus luteum.**

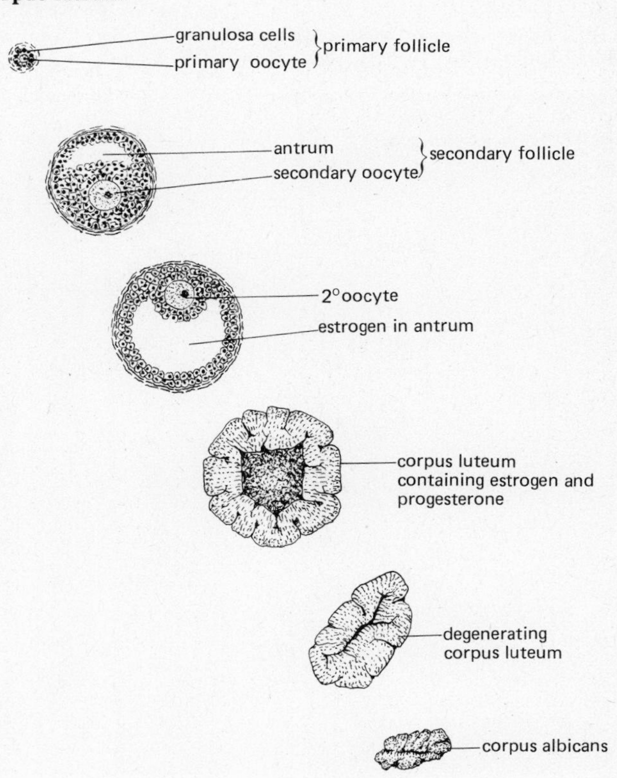

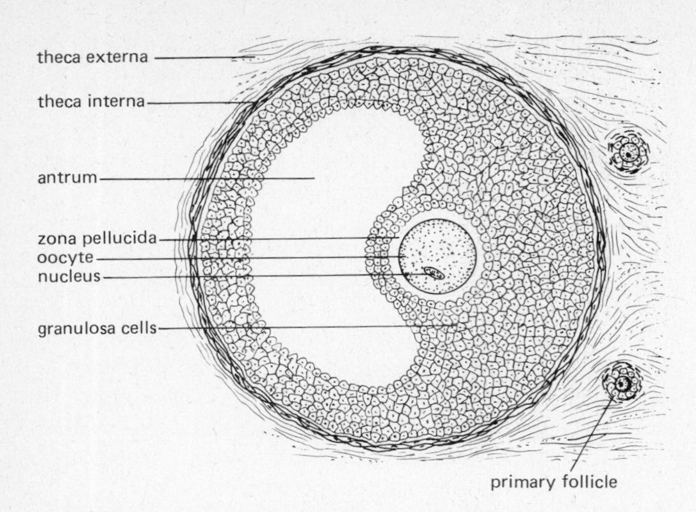

Figure 26-15. **The secondary (Graafian) follicle. Two primary follicles are also shown.**

cells. In the human female, the primary oocyte is about 100 μm in diameter. By puberty, the number of primary follicles has been reduced to about 34,000, the remainder having degenerated. Of these remaining follicles, only about 400 become mature ova in the life of a woman. Contrast this with the hundreds of millions of spermatozoa contained in a single ejaculate of the male.

Figure 26-16. **Secondary oocyte. The nucleus (nu) is large, and the nucleolus (no) has a definite structure. Many small vesicles have fused together (mv); cm = cell membrane and zp = zona pellucida. The corona radiata (cor) is beginning to form. (From J. Rhodin, *An Atlas of Ultrastructure,* W. B. Saunders Company, Philadelphia, 1963.)**

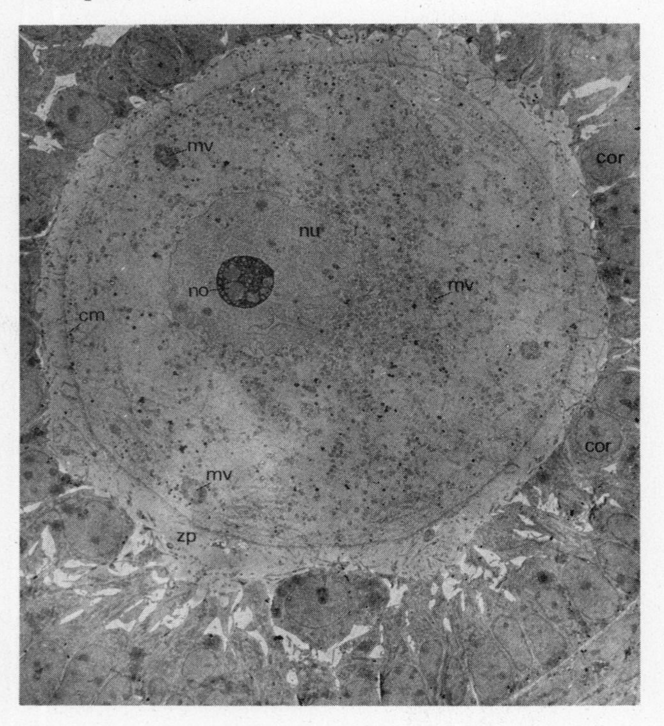

Secondary follicle (Graafian follicle)

Starting at puberty, under the influence of the gonadotropins, the primary follicle develops into the secondary follicle.

1. The *primary oocyte enlarges* to form the secondary oocyte, which is about twice the diameter of the primary oocyte. A thick viscous membrane, the *zona pellucida,* develops around it (Figs. 26-14 and 26-15).

2. The *granulosa cells proliferate* and form several layers of secretory cells.

3. The stroma of the ovary contribute another secretory layer, the *theca interna.* The granulosa and the theca interna are the cells that are capable of secreting estrogen and progesterone.

4. A connective tissue capsule, the *theca externa,* develops around the theca interna.

5. As the hormones are secreted into the follicle, a space, the *antrum,* is formed. The antrum is filled with the *sex hormones,* which surround the secondary oocyte, asymmetrically placed at one side of the follicle. These stages of follicular development are illustrated in Fig. 26-14, and the detailed structure of the secondary follicle is seen in Figs. 26-15 and 26-16.

Functions of the Secondary Follicles. The secondary follicles have the following functions:

1. They produce and expel the mature ovum, usually one each month during the reproductive years. This process is *ovulation.*
2. They produce *estrogen* during the time of growth of the ovum.
3. They are transformed into the *corpus luteum* (yellow body), which produces both *estrogen and progesterone* in the second phase of the ovarian cycle, after ovulation.

Oogenesis

Oogenesis is the growth and maturation of the oogonia into the ova. Like spermatogenesis, meiosis changes diploid germ cells to haploid gametes. However, in oogenesis, of the *four* daughter cells produced from each oogonium, *only one becomes the ovum.* The remaining *three polar cells* receive very little cytoplasm and degenerate.

If we follow the details of oogenesis a little more closely, we see that the nuclear distribution is the same as in spermatogenesis (Fig. 26-17).

Mitotic Divisions. In the eighth to twentieth weeks of intrauterine life, the oogonia divide rapidly by mitosis to form the many primary oocytes.

Meiosis Begins. In the third month of intrauterine life, the primary oocytes enter the long prophase of the first meiotic division that is going to last until puberty. The DNA content has been doubled, tetrads are formed, and crossing over may take place, but there is no distribution of the nuclear or cytoplasmic material.

Meiosis Continues. From puberty throughout the reproductive life of the female, meiosis continues. The two principal stages of meiotic division follow.

1. The first meiotic division is completed in the secondary follicle prior to ovulation. The primary oocyte divides into two daughter cells, one of which will be the *secondary oocyte;* the other is the *first polar body* or *cell.*

In this division, the tetrads are pulled apart, so that each nucleus receives two members of each tetrad. The nuclei are quantitatively the same, but differ qualitatively, depending on the source (maternal or paternal) of each chromosome.

The polar body can be seen beneath the zona pellucida of the secondary oocyte (Fig. 26-18).

2. Meiosis is completed in the oviduct, *only if the ovum* (secondary oocyte) *has been fertilized.* The response is very rapid; spindles are seen in both the secondary oocyte and the polar cell, and the chromosomes are redistributed without any duplication having occurred. Again, an unequal distribution of cytoplasm is responsible for the division of the secondary oocyte into a *mature ovum* and a *second polar body.* Two more tiny polar bodies are formed from the first one, so that *four haploid cells* result, only *one* of which is the mature ovum. As far as is known, the polar cells are never fertilized; or if they are, the zygote does not develop because the amount of cytoplasm is insufficient. It is interesting to speculate on the role of the cytoplasm and also the changes initiated by the entry of the sperm: How does the presence of a sperm nucleus in the secondary oocyte cause not only the division of the nucleus of the oocyte, but also that of the first polar body, which is on its surface?

Figure 26-17. **Oogenesis commences early in the intrauterine life (primary oocyte), continues from puberty through reproductive life (secondary oocyte), and is completed only by fertilization of the secondary oocyte by the sperm (mature ovum and the second polar body).**

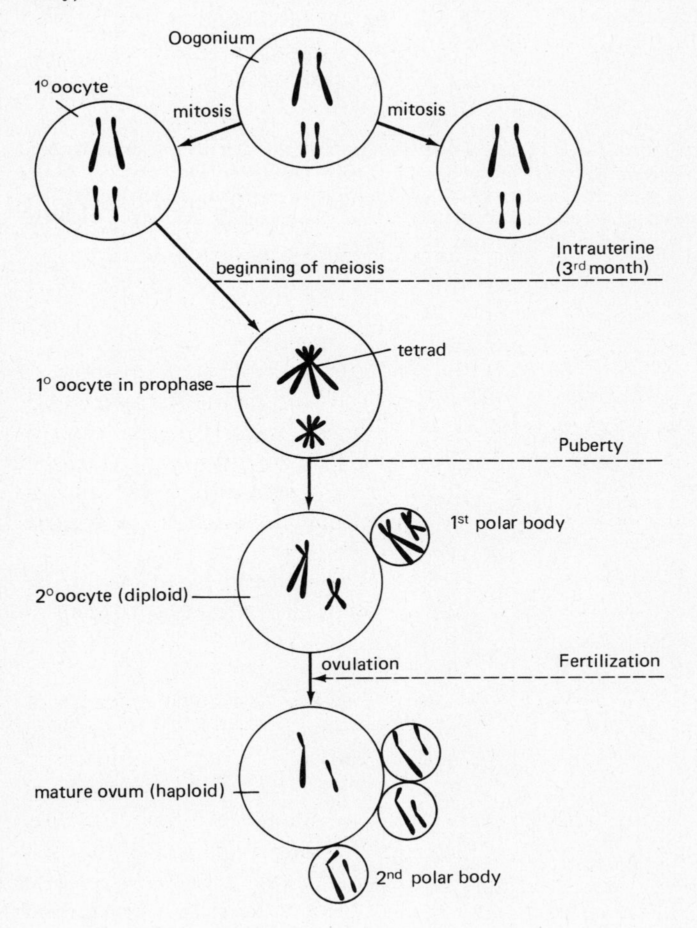

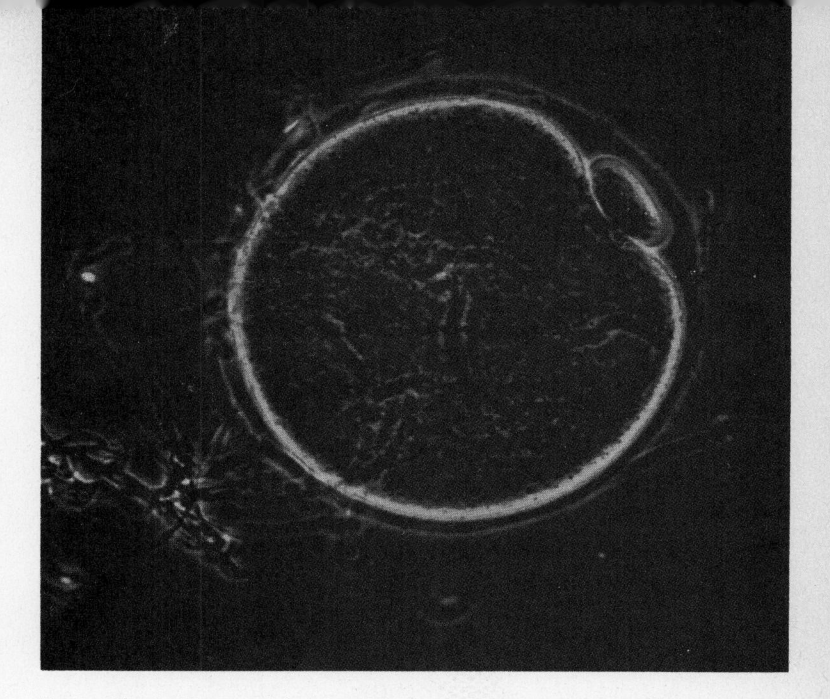

Figure 26-18. **Mouse ovulated ovum. The first polar body is clearly seen beneath the zona pellucida. Many sperm are attached to the zona pellucida. (From S. Suzuki, in *An Atlas of Mammalian Ova,* Igaku Shoin Ltd., Tokyo, Japan, 1973.)**

Preselection of Ovum

Normally, only one of the secondary follicles grows large enough to release an ovum. As the follicle grows, it can sometimes be seen as a bump on the surface of an exposed ovary. There are usually several of such follicles in various stages of development in the two ovaries, but the hormonal control that stimulates the most advanced follicle to grow and release its ovum causes at the same time a dissolution of the next largest follicle. This handicap limits ovulation to one ovum per month.

Ovulation

Ovulation has recently been observed in the human follicle. The events proceed rather slowly as the ovum is gently released from the ovary. First, the ovum and its surrounding granulosa cells, now picturesquely called the *corona radiata*

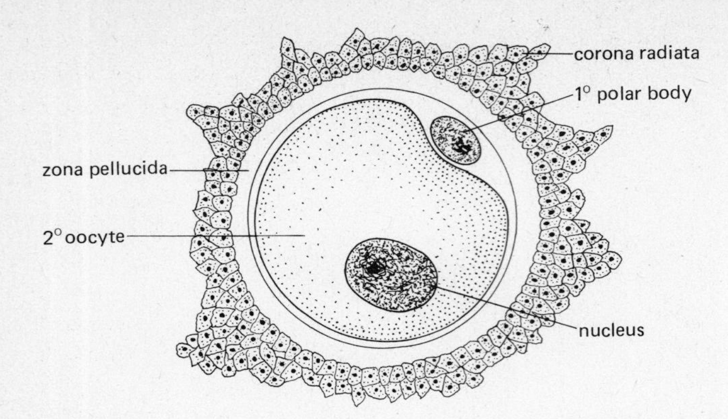

Figure 26-19. **The ovulated ovum with its corona radiata.**

(Fig. 26-19), become detached from the wall of the follicle. The fluid leaks out of the antrum, carrying the ovum to the surface of the ovary. The ovum is captured by the fringed ends of the oviduct and transported through this tube toward the uterus.

It is believed that changes in blood flow, causing necrosis, and perhaps the action of local proteolytic enzymes, are responsible for the rupture of the follicle. Another explanation is that the pressure inside the follicle builds up to burst it; but in the observed human ovulation, no explosive events have been noted.

Ovulation is dependent upon a surge in FSH and LH secretion by the anterior pituitary gland. The hormonal regulation of ovulation is discussed further in Sec. 26-8.

Corpus luteum formation

The empty follicle now undergoes a rapid change because of the growth and proliferation of the remaining, enlarged thecal cells, which become filled with yellow lipid, giving this structure its name, the *corpus luteum* (*luteus* = yellow). The corpus luteum also becomes very well vascularized and grows to about 1.5 cm. During this time, it secretes both estrogen and progesterone (Fig. 26-14).

The life span of the corpus luteum depends upon whether the ovum is fertilized or not. *If the ovum is not fertilized,* then the corpus luteum depends entirely upon the cyclic secretion of LH by the pituitary gland. In this case, the corpus luteum reaches its maximum development about 7 to 8 days after ovulation, then starts to involute, shriveling up and losing its yellow lipid. It is now the white *corpus albicans. If the ovum is fertilized* and implantation into the lining of the uterus occurs, the outer layer of cells of the early fertilized ovum, the *blastocyst,* produces a gonadotropin (*chorionic gonadotropin*) that has almost the same structure and function as pituitary LH. The chorionic gonadotropin then takes over the maintenance of the corpus luteum for the duration of the pregnancy.

26–6 Fertilization and Activation

Not only is the sperm vital for the final *maturation of the ovum,* but it has two other equally important functions. The first is to supply the genetic complement of nuclear material from the male (*fertilization*), and the second is *activation* of the egg, starting its development into the embryo.

Fertilizing Capacity of Sperm

The fertilizing ability of the sperm is increased (see Secs. 26–3, and 26–6, *capacitation*) during their stay in the oviduct, while the sperm await the ovum. It is assumed that there is some chemical secretion of the oviduct that capacitates the sperm, relatively few of which have survived the long and hazardous journey through the vagina and uterus and the lower parts of the oviduct (Fig. 26-12).

In *artificial insemination,* semen (sperm in their natural fluid) is injected with a syringe into the female reproductive tract. The sperm now avoid the hostile secretions of the cervix of the uterus; therefore only 10 per cent of the normal count of hundreds of millions of spermatozoa are needed for fertilization. This is sometimes helpful in cases of relative sterility, in which the sperm count is low.

Sperm lose their fertilizing capacity before they lose their motility. Human sperm remain fertile in the female reproductive tract from 28 to 48 hr, even though they may remain

Figure 26-20. **The cortical reaction of the egg to the penetrating sperm prevents other sperm from entering. The cortical cone engulfs the sperm head; microvilli are formed; and polysaccharides are released from granules in the cortex of the ovum, making the zona pellucida impenetrable.**

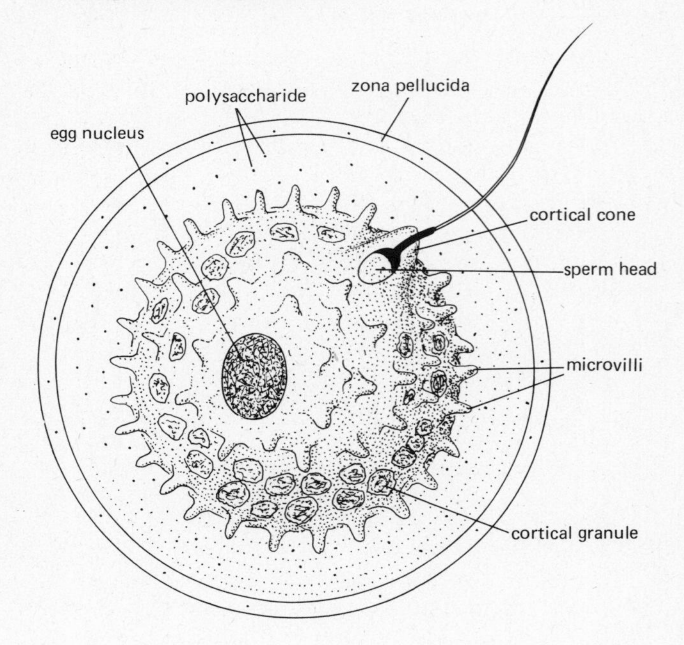

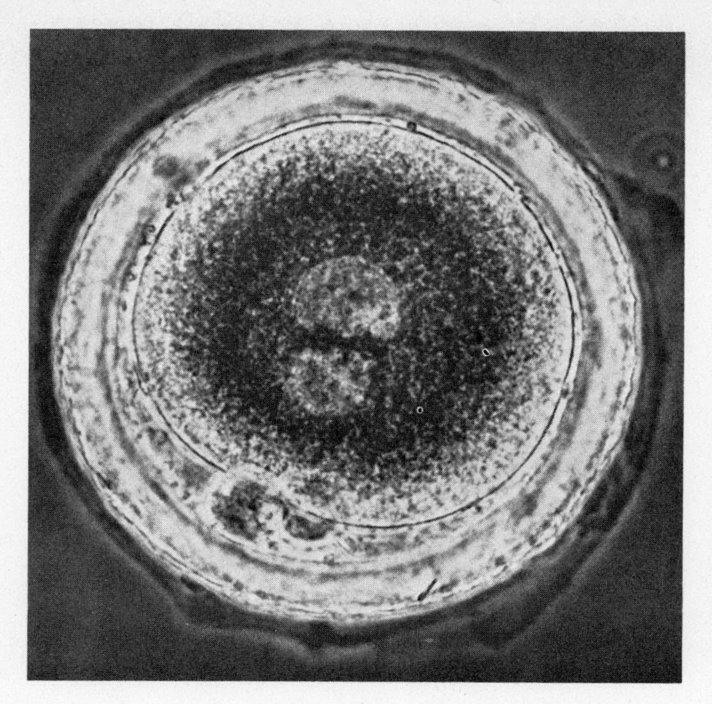

Figure 26-21. **The male and female pronuclei of a fertilized rabbit ovum are clearly seen in the cytoplasm of the egg. The polar bodies are visible beneath the zona pellucida. [From Shuetu Suzuki and Luigi Mastroianni, Jr., "In vitro fertilization of rabbit ova in tubal fluid," *Am. J. Obstet. Gynecol.* 93: 465–471 (1965).]**

motile for up to 60 hr, so that the ovum must reach the oviduct, where fertilization occurs, within 24 to 36 hr after mating.

Sperm Penetration into the Ovum

The spermatozoa must penetrate the acellular *zona pellucida* and the densely cellular *corona radiata,* which cover the mammalian ovum. When the ovum arrives in the oviduct, the high concentration of *bicarbonate* present in the oviduct secretions starts to loosen the corona cells. In vitro studies with rabbit ova, and more recently with human ova, have shown that the acrosomal part of the sperm head contains enzymes that are capable of dissolving both the acellular and the cellular components of these protective layers. The enzymes appear to be a complex consisting of *hyaluronidase* and a *trypsinlike enzyme.*

Ovum Reaction to Sperm Penetration

Cortical reaction

The sperm head is pulled into the cortical cytoplasm of the ovum by a growing cone that engulfs it. In most species, the tail is also incorporated into the egg cytoplasm. The cortical reaction includes the formation of microvilli and the release of polysaccharides from granules in the cortex (Fig. 26-20). This polysaccharide material alters the chemical composition of the zona pellucida so that no other sperm can enter, and polyspermy is thus prevented. At this time, there is a marked increase in the viscosity of the cortical cytoplasm.

Nuclear reaction

The act of penetration of the sperm into the ovum initiates the final stage of meiosis, transforming the secondary oocyte into the mature ovum and the second polar cell.

As the sperm nucleus is pulled into the ovum by the cortical cone, it loses the membrane that surrounded the sperm head. The naked nucleus now becomes hydrated, swelling to four or five times its original size; it is known as the *male pronucleus* (Fig. 26-21). It approaches the haploid nucleus of the ovum (the *female pronucleus*), and the two pronuclei fuse to form the *diploid zygote.* From now on, all the cell divisions that will result in the formation of the tissues of the embryo are mitotic.

Activation

The changes in the characteristics of the cortical cytoplasm initiated by the penetration of the sperm are called *activation.* Unless the egg is activated by the penetration of the sperm into its cytoplasm, it fails to develop further and rapidly degenerates. There appears to be a short period after ovulation when fertilization is most likely to occur successfully. Beyond this point—and it is only a matter of hours—the metabolic rate of the ovum falls so low that the penetration of the sperm cannot reverse the degenerative changes.

The activating role of the sperm can be taken over by other agents, such as pricking the egg with a needle or exposing it to cold or certain chemicals. This type of development is called *artificial parthenogenesis,* in distinction to *natural parthenogenesis,* the normal mode of development of certain members of insect societies, such as the worker ants. These different stimuli all suffice to initiate the nuclear and cytoplasmic divisions leading to the formation of the embryo, but of course this embryo will possess only the haploid number of chromosomes from the mother.

In mammals such as the rabbit, a small percentage of eggs has been induced to develop parthenogenetically, usually after exposure to cold. The resulting "fatherless" embryos develop into female rabbits, with one X chromosome. Like human females with an XO constitution (Sec. 27-4), the ovaries and reproductive tract are immature, and the parthenogenetic rabbits are sterile. Moreover, they are not as viable as normal females; one can conclude that although development of the entire organism is possible from the instructions received from only one set of chromosomes, the translation by the cytoplasm is not as dependable.

26-7 Reproductive Cycles in the Female

Unless implantation of the fertilized ovum occurs, the rhythmic development and regression of the ovarian follicle and the consequent changes in the rest of the reproductive tract reoccur each month, forming the *menstrual* (*monthly*) *cycle* characteristic of the primates. The overt sign is the bleeding associated with the degeneration and sloughing off of the uterine lining. The day on which bleeding is first noted is referred to as the first day of the menstrual cycle. Ovulation may occur any time between the tenth and twentieth day of the cycle, but usually takes place about the middle of the 28-day cycle, that is, on the fourteenth day (Fig. 26-22).

The reproductive cycle in lower mammals is called *estrus* and varies in several respects from the primate menstrual cycle. Bleeding is not associated with estrus, but an intense mating drive reappears with each estrus, and sexual receptivity is limited to this period of "heat," which usually occurs during the spring and early summer. Most wild mammals, such as wolves, squirrels, and foxes, complete an estrus cycle annually. Laboratory rodents have several estrus cycles a year, but wild rodents may not do so.

Despite the many differences between the menstrual and the estrus cycle, the basic hormonal regulation is very similar.

Figure 26-22. **Hypothalamic, anterior pituitary, and ovarian hormones during the menstrual cycle.**

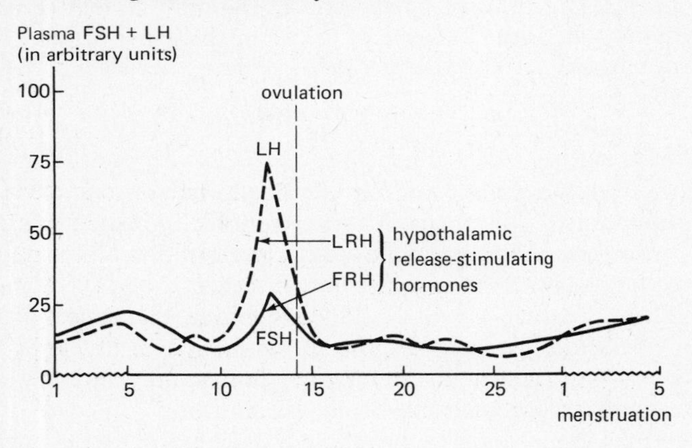

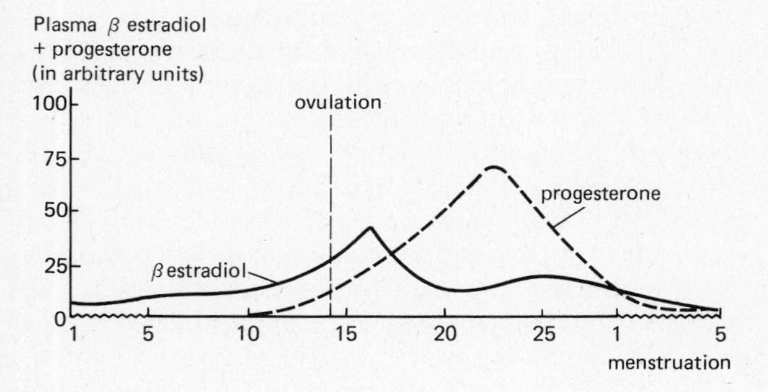

26-8 Hypothalamic and Anterior Pituitary Regulation of the Ovary

The growth of the secondary follicle, ovulation, and the secretion of the female sex hormones are all dependent upon hormonal stimulation by the anterior pituitary gonadotropins, which in turn are regulated by the hypothalamic secretions. The cyclic relationship among these three categories of hormones is shown in Fig. 26-22.

Cyclic Activity of the Female Hypothalamus

Like the male, the female hypothalamus secretes the gonadotropin-releasing hormones LRH and FRH, or a dual-acting LRH/FRH. However, the female hypothalamus has been programmed during development for cyclic secretion of the gonadotropin releasing hormones (see Sec. 11-2). This programming is reflected in the cyclic nature of the pituitary secretion of the gonadotropins.

Regulation of the Ovary by the Anterior Pituitary

First part of the ovarian cycle and estrogen secretion

The FSH and LH levels increase at the beginning of each month of the female sexual cycle—that is, the first day of menstruation. Under the influence of FSH, the follicle grows in size. The granulosa and the thecal layers thicken and secrete *estrogen,* which fills the antrum. This FSH effect is potentiated both by LH from the pituitary and by the estrogen secreted by the follicle itself.

Preovulatory surge of LH and FSH and ovulation

About 2 days prior to ovulation, there is a marked increase in gonadotropin secretion by the anterior pituitary, an increase that is especially evident in the almost tenfold increase in LH concentration. Under the influence of LH, the granulosa and thecal cells switch their secretion from estrogen to *progesterone.* Consequently, about 1 day prior to ovulation, estrogen levels drop while progesterone levels begin to rise. Ovulation is dependent upon these LH-mediated changes, and FSH probably plays a synergistic role.

It is LH, too, that limits the number of ova released to one, or at the most two, each month in the human being, by causing the dissolution of immature follicles in the ovary, while initiating the release of the most mature oocyte.

Anovulatory cycles

If the preovulatory surge of LH is not adequate, some cycles may be *anovulatory* (i.e., no ovulation occurs). Such anovulatory cycles occur most frequently at the beginning (puberty) and the end (menopause) of female reproductive life, when presumably LH production is not at its peak.

SECOND PART OF THE OVARIAN CYCLE AND PROGESTERONE SECRETION

The transformation of the ruptured follicle into the corpus luteum, as described in Sec. 26-5, is completely dependent upon LH. The secretion of progesterone by the corpus luteum also requires constant stimulation by LH. Progesterone secretion reaches its peak about 22 days after the beginning of the cycle and then declines rapidly as pituitary LH secretion diminishes. Unless the pituitary source of LH is supplanted by another source, such as the outer layer of cells of a newly implanted blastocyst, the corpus luteum will degenerate into the nonsecretory corpus albicans.

26-9 Feedback Regulation of Hypothalamic and Anterior Pituitary Secretion

The activity of the hypothalamus is affected by neural and hormonal influences. Some irregularities in the female sexual cycle can be correlated to stresses of various types, such as emotional stress, changes in altitude and time zones, and drugs of many kinds.

The normal function of the hypothalamus depends on the interplay among it, the anterior pituitary hormones, and the gonadal steroids. In the female reproductive cycle, this interplay appears to involve both a *negative* and a *positive* feedback circuit (Fig. 26-23).

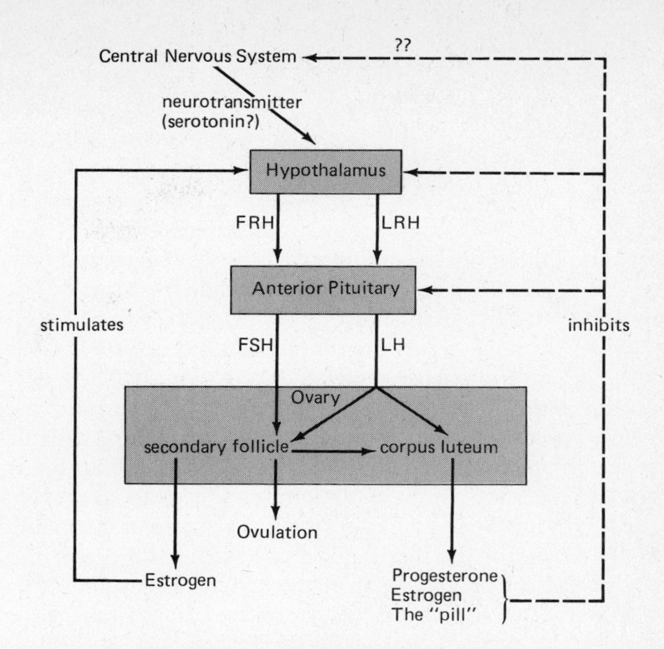

Figure 26-23. **Feedback relations among the hypothalamus, anterior pituitary, and ovary. High progesterone levels inhibit the hypothalamus and anterior pituitary (negative feedback), while high estrogen levels increase hypothalamic secretion of FRH and LRH (positive feedback). The contraceptive "pill," a combination of estrogen and progesterone, will also prevent ovulation by inhibiting LRH/FRH secretion.**

Negative Feedback

After ovulation, the rising levels of estrogen and progesterone produced by the corpus luteum feedback to *inhibit* hypothalamic secretion of LRH/FRH. There is probably also a secondary, direct inhibition by the sex steroids on the anterior pituitary. As LH levels diminish, the corpus luteum slows its secretory activity, permitting hypothalamic secretion to recommence just prior to the beginning of the next cycle.

Positive Feedback

The cause of the preovulatory surge of FSH and LH is not well understood, but it is probably due to a positive feedback mechanism. Estrogen levels peak about 2 days before ovulation and it is believed that the high estrogen level at that time *stimulates* FSH and LH secretion. Whether this positive feedback is exerted on the hypothalamic or anterior pituitary level is not known.

In some animals, like the rabbit, the act of *copulation* sets up nervous impulses to the hypothalamus, which then cause the release of LH from the pituitary gland. This rather efficiently ensures ovulation right after mating, a logical explanation for the noted fecundity of the animal.

Clinical Use of Gonadotropins to Stimulate Ovulation

In some cases, when infertility is due to failure to ovulate, the cause may be an insufficient secretion of gonadotropins. Ovulation can be induced by the administration of a gonadotropin obtained from the human placenta, *human chorionic gonadotropin* (HCG). Pituitary gonadotropins are not available on a large scale because of the great difficulty in extracting adequate amounts of the hormones from such a small organ.

Human chorionic gonadotropin has almost exactly the same effects as pituitary LH and will cause ovulation. However, unlike LH, HCG often causes the simultaneous release of several ova (*superovulation*), resulting in multiple births.

Use of Estrogen and Progesterone to Inhibit Ovulation—The "Pill"

With the synthesis in the early 1950s of orally effective steroids came the rapid development of the "pill," an *oral contraceptive.* The pill is a combination of estrogen and progesterone, which, if given in appropriate concentrations, will inhibit by negative feedback (Fig. 26-23), the preovulatory surge of LRH/FRH. Consequently, ovulation does not occur.

The most effective pill is a *combination* of estrogen and progesterone, which, taken for the first 21 days of the cycle, prevents ovulation but permits normal menstruation, as explained in the section on menstruation (Secs. 26–11 and 26–12). The *sequential pill,* in which first estrogen and then progresterone are taken, is not recommended; estrogen alone may have carcinogenic effects.

Many women experience unpleasant side effects of the pill, ranging from nausea and water retention to the dangerous formation of blood clots; therefore, new "mini-pills" are being tested which have only one third of the synthetic progesterone contained by the older type of pill. This lower amount does not inhibit ovulation, but it appears to prevent the sperm from penetrating through the mucus at the mouth of the cervix of the uterus. The mini-pill has a 3 per cent higher risk of pregnancy than the conventional pill.

Other ways of inhibiting ovulation are being tried. One of them is the development of drugs that may act as *hormone antagonists,* binding competitively to the receptors of the releasing hormones. These drugs would then act at the level of the anterior pituitary cell membrane.

There are many stages at which contraception may involve interference with reproduction. Some other methods of contraception are discussed in Sec. 26–18.

26–10 Ovarian Hormones

Estrogen: Origin, Metabolism, and Effects

ORIGIN OF ESTROGENS

The estrogens received their name in 1929, when Doisy and his co-workers found that extracts from the urine of pregnant women were able to produce estrus in immature female animals. There is a series of naturally occurring estrogens, differing slightly from one another in chemical structure and in potency. The most effective estrogen is *β-estradiol;* two somewhat less effective ovarian estrogens are *estrone* and *estriol.* Figure 26-24 shows the formulas for these three estrogens.

Figure 26-24. **Structural formulas of the natural estrogens, estradiol, estriol, and estrone. The β estradiol is the most potent of these hormones.**

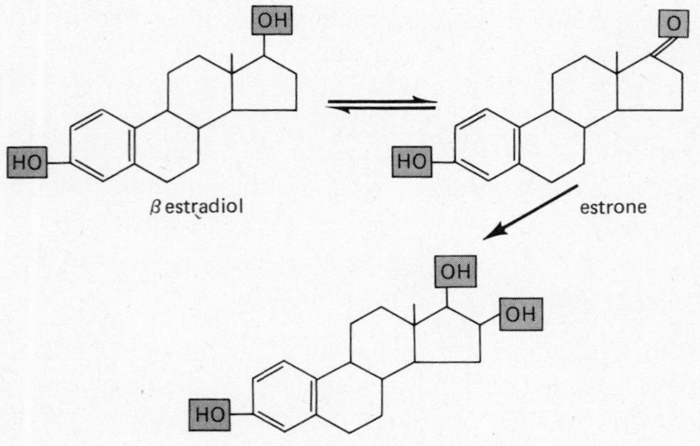

OH
HO
β estradiol
O
HO
estrone
OH
OH
HO
estriol

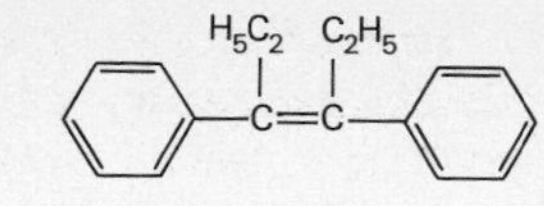

H_5C_2 C_2H_5
C=C
diethylstilbestrol (DES)

Figure 26-25. **The synthetic estrogen diethylstilbestrol is more potent than the natural estrogens, but because it is not physiologically inactivated, it has carcinogenic properties.**

The *ovary* is the main source of estrogens in the nonpregnant female, but during pregnancy the amount of estrogen formed by the *placenta* is considerably greater than that secreted by the ovaries. The *testes* of human males also secrete small quantities of estrogen, so that it is quite normal to find estrogenic metabolites in male urine. The urine of the stallion is a particularly rich source of natural estrogen. Minute amounts of estrogen are also secreted by the *adrenal cortex.* Because the sex steroids and the adrenal steroids are all synthesized from a common precursor, cholesterol (Sec. 10–7), this is not a surprising development.

SYNTHETIC ESTROGENS

Several different synthetic estrogens have been produced, some of which are more potent than the naturally occurring estrogens. One of these is *DES,* or *diethylstilbestrol* (Fig. 26-25), an extremely active estrogen even when taken by mouth, because, unlike the natural estrogens, DES is not inactivated by the liver. Because DES has been shown to increase the incidence of cancer in certain strains of animals, it is no longer permissable to use it to increase the efficiency with which farm animals are fattened for human consumption.

METABOLISM OF ESTROGENS

All the naturally occurring estrogens are steroids, similar in structure to cholesterol, and are synthesized from cholesterol or from acetate compounds (Sec. 10–7). The liver converts the active estradiol into less active forms, which are excreted in the urine. The digestive tract destroys the natural estrogens almost completely if they are administered orally, but the synthetic estrogens, such as diethylstilbestrol, can be absorbed undamaged.

EFFECTS OF ESTROGENS

The effects of the ovarian hormones are best demonstrated by the administration of such hormones to immature animals that are not producing significant amounts of these hormones themselves, or to castrated mature animals that have had the normal source of sex hormones removed. Similar but less clear results can be obtained by using hypophysectomized animals, because the removal of the pituitary gland causes the ovaries to atrophy so that very little hormone is synthesized. Clinically, the administration of hormones often supplements effectively an inadequate physiological production.

Growth of the Female Reproductive System. The estrogens stimulate tissue growth by increasing the number of *mitoses* and also by causing the accumulation of some water in the cells. These effects are most marked on the tissues of the reproductive tract, the external genitalia, and the mammary glands, so that normal growth and maturation of these structures are dependent on the secretion of adequate amounts of estrogens.

The monthly cyclic changes in the *endometrium* of the uterus reflect changes in estrogen (and progesterone) levels. These endometrial changes are discussed later in this chapter.

Stratification of the vaginal epithelium is particularly characteristic of estrogenic activity and is used as an assay technique, urine extracts being injected into test animals.

Estrogens cause the *growth of the mammary tissues* and the deposition of fat in the breasts, as effectively in males as in females. Normally, however, only the female mammary tissue is subjected to large amounts of estrogen, and the hormone is responsible for the growth of the tissue stroma, the deposition of fat, and the development of the duct system. Estrogen alone cannot cause milk production.

Fat Deposition. Estrogen regulates fat deposition; at puberty, the deposition of fat in the subcutaneous tissues increases, particularly in the breasts, the buttocks, and thighs.

Skin. Estrogens affect the epithelium of the skin, causing it to become soft and smooth, unlike androgens, which thicken and toughen it.

Skeleton. The skeleton is also affected by the growth-promoting proclivities of estrogen, and the osteoblasts in the female become remarkably active at puberty, causing a *sharp spurt in growth.* As might be expected, calcium and phosphate are retained during this period of growth. However, estrogen also stimulates *rapid union of the growth centers* of the bones, the epiphyses, with the shafts of the long bones, so that growth for girls ceases earlier than in boys. A special effect of estrogen on the bones of the *pelvis* causes them to flatten and widen, forming a broad outlet.

Mechanism of estrogen action

Like other steroid hormones, estrogen combines with specific receptors in the cytoplasm of its target organ cells (Sec. 10-8). The hormone-receptor complex enters the nucleus where it stimulates DNA transcription of RNA, rapidly increasing protein synthesis in the ribosomes as a result. A slower process of increased DNA synthesis is responsible for mitosis and cell proliferation.

Estrogens and malignant growth

Unlike androgens, which increase metabolic activity of tissues in general, estrogens affect only the metabolism of their specific target organs. The relationship between the growth-promoting activities of estrogens and their possible carcinogenic nature is a topic of much concern. Cancer can be induced by certain estrogens in specific strains of experimental animals, indicating that the genetic background and the hormonal environment interact to produce the undisciplined mitosis characteristic of malignant growth.

Progesterone: Origin, Metabolism, and Effects

Origin of progesterone

In the nonpregnant female, the *corpus luteum* is the most important site of production of the steroid hormone progesterone, although the *adrenal cortex* probably produces small amounts as well. During pregnancy, large amounts of progesterone are produced by the *placenta* after about the fourth month of gestation. Progesterone is closely related chemically not only to the estrogens and androgens but also to the steroid hormones of the adrenal cortex, so that administration of progesterone sometimes produces effects similar to those obtained from the corticoids and androgens. It may be that the tissues convert progesterone into the other similar hormones, rather than utilize the progesterone itself.

Metabolism of progesterone

The liver is responsible for the inactivation of progesterone to *pregnanediol,* which is excreted in the urine. The amounts of pregnanediol increase considerably during pregnancy and can be correlated to the amounts of progesterone produced in the body. Because this form of the hormone is quite inactive biologically, it can be measured only by chemical analysis, not by biological assay. The structural formulas of progesterone and pregnanediol are shown in Fig. 26-26.

Figure 26-26. **Structural formulas of progesterone and pregnanediol.**

progesterone

pregnanediol

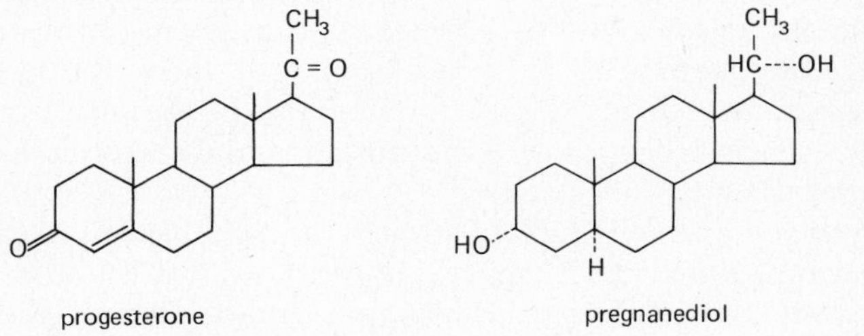

The digestive tract completely inactivates progesterone so that it must be administered clinically by injection, or the synthetic *pregneninolone* may be given orally. Its effects are not quite the same as those of the natural hormone.

EFFECTS OF PROGESTERONE

Uterine Development. The main functions of progesterone are to *increase the secretory activity and vascularization* of the uterus and to *decrease its motility.* Corner (1929) showed, by destroying the corpora lutea in pregnant animals, that progesterone was the hormone necessary for *implantation* and for the *maintenance of the fetus.* However, progesterone can act only on uterine tissue that has previously been stimulated by estrogen; estrogen and progesterone work synergistically to obtain the optimal development of the reproductive tract.

Mammary Glands. Progesterone and estrogen also function synergistically in the development of the mammary glands, progesterone continuing the development of the tissues initiated by estrogen. These changes in the mammary glands are discussed further in Sec. 26-15.

Fluid Balance. Because progesterone is so similar to the adrenocortical hormones, it also has a profound effect on fluid balance, causing the kidneys to retain water and the tissues to accumulate fluid. This retention is especially noticeable during the last part of the menstrual cycle, when the progesterone level is high, and, of course, during the latter months of pregnancy, when a considerable part of the weight gained may be caused by fluid retention.

26-11 Menstruation

If no blastocyst implants after the uterus has been so sedulously prepared first by estrogen and then by progesterone, the corpus luteum degenerates, and both *estrogen and progesterone levels drop* (Fig. 26-22). The highly coiled arteries (Sec. 26-12) become constricted, and the outer tissues they supply receive no blood and die; in the meanwhile, small hemorrhages occur in the deeper layers. These morphological changes can be observed by using a technique whereby an endometrial transplant from a monkey is placed in the transparent anterior chamber of its eye. The transplant rapidly becomes vascularized and undergoes the same cyclic changes as the rest of the uterus.

The presence of the blood and necrotic tissues inside the uterus, together with the absence of progesterone, stimulate the uterine muscles to contract and expel these discarded structures into the vagina, forming part of the menstrual flow: a disturbing or welcome indication to the woman that pregnancy has not occurred.

Blood that flows rapidly through the blood vessels of the endometrium clots normally, but the blood that has seeped slowly through the degenerating endometrium has been exposed to the action of an anticoagulant or *fibrinolysin,* and normally does not clot. If excessive bleeding of the endometrium occurs, not enough fibrinolysin is available, and blood clots appear in the menstrual flow. Together with the debris and blood are found large numbers of leukocytes, and their appearance may be responsible for the great resistance to infection that the raw endometrium has at this time.

The hormonal control of menstruation is clearly indicated by the fact that the surgical removal of both ovaries is followed by menstruation; the sudden drop in the estrogen and progesterone level is the causative factor. The menstrual cycle can be mimicked by the appropriate administration and withdrawal of estrogen and then by progesterone, but of course normal ovulatory cycles cannot be reestablished.

26-12 Hormonal Regulation of the Uterus

Menstrual Cycle

The special ability of the uterus to adapt for pregnancy begins on the first day of the menstrual cycle and may be credited to the extreme sensitivity of the tissues and arteries of the endometrium to changes in estrogen and progesterone levels.

The endometrium has a *dual arterial supply* consisting of *deep arteries* that continuously maintain the blood supply to the deeper layers, and of *coiled arteries* that supply blood to the middle and superficial endometrium (Fig. 26-27).

THE PROLIFERATIVE PHASE (ESTROGEN DOMINANT)

At the beginning of the cycle, the endometrium is less than 1 mm thick; most of it has been sloughed off (*desquamated*) during the previous menstruation. Only a few deep epithelial cells and glands remain. Under the influence of estrogen, the *glands of the endometrium increase* in number and secretory activity, and the *epithelial cells proliferate* and become pseudostratified (Fig. 26-27).

The *coiled arteries* grow rapidly in response to estrogen, twisting and coiling more as the cycle progresses. As a result of all this growth, the thickness of the endometrium increases considerably, so that at the time of ovulation it is between 3 and 5 mm thick. *Contractions* of the myometrium also increase under the influence of estrogen.

THE LUTEAL PHASE (PROGESTERONE DOMINANT)

In the second part of the cycle, following ovulation, when progesterone is the dominant sex hormone, a tremendous increase in *growth and coiling of the superficial arteries* occurs (Fig. 26-27), in preparation for the increased blood supply necessary for an implanting blastocyst. The stroma becomes loose and edematous, which facilitates implantation. The *endometrial glands* are also extremely active, pro-

resulting debris. Then, as the mesoderm joins with the trophoblast to form the shaggy, fingerlike *chorion* (Figs. 2-26 and 26-29), a real system of vessels is begun. The branches of the chorion fan out to form the twigs of the arterioles and capillaries deep in the endometrium, so that ultimately only two extremely thin membranes separate the blood of the mother from that of the embryo.

This is the beginning of the fetal part of the *placenta*, that richly vascular endocrine organ which provides:

1. The enormous *surface area* of membranes through which nutrients pass to the embryo and wastes return to the maternal circulation.
2. The *hormones* necessary for the maintenance of the embryo within the uterus.

As the extraembryonic mesoderm follows the trophoblastic projections, proper villi containing blood vessels are formed (Fig. 26-29). These are the *chorionic villi* (the chorion is formed by the double-layered trophoblast and mesoderm; see Sec. 2-5), which start producing hormones even before the blood vessels become functional.

Uterine response to implantation

Stimulated by this cannabalism, the maternal tissues react by forming a swelling around the invader. This is really a nonspecific response: tiny glass beads implanted in the uterus during this part of the cycle will also cause the same tumorlike *decidual response* (Fig. 26-28). The endometrium continues to develop both its glands and its circulation, especially in the region where the embryo lies. There is also an increase in the size and strength of the muscle fibers. This is the *maternal contribution to the placenta* (Fig. 26-30).

Figure 26-30. **Formation of the placenta by the extraembryonic membranes and the maternal endometrium.**

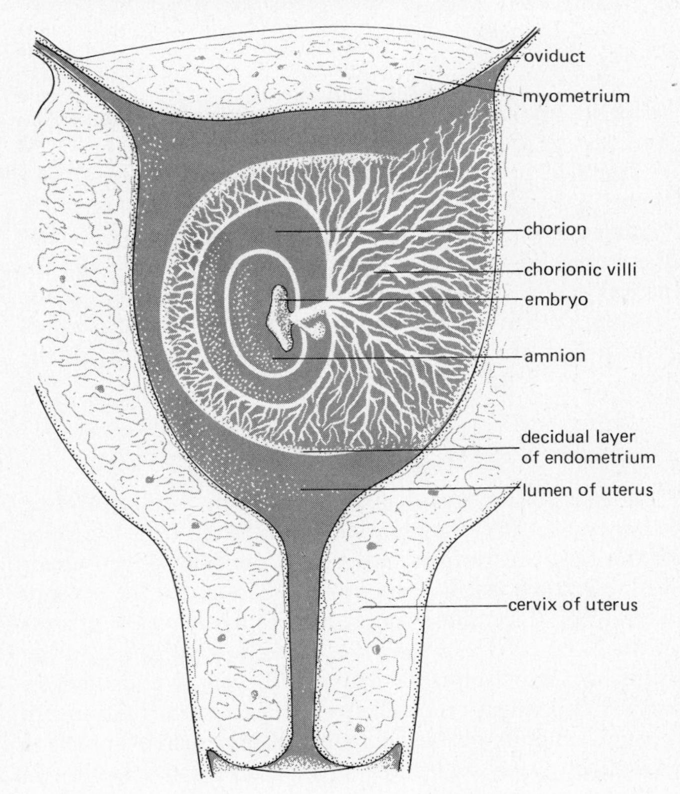

It has been stressed that implantation and the formation of the placenta are possible only in the uterus that has been prepared first by estrogen and then by progesterone. The maintenance of pregnancy also requires the continued production of these hormones, particularly progesterone. This means that the corpus luteum must remain active or that another source for hormonal production must be found. The very early *chorion* is responsible for secreting *gonadotropins* that continue the stimulation of the corpus luteum, because by this time the pituitary supply of LH is waning (Fig. 26-31).

The maternal tissues contribute to the placenta in response to the continued hormonal stimulation. The endometrium develops rapidly; its glands and blood vessels increase in size and complexity until, toward the end of *gestation* (the time spent by the developing embryo in the uterus), the placenta is surprisingly large.

Functions of the Placenta

In most mammals, the placenta has evolved into a highly specialized endocrine organ, as well as the organ necessary for the interchange of materials between the mother and the fetus.

Interchange of materials

The fetal blood is pumped into the umbilical artery through the chorionic vessels, which then distribute it to the capillaries that penetrate the blood-filled spaces of the endometrium. The fetal circulation was described in Sec. 13-6.

Nutrients. The fetus is nourished by materials that pass from the maternal blood through the two delicate separating membranes into the fetal blood. Similarly, oxygen diffuses from the mother to the fetus, with carbon dioxide passing in the opposite direction, so that the placenta acts as a *respiratory membrane* similar to the one in the lungs.

Diffusible Wastes. The fetus passes back diffusible wastes into the maternal circulation, so that the mother's kidneys excrete both her own wastes and those of the fetus. The kidneys of the fetus do not become independently active until after birth.

Large Particles. Large particles such as proteins and bacteria usually cannot get through the placenta, which acts as a somewhat variable barrier. However, certain proteins, such as antibodies, do get through and are responsible for the buildup of maternal antiRh agglutinins, which in turn

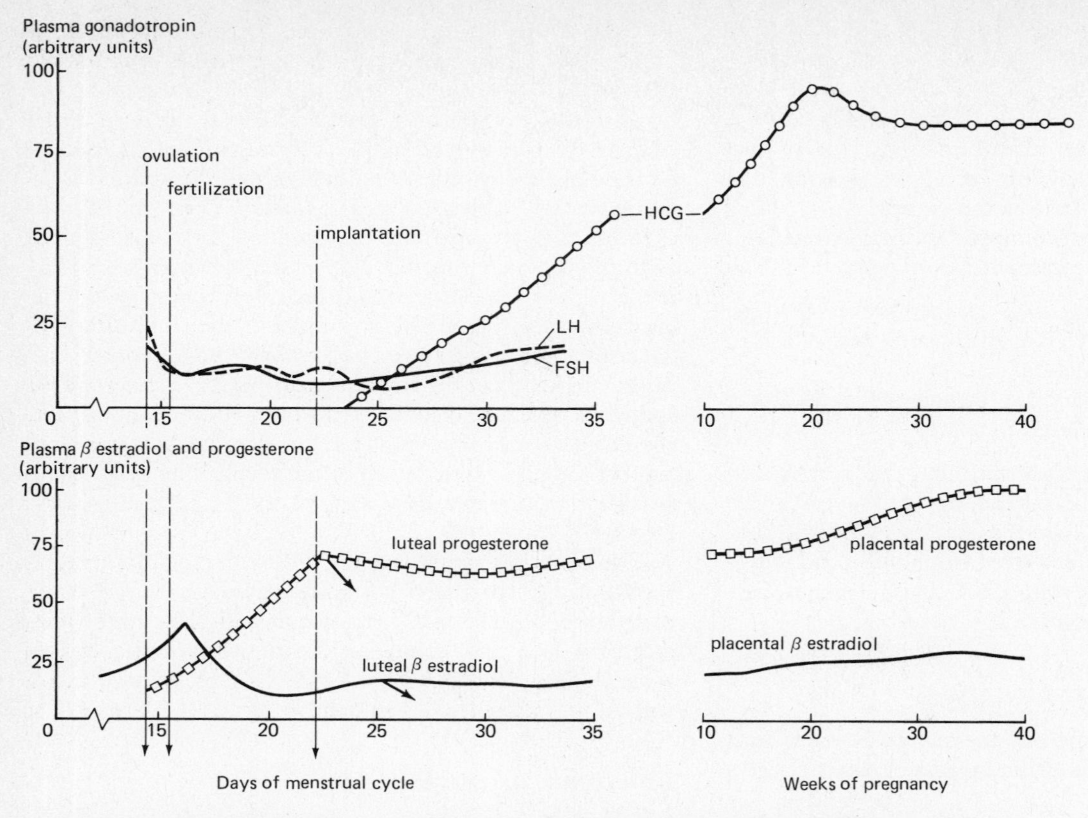

Figure 26-31. **Effect of pregnancy on the hormone levels illustrated in Fig. 26-22. Human chorionic gonadotropin (HCG) secreted by the trophoblast stimulates the corpus luteum to continue secreting progesterone, although pituitary LH has dropped. As the placenta develops, it takes over the production of progesterone and estrogen. The arrows indicate the drop in luteal hormone secretion that would have occurred if no blastocyst had been implanted. Low levels of estrogen and progesterone would then result in menstruation.**

pass back through the placenta and destroy the red blood cells of the fetus (Sec. 14-16).

The newly born mammal is also immune to certain diseases to which the mother has built up an antibody reserve, because these antibodies have passed into its circulation. This *passive immunity* only lasts for a few months after birth.

Hormones of the placenta

The placenta produces both peptide and steroid hormones. The gonadotropin produced by the chorion of the human placenta, human chorionic gonadotropin, supplements the activity of the pituitary gonadotropin to maintain the corpus luteum. In the first few weeks of pregnancy, HCG is secreted in amounts large enough, however, to maintain pregnancy even after hypophysectomy.

The human placenta can also supplant the hormonal secretions of the ovaries because the placenta produces so much *estrogen* and *progesterone* (Fig. 26-31). This is not true of all mammals, most of which remain dependent upon the pituitary gonadotropins and ovarian steroids. There seems to be an evolutionary trend in primates, in which the placenta becomes able to replace both the pituitary gonadotropins and ovarian steroids.

The rich hormonal content of human (and horse) placentas is a relatively inexpensive source of hormones for commercial use. Even the amniotic fluid (the bag of waters) that surrounds the developing fetus is being collected for incorporation in many "rejuvenating" cosmetics, on the assumption that it contains substances that will be as effective in nurturing aging tissues as it was in nurturing the fetus.

Pregnancy Tests

Chorionic gonadotropins are produced by the implanting blastocyst so early that they can be extracted from the urine of the pregnant woman and used as the basis of *pregnancy tests.* They are usually concentrated enough by the seventh week to produce blood spots (corpora lutea) on the ovaries of mice and rats (*Aschheim-Zondek test*). If pregnancy urine is injected into rabbits, ovulation will occur, a phenomenon usually dependent on copulation (*Friedman test*). Strangely enough, amphibia are much more sensitive than mammals to pregnancy urine, and a very inexpensive and reliable test depends on the release of sperm from the African toad after injection with a small amount of extracted pregnancy urine.

26–14 Parturition

The time of gestation is comparatively constant within one species but varies considerably among species. In humans and certain apes it is 38 weeks, the elephant is well known for its long gestation period of 20 months, and the cat and dog produce offspring in 9 weeks. Many factors share the responsibility for the expulsion of the uterine contents at the proper time, but physiologists still do not have a really adequate explanation for this complex phenomenon.

Distention of the Uterus

When the uterus has been distended to a certain size by the size and weight of the fetus (or fetuses), it appears to be ready to respond to the nervous and hormonal stimuli that will cause the birth contractions. The larger the litter within the uterus, the smaller will each individual be at birth and the less mature. In the human uterus, which usually contains only one fetus, this fetus is relatively large and well developed at birth. Twins are often born prematurely, since the uterine contents reach their maximum weight earlier.

Hormonal Regulation

Progesterone, which quiets the contractions of the uterus and so is essential for the maintenance of pregnancy, probably decreases toward the end of the gestation period.

Estrogen, which stimulates the myometrium to contract, is produced in increased amounts and in a more active form toward the end of pregnancy. The action of these two hormones apparently prepares the uterus to respond to two other hormones, *oxytocin* and *relaxin.*

Oxytocin, a hormone that is produced by the hypothalamus and probably also by the placenta, increases uterine contractions, *provided the uterus has been sufficiently distended and the levels of estrogen and progesterone are suitable.* Under these conditions, oxytocin has been successfully administered to speed labor that is unduly prolonged. Oxytocin is ineffective if these conditions are not met.

Relaxin is a hormone found in the blood only during pregnancy. It relaxes the pubic ligaments and also relaxes the lower part of the uterus, causing the cervix to dilate and so eases the passage of the offspring at birth. Because relaxin also increases the sensitivity of the uterus to oxytocin, it is sometimes given with oxytocin to induce labor. It is not quite clear whether the ovary or the placenta produces relaxin; it disappears quickly from the blood after parturition.

Prostaglandins are extremely effective in causing uterine contractions (Sec. 10-9). Prostaglandins are found in the blood of women during parturition and appear to act synergistically with oxytocin. Karim (1970) was the first to demonstrate the efficiency of prostaglandins in inducing labor and has recently reported that these hormones are also effective orally.

Prostaglandins are also used for second-trimester abortions, but much higher dosages are required than at the end of the normal gestation period, probably because there is no synergistic action with oxytocin until this time.

Positive Feedback in Uterine Contractions

Under the influence of the hormones we have discussed, and when the uterus is fully distended by the full-term fetus, sudden strong uterine contractions occur that are called *labor contractions.* These contractions, which stretch the cervix and push the baby through the birth canal, are accompanied by severe pain.

It is believed that once the contractions start, they elicit a series of stronger contractions until they are strong enough to expel the baby. There are at least two ways through which this positive feedback may be operating. First, the increased pressure of the baby's head against the cervix with each contraction excites the fundus or body of the uterus to contract. The second possibility is that there is a reflex secretion of oxytocin in response to cervical irritation. The oxytocin released would stimulate uterine contractions.

Labor

Once the *uterine contractions* have begun, they gradually increase in both frequency and force, until they occur every 1 or 2 min in the final stages of labor. Before this, however, there is a period of from 12 to 18 hr when the cervix is being dilated. *Dilation of the cervix* is probably helped by the force of the head of the baby pushing against it, especially after the *fetal membranes* (*bag of waters*) *have been ruptured* by the increased pressure within the uterus. The fluid that pours out is the *amniotic fluid.*

After the cervix has been completely dilated, there is a short period when the uterus is quieter, and then the last stages of labor begin (Fig. 26-32). The forceful contractions at this time are accompanied by severe pain, because the uterine musculature goes into spasm, and the birth canal is being torn. This pain reflexly causes the *abdominal muscles* to contract, adding to the tremendous pressure downward on the baby (about 25 lb during each contraction).

Ultimately, the baby is forced through the vagina and "delivered," but the placenta still remains behind, attached to the baby by the umbilical cord. For a few minutes after the birth, blood still pulses through the umbilical vein, but then the pulsation stops, and because no more blood will be delivered to the baby, the cord is tied and severed. The stump shrivels up after about a week, leaving a depressed scar, the *umbilicus* or *navel.*

About half an hour after the birth of the baby, a second series of uterine contractions expels the placenta, which has been separated from the rest of the uterus. It is a thick disk, weighing more than a pound, and is called the *afterbirth.*

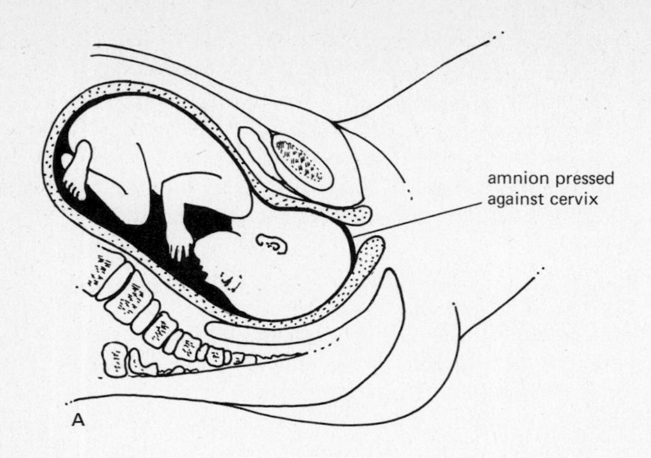

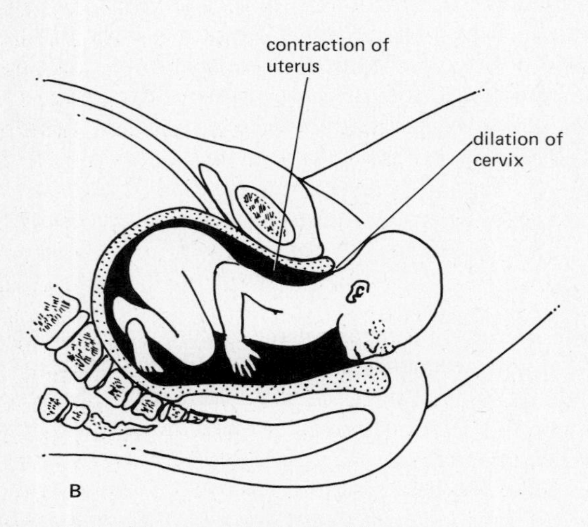

Figure 26-32. **Two stages in parturition.**

26–15 The Mammary Glands and Lactation

The mammary glands or breasts are present in a rudimentary form in both sexes but develop normally in females under the influence of the female sex hormones estrogen and progesterone. Each gland is composed of about 20 *lobules* that radiate from the nipple and are drained by *lactiferous ducts.* These ducts extend toward the *areola,* the pigmented area around the nipple, and open at the tip of the *nipple* (Fig. 26-33*A*).

Puberty

At puberty, with the sudden increase in estrogen and progesterone, there is a rapid deposition of fat in the breasts, giving the developing girl the contours characteristic of her sex, but there is no extensive change in the duct system until pregnancy occurs.

During Normal Female Cycle

During the normal female cycle, some swelling of the breasts and tenderness in the second part of the cycle are often associated with the increased amounts of *progesterone* produced by the corpus luteum. Progesterone increases the amount of water in the tissues, and turgidity of the breasts and "bloatedness" of the abdominal region are the result. As much as 3 to 4 lb of water may be accumulated at this time. When the corpus luteum involutes, if pregnancy does not occur, the amount of progesterone produced falls sharply, and the swollen tissues lose their water. (The mammary glands in the male can be made to develop very similarly with the appropriate administration first of estrogen, then progesterone.)

Pregnancy

If pregnancy does occur, the continuously increasing amounts of estrogen and progesterone cause the *ducts* to increase greatly in development and, in addition, transform the breasts into glands capable of secretion [Fig. 26-33(B)]. The ends of the ducts develop into little *sacs* or *alveoli,* lined by secretory epithelium [Fig. 26-33(C)], but the secretion of milk does not begin until parturition has occurred.

Figure 26-33. **Changes in a mammary gland during pregnancy and lactation. A sagittal section of the breast is shown in (A). The development of the alveoli during pregnancy is shown in (B). Secretion of milk by the alveolar cells (C) occurs after parturition and is under hormonal control, as illustrated in Fig. 26-34.**

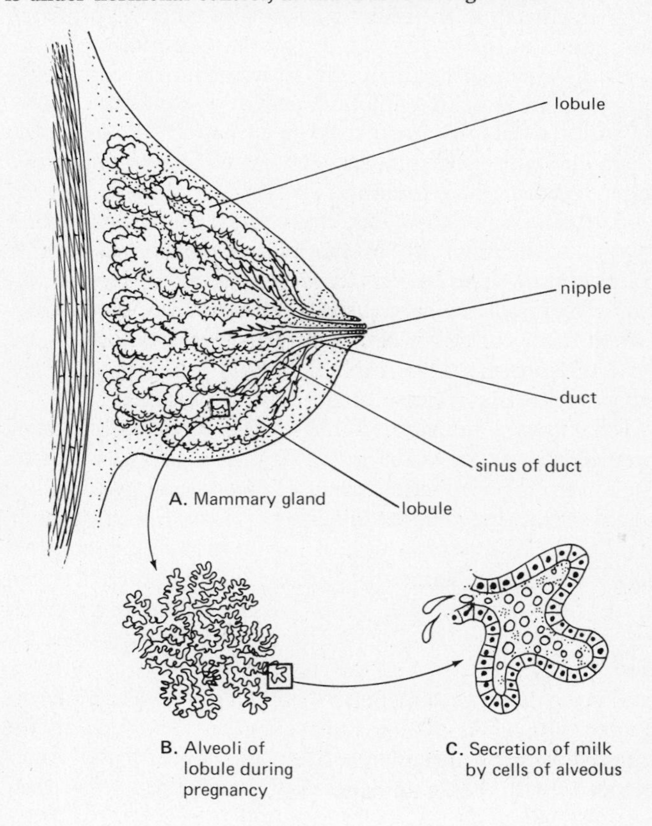

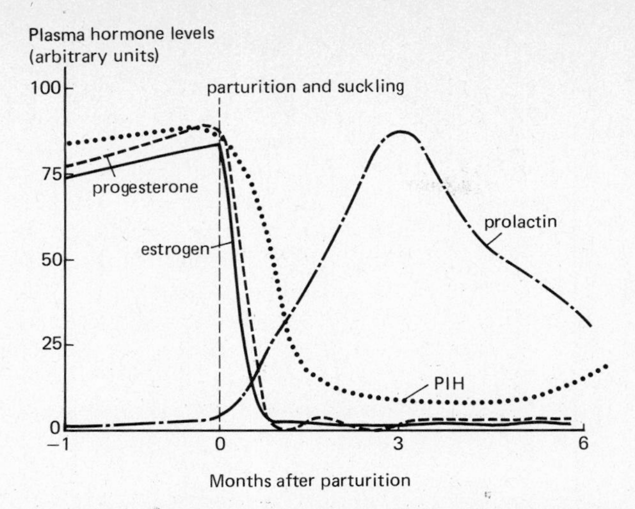

Figure 26-34. **Postparturition hormonal changes that result in lactation. As the levels of estrogen and progesterone drop, so does the secretion of the hypothalamic prolactin release-inhibiting hormone (PIH). This permits the anterior pituitary gland to secrete prolactin, which results in lactation.**

Hormonal Regulation of Lactation

Lactation, or milk production, requires a complex of hormones, including estrogen, progesterone, prolactin, thyroxine, insulin, and the adrenal cortical hormones. The last three hormones are indirectly involved through their effects on metabolism. Estrogen, progesterone, and prolactin are directly involved in lactation.

Hypothalamic inhibition of prolactin secretion

The anterior pituitary normally produces very little prolactin because the pituitary is continuously inhibited by the prolactin release-inhibiting hormone (PIH) from the hypothalamus (Sec. 11-3). If the stalk between the anterior pituitary and the hypothalamus is cut, or if the anterior pituitary is transplanted, most pituitary secretions cease, except for prolactin secretions, which rise.

The removal of this inhibitory hypothalamic action occurs just after parturition and is probably caused by nervous stimulation of the hypothalamus induced by suckling. This nervous inhibition of PIH release has to be superimposed upon a well-developed secretory system in the mammary glands *and* a drop in estrogen and progesterone secretion.

Relationship between prolactin secretion and estrogen and progesterone

After the expulsion of the placenta, the level of estrogen and progesterone immediately drops. Apparently, the sustained high levels of these hormones during pregnancy have caused the marked growth and development of the secretory structures of the mammary glands but at the same time inhibited the production of milk. Once this *inhibition is removed,* and if *suckling* begins, the anterior pituitary starts to secrete *prolactin,* which results in lactation a few days after parturition. Figure 26-34 shows the hormonal changes that result in lactation following parturition.

During lactation, the continued production of prolactin diminishes the secretion of the gonadotropins, so that little estrogen or progesterone is produced, causing the rapid return of the uterus to its nongravid size and form. It is unusual for the sexual cycle to reestablish itself during the nursing period, so that conception is unlikely to occur during this time, although it is by no means unknown.

The Composition of Milk

Human milk is made up of about 88 per cent water, 6.8 per cent milk sugar (lactose), 3 per cent protein, and a small amount of salts. The fat droplets that make up about 3 per cent of its composition are suspended in this watery fluid. Human milk contains considerably more lactose than cow's milk and much less protein. Because almost $1\frac{1}{2}$ liters of milk may be produced by a lactating woman each day, the drain on the metabolism is acute and must be replaced by a diet high in calcium and milk.

Milk ejection is the emptying of the milk from the alveoli into the large ducts, a process that requires the actual contraction of the myoepithelial cells in the walls of the alveoli and the smaller ducts. Unless the breasts are emptied regularly, the secretion of milk stops, and the mammary glands dry up. *Suckling* by the baby is an important stimulus for milk ejection; not only does the baby remove the milk, it also sets up a reflex secretion of the hypothalamic hormone *oxytocin.* Vasopressin is also secreted, but the most important hormone for milk secretion is oxytocin. The receptors for this *milk ejection reflex* are in the nipple and the nerve pathway runs to the hypothalamus. Oxytocin is released from the supraoptic and paraventricular nuclei of the hypothalamus, by way of the posterior pituitary, into the circulation to reach the mammary glands (Fig. 26-35). Oxytocin causes the contraction of the smooth muscles around the alveoli and ducts, squeezing out the milk. This reflex is

Figure 26-35. **Reflex regulation of milk ejection.**

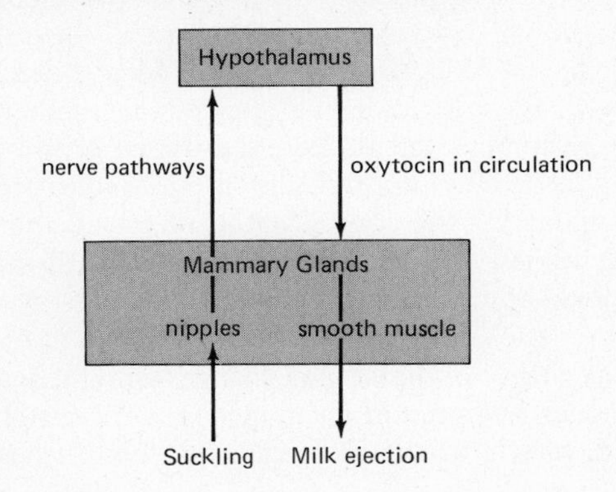

easily inhibited by psychological disturbances that result in the drying up of the breasts.

26–16 The Female Sexual Act

Penetration of the penis into the vagina is greatly facilitated by the secretions of the mucous glands beneath the labia minora. This lubricating action is under parasympathetic control, from nerves in the sacral plexus that pass to the external genitalia.

The parasympathetic nerves are also responsible for the engorgement with blood of the erectile tissue of the clitoris and the area around the opening of the vagina. Stimulation of these nerves occurs both from reflexes set up as a result of *local stimulation* of the erectile tissue and from *psychic stimulation.* It is very likely that these two types of sexual stimuli act synergistically to maximally arouse the female. In addition, local stimulation of the breasts, especially the nipples, will enhance sexual sensations and vaginal lubrications.

Orgasm in the Female

Although no phenomenon in the female is comparable to ejaculation in the male, the other physiological changes characteristic of extreme sexual excitement are essentially the same. The rhythmic movements of the erect penis inside the vagina, and the friction of the penis against the clitoris, set up a series of positive feedback reactions that enhance sexual response. When this local stimulation is reinforced by appropriate psychological stimulation, the same visceral and muscular reflexes are evoked as in the male, culminating in the *orgasm,* with its heightened sexual sensations.

The reflex orgasm is followed by feelings of pleasant fatigue and relaxation. Orgasm in women is apparently a much more graded and varied phenomenon than in men, and probably more dependent upon psychic factors.

26–17 Menopause (Female Climacteric)

Menopause literally means the cessation of the menses, or the end of the menstruation, and signifies that the rhythmic secretion of the hormones by the ovary is slowly coming to a halt. By the age of 45 or 50 years, almost all the follicles of the ovary have produced ova that have been expelled from the ovary, or the follicles have degenerated. In other words, there are no follicles left to secrete estrogen or to be transformed after ovulation into the progesterone-producing corpus luteum. As the concentration of estrogen and progesterone wanes, the pituitary gland puts forth more and more follicle-stimulating hormone but to no avail (Fig. 26-36).

The physiology of the woman changes with the loss of the sex hormones. A common phenomenon is the instability of the blood vessels of the skin, causing "hot flashes." The nervous system is affected, and the menopause may be accompanied by symptoms ranging from increased irritability all the way to psychotic states. Adaptation to the changed hormonal state usually occurs after a few years, except in extreme cases. The menopause shows clearly that the endocrine and nervous systems are closely interwoven in their integrative action in the body.

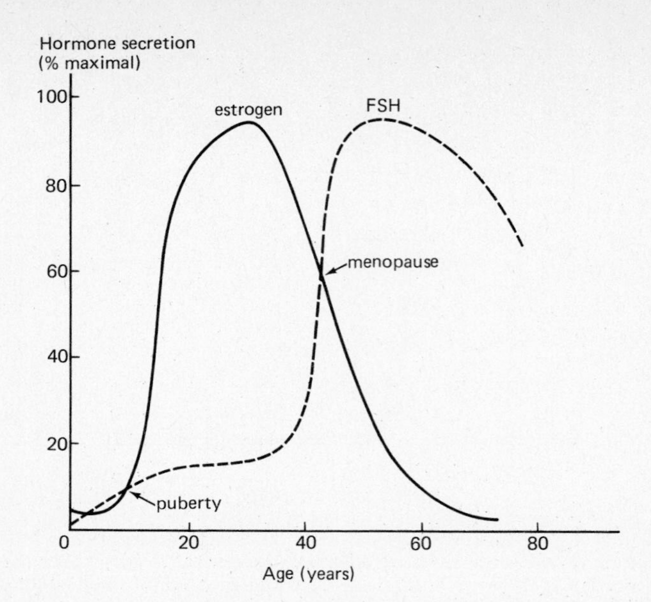

Figure 26-36. **Secretion of FSH and estrogen during the female life cycle.**

26–18 Contraception

The pioneering work of Margaret Sanger of the Planned Parenthood Federation, in cooperation with Gregory Pincus (1965) of the Worchester Foundation for Experimental Biology, laid the basis for many of the advances in methods of contraception. Although the perfect contraceptive has not yet been found, there are a number of sites at which the reproductive process may be interrupted.

Inhibition of Ovulation

Contraceptive devices such as the "pill," which are based on hormonal inhibition of ovulation, were discussed in Sec. 26-9. The administration of progesterone by nasal spray may inhibit ovulation in much smaller doses than the "pill."

Blocking Sperm Entry into the Cervix

The age-old way to prevent pregnancy has been for the male to *withdraw* prior to ejaculation. This method is psychologically unsatisfactory to both the male and female.

Sperm may be prevented from entering the cervix by a *condom,* worn by the male, or by a *diaphragm,* inserted in the female. The condom has an advantage over other contraceptive techniques in that it also protects against venereal disease (Fig. 26-37).

Surgical interruption of either the oviducts or the vas deferens (*sterilization*) is the most successful contraceptive method, but it is usually irreversible. Attempts to perform a *reversible vasectomy,* with a valve in the vas that can be opened or closed, have had variable success.

Avoiding the Fertile Period: The Rhythm Method

The rhythm method depends upon periodic abstinence in order to avoid coitus around the time of ovulation. The unreliability of this method (it fails about 25 per cent of the time) is caused by the inadequacy of knowledge of the exact time of ovulation. This method could possibly be improved by the oral administration of LRH, which induces ovulation and thus would permit a woman to control the time of her ovulation.

Prevention of Implantation: IUD

The intrauterine device (IUD) is a plastic or copper form (T, Y, or ring) that is inserted into the uterus. It probably prevents implantation of the fertilized ovum, although this assumption is not certain. The IUD is not tolerated by a fairly large number of women, especially those who have never had a baby.

Figure 26-37. **The Swedish Association for Sex Education, a non-governmental organization, uses the Kondom symbol to publicize the advantages of the condom.**

Combination Action: IUDs with Progesterone

An IUD can be coated with a natural progesterone, which appears to either prevent sperm capacitation or to prevent implantation. The natural progesterone is used because it is rapidly inactivated by the uterine lining and so cannot cause disturbing changes elsewhere in the body.

Postcoital Contraceptives: The "Morning-After Pill"

Estrogens

Estrogens are effective as postcoital contraceptives, probably because they speed the passage of the fertilized ovum through the oviduct so that it reaches the uterus before the endometrium has been prepared in the luteal phase of the cycle. However, the very large doses of estrogen that are needed for this effect produce unpleasant side effects. DES, the synthetic diethylstilbestrol, is the most effective for this purpose, but because of its carcinogenic properties cannot be used routinely. Its use is confined to emergencies, such as rape.

Luteolytic agents

Agents that will destroy the corpus luteum needed to maintain early pregnancy, are possible "morning-after" candidates. Prostaglandins may be potential postcoital contraceptives but at the moment are successfully used only to induce abortions fairly late in pregnancy, because they are potent stimulators of uterine contractility.

Immunological Methods

Preliminary experiments indicate that antigens specific to sperm can be used to immunize both males and females against the sperm that carry the antigen. This effect appears to be reversible.

Oral Contraceptives for Males

The synthetic compounds that suppress spermatogenesis also suppress male libido, a side effect pleasing to neither males nor females.

Cited References

Corner, C. W., and W. M. Allen. "Physiology of the Corpus Luteum: Production of a Special Uterine Reaction (Progestational Proliferation) by Extracts of the Corpus Luteum." *Amer. J. Physiol.* **88:** 326 (1929).

Karim, S. M. M., and G. M. Filshie. "Therapeutic Abortion Using Prostaglandin F_2." *Lancet* **1:** 157 (1970).

Pincus, G. *The Control of Fertility.* New York, Academic Press, Inc., 1965.

Additional Readings

BOOKS

AUSTIN, C. R., and R. V. SHORT, eds. *Reproduction in Mammals Series,* Books 1-5. Cambridge University Press, New York, 1970.

BARRINGTON, E. J. W. *An Introduction to General and Comparative Endocrinology,* 2nd ed. Clarendon Press, Oxford, 1975. (Paperback.)

BRANDES, D., ed. *Male Accessory Sex Organs.* Academic Press, Inc., New York, 1974.

BRIGGS, M. H., and M. BRIGGS. *Biochemical Contraception.* Academic Press, Inc., New York, 1975.

COUTINO, E. M., and F. FUCHS, eds. *Physiology and Genetics of Reproduction.* Basic Life Sciences Series, Vol. 4. Plenum Publishing Corporation, New York, 1974.

DUCKETT-RACEY, J. D., ed. *The Biology of the Male Gamete.* Academic Press, Inc., New York, 1975.

DUNCAN, E. W., E. J. HILTON, P. KREAGER, and A. A. LUMSDAINE, eds. *Fertility Control Methods.* Academic Press, Inc., New York, 1973.

FRIEDEN, E., and H. LIPNER. *Biochemical Endocrinology of the Vertebrates.* Foundations of Modern Biology Series. Prentice-Hall, Inc., Englewood Cliffs, N. J., 1971. (Paperback.)

GREEP, R. O., and E. B. ASTWOOD, eds. *Handbook of Physiology.* Sec. 7, Vol. 4, Part 2. The Williams & Wilkins Company, Baltimore, 1974.

GUYTON, A. C., ed. *MTP International Review of Science: Physiology,* Vol. 8. University Park Press, Baltimore, 1974.

HAFEZ, E. S. E., and T. N. EVANS. *Human Reproduction.* Harper & Row, Publishers, New York, 1973.

HARRISON, R. G. *A Textbook of Human Embryology.* Blackwell Scientific Publications, Ltd., Oxford, 1963.

JAMES, V. H. T., and L. MARTINI, eds. *The Endocrine Function of the Human Testis.* Academic Press, Inc., New York, 1974.

KAWAKAMI, M., ed. *Biological Rhythms in Neuroendocrine Activity.* Igaku Shoin, Tokyo, Japan, 1974.

LARSON, B. L., and V. R. SMITH. *Lactation.* Academic Press, Inc., New York, 1974.

LLOYD, C. W. *Human Reproduction and Sexual Behavior.* Lea & Febiger, Philadelphia, 1964.

MASTERS, W. H., and E. V. JOHNSON. *Human Sexual Response.* Little, Brown and Company, Boston, 1966.

MOGHISSI, K. S., and E. S. E. HAFEZ, eds. *Biology of Mammalian Fertilization and Implantation.* Charles C Thomas, Publisher, Springfield, Ill., 1972.

ODELL, W. D., and D. L. MOYER. *Physiology of Reproduction.* C. V. Mosby Co., St. Louis, Mo., 1971.

PAGE, E. W., C. A. VILLEE, and D. B. VILLEE. *Human Reproduction.* W. B. Saunders Company, Philadelphia, 1972.

PHILLIPS, D. M. *Spermiogenesis.* Academic Press, Inc., New York, 1974.

SANDLER, M., and G. L. GESSA, eds. *Sexual Behavior—Pharmacology and Biochemistry.* Raven Press, New York, 1975.

SEGAL, S. J., R. CROZIER, P. A. CORFMAN, and P. G. CONLIFFE, eds. *The Regulation of Mammalian Reproduction.* Charles C Thomas, Publisher, Springfield, Ill., 1973.

TEPPERMAN, J. *Metabolic and Endocrine Physiology.* Year Book Medical Publishers, Chicago, 1962.

TIMIRAS, P. S. *Developmental Physiology and Aging.* Macmillan Publishing Co., Inc., New York, 1972.

TURNER, C. D. *General Endocrinology,* 5th ed. W. B. Saunders Company, Philadelphia, 1971.

WHEELER, R. G., G. W. DUNCAN, and J. J. SPEODEL, eds. *Intrauterine Devices.* Academic Press, Inc., New York, 1974.

ARTICLES

ALLEN, R. D. "The Moment of Fertilization." *Sci. Am.,* July 1959.

ARMSTRONG, D. T. "Reproduction." *Ann. Rev. Physiol.* **32:** 439 (1970).

BEER, A. E., and R. E. BILLINGHAM. "The Embryo as a Transplant." *Sci. Am.,* Apr. 1974.

BRACKETT, B. G. "Mammalian Fertilization in Vitro." *Fed. Proc.* **32:** 2065 (1973).

CLERMONT, Y. "Kinetics of Spermatogenesis in Mammals: Seminiferous Epithelium Cycle and Spermatogonial Renewal." *Physiol. Rev.* **52:** 198 (1972).

CSAPO, A. "Progesterone." *Sci. Am.,* Apr. 1958.

DONOVAN, B. T. "The Role of the Hypothalamus in Puberty." *Prog. Brain Res.* **41:** 239 (1974).

FINN, C. A., and L. MARTIN. "The Control of Implantation." *J. Reprod. Fertil.* **39:** 195 (1974).

GOLDBERG, V. J., and P. W. RAMWELL. "Role of Prostaglandins in Reproduction." *Physiol. Rev.* **55:** 325 (1975).

GOMES, W. R., and N. L. VAN DEMARK. "The Male Reproductive System." *Ann. Rev. Physiol.* **36:** 307 (1974).

JACKSON, H. "Chemical Methods of Male Contraception." *Amer. Sci.* **61:** 188 (1973).

KEYE, W. R., JR., B. H. YUEN, and R. B. JAFFE. "New Concepts in the Physiology of the Menstrual Cycle." *Clin. Endocrinol. Metabol.* **2:** 451 (1975).

KISTNER, R. W. "The Menopause." *Clin. Obstet. Gynecol.* **16:** 106 (1973).

OWEN, J. A., JR. "Physiology of the Menstrual Cycle." *Amer. J. Clin. Nutr.* **28:** 333 (1975).

SCHALLY, A. V., A. J. KASTIN, and A. ARIMURA. "The Hypothalamus and Reproduction." *Amer. J. Obstet. Gynecol.* **114:** 423 (1972).

SULMAN, F. G. "Hypothalamic Control of Lactation." *Monogr. Endocrinol.* **3:** 1 (1970).

TAYMOR, M. L. "Induction of Ovulation with Gonadotropins." *Clin. Obstet. Gynecol.* **16:** 201 (1973).

WILSON, J. D. "Recent Studies on the Mechanism of Action of Testosterone." *New Engl. J. Med.* **287:** 1284 (1972).

Chapter 27

Regulation of Inheritance

I am the family face;
Flesh perishes, I live on,
Projecting trait and trace
Through time to times anon,
And leaping from place to place
Over oblivion.

The years-heired feature that can
In curve and voice and eye
Despise the human span
Of durance—that is I;
The eternal thing in man,
That heeds no call to die.

Thomas Hardy, "Heredity"

27-1 Gene Regulation

The gene is the fundamental unit of heredity. Genes are arranged in linear order in the chromosomes, each gene having its own position or *locus* on the chromosome. The structure of the gene, as first proposed by Watson and Crick in 1953, accounts for three important characteristics of genetic material: gene *specificity*, gene *replication*, and gene *mutation*.

A gene is a length of several hundred or thousand base pairs of DNA. These base pairs, held together by phosphate-sugar bonds, form two chains, which are coiled into a double helix. The sequence of the base pairs in the DNA molecule is the *genetic code* by which information is relayed by way of RNA to the cytoplasmic ribosomes, where protein synthesis occurs (see Sec. 3-3.2). Because the genetic code is the blueprint for this activity, the genes determine the *specific enzymes and proteins* that will be synthesized by a cell at a particular time.

Replication of the gene is achieved by the unwinding of the two chains, which permits each chain to act as a template for the synthesis of the other. In this way, an exact replica of each chain is produced.

A gene mutation may be as slight as an alteration in the base pair sequence of a particular gene, or it may be as drastic as a deletion or duplication of part of the gene sequence. Mutant genes at the same loci are *alleles*. Some mutations may occur without a corresponding, overt (phenotypic) change; on the other hand, some gene mutations are lethal.

Inherited differences between individuals of the same species may be due to different combinations of alleles. When the differences are expressed as abnormalities, however, in physical, physiological, or mental characteristics, they are described as an *inherited disease*. Some inherited diseases, such as sickle cell anemia and hemophilia, have been mentioned in Secs. 14-7 and 14-15; these and other inherited diseases are discussed in more detail in this chapter. More than 2000 genetic diseases are now known, and more are discovered each year. Fortunately, most of these disorders affect a very small proportion of the population.

Social Implications of Gene Regulation

New genetic techniques, developed in the last decade, can be applied to improve human genetic qualities. We are suddenly faced with decisions over which we previously had no control. Should parents have a right to produce a defective child knowingly? Should a genetically defective fetus be aborted? Do parents have the right to select the sex of their child?

Genetic engineering has resulted in the *transfer of genes* from animals into bacteria, which then reproduce the foreign DNA as well, and as rapidly, as their own DNA. The potential benefits from this transfer of genetic material into the efficient biochemical factory of the bacterium include the large-scale production of scarce hormones, such as insulin and growth hormone. The future dangers include the transformation of innocuous bacteria into new and potentially virulent strains, against which there is no protection.

Even more revolutionary is the *formation of totally new combinations of DNA* segments, which are assembled from different organisms and which can give rise to completely unknown characteristics. A recent exciting development is the ability to *synthesize a complete gene,* which functions in living cells as does the natural gene. It is completely feasible, therefore, that missing genes may be supplied to those individuals lacking them in their inherited genetic endowment. These aspects of gene regulation are discussed in more detail at the end of this chapter.

27-2 The Distribution of Genes and the Formation of a New Individual

The Formation of a New Individual

The instant of fusion of the sperm nucleus with that of the ovum initiates a fascinatingly complex series of reactions which ultimately are designed to bring to full development an *individual* of the same *species* as the parents, bearing all the characteristics of his progenitors but individually different in a multitude of small ways. At this important instant, the genetic sex has also been determined.

Although the zygote has received chromosomes from both parents, it has the same number as they, because in the formation of the gametes, each gamete received only *half the number* of chromosomes possessed by the parent cell. The distribution of the members of the homologous chromosome pairs is on an unprejudiced, random basis, so that the *gametes vary qualitatively* from each other. This process of cell division, by which the gametes receive half the number of chromosomes, distributed at random in terms of parental origin, is called *meiosis* (Secs. 26-3 and 26-5).

Fertilization is the means of restoring the full complement of chromosomes to the young zygote and at the same time encouraging a random assortment of characteristics from both parents, leading to a great deal of diversity in the offspring. It has been calculated that the possible genetic combinations from one pair of human parents is somewhere in the neighborhood of 8 million.

Once the fate of the zygote has been cast by the particular pair of gametes that have united in its formation, the delicately controlled steps of embryonic development commence. As the one-celled fertilized egg divides into two, then into four, and steadily on until a mass of cells has been formed, each new daughter cell receives the *same number* and the *same quality* of chromosomes as its parent cell. This process of cell division is *mitosis.* There is no more assortment of characteristics; these cells are replicas of one another. Yet some areas of the embryo develop into the nervous system, others into muscle or bone, others into the digestive system or the excretory system. What has caused this diversity of paths? What are the factors responsible for the specific differentiation of the cells into tissues and organs after a certain stage of development has been reached? What regulates the time at which these cells become functional? These are some of the timeless riddles that experimental embryology is slowly solving. We trace the influence of the information, coded into the structure of DNA within the chromosomes, upon the metabolism of the cell; the influence of this cell upon its neighboring cells; the diffusion of embryonic organizers and hormones that direct the development of distant zones in the embryo; and the gradual increase in the stability of the cells as they settle down as permanent tissues, having lost their youthful exuberance and ability to develop in any direction. The cell has formed a new, mature individual, and we have come full cycle.

The Formation of a New Species: Evolution

In the early nineteenth century, August Weismann emphasized the importance of the *germ plasm* in heredity. The germ plasm consists of those cells destined to give rise to the gametes; only these cells will transmit the characteristics of the parent to the offspring, and only changes to these cells will be reflected in subsequent generations. This theory is in direct contrast to other theories, popular at that time, which were based on the concepts of the French biologist Lamarck. Lamarck believed that organisms evolved because of their ability to adapt themselves to their needs in response to changes in the environment; these changed characteristics are called *acquired characteristics* and certainly develop in a wide variety of circumstances. Sustained use of the muscles builds them up, whereas disuse results in their atrophy; but it does not necessarily follow that the son of a lumberjack will be born with well-developed muscles. If he takes up his father's occupation, he will develop his musculature similarly; if he becomes a pianist, he will improve the control and strength of his fingers; if he becomes a sedentary businessman, he will probably develop a paunch. Hundreds of generations of Jewish boys have been circumcised, yet Jewish boys are still born with a prepuce. Generations of Chinese girls have had their feet bound according to tradition, but this ordeal had to be endured anew by each generation; the daughter did not inherit the tiny feet of her mother.

Weismann crudely but dramatically—by cutting off the tails of many succeeding generations of mice—showed that acquired characteristics are not inherited; baby mice still continued to be born with tails of normal length. How then do organisms adapt to their environment and transmit these adaptations to their progeny in such a way as to account for the evolution of different groups and species and families?

Darwin's *Origin of Species,* published in 1859, explained the development of the vast number of different plant and animal species on the basis of *variation* and the *selection* of the most suitable variants for survival. In other words, those organisms whose variations made their survival easier would survive to transmit these beneficial variations to their offspring. When we combine "the survival of the fittest" with the realization that acquired characteristics are not inherited, it is obvious that a selection is being made from chance mutations in the germ plasm. A *mutation* is a change in the structure of the gene, and it may be beneficial to the organism, harmful, or even lethal, or else of no certain significance. The mutations that are of little significance will continue to be transmitted without affecting the offspring particularly; the more harmful ones will be eradicated through their weakening or lethal effect on the next generations. Sometimes harmful genes survive when their effects are masked by the presence in the zygote of other dominant genes.

Another very important factor in the evolution of species is that the organisms that are undergoing mutation must be separated from one another for a long enough period of time for a wide spectrum of different genes to be produced. Otherwise, interbreeding will distribute the genes equally so that only one common species results. This type of *isolation* may be *geographic* (the most dramatic example of this is the development of the marsupials in Australia), or it may be *genetic,* when a mutation results in two different groups that are sterile when intermated. The *races of human beings* have resulted from many small mutations and *geographic and cultural isolation* that prevent interbreeding, but these genetic changes have not resulted in interracial sterility.

27–3 Genetics—The Science of Inheritance—Mendel's Laws

Strangely enough, it is this very complicated field of biological investigation that perhaps deserves most clearly to be called a science. There are laws and rules that can be applied to many situations, and we even have a fairly good idea as to how these rules are established and how they affect development—a great deal more than can be said for many other aspects of biology. In genetics the effects of the laws were discovered first by extremely disciplined experimentation by the Austrian monk Gregor Mendel in 1866. Mendel did his famous series of cross-matings in peas, *controlling one variable factor at a time,* until he was able to get mathematic ratios from which he deduced his two laws of inheritance.

At that time, there was no knowledge of chromosomes and their distribution by meiosis to the gametes, but Mendel deduced from his results that the only logical explanation was that certain inherited factors must be transmitted to the gametes in such a way that each gamete received only half the number of factors present in the parent cell. Second, he deduced that these factors were distributed at random to the gametes, so that the gametes varied in the type of factor they received. And, very important, the union of these factors once again by fertilization resulted in the formation of the new individual with the original number of factors but with no blending of the factors themselves. The factors interact but can be separated cleanly, uncontaminated, to be distributed at random once more to the next generation of gametes.

The brilliance of Mendel's work has been supported by later discoveries that showed that these factors of inheritance are on the chromosomes within the nucleus, and the distribution pattern of the chromosomes in oogenesis and spermatogenesis is responsible for the nature of the laws of inheritance. Then detailed studies of the chromosomes showed that these factors occupy specific sites, in a definite linear order along the chromosome. These sites are the genes.

Following the Inheritance of a Single Characteristic

Mendel's procedure was to cross plants that differed in one readily apparent characteristic, such as flower color or seed shape. These original plants are referred to as the parent generation (P_1); if self-fertilized, they always produce the same kind of plant—they "breed true." However, when these plants are cross-fertilized, they produce plants (the first filial generation or F_1) that may show the *characteristics of only one of the parent plants or characteristics intermediate to those of the parents.*

Cross-fertilization between plants of the F_1 generation produces the really interesting results, because it is here that the effects of *independent assortment* of the genes during meiosis and the *lack of blending* of genes in the zygote become apparent. This generation is called the second filial generation (F_2).

Let us take two examples of plant crosses:

1. Mendel's classic experiment with pea plants in which he crossed a pure-breeding, red-flowered pea with a pure-breeding, white-flowered pea.
2. A cross between pure-breeding red and white snapdragons.

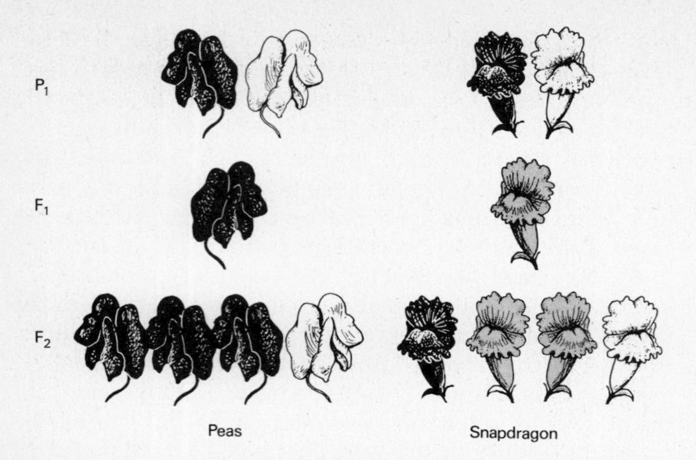

Figure 27-1. **Complete and incomplete dominance. A comparison of the results obtained by crossing pure red with pure white flowers in the pea (complete dominance of red over white) with results obtained in the snapdragon (incomplete dominance).**

Figure 27-1 shows the results of these experiments. The F_1 pea plants are all red; the F_1 snapdragons all pink. The F_2 peas are produced in a ratio of three red-flowering plants to one white-flowering plant, whereas the ratios in the F_2 snapdragons consist of one red-flowering plant, one white-flowering plant, and two pink-flowering plants.

Despite the apparent difference in the ratios of the plants in the F_2 generation, exactly the same *genetic laws* are illustrated:

1. In the formation of the gametes, half the number of factors are distributed to each gamete at random, so that the gametes vary qualitatively from one another, and quantitatively from the parents.
2. These factors can be separated uncontaminated after they have been brought together by fertilization, to be redistributed in the formation of the next generation of gametes.

Complete and incomplete dominance

Genes occupying the same location on homologous chromosomes affect the same characteristic by their influence on the chemical reactions of the cells. Such genes are *alleles,* and one allele may activate enzyme reactions necessary for the synthesis of pigment, for example, whereas another may suppress these reactions. When such *heterozygous* alleles are present in the same cell, if the effect of one overrides that of the other, the effective gene is *dominant,* and the repressed gene is *recessive.* The genetic composition is called the *genotype;* the resulting characteristics form the *phenotype.*

In the pea, the gene for the development of the red flower is completely dominant over the gene for white. In the snapdragon, however, the two allelic genes modify one another in their effect on flower color, so that when both are present, they produce an intermediate color, pink. This is *incomplete dominance.* This is seen clearly in the F_1 generation in Fig. 27-1.

In neither case is the gene itself altered by its allele, because in the next cross, the F_2 generation shows that the genes for red and white come through uncontaminated (Fig. 27-1). The only logical explanation for the ratios obtained is as follows.

The Pea. Using the symbol *R* for the dominant gene for red and *r* for the recessive gene for white, we find that the pure-red parent generation must be *homozygous* (both alleles are the same) for R. The pure-white parent generation must be *homozygous* for r. (Fig. 27-2)

The Snapdragon. The distribution of genes to the gametes and the F_1 and F_2 generations is exactly the same as for the pea; the only difference lies in those cases in which there is a heterozygous genotype, that is, in which red and white alleles are both present. This gives rise to an intermediate amount of pigment formation so that a third phenotype develops, pink flowers (Fig. 27-3).

Following the Inheritance of More Than One Characteristic

The inheritance of eye color in human beings is regulated chiefly by one pair of genes responsible for the amount of

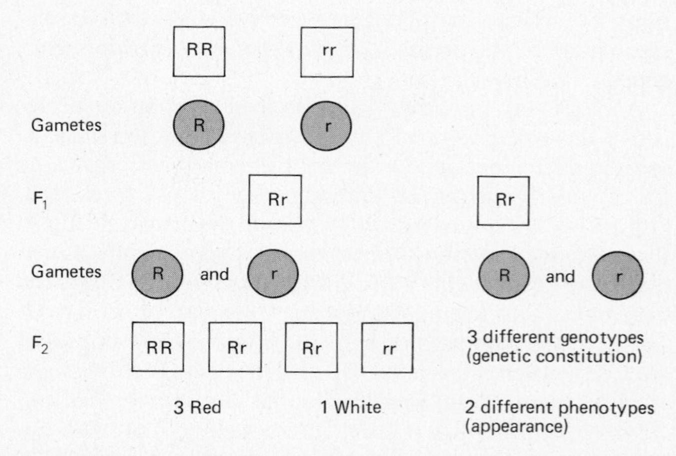

Figure 27-2. **Genotypes and phenotypes in an example of complete dominance in the pea.**

Figure 27-3. **Genotypes and phenotypes in incomplete dominance in the snapdragon.**

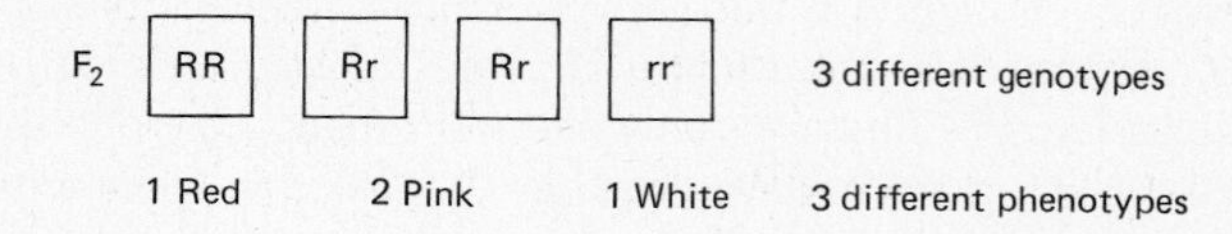

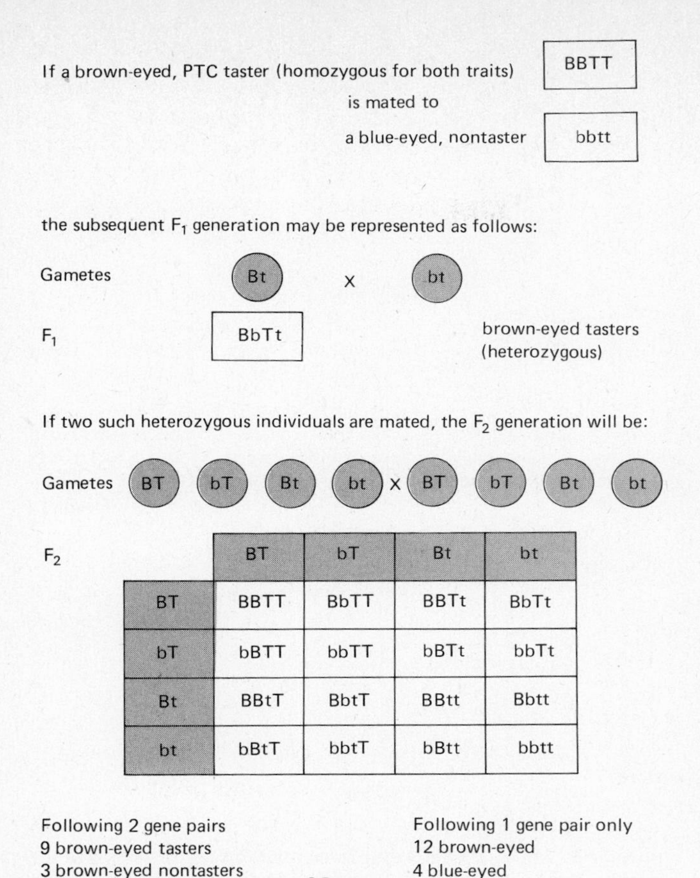

Figure 27-4. **The inheritance of more than one characteristic: eye color and the ability to taste phenylthiocarbamide.**

the pigment melanin, which is formed and deposited in the iris. Brown-eyed people have far more melanin than blue- or gray-eyed people; albinism, however, is due to a recessive gene that prevents any melanin from being formed at all. The gene for brown eyes (B) is dominant over the gene for blue eyes (b).

Similarly, the ability to taste a bitter substance, phenylthiocarbamide (PTC), is brought about by the presence of a dominant gene (T). Inability to taste this substance is caused by the presence of the homozygous recessive gene (tt).

These genes are on *different chromosomes,* so that if we follow one trait at a time, we see that each is inherited in a simple three-to-one ratio. If we concern ourselves with both traits at the same time, the ratio becomes 9:3:3:1 for the various possible combinations of tasters and eye color. Obviously, if we try to follow several characteristics at one time, the ratios appear more and more complex, and this is where Mendel's contribution was of such great importance. He was able to formulate the basic laws of inheritance by painstakingly recording the inheritance of one characteristic at a time. Once this was understood, it became apparent that the pattern for inheritance of many characteristics was often on the basis of simple mathematics (Fig. 27-4).

Interaction of Genes

Many complex characteristics are developed as a result of the interaction of different genes on the same or different chromosomes within the cell, resulting in great variety in the phenotypes produced.

Human skin color is an example of the *additive effect* of two pairs of genes, AA and BB, on two different pairs of chromosomes. If the individual inherits the four dominant alleles, the effects of these genes will be to develop the maximum amount of melanin (dark-brown) pigment in the skin and, the color will be that of the true black. If he has all four recessives, there will be relatively little pigment formed, and the skin color will be "white". Depending on the number of dominant genes, there will be gradations in skin color from "dark" with three dominants, through "mulatto" with two dominants, and "light" with one dominant (Fig. 27-5).

If the first parental generation is black and white, the children of the F_1 generation will be mulatto. The offspring of two mulatto parents, however, can range in skin color from black through to white (Fig. 27-6).

Figure 27-5. **The additive effect of genes determining skin color. The figure shows the distribution of genotypes in the children of two mulatto parents.**

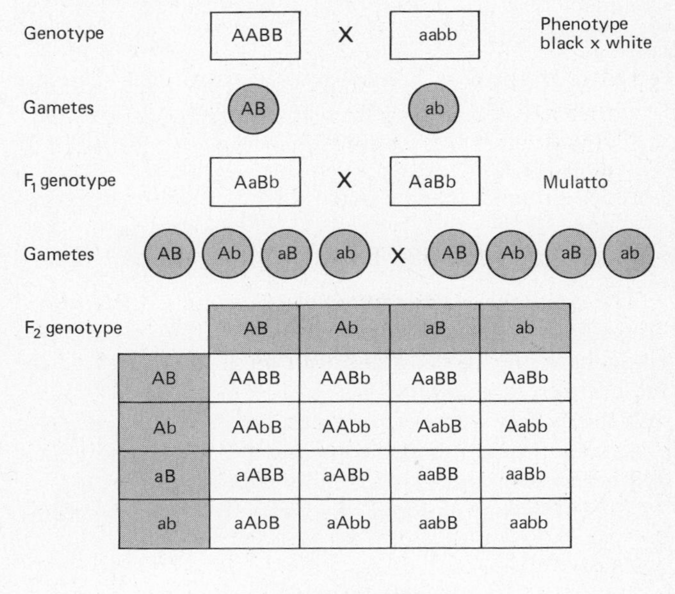

Figure 27-6. **Additive effects of genes as shown by the inheritance of skin color. The figure shows diagrammatically the expected range of skin color in the children of two mulatto parents.**

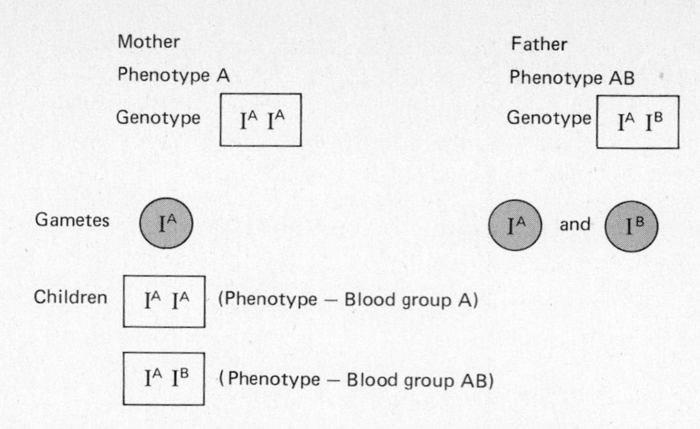

Figure 27-7. **Multiple alleles: the inheritance of blood groups from a mating of a homozygous female (blood type A) and a heterozygous male (blood type AB).**

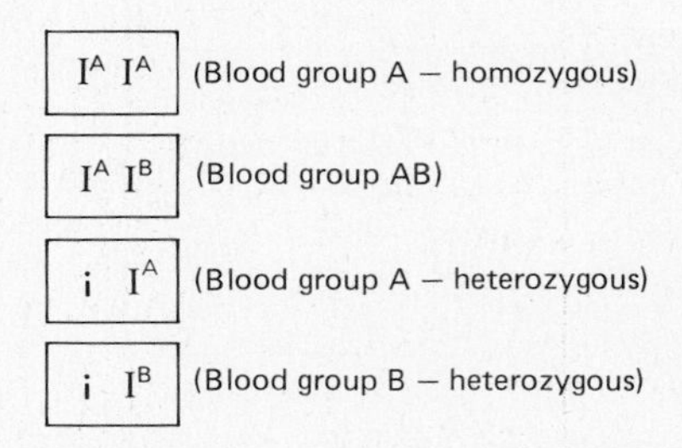

Figure 27-8. **Multiple alleles: the inheritance of blood groups in the offspring from a mating of a heterozygous female (blood group type A) and a heterozygous male (blood type AB).**

Multiple Alleles

The main blood groups in humans have been shown to result from various combinations of a series of allelic genes that affect the development of the agglutinogens in the red blood cells and the agglutinins in the serum. These blood groups, A, B, AB, and O, are determined by the three alleles of the gene I, which can be designated as I^A, I^B, and i. When I^A and I^B are present together, each exerts its own effect to the full so that the cell contains both agglutinogens; each is dominant over i, the lack of both agglutinogens (see also Sec. 20-4).

Blood group O is homozygous recessive and therefore must be ii.
Blood group A may be homozygous, $I^A I^A$, or heterozygous, $I^A I^O$.
Blood group B may be homozygous, $I^B I^B$, or heterozygous, $I^B I^O$.
Blood group AB must be heterozygous and is $I^A I^B$.

The children of a mother of blood group A (determined on the basis of agglutination tests) and a father of group AB could have the blood groups shown in Fig. 27-7 if the mother were homozygous for A.

If the mother were heterozygous, possessing the recessive gene i, she would produce equal numbers of I^A and i eggs, and her possible offspring by this man are shown in Fig. 27-8. No O-group children could result from these matings.

Linkage of Genes and Crossing Over

Because the entire chromosome is normally distributed to the daughter cell in meiosis, those characteristics regulated

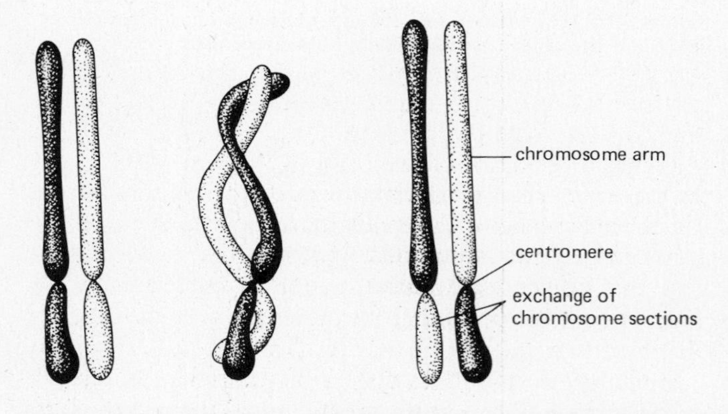

Figure 27-9. **Crossing over of homologous chromosomes during synapsis. At this stage of meiosis, each chromosome is really double so that crossing over occurs in the tetrad.**

by the genes on the same chromosome tend to be inherited together. Sometimes, however, in the close twining of homologous chromosomes in synapsis, pieces of the chromosomes may be exchanged, so that allelic genes are interchanged. This is called *crossing over* and will break the linkage between some genes on the chromosome (Fig. 27-9). Crossing over is more frequent in some chromosome pairs than in others, and the linkage between some genes is much firmer than between others on the same chromosome. The results of crossing over will be seen in the frequency with which new combinations appear in the offspring. These studies have been of great help in mapping the position of the genes along a chromosome, because those genes that never cross over seem to be very close together. The more frequent the rate of crossing over, the farther away from one another are two genes likely to be.

27–4 The Genetic Determination of Sex

The Sex Chromosomes

Of the 23 pairs of chromosomes normally found in human cells (diploid number 46), 22 are called *autosomes,* and one pair is the pair of *sex chromosomes.* The sex chromosomes are quite different from one another in appearance and also in the number of genes they carry. One of the pair is fairly large and is called the X chromosome; the other smaller one appears to be a fragment of a chromosome and is the Y chromosome. The somatic cells of normal females contain two X chromosomes, XX; normal males possess one X and one Y chromosome, XY. In some species—the birds and butterflies, for example—females have an XY pair and males an XX, but in human beings it is the presence of the Y chromosome that will produce a male.

For many years it was believed that the Y chromosome carried no genes at all, but studies of human pedigrees have located several genes upon this small, important chromosome. It is the genes upon the Y chromosome that determine the differentiation of the primary gonad into a testis rather than an ovary and thereby direct the production in the fetal testis of the androgens that are responsible for the development of the male reproductive tract. This does not mean that the autosomes have nothing to do with the development of the tissues and organs characteristic of each sex; but their influence is probably dependent on the instructions initially generated by the sex chromosomes.

Genes on the Y chromosome have also been associated with differences in the rate of maturation of the two sexes. Girls are physically more mature than boys at all ages from birth to adulthood. These differences are quite marked early in embryonic life, beginning at the time when the embryonic testis differentiates.

Studies on children with chromosomal abnormalities indicate that the genes on the Y chromosome are responsible for the slower but longer-lasting growth of the bones of the male, so that ultimately the skeleton is considerably larger and stronger than that of the female.

In the random distribution of one of each pair of homologous chromosomes to the gametes formed by meiosis, the pair of sex chromosomes is separated also. In the formation of the sperm, the XY chromosomes of the male will be separated, and half the spermatozoa will receive the X

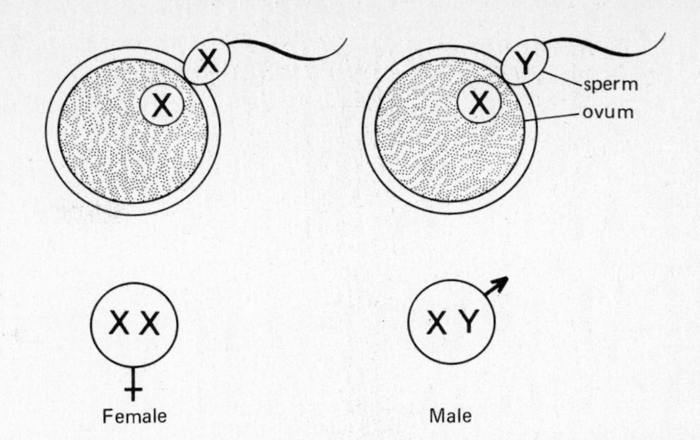

Figure 27-10. **The determination of genetic sex occurs when the sperm fertilizes the ovum.**

chromosome plus 22 autosomes, whereas the other half will receive the Y chromosome and 22 autosomes. When an X sperm fertilizes an egg (which can have only an X chromosome), it will produce an XX, or female individual. An egg fertilized by a Y sperm will result in an XY constitution, that is, a male (see Fig. 27-10).

The two kinds of sperm differ slightly in weight, because of the smaller size of the Y chromosome; they also differ in their electric charge and even in their antibody reactions. These differences have led to strenuous experimentation to separate the X-bearing sperm from the Y-bearing sperm. Economically, the feasibility of being able to separate male- and female-determining sperm is of great interest to dairy farmers and other animal breeders and perhaps not uninteresting to human parents.

Variations in the Number of Sex Chromosomes

NUCLEAR SEX

In 1949 Barr found that attached to the nuclear membrane of nerve cells in some cats was a deeply staining body. Strangely enough, not all cats showed the presence of this body in their cells. Then it was discovered that only cells from female cats possessed this *sex chromatin* or Barr body.

Sex chromatin is quite clearly seen in the cells of human

Figure 27-11. **Human female blood cell showing "drumstick" or sex chromatin attached to the nucleus. This represents the second X chromosome and is absent from cells of normal males.**

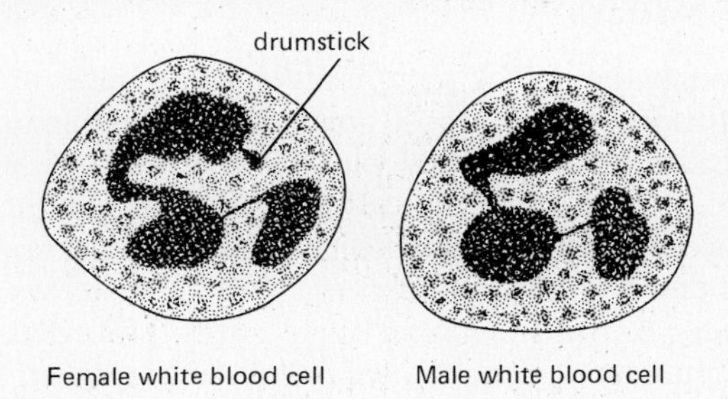

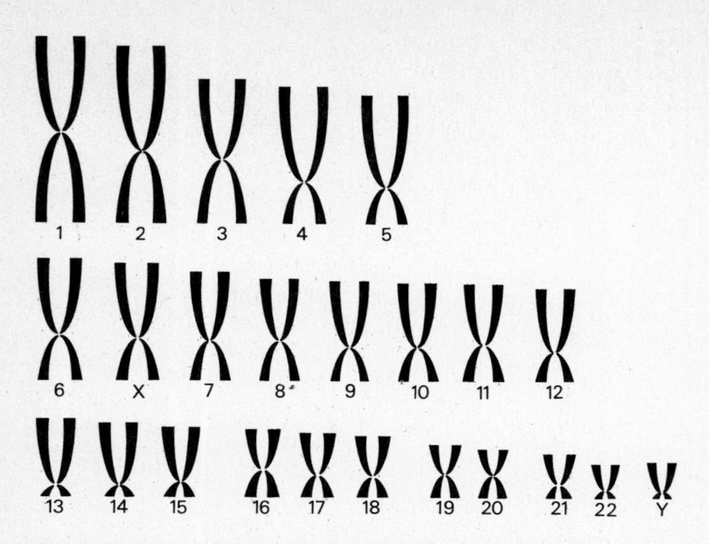

Figure 27-12. **An idealized chromosome set (karyotype) of the human male, numbered according to the international Denver system. Only one member of each pair is represented, but each chromosome has already reproduced itself, with the mirror images remaining attached at the centromere. [Redrawn from B. Lennox, *Lancet,* 1046–1051 (May 13, 1961).]**

females, but it is rarely found in the cells of normal males (Barr, 1960). If sections are made from human skin, or more conveniently, if cells are taken from an oral smear and studied, the genetic or nuclear sex can be determined. Another valuable indicator of nuclear sex has since been discovered in the leukocytes of human blood. In stained blood smears from human females, these white blood cells show a small protrusion attached by a thin thread to the main lobules of the nucleus. These protrusions look like *drumsticks* and are very rarely found in the blood cells of normal males (Fig. 27-11).

The sex chromatin bodies and the drumsticks are indicators of the number of X chromosomes in the nucleus; if more than one X chromosome is present, it is likely to attach itself to the nuclear membrane. In the normal female, therefore, one X chromosome will be indistinguishable in the chromatin network of the interphase nucleus, whereas the other one will form the sex chromatin. The normal male has only the indistinguishable X chromosome. If more than two X chromosomes are present, the additional ones will be found as extra sex chromatin bodies on the nuclear membrane.

Sex anomalies

Although the sex of most individuals is quite obvious, there are some people whose apparent sex is different from their genetic sex. Associated with this confusion are hormonal deficiencies that have placed these disorders in the realm of endocrine "diseases." However, the basic cause has been tracked down to abnormalities in the number of sex chromosomes. A nearly normal male, with underdeveloped testes, immature genitalia, and some development of the breasts (Klinefelter's syndrome) can be shown on the basis of nuclear sexing to have a chromatin body in most of his cells. This indicates the presence of an extra X chromosome; an analysis of the chromosome count (karyotype) is not 46, like that of a normal individual, but 47, because this male also has a Y chromosome making him XXY (Fig. 27-12 shows a normal karyotype). On the other hand, an individual with only one X and no Y chromosome (XO) is a female with immature but properly formed female reproductive tract and genitalia (Turner's syndrome). Her total chromosomal count is 45. For the appearance of male characteristics, the presence of the Y chromosome is essential, and even persons with four X chromosomes to one Y chromosome have masculine characteristics, a potent argument for the existence of genes on the Y chromosome.

Sex-Linked Characteristics

Certain characteristics that have nothing to do with the gonads or reproductive tract are found to occur far more frequently in males than in females. These are called sex-linked characteristics, the most well known of which is probably hemophilia, the "royal" disease transmitted through the descendants of Queen Victoria to many of the members of the ruling houses of Europe.

The responsible genes in sex-linked characteristics are usually recessive genes on the X chromosome. Since the male has only one X chromosome, and no allele on the Y chromosome, the recessive gene can exert its effect in the male, whereas in the female the likelihood is far greater that it will be repressed by the activity of the normal dominant allele on the second X chromosome.

Hemophilia, the failure of the blood to clot properly, is

Figure 27-13. **The inheritance of a sex-linked characteristic hemophilia; H represents the dominant normal gene and h indicates the recessive gene for hemophilia.**

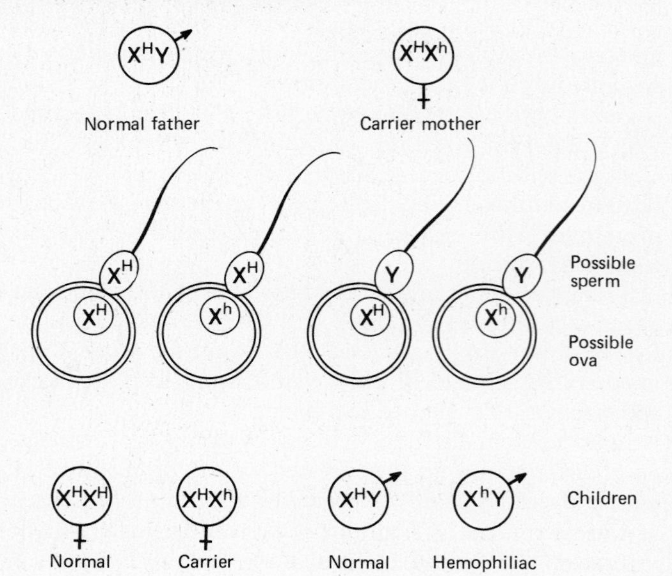

due to the absence of a clotting factor which is essential for the normal coagulation process. The absence of this factor, in turn, is caused by a mutation of a gene on the X chromosome. The mutant gene, a recessive, is designated as h, and the normal, dominant allele is H. A normal male will have the H gene on his X chromosome but no comparable allele on the Y. This can be represented as $X^H Y$. A normal female is $X^H X^H$. If the female posseses the mutant gene h, her genotype is $X^H X^h$, but her blood will coagulate normally. She is a *carrier* of this defect, however, since she will pass the mutant to half of her children. Of this half, the girls will become carriers, and the boys will suffer from the disease (Fig. 27-13).

Females can be afflicted with hemophilia too, but this condition is far less usual because it requires the presence of two recessive genes, one from a "carrier" mother and the other from a hemophiliac father. However, several such cases have been reported. The genotype of this female would be $X^h X^h$.

Other traits in human beings are sex-linked—*color blindness,* for example. Again, the mother is the carrier, and half her sons may inherit this deficiency. Obviously, all sex-linked characteristics should show some linkage with one another, if they are caused by mutations of genes on the X chromosome. Some unfortunate males therefore may inherit both color-blindness and hemophilia from their apparently normal mothers.

Sex Hormones and the Expression of the Gene

The rate of mutation of the gene and its ability to evoke its effect in the organism often result from the presence of either the male or the female sex hormone. The mutation rate for hemophilia is almost 10 times higher in the male X chromosome than in the female, so that not all the blame for the transmission of this disease can be placed on women. Some characteristics are *sex-influenced* rather than sex-linked, because they develop under the direction of genes on the autosomes, not the sex chromosomes. The expression of the gene is altered, however, by the amount of male or female sex hormone present. *Baldness* in males is sex-influenced; one gene for baldness, working in an environment in which sufficient androgens are present, will cause a man to lose his hair; a woman retains her hair unless she has two genes for baldness. (Of course, certain diseases may also cause baldness.) This fact is particularly interesting because it indicates once more the importance of the cytoplasmic environment in permitting or repressing the activities of genes.

27–5 Twinning and Multiple Births

Monovular (identical) twins

When one ovum is fertilized by one sperm, there may be a separation of the zygote into two separate masses some time after the first cell division (Fig. 27-14). These masses then develop into a complete embryo. Partial separation may account for conjoined or Siamese twins.

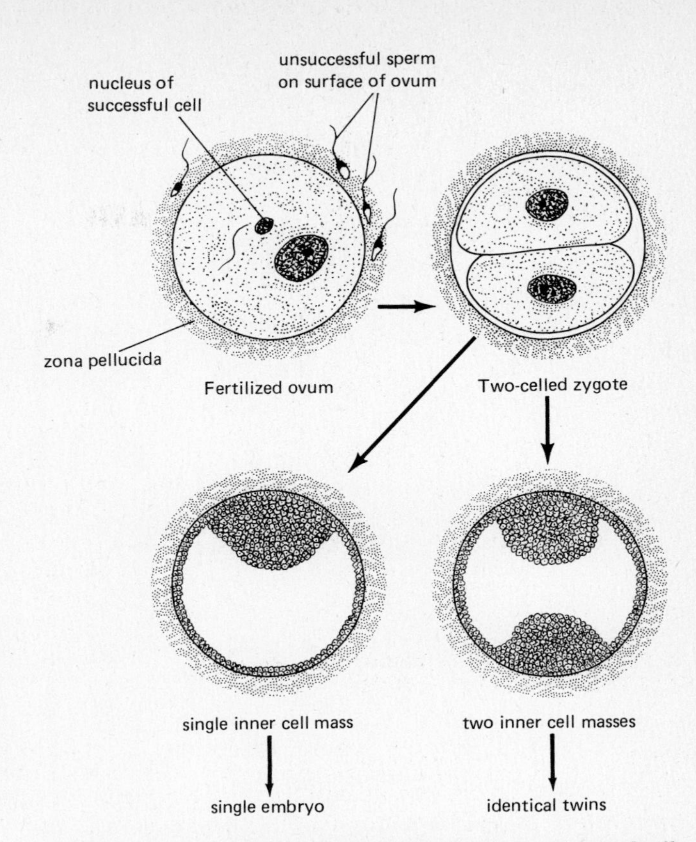

Figure 27-14. **Development of identical twins from a single, fertilized ovum. The two-celled zygote resulting from the first cleavage of the fertilized egg usually develops into a single embryo but sometimes may develop into two identical embryos.**

Monovular twins have exactly the same genetic complement and so are genetically identical, although slight variations in the environment of each growing fetus may cause minor differences between the twins. Identical twins must always be of the same sex.

Biovular (fraternal) twins

If two ova are present at the same time in the oviduct, and both are fertilized, then fraternal twins are produced that have no more genetic similarity than any other siblings of the same parents. The ova may have both come from one ovary or one from each ovary. Fraternal twins may be of the same or different sexes.

Monovular multiple births

Monovular multiple births in humans are rare, but identical triplets, quadruplets, and quintuplets are known. Figure 27-15 shows the probable division of a single fertilized ovum to produce five identical children.

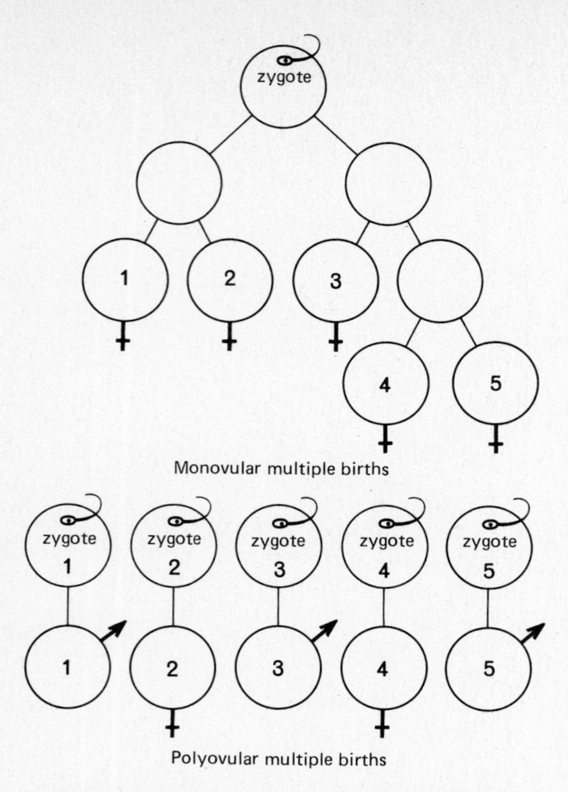

Figure 27-15. **Monovular and polyovular multiple births. The upper diagram shows that when all the offspring are derived from one zygote, they will all be of the same sex. In polyovular multiple births, many ova are fertilized and the determination of sex is random.**

Polyovular multiple births

Polyovular multiple birth in humans has become much more common since the use of fertility drugs—gonadotropin-like compounds that induce multiple ovulation (Sec. 26-9). The resulting siblings are fraternal and may be of both sexes (Fig. 27-15).

27–6 Variations in the Number of Chromosome Sets

The normal diploid number of chromosomes within a species is a rather constant figure. In plants, however, three representatives of each chromosome may be found (triploidy), or even many more (polyploidy), and this has resulted in great variation in form and vigor in the different plants. Many of the cultivated varieties of corn and tomatoes are polyploids, and they are considerably larger and stronger than their diploid relatives. In animals, however, this polyploid formation does not appear to have been successful, apart from a few groups such as some shrimp and moths that reproduce by parthenogenesis. In higher animals an extra chromosome is sometimes present, or one may be missing, but entire sets do not seem to be added.

27–7 Variations in the Number of Chromosomes

Many abnormalities in humans have been shown to result from variations in the number of chromosomes, presumably originating from some fault in the formation of the gametes. Those abnormalities that arise from variations in the number of sex chromosomes have been discussed in Sec. 27-4. If one of the other chromosomes, the autosomes, is missing or present in triplicate, other syndromes appear, including some acute *leukemias* (cancer of the blood). The best-documented syndrome is *mongolism,* which is caused by an extra representative of chromosome number 21 (total chromosome count is 47). This karyotype is shown in Fig. 27-16.

Mongolism occurs in families that have very often produced several normal children. The afflicted child may be severely retarded mentally and physically, with abnormalities in the structures of the face, hands, and other parts of the body. The name *mongolism* comes from the somewhat oriental appearance of the eyes, resulting from a typical fold of the eyelid, but the terminology is not very apt. The chances of a mongoloid baby being born increase greatly with the age of the mother but are quite independent of the number of preceding children or the age of the father. This fact must mean that abnormalities in meiosis increase in frequency as the mother grows older; probably in these cases the distribution of chromosome number 21 to the oocytes is impeded in some way, so that some receive 24 chromosomes and the polar cells 22. It is also possible that some polar cells receive 24, and the oocyte only 22, but presumably these deficient oocytes are nonviable.

27–8 Genes and Disease

Many diseases, including mental diseases, have been shown to be caused by the mutation of a specific gene, or the interaction between the genetic constitution of an individual and the environment in which he is reared. This is the social, rather than immediate biological, environment. Hemophilia (discussed in Sec. 27-4) is caused by a mutation of a gene on an X chromosome, but many other inherited diseases are not sex-linked and are brought about by genic mutations on the autosomes.

Sickle Cell Anemia. The physiological significance of sickle cell anemia was discussed in Sec. 14-7. It is a severe form of anemia that is usually fatal in early life. It exists in two forms, *sickle cell trait* and *sickle cell anemia.*

Those individuals who are heterozygous for the condition show sickling of their red blood cells when exposed to hypoxia, but are quite healthy. These heterozygotes have the sickle cell trait, and their hemoglobin is normal (HbA).

Homozygotes suffer from the serious anemia as well as the sickling phenomenon. Their red cells contain sickle cell hemoglobin (HbS). These individuals have received an ab-

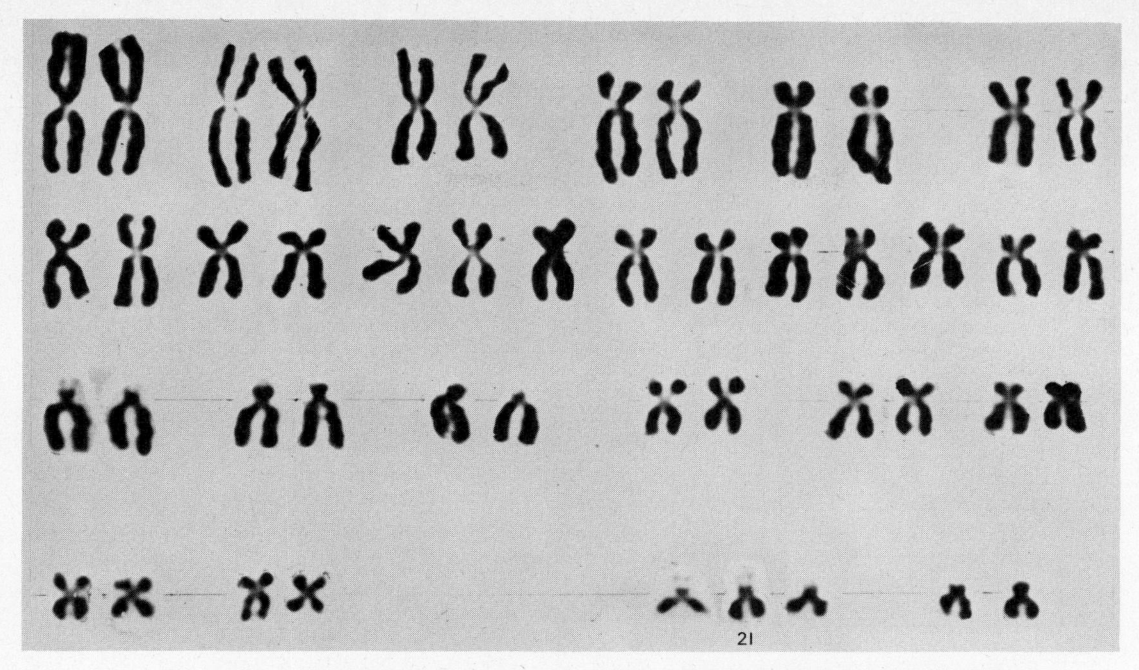

Figure 27-16. **Karyotype in mongolism (47 chromosomes). The extra chromosome is a third representative of chromosome 21. [From K. Hirschhorn and H. L. Cooper, *Amer. J. Med.* 31:442 (1961).]**

normal autosomal gene from each parent and consequently have sickle cell anemia. This disease afflicts black African races almost exclusively. This mutant gene also confers additional protection against malaria in heterozygotes.

Phenylketonuria. Phenylketonuria is a disease caused by a recessive gene that interferes with the proper enzymatic metabolism of the important amino acid *phenylalanine.* As a result, ketones accumulate in the blood and tissues. The nervous system is especially sensitive to the toxic action of these ketones, and the development of the brain, even after birth, is seriously impeded, resulting in severe mental retardation. For the disease to develop, the individual must be homozygous for this pair of recessive genes; heterozygous persons metabolize phenylalanine more slowly than normal persons but do not show symptoms of the disease. They are, of course, carriers, but because the gene is located on an autosome, both sons and daughters are equally likely to inherit it. A routine test of the newborn for this disease is now mandatory in several states. Early diagnosis enables the physician to keep the infant on a diet low in phenylalanine, so that most of the damage can be avoided.

Schizophrenia. In the complex mental disorder known as schizophrenia, the patient at times loses touch with reality and as a result seems to have two personalities. This disease is found to be more frequent in certain families than in the rest of the population and apparently is determined by several pairs of genes. These genes "predispose" the individual to this disease, but the actual development of schizophrenia depends greatly on the environment in which the person develops. From studies in identical twins, statistics indicate that if one twin develops schizophrenia, the chances are 68 in 100 that the other twin also will become ill.

27–9 Radiation and Chromosomal Damage

Radioactive substances emit high-energy radiations that penetrate living tissues with varying ease, depending on the type of radiation. Once they have penetrated and become absorbed, however, they all produce the same basic damage to the cell. These high energy radiations, including X rays and the radiations from atomic explosions, are very often called *ionizing radiations* because their enormous energy displaces electrons from atoms, breaking chemical bonds and leaving charged particles, or ions. These ionizing radiations also produce a host of new substances in the cell, most of which are toxic.

The primary site of radiation damage to the cell appears to be the *chromosomes.* The DNA of the gene may be structurally altered, resulting in a mutation, or the chromosomes may be broken, and the odd fragments sometimes join up with different chromosomes and sometimes are lost entirely during mitosis. The damage may be even more radical: the entire process of cell division may be impeded so that the tissues die because of lack of cell replacement. Those tissues that normally have the highest rate of cell division, such as the blood-cell-producing bone marrow, are most sensitive to radiation. It may be simply that the damage is made apparent sooner, for the number of white blood cells falls dramatically.

The *embryo* is particularly radiosensitive. A cell undergoing differentiation is even more sensitive than the same cell would be later in mitosis. Metabolism during differentiation must therefore be even more highly organized than during mitosis, so that any slight disturbance sets up a train of deranged reactions. Because the organs start to differentiate very early, any exposure of the embryo at this time should be avoided. For this reason, Rugh at Columbia University recommends that no woman should have pelvic irradiation, even for diagnostic reasons unless it is imperative, during the last 2 weeks of the menstrual cycle. It is during the first 38 days of pregnancy, when the woman may not even know that she is pregnant, that the most damage can be done to the embryo (Fig. 27-17).

Figure 27-17. **Anomalies of development in the mouse embryo following low levels of X-irradiation (5 to 100 r) during the gestation period (top time scale). The bottom time scale indicates the corresponding day of gestation for the human embryo. The large dots show the period of maximum susceptibility. [From R. Rugh, *Radiology* 82:917 (1964).]**

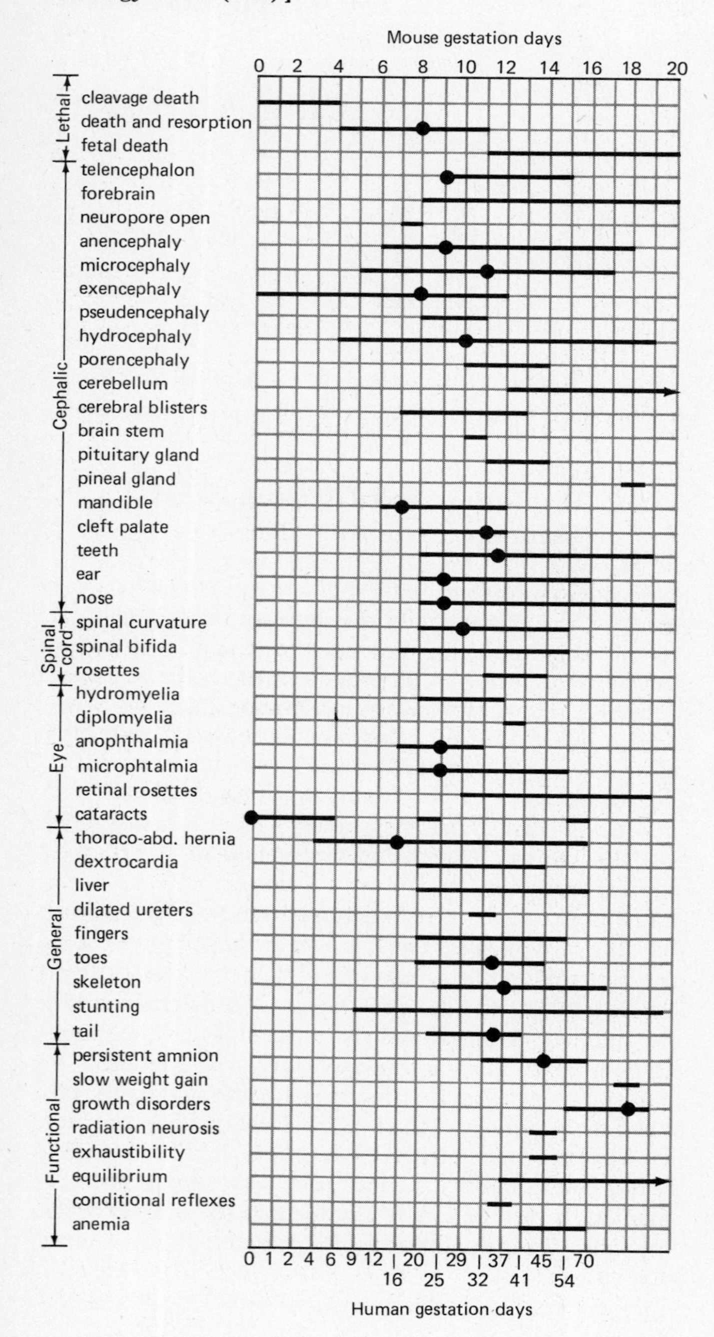

27–10 Environmental Regulation of the Gene

An inborn error of metabolism that manifests itself as a disease may become apparent only when the environment permits its expression. An example of this interaction between genetic and environmental factors is seen in *favism,* a severe form of hemolytic anemia induced by eating fava beans (a Mediterranean favorite). Not all people who eat the beans develop favism, but those who do have been shown to have a deficiency of the enzyme *glucose-6-phosphate dehydrogenase* (*G6PD*). If these individuals never eat the beans, they never develop the disease.

Another good example of the way in which the environment affects the expression of the gene is found in the inherited disease *galactosemia,* an inborn error of carbohydrate metabolism in which the hexose sugar galactose cannot be metabolized. This is caused by the genetically determined deficiency of the enzyme *galactose-1-phosphate uridyl transferase.* An infant's diet usually contains considerable amounts of galactose, derived from lactose, the principle disaccharide in milk. The block in metabolism leads to the accumulation of galactose in the blood and galactose-1-phosphate in the cells, which can have fatal consequences. However, if a diet completely free of galactose, but otherwise nutritionally adequate, is fed the child before irreversible damage has occurred, normal development is possible.

Time, Age, and Hormones

Another important aspect of environmental regulation of gene activity is the interdependence of many different genes functioning together or at well-defined sequences of time. It is obvious importance, for example, that the genes that induce the development of the neural tube be active before those that cause the mesoderm to develop into the bony vertebral column that surrounds it. Age is of extraordinary importance in the regulation of the genes, because the activity of many cells varies enormously as the organism progresses from intrauterine life through infancy, puberty, maturity, and senescence.

It has even been suggested that there is a *"death-clock"* in the hypothalamus, which gradually loses its sensitivity to feedback suppression and so permits degenerative, aging changes to occur. Is it the gene that is slowly aging or is it the cytoplasmic environment in which it is placed? Although we know that some genes are greatly affected in their expression by the amount of sex hormone present (the reproductive tract and the hypothalamus, for example, at critical times in their development), we know very little about the specific ways by which genes are regulated. Gene induction and repression, topics that were considered in Chapter 3, are almost certainly involved in this type of regulation.

Gene puffs

Work with lower organisms has exciting implications in terms of *hormonal regulation of the time at which genes become active.* The fruit fly *Drosophila* has been the favorite subject of genetic experimentation for many years. Its salivary gland cells are enormous, and the giant chromosomes within them are large enough to permit definite bands along their length to be seen under the light microscope. Because fruit flies breed so rapidly, sufficient information could be gathered for geneticists, like Morgan, Bridges (1925), and other eminent investigators, to be able to correlate these bands with the position of specific genes and make *chromosome maps* from this information.

Using the electron microscope, Clever and his associates from the Max Planck Institute in Tübingen have found that they can actually see the difference between *active and inactive genes* on the giant chromosomes of the midge (*Chironomus*). When a gene is active, it puffs up, surrounding itself with a mass of material which is RNA. The inactive gene consists mainly of DNA, which stains a clear blue with toluidine blue, and because the RNA takes a bright red violet with this dye, not only is there a large *puff* around an active gene, but the dramatic difference in color shows a difference in the chemical constitution of the gene and its product.

Figure 27-18. **Chromosome puff formed by active gene on the salivary chromosome of the midge in preparation for metamorphosis. No puff is formed and metamorphosis does not occur if the gland that produces the hormone essential for metamorphosis is removed.**

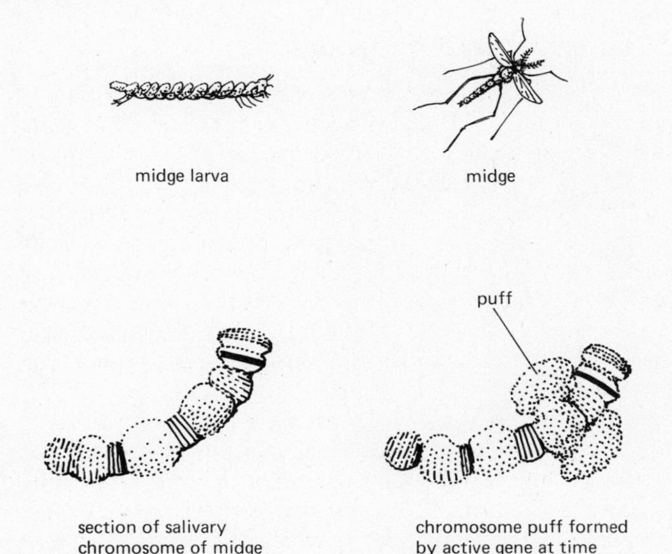

Beermann and Clever (1972) noticed that certain genes puffed up when the midge was getting ready to undergo metamorphosis (Fig. 27-18). If they removed the gland that produced the hormone (ecdysone) essential for metamorphosis, these puffs were not formed, and the insect did not molt. Injections of this hormone cause just these particular genes to produce puffs. If this information can be applied to other organisms, it gives us far more insight as to how cells differentiate and leads logically to the way in which genes exert their action at different times in the life of an organism. However, one point is equally clear: just as hormones may control the activity of the gene, which in turn will regulate metabolic activities of the cells, so the cells themselves must influence genic activity. Although we see that specific genes become active in certain cells and thereby regulate the differentiation and metabolism of that cell in a particular direction, there must be a reason for these genes becoming active in the first place. We are back again to the cytoplasm and the interaction between cytoplasm and environment and cytoplasm and gene—an interaction so complex that despite the similarities in living organisms throughout the world, each individual is unique.

27–11 Prevention of Birth Defects

The use of genetic techniques to improve genetic qualities has two quite different facets. One facet is the reduction in the number of genetically defective children produced; the other facet is selective breeding to improve genetic qualities. Both goals are laudable; both carry with them serious possibilities for misuse. Should society selectively sterilize, or prevent from reproducing, those individuals considered to have unfavorable genes? Who is to decide what is optimal with reference to fitness, beauty, personality, and other values?

Genetic Screening

Prevention of most defects that are inherited requires an understanding of the frequency of occurrence and modes of inheritance of these diseases. This information can only be obtained by genetic screening of the population, including normal individuals and those affected by the disease. Tracing human pedigrees is expensive and not always feasible; many inherited diseases cannot be identified in heterozygotes.

Screening for genetic disease also may cause racial problems. Some diseases, such as PKU, are so rare in black populations that it makes little sense to screen black children; few whites are sickle cell trait carriers. *Tay-Sachs disease,* a disease that causes blindness, severe mental retardation, and early death, is far more common in Jews of Northern European origin than in the general population. It makes little economic sense to screen the entire population

Table 27–1. **Percentage Probabilities of Successive Child Having Same Abnormality as Previous Child**

Magnitude of Risk		Genetic Basis
Total	100	Both parents are homozygous recessives. (No progeny test is needed.)
High	75	Both parents are heterozygous for an autosomal dominant with full penetrance.
	50	One parent is heterozygous for an autosomal dominant. For sons in case of a sex-linked gene carried by the mother.
Moderate	25	Recessives with full penetrance. Incompletely penetrant autosomal dominants.
Low	5 or less	Congenital malformations with unknown causes. Trisomy 21.

Source: Data from A. G. Motulsky and F. Hecht, as prepared by I. M. Lerner. From *Heredity, Evolution, and Society,* First Edition, by I. Michael Lerner. W. H. Freeman and Company, San Francisco. Copyright © 1968.

for all these diseases, yet some object to a program in which certain races are eliminated as being discriminatory.

Despite these problems, the importance of genetic screening is obvious. Early detection of medically treatable diseases, such as PKU, can prevent mental retardation, and the tremendous emotional and financial cost to the family and to society. Provided that screening programs follow ethical guidelines for confidentiality, their ultimate benefit to society should be considerable.

Genetic Counseling

One cannot select the genes with which one was born, but it is possible to select *against* genes that cause hereditary diseases in one's children. Genetic counselors can provide information about the history of genetic disease and about the risks of these diseases occurring in offspring, and thus permit couples to make responsible decisions about having or not having children. Freedom of action on the advice is essential, but decisions should be made on the basis of both personal and social concern.

Problems in genetic counseling arise under many circumstances. Should a clinically unaffected individual be told that there is a 50 per cent chance of later development of a genetic disease, for which there is no known treatment? This problem arises in *Huntington's chorea,* a disease in which there is progressive degeneration of the nervous system, development of involuntary, jerky movements, and death. It is caused by a single, dominant gene that exerts its effect in the 40- to 45-year-old individual. Would early knowledge of this inheritance decide the victim not to reproduce? Would most people want to know many years ahead that they will die prematurely of an incurable disease?

Alternatives to Bearing Genetically Defective Children

If a couple has been properly informed about the risks of their producing abnormal children, they can decide to adopt children or to attempt to have their own child, relying on intrauterine diagnosis to determine whether the fetus is normal or defective. If the fetus is abnormal, they may then decide to have it aborted.

Estimation of Risk. Genetic counseling is particularly valuable to parents who have already produced abnormal children and are concerned about the probability that their next child could have the same defect. Recurrence risk figures are calculated from knowledge of the genetic basis of a disease and its incidence in the population. Table 27-1 shows a general form of such probabilities, but for many genetic disorders, specific values can be given. A pregnant woman 40 years of age is 10 times more likely to have a child affected with mongolism than a 25-year-old woman.

Figure 27-19. **Amniocentesis, the withdrawal of amniotic fluid from the pregnant uterus.**

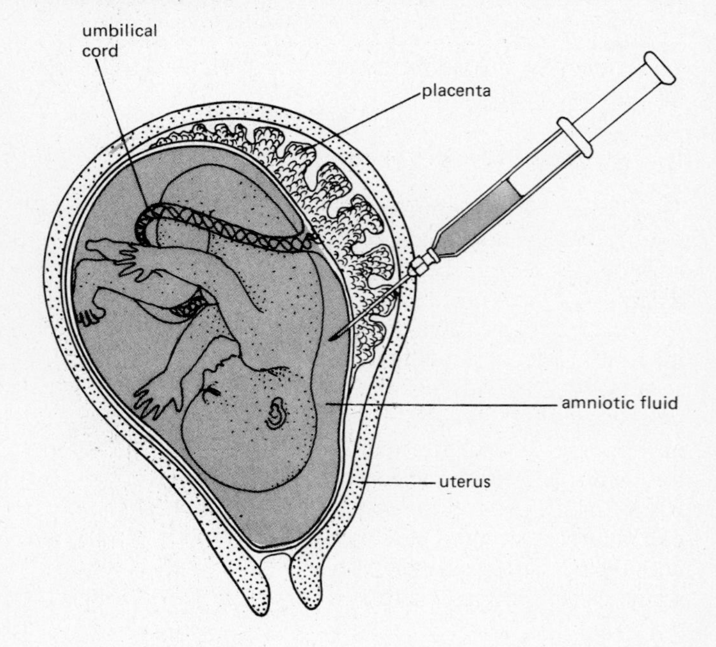

This risk factor is, however, quite independent of the number of previous children.

Intrauterine Diagnosis: Amniocentesis

Prenatal genetic diagnosis is based on studies of the amniotic fluid that surrounds the developing fetus in the uterus. Cells from the skin and respiratory tract of the fetus are shed into this fluid. *Amniocentesis* is the withdrawal of a small amount of amniotic fluid by needle puncture (Fig. 27-19) and its subsequent analysis for genetic defects. Both the fluid and the cells contain valuable clues concerning the genetic composition of the fetus. This technique does, however, carry a slight possibility of damage to the fetus.

AMNIOTIC FLUID

The first studies on amniotic fluid in the 1960s demonstrated that it was possible to identify Rh babies (Sec. 14-16) on the basis of the concentration of hemoglobin breakdown products in the amniotic fluid. Other diseases can be identified by abnormalities in metabolites in the amniotic fluid.

CHROMOSOMAL NUMBER AND ARRANGEMENT

At a very early stage in development, it is possible to distinguish the sex of the fetus and to determine abnormalities in chromosome number and type. Determination of fetal sex will rule out hemophilia in a female, although 50 per cent of the females will be carriers. A male fetus has a 50 per cent chance of being affected.

Chromosomal analysis of cells obtained by amniocentesis can indicate mongolism if fetal cells show 47 chromosomes, with the extra chromosome a third number 21. Similarly, the 47 chromosomes of Klinefelter's syndrome, with the extra X chromosome, are diagnostic of this disease. Turner's syndrome (45 chromosomes, only one X chromosome) can also be predicted.

BIOCHEMICAL AND ENZYMATIC STUDIES

Cells taken from amniotic fluid can be cultured and tested for their enzymatic characteristics. Deficiencies of specific enzymes indicate genetic disease. A disease that can be diagnosed in this manner is the *Lesch-Nyhan syndrome,* a severe neurological disease in males. It is characterized by mental retardation, involuntary writhing movements, and self-mutilation by biting. A defective gene on the X chromosome results in a lack of the enzyme hypoxanthine-guanine phosphoribosyl transferase (HGPRT), which is needed for the formation of nucleotides that are incorporated into RNA and DNA. Interference with this synthetic pathway particularly affects the cells of the basal ganglia. Fetal cells with the Lesch-Nyhan syndrome will not incorporate radioactive hypoxanthine, whereas normal cells will do so.

27–12 Genetic Engineering

Genetic engineering is the changing of undesirable genes to more desirable forms. There are at least four possible approaches to this goal:

1. *Directed mutation.* Mutations are induced in genes at specific loci.
2. *Transduction.* Genetic material is transferred from one cell to another via a virus.
3. *Transformation.* A segment of DNA from one cell is incorporated into the genetic material of another cell.

Transformation is accomplished by using newly discovered enzymes, DNA restriction endonucleases, which cleave the DNA molecule at specific base sequences. The exposed sticky ends of the DNA fragments can be made to stick to other DNA segments and can be sealed together by treatment with DNA ligase. This means that a DNA fragment from a toad can be joined to DNA from a mouse and introduced into a bacterium. Entirely new types of biologically active DNA can be constructed in this manner.

4. *Gene synthesis.* A complete gene is chemically synthesized.

The first artificial gene was synthesized in 1976 by Nobel laureate Khorana and his associates at the Massachusetts Institute of Technology (see report by Maugh, 1976), who linked 39 chemically synthesized fragments, each containing 10 to 15 nucleotides. The completed gene contains 126 nucleotides and is active in living cells (bacteria), coding for tyrosinase transfer RNA.

Not only can genes normally present in cells be synthesized, but changes in gene sequences can be made and the resulting alterations in cell function studied. This remarkable achievement should make it possible to learn more about the intricate mechanisms by which genes function.

PRACTICAL APPLICATIONS

One possible application would be the insertion of nitrogen-fixing genes into crop plants, so that fertilizer becomes unnecessary. Another potential application would be the construction of microorganisms that could synthesize some of the products now obtained from oil.

An ingenious suggestion is that a gene for the production of a hormone be synthesized and inserted into a bacterium so that the hormone could be obtained inexpensively in large quantities from bacterial cultures. Success with this technique has already been reported for insulin.

BIOLOGICAL HAZARDS

This technique of recombinant DNA and the ability to synthesize new genes gives humans the power to design and create new gene combinations.

There is serious concern, consequently, about the creation

of new types of infectious DNA, the biological properties of which cannot be anticipated. New drug-resistant or disease-producing bacteria might be created and escape from the laboratory.

To lessen the possible risks of such experiments, a group of eminent American scientists first put all genetic manipulation work under an almost complete embargo for 18 months. During this time a strict set of guidelines was established for experimentation. Subsequently, the Secretary of the Department of Health, Education and Welfare requested Congress to impose federal restrictions on research with recombinant DNA. This unusual procedure emphasizes the sense of scientific responsibility and cooperation that is essential to control this new genetic technology.

Cited References

BARR, M. L. "Sexual Dimorphism in Interphase Nuclei." *Amer. J. Human Genetics* **12:** 118 (1960).

BEERMAN, W., ed. *Developmental Studies on Giant Chromosomes.* Springer-Verlag New York, Inc., New York, 1972.

DARWIN, C. *The Origin of Species by Means of Natural Selection, or the Preservation of Favoured Races in the Struggle for Life.* Philosophical Library, New York, reprint of 1859 ed., 1951.

MAUGH, T. H. "The Artificial Gene: It's Synthesized and It Works in Cells." *Science* **194:** 44 (1976).

MENDEL, G. "Versuche über Pflanzen-hybriden." *J. Hered.* **42:** 1. English translation under the title "Experiments in Plant Hybridization," Harvard University Press, Cambridge, Mass., 1866.

MORGAN, T. H., C. B. BRIDGES, and A. H. STURTEVANT. "The Genetics of Drosophila." *Bibliographia Genetica* **2:** 1. The Hague, Netherlands, 1925.

WEISMANN, A. *Germplasm.* Charles Scribner's Sons, New York, 1893.

Additional Readings

BOOKS

BRESLER, J. B., ed. *Genetics and Society.* Addison-Wesley Publishing Co., Inc., Reading, Mass., 1973.

ETZIONI, A. *Genetic Fix.* Macmillan Publishing Co., Inc., New York, 1973.

GARDNER, E. J. *Principles of Genetics.* John Wiley & Sons, Inc., New York, 1975.

HARRIS, H. *The Principles of Human Biochemical Genetics,* 2nd ed., North Holland/American Elsevier Publishing Co., New York, 1976. (Paperback.)

LERNER, I. M., and W. J. LIBBY. *Heredity, Evolution and Society,* 2nd ed. W. H. Freeman and Company, San Francisco, 1976.

ARTICLES

COHEN, S. N. "The Manipulation of Genes." *Sci. Am.,* July 1975.

COHEN, S. N. "Recombinant DNA: Fact and Fiction." *Science* **195:** 654 (1977).

FRIEDMANN, T. "Prenatal Diagnosis of Genetic Disease." *Sci. Am.,* Nov. 1971.

MITTWOCH, U. "Sex Differences in Cells." *Sci. Am.,* July 1963.

Appendix I

Additional Readings

GENERAL REFERENCE TEXTS

BRESNICK, E., and A. SCHWARTZ. *Functional Dynamics of the Cell.* Academic Press, Inc., New York, 1968.

DAVSON, H. *A Textbook of General Physiology,* 4th ed. Churchill Livingstone, Division of Longman, Inc., New York, 1970.

FLOREY, E. *Introduction to General and Comparative Animal Physiology.* W. B. Saunders Company, Philadelphia, 1966.

GUYTON, A. *Textbook of Medical Physiology,* 5th ed. W. B. Saunders Company, Philadelphia, 1976.

MOGENSON, G. J., and F. R. CALARESU, eds. *Neural Integration of Physiological Mechanisms and Behaviour.* University of Toronto Press, Toronto, 1975.

QUARTON, G. C., T. MELECHUK, and F. O. SCHMITT, eds. *The Neurosciences: A Study Program.* The Rockefeller University Press, New York, 1967.

RUCH, T. C., and H. D. PATTON. *Medical Physiology and Biophysics,* 20th ed. W. B. Saunders Company, Philadelphia, 1973.

RUGH, R. *Vertebrate Embryology.* Harcourt Brace Jovanovich, Inc., 1964.

SCHMITT, F. O., ed. *The Neurosciences: Second Study Program.* The Rockefeller University Press, New York, 1970.

SCHMITT, F. O., and F. G. WORDEN, eds. *The Neurosciences: Third Study Program.* The M.I.T. Press, Cambridge, Mass., 1974.

TIMIRAS, P. *Developmental Physiology and Aging.* Macmillan Publishing Co., Inc., New York, 1972.

WILSON, J. A. *Principles of Animal Physiology.* Macmillan Publishing Co., Inc., New York, 1972.

REVIEWS AND HANDBOOKS

The American Physiological Society publishes a series of *Handbooks of Physiology,* written in great detail, but with enormous coverage of the field. These handbooks are arranged in Sections, e.g., Section I is Neurophysiology, Section 2 is Circulation. Many sections contain more than one volume. There are sections also on Endocrinology, Respiration, Adaptation to the Environment, Renal and Electrolyte Physiology, Alimentary Canal, and Adipose Tissue.

Physiological Reviews is also published by the American Physiological Society. This journal appears quarterly, and each issue contains four to six excellent review articles.

The Annual Review of Physiology provides yearly reviews of specific physiological areas. These reviews are very up-to-date but highly detailed.

MTP International Review of Science. Series One is devoted to physiology. Volume 1 is cardiovascular physiology; Vol. 2, respiratory physiology; Vol. 3, neurophysiology; Vol. 4, gastrointestinal physiology; Vol. 5, endocrine physiology; Vol. 6, kidney and urinary tract physiology; Vol. 7, environmental physiology; and Vol. 8, reproductive physiology. (University Park Press, Baltimore.)

Scientific American publishes excellent articles for the nonspecialist, articles that nevertheless give insight into research techniques and concepts. Many of those articles can be purchased as individual offprints from W. H. Freeman and Co., San Francisco, California 94104. Their catalogue of offprints also includes several very good books of selected readings on one topic, for example, *From Cell to Organism,* 1967, and *Human Physiology and the Environment in Health and Disease,* 1976.

Appendix

Normal Physiological Values (Averages)

Whenever available, values for both sexes, at different ages, have been included.

Blood: Cellular Elements

	Erythrocyte (RBC) Count (millions/mm^3 blood)	Reticulocyte Count (per 1000 RBC)
At birth	5.5	39.2
6–10 years	4.7	7.6
Adult male	5.4	6.0
Adult female	4.8	9.0

	Hematocrit—Packed Cell Volume (volume % RBC in blood)	Hemoglobin Content (g/100 ml blood)
At birth	52.6	19.5
6–10 years	38.5	12.9
Adult male	46.2	16.0
Adult female	40.6	14.0

Thrombocytes (platelets)

	Thrombocyte Count (thousands/mm^3 blood)
At birth	227
Adult	250

Leukocytes

First line: number/mm^3 blood
Second line: % total leukocytes

	Total Leukocytes	Neutrophils	Eosinophils	Basophils	Lymphocytes	Monocytes
At birth	17,990	11,000	400	100	5,500	1,050
	100	60.4	2.2	0.6	31	5.8
6–10 years	8,260	4,360	200	50	3,300	350
	100	53	2.4	0.6	39	4.2
Adult	7,420	4,420	200	40	2,500	300
	100	59	2.7	0.5	34	4

Blood: Normal Hematological Values

	Blood Volume (ml/kg body wt)	Plasma Volume (ml/kg body wt)	RBC Volume (ml/kg body wt)	RBC Survival Time (days)
Males	75	44	30	120
Females	67	43	24	109

Blood Coagulation Tests

Bleeding time: time elapsed before the cessation of bleeding after puncture of ear lobe or finger with a lancet
Normal value   Less than 4 min

Coagulation time: time required for complete coagulation in vitro, of blood drawn with a silicone-coated hypodermic needle (37°C)
Normal value   19–60 min (silicone-treated tube)
6–17 min (ordinary test tube)

Physical and Chemical Characteristics of Blood

Color   Arterial blood: scarlet, from oxyhemoglobin
Venous blood: dark red, from hemoglobin
Plasma: transparent yellow

Specific gravity		Sedimentation rate (RBC)
Male	1.059	up to 5 mm in 1 hr
Female	1.056	up to 8 mm in 1 hr

Surface tension of serum at 37°C	47 dynes/cm
Osmolality of serum	285-295 mOsm/liter

Hydrogen Ion Concentration—pH

	Arterial	Venous
At birth	—	7.38
6–10 years	7.40	7.38
Adult male	7.39	7.35
Adult female	7.40	7.37

Viscosity

Male	5.0	(water = 1)
Female	4.5	

Chemical Components of Blood
(values in mg/100 ml unless otherwise indicated)

Glucose (fasting)

At birth	60
Adult	83

Lipids

Total lipids

At birth	591		
Males, 18–30 years	743	Females, 18–30 years	703
31–45 years	793	31–45 years	757
46–60 years	806	46–60 years	812

Total cholesterol (free and bound)

At birth	138		
Males, 18–30 years	177	Females, 18–30 years	180
31–45 years	199	31–45 years	192
46–60 years	205	46–60 years	204

***Chemical Components of Blood* (*cont.*)**

Phospholipids			
At birth	131		
Males, 18–30 years	177	Females, 18–30 years	180
31–45 years	191	31–45 years	189
46–60 years	190	46–60 years	189
Triglycerides	10–150		
Plasma proteins			
Total	7.0 g/100 ml		
Albumin	4.04 g/100 ml		
Globulins	2.34 g/100 ml		
α globulins	0.79 g/100 ml		
β globulins	0.81 g/100 ml		
γ globulins	0.74 g/100 ml		
Fibrinogen	300		
Immunoglobulins			

Immunoglobulins	IgG	IgA	IgM	IgE
At birth	900–1500	0–5	5–20	very low
6–10 years	550–1200	60–170	40–95	very low
Adult	550–1900	60–330	45–145	very low

	Males	Females
Nonprotein nitrogen (NPN)—plasma		
Total		
Components		
Urea	27	26
Free amino acids	40	39
Creatine	33	64
Creatinine	1.03	0.79
Uric acid	5.3	4.3
ATP (whole blood)	46	39
Bilirubin		
At birth	2.68	
5th day	6.08	
Adult	0.60	
Intermediary metabolites (whole blood)		
α keto acids		
At birth	0.8	
Adult	1.3	
Pyruvic acid	0.76	
Lactic acid	9.9	
Citric acid	1.9	
Electrolytes mEq/liter plasma		
Sodium	144.7	
Potassium	4.3	
Calcium	9.5	
Magnesium	1.99	
Iron (μg/100 ml)	80–160	60–135
Phosphorus (mg/100 ml)	4.0	
Chloride	103.0	
Trace elements		
Copper	0.105	0.116
Fluorine	0.028	
Iodine		
Total	6.9	
Protein-bound iodine		
At birth	9.0	
Adult	0.0006	

Blood gases (ml gas in 100 ml whole blood)	Male	Female
Oxygen		
Capacity	20	18
Total content		
Arterial	20	18
Venous	15	14

Chemical Components of Blood (cont.)

In solution ("free" O_2)		
Arterial	0.3	0.3
Venous	0.1	0.1
Combined (oxyhemoglobin)		
Arterial	20	18
Venous	15	14
Carbon dioxide		
Total content		
Arterial	49	48
Venous	53	51
In solution (free and carbonic acid CO_2)		
Arterial	3.4	2.5
Venous	3.0	2.8
Combined (carbamino-CO_2 and bicarbonate CO_2)		
Arterial	46	46
Venous	50	49
Nitrogen (in solution)		
Arterial	0.98	0.97
Venous	0.98	0.97

Heart and Blood Pressure (Resting Values)

Heart Rate (beats/min)		Arterial Blood Pressure (mm Hg) Systolic	Diastolic
In fetal life	156		
At birth	143	69	38
Infant	120		
6–10 years			
Male	72	100	67
Female	80	100	68
Adult			
Male (22 years)	68	123	76
Male (72 years)	63	145	82
Female (22 years)	75	116	72
Female (72 years)	70	159	82

Effect of Training on Heart Rate (resting rate in beats/min)

Untrained	68
Sprinter	61
Skier	52
Marathon runner	43

Cardiac Output (liters/min)	
Adult	
Male (22 years)	6
Male (60 years)	5
Female (22 years)	5
Female (60 years)	4

Venous Blood Pressure (mm Hg) (median basilic vein, near elbow)	
3–5 years	3.4
6–10 years	4.3
Adult	
Male	7.4
Female	6.9

Distribution of Blood to Various Tissues

	% cardiac output
Head	15–20
Heart	3–7
Kidney	20–25
Liver	30–35
Muscle	10–20
Other organs	10–20

Urine

Synopsis of Urine[a]

Normal values in milligrams per 24-hr urine in adults on a normal balanced diet (unless otherwise stated).

Appearance. At the moment of voiding the urine is clear and transparent; after a rich meal which turns the urine alkaline, however, it can sometimes be more or less turbid. If clear urine is allowed to stand for some time *nubeculae* appear, that is, cloudy opacities formed by mucin from the urinary passages and, in alkaline urine, by various crystals (phosphates of alkaline-earth metals). The urine can also be turbid when fats are present.

Odor. The faint and usually aromatic odor is due to unidentified substances. Following ingestion of thymol or asparagus, for instance, the odor changes completely.

Color. Normally, the urine is more or less deep yellow in color, due mainly to the presence of urochrome. In certain diseases and after ingestion of various substances (drugs, foodstuffs, etc.) it can assume many shades of red, dark brown, or blue as shown in the following scheme.

Red urine:
  Blood, aminopyrine, aniline dyes (in candies), etc.
Dark brown urine:
  Hemoglobinuria, poisoning by phenol or cresol, melanin, alkaptonuria, etc.
Blue urine:
  Methylene blue, indigo blue.
Almost colorless:
  diabetes insipidus

	Mean	Range
Daily volume		
Newborn (1–2 days old)	—	30–80ml/24 hr
Infants of		
3–10 days	—	100–300ml/24 hr
10–60 days	—	250–450ml/24 hr
60–365 days	—	400–500ml/24 hr
Children of		
1–3 years	—	500–600ml/24 hr
3–5 years	—	600–700ml/24 hr
5–8 years	—	650–1000ml/24 hr
8–14 years	—	800–1400ml/24 hr
Adults	—	600–1600ml/24 hr
Old people	853 ml/24 hr	250–2400 ml/24 hr

	Mean	Range
Specific gravity		
Newborn (first few days)	1.012	—
Infants	—	1.002–1.006
Adults	—	1.001–1.030
pH		
Newborn	6.2	—
Men	5.7	4.8–7.5
Women	5.8	4.8–7.5

[a] Source: Geigy, *Scientific Tables,* 7th ed., 1970, by permission of J. R. Geigy. S. A. Basle, Switzerland.

Kidney Function

Glomerular Filtration Rate (GFR)	
Females	117 ml/min
Males	131
Maximum Tubular Reabsorption of Glucose (Tm glucose)	
Females	303 mg/min
Males	375 mg/min
Maximal Excretory Capacity (Diodrast)	
Females	42 mg/min
Males	51 mg/min

Respiration

Respiratory rate (respirations/min)	
At birth	40
Infant	30
6–10 years	22
Adult	16
Vital capacity (liters)	
5 years	
Male	1.3
Female	1.0
Adult	
Male	4.6
Female	4.0

Hormones in Blood and Urine

	Blood μg/ml	Urine Units as indicated/ 24 hrs
Androgens (testosterone)		
Male		
At birth	0.7	
2 weeks	0.01	
16 weeks	0.1	
Adult	0.7	70 μg
Female, adult	0.04	5 μg
17-ketosteroids (Most androgens are excreted as ketosteroids)		
Males		
6 years		0–2 mg
16 years		2–18 mg
Adult		8–18 mg
Female, adult		5–15 mg
Estrogens		
Female		
Onset of menstruation	Very low in plasma	4–25 μg
Ovulation peak		30–79 μg
Luteal maximum	> 0.2 μg	22–105 μg
Pregnancy	9.7 μg	20–46 mg[a]
Male, adult	Very low	4–25 μg
Corticosteroids		
Cortisol		
Male	10 μg	35 μg[b]
Female	9 μg	30 μg[c]
Corticosterone		
Male	1 μg	5.8 μg[b]
Female	1 μg	5.8 μg[c]

[a] Note tremendous increase.

[b] At least 50% excreted as 17-ketosteroids.

[c] Almost all excreted as 17-ketosteroids.

Total Body Water (as % body weight)

First fetal month	90
At birth	76
6–10 years	60
Adult	
Male (22 years)	61
Male (72 years)	52
Female (22 years)	50
Female (72 years)	46

Body Temperature (Resting Values)

Body temperature varies not only with age and sex and time of day, but the temperature range for repeated measurements on the same individual at the same time of day is about 1.2°F for males, 1.5°F for females.

Female body temperature is also affected by ovulation and pregnancy (Fig. 21-4). Temperature fluctuations are even more marked in infancy and early childhood.

Rectal temperatures

	°F	°C
1 month		
Males	98.97	37.21
Females	98.96	37.20
7 months		
Males	99.88	37.70
Females	99.58	37.54
36 months		
Males	99.22	37.34
Females	98.72	37.61
Adults		
Males	99.60	37.56
Females	99.80	37.67

Appendix III

Measurements

THE METRIC SYSTEM

The metric system is internationally used in scientific work. Because it is based on the decimal system, it avoids the complex fractions required by the systems common to the English-speaking countries in the conversion of measurements to larger or smaller units.

The fundamental unit of length in the metric system is the *meter* (m), which is equivalent to 39.37 in. The basic unit of weight is the *gram* (g).

Dividing or multiplying the meter or the gram by 10, 100, and so forth, gives the units that are distinguished by the appropriate Latin prefixes, as listed below:

Factor	Name x m (length) x g (weight)	Symbol
10^{12}	tera	T
10^{9}	giga	G
10^{6}	mega	M
10^{3}	kilo	k
10^{2}	hecto	h
10^{1}	deca	da
10^{0}		
10^{-1}	deci	d
10^{-2}	centi	c
10^{-3}	milli	m
10^{-6}	micro	μ
10^{-9}	nano	n
10^{-12}	pico	p
10^{-15}	femto	f
10^{-18}	atto	a

1 Ångstrom unit (A) = 10^{-10} m
= 10^{-1} nm (preferred term)

UNITS OF VOLUME

1 liter	(1) = standard unit of volume = 1.06 quarts
1 milliliter	(ml) = 0.001 1 (by weight)
1 cubic centimeter	(cc) = 0.001 1 (by volume)
	For ordinary purposes, ml and cc may be considered identical.

UNITS OF TEMPERATURE

The centigrade scale is universally used in scientific work. The temperature of boiling water is arbitrarily assigned a value of 100 degrees, whereas the freezing point of water is 0 degrees on this scale. The absolute zero (the temperature at which all molecular and atomic motion stops completely) is −273°C.

The Fahrenheit scale, like the other English and American systems of weights and measures, is cumbersome and more difficult to use. To convert centigrade to Fahrenheit, multiply by $\frac{9}{5}$ and add 32 (e.g., 37°C = 98.6°F). To convert Fahrenheit to centigrade, multiply by $\frac{5}{9}$ and subtract 32.

Comparison of Some Centigrade and Fahrenheit Temperatures

Boiling point of water	100°C	212°F
Normal room temperature	22	71.6
Normal body temperature	37	98.6
Freezing point of water	0	32
Absolute zero	−273	−460

Author Index

Italic numbers indicate pages with one or more bibliographical entries of the author's work.

Subject Index